POWER ELECTRONICS FOR RENEWABLE ENERGY SYSTEMS, TRANSPORTATION AND INDUSTRIAL APPLICATIONS

POWER ELECTRONICS FOR RENEWABLE ENERGY SYSTEMS, TRANSPORTATION AND INDUSTRIAL APPLICATIONS

Edited by

Haitham Abu-Rub

Texas A&M University at Qatar, Doha, Qatar

Mariusz Malinowski

Warsaw University of Technology, Warsaw, Poland

Kamal Al-Haddad

École de Technologie Supérieure, Montreal, Canada

IEEE

IEEE PRESS

WILEY

A co-publication of IEEE Press and John Wiley & Sons Ltd

This edition first published 2014

Registered office
John Wiley & Sons Ltd, The Atrium, Southern Gate, Chichester, West Sussex, PO19 8SQ, United Kingdom

For details of our global editorial offices, for customer services and for information about how to apply for permission to reuse the copyright material in this book please see our website at www.wiley.com.

Library of Congress Cataloging-in-Publication Data

Power electronics for renewable energy systems, transportation, and industrial applications / edited by Haitham Abu-Rub, Mariusz Malinowski, Kamal Al-Haddad.
pages cm
Author's surname spelled "Haitham" on title page.
Includes bibliographical references and index.
ISBN 978-1-118-63403-5 (cloth)
1. Power electronics. 2. Industries – Power supply. I. Abu-Rub, Haithem, editor of compilation. II. Malinowski, Mariusz (Electrical engineer), editor of compilation. III. Al-Haddad, Kamal, editor of compilation.
TK7881.15.P6725 2014
621.31′7 – dc23

2014001834

A catalogue record for this book is available from the British Library.

ISBN: 9781118634035

Typeset in 9/11pt TimesLTStd by Laserwords Private Limited, Chennai, India

1 2014

This book is dedicated to our families and parents.

Contents

Foreword

It is my great honor and pleasure to write the foreword for this state-of-the-art book *Power Electronics for Renewable Energy Systems, Transportation, and Industrial Applications*. Power electronics and drives control is an extremely complex field with multiple disciplines throughout the field of electrical engineering. It is virtually impossible to write a book covering the entire area by one individual specialist, particularly witnessing the recent developments in neighboring fields such as control theory, signal processing, and applications in renewable energy systems, as well as electric and plug-in hybrid vehicles, all of which strongly influence new solutions in power electronic systems. For this reason, the book has been written by the key specialists in these areas.

This book comprising 24 chapters is divided into three parts: (1) *Impact of Power Electronics for Emerging Technologies* (Chapters 1–5), (2) *Power Electronics for Distributed Power Generation Systems* (Chapters 6–11), and (3) *Power Electronics for Transportations and Industrial Applications* (Chapters 12–24). The first chapter is written by the world-renowned power electronics expert Professor Bimal K. Bose, and is followed by a review of power electronics in high-voltage direct current (HVDC) and flexible AC transmission systems (FACTS), a chapter on smart grid concepts and technologies, and then a chapter on recent advances in power semiconductor technology. Chapter 5 is the last in Part 1 and this presents a new class of AC-link universal power converters. The second part of the book begins with two chapters (Chapters 6 and 7) that provide an overview of renewable technology, both of which are co-authored by the world-known specialist Professor Frede Blaabjerg; these chapters deal with power electronics for wind turbines and photovoltaic (PV) energy. The next four chapters (Chapters 8–11) cover controllability analysis, distributed power generation, variable speed doubly fed induction machine (DFIM), and AC–DC–AC converters in renewable energy systems. The third and largest part of the book begins with two chapters (Chapters 12 and 13) on transportation, including one on modern power electronic solutions for aircrafts and the other on electric and plug-in hybrid vehicles. These two chapters are followed by a discussion and a presentation of multilevel converters (Chapter 14), multiphase matrix converters (Chapter 15), high-power factor rectifiers (Chapter 16), active power filters (Chapter 17), hardware-in-the-loop power electronic systems (Chapter 18), predictive control of converter-fed electric machines (Chapter 19), current source converters for drives (Chapter 20), reduction of common-mode (CM) voltage and bearing currents in pulse-width modulation (PWM)-fed drives, high-power industrial drives, modulation and control of single-phase grid-connected converters (Chapter 23), and, finally, impedance Z-source inverters (ZSI) and quasi-Z-source inverters (qZSI) (Chapter 24).

The work has typical attributes of a contemporary book and discusses several aspects of the authors' current research in an innovative and original way. Easy description and good illustrations make the book attractive for researches, engineering professionals, graduate students of electrical engineering, and power systems faculties.

Finally, I would like to applaud the initiative taken by editors in this timely book to cover a wide area of power electronic applications in renewable energy systems, smart grids, distributed generation, transportation, and other industrial areas. This work perfectly fills the current gap and contributes to a better understanding and further applications of power electronic systems.

Marian P. Kazmierkowski, IEEE Fellow
Institute of Control and Industrial Electronics
Warsaw University of Technology, Poland

Preface

It is our pleasure to present this book on up-to-date power electronics technologies and advancements in their use in renewable energy, transportation systems, and various industrial applications.

We have written this book in response to the current lack of relevant research available to researchers, professionals, and students. It is our hope that we successfully convey our passion for this field in a manner that is easy to follow textually and visually. We have chosen to write this as a joint initiative because of the expertise needed in an all-encompassing research on power electronic systems.

In this book we cover a wide range of power electronic components, renewable energy systems, smart grids, distributed generations, transportation systems, and other industrial areas. This work fills a gap in engineering literature and contributes to a better understanding and further application of power electronic systems. Power electronic components and applications are among the fastest growing engineering areas today and are key in responding to our current environmental constraints and energy demands. This book integrates material in order to answer current problems and offer solutions for the growing commercial and domestic power demands.

The book discusses several aspects of current research, and the participation of the world's top scientists solidifies the book's credibility, including *IEEE* life fellows Prof. Bimal K. Bose and Prof. Joachim Holtz. Other scientists who participated in the writing of this book include Professors Frede Blaabjerg, Leopoldo G. Franquelo, Carlo Cecati, Hamid A. Toliyat, Bin Wu, Fang Zheng Peng, Ralf M. Kennel, and Jose Rodriguez.

The book is divided into three main parts: (1) The Impact of Power Electronics for Emerging Technologies (Chapters 1–5), (2) Power Electronics for Distributed Power Generation Systems (Chapters 6–11), and (3) Power Electronics for Transportations and Industrial Applications (Chapters 12–24).

Chapter 1 offers a brief but comprehensive review of the world's energy resources and climate change problems because of fossil fuel burning, along with possible solutions or mitigation methods. The author concludes with a discussion of the impact of power electronics that have on energy conservation, renewable energy systems, bulk storage of energy, and electric/hybrid vehicles in the present century.

Chapter 2 focuses on the contribution of power electronics to achieve efficient energy transmission and distribution, enable a high penetration of renewables in the power system, and develop more electrical transportation systems. This chapter also addresses flexible AC transmission system (FACTS) devices; high voltage direct current (HVDC) transmission systems; power electronics converters for wind, photovoltaic (PV), and ocean sources; power conversion for electric vehicles; and energy storage systems.

Chapter 3 gives an overview of the main technologies, features, and problems of distributed generation and smart grids. This chapter gives a short but comprehensive overview of these emerging topics.

Chapter 4 presents recent advances in power semiconductors technology, focusing specifically on wide bandgap transistors. The authors offer a short introduction to state-of-the-art silicon power devices and the characteristics of the various SiC power switches. Design considerations of gate and base drive circuits for various SiC power switches, along with experimental results of their switching performance, are presented in details alongside a discussion of their applications.

In Chapter 5, the authors categorize AC-link universal power converters within a new class of power converters, and demonstrate how they can interface multiple loads and sources while remaining a single-stage converter.

Chapter 6 expands on technological developments and market trends in wind power application. The authors review a variety of wind turbine concepts, as well as power converter solutions, and offer an explanation of control methods, grid demands, and emerging reliability challenges.

Chapter 7 presents a comprehensive overview of grid-connected PV systems, including power curves, grid-connected configurations, different converter topologies (both single and three phases), control schemes, maximum power point tracking (MPPT), and anti-islanding detection methods. The chapter focuses on the mainstream solutions available in the PV industry, in order to establish the current state of the art in PV converter technology. In addition, the authors offer a discussion of recently introduced concepts on multilevel converter-based PV systems for large-scale PV plants, along with trends, challenges, and possible future scenarios of PV converter technology.

In Chapter 8, the authors demonstrate that the components of renewable energy systems, including interfacing filters, are first selected to ensure steady optimum performance operation, after which controllers are designed and implemented to ensure stability, high dynamic performance, and robustness to disturbance and parameter variations. The controllability analysis of an interior permanent magnet (IPM) wind turbine generator connected to the grid through a filter interface is analyzed, and the stability of the nonlinear system and the study of the zero dynamics provide insights into potential constraints on controller structure and dynamics.

Chapter 9 points out that the role of the power converter's control is fundamental and involves a number of issues: power flow control, synchronization with the main grid, reactive power capability, voltage regulation at the point of common coupling and power quality constraints. In addressing these matters, the authors focus on PV and small wind turbine systems, as well as the management of the transition among grid connection, stand-alone operation, and synchronization.

Chapter 10 describes the main properties and control methods of the doubly fed induction machine, which are related to both grid-connected and stand-alone operation modes. The chapter presents the properties of a wind turbine equipped with a doubly fed induction machine, and offers a short description of wind turbine aerodynamics, wind turbine superior control, and steady-state performance of wind turbine.

Chapter 11 is devoted to various topologies of AC–DC–AC converters and their design. It offers an in-depth discussion of classical three-phase/three-phase transistor-based AC–DC–AC converters (two-level and three-level diode-clamped converters (DCCs) and flying capacitors converters (FCC)) and simplified AC–DC–AC converters (two-level and three-level three-phase/one-phase and three-phase/three-phase DCC).

Chapter 12 describes how More Electric Aircraft (MEA) technology is continually evolving and being widely recognized as the future technology for the aerospace industry. This chapter provides a brief description of the electrical power generation, conversion, and distribution in conventional aircrafts and in more electric aircrafts, such as Airbus 380 and Boeing 787. The author also discusses more electric architectures, power distribution strategies, more electric engine concepts, and the effect of high-voltage operation at high altitudes.

Chapter 13 presents the structure and basic design aspects of electric vehicles (EVs) and plug-in hybrid electric vehicles (PHEVs), as well as future trends in EV manufacturing. The authors also discuss the integration of EVs with green, renewable energy sources and introduce the design of such systems.

Chapter 14 is dedicated to explaining multilevel converters/inverters and describing their pros and cons regarding their most suitable applications. The chapter presents how multilevel inverters are applied to static var generation (SVG), static synchronous compensator (STATCOM), and FACTS devices. The authors further explore magnetic-less multilevel DC–DC converters and analyze the multilevel converters' fault tolerance and reliability.

Chapter 15 elaborates on the theoretical and analytical aspects of multi-phase matrix converters, encompassing existing and emerging topologies and control. The authors also discuss various control algorithms for efficient operation.

Chapter 16 presents a detailed analysis of three boost-type preregulators commonly used for power factor correction in single-phase rectifiers: the single-switch basic boost, the two-switch asymmetric half-bridge boost, and the interleaved dual-boost topology. The authors also illustrate the mathematical modeling approach, applying it to the first two topologies. In so doing, the authors are able to demonstrate the effectiveness of these converters associated with their respective control systems.

Chapter 17 looks at how power electronics applications have penetrated multiple areas of modern life, thereby increasing nonlinear loads in comparison with linear loads. Simultaneously, power electronics-based loads are sensitive to harmonic distortion, which leads to a discussion of active power filters that can be employed to cancel or mitigate harmonics and their effects.

In Chapter 18A, the discussion provided proves that the so-called virtual machine (VM) is a hardware-in-the-loop (HiL) system allowing an inverter to be tested at real power levels without the need for installing and operating real machines as the VM has the same characteristics as a real induction motor or even a synchronous motor. Different machines and their respective load conditions can be emulated by software, meaning that the drive inverter under test can operate in its normal mode.

Chapter 18B also relates to the HiL systems, with a thorough presentation of the modular multilevel converter (MMC). The authors explain the limitations of standard simulation methods and propose more suitable control techniques. Issues raised by the converter topology are discussed with regard to the choice of hardware to achieve real-time simulation, and examples of implementation for real-time application using OPAL-RT real-time simulator are given for the different techniques previously discussed.

Chapter 19 describes the use of model predictive control (MPC) for speed control in electrical machines. The authors also show how the MPC is a conceptually different control technique that offers a high flexibility to control different power electronics topologies and manages several control objectives, without adding significant complexity to the system.

Chapter 20 presents two approaches used to control electric machines supplied by the current source inverter. The first approach is based on the current control and the second approach contains the voltage control with multiscalar model approach. The topologies are analyzed for controlling a supply squirrel-cage induction motor, doubly fed machine, and permanent magnet synchronous machine.

In Chapter 21, the author shows how the high dv/dt and the common-mode voltage generated by the inverter pulse-width modulation (PWM) control result in the appearances of bearing currents, shaft voltages, motor terminal overvoltages, the decrease in motor efficiency, and electromagnetic interference. A common-mode motor equivalent circuit is analyzed, with an emphasis on the bearing currents and various aspects of currents' limitation. The author dedicates much of the chapter to analyzing the active methods on the limitation of common-mode currents based on PWM modifications.

In Chapter 22, the impacts of megawatt variable frequency drives (VFDs) for liquefied natural gas (LNG) plants are discussed. This chapter presents few examples of actual high-power VFDs that can realize up to 100 MW systems using four sets of 25 MW neutral-point-clamped (NPC)-based multilevel voltage source inverters (VSIs). The chapter starts with an overview of LNG plants, outlines conventional gas turbine (GT), drives techno-economic and environmental implications, and introduces various electric drive technologies used for LNG plants, highlighting their limitations, technological problems, and their impact on future LNG plants.

Chapter 23 is devoted to the modulation and control of single-phase, active front-end converters. The first part of the chapter presents a literature review and analysis of PWM techniques with unipolar switching for three main multilevel converter topologies. The second part of the chapter is devoted to current control of single-phase voltage source converters.

The final chapter offers a comprehensive and systematic reference for the current and future development of the high-performance Z-source inverter (ZSI)/quasi-Z-source inverter (qZSI), and provides a detailed explanation of the impedance parameter design. It looks at ZSI/qZSI, otherwise known as an impedance source inverter. This inverter has attracted increasing interest because of a single-stage power conversion with a step-up/step-down function, handling the DC voltage variations in a wide range without overrating the inverter and allowing switches on the same bridge leg to simultaneously turn on. The authors present the operation principle and control methods of conventional ZSI/qZSI, and offer a discussion of the advantages of novel extended topologies, such as qZSI with battery and qZSI-based cascade multilevel systems.

Haitham Abu-Rub
Mariusz Malinowski
Kamal Al-Haddad

Acknowledgments

We would like to take this opportunity to express our sincere appreciation to everyone who directly or indirectly helped in making this book a reality. Our special thanks go to Mrs. Suzan Nasser and Mr. Wesam Mansour for assisting us in this work. We are also very grateful to Mrs. Amy Hamar for her help in revising the language of specific chapters. Special thanks also go to Texas A&M University, Qatar, for funding the language revision, editing, and other related work.

We are indebted to our families for their continuous support, patience, and encouragement without which this project would not have been completed. We would also like to express our appreciation and sincere gratitude to the staff of Wiley, especially Miss Laura Bell and Mr. Richard Davies, for their help and cooperation.

Haitham Abu-Rub
Mariusz Malinowski
Kamal Al-Haddad

List of Contributors

Gonzalo Abad, Electronics and Computing Department, Mondragon University, Mondragon, Spain

Ayman S. Abdel-Khalik, Electrical Engineering Department, Alexandria University, Egypt

Haitham Abu-Rub, Department of Electrical and Computer Engineering, Texas A&M University at Qatar, Doha, Qatar

Shehab Ahmed, Electrical and Computer Engineering Department, Texas A&M at Qatar, Doha, Qatar

Kamal Al-Haddad, Department of Electrical Engineering, Ecole de Technologie Supérieure, Montreal, Canada

Mahshid Amirabadi, Department of Electrical and Computer Engineering, University of Illinois at Chicago, Illinois, USA

Jean Bélanger, Opal-RT Technologies Inc, Montréal, Canada

Lazhar Ben-Brahim, Department of Electrical Engineering, Qatar University, Doha, Qatar

Frede Blaabjerg, Department of Energy Technology, Aalborg University, Aalborg, Denmark

Handy. F. Blanchette, Ecole de Technologie Supérieure, Montreal, Canada

Till Boller, Institute for Electrical Machines and Drives, University of Wuppertal, Wuppertal, Germany

Bimal K. Bose, Department of Electrical Engineering and Computer Science, The University of Tennessee, Tennessee, USA

Concettina Buccella, Department of Industrial and Information Engineering and Economics, and DigiPower Ltd., University of L'Aquila, L'Aquila, Italy

Giampaolo Carli, Department of Electrical and Computer Engineering, Concordia University, Montreal, Canada

Carlo Cecati, Department of Industrial and Information Engineering and Economics, and DigiPower Ltd., University of L'Aquila, L'Aquila, Italy

Christian Dufour, OPAL-RT TECHNOLOGIES Inc, Montréal, Canada

Leopoldo G. Franquelo, Department of Electronic Engineering, University of Seville, Seville, Spain

Baoming Ge, School of Electrical Engineering, Beijing Jiaotong University, Beijing, China; Department of Electrical Engineering, Texas A&M University, Texas, USA

Luc A. Grégoire, Electrical engineering department, Ecole de Technologie Supérieure, Montréal, Canada

Jaroslaw Guzinski, Faculty of Electrical and Control Engineering, Gdansk University of Technology, Gdansk, Poland

Joachim Holtz, Institute for Electrical Machines and Drives, University of Wuppertal, Wuppertal, Germany

Atif Iqbal, Department of Electrical Engineering, Qatar University, Doha, Qatar; Aligarh Muslim University, Aligarh, India

Grzegorz Iwanski, Institute of Control and Industrial Electronics, Warsaw University of Technology, Warsaw, Poland

Marek Jasinski, Faculty of Electrical Engineering, Warsaw University of Technology, Warsaw, Poland

Hadi Y. Kanaan, Department of Electrical and Mechanical Engineering, Saint-Joseph University – ESIB, Mar Roukoz, Lebanon

Hossein Karimi-Davijani, Department of Electrical and Computer Engineering/Center for Energy Systems Research, Tennessee Technological University, Tennessee, USA

Ralph M. Kennel, Institute for Electrical Drive Systems and Power Electronics, Technische Universitaet Muenchen, Munich, Germany

Samir Kouro, Department of Electronics, Universidad Tecnica Federico Santa Maria, Valparaiso, Chile

Zbigniew Krzeminski, Department of Automatic Control of Electrical Drives, Gdansk University of Technology, Gdansk, Poland

Jose I. Leon, Department of Electronic Engineering, University of Seville, Seville, Spain

Yongdong Li, Department of Electrical Engineering, Tsinghua University, Beijing, China

Marco Liserre, Christian-Albrechts-University of Kiel, Kaiserstr, Germany

Yushan Liu, Department of Electrical and Computer Engineering, Texas A&M University at Qatar, Doha, Qatar; School of Electrical Engineering, Beijing Jiaotong University, Beijing, China

Ke Ma, Department of Energy Technology, Aalborg University, Aalborg, Denmark

Mariusz Malinowski, Faculty of Electrical Engineering, Warsaw University of Technology, Warsaw, Poland

Ahmed M. Massoud, Electrical Engineering Department, Qatar University, Doha, Qatar

Rosa A. Mastromauro, Department of Electrical and Information Engineering, Politecnico di Bari, Bari, Italy

Pawel Mlodzikowski, Faculty of Electrical Engineering, Warsaw University of Technology, Warsaw, Poland

SK. Moin Ahmed, Department of Electrical and Computer Engineering, Texas A&M University at Qatar, Doha, Qatar; Department of Electrical Engineering, Universiti Teknologi Malaysia, Johor Bahru, Malaysia

Marcin Morawiec, Department of Automatic Control of Electrical Drives, Gdansk University of Technology, Gdansk, Poland

Antonella Nagliero, Department of Electrical and Information Engineering, Politecnico di Bari, Bari, Italy

Hans-Peter Nee, KTH Royal Institute of Technology, Stockholm, Sweden

Olorunfemi Ojo, Department of Electrical and Computer Engineering/Center for Energy Systems Research, Tennessee Technological University, Tennessee, USA

Fang Z. Peng, Department of Electrical & Computer Engineering, Michigan State University, Michigan, USA

Dimosthenis Peftitsis, KTH Royal Institute of Technology, Stockholm, Sweden

Marcelo A. Pérez, Department of Electronics, Universidad Técnica Federico Santa María, Valparaíso, Chile

Jacek Rąbkowski, Emissions Research and Measurement Division, Warsaw University of Technology, Warsaw, Poland; KTH Royal Institute of Technology, Stockholm, Sweden

Kaushik Rajashekara, Department of Electrical Engineering, University of Texas at Dallas, Texas, USA

José Rodríguez, Department of Electronics, Universidad Técnica Federico Santa María, Valparaíso, Chile

Arash Shafiei, Department of Electrical and Computer Engineering, Concordia University, Montreal, Canada

Sebastian Stynski, Faculty of Electrical Engineering, Warsaw University of Technology, Warsaw, Poland

Hamid A. Toliyat, Department of Electrical and Computer Engineering, Texas A&M University, Texas, USA

Sergio Vazquez, Department of Electronic Engineering, University of Seville, Seville, Spain

Sheldon S. Williamson, Department of Electrical and Computer Engineering, Concordia University, Montreal, Canada

Bin Wu, Department of Electrical and Computer Engineering, Ryerson University, Ontario, Canada

Teruo Yoshino, Power Electronics Systems Division of Toshiba Mitsubishi-Electric Industrial Systems Corporation, Tokyo, Japan

Héctor Young, Department of Electronics, Universidad Técnica Federico Santa María, Valparaíso, Chile

1

Energy, Global Warming and Impact of Power Electronics in the Present Century

Bimal K. Bose
Department of Electrical Engineering and Computer Science, The University of Tennessee, Tennessee, USA

1.1 Introduction

Power electronics concerns the conversion and control of electrical power with the aid of power semiconductor devices, which include diodes, thyristors, triacs, gate turn-off thyristors (GTO), power metal oxide silicon field effect transistors (MOSFET), insulated gate bipolar transistors (IGBT) and integrated gate-commutated thyristors (IGCT). Their applications include regulated power supplies (DC and AC), uninterruptible power supply (UPS) systems, electrochemical processes (such as electroplating, electrolysis, anodizing and metal refining), control of heating and lighting, electronic welding, power line static var compensators (SVC or STATCOM) and flexible AC transmission systems (FACTS), active harmonic filters (AHFs), high-voltage direct current (HVDC) systems, photovoltaic (PV) and fuel cell (FC) converters, electronic circuit breakers, high-frequency heating, energy storage and motor drives. The extensive applications of power electronics in global industrialization have been somewhat unprecedented in history. We now live in a global society in which the nations in the world are highly interdependent. In the present trend, it is expected that future wars in the world will be fought on an economic front rather than military. In future, all the nations in the world will face severe industrial competitiveness for their survival and prosperity. In such an environment, power electronics and motion control will play a very important role. In addition, as the price of energy increases and environmental regulations become more stringent, power electronics applications will spread practically everywhere. The role of power electronics in this century will be as, if not more, important as computer, communication and information technologies.

It should be mentioned here that power electronics technology has now gained significant maturity after several decades of the dynamic technology evolution of power semiconductor devices, converters, pulse width modulation (PWM) techniques, electrical machines, variable-frequency drives, advanced controls and computer simulation techniques. According to the estimate of the Electric Power Research Institute

Power Electronics for Renewable Energy Systems, Transportation and Industrial Applications, First Edition.
Edited by Haitham Abu-Rub, Mariusz Malinowski and Kamal Al-Haddad.

(EPRI) of the United States, roughly 70% of electrical energy in the United States is now processed through power electronics, which will eventually expand to 100%. In the present century, we expect to see the significant impact of power electronics not only in global industrialization and general energy systems, but also in energy conservation, renewable energy systems, bulk energy storage and electric vehicles (EVs) and hybrid electric vehicles (HEVs). The corresponding impact in solving or mitigating climate change problems is expected to be significant. This chapter will start with a review of the global energy perspective, environmental pollution because of the burning of fossil fuels, the corresponding climate change or global warming problems and the methods for their mitigation. Then, it will review the impact of power electronics in energy saving, renewable energy systems, the bulk storage of energy and EVs and HEVs. Finally, the future scenario of power electronics technology evolution will be reviewed in the conclusion.

1.2 Energy

Let us first discuss the global energy perspective [1–6]. Figure 1.1 shows the global energy generation and the generation of US energy in the same perspective. About 84% of global energy is generated by fossil fuels, 3% from nuclear plants and the remaining 13% comes from renewable sources, such as hydro, wind, solar, biofuels, geothermal, wave and tidal power. US energy generation follows a similar pattern. About 41% of US energy comes from oil, of which a significant part is imported. The present availability of shale oil (and natural gas) is decreasing this dependence, and according to the forecast of the International Energy Agency (IEA), the United States will be energy independent in the future. The global per capita energy consumption is highest in the United States. With nearly 4.5% of world's population (313 million out of 7 billion), the United States consumes nearly 28% of global energy and this reflects a very high standard of living. In comparison, China, now the world's second largest economy, with nearly 19% of the world's population (1.34 billion), consumes nearly half the total energy consumed by the United States. Of course this scenario is changing fast because of the rapid industrialization of China.

Figure 1.2 shows the idealized energy depletion curves of fossil and nuclear fuels throughout the world, which are somewhat Gaussian in nature. The world has enormous reserves of coal, and at the present consumption rate it is expected to last approximately 200 years. From the oil depletion curve, it appears that we are presently near the peak and is expected to become exhausted in 100 years. The oil price has been increasing recently because of the rising demand and dwindling supply. The natural gas reserves are expected to last around 150 years.

The recent availability of large amounts of shale oil and gas is creating an economic boom in some countries, particularly in the United States, as mentioned previously. Uranium (U-235) has a very low reserve and is expected to become exhausted in around 50 years.

With adequate conservation, the curves in the figure can be flattened in order to last for a longer period. Exploration of new fuel sources, particularly offshore resources, can provide new sources of oil and gas. Renewable energy resources (not included in the figure) will theoretically extend the energy depletion curve to infinity. It is no wonder that due to their competitive costs, extensive availability and environmentally clean nature, renewable energy sources are now being extensively explored all over the world. Recent studies show that renewable energy alone (with adequate storage) can supply all the energy needs of the world.

Figure 1.3 shows electricity generation according to different fuel types for the United States, Japan, China and India. For example, in the United States, 40% of the total energy is consumed in electrical form, of which typically 50% comes from coal, 2% from oil, 18% from natural gas, 20% from nuclear plants and the remaining 10% comes from renewables (mainly hydroelectricity). The electricity generation from natural gas is more favored (with the corresponding decrease in coal) because of the availability of cheap and abundant shale gas. Japan produced 31% of its electricity from nuclear resources, but the recent accident at the Fukushima Daiichi nuclear plant is changing the scenario, with the emphasis now

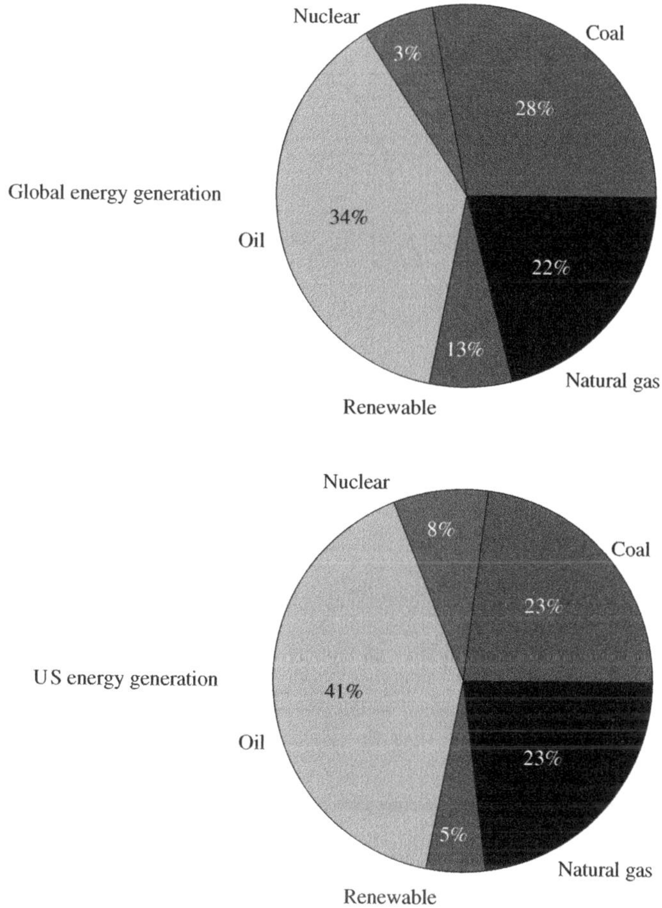

Figure 1.1 Global and US energy generation scenarios (2008) [2]

more on renewable resources. It is interesting to note that the world's two fastest developing economies (China and India) generate more than 70% of their electricity from coal.

1.3 Environmental Pollution: Global Warming Problem

Unfortunately, fossil fuel burning generates gases (SO_2, CO, NO_X, HC and CO_2) that cause environmental pollution. The most dominant effects of fossil fuel burning are climate change or global warming problems [3, 4], which is mainly caused by CO_2 and other gases – called greenhouse gases (GHG). GHG trap solar heat from the atmosphere (called the greenhouse effect). The United Nations (UN) Intergovernmental Panel on Climate Change (IPCC) has ascertained that the man-made burning of fossil fuels causes the global warming problem. It may be mentioned here that nuclear power does not have the traditional environmental pollution problem, but the safety of nuclear plants with regard to the radiation hazard is of serious concern. Another problem with nuclear power is that the waste from nuclear plants remains radioactive for thousands of years, and we do not know how to safely dispose of such waste. It is possible that in the future this waste will cause considerable damage to our society.

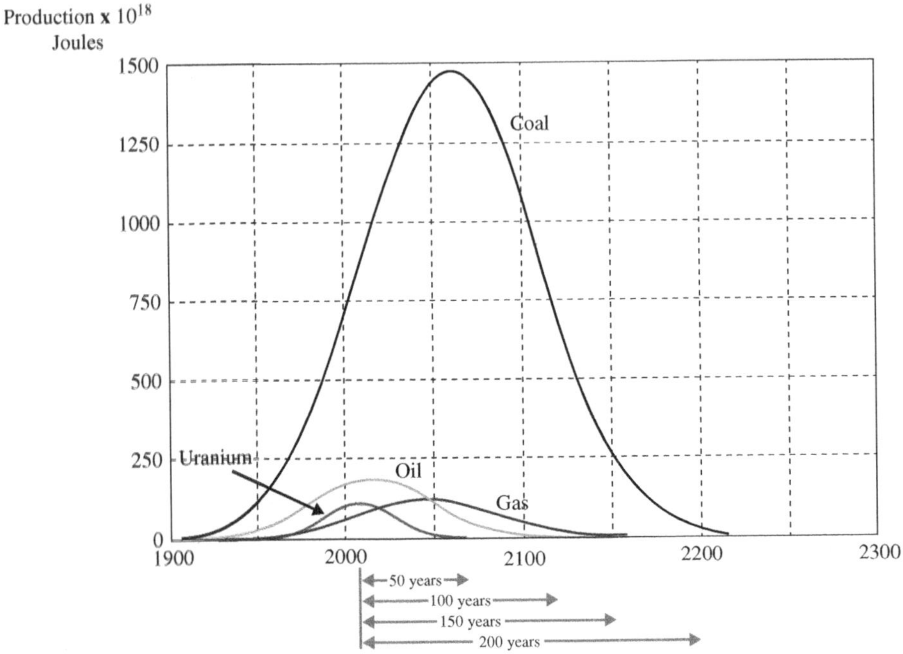

Figure 1.2 Idealized fossil and nuclear energy depletion curves for the world [2]

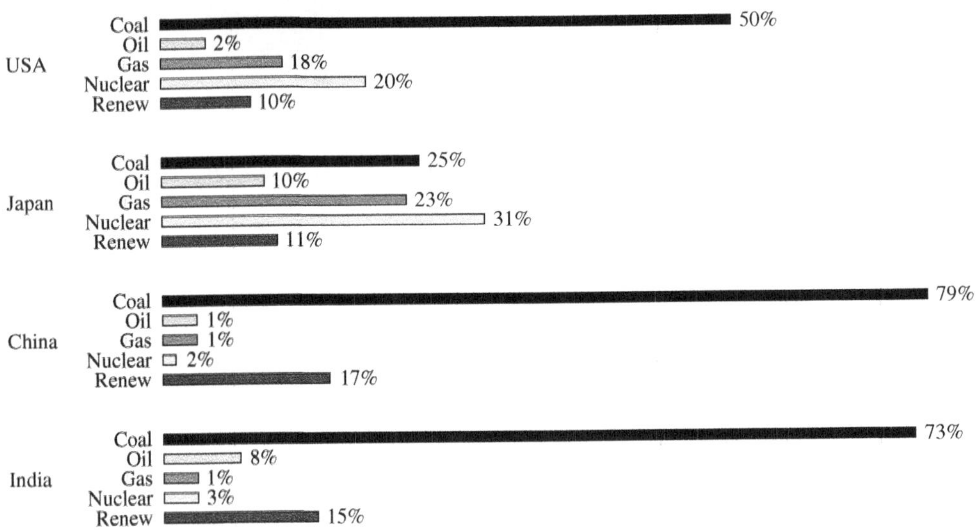

Figure 1.3 Electricity generation by fuel types of some selected countries (2008) [2]

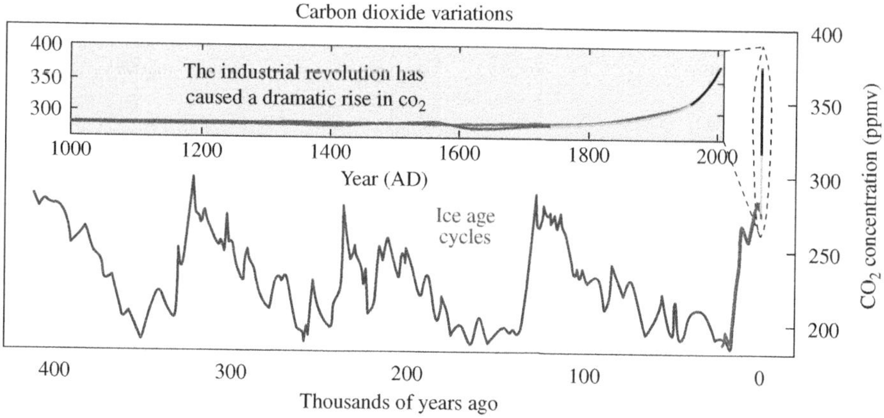

Figure 1.4 CO_2 variation in atmosphere thousands of years ago [4]

Scientists have studied carbon (CO_2) concentration variations in the atmosphere over a long period of time by analyzing ice samples from great depths in Antarctic regions. Figure 1.4 shows the CO_2 concentration curves in the atmosphere over thousands of years. The causes for the cyclic variation in CO_2 (ice age cycles) in the atmosphere is not well understood. However, one thing we are sure of is that in the past 1000 years (including the postindustrial revolution), the CO_2 concentration has increased dramatically (see the upper curve in Figure 1.4), and is much greater than the upper limit of normal cyclic variation. Scientists believe that this peaking is due to the man-made burning of fossil fuels and is irreversible, and therefore the consequences are serious.

It has been estimated that nearly 80% of atmospheric CO_2 is generated by the man-made fossil fuel burning, of which typically 50% is due to electric power generation. Figure 1.5 shows the per capita CO_2 emissions versus population for several selected countries in the world, where the horizontal axis shows the population and the vertical axis shows the CO_2 emission per person (in tons/year). The United States has the highest per capita emission in the world and Canada is very close. Next comes Australia and the European nations, as well as Russia and Japan, where the emissions are typically less than 50% of that of the United States. The total emission of a country, given by the area of the rectangle, is very important with respect to the global warming problem.

The standard of living in China is much lower than that of the United States; therefore, its per capita emission is very low. However, because of its large population the total emissions in China are large and, in fact, have exceeded that of the United States since 2006. Brazil has a good standard of living, but its per capita emission is low. In Brazil, typically 90% of energy (in electrical form) comes from hydroelectricity. In addition, it has a large Amazon rain forest that reduces CO_2, and 50% of its automobiles run on renewable sugarcane-based biofuel. Biofuels are said to have a carbon neutralization effect because they absorb CO_2 during the plant's growth but emit CO_2 at burning.

At the present rate of growth of the world's energy consumption (if no remedial actions are taken), the estimate of the potential CO_2 rise between 2002 and 2030 is provided in Figure 1.6. In 2002, the total energy consumption in the world was typically 410 quadrillion Btu (1 quadrillion = 10^{24} units), of which the share between fossil fuels and alternative energy sources is shown in the figure. The alternative energy sources consist of nuclear and renewable sources. The corresponding generation of CO_2 by fossil fuels was typically 2.6 billion tons/year. The total energy consumption in the world is expected to grow by 60% over the next 28 years, and the corresponding CO_2 level will rise by 62% (4.2 billion tons/year) because of an increased share of fossil fuels. Evidently, as a result, the global warming effect will be significant.

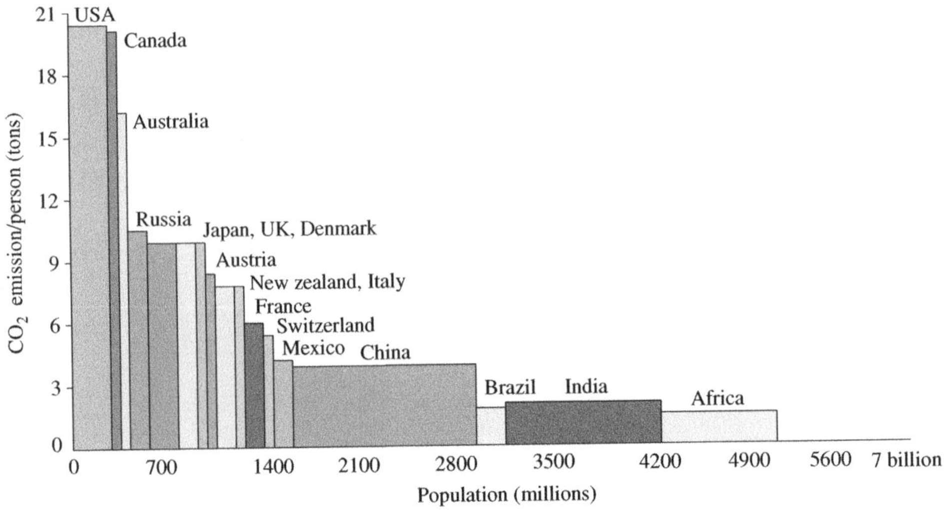

Figure 1.5 Per capita CO_2 emissions versus population of selected countries [2]

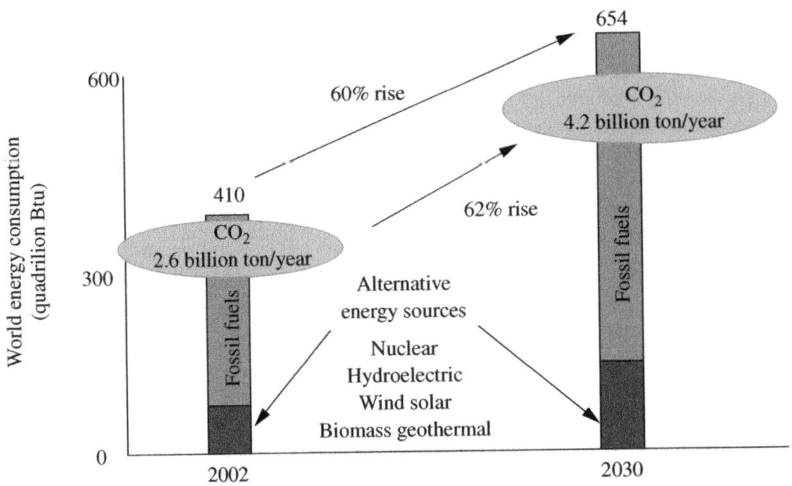

Figure 1.6 Potential CO_2 rise during 2002–2030 without remedial measures [4]

1.3.1 Global Warming Effects

As mentioned previously, GHG cause the earth's atmosphere to accumulate solar heat and raise its temperature, which may be by a few degrees in 100 years. Climate scientists are trying to model the climate system (extremely complex) and predict the atmospheric temperature rise by extensive simulation studies on supercomputers. Figure 1.7 shows global warming over the past hundred years (1900–2000) and its future projection for the next hundred years (2000–2100) by various research agencies throughout the world. The large error in the projection is due to the inaccuracy of the climate model, which

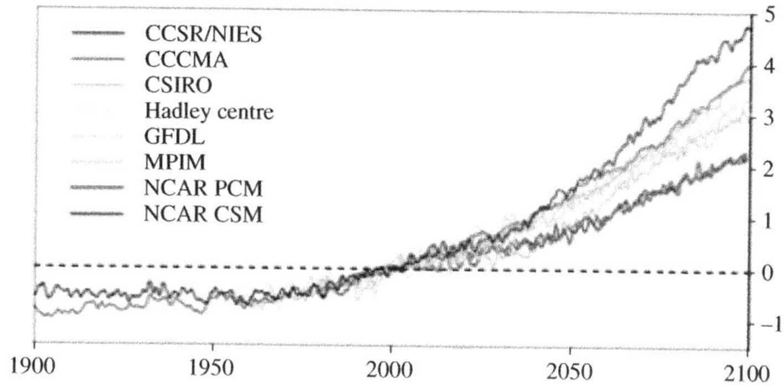

Figure 1.7 Global warming projections by different research agencies [3]

is extremely complex. However, note that none of the studies argue that there will not be any global warming.

The most harmful effect of global warming is the melting of the ice in the Arctic, the Antarctic, Greenland, the Himalayas and a large number of glaciers around the world. In fact, the study indicated that 500 000 mi^2 of ice in the Arctic ocean melted in the year 2006 alone, which is much higher than the computer simulation studies by climate scientists. This is causing a great deal of confusion to climate scientists. The melting of ice is raising the levels of the oceans, with possible flooding to low-lying areas. It is estimated that about 100 million people of the world live within 3 ft of sea level. Again, with the predicted rise in the sea level, the estimate is that half of Bangladesh will be submerged in 300 years, which is expected to displace 75 million people. Again, the prediction is that should all the ice in Greenland and Antarctica melt, the sea level will rise by 200 ft. The Arctic region is expected to be free of ice (in the summer) by 2070.

The melting of the polar ice caps is destroying the habitats of polar bears and penguins, with the expected future extinction of these species. Sensitive corals in the sea are dying due to higher water temperatures and the acidity of dissolved CO_2. The higher acidity of seawater will also adversely affect marine life (which is being studied extensively). Besides the rise in the ocean level, global warming will cause adverse effects to the world climate that will result in severe droughts in tropical countries. This will damage agriculture and vegetation; bring about hurricanes, tornados, heavy rains and floods; and cause the spread of diseases. These effects of global warming have been very evident in the United States in 2012. According to a UN estimate, if fossil fuel burning is completely stopped today, the ocean level will rise by nearly 5 ft over the next 1000 years. These climate change effects will result in serious unrest and instability in the world.

Considering these serious consequences, the UN Kyoto Protocol (an international agreement) was developed in 1997. According to this treaty, the participating countries are required to have a mandatory emission reduction within an assigned quota. If a country cannot meet the quota, it can purchase credits from a country that has a lower than assigned emission quota (called cap-and-trade policy). For example, the United Kingdom, with a higher than quota emission, can purchase credits from Brazil, which has an emission below the quota. Again, within a country a quota is assigned to each industry, which can buy or sell credits from another industry to satisfy the quota. The same principle is also applicable to individuals or families. There are trading organizations that control the buying and selling of credits, which are similar to transactions of shares. European countries took the leadership for the implementation of the Kyoto Treaty, but unfortunately the implementation is not being very successful in recent years. The United States did not sign up to the mandatory reduction formula of the Kyoto Treaty, although unofficial attempts are being made to reduce GHG emissions.

1.3.2 Mitigation of Global Warming Problems

The possible solutions or mitigation methods to global warming problems can be summarized as follows:

- Promote all energy consumption in electrical form. Centralized fossil fuel power stations can effectively use emission control strategies.
- Cut down or eliminate coal-fired power generation. Otherwise, develop clean coal technology with CO_2 capture and underground sequestration, or an integrated gasification combined cycle (IGCC) process. As mentioned earlier, coal-based electricity generation is decreasing recently with a corresponding increase in natural gas-based generation that is cheaper and cleaner.
- Increase nuclear power. Nuclear power has the usual safety and radioactive waste problems. (The trend toward increasing nuclear power is reversing after the recent Fukushima Daiichi nuclear power plant accident in Japan.)
- Since trees absorb CO_2, preserve rain forests throughout the world and promote widespread forestation.
- Control human and animal populations since they exhale GHG. In addition, a larger population means greater energy consumption. This method is not possible to implement.
- Extensively promote environmentally clean renewable energy systems.
- Replace internal combustion engine (ICE) vehicles by EVs and HEVs.
- Promote mass transportation, particularly electric trains and buses.
- Save energy by the more efficient generation, transmission, distribution and utilization of electricity, which is the goal of the future smart grid (discussed later).
- Finally, prevent energy wastage. Energy consumption should be economized to simplify life styles. Almost one-third of the total energy consumption in the world can be saved by this method.

There are, of course, a few beneficial effects of climate change. These are the exploration of offshore Arctic oil and gas, opening of shipping routes through Arctic Ocean in the summer, recovery of new land and the economy of space heating in cold climates.

1.4 Impact of Power Electronics on Energy Systems

Now let us discuss why power electronics is so important today with regard to energy systems, particularly from the viewpoint of mitigating global warming problems.

1.4.1 Energy Conservation

The improved efficiency of power electronics-based energy systems has been discussed previously. The saving of energy gives financial benefit, particularly where the energy cost is high. In addition, lower energy consumption means reduced generation, which mitigates environmental pollution or global warming problems. Power electronic control is obviously more efficient because of the switching mode operation of power semiconductor devices. According to EPRI of the United States, nearly 60–65% of grid-generated energy in the United States is consumed by motor drives and 75% of these are pumps, fans and compressor-type drives for industrial applications. For these applications, the conventional method of flow control is by throttle or damper opening, where a coupled induction motor runs at a constant speed. This method causes a lot of energy wastage due to the fluid vortex. In such applications, a power electronics-based variable-frequency motor speed control with fully open throttle can save around 20% of energy under a light load. In addition, a converter-fed machine's efficiency can be improved further by flux programming control at light load. Power electronics-based load-proportional speed control of air conditioner/heat pump can save energy by up to 20%. For example, the cost of electricity is high in Japan. For this reason, most Japanese homes use variable-speed air-conditioning to save energy. The extra cost of power electronics is recovered within a short period. One popular application of power electronics in recent years is variable-frequency multi-megawatt drives in diesel–electric ship propulsion, which can save a considerable amount of fuel compared with that of the traditional diesel–turbine drive. The flexibility of equipment installation in a ship is an additional advantage. It is estimated that about 20% of grid energy in the United States is consumed in lighting. Power electronics-based compact fluorescent lamps (CFLs) are typically four times more efficient than incandescent lamps, besides having a 10 times longer life. Some countries in the world, such as Australia and Cuba, have already banned incandescent lamps. In the United States, they will be banned in the near future. Currently, solid-state LED lamps are being increasingly used, they consume 50% less energy than CFLs and have a five times longer life, but are more expensive. High-efficiency induction and microwave cooking also saves a lot of energy. The smart or intelligent grid of tomorrow will extensively use state-of-the-art power electronics, computers, and communication technologies, and will permit optimum resource utilization, economic electricity to customers, higher energy efficiency, higher reliability and improved system security. It has been estimated that the widespread efficiency improvement in power electronics and other methods with existing technologies can save 20% of the global energy demand. Unfortunately, approximately one-third of global energy is simply wasted (particularly in the United States) because of energy being very cheap and people being affluent and have bad consumer habits. Eliminating such wastage by regulation or price increases can have a significant impact on climate change problems.

1.4.2 Renewable Energy Systems

It was mentioned earlier that renewable energy resources [7–9], such as hydro, wind, solar, biofuels, geothermal, wave and tidal powers, are environmentally clean (called "green energy") and abundant in nature, and are therefore receiving very strong emphasis throughout the world. *Scientific American* recently published a paper by two Stanford University professors [10], which predicts that renewable energies alone (with adequate storage) can supply all the energy needs of the world. Another study by the UN IPCC reports that 50% of total world energy can be met by renewable resources by 2050. After the recent nuclear accident in Japan, both Japan and Germany are heavily emphasizing renewable energy. Wind and solar resources, which are heavily dependent on power electronics for conversion and control, are particularly important to meet our growing energy needs and mitigate global warming problems.

1.4.2.1 Wind Energy

In wind energy systems, the energy from the wind is converted into electricity through a generator that is coupled to a variable-speed wind turbine. Many old installations in the world use fixed-speed, cage-type induction machines. Figure 1.8 shows the torque developed in wind turbine as a function of speed at different wind velocities. If the wind velocity remains constant and the turbine speed is increased, the torque increases, reaches a peak value and then decreases. Superimposed on the family of curves is a set of peak constant power curves that are tangential to the respective torque–speed curves. For a particular wind velocity, the turbine speed can be varied so that it can deliver the maximum power (maximum aerodynamic efficiency) by maximum power point tracking (MPPT) control. The torque follows the traditional square-law characteristics with speed and the output power follows the cube law, as indicated in the figure. Figure 1.9 shows a typical variable-speed wind power generation system with an induction generator and two-sided PWM converters. The generator excitation is supplied by the PWM rectifier in order to maintain the excitation current or flux constant (or programmable, as desired). The vector control within the speed control loop (not shown) controls the excitation and active currents. The speed is programmed with the wind velocity by an MPPT control. The variable-frequency variable-voltage power is then converted to a constant voltage and frequency through a DC-link converter system before feeding to the grid. The line-side converter is responsible for maintaining the constant DC-link voltage,

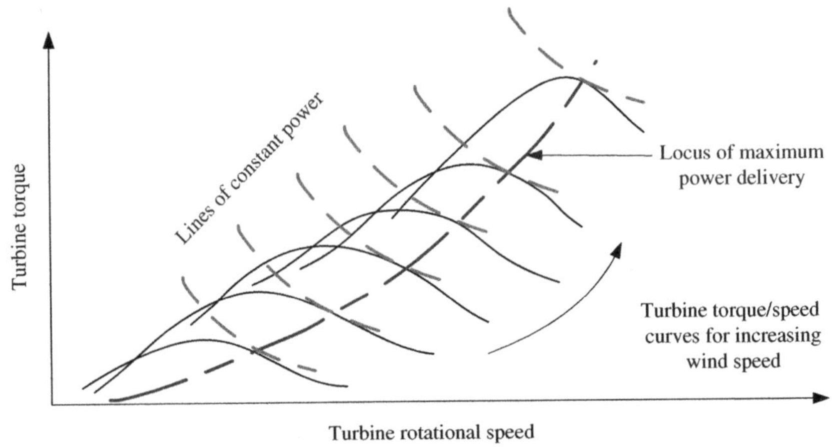

Figure 1.8 Family of torque–speed characteristics for a variable-speed wind turbine at different wind velocities

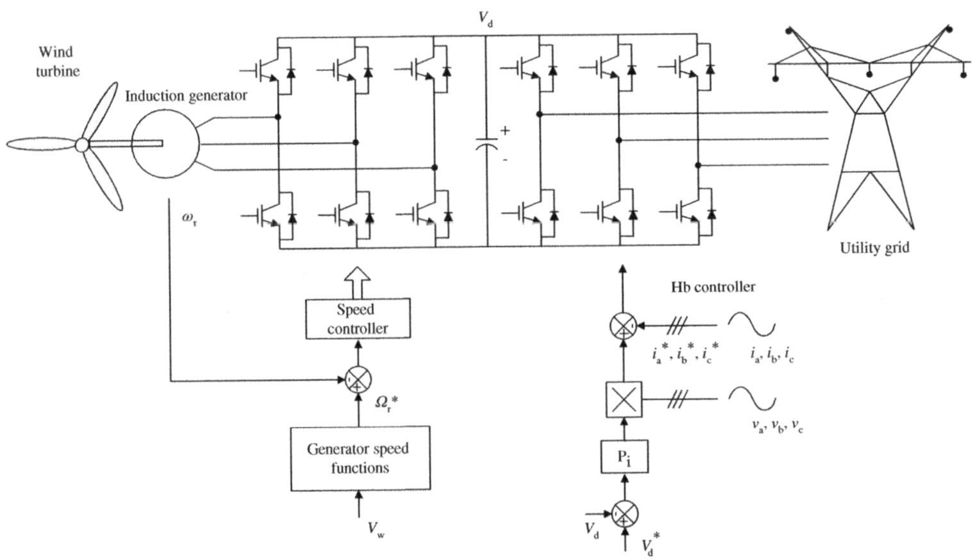

Figure 1.9 Typical wind power system with two-sided PWM converters

as shown. A simplified control using hysteresis-band current control is shown, although synchronous current control can also be used. A Scherbius type of drive with a doubly fed induction generator (DFIG) and slip power control can be used. Alternatively, a permanent magnet synchronous generator (PMSG) with diode rectifier/boost DC–DC converter/PWM inverter or a two-sided PWM converter system can be used.

The world has enormous wind energy resources, and currently they are the most economic "green" energy. Stanford University estimates that only 20% of the available wind energy can meet all the electricity needs of the world. Recent technology advances in variable-speed wind turbines, power electronics, and machine drives have made wind energy very competitive, almost equal to that of fossil fuel power. Wind and PV energy are particularly attractive to the one-third of the world's population that lives outside electric grids. Among the developing countries, for example, China and India have large expansion programs for wind energy. Currently, in terms of percentage energy consumption, Denmark is the leader with 25% wind energy, which is expected to rise to 40% by 2030. In terms of installed capacity, China is now the leader and the United States occupies second place (close to Germany and Spain), with a total penetration of barely more than 3% (this is because the US energy consumption is the highest in the world). The United States has the ambitious goal of increasing it to 20% by 2030. One drawback of wind energy is that its availability is sporadic in nature and may require backup power from fossil or nuclear power plants. Of course, surplus wind energy at high wind can be stored (storage will be discussed later) for lean-time utilization. Offshore wind farms generally offer higher energy output than onshore farms, although their installation and maintenance are more expensive.

1.4.2.2 PV Energy

PV devices (crystalline or amorphous Si, CdTe and copper indium gallium selenide) directly convert sunlight into electricity. The generated DC is then converted to AC and fed to the grid, or used in isolated load. In an autonomous system, there is usually a battery backup energy source. Figure 1.10 shows a typical configuration of a PV system, where a PV string consists of series–parallel combination of cells. The generated DC voltage is boosted by a DC–DC converter before inversion to AC by a PWM sinusoidal inverter. The DC–DC converter controls the maximum power output of the string by MPPT control. The channels are paralleled on the AC side and boosted in voltage by a transformer before connecting to the AC grid. A multilevel cascaded H-bridge (CHB) transformerless connection has also been proposed.

The PV devices have the advantages that they are static, safe, reliable, environmentally clean (green) and hardly require any repair and maintenance like wind power systems. However, in the current state of the technology, PV energy is more expensive than wind energy (see Figure 1.14), but it truly depends on the installation cost and utilization factor. The cost of power electronics is less compared to that of the PV cell cost. Although, currently, PV energy is more expensive than that of solar thermal, with the

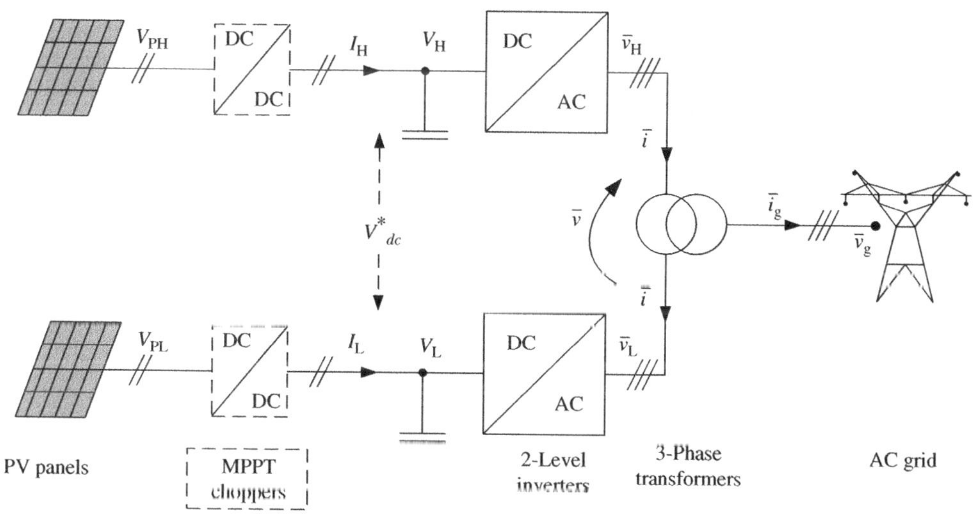

Figure 1.10 Typical configuration of photovoltaic system

present trend of aggressive research the price is falling sharply to be more competitive in the future. PV energy is expected to experience a significant expansion around the world. The IEEE asserts the ambitious prediction that by 2050 PV will supply 11% of the global electricity demand. With the present trend, it may not be surprising that PV power may exceed that from the wind in the future. The lifetime of a PV panel is typically 20 years, and conversion efficiency is typically 16% with the commonly used thin-film amorphous silicon. PV power has been extensively used in space applications (where cost is not a concern). Currently their applications are expanding to rooftop installations, off-grid remote applications and grid-connected installations. Japan is very determined with regard to PV research and applications, because it does not have much indigenous energy resources, and there energy is expensive. The recent accident at the Fukushima Daiichi nuclear reactors is now putting Japan's shift of emphasis from nuclear to PV resources. Recently, China has been manufacturing PV cells very economically. Unfortunately, like wind, PV power is also statistical in nature and therefore requires backup energy sources or bulk storage of surplus power. The recent smart grid concept can shift the power demand curve to match the available power curve, as will be discussed later. Currently there are ambitious plans to explore solar energy from the African deserts such as the Sahara and the Kalahari through extensive PV installations, and then tying this to the European grid through HVDC transmission.

1.4.2.3 Wave Energy

Currently wave and tidal energies (often called ocean energy) [11] are receiving an emphasis, although these resources are considerably small. Wave energy tends to explore energy from the ocean surface waves that are induced by wind. The technology is yet in an infant stage. The ocean surface goes through up-and-down oscillatory movements due to waves, and the mechanical energy is converted into electrical energy by a wave energy converter (WEC). There are more than 1000 patents for WEC. One popular type is the Pelamis converter, shown in Figure 1.11, which was installed offshore Scotland (2004). In this scheme, a number of cylindrical sections of the Pelamis converter (anchored at the bottom of the ocean) are linked by hinged joints, which produce the up-and-down movement in order to pump oil by piston that drives a hydraulic motor. The motor is connected to a permanent magnet (PM) synchronous generator. The generated AC of an irregular waveform is rectified to DC, and a number of units are series-connected to boost the DC voltage. The DC is then inverted to AC for local supply or boosted by transformer for grid connection.

However, the total available wave power is low and expensive (typically \$4–\$15/W), but the cost per kilowatt-hour is generally low. Wave power has the advantage that it is consistent and predictable, unlike wind and PV, but is vulnerable to bad weather conditions. The first experimental wave farm was installed in Portugal in 2008, with a capacity of 2.25 MW. The United States does not have wind power except

Figure 1.11 Principle of wave energy by Pelamis wave energy converter [11]

for a small installation in Hawaii. Recently Oregon State University has been taking a lot of research initiative in this area.

1.4.2.4 Tidal Energy

Tidal energy [12] is another form of renewable energy that explores energy from tides due to the earth–moon gravitational attraction and the earth's rotation. Lunar attraction has a 12-h cycle period, i.e., every 12-h period the water level reaches a peak level. Tidal energy can be classified into three different types, that is, tidal stream generator, barrage generator and dynamic tidal power. The world's first commercial stream generator (1.2 MW capacity), which was installed in Ireland in 2007, is shown in Figure 1.12. The scheme uses a submerged axial flow water turbine (like a wind turbine) that rotates slowly in one direction and generates electricity through the coupled synchronous generator. In a tidal barrage scheme, the water is stored in a pond by a barrage at high tide. At low tide, the low potential head of the pond water drives a water turbine-synchronous generator to generate electricity. The world's largest tidal barrage power station (254 MW) was installed in South Korea in 2011. The dynamic tidal power combines generation from up-and-down water flows across the barrage. The construction cost of the barrage and pond is high, and therefore exploration of this type of energy becomes expensive. The United States does not have any tidal power plants.

1.4.2.5 Geothermal Energy

For the completeness of renewable energy sources, geothermal energy [13] will be covered. Geothermal energy is extracted from the thermal energy stored and generated within the earth. The evidence of this energy is volcanos, hot springs and so on. In the latter case, rain water seeps into the earth and is heated by geothermal energy. The stored energy comes from the earth and was created by the sun, and the generated energy comes from the radioactive decay of minerals within the earth. The geothermal energy source is limited and originates from the regional geothermal heat belts of the earth. There are three types of geothermal energy: one is direct use (or district heating system), where the energy flows naturally on earth's surface (an example is a hot water spring). Another is geothermal heat pumps for home heating, where water is injected in a horizontal or vertical coil of pipes to absorb heat. In Iceland, for example, 93% of homes are heated by geothermal energy. The third is geothermal electricity, where

Figure 1.12 Principle of tidal energy by tidal stream generator (1.2 MW plant in Ireland) [12]

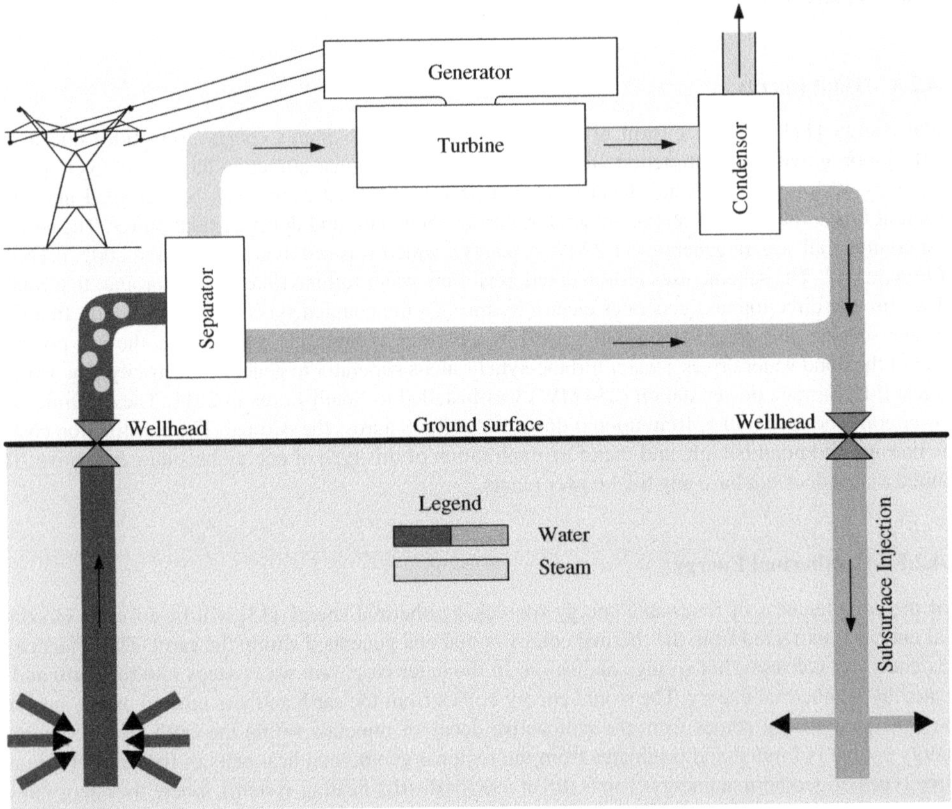

Figure 1.13 Principle of geothermal electricity by flash steam plant

water injected through a pipe transfers geothermal heat, and the generated steam runs a turbo-generator like a conventional steam power plant. There are three types of geothermal electricity: (1) dry steam plant, (2) flash steam plant and (3) binary cycle plant.

Figure 1.13 shows the principle of geothermal electric power generation by a flash steam plant (most commonly used). In this method, water is injected in a pipe (may be 1 or 2 miles deep) and steam and hot water under pressure are collected in another pipe. The water is separated from the steam through a separator and recirculated and steam drives a turbine-generator system to generate electricity, as shown in the figure. In a dry steam plant, water is fully converted to dry steam, whereas in a binary cycle plant, hot water exchanges heat to a low boiling point fluid to convert it to steam for driving a turbine. Geothermal power is economic, reliable and always available unlike wind energy and solar energy. The cost is typically \$3.00/W and \$0.10/kWh. The worldwide production of geothermal energy is 10.7 GW (2010), and the United States is the world leader (3 GW). Interestingly, Iceland generates 100% electrical energy from renewable sources, of which 70% is by hydroelectric means and 30% is geothermal.

1.4.2.6 Cost Comparison of Renewable Energy Sources

Figure 1.14 shows a cost comparison of different renewable energy sources in a decreasing order of cost, but excludes wave and tidal energy. Solar PV is the most expensive, and solar thermal is very close

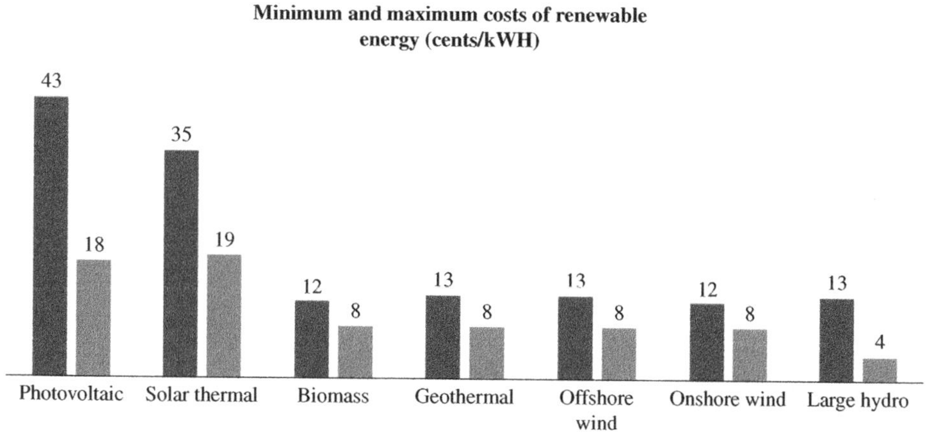

Figure 1.14 Cost comparison of renewable energy sources

to it. The minimum cost of PV is shown as 18 cents/kWh, whereas the maximum cost is 43 cents/kWh. The cost depends on various factors, such as the cost of materials, land and labor, the length of day, weather conditions and so on. Although, currently the cost of solar thermal installation is lower than that of PV, it is expected that eventually with cost reduction PV usage will be much higher than solar thermal. The minimum cost of biomass, geothermal and offshore and onshore winds is almost the same, but hydroelectricity is the cheapest (4 cents/kWh). Offshore wind installation and maintenance costs are much higher than that of onshore installation, but electricity generation is also higher because of the strong and consistent winds, which balance the high cost. The cost of power electronics and grid connection is not included in the figure.

1.4.2.7 FC Energy

An FC is an electrochemical device, in which hydrogen is the gaseous fuel that combines with O_2 to produce electricity and water. FC stacks can be considered equivalent to series-connected low-voltage batteries. The DC voltage generated by the FC is boosted by a DC–DC converter, and then converted to AC by an inverter for AC power supply or other applications. The FC has the properties of high output resistance and sluggish transient response. H_2 can be generated from water by electrolysis (reverse process of FC) or from hydrocarbon fuels (gasoline and methanol) onboard through a reformer. An FC can be defined as a clean energy source if H_2 is generated from a clean energy source. FC is safe and static, and has a high efficiency (typically 54%). FC varieties are defined by the types of their electrolytes. These can be classified as proton exchange membrane FC (PEMFC), phosphoric acid FC (PAFC), direct methanol FC (DMFC) and solid state FC (SOFC). All are commercially available, but PEMFC is the most economic with a high power density and low temperature (60–100°C), and therefore is important for FC-based electric cars. Although it is bulky and expensive in the present state of the technology, intensive R&D has recently reduced the cost of FC dramatically. Figure 1.15 shows the principle of an FC-based EV, and also summarizes the different methods of H_2 generation. In an FC vehicle, a PEMFC generates the DC power, which is boosted by a DC–DC converter (not shown) and then converted into variable-frequency variable-voltage power to drive the AC motor of the vehicle. Since FC cannot absorb the vehicle's regenerative power, a battery or ultracapacitor (UC) is needed at the FC terminal (through another DC–DC converter (not shown)). The battery/UC also supplies power during acceleration because of the sluggish FC response. The H_2 fuel is supplied from a tank, where it can be stored in the form of cryogenically cooled liquid or compressed gas. H_2 is usually generated from water using electricity from

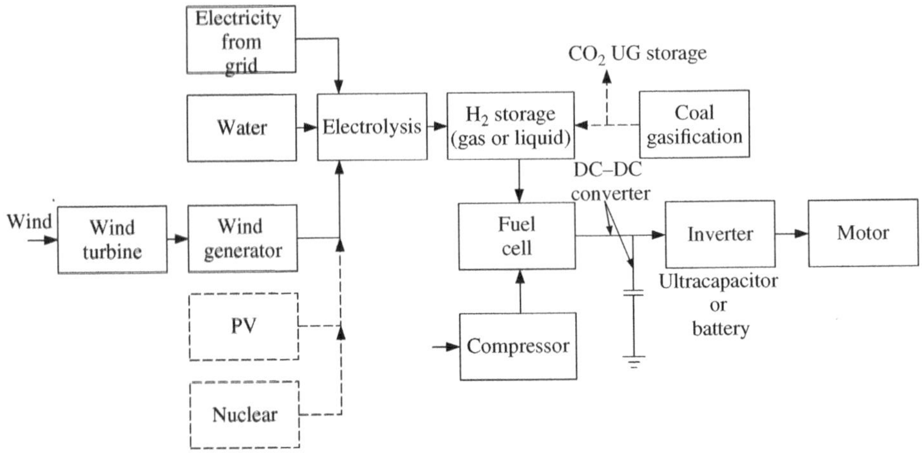

Figure 1.15 Fuel cell-based electric vehicle showing H_2 generation methods

the grid, or from environmentally clean sources such as wind, PV, or nuclear. The O_2 for the FC is obtained from air through a compressor.

1.4.3 Bulk Energy Storage

As mentioned earlier, renewable energy sources such as wind and PV are statistical in nature, because of their dependence on weather conditions (and time of the day), and therefore require bulk storage of surplus energy to match the load demand curve (which is also fluctuating) on the grid. On the other hand, fossil and nuclear power plants (and also hydroelectricity) can easily adjust their generation to match fluctuating consumption patterns, and therefore storage may not be essential. However, bulk storage may be desirable for fossil and nuclear plants, considering extra energy losses due to no load or part-load operation.

Grid energy storage [14, 15] methods can be briefly summarized as follows:

1.4.3.1 Pumped Storage in Hydro Power Plant

Hydroelectric generators can be used as motor pumps to pump water from a "tail" reservoir to a "head" reservoir and store it at high level using off-peak economic grid energy. During peak demand, the head water runs the generators, as usual, to supply the demand. It is possibly the cheapest method of energy storage, but is applicable with favorable site facilities. Otherwise it may be expensive. The typical cycle energy efficiency may be 75% and cost may be less than \$0.01/kWh. Currently, there is over 90 GW of pumped storage facility around the world.

Figure 1.16 shows an example of the 400 MW Scherbius drive for a variable-speed hydroelectric generator and pump storage system (the world's largest) of the Kansai Power Co. of Japan [16]. Normally a synchronous machine at constant speed is used as the generator.

In a hydroelectric power system, the water head in the upstream reservoir normally varies widely. It can be shown that with variable head, if a constant speed machine is replaced by a variable speed drive (speed increasing with head), an efficiency improvement of as much as 3% can be obtained. In the

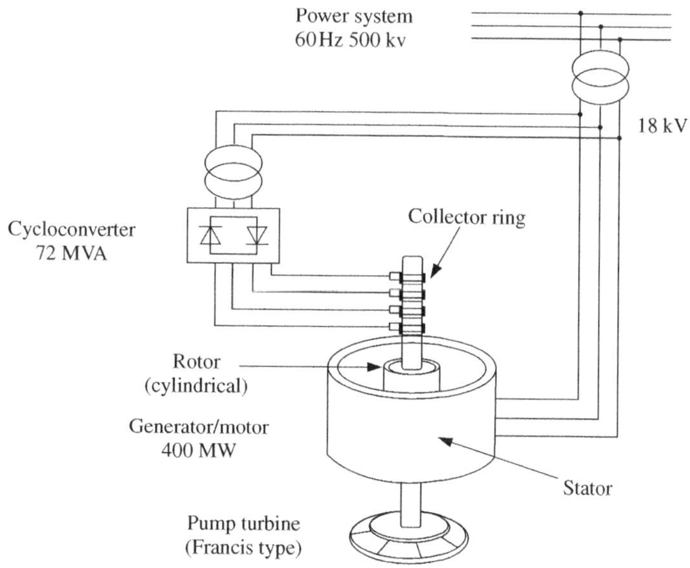

Figure 1.16 400 MW Scherbius drive for variable-speed hydroelectric generator and pump storage system [16]

daytime, the drive runs as a generator feeding power to the 500 kV grid. The DFIG speed varies from sub-synchronous (330 rpm) to super-synchronous mode (390 rpm) as the water head increases. During night time, when utility demand is low and surplus grid power is available, the machine runs as a variable speed motor to pump water from the tail to the head storage. In pumping mode, the range of speed is the same. Note that the lagging RKVA of the cycloconverter can be canceled by the leading reactive kilovolt amperes (RKVA) of the machine stator so that the total power factor can be unity. In a modern drive, the cycloconverter can be replaced by a two-sided PWM converter system.

1.4.3.2 Battery Storage

The battery has been the most common form of energy storage in the grid. In this method, electrical energy from the grid is converted to DC and stored in a battery in chemical form. Then, the stored energy is retrieved through the same converter system to feed the grid. Although very convenient, with a high cycle efficiency (typically 90%), battery storage is possibly the most expensive (typically >$0.1/kWh). Historically, lead-acid batteries have been used extensively, but recently NiCd, NaS, Li-ion and vanadium redox flow batteries are finding preference. For example, GE installed a 10-MVA lead-acid battery storage in the Southern California Edison grid in 1988. The world's largest battery storage was installed by ABB at Fairbanks, Alaska, in 2003, which uses a NiCd battery with a capacity of 27 MW for 15 min. Flow batteries have a fast response and can be more economic in large-scale storage, but the "chemical plant" is complex with continuously circulating electrolyte by pumps.

Figure 1.17 shows the battery storage system in the Southern California Edison grid [16]. Basically, it is an 18-step GTO-based phase-shifted square-wave converter system with harmonic neutralization. The system fabricates an 18-step output voltage wave by using three groups of H-bridges, where the second and third groups have a 20° and 40° phase lead, respectively, with respect to the first group. The transformer secondary connections with 15 segments fabricate the 18-step voltage wave to neutralize the harmonics (17th, 19th, etc.) and boost the output voltage, and the capacitor bank filters the

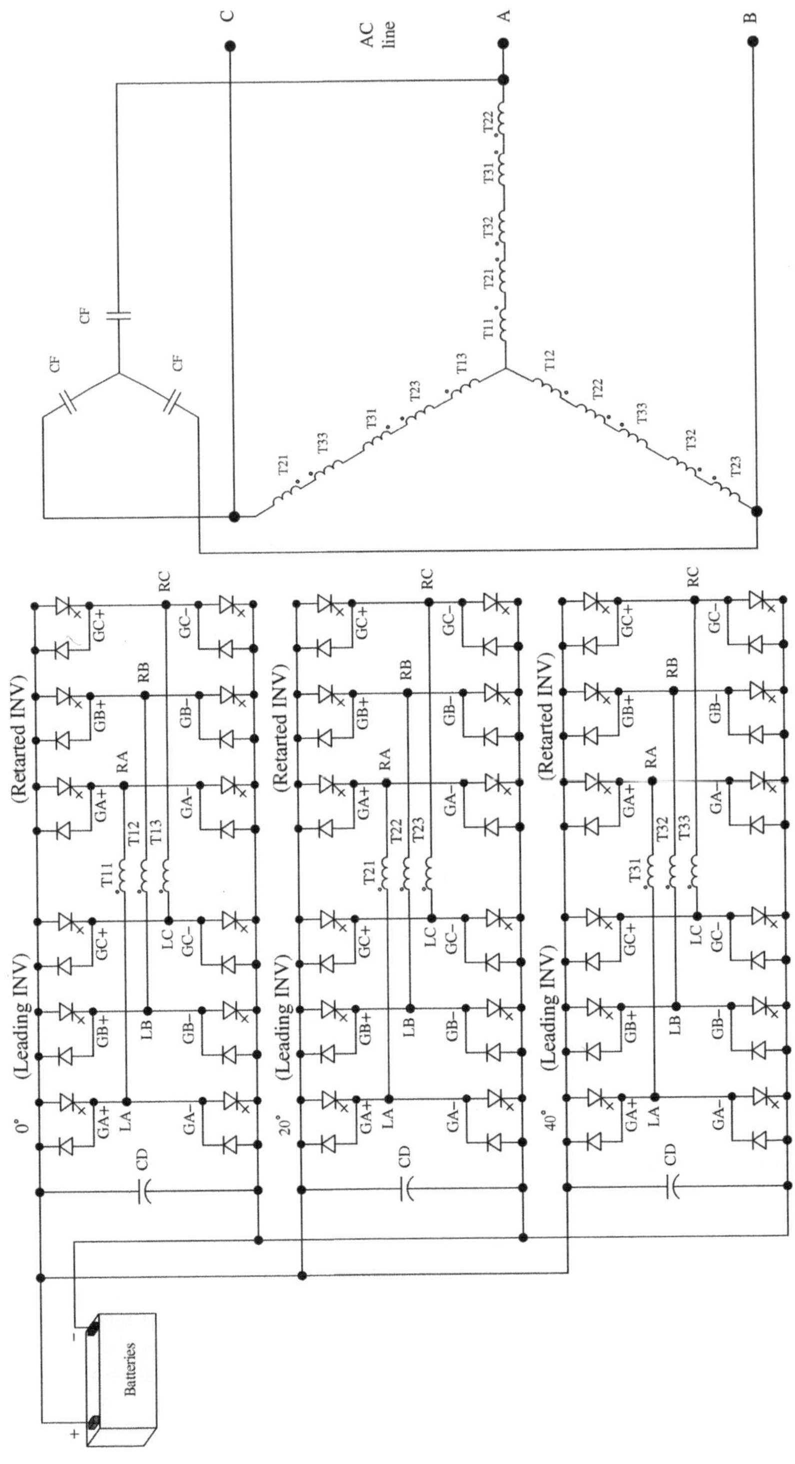

Figure 1.17 10 MVA utility system battery storage

harmonics. The converter system can operate as a SVC, static var generator (SVG), or static synchronous compensator (STATCOM) (leading or lagging) that can regulate the bus voltage and stabilize the power system. In modern installations, multilevel converters can be used.

1.4.3.3 Flywheel (FW) Storage

In this form of storage, electrical energy from the grid is converted into mechanical energy through a converter-fed drive system that charges an FW, and then the energy is retrieved by the same drive operating in generating mode. The FW can be placed in a vacuum or H_2 medium, and active bearing can be used to reduce the energy loss. Steel or composite material can be used in FW to withstand the high centrifugal force due to the high speed. FW storage is more economical than battery storage ($0.05/kWh) and has been used before, but mechanical storage has the usual disadvantages. The drive system in Figure 1.16 has been used to construct an FW storage system.

1.4.3.4 Superconducting Magnet Energy Storage

In the superconducting magnet energy storage (SMES) method, grid energy is rectified to DC that charges an SMES coil to store energy in magnetic form. Then the energy is retrieved by the reverse process. The coil is cooled cryogenically so that dissipation resistance tends toward zero, and then the energy can be stored indefinitely. Either liquid helium (0°K) or a high-temperature superconductor (HTS) material in liquid nitrogen (77°) can be used. The cycle efficiency can be higher than 95%. This type of energy storage is very expensive.

1.4.3.5 UC Storage

This is an energy storage device like an electrolytic capacitor (EC), but its energy storage density can be as much as 100 times higher than that of EC. UCs are available with a low-voltage rating (typically 2.5 V), and capacitor values can be very high (up to several thousand farads). The units are connected in series–parallel for higher voltage and higher capacitance values. However, the watt-hour per kilogram of UC is low compared to that of a battery. The power density (W/kg) of UC is very high, and a large amount of power can cycle through it without causing any deterioration. It is a good temporary storage device in an FC car, as shown in Figure 1.15. UCs are presently very expensive for bulk grid energy storage.

1.4.3.6 Vehicle-to-Grid Storage (V-2-G)

This is a new method for bulk energy storage, assuming that a large number of EV batteries are plugged into the grid. An EV can sell electricity to the grid during peak demand and then charge a battery during off-peak hours. V-2-G technology can be used, turning each vehicle with its 20–50 kWh battery pack into a distributed load-balancing device or emergency power source. However, charge–discharge cycles can shorten the battery life.

1.4.3.7 H_2 Gas Storage

H_2 can be used as a bulk energy storage medium, and then used in FC or burnt as fuel in an ICE. This idea has generated the recent concept of hydrogen economy, i.e. H_2 as the future clean energy source. H_2 can be generated from abundantly available renewable energy sources like wind and PV, and stored as compressed or liquefied gas as a high density fuel. Of course it can also be generated from hydrocarbon fuels with underground sequestration of undesirable CO_2 gas. Bulk H_2 production using biomass and its underground storage in caverns, salt domes and depleted oil and gas fields are now being investigated. The overall energy efficiency of H_2 storage cycle may be 50% to 60%, which is lower than that of battery or pumped storage systems.

1.4.3.8 Compressed Air Storage

Compressed air storage is another grid energy storage method, where electricity is used to compress air through a variable speed drive and store it underground. When the electricity demand is high, compressed air is heated with a small amount of natural gas and then burnt in a gas turbine to generate electricity. The scheme has been used in Europe.

1.5 Smart Grid

Currently, in the United States and some European countries there is a tremendous stimulus of R&D activities in smart or intelligent grid [17]. Unfortunately, our present power grid system is too old, obsolete, and inefficient and presents inadequate protection for faults. What is a smart grid? A smart grid is basically an advanced electric power grid of tomorrow, integrating the advanced state-of-the-art technologies of power electronics, computer and communication. The objectives of a smart grid can be summarized as follows:

- Optimum resource utilization
- Economic electricity distribution to customers
- Higher energy efficiency
- Higher system reliability
- Higher system security

The basic concept of the smart grid is explained in Figure 1.18. The smart grid will integrate distributed renewable energy systems, such as wind, PV, etc., along with bulk power generation plants (hydro, nuclear, and fossil fuel power sources). The future power grid will also integrate bulk energy storage systems, such as pumped storage, battery, FW, UC, H_2, etc., because renewable resources are fluctuating in nature, and due to the fluctuating nature of consumer demand. One of the most important

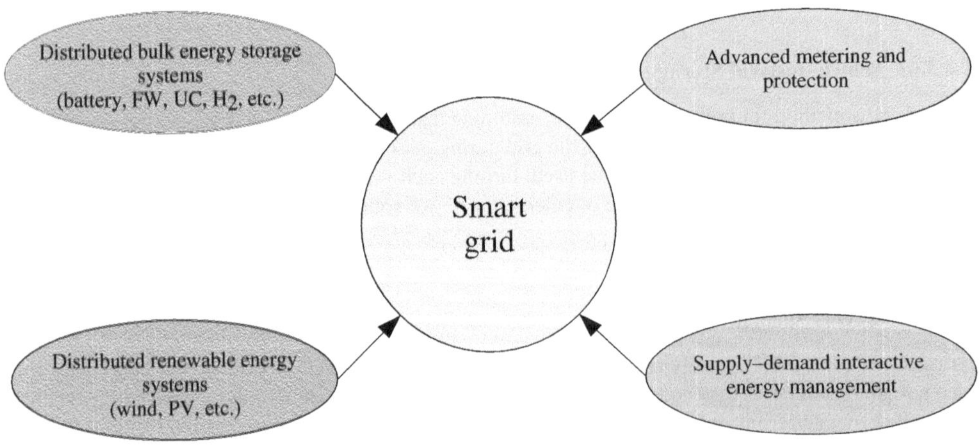

Figure 1.18 Concept of smart grid

elements of the smart grid is its supply–demand interactive energy management. Normally, in a grid the energy generation follows the fluctuating consumer's demand. This causes surplus generation capacity in off-peak hours and a shortage of generation at peak demand hours. Using advanced smart metering, the demand curve can be controlled (demand-side energy management) to match the available generation capacity. This means that some loads (such as EV battery charging, operation of washer/dryer, dish washer) can be programmed to operate in off-peak hours, with the incentive of lower tariffs to consumers. Ideally, if the demand curve matches the available supply curve and there is no fault in the system, no bulk storage or spare system capacity is needed. This results in optimum resource utilization. In a smart grid, when demands are established at certain grid points, optimum generation of the resources and the electricity flow paths can be programmed for the most efficient supply of electricity to the consumers. Of course, bus voltages and system frequency should always be maintained within tolerance limits. A smart grid can use power electronics-based HVDC transmission, FACTS, STATCOM, UPS, etc. to achieve the objectives mentioned above. Smart grids will help us gradually transition to a carbon-free society.

1.6 Electric/Hybrid Electric Vehicles

The shortage of oil and environmental pollution problems are the main reasons for the worldwide R&D activities in EV/HEV over more than the last three decades. Of course, replacing ICE-based vehicles by EV/HEV also helps solve the climate change problem.

Figure 1.19 shows a typical EV drive system [18], where a battery (lead-acid 204 V) is the energy storage device. The DC from the battery is converted to variable-frequency variable-voltage AC by a PWM inverter to drive an AC motor (induction or synchronous). The particular drive system (GE-ETXII) using interior permanent magnet (IPM) synchronous machine (NdFeB, 70 hp) and vector control, was developed by the author while employed at GE. The IPM motor has better efficiency, a reduced size and higher field-weakening speed control range. In EV, the braking energy is easily regenerated to recharge the battery, which improves the energy efficiency.

In a HEV, the battery is the energy storage device. This is assisted by a power device which is usually a gasoline ICE. While range is the main problem in pure EV, there is no such problem in HEV. Figure 1.20

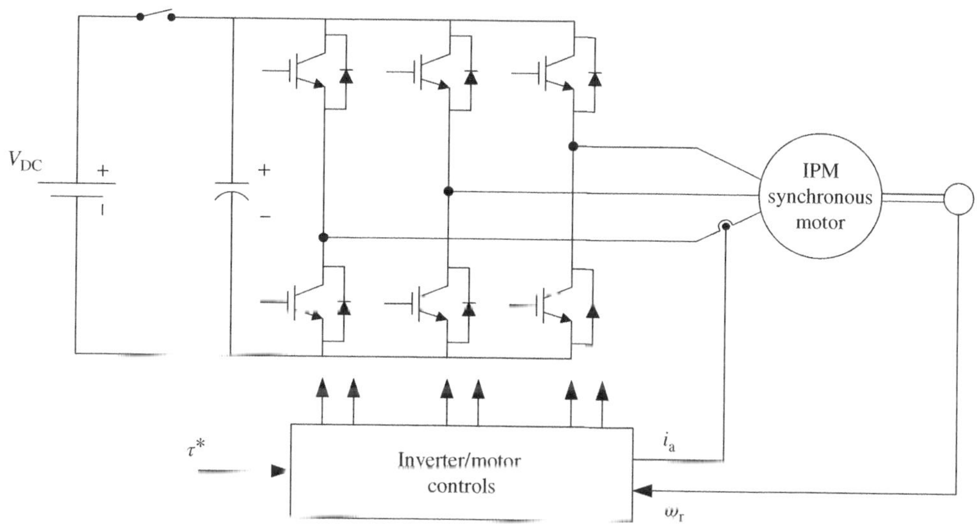

Figure 1.19 Electric vehicle drive with IPM synchronous motor (GE-ETXII)

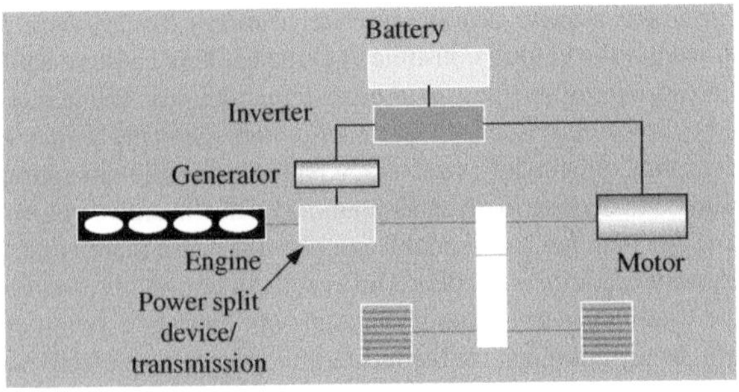

Figure 1.20 Hybrid electric vehicle drive system for Toyota Prius II (2004)

is an example configuration of the HEV drive of the popular Toyota Prius II. The axle drive power can be supplied either by an IPM synchronous motor (IPMSM) (50 kW, 500 V) or engine (57 kW at 5000 rpm), or both. The motor gets the drive power from a Ni–Mh battery (201.6 V, 1.2 kWh, 21 kW) through an IGBT inverter, which also gets the braking power from the motor. The battery has a boost DC–DC converter (not shown) to raise the DC voltage. There is no EV driving mode, but the battery assists motor acceleration and hill climbing, and (partially) absorbs the braking energy. It is charged by the engine/generator (non plug-in).

The power electronics and motor drive technology for EV/HEV is somewhat mature, with reasonably low cost. However, battery technology is not yet mature. The battery is expensive and bulky, with a large weight and limited cycle life, and charging takes several hours. Although Ni–Mh batteries have been extensively used, more recently improved Li-ion batteries are becoming available on the market. The latter has a higher storage density than the former, but is more expensive. It appears that Li-ion battery is the battery of the future, and currently there is strong emphasis on its research in the United States. HEVs are expected to disappear from the market as battery technology matures, when economic EVs with long range become available.

Currently, a number of EVs and HEVs are commercially available on the market. Among the HEVs, currently the Toyota Prius II is the most popular. Soon, a plug-in (chargeable from outlet) version with a Li-ion battery will be introduced that will give a pure EV mode of driving. Currently, Tesla Roadstar in the United States sells an EV (Li-ion battery, 215 kW, 3.5 h charging, 245 mile range) at a very high price. The battery life is typically 100 000 miles. Some examples of recent introductions to the market are the Nissan Leaf and Chevy Volt, both of which use a Li-ion battery. Leaf is a pure EV with a 100 mile range, the Volt is an HEV, for which the battery can either be charged by the ICE or from an outlet (plug-in). In pure EV mode, the range is only 40 miles, but this can extend to 360 miles with ICE charging of the battery.

1.6.1 Comparison of Battery EV with Fuel Cell EV

The R&D for both battery EV and FC EV are currently progressing simultaneously, and it is worth making some comparisons between the two types of EV. Figure 1.21 presents this comparison [19] in terms of the present technology for mass production with identical (300 mile) range, assuming that both deliver 60 kWh to the drive wheels. The battery EV is assumed to receive charging from wind energy (although currently it is mostly from coal or nuclear), which is required to supply 79 kWh with a power line efficiency of 92%, battery charging efficiency of 89%, battery efficiency of 94% and drive train efficiency of 89%, as shown in the figure. Typically, 6 kWh of regenerated energy is assumed in this calculation. The total energy efficiency of battery EV is calculated as 68%. The estimated cost of the

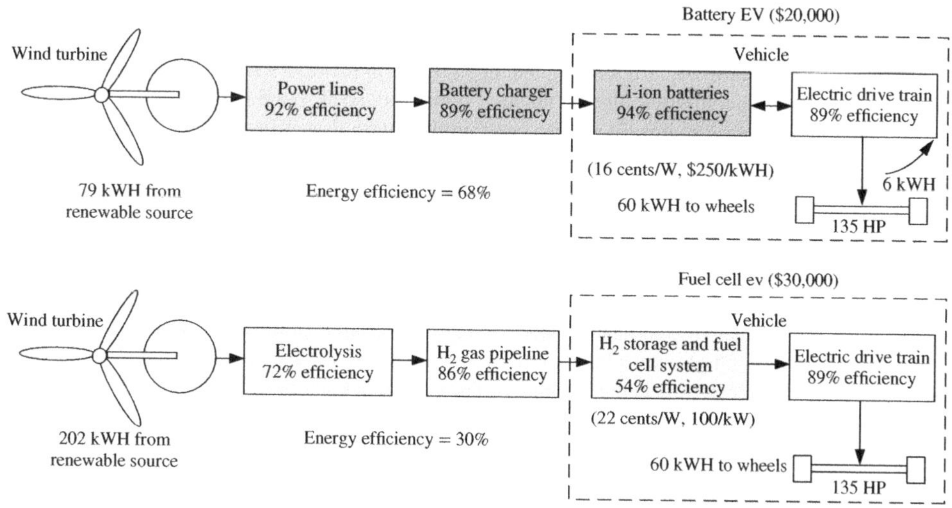

Figure 1.21 Comparison of battery EV with fuel cell-based EV (300 miles range)

vehicle is shown as \$20 000 with a battery cost of \$0.16/W and \$250/kWh. The FC EV is also assumed to obtain its energy supply from wind turbines. Considering the efficiency figures of the FC EV line, the total energy efficiency is only 30%, which means that 202 kWh needs to be supplied from wind turbines. The corresponding cost figures for FC EV are indicated in the figure. Note that the auxiliary storage of the FC EV has been ignored for simplicity. Therefore, its cost figure is somewhat optimistic. In summary, the FC EV is 38% less efficient, has 43% more weight and is 50% more expensive compared to that of the battery EV. Due to these disadvantages, FC EV research has recently been considerably reduced in the United States.

1.7 Conclusion and Future Prognosis

This chapter gives a brief but comprehensive review of the world's energy resources and climate change problems due to fossil fuel burning, along with possible solutions or mitigation methods. Then, it discusses the impact of power electronics on energy conservation, renewable energy systems, bulk storage of energy and electric/hybrid vehicles in the present century. Although tidal and geothermal powers may not use power electronics, these are included for the completeness of the discussion on renewable energy systems.

Power electronics has now emerged as a major area in electrical engineering and considering the saturating trends of the technology, the present century will find significant emphasis on its applications. It is now tending to integrate with the "classical" power engineering domain. It appears that the role of power electronics in the future will be as, if not more, important and versatile as computers and information technology today. It also appears that computers, information technology, power electronics and power systems will integrate, creating a complex interdisciplinary technology. This trend is becoming evident in the emergence of smart grid technology.

What are the future trends of power electronics technology? As the technology is maturing, we will find an increasing emphasis on application-oriented R&D in converter system modularization, system modeling, system analysis, computer simulation, design and experimental evaluations. This trend is already evident in recent literature publications. In general, some advances and trends in power electronics can be mentioned as follows. In power semiconductor devices, IGBT has now emerged as the dominant device in medium- to high-power applications, whereas power MOSFET is universal in low-power high-frequency applications. The IGCT is tending to lose competition with IGBT in the high power area. The traditional silicon-based bipolar junction transistors (BJTs) and GTOs are already obsolete and thyristors indicate a trend toward obsolescence in the future. Large bandgap devices (such as SiC and GaN) are expected to bring major evolution in the usage of power electronics, particularly in high power for large drives and utility system applications. SiC-based Schottky diodes and power MOSFET half-bridges (with freewheeling diodes) at higher power are already available on the market. In fact, SiC MOSFETs with voltage ratings up to 6 kV, when available, will make most Si-based power devices obsolete. High-voltage high-power SiC MOSFET (up to 10 kV), IGBT (up to 25 kV), GTO (up to 40 kV), junction barrier Schottky (JBS) diode and P-I-N diode (up to 10 kV) are presently in the R&D stage and their emergence will create a significant impact on high-power applications of power electronics. Currently, the replacement of high-power bulky 60-Hz transformers by solid-state high-frequency-link power transformers with SiC devices is being attempted. The emerging GaN-on-Si power devices have all the advantages of SiC devices, but show significant potential for cost reductions.

Power quality and lagging displacement power factor (DPF) problems are gradually making the classical phase-controlled power electronics, which reigned for so long, obsolete and promoting PWM line-side active converters. Of course, active harmonic filters (AHF) and static var compensators tend to mitigate the problems of classical power electronics with diodes and thyristors. In the author's view, harmonic problems will be nonexistent in future power electronic systems and thus AHF will become obsolete. Among all the classes of converters, the voltage-fed type will become universal, replacing the present current-fed and cycloconverter classes. The Z-source class of inverters, which are recently showing a lot of visibility in the literature, is expected to have a doubtful future. Multilevel voltage-fed converters are showing increasing popularity in high-voltage high-power utility systems and drive applications. CHB topology has the advantages of modularity and fault-tolerant applications. Traditional matrix converters (also known as Venturini converters) using AC switches have been on and off many times since their invention in the 1980s and in the author's view its future promise appears to be bleak. The space vector modulation (SVM) technique is becoming increasingly popular over sinusoidal PWM and currently there is a trend toward SVM algorithm simplification for multilevel converters. The traditional resonant-link-based soft-switched converters for motor drives and other high-power applications have evidently lost all their promise, despite their technology evolution for a prolonged period of time. The future emphasis on converters will be mainly on modularization and system integration – similar to the trend of (very large-scale integration (VLSI) technology of microelectronics. Power electronics will evidently play a very important role in the smart grid, as discussed previously. With the increasing emphasis on distributed renewable energy sources and the installation of bulk energy storage devices, maintenance of system frequency and bus voltages with optimum resource utilization, economic electricity supply to consumers, high system energy efficiency, high system reliability and fault-tolerant operation, extensive system studies will be required.

With regard to electrical machines, although nearly a mature technology, incremental research will continue on performance optimization, precision parameter estimation, fault diagnosis and fault-tolerant control. With increasing energy costs, permanent magnet synchronous motors (PMSMs) (with NdFeB magnet) will find increasing acceptance, although they are more expensive than induction machines. IPMSM is principally more attractive because of its field-weakening capability for extended speed operation and efficiency improvement (by flux programming) under light load operation in a constant torque region. If magnet costs become sufficiently low, PMSMs will dominate over induction motors in general industrial applications. Axial flux PMSMs are expected to find applications in direct drives, particularly in wind generation systems and electric vehicles. Again, for very high-power applications, wound-field synchronous motors (WFSMs) will remain popular. In the author's view, switched reluctance motor (SRM) drives do not show any future promise in the majority of industrial applications and has a trend toward obsolescence. The majority of electrical machines are expected to have converters in the front end in the present trend of decreasing converter cost and modularization, and integrated machine-converter-controller (particularly in lower end of power) remains a definite trend. Again, slip power recovery drives that use DFIG with slip rings and reduced converter ratings for partial range speed control range will tend toward obsolescence. Currently, they are popular in wind generation systems. Among all the drive control techniques, the scalar control techniques (including direct torque control) will tend to become obsolete and vector control will emerge as the universal controller. The cost difference in the complex vector drive and simple scalar control is barely noticeable because only the software is more complex in the former, whereas the control hardware essentially remains the same. MATLAB/Simulink-based simulation, particularly real-time simulation with hardware-in-the-loop (HIL), is recently receiving more emphasis. Although the sensorless vector drive is already commercially available, zero-speed (or zero-frequency) precision speed or position estimation remains a challenge because of the need for machine saliency, complex signal processing with externally injected signal and precision machine parameter estimation problem. This estimation is more complex for induction machines compared to that of PMSM (which has built-in saliency). However, zero-frequency sensorless PMSM drives have recently been used in commercial drives. Online fault diagnosis and the fault-tolerant control of converter-machine systems is an important R&D area, which will have increasing emphasis in the future. With the present trend for digital signal processor (DSP) and field programmable gate array (FPGA) (or application-specific integrated circuit (ASIC)), the single chip control of sensorless vector drive with fault-tolerant control is not far away. As artificial intelligence technology matures, intelligent control and estimation (particularly based on neural networks) will find increasing acceptance in power electronic systems. With the maturing DSP and FPGA technologies, the predictive control of power electronic systems based on plant models and system variables with the well-known developed theory are showing a comeback for enhanced system performance. Finally, R&D in FCs, PV cells, batteries, passive circuit components, high-temperature superconductivity (HTS), DSPs and ASIC chips will significantly impact on power electronics evolution in this century.

References

1. Bose, B.K. (2011) Energy scenario and impact of power electronics in 21st. century. Proceedings of Qatar Workshop of Power Electronics for Industrial Applications and Renewable Energy Conversion, held in Doha, Qatar on November 3–4, 2011, pp. 10–22.
2. Bose, B.K. (2013) Global energy scenario and impact of power electronics in 21st century. *IEEE Transactions on Industrial Electronics*, **60**, 2638–2651.
3. Global Warming http://en.wikipedia.org/wiki/Global_warming (accessed 20 September 2011).
4. Bose, B.K. (2010) Global warming. *IEEE Industrial Electronics Magazine*, **4**, 1–17.
5. Bose, B.K. (2010) Energy, global warming and power electronics. Proceedings of the 4th National Power Electronics Conference, Roorkee, India, June 11, 2010.
6. Roth, J.R. (1995) *Long Term Global Energy Issues in Industrial Plasma Engineering*, vol. **1**, Institute of Physics Publishing, Philadelphia, PA.
7. IPCC Summary for Policy Makers (2011) Special Report Renewable Energy Sources, UAE, May 5–8, 2011.
8. Simoes, M.G. and Farret, F.A. (2004) *Renewable Energy Systems*, CRC Press, Boca Raton, FL.
9. Farret, F.A. and Simoes, M.G. (2006) *Integration of Alternate Sources of Energy*, John Wiley/IEEE Press, Piscataway, NJ.
10. Jacobson, M.Z. and Delunocchi, M.A. (2009) A path to sustainable energy by 2030. *Scientific American*, **282**, 58–65.
11. Ocean Energy Regulatory Development, Policy and Guidelines, http://ocsenergy.anl.gov/guide/wave/index.cfm (accessed 10 July 2011).
12. Tidal Power http://en.wikipedia.org/wiki/Tidal_power (accessed 20 September 2011).
13. Geothermal Energy http://en.wikipedia.org/wiki/Geothermal_energy (accessed 22 June 2011).
14. Grid Energy Storage http://en.wikipedia.org/wiki/Grid_energy_storage (accessed 10 June 2011).
15. Vazquez, S., Lukic, S., Galvan, E. *et al.* (2010) Energy storage systems for transport and grid applications. *IEEE Transactions on Industrial Electronics*, **57**, 3881–3895.

16. Bose, B.K. (2006) *Power Electronics and Motor Drives – Advances and Trends*, Academic Press, Burlington, MA.
17. Ipakchi, A. and Albuyeh, F. (2009) Grid of the future. *IEEE Power & Energy Magazine*, **7** (2), 52–62.
18. Bose, B.K. (1988) A high performance inverter-fed drive system of an interior permanent magnet synchronous machine. *IEEE Transactions on Industry Applications*, **24**, 987–997.
19. Eaves, S. and Eaves, J. (2003) A cost comparison of fuel-cell and battery electric vehicles. *Journal of Power Sources*, 24–30.
20. Bose, B.K. (2000) Energy, environment and power electronics. *IEEE Transactions on Power Electronics*, **15** (4), 688–701.
21. Wu, B., Lang, Y., Zargari, N., and Kouro, S. (2011) *Power Conversion and Control of Wind Energy Systems*, John Wiley & Sons, Ltd/IEEE Press, Piscataway, NJ.
22. Teodorescu, T., Liserre, M., and Rodriguez, P. (2011) *Grid Converters for Photovoltaic and Wind Power Systems*, John Wiley & Sons, Ltd/IEEE Press, Piscataway, NJ.
23. Bose, B.K. (2009) Power electronics and motor drives – recent progress and perspective, *IEEE Transactions on Industrial Electronics*, **56** (2), 581–588.

2

Challenges of the Current Energy Scenario: The Power Electronics Contribution

Leopoldo G. Franquelo, Jose I. Leon and Sergio Vazquez
Department of Electronic Engineering, University of Seville, Seville, Spain

2.1 Introduction

The current energy scenario is becoming increasingly complex, and the need for an intelligent grid is ever more urgent. This has introduced several challenges to be solved, such as the interconnection of different sub-grids, the requirement of maintaining power quality above some limits, the improvement of grid stability, the integration of clean energy generation and the use of efficient high power and energy storage systems (ESSs).

Currently, the power grid is changing continuously, being adapted to fulfill minimum requirements in terms of nominal power, stability and reliability [1–3]. The increasing demand for energy by industry and housing requires the commissioning of new power plants or the purchase of energy from neighboring regions or countries. This necessitates the implementation of a highly meshed decentralized energy distribution grid, which allows the interconnection of power systems with different natures and from different regions. Multiple interconnections between different grids are currently required but this leads to some challenges, such as the stability and reliability of the overall energy generation and distribution system.

In addition, hazardous nuclear waste, the effects of global warming and CO_2 emissions from the burning of fossil fuels are factors against the installation of such power systems, which drives the development of new sources of energy generation. The main objective of these new systems is to generate electricity efficiently with a minimum impact on waste. In this sense, renewable energy sources have contributed in a fundamental way at the expense of introducing a stochastic behavior into the power grid owing to the inconsistent nature of the energy source. Thus, the integration of new renewable energy sources into the grid (mainly wind power and photovoltaic (PV) plants during recent decades) has led to the development of a highly decentralized distributed energy system with a large number of energy generation plants. The modern distributed structure of a power grid is shown in Figure 2.1. This structure presents a highly meshed network, medium-to-high integration of renewable energies and an adequate connection

Power Electronics for Renewable Energy Systems, Transportation and Industrial Applications, First Edition.
Edited by Haitham Abu-Rub, Mariusz Malinowski and Kamal Al-Haddad.

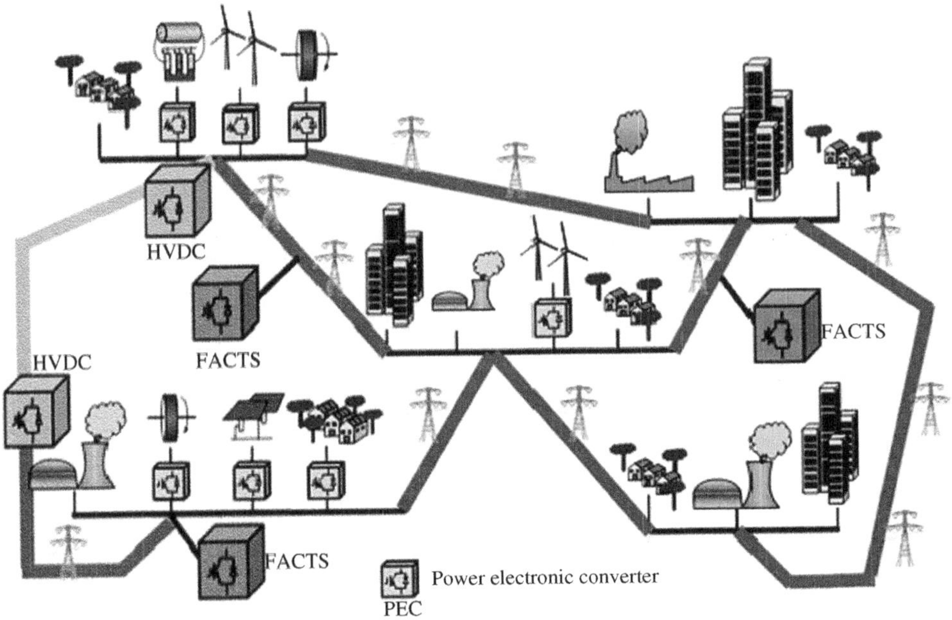

Figure 2.1 Modern scenario of the electrical grid

between different sub-grids. In this chapter, the concept of a new distribution grid is presented. This new structure solves the problems typical of the classical power grid through the extensive use of power electronic converters.

Power electronic converters offer efficient solutions that can be applied to the new smart distributed energy grid concept. This chapter reviews briefly the current scenario where power electronic converters are being applied. The focus moves toward energy transportation and distribution systems, renewable energy sources, the transportation sector and ESSs, reviewing the different alternatives and topologies already in existence as industrial products or still as lines of research in academia.

2.2 Energy Transmission and Distribution Systems

Demand for electrical energy has grown in such a way that it is pushing the existing energy transmission and distribution systems to their limits. In this scenario, new challenges have appeared to electrical and electronics power system engineers. In order to ensure static and dynamic power system stability, new technologies have been adopted or are being developed. These new systems include flexible AC transmission systems (FACTS) devices and high-voltage DC (HVDC) transmission lines.

2.2.1 FACTS

FACTS are defined by the IEEE as "Alternating Current Transmission Systems incorporating power electronics-based and other static controllers to enhance controllability and power transfer capability" [4]. Over the years, FACTS devices have been used conventionally in transmission and distribution for solving steady-state control problems in the power system, such as voltage regulation and the optimization of

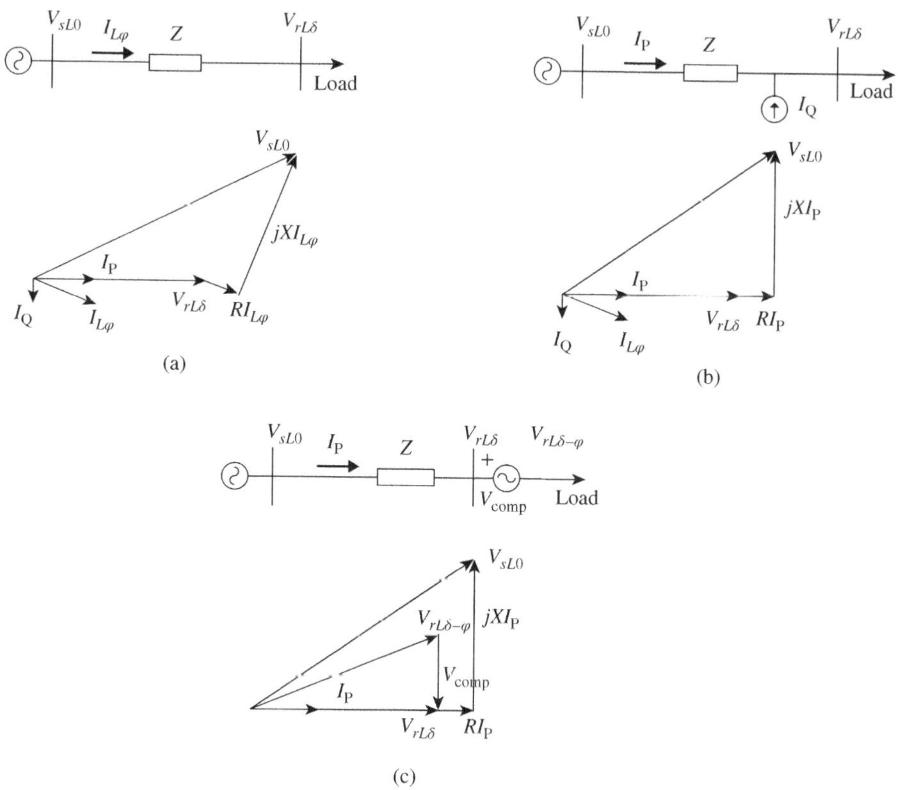

Figure 2.2 Reactive power compensation principles: (a) system without compensation, (b) shunt compensation principle, (c) and series compensation principle

power loading in transmission lines by power flow control over specific transmission routes and the enhancement of transfer capability of the lines. However, new developments in power semiconductors and converter topologies have led to the use of FACTS controllers for solving dynamic stability problems of the power systems, in addition to the static stability and power flow control tasks [5–8].

Figure 2.2 shows the effects of shunt and series reactive power compensation in a basic transmission line. The system without compensation together with its phasor diagram is represented in Figure 2.2a. The power system is formed by a voltage source (V_s), the equivalent transmission line modeled with an impedance ($Z = R + jX$) and an inductive load. From this phasor diagram, it is clear that the line current (I) can be reduced or minimized only if the active current (I_P) flows through the transmission line. The main benefits are a reduction of power losses in the transmission line, an improvement of voltage regulation at load terminals and an increase in the active power transmission capability through the transmission line. The reactive power compensation can be performed by using passive components (capacitors or reactors) or, more recently, by using voltage source converters (VSCs). However, shunt compensation can be represented by a current source connected at the load terminals, as shown in Figure 2.2b and series compensation can be described by using a voltage source added between the line and the load, as shown in Figure 2.2c. The shunt compensation is performed by the FACTS device by supplying the reactive current (I_Q) near the connection point of the load. On the other hand, the series compensation is accomplished by the FACTS device by modifying the equivalent impedance of a power line and, therefore, by changing the voltage at the load side [9].

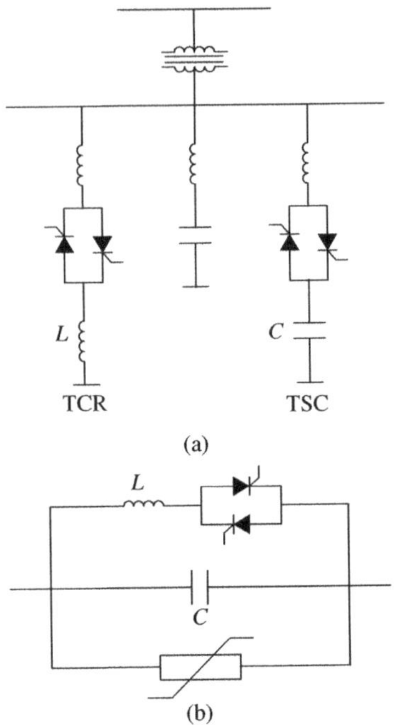

Figure 2.3 Thyristor-based FACTS devices: (a) SVC and (b) TCSC

Regarding the power semiconductor technology used for their implementation, FACTS devices can be classified as thyristor-based and gate-controlled-based var compensators [9]. The thyristor-based var compensators use fixed capacitor and reactor banks working together with thyristor switches in shunt or series circuit arrangements. The switching of the power semiconductor device is controlled in order to achieve variable reactive power exchange, which includes static var compensators (SVCs) formed by groups of thyristor-switched capacitors (TSCs), thyristor-controlled reactors (TCRs) and thyristor-controlled series capacitors (TCSCs). The schematics of these devices are shown in Figure 2.3 [9]. These FACTS devices can be considered a mature technology because they have been analyzed and improved in real power system applications since their introduction. For instance, it is possible to find references to them from as early as 1971 [10].

The gate-controlled-based var compensators use VSCs built with gate-controlled power switches, such as gate turn-off (GTO) thyristors, insulated-gate bipolar transistors (IGBT), or integrated gate-commutated thyristors (IGCT). The firing pulses of the power semiconductors are defined in order to control the amount of reactive power generated or absorbed by the power converter. In this way, they can provide capacitive or inductive behavior without requiring large banks of AC capacitors or reactors; therefore, there is no necessity for extra devices. The most common VSC-FACTS are the static synchronous compensator (STATCOM), the static synchronous series compensator (SSSC), the dynamic voltage restorer (DVR) and the unified power flow controller (UPFC). Representations of these devices are shown in Figure 2.4.

The VSC-FACTS devices are technology in development. One of their most significant advantages compared with thyristor-based compensators is the reduction of size and cost because of the reduction or elimination of large passive components. In addition, their most interesting characteristic is that

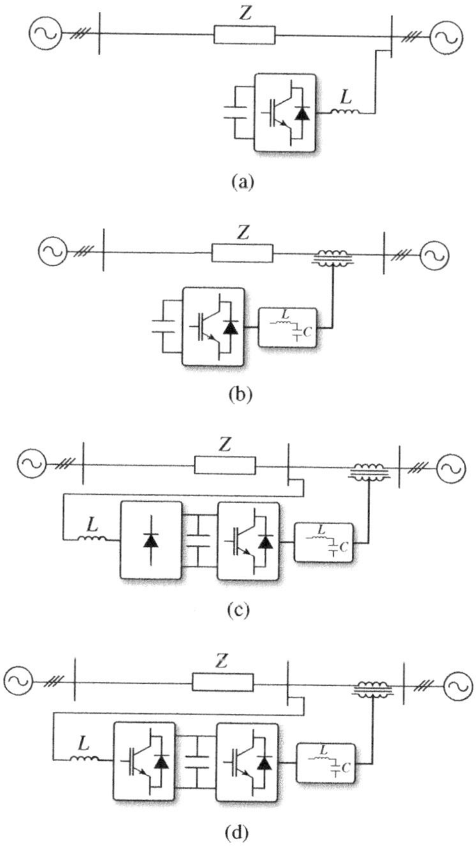

Figure 2.4 VSC-FACTS devices: (a) STATCOM, (b) SSSC, (c) DVR and (d) UPFC

VSC-FACTS present faster dynamic response compared with their thyristor-based counterparts. Thus, VSC-FACTS devices can be applied to existing problems in the power system that cannot be solved by using conventional technology, such as dynamic stability problems, or low-voltage ride-through (LVRT) capability for wind turbines or PV plants [11, 12]. Much research effort has been expended on establishing a proper power converter topology for each application and on developing high-performance control algorithms and modulation techniques suitable for these applications [13]. Despite this, several challenges still require study in order for this technology to achieve maturity. For instance, the connection of VSCs to high-voltage systems is not straightforward owing to the limited reverse voltage blocking capability of the current silicon-based switches (GTO, IGBT, or IGCT). In the near future, switches based on wide-bandgap materials could alleviate this problem. However, when using the available silicon-based switches, the problem can be solved by using multilevel power converter topologies, which appear to be the most suitable solution [14–16]. In addition, grid connection of VSC-FACTS is not a trivial issue from the perspective of control and modulation strategies. Thus, new modulation techniques have appeared to deal with the problem of DC-link voltage balance for multilevel converters [17–19], and new control algorithms have been proposed to achieve precise active and reactive power command tracking under balanced or unbalanced input grid conditions [20–23].

2.2.2 HVDC

HVDC transmission systems can be considered a mature technology because the first commercial installation appeared in 1954 [24]. Although AC transmission lines are most commonly used, HVDC systems offer several advantages over AC electrical energy transmission under certain conditions. From the technical point of view, the following issues can be identified:

- The amount of power that can be transferred in an HVDC system is constant with distance, whereas in high-voltage AC (HVAC) lines, the power transmission capability is reduced as the length of the line increases.
- It is not possible to connect directly two AC systems with different fundamental frequencies.
- When economic costs are considered, there is a critical distance beyond which the investment cost for building new HVDC transmission lines is less than that for its AC counterpart. Although this length depends on the rated power of the transmission lines, typical values are 600–800 km for overhead lines and 25–50 km for undersea installations.
- There are environment aspects of HVDC that are also better compared with AC transmission systems, such as lower visual impact and space requirements [25].

HVDC installations include several devices. Among them, one of the most important is the power electronic converter. As for FACTS devices, the power semiconductor technology used for building the power converter can be thyristor-controlled or gate-controlled power switches [26].

Thyristor-based HVDC systems use line-commutated current source converters (CSCs) as the preferred technology, as presented in Figure 2.5a. This technology has been used since the earliest installation and it is considered mature and well established for high power developments, typically over 1000 MW.

Gate-controlled-based HVDC power converters are built using gate-controlled power switches, such as GTO thyristors, IGBT, or IGCT. They adopt force-commutated VSCs as the key technology, as shown in Figure 2.5b. At the current state of development, VSC-HVDC is used for a power range of around 300–500 MW. For this reason, its applications are the in-feed to small isolated remote loads and city centers, and the grid connection of remote small-scale and offshore generation [27].

Depending on the rated power range of the transmission line, different HVDC technologies can be chosen. In addition, the application and location of the transmission system define the most suitable configuration of the HVDC system. Different valid configurations can be described for systems built with CSC-HVDC or VSC-HVDC technologies [25, 26]. The simplest configuration is a back-to-back system, as presented in Figure 2.6a. This is the choice when both HVDC power converters are located in the same substation and when there is no need for a DC-link cable over a long distance [28, 29]. When it

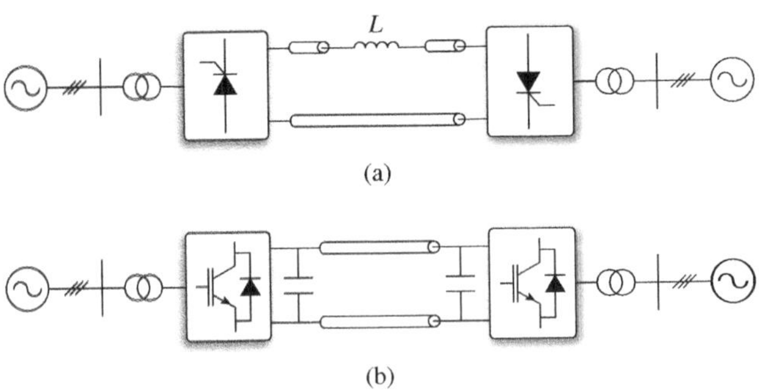

Figure 2.5 HVDC technologies: (a) CSC-HVDC and (b) VSC-HVDC

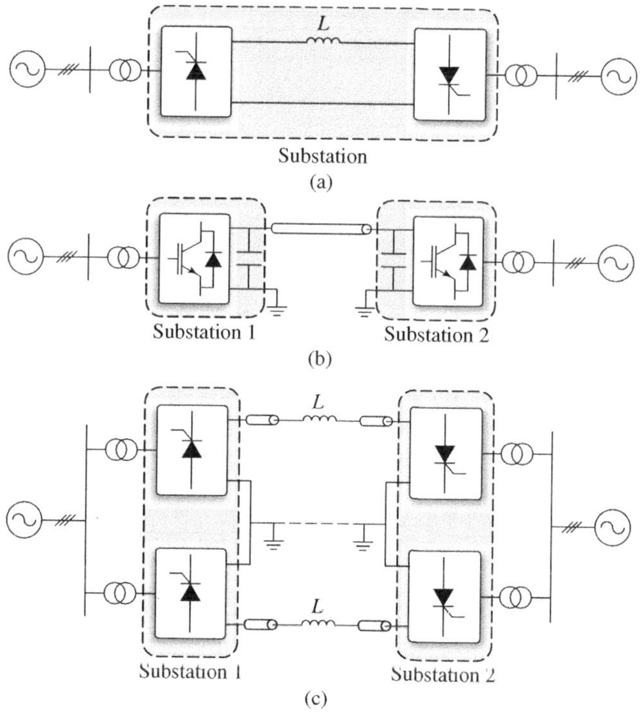

Figure 2.6 HVDC line configurations: (a) back-to-back, (b) monopolar and (c) bipolar

is necessary to transport the energy over a long distance, there are two possible choices: the monopolar and bipolar configurations. In monopolar HVDC systems, the power converters are connected using a single conductor line, as shown in Figure 2.6b. The polarity can be positive or negative and the returned current can be transmitted using the ground or a small metallic conductor. This configuration is very suitable for undersea energy transmission as in the connection of offshore wind farms to the grid [27]. The bipolar HVDC configuration is the preferred choice for the transportation of bulk power. It is the most common configuration used in overhead HVDC transportation lines. Its main advantage is reliability. In this case, the power converters are connected using two conductor lines, as illustrated in Figure 2.6c. Each pole has a positive polarity and a negative polarity; thus, the system can be viewed as the union of two monopolar systems. If one cable runs out of service, the other pole can continue to transmit power independently [30, 31].

Comparing VSC-HVDC with CSC-HVDC, the following characteristics can be noticed:

- In the power converter of the VSC HVDC, the active and reactive power can be controlled indepen dently; therefore, no reactive compensation equipment is needed at the converter stations.
- Using VSC-HVDC, if a suitable converter topology is chosen, the use of converter transformers can be avoided. Much research effort has been expended on analyzing different multilevel converter topologies, such as neutral point clamped (NPC), active NPC (ANPC), or multimodular converter (MMC) [26]. The VSC-HVDC system based on the MMC topology, represented in Figure 2.7, has several advantages over conventional VSC topologies [32].
- VSC-HVDC presents a low short-circuit level; thus, after major blackouts, it can be used to start up weak AC networks that lack generation.

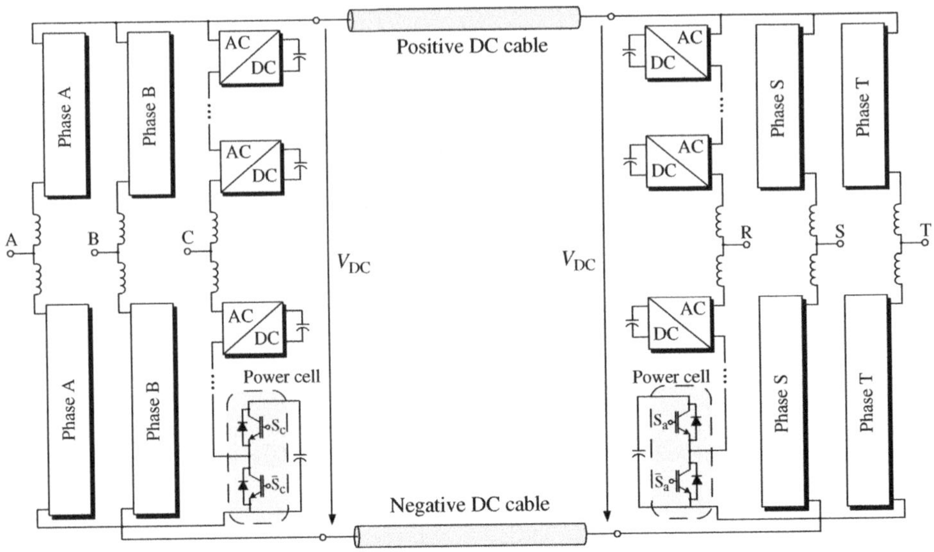

Figure 2.7 VSC-HVDC based on MMC topology

- Using VSC-HVDC, active power in the DC line can be controlled in the full range between 0% and 100%. This a major advantage over CSC-HVDC, which has a minimum power level that can be transferred.
- VSC-HVDC needs smaller AC filters for the elimination of harmonics than CSC-HVDC [25].

2.3 Renewable Energy Systems

The effects of global warming, the necessary reduction in contaminant emissions and the huge dependence on fossil fuels (becoming increasingly expensive) have led, over recent decades, to the development of new energy sources based on renewable systems. In fact, the European Union presented a plan for the period 2011–2020 to establish the lines for the development of renewable energy sources, emissions reduction and energy efficiency. The targets addressed by this plan to be achieved by 2020, usually called 20/20/20, are as follows:

- 20% reduction in greenhouse gases to below the values of 1990
- 20% of energy consumption must be generated by renewable energy sources
- 20% reduction in consumption of primary energy

It must be noticed that in 2010, these objectives were far from being achieved and, thus, remain a significant challenge for the European governments. In fact, in 2010, the share of renewable energy in the gross final energy consumption was 12.5% and the economic crisis in Europe will make it even more difficult for the proposed targets to be achieved.

The main renewable energy sources are as follows:

- Hydroelectric energy
- Geothermal energy
- Wind energy

- Solar energy (PV or thermal)
- Wave energy
- Tidal energy
- Biomass
- Other alternative sources

A classification of the current renewable energy sources is presented in Figure 2.8, where they are classified depending on their nominal power, the usual power converters and corresponding power semiconductors required for the grid integration (if necessary), the availability of the renewable energy source and the corresponding technology trend.

Currently, and steadily gaining in prominence, renewable energy sources play an important role in the energy generation market, even considering the sudden braking effect caused by the global economic crisis. Illustrating this fact, the Energy Information Administration (EIA) (Official Energy Statistics from the US Government) estimates that in 2011 about 10% of the worldwide marketed energy consumption was from renewable energy sources (hydropower, biomass, biofuels, wind, geothermal and solar) with a projection of it reaching 14% by 2035. In addition, the EIA estimates that in 2011 about 19% of worldwide electricity generation was from renewable energy with a projection of it being nearly 23% by 2035. For instance, in Spain, 32.4% of the energy consumed during 2011 was generated by renewable energy sources, constituting the most important energy source of the year.

Among the renewable energy sources, hydroelectric energy plays the most important role, providing an important percentage of the renewable power generation. The contribution of renewable energy sources, such as wind energy, PV energy and wave energy, is currently limited compared with hydroelectric power. However, worldwide wind power capacity increased by 20% in 2011, achieving around 238 GW. China was the leader and accounted for almost 44% of the global market, followed by the United States and India, while Germany remained the largest market in Europe. However, it should be stated that although its market share remained relatively small, the offshore wind sector continued to expand with the use of larger turbines, going into deeper water, further from shore. On the other hand, the use of small-scale turbines is also increasing.

The global total of installed solar PV power was roughly 70 GW at the end of 2011 compared with just 1.5 GW in 2000. Over the past 5 years (2006–2011), solar PV has averaged an annual growth rate of over 50% (for instance, from 40 to 69.3 GW during 2011); however, PV still generates just a small percentage of the total yearly electricity production. Germany is the leader of this energy sector, accounting for almost 36% of the global market with 25 GW installed, followed by Italy (12.7 GW), Japan (4.9 GW), the United States (4.4 GW), Spain (4.2 GW) and China (3 GW). In 2011, the growth has been mostly concentrated in a few countries, such as Italy (from 3.5 to 12.7 GW), China, the United Kingdom, France and the United States. Finally, wave and tidal energies are based on electricity generated from the movement of wave and tidal flows. Tidal and wave energies are more predictable than wind power and usually increase during winter when electricity demand is usually high. However, currently, wave and tidal energies cannot be considered a mature technology and only a few prototypes have been built.

Taking into account the differing nature of renewable energy sources, their stochastic behavior and the requirement of maintaining a minimum power quality, power conversion plays a fundamental role in the integration of renewable energy sources into the electrical grid. On the other hand, the widespread locations of the increasing number of renewable energy plants require specific operational and management procedures in order to maintain, or even to improve, the power reliability, quality and stability of the grid [33].

2.3.1 *Wind Energy*

As has been commented previously, wind energy is nowadays a mature technology, which has become viable in recent decades achieving a non-negligible impact on the energy market [34]. For instance,

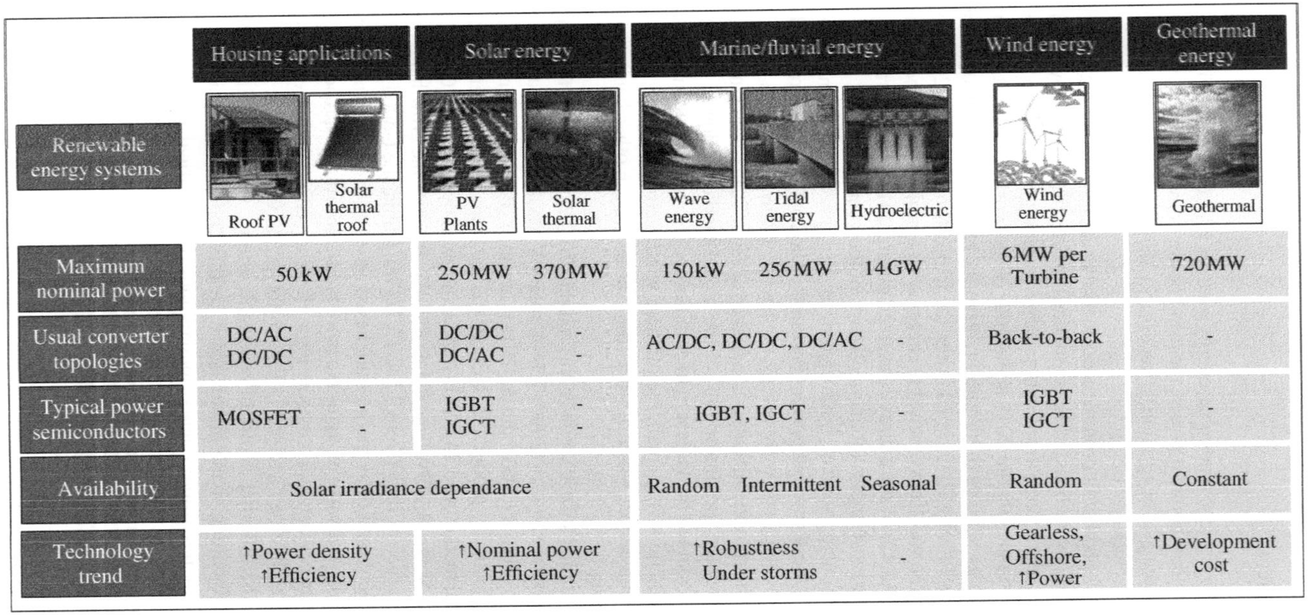

Figure 2.8 Classification of renewable energy sources

in 2010, wind energy integration attained values equal to 21% in Denmark, 18% in Portugal, 16% in Spain and 9% in Germany. However, in China and the United States, although their current wind power integration is around only 1.5%, over the last 2 years the wind power installations in both countries have increased substantially.

From a historical point of view, following the invention of windmills thousands of years ago, the first wind turbines to generate electricity were designed in the 19th century. Already in the 20th century, the first modern wind turbines were designed to work with fixed speed and stall regulated or active stall. In this case, the wind generator is connected directly to the grid through a transformer without using any power converter (only using a gearbox, a soft starter stage, a reactive power compensator and a main switch). This configuration is simple and cheap, but it is not able to extract the maximum power from the wind, suffers mechanical stress in large wind turbines and does not meet current regulations.

In order to solve these problems, in the 1990s, wind turbine technology evolved to operate with variable speed and pitch regulation. Initially, the most successful configuration of this family of wind generators was the doubly fed induction generator (DFIG) with a reduced-capacity back-to-back converter, as shown in Figure 2.9a. This configuration permits the variation of the turbine speed by 60% making the maximum power point tracking (MPPT) target possible. Although this configuration has been very successful over the last 20 years, recently, its leading role is being lost because this kind of wind turbine has problems in dealing with the voltage ride-through regulation, imposed by the regulating authorities in every country, to improve grid stability when a grid fault occurs.

Currently becoming very competitive (and in the near future expected to lead the wind turbine market), the new family of wind turbines is shown in Figure 2.9b. It is based on the connection of the turbine to the grid by using an electrical machine, mainly a permanent magnet synchronous generator (PMSG) and a full-capacity back-to-back power converter topology. In this case, the turbine speed and the active and reactive power are fully controllable and, thus, meet the regulations at the expense of an increment of the power converter cost.

Finally, other new configurations have been designed and tested, which improve some features of the wind generators. For example, a multi-winding PMSG connected to multiple distributed back-to-back converters can be used, as shown in Figure 2.9c. The advantages of this topology are that the power is shared, and that the system is inherently fault tolerant. Another possibility is shown in Figure 2.9d, where multiple PMSGs are fed by a distributed gearbox. The overall system has the same properties as the multi-winding PMSG, but at the expense of using a complex gearbox. Finally, a third solution, shown in Figure 2.9e, is based on a gearless wind turbine with a multipole machine connected to the grid using a full-capacity back-to-back converter. This gearless topology leads to a reduction in weight and maintenance, but the generator has a larger diameter.

2.3.2 *Photovoltaic Energy*

The continuously decreasing price of PV modules and the strong support of governments in pushing this technology have led to the consideration of PV and solar thermal energy as important factors in the present and future renewable energy market. One important part of these installations is based on housing applications (<5 kW), but as part of the great evolution of the PV sector in recent years, larger PV power plants are steadily being constructed, which currently achieve a nominal power up to 250 MW (Agua Caliente Solar Project in the United States). Today, the main installed PV capacity is grid-connected with the off-grid sector accounting for an estimated 2% of global capacity.

The output of PV panels is a DC voltage, and the existence of a power converter is required to generate an output AC voltage. Therefore, there is a big market for PV inverters with different topologies [35]. In general, there are four possible configurations, as shown in Figure 2.10:

- Series and parallel connections of PV panels and a three-phase central inverter (see Figure 2.10a). This configuration is common for PV plants (power range 10–250 kW or even more) and gives high efficiency and low cost at the expense of low reliability and poor MPPT.

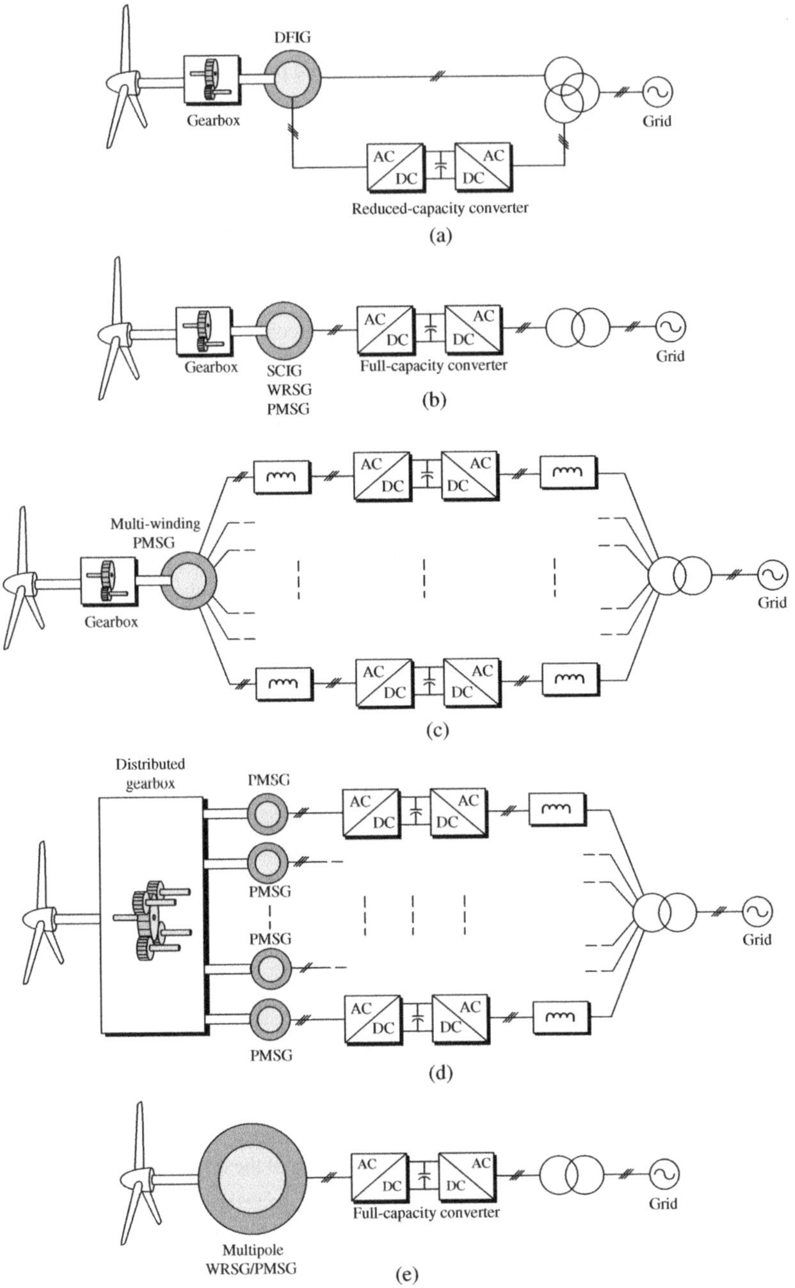

Figure 2.9 Configurations for the integration of a wind turbine in the electrical grid [5]: (a) doubly fed induction generator (DFIG) and a reduced-capacity back-to-back converter, (b) electrical machine (wound rotor induction generator (WRIG), squirrel-cage induction generator (SCIG) or permanent magnet synchronous generator (PMSG)) and a full-capacity back-to-back converter, (c) multi-winding PMSG with distributed back-to-back converters, (d) multiple PMSGs with distributed drive train, (e) multipole electrical machine (WRIG or PMSG) with full-capacity back-to-back converter

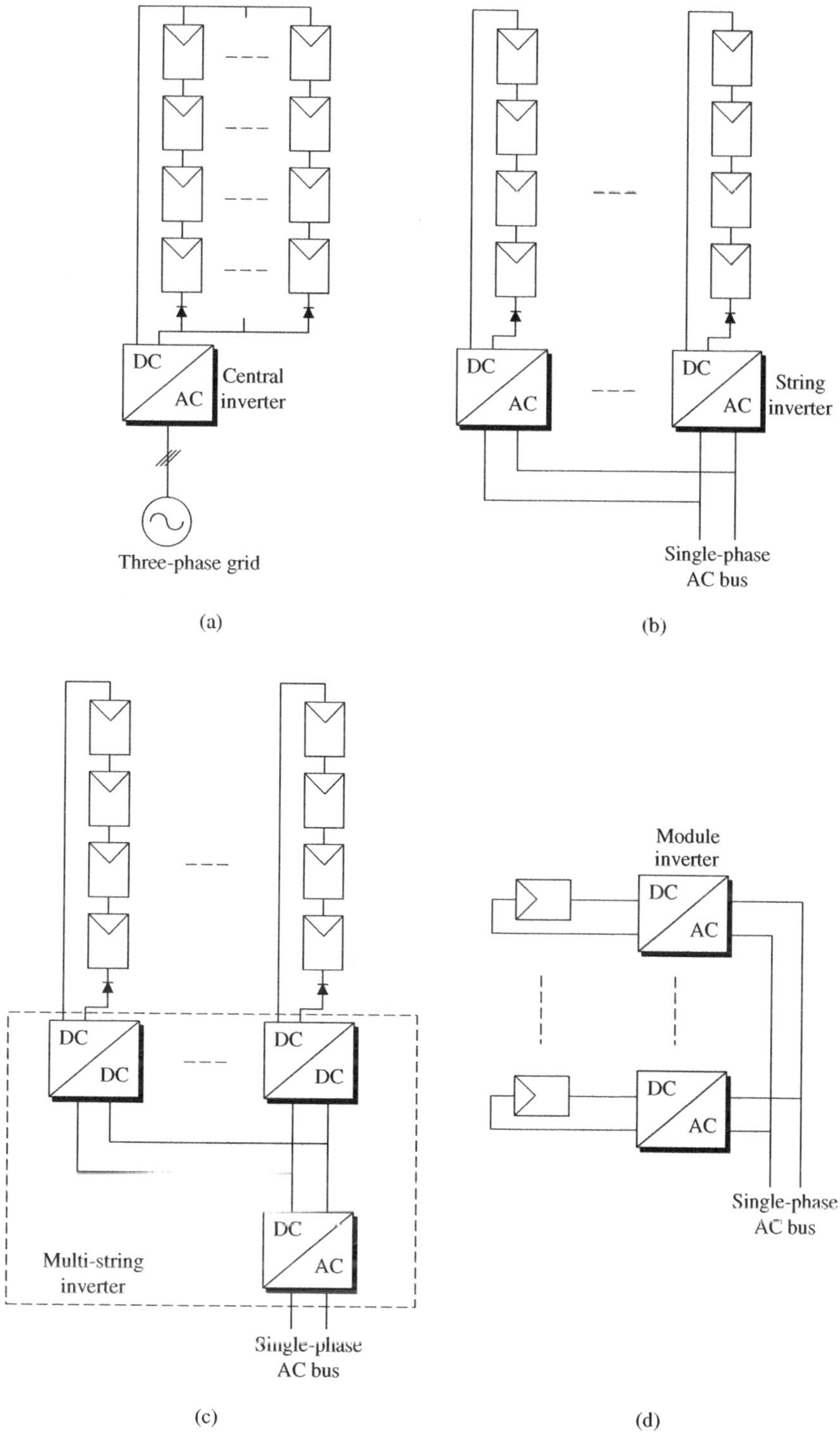

Figure 2.10 Configurations for the integration of a PV generation system: (a) series and parallel connections of PV panels and a three-phase central inverter, (b) series connection of the PV panels forming a string and individual inverters per string, (c) multi-string inverter formed by several DC/DC converters and a central inverter, and (d) module inverter dedicated to each PV panel

- String or multi string inverters usually for residential applications (<5 kW) (see Figure 2.10b and c). In these cases, each string has its own converter such that the MPPT can be obtained in a better way. In addition, the modularity is higher and therefore the reliability and fault tolerance capability are improved.
- PV systems where each PV panel has its own module inverter (see Figure 2.10d) commonly used for low-power applications (<500 W). This configuration enables optimal MPPT at the expense of increasing cost.

It must be noticed that each country has its own regulations and, for instance, the presence of a transformer is sometimes a requirement. In addition, other regulations are common, such as voltage and frequency requirements and active and reactive power limitations. In addition, an islanding effect for grid-connected PV systems takes place when the PV inverter does not disconnect immediately the grid is tripped. This fact causes problems when the grid is retripped or creates a safety hazard for utility line workers who might assume non-energized lines during a fault. In order to avoid this, safety measures called anti-islanding requirements have been included in the regulations.

Therefore, the power conversion stage (the topology for the DC/DC converters and the inverters) and the control strategy have to take these facts into account. Usually, topologies, such as the conventional H-bridge, the H5 and H6 inverters and the Heric inverter, are used and the transformer can be added in the output of the inverter, or the galvanic isolation can be achieved by the previous DC/DC stage [35, 36].

2.3.3 *Ocean Energy*

Other renewable energy sources such as ocean energy are the focus of researchers all over the world. Ocean energy includes different ways to obtain energy from water movement, such as waves (kinetic and potential energies), current (kinetic energy of flowing ocean and river currents), tidal (from the rise and ebb of tides), thermal (using the natural temperature gradient as a function of depth in tropical oceans) and salinity (chemical electric potential formed by the difference in salt concentration between freshwater and seawater) [37].

Wave and tidal energy sources have been the focus of researchers during the last decade, and currently, there are several prototypes and power plants installed. The efficiency and the feasibility of these technologies will be proven in the coming years. These kinds of devices have to fulfill the following requirements:

- The overall system has to be prepared to operate under harsh weather conditions (usually offshore). This leads to the design of over-dimensioned wave energy prototypes in order to withstand these severe conditions. However, ongoing maintenance operations of the system remain constant and are usually difficult to perform, and this important issue needs to be solved in the near future.
- The wave power conversion system has to capture the very low-frequency energy from the waves and be able to transform the power to the grid requirements. This issue can be solved by using an energy storage device (batteries, ultracapacitors, superconducting magnetic energy devices, or hydraulic accumulators) or by connecting the wave energy devices in an array to smooth the power obtained from the waves.

Among the multiple possibilities to develop a wave energy conversion system, the most common methods to capture the energy can be summarized as follows:

- Vertical oscillating systems: The device obtains the energy from the bidirectional movement of air inside a chamber filled with seawater and air. When the wave reaches the device, the seawater compresses the air inside the chamber moving it through a turbine. The reverse phenomenon occurs when the wave leaves the device.

- Overtopping systems: The energy of the waves is captured because the water overtops the edge of the device. The water is several meters above sea level, and the potential energy is used by the turbines to create the electricity.
- Wave activated bodies: The structure of the system is formed by several units, usually called bodies, which are able to move and oscillate around a reference point. When the system is placed in the water, the waves excite the system by moving these bodies and the energy is derived from this movement.

All these ocean energy conversion systems are based on electrical turbines generating an AC voltage that has to be transformed to fulfill the amplitude and frequency requirements of the power grid. Usually, a power converter in a direct drive back-to-back configuration is used, where an independent AC connection of each system to the AC grid is considered, or a common DC bus is used, where the first power stage only of each system is connected, using a central inverter to carry out the connection with the grid. In addition, an energy storage device is also needed in order to smooth the pulsating power from the waves [38].

Currently, several prototypes have been built and are in operation. Pelamis, Wave Dragon, WaveStar, Limpet, Archimedes Wave Swing and PowerBuoy, among others, have attracted increasing interest with regard to demonstrating the potential of this renewable energy source. Countries such as the United Kingdom, Portugal, the United States and Australia have built some wave energy plants, which are currently under evaluation.

However, it is expected that tidal energy will become an important source because tides are more predictable than wind energy and solar power. Sihwa Lake Tidal Power Station in South Korea is currently the most powerful tidal power plant in operation (254 MW), but the South Korean Government is planning to build a power plant (Incheon Tidal Power Station) in the next years with a capacity in the gigawatt range. In addition, in Russia, there are plans to build the Penzhin Tidal Power Plant with a capacity in the range of 100 GW, although this is still in the proposal stage.

2.4 Transportation Systems

Transportation (mainly vehicles, trains and airplanes) is nowadays one of the fundamental needs of daily life. For instance, the air traffic market is estimated to grow at 5% per year until 2020. During recent decades, the development of transportation systems has been huge. Their evolution has been focused on improvements in their efficiency, reliability, availability, fault tolerant operation and the reduction of fuel consumption and maintenance costs. Many of these challenges are being achieved by introducing power electronics in the drive train, reducing the volume and weight of the system, increasing the fault tolerance capability and finally, decreasing the fuel consumption and, thus, reducing the emission of noxious gases and the environmental impact.

A conventional aircraft has four different integrated energy vectors: pneumatic power (air conditioning, ice protection and engine start-up), electric power (avionic systems and commercial loads), hydraulic power (pumps and flight control actuators) and mechanical power (to drive hydraulic or fuel pumps). The more electric aircraft (MEA) concept is based on the replacement of these conventional devices by electrical systems. With the A380 and B787, Airbus and Boeing have caused a major upset in the aeronautical world by speeding up the transition toward an MEA. However, some challenges remain, such as improving the integration of the power systems, simplifying the electric structure of the aircraft, designing a high-voltage DC standard architecture and introducing efficient and advanced ESSs in the power train of the aircraft [39].

On the other hand, the increasing cost of fuel and the concern over noxious gas emissions are the main reasons for the recent significant development of electric vehicles (EVs). Although the EV market is relatively new, an optimistic scenario foresees 8–10 million EVs in global sales by 2020, assuming higher oil prices, accelerated reduction of battery cell costs and a stronger political support from governments.

This forecast will only be achieved by means of improving the vehicle drive train using efficient power electronics and ESSs [40, 41].

The term EV actually includes several different vehicle technologies, which are summarized in Figure 2.11 [42]:

- Pure EV: EVs run on a pure electric motor powered by batteries that are recharged by plugging in the vehicle.
- Hybrid EV (HEV): HEVs run on gasoline with a motor and use batteries to improve fuel efficiency (usually at low speed). They do not use electricity from any external source.
- Plug-in HEV (PHEV): PHEVs can be charged with electricity like EVs and run under engine power like HEVs. The combination offers increased driving range with potentially large fuel and cost savings and emission reductions. There are three types of PHEV. Parallel hybrids are PHEVs in which both the electric motor and the combustion engine are mechanically coupled to the wheels through a transmission. Series hybrids, also known as extended range EVs, are PHEVs in which the electric motor is coupled directly to the wheels and the combustion engine is only used to charge the batteries when needed. Series–parallel hybrids are a combination of the series and parallel hybrids making it possible to combine the advantages of both configurations.

Considering the growing environmental concerns, increasing fuel cost and more stringent government regulations, the sector is currently forced to optimize fuel consumption by improving the drive trains via downsizing or hybridization. Firstly, the vehicles will follow the hybridization path mainly powered by internal combustion engines, but in the near future, the solution will be to develop almost zero-emission pure electric propulsion with pure EVs and PHEVs.

2.5 Energy Storage Systems

ESSs include technologies where power electronic converters are needed in order to be able to interconnect different kinds of energies. ESSs are very important systems because they permit efficient grid management in the transmission and distribution systems. They also facilitate the accommodation of the discontinuous character of renewable energies or the charging electrical vehicles [43, 44]. These are only a few examples that have been addressed previously. In this section, the most common ESSs are addressed.

2.5.1 Technologies

A summary of the different technologies of ESSs is introduced in Figure 2.12, a classification of the ESSs depending on the power density is shown in Figure 2.13, and the main characteristics of these ESSs are summarized in Table 2.1.

- Batteries: Batteries are the oldest ESSs (lead acid) that have been used for more than 150 years, well before power electronic converters were introduced. Lead-acid batteries are still used in applications where the cost is very important, but tend to be replaced by other alternatives with better energy and power density, as can be observed in Table 2.1.
- Li-ion batteries are among those most frequently used because of their high energy density. These batteries are becoming very popular in low-power consumer applications (cameras, phones) and also in electric or hybrid vehicles. Their good energy efficiency and the absence of memory effect have also contributed highly to the expansion of their use, but their high price still renders them unaffordable for low-cost applications.

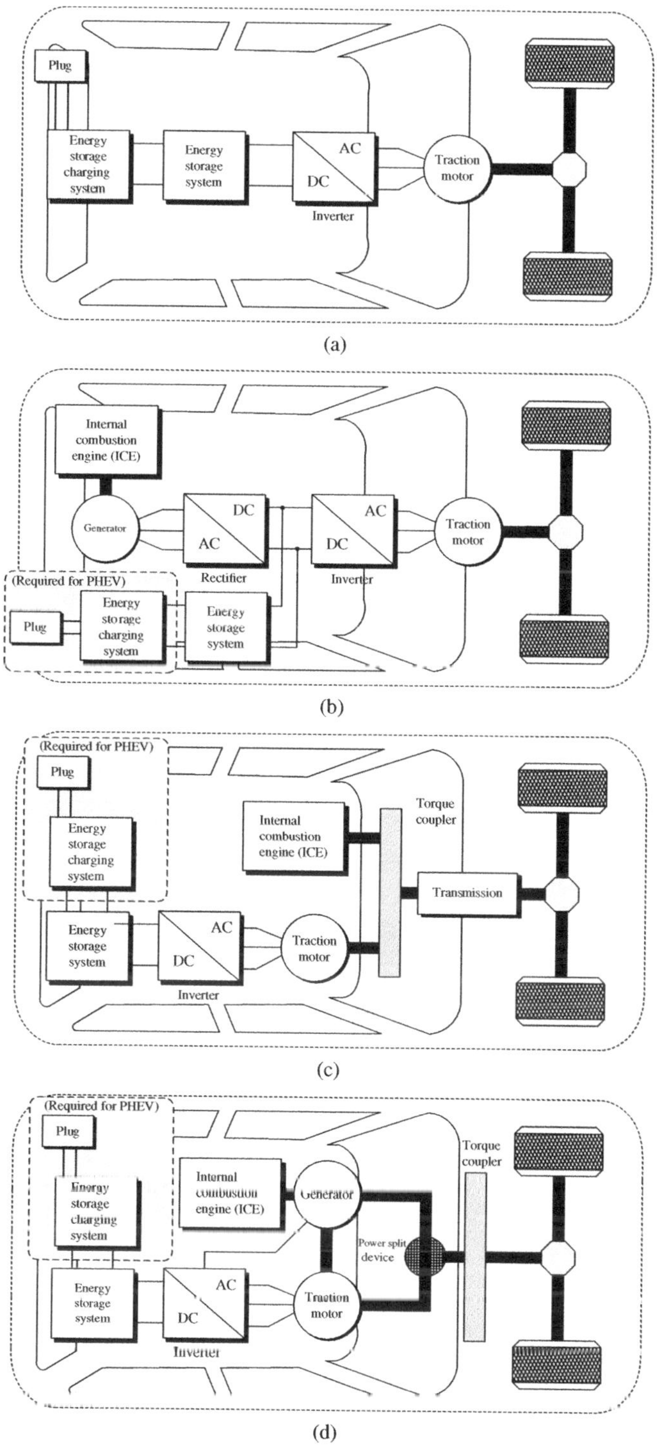

Figure 2.11 Typical layouts of the drive train of electric and hybrid electrical vehicles: (a) drive train of a pure electric vehicle, (b) drive train of a (plug-in) series hybrid electric vehicle, (c) drive train of a (plug-in) parallel hybrid electric vehicle and (d) drive train of a (plug-in) series–parallel hybrid electric vehicle

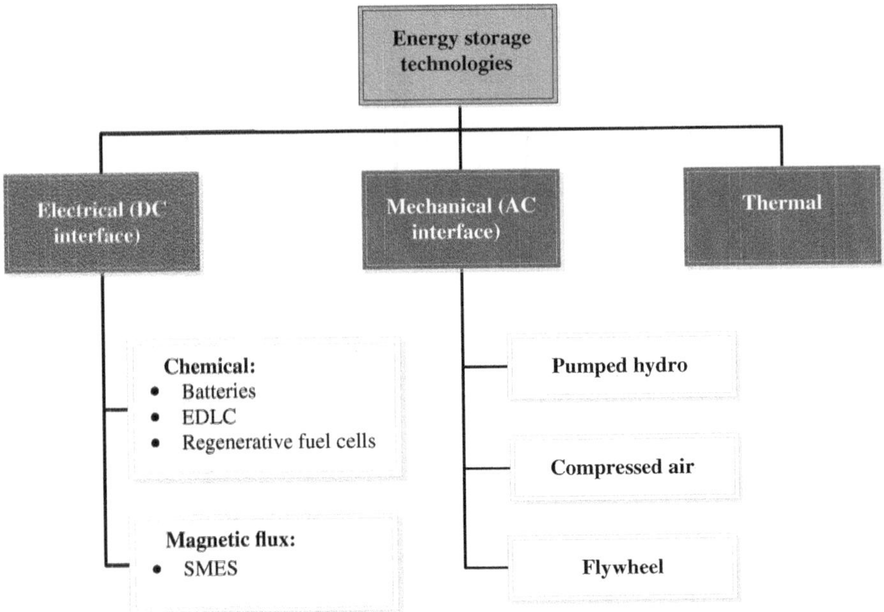

Figure 2.12 Technology of the energy storage devices

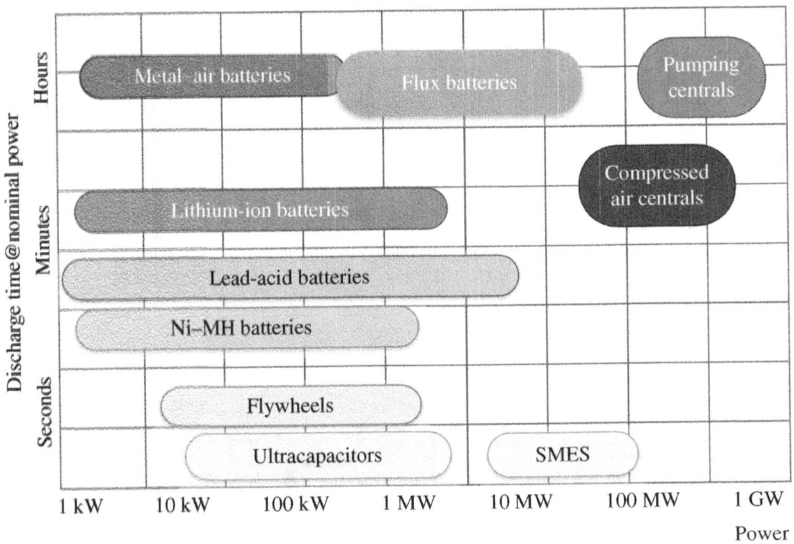

Figure 2.13 Classification of the ESSs depending on the power density [42]

Table 2.1 Characteristic parameters of different energy storage technologies [42]

Type	Energy efficiency (%)	Energy density (Wh = kg)	Power density (W = kg)	Cycle life (cycles)	Self-discharge
Pb-acid	70–80	20–35	25	200–2 000	Low
Ni–Cd	60–90	40–60	140–180	500–2 000	Low
Ni–MH	50–80	60–80	220	<3000	High
Li-ion	70–85	100–200	360	500–2000	Medium
Li-polymer	70	200	250–1 000	>1 200	Medium
NaS	70	120	120	2 000	–
VRB	80	25	80–150	>16 000	Negligible
EDLC	95	<50	4000	>50 000	Very high
Pumped hydro	65–80	0.3	–	>20 years	Negligible
CAES	40–50	10–30	–	>20 years	–
Flywheel (steel)	95	5–30	1 000	>20 000	Very high
Flywheel (composite)	95	>50	5 000	>20 000	Very high

- Other batteries, such as NiMH or NiCd, were very popular several decades ago, but are now being replaced by Li-ion or Li-polymer because of their high power and energy densities. These batteries are used mainly in massive consumer applications (mobile phones, cameras, etc.) and are currently the best option for hybrid and electric cars. Unfortunately, their high cost is preventing the spread of their use in cost-sensitive applications.
- There are also other promising battery technologies, such as sodium-sulfur battery, which has good performances, but is only able to work at high temperatures. Another promising technology is flow batteries (FB), and their main advantage is the decoupling of the power delivered and energy stored, in other words, the power is rated by the size of the reactor chamber, but the stored energy depends on the volume of fluids stored in low-cost reservoirs (plastic). There are two deposits: one containing the fresh reactive and a second with the exhausted reactive; thus, no self-discharge is observed. The reactives can be replaced, which allows a quick battery recharge.
- Electrochemical double-layer capacitors (EDLCs): EDLCs store energy in the same way as a classical capacitor does; therefore, no chemical/ionic reaction is held. This is an advantage with respect to batteries because better efficiency and durability can be expected. The power density is also very high. In order to obtain high capacity, the capacitor is built using a very porous (large surface) material, but one that is very thin. This implies that only very low-voltage devices can be used (typically 2–3 V). The capacity can be as high as kilofarads, but for practical use, series connection is needed in order to be able to connect to higher voltages.
- Regenerative fuel cells (FCs): To some extent, it can be said that FCs work in a similar way to FBs; however, in this case, the reactives are gases (oxygen and hydrogen) or gas–liquid (oxygen and methanol). Research efforts are focused on improving the durability of the membranes.
- Compressed air energy systems (CAESs): CAES is a technology that stores energy as compressed air. The energy is recovered by means of a gas turbine where the air is expanded. Large plants are possible if large reservoirs are available (usually caverns). The process of compressing air is exothermic and the expansion is endothermic, this implies an extra complication in the system design necessary to deal with this extra heat.
- Flywheel energy storage systems (FESSs): FESSs store energy as kinetic energy in a rotating mass. As the energy depends on the square of the rotating speed, this is the most important parameter to be extended. New materials permit speeds of over 10 000 rpm. The rotational speed is a simple and accurate measure of the "state of charge." The interface is an electrical AC machine with variable speed, and a frequency converter is needed to connect the system to the power grid.

- Superconductive magnetic energy storage systems (SMESs): SMESs store the energy as magnetic flux in a superconducting coil. The main difficulty is to maintain the coil at a very low temperature. SMESs are one of the systems with a high dynamic response with the highest power density among all of the other ESSs.
- Thermoelectric energy storage (TEES): TEES stores energy by increasing the temperature in synthetic oil or molten salt, maintaining it in a thermally isolated tank to be used later.

2.5.2 Application to Transmission and Distribution Systems

Traditionally, the electrical utilities have based their needs for storing excess generated energy in a few concentrated pumped hydro stations. In this way, when the generated power exceeds consumption, the water is pumped from the lower reservoir to the higher reservoir, providing a use for the extra energy produced. When the situation is the opposite, the utilities have two alternatives: to use the high-cost plants or to reverse the flow in the pumped station to produce the power needed. This model has been working well for years, but has a limitation; finding good sites is not easy.

Modern "Smart Grids," as shown in Figure 2.1, tend to have a mixed distributed generation where smaller plants closer to the consumers can be associated with smaller distributed storage facilities. In this way, different storage technologies contribute to the grid control, complementing the generators by shaving peaks and load leveling. Basically, the system will store the energy when it is cheap and the energy is returned to the grid when its generation is more expensive.

2.5.3 Application to Renewable Energy Systems

The discontinuous nature of renewable energy systems (RESs), depending on the wind speed or irradiance of the sun, makes its connection to a power grid difficult, because the network stability can be compromised when an abrupt change in the injected power occurs. This is a growing problem as the penetration percentage of RESs reaches significant values, implying that at some moments over 50% of the total generated power is produced from RESs. An example of this can be observed in Figure 2.14,

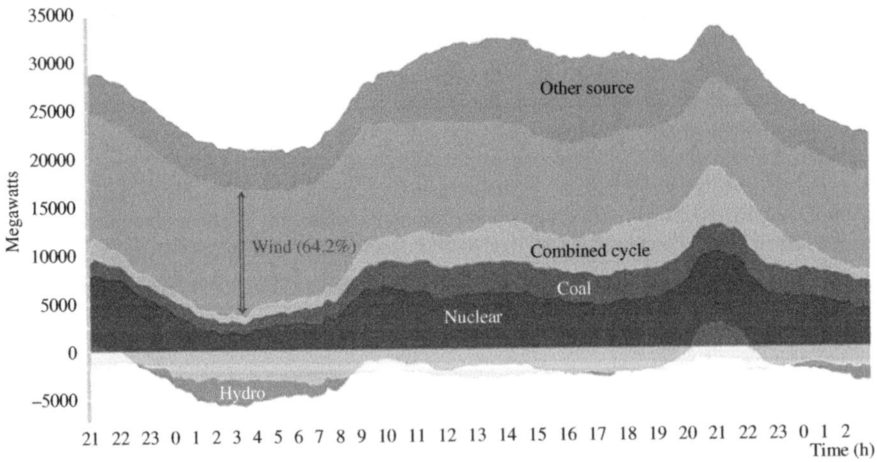

Figure 2.14 Spanish power consumption share on 24 September 2012. It shows a wind generation record of 13.285 MW out of 20.677 MW (64.28%). Source: https://demanda.ree.es/generacion_acumulada.html

which shows the Spanish power consumption on 24 September 2012. On that day, the energy provided by wind energy plants achieved a peak of around 64% of the energy production. This shows how the grid has become dependent on renewable energy sources. ESSs allow the decoupling of generation and consumption and they benefit from different technologies, such as batteries or compressed air.

2.5.4 Application to Transportation Systems

In the transportation systems, ESSs are being used in autonomous electric or hybrid vehicles (based mainly on lithium-based batteries), but grid-connected vehicles (trains) can also benefit from the added advantages of using an ESS.

The braking energy of vehicles is an important amount of energy that can be stored for later use, in particular, this energy, if not stored, causes an increase in the grid voltage. When a train connected to the grid is accelerating, the stored energy is recovered and this energy can be stored in the locomotive (mainly batteries or EDLC) or in the substation feeder.

2.6 Conclusions

The multiple interconnections between different regional grids, the integration of renewable energy sources and the matching between local consumption and generation are leading to the development of a new more distributed electrical grid. In addition, the transportation sector has to deal with the increasing cost of fossil fuels and stricter environmental requirements. Power electronics offer alternatives to face the challenges of this new distributed grid.

The integration of renewable energy sources can be achieved by efficient power converters, where the power flow and the random or intermittent nature of the energy source have to be fully controlled. New power electronic converters, such as FACTS and HVDC systems, are being installed in order to achieve efficient power distribution with the required stability margins and power quality demanded by the grid codes; for instance, facilitating the inland connection of offshore wind farms. Finally, last generation ESSs collaborate on grid stability, on the matching between generation and consumption, on the efficient integration of renewable energy sources and on the extensive use of EVs. In summary, power electronics is the key to the present and future energy sector.

References

1. Bose, B.K. (2009) Power electronics and motor drives – recent progress and perspective. *IEEE Transactions on Industrial Electronics*, **56** (2), 581–588.
2. Bouhafs, F., Mackay, M., and Merabti, M. (2012) Links to the future: communication requirements and challenges in the smart grid. *IEEE Power and Energy Magazine*, **10** (1), 24–32.
3. Liserre, M., Sauter, T., and Hung, J.Y. (2010) Future energy systems: integrating renewable energy sources into the smart power grid through industrial electronics. *IEEE Industrial Electronics Magazine*, **4** (1), 18–37.
4. Edris, A.A., Aapa, R., Baker, M.H. *et al.* (1997) Proposed terms and definitions for flexible AC transmission system (FACTS). *IEEE Transactions on Power Delivery*, **12** (4), 1848–1853.
5. Hingorani, N. and Gyugyi, L. (1999) *Understanding FACTS: Concepts and Technology of Flexible AC Transmission Systems*, John Wiley & Sons, Ltd-IEEE Press, December 1999.
6. Jonsson, H.-A. (2000) FACTS: transmission solutions in a changing world. International Conference on Power System Technology (PowerCon 2000), Vol. 1, pp. 375–380.
7. Tyll, H.K. and Schettler, F. (2009) Power system problems solved by FACTS devices. IEEE/PES Power Systems Conference and Exposition (PSCE '09), March 15–18, 2009, pp. 1–5.
8. Tyll, H.K. and Schettler, F. (2009) Historical overview on dynamic reactive power compensation solutions from the begin of AC power transmission towards present applications. IEEE/PES Power Systems Conference and Exposition (PSCE '09), March 15–18, 2009, pp. 1–7.

9. Dixon, J., Moran, L., Rodriguez, J., and Domke, R. (2005) Reactive power compensation technologies: state-of-the-art review. *Proceedings of the IEEE*, **93** (12), 2144–2164.
10. Frank, H. and Landstrom, B. (1971) Power factor correction with thyristor-controlled capacitors. *ASEA Journal*, **44** (6), 180–184.
11. Molinas, M., Vazquez, S., Takaku, T. *et al.* (2005) Improvement of transient stability margin in power systems with integrated wind generation using a STATCOM: an experimental verification. International Conference on Future Power Systems, November 18, 2005, pp. 6.
12. Molinas, M., Suul, J.A., and Undeland, T. (2008) Low voltage ride through of wind farms with cage generators: STATCOM versus SVC. *IEEE Transactions on Power Electron*, **23** (3), 1104–1117.
13. Abido, M.A. (2009) Power system stability enhancement using facts controllers: a review. *The Arabian Journal for Science and Engineering*, **34**, 153–172.
14. Franquelo, L.G., Rodriguez, J., Leon, J.I. *et al.* (2008) The age of multilevel converters arrives. *IEEE Industrial Electronics Magazine*, **2** (2), 28–39.
15. Rodriguez, J., Franquelo, L.G., Kouro, S. *et al.* (2009) Multilevel converters: an enabling technology for high-power applications. *Proceedings of the IEEE*, **97** (11), 1786–1817.
16. Kouro, S., Malinowski, M., Gopakumar, K. *et al.* (2010) Recent advances and industrial applications of multilevel converters. *IEEE Transactions on Industrial Electronics*, **57** (8), 2553–2580.
17. Leon, J.I., Vazquez, S., Portillo, R. *et al.* (2009) Three-dimensional feedforward space vector modulation applied to multilevel diode-clamped converters. *IEEE Transactions on Industrial Electronics*, **56** (1), 101–109.
18. Leon, J.I., Vazquez, S., Sanchez, J.A. *et al.* (2010) Conventional space-vector modulation techniques versus the single-phase modulator for multilevel converters. *IEEE Transactions on Industrial Electronics*, **57** (7), 2473–2482.
19. Leon, J.I., Kouro, S., Vazquez, S. *et al.* (2011) Multidimensional modulation technique for cascaded multilevel converters. *IEEE Transactions on Industrial Electronics*, **58** (2), 412–420.
20. Vazquez, S., Sanchez, J.A., Carrasco, J.M. *et al.* (2008) A model-based direct power control for three-phase power converters. *IEEE Transactions on Industrial Electronics*, **55** (4), 1647–1657.
21. Cortes, P., Ortiz, G., Yuz, J.I. *et al.* (2009) Model predictive control of an inverter with output filter for UPS applications. *IEEE Transactions on Industrial Electronics*, **56** (6), 1875–1883.
22. Reyes, M., Rodriguez, P., Vazquez, S. *et al.* (2012) Enhanced decoupled double synchronous reference frame current controller for unbalanced grid-voltage conditions. *IEEE Transactions on Power Electronics*, **27** (9), 3934–3943.
23. Portillo, R., Vazquez, S., Leon, J.I. *et al.* (2013) Model based adaptive direct power control for three-level NPC converters. *IEEE Transactions on Industrial Informatics*, **9** (2), 1148–1157.
24. Rudervall, R., Charpentier, J.P., and Sharma, R. (2000) High voltage direct current (HVDC) transmission systems technology review paper. Energy Week 2000, Washington, DC, March 7–8, 2000.
25. Setreus, J. and Bertling, L. (2008) Introduction to HVDC technology for reliable electrical power systems. Proceedings of the 10th International Conference on Probabilistic Methods Applied to Power Systems (PMAPS '08), May 25–29, 2008, pp. 1–8.
26. Flourentzou, N., Agelidis, V.G., and Demetriades, G.D. (2009) VSC-based HVDC power transmission systems: an overview. *IEEE Transactions on Power Electronics*, **24** (3), 592–602.
27. Glasdam, J., Hjerrild,J., Kocewiak, L.H., and Bak, C.L. (2012) Review on multi-level voltage source converter based HVDC technologies for grid connection of large offshore wind farms. IEEE International Conference on Power System Technology (POWERCON 2012), October 30–November 2, 2012, pp. 1–6.
28. Bagen, B., Jacobson, D., Lane, G., and Turanli, H.M. (2007) Evaluation of the performance of back-to-back HVDC converter and variable frequency transformer for power flow control in a weak interconnection. IEEE Power Engineering Society General Meeting, June 24–28, 2007, pp. 1–6.
29. Reynolds, M., Stidham, D., and Alaywan, Z. (2012) The golden spike: advanced power electronics enables renewable development across NERC regions. *IEEE Power and Energy Magazine*, **10** (2), 71–78.
30. Shao, Z., Lu, J., Lu, Y. *et al.* (2011) Research and application of the control and protection strategy for the ±500kV Xiluodu-Guangdong double bipole HVDC project. 4th International Conference on Electric Utility Deregulation and Restructuring and Power Technologies (DRPT 2011), July 6–9, 2011, pp. 88–93.
31. Al-Haiki, Z.E. and Shaikh-Nasser, A.N. (2011) Power transmission to distant offshore facilities. *IEEE Transactions on Industry Applications*, **47** (3), 1180–1183.
32. Chuco,B. and Watanabe, E.H. (2010) A comparative study of dynamic performance of HVDC system based on conventional VSC and MMC-VSC. iREP Symposium Bulk Power System Dynamics and Control (iREP 2010), August 1–6, 2010, pp. 1–6.

33. Carrasco, J.M., Franquelo, L.G., Bialasiewicz, J.T. *et al.* (2006) Power-electronic systems for the grid integration of renewable energy sources: a survey. *IEEE Transactions on Industrial Electronics*, **53** (4), 1002–1016.
34. Wu, B., Lang, Y., Zargari, N., and Kouro, S. (2011) *Power Conversion and Control of Wind Energy Systems*, 1st edn, John Wiley & Sons, Ltd-IEEE Press, August 2011.
35. Teodorescu, R., Liserre, M., and Rodriguez, P. (2011) *Grid Converters for Photovoltaic and Wind Power Systems*, 1st edn, John Wiley & Sons, Ltd-IEEE Press, February 2011.
36. Romero-cadaval, E., Spagnuolo, G., Franquelo, L.G., Ramos-Paja, C.A., Suntio, T., and Xiao, W.M., (2013) Grid-Connected Photovoltaic Generation Plants: Components and Operation, *Industrial Electronics Magazine, IEEE*, **3** (7), pp. 6–20.
37. Czech, B. and Bauer, P. (2012) Wave energy converter concepts: design challenges and classification. *IEEE Industrial Electronics Magazine*, **6** (2), 4–16.
38. Igic, P., Zhou, Z., Knapp, W. *et al.* (2011) Multi-megawatt offshore wave energy converters – electrical system configuration and generator control strategy. *IET Renewable Power Generation*, **5** (1), 10–17.
39. Roboam, X., Sareni, B., and Andrade, A.D. (2012) More electricity in the Air: toward optimized electrical networks embedded in more-electrical aircraft. *IEEE Industrial Electronics Magazine*, **6** (4), 6–17.
40. Michael, V.-U. and Wolfgang, B. (2009) *Powertrain 2020 – The Future Drives Electric*, Roland Berger Strategic Consultants.
41. Dickerman, L. and Harrison, J. (2010) A new car, a new grid. *IEEE Power and Energy Magazine*, **8** (2), 55–61.
42. Emadi, A., Williamson, S.S., and Khaligh, A. (2006) Power electronics intensive solutions for advanced electric, hybrid electric, and fuel cell vehicular power systems. *IEEE Transactions on Power Electronics*, **21** (3), 567–577.
43. Vazquez, S., Lukic, S.M., Galvan, E. *et al.* (2010) Energy storage systems for transport and grid applications. *IEEE Transactions on Industrial Electronics*, **57** (12), 3881–3895.
44. Lukic, S. (2008) Charging ahead. *IEEE Industrial Electronics Magazine*, **2** (4), 22–31.

3

An Overview on Distributed Generation and Smart Grid Concepts and Technologies

Concettina Buccella[1], Carlo Cecati[1] and Haitham Abu-Rub[2]
[1]*Department of Information Engineering, Computer Science and Mathematics, University of L'Aquila, and DigiPower Ltd. L'Aquila, Italy*
[2]*Electrical and Computer Engineering, Texas A&M University at Qatar, Doha, Qatar*

3.1 Introduction

The existing power grid can be considered as a hierarchical system where power plants are at the top of the chain and loads are at the bottom, resulting in a unidirectional electrical pipeline managed with limited information about the exchange among sources and end points.

This situation has severe drawbacks, including the following:

- The system is sensitive to voltage and frequency instabilities as well as to power security issues caused by load variations and dynamic network reconfigurations.
- The implementation of demand side management strategies, which would be very useful for reducing the risk of failures and blackouts and for increasing system efficiency, is not allowed, moreover.
- It is not suitable for the integration of renewable energy.

During the last decade, the electrical energy market has been characterized by a growing demand for energy and two important innovations: the quick growth and massive diffusion of renewable energy systems (RESs) and the subsequent rapid development of distributed generation (DG) systems and smart grids (SGs) [1–5]. Conventional unidirectional power systems consisting of few very high-power generators producing hundreds of megavolt-amperes and connected with substations through transmission lines and feeding distribution systems supplying loads through a tree are being replaced by wide, complex, and heterogeneous meshes, embedding a huge number of loads and generators, the latter supplied by different sources (fossil, nuclear, gas, wind, sunlight, biomass, etc.) and operating at different power

Power Electronics for Renewable Energy Systems, Transportation and Industrial Applications, First Edition.
Edited by Haitham Abu-Rub, Mariusz Malinowski and Kamal Al-Haddad.

levels (from a few kilovolt-amperes up to hundreds of megavolt-amperes) and voltages, each with its own characteristics (e.g., electrical, mechanical, thermal, photovoltaic (PV) surface) and dynamics.

According to many analyses, future electrical systems should have the following parameters [6–12]:

- High power capability: with the increasing demand for energy in industrial, residential and civil applications and the incoming large-scale diffusion of electric vehicles, electricity is becoming the main power source of the modern world and hence the need for it will increase significantly during the next years; this trend is expected to remain positive for many decades and will be marginally influenced by external perturbations such as economic or political crises.
- High power quality and reliability: electricity must be available whenever it is needed with the lowest or no latency, stable voltage and frequency and low harmonic distortion.
- High efficiency: electricity should not be dispersed during production, transportation and distribution processes; the grid and the loads should be managed to achieve maximum system efficiency.
- High flexibility: the power system should be highly configurable and should allow smooth integration among different power sources; moreover, dynamic changes of loads and power sources should not influence power quality.
- Low environmental impact (i.e., sustainability): renewable energy sources should progressively replace traditional sources; moreover, even in the short term they should be fully integrated into the existing power system.

Previous requirements cannot be satisfied by traditional power systems; therefore, during the next years a huge revision of the present systems is expected with the introduction of many new functions, systems and modus operandi, commonly referred to as the distributed generation (DG) and smart grid (SG) revolution. This is changing the way in which next-generation power systems have to be designed, operated and maintained, and can be achieved only by introducing new technologies, functionalities and operational approaches, which are as follows [13–18]:

- full exploitation of all renewables
- technological enhancements and large-scale diffusion of energy storage systems
- massive introduction of Information and Communication Technologies (ICTs)
- implementation of high-granularity self-healing and resiliency against unwanted situations, such as blackouts or natural disasters
- consumers' active participation to the electricity market
- introduction of new products, services and markets.

With reference to the last point, the same ICT infrastructure could be shared among different services (water, gas, electricity, heat, etc.) or different service providers. This should allow the hardware (HW) and software (SW) resource optimization necessary to sustain large investments in the realization of the communication infrastructure.

The following sections give an overview of the main technologies, features and problems of DG and SGs. Due to the breadth of topics, this chapter gives a short but comprehensive overview of these emerging topics.

3.2 Requirements of Distributed Generation Systems and Smart Grids

In order to fully satisfy energy demand, conventional power systems are operated taking into account the predicted power needs, thus producing a corresponding quantity of high-quality energy, that is, with stable output voltage and frequency. Because of the high thermal inertia of steam generators, peak demands are satisfied by producing amounts of electrical energy close to peak values, and hence significant energy surpluses are often available during periods of light load operations. When possible, the energy surplus is employed to increase the quantity of stored energy, typically in hydroelectric reserves.

However, the massive introduction of renewable energies does not solve the last problem and introduces several additional problems, mainly related to the fluctuating quantity and quality of the energy injected in the grid. Hence, in general, power system stability deteriorates instead of improving.

With this situation, intelligent management of produced and absorbed energy, together with the introduction of small-to-medium fully controllable generators, for example, turbo gas (TG) and combined heat and power (CHP) systems, which, being fully operative within few minutes or seconds, can be activated and deactivated according to short-term predictions, represents a viable if not unique short-term to medium-term solution.

In fact, at the current stage of technologies, energy storage systems, except hydro power plants, are unable to store and release the amount of energy often needed to satisfy peak demands, while static compensators (STATCOMs) and other high-power electronic devices, even if they are efficient under some kinds of transient disturbances, could be insufficient to guarantee stability in the case of large and long-term variations of electrical quantities. In addition, some types of energies such as nuclear or fossil are becoming critical either because of their impact on health and the environment or because they are potentially dangerous.

Another fundamental issue motivating the diffusion of SG is that the increasing demand for energy requires permanent enhancements or a redesign of transmission lines, and thus requires huge investments at national and international levels. Hence, several countries are changing their energy policies in favor of full exploitation of DG, the combination of RESs and clean traditional sources, improvement of existing power plants and user awareness about saving energy [19].

The new power systems will include a huge number of power lines often operating at different power and voltage levels, connected through intelligent routing nodes and capable of sustaining, monitoring, controlling and billing energy in a fully bidirectional power flow environment.

Other key elements of modern power systems are the electronic static converters necessary for the connection of RESs to the grid and storage systems, which is necessary to reduce the effects of energy fluctuations and power losses along lines. Their operations ensure power quality enhancements [20].

In short, DG and SGs are very complex systems requiring many technologies including, but not limited to, power systems, power electronics, communications, computer science, computational intelligence and so on, which are necessary for full and optimized integration among generators, loads and lines [21].

Moreover, tools based on previous technologies are necessary for improving overall energy management according to economic criteria and for creating sensibility among users concerning the rational use of energy.

At the current state of technology, the most diffused RES generators are PV, wind and hydro generators. Other possible sources suitable for DG are CHP systems, combustion motors, geothermal systems and fuel cells (Figure 3.1).

3.3 Photovoltaic Generators

PV cells are direct current (DC) generators. Their voltage level depends on the intrinsic cell characteristics, the number of cascaded cells and their temperature. The available current depends on cell characteristics, the number of parallel strings (a string is a group of cascaded cells) and sunlight intensity. With present technology, they are arranged in panels providing up to 220–250 W at a voltage rating of 48–60 V. In fact, typical panels do not exceed 1.5–2 m^2 due to the need for easy handling. New technologies are emerging that improve PV energy conversion by means of cell efficiency enhancement and/or optical concentrators capable of increasing the quantity of energy applied to the cell surface [22]. This solution increases the power produced per surface unit but at the cost of very high temperatures, and hence suitable heat transfer systems are needed to maintain safe and efficient operations. Currently, some experimental systems propose to combine electricity and heat production (Figure 3.2).

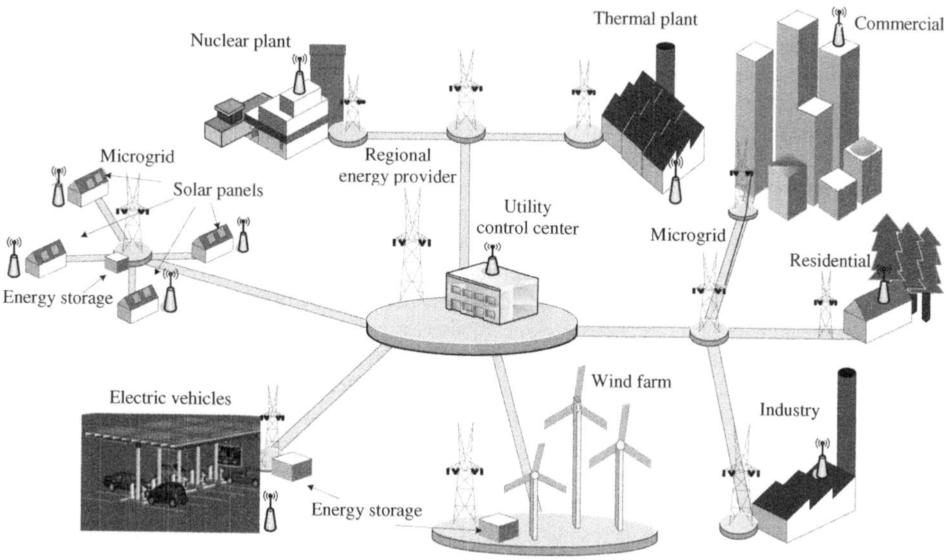

Figure 3.1 Smart grid architecture

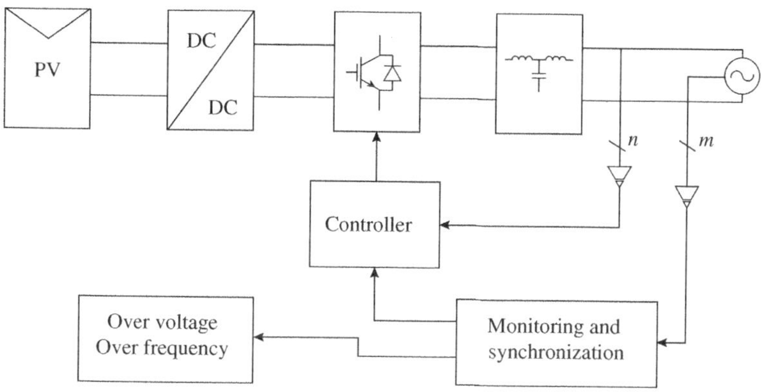

Figure 3.2 Schematic diagram of a photovoltaic system

In order to supply standard loads operating with alternating current (AC) – AC motors, lamps, air conditioners, electronic equipment, domestic and industrial loads and so on – panels are connected in series and/or in parallel, thus reaching the desired level of voltage and current; hence they supply an inverter [23–24]. The latter has to guarantee a stable and standard output (e.g., 400 V, 50 Hz), high conversion efficiency, grid synchronization (in the case of grid connected systems), low voltage and current distortion and low cost. Moreover, it might be in charge of power factor control and harmonic mitigation. Usually, inverter operations are based on feedback control and two-level pulse width modulation (PWM); moreover, they are connected with the grid through suitable passive output filters.

A common solution includes inductive/capacitive filter (LC or LCL) configuration. In order to maximize the extracted energy and to adapt the output voltage amplitude to the required level, a DC/DC converter with maximum power point tracking (MPPT) capability is usually interposed between cells and the single-phase or three-phase inverter. In grid-connected systems, a further requirement for the converter is the implementation of a suitable synchronization algorithm, needed to ensure proper operations and the injection of the maximum amount of energy while avoiding instabilities and/or failures. It is usually based on phase-locked loop (PLL) or its variants: synchronous reference frame (SRF-PLL), enhanced phase-locked loop (EPLL), quadrature phase-locked loop (QPLL) and dual second-order generalized integrator (SOGI) [25–26].

When dealing with PV generators in DG, there are many issues that depend on both the power level and the input and output voltage. The choice of plant as well as converter topology is important for extracting the highest amount of power by PV cells, ensuring the optimum connection between the converter and the grid and implementing protective functions. It is necessary to balance electrical considerations such as voltage and current levels, power losses, robustness, reliability, expandability, complexity, cost and actual operating conditions (e.g., sunlight level and its daily/monthly distribution, shadowing effects and so on), which play a role often more important than technical and/or technological issues. Often large PV plants are microgrids (MGs), that is, small autonomous systems interconnected with others and/or with higher hierarchical level power grids. In low-power applications (few kilovolt-amperes), typical systems include a simple high-frequency DC/DC converter (boost or buck-boost) with or without a high-frequency transformer and a single-phase PWM inverter.

New converter topologies such as resonant DC/DC converters are being considered, with the aim of increasing efficiency and reducing costs and size. Cells are arranged such that the highest efficiency is achieved: due to the relatively high number of cells, the PV field surface is quite limited (some tens of square meters) and hence shadowing effects caused by clouds or other random obstacles interposed between the sun and the panels are almost negligible if certain precautions are taken during the positioning of panels. On the other hand, in order to avoid performance drops, cascaded cells require higher consistency of their electric characteristics compared to a parallel arrangement. Hence, the choice of series or parallel connection is not unique and always represents a trade-off.

Recently, "plug-and-play" solutions have been proposed: each single panel has its own complete power converter, and thus it can be directly interconnected with the grid [27].

This solution has some advantages: high modularity, quick installation, high reliability, optimized performance and reduced losses because of the direct availability of AC output voltage and simple maintenance (in the case of failure of a panel, the others and hence the whole system, continue operation and the faulty panel can be easily replaced without service interruption). Alternatively, each panel can incorporate its own DC/DC converter, thus allowing its accurate control and optimization, while the DC/AC conversion is provided by a system-level inverter [28].

This approach seems to be very attractive as the cost of the low-power DC/DC converter is not high, while accurate MPPT and fixed DC voltage are ensured at panel level.

Medium- and high-power systems require some choices imposed by a number of distinct factors including the following:

- PV field electrical arrangement (i.e., series or parallel connection of panels) due to electrical and shadowing factors
- the limited power achievable by insulated DC/DC converters due to the unavailability of large ferrite cores
- the choice of output current/voltage
- the choice of the inverter (single inverter or multiple inverters in parallel)
- the intrinsic problems of MPPT in large systems
- the presence of low-frequency transformers between the PV plant and the grid.

Also in this field, plant technology is moving toward modular solutions, which basically consist of a double-stage power converter (high-frequency DC/DC converter and inverter) for each string of the PV plant, the latter producing a relatively high amount of power.

Single-phase inverters may be considered for each string, achieving three phases at the grid level. In this case, blocking diodes are not requested, MPPT algorithms are applied separately to each string and accurate power control can be achieved.

Alternatively, several DC/DC converters may be connected to a high-voltage DC bus, the latter linked with a single or multiple three-phase inverters. The DC/DC units can be current source converters or voltage source converters.

The first type is preferred due to its high robustness; the other requires large electrolytic capacitors, which are bulky and have low reliability due to their intrinsic degradation because of aging. Alternatively, numerous expensive polypropylene capacitors might be used, guaranteeing constant performance without any degradation or influence by ambient temperature. Whatever the choice, modular designs improve reliability and performance, but at the cost of higher complexity.

New topologies have been recently proposed in order to increase efficiency or reduce cost or both. Among them, a very interesting solution is the use of cascaded H-bridge multilevel converters, consisting of a number of H-bridges connected in series, each one supplied by a separate PV generator connecting a suitable group of cells in series and/or in parallel, thus reaching the desired level of input voltage without the use of additional converters. This architecture, also known as cascaded multilevel inverters (CML), allows better utilization of the available input energy because each H-bridge can embed its own MPPT algorithm; moreover, suitable voltage levels can be chosen, thus extracting the highest amount of solar energy without the need for a DC/DC converter, while conversion efficiency is higher due to the lower switching losses resulting from lower switching frequencies.

Faster and more efficient power devices can also be employed due to the lower operating voltage of each H-bridge, further improving overall performance. In high-power high-voltage (kV) applications, direct operations are allowed without the need for line frequency transformers or high-voltage DC sources. Another significant advantage of CML inverters is their better output waveforms, which achieve a significant reduction of the output filter and an increase in the efficiency of PV energy conversion.

3.4 Wind and Mini-hydro Generators

Wind and mini-hydro systems use AC synchronous or asynchronous generators that convert the mechanical energy generated by wind or water and available at the machine shaft into electrical energy.

The simplest high-power wind generator consists of a squirrel-cage induction generator (SCIG) connected with the grid through a transformer. Owing to its simplicity and the intrinsic induction machine behavior, it can operate at an almost fixed speed and in the presence of a stiff grid. Its limited control capabilities are often achieved at a mechanical level (e.g., blade pitch control), while reactive power control requires the availability of a STATCOM.

Current power electronic technologies and devices offer more sophisticated solutions, such as multipole permanent magnet or induction generators connected with an AC/AC or an AC/DC/AC power converter capable of stabilization of both voltage amplitude and frequency at the desired levels. Generator sets operate at voltage levels starting from 400 V, 50 or 60 Hz, up to some kilovolts, the latter being typical of multi-megawatt systems. However, typical generators operate at voltage levels lower than the grid and therefore their interconnection requires the use of a line-frequency step-up transformer. It is worth noticing that the use of multilevel converters might allow the elimination of such a transformer.

Depending on the power level and the required performance, the power converter consists of a simple unidirectional diode rectifier coupled with a PWM inverter (or any other topology) through a DC link or the more sophisticated back to back converter, providing bidirectional power flow. The inverter is coupled with the grid through a low-pass filter, needed to ensure the elimination of high-frequency harmonics and a line frequency transformer.

In the range of very high-power generators (hundreds of kilovolt-amperes up to several megavolt-amperes), doubly fed induction generators (DFIGs) represent a very common and interesting

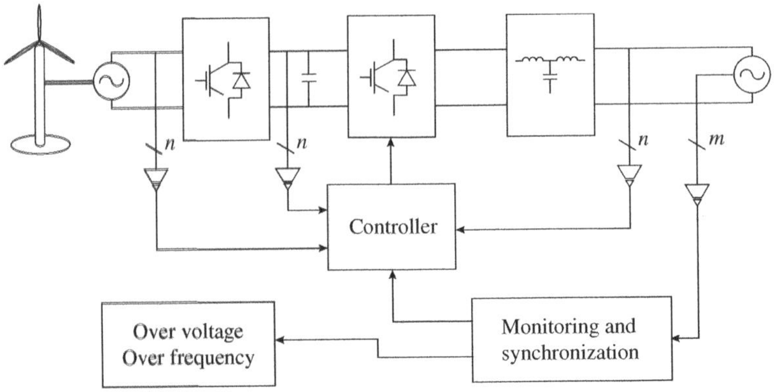

Figure 3.3 Schematic diagram of a wind generation system

solution because of their high efficiency, full control of active and reactive power and the need for a converter with a rating of about 20–30% of the generator power. In fact, the machine stator windings are directly connected with the grid or with each interposed line frequency transformer, while full active/reactive control is achieved by modifying the supply characteristics of wound rotor windings through a power converter. This reduces power converter cost, saves energy (reducing power converter losses) and extends the power range (limited by the power devices rating). DFIGs may operate in the kilovolt range and the power converter could consist of a matrix or a multilevel converter [29–31].

In low-power applications, a DC/DC converter might be included in a DC link for MPPT, but the available technology of magnetic materials limits their use to a few tens of kilowatts (Figure 3.3).

3.5 Energy Storage Systems

The intermittent or fluctuating operations of RESs can be mitigated by energy storage systems (ESS) [32].

Traditional systems consist of hydro and pumped storages. Their capabilities are often enormous, but usually they are very far from loads and/or wind or sunlight generators. Consequently, their effects in the event of voltage sags, swell, or other transient disturbances are negligible, while they are important in balancing long-term predictable fluctuations, which can be eliminated through suitable accumulate–release energy strategies. Recently, mini-hydro and micro-hydro have been gaining diffusion. Their localization is intrinsically less critical, but their operations cannot satisfy fast peak demands, except when they are controlled by demand prediction [33].

Typical energy storage systems for fast transient operations are batteries, supercapacitors, freewheels and fuel cells.

Batteries appear to be the most effective technology available today in terms of performance and costs, even if their reliability and duration are not outstanding. Traditional lead acid batteries are leaving space to other technologies such as nickel–metal hydride (Ni-HM) and lithium-ion batteries (in some variants), which present higher capacity and faster recharge time. Batteries allow a fast transient response, satisfying the most demanding applications, but their drawback is a long recharge time.

Supercapacitors have an energy density hundreds of times greater than that of conventional electrolytic capacitors and a power density much higher than that of batteries or fuel cells; their speed in supplying current is very high but their voltage level is too low for requirements so a lot of cells must be connected in series in order to reach the desired voltage level; hence, their cost is not yet attractive.

ESSs are needed for quick delivery of energy, and thus heat-based sources are needed for long-term disturbances caused by predictable meteorological variations.

ESSs can be either DC or AC sources. In the first case, they are usually connected with a DC link shared with the existing power converters. The cost of this solution is attractive but the achievable power is often limited; moreover, it has low flexibility and low stiffness to failures, sharing the DC/AC converter with RES. A better solution consists of a fully independent ESS directly connected with the AC grid, as each element can be sized independently and easily modularized.

Local or residential micro storage systems (MSSs) will be integrated with the end-users of future SGs, allowing utilities to dispatch stored energy produced earlier by local generators. To perform data gathering and control, the ESS should communicate with other devices, such as the power meter, to monitor the power consumed.

3.6 Electric Vehicles

Electric vehicles (EVs), hybrid electric vehicles (HEVs) and plug-in hybrid electric vehicles (PHEVs) use electricity either as their primary fuel or to improve the efficiency of conventional combustion engines [34]. All of them use a lithium battery (or other recent technology) to store the electrical energy needed to power the motor. Their diffusion is expected to grow very quickly during the next few years, as their cost, considering a balancing between various costs (purchase, use, maintenance, disposal), will become similar to or better than that of traditional combustion engine cars. For this reason, they are being considered as a viable solution for energy storage in SGs through the so-called vehicle-to-grid (V2G) technology: during recharge operations, the vehicle battery and recharge system are in parallel with the grid; therefore, using a bidirectional converter for recharge and following a suitable demand side management strategy, the vehicle can supply energy to the grid like any other storage system [35]. The main drawback of this strategy is represented by the limited quantity of available electricity, which is strictly dependent on the capacity of the installed batteries. In fact unlike "static" ESS, battery recharge is very common in V2G because of the main use of the vehicle. On the other hand, using proper recharge operations at high power, a typical car could be recharged within 20–30 min. With present technology, this is only possible when operating connected with a 400 V three-phase line sustaining currents of around 150–200 A. These kinds of recharge operations require accurate control of both electrical quantities and main battery parameters; otherwise, the battery life may be very short or the battery might be seriously damaged [36–37]. Moreover, the huge quantity of electricity necessary to recharge EV fleets is evident. This last aspect is crucial and requires large investments in the power system infrastructure. Alternatively, relatively small electric generators (some megavolt-amperes), for instance CHP, might be installed in the vicinity of parking with the dual task of recharging batteries and providing heat to nearby buildings. An interesting solution is represented by use of PV energy: large PV plants can be realized by exploiting the surfaces of roofs of shopping centers or by creating covered parking.

3.7 Microgrids

With the advent of DG and SG, power systems are moving toward a different scheme with a huge number of reduced scale systems, called microgrids (MGs), interconnected by meshes of complex transmission lines, whose nodes have a high level of intelligence. MGs could be interconnected among them at a distribution line level [38–39].

Each MG is autonomous and includes electricity generators, energy storage, distribution lines, loads and interfaces with the higher level grid called common coupling point (CCPs), where the higher level grid is often referred to as the macrogrid. MG operates at low- and medium-voltage levels. This operation

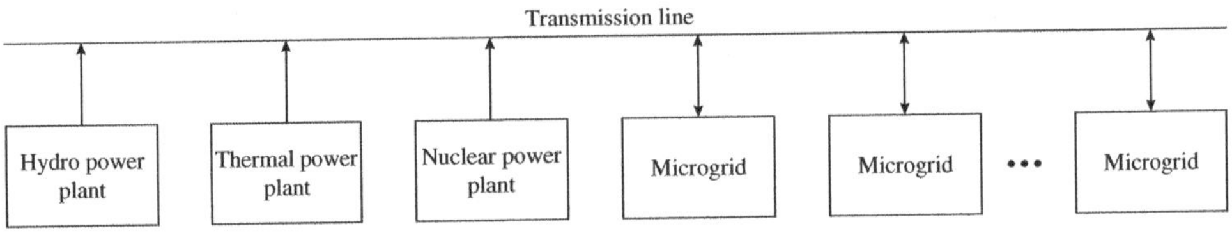

Figure 3.4 Power system with microgrids

mode is known as "islanding." CCPs include line frequency transformers and protection and intelligent devices with communication capability (Figure 3.4).

This approach has several advantages. In fact, owing to the high modularity and local use of energy, it simplifies the system; eliminates transmission losses; improves efficiency; provides active load control; isolates MGs from foreign disturbances, failures and outages; and simplifies self-healing, thus increasing the resilience of the overall system [40–41].

The quantity of energy generated in an MG usually corresponds to local demand, but the concept is general and therefore it could be applied to wind farms or other similar systems, where most elements are generators. Connections with higher hierarchical levels occur only to satisfy peak demands that cannot be handled locally, to buy energy produced in other MGs, or to sell energy. Generators are supplied by RESs, but in order to limit the drawbacks of RESs (energy fluctuations and daily variations), they are integrated with fuel-based generators, typically TG eventually supplied by biomass, or CHP systems (Figure 3.5).

In the future fuel cells could replace integrated thermal power plants but nowadays their cost is prohibitive.

Figure 3.5 Example of microgrid

The use of no-RES generators depends on the level of intelligence of the system, its predictive capability and the location of the MG. Typical TG generators reach their full power within a few seconds, and hence they might be operated on a prediction basis, thus supplying electricity only when RESs are not sufficient to cover peak demand or during the night and more generally during periods with limited sunlight or wind due to adverse weather conditions. This could allow the use of "foreign" energy to be avoided or limited to exceptional circumstances [42].

In order to ensure stable operations and the necessary power quality, energy storage systems must be included in an MG. Their quantity and dimensions depend on several factors, including the number and type of loads, the quality of energy available in the MG and the possibility of operations tightly connected with the grid.

3.8 Smart Grid Issues

Connecting RESs with the grid and managing a MG or a SG may become a challenging task if the number of generators and users is very large and their power quality and/or that of the grid is not high. Therefore, optimal operations can be achieved only by addressing stability and reliability issues. Moreover, the intrinsic nature of DG imposes flexibility and expandability.

Dynamic perturbations caused by interactions among fluctuating power sources and variable loads often lead to instabilities, which, if not adequately controlled, might cause failures and blackouts. These situations can be reduced by introducing large storage systems and in a well-designed MG, they are intrinsically less frequent and easier to manage. Instability should be taken into account while designing power system and power converters in order to either eliminate it at source or mitigate its effects.

Suitable protection policies and control algorithms can be implemented using both HW and SW, eliminating the risk of network instabilities and failures of power sources or power converters. This approach requires real-time sensing of output quantities (voltage, current, frequency) and, eventually, fast breakers capable of real-time disconnection of elements from the grid, thus preserving its integrity. Well-designed inverters can intrinsically avoid the injection of additional energy in an already unstable system and, in some circumstances, operate similar to that of a STATCOM or a Unified Power Factor Controller (UPFC). For instance, they can provide active and reactive power. Such systems usually adopt PLL techniques (already listed in an earlier section) and zero-crossing grid voltages detection (ZCD) implemented through a group of combined filters coupled with a nonlinear transformation. The latter approach may be negatively influenced by poor power quality determined by electromagnetic interferences (EMIs) and other disturbances, such as random voltage sags, swell and dips, which may lead to erroneous triggering and time delays. These drawbacks can be mitigated using appropriate design techniques and filters.

STATCOMs and UPFC, eventually coupled with supercapacitors, are typical power converters specific for dynamic stabilization.

Instabilities caused by long-term fluctuations or other disturbances as well as by the effects of remote failures can only be managed through the availability of energy storage systems (e.g., batteries) and converters with appropriate capability, and, if this is not successful, through the use of hydropump plants or, at worst, disconnection.

Wide-area situational awareness (WASA) systems integrate technologies for effective power system management and monitoring. They represent one of the key functions of SGs, since they provide reliability, security and interoperability among so many interconnected systems and devices. Synchrophasors are counted as new wide-area measurement technologies. Their primary task is to measure the different portions of the power system and to put these measurements on the same time basis, enabling a view of the whole power system at the same time and thus simplifying the comparison of different portions of the power system in real time.

Reliability is another important requirement in DG and SGs. Similar to other fields, it can be achieved by adopting best practices during design and realization of single elements and of the whole power system. The availability of a network of sensors and actuators, typical of SG, is commonly exploited for implementation of early prediction, fault detection and diagnosis. Power converters and the other intelligent systems connected with an SG usually include sensors and feedback control, and therefore they embed self-protection mechanisms and, eventually, self-healing.

In conclusion, DG and SGs intrinsically address the problems of energy flow optimization, flexibility and expandability. However, these performances can be greatly improved by using new HW systems as well as sophisticated and high-performance SW techniques capable of addressing the high complexity of the problem. For instance, the approach based on smart agents, recently proposed for control and management of SGs, reduces problem complexity by splitting the complex system into many simpler elements (agents) and defining a set of behavioral rules. Other ongoing researches use different approaches to reach the expected results.

3.9 Active Management of Distribution Networks

Active networks (ANs) consist of local areas with full integration, at control level, of distributed generators, network devices and customers, with the aim of balancing the electrical energy supply and demand, thus obtaining a better utilization of the available energy and more reliable operation. This enhances overall performance, offering new system services such as flexible voltage regulation, optimized power flow paths, congestion management, fault isolation, situational awareness, outage management, distribution automation, distribution management, asset management, smart metering, demand response management, management of other individual customers, and full network security [43–45].

ANs could be defined at an MG level, or in the case of high complexity, an MG could include a distinct AN area.

By virtue of the extensive use of ICT and an advanced metering infrastructure, the distribution network fully connects (power + data paths) suppliers, distributed system operators (DSOs) and customers; hence, ANs permit full coordination among generators, voltage regulators, reactive power compensators, on-load tap changers, loads and all other network devices. The information exchanges are used both for network management, thus to ensure optimum operations and a high level of security and for sophisticated billing management, thus for full exploitation of energy market operations [46–50].

This approach allows significant advantages in terms of efficiency of the power system and better utilization of physical infrastructure, in particular the distribution lines, resulting in a reduction or deferral of network investments. Some researches have demonstrated the economic benefits deriving from active management as compared with traditional network reinforcement strategies.

Ongoing research projects on the management concepts of ANs are focused on the resolution of interconnection issues of DGs. At the entry-level stage, AN management consists of monitoring and remote control of DG. At the intermediate stage, it permits the optimum allocation of a significant amount of DG, once the local and global services and trading issues have been defined. Full exploitation of ANs relies on overall network management.

However, due to the huge investments, new market rules are still required in order to balance the economic contribution among players in order to achieve government targets for the utilization of renewable sources and to offer economic benefits to DNOs. Therefore, in most countries, the present situation can be considered far from full exploitation of the current technical possibilities.

3.10 Communication Systems in Smart Grids

The communication infrastructure connecting energy generation, transmission, distribution systems and consumption points is an essential component for the development of an SG and to ensure full operations, security, reliability, flexibility, response demand management and other important features. The adopted technologies are fully digital, thus enabling data digitization, intelligent self-awareness and increased reliability.

SG communication infrastructures can be public or private. Public networks like the Internet and the 2G, 3G and 4G mobile networks are excellent and are at low cost, but due to the existing communications, their use may lead to several problems. Among them the most evident are the difficulty of achieving effective real-time and reliable communication and security issues. These issues can be addressed by defining suitable virtual private networks (VPNs). Power distribution lines themselves act as communication carriers by using Power Line Communication (PLC) technology, coupled with a combination of wired and wireless technologies, thus creating a complex but reliable communication infrastructure [51].

Due to the complexity and critical nature of problems, standardization of HW and SW communication technologies specific for SG applications is fundamental for successful operations.

The selection of a suitable communication architecture is subject to many challenging technical restrictions regarding the needs of bidirectional and real-time communications, interoperability between applications and reliable communications with low latencies and sufficient bandwidth [25]. Bandwidth, speed and real-time requirements depend on the specific application; for instance, billing information requires a very low bandwidth, while fast information exchanges are fundamental for early prevention of or system recovery from outages. Moreover, high system security has to be ensured to prevent cyber attacks and to provide power system stability and reliability.

A first element in the design of an SG communication infrastructure is the consideration of the system's scalability, as this greatly simplifies the system design, operations and maintenance.

Another fundamental element is usability, which can be easily achieved by using common interfaces such as a Web server, which is easily reachable and configurable using traditional and widely known browsers at SW level and widely diffused interfaces and protocols (e.g., Ethernet, Wi-Fi, ZigBee) for interconnection of nodes. However, due to security issues, they require some enhancements over general purpose applications.

Interoperability, that is, the ability of SG components to work cooperatively, exchanging information is another requirement. This requires adoption of some standards across the communication network.

There are two IEC standards (61970–301 and 61968–11) that are capable of describing different components and their interrelationships within a hierarchical architecture. Moreover, the IEC 61850 standard aims to improve the interoperability between Intelligent Electronic Devices (IEDs) for substation automation systems, while the IEEE P2030 standard provides interoperability between the electric power system (EPS) and customer-side applications, and the ANSI C12.22 standard is for communication modules and smart meters [19].

Quality of service (QoS) is another important requirement. Performance degradation such as delays or outages may compromise stability; therefore, QoS mechanisms must be provided to satisfy communication requirements. Usually they include specifications such as average delay, jitter and connection outage probability. To derive the QoS requirement, it is important to describe the probabilistic dynamics of the power system needed to evaluate the impact of different QoS specifications on the SG system and to derive the QoS requirement from the corresponding SG system.

In both wired and wireless networks, a very important problem is the availability of robust routing infrastructures, which are fundamental for reducing the influence of the above problems. Hybrid routing protocols that combine local agility with centralized control are still under development.

Other problems are the remote location of nodes, requiring very low-power electronic devices and lossy environments, where the presence of high interference requires adaptive and reconfigurable network operations.

Communications can be divided into three distinct categories or layers [19]:

- Wide area networks (WANs)
- Field area networks (FANs)
- Home area networks (HANs).

The three main tiers that are located between these three networks are the core backbone, back-haul distribution and access point.

The communication between back-haul aggregation points and the core backbone utility center is carried over different types of communication networks, such as star networks and fiber or wireless networks. In the following, a brief description of transmission categories is given:

- WANs provide high-bandwidth communications between electric utilities and substations for sensing, monitoring and control of the SG. Cellular networks, WiMAX and wired communications can be used for WAN, but fiber and microwave communications are preferred for their higher bandwidth and reliable communications. PLC is also possible but with reduced bandwidth [54]. Wired communications are better from a security point of view because of their intrinsically low possibility of cyber attacks.
- FANs act as bridges between customer premises and substations. Wireless communications such as Wi-Fi, WiMAX and Radio Frequency (RF) mesh technologies are suitable for FAN communications. Smart meters, advanced distribution automation and integration of distributed energy resources are some of the areas of application of FANs.
- HANs are connected with FANs by smart meters. HANs are low bandwidth and support communications among home electrical appliances and the smart meter. Many functions are implemented on HANs: smart metering, customers' information about their own consumption and electricity market opportunities (e.g., prediction of energy cost during the day) for optimum load management and communication with the FAN if the customer produces energy, for instance, with PV panels. Many interesting functions will be available in the future at HAN level for implementing intelligent load management. ZigBee, Wi-Fi, HomePlug, Z-wave and M-Bus are suitable for the HAN category. ZigBee has the ability to operate in a mesh network topology, which offers some advantages; that is, some devices in a ZigBee mesh can remain in sleep mode when they are not active in the network, which results in energy conversion. Wi-Fi cannot support mesh networking and is more expensive and more energy hungry. On the other hand, Z-wave is an interference-free wireless standard that is specifically designed for the remote control of appliances and widespread for HANs.

3.11 Advanced Metering Infrastructure and Real-Time Pricing

Smart metering and advanced metering infrastructures (AMIs) are fundamental in SGs. They create a bidirectional communication channel between provider and user, thus allowing cooperation among remote advanced sensors, monitoring systems and all computers, SW and data management systems available in a modern smart home connected with the SG, thus allowing consumers' participation in the energy market. The AMI includes a meter data management system (MDMS), located at FAN level, necessary for handling the huge amount of data and managing the raw data, creating meaningful information and messages for customers and assisting them in using energy intelligently. The same infrastructure can be used for managing information generated by different technological networks, such as electricity, gas,

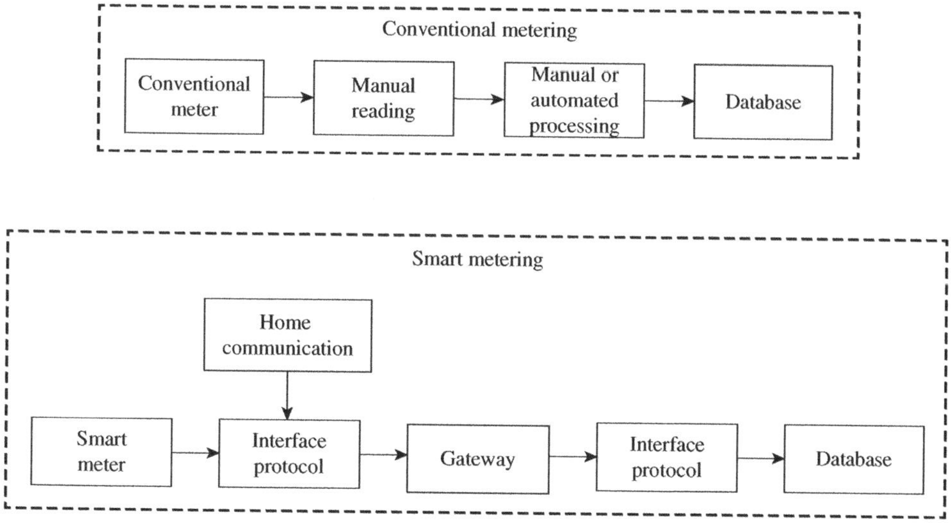

Figure 3.6 Smart metering

water and – if in the vicinity of a CHP thermal plant – heating. The most known advantage of AMI is that customers can participate in the energy market by changing their energy consumption instead of being passively exposed to fixed prices, resulting in profits for both service providers and end-users, but it also ensures the best SG performance (Figure 3.6) [55].

AMI can use different communication technologies, whose choice depends on the number of customers and the coverage per area, the availability of Internet connection, the scalability, the required data rate and expected communication delay, the expected energy efficiency and so on.

Low latency and high bandwidth are essential for some AMI critical applications, such as recovery from remote failures or elimination of incipient failures detected early, thanks to the available sensors network. PLC is widely diffused in urban areas, but may be insufficient for some real-time applications requiring bandwidths up to 100 kbps per device. The availability of optical fibers could solve the problem while increasing the number of services, too. Instead, RF meshes, GPRS and other cellular transmissions are the only viable technology to connect low-density areas. Therefore, in rural areas, smart meters are connected with MDMS by using wireless communications.

In order to successfully implement home energy management (HEM), consumption data gathered from each appliance are measured and transferred to a data concentrator, which could be the smart meter itself. Hence, data can be used locally to inform customers about their consumption behavior or for their active control and are also sent to the utility company for accounting and to enable remote response demand management. As already pointed out, ZigBee is currently considered an excellent solution for HEM networks. HomePlug and Bluetooth technologies could also be considered, but their cost is not dissimilar to that of ZigBee, while their performances are often lower (Figure 3.7).

3.12 Standards for Smart Grids

Full standardization of DG and SG devices and protocols is necessary to guarantee proper operations with high efficiency, reliability and security. Indeed, efficient and reliable operations can be achieved only by adhering to standards.

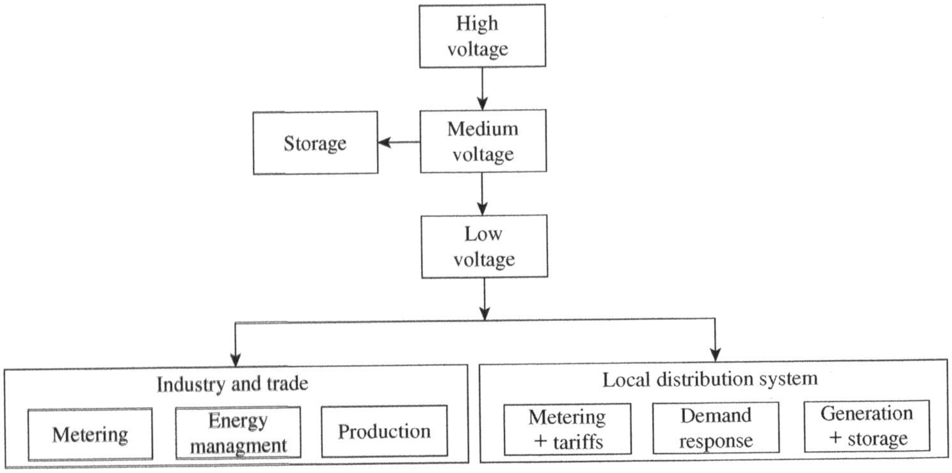

Figure 3.7 Advanced metering infrastructure

Because of the complexity and the wide spectrum, numerous distinct standards have been proposed and often revised to fit with the new developments for SG while others are under study.

Important standards with regard to the characteristics of DG and SGs are IEEE 1547, IEC 61850-7-420, IEC 61400–25, IEEE 1379, IEEE 1344 – IEEE C37.118 and IEEE 519.

IEEE 1547 is the standard for interconnection of distributed energy resources (DERs) and was published in 2003. It gives a set of criteria and requirements for the interconnection of DG resources into the power grid. Currently, there are six complementary standards designed to expand upon or clarify the initial standard, two of which are published, while the other four are still drafts.

IEC 61850-7-420 represents the standard related to communication and control interfaces for all DER devices. This standard defines the information models to be used in data exchange among DERs, which comprise DG devices and storage devices, including fuel cells, microturbines, PVs and combined heat and power. Where possible, it utilizes existing IEC 61850-7-4 logical nodes and defines DER-specific logical nodes where needed. Such standards allow significant simplifications in implementation, reduction of installation costs, sophisticated market-driven operations and simplification of maintenance, thus improving the overall reliability and efficiency of power system operations.

IEC 61400–25 defines the communication needed for monitoring and control of wind power plants. It is a subset of IEC 61400, a set of standards for the design of wind turbines. The standard allows control and monitoring of information from different wind turbine vendors in a homogeneous manner. The information is hierarchically structured and covers, for example, common information regarding the rotor, generator, converter, grid connection and the like. It covers all components required for the operation of wind power plants including the meteorological subsystem, the electrical subsystem and the wind power plant management system.

IEEE 1379 is the recommended practice for data communications between remote terminal units and intelligent electronic devices in a substation. This standard, published in 2000, provides a set of guidelines for communications and interoperations of remote terminal units (RTUs) and intelligent electronic devices (IEDs) in a substation. It does cover two widely used protocols for Supervisory Control and Data Acquisition (SCADA) systems: IEC 60870–5 and DNP3.

IEC standard 60870–5 deals with telecontrol, teleprotection and the associated telecommunications for electric power systems. It provides a communication profile for sending telecontrol messages between two power substations and defines operating conditions, electrical interfaces, performance requirements and data transmission protocols.

DNP3 (Distributed Network Protocol) is a set of communication protocols used between components in process automation systems. Recently, IEEE adopted DNP3 as IEEE standard 1815–2010 in 2010 and then modified it in the present standard 1815–2012.

Standard IEEE C37.118 is the current standard for measurement systems of synchronized phasors [56]. Synchronization is fundamental for ensuring proper operations and for eliminating potential faults and failures. A phasor measurement unit (PMU), which can be a stand-alone physical unit or a functional unit within another physical unit that estimates an equivalent synchrophasor for an AC waveform, is introduced. The total vector error compares both the magnitude and the phase of the PMU estimate with the theoretical phasor equivalent signal for the same instant of time. It provides an accurate method for evaluating the PMU measurement and establishes compliance requirements under steady-state conditions. The latter define the levels for phasor frequency, magnitude and angle measurements, harmonic distortion and out-of-band interference. It is worth noticing that IEEE C37.118 does not establish compliance requirements under dynamic conditions or other tests during which the amplitude or frequency of the signals varies, even if several dynamic tests are described in this standard.

IEEE 519 establishes the limits on harmonics amplitudes for currents and voltages at the PCC or at the point of metering in an EPS. The limits assure that the electric utility can deliver relatively clean power to all customers and protect its electrical equipment from overheating, avoid loss of life from excessive currents harmonics and prevent excessive voltage stress because of excessive voltage harmonics.

IEEE P2030 deals with interoperability of energy technology and information technology operation with the EPS and customer-side applications. It is responsible for bidirectional data transfer for electricity generation and reliable power delivery.

IEC 62351 is a standard for cyber security and protection of communication protocols from hackers' attacks.

Other standards have been defined for communications in WANs, FANs and HANs, for example, G3-PLC, HomePlug, PRIME, U-SNAP, IEEE P1901, Z-Wave, IEC 61970 and IEC 61969 and IEC 60870–6.

Finally, V2G operations are regulated by SAE J2293, which provides the requirements for EVs and electric vehicle supply equipment and SAE J2836 with regard to the communication between plug-in electric vehicles (PEV) and power grid, while SAE J2847 is specific for communications between PEVs and grid components.

In conclusion, numerous international standards regulate almost all aspects of DG systems and SGs, thus obliging equipment producers and users (service providers, generator builders and final users) to fulfill standard requirements to guarantee smooth and correct operation of the whole SG.

References

1. Xinghuo, Y., Cecati, C., Dillon, T., and Simões, M.G. (2011) The new frontier of smart grids. *IEEE Industrial Electronics Magazine*, **5** (3), 49–63.
2. Farhangi, H. (2010) The path of the smart grid. *IEEE Power and Energy Magazine*, **8** (1), 18–28.
3. ABB Toward a Smarter Grid: ABB's Vision for the Power System of the Future, http://www02.abb.com /db/db0003/db002698.nsf/0/36cc9a21a024dc02c125761d0050b4fa/$file /Toward_a_smarter_grid_Jul+09.pdf.
4. U.S. Department of Energy The Smart Grid: An Introduction, http://www.oe.energy.gov/DocumentsandMedia /DOE_SG_Book_Single_Pages.pdf (accessed 17 December 2013).

5. Smart Grids Smart Grids European Technology Platform SmartGrids, http://www.smartgrids.eu (accessed 17 December 2013).
6. European SmartGrids Technology Platform (EC) Vision and Strategy for Europe's Electricity Networks of the Future, ftp://ftp.cordis.europa.eu/pub/fp7/energy/docs/smartgrids_en.pdf (accessed 17 December 2013).
7. World Economic Forum Accelerating Smart Grid Investments, http://www.weforum.org/pdf/SlimCity/ Smart-Grid2009.pdf (accessed 27 December 2013).
8. The European Renewable Energy Council & Greenpeace [R]enewables 24/7: Infrastructure Needed to Save the Climate, http://www.greenpeace.org/raw/content/international/press/reports/renewables-24-7.pdf (accessed 27 December 2013).
9. Garrity, T.F. (0000) Getting smart. *IEEE Power and Energy Magazine*, https://w3.energy.siemens.com /cms/us/whatsnew/Documents/Getting%20Smart_Garrity.pdf.
10. Flynn, B. (2007) What is the real potential of the Smart Grid. AMRA International Symposium, 2007, http:// www.gepower.com/prod_serv/plants_td/en/downloads/real_potential_grid.pdf (accessed 27 December 2013).
11. Fabro, M., Roxey, T., and Assante, M. (2010) No grid left behind. *IEEE Security and Privacy*, **8** (1), 72–76.
12. Peças Lopes, J.A., Hatziargyriou, N., Mutale, J. *et al.* (2007) Integrating distributed generation into electric power systems: a review of drivers, challenges and opportunities. *Electric Power Systems Research*, **77** (9), 1189–1203.
13. Spagnuolo, G., Petrone, G., Araujo, S.V. *et al.* (2010) Renewable energy operation and conversion schemes: a summary of discussions during the seminar on renewable energy systems. *IEEE Industrial Electronics Magazine*, **4** (1), 38–51.
14. Zobaa, A.F. and Cecati, C. (2006) A comprehensive review on distributed power generation. Proceedings of SPEEDAM 2006, pp. 514–518.
15. Cecati, C., Mokryani, G., Piccolo, A., and Siano, P. (2010) An overview on the smart grid concept. Proceedings of 36th Annual Conference IEEE IECON, pp. 3322–3327.
16. Zhang, P., Li, F., and Bhatt, N. (2010) Next-generation monitoring, analysis, and control for the future smart control center. *IEEE Transactions on Smart Grid*, **1** (2), 186–192.
17. Blaabjerg, F., Teodorescu, R., Liserre, M., and Timbus, A.V. (2006) Overview of control and grid synchronization for distributed power generation systems. *IEEE Transactions on Industrial Electronics*, **53** (5), 1398–1409.
18. Rodriguez, P., Timbus, A.V., Teodorescu, R. *et al.* (2007) Flexible active power control of distributed power generation systems during grid faults. *IEEE Transactions on Industrial Electronics*, **54** (5), 2583–2592.
19. Gungor, V.C., Sahin, D., Koçak, T. *et al.* (2011) Smart grid technologies: communication technologies and standards. *IEEE Transactions on Industrial Informatics*, **7** (4), 529–539.
20. Gungor, V.C., Sahin, D., Koçak, T. *et al.* (2013) A survey on smart grid potential applications and communication requirements. *IEEE Transactions on Industrial Informatics*, **9** (1), 28–42.
21. Gungor, V.C., Sahin, D., Koçak, T. *et al.* (2012) Smart grid and smart homes: key players and pilot projects. *IEEE Transactions on Industrial Electronics Magazine*, **6** (4), 18–34.
22. Cecati, C., Ciancetta, F., and Siano, P. (2010) A multilevel inverter for photovoltaic systems. *IEEE Transactions on Industrial Electronics*, **57** (12), 4115–4125.
23. Vaccaro, A., Velotto, G., and Zobaa, A. (2011) A decentralized and cooperative architecture for optimal voltage regulation in smart grids. *IEEE Transactions on Industrial Electronics*, **58** (10), 4593–4602.
24. Carrasco, J.M., Franquelo, L.G., Bialasiewicz, J.T. *et al.* (2006) Power-electronic systems for the grid integration of renewable energy sources: a survey. *IEEE Transactions on Industrial Electronics*, **53** (4), 1002–1016.
25. Liserre, M., Sauter, T., and Hung, J.Y. (2010) Future energy systems: integrating renewable energy sources into the smart power grid through industrial electronics. *IEEE Transactions on Industrial Electronics Magazine*, **4** (1), 18–37.
26. Balaguer, I.J., Lei, Q., Yang, S. *et al.* (2010) Control for grid-connected and intentional islanding operations of distributed power generation. *IEEE Transactions on Industrial Electronics*, **58** (1), 147–157.
27. Buccella, C., Cecati, C., and Latafat, H. (2012) Digital control of power converters - a survey. *IEEE Transactions on Industrial Informatics*, **8**, 437–447.
28. Buccella, C., Cecati, C., Latafat, H., and Razi, K. (2013) A grid-connected PV system with LLC resonant DC-DC converter, Clean Electrical Power (ICCEP), 2013 International Conference on, pp. 777-782.
29. Portillo, R.C., Martín Prats, M.A., León, J.I. *et al.* (2006) Modeling strategy for back-to-back three-level converters applied to high-power wind turbines. *IEEE Transactions on Industrial Electronics*, **53** (5), 1483–1491.

30. Cárdenas, R., Peña, R., Tobar, G. *et al.* (2009) Stability analysis of a wind energy conversion system based on a doubly fed induction generator fed by a matrix converter. *IEEE Transactions on Industrial Electronics*, **56** (10), 4194–4206.
31. Cecati, C., Citro, C., Piccolo, A., and Siano, P. (2010) Smart operation of wind turbines and diesel generators according to economic criteria. *IEEE Transactions on Industrial Electronics*, **58** (10), 4514–4525.
32. Whittingham, M.S. (2012) History, evolution, and future status of energy storage. *Proceedings of the IEEE*, **100**, S.I., 1518–1534.
33. Li, X., Hui, D., and Lai, X. (2013) Battery Energy Storage Station (BESS)-based smoothing control of photovoltaic (PV) and wind power generation fluctuations. *IEEE Transactions on Sustainable Energy*, **4** (2), 464–473.
34. Chan, C.C. (2007) The state of the art of electric and hybrid, and fuel cell vehicles. *Proceedings of the IEEE*, **95** (4), 704–718.
35. Emadi, A., Lee, Y.J., and Rajashekara, K. (2008) Power electronics and motor drives in electric, hybrid electric, and plug-in hybrid electric vehicles. *IEEE Transactions on Industrial Electronics*, **55** (6), 2237–2245.
36. Saber, A.Y. and Venayagamoorthy, G.K. (2011) Plug-in vehicles and renewable energy sources for cost and emission reductions. *IEEE Transactions on Industrial Electronics*, **58** (4), 1229–1238.
37. Yilmaz, M. and Krein, P.T. (2013) Review of the impact of vehicle-to-grid technologies on distribution systems and utility interfaces. *IEEE Transactions Power Electronics*, **28** (12), 5673–5689.
38. Hatziargyriou, N., Asano, H., Iravani, R., and Marnay, C. (2007) Microgrids. *IEEE Power and Energy Magazine*, **5**, 78–94.
39. Katiraei, F. and Iravani, M.R. (2005) Power management strategies for a microgrid with multiple distributed generation units. *IEEE Transactions on Power Apparatus and Systems*, **21**, 1821–1831.
40. Katiraei, F., Iravani, R., Hatziargyriou, N., and Dimeas, A. (2008) Microgrids management. *IEEE Power and Energy Magazine*, **6**, 54–65.
41. Colson, C.M. and Nehrir, M.H. (2009) A review of challenges to real-time power management of microgrids. *Proceedings of the IEEE Power & Energy Society General Meeting*, pp. 1–8.
42. Guerrero, J.M., Vasquez, J.C., Matas, J. *et al.* (2011) Hierarchical control of droop-controlled AC and DC microgrids – A general approach toward standardization. *IEEE Transactions on Industrial Electronics*, **58** (1), 158–172.
43. Watanabe, E.H., Aredes, M., Barbosa, P.G. *et al.* (2007) Flexible AC transmission systems, in Power Electronics Handbook, 2nd edn, Elsevier, pp. 822–794.
44. Palensky, P. and Dietrich, D. (2011) Demand side management: demand response, intelligent energy systems, and smart loads. *IEEE Transactions on Industrial Informatics*, **7** (3), 381–388.
45. Chen, Z., Wu, L., and Fu, Y. (2012) Real-time price-based demand response management for residential appliances via stochastic optimization and robust optimization. *IEEE Transactions on Smart Grid*, **3** (4), 1822–1831.
46. Lu, D., Kanchev, H., Colas, F. *et al.* (2011) Energy management and operational planning of a microgrid with a PV-based active generator for smart grid applications. *IEEE Transactions on Industrial Electronics*, **58** (10), 4583–4592.
47. Harrison, G.P., Siano, P., Piccolo, A., and Wallace, A.R. (2007) Exploring the trade-offs between incentives for distributed generation developers and DNOs. *IEEE Transactions on Power Apparatus and Systems*, **22** (2), 821–828.
48. Albadi, M.H. and El-Saadany, E.F. (2007) Demand response in electricity markets: an overview. *IEEE Power Engineering Society General Meeting*, pp. 1–5.
49. Paudyal, S., Canizares, C., and Bhattacharya, K. (2011) Optimal operation of distribution feeders in smart grids. *IEEE Transactions on Industrial Electronics*, **10** (58), 4495–4503.
50. Shafiu, A., Bopp, T., Chilvers, I., and Strbac, G. (2004) Active management and protection of distribution networks with distributed generation. Power Engineering Society General Meeting, **1**, pp. 1098–1103.
51. Yang, Q., Barria, J.A., and Green, T.C. (2011) Communication Infrastructures for distributed control of power distribution networks. *IEEE Transactions on Industrial Informatics*, **2** (7), 316–327.
52. McMorran, A.W. (2007) An Introduction to IEC 61970-301 & 61968-11: The Common Information Model, Institute for Energy and Environment, Department of Electronic and Electrical Engineering, University of Strathclyde, Glasgow.
53. Southern Company Services, Inc (2010) Comments Request for Information on Smart Grid Communications Requirements, http://www.alvarion.com/index.php/(accessed 11 February 2014).

54. Paruchuri, V., Durresi, A., and Ramesh, M. (2008) Securing powerline communications. *Procedure of the IEEE International Symposium on Power Line Communications and its Applications (ISPLC)*, pp. 64–69.
55. Benzi, F., Anglani, N., Bassi, E., and Frosini, L. (2011) Electricity smart meters interfacing the households. *IEEE Transactions on Industrial Electronics*, **10** (58), 4487–4494.
56. Martin, K. (2011) Synchrophasor Standards Development – IEEE C37.118 & IEC 61850. *Procedure of the 44th Hawaii International Conference on System Sciences*, Hawaii.

4

Recent Advances in Power Semiconductor Technology

Jacek Rąbkowski[1,2], Dimosthenis Peftitsis[2] and Hans-Peter Nee[2]
[1]*Emissions Research and Measurement Division, Warsaw University of Technology, Warsaw, Poland*
[2]*Department of Electrical Energy Conversion, KTH Royal Institute of Technology, Stockholm, Sweden*

4.1 Introduction

Rapid development in power semiconductor technology is one of the driving forces behind the dramatic growth of the power electronics field. There is no doubt that traditional silicon technology dominates all types of devices, in applications ranging from tiny transistors in mobile and smartphones to 100–200 mm sized power transistors and thyristors in grid applications such as high-voltage direct current (HVDC) transmission. Silicon technology has established itself in this dominant position because of a mature manufacturing process, which results in power devices with application-suited parameters at a reasonable cost. It is very hard to imagine a situation in which devices made of any wide-bandgap (WBG) material can replace silicon power devices in low-cost applications, such as standard switch-mode power supplies or general-purpose industrial motor drives. The situation may be different in applications where specific device properties become more important than the cost. However, silicon devices are still improving despite the common opinion that this technology has reached its physical limits. The never-ending development of silicon technology can be illustrated by a multitude of examples. These include low-voltage metal-oxide semiconductor field-effect transistors (MOSFET)s reaching the on-state resistances of single mΩ, superjunction MOSFETs with breakdown voltages close to 1 kV with reasonable figures of merit (on-state resistance × gate charge), and insulated gate bipolar transistors (IGBTs) with improved thermal properties (maximum junction temperature of 175 °C).

At the same time, a big step has been made in silicon carbide (SiC) power electronics from being a promising future technology to being a potent alternative to state-of-the-art silicon technology in applications in which high efficiency, high frequency and high temperature are required. The reasons for this are that SiC power electronics may offer better performance: higher voltage ratings and lower voltage drops, higher maximum temperatures and higher thermal conductivities. Today, several manufacturers have both the knowledge and technology to process high-quality transistors at cost levels that facilitate the introduction of new products in applications where the benefits of the SiC technology can provide significant system advantages. The additional cost of the SiC transistors in comparison with their Si counterparts is

Power Electronics for Renewable Energy Systems, Transportation and Industrial Applications, First Edition.
Edited by Haitham Abu-Rub, Mariusz Malinowski and Kamal Al-Haddad.

Figure 4.1 40 kVA SiC inverter (300 × 370 × 100 mm) shown together with a copy of the *IEEE Industrial Electronics Magazine* [1]

today seriously smaller than the reduction in cost or increase in value seen from a system's perspective in many applications. In all the cases mentioned earlier, the SiC transistors are unipolar devices, such as junction field-effect transistors (JFET), MOSFET and bipolar junction transistors (BJT). At first sight the BJT appears to be a bipolar device. From experiments, however, it is found that presently available 1.2 kV SiC BJTs have properties very close to unipolar devices. At turn-on and turn-off, very small amounts of charge have to be injected and extracted, respectively. The reason for this is that the doping levels of 1200 V SiC transistors are so high that any considerable carrier injection is superfluous for the conduction mechanism. It is likely that true bipolar devices will be necessary for voltage ratings beyond 4.5 kV. The high-voltage high-power area in applications such as HVDC transmission seems to be an excellent field of application for IGBTs and BJTs in SiC, as very high number of series-connected devices would be necessary to withstand the system voltages. However, since the trend in voltage source converter (VSC) based HVDC is focused on modular converters, the need for devices with voltage ratings over 10 kV is not certain. The problem may be the relatively high voltage drop because of a SiC built-in potential of more than 3 V. In addition, since the fabrication of SiC IGBTs is far more complex than, for instance, that of an SiC JFET or BJT, it makes sense to first fully exploit the benefits of these devices. It is possible to build switch-mode inverters in the 10–100 kW range with high efficiencies (in excess of 99.5%). A fitting example may be a 40 kVA three-phase inverter built with 10 paralleled SiC JFETs on each switch position, shown in Figure 4.1. Reduced on-state resistances below 10 mΩ together with fast switching result in an efficiency of approximately 99.7%.

Finally, it must be noticed that devices made of another WBG material – gallium nitride (GaN) – has offered a voltage range up to 600 V. The advantage is that the fabrication cost is not that high in comparison with SiC transistors. At the same time, the parameters of the GaN unipolar devices outperform their silicon counterparts and may be interesting alternatives. But still many issues have to be solved to consider GaN power devices in well-established technology.

4.2 Silicon Power Transistors

As a background to the description of WBG semiconductor power transistors, a brief overview of state-of-the-art silicon-based power transistors is given. The presentation is focused on MOSFETs and

IGBTs because these two types of power transistors match the currently available WBG semiconductor power transistors with respect to voltage and power ratings. High-power devices such as thyristors and integrated-gate commutated thyristors (IGCTs) are not considered.

4.2.1 *Power MOSFETs*

The MOSFET was first described by Hofstein and Heiman in 1963 [2]. The operation of this device relies on the formation of a thin insulating oxide layer on top of the silicon surface. If a metallization is deposited on top of the oxide layer, the conductivity of the silicon beneath the oxide can be controlled by applying a voltage to the metallization on the surface of the oxide layer. If the silicon beneath the oxide layer is of p-type, a positive voltage on the metallization attracts electrons to the p-type silicon beneath the oxide. If a sufficient quantity of electrons is attracted, the p-type silicon appears to be an n-type material. This phenomenon can be used to create an n-type conduction channel through a p-type region. Then, the conduction channel can connect two n-type regions on each side of the p-type region beneath the oxide layer. The basic structure of a typical vertical power MOSFET is shown in Figure 4.2.

If a positive voltage is applied across the drain and source terminals in Figure 4.2 (with the higher potential on the drain terminal), the current through the channel can be controlled by means of the gate voltage; the higher the voltage, the higher the current. For gate voltages below a certain threshold voltage (about 5 V), the current is blocked because a sufficient amount of electrons cannot be attracted to the p-region below the oxide layer. For high values of gate voltage, the current saturates to a value determined by Rds(on), that is, the resistance from the source to drain when the device is saturated. Note that this conduction mechanism relies entirely on the drift of the majority carriers, which means that no dynamic effects are as such associated with the conduction mechanism. In order to charge the gate, however, a sufficient amount of charge has to be provided. This charge transport requires a certain time, both at turn-on and turn-off, but this period can be chosen with the gate driver design.

In the conduction state, the device is saturated with an on-state resistance Rds(on). The value of Rds(on) depends on the conductivity of the various regions in the structure and on the total cross-sectional area of the current flow from drain to source. In this context, it must be stated that the conductivity of the n^-

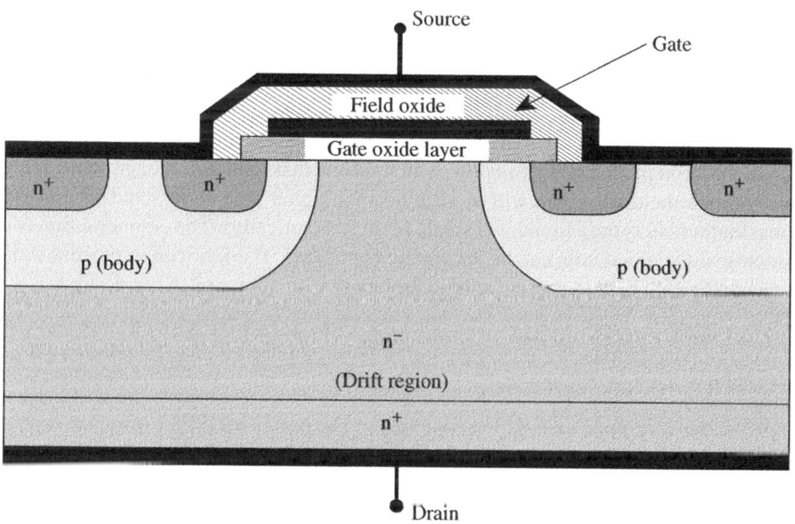

Figure 4.2 Basic structure of a vertical power MOSFET

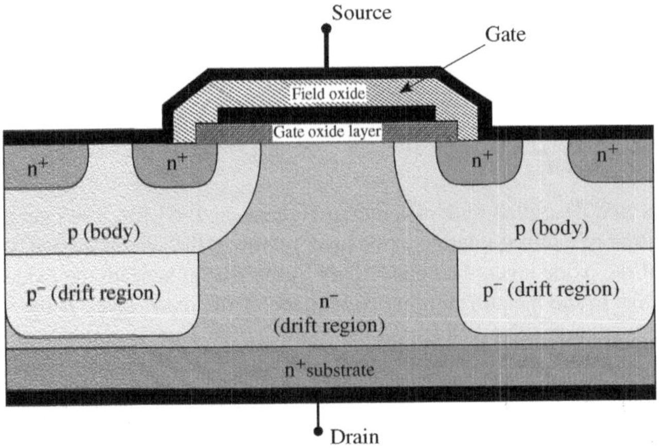

Figure 4.3 Basic structure of the superjunction MOSFET

drift region depends on the voltage rating of the device. For high blocking voltages, the doping must be low in order to keep the electric field in the depletion layer below the maximum allowable value. When the doping level is reduced, the conductivity of the material is also reduced. In addition, the depletion layer extends further down in the structure, which means that the thickness of the n^- drift region must be increased. The combination of reduced conductivity and increased distance gives a dramatic increase in Rds(on) if the voltage rating of the device is increased. As a rule of thumb, Rds(on) is approximately proportional to the voltage rating to the power of 2.5. For voltage ratings above 300 V, power transistors other than the standard MOSFET are preferred.

In order to increase the maximum useful blocking voltage of MOSFETs, the superjunction MOSFET was introduced by Deboy *et al.* in 1998 [3]. The new structure is shown in Figure 4.3. This design permits much lower values of Rds(on) than those of a standard MOSFET for voltage ratings above 300 V. For a 600 V design, Rds(on) may be more than three times lower than for a standard power MOSFET. For higher blocking voltages, the reduction in Rds(on) is even higher. Voltage ratings up to 900 V are presently available.

The basic idea behind the reduction in Rds(on) is a p^- column that penetrates deep into the structure from the p-body. This arrangement is often referred to as superjunction. The donors of the n^- drift region are compensated by acceptors in the p^- column. With effective compensation, the vertical electric field is almost constant throughout the drift region, which means that the total voltage when integrating the electric field through the drift region will be significantly higher than for a standard power MOSFET, where the maximum field is only found in a single point (theoretically). The compensation is only effective if the doping of the p^- column can be accurately controlled. The fabrication process is, therefore, a significant technological challenge, especially if structures with a finer pitch are desired.

4.2.2 IGBTs

In the late 1970s, the very good switching properties of the power MOSFET-stimulated researchers to locate modifications in the power MOSFET would permit higher blocking voltages without significantly increasing the on-state voltage drop. The key to this development was conductivity modulation, which means that the conduction process involves not only electrons as in an n-channel power MOSFET but also holes. For this modulation to take place, a highly doped p^+ region, which would inject the necessary holes into the structure, had to be added. In 1980, Becke and Wheatley filed a patent [4] describing the structure which today is called the IGBT. Basically, the IGBT is just a MOSFET with a highly doped p^+

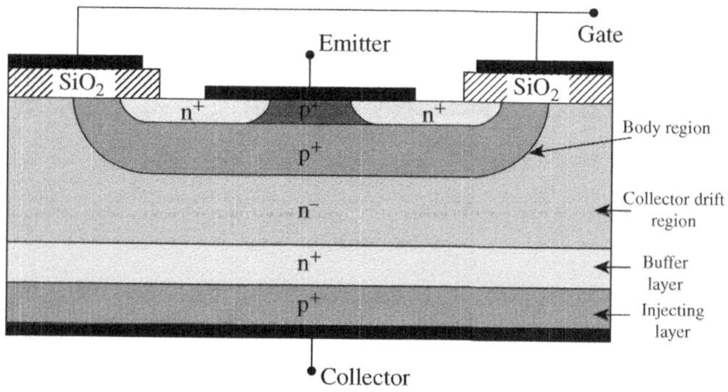

Figure 4.4 Basic structure of the punch-through IGBT

region added outside the drain of a power MOSFET. The first description of the benefits of the IGBT was presented by Baliga *et al.* [5]. Here, the n^+ drain from the power MOSFET had been replaced with the p^+ collector. The first commercial devices, however, re-introduced a more highly doped n-region between the collector and the drift region. This was called the buffer layer, and because this region effectively stopped the electric field the resulting device structure was later called a punch-through (PT) IGBT. The reason for re-introducing the buffer layer was to prevent the activation of a parasitic thyristor in the four-layered structure of the IGBT. The basic device structure of the PT-IGBT is shown in Figure 4.4.

As already indicated above, the IGBT is similar to the power MOSFET both with respect to conduction and switching. By controlling the gate voltage, the n-channel through the body region could be controlled in the same way as in the power MOSFET. An important difference, however, is that the conduction process also involves minority carriers injected from the collector into the lowly doped drift region. The process of generating the high level of charges in the drift region introduces a new dynamic property of the device structure because it takes a finite time to build up the charges. As a consequence, the saturation voltage in the on-state is only reached after a certain forward recovery process, and during turn-off it may have a considerable tail current. These effects have been described in numerous publications, for instance [6].

In 1988, Tihanyi introduced the non-punch-through (NPT) IGBT [7]. Owing to the absence of the buffer layer, the drift region of the NPT-IGBT must be thicker than for a PT-IGBT with the same blocking voltage in order to accommodate for the depletion region in the blocking state. This is of course not a benefit, but the on-state voltage drop does not necessarily have to be higher because the thickness of the substrate can be thinner for the NPT-IGBT. One benefit of the NPT-IGBT is the increased controllability of the manufacturing process compared to the PT-IGBT. As a result of this, the latch-up problem could also be solved without the need for a buffer layer. Another difference with the NPT-IGBT compared to the PT-IGBT is that the NPT-IGBT has a lower carrier plasma density on the collector side and a higher carrier lifetime. This means that the tail current is lower in magnitude but has a considerably longer duration. This may be a drawback in soft switching applications [6]. The basic device structure of the NPT-IGBT is shown in Figure 4.5.

Later, this type of IGBT was developed for ever higher voltage and current ratings, and for several years IGBTs with voltage ratings of up to 6 kV have been available. These devices typically have bipolar characteristics with significant conductivity modulation lags and tail currents. The high-power devices are also available in presspack modules, where bond wires and solderings are avoided in order to increase the reliability of the device.

Many modern IGBT designs make use of a trench gate [8]. The main objective of the trench-gate structure is to increase the carrier plasma density on the emitter side of the n^- drift region. If this is achieved, the voltage drop across the drift region can be reduced [9]. At first sight this may seem problematic, since an increased carrier plasma density could be associated with an increased tail current during

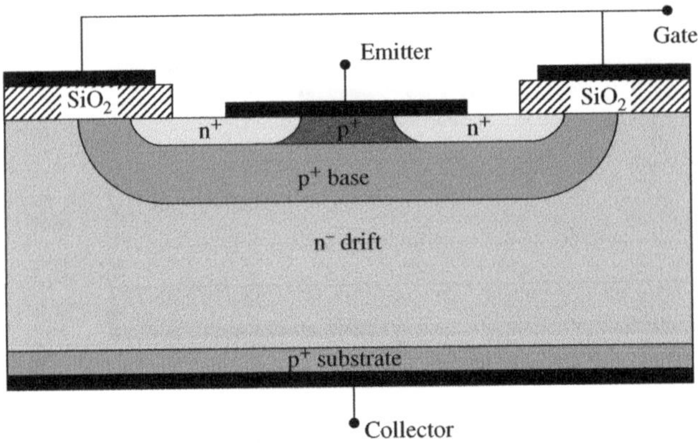

Figure 4.5 Basic structure of the non-punch-through IGBT

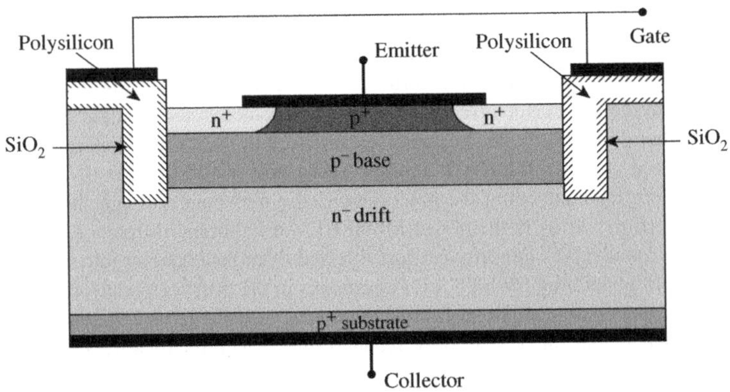

Figure 4.6 Basic structure of a trench-gate IGBT

turn-off. From experience, however, it is found that this is not a significant problem. On the contrary, the plasma on the collector side dominates the influence on the tail current. Especially (but not only) for high-voltage designs, trench-gate IGBTs have been successfully commercialized due to their low on-state voltage drop, good switching performance and good non-latch-up property. In Figure 4.6, the basic structure of a trench-gate IGBT is illustrated.

As already mentioned, the NPT-IGBT requires a comparably thick n^- drift region because of the triangular electric field shape. This means that there is the potential to reduce both the on-state losses and switching losses if the drift region can be made thinner. By implementing a field-stop layer below the p^+ emitter, the field can be stopped without influencing the p^+ emitter [10]. In this way, a more trapezoidal electric field shape is achieved, which results in a reduction in thickness of approximately one-third. It is essential that the field-stop layer has a low dose, such that the comparably low doping of the NPT p^+ emitter can be maintained. Otherwise, the magnitude of the tail current could increase. Today, the field-stop technology is state of the art, often in combination with trench-gate structures. Sometimes, also an integrated antiparallel diode is implemented by modifying the structure at the collector side. All these new devices have both low on-state losses and low switching losses. As such, they are serious competitors to all new WBG power transistors.

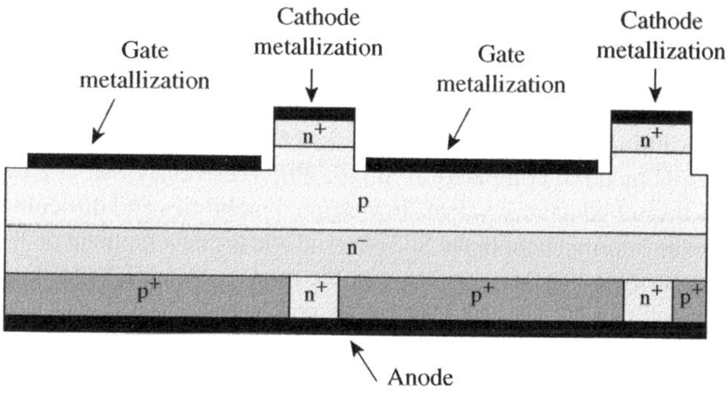

Figure 4.7 Basic structure of the IGCT

4.2.3 High-Power Devices

Although high-voltage IGBTs can be designed for very high-power devices by the parallel connection of multiple chips, the truly tailor-made power switch in this field is the IGCT. This makes significantly better use of the silicon area than the high-voltage IGBT because the IGCT is a whole-wafer component where the edge terminations constitute only a tiny fraction of the area, whereas in high-voltage IGBTs the edge terminations considerably reduce the effective silicon area with chip sizes of approximately 12 mm times 12 mm. The IGCT is a refined development of the gate turn-off thyristor (GTO), where an optimized gate driver is integrated with the thyristor. The IGCT exhibits the following characteristics when compared with those of the GTO [11]:

- Reduced on-state and turn-off losses by the minimization of the silicon thickness.
- Eliminated need for d*v*/d*t* snubber.
- Reduced gate-power requirements, especially during conduction.
- Dramatically reduced storage time at turn-off (from 20 μs to 1 μs).

Altogether, the IGCT is close to the ultimate power switch and it is therefore suitable for various high-power applications. The only thing that can be put against the IGCT is that it is an inherently slow power switch. This is a result of an optimization of the conduction losses. From an application viewpoint it is also questionable whether fast switching speeds are desirable, because this would result in additional stresses on the insulation systems of the surrounding equipment (Figure 4.7).

4.3 Overview of SiC Transistor Designs

The combination of great quality improvement in the SiC material [12] together with excellent research and development has resulted in a breakthrough in the design and fabrication of SiC devices. This has paved the way for a strong commercialization of SiC switch-mode devices [12, 13]. Nevertheless, the SiC device market is still in an early stage, and, today, the majority of available SiC transistors (the JFET [13, 14], the BJT [15] and the MOSFET [16, 17]) are still not in mass production. Despite the visible downward trend, the cost of these components is still higher than their silicon counterparts. In addition, currently available discrete devices are only suitable for low-power applications because of the low voltage and current ratings of these SiC devices. In particular, voltage ratings in the range of 1200 V and current ratings of few tens of amperes are available. Regardless of the type of SiC transistor, the driver counts as a vital part when the devices operate in a power electronics converter. The driver requirements differ among devices and they should be designed in such a way that they ensure reliable operation. Finally, it is also worth mentioning the research progress on the SiC IGBT [13].

4.3.1 SiC JFET

The first attempts to design and fabricate an SiC JFET were made in the early 1990s, when the main research issues were dealing with design optimization in order to realize high-power and high-frequency SiC devices [18]. It was during these years that a few research groups started mentioning the advantageous characteristics of the SiC material compared to Si [19, 20]. These early SiC JFETs, however, suffered from comparably low transconductance values, low channel mobilities and difficulties in the fabrication process [19, 20]. Both the improvement in the SiC material and the development of 3″ and 4″ wafers over the last decade have contributed to the fabrication of the modern SiC JFETs [13], and it was in around 2005 when the first prototype samples of SiC JFETs were released.

One of the modern designs of the SiC JFET is the so-called lateral channel JFET (LCJFET), as shown in Figure 4.8.

The drain current can flow in both the positive and negative directions depending on the external voltage applied, and it is controlled by a buried p^+ gate and an n^+ source p–n junction. This SiC JFET is a normally-on device, and a negative gate–source voltage must be applied in order to turn the device off. By applying a negative gate–source voltage, the channel width is decreased due to the creation of a certain space charge region and a reduction in the current is obtained. Below a certain negative value of the voltage applied to the gate–source junction no drain current can flow. This gate–source voltage is named the "pinch-off" voltage. The range of the pinch-off voltages of this JFET type is usually between −16 V and −26 V. Another important feature of this structure is that the p^+ source side, the n^- drift region and the n^{++} drain form the antiparallel body diode. However, the forward voltage drop of the body diode is higher compared to the on-state voltage of the channel [13, 21] at rated (or lower) current densities. From an on-state power losses point of view, the use of the channel is preferable for negative currents. Thus, the body diode may only be used for safety for short-time transitions [22]. This type of SiC JFET was released by SiCED (Infineon Technologies) a few years ago.

The second commercially available SiC JFET was the vertical trench JFET (VTJFET) (released in 2008 by SemiSouth Laboratories) [23, 24]. A cross-sectional schematic of its structure is shown in Figure 4.9. The SiC VTJFET can be produced either as a normally-off (enhancement-mode vertical trench JFET, EMVTJFET) or as a normally-on (depletion-mode vertical trench JFET, DMVTJFET) transistor. The difference in the device characteristics is made by the change in the vertical channel thickness and the doping levels of the structure. A negative gate–source voltage is required in order to keep the normally-on depletion-mode JFET in the off-state. On the other hand, a significant gate current (approximately 200 mA for a 30 A device) is necessary for the normally-off enhanced-mode JFET in order to

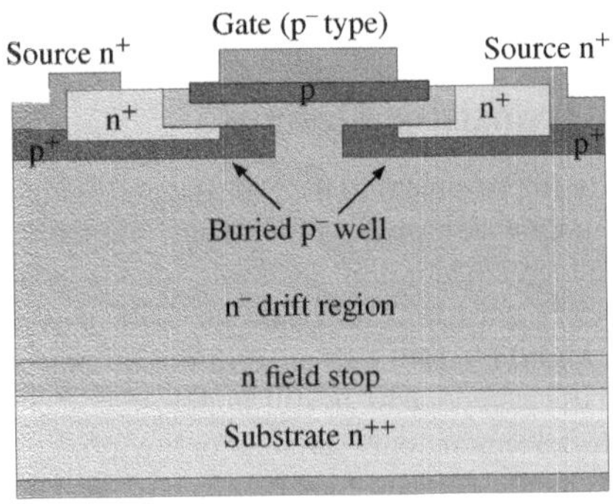

Figure 4.8 Cross-section of the normally-on SiC LCJFET [1]

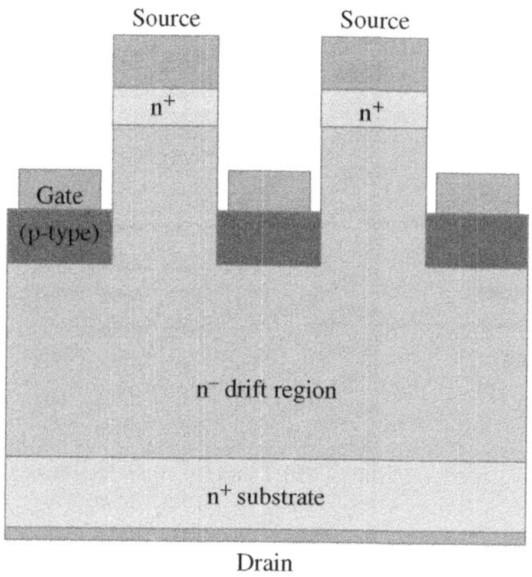

Figure 4.9 Cross-section of the SiC VTJFET [1]

keep it in the conduction state. The value of the pinch-off voltage for the DMVTJFET is negative (in the level of −6 V), while the value of the pinch-off voltage for the EMVTJFET is slightly positive, around 1 V. Comparing this type of SiC JFET with the LCJFET, the absence of the antiparallel body diode in the DMVTJFET makes the LCJFET design more attractive in numerous applications. On the other hand, an external, antiparallel SiC Schottky diode seems to very interesting solution for the VTJFET. Again, this diode may only be utilized for short-time transients, which is similar to the internal diode of the LCJFET. Except during these short-time transients, the reverse current should flow through the channel. When VTJFETs are operating in parallel connection, the antiparallel SiC Schottky diode is an especially attractive solution because the voltage drop across the transistors is lower than the threshold voltage of the diode. It must be noted that a single diode would be adequate for all parallel JFETs due to the short (<500 ns) conduction time of the diode.

Two more designs of SiC JFET were demonstrated, as depicted in Figure 4.10a and b [25]. The first is a buried grid JFET (BGJFET), which is shown in Figure 4.10a. Both low-specific on-state resistance and high-saturation current densities are achieved due to the small cell pitch of the BGJFET. However, difficulties in the fabrication process may be considered a basic drawback of this design compared to the LCJFET. Figure 4.10b presents the double-gate vertical channel trench JFET (DGVTJFET) design, proposed by Denso [25], which combines the design characteristics of the LCJFET and the BGJFET. In [25], it is claimed that fast switching performance can be achieved due to the low gate–drain capacitance. Moreover, the small cell pitch and the double-gate control contribute to the low-specific on-state resistance.

Toward the end of 2012, the commercially available SiC JFETs had voltage ratings of mainly 1200 V and 1700 V. Current ratings of 48 A and 30 A for normally-on JFETs and normally-off counterparts, respectively, were also available.

4.3.2 Bipolar Transistor in SiC

The SiC bipolar transistor is a normally off device that combines both a low saturation voltage (0.32 V at 100 A cm^{-2}) [15] and very fast switching performance. Figure 4.11 illustrates a typical cross-section

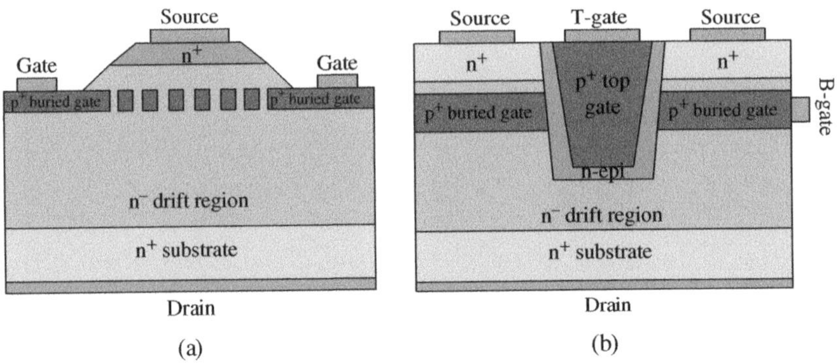

Figure 4.10 (a) Cross-section of the BGJFET and (b) cross-section of the SiC DGTJFET [1]

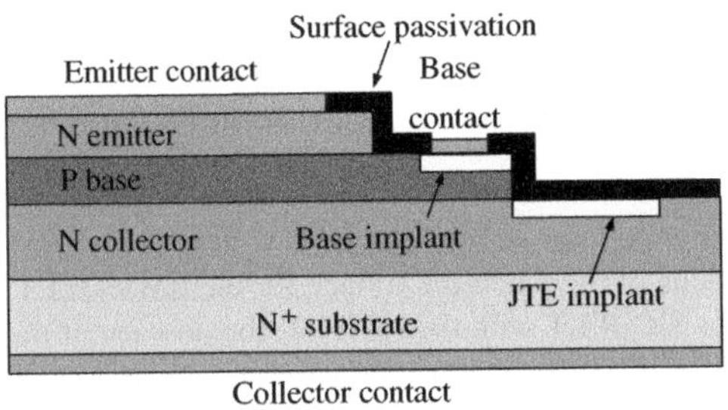

Figure 4.11 Cross-section of the SiC BJT [1]

of this device (NPN BJT). The cancelation of the base–emitter and base–collector junction voltages results in very low-saturation voltage values. However, since the SiC BJT is a current-driven device, a substantial continuous base current is required to keep the device in deep saturation. SiC BJTs of 1200 V with current ratings in the range of 6–20 A and current gains higher than 70 were reported [26]. The fabrication of SiC BJTs with improved surface passivation leads to current ratings of 50 A at 100 °C and current gains higher than 100. A serious drawback is that the current gain is significantly dependent on the temperature (is reduced by more than 50% when the junction temperature is increased by 250 °C). Despite the base current requirement, SiC BJTs with a competitive performance in the kV range are expected in the future.

4.3.3 SiC MOSFET

Research and development into the MOSFET in SiC was a challenging issue, especially when the fabrication and stability of the oxide layer is considered. Figure 4.12 presents a cross-section of a typical SiC MOSFET design, which is very attractive to designers of power electronic converters due to the normally-off characteristic. Unfortunately, low channel mobility results in an increased on-state resistance of the device, and thus additional conduction power losses also ensue. Moreover, the reliability and the stability of the gate oxide layer and the body diode, especially over long time periods and at elevated temperatures, have not yet been confirmed. Fabrication issues also contribute to the deceleration in SiC MOSFET development. However, the currently available SiC MOSFETs show a quite

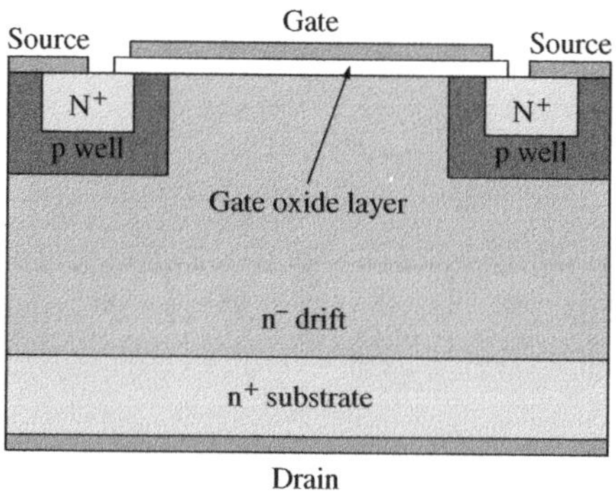

Figure 4.12 Cross-section of the SiC MOSFET [1]

promising performance in terms of switching behavior and on-state resistance values. It is believed that these devices will be massively produced in a first place by various manufacturers.

The first commercial SiC MOSFET was released during the end of 2010 by Cree, while other manufacturers (e.g., ROHM, ST Microelectronics) are also close to having their devices on the market [12, 16]. Currently, 1200 V SiC MOSFETs with current ratings of 20–48 A (on-state resistances of 80 mΩ and 160 mΩ) are also available, as well as 25 mΩ bare dies. Furthermore, SiC MOSFET chips rated at 10 A and 10 kV have also been reported by Cree as a part of a 120 A half-bridge module [27]. The 10 kV SiC MOSFETs offer very good performance than the state-of-the-art 6.5 kV Si IGBT, but the commercialization of such a unipolar SiC device is not foreseen in the near future.

4.3.4 SiC IGBT

Owing to its excellent performance, the silicon-based IGBT dominates in many application fields for a wide range of voltage and current ratings. The fabrication of a silicon n-type IGBT is started on a p-type substrate, and such substrates are also available in SiC. Unfortunately their resistivity was found to be unacceptably high, which prevents these components from being applied in power electronics converters because of large conduction losses. Furthermore, the gate oxide layer issue also results in high channel resistivities. These problems have already been reported by many research teams, and it is believed that such SiC devices will not be commercialized within the next 10 years [13, 28]. In contrast, even if high-voltage SiC IGBTs are released into the market in the future, it is not clear that they will provide power losses as low as a high-voltage SiC JFET (e.g., 3.3 kV SiC JFET). This may be an issue if converters based on modular concepts (for instance modular multilevel converters – M2C) will be used for high-voltage high-power applications [29].

4.3.5 SiC Power Modules

In order to facilitate the easy and compact design of main circuits for high-power converters with SiC switches, power modules must be developed. Recently, a number of attempts have been made to achieve both high current ratings and high temperatures. The highest current ratings presented so far are the 880 A SiC MOSFET module in [12] and the 800 A SiC MOSFET module in [13]. In [12], plans for 1600 A modules are also presented. Another interesting example is the 1200 V/770 A module shown in Figure 4.13. This module is a half-bridge with 14 parallel normally-on SiC JFETs with a total current

Figure 4.13 Photograph of the APEI power module

rating of 378 A. The module has been designed by APEI using SiC 85 mΩ JFET chips from SemiSouth Laboratories.

4.4 Gate and Base Drivers for SiC Devices

The advantageous performance of SiC transistors, especially during the switching transients, may be only utilized with suitable driver circuits. These circuits must be capable of providing rapid switching, but should also be accompanied by optimized power consumption. In addition, operation at high temperatures may also be important for these drivers due to the high-temperature capability of SiC devices. Various driver circuits for SiC JFETs, SiC BJTs and SiC MOSFETs have been proposed since these devices have begun to be available on the market, and selected concepts are presented in this section.

4.4.1 Gate Drivers for Normally-on JFETs

The normally-on SiC JFET is a voltage-controlled device, which is in on-state without any gate bias. A negative gate–source voltage, lower than the pinch-off, must be provided in order to turn-off the drain current. The pinch-off voltage varies between different designs: in the range of −16 V and −26 V for LCJFET and around −6 V for DMVTJFET. The simplest method of the normally-on JFET driving is the use of the standard totem-pole driver without a positive supply (see Figure 4.14a). When the upper transistor is in the on-state the gate potential is the same as the source and the JFET is turned on, otherwise the negative supply V_{EE} turns the device off. The switching speed can be controlled by means of a gate resistor R_G. According to JFETs features, a slight positive voltage applied to the gate–source junction during the on-state may reduce the device's resistance by 10–15%. Thus, a positive supply of 2.5 V can be added to the driver configuration in Figure 4.14a. An example of the switching performance of the simple totem-pole driver is presented in Figure 4.15a, in which the turn-on process of the 1200 V/85 mΩ JFET can be observed.

In order to ensure the safe operation of the driver, the negative supply voltage V_{EE} should be chosen to be lower than the pinch-off but higher than the reverse breakdown voltage of the gate–source diode. A low margin between the pinch-off and the breakdown voltages in conjunction with the parameters' spread of

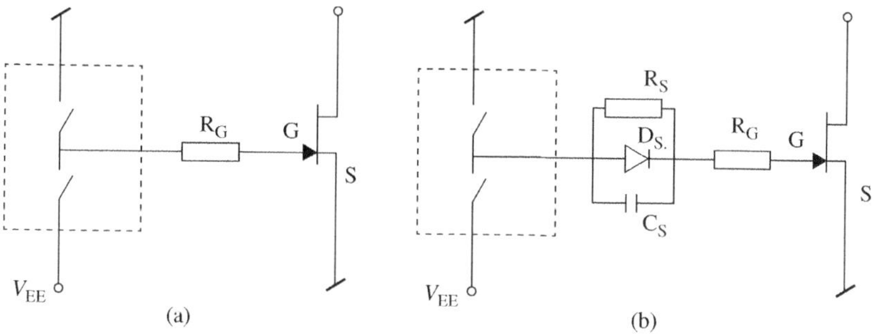

Figure 4.14 Gate drivers of the normally-on SiC JFET: (a) simple totem-pole driver and (b) driver with DRC network presented in [30]

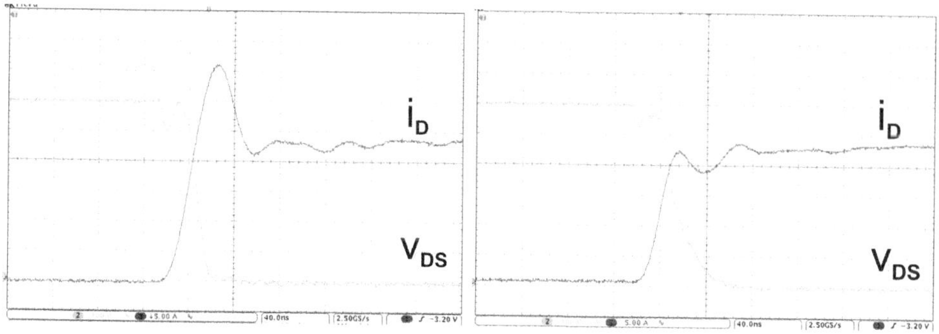

Figure 4.15 The turn-on process of the SiC JFET with (a) the simple totem-pole 2.5 V/–15 V driver, (b) DRC driver ($C_S = 22$ nF, $R_G = 10\,\Omega$, $V_{EE} = 30$ V) – all 40 ns/div, 100 V/div/5 A/div

the first-released LCJFETs made the selection of the appropriate negative supply very hard. One of the most popular gate drivers for the LCJFET was proposed in [30], which seems to be a very good solution in order to overcome this problem. The main part of the driver (Figure 4.14b) is a parallel-connected network with a diode D_S, a capacitor C_S and high-value resistor R_S, while a gate resistor R_G is connected in series. The negative supply V_{EE} may be lower than the reverse breakdown voltage of the gate–source diode as the gate current is limited by R_S. In order to keep the JFET in the on-state, the totem-pole supplies 0 V by shorting the gate to the source potential. However, when the device is expected to be turned off, the totem-pole voltage jumps from zero to the negative supply. As a consequence, a current peak is supplied to the gate of the JFET through the gate resistor R_G and the capacitor C_S. In the steady state, the voltage drop across the capacitor C_S equals the voltage difference between $-V_{EE}$ and the reverse breakdown voltage of the gate. A small amount of the gate current flows through the resistor R_S but the associated power loss can be neglected as the current value is very low (a few milli-amperes). The switching speed can be adjusted to any value by selecting a suitable value of the gate resistor R_G and the capacitor C_S. An example of the switching waveforms of SiC JFET driven by the gate driver shown earlier is shown in Figure 4.15b.

Despite the various advantages of the normally-on SiC JFET, a serious issue is that in case of the power supply for the gate driver being lost a possibly shoot-through may destroy the device. Therefore, a “smart” gate driver is required in order to deal with this problem in practical applications. A successful example of a “smart” gate driver for normally-on SiC JFETs is the self-powered gate driver (SPGD), which has been presented in [31]. It is basically not only a circuit solution to the shoot-through problem

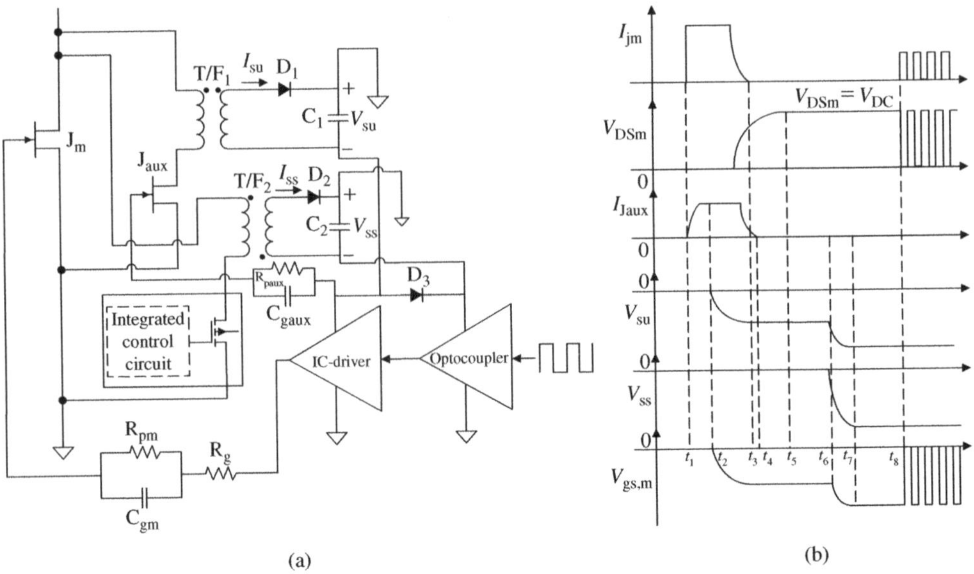

Figure 4.16 Detailed schematic (a) and the start-up sequence (b) of the self-powered gate driver [31]

(start-up process), but it is also able to supply the required power to the gate during steady-state operation. A circuit schematic of the SPGD is shown in Figure 4.16a. The SPGD consists of two DC/DC converters: start-up and steady-state converters. Both of these converters are connected in parallel to the normally-on SiC JFET. It must also be noted that an optocoupler and an integrated circuit driver (IC driver) are also necessary. The first isolates the controller signals from the gate driver circuit, while the IC driver is a totem-pole circuit consisting of MOSFETs.

The start-up converter is basically a modified forward DC/DC converter. When the normally-on SiC JFET (J_m) is subjected to a DC-link voltage, a shoot-through current is flowing and the energy of the shoot-through current basically feeds the start-up converter and an adequately negative voltage V_{su} is supplied for a certain time in order to turn-off the main transistor J_m. The vital parts of this converter are the auxiliary normally-on SiC JFET, J_{aux}, the high-turns ratio transformer T/F$_1$, the diode D_1 and the output capacitor C_1. Similar to J_m, J_{aux} is also turned off when the output voltage V_{su} is more negative than its pinch-off voltage.

The steady-state converter is a DC/DC flyback converter that converts the high-value blocking voltage across J_m to a low and negative voltage V_{ss}, which supplies the gate driver. The MOSFET with the integrated controller is able to start operating when the drain-to-source voltage across it exceeds a certain value. Thus, the steady-state converter also starts operating and supplies both the optocoupler and the IC driver with a negative voltage. It is clear from Figure 4.16a that the IC driver only supplies the gate of the main JFET J_m, while the gate of the auxiliary SiC JFET J_{aux} is directly connected to the output of the converters. Thus, it means that J_{aux} is always kept in the off-state during the steady-state operation of the gate driver, while J_m is controlled by the IC driver.

Figure 4.16b shows the start-up sequence of the SPGD. It is worth looking at the last trace in this figure, where the gate–source voltage of the main SiC JFET $V_{gs,m}$ is illustrated. This voltage equals the output voltage of the start-up converter until t_6, while after this time (when the steady-state converter has started operating) it is equal to V_{ss}. If it is assumed that after t_8 the pulse-width modulation (PWM) operation starts, it is clear that $V_{gs,m}$ jumps from 0 to each negative value and vice versa.

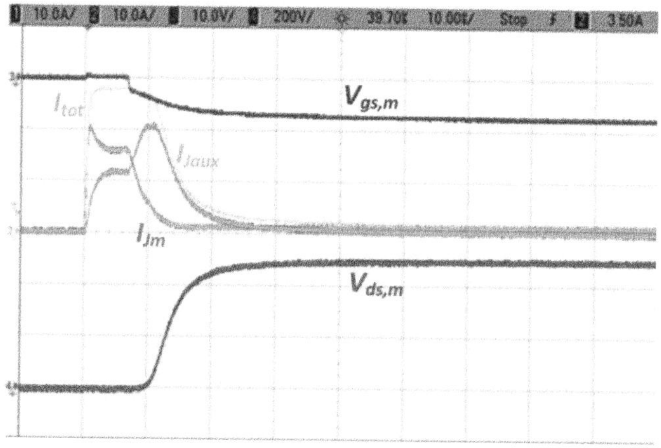

Figure 4.17 Measured gate–source voltage of the main SiC JFET, J_m (purple line, 10 V/div), drain–source voltage of J_m (dark-pink line, 200 V/div), shoot-through current I_{Jm} (green line, 10 A/div), short-circuit current flows through J_{aux}, I_{Jaux} (yellow line, 10 A/div) and the sum of the shoot-through currents (light-pink line, 10 A/div; timebase 10 μs/div) [31]

The SPGD has been tested on both a stand-alone test circuit and a half-bridge converter consisting of two normally-on SiC JFETs, J_{m1} and J_{m2}. Experimental results showing the performance of the SPGD when it is employed on the stand-alone circuit is presented in Figure 4.17. In particular, this figure shows the shoot-through currents I_{jm} and I_{jaux} which are cleared after approximately 20 μs, while the blocking voltage of J_m is also illustrated. Finally, Figure 4.18 shows the performance of the SPGD when it is operating on the half-bridge converter. The time intervals where the start-up and the steady-state converters are operating are clearly defined in this figure. It is also observed from Figure 4.18 that after a certain time of steady-state converter operation, the PWM signal is provided and the switching process starts.

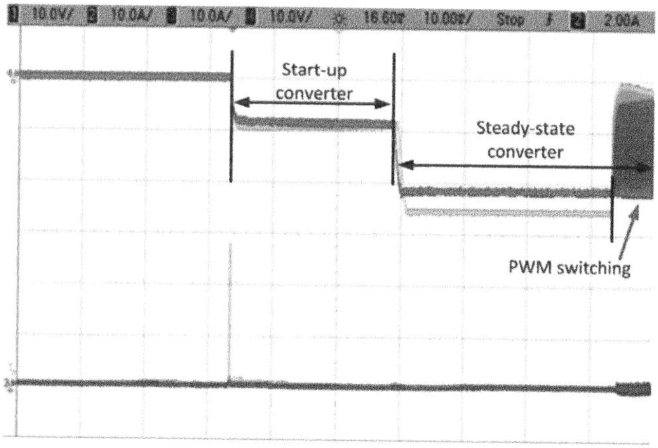

Figure 4.18 Measured gate–source voltage of the upper SiC JFET, J_{m1} (pink line, 10 V/div), gate–source voltage of the lower SiC JFET, J_{m2} (yellow line, 10 V/div), shoot-through current I_{Jm} measured on the drain of J_{m1} (green line, 10 A/div) and on the drain of J_{m2} (purple line, 10 A/div; timebase 10 ms/div) [31]

Apart from the solution analyzed above, there are a few more solutions to the normally-on problem, which have been presented so far. A well-known solution is based on a cascode connection of the SiC JFET with a low-voltage silicon MOSFET [32]. Thus, the normally-on SiC JFET performs as a normally-off switch. However, a serious drawback of this solution is the additional voltage drop across the MOSFET, while by introducing a silicon device high-temperature operation is prohibited. A second solution that is specialized for VSCs has been studied in [33].

The protection scheme proposed in this paper is based on a linear regulator consisting of a normally-on SiC JFET, which is connected to the DC link. During a shoot-through condition the protection scheme is enabled, and an ultrafast converter supplies a negative voltage to the gates of the lower SiC JFETs of the VSC. An external supply, however, is necessary to supply the steady-state power to the gates. Another protection scheme for VSCs built with normally-on SiC JFETs has also been presented in [34]. The Si IGBT is connected in series with a relay while both are connected in parallel with a charging resistor. Using this protection scheme, which is applied in the mid-point between the DC-link capacitors, short circuits can be cleared within very short periods. However, the steady-state power to the gate drivers of the SiC JFETs also needs to be externally supplied. A protection circuit for normally-on SiC JFETs employed in a switch-mode power supply (SMPS) fed by the grid has been presented in [35]. In this case, the inrush current during the start-up process is utilized in order to turn-off the JFET. Regardless of whether the sinusoidal input voltage starts at zero or at any other value, the safe operation of the JFET is ensured. As for most of the previous ideas described above, the drawback of an external power supply for the gate driver under steady-state operation is also met in this case.

4.4.2 *Base Drivers for SiC BJTs*

The SiC BJT is a current-driven device, which requires a substantial base current during the on-state in order to maintain the device in the saturation region. Then, owing to voltage drops cancelation (v_{BE} and v_{BC}), a collector–emitter voltage is significantly lower than the built-in voltage of the SiC P–N junction. An amount of the base current that needs to be delivered to the base–emitter junction from the base driver units depends on the power circuit conditions (collector current, junction temperature) and the BJT parameters (current gain β). As an example, a 1200 V/6 A BJT employed in a 2 kW DC/DC boost converter [36] can be taken, where 320 mA of steady-state base current leads to collector–emitter voltage being lower than 1 V. The power loss becomes a critical issue for the base-drive unit as the substantial steady-state base current flows through the base–emitter junction (voltage drop around 3 V). In applications where the collector current is varying versus time, proportional drivers could be interesting solutions in order to limit the driving power loss of the BJT. Since SiC BJTs are definitely faster devices than their Si counterparts, the well-known proportional base driver circuits with the use of transformers might not be the best option. The most problematic issue is the introduction of the parasitic inductances and capacitances to the main circuit, which could affect the transistor switching performance. Doubtless, more suitable solutions can be found with the use of modern electronic components. All in all, the necessary condition is a fair trade-off between the base driver complexity and expected power loss savings. In the following section, different concepts of the SiC BJT base driver units are discussed, all of which operate with a fixed steady-state base current.

Figure 4.19a shows a simple base driver for SiC BJTs utilizing the totem-pole circuit and a series base resistor R_B, which is selected such that a suitable steady-state base current is adjusted. As a consequence, the switching speed of the BJT with such a driver is rather low. The switching performance may be improved by connecting a parallel speed-up capacitor, C_B, as shown in Figure 4.19b [15]. After every change in the totem-pole output state this capacitor is charged/discharged, which results in a base current peak. More rapid peaks lead to higher switching speeds, which may be adjusted by the supply voltage V_{CC} and the capacitance of C_B. The higher the supply voltage and the capacitance, the faster the switching transients. On the other hand, the power consumption of the base driver is also increased and, therefore, it seems that the switching speed and the power consumption is a trade-off.

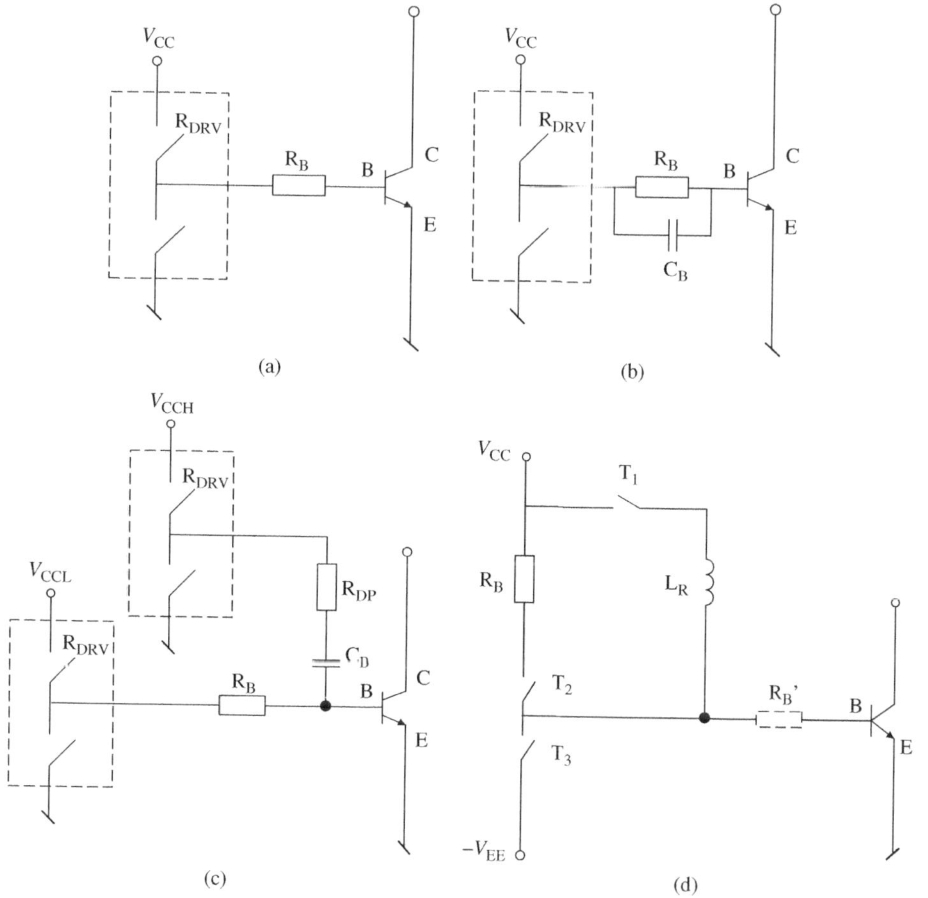

Figure 4.19 Base-drive units for SiC BJT: (a) simple totem-pole driver, (b) driver with speed-up capacitor, (c) dual source with resistors and capacitors (2SRC) and (d) current source driver [36, 37]

Another base driver circuit, with two voltage sources, can be employed in order to combine fast switching performance and low power consumption (dual source, see Figure 4.19c) [36]. The high-voltage supply V_{CCH}, which delivers dynamic current peaks through the capacitor branch, contributes to the fast switching speed. The power consumption of the driver is reduced by using the low-voltage supply V_{CCL} connected to a carefully chosen base resistor. An example of the waveforms during the turn-on process is presented in Figure 4.20, where the high current peak provided to the base may also be observed (see Figure 4.20a). Turn-on times of the 1200 V/6 A SiC BJT in the range of 30 ns can be observed on the waveforms in Figure 4.20b. The steady-state base current is supplied from a low-voltage source, and the driver losses during the on-state can be kept fairly low. The dual-source circuit drives the 6 A SiC BJT at a switching frequency of 100 kHz, and the 50% duty ratio consumes 1.6 W from the auxiliary supply while the corresponding single source design is 2.5 W [36]. Furthermore, the efficiency of the DC/DC boost converter with dual source was close to 99%, in contrast with 98.4% measured for the single-stage solution.

The very good performance of the dual-source driver is accomplished by features that could be recognized as a drawback in some applications. First, the lack of a negative voltage supply could be a problem

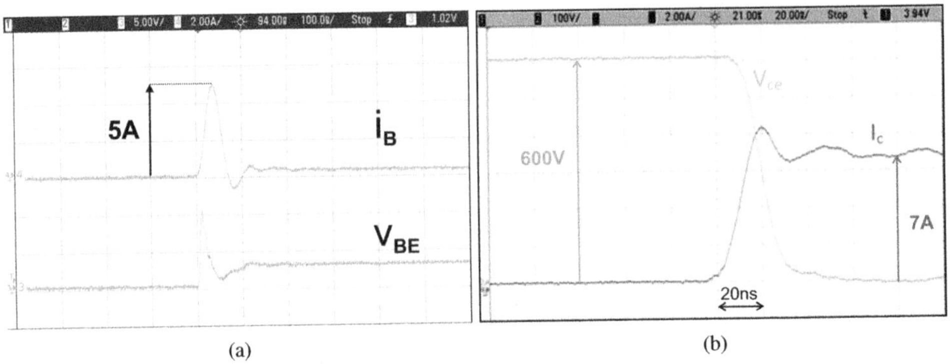

Figure 4.20 Turn-on process of the 6 A SiC BJT with dual-source base driver: (a) base current (2 V/div, 100 ns/div) and base–emitter voltage (5 V/div) and (b) V_{ce}, collector–emitter voltage – 100 V/div; and I_c, collector current – 2 A/div, timebase 20 ns/div [36]

when the devices operate in a phase-leg configuration and undesired turn-on may occur. Second, the base capacitor must be discharged in order to make the driver ready for the next turn-on process, and this is the reason to an associated duty-cycle limitation. Furthermore, C_B is a source of high-frequency oscillations with parasitic inductances of the transistors package, which must be eliminated as much as possible.

A gate base driver that has got rid of the mentioned drawbacks above was presented in [37]. Similar to the previous solution, a steady-state current is delivered from the low-voltage source V_{CC} through the base resistor R_B when T_2 is on (see Figure 4.19d). The steady-state power loss may therefore be easily optimized. The dynamic current peak that is necessary to turn-on the BJT rapidly is generated by the special current source, also fed from V_{CC}. In contrast to the dual-source driver, the sole component of the dynamic stage is the inductor L_R, which minimizes the high-frequency oscillations problem. As can be seen in Figure 4.21, the turn-on of T_1 while T_3 is also in the on-state leads to the inductor current rise at the required value. Then, the switching between T_3 and T_2 breaks the current path to the negative source V_{EE} and the inductor current is conveyed to the base of the BJT. It should be highlighted that the current

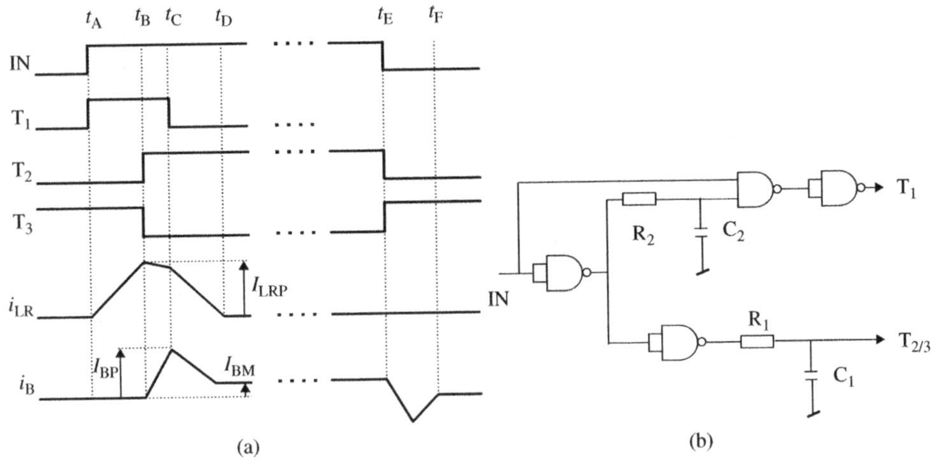

Figure 4.21 Operating principles of the current source driver for SiC BJT (a) and simple logic to obtain the control sequence (b) [37]

peak value can be easily adjusted by properly adjusting the inductance and/or the time values. A fast turn-off process is ensured by supplying a negative voltage bias using V_{EE} and could also be adjusted by changes of the resistance R_B. At the first sight, the current source driver seems to be a complex design, especially because of the inductor and control sequence. However, the point is that the inductor can be just an air-core inductor (in the range of hundreds of nH) and in order to obtain the suitable logic sequence a simple circuit (Figure 4.21b) with one IC (4×2 input NANDs) is relatively sufficient. Measurements presented in [37] show slightly higher power consumption (1.9 W at 100 kHz) than the dual-stage design, but this seems rather to be a result of the non-optimized base driver circuit.

4.4.3 Gate Drivers for Normally-off JFETs

At first sight, the SiC EMVTJFET seems to be a voltage-controlled device (the threshold voltage is close to 1 V). However, a steady-state gate current is required in order to conduct the drain current, especially if a reasonable on-state resistance is targeted [38]. In addition, high current peaks should be provided to the gate so that the recharge of the input capacitance of the device will be faster. This means that the requirements for the gate driver of the normally-off SiC JFET are very close to those of the SiC BJT, which have been discussed in the previous section. Thus, all the designs for the bipolar transistor presented above may be utilized after introducing slight changes. A very good proof of this is the current source driver case. The same driver board (scheme in Figure 4.19d) was employed to drive the normally-off SiC JFET. The only difference was the value of the base (gate) resistor as the JFET requires a lower steady-state current. Measurements performed using a double-pulse test circuit are presented in Figure 4.22. As can be seen from this figure, a very good dynamic performance of the normally-off SiC JFET (1200 V/100 mΩ) is observed as switching times counted in tens of ns were achieved.

The normally-off JFET was the first commercially available and, therefore, became the most widely tested and used SiC transistor. Thus, a number of gate driver concepts can be found in the literature and especially in manufacturer application notes [39]. Besides RC-coupled drivers, discussed in the previous section for SiC BJT, a two-stage gate driver with resistors has been also proposed for a 1200 V/30 A normally-off JFET as shown in Figure 4.23a. This driver is built with two operating stages. A dynamic stage, which consists of a standard driver and a resistor R_{B2}, provides high current peaks to turn the transistor on and off very fast. A DC/DC step-down converter, a BJT and a resistor R_{B1} constitute the second, static operating stage. The auxiliary BJT is turned on when the turn-on transition is finished in order to supply a gate current of approximately 200 mA. The absence of the speed-up capacitor results in elimination of problems related to charging and discharging times. To some extent, this, may count as an advantage of this driver concept, which was also recommended for use in power modules populated with SiC JFETs [40].

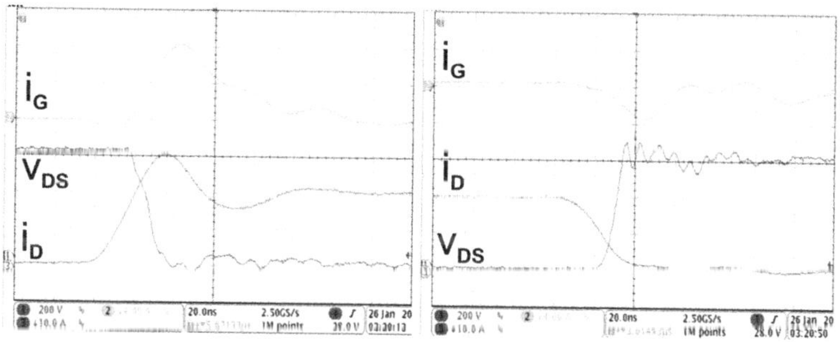

(a) (b)

Figure 4.22 Turn-on (a) and turn-off (b) of the SiC EMVTJFET when current source driver is applied during double-pulse test (base current, 4 A/div; drain current, 10 A/div; drain–source voltage, 200 V/div, timescale 20 ns/div)

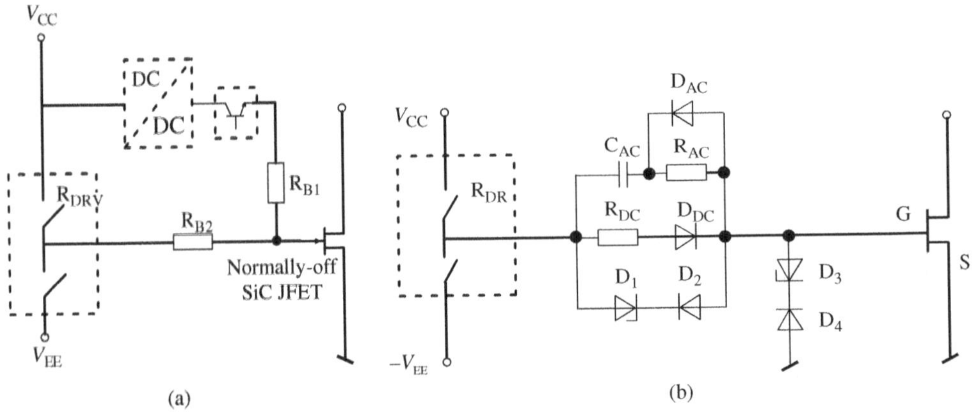

Figure 4.23 (a) Two-stage gate drive unit for normally-off SiC JFETs [20] and (b) AC-coupled gate driver from [41]

Finally, another AC-coupled gate driver for normally-off SiC JFETs has been presented in [41] in order to utilize the fast switching performance of this device (see Figure 4.23b). The steady-state current flows to the GS junction via a totem-pole driver, a resistor R_{DC} and a diode D_{DC} during the on-state. Power loss is limited as the positive supply V_{CC} is very low, close to 3 V. The speed-up capacitor C_{AC} supports the turn-on process together with the positive supply V_{CC}, while the switching speed can be adjusted by selecting an adequate R_{AC} value. At the end of the turn-on process V_{CAC} is close to zero and, therefore, the duty-cycle limitation problem is avoided. During the turn-off transient, the low impedance path with C_{AC} is utilized to achieve the very fast discharge of the input capacitance of the normally-off SiC JFET. Afterward the device is kept in the off-state with the use of the diodes D1–D4, which are intended to bypass any currents caused by the Miller effect. This feature of the AC-coupled driver is especially important due to low pinch-off voltage and significant gate-to-drain parasitic capacitance of the SiC EMVTJFET.

4.4.4 *Gate Drivers for SiC MOSFETs*

The MOSFET, among all SiC transistors, seems to be most similar to the state-of-the-art Si counterparts or IGBTs when the gate driver is taken into consideration. However, some specific features of the SiC MOSFET must be pointed out. Similar to the normally-off SiC JFET, the pinch-off voltage is rather low (around 2.5 V) and it decreases with junction temperature. This makes the device more immune to d*v*/d*t* noise. Furthermore, the transconductance is modest, which can also be counted as a drawback from a noise immunity point of view. During rapid changes in the main circuit, the gate circuitry can be easily distorted. Another important feature, which is a result of the low transconductance, is that the voltage required to turn-on the MOSFET is higher: 20 V is the manufacturer's recommended value. At the same time, the minimum negative voltage is only 5 V. All in all, the driver topology can be, for instance, a well-known totem-pole configuration (Figure 4.24a), but special attention must be paid in order to make a careful design and eliminate the parasitics.

Switching test results of 1200 V/160 mΩ SiC MOSFET with 1200 V/30 A SiC Schottky with the use of a simple totem-pole driver (Figure 4.24a) are presented in Figure 4.25. The positive supply voltage of the totem-pole driver equals 24 V, resulting in a turn-on time of approximately 30 ns. The turn-off, on the other hand, takes around 50 ns as the negative supply is only 5 V and the gate current dynamics are much lower.

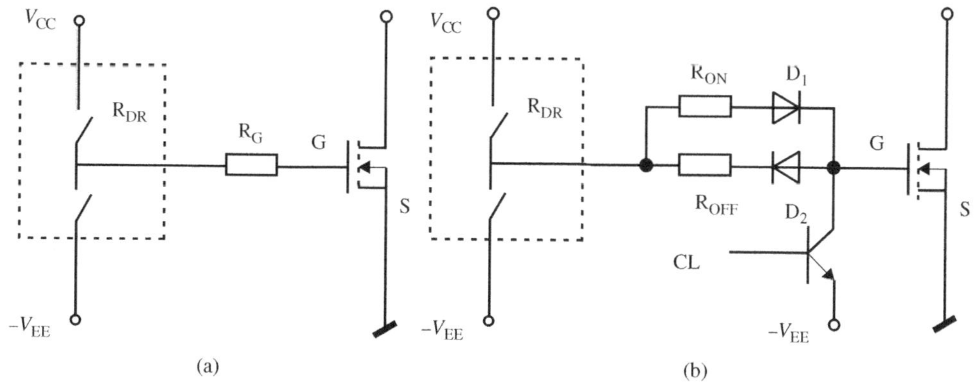

Figure 4.24 Totem-pole driver (a) for SiC MOSFETs and modified version with Miller clamp [42] (b)

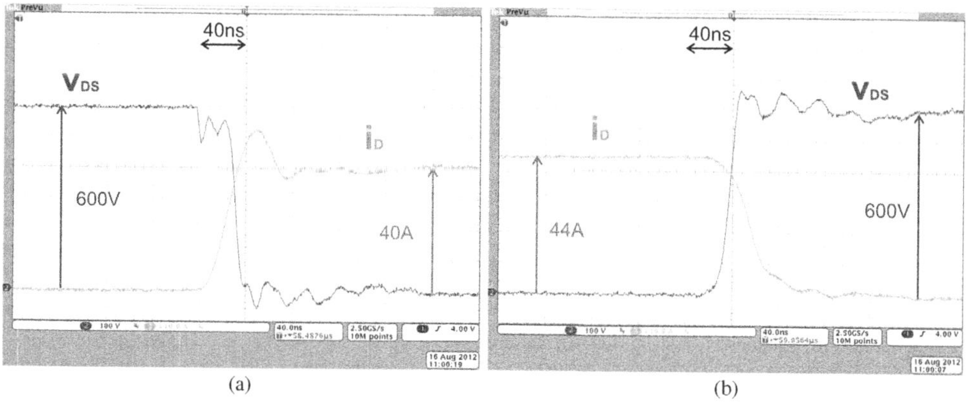

Figure 4.25 Waveforms of the SiC MOSFET (CMF10120D) drain–source voltage v_{DS}, drain current i_D recorded during turn-on (a) and turn-off (b) processes (totem-pole driver)

In order to avoid any noise problems caused by the Miller effect in the bridge-leg configuration, a modified totem-pole driver was employed as shown in [42]. Besides the main part with separate turn-on and turn-off resistors (Figure 4.24b), an extra bipolar transistor was added to create a low impedance path between the MOSFET gate and the negative supply. When the main transistor is in the off-state the clamping transistor is turned on and the current, generated after rapid discharging of the DG capacitance, flows directly to V_{EE}. The gate resistor and the driver resistance are bypassed, and therefore the slight voltage drop caused by the Miller current is still lower than the difference between the negative supply V_{EE} (−5 V) and the pinch-off voltage (+2.5 V). Thanks to this driver configuration, even very fast changes of the drain potential are not able to turn-on the MOSFET, as shown in [42].

4.5 Parallel Connection of Transistors

The chip size and the associated current ratings of available SiC transistors will not always be sufficient in several power electronics applications. Consequently, it is necessary either to build multichip modules or to connect discrete components in parallel. In both cases, it is important to examine both steady-state and transient current sharing of the paralleled chips. If the transient current is not uniformly distributed among the transistors, the switching losses caused in the most loaded one will result in an increased junction temperature. An unequal temperature distribution may also be caused due to non-uniform steady-state current sharing.

The problems of parallel-connected normally-off SiC JFETs are discussed in [43]. Similar research dealing with discrete SiC BJTs has also been presented by the same authors. It has been shown that the transient currents of the transistors might suffer from mismatches, especially when the switching speed is high. On the contrary, they have shown excellent current sharing during the steady-state operation. Even though these two SiC devices seem to be preferred from the normally-off characteristics point of view, it has been shown that the on-state resistance of the normally-on SiC JFET is comparably lower [44]. It is, therefore, worth deeply investigating the parallel connection of this type of SiC JFET.

As has already been mentioned previously, there are two types of normally-on SiC JFETs available as either engineering samples or commercial products. An investigation into parallel-connected normally-on LCJFETs and DMVTJFETs has been presented in [45]. Regardless of the type of the normally-on SiC JFET, the device parameters that affect the performance of parallel connection are the same for both JFET designs. In particular, there are four device parameters that count as the most crucial parameters affecting both the transient and the steady-state current sharing. The first crucial parameter is the on-state resistance of the parallel-connected JFETs, which must have a positive temperature coefficient and a low spread. Figure 4.26 illustrates the on-state resistance of four different samples of DMVTJFETs at elevated temperatures. It is clear from this figure that there might be significant differences in the on-state resistance among the JFETs. However, the spread of the on-state resistance as such remains constant as the temperature increases.

The second crucial device parameter is the variations in static transfer characteristics (I–V transfer characteristics) of the parallel-connected JFETs. Figure 4.27 shows the static transfer characteristics at various gate–source voltages for two different samples of DMVTJFETs. Ideally, the I–V characteristics of the parallel-connected devices must be identical when looking at the same gate–source voltage. Nevertheless, it is obvious from Figures 4.27 and 4.28 that there is a spread in the I–V curves of these two devices. For instance, for the gate–source voltage of $V_{gs} = -5$ V the device shown in Figure 4.27 still conducts the current, while the other device (Figure 4.28) is kept in the off-state. A similar situation

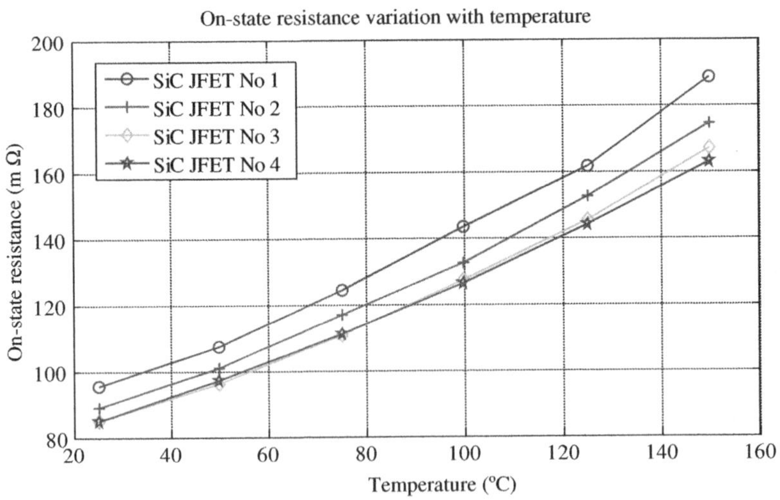

Figure 4.26 On-state resistance variation with temperature for four different DMVTJFET samples [45]

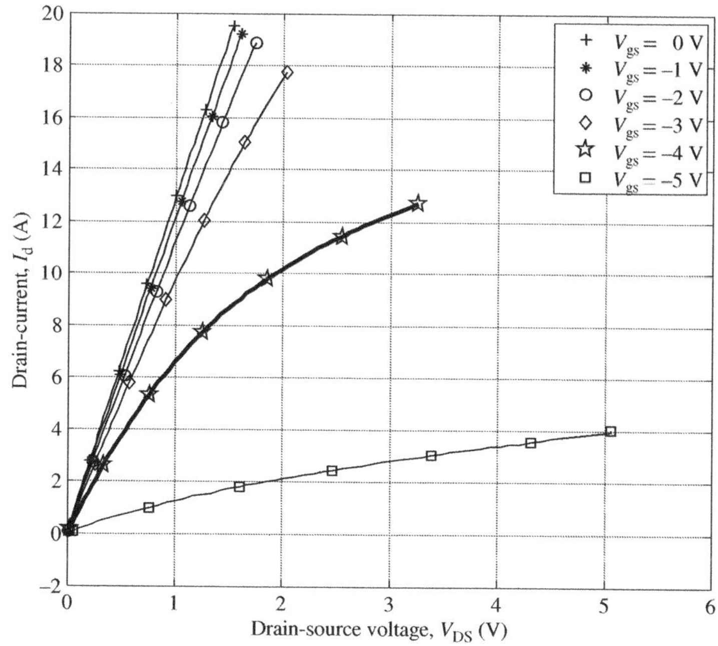

Figure 4.27 Static transfer characteristics at various gate–source voltages for a specific DMVTJFET sample [45]

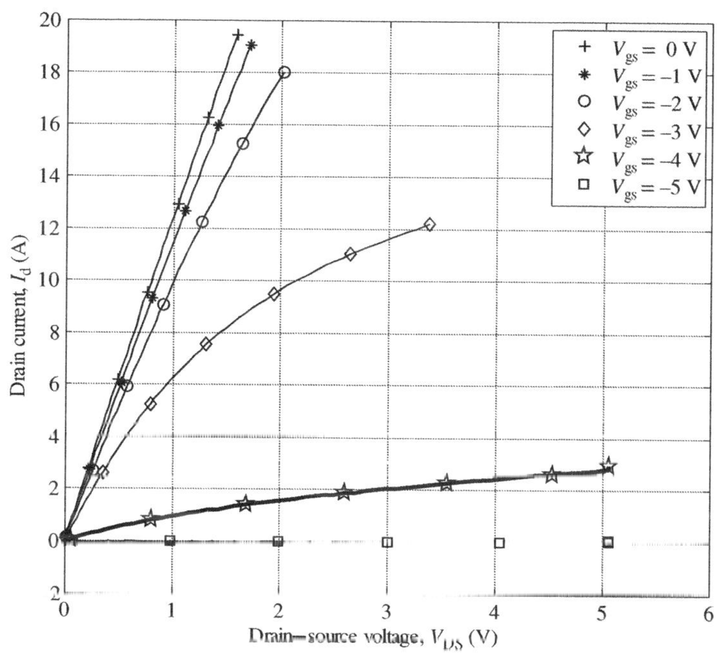

Figure 4.28 Static transfer characteristics at various gate–source voltages for a specific DMVTJFET sample [45]

can be observed if a gate–source voltage of $V_{gs} = -4$ V is considered. By investigating the I–V curves corresponding to this certain gate–source voltage, a spread can be shown. The differences in the I–V transfer characteristics result in different values of transconductance for the parallel-connected SiC JFETs. Thus, there might be current mismatched during the switching transitions. At turn-on, for example, the JFET having the lowest transconductance turns on slightly more slowly than the others.

The third and the fourth device parameters are the pinch-off voltage and the reverse breakdown voltage of the gate, $V_{br,g}$, of the parallel-connected JFETs, respectively. Since the SiC devices are aimed at high-temperature operation, it makes sense to investigate their temperature dependences. It is experimentally found that the pinch-off voltage of both JFET designs is almost temperature independent. This has been experimentally shown for the LCJFET in [45], where the static transfer characteristics of a certain JFET at room temperature and at 150 °C are shown and it is clear that the pinch-off voltage is unchanged. Even though the pinch-off voltage is constant at elevated temperatures for both JFET structures, the performance of $V_{br,g}$ with respect to the temperature differs between LCJFET and DMVTJFET. For the DMVTJFET $V_{br,g}$ is almost unaffected from temperature variations, as shown in Figure 4.29, whereas for LCJFET the $V_{br,g}$ increases as the temperature increases [45]. In particular, after experimentation, it has been shown that the temperature dependence of $V_{br,g}$ for two different samples of the LCJFET design might be different. Thus, the margin between the pinch-off voltage and $V_{br,g}$ might also vary among the parallel-connected LCJFETs. Finally, at elevated temperatures, it is clear that $V_{br,g}$ is increasing (or, in other words, its absolute value is decreasing) and the margin between $V_{br,g}$ and the pinch-off voltage is also becoming narrow.

A serious problem might be faced when two or more LCJFETs with different values of $V_{br,g}$ or even different pinch-off voltages are parallelly connected. Especially at high temperatures, the margin between these two parameters is so narrow or even negative, so that an unstable operation might occur. A solution to this problem [45] is to employ the gate driver as shown in Figure 4.14b, which is supplied by a more negative voltage than the most negative $V_{br,g}$ of all parallel-connected devices and limits the gate leakage current by means of a high-value resistor R_p. Thus, all the devices are forced to PT, but none of them will suffer from destruction of the gate–source junction. On the contrary, in the case of the DMVTJFET design both the pinch-off voltage and $V_{br,g}$ are temperature independent, and thus problems

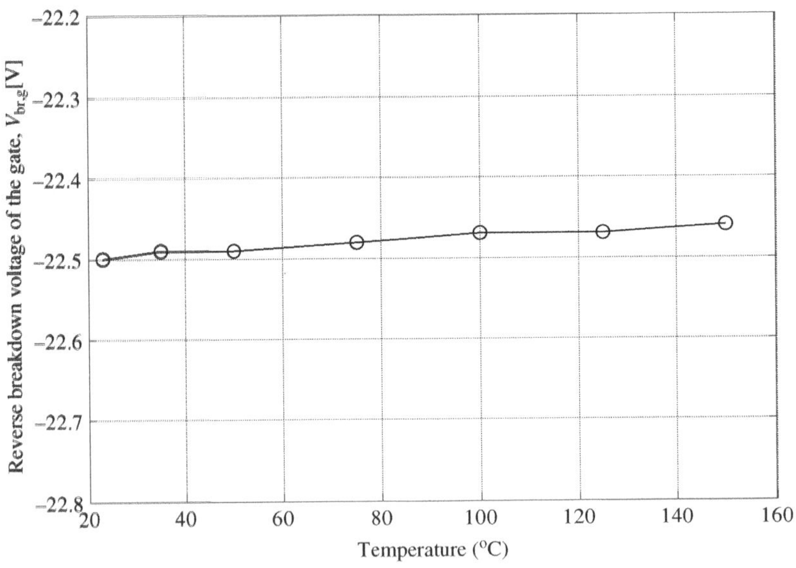

Figure 4.29 Measurements of the temperature dependence of the reverse breakdown voltage of the gate of a DMVTJFET [45]

related to the safety margin between these two parameters are eliminated. However, there might still be transient current mismatches caused by the different values of the pinch-off voltage and $V_{br,g}$ among the parallel-connected devices.

Apart from the investigation into the static parameters of the parallel-connected SiC JFETs, it also makes sense to study the switching performance. Before going into more details about this it is necessary to mention that in order to study the switching performance, the DMVTJFET design was chosen. This choice is mainly based on the temperature independent margin between the pinch-off voltage and $V_{br,g}$, which has been observed for the DMVTJFET design. Furthermore, the unavailability of new versions of the LCJFET design, which in the future might have comparable performance to the DMVTJFET, also contributes to the choice of the SiC JFET design.

The switching performance is studied assuming two pairs of two parallel-connected DMVTJFETs. The first pair has different values of $V_{br,g}$ while the other pair of JFETs has the same (−28.1 V/−19.2 V and −28.1 V/−28.0 V, respectively) [45]. It is experimentally shown that a fifth parameter also affects the performance of the parallel-connected SiC JFETs. This parameter deals with the positioning of the devices on the circuit layout. In particular, different positioning results in different values for the stray inductances between the pins of the discrete JFETs. Thus, a different switching performance for each parallel-connected device is anticipated. Graphical illustrations for the three different positionings considered are shown in Figure 4.30. The circuit layout shown in Figure 4.30c is based on the symmetrical placement of the components (the diode is mounted between the JFETs).

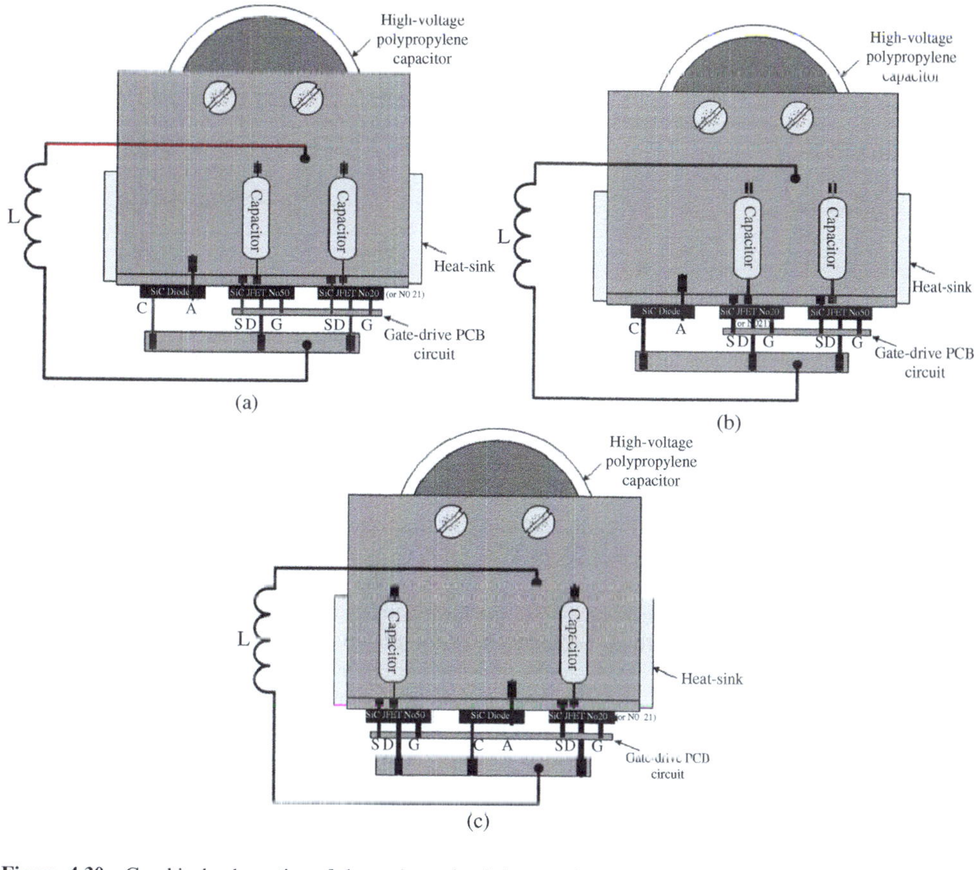

Figure 4.30 Graphical schematics of the various circuit layouts (a) circuit layout L1, (b) circuit layout L2 and (c) circuit layout L3 [45]

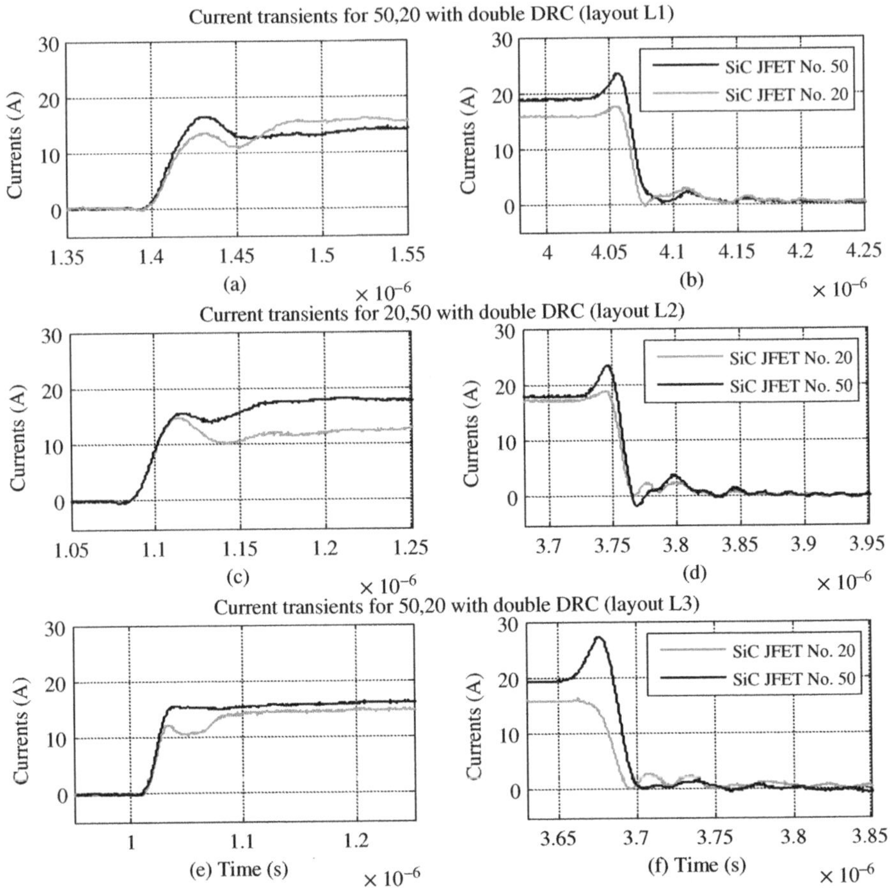

Figure 4.31 Switching transients for the SiC JFETs with the same reverse breakdown voltage of the gates when various circuit layouts are used (a) turn-on and (b) turn-off transients using circuit layout L1; (c) turn-on and (d) turn-off transients using circuit layout L2; and (e) turn-on and (f) turn-off transients using circuit layout L3 [45]

Turn-on and turn-off switching transients for the parallel-connected SiC JFETs with the same and different values of $V_{br,g}$ are shown in Figures 4.31 and 4.32, respectively. From these two figures, it is clear that an improved switching performance is observed for devices with the same $V_{br,g}$ compared to those with different $V_{br,g}$. Moreover, the placement of the JFETs on the circuit layout also affects the current transient. Considering for instance Figure 4.31, the lowest traces that correspond to the symmetrical placement (L1) are somehow improved compared to the other two circuit layouts. However, this is not the rule. In fact, if Figure 4.32 is considered the transient current mismatches have not been eliminated due to the symmetrical placement of JFETs. Thus, it must be concluded that the switching performance of the parallel-connected SiC JFETs is governed by a combination of various parameters related to not only the devices but also the circuit layout.

In order to get rid of the stray inductances, in the future modules populated with several chips must be built. In such a case, the question that arises deals with the sorting criterion of the chips. For instance, the chips might be sorted with respect to either their pinch-off voltages or $V_{br,g}$.

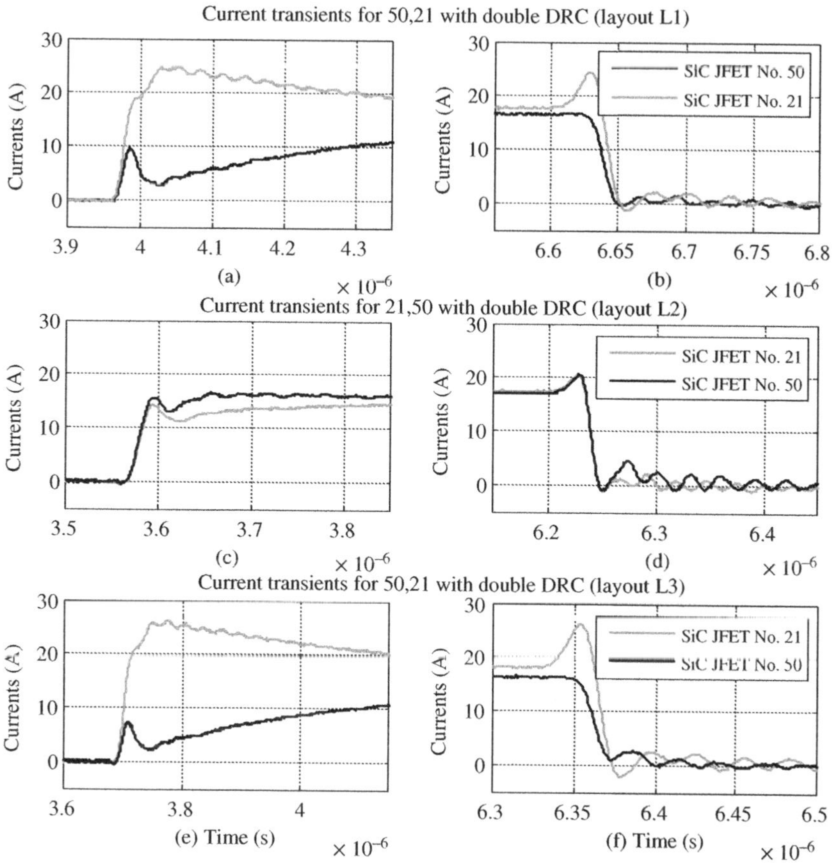

Figure 4.32 Switching transients for the SiC JFETs with different reverse breakdown voltage of the gates when various circuit layouts are used (a) turn-on and (b) turn-off transients using circuit layout L1; (c) turn-on and (d) turn-off transients using circuit layout L2; and (e) turn-on and (f) turn-off transients using circuit layout L3 [45]

A successful example of the parallel connection of 10 JFETs in each switch position is the 40 kVA inverter [46], as shown in Figures 4.1 and 4.33. The on-state resistance of a single switch is lower than 10 mΩ, but because of the relatively low switching frequency (10 kHz) most of the power losses are still caused by conduction. Thus, uniform switching of 10 discrete devices is not crucial in this case. However, special gate drivers were designed in order to achieve similar driving conditions for all JFETs. The number of parallel-connected devices is high (60 transistors and 6 diodes) and the total semiconductor cost was around 37.5 USD/kW of rated power, which is higher than the corresponding cost of the inverter with Si IGBTs. Nevertheless, the outstanding switching performance of SiC JFETs, as shown in Figure 4.33b, and the very high efficiency (over 99.5%, close to 99.7%) constitute the two driving factors encouraging investment in such a converter.

Another interesting example of the parallel-connected SiC devices is a DC/DC boost converter built with four SiC BJTs (see Figure 4.34a). For the parallel connection of bipolar transistors, the most important parameters are the on-state voltage drop, which must have a positive temperature coefficient, the current gain, the static transfer characteristics and the gate charge of the device, which is required to

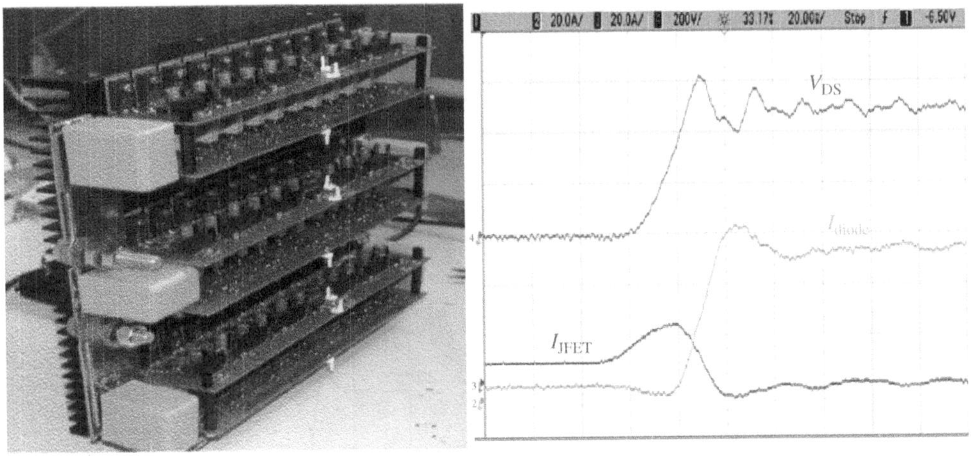

(a) (b)

Figure 4.33 Photograph of the three-phase inverter built with 10 parallel JFETs for each switch position (a) and turn-off process of the 10 parallel SiC JFETs (b) (V_{DS}, drain–source voltage – 200 V/div; and IJFET, drain current – 20 A/div of one single JFET) with antiparallel SiC Schottky diode (I_{diode}, diode current – 20 A/div, timebase 20 ns/div)

(a) (b)

Figure 4.34 Photograph of the DC/DC boost converter with four parallel SiC BJTs (a) and the dual-source driver (b) [47]

be supplied by the base-drive circuit in order to turn it on. The current gain must be approximately the same for all parallel-connected devices in order to ensure equally steady-state current distribution if all devices are driven from the same drive unit. Transient current mismatches might be caused if the parallel-connected BJTs have different static transfer characteristics and/or different required gate charges of the gates. Moreover, as has been demonstrated, the circuit layout might also affect the switching performance of the parallel-connected devices. The stray inductances between the pin connections of the BJTs and the diode in a DC/DC boost converter, for instance, might contribute to further transient current mismatches. But the key issue is to design a suitable base driver that ensures the similar driving conditions of all four devices including both dynamic and steady-state currents. For the present DC/DC

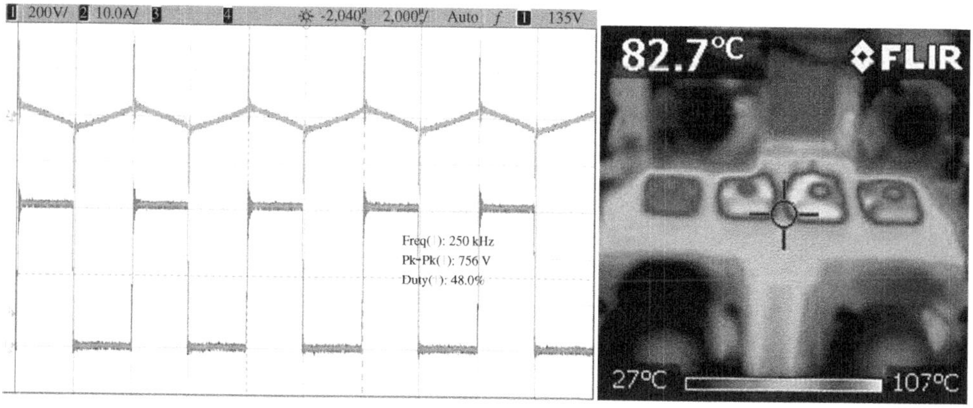

(a) (b)

Figure 4.35 Inductor current and middle-point voltage of the boost converter (a) and heat camera image during full power operation (250 kHz, 6 kW) [47]

converter, a dual-source concept was employed to drive four 1200 V/6 A SiC BJTs (Figure 4.34b). The measured power consumption of the driver was 8.5 W at a switching frequency of 250 kHz and 50% duty ratio (>2 W/device). Tests conducted at nominal conditions (300 V/600 V, 6 kW, 250 kHz) have shown a very good performance of the DC/DC boost converter (see Figure 4.35). Owing to difficulties with electrical measurements, especially the BJT currents (the devices and the driver are close to each other interfering measurements with Rogowski coils), a heat camera was used to verify the current/power sharing. The temperatures of the transistor packages are obviously not the same, but the differences are on an acceptable level and thermal runaway effect was not observed. This proves the very good performance of the parallel connection as most of the total power loss (115 W including passive elements – efficiency 98.1%) is caused by switching (conduction loss is around 2.1 W/transistor).

4.6 Overview of Applications

On the basis of a brief analysis of the available SiC power devices performance presented in earlier sections, it is clearly shown that their features are moving power electronics into new areas. When the design of power electronic converters is considered with SiC devices instead of their classic Si counterparts, three different design directions may be chosen as shown in Figure 4.36.

Doubtless, the most obvious direction is the increase in the switching frequency up to a few hundreds of kilohertzs as the SiC devices may be switched in the range of nanoseconds. This means that SiC converters may reach switching frequencies that are not favorable to Si devices (for voltage ratings above 600 V) unless soft switching techniques are employed [48]. The increase in the switching frequency contributes to the further reduction of volume and weight of passive elements (e.g., inductors, capacitors), which may be counted as a major advantage from a systems perspective [49]. Consequently, more compact converters, especially DC/DC converters and inverters with passive filters, might be constructed. Finally, owing to the faster switching speeds and higher switching frequencies, the voltage and current harmonics are moved up higher in frequency spectrum. This also leads to the volume reduction of the EMI (electromagnetic interference) filters [50].

The switching times and voltage drops across the SiC devices are definitely lower than their silicon counterparts, which result in serious reductions in switching and conduction power losses. In applications where the rise in the switching frequency is not a priority, one of the rest two remaining design directions

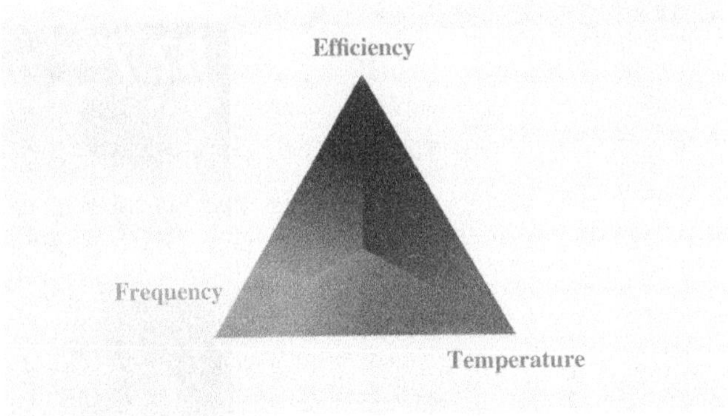

Figure 4.36 Three main directions in design of power electronics converters with SiC devices

may be chosen. The reduction of the converter power losses is reflected in the decrease in the volume and weight of the cooling equipment [46]. Moreover, either liquid or air-forced cooling systems might be avoided due to the fact that the amount of dissipated heat is a significantly lower. High efficiency is also valuable when power delivery and distribution systems as well as photovoltaic applications are considered. In such cases, the lower power loss is directly recalculated into profit.

The last possible design direction deals with high-temperature converters because of high-temperature characteristics of SiC devices (junction temperatures over 450 °C). This is a roughly new area of power electronics aimed at automotive, space, or drill-hole applications. However, the main obstacle at high temperatures is not the device itself, but rather the immature packaging technology, lack of high-temperature passive components and logic electronics. Small steps may be made with existing SiC transistors before totally high-temperature converters will be released. For example, assuming junction temperatures between 200 and 250 °C, a reduction in volume and weight may be achieved due to the potentially higher thermal resistance of the heat sink. Recently successful examples of high-temperature packaging for power modules have been published [51]. Even though the SiC devices were operating at junction temperature of 250 °C (case temperature below 200 °C), there are still questions on the reliability and long-term stability requiring answers. High-temperature gate drivers based on the silicon-on-insulator technology are currently available [52, 53]. Simultaneously, SiC-based control electronics are under development and exhibit very promising characteristics [54].

It cannot be denied that SiC power device technology is currently thriving, if not to say in an explosive stage, as new things such as new devices, driver concepts, or application examples continuously appear. Undoubtedly, only a small portion of all new achievements is shown. It is likely that the most attractive examples are kept hidden. This means that it is hard to present the real situation of current SiC technology, but, on the other hand, it is still possible to show an overview of what has previously demonstrated. SiC Schottky diodes are well-established and mass-produced devices available in either discrete or module packages. In contrast, real applications of SiC transistors still seem to be in an early development stage, excluding a few low-power application examples. However, using the data of available SiC transistors, reasonable forecasts of the performance of future SiC converters are possible. That is why numerous laboratory prototypes and demonstrators have been developed and experimentally tested by several research groups around the world.

4.6.1 Photovoltaics

The advantageous characteristics of SiC devices match perfectly in order to meet the expectations of the photovoltaic industry, especially when referring to an efficiency increase and integration of the inverter with the photovoltaic panel [55]. Efficiencies above 99% can be reached by replacing Si with SiC components in the power range of single kilowatt. Even if the device cost is still higher, the system benefits are significant. Furthermore, small inverters integrated on the backside of photovoltaic panels may employ SiC transistors in the future. In photovoltaic applications, harsh environmental conditions are usually met and high reliability and long lifetime requirements may not be achieved with existing Si technology. Nevertheless, the remaining argument against SiC devices is the high cost. Most likely, the overall system benefits as well as the expected reduction in the device cost may contribute in solving this issue in the future.

4.6.2 AC Drives

The chances for the mass introduction of SiC electronics in the field of AC drives are limited. The main reason for this is the higher cost than the state-of-the-art Si devices (especially IGBTs), which currently almost fulfill the requirements for inverter-fed AC drives. Nevertheless, it is very likely that features of the SiC devices could be utilized in many niches. An additional advantage of using SiC devices may be, for example, utilized in applications of AC drives that require an LC filter. The increase in switching frequency may lead to a significant reduction in the volume and weight of such systems. SiC transistors can also be beneficial when high efficiencies and high-power densities are required [48–50]. Finally, they may be considered in applications where a high switching frequency is necessary in order to feed a high-speed motor.

4.6.3 Hybrid and Plug-in Electric Vehicles

The possible system gains when SiC transistors are employed in inverters for hybrid electric vehicles are extremely high [56–59]. The integration of power electronic converters with the combustion engine using a common cooling system at a temperature above 100 °C counts as a current trend. It is not likely that all these requirements can be fulfilled using Si electronics. Thus, SiC electronics with much higher temperature limits seem to be the best choice. Moreover, very high-efficient converters and inverters may be constructed. However, the lack of reliable high-temperature packaging still counts as a serious problem. Finally, auxiliary components such as gate drivers, capacitors and so on must also be capable of high-temperature operation, which is a great challenge.

4.6.4 High-Power Applications

The much higher blocking voltages of the SiC devices make them attractive in the area of high-power converters for grid applications such as HVDC. Unipolar devices like JFETs will be the best choice in a voltage range up to 4.5 kV according to [27]. Future bipolar devices (BJT, IGBTs) may be superior for higher voltage ranges. However, the availability of SiC devices with high voltage ratings is currently limited, even if a few interesting examples can be found in the literature. The case of a 300 MVA modular multilevel converter for 300 kV HVDC transmission is discussed in [60]. As the switching frequency is only 150 Hz, the switching losses are very low. Together with the low on-state resistance of the SiC JFETs, it is possible to achieve an efficiency increase by 0.3% (900 kW) with respect to Si IGBTs case [61]. The benefit is not only the lower amount of heat that needs to be dissipated but also the lower energy cost. The possible application of 20 kV SiC GTOs in a 120 kV/1 kA HVDC interface has been discussed in [62]. These devices are compared with 4 × 5 kV Si counterparts by using an analytical device model and by performing system simulations. Employing the 20 kV SiC GTOs, a significant efficiency increase is expected at higher junction temperatures. Another interesting example of the high-voltage capability of SiC JFETs is shown in [48], where six 1.2 kV transistors connected in a super cascode configuration were tested up to 5 kV in a DC/DC converter. With 6.5 kV JFETs, this converter would be able to operate at a voltage level over 20 kV in a distribution system. Moreover, the 10 kV SiC MOSFET is often considered for high-voltage applications such as solid-state transformers [63]. In this case, while the power losses are a few times lower than for 6.5 kV IGBTs, the switching frequency of such a converter is increased from 1 kHz to 20 kHz in order to reduce the size of the high-voltage transformers. The 15 kV n-channel SiC IGBT has been considered in a study regarding smart grid applications [64]. Despite the very promising simulation results of the n-channel SiC IGBT (low on-state voltage, switching energies >6.5 kV Si counterpart, etc.), experimental verification has only been done with a 12 kV/10 A SiC IGBT in the laboratory [27].

4.7 Gallium Nitride Transistors

GaN is another interesting WBG semiconductor material that offers a great potential in the area power semiconductor devices. A brief analysis of the material properties, presented in Table 4.1, shows that both WBG materials (SiC and GaN) outperform silicon in most parameters. Possible improvements of the power semiconductor devices may also be found in this table. The higher critical field strength of WBG semiconductors allows devices to be built with higher blocking voltages and lower leakage currents. The electron mobility, which is much higher for GaN than for silicon or SiC, is an important factor when the operating frequency is considered. Thus, GaN devices are superior at high frequencies, and this feature has already been utilized in many radio frequency (RF) applications. On the other hand, thermal conductivity, which is comparable to Si, reduces the expectations for high-temperature operation.

Nowadays, mass-produced blue laser diodes are built on bulk GaN substrates, but for the manufacturing of RF or power devices a high-quality material (low number of defects) is required. It seems that heteroepitaxy wafers are the best choice and GaN on Si is advantageous over GaN on SiC because of the expectations of lower costs in the future. On the other hand, the thermal expansion coefficient of GaN (5.6) is more than two times higher than that of Si (2.6), and therefore the thermal cycling of the devices may be an issue. Currently, solutions based on GaN on sapphire substrate are also considered, but in most cases GaN-on-Si substrate are used. Wafers with diameters of 6″ are currently commercially available, but 8″ wafers have also been demonstrated.

There is no doubt that the technological development of GaN power semiconductor devices is a few years behind that of SiC, but the progress is very rapid, and in addition to engineering samples, commercial devices are also available.

The most common device design in GaN is the high-electron-mobility transistor (HEMT). The basic structure of this transistor (Figure 4.37a) suffers from the same drawback as the first designs of the SiC JFET: that is, the normally-on property. A negative voltage must be supplied to the gate in order to block the current flow between the drain and the source contacts as can be seen in Figure 4.37a [65]. The two-dimensional electron gas with high mobility existing between the GaN and AlGaN layers results in a very high conductivity and low on-state resistance. The high critical field leads to very small device

Table 4.1 Material properties of silicon, 4H-SiC and GaN

Material property	Si	4H-SiC	GaN
Bandgap (eV)	1.1	3.2	3.4
Critical field (106 V cm^{-1})	0.3	3	3.5
Electron mobility ($cm^2\ V^{-1}\ s^{-1}$)	1450	900	2000
Electron saturation velocity (106 cm s^{-1})	10	22	25
Thermal conductivity (W cm^{-2} K)	1.5	5	1.3

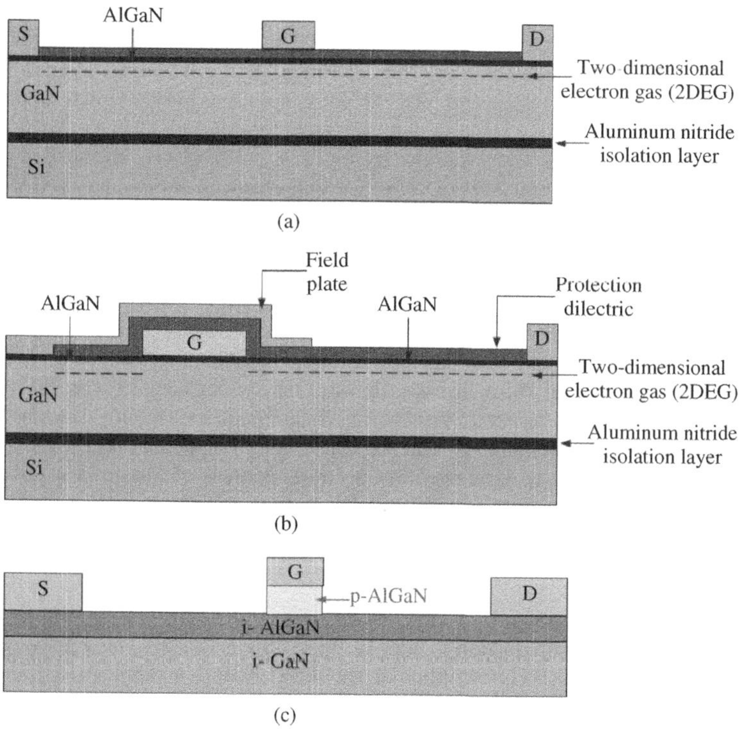

Figure 4.37 Structure of GaN transistors (a) normally-on HEMT, (b) normally-off HEMT and (c) GIT

sizes and, therefore, very low capacitance. All in all, GaN HEMTs show much higher RonxC figures of merit than silicon FETs.

In order to overcome potential difficulties with the normally-on characteristic a cascode circuit can be used [66], and also enhanced-mode transistors with normally-off characteristics have been developed [65]. An example of a device structure that is commercially available and has ratings of 200 V/12 A [66] is presented in Figure 4.37b. This device offers low on-state resistance with a slightly lower temperature dependence than that of Si MOSFETs and low capacitances because of the lateral structure. On the other hand, the low threshold voltage (only 1.6 V) may, unfortunately, result in gate mistriggering problems. Owing to the expected difficulties with the parasitic components, the devices are delivered in an land grid array (LGA) package where the parasitic inductance of the electrodes is significantly lower compared to other packaging technologies.

Other designs based on HEMT concepts have also been presented [66–68]. Definitely the 600 V class devices [67, 68] made on SiC substrates may be recognized as the most interesting. For instance, on-state resistances of 150 mΩ and pulsed drain currents of 70 A have been reported for the normally-on HEMT in cascade configurations [67]. Furthermore, similar data can be found for the device presented in [68].

Another device design, the so-called gate injection transistor (GIT) – see cross-section in Figure 4.37c – also targets higher blocking voltages [69]. The p-type AlGaN gate decreases the electron concentration in the electron gas under the gate and allows designs with normally-off characteristics. At appropriate gate voltages, holes can also be injected into the channel region and channel-conductivity modulation occurs. This will result in higher drain currents, but this, unfortunately, also means that a certain gate current is required. According to [70], a 600 V/15 A GIT with 58 mΩ and 10 nC is available.

The competitive performance of the GaN transistors has been proven by a number of demonstrators. According to [71], a DC/DC isolated converter (1.5 MHz 42 V/12 V) that employs HEMTs offers up to 5–25% better efficiency than a comparable system consisting of Si MOSFETs. Similar improvements can be found in [66] where, for instance, the DC/DC LCC converter (400 kHz 300 V/30 V) is 3–17% more efficient than a corresponding solution with superjunction MOSFETs in Si. Apart from DC/DC converters, three-phase inverters in GaN were also presented. A 2 kW/100 kHz inverter operating at 350 V DC has been reported in [67]. An efficiency of 98.5% was achieved with the use of the 600 V HEMTs. The same application example, basically a low-voltage motor drive, can also be found for GITs [72]. Other concepts such as bidirectional switches can also be obtained by considering this device. Photovoltaic inverters count as another interesting example of GaN GITs for possible future applications [70]. A boost converter, presented in [70], with 600 V/15 A GIT and GaN Schottky reaches an efficiency of 98.6% at a power level of 1.2 kW (250 V/350 V conversion).

It is very likely that the fast growth of GaN power transistor technology will play an important role within the semiconductor market. The main expansion area may be placed below the 600 V class. Doubtless, GaN devices offer better performance than their Si counterparts and the only necessary requirement for a breakthrough is a cost reduction to an acceptable level. The competition against Si and SiC in the 1200 V class is also possible, but the parameters of GaN transistors are still uncertain at this voltage level, and, therefore, their impact on applications still remains questionable.

4.8 Summary

This chapter aims to give an introduction to state-of-the-art SiC and GaN power transistors and their applications. As background, a short description of MOSFETs, IGBTs and IGCTs in silicon is first presented. After this, a presentation of power transistors in SiC is given. Even though device structures are shown, the focus of the presentation is the use of SiC power transistors in power electronic converters in the kilowatt to megawatt power range. The benefits, drawbacks and specific properties of the different transistor designs are treated in detail.

Special attention is also given to gate and base drivers of the SiC power transistors presented. This section is especially important to readers who want to start new experimental activity on SiC power electronics, because knowledge of gate and base drivers is one of the most important issues for the successful operation of power electronic converters. From the presentation, it is clear that standard silicon MOSFET drivers will typically not yield fruitful results when applied to currently available SiC power transistors. A dedicated section is devoted to solutions for the normally-on problem of some SiC JFET deigns.

Another important issue is parallel connection as the available SiC power transistors usually have comparably low current ratings. Both theoretical foundations and experimental results for specific cases are presented in order to guide the reader in this matter. Converter designs with up to 10 parallel devices are presented as examples.

In the next section, an overview of the applications for SiC power transistors is given. First, how the properties of SiC power transistors can be utilized in order to achieve specific benefits is demonstrated. Next, different applications such as photovoltaics, AC drives, hybrid electric vehicles and high-power applications, are presented.

Finally, an overview of GaN power transistors is given. The presentation is focused on HEMT and GIT designs.

References

1. Rabkowski, J., Peftitsis, D., and Nee, H.-P. (2012) SiC power transistors – a new era in power electronics is initiated. *IEEE Industrial Electronics Magazine*, **6** (2), 17–26.

2. Hofstein, S.R. and Heiman, F.P. (1963) The silicon insulated-gate field-effect transistor. *Proceedings of the IEEE*, **51** (9), 1190–1202.
3. Deboy, G., März, M., Stengl, J.P. *et al.* (1998) A new generation of high voltage MOSFETs breaks the limit line of silicon. Procedings of International Electron Devices Meeting 1998 (IEDM 1998), pp. 683–685.
4. Becke, H.W. and Wheatley, C.F. Jr, (1980) Power MOSFET with an anode region. US Patent 4,364,073, Dec. 14, 1982, filed Mar. 25, 1980.
5. Baliga, B.J., Adler, M.S., Gray, P.V. *et al.* (1983) The insulated gate rectifier (IGR): a new power switching device. *IEEE Electron Device Letters*, **4** (12), 452–454.
6. Ranstad, P. and Nee, H.-P. (2011) On dynamic effects influencing IGBT losses in soft-switching converters. *IEEE Transactions on Power Electronics*, **26** (1), 260–271.
7. Tihanyi, J. (1988) MOS-Leistungsschalter. ETG-Fachtagung Bad Nauheim. Fachbericht Nr. 23, VDE-Verlag, May 1988, pp. 71–78.
8. Chang, H.-R. and Baliga, B.J. (1989) 500-V n-channel insulated-gate bipolar transistor with a trench gate structure. *IEEE Transactions on Electron Devices*, **36** (9), 1824–1829.
9. Kitagawa, M., Omura, I., Hasegawa, S. *et al.* (1993) A 4500 V injection enhanced insulated gate bipolar transistor (IEGT) operating in a mode similar to a thyristor. Proceedings of International Electron Devices Meeting 1993 (IEDM 1993), pp. 679–682.
10. Laska, T., Münzer, M., Pfirsch, F. *et al.* (2000) The field stop IGBT (FS IGBT) – a new power device concept with a great improvement potential. Proceedings of the 12th International Symposium on Power Semiconductor Devices and ICs (ISPSD 2000), pp. 355–358.
11. Steimer, P.K., Gruning, H.E., Werninger, J. *et al.* (1997) IGCT – a new emerging technology for high power, low cost inverters. Proceedings of IEEE Industry Applications Society Annual Meeting, New Orleans, LA, October 5–9, 1997, pp. 1592–1599.
12. Agarwal, A.K. (2010) An overview of SiC power devices. International Conference on Power, Control and Embedded Systems (ICPCES 2010), November 29–December 1, 2010, pp. 1–4.
13. Friedrichs, P. and Rupp, R. (2005) Silicon carbide power devices – current developments and potential applications. Proceedings European Conference on Power Electronics and Applications.
14. Ritenour, A., Sheridan, D.C., Bondarenko, V., and Casady, J.B. (2010) Saturation current improvement in 1200 V normally-off SiC VJFETs using non-uniform channel doping. 22nd International Symposium on Power Semiconductor Devices and IC's (ISPSD 2010), June 6–10, 2010, pp. 361–364.
15. Lindgren, A. and Domeij, M. (2010) 1200V 6A SiC BJTs with very low VCESAT and fast switching. 6th International Conference on Integrated Power Electronics Systems (CIPS 2010), March 16–18, 2010, pp. 1–5.
16. Cooper, J.A. Jr.,, Melloch, M.R., Singh, R. *et al.* (2002) Status and prospects for SiC power MOSFETs. *IEEE Transactions on Electron Devices*, **49** (4), 658–664.
17. Stephani, D. (2001) Status, prospects and commercialization of SiC power devices. Proceedings of Device Research Conference.
18. Kelner, G., Shur, M.S., Binari, S. *et al.* (1989) High-transconductance β-SiC buried-gate JFETs. *IEEE Transactions on Electron Devices*, **36** (6), 1045–1049.
19. Kelner, G., Binari, S., Sleger, K., and Kong, H. (1987) β-SiC MESFET's and buried-gate JFET's. *IEEE Electron Device Letters*, **8** (9), 428–430.
20. McGarrity, J.M., McLean, F.B., DeLancey, W.M. *et al.* (1992) Silicon carbide JFET radiation response. *IEEE Transactions on Nuclear Science*, **39** (6), 1974–1981.
21. Tolstoy, G., Peftitsis, D., Rabkowski, J., and Nee, H.-P. (2010) Performance tests of a 4.1x4.1mm2 SiC LCVJFET for a DC/DC boost converter application. Proceedings European Conference on Silicon Carbide and Related Materials (ECSCRM 2010), Oslo, Norway, August 29–September 2, 2010.
22. Ållebrand, B. and Nee, H.-P. (2001) On the choice of blanking times at turn-on and turn-off for the diode-less SiC JFET inverter bridge. Proceedings of the European Conference on Power Electronics and Applications (EPE), August 2001.
23. Sankin, I., Sheridan, D.C., Draper, W. *et al.* (2008) Normally-off SiC VJFETs for 800 V and 1200 V power switching applications. 20th International Symposium on Power Semiconductor Devices and IC's (ISPSD '08), May 18–22, pp. 260–262.
24. Kelley, R.L., Mazzola, M.S., Draper, W.A., and Casady, J. (2005) Inherently safe DC/DC converter using a normally-on SiC JFET. Twentieth Annual IEEE Applied Power Electronics Conference and Exposition (APEC 2005), March 6–10, 2005, Vol. 3, pp. 1561–1565.

25. Malhan, R.K., Bakowski, M., Takeuchi, Y. *et al.* (2009) Design, process, and performance of all-epitaxial normally-off SiC JFETs. *Physica Status Solidi A*, **206** (10), 2308–2328.
26. Lindgren, A. and Domeij, M. (2011) Degradation free fast switching 1200 V 50 a silicon carbide BJT's. Twenty-Sixth Annual IEEE Applied Power Electronics Conference and Exposition (APEC), March 6–11, 2011, pp. 1064–1070.
27. Palmour, J.W. (2009) High voltage silicon carbide power devices. Presentation at the ARPA-E Power Technologies Workshop, February 9, 2009.
28. Zhang, Q., Wang, J., Jonas, C. *et al.* (2008) Design and characterization of high-voltage 4H-SiC p-IGBTs. *IEEE Transactions on Electron Devices*, **55** (8), 1912–1919.
29. Peftitsis, D., Tolstoy, G., Antonopoulos, A. *et al.* (2011) High-power modular multilevel converters with SiC JFETs. *IEEE Transactions on Power Electronics*, **27** (1), 28–36.
30. Round, S., Heldwein, M., Kolar, J.W. *et al.* (2005) A SiC JFET driver for a 5 kW, 150 kHz three-phase PWM converter. Conference Record of the Fortieth IAS Annual Meeting – Industry Applications Conference, Vol. 1, pp. 410–416.
31. Peftitsis, D., Rabkowski, J., and Nee, H.-P. (2013) Self-powered gate driver for normally-on silicon carbide junction field-effect transistors without external power supply. *IEEE Transactions on Power Electronics*, **28** (3), 1488–1501.
32. Siemieniec, R. and Kirchner, U. (2011) The 1200V direct-driven SiC JFET power switch. Proceedings of the 14th European Conference on Power Electronics and Applications (EPE).
33. Dubois, F., Bergogne, D., Risaletto, D. *et al.* (2011) Ultrafast safety system to turn-off normally-on SiC JFETs. Proceedings of the 14th European Conference on Power Electronics and Applications (EPE).
34. Rixin, L., Fei, W., Burgos, R. *et al.* (2010) A Shoot-through protection scheme for converters built with SiC JFETs. *IEEE Transactions on Industry Applications*, **46** (6), 2495–2500.
35. Kim, J.-H., Min, B.-D., Baek, J.-W., and Yoo, D.-W. (2009) Protection circuit of normally-on SiC JFET using an inrush current. 31st International Telecommunications Energy Conference (INTELEC 2009), October 18–22, 2009, pp. 1–4.
36. Rabkowski, J., Tolstoy, G., Peftitsis, D., and Nee, H.P. (2012) Low-loss high–performance base-drive unit for SiC BJTs. *IEEE Transactions on Power Electronics*, **27** (5), 2633–2433.
37. Rabkowski, J., Zdanowski, M., Peftitsis, D., and Nee, H.-P. (2012) A simple high-performance low-loss current-source driver for SiC bipolar transistors. Proceedings of 7th International Power Electronics and Motion Control Conference – ECCE Asia, Harbin, China,
38. Kelley, R., Ritenour, A., Sheridan, D., and Casady, J. (2010) Improved two-stage DC-coupled gate driver for enhancement-mode SiC JFET. Proceedings Twenty-Fifth Annual IEEE Applied Power Electronics Conference and Exposition (APEC), pp. 1838–1841.
39. SemiSouth AN-SS3:6A Gate Driver Reference Design and Demoboard, Application Note, www.semisouth.com (accessed 18 December 2012).
40. SemiSouth AN-SS5: Operation and Intended Use of the SGDR2500P2 Dual-Stage Driver Board, www.semisouth.com (accessed 18 December 2013).
41. Wrzecionko, B., Kach, S., Bortis, D. *et al.* (2010) Novel AC coupled gate driver for ultra fast switching of normally-off SiC JFETs. 36th Annual Conference on IEEE Industrial Electronics Society (IECON 2010), November 7–10, 2010, pp. 605–612.
42. Chen, Z., Danilovic, M., Boroyevich, D., and Shen, Z. (2011) Modularized design consideration of a general-purpose, high-speed phase-leg PEBB based on SiC MOSFETs. Proceedings of the 14th European Conference on Power Electronics and Applications (EPE 2011).
43. Chinthavali, M., Ning, P., Cui, Y., and Tolbert, L.M. (2011) Investigation on the parallel operation of discrete SiC BJTs and JFETs. Proceedings of the Twenty-Sixth Annual IEEE Applied Power Electronics Conference and Exposition (APEC), March 6–11, 2011, pp. 1076–1083.
44. Mihaila, A.P., Udrea, F., Rashid, S.J. *et al.* (2005) SiC junction FETs – a state of the art review. Proceedings of the International Semiconductor Conference, Vol. 2, pp. 349–352.
45. Peftitsis, D., Baburske, R., Rabkowski, J. *et al.* (2013) Challenges regarding parallel-connection of SiC JFETs. *IEEE Transactions on Power Electronics*, **28** (3), 1449–1463.
46. Rabkowski, J., Peftitsis, D., and Nee, H.P. (2012) Design steps towards a 40-kVA SiC inverter with an efficiency exceeding 99.5%. Twenty-Seventh Annual IEEE Applied Power Electronics Conference and Exposition (APEC), pp. 1536–1543.

47. Rabkowski, J., Peftitsis, D., and Nee, H.P. (2013) A 6kW, 200kHz boost converter with parallel-connected SiC bipolar transistors. IEEE Applied Power Electronics Conference and Exposition (APEC), pp. 1991–1998.
48. Biela, J., Schweizer, M., Waffler, S., and Kolar, J.W. (2011) SiC vs. Si – evaluation of potentials for performance improvement of inverter and DC-DC converter systems by SiC power semiconductors. *IEEE Transactions on Industrial Electronics*, **57** (7), 2872–2882.
49. Friedli, T., Round, S.D., Hassler, D., and Kolar, J.W. (2009) Design and performance of a 200-kHz All-SiC JFET current DC-link back-to-back converter. *IEEE Transactions on Industry Applications*, **45** (5), 1868–1878.
50. Lai, R., Wang, F., Ning, P. *et al.* (2010) A high-power-density converter. *IEEE Industrial Electronics Magazine*, **4** (4), 4–12.
51. Cilio, E., Homberger, J., McPherson, B. *et al.* (2007) A novel high density 100kW three-phase silicon carbide (SIC) multichip power module (MCPM) inverter. Twenty Second Annual IEEE Applied Power Electronics Conference (APEC), pp. 666–672.
52. Doucet, J.-C. (2011) Gate driver solutions for SiC switches. Proceedings of the International SiC Power Electronics Application Workshop (ISiCPEAW).
53. Greenwell, R.L., McCue, B.M., Zuo *et al.* (2011) SOI-based integrated circuits for high-temperature power electronics applications. 26th Annual IEEE Applied Power Electronics Conference and Exposition (APEC), pp. 836–843.
54. Zetterling, C.-M., Lanni, L., Ghandi, R. *et al.* (2012) Future high temperature applications for SiC integrated circuits. *Physica Status Solidi C*, **9** (7), 1647–1650.
55. Stalter, O., Kranzer, D., Rogalla, S., and Burger, B. (2010) Advanced solar power electronics. 22nd International Symposium on Power Semiconductor Devices and IC's (ISPSD), pp. 3–10.
56. Zhang, H., Tolbert, L.M., and Ozpineci, B. (2011) Impact of SiC devices on hybrid electric and plug-in hybrid electric vehicles. *IEEE Transactions on Industry Applications*, **47** (2), 912–921.
57. Wrzecionko, B., Biela, J., and Kolar, J.W. (2009) SiC power semiconductors in hevs: influence of junction temperature on power density, chip utilization and efficiency. 35th Annual Conference of IEEE Industrial Electronics (IECON '09), pp. 3834–3841.
58. Waffler, S., Preindl, M., and Kolar, J.W. (2009) Multi-objective optimization and comparative evaluation of Si soft-switched and SiC hard-switched automotive DC-DC converters. 35th Annual Conference of IEEE Industrial Electronics (IECON '09), pp. 3814–3821.
59. Bortis, D., Wrzecionko, B., and Kolar, J.W. (2011) A 120°C ambient temperature forced air-cooled normally-off SiC JFET automotive inverter system. Twenty-Sixth Annual IEEE Applied Power Electronics Conference and Exposition (APEC), March 6–11, 2011, pp. 1282–1289.
60. Allebrod, S., Hamerski, R., and Marquardt, R (2008) New transformerless, scalable modular multilevel converters for HVDC-transmission. Proceedings of Power Electronics Specialists Conference (PESC 2008), Rhodes, pp. 174–179.
61. Peftitsis, D., Tolstoy, G., Antonopoulos, A. *et al.* (2012) High-power modular multilevel converters with SiC JFETs. *IEEE Transactions on Power Electronics*, **27** (1), 28–36.
62. Chinthavali, M., Tolbert, L.M., and Ozpineci, B. (2004) SiC GTO thyristor model for HVDC interface. IEEE Power Engineering General Meeting, June 6–10, 2004, Denver, CO, pp. 680–685.
63. Wang, G., Huang, X., Wang, J. *et al.* (2010) Comparisons of 6.5kV 25A Si IGBT and 10-kV SiC MOSFET in solid-state transformer application. Proceedings Energy Conversion Congress and Exposition (ECCE), pp. 100–104.
64. Wang, J., Huang, A., Sung, W. *et al.* (2009) Smart grid technologies. *IEEE Industrial Electronics Magazine*, **3** (2), 16–23.
65. Efficient Power Conversion Corporation (2012 GaN Technology Overview. Efficient Power Conversion, http://www.epc-co.com/epc (accessed 17 December 2013).
66. Briere, M. (2012) The status of GaN power devices at international rectifier. Proceedings of Power Conversion Intelligent Motion (PCIM), Nurnberg, Germany.
67. Wu, Y., Kebort, D., Guerrero, J. *et al.* (2012) High frequency GaN diode-free motor drive inverter with pure sine-wave output. Proceedings of Power Conversion Intelligent Motion (PCIM), Nurnberg, Germany.
68. Sonmez, E., Heinie, U., Daumiller, I., and Kunze, M. (2012) Efficient power electronics for the price of silicon – 3D-GaN technology for GaN-on-silicon. Proceedings of Power Conversion Intelligent Motion (PCIM), Nurnberg, Germany.

69. Uemoto, Y., Hikita, M., Ueno, H. *et al.* (2007) Gate injection transistor (GIT) – a normally-off AlGaN/GaN power transistor using conductivity modulation. *IEEE Transactions on Electron Devices*, **54** (12), 3393–3399.
70. Hensel, A., Wilhelm, C., and Kranzer, D. (2012) Application of a new 600 V GaN transistor in power electronics for PV systems. Proceedings of International Power Electronics and Motion Control Conference (EPE-PEMC), Novi Sad, Serbia.
71. Delaine, J., Jeannin, P.-O., Frey, D., and Guepratte, K. (2012) High frequency DC-DC converter using GaN device. Proceedings of Applied Power Electronics Conference (APEC), Orlando.
72. Otsuka, N. (2012) GaN power electron devices. Proceedings of International SiC Power Electronics Applications Workshop (ISiCPEAW), Kista, Sweden.

5

AC-Link Universal Power Converters: A New Class of Power Converters for Renewable Energy and Transportation

Mahshid Amirabadi[1] and Hamid A. Toliyat[2]

[1]*Department of Electrical and Computer Engineering, University of Illinois at Chicago, Illinois, USA*

[2]*Department of Electrical and Computer Engineering, Texas A&M University, College Station, Texas, USA*

5.1 Introduction

This chapter introduces a new class of power converters, that is, ac-link universal power converters [1–12]. This category of power converters is described as "universal," because the inputs and outputs of these converters can be dc, ac, single phase, or multiphase. Therefore, they can be employed in a variety of applications including, but not limited to, photovoltaic (PV) power generation, battery–utility interfaces, battery chargers and wind power generation.

In ac-link universal power converters, the link has high-frequency alternating current and voltage; therefore, there is no need for dc electrolytic capacitors at the link. Considering all the problems associated with bulky electrolytic capacitors, such as temperature dependency, this converter is an excellent alternative to dc-link converters.

In fact, the ac-link universal power converter is an extension of the dc–dc buck–boost converter and by adding complementary switches and modifying the switching scheme, the link inductor, which is the main energy storage element in this converter, can have alternating current instead of direct current. This improves the performance of this converter and increases significantly the utilization of the link.

The following section presents a summary of the steps required to convert the dc–dc buck–boost converter to an ac-link ac–ac buck–boost converter. In Section 5.3, the soft switching version of the ac-link universal power converter is introduced. It is shown that placing a small capacitor in parallel with the link inductor enables the converter to benefit from soft switching, which significantly increases

Power Electronics for Renewable Energy Systems, Transportation and Industrial Applications, First Edition.
Edited by Haitham Abu-Rub, Mariusz Malinowski and Kamal Al-Haddad.

its efficiency. The principles of operation of the soft switching ac-link universal power converter are studied in Section 5.4. A design procedure and an analysis method are presented in Sections 5.5 and 5.6, respectively. Some of the applications of this converter are discussed in Section 5.7, and finally Section 5.8 summarizes this chapter.

5.2 Hard Switching ac-Link Universal Power Converter

Let us start from the most basic configuration, a dc–dc buck–boost converter, as shown in Figure 5.1. In a buck–boost converter, the link is first charged from the input and then discharged into the output. Clearly, this circuit has two modes of operation: mode 1, the charging of the link and mode 2, the discharging of the link. During mode 1, switch S1 conducts and during mode 2, switch S2 conducts. It is assumed that this buck–boost converter operates at the boundary of continuous and discontinuous conduction modes. Switch S2 needs to block reverse voltage. Hence, in Figure 5.1, a diode is placed in series with this switch. Voltages $V1$ and $V2$ are both assumed to be positive. In a conventional buck–boost converter, switch S2

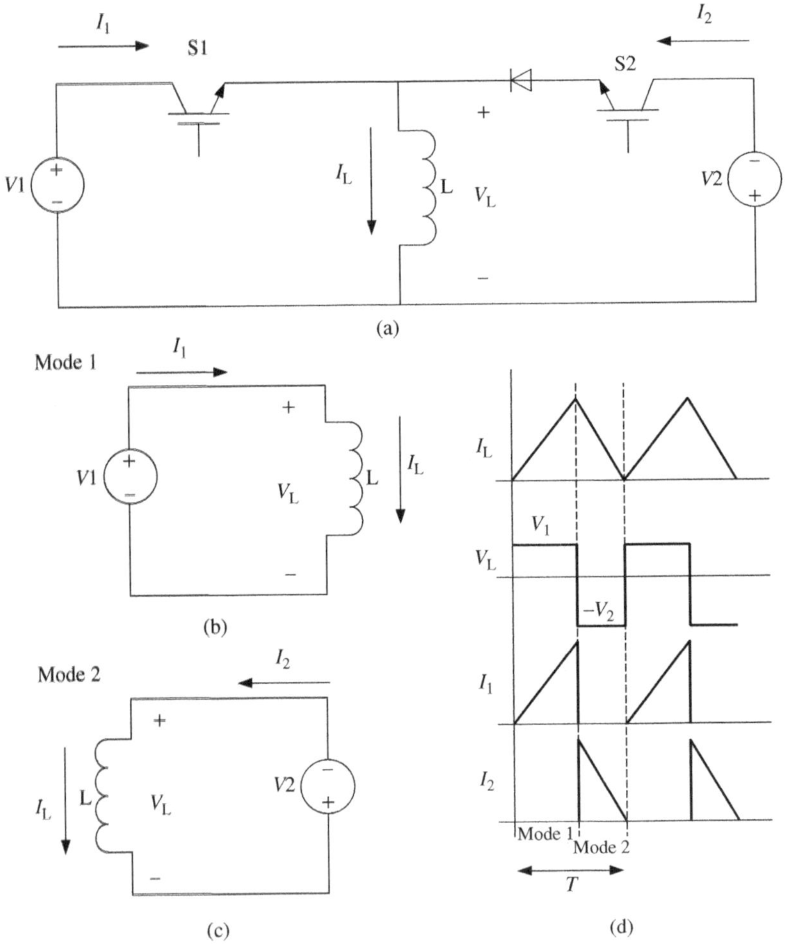

Figure 5.1 dc–dc buck–boost converter: (a) configuration, (b) energizing mode (mode 1), (c) de-energizing mode (mode 2) and (d) voltage and current waveforms

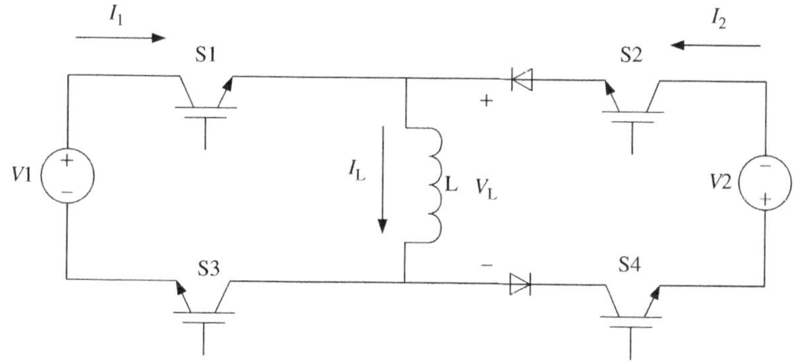

Figure 5.2 An alternative representation of the dc–dc buck–boost converter

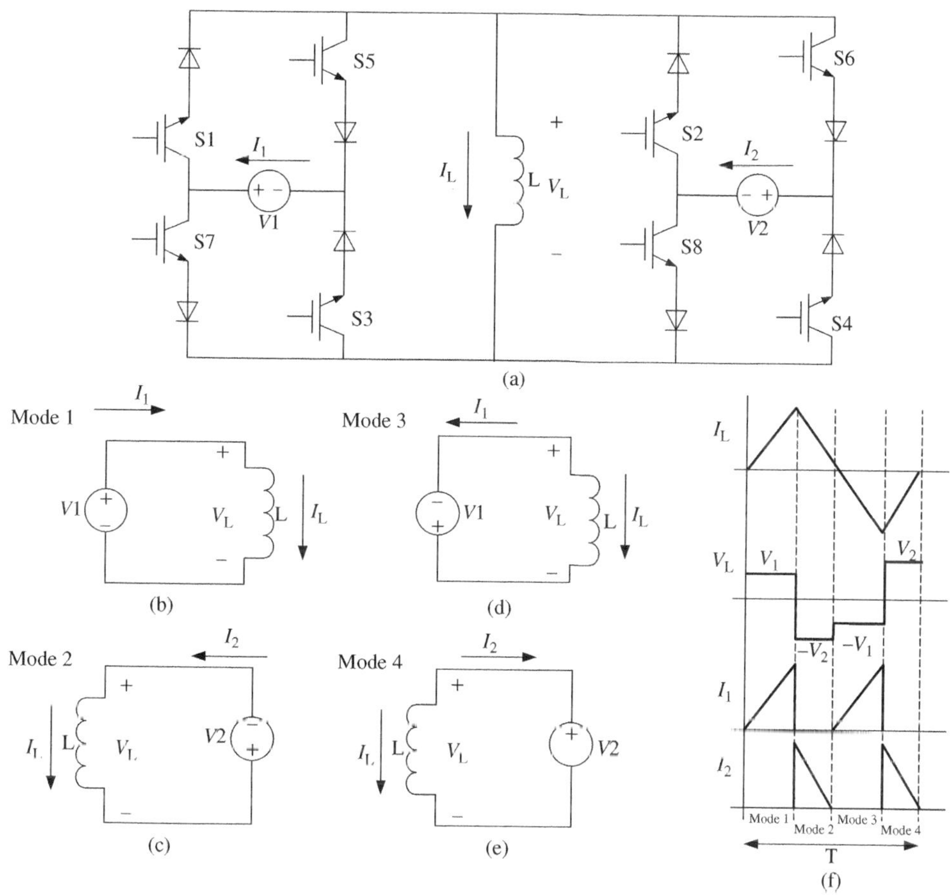

Figure 5.3 ac-link dc–dc buck–boost converter: (a) configuration, (b) energizing mode in positive direction (mode 1), (c) de-energizing mode in positive direction (mode 2), (d) energizing mode in negative direction (mode 3), (e) de-energizing mode in negative direction (mode 4) and (f) current and voltage waveforms

is usually removed and the diode conducts during mode 2. This converter can also be demonstrated, as represented in Figure 5.2. Although switches S3 and S4 are unnecessary for this case, they are required for the subsequent steps.

In order to have an alternating inductor current or, in other words, to allow the link inductor to charge and discharge in both positive and negative directions, four other switches should be added, as depicted in Figure 5.3. The link cycle can then be divided into two half cycles with the positive link current during the first half cycle and the negative link current during the second half cycle. This results in an alternating link current. This converter has four modes of operation: mode 1, the charging of the link with positive current; mode 2, the discharging of the link with positive current; mode 3, the charging of the link with negative current; and mode 4, the discharging of the link with negative current. Switches S1, S2, S3 and S4 conduct when the inductor current is positive, whereas switches S5, S6, S7 and S8 conduct in the case of a negative link current.

If bidirectional power flow is required, then the number of switches needs to be doubled, as shown in Figure 5.4. Switches S1–S8 transfer the power from *V*1 to *V*2 and switches S9–S16 are responsible for transferring the power from *V*2 to *V*1. The converter shown in Figure 5.4 is, in fact, a bidirectional ac-link dc–dc buck–boost converter.

Adding more legs to each side of this converter yields multiphase bidirectional ac-link converters, such as the dc to multiphase ac, multiphase ac to dc and multiphase ac to multiphase ac configurations shown in Figure 5.5. Dc–ac, ac–dc and ac–ac buck–boost converters were first proposed by Ngo [13]. However, in those configurations the switches were unidirectional, resulting in a dc-link current.

The principle of operation of the multiphase converter is similar to that of the dc–dc converter; however, the main challenge is to determine the contribution of each phase of the multiphase system to the charging or discharging of the link. Let us consider a three-phase inverter, as shown in Figure 5.5(a).

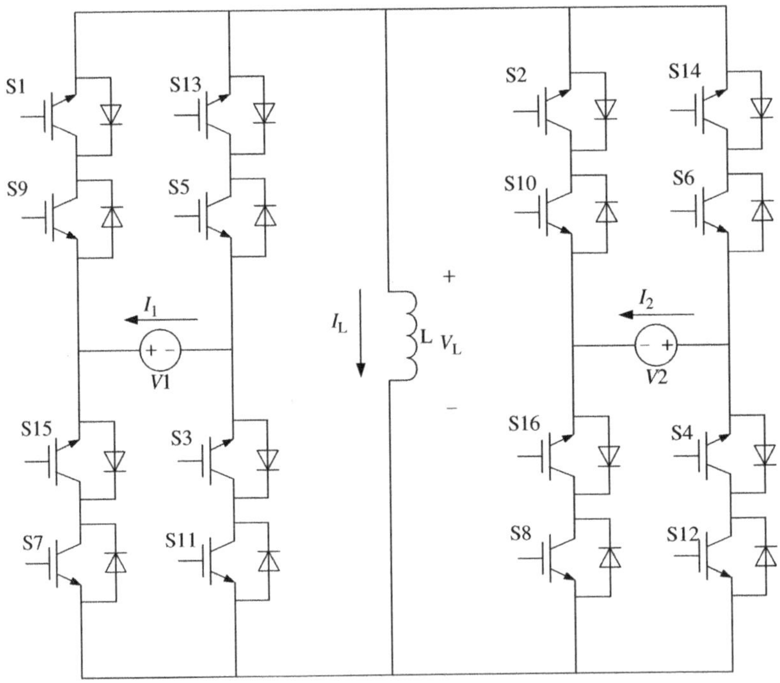

Figure 5.4 Bidirectional ac-link dc–dc buck–boost converter

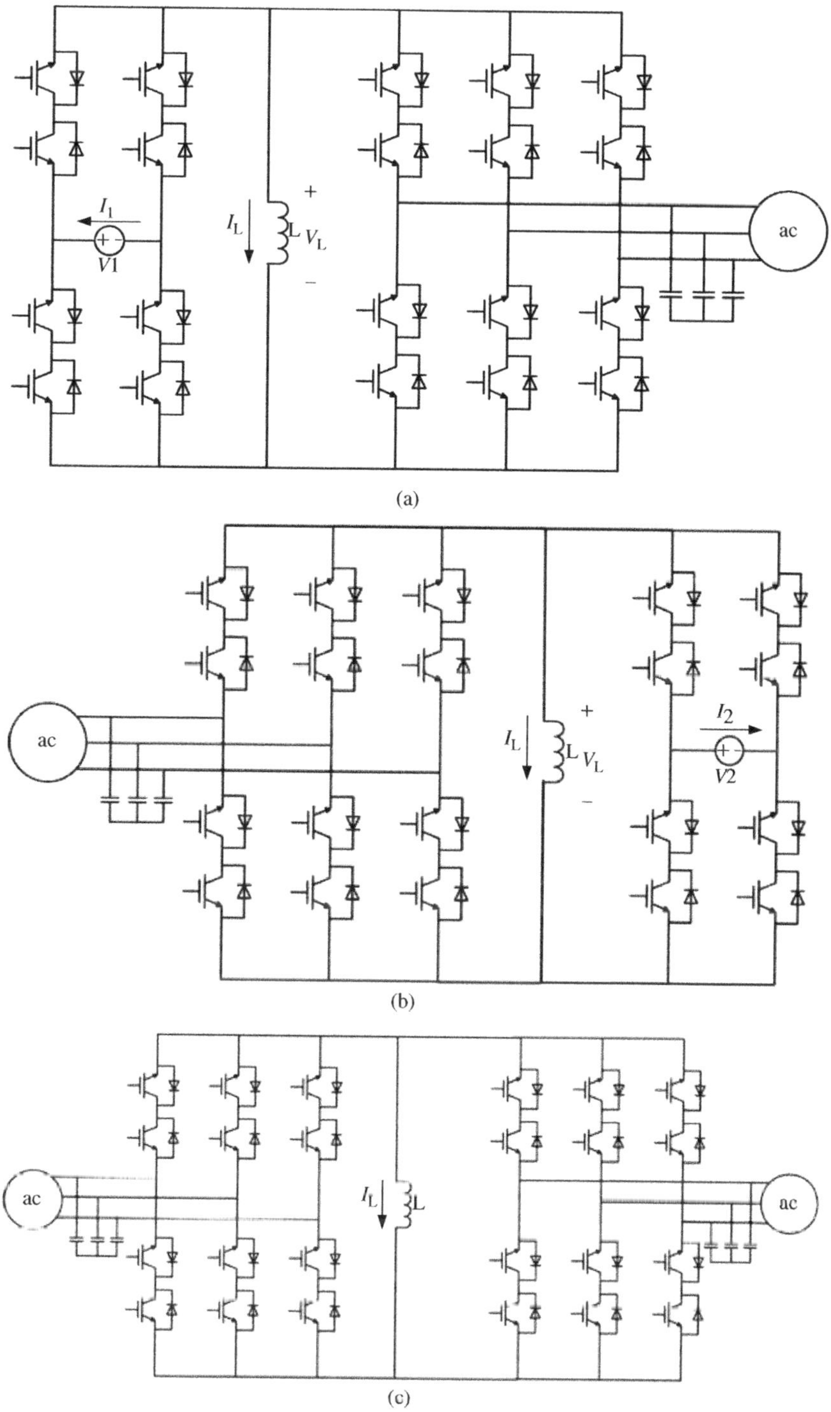

Figure 5.5 (a) Bidirectional ac-link dc–ac buck–boost converter, (b) bidirectional ac-link ac–dc buck–boost converter and (c) bidirectional ac-link ac–ac buck–boost converter

As the input of the inverter is dc, the charging of the link in this case is similar to that of the dc–dc converter. However, during the discharging mode, there is one link to be discharged and two phase pairs to be charged from the link. In a balanced three-phase system, the sum of the phase currents at any instant is zero. One of them is the highest in magnitude with a certain polarity, while the other two phase currents have lower magnitude with the opposite polarity. Although there are three phase pairs in a three-phase system, considering the polarity of the current in each phase, only two of these phase pairs can provide a path for the current when connected to the link. Therefore, the charged link transfers power to the output by discharging into two phase pairs. The two phase pairs are the one formed by the phases having the highest current and the second highest current, and the other formed by the phases having the highest current and the lowest current, where the currents are sorted as highest, second highest and lowest in terms of magnitude alone. For example, if $I_{a\_o} = -10$ A, $I_{b\_o} = 7$ A and $I_{c\_o} = 3$ are the three-phase currents, then phase pairs ab and ac are chosen to be charged from the link, because phase pair bc cannot provide any paths for the current. Therefore, in the case of dc to three-phase ac, there are six modes of operation. Similarly, in a three-phase ac–ac system, each charging and discharging mode will be split into two modes, which results in eight modes of operation.

It must be noted that, in practice, the incoming switches on the output side need to be turned on before turning off the outgoing switches on the input side. However, being reversed biased, the incoming switches do not conduct before the outgoing switches are turned off. On the other hand, the incoming switches on the input side might need to be turned on after the link current is zero and the outgoing output-side switches are turned off. This might result in operation of the converter in discontinuous conduction mode.

5.3 Soft Switching ac-Link Universal Power Converter

As seen in the previous section, the ac-link universal power converter transfers power entirely through the link inductor (L). Placing a small capacitor (C) in parallel with the link inductor allows the switches to be turned on at zero voltage, as well as to benefit from a soft turn-off. Figure 5.6 shows the configuration of an ac–ac soft switching ac-link universal power converter.

This converter is a partial resonant converter, that is, only a small time interval is allocated to resonance in each cycle. Hence, while the resonance facilitates the zero voltage turn-on and soft turn-off of the switches, the LC link has low reactive ratings and low power dissipation. This is the main difference

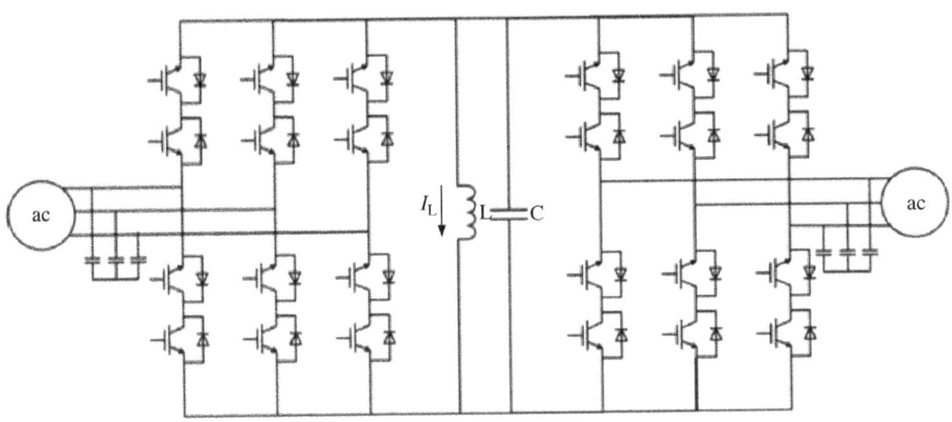

Figure 5.6 Soft switching bidirectional ac-link ac–ac buck–boost converter

between the soft switching ac-link universal power converter and the parallel resonant ac-link converter introduced in [14]. Although these converters seem similar, their principles of operation are totally different.

The first partial resonant ac–ac buck–boost converter was proposed by Kim and Cho [15]; however, it contained merely unidirectional switches and, hence, the link current had a dc component that reduced significantly the utilization of the link.

Owing to the soft switching, the switching losses in soft switching ac-link universal power converters are negligible and the frequency of the link can be very high, which results in a compact link inductor and filter components. Moreover, the zero voltage turn-on of the switches results in reduced voltage stress on the switches.

5.4 Principle of Operation of the Soft Switching ac-Link Universal Power Converter

The principle of operation of the soft switching ac-link universal power converter is similar to that of the hard switching ac-link universal power converter. The main difference is that between each charging and discharging mode there is a resonating mode during which none of the switches conduct and the link resonates to facilitate the *zero voltage turn-on* and *soft turn-off* of the switches.

To explain the principle of operation, a three-phase ac–ac converter is considered. The basic operating modes and relevant waveforms of this converter are represented in Figures 5.7 and 5.8. Each link cycle is divided into 16 modes, with 8 power transfer modes and 8 partial resonant modes taking place alternately. The link is energized from the input phase pairs during modes 1, 3, 9, and 11 and is de-energized to the output phase pairs during modes 5, 7, 13 and 15. Modes 2, 4, 6, 8, 10, 12, 14 and 16 are resonating modes. The following are the details of various operating modes:

Mode 1 (energizing): Before the start of mode 1, input switches that are supposed to conduct during modes 1 and 3 are activated (S6, S10 and S11 in Figures 5.7 and 5.8); however, they do not immediately conduct because they are reverse biased. Once the link voltage, which is resonating before mode 1, becomes equal to the maximum input line-to-line voltage that is supposed to charge the link (V_{AB} in Figures 5.7 and 5.8), the proper switches (S6 and S10) become forward biased, initiating mode 1. Therefore, the link is connected to the input voltage pair having the highest voltage via switches that charge it in the positive direction. Owing to the high frequency of the link, V_{AB} can be assumed constant during mode 1. The link current (i_{Link}) during mode 1 can be calculated using the following equations:

$$V_{\text{AB}} = L\frac{\mathrm{d}i_{\text{Link}}(t)}{\mathrm{d}t} \tag{5.1}$$

$$i_{\text{Link}}(t) = \frac{1}{L}\int_0^t V_{\text{AB}}\mathrm{d}t = \frac{V_{\text{AB}}t}{L} + i_{\text{Link}}(0) \tag{5.2}$$

In the above equations, L is the link inductance. During this mode, the link voltage is equal to V_{AB}, as shown in Figure 5.7.

The link charges until the current of phase B on the input side, when averaged over a cycle, meets its reference value. It is assumed that phase A carries the maximum input current; hence, it will be involved in charging the link during both modes 1 and 3. At the end of mode 1, switch S10 is turned off. As mentioned earlier, the link capacitor acts as a buffer across the switches during their turn-off, which results in negligible turn-off losses.

Mode 2 (partial resonance): During this mode, none of the switches conduct and the link resonates until its voltage becomes equal to that of the other input phase pair, which is supposed to charge the link (V_{AC} in Figures 5.7 and 5.8). The voltage across this phase pair is lower than the voltage across the

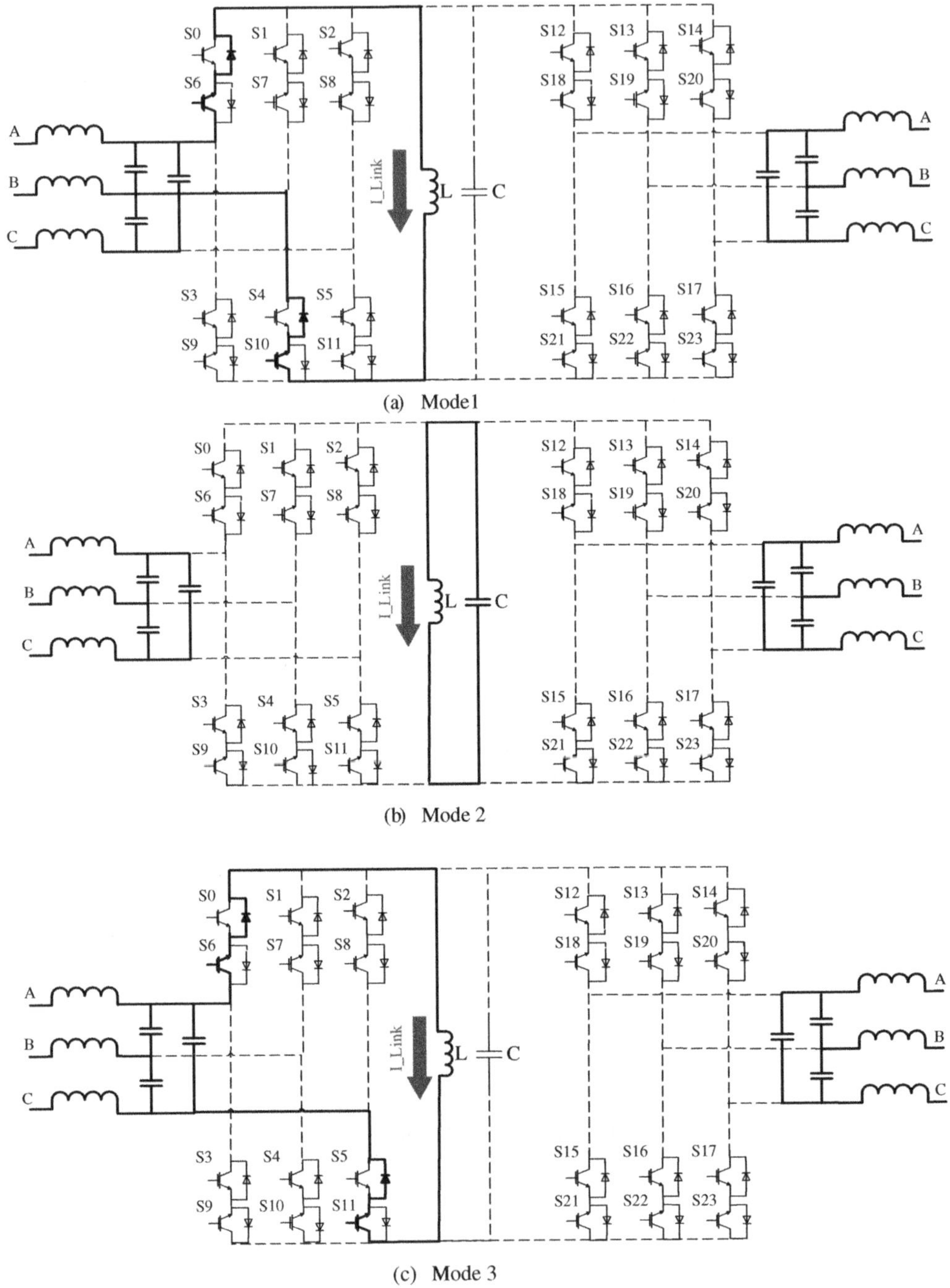

Figure 5.7 Circuit behaviors in different modes of operation [12]

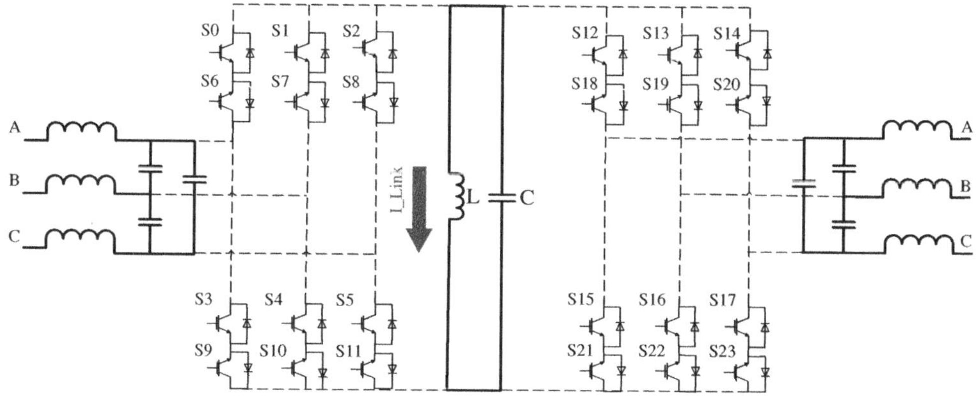

(d) Mode 4

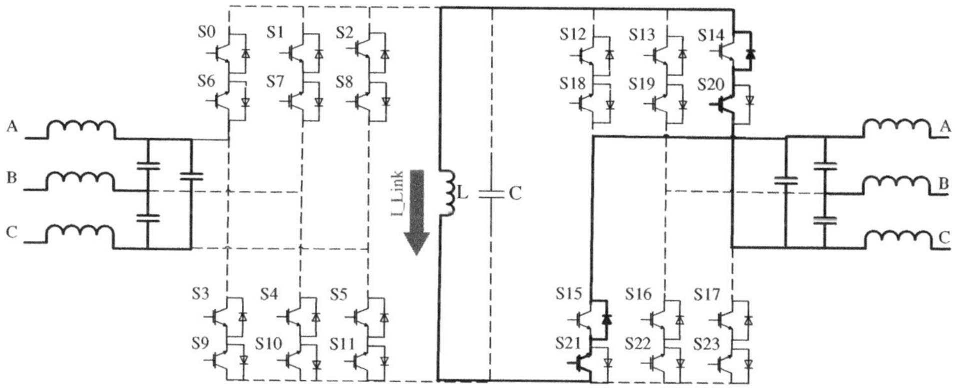

(e) Mode 5

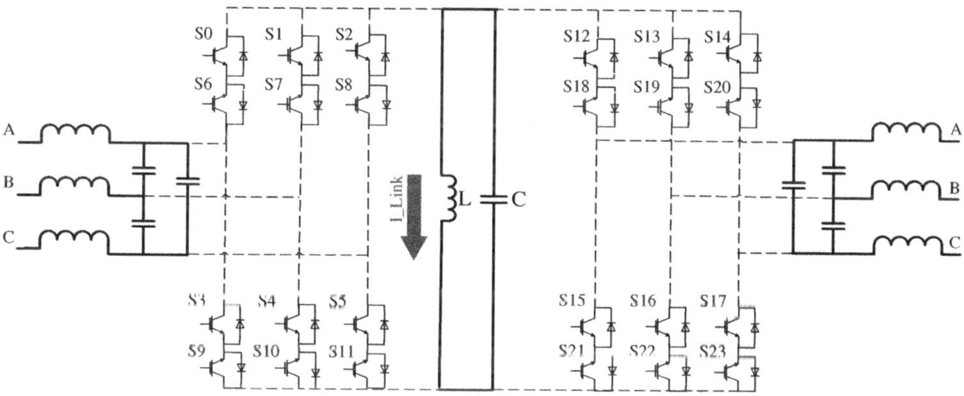

(f) Mode 6

Figure 5.7 *(continued)*

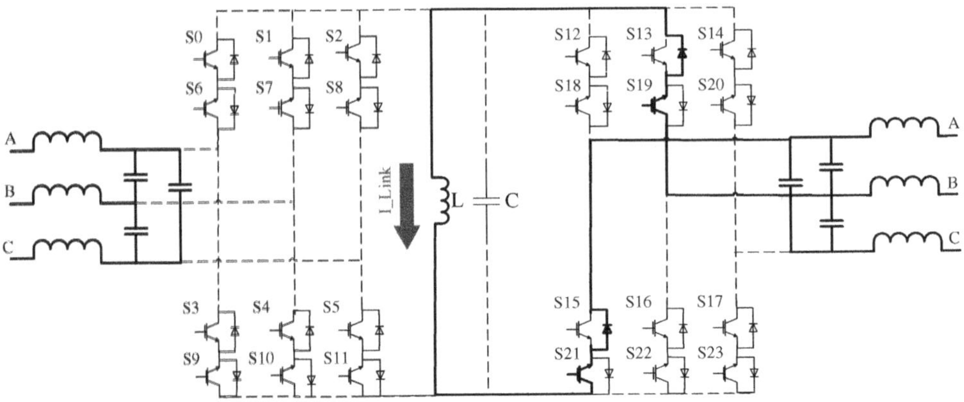

(g) Mode 7

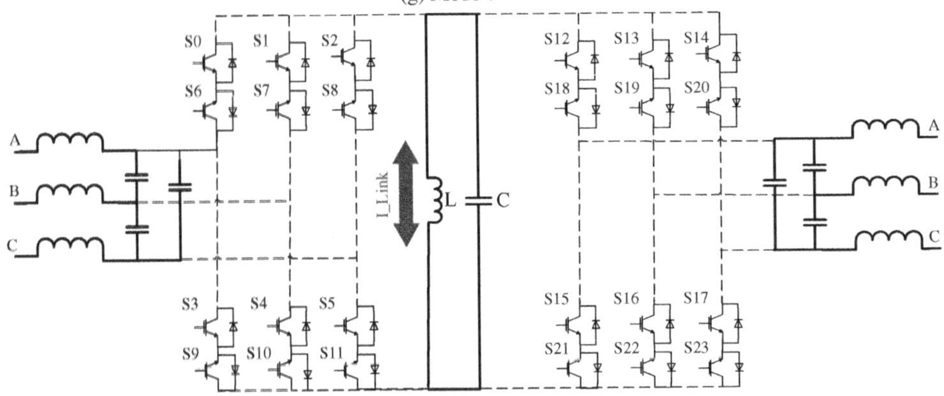

(h) Mode 8

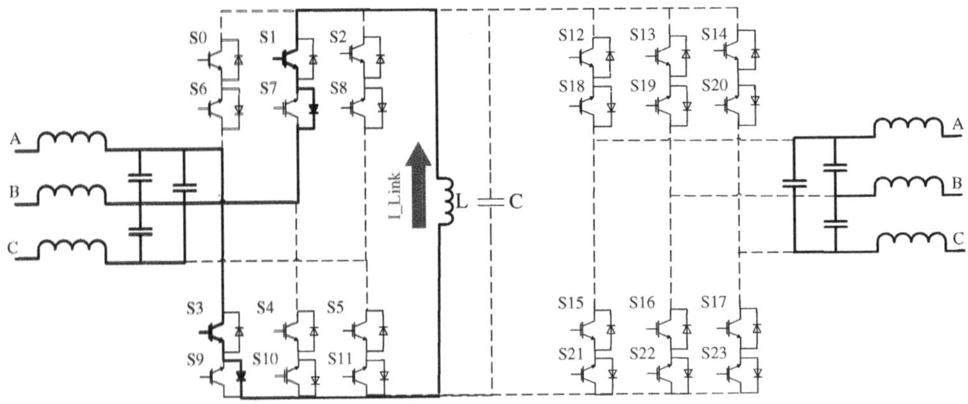

(i) Mode 9

Figure 5.7 *(continued)*

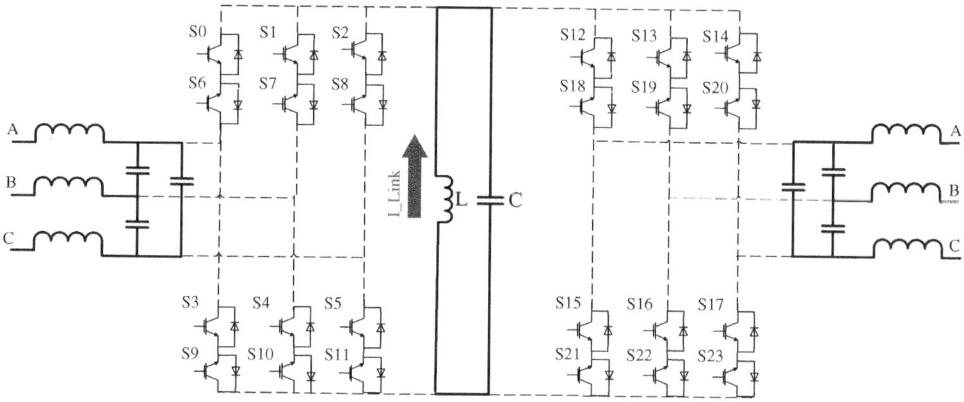

(j) Mode 10

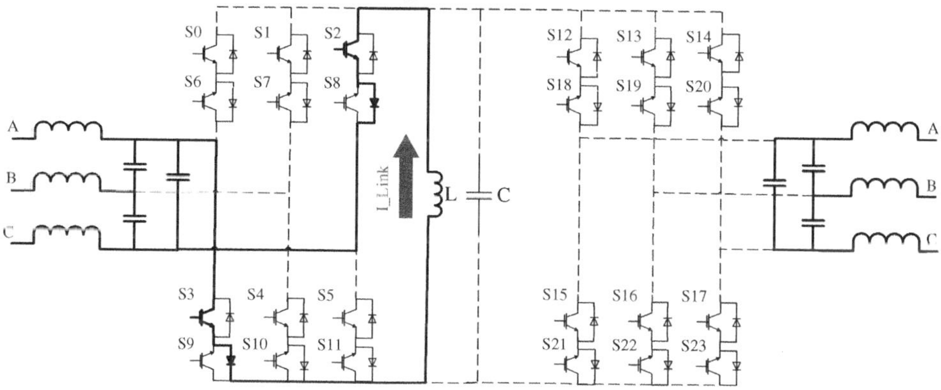

(k) Mode 11

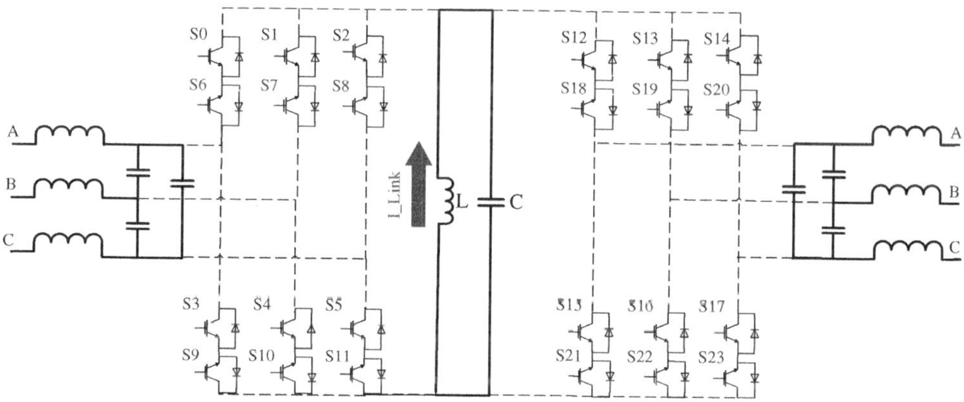

(l) Mode 12

Figure 5.7 (*continued*)

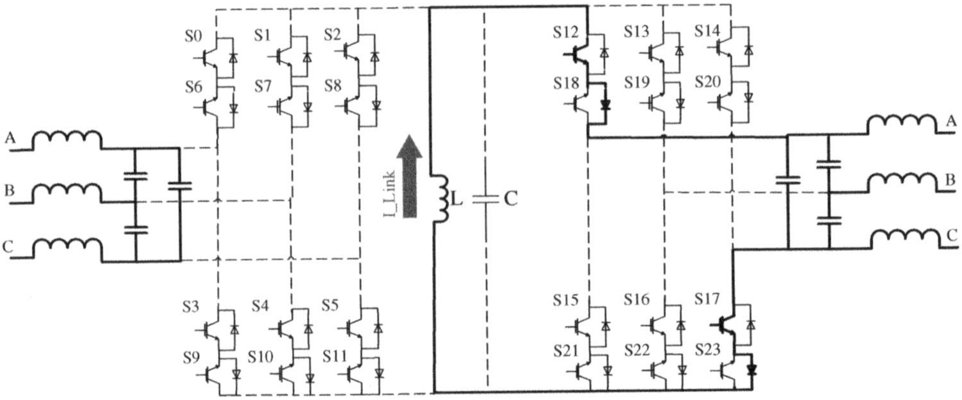

(m) Mode 13

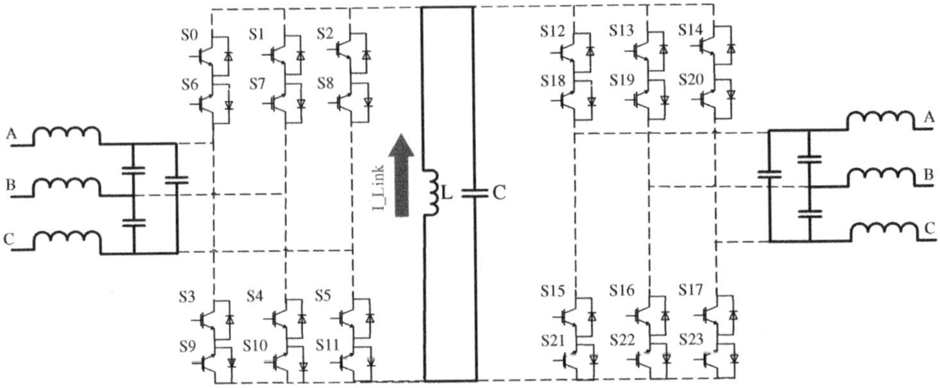

(n) Mode 14

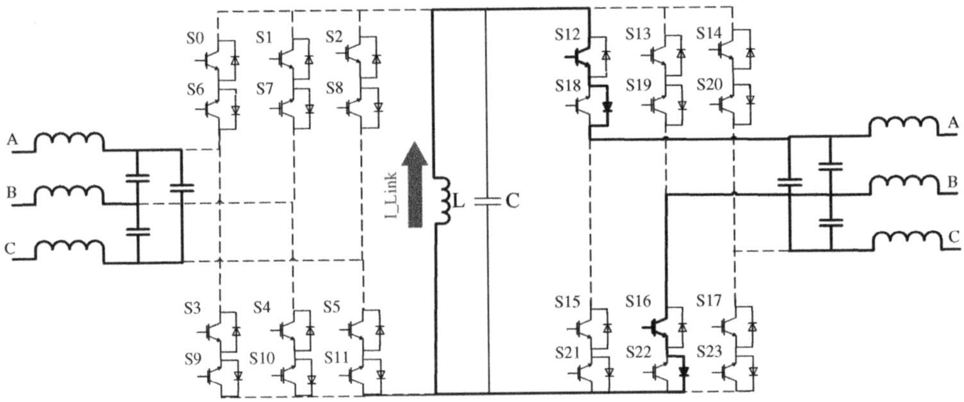

(o) Mode 15

Figure 5.7 *(continued)*

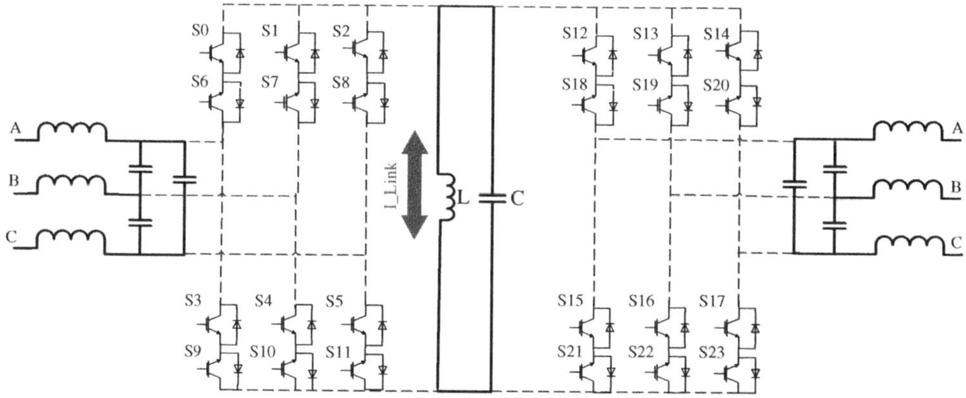

(p) Mode 16

Figure 5.7 *(continued)*

phase pair that charged the link during mode 1. In this mode, the circuit behaves as a simple LC circuit described by the following equations:

$$i_c(t) = -i_{\text{Link}}(t) = C\frac{dV_{\text{Link}}(t)}{dt} \tag{5.3}$$

$$V_{\text{Link}}(t) = L\frac{di_{\text{Link}}(t)}{dt} \tag{5.4}$$

In the above equations, $i_c(t)$, $V_{\text{Link}}(t)$ and C are the capacitor current, the link voltage and the link capacitance, respectively. As the current passing through the capacitor is equal to "$-i_{\text{Link}}(t)$" and the inductor current is positive, $(dV_{\text{Link}}(t))/dt$ is negative, implying that the link voltage is decreasing.

Mode 3 (energizing): Once the link voltage reaches the voltage across the input phase pair AC, switches S6 and S11 are forward biased, initiating mode 3, during which the link continues to charge in the positive direction from the input phase pair having the second highest voltage (V_{AC}). At the end of mode 2, the link voltage is equal to V_{AC}. Hence, at the instant of switch turn-on, the voltage across the corresponding switches is zero. This implies that the turn-on occurs at zero voltage as the switches transition from reverse to forward bias. For the case shown in Figure 5.7, during mode 3, the link charges until the current of phase A on the input side, averaged over a cycle, meets its reference value. All the input switches are then turned off, initiating another resonating mode.

Mode 4 (partial resonance): During mode 4, the behavior of the circuit is similar to that of mode 2 and the link voltage decreases until it reaches zero. At this point, the output switches that are supposed to conduct during modes 5 and 7 are activated (S19, S20 and S21 in Figures 5.7 and 5.8); however, being reverse biased, they do not conduct immediately.

The sum of the output currents at any instant is zero. One of them is the highest in magnitude and of one polarity, while the two lower magnitude ones are of the other polarity. As mentioned earlier, the charged link transfers power to the output by being discharged into two output phase pairs. The two phase pairs are the one formed by the phases having the highest current and the second highest current, and the one formed by the phases having the highest current and the lowest current, where the currents are sorted as highest, second highest and lowest in terms of magnitude alone. The phase pair with the lower line-to-line voltage is chosen as the first one into which the link is discharged. In Figures 5.7 and 5.8, it is assumed that phase A carries the maximum output current.

Once the link voltage reaches V_{ACO} (assuming V_{ACO} is lower than V_{ABO}), switches S21 and S20 will be forward biased and will start to conduct, initiating mode 5.

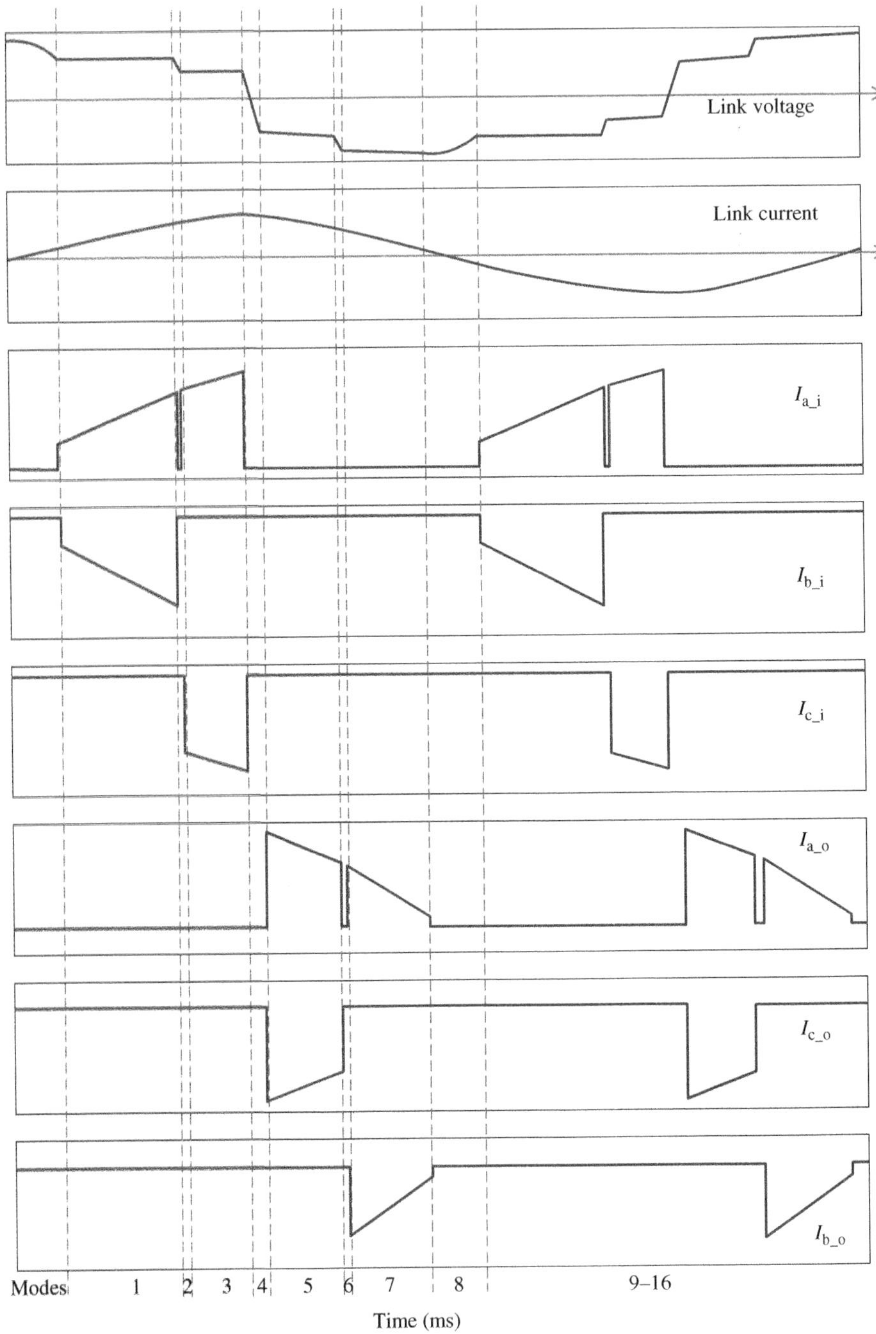

Figure 5.8 Voltage and current waveforms showing the behavior of the circuit during different modes of operation

Mode 5 (de-energizing): The output switches (S20 and S21) are turned on at zero voltage to allow the link to discharge into the chosen phase pair until the current of phase C on the output side, averaged over the link cycle, meets its reference. At this point, S20 will be turned off, initiating another resonating mode.

Mode 6 (partial resonance): The link is allowed to swing to the voltage of the other output phase pair chosen during mode 4; for the cases shown in Figures 5.7 and 5.8, the link voltage swings from V_{ACO} to V_{ABO}.

Mode 7 (de-energizing): During mode 7, the link is discharged into the selected output phase pair until there is just sufficient energy left in the link to swing to a predetermined voltage (V_{max}), which is slightly higher than the maximum input and output line-to-line voltages. At the end of mode 7, all the switches are turned off allowing the link to resonate during mode 8.

Mode 8 (partial resonance): The link voltage swings to V_{max} and then it starts to increase (because it is negative, its absolute value decreases). At this moment, the input switches that are supposed to conduct during modes 9 and 11 are turned on; however, they do not conduct immediately because they are reverse biased. Once the absolute value of the link voltage becomes equal to V_{AB}, switches S1 and S3 will be forward biased.

Modes 9 through 16 are similar to modes 1 through 8, except that the link is charged and discharged in the reverse direction. For these modes, the complimentary switch in each leg is switched when compared with the ones switched during modes 1 through 8.

Choosing the proper input-side switches during modes 8 and 16 is similar to choosing the proper output-side switches during modes 4 and 12. The sequence and the pairs are selected in order to minimize the partial resonance times while meeting the desired harmonic levels. Therefore, first the two phase pairs that can be involved in the energizing process need to be determined. These are the phase pairs formed by the phase with the maximum current and the two other phases. The switches connected to the input phase pair that has the higher line-to-line voltage conduct during the first energizing mode (mode 1 or 9). Whereas the switches connected to the output phase pair that has the lower line-to-line voltage conduct during the first de-energizing mode (mode 5 or 13). This will result in minimized partial resonance time and guarantees the zero voltage turn-on of the switches.

As observed, the input-side and output-side switches never conduct simultaneously, which implies that the input and output are isolated. However, if galvanic isolation is still required, a single-phase high-frequency transformer can be added to the link. In order to have a more compact converter, the transformer can be designed such that its magnetizing inductance plays the role of the link inductance, as depicted in Figure 5.9. In practice, the link capacitor needs to be split into two capacitors placed at the primary and the secondary of the transformer. The principle of operation of the converter with galvanic isolation is similar to that of the original converter.

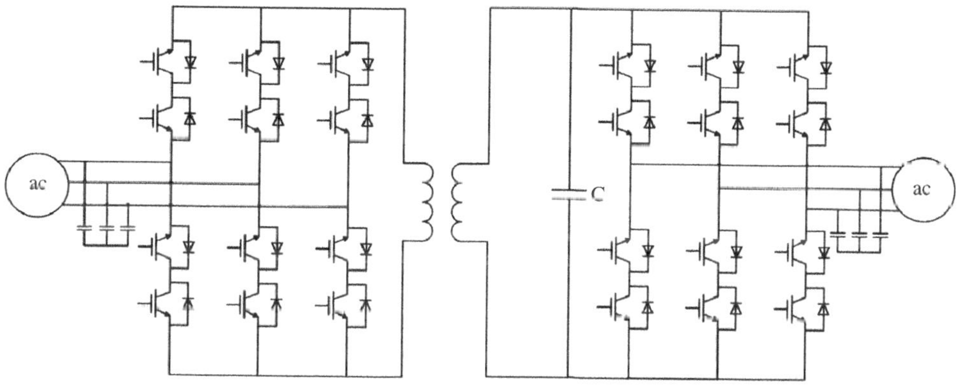

Figure 5.9 Soft switching bidirectional ac-link ac–ac buck–boost converter with galvanic isolation

5.5 Design Procedure

To simplify the design procedure, the resonating time, which is much shorter than the power transfer time at full power, will be neglected. Moreover, charging and discharging are each assumed to take place in one equivalent mode instead of two modes during each power cycle. For this, the link is assumed to be charged through a virtual input phase with the input equivalent current and voltage, and similarly the link is assumed to be discharged into a virtual output phase with the output equivalent current and voltage. Considering that the phase carrying the maximum input current is involved in the energizing of the link during both modes 1 and 3, and, similarly, that the phase carrying the maximum output current is involved in the de-energizing of the link during both modes 5 and 7, the input and output equivalent currents in a three-phase ac–ac ac-link universal power converter can be calculated as follows:

$$I_{\text{i-eq}} = \frac{3I_{\text{i,peak}}}{\pi} \tag{5.5}$$

$$I_{\text{o-eq}} = \frac{3I_{\text{o,peak}}}{\pi} \tag{5.6}$$

where $I_{\text{i,peak}}$ and $I_{\text{o,peak}}$ are the input and output peak phase currents. It can be shown that input and output equivalent voltages are equal to

$$V_{\text{i-eq}} = \frac{\pi V_{\text{i,peak}}}{2} \cos\theta_{\text{i}} \tag{5.7}$$

$$V_{\text{o-eq}} = \frac{\pi V_{\text{o,peak}}}{2} \cos\theta_{\text{o}} \tag{5.8}$$

where $V_{\text{i,peak}}$, $V_{\text{o,peak}}$, $\cos(\theta_{\text{i}})$ and $\cos(\theta_{\text{o}})$ are the input peak phase voltage, output peak phase voltage, input power factor and output power factor, respectively. It should be noted that for the dc–dc, dc–ac, or ac–dc cases, we do not need to consider any virtual phases for the dc side. Design of these cases is similar to the ac–ac case, except that the equivalent current and voltage of the dc side are, in fact, the average current and voltage of that side.

Figure 5.10 represents one cycle of the link current. The following equations describe the behavior of the circuit during the charging and discharging of the link:

$$I_{\text{Link,peak}} = \frac{V_{\text{i-eq}} t_{\text{charge}}}{L} \tag{5.9}$$

$$I_{\text{Link,peak}} = \frac{V_{\text{o-eq}} t_{\text{discharge}}}{L} \tag{5.10}$$

In the above equations, the following parameters $I_{\text{Link,peak}}$, t_{charge} and $t_{\text{discharge}}$ represent the peak of the link current, the total charge time during modes 1 and 3, and the total discharge time during modes 5 and 7, respectively.

Equations (5.9) and (5.10) determine the relationship between the charge time and discharge time as follows:

$$t_{\text{charge}} = \frac{V_{\text{o-eq}} t_{\text{discharge}}}{V_{\text{i-eq}}} \tag{5.11}$$

Consequently, the ratio of the total duration of modes 1 and 3 to the period of the link current, which is twice the sum of the charging time and discharging time, is

$$\frac{t_{\text{charge}}}{T} = \frac{1}{2}\frac{V_{\text{o-eq}}}{V_{\text{i-eq}} + V_{\text{o-eq}}} \tag{5.12}$$

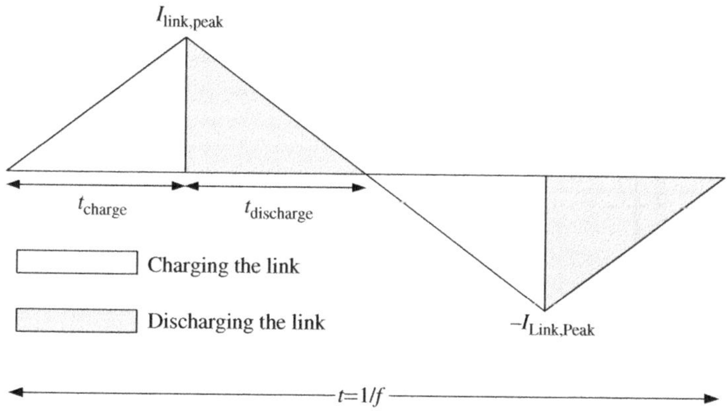

Figure 5.10 Link current when the resonating time is negligible

On the other hand, the equivalent input current can be calculated based on the link peak current, the energizing time and the link current period, which is as follows:

$$I_{\text{i-eq}} = 2 \times \frac{1}{T} \times \frac{1}{2} \times t_{\text{charge}} \times I_{\text{Link,peak}} \tag{5.13}$$

Using (5.12) and (5.13), the following equation is derived for determining the link peak current:

$$I_{\text{Link,peak}} = 2 \times I_{\text{i-eq}} \times \left(1 + \frac{V_{\text{i-eq}}}{V_{\text{o-eq}}}\right) = 2 \times (I_{\text{i-eq}} + I_{\text{o-eq}}) \tag{5.14}$$

This implies that the input and output currents determine the link peak current. The frequency of the link f can be chosen based on the power rating of the system and the characteristics of the available switches. Once the frequency is chosen, the following equation determines the inductance of the link:

$$L = \frac{P}{f(I^2_{\text{Link,peak}})} \tag{5.15}$$

where P is the rated power. Figure 5.11 shows the relationship between the link frequency and link inductance. It can be observed that choosing a higher link frequency results in a lower link inductance.

Link capacitance is chosen such that the resonating periods are retained within a small percentage of the link cycle:

$$\frac{1}{2\pi\sqrt{LC}} \gg f \tag{5.16}$$

5.6 Analysis

As mentioned earlier, the resonating time is normally negligible at full power. However, at lower power levels, the power transfer time (energizing and de-energizing time) is usually shorter than the power transfer time at full power, whereas the resonating time is almost constant. Therefore, the resonating

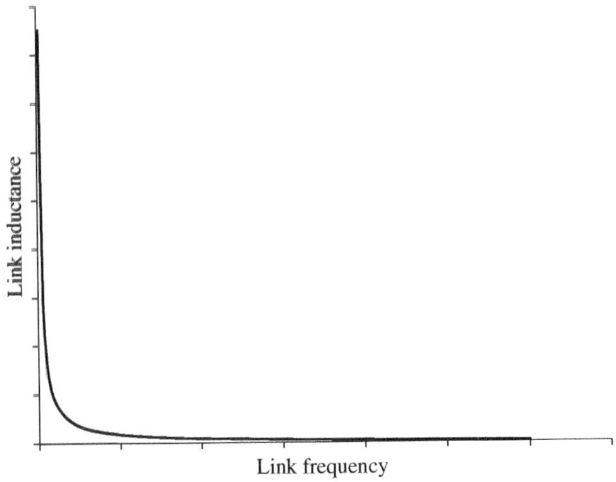

Figure 5.11 Link inductance versus link frequency

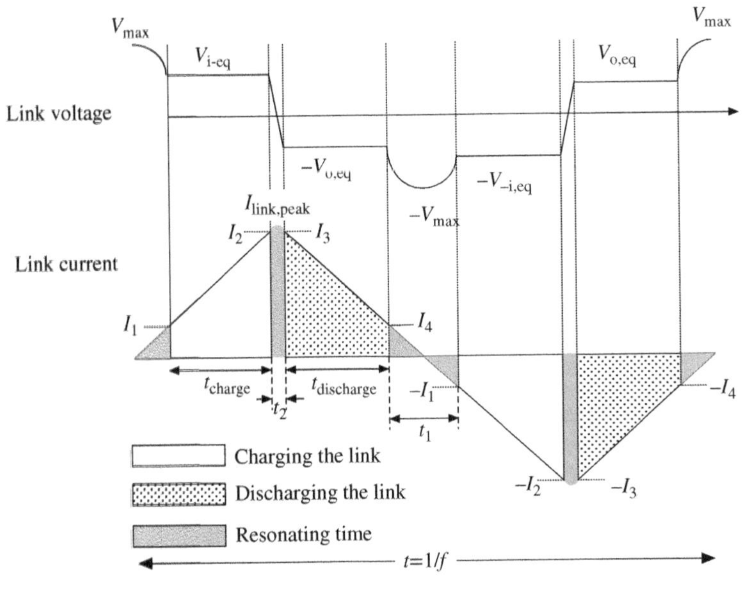

Figure 5.12 Link voltage and current when resonating time is not negligible

time cannot be neglected at lower power levels. Figure 5.12 shows the link voltage and current over one cycle, assuming that the resonating time is not negligible.

If the resonating time is negligible, then (5.14) and (5.15) may be used to calculate the link peak current and the link frequency at different power levels. However, when the resonating time cannot be neglected, the analysis will become more complicated.

Considering the principle of operation, the link current at the end of the de-energizing mode I_4 can be calculated by

$$I_4 = \sqrt{\frac{C}{L}(V_{\max}^2 - V_{o,eq}^2)} \tag{5.17}$$

Solving the resonant LC circuit during the time period t_1, I_1 and t_1 can be calculated as follows:

$$I_1 = \sqrt{\left(I_4^2 + \left(\frac{V_{o,eq}}{L\omega_r}\right)^2 - \left(\frac{V_{i,eq}}{L\omega_r}\right)^2\right)} \tag{5.18}$$

$$\begin{aligned} t_1 &= \frac{1}{\omega_r}\left(\pi + \tan^{-1}\left(\frac{I_1 L\omega_r}{V_{i,eq}}\right) - \pi + \tan^{-1}\left(\frac{I_4 L\omega_r}{V_{o,eq}}\right)\right) \\ &= \frac{1}{\omega_r}\left(\tan^{-1}\left(\frac{I_1 L\omega_r}{V_{i,eq}}\right) + \tan^{-1}\left(\frac{I_4 L\omega_r}{V_{o,eq}}\right)\right) \end{aligned} \tag{5.19}$$

In the above equations, ω_r is the resonant angular frequency, which can be calculated by

$$\omega_r = \frac{1}{\sqrt{LC}} \tag{5.20}$$

In order to find the link peak current and the link frequency, the current at the end of the energizing mode (I_2), the current at the beginning of the de-energizing mode (I_3), the resonating time between the energizing and the de-energizing mode (t_2), and the total energizing and de-energizing time should be determined first. Five other equations should be solved in order to determine I_2, I_3, t_2, t_{charge} and $t_{discharge}$. These equations are as follows:

$$V_{i,eq} = L\frac{I_2 - I_1}{t_{charge}} \tag{5.21}$$

$$V_{o,eq} = L\frac{I_3 - I_4}{t_{discharge}} \tag{5.22}$$

$$I_{i,eq} = \frac{t_{charge}}{2(t_{charge} + t_{discharge} + t_1 + t_2)}(I_2 + I_1) \tag{5.23}$$

$$I_{o,eq} = \frac{t_{discharge}}{2(t_{charge} + t_{discharge} + t_1 + t_2)}(I_3 + I_4) \tag{5.24}$$

$$t_2 = \frac{1}{\omega_r}\left(\pi - \tan^{-1}\left(\frac{I_3 L\omega_r}{V_{o,eq}}\right) - \tan^{-1}\left(\frac{I_2 L\omega_r}{V_{i,eq}}\right)\right) \tag{5.25}$$

Once these equations are solved, $I_{Link,peak}$ and f can be calculated as follows:

$$I_{Link,peak} = \sqrt{\left(I_2^2 + \left(\frac{V_{i,eq}}{L\omega_r}\right)^2\right)} \tag{5.26}$$

$$f = \frac{1}{2(t_{charge} + t_{discharge} + t_1 + t_2)} \tag{5.27}$$

It can be shown that by decreasing the power level, the link peak current decreases, whereas the link frequency increases. It should be noted that in this converter, the maximum link frequency is equal to the resonant frequency. Figures 5.13 and 5.14 represent the link peak current and link frequency variations versus power, respectively.

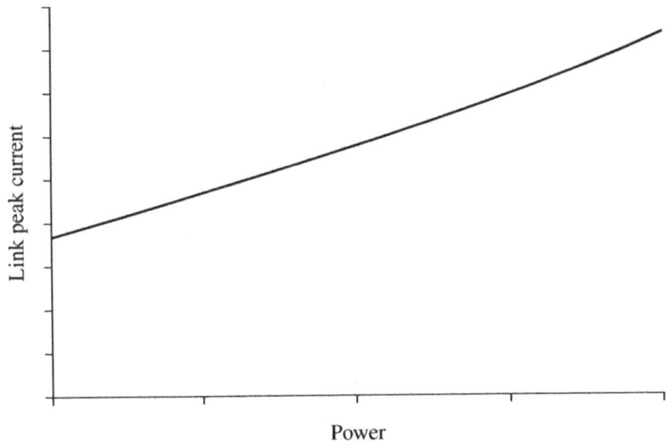

Figure 5.13 Link peak current variations versus power

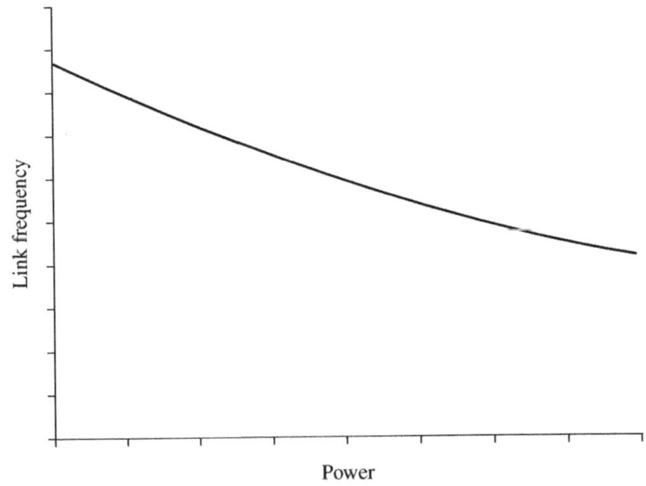

Figure 5.14 Link frequency variations versus power

5.7 Applications

5.7.1 Ac–ac Conversion (Wind Power Generation, Variable frequency Drive)

Ac–ac converters are needed in several applications, including wind power generation and variable frequency drives. Different configurations of variable speed wind turbine systems are illustrated in Figure 5.15. As seen in this figure, the line-frequency transformer is an inseparable part of these configurations. Traditionally, dc-link converters, which are formed by a rectifier and an inverter, are employed in wind power generation systems. In the simplest case, the rectifier can be formed by a diode bridge. To improve the currents, the dc-link converter can be implemented by a PWM rectifier

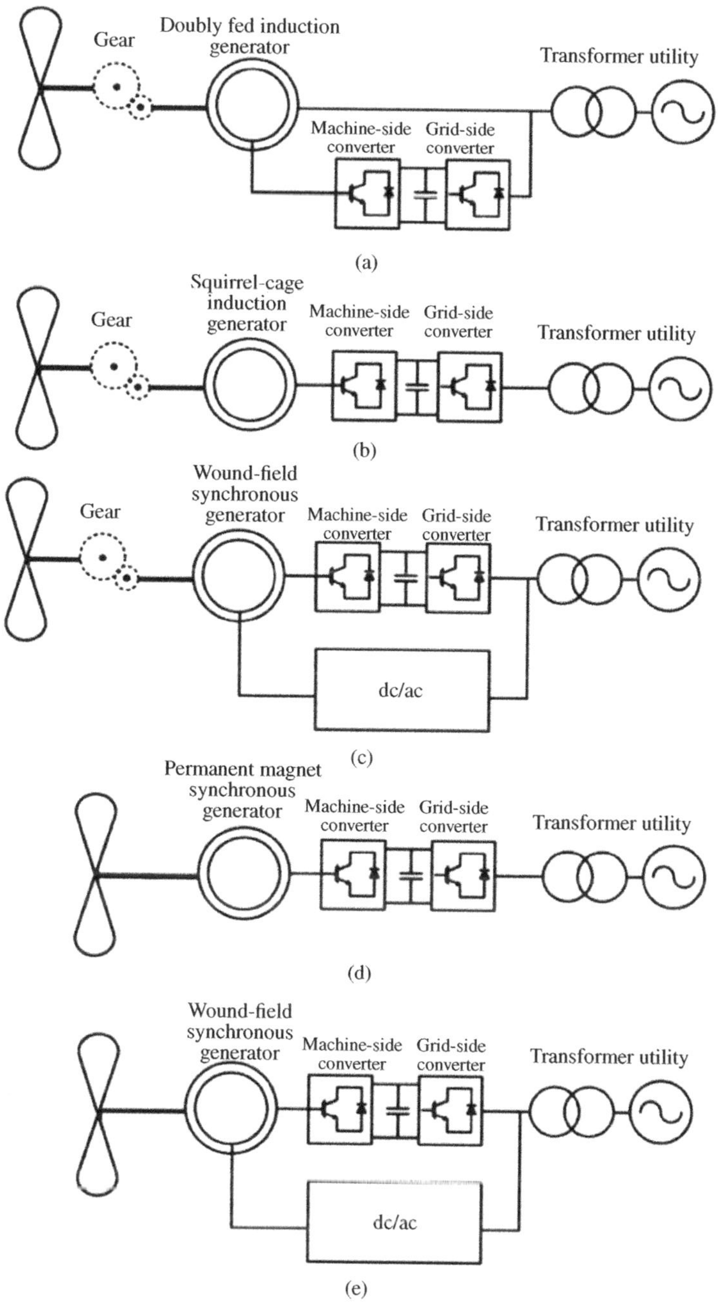

Figure 5.15 Wind turbine configurations in variable speed constant frequency systems: (a) indirect drive using doubly fed induction generator, (b) indirect drive using squirrel-cage induction generator, (c) indirect drive using wound-field synchronous generator, (d) direct drive using permanent magnet synchronous generator and (e) direct drive using wound-field synchronous generator [17]

and a PWM inverter. Regardless of the topologies of the inverter and the rectifier, a dc electrolytic capacitor, which deteriorates severely the circuit reliability and lifetime, is an integral part of dc-link converters. High switching losses and high device stresses are other important limitations of this type of converter [16, 17].

Because both the frequency and amplitude of its voltage can be changed, the ac-link universal power converter can be employed as an interface between wind-driven generators and the grid. Owing to its high reliability, compactness and low weight, this converter is an excellent candidate for wind power generation applications. In order to employ the ac-link universal power converter in such an application, the three-phase transformer used for wind power generation is replaced with a single-phase high-frequency transformer added to the link.

5.7.2 Dc–ac and ac–dc Power Conversion

5.7.2.1 PV Power Generation

In past designs, a centralized converter-based PV system was the most commonly used type of PV system. As shown in Figure 5.16, in this system, PV modules are connected to a three-phase voltage source inverter. The output of each phase of the inverter is connected to an LC filter to limit the harmonics. A three-phase transformer, which steps up the voltage and provides galvanic isolation, connects the inverter to the utility [18–26].

Low-frequency transformers are considered poor components mainly because of their large size and low efficiency. To avoid the need for low-frequency transformers, multiple-stage converters are widely used in PV systems. The most common topology, which is represented in Figure 5.17, includes a voltage source inverter and a dc-dc converter. Commonly, the dc-dc converter contains a high-frequency transformer. Despite offering a high boosting capability and galvanic isolation, this converter consists of multiple power processing stages, which lower the efficiency of the overall system. Moreover, bulky electrolytic capacitors are required for the dc link. Electrolytic capacitors, which are very sensitive to temperature, might cause severe reliability problems in inverters, and an increase by even 10 °C can halve their lifetime. Therefore, PV inverters containing electrolytic capacitors are not expected to provide the same lifetime as the PV modules. Consequently, the actual cost of the PV system involves periodic replacement of the inverter, which increases the levelized cost of energy extracted from the PV system. Considering the aforementioned problems, it is essential to support the design of alternative inverter topologies with higher reliability and lower cost [18–26].

The ac-link universal power converter can overcome most of the problems associated with existing PV inverters. As mentioned earlier, the control scheme of this converter guarantees the isolation of the input and output. However, if galvanic isolation is required, the link inductor can be replaced with

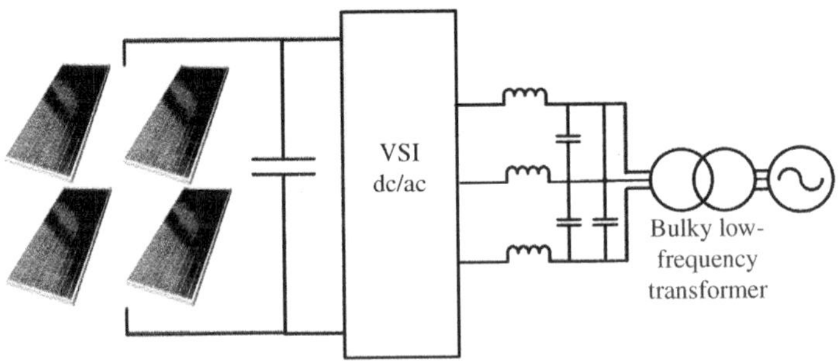

Figure 5.16 Centralized converter-based PV system [6]

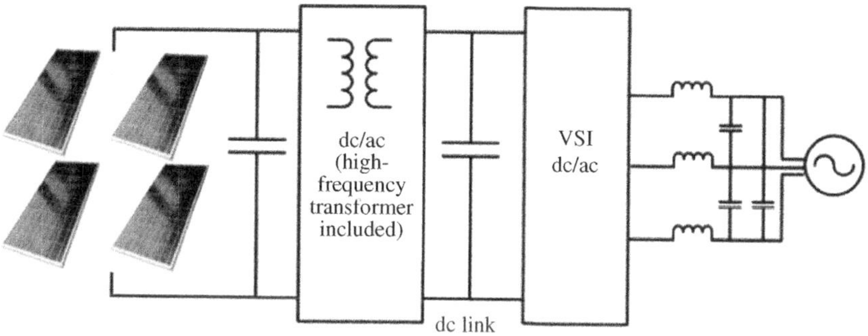

Figure 5.17 Multiple-stage conversion systems used in PV applications [6]

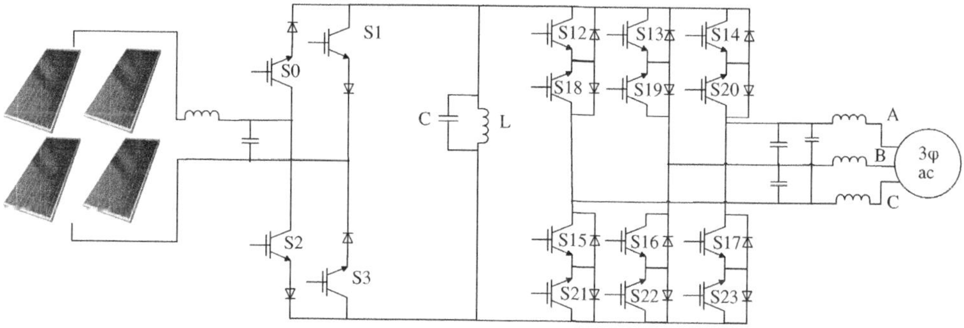

Figure 5.18 Soft stitching ac-link universal power converter for PV applications [6]

a single-phase high-frequency transformer, which eliminates the need for low-frequency transformers employed in traditional centralized converter-based PV systems. With or without a transformer, the proposed inverter is capable of stepping up or stepping down the voltage. The other merits of the proposed inverter are the elimination of the dc link and the replacement of the bulky electrolytic capacitors employed in the multiple-stage conversion systems with an LC pair having alternating current and voltage.

Because the direction of power flow in a PV inverter is always from the PV toward the load, the PV-side switches do not need to be bidirectional. Figure 5.18 shows the soft switching ac-link PV inverter.

Although this inverter does not have a dc link, it can inject reactive power into the grid during voltage sags. To provide the low-voltage ride-through (LVRT) feature, the PV-side switches should be replaced by bidirectional switches, as illustrated in Figure 5.19. The principle of operation of the inverter during the grid fault is slightly different from that of normal operation. During mode 1, the link will be charged through the PV up to a certain level. Then, similarly to normal operation, it will be discharged into the output phases; however, no net energy is taken from the link in this case. After the output-side switches are turned off, the energy stored in the link is discharged into the PV-side capacitor. In this case, the PV-side filter capacitor absorbs the energy discharged into the input. Therefore, in order to provide the LVRT feature, the PV-side filter capacitor needs to be designed based on the reactive power rating of the inverter. Once the link is completely discharged, it will be recharged from the PV with current flowing in the opposite direction. In fact, this method of control can also be used for normal operation. In the case of normal operation, the energy remaining in the link after turning off the output-side

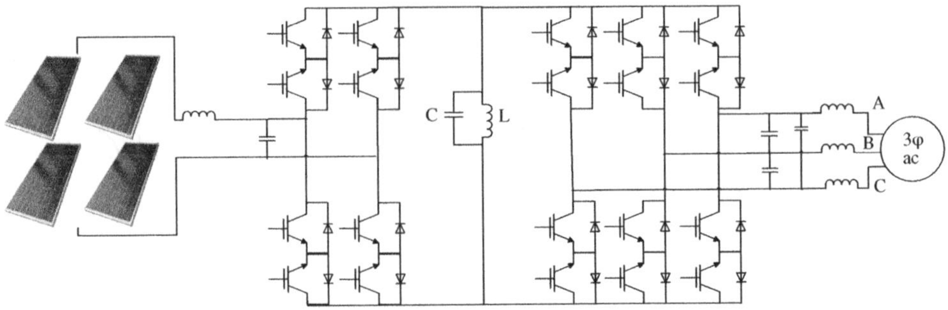

Figure 5.19 ac-link universal inverter for PV application with LVRT feature [6]

switches, which needs to be discharged into the PV-side capacitor, is much lower than that of the grid fault case.

5.7.2.2 Battery–Utility Interface

A battery–utility interface is a bidirectional dc–ac converter. In such a system, during times of excess energy, the battery is charged from the grid through the interfacing ac–dc converter, storing dc electricity in the battery. Once the stored energy is required, the battery should switch from charge mode to discharge mode such that it can feed the load. Of course, for this purpose, the dc electricity should be converted to ac through the dc–ac inverter.

Electrochemical batteries can switch between their charge and discharge modes very quickly; therefore, the main challenge is to develop low cost and robust bidirectional converters that can switch rapidly between charge and discharge modes.

The duration of the power reversal process in soft switching ac-link universal power converters is only limited by the resonance period of the input/output filters. As the frequency of the link can be as high as permitted by the switches and their associated controller, the resonance frequency of the input and output filters can be very high, resulting in a short power reversal time. This feature, along with other merits of the soft switching ac-link universal power converter, makes it an excellent candidate for battery–utility interface applications.

5.7.3 Multiport Conversion

Owing to the intermittent nature of PV and wind power, hybrid energy sources are very common in renewable energy systems. In the case of a hybrid system, both power sources need power electronics interfaces to be connected to the load. One solution is to use two separate converters, but a more efficient and reliable method is to use a multiport converter to connect several power sources to the load. Thus, it is essential to investigate multi-input/output converters.

As an example, a hybrid PV–battery source is considered in this section. Two common inverter configurations exist for hybrid distributed energy systems including dc sources. The first system is formed by a hybrid inverter that consists of two inverters operating in parallel, whose outputs are tied to the grid through a low-frequency transformer. Similar to traditional single-input/single-output inverters, the main drawback of this scheme is the existence of the low frequency and, thus, the bulky transformer. The other configuration is a multi-input dc–dc converter coupling with an inverter for feeding ac loads, which has the same disadvantages as the multi-stage PV inverters discussed in the previous section [18–26].

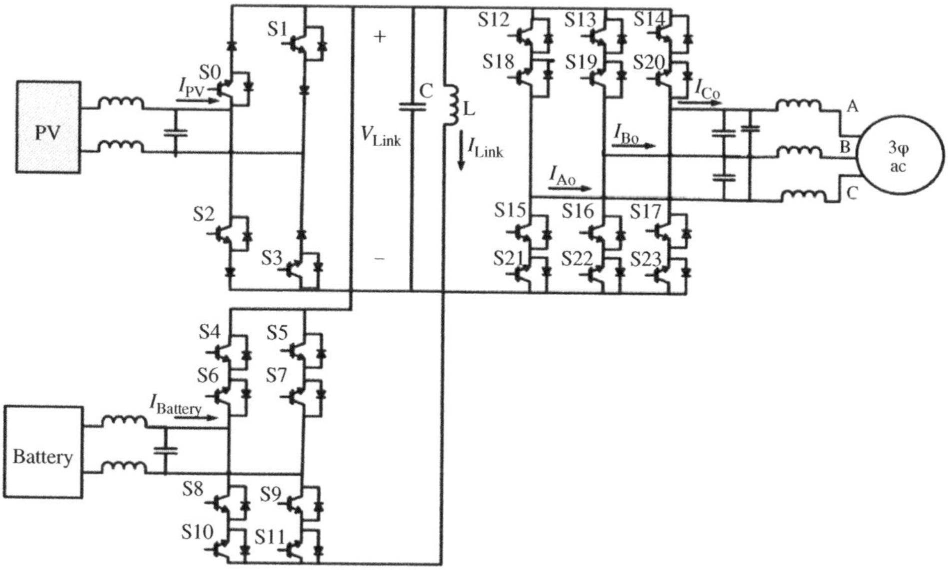

Figure 5.20 Multiport ac-link universal converter [11]

In light of these problems, a single-stage multiport high-frequency ac-link converter might be an outstanding candidate for this application. Figure 5.20 depicts the schematic of a soft switching multiport ac-link universal power converter interfacing a PV source, a battery and a three-phase ac load. There are three switch bridges in this configuration: one connected to the PV modules, one connected to the battery and one connected to the load. As seen in Figure 5.20, the PV switch bridge contains only unidirectional switches, while the other two bridges contain bidirectional switches to allow bidirectional power flow.

Depending on the power generated by the PV modules, the battery state of charge (SOC) and the load requirements, there are four possible power flow scenarios in this converter, that is, power flow:

1. From the PV modules to the load
2. From the battery to the load
3. From the PV modules to the load and the battery
4. From the PV modules and the battery to the load

If the grid-connected configuration is considered, another scenario is one in which power flows from the grid to the battery.

Depending on the power flow scenario, there might be more than one input phase pair used to charge the link or more than one output phase pair into which the link is discharged. In order to have more control over the currents and to minimize input and output current harmonics, the link charging or discharging mode can be split into two or more modes.

Figures 5.21–5.23 represent one cycle of the link current in each power flow scenario. Of course, the sequence of the charging or discharging of the sources/loads in a link cycle also depends on their voltage values. For example, in Figure 5.23, it is assumed that the voltage of the battery is higher than the voltage across the PV modules; therefore, it charges the link in mode 1 and the PV modules charge the link during mode 3. If the voltage across the PV modules is higher than the battery voltage, they charge the link during mode 1 and the battery charges the link in mode 3.

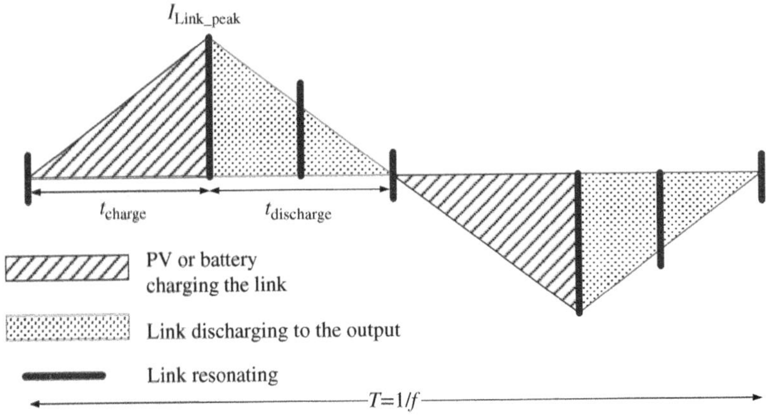

Figure 5.21 Link current of the multiport ac-link universal power inverter during the first and second power flow scenarios [11]

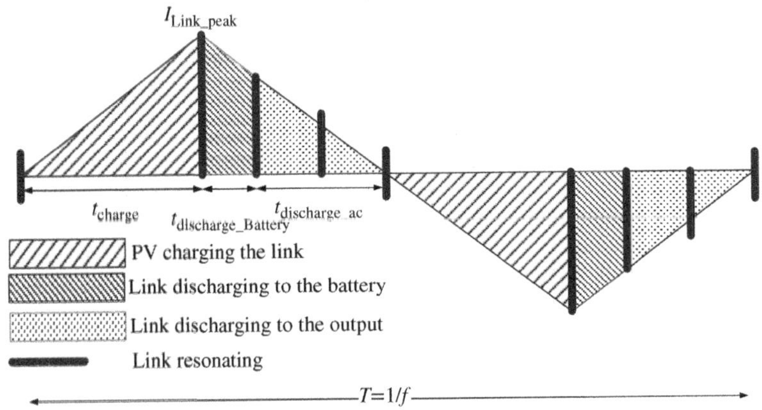

Figure 5.22 Link current of the multiport ac-link universal power inverter during the third power flow scenario [11]

In the first and second power flow scenarios, the converter behaves as a dc–ac inverter. In this case, the link cycle is divided into 12 modes with 6 power transfer modes and 6 resonating modes. The link is energized from the battery (in the second power flow scenario) or PV modules (in the first power flow scenario) during modes 1 and 7, and is de-energized into the load during modes 3, 5, 9 and 11. Modes 2, 4, 6, 8, 10 and 12 are the resonating modes.

In the third power flow scenario, the converter converts dc power to ac and dc (dc/ac+dc). In this scenario, the link cycle is divided into 16 modes with 8 power transfer modes and 8 resonating modes. The link is energized from the PV during modes 1 and 9 and is de-energized to the load and the battery during modes 3, 5, 7, 11, 13 and 15.

In the fourth power flow scenario, power flows from two dc sources to an ac load (dc+dc/ac). In this case, the link cycle is divided into 16 modes, including 8 power transfer modes and 8 resonating modes.

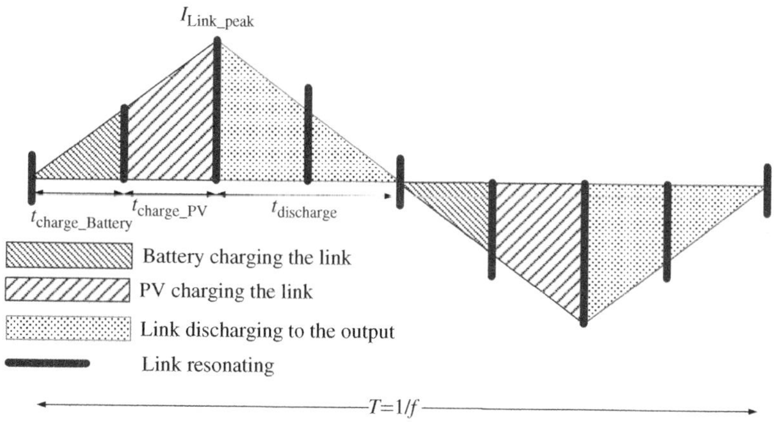

Figure 5.23 Link current of the multiport ac-link universal power inverter during the fourth power flow scenario [11]

The link is energized from the PV and the battery during modes 1, 3, 9 and 11 and is de-energized into the load during modes 5, 7, 13 and 15.

5.8 Summary

The ac-link universal power converter is a new class of power converter with several advantages. In fact, this converter is an extension of the dc-dc buck–boost converter to dc–ac, ac–dc and ac–ac converters. Hence, the main energy storage component in this converter is the link inductor, which has a high-frequency alternating current. Placing a small capacitor in parallel with the link inductor facilitates the zero voltage turn-on and soft turn-off of the switches. In this converter, the input and output are isolated; however, if galvanic isolation is required, a single-phase high-frequency transformer can be added to the link.

Considering its negligible switching losses, higher efficiency, higher reliability and compactness, this converter is an excellent candidate for several applications, including PV power generation, wind power generation, battery–utility interfaces and power electronics interfaces for hybrid energy sources.

Acknowledgment

This chapter was made possible by an NPRP Grant No. 04-080-2-030 from the Qatar National Research Fund (a member of Qatar Foundation). The statements made herein are solely the responsibility of the authors.

References

1. Alexander, W.C. (2008) Universal power converter. US Patent 2008/0013351A1, Jan. 17, 2008.
2. Balakrishnan, A., Toliyat, H.A., and Alexander, W.C. (2008) Soft switched ac link buck boost converter. Proceedings of the 23rd Annual IEEE Applied Power Electronics Conference and Exposition, APEC 2008, pp. 1334–1339.

3. Balakrishnan, A.K. (2008) Soft switched high frequency ac-link converter. MS thesis. ECE Department, Texas A&M University, College Station, TX.
4. Toliyat, H.A., Balakrishnan, A., Amirabadi, M., and Alexander, W. (2008) Soft switched ac-link AC/AC and AC/DC buck-boost converter. Proceedings of the IEEE Power Electronics Specialists Conference (PESC), pp. 4168–4176.
5. Amirabadi, M., Balakrishnan, A., Toliyat, H.A., and Alexander, W. (2008) Soft switched ac-link direct-connect photovoltaic inverter. Proceedings of the IEEE International Conference on Sustainable Energy Technologies (ICSET), pp. 116–120.
6. Amirabadi, M., Balakrishnan, A., Toliyat, H., and Alexander, W. (2014) High frequency AC-link PV inverter. *IEEE Transactions on Industrial Electronics*, **61**, 281–291.
7. Balakrishnan, A., Amirabadi, M., Toliyat, H., and Alexander, W. (2008) Soft switched ac-link wind power converter. Proceedings of the IEEE International Conference on Sustainable Energy Technologies (ICSET), pp. 318–321.
8. Amirabadi, M., Toliyat, H.A., and Alexander, W. (2009) Battery-utility interface using soft switched AC Link supporting low voltage ride through. Proceedings IEEE Energy Conversion Congress and Exposition Conference (ECCE), pp. 2606–2613.
9. Amirabadi, M., Toliyat, H.A., and Alexander, W. (2009) Battery-utility interface using soft switched AC link buck boost converter. Proceedings of the IEEE International Electric Machines and Drives (IEMDC) Conference, pp. 1299–1304.
10. Amirabadi, M., (2013) Soft-switching High-Frequency ac-Link Universal Power Converters with Galvanic Isolation, Ph.D. Dissertation, Texas A&M University, College Station, TX, USA.
11. Amirabadi, M., Toliyat, H.A., and Alexander, W.C. (2013) A multi-port AC link PV inverter with reduced size and weight for stand-alone application. *IEEE Transactions on Industry Applications*, **49** (5), 2217–2228.
12. Amirabadi, M., Toliyat, H.A., and Alexander, W.C. (2012) Partial resonant AC link converter: a highly reliable variable frequency drive. Proceedings of the IEEE Industrial Electronics Society Annual Conference (IECON), pp. 1946–1951.
13. Ngo, K. (1984) Topology and analysis in PWM inversion, rectification, and cycloconversion. PhD Dissertation. California Institute of Technology.
14. Lipo, T.A. (1988) Resonant link converters: a new direction in solid state power conversion. Presented at 2nd International Conference Electrical Drives, Eforie Nord, Romania, September 1988.
15. Kim, I.-D. and Cho, G.-H. (1990) New bilateral zero voltage switching ac/ac converter using high frequency partial-resonant link. Proceedings IEEE Industrial Electronics Society Annual Conference (IECON), pp. 857–862.
16. Kolar, J.W., Friedli, T., Rodriguez, J., and Wheeler, P.W. (2011) Review of three-phase PWM AC–AC converter topologies. *IEEE Transactions on Industrial Electronics*, **58** (11), 4988–5006.
17. Abolhassani, M.T. (2004) Integrated electric alternators/active filters. PhD Dissertation. ECE Department, Texas A&M University, College Station, TX.
18. Chakraborty, S., Kramer, B., and Kroposki, B. (2009) A review of power electronics interfaces for distributed energy systems towards achieving low-cost modular design. *Renewable and Sustainable Energy Reviews*, **13** (9), 2323–2335.
19. Kerekes, T., Teodorescu, R., Rodríguez, P. *et al.* (2011) A new high-efficiency single-phase transformerless PV inverter topology. *IEEE Transactions on Industrial Electronics*, **58** (1), 184–191.
20. Grandi, G., Rossi, C., Ostojic, D., and Casadei, D. (2009) A new multilevel conversion structure for grid-connected PV applications. *IEEE Transactions on Industrial Electronics*, **56** (11), 4416–4426.
21. Margolis, R. (2006) *A Review of PV Inverter Technology Cost and Performance Projections*, NREL/SR-620-38771, National Renewable Energy Laboratory, January 2006, http://www.nrel.gov/pv/pdfs/38771.pdf (accessed 17 December 2013).
22. Jain, S. and Agarwal, V. (2008) An integrated hybrid power supply for distributed generation applications fed by nonconventional energy sources. *IEEE Transactions on Energy Conversion*, **23** (2), 622–631.
23. Kjaer, S.B., Pedersen, J.K., and Blaabjerg, F. (2005) A review of single-phase grid-connected inverters for photovoltaic modules. *IEEE Transactions on Industry Application*, **41** (5), 1292–1306.

24. Araujo, S.V., Zacharias, P., and Mallwitz, R. (2010) Highly efficient single-phase transformerless inverters for grid-connected photovoltaic systems. *IEEE Transactions on Industrial Electronics*, **57** (9), 3118–3128.
25. Gonzalez, R., Gubia, E., Lopez, J., and Marroyo, L. (2008) Transformerless single-phase multilevel-based photovoltaic inverter. *IEEE Transaction on Industrial Electronics*, **55** (7), 2694–2702.
26. Kerekes, T., Liserre, M., Teodorescu, R. *et al.* (2009) Evaluation of three-phase transformerless photovoltaic inverter topologies. *IEEE Transactions on Power Electronics*, **24** (9), 2202–2211.

6

High Power Electronics: Key Technology for Wind Turbines

Frede Blaabjerg and Ke Ma
Department of Energy Technology, Aalborg University, Aalborg, Denmark

6.1 Introduction

To date, wind power is still the most promising renewable energy source, as reported in [1], because it has relatively low energy cost. The wind turbine system (WTS) technology started with a few tens of kilowatts of power in the 1980s, whereas nowadays, multi-megawatt (MW) wind turbines are generally installed and their size is still growing [2–5]. There is a widespread use of wind turbines in the distribution networks and more and more wind power stations are acting as traditional power plants, which are connected to the transmission networks. For example, Denmark has high wind power penetration in major areas of the country, and today, more than 30% of the country's electrical power consumption is covered by wind [6].

Initially, wind power did not have a serious impact on the power grid system. The wind turbine solution was based on a squirrel-cage induction generator (SCIG) connected directly to the grid, and thus, power pulsations in the wind were almost directly transferred into the networks. Furthermore, there was no controllability for the delivered active and reactive powers, which are important control parameters in regulating the frequency and voltage of the grid system. Nowadays, following the dramatic increase in the penetration and power level of wind turbines, wind power has a significant impact on the grid operation. Wind turbine solutions with more power electronics have become dominant [7]; this is mainly because the power electronics can change the characteristics of the wind turbines from being an unregulated energy source to being an active power generation unit. It is worth mentioning that the power electronics technology used for wind turbines is not new, but some special considerations that are valid in wind power applications have to be carefully taken into account [2–5, 8].

The scope of this chapter is to give an overview and discuss some trends in power electronics technologies used for wind turbines. First, the developments of technology and of market trends in wind power applications are discussed. Next, different wind turbine concepts are reviewed and some dominant and promising power converter solutions in wind power applications are presented. Furthermore, some control methods and grid demands, as well as the emerging reliability challenges, for future wind turbines are explained.

Power Electronics for Renewable Energy Systems, Transportation and Industrial Applications, First Edition.
Edited by Haitham Abu-Rub, Mariusz Malinowski and Kamal Al-Haddad.

6.2 Development of Wind Power Generation

The wind power has grown rapidly to a cumulative level of 282 GW worldwide with over 45 GW installed in 2012 alone; this is more than any other renewable energy source [1]. The annual market for electrical power capacity in 2011 was around 208 GW; this number indicates that wind power is really an important player in the modern electrical energy supply system. The worldwide energy penetration by wind power was 2.5% in 2010 and the prediction for 2019 is more than 8% or 1 TW. China became the largest market in 2011 with over 17.6 GW wind power installed, together with the entire European Union (EU) (9.6 GW) and the United States (6.8 GW), sharing around 85% of the global market in 2011. The cumulative capacity of wind power generation from 1996 to 2012 is shown in Figure 6.1.

Regarding the markets and manufacturers, in 2011, the Danish company Vestas was still on the top position among the largest wind turbine manufacturer in the world, closely followed by GE and Goldwind, both sharing the second position in the market. Figure 6.2 summarizes the worldwide top suppliers of

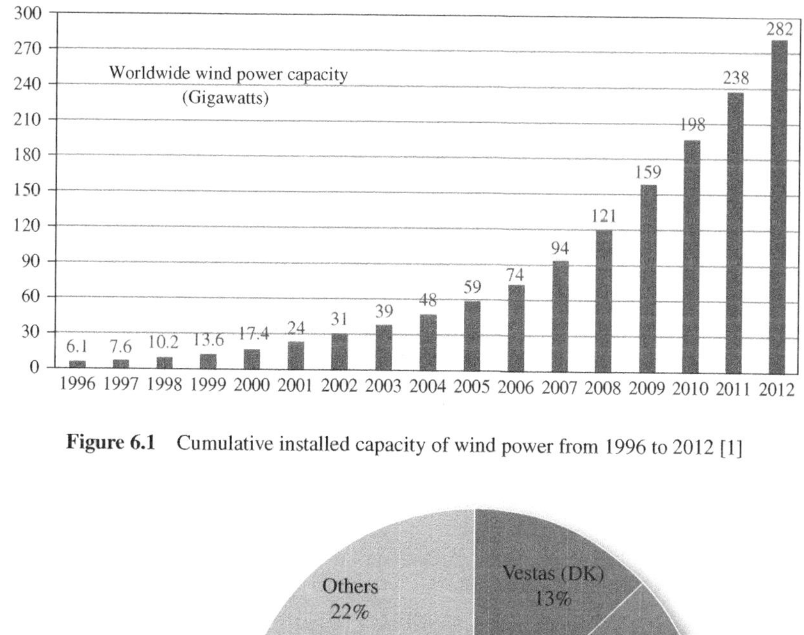

Figure 6.1 Cumulative installed capacity of wind power from 1996 to 2012 [1]

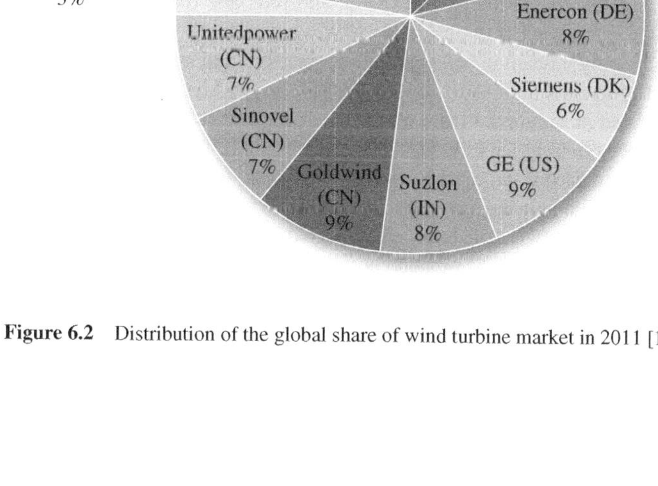

Figure 6.2 Distribution of the global share of wind turbine market in 2011 [1]

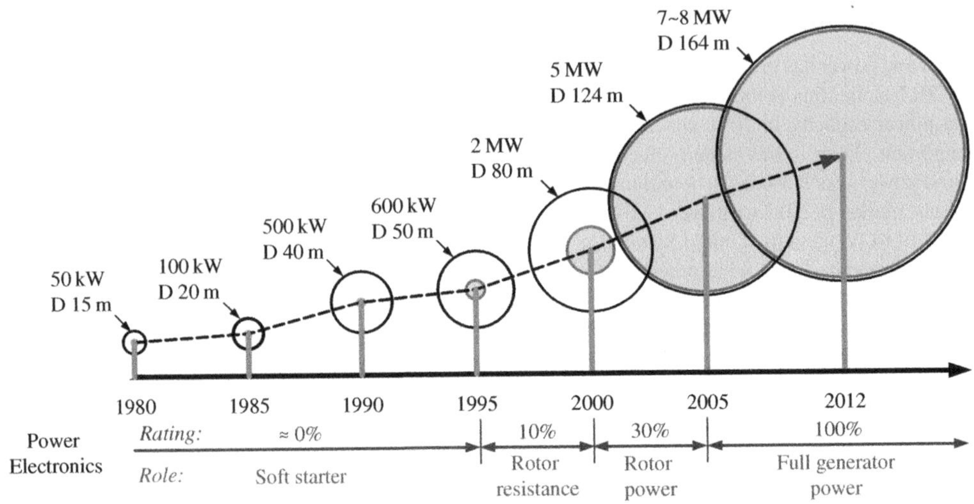

Figure 6.3 Evolution of wind turbine size and the power electronics impact in the last 30 years (shadow area inside the circle indicates the power coverage by converters)

wind turbines in 2011. It is interesting to note that four Chinese companies are in the top 10 manufacturers with a total market share of 26%.

The size of the wind turbine unit is also increasing dramatically in order to reduce the price per generated kilowatt-hour. In 2011, the average individual wind turbine size delivered to the market was 1.7 MW, among which the average offshore turbine size achieves a power level of 3.6 MW. The growing trends of emerging wind turbine size between 1980 and 2012 are shown in Figure 6.3 [7]. It can be noted that in 2012, the cutting-edge 8 MW wind turbines with diameters of 164 m had already appeared on the market [9]. Currently, most manufacturers are developing wind turbine systems in the range of 4.5–8 MW, and it is expected that the increasing number of large wind turbines with multi-MW power levels, even up to 10 MW, will be present in the next decade, driven mainly by the overall goal to lower the energy cost [10].

6.3 Wind Power Conversion

The WTS captures wind power by means of aerodynamic blades and converts it to rotating mechanical power in the shaft of the generator. In order to have proper power conversion, the tip speed of the blade should be lower than half that of the sound speed, and thus, the rotational speed will decrease as the diameter of the blade increases. For typical multi-MW wind turbines, the rotational speed ranges between 5 and 16 rpm, and this might result in bulky generator solutions and might increase installation costs. One of the most weight-efficient solutions to convert the low-speed, high-torque mechanical power is to use a gearbox, as illustrated in Figure 6.4 [6].

Today, the high-speed doubly fed induction generator (DFIG) with a large gearbox and a partial-scale power converter is dominating the market, but in the future, in order to obtain optimum overall performance, multipole permanent magnet synchronous generators (PMSGs) with simpler or no gearbox solutions using a full-scale power converter, are expected to take over. Actually, synchronous generators with either external excitation or with permanent magnets are becoming the preferred technology in the best-selling power ranges of the wind turbines [1–4]. However, the uncertain price trends of permanent

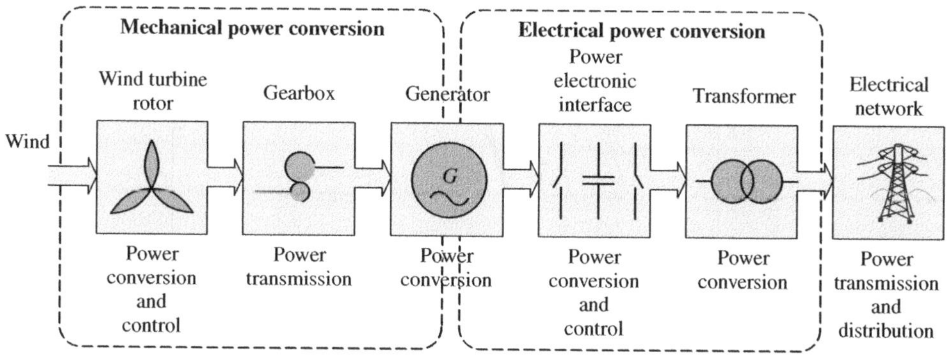

Figure 6.4 Converting wind power into electrical power in a wind turbine [11]

magnet materials might change the philosophy of the adopted generator and drive trains of WTSs in order to avoid the risk of high expense.

Between the grid and the generator, a power converter can be inserted where the transformers and filters have a pivotal role regarding the volume and losses of the entire system. All wind turbine manufacturers use a step-up transformer for connecting the converter/generator to the power grid; however, research is ongoing to replace it with more power semiconductors – leading to future high-power high-voltage transformerless solutions, which is a future technological challenge [12].

6.3.1 Basic Control Variables for Wind Turbines

For a WTS, it is essential to be able to limit the mechanical power generated by the turbines under higher wind speeds in order to prevent overloading. The power limitation can be achieved by stall control (blade position is fixed but stall of the wind appears along the blade), active stall control (blade angle is adjusted in order to create stall along the blades), or pitch control (blades are turned out of the wind). The characteristics of the output mechanical power using the three types of power limitation methods are compared in Figure 6.5 [3]. It can be seen clearly that pitch control can achieve the best power limitation performance, and it has already become the dominant technology in the most recently established wind turbines.

Another control variable of wind turbines is the rotational speed of blades. In the past, this control freedom has not been utilized and the rotational speed is fixed during the entire operational range of wind speeds. Although fixed speed turbines have the advantages of being simple and robust with low-cost electrical parts, the drawbacks are even more significant, such as uncontrollable electrical power, large mechanical stress during wind gusts, and limited output power quality.

Nowadays, variable speed wind turbines are widely used in order to achieve better aerodynamic efficiency and overall control performance. By introducing variable speed operation, it is possible to continuously adapt the rotational speed of the wind turbine to the wind speed, such that the tip speed ratio is maintained constant in order to achieve the maximum power coefficient of the blades and, thereby, the maximum power-extracting efficiency of the wind turbine. The power variations in wind will be absorbed by the speed changes of the rotor; thus, the mechanical stress and acoustic noise can be reduced. In addition, the power converters used for speed adjustment can also provide better power controllability of the wind turbines, which helps fulfill the higher technical demands imposed by the grid operators. This feature is becoming a critical determining factor in the development of future wind power technologies [12].

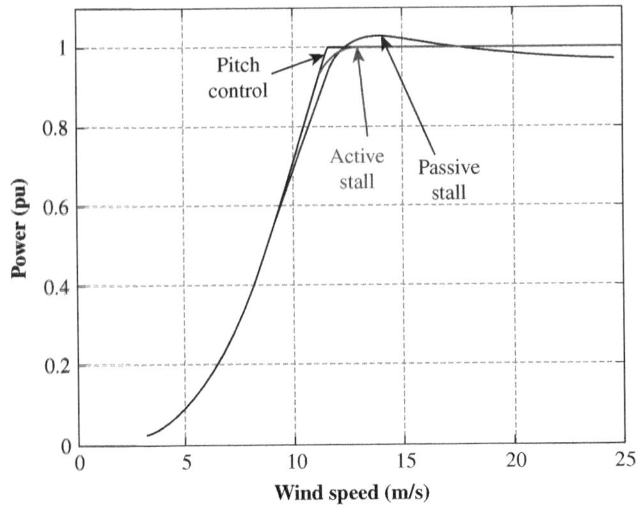

Figure 6.5 Power limitation by different methods (passive stall is based on fixed speed operation) [11]

6.3.2 *Wind Turbine Concepts*

The solutions used for WTSs have also changed dramatically during the last 30 years with four to five generations having emerged, as illustrated in Figure 6.3. Generally, existing wind turbine configurations can be categorized into four concepts [5]. The main differences between these concepts are the types of generator, the power electronics, speed controllability and the way in which the aerodynamic power is limited.

6.3.2.1 Fixed Speed Wind Turbines (WT Type A)

As shown in Figure 6.6, this configuration corresponds to the so-called Danish concept that was very popular in the 1980s. The wind turbine is equipped with an asynchronous SCIG and smoother grid connection can be achieved by incorporating a soft starter, which is bypassed during the normal generation mode.

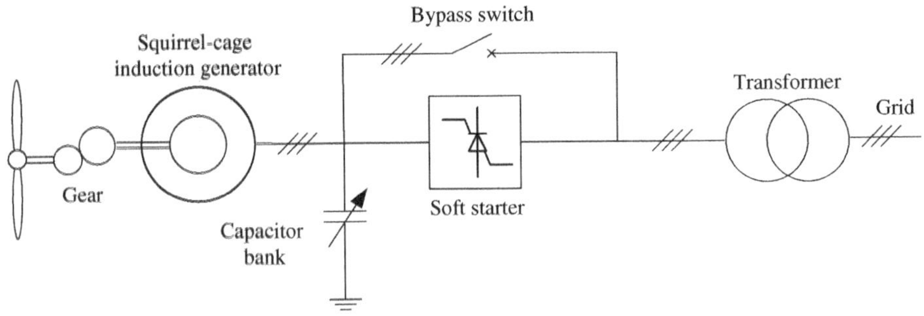

Figure 6.6 Fixed speed wind turbine with direct grid connection using a soft starter during connection

A disadvantage of this early concept is that a reactive power compensator (e.g., capacitor bank) is required to compensate the reactive power demand from the asynchronous generator (AG). Because the rotational speed is fixed without any controllability, other considerations are that the mechanical parts must be strong enough to withstand adverse mechanical torque, and the wind speed fluctuations are transferred directly into the electrical power pulsations, which could yield to unstable voltage/efficiency in the case of a weak grid.

6.3.2.2 Partial Variable Speed Wind Turbine with Variable Rotor Resistance (WT Type B)

As presented in Figure 6.7, this concept is also known as OptiSlip (Vestas™) and emerged in the mid-1990s [9, 13]. It introduces the variable rotor resistance and, thus, limits the speed controllability of wind turbines. Typically, a wound rotor induction generator (WRIG) and the corresponding capacitor compensator are used, and the generator is connected directly to the grid by a soft starter, as with the Type A concept.

A technological improvement of this concept is that the rotational speed of the wind turbine can be partially adjusted by dynamically changing the rotor resistance. This feature contributes to mechanical stress relief and smoother output of electrical power. However, the constant power loss dissipating in the rotor resistors is a significant drawback for this concept.

6.3.2.3 Variable Speed WT with Partial-Scale Frequency Converter (WT Type C)

Currently, this concept is the most established solution and it has been widely used since the start of the century. As shown in Figure 6.8, a back-to-back power electronic converter is adopted in conjunction with a DFIG. The stator windings of the DFIG are connected directly to the power grid, while the rotor windings are connected to the power grid by the power electronic converter, which normally has a capacity of 30% of that of the wind turbine [14, 15].

Through use of a power electronic converter, the frequency and current in the rotor can be regulated flexibly and, thus, the variable speed range can be extended further to a satisfactory level. Meanwhile, the power converter can regulate partially the output power of the generator, improving the power quality and providing limited grid support. The smaller converter capacity makes this concept attractive from a cost point of view. However, its main drawbacks are the use of slip rings and the challenging power controllability in the case of grid faults; these disadvantages might compromise the reliability performance and they might not satisfy future grid requirements.

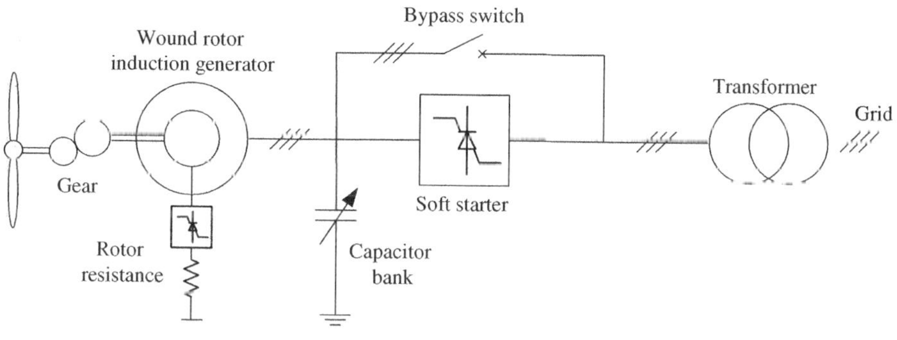

Figure 6.7 Partial-scale variable speed wind turbine with variable rotor resistance

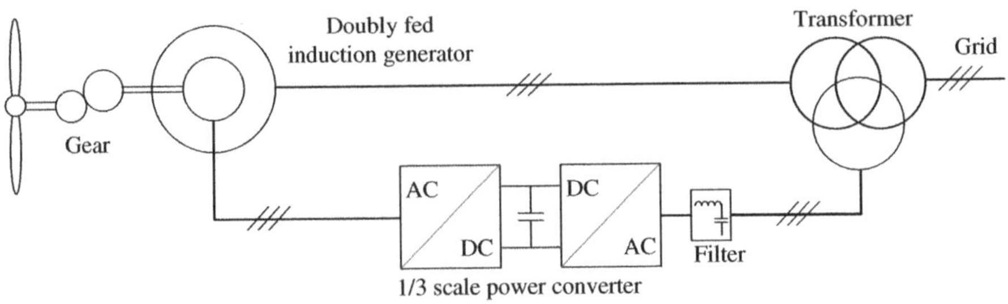

Figure 6.8 Variable speed wind turbine with partial-scale power converter and a doubly fed induction generator

6.3.2.4 Variable Speed Wind Turbine with Full-Scale Power Converter (WT Type D)

Another promising concept that is becoming popular for newly developed and installed wind turbines is shown in Figure 6.9. It introduces a full-scale power converter to interconnect the power grid and stator windings of the generator; thus, all the power generated by the wind turbine can be regulated. In this concept, the AG, wound rotor synchronous generator (WRSG) and PMSG have all been reported.

The elimination of slip rings, a simpler or even an eliminated gearbox, full power and speed controllability, as well as better grid support capability, are the main advantages of this solution compared with the DFIG-based concept. Owing to the use of a full-scale power converter, the voltage level of the power conversion stage can be flexible. In the future, the voltage might be high enough to connect directly to the power grid without need of a bulky low-frequency transformer, this advantage might be an attractive feature for future WTSs. However, the more stressed and expensive power electronics and the price of permanent magnet materials might raise some uncertainties for this concept regarding further commercialization [7].

6.3.2.5 Wind Turbine Concepts Comparison

Comparing the different wind turbine solutions reveals a contradiction between cost and performance. Table 6.1 shows a comparison of the grid-side control, cost, maintenance and internal turbine performance of the four turbine concepts [11]. It can be seen that owing to the use of power electronic converters, Type C and Type D WTSs can achieve much better power controllability regarding the rotational speed, control bandwidth and delivered active/reactive power compared with Types A and B solutions. Moreover, it is clear that wind turbine concepts of Types C and D enable many important features when connecting to the power grid; therefore, they are much more suitable to be grid integrated. Considering that the price for the power semiconductor devices has been kept decreasing during recent decades,

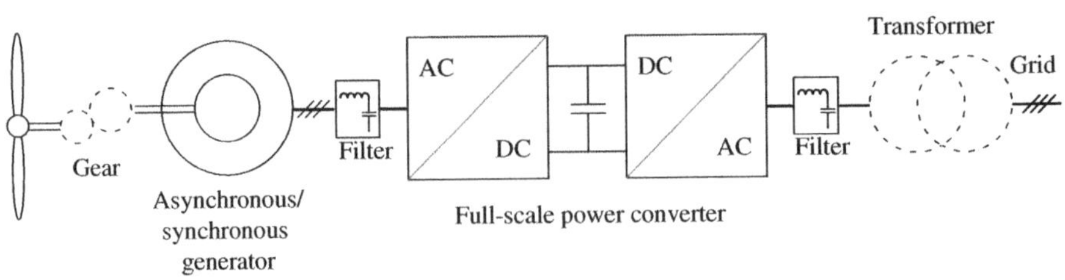

Figure 6.9 Variable speed wind turbine with full-scale power converter

Table 6.1 System comparison of wind turbine configurations

System	Type A	Type B	Type C	Type D
Variable speed	No	No	Yes	Yes
Control active power	Limited	Limited	Yes	Yes
Control reactive power	No	No	Yes	Yes
Short circuit (fault-active)	No	No	No/yes	Yes
Short circuit power	Contribute	Contribute	Contribute	Limit
Control bandwidth	1–10 s	100 ms	1 ms	0.5–1 ms
Standby function	No	No	Yes +	Yes ++
Flicker (sensitive)	Yes	Yes	No	No
Soft starter needed	Yes	Yes	No	No
Rolling capacity on grid	Yes, partly	Yes, partly	Yes	Yes
Reactive compensator (C)	Yes	Yes	No	No
Island operation	No	No	Yes/no	Yes
Investment	++	++	+	0
Maintenance	++	++	0	+

Types C and D concepts are more cost-effective compared with Types A and B; thus, they have come to dominate the current market [3–5].

6.4 Power Converters for Wind Turbines

Because of the rapid development of capacity and technology in wind power generation, the power electronic converter is becoming an increasingly important part of the entire system, as indicated in Figure 6.3. However, power electronic converters also need to satisfy requirements that are much tougher than ever before. Generally, these requirements can be categorized into the following three groups, as shown in Figure 6.10 [7]:

On the generator side: the current flowing in the generator rotor or stator should be controlled to adjust torque and, consequently, the rotating speed of the wind turbine. This will contribute to the active power

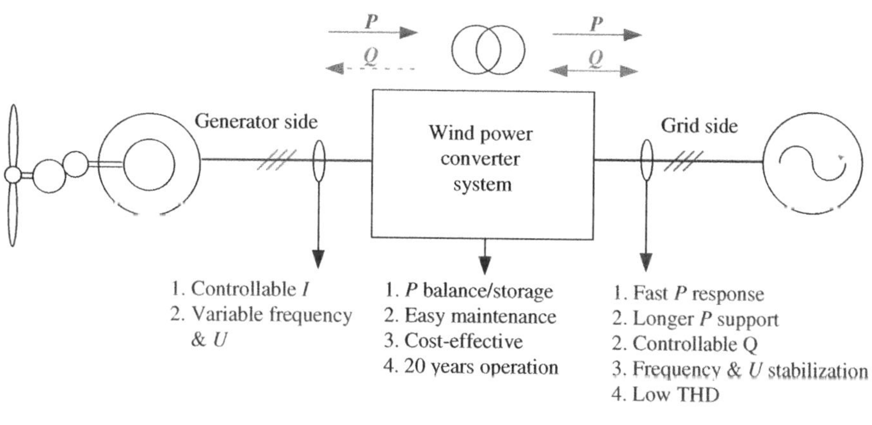

Figure 6.10 Requirements of modern wind power converters

balance not only in normal operation when extracting the maximum power from the wind turbine, but also in the case of grid faults when the generated power needs to be reduced quickly. Moreover, the converter should have the capability to handle a variable fundamental frequency and voltage amplitude from the generator outputs.

On the grid side: the converter must comply with the grid requirements regardless of the wind speed. This means that it should have the capability to control the reactive power Q delivered to the power grid and to perform a fast response on the active power P response. The fundamental frequency, as well as voltage amplitude on the grid side, should be maintained almost fixed under normal operation, and the total harmonic distortion (THD) of the current must be restrained at a low level [2, 16, 17].

Inherently: the converter system needs to be cost-effective, easy to maintain and have high reliability. This requires high power density, reliability and modularity designs for each part of the converter system. Furthermore, the wind power converter might need the ability to store some active power and boost up the voltage from the generator side to the grid side.

Considering the mission profiles for modern wind power converters, some of the dominant and promising converter topologies in wind power applications will be shown and discussed in the following.

6.4.1 Two-Level Power Converter

A pulse width modulation-voltage source converter with two-level output voltage (2L-PWM-VSC) is the most frequently used three-phase topology in wind power applications. The knowledge available in this converter topology is extensive and well established. However, the 2L-PWM-VSC topology might suffer from larger switching losses and lower efficiency at MW and medium-voltage (MV) levels. The available switching devices or converter building blocks might need to be connected in parallel or in series in order to obtain the required power and voltage [18]. Another problem of the 2L-PWM-VSC is the two-level output voltage, which introduces relatively higher dv/dt stresses to the generator and transformer windings and bulky output filters might be necessary to limit the voltage gradient and reduce the harmonics level [19]. In WTSs, the 2L-PWM-VSC can be used in different configurations:

6.4.1.1 Two-Level Unidirectional Voltage Source Power Converter (2L-UNI)

It is becoming a trend to use PMSG for wind turbine concepts with full-scale power converters. Because there is no reactive power required in such a generator and because active power flows unidirectionally from the generator to the power grid, only a simple diode rectifier need be applied on the generator side, which achieves a cost-efficient solution, as shown in Figure 6.11. However, the diode rectifier might introduce low-frequency torque pulsations that could trigger shaft resonance [20]. Semicontrolled rectifier solutions are also possible using this circuit topology [21].

In order to obtain variable speed operation and a stable DC bus voltage, a boost DC/DC converter could be inserted in the DC-link, or the DC-voltage could be controlled by using rotor excitation, as shown in Figure 6.11. It must be mentioned that for power levels in the range of MW, the DC/DC converter needs to be made by several interleaved units or by a three-level solution [22].

6.4.1.2 Two-Level Back-To-Back Voltage Source Power Converter (2L-BTB)

It is very popular to configure two 2L-PWM-VSCs in a back-to-back structure (2L-BTB) in the wind power conversion system, as shown in Figure 6.12. A technical advantage of the 2L-BTB solution is the full power controllability (four-quadrant operation) with a relatively simple structure and few components, which contributes to a well-proven robust and reliable performance. The 2L-BTB topology is a state-of-the-art solution in DFIG-based wind turbines (see, e.g., [3, 4, 23]). Several manufacturers are also using this topology for a full-scale power converter concept with an SCIG.

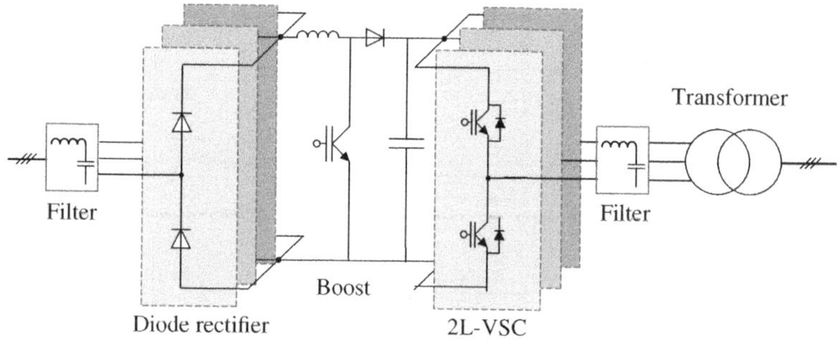

Figure 6.11 Two-level unidirectional voltage source converter for wind turbine (2L-UNI) [7]

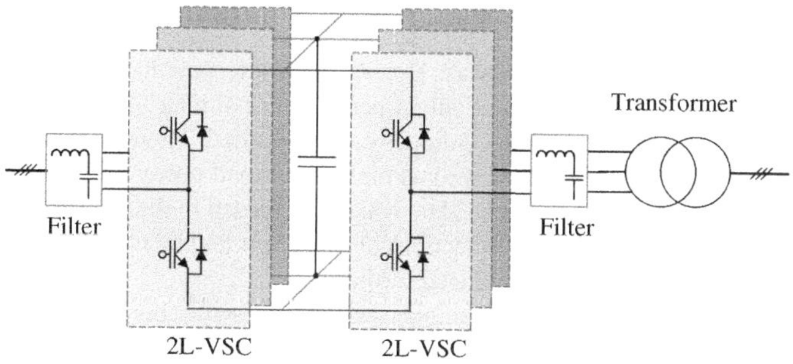

Figure 6.12 Two-level back-to-back voltage source converter for wind turbine (2L-BTB) [7]

6.4.2 Multilevel Power Converter

With the capabilities of more output voltage levels, higher voltage amplitude and larger power handling ability, the multilevel converter topologies are becoming interesting and promising candidates in the wind turbine applications [24–26]. Generally, the multilevel converters can be classified into three categories [26–29]: neutral-point diode clamped, flying capacitor clamped and cascaded converter cells. In order to achieve a cost-effective design, multilevel converters are mainly used in 3–8 MW variable speed wind turbines with full-scale power converters.

6.4.2.1 Three-Level Neutral-Point Diode Clamped Back-To-Back Topology (3L-NPC BTB)

The three-level neutral-point diode clamped topology is one of the most commercialized multilevel topologies on the market. Similar to the 2L-BTB, it is usually configured as a back-to-back structure in wind turbines, as shown in Figure 6.13, which for convenience, is called 3L NPC BTB.

It achieves one more output voltage level and less d*v*/d*t* stress compared with the 2L-BTB; thus, the filter size is smaller. The 3L-NPC BTB is also able to double the voltage amplitude compared with the 2L-BTB converter via switching devices with the same voltage rating. The mid-point voltage fluctuation of the DC bus used to be a drawback of the 3L-NPC BTB, but this problem has been researched extensively and is considered improved by the controlling of redundant switching states [30]. However, it is found that the loss distribution is unequal between the outer and inner switching devices in a switching arm, and this problem might lead to de-rated power capacity when it is practically designed.

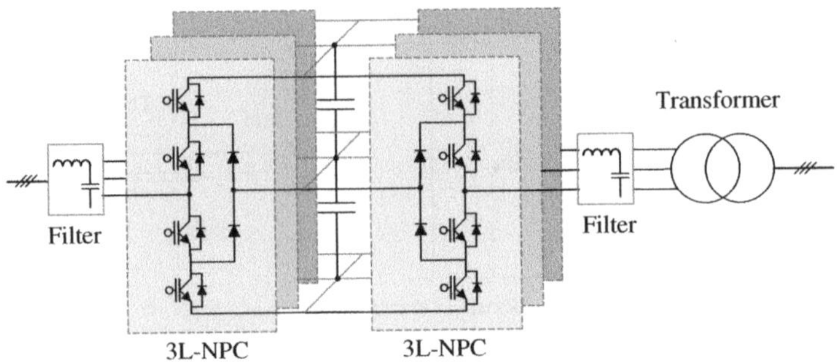

Figure 6.13 Three-level neutral-point clamped back-to-back converter for wind turbine (3L-NPC BTB) [7]

6.4.2.2 Three-Level H-Bridge Back-To-Back Topology (3L-HB BTB)

The 3L-HB BTB solution comprises two 3-phase H-bridge converters configured in a back-to-back structure, as shown in Figure 6.14. It achieves an output performance similar to the 3L-NPC BTB solution, but the unequal loss distribution and clamped diodes can be avoided. Thereby, more efficient and equal loading of the power switching devices, as well as higher designed power capacity, might be obtained [24, 25, 31]. Moreover, as only half of the DC bus voltage is needed in the 3L-HB BTB compared with the 3L-NPC BTB, there are fewer series connections of capacitors and no mid-point in the DC bus; thus, the size of the DC-link capacitors can be reduced further.

However, the 3L-HB BTB solution needs an open-winding structure both in the generator and in the transformer in order to achieve isolation among each phase. This feature has advantages and disadvantages. On the one hand, a potential fault-tolerant ability is obtained, if one or even two phases of the generator are out of operation. On the other hand, doubled cable length is needed and extra cost, weight, loss and inductance can be major drawbacks in such a converter configuration.

6.4.2.3 Five-Level H-Bridge Back-To-Back Topology (5L-HB BTB)

The 5L-HB BTB converter comprises two 3-phase H-bridge converters making use of 3L-NPC switching arms, as shown in Figure 6.15. This is an extension of the 3L-HB BTB solution and shares the same requirements for an open-winding generator and transformer. The 5L-HB BTB can achieve five-level output voltage and double the voltage amplitude compared with the 3L-HB BTB solution with the same devices. These features enable the use of smaller output filters and lower current rating in the switching

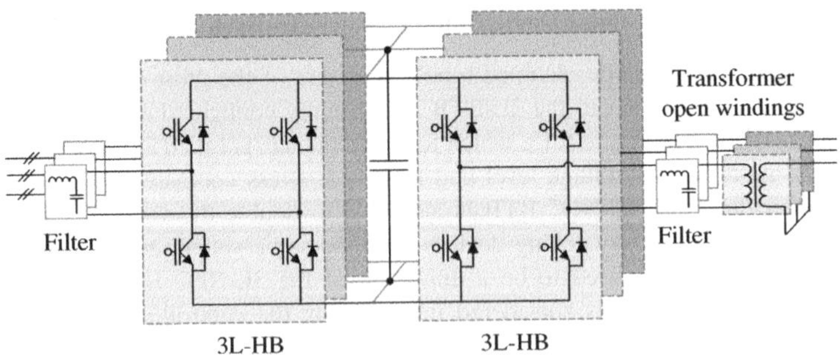

Figure 6.14 Three-level H-bridge back-to-back converter for wind turbine (3L-HB BTB) [7]

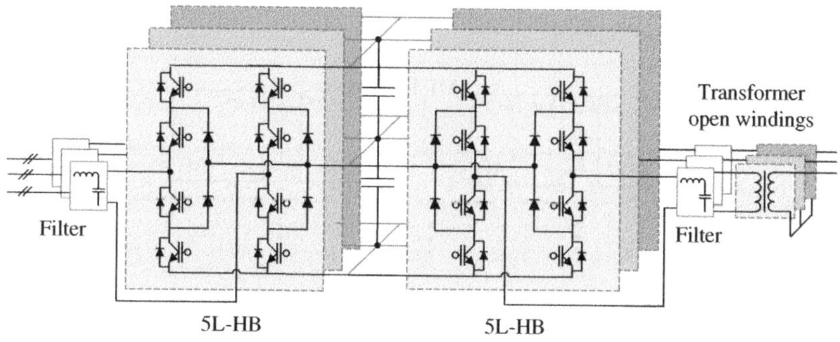

Figure 6.15 Five-level H-bridge back-to-back converter for wind turbine (5L-HB BTB) [7]

devices and cables [19, 32]. However, the 5L-HB BTB converter introduces more switching devices, which could reduce the reliability of the total system.

6.4.3 Multicell Converter

Nowadays, most of the newly established wind turbines achieve the multi-MW power level. In order to handle the fast growth in power capacity, some multicell converter configurations (i.e., parallel/series connection of converter cells) have been developed and are widely adopted by the industry.

6.4.3.1 Multicell Converter with Paralleled Converter Cells (MC-PCC)

Figure 6.16 (a) shows a multicell solution adopted by Gamesa in the 4.5-MW wind turbines [33], which have 2L-BTB single-cell converters paralleled both on the generator side and on the grid side. Siemens also introduced a similar solution in their best-selling multi-MW wind turbines, as indicated in Figure 6.16 (b) [34]. The standard and proven low-voltage converter cells, as well as the redundant and modular characteristics, are their main advantages. This converter configuration is the state-of-the-art solution in the industry for wind turbines with power levels above 3 MW.

6.4.3.2 Cascaded H-Bridge Converter with Medium-Frequency Transformers (CHB-MFT)

This configuration shares a similar idea to the next generation traction converters [35, 36], and it has been proposed in the European UNIFLEX-PM Project [37], as shown in Figure 6.17. It is based on a structure

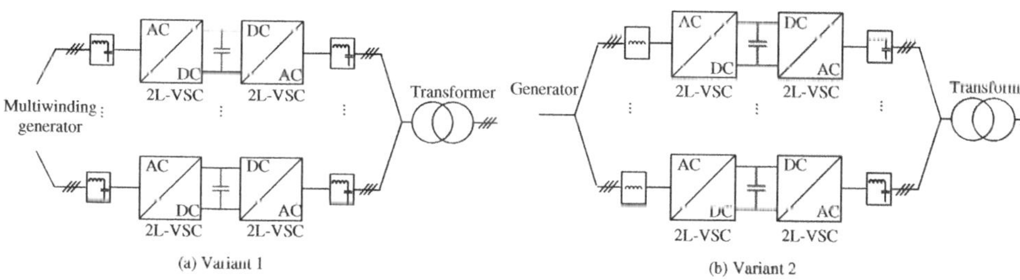

Figure 6.16 Multicell converter with paralleled converter cells (MC-PCC)

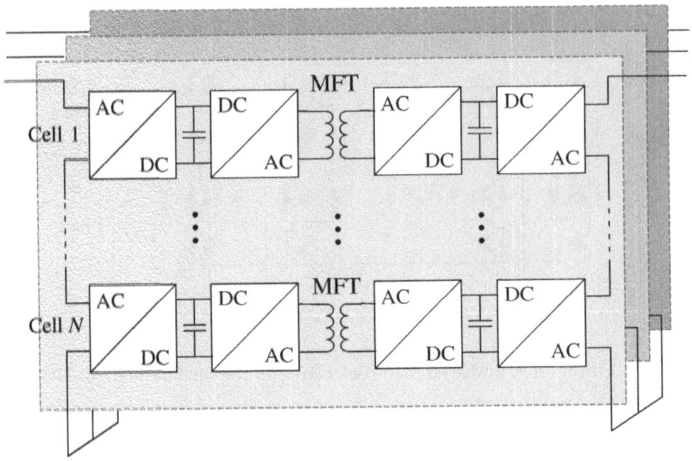

Figure 6.17 Cascaded H-bridge converter with medium-frequency transformer for wind turbine system (CHB-MFT)

of back-to-back cascaded H-bridge converters with galvanic insulated DC/DC converters as an interface. The DC/DC converters have a medium-frequency transformer (MFT) operating at several kilohertzs up to dozens of kilohertzs; thereby, the transformer size can be reduced significantly in both weight and volume. Moreover, because of the cascaded structure, it can be connected directly to the distribution power grid (10–20 kV) with high-output voltage quality, filterless design and redundant capability [35, 36]. This solution would become more attractive if it could be placed in the nacelle of wind turbines because the heavy/bulky low-frequency transformer could be replaced by the more compact and flexibly configured power semiconductor devices.

6.4.3.3 Modular Multilevel Converter (MMC)

This configuration shares a similar idea with some of the emerging converters used for high-voltage direct current (HVDC) transmission [38, 39], as shown in Figure 6.18. It is also based on a back-to-back structure with cascaded DC/AC converter cells. One advantage of this configuration is the easily scalable voltage/power capability; therefore, it can achieve very high power conversion at dozens of kilovolts with good modularity and redundancy performance. The output filter can also be eliminated because of the significantly increased voltage levels. However, the useable voltage rating in a WTS might be limited significantly by the insulation capability of the generator. Moreover, the low fundamental frequency of the generator outputs, which is the normal case for MW wind turbines, might introduce large DC voltage fluctuations on the generator-side converters, resulting in bulky DC capacitors in the converter system, which might be undesirable in wind power applications.

It can be seen that multicell converters have a modular and fault-tolerant capability, which can contribute to achieve higher reliability performance. However, all of the three configurations have significantly increased components count, which could compromise the system reliability and increase the cost significantly. The overall merits and defects of these multicell converters in wind power applications still needs further evaluation, because the technologies for power semiconductor devices are developing rapidly.

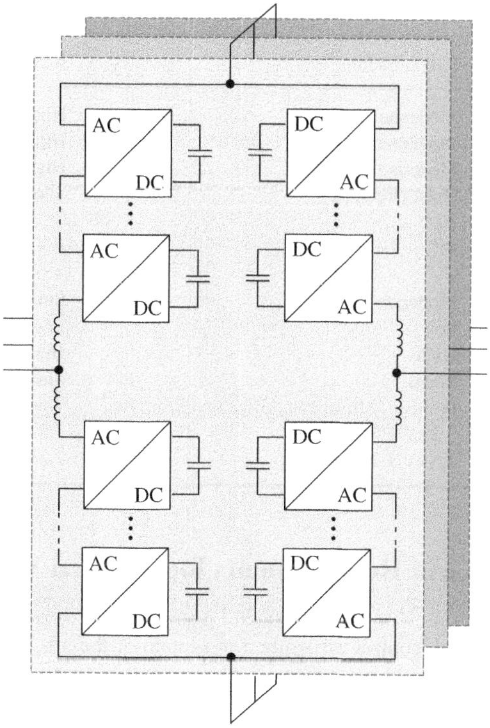

Figure 6.18 Modular multilevel converters for wind turbine (MMC)

6.5 Power Semiconductors for Wind Power Converter

As the backbone component for the converters, power semiconductor devices are also playing an important role in the development of high-performance wind turbines [40]. The dominant choices for power switching devices, as reported in the wind power industry, are based on the module packaging insulated gate bipolar transistor (IGBT), the presspack packaging IGBT and the presspack packaging integrated gate-commutated thyristor (IGCT). These three types of power semiconductor devices have quite different characteristics and they are compared in Table 6.2 [7].

The module packaging technology for IGBT has a longer record of applications and fewer hardware-mounting regulations. However, owing to the soldering and bond wire connections of the internal chips, module packaging devices might suffer from larger thermal resistance, lower power density and higher failure rates [7]. The presspack packaging technology improves the connection of the chips by direct presspack contacting, which leads to improved reliability (known from industrial experience), higher power density (easier stacking for connection) and better cooling capability, but with the disadvantage of higher cost compared with the module packaging devices. Presspack IGCTs were introduced into MV converters in the 1990s and are already becoming state-of-the-art technology used in high-power motor drives, but they have not yet been adopted widely in the wind power industry [40–42].

Table 6.2 Dominant power switching devices for wind power application

	IGBT module	IGBT presspack	IGCT presspack
Power density	Moderate	High	High
Reliability	Moderate	High	High
Cost	Moderate	High	High
Failure mode	Open circuit	Short circuit	Short circuit
Easy maintenance	+	–	–
Insulation of heat sink	+	–	–
Snubber requirement	–	–	+
Thermal resistance	Moderate	Small	Small
Switching loss	Low	Low	High
Conduction loss	High	High	Low
Gate driver	Small	Small	Large
Major manufacturers	Infineon, Mitsubishi ABB, Semikron, Fuji	Westcode, ABB	ABB
Medium voltage ratings	3.3 kV/4.5 kV/6.5 kV	2.5 kV/4.5 kV	4.5 kV/6.5 kV
Maximum current ratings	1.5 kV/1.2 kA/750 A	2.2 kA/2.4 kA	2.1 kA/1.3 kA

6.6 Controls and Grid Requirements for Modern Wind Turbines

Controlling a wind turbine involves both fast and slow control dynamics [43–53], as shown in Figure 6.19, where a general control structure for a state-of-the-art WTS, including the turbine, generator, converter and power grid are indicated. The wind turbine concept can either be the Type C shown in Figure 6.8, or Type D, as shown in Figure 6.9.

Generally, the energy flowing in and out of the generation system has to be managed carefully. The power generated by turbines should be controlled by means of mechanical parts (e.g., pitch angle of blades and yawing system). Meanwhile, the entire control system has to follow the power production commands given by the transmission system operator (TSO).

The more advanced features of wind turbine might be taken into account in the control system, such as the maximization of the generated power, riding through the grid faults and providing grid support.

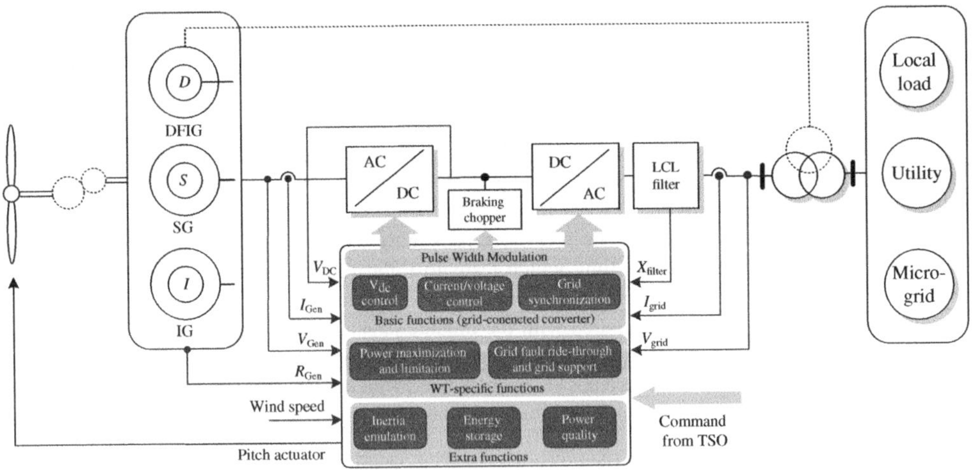

Figure 6.19 General control structure for modern wind turbines

With the concept of variable speed wind turbines, the current in the generator will typically be changed by controlling the generator-side converter and thereby, the rotational speed of the turbine rotor can be adjusted to achieve maximum power production based on the available wind power. Regarding the grid fault and support conditions, coordinated control by several subsystems of the wind turbine, such as the grid-side converter, braking chopper, pitch angle system and generator, might be necessary.

Finally, basic control, such as current regulation, DC bus stabilization and grid synchronization has to be performed quickly by the power converter; proportional-integral (PI) controllers and proportional-resonant (PR) controllers are typically used.

Most countries have dedicated grid codes for wind turbines and they are updated regularly [54–58]. In most cases, these requirements reflect the significant penetration of wind power into the grid system, and the requirements cover a wide range of voltage levels from MV to very high voltage. Basically, the grid codes are always trying to make wind farms act as conventional power plants from the point of view of the electricity network. Thus, generally, the focus is on the power controllability, power quality, fault ride-through capability and grid support capability of WTSs during network disturbances. Examples of grid codes in different countries for active and reactive power control, power quality and ride-through capabilities are given in the following, and they are regulations either for individual wind turbines or for the entire wind farm.

6.6.1 Active Power Control

According to most grid codes, individual wind turbines must be able to control the active power in the point-of-common coupling (PCC) within a given power range. Typically, the active power is controlled based on the system frequency, for example, in Denmark, Ireland and Germany, such that the power delivered to the grid is decreased when the grid frequency rises above 50.1 Hz. Typical characteristics for frequency control in the Danish and German grid codes are shown in Figure 6.20.

For larger generation units on the wind farm scale, which are normally connected at the transmission line, the wind turbines should act as conventional power plants providing a wide range of controlled active power based on the demands of the TSO. In addition, they have to participate in primary and secondary control of the power system. Several active power control requirements for wind farms by the

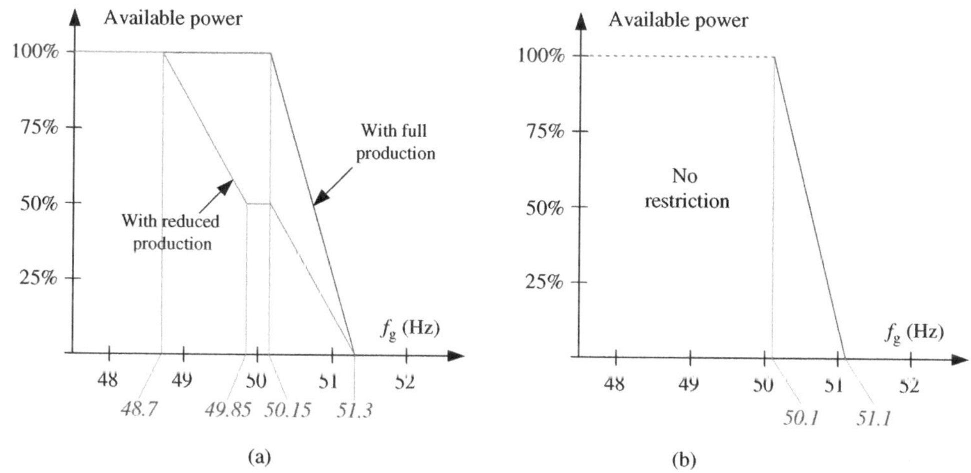

Figure 6.20 Frequency control profiles for wind turbines connected to (a) the Danish grid [57] and (b) the German grid [58]

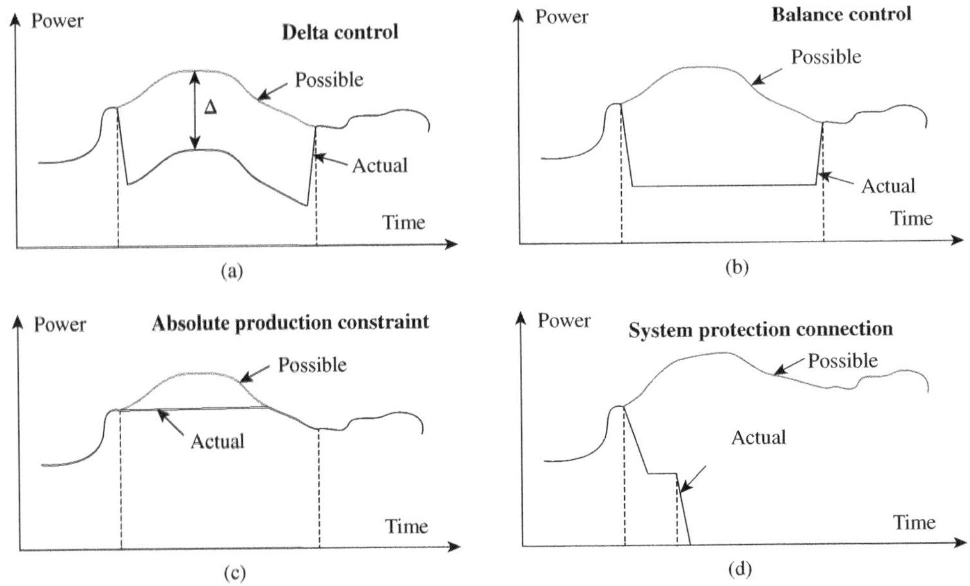

Figure 6.21 Regulations for active power control required by the Danish grid codes: (a) delta control, (b) balance control, (c) absolute production constraint and (d) system protection [56]

Danish grid codes are illustrated in Figure 6.21(a)–(d). It can be seen that these active power controls always require some reserved power capacity from the possible power generation by the wind farms, which will provide enough support in case extra active power is demanded, reducing the need for the energy storage systems.

6.6.2 *Reactive Power Control*

During normal operation, the reactive power delivered by the wind turbine or wind farm also has to be regulated by the grid codes within a certain range. The grid codes in different countries specify different reactive power control behaviors.

As shown in Figure 6.22, both the Danish and the German grid codes give a range for controlling the reactive power of the WTS against the active power output. In addition, the TSOs will normally specify the reactive power range delivered by wind farms according to the grid voltage levels, as shown in Figure 6.23, where the German grid code for offshore wind farms is given as an example. It should be noted that this basic form of reactive power control should be realized slowly on a timescale of minutes [54].

6.6.3 *Total Harmonic Distortion*

Power quality issues are addressed especially for wind turbines connected to MV networks. However, some grid codes, for example, in Denmark and Ireland, also have requirements at a transmission level. Generally, two standards are used for defining the power quality parameters, namely, IEC 61000-x-x and EN 50160. Specific values are given for fast variations in voltage, short-term flicker severity, long-term flicker severity and THD. The limits for individual orders of harmonic distortion are also given based on

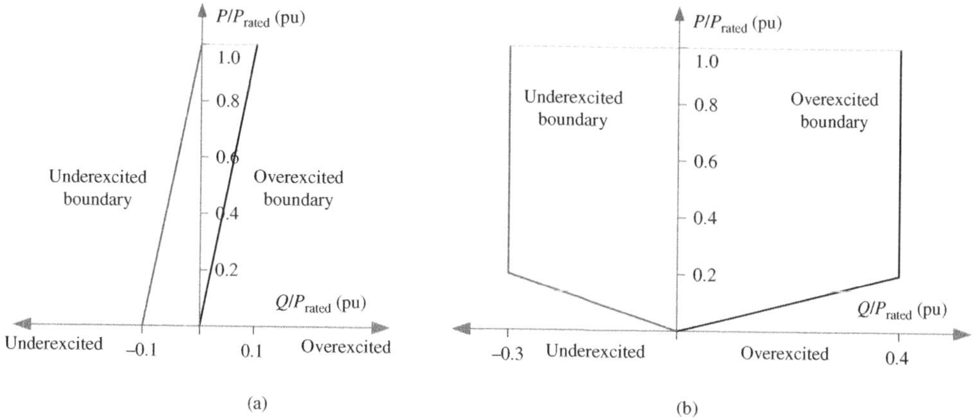

Figure 6.22 Reactive power ranges under different generating powers: (a) for a wind turbine specified by Danish grid codes [56] and (b) for a wind farm specified by German grid codes [58]

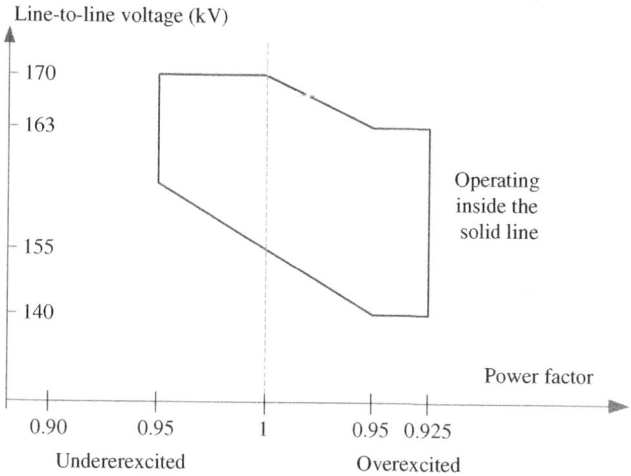

Figure 6.23 Reactive power ranges under different voltage levels by German grid codes [58]

standards, or in some cases, for example, Denmark, on custom-defined harmonic compatibility levels. Interharmonics might also be considered by the grid codes.

6.6.4 *Fault Ride-Through Capability*

In addition to normal operation, TSOs in different countries have issued strict low-voltage ride through (LVRT) codes for wind turbines/wind farms. Figure 6.24 [54–58] shows the boundaries with various grid voltage dipping amplitudes and the allowable disturbance time for a wind farm. One of the uncertainties still under discussion is the definition of the voltage amplitude during unsymmetrical grid faults; this is not clearly specified in most grid codes.

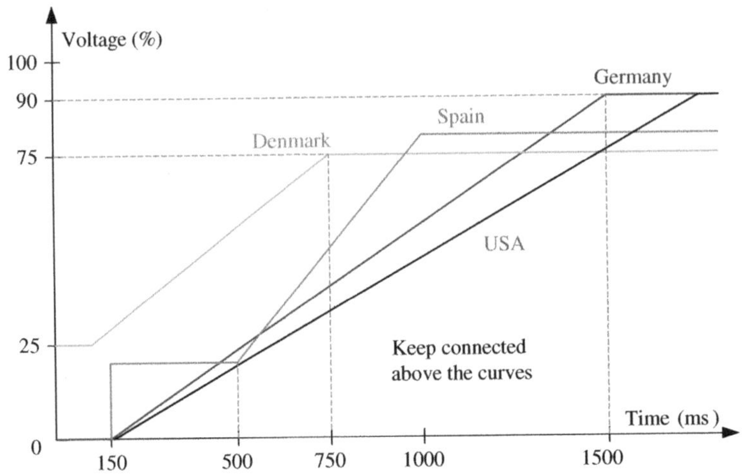

Figure 6.24 Voltage profile for low-voltage fault ride-through capability of wind turbines [7]

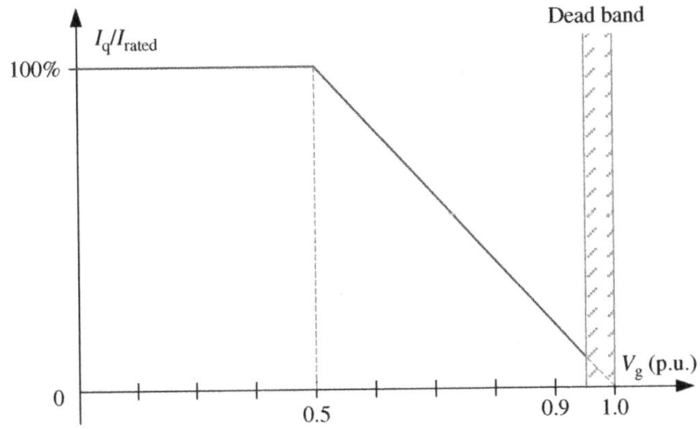

Figure 6.25 Reactive current requirements for a wind farm during grid sags by the German and Danish grid codes [57, 58]

In addition to the fault ride-through capability, it is becoming necessary that wind power generation systems should provide reactive power (up to 100% current capacity), in order to contribute to the voltage recovery when grid faults are present. Figure 6.25 shows the required amount of reactive current for wind farms against the grid voltage amplitude specified by the German [58] and Danish grid codes [57]. This demand is relatively difficult to meet by some of the wind turbine concepts, such as Type A (Figure 6.6), Type B (Figure 6.7) and even Type C (Figure 6.7).

The grid codes have raised great challenges for WTSs during the last decade and are continuing to push technological developments in power electronics for wind power applications. On the one hand, these requirements have increased the cost per produced kilowatt-hour but, meanwhile have made wind power technology much more suitable to be utilized and integrated into the power grid. It can be predicted that stricter grid codes in the future will keep challenging WTSs and push forward the technology of power electronics.

6.7 Emerging Reliability Issues for Wind Power System

The dramatic growth in the total number of installations and of the individual capacity of wind turbines makes their failure harmful or even unacceptable from the TSO point of view. Failures of WTSs will not only cause stability problems for the power grid owing to the sudden absence of large amounts of power, but also result in very high costs for repair and maintenance, especially for those turbines that are large and remote. Therefore, reliability performance is a critical design consideration for the next generation of wind power converter systems.

Unfortunately, former field feedback has shown that the larger wind turbines seem more prone to failure, as indicated in [59]. When looking at the failure rates and downtime distribution of individual WTSs, as shown in Figure 6.26, it is obvious that the control and power electronic parts tend to have higher failure probability than other subsystems by a factor of 2–4 [60]. It is noted that although the generator and gearbox have the largest downtime (i.e., time needed for repair), their probability of failure is lower than that of the electrical and control parts. Thus, understanding and improving the reliability performance of the power electronic converters will be crucial for wind turbines in the future, especially for the larger ones at multi-MW level.

Research into the reliability of power electronics has been carried out for decades and is now moving from a solely statistical approach that has been proven unsatisfactory in the automotive industry to a more physical approach, which not only involves statistics but also investigation and modeling of the root cause behind the failures [61, 62]. As shown in Figure 6.27, in order to achieve more cost-efficient and reliable power electronics, multidisciplines are necessary, which involve stress analysis, strength modeling, statistical considerations and also the online monitoring/control/maintenance of the converter system.

Stress analysis might focus on the complete mission profile definition, converter design and stress estimation and measurement. This group of disciplines will target the accurate determination of the converter's loading profile, which can trigger failure mechanisms of critical components, such as thermal cycling in power devices [63], voltage increase in the DC bus [64], vibration and humidity [65, 66]

Strength modeling might involve the identification, modeling and accelerating tests of failure mechanisms in the converter system, for example, the bond wire lift-off and soldering cracks inside the power devices [63]. The goal of this group of disciplines is to seek correlations between the established/measured stresses and the quantified fatigues/failures of critical components.

The monitoring and control approach might relate to lifetime monitoring [67, 68], stress relief controls [69, 70] and intelligent maintenance. This group of disciplines will target the monitoring and control of

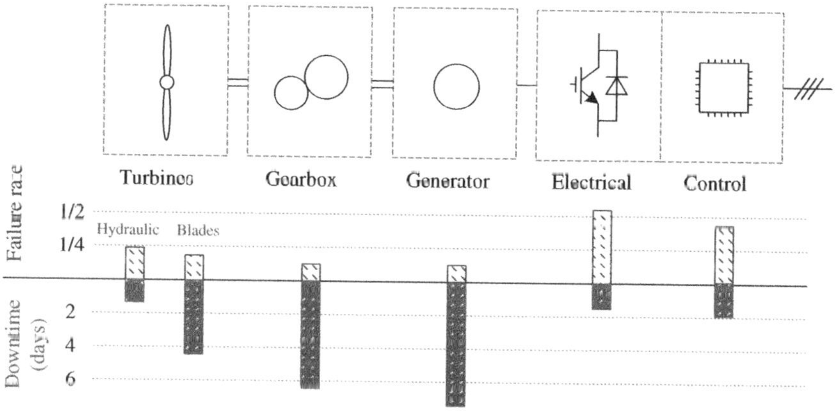

Figure 6.26 Failure rate and downtime for different parts of wind turbine system [7]

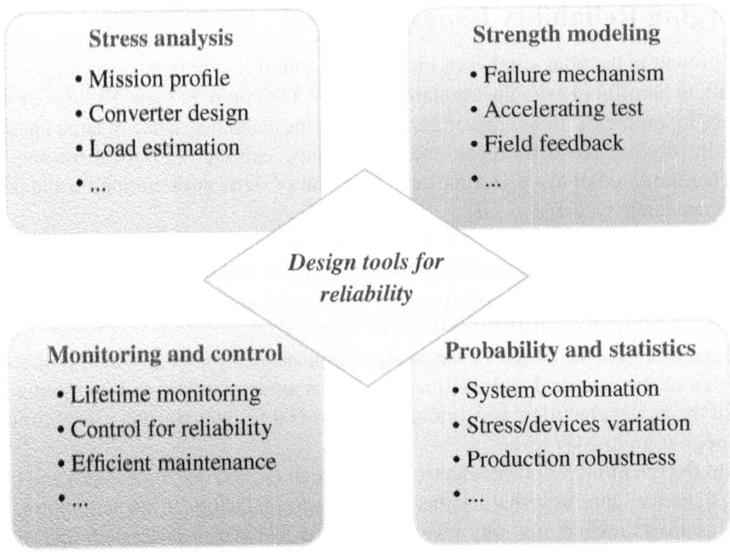

Figure 6.27 Multidiscipline approaches for more reliable power electronics

the converter's lifetime during operation. For example, the collector–emitter voltage V_{CE} of an IGBT, which is subject to an accelerated test, experiences a sudden increase just before IGBT failure [67], and it could be used for predictive maintenance in a WTS.

The probability and statistics might add the statistical distribution and correlation to the acquired stress, strength and component configuration. This group of disciplines will target the enhancement of the robustness of the designed converter and take into account the severe usage, as well as quality variations of the components.

6.8 Conclusion

This chapter discusses the power electronics used in WTSs. The configurations and roles of power electronics in various wind turbine concepts are illustrated showing that the wind turbine behavior/performance is significantly improved by introducing power electronics. It is possible for wind turbines to act as an active contributor to the frequency and voltage control in the grid system by means of power control using power electronic converters.

As the WTS capacity and voltage are increasing, a trend is to use multilevel topologies or parallel the converter cells in the wind power converter, and a number of these are shown in this chapter. Different layers of control are discussed, as are the state-of-the-art requirements by grid codes. Finally, the emerging challenges regarding reliability and the approaches for more reliable wind power converters are also highlighted.

References

1. REN21 (2012) REN21 – Renewables 2012 Global Status Report, June, 2012, http://www.ren21.net (accessed 17 December 2013).
2. Liserre, M., Cardenas, R., Molinas, M., and Rodriguez, J. (2011) Overview of Multi-MW wind turbines and wind parks. *IEEE Transactions on Industrial Electronics*, **58** (4), 1081–1095.

3. Chen, Z., Guerrero, J.M., and Blaabjerg, F. (2009) A review of the state of the art of power electronics for wind turbines. *IEEE Transactions on Power Electronics*, **24** (8), 1859–1875.
4. Blaabjerg, F., Chen, Z., and Kjaer, S.B. (2004) Power electronics as efficient interface in dispersed power generation systems. *IEEE Transactions on Power Electronics*, **19** (4), 1184–1194.
5. Hansen, A.D., Iov, F., Blaabjerg, F., and Hansen, L.H. (2004) Review of contemporary wind turbine concepts and their market penetration. *Journal of Wind Engineering*, **28** (3), 247–263.
6. Green Growth Leaders (2010) "Green Energy – the Road to a Danish Energy System without Fossil Fuels," Report of Danish Commission on Climate Change Policy, September 2010. http://www.ens.dk/en/policy/danish-climate-energy-policy/danish-commission-climate-change-policy/green-energy
7. Blaabjerg, F., Liserre, M., and Ma, K. (2012) Power electronics converters for wind turbine systems. *IEEE Transactions on Industry Application*, **48** (2), 708–719.
8. Kazmierkowski, M.P., Krishnan, R., and Blaabjerg, F. (2002) *Control in Power Electronics-Selected Problems*, Academic Press. ISBN: 0-12-402772-5.
9. Website of Vestas Wind Power (2011) Wind Turbines Overview, April 2011, http://www.vestas.com/ (accessed 17 December 2013).
10. EWEA (2011) UpWind Project, Design Limits and Solutions for Very Large Wind Turbines, March 2011, http://www.ewea.org/fileadmin/ewea_documents/documents/upwind/21895_UpWind_Report_low_web.pdf (accessed 17 December 2013).
11. Blaabjerg, F., Iov, F., Chen, Z., and Ma, K. (2010) Power electronics and controls for wind turbine systems. Proceedings of EnergyCon' 2010, pp. 333–344.
12. Ng, C.H., Parker, M.A., Ran, L. *et al.* (2008) A multilevel modular converter for a large, light weight wind turbine generator. *IEEE Transactions on Power Electronics*, **23** (3), 1062–1074.
13. Sun, T., Chen, Z., and Blaabjerg, F. (2003) Voltage recovery of grid-connected wind turbines after a short-circuit fault. Proceedings of IECON'2003, pp. 2723–2728.
14. Muller, S., Deicke, M., and De Doncker, R.W. (2002) Doubly fed induction generator systems for wind turbines. *IEEE Industry Applications Magazine*, **8** (3), 26–33.
15. Xiang, D., Ran, L., Tavner, P.J., and Yang, S. (2006) Control of a doubly fed induction generator in a wind turbine during grid fault ride-through. *IEEE Transactions on Energy Conversion*, **21** (3), 652–662.
16. Teodorescu, R., Liserre, M., and Rodriguez, P. (2011) *Grid Converters for Photovoltaic and Wind Power Systems*, John Wiley & Sons, Ltd/IEEE Press.
17. Blaabjerg, F., Teodorescu, R., Liserre, M., and Timbus, A.V. (2006) Overview of control and grid synchronization for distributed power generation systems. *IEEE Transactions on Industrial Electronics*, **53** (5), 1398–1409.
18. Rodriguez, J., Bernet, S., Bin, W. *et al.* (2007) Multilevel voltage-source-converter topologies for industrial medium-voltage drives. *IEEE Transactions on Industrial Electronics*, **54** (6), 2930–2945.
19. Kouro, S., Malinowski, M., Gopakumar, K. *et al.* (2010) Recent advances and industrial applications of multilevel converters. *IEEE Transactions on Power Electronics*, **57** (8), 2553–2580.
20. Faulstich, A., Stinke, J.K., and Wittwer, F. (2005) Medium voltage converter for permanent magnet wind power generators up to 5 MW. Proceedings of EPE 2005, pp. 1–9.
21. Oliveira, D.S., Reis, M.M., Silva, C. *et al.* (2010) A three-phase high-frequency semicontrolled rectifier for PM WECS. *IEEE Transactions on Power Electronics*, **25** (3), 677–685.
22. Wu, B., Lang, Y., Zargari, N., and Kouro, S. (2011) *Power Conversion and Control of Wind Energy Systems*, John Wiley & Sons, Ltd.
23. Pena, R., Clare, J.C., and Asher, G.M. (1996) Doubly fed induction generator using back-to-back PWM converters and its application to variable speed wind-energy generation. *Electric Power Application*, **143** (3), 231–241.
24. Ma, K., Blaabjerg, F., and Xu, D. (2011) Power devices loading in multilevel converters for 10 MW wind turbines. Proceedings of ISIE 2011, June 2011, pp. 340–346.
25. Ma, K. and Blaabjerg, F. (2011) Multilevel converters for 10 MW wind turbines. Proceedings of EPE'2011, Birmingham, UK, pp. 1–10.
26. Carrasco, J.M., Franquelo, L.G., Bialasiewicz, J.T. *et al.* (2006) Power-electronic systems for the grid integration of renewable energy sources: a survey. *IEEE Transactions on Industrial Electronics*, **53**, 1002–1016.
27. Krug, D., Bernet, S., Fazel, S.S. *et al.* (2007) Comparison of 2.3-kV medium-voltage multilevel converters for industrial medium-voltage drives. *IEEE Transactions on Industrial Electronics*, **54** (6), 2979–2992.
28. Rodriguez, J., Bernet, S., Steimer, P.K., and Lizama, I.E. (2010) A survey on neutral-point-clamped inverters. *IEEE Transactions on Industrial Electronics*, **57** (7), 2219–2230.
29. Teichmann, R. and Bernet, S. (2005) A comparison of three-level converters versus two-level converters for low-voltage drives, traction, and utility applications. *IEEE Transactions on Industry Applications*, **41** (3), 855–865.

30. Bruckner, T., Bernet, S., and Guldner, H. (2005) The active NPC converter and its loss-balancing control. *IEEE Transactions on Industrial Electronics*, **52** (3), 855–868.
31. Senturk, O.S., Helle, L., Munk-Nielsen, S. *et al.* (2009) Medium voltage three-level converters for the grid connection of a multi-MW wind turbine. Proceedings of EPE'2009, pp. 1–8.
32. Hosoda, H. and Peak, S. (2010) Multi-level converters for large capacity motor drive. Proceedings of IPEC' 10, pp. 516–522.
33. Andresen, B. and Birk, J. (2007) A high power density converter system for the Gamesa G10x 4.5 MW Wind turbine. Proceedings of EPE'2007, pp. 1–7.
34. Jones, R. and Waite, P. (2011) Optimised power converter for multi-MW direct drive permanent magnet wind turbines. Proceedings of EPE'2011, pp. 1–10.
35. Engel, B., Victor, M., Bachmann, G., and Falk, A. (2003) 15 kV/16.7 Hz energy supply system with medium frequency transformer and 6.5 kV IGBTs in resonant operation. Proceedings of EPE'2003, Toulouse, France, September 2–4, 2003.
36. Inoue, S. and Akagi, H. (2007) A bidirectional isolated DC–DC converter as a core circuit of the next-generation medium-voltage power conversion system. *IEEE Transactions on Power Electronics*, **22** (2), 535–542.
37. Iov, F., Blaabjerg, F., Clare, J. *et al.* (2009) UNIFLEX-PM-A key-enabling technology for future European electricity networks. *EPE Journal*, **19** (4), 6–16.
38. Davies, M., Dommaschk, M., Dorn, J. *et al.* (2008) *HVDC PLUS – Basics and Principles of Operation*, Technical articles,, Siemens Energy Sector.
39. Lesnicar, A. and Marquardt, R. (2003) An innovative modular multilevel converter topology suitable for a wide power range. Proceedings of IEEE Bologna PowerTech Conference, pp. 1–6.
40. Ma, K. and Blaabjerg, F. (2012) The impact of power switching devices on the thermal performance of a 10 MW wind power NPC converter. *Energies*, **5** (7), 2559–2577.
41. Jakob, R., Keller, C., and Gollentz, B. (2007) 3-Level high power converter with press pack IGBT. Proceedings of EPE' 2007, September 2–5, 2007, pp. 1–7.
42. Alvarez, R., Filsecker, F., and Bernet, S. (2011) Comparison of press-pack IGBT at hard switching and clamp operation for medium voltage converters. Proceeding of EPE'2011, pp. 1–10.
43. Prasai, A., Jung-Sik, Y., Divan, D. *et al.* (2008) A new architecture for offshore wind farms. *IEEE Transactions on Power Electronics*, **23** (3), 1198–1204.
44. Iov, F., Soerensen, P., Hansen, A., and Blaabjerg, F. (2006) *Modelling, Analysis and Control of DC-connected Wind Farms to Grid*, International Review of Electrical Engineering, Praise Worthy Prize, p. 10, February 2006, ISSN: 1827–6600.
45. Lima, F.K.A., Luna, A., Rodriguez, P. *et al.* (2010) Rotor voltage dynamics in the doubly fed induction generator during grid faults. *IEEE Transactions on Power Electronics*, **25** (1), 118–130.
46. Santos-Martin, D., Rodriguez-Amenedo, J.L., and Arnaltes, S. (2009) Providing ride-through capability to a doubly fed induction generator under unbalanced voltage dips. *IEEE Transactions on Power Electronics*, **24** (7), 1747–1757.
47. El-Moursi, M.S., Bak-Jensen, B., and Abdel-Rahman, M.H. (2010) Novel STATCOM controller for mitigating SSR and damping power system oscillations in a series compensated wind park. *IEEE Transactions on Power Electronics*, **25** (2), 429–441.
48. Dai, J., Xu, D.D., and Wu, B. (2009) A novel control scheme for current-source-converter-based PMSG wind energy conversion systems. *IEEE Transactions on Power Electronics*, **24** (4), 963–972.
49. Yuan, X., Wang, F., Boroyevich, D. *et al.* (2009) DC-link voltage control of a full power converter for wind generator operating in weak-grid systems. *IEEE Transactions on Power Electronics*, **24** (9), 2178–2192.
50. Rodriguez, P., Timbus, A., Teodorescu, R. *et al.* (2009) Reactive power control for improving wind turbine system behavior under grid faults. *IEEE Transactions on Power Electronics*, **24** (7), 1798–1801.
51. Timbus, A., Liserre, M., Teodorescu, R. *et al.* (2009) Evaluation of current controllers for distributed power generation systems. *IEEE Transactions on Power Electronics*, **24** (3), 654–664.
52. Liserre, M., Blaabjerg, F., and Hansen, S. (2005) Design and control of an LCL-filter-based three-phase active rectifier. *IEEE Transactions on Industry Applications*, **41** (5), 1281–1291.
53. Rodriguez, P., Timbus, A.V., Teodorescu, R. *et al.* (2007) Flexible active power control of distributed power generation systems during grid faults. *IEEE Transactions on Industrial Electronics*, **54** (5), 2583–2592.
54. Altin, M., Goksu, O., Teodorescu, R. *et al.* (2010) Overview of recent grid codes for wind power integration. Proceedings of OPTIM'2010, pp. 1152–1160.
55. Tsili, M. (2009) A review of grid code technical requirements for wind farms. *IET Journal of Renewable Power Generation*, **3** (3), 308–332.

56. Energinet (2003) Energinet – Wind Turbines Connected to Grids with Voltages Below 100 kV, January 2003.
57. Energinet (2010) Energinet – Technical Regulation 3.2.5 for Wind Power Plants with a Power Output Greater than 11 kW, September 2010.
58. E.ON-Netz – Grid Code (2008) Requirements for Offshore Grid Connections in the E.ON Netz Network, April 2008.
59. Faulstich, S., Lyding, P., Hahn, B., and Tavner, P. (2009) Reliability of offshore turbines–identifying the risk by onshore experience. Proceedings of European Offshore Wind, Stockholm, Sweden.
60. Hahn, B., Durstewitz, M., and Rohrig, K. (2007) Reliability of wind turbines – experience of 15 years with 1500 WTs, in *Wind Energy*, pp. 330–332, Spinger, Berlin.
61. Wolfgang, E., Amigues, L., Seliger, N., and Lugert, G. (2005) Building-in reliability into power electronics systems. The World of Electronic Packaging and System Integration, pp. 246–252.
62. Hirschmann, D., Tissen, D., Schroder, S., and De Doncker, R.W. (2005) Inverter design for hybrid electrical vehicles considering mission profiles. IEEE Conference on Vehicle Power and Propulsion, September 7–9, 2005, pp. 1–6.
63. Busca, C., Teodorescu, R., Blaabjerg, F. *et al.* (2011) An overview of the reliability prediction related aspects of high power IGBTs in wind power applications. *Microelectronics Reliability*, **51** (9–11), 1903–1907.
64. Kaminski, N. and Kopta, A. (2011) Failure Rates of HiPak Modules Due to Cosmic Rays, ABB Application Note 5SYA 2042–04, March 2011.
65. Wolfgang, E. (2007) Examples for failures in power electronics systems. Presented at ECPE Tutorial on Reliability of Power Electronic Systems, Nuremberg, Germany, April 2007.
66. Yang, S., Bryant, A.T., Mawby, P.A. *et al.* (2011) An industry-based survey of reliability in power electronic converters. *IEEE Transactions on Industry Applications*, **47** (3), 1441–1451.
67. Yang, S., Xiang, D., Bryant, A. *et al.* (2010) Condition monitoring for device reliability in power electronic converters: a review. *IEEE Transactions on Power Electronics*, **25** (11), 2734–2752.
68. Due, J., Munk-Nielsen, S., and Nielsen, R. (2011) Lifetime investigation of high power IGBT modules. Proceedings of EPE'2011 –Birmingham, UK.
69. Ma, K. and Blaabjerg, F. (2012) Thermal optimized modulation method of three-level NPC inverter for 10 MW wind turbines under low voltage ride through. *IET Journal on Power Electronics*, **5** (6), 920–927.
70. Ma, K., Blaabjerg, F., and Liserre, M. (2012) Reactive power control methods for improved reliability of wind power inverters under wind speed variations. Proceedings of ECCE' 2012, pp. 3105–3122.

7

Photovoltaic Energy Conversion Systems

Samir Kouro[1], Bin Wu[2], Haitham Abu-Rub[3] and Frede Blaabjerg[4]
[1]*Department of Electronics, Universidad Tecnica Federico Santa Maria, Valparaíso, Chile*
[2]*Department of Electrical and Computer Engineering, Ryerson University, Ontario, Canada*
[3]*Department of Electrical Engineering, Texas A&M University, Doha, Qatar*
[4]*Department of Energy Technology, Aalborg University, Aalborg, Denmark*

7.1 Introduction

Grid-connected photovoltaic (PV) energy is one of the fastest growing and most promising renewable energy sources in the world. In fact, since 2007, it has increased over 10 times (from 10 to 100 GW of installed capacity), with 60 GW installed in the last two years alone, as can be seen in Figure 7.1 [1]. Furthermore, PV power has reached 35% of the installed wind power capacity, although in 2007 it accounted only for 10%. Following this trend, PV energy penetration is currently around five years behind the wind energy industry. Nevertheless, grid-connected PV systems are experiencing an accelerated version of the trend experienced by wind energy during the last two decades, which offers a promising glimpse of the future ahead for the solar industry.

The main reason for this remarkable development is the increased competitiveness of PV energy because of the cost reduction of PV modules and the introduction of economic incentives or subsidies. The latter is a reaction to the continuously rising fossil fuel prices, their limited reserves and geopolitical concentration and growing environmental concerns. This has made PV-generated electrical energy cost-effective and competitive in some regions of the world with good solar irradiation conditions. It is expected that PV technology costs will continue to decline in the next decade, making large-scale PV systems more and more attractive. In addition, PV systems can range from small scale to large scale, making it possible to implement PV plants by individuals compared to other renewable energy sources (wind, marine, geothermal, etc.) that have higher capital costs.

Although the core of a PV system is the PV cell (also known as a *PV generator*), power electronics plays a fundamental role as an enabling technology for an efficient PV system control and interface to transfer the generated power to the grid [2]. The functions of the power converter stage of a PV system include maximum power point tracking (MPPT), DC-to-AC power conversion, grid synchronization, grid code compliance (power quality), active and reactive power control and anti-islanding detection.

Power Electronics for Renewable Energy Systems, Transportation and Industrial Applications, First Edition.
Edited by Haitham Abu-Rub, Mariusz Malinowski and Kamal Al-Haddad.

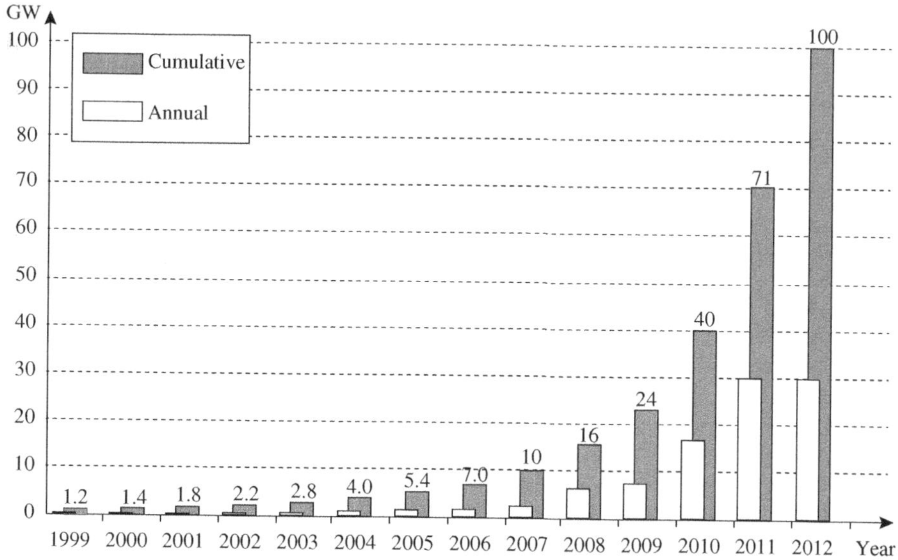

Figure 7.1 World annual and cumulative installed photovoltaic capacity [1]

An overview of a generic power converter interface for grid-connected PV systems is shown in Figure 7.2. The system includes a PV generation system, which can be a single module, a string of series-connected modules, or an array of parallel-connected strings. The PV system is followed by a passive input filter, generally a capacitor, which is used to decouple the input voltage and current from the subsequent power stages by reducing current and voltage ripple (and hence power) at the PV side. The input filter can be followed by a DC–DC stage, which is generally used to perform the MPPT of the PV system, elevate its output voltage and in some occasions also provide galvanic isolation (when using DC–DC converters with high-frequency (HF) transformers). As will be discussed later, some PV systems include several DC–DC power converters to distribute the conversion and control on the DC side. The DC–DC stage (or the input filter if no DC–DC stage is used) is connected through a DC link to the grid-tied DC–AC converter, commonly referred to as the *PV inverter*. In PV systems where no DC–DC stage is used, the input filter is equivalent to the DC-link capacitor as shown in Figure 7.2.

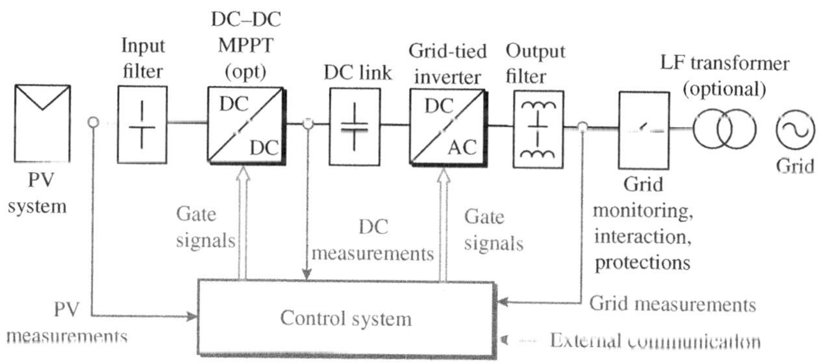

Figure 7.2 Overview of a generic grid-connected PV energy conversion system

The PV inverter is connected to the grid via an output filter, usually made of a combination of inductors (L) and capacitors (C), typically in L, LC, or LCL configurations. The AC filter enables harmonic mitigation and assists the converter–grid interface control. Depending on the PV system requirements and the grid connection available, a low-frequency (LF) transformer is used to elevate the voltage and provide isolation (this is not necessary if an isolated DC–DC stage is used).

The grid-tied converter also includes a grid monitoring and interaction unit (with circuit breaker and fuse) to disconnect the system when required, such as during the night, grid faults, or islanding operation. The feedback for the control system is composed of several current and voltage sensors at the PV input side (for MPPT), DC-link stage (for DC-link voltage control) and grid side (for grid synchronization and active/reactive power control). The control system is further composed of analog-to-digital signal converters, digital microprocessors (or equivalent) and gate drive units to control the semiconductor devices of the different power stages.

Not all grid-connected PV systems are the same; they can vary significantly in size and power from a small-scale (a single module of a few hundred watts) to a large-scale power plant (currently up to 290 MW). They can also be arranged in different string configurations and connected to different available grids (single phase or three phase, 50 or 60 Hz, low voltage (LV) or medium voltage (MV) at point of common coupling (PCC), etc.). Therefore, several power conversion configurations have been developed for better adjustment to the needs of each PV system. This chapter presents the most widely used grid-connected PV system configurations, DC–DC and DC–AC power converter topologies, including their operating principles and control schemes. Additional PV system concepts, such as anti-islanding detection, maximum power point (MPP) and different MPPT methods are introduced. Finally, a brief discussion on the recent developments in multilevel converter-based PV systems is also included.

7.2 Power Curves and Maximum Power Point of PV Systems

PV modules are composed of series-connected PV cells, which are essentially made of two layers of crystalline silicon semiconductor devices (monocrystalline and polycrystalline), or the so-called *thin-film devices* (cadmium telluride, copper indium gallium selenide and amorphous silicon), forming a p–n type junction field [3]. The underlying operating principle of all these technologies is the photoelectric effect, by which photons of equal or greater energy than the bandgap of the semiconductor material can excite and free electrons. When the PV circuit is closed (by connecting a load or converter), the freed electrons generate a DC current from the positive layer to the negative layer to fill the "positive holes." Therefore, the current generated by a PV module is directly dependent on the number of incoming photons and, thus, the solar irradiation: higher irradiation means more photons, hence more free electrons and thus higher currents.

7.2.1 *Electrical Model of a PV Cell*

The electrical characteristics of a PV module or cell are nonlinear and are highly dependent on solar irradiation and temperature [4]. The PV cell can be electrically modeled by an equivalent circuit such as the one shown in Figure 7.3, with a photocurrent source in parallel with a diode, a shunt resistance R_{sh} and a series resistance R_s.

The model of Figure 7.3 can be mathematically described by Kishor *et al.* [4]

$$i_{PV} = i_L - i_0 \left[e^{\frac{q(v_{PV}+i_{PV}R_s)}{nKT}} - 1 \right] - \frac{v_{PV} + i_{PV}R_s}{R_{sh}} \tag{7.1}$$

where i_{PV} is the output current of the PV cell, which is a function of its output voltage v_{PV}, and a number of variables as defined in Table 7.1.

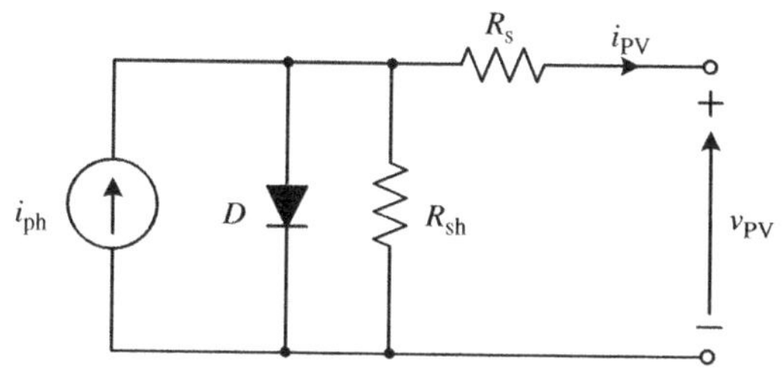

Figure 7.3 PV cell electrical model

Table 7.1 PV cell parameters

Variable	Parameter
v_{PV}	Cell output voltage (V)
i_{PV}	Cell output current (A)
R_s	Cell series parasitic resistance (Ω)
R_{sh}	Cell shunt parasitic resistance (Ω)
q	Electronic charge: 1.6×10^{-19} (Coulombs)
K	Boltzmann constant: 1.38×10^{-23} (J/K)
T	Absolute temperature (K)
n	Diode ideality factor: ideally $n - 1$
i_o	Cell reverse saturation current: 10^{-12} (A/cm^2)
i_{ph}	Cell photocurrent: 35–40 mA/cm^2/Sun for Si cells

7.2.2 *Photovoltaic Module I–V and P–V Curves*

Typical PV module's current–voltage ($I-V$) curves and power–voltage ($P-V$) curves can be observed in Figure 7.4. From the $I-V$ curves, it can be seen that the PV module behaves as a DC current source that remains almost constant even if the PV module is connected to different output voltage potentials. When the voltage across the module gets higher, the free electrons start recombining and do not generate current. This effect is not linear with respect to the voltage and it cuts off abruptly. When all electrons are recombined and no current is produced, the module is operating at an open-circuit voltage v_{oc}. This voltage varies slightly for different levels of solar irradiation (at the same temperature), as it can be seen in Figure 7.4(a). In contrast, when short circuited, the module produces the largest current (short-circuit current i_{sc}), which is very linearly dependent on the solar irradiation (the more photons, the more free electrons). Consequently, the $P-V$ curves (obtained by multiplying the axes of the $I-V$ curve) of the PV module will be composed of three segments: a constant positive $dp/dv > 0$ slope equal to the generated constant DC current, a negative $dp/dv < 0$ slope when operating close to the open-circuit voltage, and a $dp/dv = 0$ segment corresponding to the MPP for that given irradiation [5]. The voltage at which the MPP is obtained is known as the *maximum power voltage* v_{mp}. The slopes of the $I-V$ curve can be adjusted by the shunt resistance R_{sh} and series resistance R_s in the PV cell model to better represent the real PV cell.

The open-circuit voltage, although not much affected by different levels of solar irradiations, does vary significantly when the module is operating at different temperatures, as can be seen in Figure 7.4(b). In summary, the higher the temperature, the lower the open-circuit voltage, reducing the maximum power that can be generated by the module.

Both solar irradiation levels and temperature affect the maximum power that can be drawn from the modules. When connecting a passive load, the $I-V$ curve of the load will intersect the $I-V$ curve of the

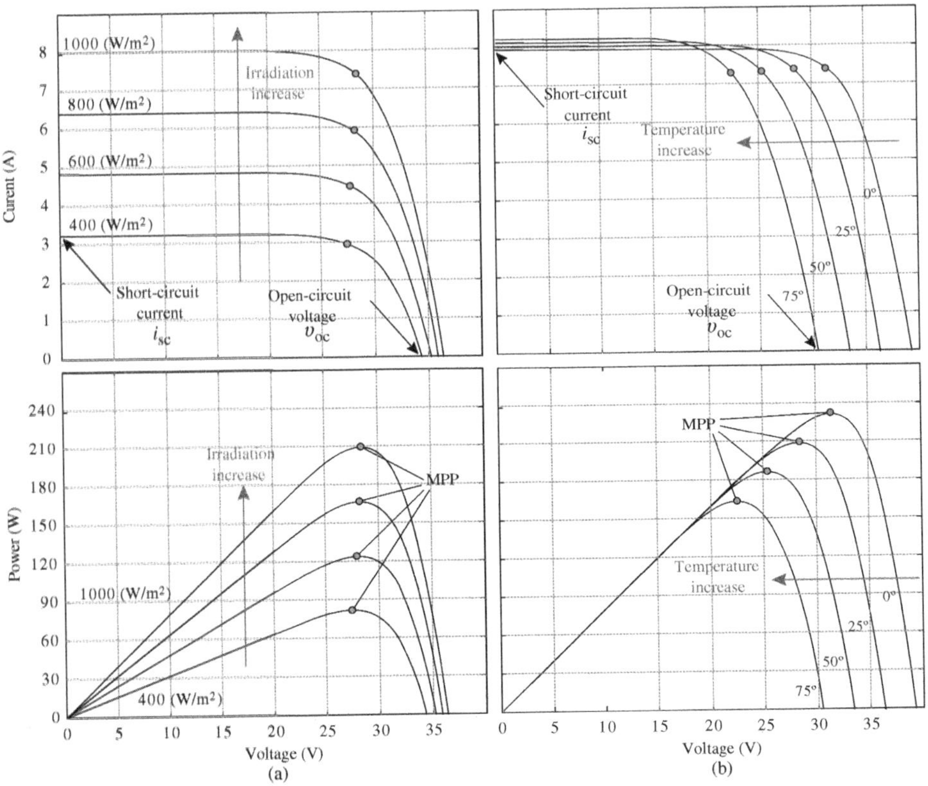

Figure 7.4 *I–V* and *P–V* curves of a PV module under: (a) different levels of solar irradiation at 25 °C and (b) different temperatures at 1000 W/m^2

module and define the power generated by the module. However, a controlled power converter, whether DC–DC or DC–AC, can control the load curve characteristic and intersect the PV curve at the MPP. This is achieved by controlling the PV output voltage at v_{mp}. As the operating conditions of irradiation and temperature are inherently time variant, the instantaneous v_{mp} is unknown (unless irradiation and temperature are measured, which is expensive). The search for the MPP is then performed by the power converter, hence the importance of power electronics for this particular application. The control techniques used for such task are known as *maximum power point tracking (MPPT) methods* and will be discussed later in this chapter. The MPPT algorithm generates the voltage reference to be controlled by the converter.

It is worth mentioning that the power curves shown in Figure 7.4 correspond to a single PV module; hence series-connected modules forming a PV string will have a different combined MPP, depending on module mismatch, temperature differences and partial shading of the modules. The same applies to PV arrays formed by parallel-connected strings. In these cases, the *I–V* curve has several local MPPs and one global MPP, making it more difficult to achieve a true MPPT.

7.2.3 MPP under Partial Shading

Partial shading and module mismatch, because of differences in fabrication or differences caused by usage and aging, have been identified as the main causes of reduced energy yield of PV strings and

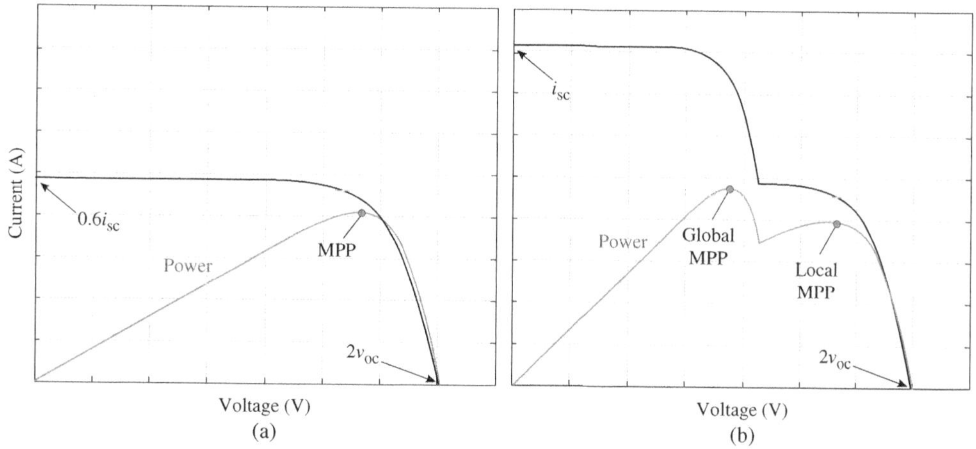

Figure 7.5 I–V and P–V curves of a two-module string, with one cell shaded by 40%: (a) without bypass diode and (b) with bypass diode

arrays. They are also responsible of the creation of hot spots in cells of a module, which reduce not only energy production but also the lifespan of the module.

In a string of two or more series-connected modules, a shaded module will have a very low or even zero photocurrent, and thus the current generated by the other modules passes through the shunt resistor, resulting in a negative voltage potential. Since R_{sh} is usually large, the shaded module reduces the overall voltage of the string instead of adding to it. To overcome this effect, a diode is usually added in parallel to the modules [5], so that in the case of partial shading the current passes through the diode resulting in only the semiconductor voltage drop. Similarly, in parallel-connected modules or strings, a diode is connected in series to each string to prevent reverse currents flowing into the lower voltage string [5].

Figure 7.5 illustrates a qualitative example of the effect of partial shading on the I–V curves of a PV string of two series-connected modules with and without a parallel-connected bypass diode. Connecting a diode in parallel to each module avoids the decrease in the short-circuit current i_{sc} during partial shading and thus increases the possible power yield of the whole PV string.

7.3 Grid-Connected PV System Configurations

Grid-connected PV energy conversion systems can be grouped into four different types of configurations: centralized configuration for large-scale PV plants (three phase), string configuration for small- and medium-scale PV systems (single phase and three phase), multi-string configuration for small- to large scale systems (single phase and three phase) and AC-module configuration for small-scale systems (commonly single phase) [2]. Simplified diagrams of each of these configurations are given in Figure 7.6.

The centralized topology (Figure 7.6(a)) has, as its main characteristic, the use of a single three-phase voltage source inverter (VSI) to connect the whole PV plant to the grid [6], or a portion of it, in case the PV plants exceed the power rating of the existing central inverters. The PV system is formed by the series connection of modules (string) to reach the desired DC-link voltage and by several strings in parallel to reach the power rating of the inverter. The advantages of this configuration are its simple structure, single LF transformer and single control system (single set of sensors, control platform and grid monitoring unit). This comes at the expense of reduced power generation because of a single MPPT algorithm for the whole PV system. Also, diode conduction losses are introduced by the string series blocking diodes. Currently, the central configuration is the most widely used topology for large-scale PV plants.

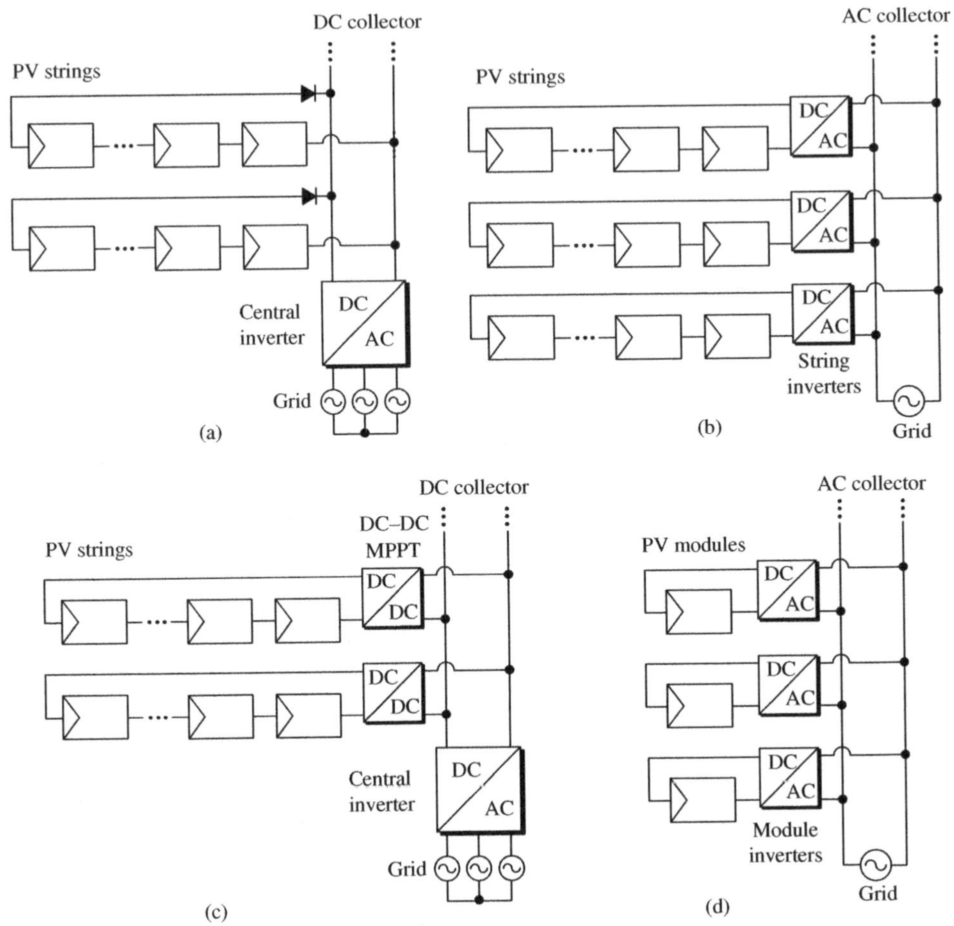

Figure 7.6 Grid-connected PV system configurations: (a) centralized configuration, (b) string configuration, (c) multi-string configuration and (d) AC-module configuration

The string configuration (Figure 7.6(b)) uses one inverter per PV string; hence there is no need for a series blocking diode. In addition, for a PV plant consisting of several string inverters instead of a single central inverter, there will be more individual MPPT available, increasing the total energy yield [7]. Partial shading and mismatch are reduced at a string level and not at an array level. The string configuration also increases modularity, since additional PV strings and string inverters can be added to the power plant without affecting existing strings. On the downside, compared to a central inverter, the string configuration has a higher component count, multiple LF or HF transformers if isolation is required and a need for several individual grid control systems (sensors, control platform, grid monitoring units, etc.) for a power plant of the same size. For large PV plants, the investment cost of a string inverter configuration can reach up to 60% higher than that of a central inverter configuration [7]. Therefore, the string topology is widely adopted as a solution for small- and medium-scale PV systems such as rooftop and residential systems.

The multi-string configuration (Figure 7.6(c)) merges the benefits of the centralized and string systems [8]. It introduces the distributed MPPT capability of the string configuration through individual

DC–DC converters interfacing each string to the centralized inverter. The DC–DC stage may also serve for voltage elevation and isolation if HF-isolated DC–DC converters are used. In this way, the system gains higher energy yield and modularity compared to the central topology, while retaining its major benefits (simple structure and single grid-side control system). In terms of component count, the multi-string configuration is above the central topology, because of additional DC–DC converters, and below the string inverter, since the DC–DC stage requires fewer components compared to the grid-tied inverters. Among the disadvantages are higher DC-cable losses, which are necessary to connect smaller parts of the PV system and DC–DC converters to the central inverter. The multi-string configuration is popular for small- and medium-scale PV systems such as rooftop ones. More recently, it has also been introduced for large- or utility-scale PV plants.

Finally, the AC-module or module-integrated topology, commonly referred to as *micro-inverter* (Figure 7.6(d)), is the most distributed power converter architecture for grid-connected PV systems, since it features one inverter per PV module [9]. Therefore, it has the best MPPT capability of all configurations. Because PV modules usually generate an LV (<50 V), voltage elevation is required for grid connection of this configuration. Therefore, AC-module topologies usually include a DC–DC boosting stage to elevate the voltage of the module. In most cases, the DC–DC stage also includes a HF isolation transformer to provide galvanic isolation. In this configuration, all power electronic devices, components, filters, control system and so on are distributed among all modules of the plant, and therefore it may result in higher cost and reduced converter efficiency (excluding MPPT efficiency, which is higher). Therefore, this topology is intended for small-scale PV systems and more domestic use.

A summary of the main characteristics of the four configurations is given in Table 7.2.

7.3.1 Centralized Configuration

The central configuration is the most widely adopted PV system for large-scale or utility-scale PV plants. Currently, utility-scale PV plants are experiencing an explosive growth and paving the way for large-scale PV energy penetration around the world. Just five years ago, there were a handful of PV plants with over 20 MW of installed capacity, with the biggest being the Olmedilla de Alarcon Solar Park in Spain with 60 MW of peak installed capacity (commissioned in 2008). Currently, more than 50 plants surpass 40 MW [10], with the biggest reaching 290 MW (Agua Caliente Solar Project, Yuma, United States) [11]. Larger plants, in the gigawatt range, are currently under development in the United States, China, Greece and Dubai, and there are more than 25 plants over the 250 MW range worldwide all to be completed within this decade [12]. Figure 7.7 shows the Sarnia PV plant located in Ontario, Canada,

Table 7.2 Summary of grid-connected PV system configuration characteristics

Configuration	Power range	Cost/W	Devices	Advantages	Disadvantages
Central inverter	<1.6 MW	Low	IGBT	Simple design, single control system	Mismatch losses, reduced energy yield, string blocking diodes
String inverter	<10 kW	Medium	MOSFET/IGBT	Separate MPPT, reduced DC wiring	High component counts, several grid control systems
Multi-string inverter	<500 kW	Medium/low	MOSFET/IGBT	Separate MPPT, single-grid control system	Medium component counts, two-stage power conversion
AC-module inverter	<300 W	High	MOSFET	No DC wiring, high-energy yield	Highest component count, one control system per module

Figure 7.7 Sarnia PV power plant, Ontario, Canada, 97 MW peak, 1.3 million modules (photo courtesy of First Solar) [13]

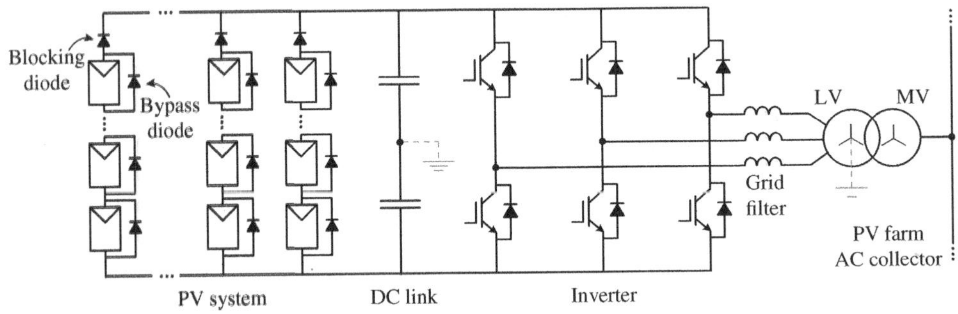

Figure 7.8 Typical central inverter configuration based on two-level voltage source inverter

with a peak installed capacity of 97 MW, which was the biggest PV plant operating in 2011 (1.3 million PV modules).

Figure 7.8 shows a typical configuration of a grid-connected central inverter PV system. The PV modules are connected in series to form strings, which in turn are connected in parallel to form an array connected to the central inverter. The first versions of the central topology were line-commutated inverters with a power factor of 0.7. The current central inverter technology is typically a two-level three-phase voltage source converter and is shown in Figure 7.8. It is equipped with self-commutated insulated gate bipolar transistor (IGBT) technology, controlled by digital signal processor (DSP) and modulated with carrier-based pulse width modulation (PWM) or space vector-based PWM modulation (SVM). The basic function of the inverter is to perform the DC–AC conversion from DC PV power to the three-phase AC grid. In addition, the inverter is in charge of performing the MPPT of the PV array by controlling the DC-link voltage to v_{mp}, and synchronizing the AC grid currents with the grid voltage for active and reactive power control. The inverter is connected to the grid via an inductive grid filter and a LF transformer to go from LV to MV of a few thousand volts (LV/MV) in order to reduce losses in transmitting PV energy to the grid.

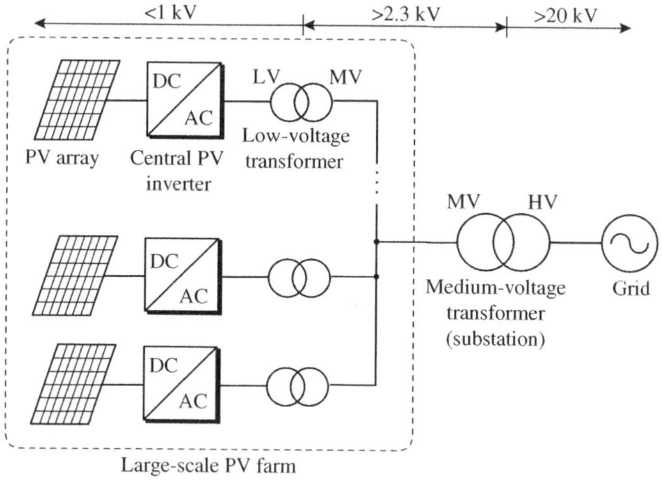

Figure 7.9 Typical utility-scale PV plant based on central inverter configuration

The typical DC-link voltage in the central inverter PV system shown in Figure 7.8 is between 550 and 850 V. The maximum operating DC-link voltage of the converter is limited to 1000 V, since this is the insulation capability of most commercial PV modules (a string may not surpass 1000 V). This means that if a typical module of 30 V v_{mp} and 250 W peak power is considered, a string could be made of around 15–25 series-connected modules, with a total peak power of 3.75–6.25 kW. Since the current IGBT technology limits the two-level VSI operating at LV to <800 kW, more than 120 strings (over 3000 modules) can be concentrated into a single central inverter. Therefore the chance of partial shading and module mismatch is highly likely, which requires a parallel connection to each module of a bypass diode (commonly already integrated in the module), and a series connection of a blocking diode to prevent weaker modules and strings from behaving as loads.

As the central inverter is the sole power converter stage of the system, high-power conversion efficiency, usually 97% can be achieved from the PV system to the grid. However, the fact that a single inverter has only one controlled DC-link voltage limits the MPPT capability of the central inverter to a single degree of freedom, reducing the generated energy output, compared to the distributed MPPT systems. Another disadvantage is the long DC cables from the strings located faraway from the inverter. Nevertheless, the low cost, capability of concentrating a large PV plant and also the simple structure have made this configuration the most adopted in utility-scale PV plants.

Figure 7.9 shows a typical utility-scale PV plant like the Sarnia solar plant shown in Figure 7.7. These large PV farms, rated at over tens and even hundreds of megawatts, surpass the power rating of the largest central inverters (800 kW) and therefore require the connection of several central inverters. The central inverters are located next to an LV to MV transformer to elevate the voltage of the AC collector busbar of the PV plant (usually ranging from 2.3 to 35 kV). Then, the power of the PV plant is transmitted to a substation where a transformer elevates the voltage from medium to high voltage for power transmission.

As PV farms become larger and larger, the PV inverter industry has introduced dual-central inverters, shown in Figure 7.10, which are basically two central inverters connected to the PV AC collector busbar through a 12-pulse transformer to reduce LF and some switching harmonics. This is why some manufacturers offer central inverters rated up to 1.6 MW, actually consisting of two 800 kW central inverters. Costs are reduced by using a single-grid monitoring system for both inverters and concentrating

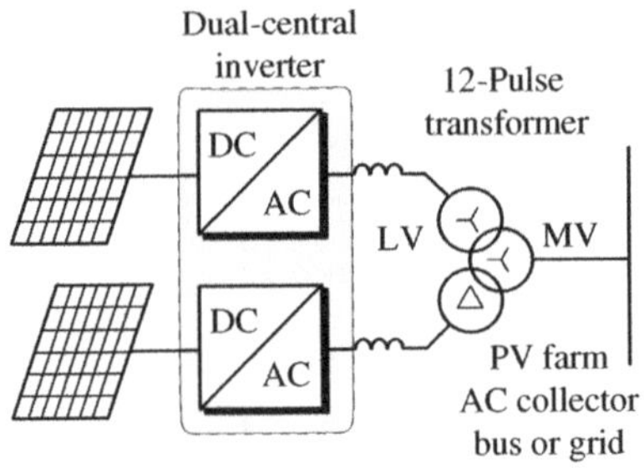

Figure 7.10 Dual-central inverter configuration with multi-winding transformer

Table 7.3 Examples of commercial dual-central inverters

Parameter	Satcon Prism Platform Equinox	ABB PVS800-MWS	SMA MV Power Platform 1.6
Input voltage range (V)	550–850	525–825	641–820
Power (MW)	1.5	1.25	1.6
Maximum DC voltage (V)	1000	1000	1000/1100
Input current (A)	2820 (2 × 1410)	2480 (2 × 1240)	2800 (2 × 1400)
Efficiency (%)	98.5	97.8	98.6
Isolation	Integrated MV transformer up to 35 kV	Integrated MV transformer up to 20 kV	Integrated MV transformer up to 34.5 kV
Independent MPPT	2	2	2
Photo			

the control system, power converter, filters and even the transformer into a single housing, generally a container. In addition, the 12-pulse transformer is capable of canceling or reducing certain harmonics. Table 7.3 shows some commercial examples of these inverters and their ratings.

The two-level VSI is currently on the edge of its capability as an optimal solution for the central inverter of large-scale PV plants. To increase converter efficiency, reduce filter size, improve power quality and be able to fulfill the increasingly demanding grid codes, new central inverter topologies have been developed. Among them are the three-phase three-level neutral point clamped (3L-NPC) converter and the three-phase three-level T-type converter (3L-T), as shown in Figure 7.11(a) and (b), respectively. These two central inverter topologies belong to the multilevel converter family.

Multilevel converters were originally conceived for high-power (>1 MW) and MV applications because of their capability to surpass classic semiconductor voltage blocking limits. However, they have recently been introduced for LV, low-power and high-performance/high-efficiency/high-power quality demanding applications, such as uninterruptible power supplies (UPSs) and PV energy conversion systems.

The extra level in the phase output voltage of these three-level inverters offers a reduction to half the original dv/dt generated by the 2L-VSI. This allows a reduction in the device's average switching frequency, hence switching losses, which improves the power conversion efficiency without compromising power quality. Several semiconductor manufacturers (Infineon, Semikron, Fuji, Mitsubishi, to name a few) offer power converter modules with these device arrangements specifically designed for UPS and PV systems.

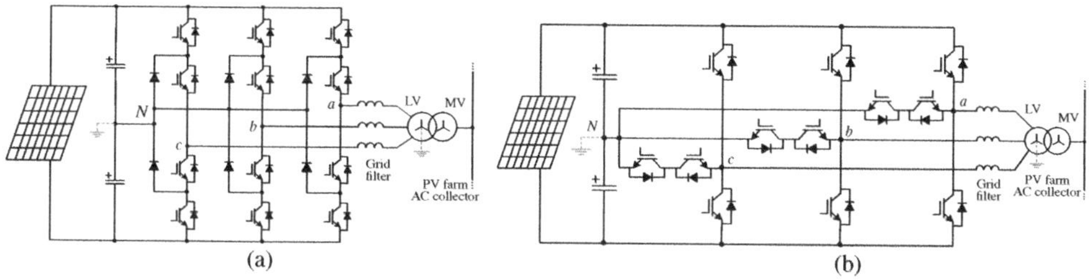

Figure 7.11 Multilevel central inverter PV systems: (a) 3L-NPC inverter PV system and (b) 3L-T inverter PV system

7.3.2 String Configuration

String inverters are reduced versions of the central inverter, in the sense that instead of PV arrays with parallel-connected strings, only single strings (with no need for a blocking diode) are connected through a dedicated grid-tied inverter with an independent MPPT as shown in Figure 7.6(b). This technology results in an increase in energy yield by 1–3% compared with that of the central inverter topology because of the reduction in mismatch and partial shading losses [7]. As they only interface a single PV string, they are aimed at low-power and single-phase grid connection and they are very popular for small-scale domestic systems (typically rooftop).

Depending on the country and grid connection standard, string inverters may or may not require galvanic isolation. When galvanic isolation is required, LF transformers on the grid side or HF transformers in the DC–DC converter stage are used. Both introduce galvanic isolation, but they can also be used for voltage elevation purposes. The LF (grid frequency) transformer has significantly lower power density compared to HF transformers. Nevertheless, DC–DC converters with HF isolation require very high switching frequencies of 20–100 kHz, which may result in more losses. To reduce this effect, fast and efficient switching devices are used. The HF isolation stage also turns out in a more complex circuit structure, more complex control and electromagnetic interference (EMI)/electromagnetic compatibility (EMC) design considerations compared with that of LF transformer or transformerless solutions.

In some countries, including Spain and Germany, isolation is not required, which can lead to cost reduction, more compact and more efficient string inverters. However, transformerless operation also requires a more careful design of the converter topology and modulation schemes, since a switched common-mode voltage introduces the risk of high leakage ground currents, because of the parasitic capacitance of the PV string [14]. This capacitance formed between the conductive layers, polymers, glass and frame of the PV modules (shown in Figure 7.12(a)) finds a current path since it is directly connected to the AC grid that is grounded for potential equalization as well as for the PV module frame (shown in Figure 7.12(b)). This capacitance increases with the size of the PV system and can be quite large, hence the risk of high leakage ground currents by the dv/dt's in the switched common-mode voltage. Since in these cases no transformer is used, an additional DC–DC stage could also be necessary to elevate the voltage for grid connection, depending on the size of the PV string (number of modules in series). If a DC–DC stage is added, they are known as *two-stage string inverters*.

Table 7.4 shows a summary of all combinations between the DC–DC stage and isolation options that are found in practical string configurations. Several manufacturers have developed and offered proprietary topologies or combinations of different DC–DC and DC–AC stages, including different types of isolation, even for single- or three-phase grid connection [15–18]. Therefore, there is a wide selection of string inverters, which cannot all be covered in this chapter.

The most popular string inverter is the single-phase full-bridge inverter, also known as *H-bridge inverter*. Many string inverter topologies are derived from this topology or used it as the basic building block with a few additional components to improve efficiency and performance. Different combinations of isolation and two-stage conversion using the H-bridge string inverter are shown in Figures 7.13

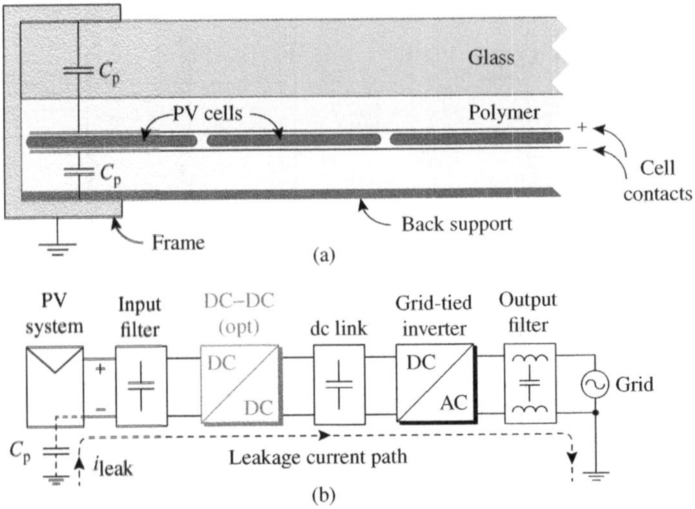

Figure 7.12 Parasitic capacitance and leakage current in transformerless PV systems: (a) parasitic capacitance (cross section of PV module) and (b) leakage current path

Table 7.4 String configurations depending on isolation and DC–DC stage

	With DC–DC stage	Without DC–DC stage
No Isolation	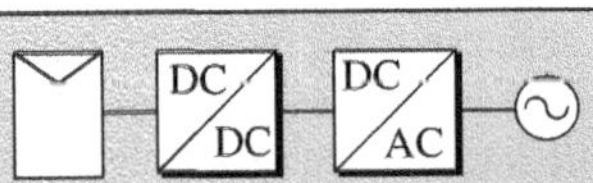	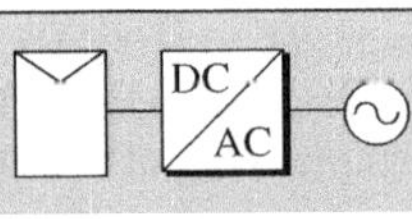
Low-frequency isolation		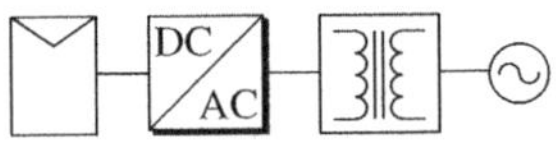
High-frequency isolation	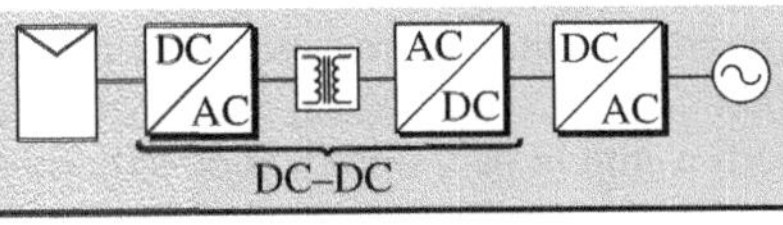	

and 7.14. Note that transformerless topologies must have the grid filter divided into two symmetric parts: one in the phase and the other in the neutral wires in order to reduce the common-mode voltage changes that otherwise can lead to higher leakage currents.

Figure 7.13(a) shows the most basic H-bridge string inverter configuration with a grid-side LF transformer and no DC–DC stage. The advantage of this configuration is the simple power circuit, galvanic isolation and voltage elevation provided by the transformer, which allows a larger range of input voltages. Unipolar PWM is used to generate a three-level voltage waveform with a harmonic content at double the device switching frequency in order to obtain higher power quality. A hybrid modulation is also used, in which the output phase leg of the H-bridge is controlled by a carrier-based PWM while the neutral leg is controlled by a fundamental frequency square waveform in order to reduce the switching losses compared to that of the unipolar PWM. However, it produces higher grid current ripple because of the lower equivalent switching frequency of the output voltage. Since both are three-level modulation schemes,

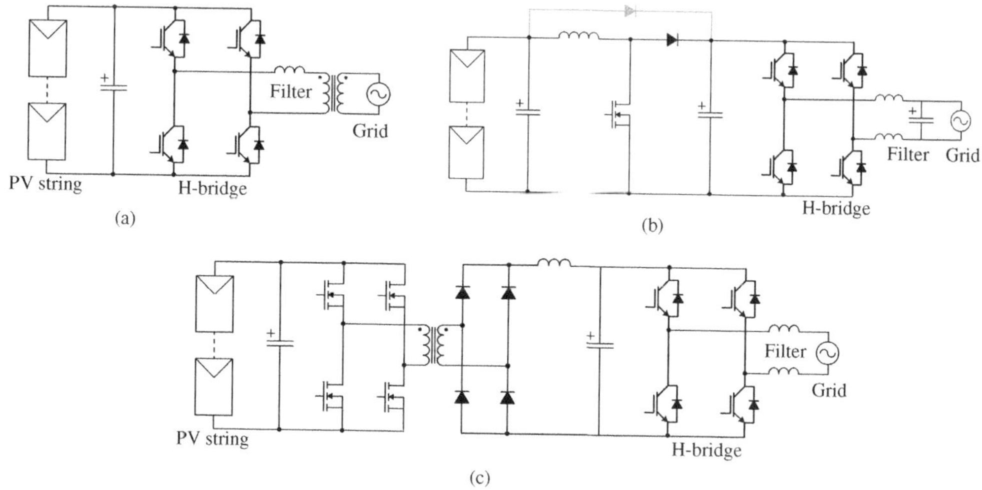

Figure 7.13 String inverters based on the H-bridge topology: (a) with low-frequency isolation transformer, (b) transformerless with DC–DC boost stage and (c) with high-frequency isolation transformer in DC–DC stage

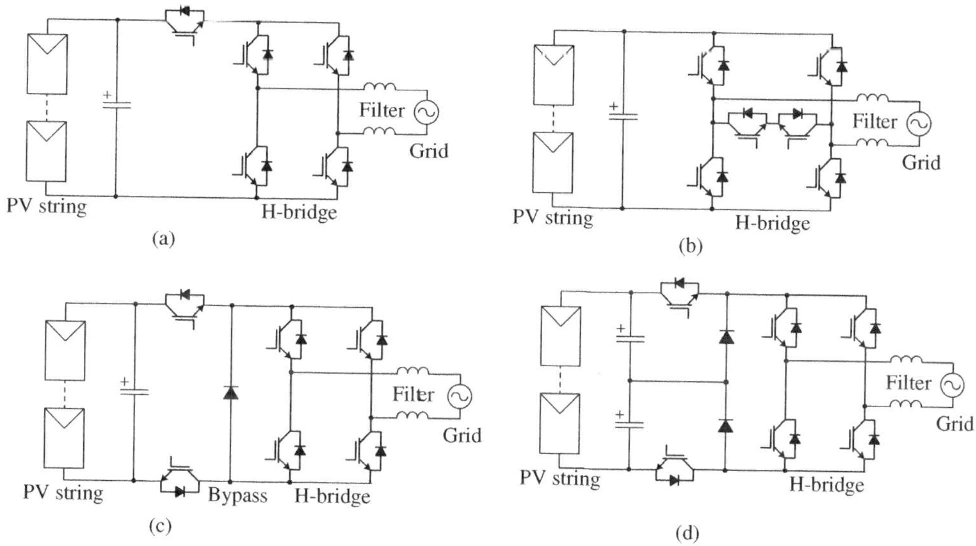

Figure 7.14 Variants of the transformerless H-bridge string inverter: (a) H5, (b) HERIC, (c) H6D1 and (d) H6D2

the bypass switching state (zero voltage level) prevents a reactive current flow between the filter inductor and the DC-link capacitor. The disadvantages of this topology are mainly introduced by the LF transformer: higher cost, lower power density, non-unity power factor at lower than-rated operation because of the magnetizing current and reduced efficiency (2% extra losses). This topology was commercialized by several manufacturers, but is continually losing popularity because of its low efficiency and large size.

The transformerless version of the H-bridge string inverter, also known as *H4 topology*, is only possible with a symmetric filter (components equally distributed between phase and neutral wires) at the grid side, and unlike the isolated version it cannot be controlled using unipolar PWM or hybrid modulation. This is due to the square-wave components in the common-mode voltage, leading to high leakage currents and corresponding safety hazards [15]. Instead, bipolar PWM (two-level output voltage) is used, since it produces a constant common-mode voltage together with the split symmetric filter, resulting in a line frequency voltage across the parasitic capacitance, which is safe for transformerless operation. However, it has lower efficiency since there is a reactive current flow between the passive elements of the circuit at zero voltage through the freewheeling diodes when using this modulation, as the DC-link capacitor is not isolated from the grid at any time [16]. This is the main reason why the H4 topology is not a popular configuration in modern string inverters.

Both versions of the H-bridge string inverter, with and without an LF transformer, can be found with an additional boost DC–DC stage. The main purpose of the boost converter is to elevate the PV string voltage, since the MPP voltage can vary depending on radiation and temperature. Hence, the DC–DC stage decouples the PV DC voltage from the inverter DC-link voltage. This is a desirable feature considering that the grid-tied inverter requires a DC-link voltage higher than the peak grid voltage to work properly (to be able to control the grid current). This extends the input voltage range of the system and also allows a more steady DC-link voltage, which improves the grid control performance. This configuration is shown for the transformerless H-bridge string inverter in Figure 7.13(b), although the same DC–DC converter can be added to the LF-isolated version. Note that there is an optional bypass diode (shown in gray), which disables the DC–DC stage when the PV system DC voltage is greater than the inverter DC-link voltage, to improve efficiency when operating close to the nominal design conditions, avoiding the switching losses and also the boost diode reverse recovery. Despite the additional DC–DC stage, this configuration retains the same drawbacks that are inherent in the use of a grid-tied H-bridge inverter addressed earlier.

Finally, Figure 7.13(c) shows an HF-isolated two-stage H-bridge string inverter. In this particular example, the isolated DC–DC stage is composed of a MOSFET full-bridge inverter and a diode full-bridge rectifier. Although the MOSFET H-bridge is most commonly used in this configuration, many HF-isolated DC–DC stages could be used instead, including the half-bridge, push–pull, forward and flyback converters. This approach greatly reduces the size of the converter compared to that of the LF isolation topologies, thus improving the power density (volume and weight). It also allows the use of unipolar PWM and extends the input voltage range of the PV system compared to that of the transformerless topologies, thus allowing a more flexible design of the PV string and a wide range for MPPT. However, the additional converter stages introduce design complexity, higher losses, higher cost and lower reliability.

To overcome the problem of the reactive current transfer between the grid filter and the DC-link capacitor in transformerless H-bridge string inverters, several proprietary solutions have been introduced by different manufacturers [15–17]. Figure 7.14(a) shows the H5 string inverter introduced by SMA, in which a switch has been added between the DC-link and the H-bridge inverter to open the current path between the passive components when freewheeling in order to increase the efficiency. The same objective can be accomplished by introducing a bidirectional switch that bypasses the whole H-bridge inverter, separating the grid filter from the converter during freewheeling, as shown in Figure 7.14(b). This converter is also known as the *Highly Efficient and Reliable Inverter Concept* (HERIC) converter, introduced by Sunways. Although the HERIC converter has one more switch compared to that of the H5, it has only two switches in the current path at any time, while the H5 has three switches when it is not freewheeling, which results in higher conduction losses.

These converters enable the generation of a three-level voltage output, including the zero voltage level, without producing a switched common-mode voltage and therefore are suitable for transformerless operation. In this way, very efficient, low-cost and high-power density string inverters are obtained compared to those in Figure 7.13 featuring the H-bridge inverter only. In addition, these converters may also include an additional DC–DC stage in order to extend the input voltage flexibility and decouple the PV system DC voltage from the inverter DC-link voltage for better grid converter control performance.

An alternative way to isolate the DC side from the AC side during freewheeling is to add an additional switch in the negative DC bar to the H5 topology, forming the H6 topology as shown in Figure 7.14(c). The advantage is that the freewheeling is evenly distributed among the H-bridge devices over a fundamental cycle, ensuring an even usage of all switches [17]. The disadvantage is that four switches conduct during normal operation, causing more conduction losses. If a diode is connected in parallel to the DC side of the H-bridge of the H6 topology, an auxiliary freewheeling current path is added, allowing the use of bipolar PWM in the H-bridge, resulting in a unipolar output voltage. This converter is also known as *H6D1*. A disadvantage is that an additional diode is conducting during freewheeling, compared to that of the H6, and hence there are more losses. By adding two auxiliary freewheeling diodes instead of one diode, the H6D2 string inverter as shown in Figure 7.14(d) is formed, with the additional diode. The blocking voltage across the DC switches can be fixed to half the DC voltage, reducing the losses compared to that of the H6D1. Both the H6D1 and H6D2 inverters were introduced by Ingeteam [19].

Other string inverters introduced in practical PV applications are the single-phase three-level NPC (3L-NPC) [20], the transistor clamped converter, also known as the *T-type converter and Conergy NPC* [20] and the five-level H-bridge NPC (5L-HNPC) [21], as shown in Figure 7.15.

The 3L-NPC string inverter, also known as the *diode-clamped converter*, is connected to the phase and neutral of the grid through the output phase node and the DC-link neutral point, respectively. Depending on the switching state, the inverter clamps the grid phase to the positive, negative and neutral potentials of the DC link, hence the three levels. The connection to the neutral point is achieved by switching on the two internal devices and by freewheeling through the clamping diodes. This switching state generates a zero voltage level at the output, since the neutral point is grounded. The three voltage levels have the same power quality as the unipolar voltage generated by the H-bridge, but in this case without the problems of HF commutations in the common-mode voltage, since the neutral of the grid is permanently connected to the neutral point of the DC link and does not change its potential with respect to the PV system for any switching state. Since the common-mode voltage is, in fact, constant (DC voltage), there are no problems with leakage currents and EMI, making the 3L-NPC a suitable topology for transformerless applications. The main drawback is that the DC-link voltage is the double of the H-bridge topologies, which might require an additional boost stage that introduces losses, particularly when a large boost effort is needed [15]. Note that in this case the grid filter does not need to be symmetrically divided in both current paths of the grid connection, as with H-bridge topologies. The 3L-NPC string inverter has been introduced in practice by Danfoss.

An alternative way to clamp the phase of the grid to the neutral point of a split DC link is through a bidirectional switch between the two nodes, hence the name "transistor clamped inverter." Since it forms a T with the half-bridge converter leg that is connected to the positive and negative bars of the DC link, it is also called *T-type converter*. Particularly in PV applications, it is more commonly referred to as the *Conergy NPC*, since it was introduced by the company of the same name. As with the previous topology, the converter generates three voltage levels but does not require the two additional clamping

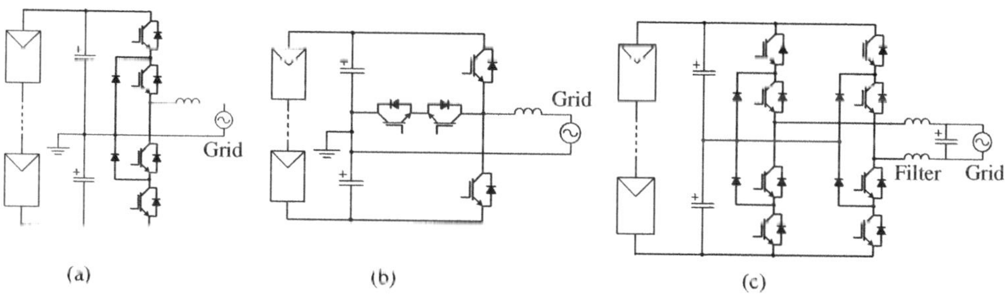

Figure 7.15 Neutral point clamped string inverters without isolation: (a) 3L-NPC, (b) T-type and (c) 5L-HNPC

diodes. Nevertheless, the phase leg switches have to block double the usual voltage if connected to the same grid, which is not the case for the clamping switches, which share the blocking voltage. Since the neutral of the grid is permanently connected to the neutral of the DC link (which is grounded), it causes a fixed common-mode voltage, and hence there is no risk of leakage current, which therefore is suitable for transformerless operation. For the same reason, no symmetrical division of the grid filter is necessary.

More recently the 5L-HNPC string inverter has been introduced by ABB [21]. Unlike the two previous converters, this topology does not connect the grid neutral to the midpoint of the DC link and to ground. In fact, it is a full-bridge made of two 3L-NPC phase legs capable of generating a five-level output voltage between both phase leg terminals. Like the H-bridge, since both converters are full-bridge topologies, it requires a split symmetrical grid filter for transformerless operation together with a particular modulation technique. Although the converter is capable of generating five levels, only three levels are generated in order to avoid switched common-mode voltages.

The cascaded H-bridge (CHB) multilevel converter has also found practical application for string inverters, and was introduced by Mitsubishi [22]. The converter consists of three series-connected H-bridge cells operating with unequal DC-voltage ratios, as shown in Figure 7.16. The PV system is connected through a boost DC–DC stage to only one of the H-bridge cells, which is the only one processing active power to the grid. The other two cells use floating DC-link capacitors and act as auxiliary series active filters for power quality improvement through the generation of more voltage levels. The voltage ratio has been chosen to be 1 : 2 : 4 among the H-bridges, since it prevents regenerating operation of the floating cells, making the voltage balancing mechanism feasible. With this asymmetry ratio, 13 voltage levels are obtained at the grid terminals, allowing an important reduction in devices switching frequency without compromising power quality and reducing filtering requirements. Moreover, the large power cell connected to the PV system operates at fundamental switching frequency, further improving efficiency. However, the topology requires a bidirectional bypass switch connected to the large cell, in order to reduce the changing potential between the PV system and ground to reduce the possibility of leakage currents and enable transformerless operation. The main advantage of this converter is the high power quality and efficiency. On the other hand, the main drawbacks are the asymmetry in the blocking voltages among devices and the more complex control scheme, which includes voltage balance

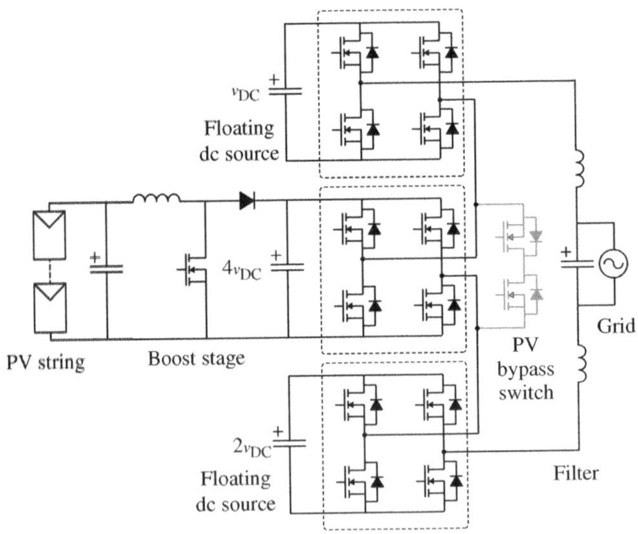

Figure 7.16 String inverter based on cascaded H-bridge with unequal DC-link voltages

Table 7.5 Examples of commercial string inverters

Parameter	Sunways NT 5000	Danfoss DLX 4.6	ABB PVS 300 TL8000	Mitsubishi PV-PN40G
Maximum input voltage (V)	900	600	900	380
Rated AC power (kW)	4.6	4.6	8	4
Grid connection	Single phase	Single phase	Single phase	Single phase
Topology	HERIC		HNPC	1:2:4-CHB
Efficiency (%)	97.8	97.3	97	97.5
Isolation	Transformerless	HF transformer	Transformerless	Transformerless
Independent MPPT	1 (2 strings)	1 (3 strings)	1 (4 strings)	1
Photo	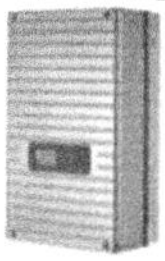			

of floating capacitors. Table 7.5 shows some examples of commercial string inverters discussed in this section and their main characteristics.

7.3.3 Multi-string Configuration

Central and string inverter configurations interface a PV array and a single PV string to the grid, respectively. Hence, only one MPPT is possible for the respective array or string. In order to introduce more flexibility in the PV system design, while mitigating the effects of partial shading and module mismatch, the multi-string concept was introduced [8]. The multi-string configuration is basically a two-stage central configuration with the peculiarity that it has more than one DC–DC stage (usually from 2 to 4). Hence, it combines the advantages of string inverter (high-energy yield by individual MPPT) and central inverter (low cost).

Each DC–DC converter is used to connect a PV array or string to a central inverter that concentrates all DC–DC stages into one grid connection. This configuration decouples the PV system from the inverter DC link, enabling a robust grid-tied converter control and extends the PV input voltage operating range. The flexibility of having a decoupled PV system from the grid connection enables the integration of even different module technologies (e.g., crystalline or thin film) and orientations (e.g., south, east and west) in the same PV plant. The modular nature of this inverter makes it more flexible and more robust since the failure of one string does not affect the entire PV plant.

Multi-string inverters can be found for single- and three-phase grid connections and can range from small systems of a few kilowatts to medium systems of several tens of kilowatts. Examples of commercial multi-string inverters and their main characteristics are given in Table 7.6.

The main difference between the multi string configuration and the central and string configuration is the multiple DC–DC stages connected in parallel to the DC link. Hence, all inverter topologies seen in the previous sections can be potentially used as multi-string inverters by connecting two or more DC–DC stages to the DC link and therefore they will not be analyzed again.

One of the first multi-string inverters introduced in practice was the half-bridge inverter with boost converters in the DC–DC stage introduced by SMA [8]. Others that have followed include the H-bridge, the H5 converter, the three-phase two-level VSI, the 3L-NPC and the three-phase three-level T-type converter (3L-T) [15]. Figure 7.17 shows some examples of practical multi-string configurations.

The most common DC–DC stages used for multi-string configurations are the boost converter (as shown in Figure 7.17(a) and (b)) and the HF-isolated DC–DC switch mode converter based on an H-bridge, HF transformer and diode rectifier (shown in Figure 7.17(c)). The boost is very simple in

Table 7.6 Examples of commercial multi-string inverters

Parameter	SMA Sunny Boy 5000TL	Danfoss TLX 15	SATCON Solstice
Nominal input voltage (V)	750	700	600
Maximum input power	5250 W	15 kW	100 kW
Grid connection	Single phase	Three phase	Three phase
Maximum input current (A)	15	3 × 12	182
Maximum efficiency (%)	97	98	96.7
Isolation	Transformerless	Transformerless	LF transformer
Independent MPPT	2 (2 strings each)	3	4–6 (9 strings each)
Photo			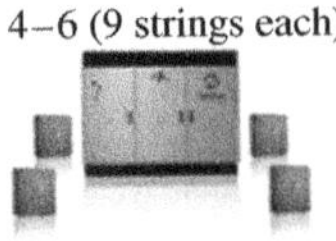

structure, design and control and is capable of elevating voltage and controlling the MPPT. The boost converter can be found implemented with the inductor and diode in the positive DC bar as well as in the return path in the negative DC bar, depending on the PV module technology. In particular, thin-film PV modules benefit more from having the inductor and diode in the negative bar [15].

On the other hand, the HF-isolated DC–DC converter offers additional input–output voltage flexibility, thanks to the transformer turns ratio between the primary and secondary sides. Nevertheless, it has a high component count compared to that of boost converters and reduced efficiency because of the high switching frequencies and transformer. However, since isolation is required in some countries, this is a viable option with efficiencies comparable to that of LF transformers but with higher power density. The development of new, faster and more efficient semiconductor devices in the future will make this DC–DC converter more competitive compared to that of other DC–DC converter stages.

7.3.4 *AC-Module Configuration*

The AC-module configuration uses a dedicated grid-tied converter for each PV module of the system. Therefore, this configuration is also known as *module-integrated inverter* and *micro-inverter* because of the small size and low power rating of the converter. The LV rating of PV modules (generally around 30 V) requires voltage elevation for grid connection. This can be achieved by adding a DC–DC boost stage before the micro-inverter or by using a line frequency step-up transformer at the grid side. The latter is not a practical solution, since the size and cost of LF transformers for a single PV module, with a power rating ranging from 150 to 250 W, is not competitive compared to that of other configurations. Therefore, all AC-module configurations that are commercially available have a DC–DC stage prior to the micro-inverter. If isolation is required, HF transformers are included in the DC–DC stage, which is the case in most commercial AC-module topologies.

Figure 7.18 shows one of the most commonly used AC-module configurations found in the literature and in practice: the flyback/H-bridge topology. A practical version of this topology was developed by Enphase Energy [23], currently commercialized by Siemens. The flyback converter performs MPPT and voltage elevation and provides galvanic isolation, while the H-bridge inverter controls the DC-link voltage (output voltage of the flyback), grid synchronization and active/reactive power control. The topology shown in Figure 7.18 features interleaved flyback converters connected in parallel at the input and output, which are controlled by phase-shifted carriers. This concept divides the power among the different flyback converters, enabling higher switching frequencies of the flyback semiconductor due to the lower current, reducing the rating of the device while enabling the use of smaller HF

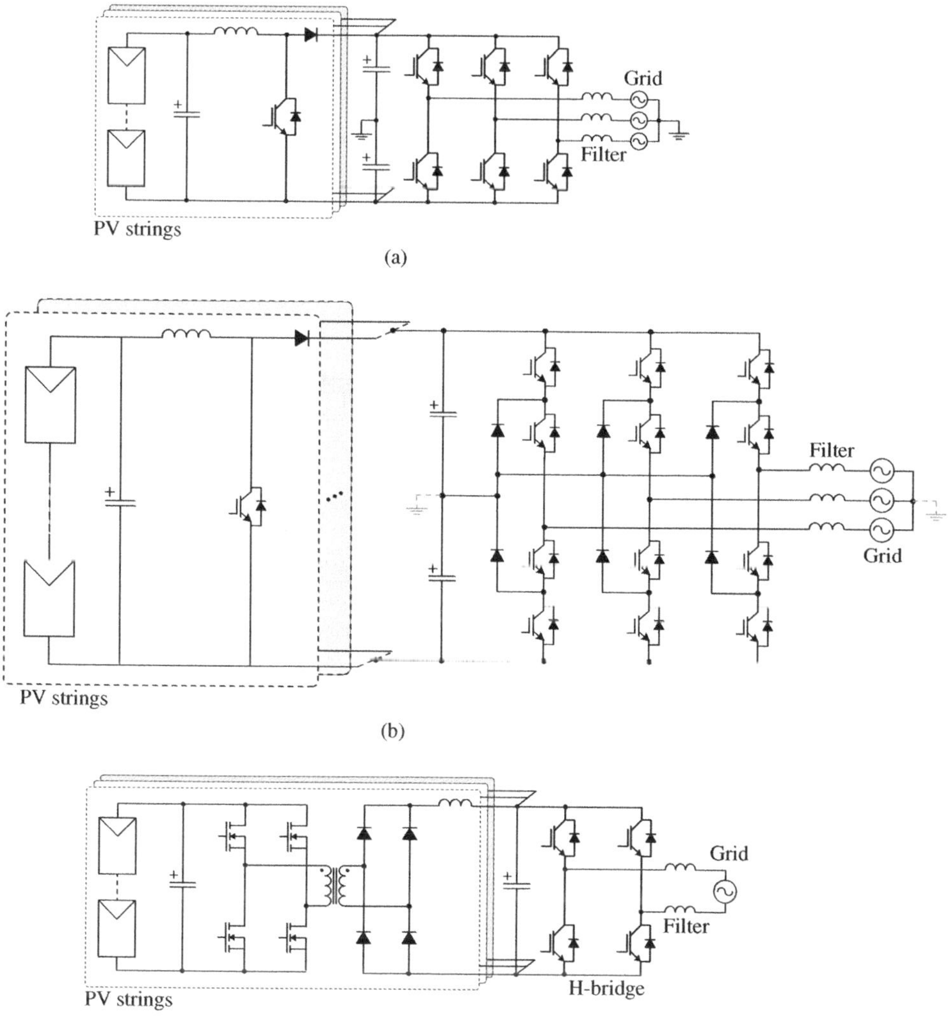

Figure 7.17 Examples of multi-string topologies: (a) three-phase 2L-VSI with boost proposed by Steca, (b) three-phase 3L-NPC with boost used by Danfoss and (c) H-bridge inverter with high-frequency DC–DC stage used by Fronius and Danfoss

transformers. This configuration not only reduces the size of the DC–DC stage but, more importantly, allows a reduction in the current ripple at both input and output because of the phase-shifted carrier modulation, and thereby extending the lifespan of the capacitors. Since capacitors are the most unreliable component of the circuit, the use of an interleaved DC–DC stage improves the lifecycle of the whole converter, making it possible to extend the manufacturer's warranty up to the lifetime of PV modules (usually 25 years). The small size of the converter allows a very compact enclosure design that can be attached to the back of each PV module hence the name "module-integrated inverter."

Another commercial AC-module integrated converter, shown in Figure 7.19, includes a resonant H-bridge stage with an HF isolation transformer and a diode bridge rectifier as the DC–DC converter

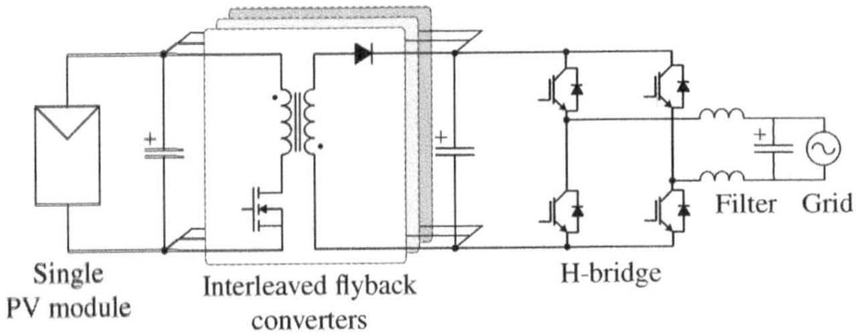

Figure 7.18 AC-module PV system based on interleaved flyback DC–DC converter and H-bridge inverter

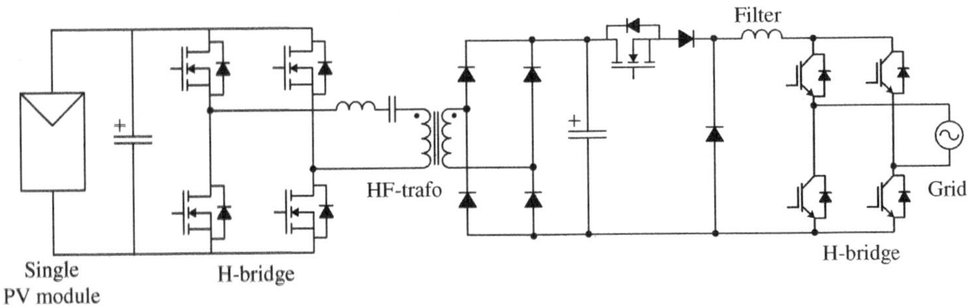

Figure 7.19 AC-module PV system based on resonant H-bridge HF DC–DC converter and H-bridge inverter

Table 7.7 Examples of commercial AC-module inverters

Parameter	Power One Aurora MICRO-0.3-I	Siemens Micro-inverter System	Enecsys Single micro-inverter
Maximum input voltage (V)	60	45	44
Input power (W)	200 and 300	190–260	240
AC connection	Single phase	Single phase, three phase	Single phase
Maximum input current (A)	10.5	10.5	12
Peak efficiency (%)	96.5	96.3	95
Isolation	HF H-bridge DC–DC	HF flyback DC–DC	HF H-bridge DC–DC
Independent MPPT	1 per module	1 per module	1 per module
Photo			

instead of the flyback. The H-bridge has better power conversion properties compared to that of the flyback (reduced ripple and higher power capacity). Therefore, interleaved operation is not necessary, although it requires more devices at both sides of the HF transformer. This converter has been developed by Enecsys [24]. An alternative topology patented by the same manufacturer includes a DC–DC stage with a half-bridge and two capacitors before the HF transformer instead of the H-bridge.

Generally, MOSFET devices are used in the DC–DC conversion stage because of the higher switching frequencies required by the converter and the LV rating, while IGBT devices are used in the grid-tied inverter circuit. Examples of AC-module integrated inverter configurations are given in Table 7.7 along with the most important parameters.

Although the concept of module-integrated inverters has been around for decades, this configuration has not been widely adopted even for small- or medium-scale PV plants. Among the reasons are that the cost per watt (and even volume per watt) of distributing the power electronics and the control system among all modules is higher (more sensors, control platforms, gate drives, semiconductors, etc.). Furthermore, despite having the best MPPT capability of all configurations (one module, one MPPT), it has lower power converter efficiency because of the high switching frequency and larger voltage elevation ratio required for grid connection. Nevertheless, this concept might benefit from new, faster and more efficient semiconductor devices (SiC and GaN) and gain more importance in the future.

7.4 Control of Grid-Connected PV Systems

According to the previous sections, grid-connected PV systems can be composed of one or two power conversion stages (DC–DC and DC–AC) and different converter topologies. Therefore, the control system and modulation schemes depend on the topology and vary from one topology to another. Nevertheless, the main control objectives are universal and include MPPT of the PV system, grid synchronization, DC-link voltage control, active and reactive power control and grid monitoring including anti-islanding detection.

7.4.1 Maximum Power Point Tracking Control Methods

Following the analysis of the power curves of PV modules presented in Section 7.2, it is clear that to maximize the energy output of a PV system, it should be operating around the MPP. The MPP depends on the solar irradiation and temperature of the module, which are variable in time. The MPP is achieved by controlling the load curve of the PV module or string, making it to intersect the I–V curve where $i_{PV} \times v_{PV}$ is maximal. As defined in Section 7.2, the voltage at which this occurs is known as the *maximum power voltage* v_{mp} which should be considered as the reference voltage for the control system $v_{PV}^* = v_{mp}$. This voltage could be obtained through direct measurement of the solar irradiation and the module temperature, which are used to compute v_{mp} from the PV module physical model. However, solar irradiation sensors (pyranometers), in particular, are very expensive. Therefore, numerical methods based on electrical variable measurements (already needed for the converter control) are used to search for the MPP. Since this task is performed online and updated periodically over time, they are known as *MPP tracking (MPPT) methods*.

There are several MPPT methods found in practice and in the literature that can reach up to 99% of the MPP efficiency [25, 26]. A classification of the most common MPPT methods is given in Figure 7.20. Hill-climbing and fractional methods are widely adopted in commercial PV systems.

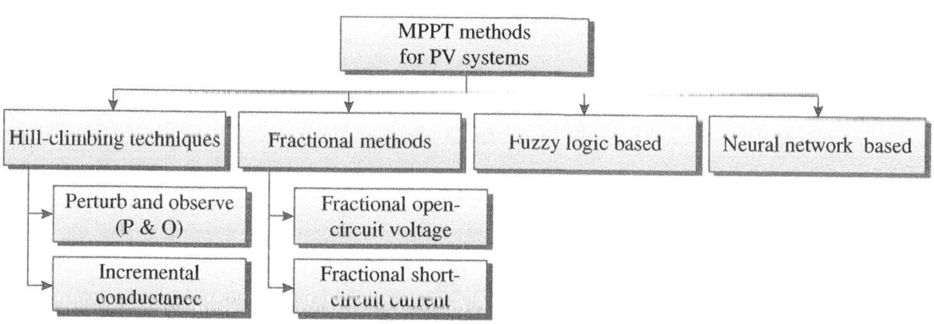

Figure 7.20 MPPT algorithm classification

7.4.1.1 Perturb and Observe Algorithm

Hill-climbing algorithms are the most popular methods used in practice. Their popularity is due to the implementation simplicity and effective tracking of the MPP. Perturb and Observe (P&O) is a type of hill-climbing method consisting of a continuous reference voltage search process to reach the MPP [27]. The search is performed by perturbing the reference voltage and then measuring the system response (observing) to determine the direction of the next perturbation. The reference voltage perturbations are performed in the direction in which the power should increase.

Increasing the reference voltage v^*_{PV} of the PV system when it operates at the left side of the MPP would increase the power output of the system (refer to Figure 7.4 for details), whereas such perturbation would decrease the power output if the system was operating at the right side of the MPP of the $P-V$ curve for a given solar irradiation and temperature. The opposite reactions are obtained when decreasing the reference voltage on both sides of the MPP. Therefore, the MPP can be tracked by introducing a perturbation in the reference voltage (Δv). The perturbation in the reference voltage is always of the same magnitude while the sign is determined by the power variation. If the power has increased from the last sample, the sign of the perturbation is maintained. If the power has decreased, the sign is reversed. Figure 7.21 shows a diagram of the algorithm implementation and execution.

This process is repeated until the system reaches the MPP and oscillates around it, locked in three voltage references in steady state, as can be seen in Figure 7.22. When a change occurs in solar irradiation or temperature, the system will change to a different $P-V$ curve, which will be detected by the power calculation and compared to that of a previous value and will introduce the corresponding perturbation to the voltage reference. This is shown in Figure 7.22, where a step in temperature from 50 to 25 °C has been introduced at Δt_5. Note how the voltage reference drifts to a new set of three voltage levels around the new MPP.

The magnitude of the oscillation around the MPP depends on the magnitude of the perturbation $|\Delta v|$. The magnitude of the perturbation is designed to be at least as large as the voltage ripple in the input capacitor of the converter (or the output of the PV array). Otherwise, the ripple in the voltage will lead to poor power calculations, which may result in inadequate directions for the

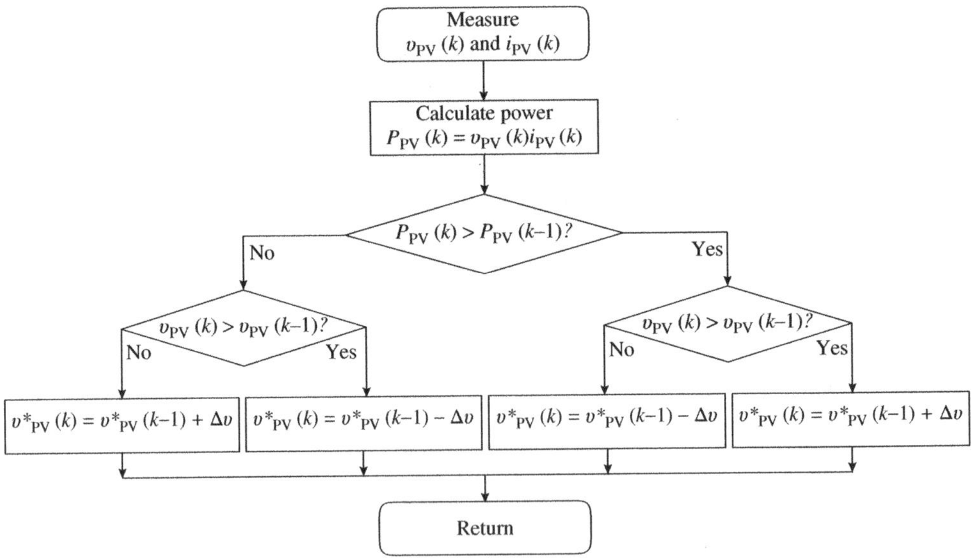

Figure 7.21 P&O algorithm

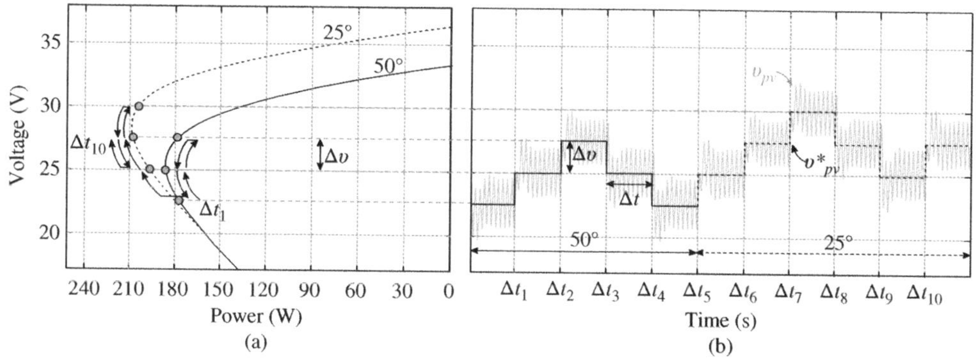

Figure 7.22 P&O under step change in temperature: (a) operating principle in the $P-V$ curve (rotated for common voltage axis) and (b) voltage reference and voltage output of PV system

perturbation and prevent the system from locking in the three voltage levels, which is around the MPP in steady state.

The other parameter that needs to be designed in P&O algorithms is the time interval in which the reference is updated Δt. This time period needs to be large enough to allow the power converter controller to reach the steady state around the input voltage reference before a new perturbation is introduced.

One of the drawbacks of this method is that it fails under rapid variation in solar irradiance and other environmental conditions. This happens when the change in power because of operating conditions is larger than the changes because of the perturbation produced by the algorithm. This condition may result in shifting the operating point in the opposite direction. This problem can be avoided by estimating the variation in power caused by the change of operating conditions separately and then perturbing the system to decide the direction of the next perturbation. The algorithm is also sensitive to measurement noise and to large ripples in the PV output voltage. The latter occurs for low capacitances at the PV output, causing a large oscillation around the MPP. Since the $P-V$ curve falls very steeply on the left side of the MPP, large oscillations in the PV voltage may produce large drops in the generated power.

7.4.1.2 Incremental Conductance Algorithm

This method is similar to the P&O algorithm and was proposed for rapidly changing atmospheric conditions [25, 26]. This method is derived from the power curves of Figure 7.4, where the slope of the curve is positive on the left side, negative on the right side and zero at the peak. The slope of the power curve can be found from

$$\frac{dP_{PV}}{dv_{PV}} = \frac{d(i_{PV}v_{PV})}{dv_{PV}} = i_{PV} + v_{PV}\frac{di_{PV}}{dv_{PV}} \tag{7.2}$$

The maximum power is then obtained when the power derivative is zero; hence

$$\frac{dP_{PV}}{dv_{PV}} = 0 \Rightarrow \frac{di_{PV}}{dv_{PV}} = -\frac{i_{PV}}{v_{PV}} \tag{7.3}$$

where i_{PV}/v_{PV} represents the conductance, while di_{PV}/dv_{PV} represents the incremental conductance. The derivative of the current with respect to the derivative of voltage can be approximated as the difference between the actual values and the previous instant values in that iteration process. Hence, by comparing

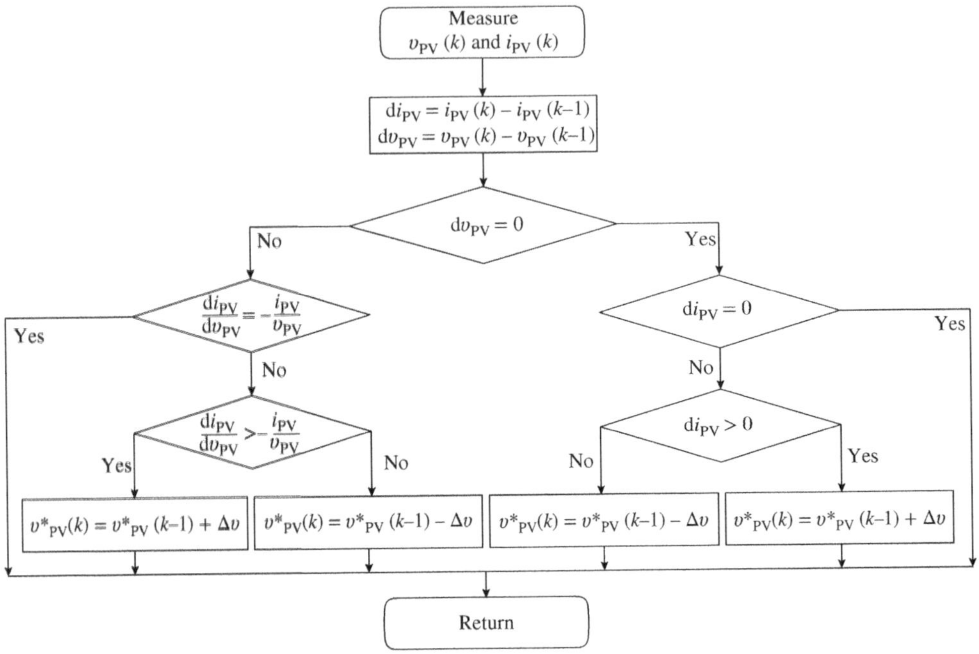

Figure 7.23 Incremental conductance MPPT algorithm

the conductance i_{PV}/v_{PV} to the incremental conductance di_{PV}/dv_{PV} as shown in Figure 7.23, the algorithm can track the MPP and stay there until a change of di_{PV} or dv_{PV} occurs as a result of a change in atmo spheric condition.

7.4.1.3 Fractional Open-Circuit Voltage Method

The $I-V$ curves of a PV module, string, or array, such as the one shown in Figure 7.4, suggest a linear relation between the open-circuit voltage (v_{oc}) and the MPP voltage (v_{mp}) for different irradiance and temperature conditions. This linear relation can be described by

$$v_{mp} = k_1 \, v_{oc} \quad \text{for } k_1 < 1 \tag{7.4}$$

The proportionality constant k_1 depends on the characteristics of the PV system and is determined by measuring the v_{mp} and v_{oc} under different irradiance and temperature conditions. Although the constant k_1 differs according to the type of the PV system, the range of k_1 is reported in the literature to be between 0.71 and 0.8. In order to find the MPP using Equation 7.4, the value of v_{oc} must be measured for a given irradiance and temperature. This can be achieved online by shutting down the power converter – either the DC–DC or the DC–AC stage depending on the configuration connected to the PV system – to simulate no load conditions, in order to measure v_{oc}. In order to track the MPP under variable conditions, this process is repeated periodically every Δt (usually around 15 ms).

This method is simple and straightforward to implement [26]. However, there is an error of approximation in the linearity of PV modules and the frequent shutdown of the system necessary to measure v_{oc} causes power losses. The latter issue can be avoided by using a separate PV cell to measure v_{oc}. This solution comes at an additional cost and mismatch between the characteristics of the reference PV cell and the actual PV system can cause a steady-state error in the MPPT.

7.4.1.4 Fractional Short-Circuit Current Method

This method is the current counterpart to the open-circuit voltage method [26]. Instead, the MPP current i_{mp} is computed through the linear relation with the short-circuit current i_{sc} under different environmental conditions. This can be expressed by

$$i_{mp} = k_2\, i_{sc} \quad \text{for } k_2 < 1 \tag{7.5}$$

The proportionality constant k_2 depends on the characteristics of the PV system and has been reported to be typically between 0.78 and 0.92. Measuring the short-circuit current i_{sc} online is more complex than measuring the open-circuit voltage and may require additional components in the power circuit such as a bypass switch. As with the open-circuit voltage method, this method produces additional losses during the measurement of i_{sc}.

7.4.2 DC–DC Stage Converter Control

According to Section 7.2, some string configurations all multi-string configurations and most AC-module integrated configurations include a DC–DC stage between the PV system and the grid-tied inverter. The main function of the DC–DC stage is to introduce a decoupling between the two DC voltages, hence decoupling the MPP voltage reference at the output of the PV system from the DC-link voltage reference at the input of the grid-tied inverter. Additional functions of the DC–DC stage include voltage elevation to adapt LV levels from the PV system to DC-link voltage requirements for grid connection and also galvanic isolation for those DC–DC converters with HF transformers.

The control scheme may vary from one DC–DC converter topology to another and also be different depending on the control method (linear, hysteresis, fuzzy, etc.). Nevertheless, most control structures found in practice are based on cascaded linear control loops, with a slower outer DC-voltage control loop and a faster inner current control loop. As an example, without loss of generality, the boost converter control will be analyzed, since it is the most representative of all DC–DC converters. Only slight modifications in the modulation stage are required for the other DC–DC converters.

Figure 7.24 shows a block diagram of a classic control system used to perform MPPT of the PV system with a boost converter. Any voltage-based MPPT algorithm may be used to generate the outer loop reference voltage v^*_{PV}. The PV system current i_{PV} and voltage v_{PV} measurements are required for the MPPT algorithm (e.g., P&O). The outer voltage loop controls the input capacitor voltage of the boost converter, that is, the PV system output voltage. Therefore, the reference for this loop is given by the MPPT algorithm. The voltage error is controlled using a linear proportional-integral (PI) controller whose output is the current through the input capacitor i_c. As the DC–DC converter does not directly control the capacitor current, a feedforward compensation can be included to compute the DC–DC converter current reference; in the case of the boost converter, this is the inductor current i_L.

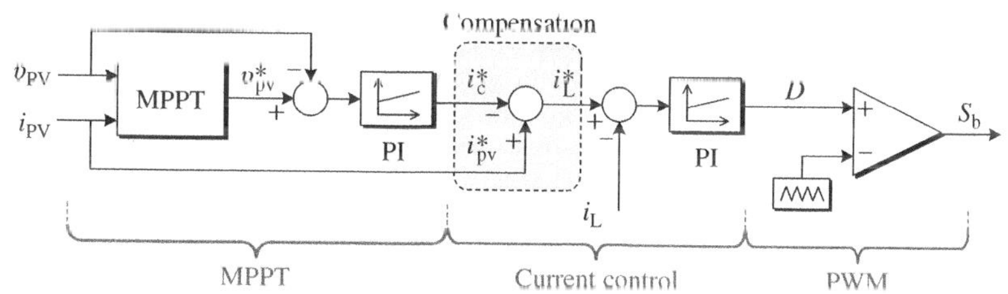

Figure 7.24 DC–DC converter stage control block diagram (adjusted here for a boost converter)

The boost inductor current i_L is measured for feedback to compute the current error, which is also controlled with a PI regulator (hysteresis controllers may be used here but introduce variable switching frequency). The output of the current control loop is the duty cycle reference of the boost converter, which is modulated using carrier-based PWM. The resulting gate signal (S_b) is used to control the boost semiconductor.

Note that a necessary condition, by design, for the proper operation of the boost converter and to effectively control the boost inductor current is that the output voltage is greater than the input voltage ($v_{DC} > v_{PV}$), which is the case for most DC–DC stages used in PV systems. The output voltage of the boost converter v_{DC} is controlled by the grid-tied inverter and can be considered as a fixed voltage source for the boost control purposes.

The boost converter control system essentially adjusts the duty cycle to achieve the desired input–output voltage ratio imposed by the MPPT algorithm reference and the DC-link voltage reference. In the steady state, the boost controller should settle the duty cycle of the boost converter D, fulfilling the voltage conversion condition

$$\frac{v_{DC}}{v_{PV}} = \frac{1}{1-D} \tag{7.6}$$

7.4.3 Grid-Tied Converter Control

The main function of the grid-tied inverter is to provide grid synchronization and to control active and reactive power flow or, seen differently, to control the grid currents and the DC-link capacitor voltage. In PV configurations where no DC–DC stage is included, as in the central topology and some string inverters, the DC-link voltage control also performs the MPPT of the PV system.

Among the most commonly used control schemes for three-phase grid-tied inverters (also known in the motor drive industry as active front-end rectifiers) are voltage-oriented control (VOC) and direct power control (DPC) [28]. Nevertheless, in PV applications, most grid-tied inverters are controlled using VOC. As the name suggests, VOC uses a rotational dq reference frame transformation oriented with the grid voltage vector to transform all AC quantities to DC values to simplify the control system design and enable the use of PI controllers. The grid current in the dq frame can be decomposed into the real part on the d-axis (i_{sd}), which is aligned with the grid voltage vector and hence it is proportional to the active power P and the imaginary part on the q-axis (i_{sq}), which is perpendicular to the grid voltage vector and hence proportional to the reactive power Q.

VOC is based on a cascaded voltage and current control loop as shown in Figure 7.25. The outer voltage loop controls the DC-link voltage v_{DC} using a PI controller whose output is proportional to the active power and hence to the desired value of i_{sd}, while reactive power is controlled with i_{sq}. Both current control loops are regulated using PI controllers, whose outputs are the dq reference voltages. These are converted to phase values to be modulated by the inverter using PWM. Note that the q-axis current reference can be arbitrarily adjusted. Normally, i_{sq} is set to zero, although during voltage dips the grid operator might request the injection of reactive power to help support the grid, as it is currently the case in wind power plants.

For feedback purposes, the three phase currents (i_a, i_b, i_c) are measured and transformed to dq values. Grid synchronization is achieved by extracting the grid voltage vector angle θ_s, which is necessary in order to have the dq frame correctly aligned with the grid voltage vector. Usually a phase-locked loop (PLL) is considered for such purposes, for which the grid phase voltages are measured (v_a, v_b, v_c). Note that the dq current control loops shown in Figure 7.25 include feedforward crossed compensation components for decoupling between the dq current axes. This can be explained using the equivalent space vector circuits in the static and rotational frames as shown in Figure 7.26. From Figure 7.26(b), the grid-side voltage equation can be obtained:

$$\mathbf{v}_s = \mathbf{i}_s R_s + L_s \frac{d\mathbf{i}_s}{dt} + j\omega_s L_s \mathbf{i}_s + \mathbf{v}_c \tag{7.7}$$

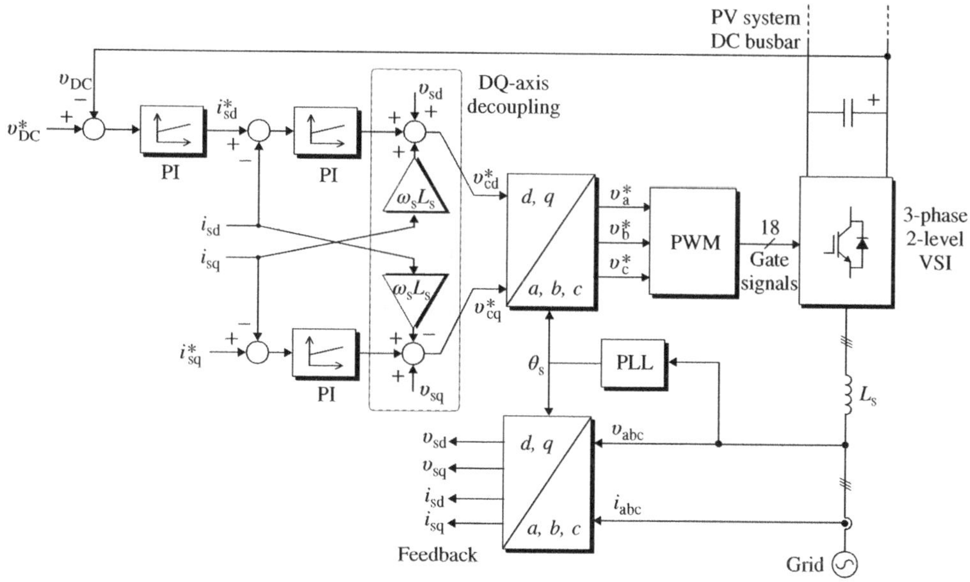

Figure 7.25 Voltage-oriented control (VOC) for three-phase grid-tied PV inverters

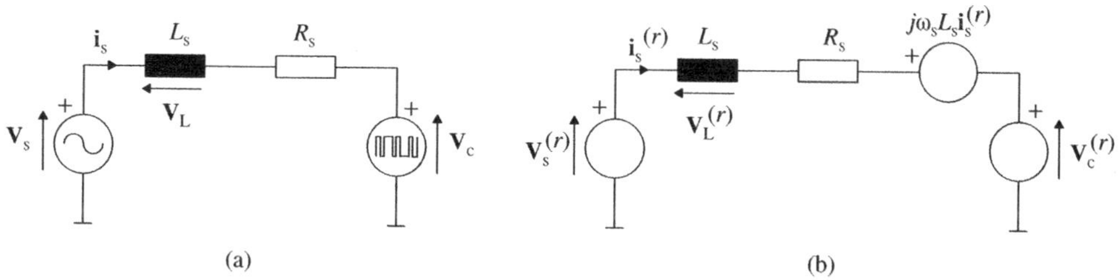

Figure 7.26 Space vector equivalent circuit of grid-tied converter: (a) stationary frame and (b) synchronous rotating frame

Then, decomposing Equation 7.7 into the *d*- and *q*-components yields

$$
\begin{aligned}
v_{sd} &= i_{ds}R_s + L_s\frac{di_{sd}}{dt} - j\omega_s L_s i_{sq} + v_{cd} \\
v_{sq} &= i_{dq}R_s + L_s\frac{di_{sq}}{dt} + j\omega_s L_s i_{sd} + v_{cq} = 0
\end{aligned}
\tag{7.8}
$$

Note that v_{sq} is zero, since the *dq* frame is aligned with the voltage space vector v_s and hence there is no projection over the *q* axis. From Equation 7.8, it is clear there is a coupling between the *dq* currents. In order to reduce the effects of one variable on the other, the coupling terms can be fed forward since the parameters are known, and the involved variables have already been measured. Since the voltages to be modulated are the voltages generated by the converter (v_{cd} and v_{cq}), solving Equation 7.8 will deliver the decoupling term included in the block diagram of Figure 7.25. The active and reactive power can be computed by

$$P = \frac{3}{2}Re\{v_s i_s^*\} = \frac{3}{2}v_{sd}i_{sd}$$
$$Q = \frac{3}{2}\text{Im}\{v_s i_s^*\} = -\frac{3}{2}v_{sd}i_{sq} \tag{7.9}$$

Note that Equation 7.9 can be used to compute the q-axis current reference from the desired reactive power reference in the case it is not zero.

Finally, the only design constraint left is the minimum DC-link voltage v_{DC} required to enable the appropriate control of the grid currents. This limit can be deduced from Figure 7.27, where the first quadrant of the space vector diagram of the grid-tied converter system is shown. Considering Equation 7.7 in steady state and neglecting the drop in the resistance result in the space vector v_c given by

$$\mathbf{v}_c = \mathbf{v}_s - j\omega_s L_s \mathbf{i}_s \tag{7.10}$$

According to the vector diagram of Figure 7.27, the converter space vector must be included within the circumference of radius h for its generation to be feasible. The radius h depends on the converter DC-link voltage and it is equal to

$$h = \frac{v_{DC}}{3} \tag{7.11}$$

Replacing this limit ($h > |v_c|$) in Equation 7.10 and solving v_{DC} in the magnitude of the resulting vector yields

$$v_{DC} \geq \sqrt{3\left[\hat{v}_s^2 + \left(\omega_s L_s v \hat{i}_s\right)^2\right]} \tag{7.12}$$

The result of Equation 7.12 establishes that the DC-link voltage reference will depend on the grid voltage amplitude, grid frequency, grid filter inductance and rating of the converter given by the peak grid current. Note that in the case if no DC–DC stage is included in the PV system, this value will also set a limit for the MPP voltage of the PV array. Therefore, the number of series-connected PV modules in the string needs to be carefully designed in order to provide an MPP voltage above this limit even under the most adverse environmental conditions (minimum radiation and maximum temperature at which the plant operates).

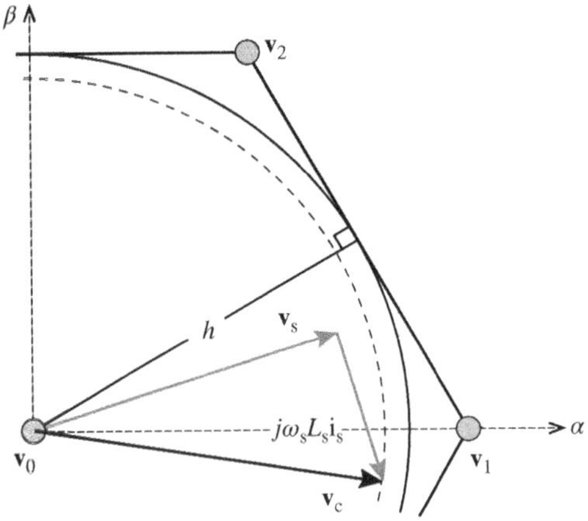

Figure 7.27 Space vector diagram of the grid-tied converter system

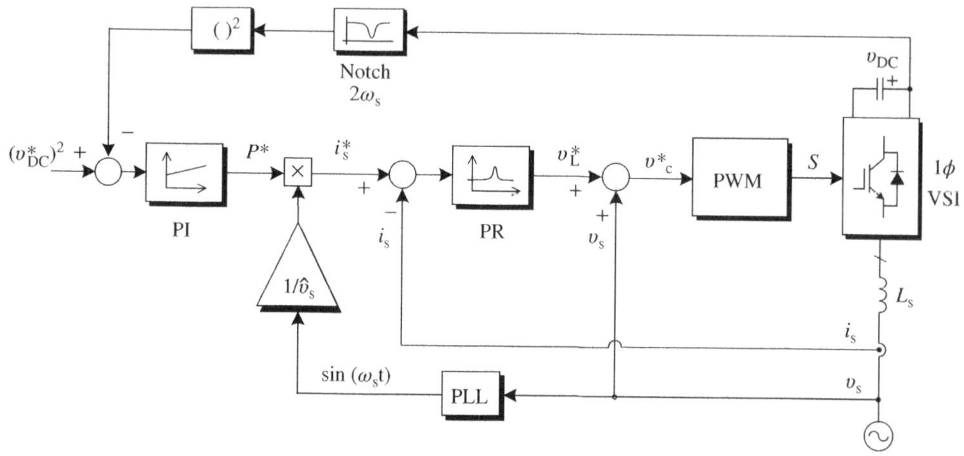

Figure 7.28 Voltage-oriented control (VOC) for single-phase grid-tied PV inverters

For single-phase grid-connected PV systems, such as some string, multi-string and AC-module integrated configurations, the previous VOC scheme in a dq rotating reference frame is not possible. Instead VOC is adapted by synchronizing the grid current reference directly with the single-phase grid voltage. The single-phase VOC block diagram is shown in Figure 7.28. The same cascaded control structure remains, in which the outer DC link voltage loop is controlled using a PI regulator [16]. The output of the voltage controller is the active power, which is divided by the grid voltage amplitude to obtain the grid current reference. In addition, a sine wave function reconstructed with a PLL to extract the angle of the grid voltage is multiplied to synchronize the reference current with the grid for unity power factor operation. The PLL is preferred instead of using the direct measurement of the grid voltage, in order to avoid the introduction of grid harmonics and other perturbations to the grid current reference.

Since the grid current reference is sinusoidal, PI controllers would generate steady-state error and phase displacement. Therefore, proportional-resonant controllers are used to tune to the grid frequency ω_s. The output of the gird current controller is the grid filter inductor voltage reference. Neglecting the voltage drop in the resistance and considering that the grid voltage is already measured, a feedforward compensation can be used to compute the converter voltage reference by

$$v_c^* = v_L^* + v_s \tag{7.13}$$

The converter voltage reference v_c^* is then modulated using the appropriate PWM strategy depending on the converter topology.

Single-phase grid-connected systems have a strong second harmonic ($2\omega_s$) in the DC-link voltage because of rectification. This harmonic must be filtered during feedback to prevent a third harmonic from appearing in the grid current reference because of frequency convolution when multiplying the active power (which would include the second harmonic) with the synchronization sine wave with fundamental frequency. This is achieved with a notch filter tuned to $2\omega_s$. If this step is not performed, the grid current will have an important third harmonic. Another way to compensate this is to greatly reduce the outer voltage loop bandwidth, so the second harmonic is rejected. However, this would degrade the overall performance of the system.

7.4.4 *Anti-islanding Detection*

According to IEEE standard 1547-2008 [29], a power system "island" is defined as "a condition in which a portion of an area of the electric power system is energized solely by one or more local electric power

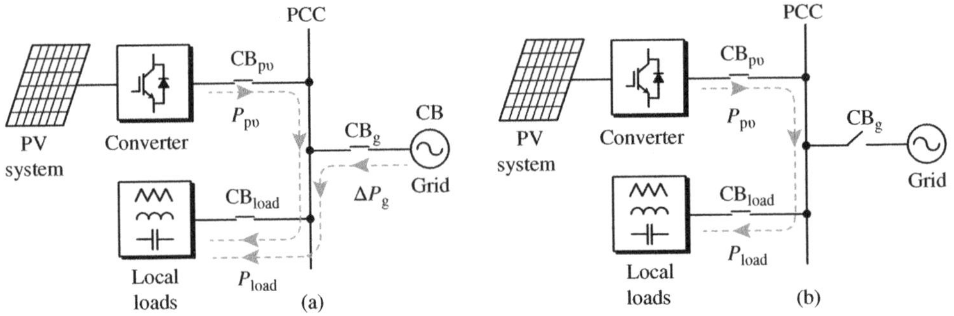

Figure 7.29 Grid-connected PV distributed generation system with local loads: (a) normal operation and (b) island operation

systems through the associated PCC while that portion of the area of the electric power system is electrically separated from the rest." Grid-connected PV systems can become islanded if they are disconnected from the grid, for example, during a grid fault. If this situation is not detected, the PV system will continue generating power, feeding the local loads. This situation is more likely to occur in distributed generation power systems such as the one shown in Figure 7.29.

Islanding operation can pose a serious risk to utility workers and maintenance personnel, since the lines remain energized [30]. It can also damage local loads as the voltage and frequency at PCC may drift from rated values. In addition, once islanding operation is over and the grid is reconnected, the PV system inverter will very likely be out of synchronization, causing damage to the converter and local loads because of the phase difference between the grid and inverter voltage. For this reason, grid connection standards of distributed generation systems mandate fast disconnection and cease of generation when island condition is detected.

Islanding detection methods can be classified into three categories, as shown in Figure 7.30, depending on the location in which the detection is performed: by the power converter (converter-resident), by additions to the power system (external impedance insertion), or by the grid operator (grid-resident) [31, 32].

Converter-resident methods can be further divided into passive and active methods. Passive methods use grid parameters and measurements (voltage, frequency, harmonic content, etc.) to detect islanding operation by comparing the measurements with boundary limits. These limits define the non-detection zone (NDZ), in which the islanding is not detected, particularly when the local loads have similar power capacity to the PV system, and all the generated power is consumed locally, producing only slight variations in voltages and currents at PCC when islanding occurs. Therefore, passive methods have a

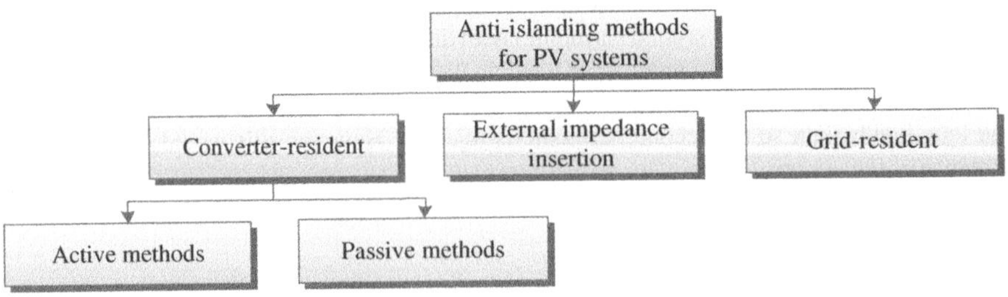

Figure 7.30 Anti-islanding detection methods for grid-connected PV systems

large NDZ. Nevertheless, passive methods are conceptually simple, incur little or no additional cost, are easy to implement and do not introduce any change in the power quality of the system. These methods are used in small- and medium-scale PV converters. Typical passive methods are under- and overvoltage detection, under- and overfrequency detection and voltage phase jump detection.

Note that according to the definitions of power flow in the distributed generation system in Figure 7.29(a), during normal operation, the following holds

$$P_{\text{load}} = P_{\text{PV}} + \Delta P_g$$
$$Q_{\text{load}} = Q_{\text{PV}} + \Delta Q_g \tag{7.14}$$

Active power is only related to voltage; hence during island operation, according to Figure 7.29(b), the voltage at the PCC should change to

$$\tilde{v}_{\text{pcc}} = v_{\text{pcc}}\sqrt{\frac{P_{\text{PV}}}{P_{\text{load}}}} \tag{7.15}$$

However, if the power consumed by the local loads is similar to that of the power generated by the PV system, then $P_{\text{PV}} \approx P_{\text{load}}$, which substituted in Equation 7.15 shows that there will be little or no variations in the PCC voltage and remain within the NDZ.

Active converter-resident methods were developed to reduce the NDZ of passive methods. This is achieved by injecting perturbations to the system, hence the name "active," to accelerate the drift of variables or drive the system toward the boundary limits. Usually, the perturbation is injected to the grid current reference waveform. One of these methods is active frequency drift (AFD) [33], in which a phase perturbation is injected to the current reference waveform, which causes the inverter frequency to drift in case of islanding operation, which does not happen when the grid is available. Then, the induced frequency drift can then easily be detected with some boundary limits. Figure 7.31 shows the injected perturbation and the resulting current reference waveform. As can be observed, the smaller NDZ obtained

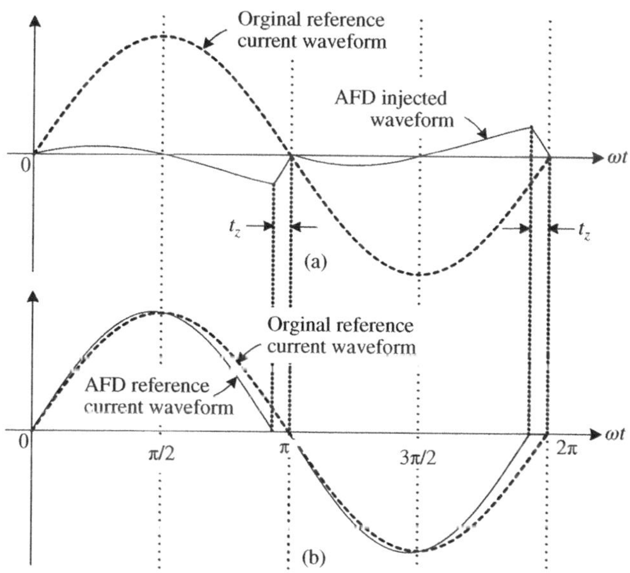

Figure 7.31 Active frequency drift (AFD) anti-islanding detection method: (a) AFD injected waveform and (b) modified current reference waveform [30]

with AFD comes at the expense of increased THD, which degrades the power quality provided by the grid-tied converter compared to that of the passive methods. Active converter-resident methods are also mainly used in small- to medium-scale PV converters.

External impedance insertion can be implemented by connecting a low-value impedance, usually a capacitor bank, at the utility side of the PCC when the grid is disconnected, which creates a step change in the phase between the current and the voltage of the converter at the PCC. This phase jump is then easily detected by passive island detection methods resident in the converter. However, the implementation of this method requires additional switching equipment and capacitor banks, making it less attractive for small PV systems.

Grid-resident methods are based on direct communication between the grid operator and the power converter such as PLCC (power line carrier communication) or SCADA (supervisory control and data acquisition). The grid operator directly informs the converter of the island condition, triggering the cease of power generation by the PV system. These methods require the involvement of the utility and are therefore not suitable for small- and medium-scale PV systems. Nevertheless, they are the mainstream solution for large-scale PV converters, where island operation is more critical because of the larger hazard that can be produced by a larger PV system.

There are several other anti-islanding detection methods, many of which are proprietary technology. A review of mainstream anti-islanding detection methods for PV systems is given in [32].

7.5 Recent Developments in Multilevel Inverter-Based PV Systems

Multilevel converters can reach higher voltages and power ratings and operate with lower switching frequency (below 1 kHz) compared to that of 2L-VSC because of an inherent increase in the output voltage waveform quality. This has made them very popular over the last two decades for high-power MV motor drive applications such as fans, pumps, conveyors, traction and propulsion drives [34]. They have these qualities since they are able to generate a stepped voltage waveform that better approximates a desired reference (usually sinusoidal). This is achieved by arranging the power switches and capacitive DC sources in such a way that they can be interconnected, generating different voltage levels through a proper modulation. Currently there are a wide variety of multilevel converter topologies as shown in the classification of Figure 7.32, many of which are commercialized and found in industrial applications [34].

As shown in Figure 7.32, multilevel converters belong to the indirect conversion category, with a capacitive DC link, making them a voltage source type. The classification also includes direct AC–AC converters and current source converters, which are currently the main competitors of the multilevel voltage source converter technology: cyclo-converters load commutated inverters and PWM current source inverters. However, these power converters are developed mainly for high-power variable-speed motor drives and therefore they are not really suitable for PV power generation.

Although the multilevel converter concept has been around since the 1960s, the earliest multilevel converters to find industrial application were the NPC, CHB and flying capacitor (FC) converters. Currently, the three-level NPC and 7- to 13-level CHB converters are the most commonly found multilevel converters in practice. The voltage and power ratings of these converters go from 2.3 to 6.6 kV and from 1 to 40 MW, respectively. These converters are mainly used in pumps, fans, compressors, conveyors and rolling mill applications. Nevertheless, in the past few years, multilevel converters have expanded their industrial presence with several new topologies and industrial applications. Among them are the three-level active NPC (3L-ANPC), five-level H-bridge NPC (HNPC), five-level hybrid between FC and ANPC (5L-ANPC), three-level neutral point piloted (NPP) and modular multilevel converter (MMC) [34, 35]. New applications include train traction, ship propulsion, wind energy conversion,

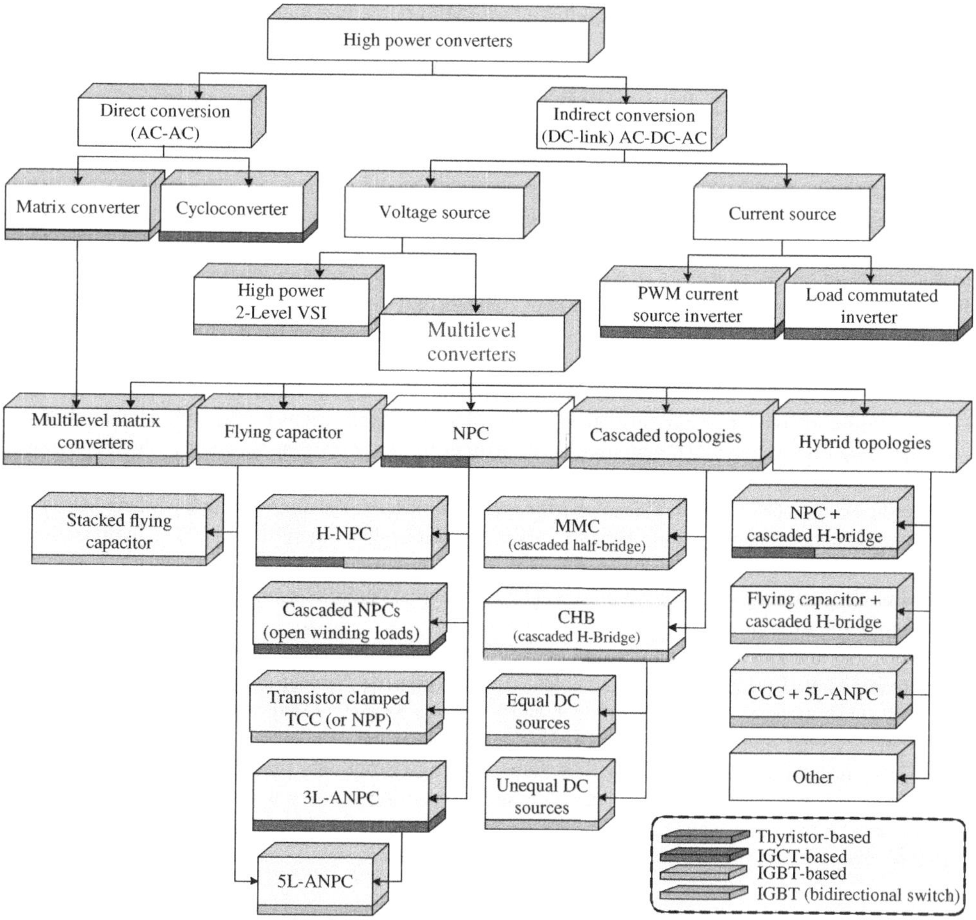

Figure 7.32 Multilevel converter topologies classification [34]

HVDC transmission and hydro-pumped storage, to name a few. Many of them are addressed in Ref. [35] and the references therein.

Because of the previously mentioned advantages, multilevel converters have also been proposed for interfacing PV systems to the grid [36], most of them based on the NPC [37–42] and CHB [43–51] topologies. In Refs. [37–41], NPC-based multilevel converter topologies are used to interface a single PV string per DC-link capacitor of the corresponding NPC topology. This approach does not need a DC–DC stage like the multi-string topology and effectively increases the voltage and enables MPPT of each string. However, it allows only a few string connections, which for a three-level NPC (3L-NPC) would be only two strings. This severely limits the size of the PV power plant managed by the multilevel converter, underutilizing the multilevel converter power capacity. In fact, commercial NPC converters are available from 0.8 to 44 MW for motor drive applications [34]. Thus, to fully use a 0.8 MW converter for PV power generation, several PV strings should be connected in parallel to both DC-link capacitors. In addition, the lack of a voltage boost stage would require the connection of a high number of modules in series in each string to reach the medium voltage (MV) range in which the NPC converter operates (3.3 and 6.6 kV).

The series connection of PV panels and parallel connection of PV strings introduce the same problem of the centralized topology. Panel mismatch and partial shading will reduce the power output. An additional problem is that the MPPT performed for each DC-link capacitor introduces an inherent DC-link imbalance (they have to operate at different voltages to ensure MPPT), which introduces distortion at the AC side. Furthermore, if the string output voltage reference given by the MPPT is low because of solar radiation and temperature conditions, the DC side could have a reduced voltage that affects or even disables the proper control of the AC side of the converter. This problem has been preliminarily addressed in Ref. [42], where an NPC multi-string configuration is presented, as shown in Figure 7.33. In this case, two DC buses are kept constant and separate the inverter from the MPPT side, which is performed by DC–DC converters. This allows an independent MPPT for each string without affecting the inverter and reduces the number of series-connected modules per string since voltage elevation is achieved by the two stacked DC buses and the DC–DC converters.

CHB multilevel converters have also been proposed for PV applications by exploiting a feature that is considered as a drawback when operating as an inverter in motor drive applications, which is the need for several isolated DC sources connected to each cell. However, in PV systems, the PV strings naturally provide the isolated DC sources. This allows the inclusion of several strings with independent MPPT and provides inherent voltage elevation because of series connection of the H-bridges. The main challenge in CHB-based PV systems is the inherent power imbalance among cells in each converter leg and the power imbalance among the different phases. Therefore, most contributions in the literature are centered on the single-phase solution.

In Ref. [49], the multilevel multi-string concept was applied to the CHB, as shown in Figure 7.34. In this case, each H-bridge cell provides a DC busbar for string connection through DC–DC converters. This concentrates a very large number of PV strings in a single converter up to several tens of megawatt. The series connection of the H-bridge cells elevates the voltage to 3.3 and 6.6 kV for three and six series-connected cells, respectively. The main challenge is not only that there is a power imbalance among cells of one phase, but that the three phases are also inherently unbalanced. The cell imbalance is dealt with in the same way as with the single-phase CHB PV system presented in Ref. [45], while

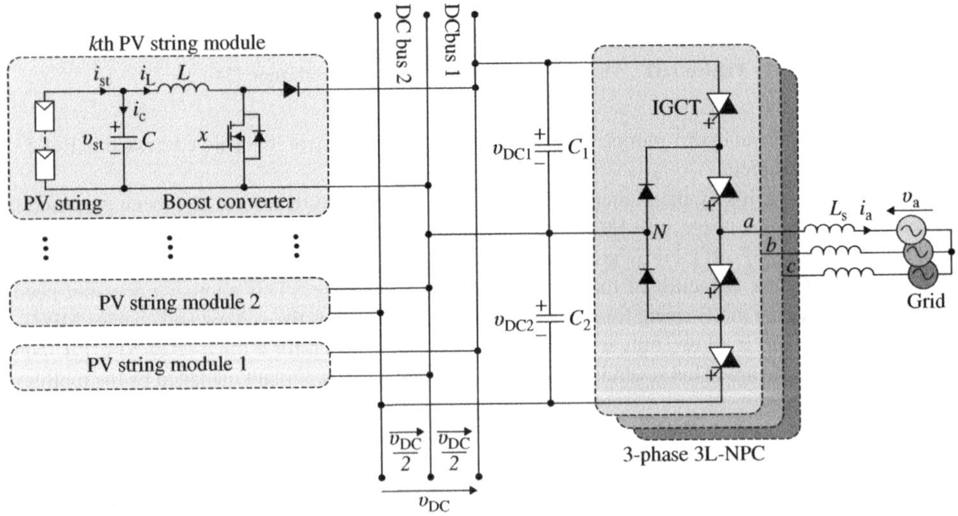

Figure 7.33 Three-phase NPC multi-string topology for multi-megawatt PV application [42]

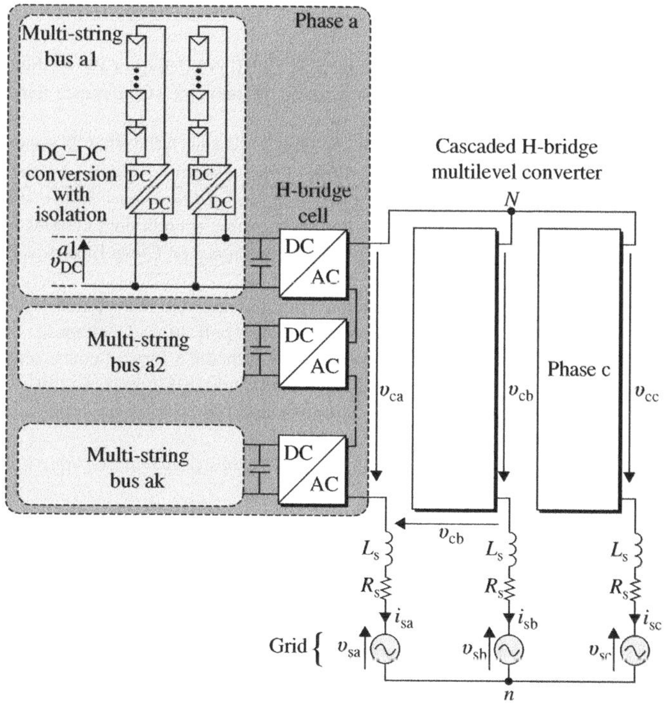

Figure 7.34 Three-phase CHB multi-string topology for multi-megawatt PV application [49]

the phase imbalance is solved by shifting the neutral of the converter voltage in such a way that the grid currents become balanced. This has been achieved in the modulation stage with a simple feedforward mechanism that compensates the voltage reference inversely to the imbalance so that the currents are balanced [49].

7.6 Summary

This chapter presents a comprehensive overview of grid-connected PV systems, including power curves, grid-connected configurations, different converter topologies (both single and three phases), control schemes, MPPT and anti-islanding detection methods. The focus of the chapter has been on the mainstream solutions available in the PV industry, in order to establish the current state of the art in PV converter technology. Some examples of commercial PV converters have been included for this purpose. In addition, some recently introduced concepts on multilevel converter-based PV systems for large scale PV plants have been discussed, along with trends, challenges and possible future scenarios of PV converter technology.

References

1. REN21 (2013) Renewable Energy Policy Network for the 21st Century, Renewables 2013 Global Status Report, http://www.ren21.net (accessed 18 December 2013).
2. Blaabjerg, F., Chen, Z., and Kjaer, S.B. (2004) Power electronics as efficient interface in dispersed power generation systems. *IEEE Transactions on Power Electronics*, **19** (5), 1184–1194.

3. Luque, A. and Hegedus, S. (2011) *Handbook of Photovoltaic Science and Engineering*, 2nd edn, John Wiley & Sons, Inc..
4. Kishor, N., Villalva, M.G., Mohanty, S.R., and Ruppert, E. (2010) Modeling of PV module with consideration of environmental factors. IEEE PES Innovative Smart Grid Technologies Conference Europe (ISGT Europe), October 2010.
5. Haberlin, H. (2012) *Photovoltaics System Design and Practice*, 1st edn, John Wiley & Sons, Ltd.
6. Carrasco, J., Franquelo, L., Bialasiewicz, J. *et al.* (2006) Power-electronic systems for the grid integration of renewable energy sources: a survey. *IEEE Transactions on Industrial Electronics*, **53** (4), 1002–1016.
7. Pavan, A.M., Castellan, S., Quaia, S. *et al.* (2007) Power electronic conditioning systems for industrial photovoltaic fields: centralized or string inverters? International Conference on Clean Electrical Power (ICCEP '07), May 2007, pp. 208–214.
8. Meinhardt, M. and Cramer, G. (2001) Multi-string-converter: the next step in evolution of string-converter technology. Proceedings of the 9th European Power Electronics and Applications Conference.
9. Carbone, R. and Tomaselli, A. (2011) Recent advances on AC PV-modules for grid-connected photovoltaic plants. International Conference Clean Electrical Power (ICCEP 2011), June 2011, pp. 124–129.
10. pvresources.com (2013) Large-Scale Photovoltaic Power Plants: Top 50 Ranking, http://www.pvresources.com/PVPowerPlants/Top50.aspx (accessed 18 December 2013).
11. First Solar Agua Caliente Solar Project, http://www.firstsolar.com/Projects/Agua-Caliente-Solar-Project (accessed 18 December 2013).
12. Wikipedia List of Photovoltaic Power Stations, http://en.wikipedia.org/wiki/List_of_photovoltaic_power_stations (accessed 18 December 2013).
13. First Solar Sarnia PV Power Plant, Ontario, Canada, http://www.firstsolar.com/en/Projects/Sarnia-Solar-Project (accessed 19 December 2013).
14. Kerekes, T., Teodorescu, R., and Liserre, M. (2008) Common mode voltage in case of transformerless PV inverters connected to the grid. IEEE International Symposium on Industrial Electronics (ISIE 2008), pp. 2390–2395.
15. Burger, B. and Kranzer, D. (2009) Extreme high efficiency PV-power converters. 13th Europe Conference Power Electronics and Applications (EPE '09), September 2009.
16. Teodorescu, R., Liserre, M., and Rodríguez, P. (2011) *Grid Converters for Photovoltaic and Wind Power Systems*, IEEE Press/John Wiley & Sons, Ltd.
17. Dreher, J.R., Marangoni, F., Schuch, L. *et al.* (2012) Comparison of H-bridge single-phase transformerless PV string inverters. IEEE/IAS International Conference Industry Applications (INDUSCON 2012), November 2012.
18. Kjaer, S., Pedersen, J., and Blaabjerg, F. (2005) A review of single-phase grid-connected inverters for photovoltaic modules. *IEEE Transactions on Industry Applications*, **41** (5), 1292–1306.
19. Gonzalez, R., Coloma, J., Marroyo, L. *et al.* (2009) Single-phase inverter circuit for conditioning and converting DC electrical energy into AC electrical energy. European Patent EP 2 053 730 A1.
20. Saridakis, S., Koutroulis, E., and Blaabjerg, F. (2013) Optimal design of modern transformerless PV inverter topologies. *IEEE Transactions on Energy Conversion*, **28** (2), 394–404.
21. Karraker, D.W., Gokhale, K.P., and Jussila, M.T. (2011) Inverter for solar cell array. US Patent 2011/0299312 A1.
22. Urakabe, T., Fujiwara, K., Kawakami, T., and Nishio, N. (2010) High efficiency power conditioner for photovoltaic power generation system. International Power Electronics Conference (IPEC 2010), June 2010, pp. 3236–3240.
23. Fornage, M. (2010) Method and apparatus for converting direct current to alternating current. US Patent 7,796,412B2.
24. Garrity, P. (2013) Solar photovoltaic power conditioning unit. US Patent 8,391,031B2.
25. Hohm, D.P. and Ropp, M.E. (2003) Comparative study of maximum power point tracking algorithms, *Progress in Photovoltaics: Research and Applications*, **11**, 1, 47–62.
26. Esram, T. and Chapman, P. (2007) Comparison of photovoltaic array maximum power point tracking techniques. *IEEE Transactions on Energy Conversion*, **22** (2), 439–449.
27. Femia, N., Granozio, D., Petrone, G. *et al.* (2007) Predictive and adaptive MPPT perturb and observe method. *IEEE Aerospace and Electronic Systems Magazine*, **43** (3), 934–950.
28. Malinowski, M., Kazmierkowski, M.P., and Trzynadlowski, A.M. (2003) A comparative study of control techniques for PWM rectifiers in AC adjustable speed drives. *IEEE Transactions on Power Electronics*, **18** (6), 1390–1396.
29. IEEE Standard (2003) 1547-2003. *Standard for Interconnecting Distributed Resources with Electric Power Systems*, IEEE.

30. Yafaoui, A., Wu, B., and Kouro, S. (2012) Improved active frequency drift anti-islanding detection method for grid connected photovoltaic systems. *IEEE Transactions on Power Electronics*, **27** (5), 2367–2375.
31. Singam, B. and Hui, L.Y. (2006) Assessing SMS and PJD schemes of anti-islanding with varying quality factor. IEEE International Power and Energy Conference (PECon '06).
32. Teoh, W.Y. and Tan, C.W. (2011) An overview of islanding detection methods in photovoltaic systems. *World Academy of Science, Engineering and Technology*, **58**, pp. 577–585.
33. Lopes, L. and Sun, H. (2006) Performance assessment of active frequency drifting islanding detection methods. *IEEE Transactions on Energy Conversion*, **21** (1), 171–180.
34. Kouro, S., Rodriguez, J., Wu, B. *et al.* (2012) Powering the future of industry: high-power adjustable speed drive topologies. *IEEE Industry Applications Magazine*, **18** (4), 26–39.
35. Kouro, S., Malinowski, M., Gopakumar, K. *et al.* (2010) Recent advances and industrial applications of multilevel converters. *IEEE Transactions on Industrial Electronics*, **57** (8), 2553–2580.
36. Calais, M. and Agelidis, V.G. (1998) Multilevel converters for single-phase grid connected photovoltaic systems-an overview. Proceedings IEEE International Symposium Industrial Electronics (ISIE '98), July 7–10, 1998, Vol. 1, pp. 224–229.
37. Busquets-Monge, S., Rocabert, J., Rodriguez, P. *et al.* (2008) Multilevel diode-clamped converter for photovoltaic generators with independent voltage control of each solar array. *IEEE Transactions on Industrial Electronics*, **55** (7), 2713–2723.
38. Gonzalez, R., Gubia, E., Lopez, J., and Marroyo, L. (2008) Transformerless single-phase multilevel-based photovoltaic inverter. *IEEE Transactions on Industrial Electronics*, **55** (7), 2694–2702.
39. Kerekes, T., Liserre, M., Teodorescu, R. *et al.* (2009) Evaluation of three-phase transformerless photovoltaic inverter topologies. *IEEE Transactions on Power Electronics*, **24** (9), 2202–2211.
40. Ozdemir, E., Ozdemir, S., and Tolbert, L.M. (2009) Fundamental-frequency modulated six-level diode-clamped multilevel inverter for three-phase stand-alone photovoltaic system. *IEEE Transactions on Industrial Electronics*, **56** (11), 4407–4415.
41. Ma, L., Jin, X., Kerekes, T. *et al.* (2009) The PWM strategies of grid-connected distributed generation active NPC inverters. IEEE Energy Conversion Congress and Exposition (ECCE), September 20–24, 2009, pp. 920–927.
42. Kouro, S., Asfaw, K., Goldman, R. *et al.* (2010) NPC multilevel multistring topology for large scale grid connected photovoltaic systems. 2nd IEEE International Symposium Power Electronics for Distributed Generation Systems (PEDG 2010), pp. 400–4005.
43. Alonso, O., Sanchis, P., Gubia, E., and Marroyo, L. (2003) Cascaded H-bridge multilevel converter for grid connected photovoltaic generators with independent maximum power point tracking of each solar array. IEEE 34th Annual Power Electronics Specialist Conference (PESC'03), June 2003, Vol. 2, pp. 731–735.
44. Villanueva, E., Correa, P., Rodriguez, J., and Pacas, M. (2009) Control of a single-phase cascaded H-bridge multilevel inverter for grid-connected photovoltaic systems. *IEEE Transactions on Industrial Electronics*, **56** (11), 4399–4406.
45. Kouro, S., Moya, A., Villanueva, E. *et al.* (2009) Control of a cascaded H-bridge multilevel converter for grid connection of photovoltaic systems. 35th Annual Conference IEEE Industrial Electronics Society (IECON09), pp. 1–7.
46. Negroni, J., Guinjoan, F., Meza, C. *et al.* (2006) Energy sampled data modeling of a cascade H-bridge multilevel converter for grid-connected PV systems. 10th IEEE International Power Electronics Congress, October 2006, pp. 1–6.
47. Cecati, C., Ciancetta, F., and Siano, P. (2010) A multilevel inverter for photovoltaic systems with fuzzy logic control. *IEEE Transactions on Industrial Electronics*, **57** (12), 4115–4125.
48. Brando, G., Dannier, A., and Rizzo, R. (2007) A sensorless control of H-bridge multilevel converter for maximum power point tracking in grid connected photovoltaic systems. International Conference on Clean Electrical Power (ICCEP '07), May 2007, pp. 789–794.
49. Rivera, S., Kouro, S., Leon, J.I. *et al.* (2011) Cascaded H-bridge multilevel converter multistring topology for large scale photovoltaic systems. 20th IEEE International Symposium Industrial Electronics (ISIE 2011), Gdansk, Poland, June 27–30, 2011.

50. Kouro, S., Fuentes, C., Perez, M., and Rodriguez, J. (2012) Single DC-link cascaded H-bridge multilevel multistring photovoltaic energy conversion system with inherent balanced operation. 38th Annual Conference IEEE Industrial Electronics Society (IECON 2012), Montreal, Canada, October 25–28, 2012.
51. Cortes, P., Kouro, S., Barrios, F., and Rodriguez, J. (2012) Predictive control of a single-phase cascaded H-bridge photovoltaic energy conversion system. 7th International Power Electronics and Motion Control Conference (IPEMC 2012), Harbin, China, June 2–5, 2012.

8

Controllability Analysis of Renewable Energy Systems

Hossein Karimi-Davijani and Olorunfemi Ojo
Department of Electrical and Computer Engineering/Center for Energy Systems Research, Tennessee Technological University, Tennessee, USA

8.1 Introduction

In recent years, there has been an accelerated development in renewable energy sources, such as photovoltaic (PV) and wind energy, to meet the increasing energy demands of growing populations. Most installed wind turbines are variable speed turbines, actuating doubly fed induction or synchronous generators in the kilowatt and megawatt ranges [1–3]. As many of the large wind turbines are located offshore, there is a need for robust machine and converter technologies that will require little maintenance. Since doubly fed induction and synchronous machines use brushes and commutators, which require frequent maintenance and replacement, new electric machine structures to replace wind generators are under investigation. The interior permanent magnet (IPM) machine is one such machine, with a considerable potential for wind power generation. This is because brushes and commutators are not required and there is no field winding copper loss since the excitation of permanent magnet is buried in the rotor. The IPM also offers better efficiency compared to other machines, although it is more expensive.

It is anticipated that microgrid systems consisting of many renewable energy sources, including IPM generators, serving autonomous loads or connected to the grid will become prevalent. The IPM generator will be required to supply an active power to the grid under minimum generator loss conditions at a controllable grid power factor. For a desired power rating, the IPM generator, rectifier, inverter and the topology of the filter connecting the inverter output to the grid must also be selected as well as the values of its inductors and capacitors. Then controllers will be designed to ensure system stability, robustness and high dynamic performance under all operating conditions. This sequential method of system design (structural (plant) design, control system design) has long been established to problematize the optimal static and dynamic operations of complex and nonlinear engineering and biological systems. In particular, controllers can be limited by system characteristics induced by nonlinearity, such as input/output multiplicities, open-loop instability and right half-plane zeros or non-minimum phase phenomenon that limits the achievable controller bandwidth.

A microgrid consisting of different types of renewable sources and loads is best conceived as a nonlinear system. The dynamic models of the sources are nonlinear and some of the loads are modeled as

Power Electronics for Renewable Energy Systems, Transportation and Industrial Applications, First Edition.
Edited by Haitham Abu-Rub, Mariusz Malinowski and Kamal Al-Haddad.

nonlinearly voltage and/or current dependent. The main focus of research to date has been on the control of microgrid systems, aimed at achieving maximum power extraction from renewable sources, the sharing of real and reactive powers between parallel sources and the seamless transition between autonomous and grid-connected modes of operation. Another area of great exertion is the design of the interface filters between the grid and the energy sources to meet steady-state filtering requirements in order to prevent converter-induced harmonics from flowing into the grid, while not provoking instabilities because of the variations in operating conditions and system parameters. Although there is an emerging consensus that the inductor-capacitor-inductor LCL filter is the best, research is ongoing to determine the controller structure for the system, the state variables to be fed back and the selection of filter parameters [4–7].

In order to elucidate the apparent difficulties in arriving at the optimal control structure for the LCL filter, this chapter refers to the concept of system controllability to highlight the inherent characteristics of the system. Controllability is originally defined as the ability of a system to move quickly and smoothly from one operating condition to another and deal effectively with disturbances [7]. Controllability analysis gives insights into the inherent characteristics and phase behavior of the system, which should inform the system design and the selection of controller architecture [8–10].

From the control viewpoint of a linear system, in a stable system all the real parts of the poles should be on the left-hand side of the s-plane. In addition, a minimum phase system has all the real parts of the zeros of its characteristic equation transfer function in the left-hand plane. Therefore, using the poles and zeros, the response of the system to any input can be studied or predicted; moreover, the type of controller needed to improve its response can be selected more clearly.

PV system is one of the most popular methods for directly producing electricity from the irradiation of the sun. It is environmentally friendly, renewable and free after initial production. Generally, the goal is to obtain the maximum possible power from PVs. For the maximum power point (MPP) tracking of solar cells, two main algorithms are suggested: open-loop and closed-loop controls. Open-loop control is the simplest, but the effect of solar irradiance variations is not taken into account and the efficiency may not be satisfactory.

In closed-loop control, owing to the nonlinear characteristic of the solar cells and unpredictable environmental conditions and their effect on the output of the PV panel, designing a precise controller is very complex and sometimes not completely applicable. The most popular methods for the close MPP control of PV systems are incremental conductance (INC) and the perturbation and observation (P&O) algorithm.

Three techniques have been proposed in the literature for implementing the P&O algorithm: reference voltage perturbation [11–13], reference current perturbation [14, 15] and direct duty ratio perturbation [12, 16, 17]. Reference current perturbation has a slow transient response to irradiance changes and a high susceptibility to noise and proportional-integral (PI) controller oscillation. Reference [18] compares voltage perturbation and direct duty ratio perturbation on the basis of system stability, performance characteristics and energy utilization for standalone PV pumping systems. In reference to voltage perturbation, the system has a faster response to irradiance and temperature transients; however, it loses stability if operated at a high perturbation rate or if low-pass filters are used for noise rejection. Direct duty ratio control offers better energy utilization and better stability characteristics at a slower transient response and worse performance at a rapidly changing irradiance. System stability is not affected by the use of low-pass feedback filters and it allows the use of high perturbation rates.

Most electrical power systems and the microgrid are highly complex and nonlinear systems. They are also multi-input multi-output (MIMO) systems, which makes their analysis even more complex. Unlike linear systems their characteristic equations cannot be defined and in many cases their stability can only be defined locally over the steady state operation points using the eigenvalues of the linearized characteristic equations. In order to define the minimum/non-minimum phase of the nonlinear system, the concept of zero dynamics has been used [10]. The unstable zero dynamics may cause problems for both the dynamic and steady-state responses. From a dynamic point of view, it causes an inverse response in which the response spends part of its time going in the wrong direction. Hence, it reduces the bandwidth of the controller, causes a delay in reaching the steady-state value and limits the degree of achievable control quality [19–21]. The problem of a non-minimum phase system in steady state is called input multiplicity, in which the same output can be obtained from combinations of different inputs and output variables. It is therefore possible to have an unobservable transition from one steady-state output to another.

This chapter therefore investigates the controllability of an IPM wind turbine generator connected to the grid using the analysis of its open-loop stability and zero phase behaviors. The zero dynamics of the grid-connected IPM generator using either an L or LCL filter are investigated. The wind turbine is working at the MPP tracking scenario, and other operational objectives include generator loss minimization, the control of the grid reactive power and the converter DC-link voltage. The effect of filter structure on the inherent characteristics of the grid-connected generator determined through the study of zero dynamics constitutes the main theme of this chapter. It is also observed that the use of the feedback linearization methodology for controller design, where the stability of the zero dynamics permits the use of static controller gains, is becoming advantageous in the design of controllers for energy systems. This chapter is an extension of the work reported in [22].

8.2 Zero Dynamics of the Nonlinear System

The definition of a nonlinear system with m number of inputs and outputs is

$$\begin{cases} \dfrac{\partial x}{dt} = f(x) + \sum\limits_{j=1}^{m} g_j(x)u_j(x) \\ y_i = h_i(x), \qquad i = 1, \dots, m \end{cases} \tag{8.1}$$

The number of derivations (differentiations) of the output needed to see the input in its equation (explicit relation between input and output) is called the relative order of the output. In an MIMO, the relative order is the sum of all relative orders of the outputs.

Another way to define the relative order of any output is to fulfill the following condition

$$L_g L_f^{r_i-1} h_i(x) = \left[L_{g1} L_f^{r_i-1} h_i(x), L_{g2} L_f^{r_i-1} h_i(x), \dots, L_{gm} L_f^{r_i-1} h_i(x)\right] \neq [0, 0, \dots, 0] \tag{8.2}$$

where the lie derivative definition is

$$L_f h(x) = \frac{\partial h}{\partial x} f(x) \tag{8.3}$$

$$L_f^2 h(x) = L_f L_f(x), \quad L_g L_f h(x) = L_g\left(\frac{\partial h}{\partial x} f(x)\right) \tag{8.4}$$

The relative order of the MIMO system is

$$r = r_1 + r_2 + \cdots + r_m \tag{8.5}$$

If the relative order of the system is less than the order of the system (number of dynamic states), part of the system is unobservable and there are internal dynamics. Internal dynamic states are not reflected during the process of feedback linearization and control design of the system; therefore, their stability must be ascertained. Zero dynamics is a special case of the internal dynamics and is defined as the internal dynamics of the system when the selected system outputs are set to zero by the input. There are two methods to isolate the zero dynamics.

8.2.1 First Method

In the first method [23] for finding the zero dynamics, after defining the number of internal dynamics ($n - \sum r_i$) all the outputs should be made equal to zero. In this case, some dynamic states will be removed, leaving the internal dynamics which are independent of the input variables.

It should be noted that if the purpose for checking the stability of the zero dynamics is to use the input/output linearization method for controller design, the characteristic matrix can be defined as

$$C_{\mathrm{h}}(x)=\begin{bmatrix} L_{\mathrm{g1}}L_{\mathrm{f}}^{r_1-1}h_1 & \cdots & L_{\mathrm{gm}}L_{\mathrm{f}}^{r_1-1}h_1 \\ \vdots & & \vdots \\ L_{\mathrm{g1}}L_{\mathrm{f}}^{r_{\mathrm{m}}-1}h_{\mathrm{m}} & \cdots & L_{\mathrm{gm}}L_{\mathrm{f}}^{r_{\mathrm{m}}-1}h_{\mathrm{m}} \end{bmatrix} \tag{8.6}$$

The non-singularity of the characteristic matrix in an MIMO system is a sufficient condition for input/output linearization to be achievable through static state feedback. For systems for which this condition is not satisfied, controllers with dynamic contents can be employed to transform them to new equations satisfying the characteristic matrix [20].

8.2.2 *Second Method*

In the second method based on [24], a new coordinate system with $t(x)$ is generated in which the elements of the following matrix are linearly independent

$$\eta=\begin{bmatrix} \eta^{(0)} \\ \eta^{(1)} \\ \vdots \\ \eta^{(\mathrm{m})} \end{bmatrix}=\left[t_1(x)\ \cdots\ t_{n-\sum r_i}(x)\ h_1(x)\ L_{\mathrm{f}}h_1(x)\ \cdots\ L_{\mathrm{f}}^{r_1-1}h_1(x)\ \cdots\ h_{\mathrm{m}}(x)\ L_{\mathrm{f}}h_{\mathrm{m}}(x)\ \cdots\ L_{\mathrm{f}}^{r_{\mathrm{m}}-1}h_{\mathrm{m}}(x)\right]^T \tag{8.7}$$

The original dynamic equations of the system are now converted into the new coordinates in which the input variables have been eliminated. For an MIMO system, the following can be defined

$$\begin{cases} F_i=L_{\mathrm{f}}t_i(x) & i=1,\ldots,(n-\sum_{i=1}^{m} r_i) \\ G_i=\left[L_{\mathrm{g1}}t_i\cdots L_{\mathrm{gm}}t_i\right] & i=1,\ldots,(n-\sum_{i=1}^{m} r_i) \\ C_i=\left[L_{\mathrm{g1}}L_{\mathrm{f}}^{r_i-1}h_i(x)\cdots L_{\mathrm{gm}}L_{\mathrm{f}}^{r_i-1}h_i(x)\right] & i=1,\ldots,m \\ W_i=L_{\mathrm{f}}^{r_i}h_i(x) & i=1,\ldots,m \end{cases} \tag{8.8}$$

The zero dynamics after converting the matrices to the new coordinates (the output variables will be either zero or at their steady-state values) are defined as follows

$$\frac{\mathrm{d}\eta^{(0)}_{n-\sum r_i}}{\mathrm{d}t}=F_{n-\sum r_i}(\eta)-G_{n-\sum r_i}(\eta)\begin{bmatrix} C_1(\eta) \\ \vdots \\ C_{\mathrm{m}}(\eta) \end{bmatrix}^{-1}\left[W_1(\eta)\ \cdots\ W_{\mathrm{m}}(\eta)\right] \tag{8.9}$$

8.3 Controllability of Wind Turbine Connected through L Filter to the Grid

As illustrated in Figure 8.1, in the study system, an IPM generator driven by a wind turbine supplies power to the grid. A back-to-back AC/DC/AC inverter converts the AC generator output voltage to a DC voltage and then to another three-phase AC voltage. The grid-side inverter is interfaced to the grid through the L-type filter.

In the IPM machine, the dynamics of the rotor can be neglected because the permanent magnet is a poor electrical conductor. Ignoring the core loss of the machine, the dynamic equations of the generator

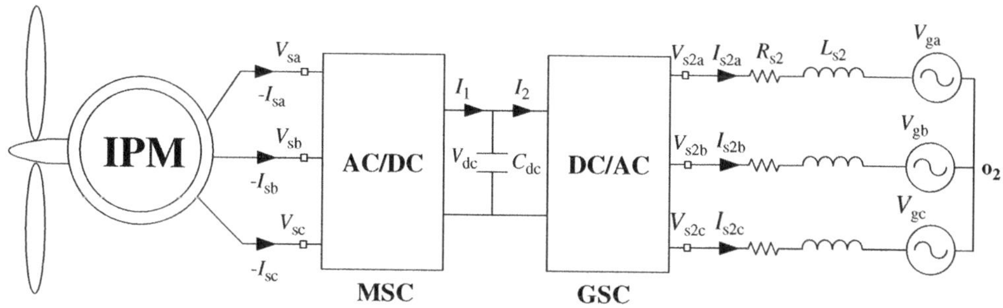

Figure 8.1 Schematic diagram of IPM wind turbine connected to the main grid with L-type filter [23]

are as follows

$$M_{q1}\frac{V_{dc}}{2} = V_{sq} = R_s I_{sq} + L_{sq} p I_{sq} + \omega_r L_{sd} I_{sd} + \omega_r \lambda_m \tag{8.10}$$

$$M_{d1}\frac{V_{dc}}{2} = V_{sd} = R_s I_{sd} + L_{sd} p I_{sd} - \omega_r L_{sq} I_{sq} \tag{8.11}$$

The converter DC capacitor equation is given as

$$C_{dc} p V_{dc} = I_1 - I_2 = -\frac{3}{4}(M_{q1} I_{sq} + M_{d1} I_{sd}) - I_2 \tag{8.12}$$

where $V_{sqd}$and I_{sqd} are the complex qd-axis voltages and currents of the stator, M_{qd1} and V_{dc} are the complex qd-axis modulation indexes of machine-side converter and the DC-link voltage, respectively. R_s is the stator resistance and L_{sqd} and λ_m are qd-axis stator inductance and magnetization flux of the generator.

The dynamic equations of the main grid with an L-type filter and grid-side converter are

$$\frac{M_{q2}}{2} V_{dc} - V_{gq} = R_{s2} I_{s2q} + L_{s2} p I_{s2q} + \omega_e L_{s2} I_{s2d} \tag{8.13}$$

$$\frac{M_{d2}}{2} V_{dc} - V_{gd} = R_{s2} I_{s2d} + L_{s2} p I_{s2d} - \omega_e L_{s2} I_{s2q} \tag{8.14}$$

The parameters of the IPM machine and the grid are presented in Table 8.1. For generation of copper loss minimization, the following requirement should be met [25]

$$\gamma = \begin{vmatrix} \dfrac{\partial T_e}{\partial I_{sq}} & \dfrac{\partial T_e}{\partial I_{sd}} \\ \dfrac{\partial P_{loss}}{\partial I_{sq}} & \dfrac{\partial P_{loss}}{\partial I_{sd}} \end{vmatrix} = f(I_{sq}, I_{sd}) = 0 \tag{8.15}$$

$$\gamma = \left(\frac{9P}{4}\right)(R_s(L_{sd} - L_{sq})I_{sq}^{\ 2} - R_s(L_{sd} - L_{sq})I_{sd}^{\ 2} - R_s \lambda_m I_{sd}) \tag{8.16}$$

8.3.1 Steady State and Stable Operation Region

Under steady state operating conditions, the derivatives of the states in (8.10)–(8.14) are set to zero. The resulting equations with the algebraic equation defining the reference grid reactive power and the

Table 8.1 Parameters of the IPM machine for controllability analysis

$V_{\text{m-rated}}$	180 V	$I_{\text{m-rated}}$	5.5 A
$\omega_{\text{rm-rated}}$	1800 rpm	P (no. of poles)	4
R_s	1.5 Ω	λ_m	0.21 Wb
L_{sd}	0.035 H	L_{sq}	0.11 H
C_{dc}	$600e^{-6}$ F	J	0.089 kg/m^2
V_{gq}	170 V	V_{gd}	0 V
R_{s2}	0.01 Ω	L_{s2}	0.4 mH

condition for minimum generator loss (8.16) are used to determine the steady-state operation. The known variables are

$$\begin{bmatrix} V_{gq} & V_{gd} & Q_g & (\omega_r \Leftrightarrow T_t) & V_{dc} \end{bmatrix} \tag{8.17}$$

and the unknown variables are

$$\begin{bmatrix} I_{sq} & I_{sd} & M_{q1} & M_{d1} & I_{s2q} & I_{s2d} & M_{q2} & M_{d2} \end{bmatrix} \tag{8.18}$$

The rotor speed and the DC capacitor voltage are varied (ω_r and V_{dc}) to solve the steady-state equations given as

$$\begin{cases} -M_{q1}\dfrac{V_{dc}}{2} + R_s I_{sq} + \omega_r L_{sd} I_{sd} + \omega_r \lambda_m = 0 \\ -M_{d1}\dfrac{V_{dc}}{2} + R_s I_{sd} - \omega_r L_{sq} I_{sq} = 0 \\ -M_{q2}\dfrac{V_{dc}}{2} + V_{gq} + R_{s2} I_{s2q} + \omega_e L_{s2} I_{s2d} = 0 \\ -M_{d2}\dfrac{V_{dc}}{2} + V_{gd} + R_{s2} I_{s2d} - \omega_e L_{s2} I_{s2q} = 0 \\ -\dfrac{3}{4}\left(M_{q1} I_{sq} + M_{d1} I_{sd}\right) - \dfrac{3}{4}(M_{q2} I_{s2q} + M_{d2} I_{s2d}) = 0 \\ T_t - \left(\dfrac{3P}{4}\right)(\lambda_m I_{sq} + (L_{sd} - L_{sq}) I_{sq} I_{sd}) = 0 \\ \gamma = \left(\dfrac{9P}{4}\right)(R_s(L_{sd} - L_{sq}) I_{sq}{}^2 - R_s(L_{sd} - L_{sq}) I_{sd}{}^2 - R_s \lambda_m I_{sd}) = 0 \\ -Q_g + \left(\dfrac{3}{2}\right)(V_{gq} I_{s2d} - V_{gd} I_{s2q}) = 0 \end{cases} \tag{8.19}$$

The steady-state operation regions of the generator system are illustrated in Figure 8.2. In this figure, the output active and reactive powers of the IPM generator are shown. The stator current and voltage of the machine are presented in Figure 8.2b and c, respectively. The steady-state parameters of the machine are simply related to the input wind power or rotor speed that is directly related to it by the MPP algorithm. The modulation index of the machine-side converter is related to both rotor speed and DC-link voltage and it is always in the linear region below unity. Because all the power coming from the wind on the other side is transmitted to the grid and the voltage of the grid is constant, the modulation index of the grid-side converter in Figure 8.2 is simply related to the DC-link voltage magnitude.

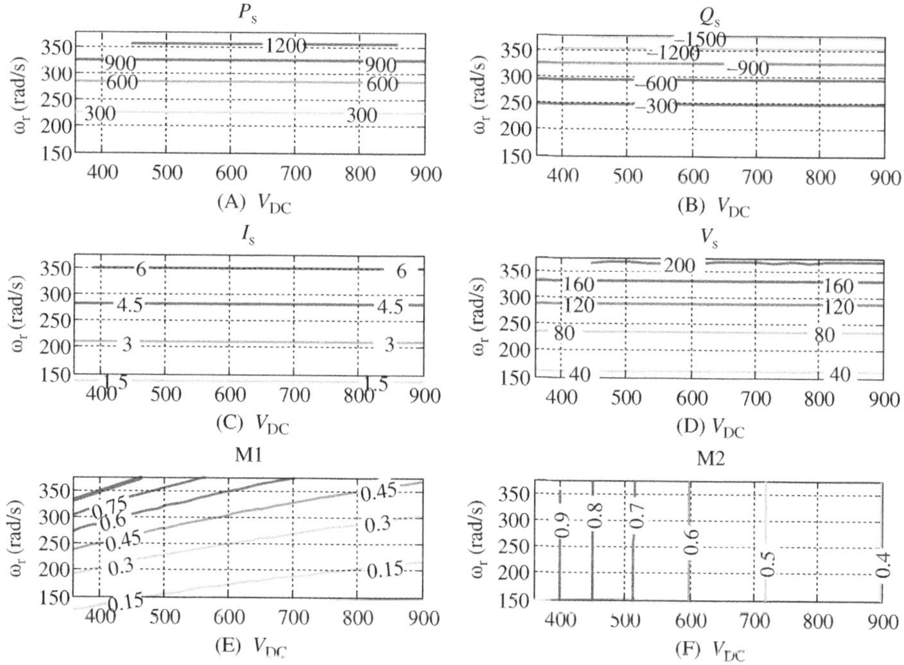

Figure 8.2 Steady-state operation region of the system: (a) output active power of the generator (W), (b) output reactive power of the generator (var), (c) stator current (A), (d) stator voltage (V), (e) modulation index magnitude of machine-side converter and (f) modulation index magnitude of grid-side converter [23]

To study the stability region of the system in all feasible operation regions, the model equations of the electrical subsystem is linearized and its characteristic equation is defined as a sixth-order equation.

$$\begin{cases} p\Delta I_{sq} = \dfrac{1}{L_{sq}}\left(M_{q1}\dfrac{\Delta V_{dc}}{2} - R_s\Delta I_{sq} - \omega_{r0}L_{sd}\Delta I_{sd} - L_{sd}I_{sd0}\Delta\omega_r - \Delta\omega_r\lambda_m\right) \\ p\Delta I_{sd} = \dfrac{1}{L_{sd}}\left(M_{d1}\dfrac{\Delta V_{dc}}{2} - R_s\Delta I_{sd} + \omega_{r0}L_{sq}\Delta I_{sq} + L_{sq}I_{sq0}\Delta\omega_r\right) \\ p\Delta I_{s2q} = \dfrac{1}{L_{s2}}\left(M_{q2}\dfrac{\Delta V_{dc}}{2} - R_{s2}\Delta I_{s2q} - \omega_e L_{s2}\Delta I_{s2d}\right) \\ p\Delta I_{s2d} = \dfrac{1}{L_{s2}}\left(M_{d2}\dfrac{\Delta V_{dc}}{2} - R_{s2}\Delta I_{s2d} + \omega_e L_{s2}\Delta I_{s2q}\right) \\ p\Delta V_{dc} = -\dfrac{3}{4C_{dc}}(M_{q1}\Delta I_{sq} + M_{d1}\Delta I_{sd} + M_{q2}\Delta I_{s2q} + M_{d2}\Delta I_{s2d}) \\ p\Delta\omega_r = \left(\dfrac{3p^2}{8J}\right)(\lambda_m\Delta I_{sq} + (L_{sd} - L_{sq})I_{sq0}\Delta I_{sd} + (L_{sd} - L_{sq})I_{sd0}\Delta I_{sq}) \end{cases} \tag{8.20}$$

$$\begin{bmatrix} p\Delta I_{sq} \\ p\Delta I_{sd} \\ p\Delta I_{s2q} \\ p\Delta I_{s2d} \\ p\Delta V_{dc} \\ p\Delta \omega_r \end{bmatrix} = \begin{bmatrix} -\frac{R_s}{L_{sq}} & -\frac{\omega_{r0}L_{sd}}{L_{sq}} & 0 & 0 & \frac{M_{q1}}{2L_{sq}} & -\frac{L_{sd}I_{sd0}+\lambda_m}{L_{sq}} \\ \frac{\omega_{r0}L_{sq}}{L_{sd}} & -\frac{R_s}{L_{sd}} & 0 & 0 & \frac{M_{d1}}{2L_{sd}} & \frac{L_{sq}I_{sq0}}{L_{sd}} \\ 0 & 0 & -\frac{R_{s2}}{L_{s2}} & -\omega_e & \frac{M_{q2}}{2L_{s2}} & 0 \\ 0 & 0 & \omega_e & -\frac{R_{s2}}{L_{s2}} & \frac{M_{d2}}{2L_{s2}} & 0 \\ -\frac{3M_{q1}}{4C_{dc}} & -\frac{3M_{d1}}{4C_{dc}} & -\frac{3M_{q2}}{4C_{dc}} & -\frac{3M_{d2}}{4C_{dc}} & 0 & 0 \\ \frac{3p^2\left(\lambda_m+\left(L_{sd}-L_{sq}\right)I_{sd0}\right)}{8J} & \frac{3p^2(L_{sd}-L_{sq})I_{sq0}}{8J} & 0 & 0 & 0 & 0 \end{bmatrix} \times \begin{bmatrix} \Delta I_{sq} \\ \Delta I_{sd} \\ \Delta I_{s2q} \\ \Delta I_{s2d} \\ \Delta V_{dc} \\ \Delta \omega_r \end{bmatrix} \tag{8.21}$$

$$|\lambda I - A| = a_6\lambda^6 + a_5\lambda^5 + a_4\lambda^4 + a_3\lambda^3 + a_2\lambda^2 + a_1\lambda + a_0 = 0 \tag{8.22}$$

Using the Routh–Hurwitz criteria

$$\begin{array}{c|ccc} \lambda^6 & a_6 & a_4 & a_2 \; a_0 \\ \lambda^5 & a_5 & a_3 & a_1 \\ \lambda^4 & b_1 = \dfrac{a_4a_5 - a_3a_6}{a_5} & b_2 = \dfrac{a_2a_5 - a_1a_6}{a_5} & a_0 \\ \lambda^3 & c_1 = \dfrac{a_3b_1 - a_5b_2}{b_1} & c_2 = \dfrac{a_1b_1 - a_5a_0}{b_1} & \\ \lambda^2 & d_1 = \dfrac{b_2c_1 - b_1c_2}{c_1} & a_0 & \\ \lambda^1 & e_1 = \dfrac{c_2d_1 - c_1a_0}{d_1} & & \\ \lambda^0 & a_0 & & \end{array} \tag{8.23}$$

the requirements for the stability of the system are

$$\begin{cases} a_6 > 0,\ a_5 > 0,\ a_4 > 0,\ a_3 > 0,\ a_2 > 0,\ a_1 > 0,\ a_0 > 0 \\ b_1 > 0,\ c_1 > 0,\ d_1 > 0,\ e_1 > 0 \end{cases} \tag{8.24}$$

As shown in Figure 8.3, only the coefficient a_0 affects the stability region in which the system is unstable when operated above a particular rotor speed.

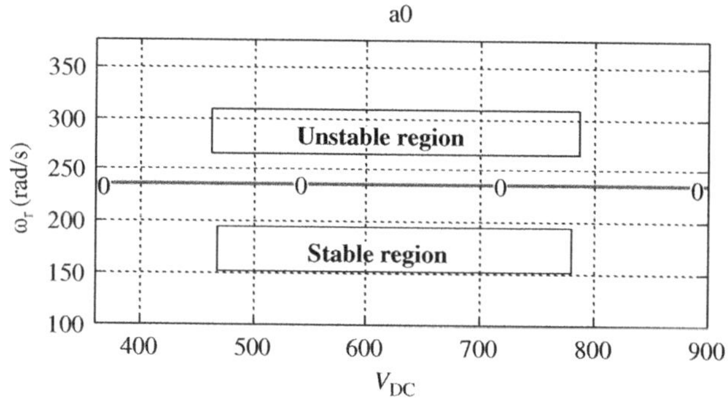

Figure 8.3 Stability region based on Routh–Hurwitz criteria. The final coefficient of the characteristic equation [23]

8.3.2 *Zero Dynamic Analysis*

To analyze the zero dynamics of the system, the first method is adopted. The four inputs and four outputs are

$$Y = \begin{bmatrix}\gamma & \omega_r & V_{dc} & Q_g\end{bmatrix}^T, \quad U = \begin{bmatrix}M_{q1} & M_{d1} & M_{q2} & M_{d2}\end{bmatrix}^T \tag{8.25}$$

With differentiating of the outputs to define the relative orders

$$\begin{cases} r_1 = 1, & r_2 = 2 \\ r_3 = 1, & r_4 = 1 \end{cases} \quad \text{and} \quad r = \sum_{i=1}^{4} r_i = 5 \tag{8.26}$$

Therefore, the system has one $(n-\sum r_i)$ internal dynamic. The zero dynamics is obtained when all the output variables are set to zero. With the grid voltage taken as the reference for the *qd* synchronous reference frame transformation and aligning the grid voltage in the *q*-axis ($V_{gq} = V_g, V_{gd} = 0$)

$$\begin{cases} C_{dc}pV_{dc} = 0 = -\frac{3}{4}\left(M_{q1}I_{sq} + M_{d1}I_{sd}\right) - \frac{3}{4}(M_{q2}I_{s2q} + M_{d2}I_{s2d}) \\ p\omega_r = 0 = T_t - \left(\frac{3P}{4}\right)(\lambda_m I_{sq} + (L_{sd} - L_{sq})I_{sq}I_{sd}) \\ Q_g = 0 = \frac{3}{2}(V_{gq}I_{s2d} - V_{gd}I_{s2q}), \omega_r = 0, \ V_{dc} = 0 \\ \gamma = 0 = \left(\frac{9P}{4}\right)(R_s(L_{sd} - L_{sq})I_{sq}{}^2 - R_s(L_{sd} - L_{sq})I_{sd}{}^2 - R_s\lambda_m I_{sd}) \end{cases} \tag{8.27}$$

then

$$I_{sq} = -\frac{M_{d1}I_{sd}}{M_{q1}} \tag{8.28}$$

and the only remaining dynamic, the zero dynamic of the system, is

$$R_s I_{sd} + L_{sd}pI_{sd} = 0 \tag{8.29}$$

Hence since the resistances and inductances of the generators have positive values, the zero dynamics is always stable, indicating that the nonlinear generator connected to the grid system through L-type filter has a minimum phase characteristics.

8.4 Controllability of Wind Turbine Connected through LCL Filter to the Grid

The LCL filter shown in Figure 8.4 is a third-order filter that provides a much better ripple and harmonic attenuation over the higher frequency range using smaller passive elements. Therefore, they are more suited for high-power conversion systems and have already been widely employed in wind farms of over hundreds of kilowatts [26–29]. The second inductance represents the inductance of the filter, the grid and the isolating transformer. The resistances also include the copper loss of the grid, transformer and switching loss of the grid-side converter.

8.4.1 Steady State and Stable Operation Region

The total dynamic equations of the system with the LCL filter are

$$\begin{cases} M_{q1}\dfrac{V_{dc}}{2} = R_s I_{sq} + L_{sq} p I_{sq} + \omega_r L_{sd} I_{sd} + \omega_r \lambda_m \\ M_{d1}\dfrac{V_{dc}}{2} = R_s I_{sd} + L_{sd} p I_{sd} - \omega_r L_{sq} I_{sq} \\ M_{q2}\dfrac{V_{dc}}{2} - V_{cq} = R_{s2} I_{s2q} + L_{s2} p I_{s2q} + \omega_e L_{s2} I_{s2d} \\ M_{d2}\dfrac{V_{dc}}{2} - V_{cd} = R_{s2} I_{s2d} + L_{s2} p I_{s2d} - \omega_e L_{s2} I_{s2q} \\ V_{cq} - V_{gq} = R_g I_{gq} + L_g p I_{gq} + \omega_e L_g I_{gd} \\ V_{cd} - V_{gd} = R_g I_{gd} + L_g p I_{gd} - \omega_e L_g I_{gq} \\ pV_{cq} = \dfrac{1}{C_g}\left(I_{s2q} - I_{gq} - \omega_e C_g V_{cd}\right) \\ pV_{cd} = \dfrac{1}{C_g}(I_{s2d} - I_{gd} + \omega_e C_g V_{cq}) \\ C_{dc} p V_{dc} = I_1 - I_2 = -\dfrac{3}{4}(M_{q1} I_{sq} + M_{d1} I_{sd}) - \dfrac{3}{4}(M_{q2} I_{s2q} + M_{d2} I_{s2d}) \\ p\omega_r = \left(\dfrac{P}{2J}\right)(T_t - T_e) = \left(\dfrac{P}{2J}\right) T_t - \left(\dfrac{P}{2J}\right)\left(\dfrac{3P}{4}\right)(\lambda_m I_{sq} + (L_{sd} - L_{sq}) I_{sq} I_{sd}) \end{cases} \tag{8.30}$$

In steady state, the derivative parts are zero so eight equations including six dynamic equations in the steady-state condition along with the two algebraic equations of reactive power and loss minimization can be solved. Therefore, the known variables are

$$\begin{bmatrix} V_{gq} & V_{gd} & Q_2 & (\omega_r \Leftrightarrow T_t) & V_{dc} \end{bmatrix} \tag{8.31}$$

and the unknown variables are

$$\begin{bmatrix} I_{sq} & I_{sd} & M_{q1} & M_{d1} & I_{s2q} & I_{s2d} & M_{q2} & M_{d2} & I_{gq} & I_{gd} & V_{cq} & V_{cd} \end{bmatrix} \tag{8.32}$$

So ω_r and V_{dc} can be varied to calculate the other 12 variables. The results are illustrated in Figures 8.5–8.11.

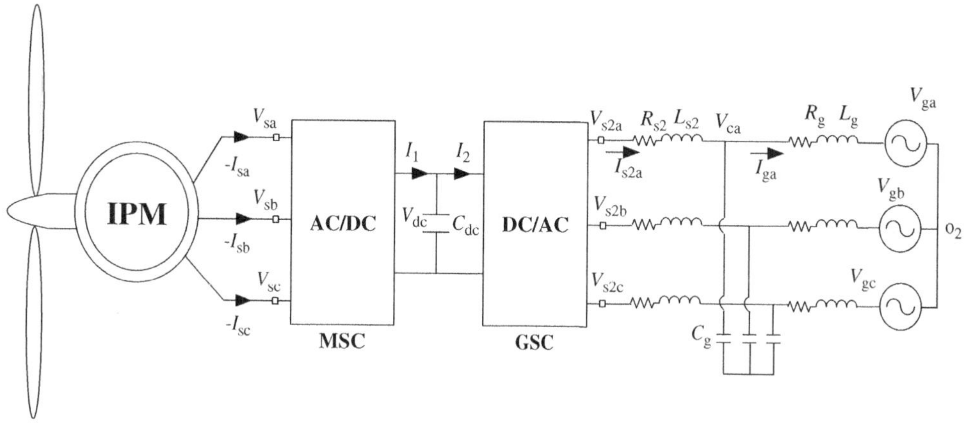

Figure 8.4 The IPM machine as a wind turbine generator connected through LCL filter

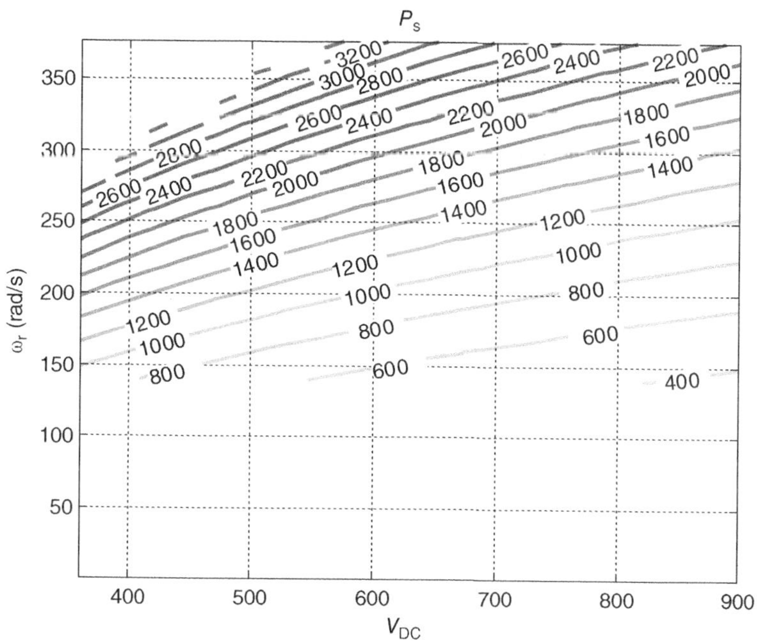

Figure 8.5 Output active power of IPM generator

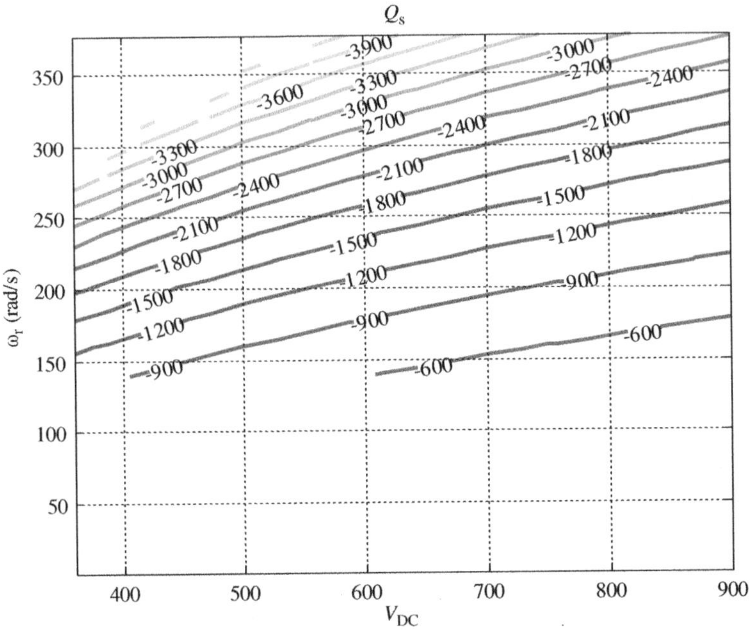

Figure 8.6 Output reactive power of IPM generator

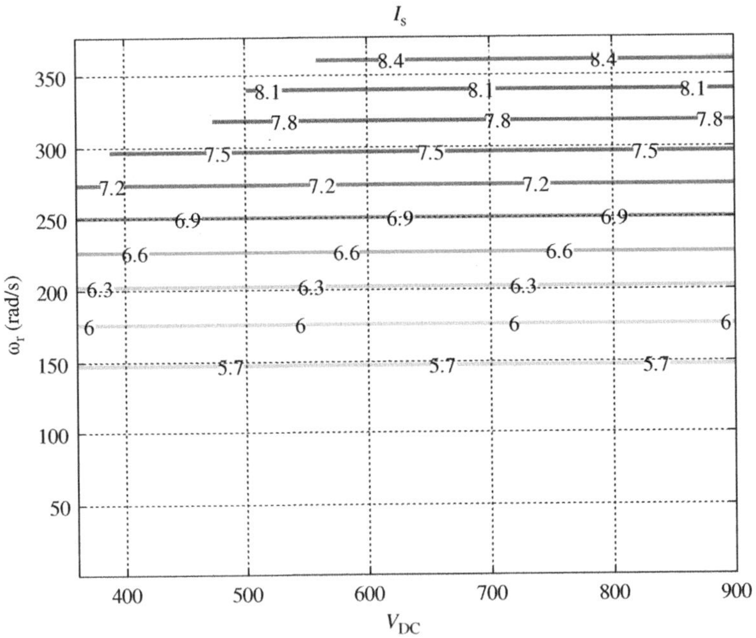

Figure 8.7 Stator current magnitude of IPM generator

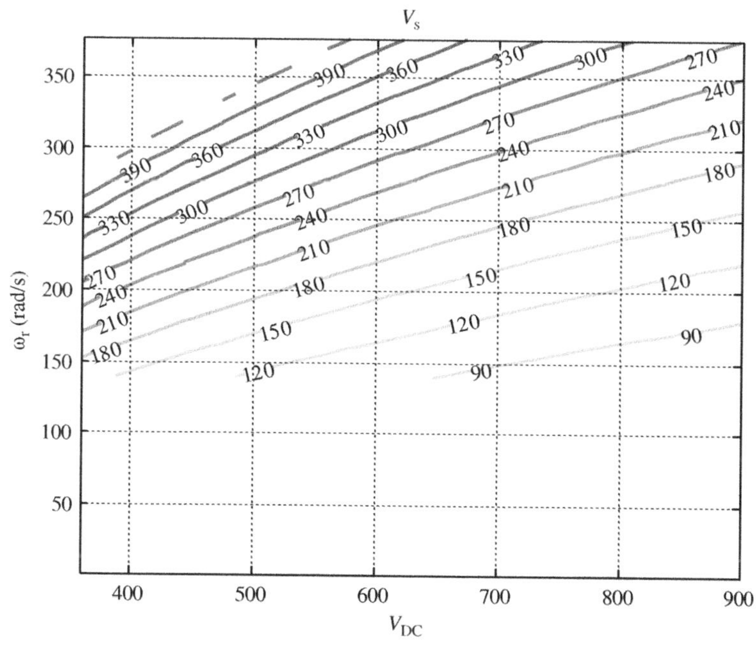

Figure 8.8 Stator voltage magnitude of IPM generator

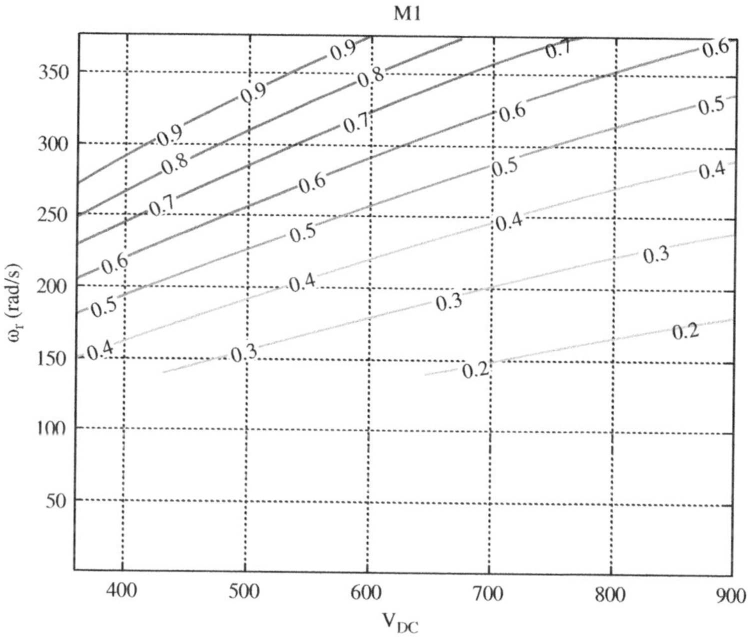

Figure 8.9 Modulation index magnitude of the machine-side inverter

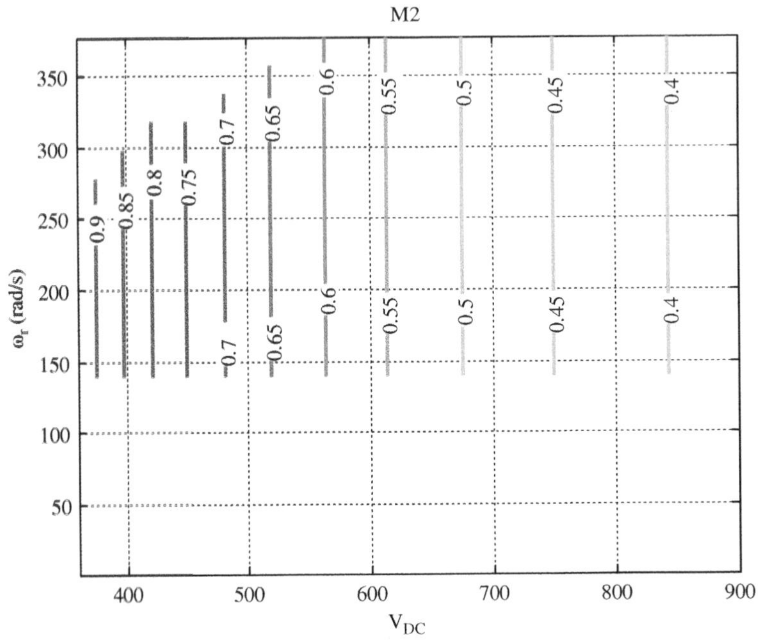

Figure 8.10 Modulation index magnitude of the grid-side inverter

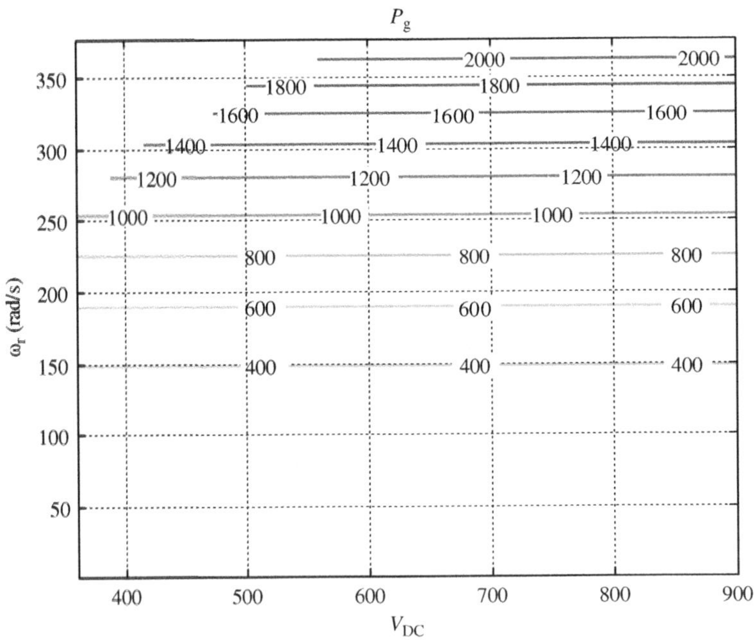

Figure 8.11 Active power flowing into the grid

The stability region of this system can be studied using the small signal analysis in any steady-state point. The eigenvalues of the system at that point can be defined and checked for stability.

$$\begin{cases} p\Delta I_{sq} = \dfrac{1}{L_{sq}}\left(M_{q1}\dfrac{\Delta V_{dc}}{2} - R_s\Delta I_{sq} - \omega_{r0}L_{sd}\Delta I_{sd} - L_{sd}I_{sd0}\Delta\omega_r - \Delta\omega_r\lambda_m\right) \\ p\Delta I_{sd} = \dfrac{1}{L_{sd}}\left(M_{d1}\dfrac{\Delta V_{dc}}{2} - R_s\Delta I_{sd} + \omega_{r0}L_{sq}\Delta I_{sq} + L_{sq}I_{sq0}\Delta\omega_r\right) \\ p\Delta I_{s2q} = \dfrac{1}{L_{s2}}\left(M_{q2}\dfrac{\Delta V_{dc}}{2} - \Delta V_{cq} - R_{s2}\Delta I_{s2q} - \omega_e L_{s2}\Delta I_{s2d}\right) \\ p\Delta I_{s2d} = \dfrac{1}{L_{s2}}\left(M_{d2}\dfrac{\Delta V_{dc}}{2} - \Delta V_{cd} - R_{s2}\Delta I_{s2d} + \omega_e L_{s2}\Delta I_{s2q}\right) \\ p\Delta I_{gq} = \dfrac{1}{L_g}(\Delta V_{cq} - R_g\Delta I_{gq} - \omega_e L_g\Delta I_{gd}) \\ p\Delta I_{gd} = \dfrac{1}{L_g}(\Delta V_{cd} - R_g\Delta I_{gd} + \omega_e L_g\Delta I_{gq}) \\ p\Delta V_{cq} = \dfrac{1}{C_g}(\Delta I_{s2q} - \Delta I_{gq} - \omega_e C_g\Delta V_{cd}) \\ p\Delta V_{cd} = \dfrac{1}{C_g}(\Delta I_{s2d} - \Delta I_{gd} + \omega_e C_g\Delta V_{cq}) \\ p\Delta V_{dc} = -\dfrac{3}{4C_{dc}}(M_{q1}\Delta I_{sq} + M_{d1}\Delta I_{sd} + M_{q2}\Delta I_{s2q} + M_{d2}\Delta I_{s2d}) \\ p\Delta\omega_r = \left(\dfrac{3p^2}{8J}\right)(\lambda_m\Delta I_{sq} + (L_{sd} - L_{sq})I_{sq0}\Delta I_{sd} + (L_{sd} - L_{sq})I_{sd0}\Delta I_{sq}) \end{cases} \tag{8.33}$$

$$\begin{bmatrix} p\Delta I_{sq} \\ p\Delta I_{sd} \\ p\Delta I_{s2q} \\ p\Delta I_{s2d} \\ p\Delta I_{gq} \\ p\Delta I_{gd} \\ p\Delta V_{cq} \\ p\Delta V_{cd} \\ p\Delta V_{dc} \\ p\Delta\omega_r \end{bmatrix} = A \begin{bmatrix} \Delta I_{sq} \\ \Delta I_{sd} \\ \Delta I_{s2q} \\ \Delta I_{s2d} \\ \Delta I_{gq} \\ \Delta I_{gd} \\ \Delta V_{cq} \\ \Delta V_{cd} \\ \Delta V_{dc} \\ \Delta\omega_r \end{bmatrix} \tag{8.34}$$

With the characteristic equation

$$|\lambda I - A| = a_{10}\lambda^{10} + a_9\lambda^9 + a_8\lambda^8 + a_7\lambda^7 + a_6\lambda^6 + a_5\lambda^5 + a_4\lambda^4 + a_3\lambda^3 + a_2\lambda^2 + a_1\lambda + a_0 = 0 \tag{8.35}$$

The stability can be defined using the Routh–Hurwitz criteria. Because all eigenvalues in all operation regions are negative, the system is always stable.

8.4.2 Zero Dynamic Analysis

The dynamic equation of the system with the LCL filter in the nonlinear format is

$$\dot{X} = f(x) + g_1u_1 + g_2u_2 + g_3u_3 + g_4u_4 \tag{8.36}$$

where

$$f(x)=\begin{bmatrix} -\dfrac{R_s x_1}{L_{sq}}-\dfrac{x_{10}L_{sd}x_2}{L_{sq}}-\dfrac{x_{10}\lambda_m}{L_{sq}} \\ -\dfrac{R_s x_2}{L_{sd}}+\dfrac{x_{10}L_{sq}x_1}{L_{sd}} \\ -\dfrac{x_7}{L_{s2}}-\dfrac{R_{s2}x_3}{L_{s2}}-\dfrac{\omega_e L_{s2}x_4}{L_{s2}} \\ -\dfrac{x_8}{L_{s2}}-\dfrac{R_{s2}x_4}{L_{s2}}+\dfrac{\omega_e L_{s2}x_3}{L_{s2}} \\ \dfrac{x_7}{L_g}-\dfrac{V_{gq}}{L_g}-\dfrac{R_g x_5}{L_g}-\dfrac{\omega_e L_g x_6}{L_g} \\ \dfrac{x_8}{L_g}-\dfrac{V_{gd}}{L_g}-\dfrac{R_g x_6}{L_g}+\dfrac{\omega_e L_g x_5}{L_g} \\ \dfrac{x_3}{C_g}-\dfrac{x_5}{C_g}-\dfrac{\omega_e C_g x_8}{C_g} \\ \dfrac{x_4}{C_g}-\dfrac{x_6}{C_g}+\dfrac{\omega_e C_g x_7}{C_g} \\ 0 \\ \left(\dfrac{p}{2J}\right)T_t-\left(\dfrac{3P^2}{8J}\right)(\lambda_m x_1+(L_{sd}-L_{sq})x_1x_2) \end{bmatrix} \tag{8.37}$$

$$g_1(x)=\begin{bmatrix} \dfrac{x_9}{2L_{sq}} \\ 0 \\ 0 \\ 0 \\ 0 \\ 0 \\ 0 \\ 0 \\ -\dfrac{3x_1}{4C_{dc}} \\ 0 \end{bmatrix},\ g_2(x)=\begin{bmatrix} 0 \\ \dfrac{x_9}{2L_{sd}} \\ 0 \\ 0 \\ 0 \\ 0 \\ 0 \\ 0 \\ -\dfrac{3x_2}{4C_{dc}} \\ 0 \end{bmatrix},\ g_3(x)=\begin{bmatrix} 0 \\ 0 \\ \dfrac{x_9}{2L_{s2}} \\ 0 \\ 0 \\ 0 \\ 0 \\ 0 \\ -\dfrac{3x_3}{4C_{dc}} \\ 0 \end{bmatrix},\ g_4(x)=\begin{bmatrix} 0 \\ 0 \\ 0 \\ \dfrac{x_9}{2L_{s2}} \\ 0 \\ 0 \\ 0 \\ 0 \\ -\dfrac{3x_4}{4C_{dc}} \\ 0 \end{bmatrix} \tag{8.38}$$

The output equations are

$$Y=\begin{bmatrix} h_1(x) \\ h_2(x) \\ h_3(x) \\ h_4(x) \end{bmatrix}=\begin{bmatrix} \left(\dfrac{9P}{4}\right)(R_s(L_{sd}-L_{sq})x_1{}^2-R_s(L_{sd}-L_{sq})x_2{}^2-R_s\lambda_m x_2) \\ x_{10} \\ x_9 \\ \dfrac{3}{2}(V_{gq}x_6-V_{gd}x_5) \end{bmatrix} \tag{8.39}$$

where

$$X=\left[I_{sq}\ I_{sd}\ I_{s2q}\ I_{s2d}\ I_{gq}\ I_{gd}\ V_{cq}\ V_{cd}\ V_{dc}\ \omega_r\right] \tag{8.40}$$

The new states and inputs are

$$U=\left[M_{q1}\ M_{d1}\ M_{q2}\ M_{d2}\right] \tag{8.41}$$

With the same output variables in the new 10th-order system as the LCL filter defined in (7.39), the relative orders are defined as

$$\begin{cases} r_1 = 1, & r_2 = 2 \\ r_3 = 1, & r_4 = 3 \end{cases} \quad \text{and} \quad r = \sum_{i=1}^{4} r_i = 7 \tag{8.42}$$

Therefore, there are three $(n - \sum r_i)$ internal dynamics. In this system, the second method has been used for defining the zero dynamics. In the first step, based on (8.7), the new coordinates of the system are

$$\eta = \begin{bmatrix} \eta^{(0)} \\ \eta^{(1)} \\ \vdots \\ \eta^{(m)} \end{bmatrix}$$

$$= \begin{bmatrix} t_1(x) \\ t_2(x) \\ t_3(x) \\ \left(\frac{9P}{4}\right)(R_s(L_{sd} - L_{sq})x_1{}^2 - R_s(L_{sd} - L_{sq})x_2{}^2 - R_s\lambda_m x_2) \\ x_{10} \\ \left(\frac{P}{2J}\right)T_t - \left(\frac{3P^2}{8J}\right)(\lambda_m x_1 + (L_{sd} - L_{sq})x_1 x_2) \\ x_9 \\ \frac{3}{2}(V_{gq}x_6 - V_{gd}x_5) \\ \left(\frac{3}{2}\frac{V_{gd}R_g + V_{gq}\omega_e L_g}{L_g}\right)x_5 + \left(\frac{3}{2}\frac{V_{gd}\omega_e L_g - V_{gq}R_g}{L_g}\right)x_6 - \frac{3}{2}\frac{V_{gd}}{L_g}x_7 + \frac{3}{2}\frac{V_{gq}}{L_g}x_8 \\ \left\{ \begin{aligned} & -\frac{3}{2}\frac{V_{gd}}{L_g C_g}x_3 + \frac{3}{2}\frac{V_{gq}}{L_g C_g}x_4 + \frac{3}{2}\left(\frac{V_{gd}}{L_g C_g} - \frac{V_{gd}R_g^2 + V_{gq}\omega_e L_g R_g}{L_g{}^2} + \frac{V_{gd}\omega_e^2 L_g^2 - V_{gq}\omega_e L_g R_g}{L_g{}^2}\right)x_5 \\ & +\frac{3}{2}\left(-\frac{V_{gq}}{L_g C_g} - \frac{V_{gd}\omega_e L_g R_g + V_{gq}\omega_e^2 L_g^2}{L_g^2} - \frac{V_{gd}\omega_e L_g R_g - V_{gq}R_g^2}{L_g^2}\right)x_6 \\ & +\frac{3}{2}\left(\frac{V_{gd}R_g + 2V_{gq}\omega_e L_g}{L_g^2}\right)x_7 + \frac{3}{2}\left(\frac{2V_{gd}\omega_e L_g - V_{gq}R_g}{L_g^2}\right)x_8 \\ & +\left(\frac{3}{2}\frac{V_{gd}R_g + V_{gq}\omega_e L_g}{L_g}\right)\left(-\frac{V_{gq}}{L_g}\right) + \left(\frac{3}{2}\frac{V_{gd}\omega_e L_g - V_{gq}R_g}{L_g}\right)\left(-\frac{V_{gd}}{L_g}\right) \end{aligned} \right\} \end{bmatrix} \tag{8.43}$$

$$\eta = \begin{bmatrix} \eta^{(0)} \\ \eta^{(1)} \\ \vdots \\ \eta^{(m)} \end{bmatrix} = \begin{bmatrix} \eta_1{}^{(0)} \\ \eta_2{}^{(0)} \\ \eta_3{}^{(0)} \\ \eta_1{}^{(1)} \\ \eta_1{}^{(2)} \\ \eta_2{}^{(2)} \\ \eta_1{}^{(3)} \\ \eta_1{}^{(4)} \\ \eta_2{}^{(4)} \\ \eta_3{}^{(4)} \end{bmatrix} = \begin{bmatrix} t_1(x) \\ t_2(x) \\ t_3(x) \\ a_1 x_1{}^2 + b_1 x_2{}^2 + c_1 x_2 \\ x_{10} \\ a_2 x_1 + b_2 x_1 x_2 + c_2 \\ x_9 \\ a_3 x_6 + b_3 x_5 \\ a_4 x_5 + b_4 x_6 + c_4 x_7 + d_4 x_8 \\ a_5 x_3 + b_5 x_4 + c_5 x_5 + d_5 x_6 + e_5 x_7 + h_5 x_8 + i_5 + j_5 \end{bmatrix} \tag{8.44}$$

where $t(x)$ is chosen as

$$t_1(x) = x_5, t_2(x) = x_4, t_3(x) = x_3 \tag{8.45}$$

With $t(x)$ given in (8.45), the linearized matrix of η will be

$$\Delta\eta = \begin{bmatrix} \Delta\eta^{(0)} \\ \Delta\eta^{(1)} \\ \vdots \\ \Delta\eta^{(m)} \end{bmatrix} = \begin{bmatrix} 0 & 0 & 0 & 0 & 1 & 0 & 0 & 0 & 0 & 0 \\ 0 & 0 & 0 & 1 & 0 & 0 & 0 & 0 & 0 & 0 \\ 0 & 0 & 1 & 0 & 0 & 0 & 0 & 0 & 0 & 0 \\ 2a_1x_{1o} & 2b_1x_{2o} + c_1 & 0 & 0 & 0 & 0 & 0 & 0 & 0 & 0 \\ 0 & 0 & 0 & 0 & 0 & 0 & 0 & 0 & 0 & 1 \\ a_2 + b_2x_{2o} & b_2x_{1o} & 0 & 0 & 0 & 0 & 0 & 0 & 0 & 0 \\ 0 & 0 & 0 & 0 & 0 & 0 & 0 & 0 & 1 & 0 \\ 0 & 0 & 0 & 0 & b_3 & a_3 & 0 & 0 & 0 & 0 \\ 0 & 0 & 0 & 0 & a_4 & b_4 & c_4 & d_4 & 0 & 0 \\ 0 & 0 & a_5 & b_5 & c_5 & d_5 & e_5 & h_5 & 0 & 0 \end{bmatrix} \tag{8.46}$$

in which the determinant of the linearized matrix $\eta(x)$ is not equal to zero

$$\det(\Delta\eta) = -(2b_2x_{1o}{}^2a_1 - 2b_1x_{2o}a_2 - 2b_1x_{2o}{}^2b_2 - c_1a_2 - c_1b_2x_{2o})(c_4h_5 - e_5d_4)a_3 \neq 0$$

$$c_4h_5 - e_5d_4 = -\frac{9}{2}\frac{V_{gd}{}^2\omega_e L_g + V_{gq}{}^2\omega_e L_g}{L_g{}^3} \neq 0 \tag{8.47}$$

Therefore, all the elements of the matrix are linearly independent. The system equations in the new coordinates are given as

$$\begin{bmatrix} x_1 \\ x_2 \\ x_3 \\ x_4 \\ x_5 \\ x_6 \\ x_7 \\ x_8 \\ x_9 \\ x_{10} \end{bmatrix} = \begin{bmatrix} f_{x1}(\eta_1{}^{(1)}, \eta_2{}^{(2)}) \\ f_{x2}(\eta_1{}^{(1)}, \eta_2{}^{(2)}) \\ \eta_3{}^{(0)} \\ \eta_2{}^{(0)} \\ \eta_1{}^{(0)} \\ f_{x6}(\eta_1{}^{(0)}, \eta_1{}^{(4)}) \\ f_{x7}(\eta_1{}^{(0)}, \eta_2{}^{(0)}, \eta_3{}^{(0)}, \eta_1{}^{(4)}, \eta_2{}^{(4)}, \eta_3{}^{(4)}) + C_{x7} \\ f_{x8}(\eta_1{}^{(0)}, \eta_2{}^{(0)}, \eta_3{}^{(0)}, \eta_1{}^{(4)}, \eta_2{}^{(4)}, \eta_3{}^{(4)}) + C_{x8} \\ \eta_1{}^{(3)} \\ \eta_1{}^{(2)} \end{bmatrix} \tag{8.48}$$

The zero dynamics are determined by setting the output variables to either zero or the steady-state values

$$\begin{cases} h_1(x) = \eta_1{}^{(1)} = \gamma^{ss} = 0 \\ h_2(x) = \eta_1{}^{(2)} = \omega_{r\_ref} \\ h_3(x) = \eta_1{}^{(3)} = V_{dc\_ref} \\ h_4(x) = \eta_1{}^{(4)} = Q_{2\_ref} \end{cases} \tag{8.49}$$

and their derivatives

$$\begin{cases} \eta_2{}^{(2)} = 0 \\ \eta_2{}^{(4)} = 0 \\ \eta_3{}^{(4)} = 0 \end{cases} \tag{8.50}$$

With Equation 8.50, the new equations are given in (8.51) which are constant.

$$\begin{cases} x_1 = a_{x1} \\ x_2 = a_{x2} \end{cases} \tag{8.51}$$

Using (8.8), the following can be calculated

$$\begin{cases} F_1 = L_f t_1 = \dfrac{x_7}{L_g} - \dfrac{V_{gq}}{L_g} - \dfrac{R_g x_5}{L_g} - \dfrac{\omega_e L_g x_6}{L_g} \\ F_2 = L_f t_2 = -\dfrac{x_8}{L_{s2}} - \dfrac{R_{s2} x_4}{L_{s2}} + \dfrac{\omega_e L_{s2} x_3}{L_{s2}} \\ F_3 = L_f t_3 = -\dfrac{x_7}{L_{s2}} - \dfrac{R_{s2} x_3}{L_{s2}} - \dfrac{\omega_e L_{s2} x_4}{L_{s2}} \end{cases} \tag{8.52}$$

The other matrices are defined as

$$\begin{cases} G_1 = \begin{bmatrix} L_{g1}t_1 & L_{g2}t_1 & L_{g3}t_1 & L_{g4}t_1 \end{bmatrix} = \begin{bmatrix} 0 & 0 & 0 & 0 \end{bmatrix} \\ G_2 = \begin{bmatrix} L_{g1}t_2 & L_{g2}t_2 & L_{g3}t_2 & L_{g4}t_2 \end{bmatrix} = \begin{bmatrix} 0 & 0 & 0 & \dfrac{x_9}{2L_{s2}} \end{bmatrix} \\ G_3 = \begin{bmatrix} L_{g1}t_3 & L_{g2}t_3 & L_{g3}t_3 & L_{g4}t_3 \end{bmatrix} - \begin{bmatrix} 0 & 0 & \dfrac{x_9}{2L_{s2}} & 0 \end{bmatrix} \end{cases} \tag{8.53}$$

To substitute the input variables with the states, the equations of $C(x)$ and $W(x)$ should be defined as

$$\begin{cases} C_1 = \left(\dfrac{9P}{4}\right)\begin{bmatrix} \left(R_s\left(L_{sd}-L_{sq}\right)x_1\right)\dfrac{x_9}{2L_{sq}} & (-2R_s(L_{sd}-L_{sq})x_2 - R_s\lambda_m)\dfrac{x_9}{2L_{sd}} & 0 & 0 \end{bmatrix} \\ C_2 = -\left(\dfrac{3P^2}{8J}\right)\begin{bmatrix} (\lambda_m + \left(L_{sd}-L_{sq}\right)x_2)\dfrac{x_9}{2L_{sq}} & (L_{sd}-L_{sq})x_1\dfrac{x_9}{2L_{sd}} & 0 & 0 \end{bmatrix} \\ C_3 = -\dfrac{3}{4C_{dc}}\begin{bmatrix} x_1 & x_2 & x_3 & x_4 \end{bmatrix} \\ C_4 = \begin{bmatrix} 0 & 0 & -\dfrac{3}{4}\dfrac{V_{gd}}{L_g C_g L_{s2}}x_9 & \dfrac{3}{4}\dfrac{V_{gq}}{L_g C_g L_{s2}}x_9 \end{bmatrix} \end{cases} \tag{8.54}$$

$$\begin{cases} W_{11} = \left(\dfrac{9P}{4}\right)\begin{pmatrix} 2R_s\left(L_{sd}-L_{sq}\right)x_1\left(-\dfrac{R_s x_1}{L_{sq}} - \dfrac{x_{10}L_{sd}x_2}{L_{sq}} - \dfrac{x_{10}\lambda_m}{L_{sq}}\right) \\ -(2R_s(L_{sd}-L_{sq})x_2{}^2 + R_s\lambda_m)\left(-\dfrac{R_s x_2}{L_{sd}} + \dfrac{x_{10}L_{sq}x_1}{L_{sd}}\right) \end{pmatrix} \\ W_{21} = -\left(\dfrac{3P^2}{8J}\right)\begin{pmatrix} \left(\lambda_m + \left(L_{sd}-L_{sq}\right)x_2\right)\left(-\dfrac{R_s x_1}{L_{sq}} - \dfrac{x_{10}L_{sd}x_2}{L_{sq}} - \dfrac{x_{10}\lambda_m}{L_{sq}}\right) \\ +(L_{sd}-L_{sq})x_1\left(\dfrac{R_s x_2}{L_{sd}} + \dfrac{x_{10}L_{sq}x_1}{L_{sd}}\right) \end{pmatrix} \\ W_{31} = 0 \\ W_{41} = a_5 f_{31} + b_5 f_{41} + c_5 f_{51} + d_5 f_{61} + e_5 f_{71} + h_5 f_{81} \end{cases} \tag{8.55}$$

The above equations are transformed into the new coordinates to yield the equations of the zero dynamics which are as follows

$$\begin{cases} \dot{\eta}_1^{(0)} = A_{z1}\eta_1^{(0)} + B_{z1}\eta_2^{(0)} + C_{z1}\eta_3^{(0)} + D_{z1} \\ \dot{\eta}_2^{(0)} = \left(\begin{array}{c} A_{z2}\eta_1^{(0)} + B_{z2}\eta_2^{(0)} + C_{z2}\eta_3^{(0)} + M_{z2} \\ \dfrac{(D_{z2}\eta_1^{(0)3} + E_{z2}\eta_1^{(0)2} + F_{z2}\eta_1^{(0)} + G_{z2}\eta_1^{(0)}\eta_2^{(0)} + H_{z2}\eta_3^{(0)}\eta_1^{(0)})}{\eta_1^{(0)}V_{gq} + \eta_2^{(0)}V_{gd}} \end{array} \right) \\ \dot{\eta}_3^{(0)} = \left(\begin{array}{c} A_{z3}\eta_1^{(0)} + B_{z3}\eta_2^{(0)} + C_{z3}\eta_3^{(0)} + M_{z3} \\ + \dfrac{(D_{z3}\eta_1^{(0)2}\eta_2^{(0)} + E_{z3}\eta_1^{(0)}\eta_2^{(0)} + F_{z3}\eta_2^{(0)2} + G_{z3}\eta_3^{(0)}\eta_2^{(0)} + H_{z3}\eta_2^{(0)})}{\eta_1^{(0)}V_{gq} + \eta_2^{(0)}V_{gd}} \end{array} \right) \end{cases} \tag{8.56}$$

The equations of the zero dynamics are nonlinear and hence the stability can be checked using the corresponding linearized equations. The characteristic equation of the zero dynamics is a cubic equation given as

$$z^3 + n_2 z^2 + n_1 z + n_0 = 0 \tag{8.57}$$

The Routh–Hurwitz criteria for the stability of this characteristic equation is given as

$$\begin{cases} n_2 > 0, n_1 > 0, n_0 > 0 \\ (n_1 n_2 - n_0) > 0 \end{cases} \tag{8.58}$$

Figure 8.12 shows how the signs of the two coefficients of (8.57) and (8.58) are influenced by the values of the rotor speed and the DC capacitor voltage. The zero dynamics is unstable for the operating regions considered, which leads to the conclusion that the IPM generator system has unstable zero dynamics, and is in non-minimum phase for all operating conditions. For the selected controlled variables, there is an inherent structural constraint that adversely affects the controller design.

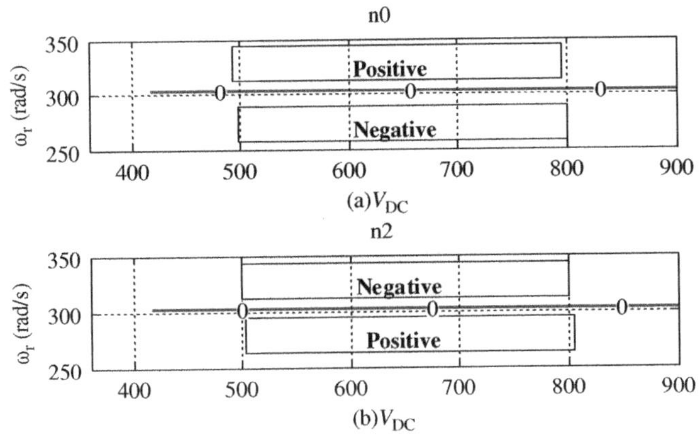

Figure 8.12 Two of the coefficients of the linearized zero dynamics characteristic equation: (a) the last coefficient (n_0) and (b) second coefficient (n_2) [23]

8.5 Controllability and Stability Analysis of PV System Connected to Current Source Inverter

As illustrated in Figure 8.13, in the study system, a PV system is connected to the grid through a current source inverter. The nonlinear characteristic equation of the PV system relating the PV voltage to its output current is as follows

$$I_{pv} = n_p I_{ph} - n_p I_{rs}\left[\exp\left(\frac{q}{k\theta A}\frac{V_{pv}}{n_s}\right) - 1\right] \tag{8.59}$$

$$I_{ph} = [I_{src} + k_\theta(\theta - \theta_r)]\frac{S}{100} \tag{8.60}$$

Equations (8.59) and (8.60) define the steady-state equations relating the current flowing out of the PV and the terminal voltage. In the equations, I_{rs} and θ are the reverse saturation current and the temperature of the $p-n$ junction. I_{src} is the short-circuit current of one PV cell, n_s and n_p are the number of series and parallel PV cells in a string, respectively. A is the ideality factor, S is the solar irradiation level, θ_r is the cell reference temperature and k_θ is the temperature coefficient. $q(=1.602e^{-19}$ C) and $k(=1.38e^{-23}$ J/K) are the unit electric charge and Boltzmann's constant. The dynamic equations of the grid are

$$\begin{cases} V_{mq} - V_{gq} = R_g I_{gq} + L_g p I_{gq} + \omega_e L_g I_{gd} \\ V_{md} - V_{gd} = R_g I_{gd} + L_g p I_{gd} - \omega_e L_g I_{gq} \end{cases} \tag{8.61}$$

where V_{gqd} and I_{gqd} are the qd-axis grid voltage and current. The inverter and the DC-side dynamic equations are as follows

$$\begin{cases} I_{slq} = \dfrac{M_q}{2} I_{dl} \\ I_{sld} = \dfrac{M_d}{2} I_{dl} \\ V_{dc} = \dfrac{3}{4}\left(M_q V_{mq} + M_d V_{md}\right) \end{cases} \tag{8.62}$$

$$\begin{cases} \dfrac{M_q}{2} I_{dl} - I_{gq} = C_m p V_{mq} + \omega_e C_m V_{md} \\ \dfrac{M_d}{2} I_{dl} - I_{gd} = C_m p V_{md} - \omega_e C_m V_{mq} \end{cases} \tag{8.63}$$

$$V_{cl} - \frac{3}{4}(M_q V_{mq} + M_d V_{md}) = R_{dl} I_{dl} + L_{dl} p I_{dl} \tag{8.64}$$

$$C_l p V_{cl} = I_{pv} - I_{dl} \text{ and } V_{cl} = V_{pv} \tag{8.65}$$

To control the input current of the inverter and the reactive power following into the grid

$$\begin{cases} R_{dl} I_{dl}^2 + \dfrac{1}{2} L_{dl} p I_{dl}^2 = V_{dl} I_{dl} - \dfrac{3}{2}\left(\dfrac{M_q}{2} I_{dl} V_{mq} + \dfrac{M_d}{2} I_{dl} V_{md}\right) \\ Q_g = \dfrac{3}{2}(V_{gq} I_{gd} - V_{gd} I_{gq}) \end{cases} \tag{8.66}$$

If the grid voltage is aligned to the q-axis reference frame ($V_{gd} = 0$) and the system operates under a unity power factor, then

$$Q_g = \frac{3}{2} V_{gq} I_{gd} = 0 \tag{8.67}$$

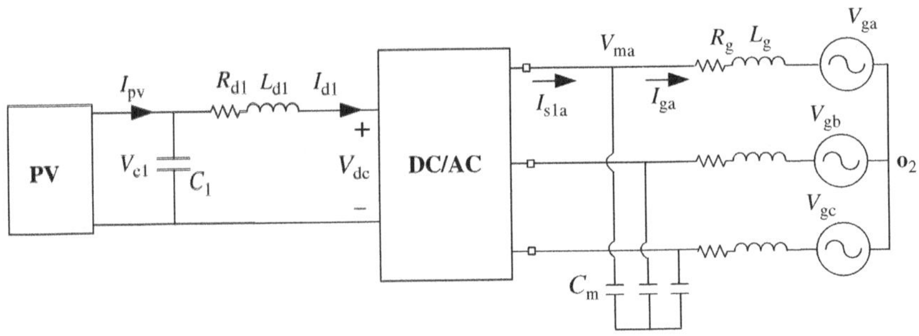

Figure 8.13 A PV generator system with a current source inverter connected through LCL filter to the grid

8.5.1 *Steady State and Stability Analysis of the System*

Under steady state operating conditions the derivatives of the states are set to zero. The resulting equations with the algebraic equation defining the reference grid reactive power and operating under the MPP of PV are used to determine the steady-state operation. They are given in Figures 8.14–8.17. The known variables are

$$\left[V_{gq}\ V_{gd}\ Q_g\ I_{pv}\ V_{pv}\right] \tag{8.68}$$

and the unknown variables are

$$\left[M_q\ M_d\ I_{gq}\ I_{gd}\ V_{mq}\ V_{md}\ I_{d1}\right] \tag{8.69}$$

The irradiation of the sun is varied (S_1) and the PV voltage and current based on the MPP operation are calculated to solve the steady-state equations, which is given as

$$\begin{cases} V_{mq} - V_{gq} - R_g I_{gq} - \omega_e L_g I_{gd} = 0 \\ V_{md} - V_{gd} - R_g I_{gd} + \omega_e L_g I_{gq} = 0 \\ \dfrac{M_q}{2} I_{d1} - I_{gq} - \omega_e C_m V_{md} = 0 \\ \dfrac{M_d}{2} I_{d1} - I_{gd} + \omega_e C_m V_{mq} = 0 \\ V_{d1} - \dfrac{3}{4}\left(M_q V_{mq} + M_d V_{md}\right) - R_{d1} I_{d1} = 0 \\ I_{pv} - I_{d1} = 0 \\ Q_g = \dfrac{3}{2} V_{gq} I_{gd} = 0 \end{cases} \tag{8.70}$$

To study the stability region of the system in all feasible operation regions, the model equations of the electrical subsystem are linearized. The output, input and state variables of the system are

$$h(x) = \begin{bmatrix} I_{d1}^{2} \\ Q_g \end{bmatrix}, U = \begin{bmatrix} M_q \\ M_d \end{bmatrix}, X = \left[V_{mq}\ V_{md}\ I_{gq}\ I_{gd}\ I_{d1}\ V_{pv}\right] \tag{8.71}$$

To define the number of internal dynamics

$$\begin{cases} p\Delta I_{\mathrm{gq}} = \dfrac{1}{L_{\mathrm{g}}}\left(\Delta V_{\mathrm{mq}} - R_{\mathrm{g}}\Delta I_{\mathrm{gq}} - \omega_{\mathrm{e}} L_{\mathrm{g}}\Delta I_{\mathrm{gd}}\right) \\ p\Delta I_{\mathrm{gd}} = \dfrac{1}{L_{\mathrm{g}}}\left(\Delta V_{\mathrm{md}} - R_{\mathrm{g}}\Delta I_{\mathrm{gd}} + \omega_{\mathrm{e}} L_{\mathrm{g}}\Delta I_{\mathrm{gq}}\right) \\ p\Delta V_{\mathrm{mq}} = \dfrac{1}{C_{\mathrm{m}}}\left(\dfrac{M_{\mathrm{qo}}}{2}\Delta I_{d1} + \dfrac{I_{\mathrm{d1o}}}{2}\Delta M_{\mathrm{q}} - \Delta I_{\mathrm{gq}} - \omega_{\mathrm{e}} C_{\mathrm{m}}\Delta V_{\mathrm{md}}\right) \\ p\Delta V_{\mathrm{md}} = \dfrac{1}{C_{\mathrm{m}}}\left(\dfrac{M_{\mathrm{do}}}{2}\Delta I_{\mathrm{d1}} + \dfrac{I_{\mathrm{d1o}}}{2}\Delta M_{\mathrm{d}} - \Delta I_{\mathrm{gd}} + \omega_{\mathrm{e}} C_{\mathrm{m}}\Delta V_{\mathrm{mq}}\right) \\ p\Delta I_{\mathrm{d1}} = \dfrac{1}{L_{\mathrm{d1}}}\left(\Delta V_{\mathrm{pv}} - \dfrac{3M_{\mathrm{qo}}}{4}\Delta V_{\mathrm{mq}} - \dfrac{3V_{\mathrm{mqo}}}{4}\Delta M_{\mathrm{q}} - \dfrac{3M_{\mathrm{do}}}{4}\Delta V_{\mathrm{md}} - \dfrac{3V_{\mathrm{mdo}}}{4}\Delta M_{\mathrm{d}} - R_{\mathrm{d1}}\Delta I_{\mathrm{d1}}\right) \\ p\Delta V_{\mathrm{pv}} = -\dfrac{qn_{\mathrm{p}}I_{\mathrm{rs}}}{C_1 k\theta A n_{\mathrm{s}}}\exp\left(\dfrac{q}{k\theta A}\dfrac{V_{\mathrm{pv}}}{n_{\mathrm{s}}}\right)\Delta V_{\mathrm{pv}} - \dfrac{1}{C_1}\Delta I_{\mathrm{d1}} \end{cases} \tag{8.72}$$

$$A = \begin{bmatrix} -\dfrac{R_{\mathrm{g}}}{L_{\mathrm{g}}} & -\omega_{\mathrm{e}} & \dfrac{1}{L_{\mathrm{g}}} & 0 & 0 & 0 \\ \omega_{\mathrm{e}} & -\dfrac{R_{\mathrm{g}}}{L_{\mathrm{g}}} & 0 & \dfrac{1}{L_{\mathrm{g}}} & 0 & 0 \\ -\dfrac{1}{C_{\mathrm{m}}} & 0 & 0 & -\omega_{\mathrm{e}} & \dfrac{M_{\mathrm{qo}}}{2C_{\mathrm{m}}} & 0 \\ 0 & -\dfrac{1}{C_{\mathrm{m}}} & \omega_{\mathrm{e}} & 0 & \dfrac{M_{\mathrm{do}}}{2C_{\mathrm{m}}} & 0 \\ 0 & 0 & -\dfrac{3M_{\mathrm{qo}}}{4L_{\mathrm{d1}}} & -\dfrac{3M_{\mathrm{do}}}{4L_{\mathrm{d1}}} & -\dfrac{R_{\mathrm{d1}}}{L_{\mathrm{d1}}} & \dfrac{1}{L_{\mathrm{d1}}} \\ 0 & 0 & 0 & 0 & -\dfrac{1}{C_1} & -\dfrac{qn_{\mathrm{p}}I_{\mathrm{rs}}}{C_1 k\theta A n_{\mathrm{s}}}\exp\left(\dfrac{q}{k\theta A}\dfrac{V_{\mathrm{pv}}}{n_{\mathrm{s}}}\right) \end{bmatrix} \tag{8.73}$$

The characteristic equation is defined as a sixth-order equation

$$|\lambda I - A| = a_6\lambda^6 + a_5\lambda^5 + a_4\lambda^4 + a_3\lambda^3 + a_2\lambda^2 + a_1\lambda + a_0 = 0 \tag{8.74}$$

Using the Routh–Hurwitz, the criteria for stability is

$$\begin{cases} a_6 > 0,\ a_5 > 0,\ a_4 > 0,\ a_3 > 0,\ a_2 > 0,\ a_1 > 0,\ a_0 > 0 \\ b_1 > 0,\ c_1 > 0,\ d_1 > 0,\ e_1 > 0 \end{cases} \tag{8.75}$$

The parameters of (8.75) are defined using (8.23). After applying the steady-state parameters, the system is stable in the whole operation region.

8.5.2 Zero Dynamics Analysis of PV

To analyze the zero dynamics of the system, the two inputs and two outputs are

$$h(x) = \left[I_{\mathrm{d1}}{}^2\ \ Q_{\mathrm{g}}\right]^T,\ U = \left[M_{\mathrm{q}}\ \ M_{\mathrm{d}}\right]^T \tag{8.76}$$

Differentiating the outputs to define the relative orders, the results are (Figures 8.14–8.17)

$$\begin{cases} r_1 = 1, \\ r_2 = 2, \end{cases} \quad \text{and} \quad r = \sum_{i=1}^{2} r_i = 3 \tag{8.77}$$

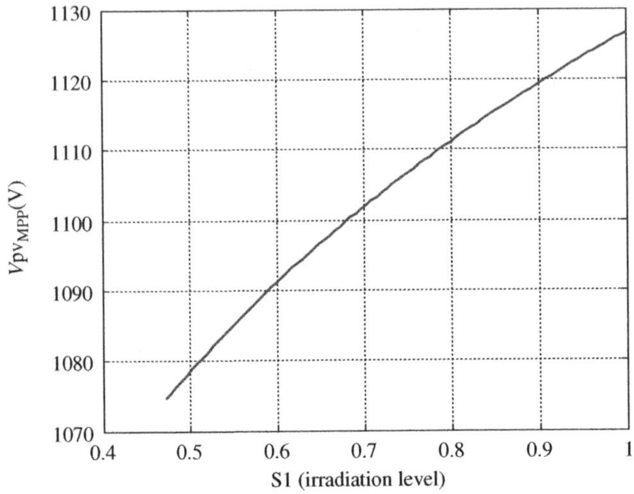

Figure 8.14 PV output voltage at MPP for different irradiation levels

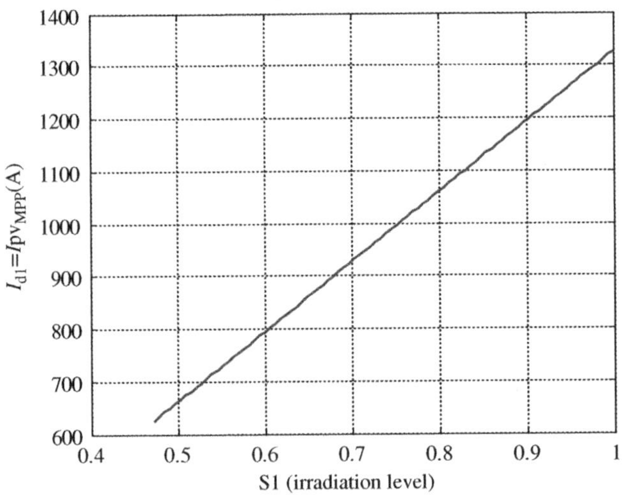

Figure 8.15 PV output current at MPP for different irradiation levels

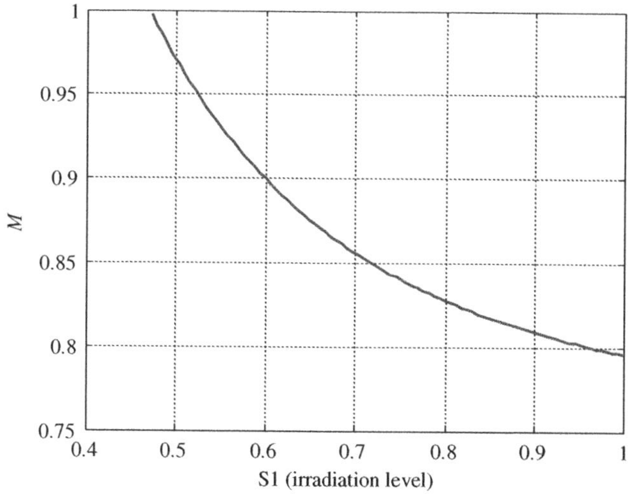

Figure 8.16 Inverter modulation index magnitude at different irradiation levels

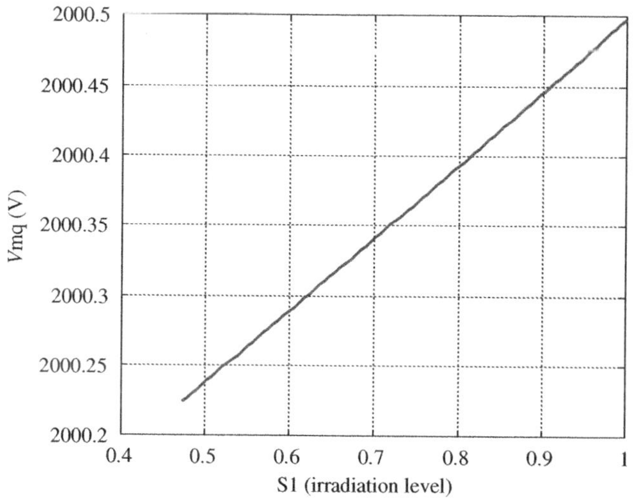

Figure 8.17 The q-axis capacitor voltage at different irradiation levels

Therefore, the system has three $(n-\sum r_i)$ internal dynamics. The dynamic equations of the system in nonlinear format are

$$\dot{X} = f(x) + g_1(x)u_1 + g_2(x)u_2 \tag{8.78}$$

$$\begin{bmatrix} \dot{x}_1 \\ \dot{x}_2 \\ \dot{x}_3 \\ \dot{x}_4 \\ \dot{x}_5 \\ \dot{x}_6 \end{bmatrix} = \begin{bmatrix} \frac{x_3}{L_g} - \frac{V_{gq}}{L_g} - \frac{R_g}{L_g}x_1 - \omega_e x_2 \\ \frac{x_4}{L_g} - \frac{R_g}{L_g}x_2 + \omega_e x_1 \\ -\frac{x_1}{C_m} - \omega_e x_4 \\ -\frac{x_2}{C_m} + \omega_e x_3 \\ \frac{x_6}{L_{d1}} - \frac{R_{d1}}{L_{d1}}x_5 \\ \frac{I_{pv}}{C_1} - \frac{x_5}{C_1} \end{bmatrix} + \begin{bmatrix} 0 \\ 0 \\ \frac{x_5}{2C_m} \\ 0 \\ -\frac{3x_3}{4L_{d1}} \\ 0 \end{bmatrix} u_1 + \begin{bmatrix} 0 \\ 0 \\ 0 \\ \frac{x_5}{2C_m} \\ -\frac{3x_4}{4L_{d1}} \\ 0 \end{bmatrix} u_2 \tag{8.79}$$

$$Y = \begin{bmatrix} h_1(x) \\ h_2(x) \end{bmatrix} = \begin{bmatrix} x_5{}^2 \\ \frac{3}{2}V_{gq}x_2 \end{bmatrix} \tag{8.80}$$

The new coordinates of the system are

$$\eta = \begin{bmatrix} \eta^{(0)} \\ \eta^{(1)} \\ \vdots \\ \eta^{(m)} \end{bmatrix} = \begin{bmatrix} t_1(x) \\ t_2(x) \\ t_3(x) \\ x_5{}^2 \\ \frac{3}{2}V_{gq}x_2 \\ \frac{3}{2}V_{gq}\left(\frac{x_4}{L_g} - \frac{R_g}{L_g}x_2 + \omega_e x_1\right) \end{bmatrix} \tag{8.81}$$

and if

$$t_1(x) = x_1,\ t_2(x) = x_3,\ t_3(x) = x_6 \tag{8.82}$$

$$\Delta\eta = \begin{bmatrix} \Delta\eta^{(0)} \\ \Delta\eta^{(1)} \\ \vdots \\ \Delta\eta^{(m)} \end{bmatrix} = \begin{bmatrix} 1 & 0 & 0 & 0 & 0 & 0 \\ 0 & 0 & 1 & 0 & 0 & 0 \\ 0 & 0 & 0 & 0 & 0 & 1 \\ 0 & 0 & 0 & 0 & 2x_{5o} & 0 \\ 0 & \frac{3}{2}V_{gq} & 0 & 0 & 0 & 0 \\ \frac{3V_{gq}\omega_e}{2} & -\frac{3V_{gq}R_g}{2L_g} & 0 & \frac{3V_{gq}}{2L_g} & 0 & 0 \end{bmatrix} \begin{bmatrix} x_1 \\ x_2 \\ x_3 \\ x_4 \\ x_5 \\ x_6 \end{bmatrix} \tag{8.83}$$

$$\det(\Delta\eta) = \frac{9V_{gq}{}^2}{2L_g}x_{5o} \neq 0 \tag{8.84}$$

With $t(x)$ given in (8.82), the determinant of the linearized matrix $\eta(x)$ is not equal to zero, in which case all the elements of $\eta(x)$ are linearly independent. The system equations in the new coordinates are given as

$$\begin{bmatrix} x_1 \\ x_2 \\ x_3 \\ x_4 \\ x_5 \\ x_6 \end{bmatrix} = \begin{bmatrix} \eta_1{}^{(0)} \\ \dfrac{2}{3V_{\mathrm{gq}}}\eta_1{}^{(2)} \\ \eta_2{}^{(0)} \\ \dfrac{2L_{\mathrm{g}}}{3V_{\mathrm{gq}}}\eta_2{}^{(2)} - \omega_{\mathrm{e}}L_{\mathrm{g}}\eta_1{}^{(0)} + \dfrac{2R_{\mathrm{g}}}{3V_{\mathrm{gq}}}\eta_1{}^{(2)} \\ \sqrt{\eta_1{}^{(1)}} \\ \eta_3{}^{(0)} \end{bmatrix} \tag{8.85}$$

The following can be calculated

$$\begin{cases} F_1 = L_{\mathrm{f}}t_1 = \dfrac{x_3}{L_{\mathrm{g}}} - \dfrac{V_{\mathrm{gq}}}{L_{\mathrm{g}}} - \dfrac{R_{\mathrm{g}}}{L_{\mathrm{g}}}x_1 - \omega_{\mathrm{e}}x_2 \\ F_2 = L_{\mathrm{f}}t_2 = -\dfrac{x_1}{C_{\mathrm{m}}} - \omega_{\mathrm{e}}x_4 \\ F_3 = L_{\mathrm{f}}t_3 = \dfrac{I_{\mathrm{pv}}}{C_1} - \dfrac{x_5}{C_1} \end{cases} \tag{8.86}$$

The other matrices are defined as

$$\begin{cases} G_1 = \begin{bmatrix} L_{\mathrm{g1}}t_1 & L_{\mathrm{g2}}t_1 \end{bmatrix} = \begin{bmatrix} 0 & 0 \end{bmatrix} \\ G_2 = \begin{bmatrix} L_{\mathrm{g1}}t_2 & L_{\mathrm{g2}}t_2 \end{bmatrix} = \begin{bmatrix} \dfrac{x_5}{2C_{\mathrm{m}}} & 0 \end{bmatrix} \\ G_3 = \begin{bmatrix} L_{\mathrm{g1}}t_3 & L_{\mathrm{g2}}t_3 \end{bmatrix} = \begin{bmatrix} 0 & 0 \end{bmatrix} \end{cases} \tag{8.87}$$

To substitute the input variables with the states, the equations of $C(x)$and $W(x)$ should be defined as

$$\begin{cases} C_1 = \begin{bmatrix} L_{\mathrm{g1}}h_1 & L_{\mathrm{g2}}h_1 \end{bmatrix} = \begin{bmatrix} -\dfrac{3x_5x_3}{2L_{\mathrm{d1}}} & -\dfrac{3x_5x_4}{2L_{\mathrm{d1}}} \end{bmatrix} \\ C_2 = \begin{bmatrix} L_{\mathrm{g1}}L_{\mathrm{f}}h_2 & L_{\mathrm{g2}}L_{\mathrm{f}}h_2 \end{bmatrix} = \begin{bmatrix} 0 & \dfrac{3V_{\mathrm{gq}}x_5}{4C_{\mathrm{m}}L_{\mathrm{g}}} \end{bmatrix} \end{cases} \tag{8.88}$$

$$\begin{cases} W_{11} = L_{\mathrm{f}}h_1 = \dfrac{2x_5x_6}{L_{\mathrm{d1}}} - \dfrac{2R_{\mathrm{d1}}}{L_{\mathrm{d1}}}x_5{}^2 \\ W_{21} = L_{\mathrm{f}}h_2 = \dfrac{3V_{\mathrm{gq}}}{2}\left(\dfrac{x_4}{L_{\mathrm{g}}} - \dfrac{R_{\mathrm{g}}}{L_{\mathrm{g}}}x_2 + \omega_{\mathrm{e}}x_1\right) \\ W_{31} = L_{\mathrm{f}}{}^2h_2 = \left(\begin{matrix} \dfrac{3V_{\mathrm{gq}}\omega_{\mathrm{e}}}{2}\left(\dfrac{x_3}{L_{\mathrm{g}}} - \dfrac{V_{\mathrm{gq}}}{L_{\mathrm{g}}} - \dfrac{R_{\mathrm{g}}}{L_{\mathrm{g}}}x_1 - \omega_{\mathrm{e}}x_2\right) \\ -\dfrac{3V_{\mathrm{gq}}R_{\mathrm{g}}}{2L_{\mathrm{g}}}\left(\dfrac{x_4}{L_{\mathrm{g}}} - \dfrac{R_{\mathrm{g}}}{L_{\mathrm{g}}}x_2 + \omega_{\mathrm{e}}x_1\right) + \dfrac{3V_{\mathrm{gq}}}{2L_{\mathrm{g}}}\left(-\dfrac{x_2}{C_{\mathrm{m}}} + \omega_{\mathrm{e}}x_3\right) \end{matrix} \right) \end{cases} \tag{8.89}$$

For calculation of zero dynamics, the outputs should be set at either zero or their steady-state values. Thus, their derivatives will be zero. Therefore,

$$\begin{cases} h_1(x) = \eta_1^{(1)} = I_{\mathrm{d1\_ref}}^{\;2} \\ h_1(x) = \eta_1^{(2)} = Q_{\mathrm{g\_ref}} \Rightarrow \eta_2^{(2)} = 0 \end{cases} \tag{8.90}$$

From the above results, during the calculation of zero dynamics

$$\begin{cases} x_2 = a_{x2} = \dfrac{2Q_{\mathrm{g\_ref}}}{3V_{\mathrm{gq}}} = cte \\ x_5 = a_{x5} = I_{\mathrm{d1\_ref}} = cte \end{cases} \tag{8.91}$$

and there is no need for their conversion to new coordinates

$$\begin{bmatrix} x_1 \\ x_2 \\ x_3 \\ x_4 \\ x_5 \\ x_6 \end{bmatrix} = \begin{bmatrix} \eta_1^{(0)} \\ a_{x2} \\ \eta_2^{(0)} \\ -\omega_{\mathrm{e}} L_{\mathrm{g}} \eta_1^{(0)} + \dfrac{2R_{\mathrm{g}} Q_{\mathrm{g\_ref}}}{3V_{\mathrm{gq}}} \\ a_{x5} \\ \eta_3^{(0)} \end{bmatrix} \tag{8.92}$$

If the matrices F_i, G_i, C_i, W_i are converted to new coordinates

$$\begin{cases} F_1 = L_{\mathrm{f}} t_1 = \dfrac{\eta_2^{(0)}}{L_{\mathrm{g}}} - \dfrac{V_{\mathrm{gq}}}{L_{\mathrm{g}}} - \dfrac{R_{\mathrm{g}}}{L_{\mathrm{g}}} \eta_1^{(0)} - \omega_{\mathrm{e}} a_{x2} \\ F_2 = L_{\mathrm{f}} t_2 = -\dfrac{\eta_1^{(0)}}{C_{\mathrm{m}}} + \omega_{\mathrm{e}}^2 L_{\mathrm{g}} \eta_1^{(0)} - \dfrac{\omega_{\mathrm{e}} 2R_{\mathrm{g}} Q_{\mathrm{g\_ref}}}{3V_{\mathrm{gq}}} \\ F_3 = L_{\mathrm{f}} t_3 = \dfrac{I_{\mathrm{pv}}}{C_1} - \dfrac{a_{x5}}{C_1} \end{cases} \tag{8.93}$$

$$\begin{cases} G_1 = \begin{bmatrix} L_{\mathrm{g1}} t_1 & L_{\mathrm{g2}} t_1 \end{bmatrix} = \begin{bmatrix} 0 & 0 \end{bmatrix} \\ G_2 = \begin{bmatrix} L_{\mathrm{g1}} t_2 & L_{\mathrm{g2}} t_2 \end{bmatrix} = \begin{bmatrix} \dfrac{a_{x5}}{2C_{\mathrm{m}}} & 0 \end{bmatrix} \\ G_3 = \begin{bmatrix} L_{\mathrm{g1}} t_3 & L_{\mathrm{g2}} t_3 \end{bmatrix} = \begin{bmatrix} 0 & 0 \end{bmatrix} \end{cases} \tag{8.94}$$

$$\begin{cases} C_1 = \begin{bmatrix} L_{\mathrm{g1}} h_1 & L_{\mathrm{g2}} h_1 \end{bmatrix} = \begin{bmatrix} -\dfrac{3a_{x5}\eta_2^{(0)}}{2L_{\mathrm{d1}}} & -\dfrac{3a_{x5}}{2L_{\mathrm{d1}}} \left(-\omega_{\mathrm{e}} L_{\mathrm{g}} \eta_1^{(0)} + \dfrac{2R_{\mathrm{g}} Q_{\mathrm{g\_ref}}}{3V_{\mathrm{gq}}} \right) \end{bmatrix} \\ C_2 = \begin{bmatrix} L_{\mathrm{g1}} L_{\mathrm{f}} h_2 & L_{\mathrm{g2}} L_{\mathrm{f}} h_2 \end{bmatrix} = \begin{bmatrix} 0 & \dfrac{3V_{\mathrm{gq}} a_{x5}}{4C_{\mathrm{m}} L_{\mathrm{g}}} \end{bmatrix} \end{cases} \tag{8.95}$$

$$\begin{cases} W_{11} = L_f h_1 = \dfrac{2a_{x5}\eta_3^{(0)}}{L_{d1}} - \dfrac{2R_{d1}}{L_{d1}}a_{x5}^{\ 2} \\ W_{21} = L_f^{\ 2}h_2 = \left(\begin{array}{c} \dfrac{3V_{gq}\omega_e}{2}\left(\dfrac{\eta_2^{(0)}}{L_g} - \dfrac{V_{gq}}{L_g} - \dfrac{R_g}{L_g}\eta_1^{(0)} - \omega_e a_{x2}\right) + \dfrac{3V_{gq}}{2L_g}\left(-\dfrac{a_{x2}}{C_m} + \omega_e\eta_2^{(0)}\right) \\ -\dfrac{3V_{gq}R_g}{2L_g}\left(\dfrac{\left(-\omega_e L_g\eta_1^{(0)} + \dfrac{2R_g Q_{g\_ref}}{3V_{gq}}\right)}{L_g} - \dfrac{R_g}{L_g}a_{x2} + \omega_e\eta_1^{(0)} \right) \end{array} \right) \end{cases} \quad (8.96)$$

$$\begin{cases} \dot{\eta}_1^{(0)} = \dfrac{\eta_2^{(0)}}{L_g} - \dfrac{V_{gq}}{L_g} - \dfrac{R_g}{L_g}\eta_1^{(0)} - \dfrac{2\omega_e Q_{g\_ref}}{3V_{gq}} \\ \dot{\eta}_2^{(0)} = a_{21} + a_{22}\eta_1^{(0)} + \dfrac{a_{23}}{\eta_2^{(0)}} + \dfrac{a_{24}\eta_1^{(0)2}}{\eta_2^{(0)}} + \dfrac{a_{25}\eta_1^{(0)}}{\eta_2^{(0)}} + \dfrac{a_{26}\eta_3^{(0)}}{\eta_2^{(0)}} \\ \dot{\eta}_3^{(0)} = \dfrac{I_{pv}}{C_1} - \dfrac{I_{d1\_ref}}{C_1} \end{cases} \quad (8.97)$$

$$\begin{cases} \dot{\eta}_1^{(0)} = a_{11} + a_{12}\eta_1^{(0)} + a_{13}\eta_2^{(0)} \\ \dot{\eta}_2^{(0)} = a_{21} + a_{22}\eta_1^{(0)} + \dfrac{a_{23}}{\eta_2^{(0)}} + \dfrac{a_{24}\eta_1^{(0)2}}{\eta_2^{(0)}} + \dfrac{a_{25}\eta_1^{(0)}}{\eta_2^{(0)}} + \dfrac{a_{26}\eta_3^{(0)}}{\eta_2^{(0)}} \\ \dot{\eta}_3^{(0)} = a_{31} \end{cases} \quad (8.98)$$

$$\begin{bmatrix} \Delta\dot{\eta}_1^{(0)} \\ \Delta\dot{\eta}_2^{(0)} \\ \Delta\dot{\eta}_3^{(0)} \end{bmatrix} = \begin{bmatrix} 0 & a_{12} & a_{13} \\ a_{22} + \dfrac{2a_{24}\eta_1^{(0)} + a_{25}}{\eta_2^{(0)}} & -\dfrac{1}{\eta_2^{(0)2}}(a_{23} + a_{24}\eta_1^{(0)2} + a_{25}\eta_1^{(0)} + a_{26}\eta_3^{(0)}) & \dfrac{a_{26}}{\eta_2^{(0)}} \\ 0 & 0 & 0 \end{bmatrix} \begin{bmatrix} \Delta\eta_1^{(0)} \\ \Delta\eta_2^{(0)} \\ \Delta\eta_3^{(0)} \end{bmatrix} \quad (8.99)$$

$$|SI - A| = s(s^2 + b_1 s + b_0) \quad (8.100)$$

$$\begin{cases} b_1 = \dfrac{1}{\eta_{2o}^{(0)2}}(a_{23} + a_{24}\eta_{1o}^{(0)2} + a_{25}\eta_{1o}^{(0)} + a_{26}\eta_{3o}^{(0)}) \\ b_0 = -\left(a_{12}a_{22} + \dfrac{2a_{12}a_{24}\eta_{1o}^{(0)} + a_{12}a_{25}}{\eta_{2o}^{(0)}}\right) \end{cases} \quad (8.101)$$

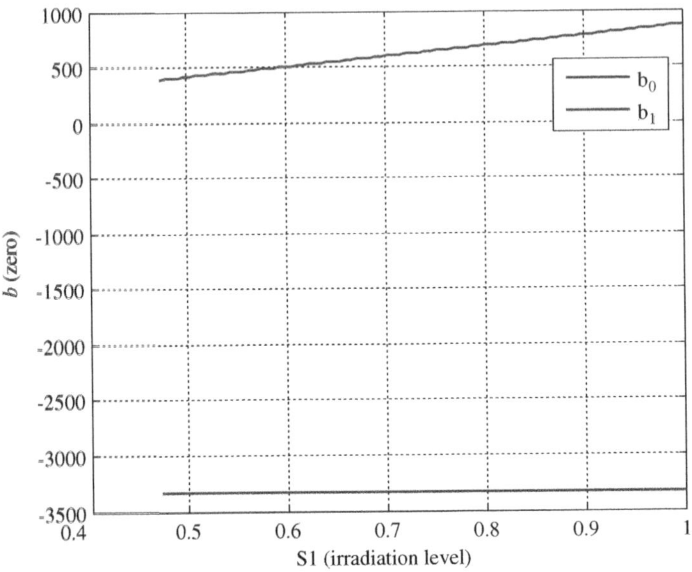

Figure 8.18 Two of the coefficients of the linearized PV zero dynamics characteristics

The Routh–Hurwitz criteria for the stability of this characteristic equation (Figure 8.18)

$$\begin{cases} b_1 > 0 \\ b_0 > 0 \end{cases} \tag{8.102}$$

$$\begin{cases} \eta_{1o}{}^{(0)} = x_1 = I_{gqo} \\ \eta_{2o}{}^{(0)} = x_3 = V_{mqo} \\ \eta_{3o}{}^{(0)} = x_6 = V_{pvo} \end{cases} \tag{8.103}$$

Therefore, besides one of the characteristic equation poles at origin (S=0), the second coefficient of the Routh–Hurwitz criteria (b_0) in all operating regions is negative as shown in Figure 8.18. This means that the PV system connected to the current source inverter is in non-minimum phase throughout its operation region.

8.6 Conclusions

This chapter investigated the controllability analysis of a wind-turbine-driven IPM generator interfaced to the grid using either an L or LCL filter. The stability regions of the operation based on the MPP tracking of the wind turbine and loss minimization of the generator were presented. With the aid of two methods used for the extraction of the zero dynamics, it was shown that the generator fitted with the L filter has a stable zero dynamics and possesses minimum phase characteristics which ensure that no structural limitation is imposed on the design of the controllers. When the generator is grid-interfaced using the LCL filter, the zero dynamics is unstable for all operating points, and in the small, the system is in non-minimum phase. There is a structural constraint (such as a limited-current regulator bandwidth and response time) on any controller to be designed for the system. Also the nonlinear controller design using the method of feedback state linearization cannot be applied to the control of the four variables specified of the IPM generator with the LCL filter. Modest dynamic performance can, however, be achieved by using higher order controllers and/or high gain observers.

The steady-state operation region and zero dynamics of the PV source connected to the current source inverter were also studied. This system is in non-minimum phase throughout its operation region.

The controllability analysis for the IPM generator system suggests that the topological structure of the interface filter and the controlled variables have significant effects on the phase behavior of the system, the agility of any controller used, and hence the static and dynamic performance of the controlled generator system. This chapter provides an explanation as to why designing controllers for renewable energy systems with LCL interfaces with the grid has continued to be challenging. The appropriate selection of controlled variables, stating to be estimated using high-gain observers and using high-order controllers, points the way for the design of controllers that will deliver a good static and dynamic performance.

References

1. Madawala, U.K., Geyer, T., Bradshaw, J.B., and Vilathgamuwa, D.M. (2012) Modeling and analysis of a novel variable-speed cage induction generator. *IEEE Transactions on Industrial Electronics*, **59** (2), 1020–1028.
2. Delli Colli, V., Marignetti, F., and Attaianese, C. (2012) Analytical and multiphysics approach to the optimal design of a 10-MW DFIG for direct-drive wind turbines. *IEEE Transactions on Industrial Electronics*, **59** (7), 2791–2799.
3. Sopanen, J., Ruuskanen, V., Nerg, J., and Pyrhonen, J. (2011) Dynamic torque analysis of a wind turbine drive train including a direct-driven permanent magnet generator. *IEEE Transactions on Industrial Electronics*, **58** (9), 3859–3867.
4. Liserre, M., Blaabjerg, F., and Hansen, S. (2005) Design and control of an LCLfilter based three phase active rectifier. *IEEE Transactions on Industry Applications*, **41** (5), 1281–1291.
5. Malinowski, M., Stynski, S., Kolomyjski, W., and Kazmierkowski, M.P. (2009) Control of three-level PWM converter applied to variable-speed-type turbines. *IEEE Transactions on Industrial Electronics*, **56** (1), 69–77.
6. Gabe, I.J., Montagner, V.F., and Pinheiro, H. (2009) Design and implementation of a robust current controller for VSI connected to the grid through an LCL-filter. *IEEE Transactions on Power Electronics*, **24** (6), 1444–1452.
7. Skogestad, S. and Postlethwaite, I. (2008) *Multivariable Feedback Control*, John Wiley & Sons, Inc.
8. Yuan, Z., Zhang, N., Chen, B., and Zhao, J. (2012) Systematic controllability analysis for chemical processes. *AICHE Journal*, **58** (10), pp. 3096–3109.
9. Yuan, Z., Chen, B., and Zhao, J. (2011) An overview on controllability analysis of chemical processes. *AICHE Journal*, **57** (5), 1185–1201.
10. Yuan, Z., Chen, B., and Zhao, J. (2011) Phase behavior analysis for industrial polymerization reactors. *AICHE Journal*, **57** (10), 2795–2807.
11. Wasynczuk, O. (1983) Dynamic behavior of a class of photovoltaic power systems. *IEEE Transactions on Power Apparatus and Systems*, **PAS-102** (9), 3031–3037.
12. Femia, N., Petrone, G., Spagnuolo, G., and Vitelli, M. (2009) A technique for improving P&O MPPT performances of double-stage grid-connected photovoltaic systems. *IEEE Transactions on Industrial Electronics*, **56** (11), 4473–4482.
13. Alonso, R., Ibáñez, P., Martinez, V. *et al.* (2009) An innovative perturb, observe and check algorithm for partially shaded PV systems. Proceedings Power Electronics and Applications Conference, Barcelona, Spain, September 8-10, 2009, pp. 1–8.
14. Tan, C.W., Green, T.C., and Hernandez-Aramburo, C.A. (2008) Analysis of perturb and observe maximum power point tracking algorithm for photovoltaic applications. Proceedings Power and Energy Conference, Johor Bahru, Malaysia, December 1-3, 2008, pp. 237–242.
15. Tan, C.W., Green, T.C., and Hernandez-Aramburo, C.A. (2007) A current mode controlled maximum power point tracking converter for building integrated photovoltaics. Proceedings Power Electronics and Applications Conference, Aalborg, Denmark, September 2-5, 2007, pp. 1–10.
16. Pandey, A., Dasgupta, N., and Mukerjee, A.K. (2008) High-performance algorithms for drift avoidance and fast tracking in solar MPPT system. *IEEE Transactions on Energy Conversion*, **23** (2), 681–689.
17. Koutroulis, E., Kalaitzakis, K., and Voulgaris, N.C. (2001) Development of a microcontroller-based, photovoltaic maximum power point tracking control system. *IEEE Transactions on Power Electronics*, **16** (1), 46–54.

18. Elgendy, M.A., Zahawi, B., and Atkinson, D.J. (2012) Assessment of perturb and observe MPPT algorithm implementation techniques for PV pumping applications. *IEEE Transactions on Sustainable Energy*, **3** (1), 21–33.
19. Panjapornpon, C. (2005) Model-based controller design for general nonlinear processes. PhD dissertation. Drexel University, USA.
20. Nazrulla, M.Sh. (2010) Control of non-minimum phase systems using extended high-gain observers. PhD dissertation. Michigan State University.
21. Kuhlmann, A. and Bogle, I.D.L. (2003) Design of nonminimum phase processes for optimal switchability. *Journal of Chemical Engineering and Processing*, **43**, 655–662.
22. Slotine, J. and Li, W. (1991) *Applied Nonlinear Control*, Prentice-Hall, Englewood Cliffs, NJ.
23. Ojo, O. and Karimi-Davijani, H. (2012) Controllability analysis of IPM generator connected through LCL filter to the grid. 38th Annual Conference on IEEE Industrial Electronics Society, Montreal, QC, October 25-28, 2012, pp. 2210–2215.
24. Isidor, A. (1995) *Nonlinear Control Systems*, 3rd edn, Springer-Verlag, New York.
25. Ojo, O., Osaloni, F., Zhiqiao, W., and Omoigui, M. (2003) A control strategy for optimum efficiency operation of high performance interior permanent magnet motor drives. Proceedings IEEE Industry Applications Conference, Vol. 45, pp. 604–610.
26. da Costa, J.P., Pinheiro, H., Degner, T., and Arnold, G. (2011) Robust controller for DFIGs of grid-connected wind turbines. *IEEE Transactions on Industrial Electronics*, **58** (9), 4023–4038.
27. Parker, M.A., Chong, N., and Ran, L. (2011) Fault-tolerant control for a modular generator–converter scheme for direct-drive wind turbines. *IEEE Transactions on Industrial Electronics*, **58** (1), 305–315.
28. Geng, H., Xu, D., Wu, B., and Yang, G. (2011) Active damping for PMSG-based WECS with DC-link current estimation. *IEEE Transactions on Industrial Electronics*, **58** (4), 1110–1119.
29. Rockhill, J.A.A., Liserre, M., Teodorescu, R., and Rodriguez, P. (2011) Grid-filter design for a multimegawatt medium-voltage voltage-source inverter. *IEEE Transactions on Industrial Electronics*, **58** (4), 1205–1217.

9

Universal Operation of Small/Medium-Sized Renewable Energy Systems

Marco Liserre[1], Rosa A. Mastromauro[2] and Antonella Nagliero[2]
[1]*Christian-Albrechts-University of Kiel, Kaiserstr, Germany*
[2]*Department of Electrical and Information Engineering, Politecnico di Bari, Bari, Italy*

9.1 Distributed Power Generation Systems

In recent years, the electrical power system (EPS) has been subject to a worldwide restructuring process. Indeed, deregulation and the advancement of technology have increased interconnection of distributed power generation units to the main grid. The constant growth of distributed power generation systems (DPGSs) presents an efficient and economical way of generating electric energy closer to the loads. Moreover, environmental concerns have encouraged governments and the scientific community to work together in order to maximize the use of renewable energy resources. The rating of DPGSs can vary between a few kilowatts to as much as 10 MW.

Various types of new distributed resources, such as microturbines, small hydro cells and fuel cells, in addition to the more traditional solar and wind power are creating significant new opportunities for the integration of DPGSs into the electricity grid. Voltage source converters (VSCs) employing pulse width modulation (PWM) techniques are often used to interface the DPGS to the grid [1, 2].

One of the most important issues introduced by the growth of DPGS is the stand-alone operation mode. Improvements in power electronic technology make it possible to design DPGSs that can switch between grid-connected and stand-alone modes without interruption to the load power supply [3, 4]. The stand-alone mode of operation is important in the case of sensitive and critical loads, as it is necessary to maintain continuous and uninterrupted AC power during planned or unplanned grid outage conditions. The IEEE standard 1547.4 states the need for implementing intentional islanding operations of DPGSs [5]. Grid-connected DPGSs should comply with the following regulations:

- regarding the power quality and protection (grid codes) [6],
- concerning the performance of DPGSs, such as energy efficiency and electromagnetic compatibility and
- concerning safety.

Power Electronics for Renewable Energy Systems, Transportation and Industrial Applications, First Edition.
Edited by Haitham Abu-Rub, Mariusz Malinowski and Kamal Al-Haddad.

During grid-connected operation, all DPGSs should operate in accordance with IEEE Standard 1547–2003 [7]. The requirements should be met at the point-of-common coupling (PCC), where monitoring and control equipment would be necessary to control islanding operations and to implement the transition from normal operational mode to islanding mode and back.

The term stand-alone DPGS, sometimes referred to as microgrids, is used to refer to intentional islanding in electric power systems that contain distributed resources and loads [8]. Stand-alone DPGSs present the following characteristics:

- to have distributed energy resources and loads
- to have the ability to disconnect from and parallel with the main grid
- to include some portions of the EPS
- are intentionally planned

The islanded DPGS needs to be designed to meet the active and reactive power requirements of the loads within the island. The microgrid should also provide frequency stability and operate within the specified voltage ranges during all expected load and resource changes. During the island mode condition, transient stability should be maintained for load steps, DPGS outage and island faults. All faults must be cleared within the island.

For the reconnection of the islanded DPGS to the grid, monitoring should indicate that the proper conditions exist for synchronizing the island with the grid. The amplitude, frequency and phase displacement between the two system voltages should be within acceptable limits in order to initiate a reconnection.

9.1.1 Single-Stage Photovoltaic Systems

In the power range of a few kilowatts, the single-stage transformerless power converter seems to be the best choice to achieve low cost and high efficiency. Small power, high-efficiency and high-reliability single-stage photovoltaic systems (PVSs) bring new control challenges; one of the most important being the interaction with the utility grid. Consequently, grid interconnection requirements, which apply to PVS distributed power generation, are continuously updated in order to maintain the power quality and the stability of the utility grid. The demanding international standards allow strict limits for the current total harmonic distortion (THD) factor [9, 7], which can be met by using inverters that feature not only reduced harmonic production, but also are controlled, in order to provide rejection capability with respect to the grid background distortion. For this purpose, current control is pivotal [10, 11]. Regarding the PV array side, a maximum power point tracker (MPPT) algorithm should be used in order to collect the maximum available power at every operating point, because solar radiation and temperature influence the characteristics of the PV modules [12]. Typical PVS control issues can be divided into two main parts: basic and ancillary. The basic control comprises mainly the maximum power extraction algorithm, PVS harmonic rejection capability and unity power factor (both fulfilled through current control) and islanding detection capability in order to stop the power production of the PVS in case of abnormal conditions. PVSs connected to low-voltage distribution grids might be designed to comply with the requirements of these standards, but they should also be designed to enhance the electrical system by providing "ancillary services", such as voltage and frequency support provided to local loads [13, 14] or to the electric power system [15]. In addition, PV inverters can be usefully adopted as reactive service providers within distribution grids. The injection of reactive power into the grid can enhance the voltage profile, which contributes to reinforcing the distribution grid, maintaining proper quality of supply and avoiding additional investment. This gives some measure of how much the perspective of PVSs has changed in recent years, especially in comparison with studies published in the past, which have suggested limits to the maximum level of PVS penetration, in order to prevent potential problems, such as voltage rise and false trips [16].

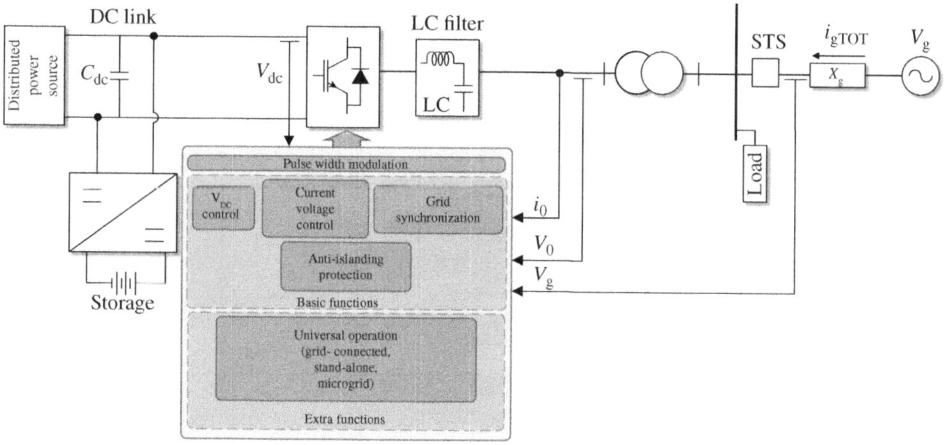

Figure 9.1 General structure of a DPGS

The general structure of a PVS comprises the power circuit and the control unit. Considering the power circuit, the converter can be a single-stage converter (DC–AC converter), as shown in Figure 9.1, or a double-stage converter (e.g., DC–DC and DC–AC converter) with or without galvanic isolation. The advantages of the single-stage converters are good efficiency, a lower price and an easier implementation [17]. The output filter is used in order to cut off the switching ripple. The current/voltage control reference signal is provided by the input power control, which comprises an MPPT algorithm and a DC voltage controller.

9.1.2 *Small/Medium-Sized Wind Turbine Systems*

Among the different sources, small/medium-sized wind turbines (power unit <200 kW) might be the preferred solution immediately outside residential areas, in order to limit the visual impact on the landscape and yet maintain profitability. In the power range of a few tens of kilowatts, the small wind turbine system (WTS) market segmentation shows that the maximum number of manufacturers in Europe corresponds to stand-alone wind applications with a maximum wind turbine size of 20 kW, but that the average rate of wind penetration is a maximum in the pumping field. It is possible to observe that the maximum wind turbine size increases in the case of wind–diesel systems with batteries for stand-alone operation.

The field of small generation has been dominated by the use of asynchronous generators connected directly to the grid/load and more recently by permanent magnet synchronous generators (PMSGs) with a diode rectifier, boost converter and inverter. The use of a high number of pole pairs allows the PMSGs to operate at low speed without decreasing efficiency, thus avoiding the necessity for a gearbox [18]. The use of a diode bridge reduces both cost and the complexity of the control algorithm. Moreover, the well-known six-pulse DC-voltage waveform allows an easy estimation of the rotor speed. The generator low-frequency harmonics (fifth and seventh) can be reduced because of the DC inductor, but, unfortunately, this has a negative effect on the power factor. Moreover, the extracted power decreases as the wind speed increases, owing to the major effect of diode commutation and at low speeds, owing to the possible discontinuous operation of the DC–DC converter. Hence, for power levels of the order of tens of kilowatts, these generators will use a back-to-back converter leading to 5–15% more power with respect to the case of a diode bridge AC–DC power conversion stage.

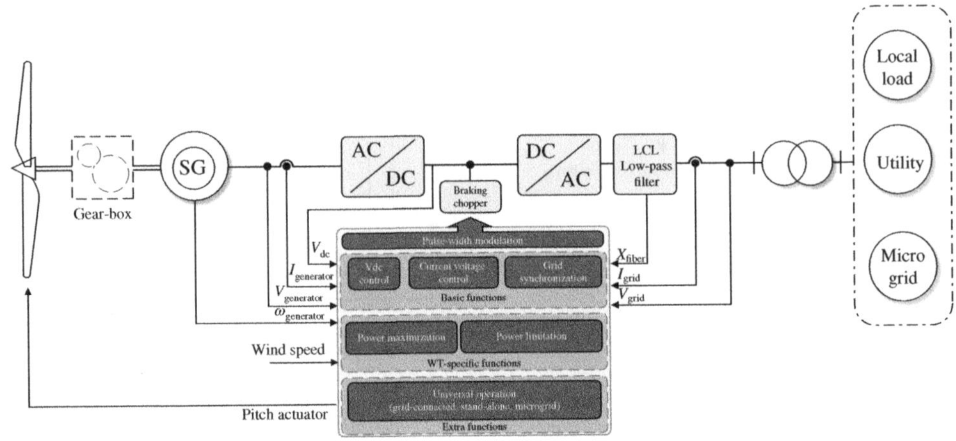

Figure 9.2 Wind turbine system with back-to-back converter [19]

On the generator side, there are two main issues present. First, the tracking of the maximum power without using an anemometer, which might not be present in this power range or could give erroneous values owing to shadow effects and second, the estimation of the rotor position and speed, because the use of a sensor is not always practical as the converter is placed outside the nacelle. As these WTSs are often integrated in rural areas characterized by weak-grid connections, it might be necessary to regulate the reactive power supplied by a variable-speed wind turbine.

Figure 9.2 shows a PMSG connected to the grid through a bidirectional converter. It consists of two voltage source PWM converters connected by a storage capacitor; the converter connected to the generator is used as a rectifier, whereas the converter connected to the grid is used as an inverter. To achieve full control of the output, the DC-link voltage must be boosted to a level higher than the amplitude of the grid voltage. The power flow of the grid-side converter is controlled in order to maintain the DC-link voltage constant, while the control of the generator side is set to suit the magnetization demand and the reference speed or torque. A technical advantage of this topology is the capacitor decoupling between the grid converter and the generator converter. In addition to affording some protection, this decoupling offers separate control of the two converters, allowing independent compensation of asymmetry on both the generator side and the grid side.

9.1.3 Overview of the Control Structure

The general structure of a DPGS consists of a power circuit and a control unit, as shown in Figure 9.3.

The power circuit comprises the following:

- the input power source (e.g., PV source or the wind generator plus the rectifier in the case of a small power wind turbine);
- the converter, which can be just a single-stage converter (e.g., DC–AC converter), or it can represent the second power conversion stage in the case of a double-stage converter (e.g., DC–DC plus DC–AC converter with or without galvanic isolation, or AC–DC plus DC–AC);
- an output filter in order to cut off the switching ripple;
- energy storage;
- the local loads.

The DPGS can be connected to local loads and/or to the main electrical grid, or it can be connected to other DPGSs.

The converter control system should be capable of providing flexible operation. The control unit comprises different control loops: the AC current loop, the AC voltage loop and the power control loop.

The inner current control loop is necessary in order to have close control on the injected current, to force the voltage source inverter to work as a current source within the current control bandwidth. The current control ensures that the inverter current is free from low-order harmonic distortion. However, the high-frequency current distortion owing to the switching frequency must be attenuated with an AC-side filter, in order to cope with the power quality standards for connection of an inverter to the grid. The current control loop is necessary in both grid-connected and stand-alone operational modes.

The voltage control is strictly necessary in stand-alone operational modes because it is responsible for maintaining the voltage profile and guaranteeing the voltage quality in the case of sensitive loads. Voltage control can also be performed in the case of grid-connected operations to regulate the voltage at the PCC and to support the voltage profile if the stiffness of the grid is low. The power control provides the reference voltage amplitude and frequency for the inner loop according to the droop characteristic, which emulates the operation of a high-power plant.

In microgrid operational mode, the power control loop is essential; hence, at least a three-level control system should be performed where the power control loop should take care of load sharing among the different units and the inner voltage loop provides the reference for the current control loop. Both the voltage and the current control loops should provide enough damping for the output LCL filter. The general structure of a grid-connected DPGS is shown in Figure 9.3. The power circuit comprises the following:

- the input power source;
- the converter;
- an output filter in order to cut off the switching ripple;
- the local loads.

For simplicity, the storage system is omitted in this representation because it is beyond the scope of the analysis. A synchronization unit is always present in the system. Inside this unit, information about the phase angle, the amplitude and the frequency of the utility voltage is used for islanding detection and protection in the case of over/under voltage (OUV) or over/under frequency (OUF).

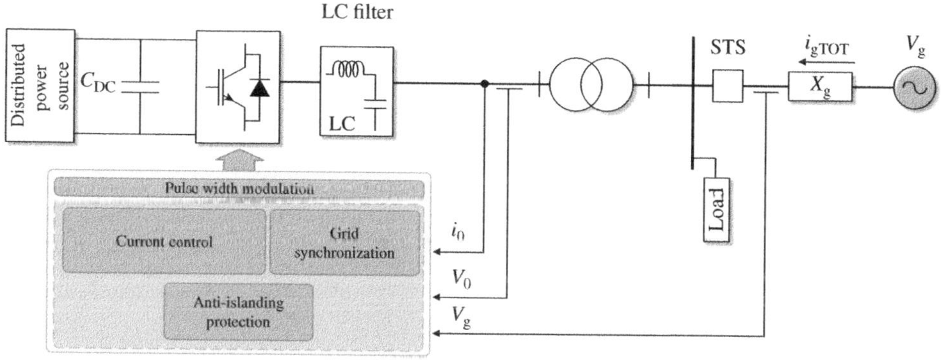

Figure 9.3 Control structure of a grid-connected DPGS

9.1.3.1 AC Current and Voltage Control

The basic control of a DPGS could be different in grid-connected mode compared with the stand-alone mode. In grid-connected mode, the DPGS is generally current controlled and delivers a pre-specified amount of active power into the distribution network. This derives from the fact that traditionally DPGSs are treated as "non-dispatchable" resources, which implies a traditional implementation of the *d-q* control strategy based on zero *q*-axis current reference. In stand-alone operational mode, or when forming a microgrid, the DPGSs are voltage controlled and are responsible for both voltage and power control. However, for the cases of both grid-connected and stand-alone operational modes, the basic control structure can be the same if using a cascade loop control. In this, the outer loop is the converter voltage control, based on a PI regulator and it provides the current reference for the inner loop; this could also be based on a PI, but it controls the LC filter current (i_{LC}) and it determines the voltage references (v_{PWM}) for the PWM module. The scheme of the primary control system is shown in Figure 9.4, which includes decoupling terms compensating for the coupling of the *d-q* dynamics, and feedforward terms used to improve transient performance [20]. It should be noted that the tuning of the current and voltage loop controllers demands special attention in order to fulfill the faster dynamics of the current inner loop with respect to the voltage loop.

It should be considered that the IEC 61727 and the IEEE 1547 standards allow a current THD limit of 5%. This requirement can be fulfilled if the adopted current controllers exhibit high rejection capability to the grid background distortion. In case the design of the current control is based on the internal model principle (IMP), the model of the harmonic disturbances is included in the controller structure in order

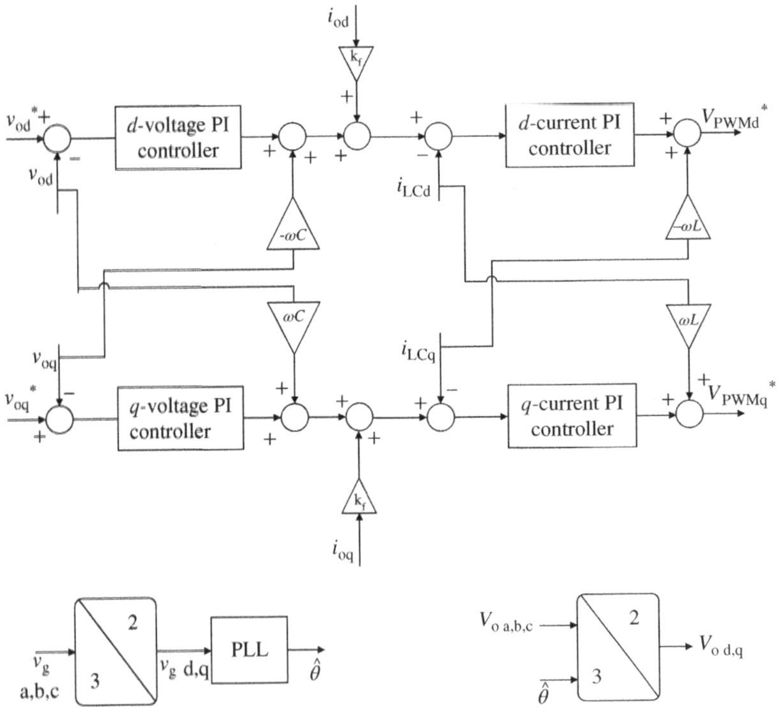

Figure 9.4 DPGS basic control system (i_{od} and i_{oq} are, respectively, the d and q-axis components of the grid current i_o) [21]

to achieve high harmonic rejection capability. Among the controllers based on IMP, both resonant and repetitive controllers provide suitable results for DPGS applications.

In the case of three-phase systems, the control system based on IMP can be easily designed in the d-q reference frame, adding a harmonic compensator τ to the conventional d-q PI controller in order to compensate some selected harmonics and improve power quality performance:

$$\tau(s) = \sum_{h=h_1,h_2,\ldots h_n} k_{\text{ih}} \frac{s}{s^2 + (h\omega)^2} \tag{9.1}$$

In the case of single-phase systems, the use of the d-q reference frame is not desirable. Here, the resonant controller based on IMP guarantees zero steady-state errors in the presence of a sinusoidal current reference. The resonant controller consists of a proportional controller plus a generalized integrator (GI):

$$C_{\text{PRes}}(s) = k_{\text{p}} + k_{\text{i}} \frac{s}{s^2 + \omega^2} \tag{9.2}$$

where $s/(s^2 + \omega^2)$ is the GI and ω is the resonance frequency. The reference tracking is ensured by setting the resonance frequency of the GI equal to the fundamental frequency. Similar performances can be reached using the repetitive controller instead of the resonant controller [10].

9.1.3.2 Synchronization Techniques

Synchronization is one of the most important issues in the control of power electronics equipment connected to the grid. The rapid development of DPGSs has resulted in the reformulation of the grid connection requirements (GCRs). The objectives of the grid codes are to ensure efficient and reliable power generation and transmission and to regulate the rights and responsibilities of the entities acting in the electricity sector. Hence, the requirements specify relevant voltage and frequency operating ranges and the corresponding trip times. In addition, requirements concerning the active power and reactive power are taken into account, which determine the control capabilities under various fault situations [22].

The phase-locked loop (PLL) technique is used in grid-connected DPGS applications for grid synchronization and for grid voltage monitoring [23]. The conventional strategy for phase locking is to estimate the difference between the phase angle of the input signal and that of the output signal and to regulate this value to zero by means of a control loop. The phase difference evaluation is in charge of a phase detector (PD) and is passed through a loop filter (LF) to generate an error signal driving a voltage-controlled oscillator (VCO). Numerous methods for grid synchronization and grid voltage monitoring have been presented in the literature regarding DPGSs. The main difference among the different PLL methods is the configuration of the PD. In particular, the PD structure of the Synchronous Reference Frame-PLL (SRF-PLL) consists of a shift of 90° with respect to the input signal fundamental frequency in order to obtain an orthogonal system. The enhanced phase-locked loop (EPLL) presents a band-pass adaptive filter (BPAF) to estimate the phase error [24]. The second-order generalized integrator PLL (SOGI-PLL) is based on frequency-adaptive quadrature signal generation by means of the second-order generalized integrator-quadrature signal generator (SOGI QSG) filter.

The tracking techniques have to provide the following:

- robustness with respect to noise, stationary and transient disturbances;
- the capability to follow amplitude variations of the grid voltage in case of voltage sags;
- accurate performance in case of grid-voltage frequency variations;
- high filtering performance in case of harmonics distortions,
- easy analog/digital and hardware/software implementations.

A set of in-quadrature signals, obtained by means of a transport delay block introducing a phase shift of 90°, is used in the SRF-PLL. The QSG based on this transport delay gives rise to errors if the grid-voltage frequency changes with respect to its rated value. In addition, it does not provide any filtering capability; hence, the presence of harmonics in the grid voltage acts as a perturbation for the PLL.

In contrast, the EPLL is based on a frequency-adaptive nonlinear synchronization approach [24]. The structure of the EPLL (Figure 9.5) consists of the BPAF, the LF and the VCO. The adaptive filter can be implemented as either a notch or a band-pass filter.

A SOGI-QSG filter is used in the SOGI-PLL structure to obtain a frequency-adaptive quadrature signal generation (Figure 9.6). The transfer function of the adaptive filter based on SOGI is

$$G_{\mathrm{SOGI}}(s) = \frac{s\omega_0}{s^2 + \omega_0^2} \tag{9.3}$$

As output signals, two sine waves (v' and qv') with a phase shift of 90° are generated. The component v' has the same phase and magnitude of the input signal (v) fundamental component. The tuning of the proposed scheme is frequency dependent; thus, problems occur when grid frequency has fluctuations. Consequently, an adaptive tuning of the structure with respect to its resonance frequency is required. Therefore, the resonance frequency value of the SOGI is adjusted by the provided frequency of the PLL. High filtering capability can be obtained by setting the gain k high. If k decreases, the band of the filter becomes narrower, but, at the same time, the dynamic response of the system becomes slower.

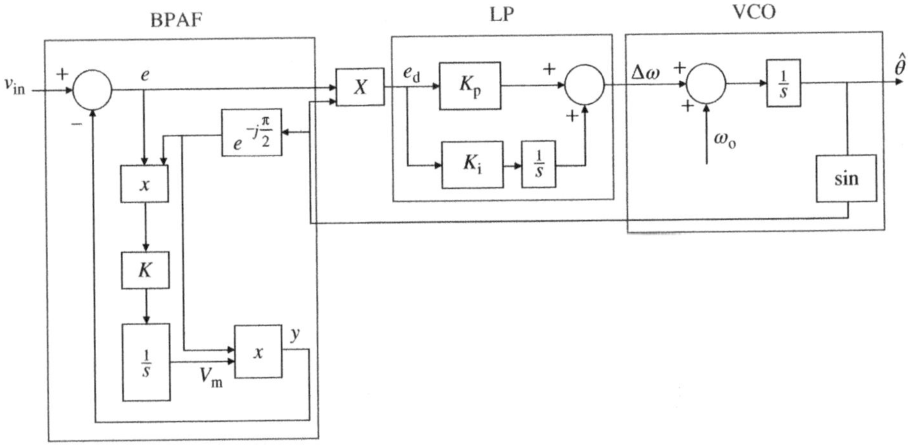

Figure 9.5 Block diagram of the EPLL [21]

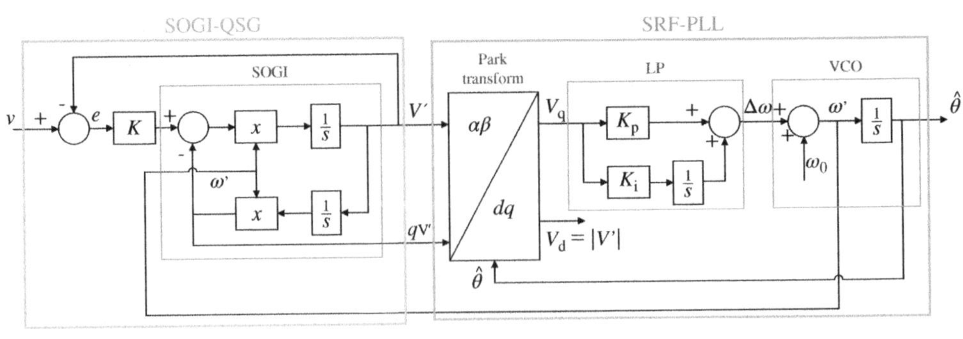

Figure 9.6 SOGI-PLL

In three-phase systems, the amplitude and phase angle of the positive-sequence voltage is used for the synchronization of the converter output variables and power flux calculations, or for the transformation of state variables into rotating reference frames. Regarding the technique used for grid monitoring, the amplitude and the phase of the positive-sequence component must be obtained quickly and accurately, even if the grid voltage is distorted and unbalanced. For three-phase systems, the synchronization technique based on SOGI is called "dual second-order generalized integrator PLL" (DSOGI-PLL). A dual SOGI-based QSG is used for filtering and obtaining the 90° shifted versions from the $\alpha\beta$ voltages. These signals act as inputs to the positive sequence calculator (PSC), which lies in the instantaneous symmetrical components (ISC) method on the $\alpha\beta$ domain:

$$v_\alpha^+ = v'_\alpha - qv'_\beta$$
$$v_\beta^+ = qv'_\alpha + v'_\beta \tag{9.4}$$

Finally, the positive-sequence $\alpha\beta$ voltages are translated to the d-q synchronous reference frame and an SRF-PLL is employed to make the system frequency adaptive. In the system presented in Figure 9.7, the feedback loop regulating the q component to zero controls the angular position of the d-q reference frame and estimates the grid frequency. The estimated grid frequency ω' is used by the outer feedback loop to adapt dynamically the DSOGI–QSG resonant frequency.

9.1.3.3 Islanding Detection Methods

Unintentional islanding of grid-connected inverter-based DPGSs can result in damage to local electrical loads, the grid-connected inverter and/or technicians during maintenance operations. Consequently, DPGSs must be operated properly when this condition occurs. Two main approaches can be employed: grid-level or inverter-side detection. In the first case, a communication system is required in order to manage the islanding condition properly, which increases both the system complexity and cost. In the second case, the controller of the grid-connected DPGS detects the islanding operation mode and acts properly on the inverter. The latter method can be divided into four categories: passive inverter-resident methods, active inverter-resident methods, active methods not resident in the inverter and methods based on the use of communications between the utility and PV inverter. Moreover, it is possible that utility breakers or circuit reclosers could reconnect the island to the utility system when out of phase, consequently causing

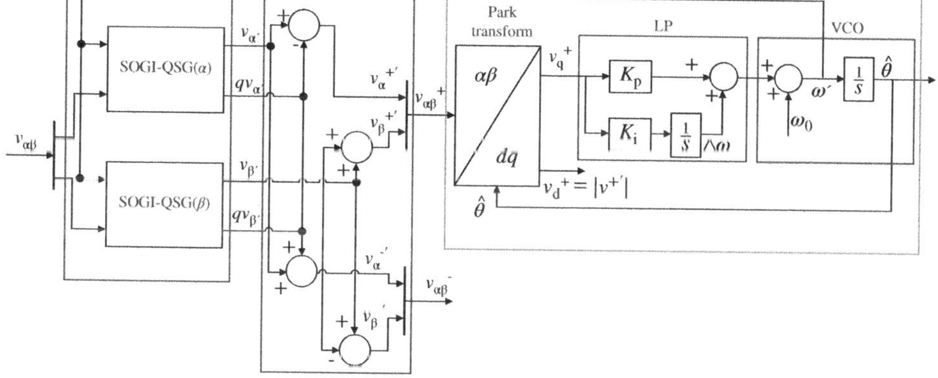

Figure 9.7 DSOGI-PLL

an overcurrent and trip out of the system. According to this scheme, the non-detection zone (NDZ) is the range of local loads inside the potential island, in which the islanding detection scheme under test fails to detect the islanding. Each detection method presents a different NDZ, and for this reason, the NDZ could be used as a performance index to evaluate the different anti-islanding algorithms. The aim of all islanding detection methods is to reduce the NDZ to near zero when detecting islanding in all cases.

Grid-level methods allow the islanding conditions to be detected accurately, but depending on the number of DPGSs and on the grid-voltage level, the number and price of sensors, communication equipment and the management complexity can vary. Consequently, inverter-resident solutions are commonly employed in cases where DPGSs are connected to low-voltage electrical grids.

Over/under Voltage (OUV) and over/under Frequency (OUF) method

All grid-connected DPGS inverters are required to have an OUF and OUV protection that causes the PV inverter to cease supplying power to the utility grid if the values in the PCC are different from set values. The behavior of the system at the time of utility disconnection will depend on the power provided at the instant before the switch opens to form an island. If the resonant frequency of the **RLC load is the same as the grid line frequency, the linear load does not absorb the reactive power. Active power is directly proportional to the voltage. After the disconnection, the load power and the power provided by the DPGS might coincide; hence, the voltage changes into

$$V' = K \cdot V \tag{9.5}$$

where

$$K = \sqrt{\frac{P_{\mathrm{DG}}}{P_{\mathrm{load}}}} \tag{9.6}$$

and P_{DG} denotes the DPGS power and P_{load} is the load power. When $P_{\mathrm{DG}} > P_{\mathrm{load}}$ there is an increase in the voltage amplitude and, conversely, when $P_{\mathrm{DG}} < P_{\mathrm{load}}$ there is a decrease in the amplitude. At the same time, reactive power is tied up to the frequency and amplitude of the voltage [25, 26].

The worst case for islanding detection is represented by a condition of balance of the active and reactive powers, which results in no change of amplitude and frequency, that is, $\Delta P=0$ and $\Delta Q=0$. It is straightforward that a small ΔP results in an insufficient change in voltage amplitude, and that a small ΔQ results in an inadequate change in frequency to disconnect the system and prevent islanding. It is possible to calculate the NDZ area from the mismatches of active and reactive powers and by setting the threshold values for the frequency and the amplitude. It should be noted that the probability that ΔP and ΔQ fall into the NDZ of OUV/OUF could be significant. Because of this concern, the OUV and OUF protective devices are generally considered insufficient anti-islanding protection.

Voltage Harmonic Monitoring Method

The voltage at the PCC during normal conditions is controlled by the grid, but in the case of islanding, the DPGS inverter controls the PCC voltage and the amplitude of the single harmonic on the RLC load change. Hence, monitoring the voltage harmonic distortion can be useful in detecting the occurrence of an islanding condition. The voltage harmonic monitoring method is based on the measurement of harmonics through the THD of the PCC voltage or on the measurement of the highest harmonic. In real power systems, a significant harmonic distortion of the PCC voltage allows the islanding condition to be revealed. It must be considered that the magnitude of the measured PCC voltage harmonics depends on the grid impedance; thus, it is not always possible to select a trip threshold that provides reliable islanding protection.

Grid Impedance Change Detection Method

The European standard EN50330-1 describes an active method for islanding detection consisting of monitoring grid impedance changes during islanding transients called ENS or MSD (abbreviations of the main monitoring units with allocated all-pole switching devices). The impedance measurement with the OUV and OUF protection is adopted in Germany and Austria in the standard DIN VDE0126. To be able to complete a measurement of impedance, a dedicated external device can be used, or it is possible to include it in the inverter control. The goal is to isolate the supply within 5 s after a change of impedance $\Delta Z = 0.5\,\Omega$ owing to a grid failure [27].

Typically, a small current at a certain harmonic order (h) is injected into the utility by the device to determine the impedance:

$$\overline{Z}_{\mathrm{h}} = \frac{V \cdot e^{j\phi_{\mathrm{V}}}}{I \cdot e^{j\phi_{\mathrm{I}}}} = Z_{\mathrm{h}} \cdot e^{j\phi_{\mathrm{Z}}} \tag{9.7}$$

or

$$\overline{Z}_{\mathrm{h}} = R_{\mathrm{g}} + j\omega_{\mathrm{h}} L_{\mathrm{g}} \tag{9.8}$$

The disadvantage of this method is the increase in the harmonic pollution-injected online by the inverter. To be able to limit this drawback, it is recommended to inject the current harmonic or interharmonic only for the time needed for the numerical elaboration of the data. Moreover, if many converters are connected in parallel, a contemporary injection could cause problems for both the effectiveness of the method and the power quality with consequent problems for the control and the stability of the system.

Active and Reactive Power Variation Methods

Figure 9.8 shows the control of a DPGS inverter with the islanding detection method based on active and reactive power injection. This method is applied to PV systems. When the inverter is connected to the grid, active power variations ΔP can happen, while it is assumed that the load absorbs constant

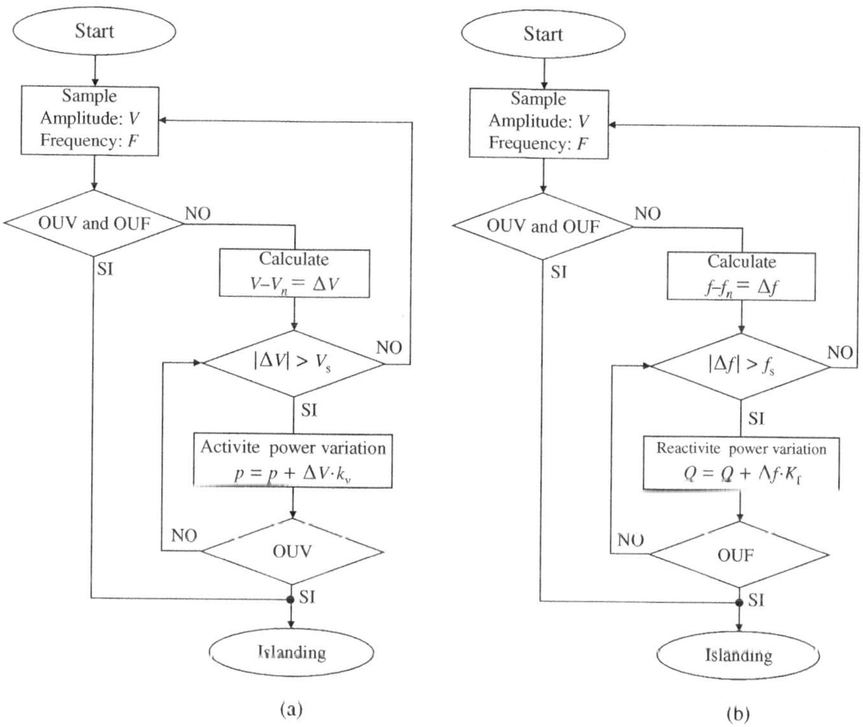

Figure 9.8 Active (a) and reactive (b) power variation islanding detection methods [26]

power P_{load}. In an islanding condition, the real power variation flows directly in the load, affecting the inverter current and the voltage in the PCC. It is possible to calculate the voltage variation versus the active power variation injected by the inverter in the load during the islanding condition:

$$P_{DG} = P_{load} = \frac{V^2}{R} \tag{9.9}$$

$$V = \sqrt{R \cdot P_{DG}} \tag{9.10}$$

Deriving P_{DG} and from (9.10):

$$\frac{\partial P_{DG}}{\partial V} = 2 \cdot \frac{V}{R} = 2 \cdot \frac{\sqrt{R \cdot P_{DG}}}{R} = 2 \cdot \sqrt{\frac{P_{DG}}{R}} \tag{9.11}$$

The voltage variation is expressed by

$$\Delta V = \frac{\Delta P_{DG}}{2} \cdot \sqrt{\frac{R}{P_{DG}}} \tag{9.12}$$

Hence, it is possible to vary the inverter's active power to carry the voltage amplitude out from the normal range of operation, and the method works just when the voltage variation exceeds a certain threshold value. The intervention time of the algorithm can be adjusted by the gain K_v, which increases or decreases the power dP proportionally to the measured voltage variation. Hence, the power variation is equal to

$$\mathrm{d}P = K_v \cdot (V - V_n) \tag{9.13}$$

Similarly, it is possible to use the strong dependence between the frequency and the reactive power to develop another islanding detection method, measuring the grid voltage frequency via the monitoring PLL. The difference is amplified by means of a gain, providing that the reactive power dQ is

$$\mathrm{d}Q = K_f \cdot (f_n - f) \tag{9.14}$$

Other Inverter-Resident Active Methods

Other islanding detection methods based on the perturbations of the system with the objective of carrying the electrical island instability have been developed in recent years [28]. One of these is called Sandia voltage shift (SVS) and it applies positive feedback to the amplitude of the voltage at the PCC (usually this is the RMS value that is measured in practice). Slip-mode frequency shift (SMS) is another method in which the current–voltage phase angle of the inverter, instead of always being controlled to be zero, is made to be a function of the frequency of the voltage at the PCC. The phase-response curve of the inverter is designed such that the phase of the inverter increases faster than the phase of the (RLC) load with a unity-power factor in the region near the utility frequency.

The active frequency drift method (AFD) operates with a waveform of the current injected by the inverter, which is slightly distorted such that there is a continuous trend to change the frequency. Hence, in an islanding condition, the frequency of the voltage at the PCC is forced to drift upward or downward. It is possible to define the chopping factor for the AFD:

$$\mathrm{cf} = \frac{2T_z}{T} \tag{9.15}$$

where T_z is the zero time of the AFD signal and T is the period of the grid voltage.

In contrast, Sandia frequency shift (SFS) applies positive feedback to the frequency, and the chopping factor for this method is

$$\mathrm{cf} = \mathrm{cf}_0 + K(f - f_n) \tag{9.16}$$

Table 9.1 Equation for active methods: SMS, AFD and SFS

Methods	Phase criterion
SMS	$\tan^{-1}\left[R\left(\omega C-\frac{1}{\omega L}\right)\right]=-\arg(G(j\omega))$
AFD	$\tan^{-1}\left[R\left(\omega C-\frac{1}{\omega L}\right)\right]=\frac{\pi\cdot \mathrm{cf}}{2}$
SFS	$\tan^{-1}\left[R\left(\omega C-\frac{1}{\omega L}\right)\right]=\frac{\pi\cdot(\mathrm{cf}_0+K(f-f_n))}{2}$

where K is the acceleration gain, cf_0 is the chopping factor when there is no frequency error and $f-f_n$ is the difference between the estimated frequency and nominal value. The SMS, AFD and SFS functions are given in Table 9.1.

9.2 Control of Power Converters for Grid-Interactive Distributed Power Generation Systems

Grid-interactive microgrids are becoming an increasingly attractive structure for the proposal of an active distribution grid. A microgrid can be conceived as a cluster of interconnected DPGS load and energy storage units coordinated in order to be treated collectively as a single controllable device [2, 17]. A preliminary analysis of the general structure of a single DPGS unit has been provided already in earlier sections, and the main control issues have been investigated. Thus, the focus is shifted to the problem of the DPGS power management, which represents the key point toward a "universal" operation of the DPGS. The goal is to allow the DPGS operation in the case of stand-alone, grid-connected and microgrid conditions by managing the disconnection and reconnection without substantial changes of the system control structure.

The basic microgrid architecture is shown in Figure 9.9, in which the components can be physically close to each other or distributed geographically, and where the loads may be critical or noncritical. Critical loads require reliable energy sources and demand stringent power quality. In the case of

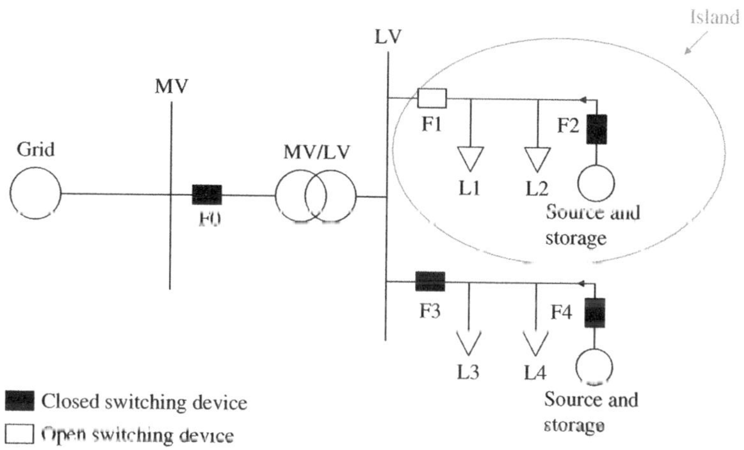

Figure 9.9 Basic microgrid architecture [29]

grid-connected operation, the static transfer switch (STS) is closed and each DPGS unit generates fixed active and reactive powers, according to the dispatched references, or according to the tracking of the maximum power that can be extracted from the source. In the case of stand-alone operation, the DPGS units should supply loads providing power sharing in accordance with the respective ratings and achieving power balance between production and consumption.

To date, DPGS power converters have been required to change their control structure with respect to grid-connected operation and stand-alone operation, because in the past, DPGSs were conceived as current source devices in the first case and as voltage source devices in the second. In this hypothesis, a current control loop is present in both cases, but for stand-alone operation, the voltage across the capacitor of the grid filter might also be controlled. As already stated in Section 9.1.3.1, the voltage control can offer a good trade-off performance in all operational modes. In particular, in grid-connected operation, the control of the voltage allows strict regulation of the voltage at the PCC and support of the voltage applied to the local loads in the case of voltage sags. In the cases of stand-alone and microgrid operational modes, the power converter's control is only responsible for the voltage regulation, because the grid is absent and the control system may guarantee high performance in terms of voltage profile and fast recovery in the case of load transients. Hence, proper operation of the microgrid in all operating conditions requires an accurate power flow management and the implementation of voltage regulation algorithms.

Two control approaches can be followed to realize the abovementioned objective, which are non-communication-based control and communication-based control. Communication-based control of a microgrid relies on the exchange of control information among the different DPGSs. This technique requires a centralized microgrid controller, which receives information from each DPGS and loads the DPGS units at different load conditions. There are many drawbacks in the communication-based approach. The high-bandwidth communication infrastructure necessary to share the dynamic sharing information is an impractical and costly solution in microgrids with long connection distances between inverters and the need to distribute control signals reduces the reliability of the microgrid system, where the control information is critical for proper operation of the DPGSs.

The implemented control algorithms should preferably have no communication links between the paralleled DPGSs. The control algorithms of each individual DPGS should be designed to use only feedback variables that can be measured locally. A simple non-communication-based power control can be realized in the sense of the frequency and voltage droop method. This concept stems from power theory, in which a generator connected to the grid droops its frequency when the power required increases. Each DPGS shares the active power by drooping the voltage frequency according to the total load demand in a manner determined by its frequency droop characteristic, and essentially utilizes the voltage frequency as a communication link between the DPGS control systems. Similarly, a drop in the voltage magnitude with the reactive power is used to ensure reactive power sharing. This control technique is known as the "droop method."

9.2.1 Droop Control

Figure 9.9 shows a microgrid composed of different types of resources, such as solar panels, fuel cells and wind turbines. Every unit has an inverter as an interface connected to the common bus through a cable that can be modeled as an impedance, as shown in Figure 9.10.

The active and reactive power flows from an inverter to the grid can be expressed as follows:

$$P = \frac{1}{Z}\left[\left(EV\cos\phi - E^2\right)\cos\theta + EV\sin\phi\sin\theta\right] \tag{9.17}$$

$$Q = \frac{1}{Z}\left[\left(EV\cos\phi - E^2\right)\sin\theta - EV\sin\phi\cos\theta\right] \tag{9.18}$$

where V is the amplitude of the inverter output voltage, E is the amplitude of the common bus voltage, ϕ is the converter voltage angle and Z and θ are the magnitude and phase of the output impedance, respectively.

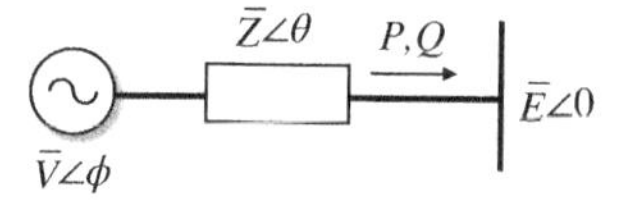

Figure 9.10 Equivalent circuit of an inverter connected to a bus [30]

Assuming that the output impedance is purely inductive $X \gg R$, then R can be neglected. Consequently, (9.17) and (9.18) can be rewritten as

$$P = \frac{EV}{X}\sin\phi \tag{9.19}$$

$$Q = \frac{EV}{X}\cos\phi - \frac{E^2}{X} \tag{9.20}$$

Considering that angle ϕ is small, then $\sin\phi \approx \phi$ and $\cos\phi \approx 1$ and thus, (9.19) and (9.20) become

$$P \approx \frac{EV}{X}\phi \tag{9.21}$$

$$Q \approx \frac{E}{X}(V - E) \tag{9.22}$$

where X is the output impedance at the fundamental frequency. From these equations, it is possible to deduce that the active power can be regulated by acting on the phase ϕ, whereas the reactive power can be regulated by controlling the converter voltage amplitude V.

These conclusions form the basis of droop control. The droop method uses the frequency instead of the phase to control active power flow. This is because the DPGS unit does not know the initial phase value of the PCC voltage, whereas the initial frequency at no load can be easily fixed. Hence, the conventional droop method can be expressed as follows:

$$\omega = \omega^* - m_p(P - P^*) \tag{9.23}$$

$$V = V^* - n_q(Q - Q^*) \tag{9.24}$$

where ω^* and V^* are the rated frequency and voltage, P^* and Q^* are the set points for active and reactive powers and m_p and n_q are the proportional coefficients.

The frequency and voltage control characteristics are shown in Figure 9.11.

The output inductance of inverter X should be selected properly. In fact, increasing the output impedance reduces the circulating current produced by the differences between the power lines or between the DPGS parameters, such as the output filter parameters or the frequency and amplitude set point.

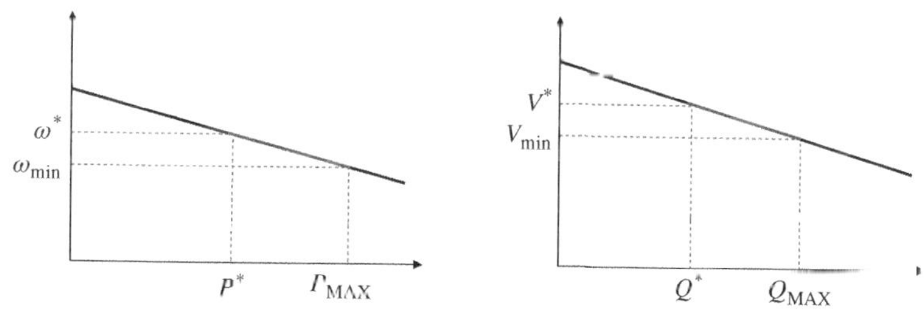

Figure 9.11 Frequency and voltage droop characteristics

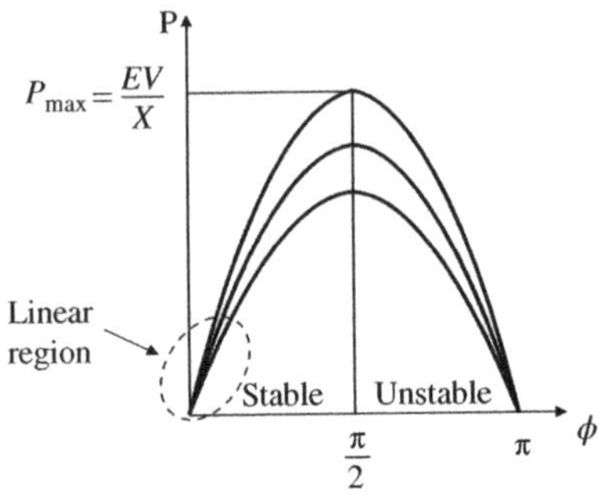

Figure 9.12 Effect of increasing the output inductance on the active power

As shown in Figure 9.12, increasing X also reduces the maximum active power that can be injected, whereas decreasing the output impedance forces angle ϕ to become smaller:

$$\phi = \sin^{-1}\left(\frac{PX}{EV}\right) \tag{9.25}$$

Thus, there is a trade-off in the output impedance design.

In the case of a low-voltage microgrid, the line resistance cannot be neglected. In this case, the droop regulation described by (9.21) and (9.22) is no longer effective [31]. Controlling the power flow based on Equations (9.17) and (9.18) will introduce coupling and the real power, and reactive power will no longer be regulated independently, but if $Z = R$ and $\theta = 0$, then

$$P = \frac{E}{X}(V - E) \tag{9.26}$$

$$Q = \frac{EV}{X}\phi \tag{9.27}$$

Hence, P and Q exchange their roles:

$$\omega = \omega^* - m'(Q - Q^*) \tag{9.28}$$

$$V = V^* - n'(P - P^*) \tag{9.29}$$

Consequently, a control scheme based on the P-ω and Q-V droops can be used for inductive impedance, whereas P-V and Q-ω droops can be used in cases of resistive impedance. In order to introduce decoupling between the active and reactive powers for every line impedance condition, it is possible to use a frame transformation matrix T to modify the active and reactive powers P and Q in P_C and Q_C:

$$\begin{bmatrix} P_C \\ Q_C \end{bmatrix} = T\begin{bmatrix} P \\ Q \end{bmatrix} = \begin{bmatrix} \sin\alpha & -\cos\alpha \\ \cos\alpha & \sin\alpha \end{bmatrix}\begin{bmatrix} P \\ Q \end{bmatrix} = \begin{bmatrix} \frac{X}{Z} & -\frac{R}{Z} \\ \frac{R}{Z} & \frac{X}{Z} \end{bmatrix}\begin{bmatrix} P \\ Q \end{bmatrix} \tag{9.30}$$

where

$$\alpha = \text{atan}(X/R) \tag{9.31}$$

Applying this transformation on (9.19) and (9.20) results in

$$P_C = \frac{EV}{X}\sin\phi \tag{9.32}$$

$$Q_C = \frac{EV\cos\phi - E^2}{Z} \tag{9.33}$$

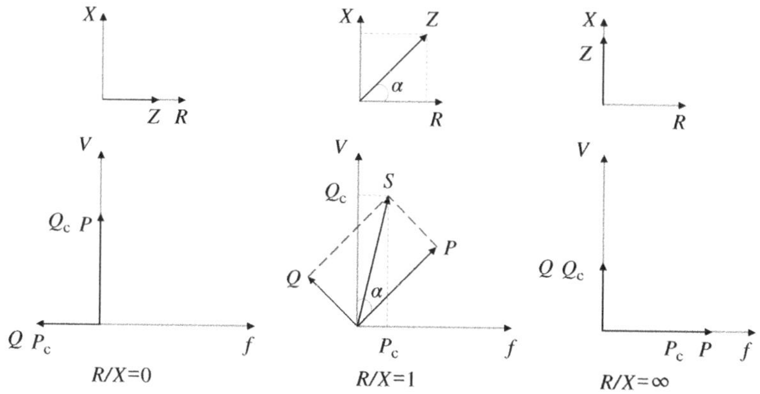

Figure 9.13 Influence of active and reactive powers on voltage and frequency for different line impedance ratios

The definition of P_C and Q_C permits independent influence on the grid frequency and amplitude (Figure 9.13). In order to obtain P_C and Q_C, it is sufficient to know the ratio *R/X*, even if, unfortunately, the transformation frame depends on the accuracy of the impedance knowledge.

9.2.2 Power Control in Microgrids

Based on droop control theory, the power management strategy of a DPGS might work both in grid-connected and in island modes with a minimum number of changes required in the control structure when moving between the different operational modes, and it might also guarantee a fast-transient response.

9.2.2.1 Power Control in Grid-Connected Operational Mode

In the grid-connected operational mode, each DPGS unit forming the microgrid or directly connected to the main electrical grid generates fixed active and reactive powers, according to the dispatched references, or as a consequence of tracking the maximum power that can be extracted from the source. The DPGS operates as a controlled current source and all loads are fed by the utility.

The difference between real-time active/reactive power and the rating active/reactive power of each unit is regulated according to the relations:

$$\omega = \omega_e - m_p(P - P^*) \tag{9.34}$$

$$\hat{V} = \hat{E} - n_p\left[(Q - Q^*) + \frac{1}{T_{iQ}}\int_{-\infty}^{t}(Q - Q^*)\mathrm{d}\tau\right] \tag{9.35}$$

where ω_e and $\hat{E}$ are the frequency and amplitude voltage references provided by a PLL that monitors the PCC voltage, P^* and Q^* are the active and reactive power reference values, m_p and n_p are the frequency and voltage proportional coefficients, respectively and T_{iQ} is the integral time-constant reactive power controller.

The active power controller is designed to dispatch the active power P^* at the grid frequency ω_e when operating in a grid-connected mode. A PI controller has been considered for the reactive power regulation, forcing the DPGS reactive power output to its desired value with zero-steady-state error.

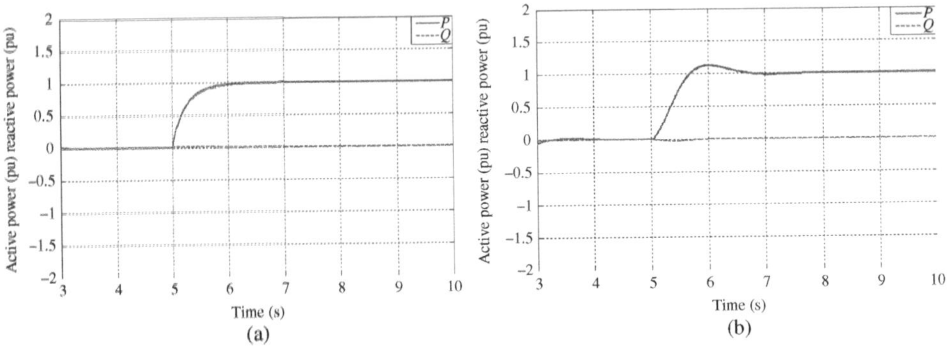

Figure 9.14 Active and reactive powers of a DPGS in the case of a step variation of the active power reference with (a) and without (b) the implementation of the decoupling transformation matrix

The active and reactive powers of the DPGS, both with and without the implementation of the decoupling transformation matrix, are shown in Figure 9.14. It is possible to notice that the settling time is about 1 s in the first case and about 2 s in the second case. The implementation of the decoupling transformation matrix makes the response more damped with a reduced settling time, as can be seen in Figure 9.15 for the output current. The output current increases with a smooth transient, maintaining the voltage and the frequency within the limits.

In the case of load variation, it is possible to observe that the active and reactive powers are not influenced (Figure 9.16) because the grid absorbs the power overproduction.

9.2.2.2 Power Control in Stand-Alone Mode

During stand-alone operation, the DPGSs supply the loads, sharing the total load demand according to the units' respective ratings, with the aim of achieving a perfect balance between production and consumption. Active and reactive powers might be modified depending on the load demand [32] and according to the following equations:

$$\omega = \omega_b - m_{island}(P - P_{MAX}) \tag{9.36}$$

$$\hat{V} = \hat{V}_{Cb} - n_{island}(Q - Q_{MAX}) \tag{9.37}$$

where $\hat{V}_{Cb}$and ω_b are the phase and amplitude voltage references set to nominal values, P_{MAX} and Q_{MAX} are the maximum active and reactive powers that the DPGSs can supply.

A "supervisory command" (SC) signal is used in the power control, as shown in Figure 9.17, to switch between grid-connected and stand-alone operations. The outputs of the control system are the voltage and frequency references for the DPGS voltage control. In the case of load power increment, the active power provided by the system matches the load demand with a fast and smooth transient, maintaining the voltage and the frequency within the limits (Figure 9.18).

9.2.2.3 Transition between the Different Operational Modes

During the island-mode operation, the DPGS can work at a frequency different from that of the main grid; however, consequently, the DPGS cannot be connected to the grid if the phase difference is non-negligible because it could lead to a high overcurrent, damaging the system or compromising stability [34]. To avoid an out-of-phase reclose, a synchronization that assures smooth transitions and unnoticeable transients

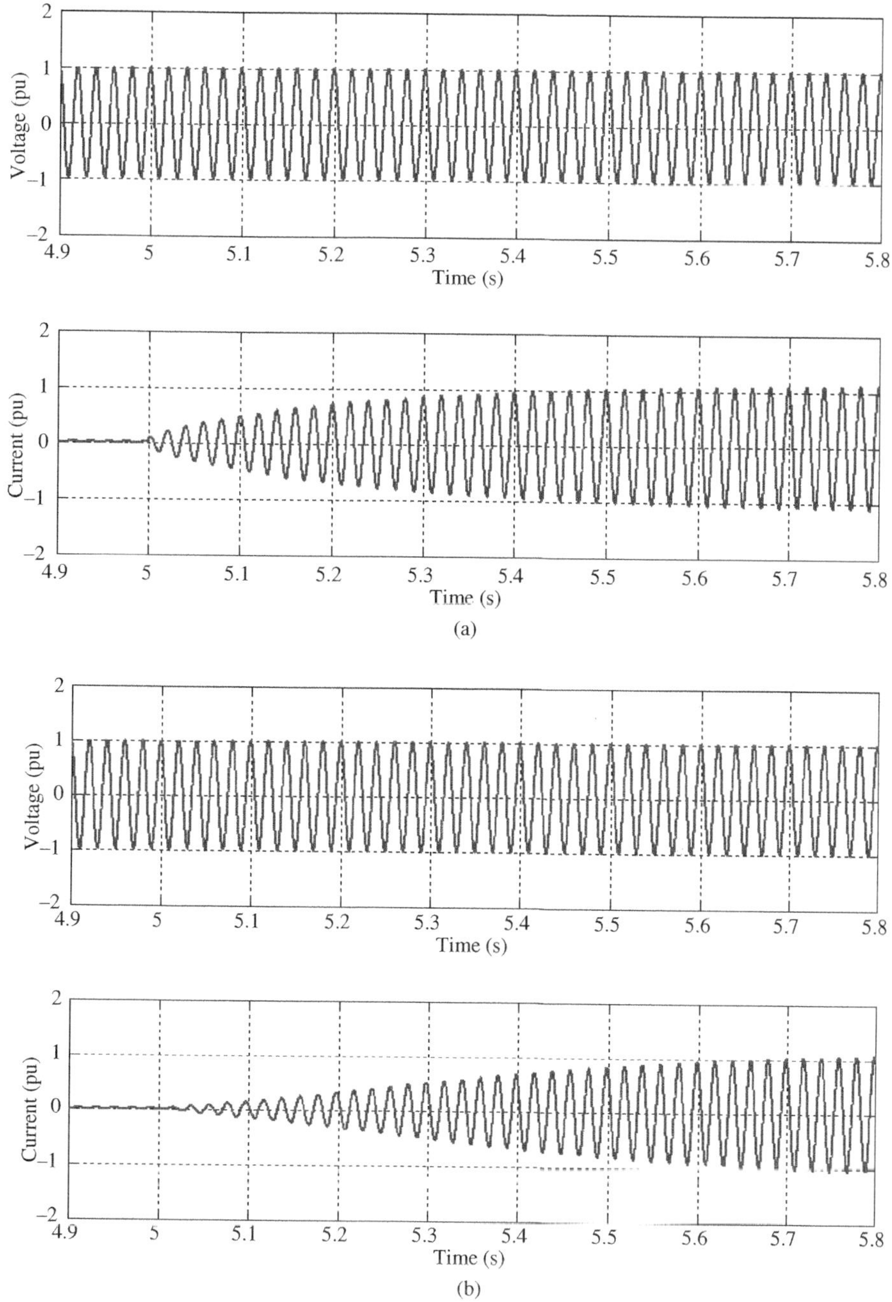

Figure 9.15 PCC voltage and output current in the case of a step variation of the active power reference with (a) and without (b) the implementation of the decoupling transformation matrix

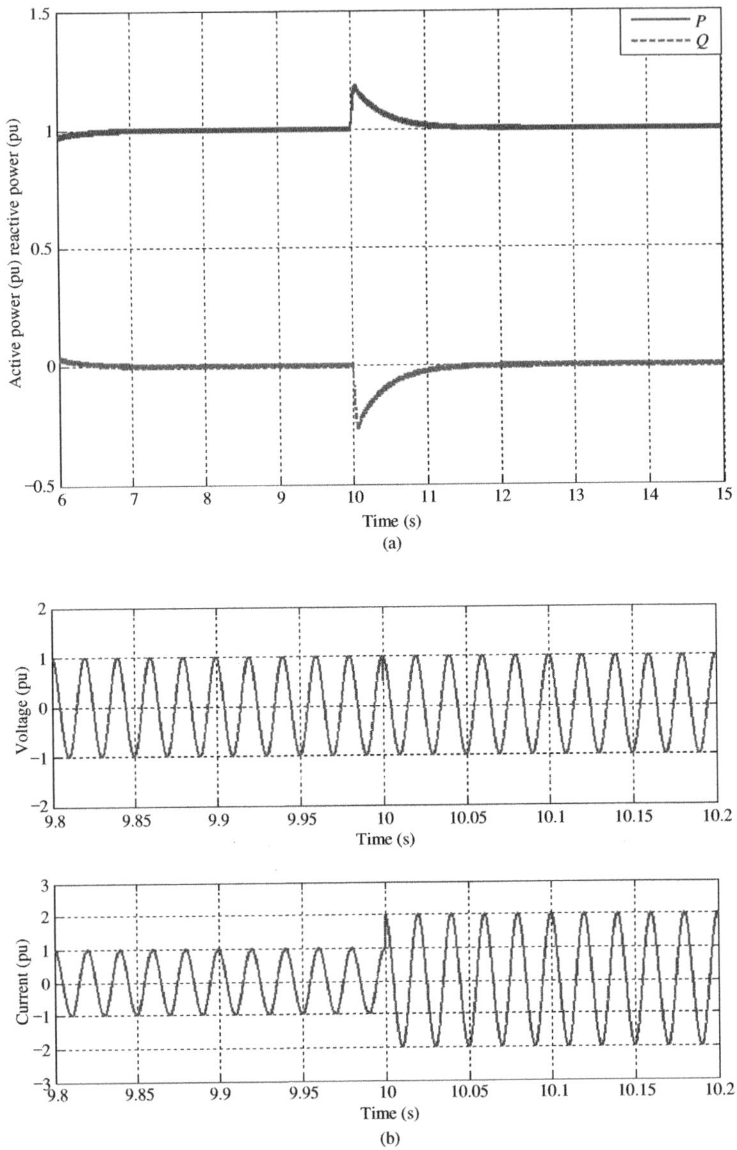

Figure 9.16 Load variation in grid-connected operational mode: (a) active and reactive powers and (b) load voltage and current

is needed. The synchronization system can be based on the comparison of the angle and amplitude information obtained by a PLL, as shown in Figure 9.19, and related to the grid voltage v_g and the DPGS voltage v_o. Two buffers are employed for monitoring the phase and amplitude errors in order that when they are lower than certain thresholds, the output signal S indicates that the synchronization can be carried out and this signal coordinates the STS. The state of the STS determines the operational mode of the DPGS, that is, grid-connected mode or stand-alone mode.

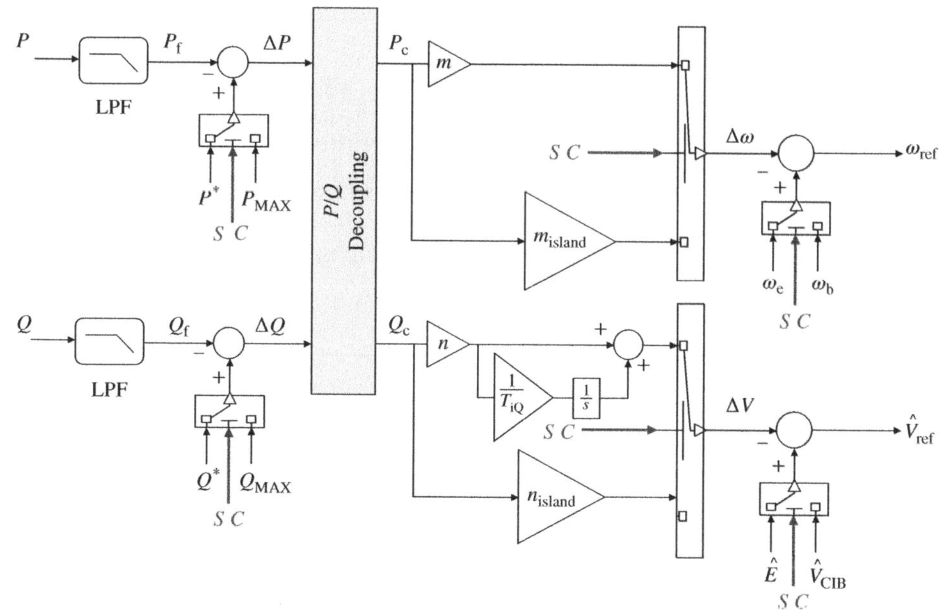

Figure 9.17 Power control based on droop control [33]

In order to define the voltage reference, verify the performance and decide on the reconnection instant: sequence detectors DSOGI-QSG and SFR-PLLs are used, as detailed in Section 9.1.3.2. The voltage reference of the DPGS converter is configured with the fundamental frequency-positive sequence (FFPS) in order to obtain a signal free of harmonics and unbalanced components. The choice of DSOGI-QSG relies on the best harmonic rejection performance with respect to the other techniques.

If the DPGS works in stand-alone mode, the problem of synchronization disappears in the case of grid failure; it is just when the main utility voltage is restored in the PCC that the synchronization system may be enabled to operating a proper connection.

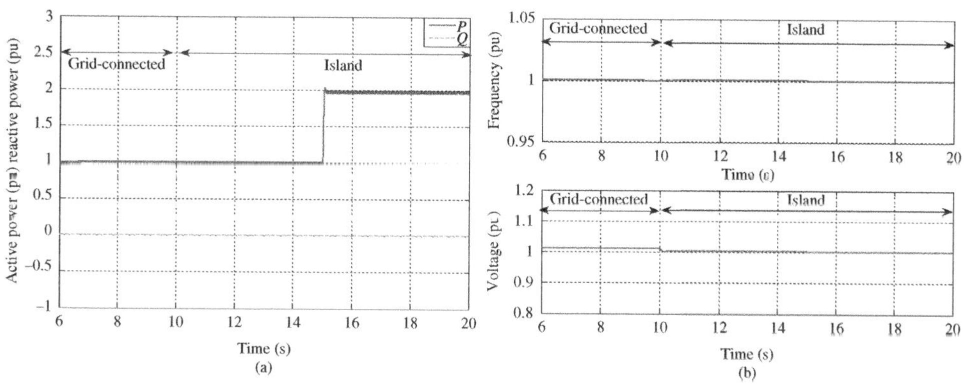

Figure 9.18 Load variation in island operational mode: (a) active and reactive powers, (b) PCC voltage amplitude and frequency and (c) load voltage and current

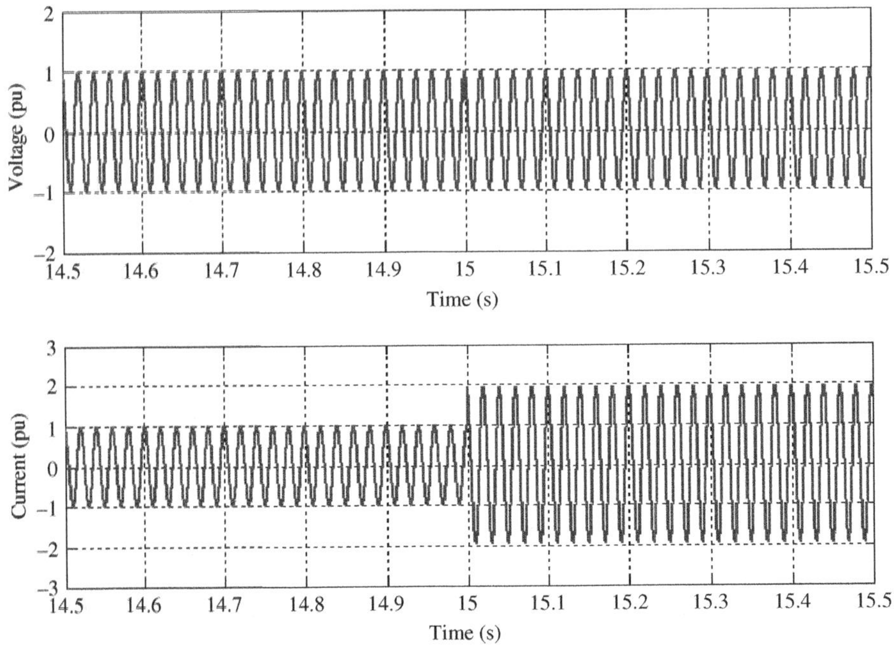

(c)

Figure 9.18 (*continued*)

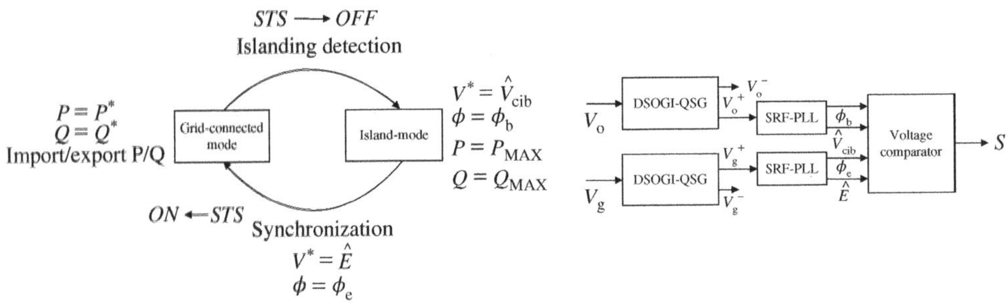

Figure 9.19 Transition between the different operational modes and synchronization system

9.2.3 Control Design Parameters

The DPGS dynamic model and the stability analysis are necessary to design properly the *P*-ω and *Q*-*V* control characteristics for both grid-connected and island operational modes. By modeling the inverter control structure with three control loops: a power-sharing control loop used to generate the amplitude and frequency of reference voltage; a voltage control loop based on a P-resonant controller providing the reference current for the innermost loop; and a P-resonant current controller, which determines the voltage reference for the PWM module (Figure 9.20), the stability analysis can be performed to design the controller parameters in the grid-connected operational mode. The outer power control loop should

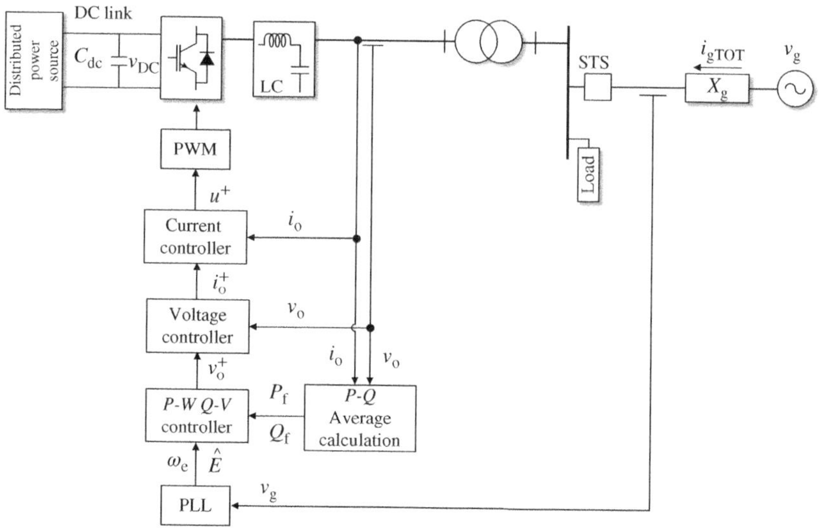

Figure 9.20 Multiloop control strategy of the DPGS

be designed with much lower bandwidth with respect to the two internal loops. The average P_f can be obtained by multiplying the output voltage v_o by the output current i_o and filtering the product using a low-pass filter. The average reactive power Q_f is obtained by delaying the output voltage by 90° and using a low-pass filter.

The low-pass filter is a second-order Butterworth filter, whose cut-off frequency should be chosen over a tenth of the fundamental frequency. The low-pass filter has a bandwidth much smaller than that of the voltage control of the inverter; thus, the performance of control will be very sensitive to this filter. The general expression of the transfer function of an nth-order Butterworth low-pass filter is given by

$$H(s) = \frac{1}{\prod_{i=1}^{n}(s - s_i)} = \frac{1}{(s - s_1)(s - s_2)\ \dots\ (s - s_n)} \tag{9.38}$$

where

$$s_i = e^{j\pi[(si+n-1)/2n]} = \cos\left(\pi\frac{2i+n-1}{2n}\right) + j\sin\left(\pi\frac{2i+n-1}{2n}\right) \tag{9.39}$$

As the pass-band of the filter is much smaller than the pass-band of the voltage control, it is reasonable to approximate with a delay, the dynamic of the voltage control loop where the power control is concerned. Linearizing (9.17) and (9.18) and assuming a P/Q decoupling, the active and reactive power control loops are obtained as follows (Figure 9.21):

The variables shown in Figure 9.21 are presented as follows:

- P_f and Q_f are the average active and reactive output powers;
- P^* and Q^* are the reference active and reactive powers;
- $G_{cP}(s)$ and $G_{cQ}(s)$ are the transfer functions of the active and reactive power controller;
- $G_{LPF}(s)$ is the transfer function of a second-order Butterworth low-pass filter;
- $G_{oP}(s)$ and $G_{oQ}(s)$ are the transfer functions of the plant of the active and reactive power control loops, respectively.

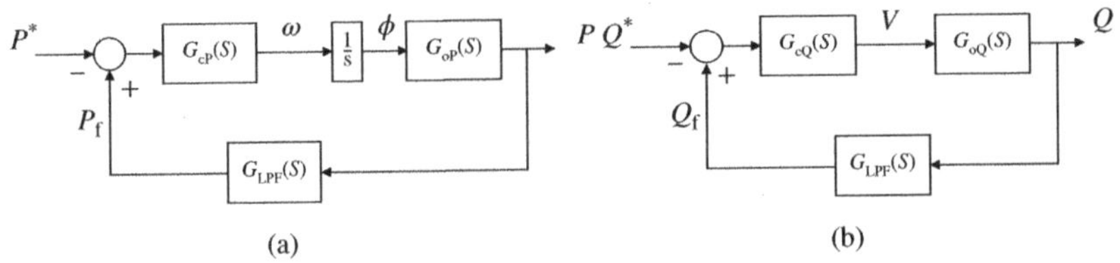

Figure 9.21 Active power: (a) reactive power and (b) control loops

Hence, the open-loop transfer function for the active and reactive control loops are given as

$$G_{\text{olP}}(s) = G_{\text{LPF}}(s)\frac{EV}{X} \cdot \frac{m(P_{\text{f}} - P^{*})}{s}$$

$$G_{\text{olQ}}(s) = G_{\text{LPF}}(s) \cdot \frac{(V-E)E}{X} \cdot n\left[(Q_{\text{f}} - Q^{*}) + \frac{1}{T_{\text{iQ}}s}(Q_{\text{f}} - Q^{*})\right] \tag{9.40}$$

The low-pass filter is a second-order Butterworth filter whose cut-off frequency should be chosen in the range of 5–10 Hz in order to extract the average value of the active and reactive powers. Considering a low-pass filter with a cut-off frequency of $f_{\text{t}} = 10$ Hz, the effect of proportional gain (m) variations on the stability and the speed response of the active power loop have been evaluated. Figure 9.22 presents the results for the active power control loop in which the proportional gain is varied. The phase margin reduces by increasing the proportional gain, but it can be varied within a limited range. The system becomes unstable for frequency proportional gain greater than 10^{-4}. Figure 9.22(c) depicts the family root locus diagrams for different values of m. It can be observed that the poles can be adjusted to obtain the desired dynamics. On the one hand, by increasing m, the poles move to imaginary axes and the system has a second-order behavior with an overdamped response; on the other hand, by decreasing the proportional gain, the system behaves as a first-order system with a response that is more damped with an increased settling time.

Figure 9.23 presents the results for the reactive power control loop in which the integrator time constant is fixed as $T_{\text{iQ}} = 0.03$, and the proportional gain is varied. From the Bode diagram, it is possible to observe that the phase margin decreases by increasing the proportional gain, but that it can be varied within a limited range. By increasing the proportional gain, the system response is faster and the settling time is reduced, but it presents an increasing overshoot, because the dominant poles are attracted to the imaginary axes.

In islanding mode, the DPGS must provide the active and reactive powers demanded by the load, maintaining the frequency within the limits allowed by the frequency-sensitive loads and a steady voltage profile. The proportional parameters of the power control can be tuned as follows:

$$m_{\text{island}} = \frac{\Delta\omega_{\text{MAX}}}{P_{\text{MAX}}} \tag{9.41}$$

$$n_{\text{island}} = \frac{\Delta V_{\text{MAX}}}{Q_{\text{MAX}}} \tag{9.42}$$

where $\Delta\omega_{\text{MAX}}$ and ΔV_{MAX} are the maximum allowed frequency and voltage deviations, respectively, and P_{MAX} and Q_{MAX} are the maximum active and reactive powers that DPGS can provide. The parameters m_{island} and n_{island} are selected in order to share the load demand among the DPGSs proportionally with their active power ratings and below the maximum allowed frequency deviation, and to guarantee both good reactive power sharing among the DPGSs and appropriate voltage regulation.

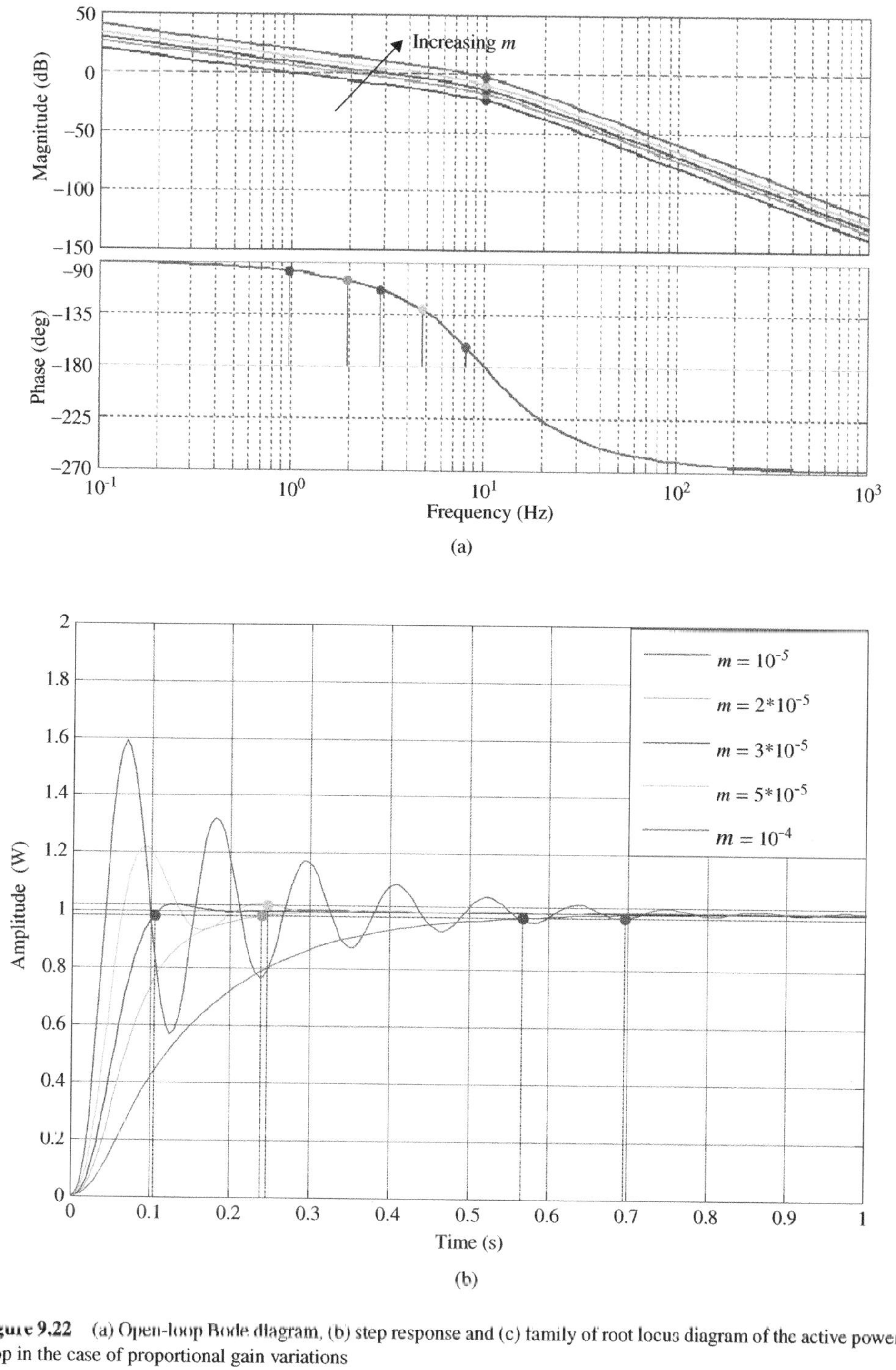

Figure 9.22 (a) Open-loop Bode diagram, (b) step response and (c) family of root locus diagram of the active power loop in the case of proportional gain variations

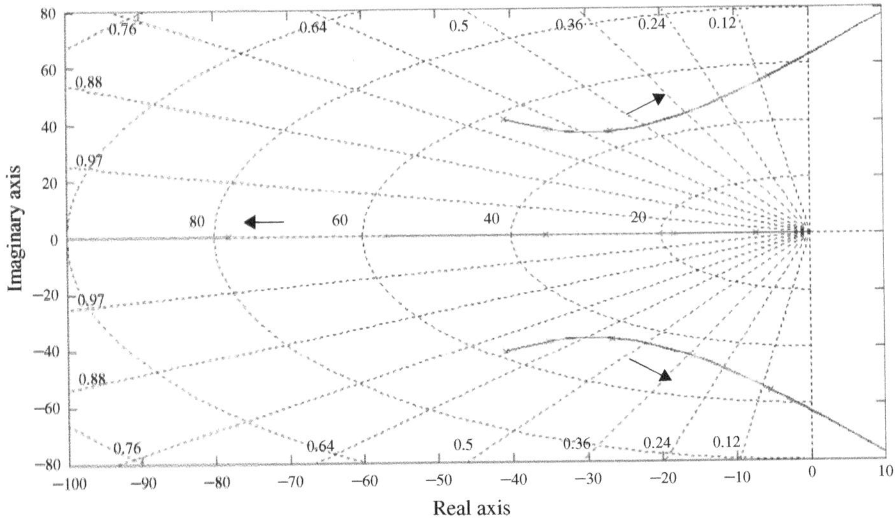

(c)

Figure 9.22 (*continued*)

9.2.4 *Harmonic Compensation*

One of the main limitations related to droop control is the poor harmonic power compensation when the DPGSs supply nonlinear loads. However, the structure of conventional droop control can be modified in order to comply with the power quality requirements; the solution results in fixing the DPGS converter output impedance presented to the single harmonic components. In the case of grid-connected operation, a proper compensation strategy can be adopted, based on a virtual impedance approach, subtracting a voltage proportional to the current harmonics from the output-voltage reference. Individual output impedances for the current harmonics of the highest values are obtained by filtering the current by means of a bank of band-pass filters. With the main aim to inject harmonic-free current into the grid, resonant filters can be employed to extract selectively the 3rd, 5th, 7th and 11th harmonics. The virtual impedance Z_H is defined by

$$Z_H(s) = \sum_{\substack{i=3 \\ \text{odd}}}^{11} \frac{2K_i s}{s^2 + 2K_i s + \omega_i^2} \tag{9.43}$$

where ω_i is the frequency of the ith harmonic and K_i are the coefficients of the filter. The coefficients K_i should be chosen properly in order not to affect the stability of the system. The voltage reference v_{ref} is modified by subtracting a voltage, which is proportional to the current harmonics, as shown in Figure 9.24.

The resulting voltage reference can be expressed as

$$v_o^* = v_{ref} - Z_H(s)\, i_o \tag{9.44}$$

where v_{ref} is the droop control output voltage reference and i_o is the inverter output current. The Bode diagram of the harmonics compensation loop, adopted in the grid-connected operational mode, in shown in Figure 9.25(a).

In contrast to the stand-alone operational mode, the main goal is to supply the local loads by providing a pure sinusoidal voltage in spite of nonlinear or unbalanced loads [13]. In the case of islanding detection,

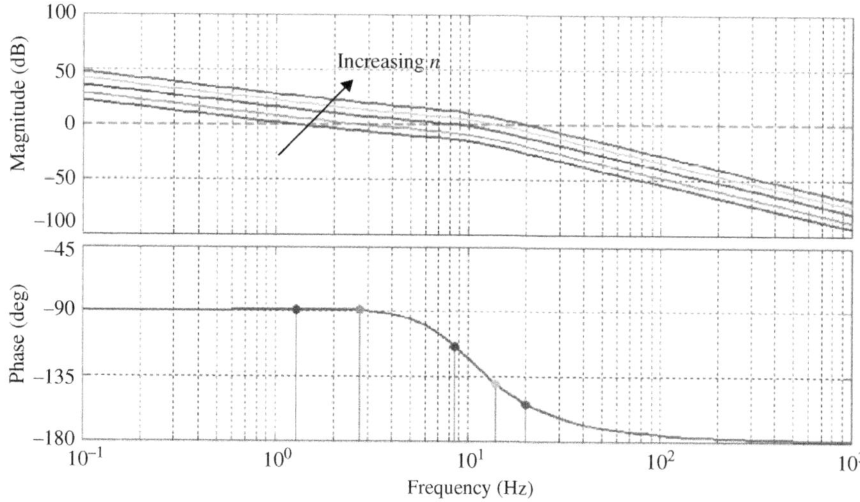

(a)

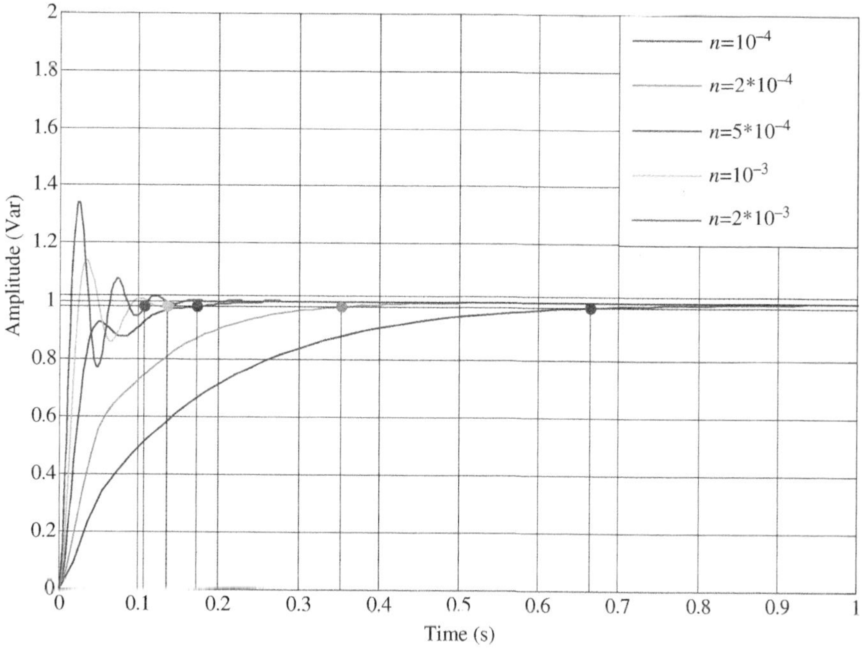

(b)

Figure 9.23 (a) Open-loop Bode diagram, (b) step response, (c) family of root locus diagram of the reactive power loop in the case of proportional gain variations fixing the integral gain at TiQ = 0.03

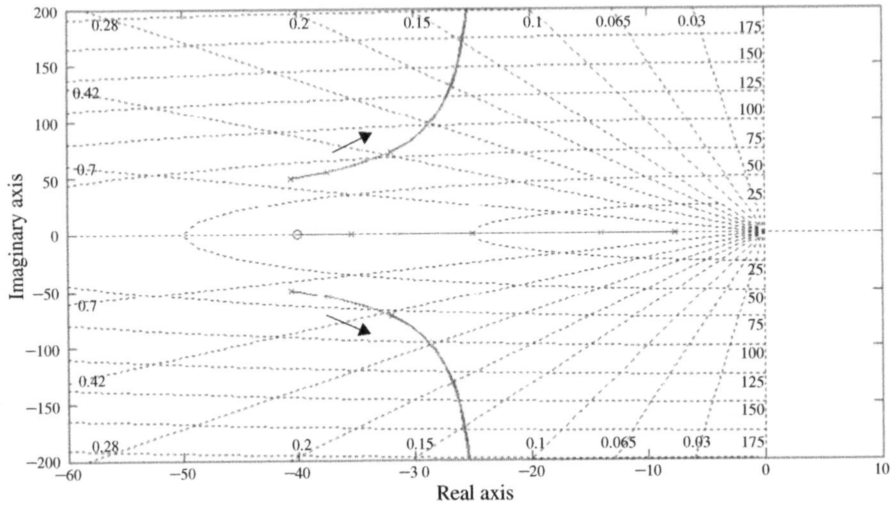

(c)

Figure 9.23 (*continued*)

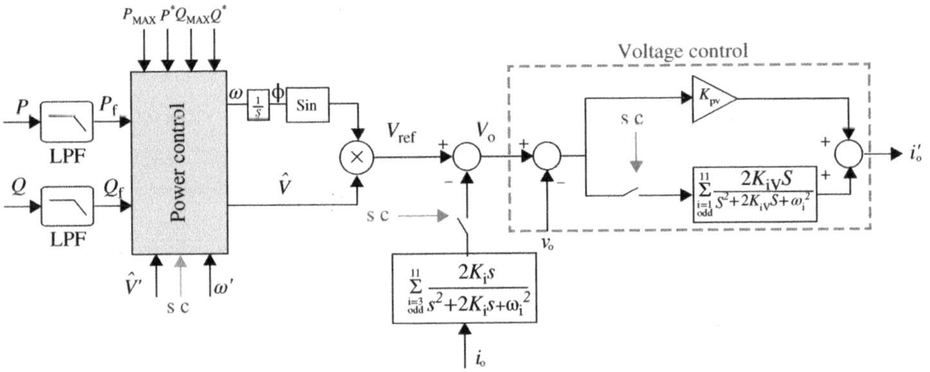

Figure 9.24 Scheme of harmonic compensation strategy

the control structure can be modified. The proposal relies in the application of resonant compensators (H_V) directly into the voltage control loop, without affecting the computation of the voltage reference v_{ref} provided by the droop control:

$$H_V(s) = \sum_{\substack{i=3 \\ \text{odd}}}^{11} \frac{2K_{iv}s}{s^2 + 2K_{iv}s + \omega_i^2} \tag{9.45}$$

Zero voltage steady-state errors and selective harmonic compensation can also be achieved in this operational mode. The Bode diagram of the voltage loop is shown in Figure 9.25(b).

In Figure 9.26, it is shown how the output current and the PCC voltage THD change when the harmonic compensation strategy of both grid-connected and stand-alone modes is enabled. In stand-alone

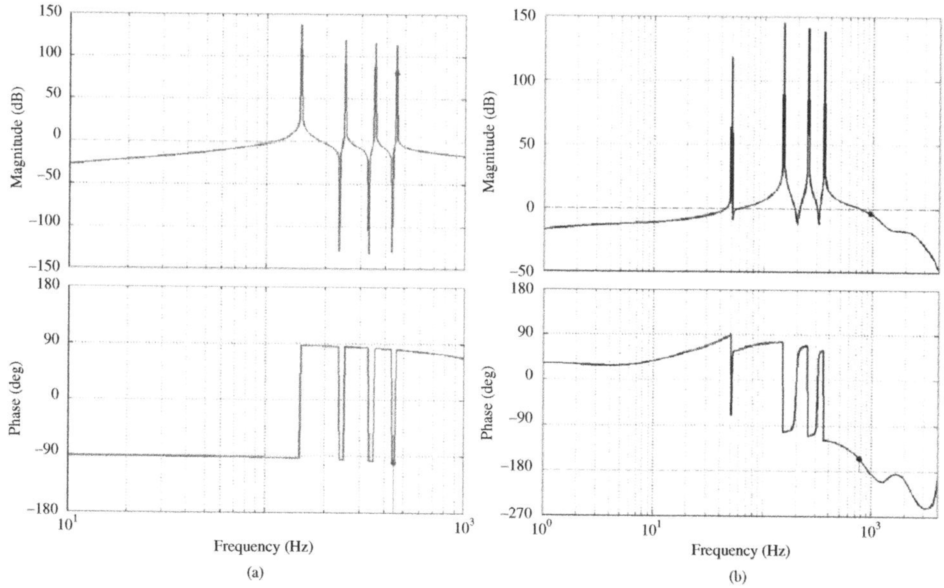

Figure 9.25 Bode diagram of (a) the current harmonics compensation loop and (b) the voltage harmonics compensation loop

operational mode, if nonlinear loads are present in the system, the voltage harmonics compensation strategy leads to a deteriorated DPGS current.

9.3 Ancillary Feature

The main ancillary services in DPGSs, as defined by the IEEE standard 1547.2 [36], are scheduling, system control and dispatch services, reactive power supply and voltage/frequency control regulation, energy imbalance service and operation of spinning reserve. In the future, ancillary services provided by DPGSs should contribute to the reinforcement of the distribution network and to the maintainence of proper quality of supply. Two different scenarios are open for PV systems and small WTSs to achieve this purpose. The first relies on a flexible control strategy providing voltage support for local loads that can be expanded to harmonic compensation functions. The second and more complex possibility is to support the voltage at the level of the electric power system area, in order to optimize the feeder voltage profile injecting reactive power under the management of the distribution system operator (DSO). These issues, together with the need for limiting the cost and size of the DPGSs such that they remain economically competitive, even when ancillary services are added, make the design problem particularly challenging.

9.3.1 Voltage Support at Local Loads Level

A DPGS grid-connected converter can be treated as a shunt controller. Generally, under power systems theory, shunt controllers have been used as static var generators for stabilizing and improving the voltage profile and for compensating current harmonics and unbalanced load currents, whereas series-converter topologies have been applied to multifunctional inverters with ride-through capability in the presence of

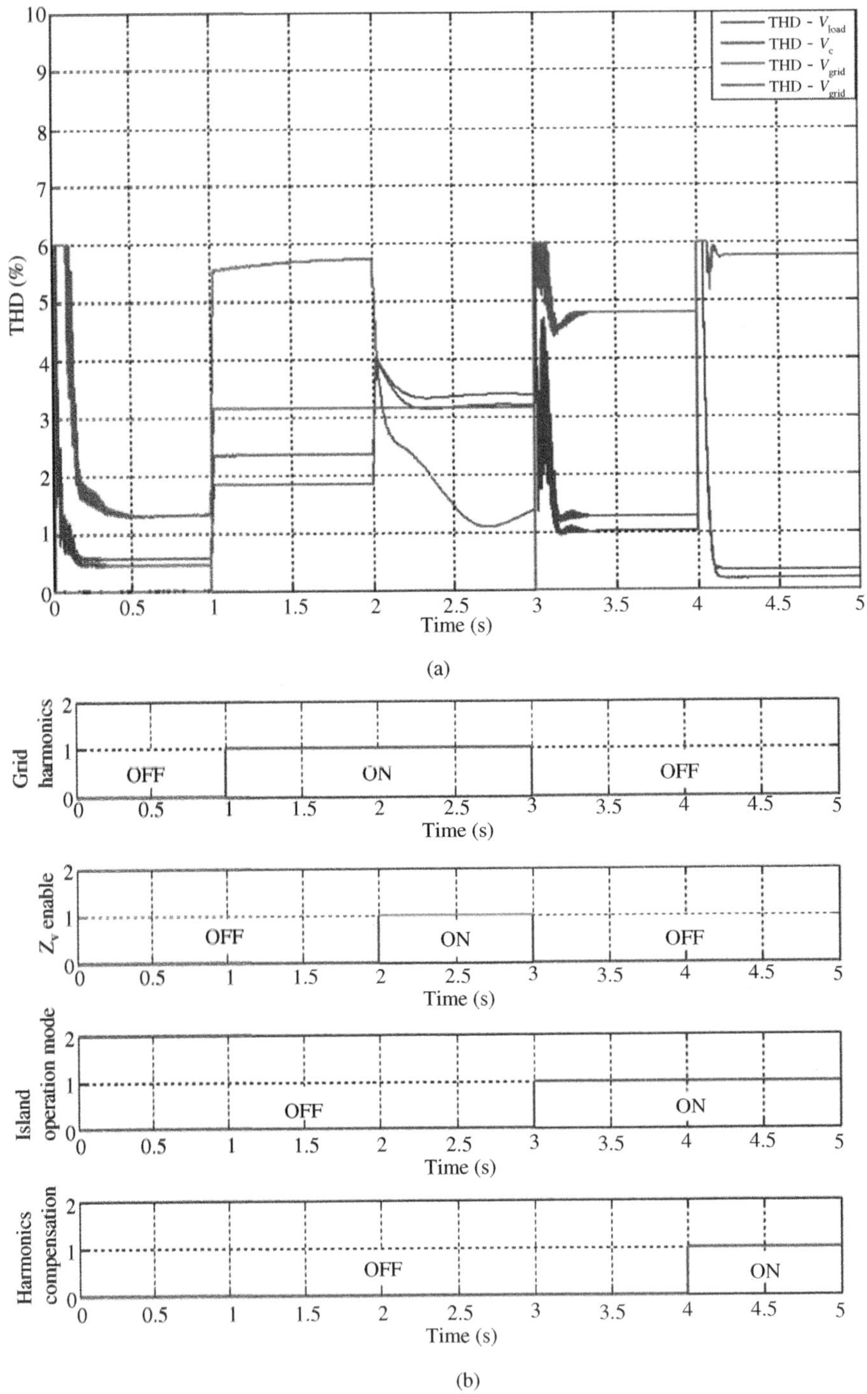

Figure 9.26 (a) PCC voltage and DPGS output current THDs, (b) time slot of different operation modes [35]

grid voltage sags. Alternatively, shunt devices are usually adopted to compensate small voltage variations, which can be controlled by reactive power injection. It is impossible to provide simultaneous control of the output voltage and of the output current harmonics. The ability to control the fundamental voltage at a certain point depends on the grid impedance and the load power factor. The compensation of a voltage dip by current injection is difficult to achieve because the grid impedance is usually low and the injected current has to be very high to increase the load voltage.

A shunt controller can be current or voltage controlled. When the converter is current controlled, it can be represented as a grid-feeding component (Figure 9.27(a)) that supports the grid voltage by adjusting its reactive output power according to grid voltage variations. When the converter is voltage controlled, it can be represented as a grid-supporting component (Figure 9.27(b)) that controls its output voltage. However, in this second case, the control action also results in injecting reactive power in order to stabilize the voltage. The vector diagrams of a shunt controller designed to provide only reactive power are presented in Figure 9.29. When the grid voltage is 1 pu, the converter supplies the reactive power absorbed by the load, and the vector diagram of the current- or voltage-controlled converter is the same. In the first case, it is controlled by the compensating current I_C and in the second case, it is controlled by the load voltage, as underlined in Figure 9.28(a) and Figure 9.28(b).

When a voltage sag occurs, the converter provides reactive power to support the load voltage and the grid current $\overline{I}_g$ has a dominant reactive component:

$$\overline{I}_g + \overline{I}_C = \overline{I}_{load} \tag{9.46}$$

The amplitude of the grid current depends on the value of the grid impedance because

$$\overline{I}_g = \frac{\overline{V}_{Lg}}{j\omega L_g} \tag{9.47}$$

where $\overline{V}_{Lg}$ is the inductance voltage drop, as shown in Figure 9.28(c). If the shunt controller supplies the load with all the requested active and reactive powers, in normal conditions it provides a compensating current $\overline{I}_c = \overline{I}_{load}$; hence, the system operates as in island mode and $\overline{I}_g = 0$.

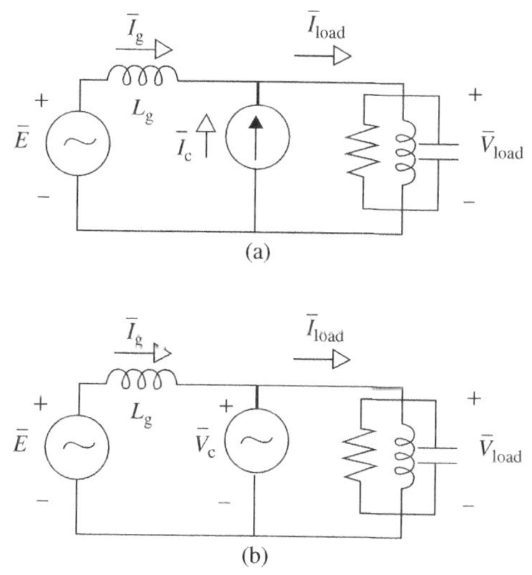

Figure 9.27 Use of a shunt controller for voltage dip compensation: (a) simplified power circuit of the current-controlled shunt controller and (b) simplified power circuit of the voltage-controlled shunt controller [13]

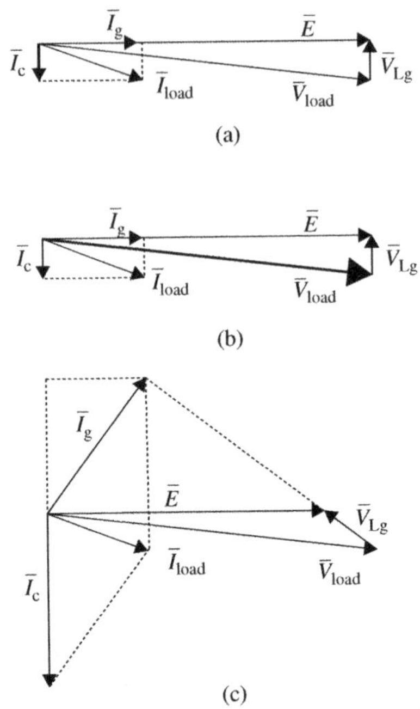

Figure 9.28 Vector diagram of the shunt controller providing only reactive power: (a) current-controlled converter in normal conditions, (b) voltage-controlled converter in normal condition and (c) vector diagram for compensation of a voltage dip of 0.15 pu [13]

In the case of voltage dips the converter has to provide the active power required by the load and it has to inject the reactive power needed to stabilize the load voltage, as shown in Figure 9.29(b). The grid current in this case is reactive. It can be seen that

$$\overline{V}_{load} = \overline{E} + \overline{V}_{Lg} \tag{9.48}$$

Hence, during a voltage sag the amount of reactive current needed to maintain the load voltage at the desired value is inversely proportional to ωL_g. This means that a large inductance will help in mitigating voltage sags, although it is not desirable during normal operation. This feature can be viewed as an ancillary service that the system could provide to its local loads.

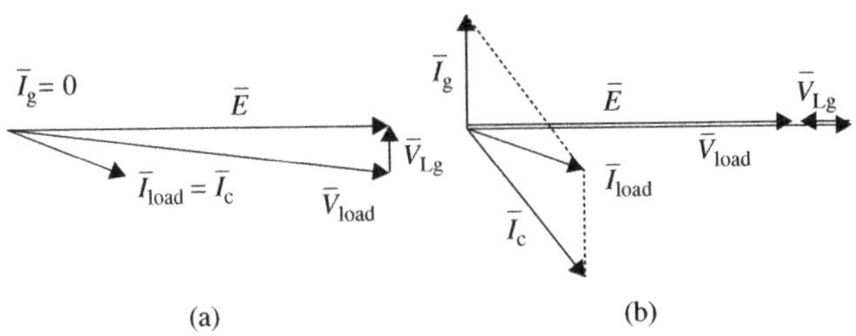

Figure 9.29 Vector diagram of the shunt controller providing both active and reactive powers: (a) normal conditions and (b) vector diagram for compensation of a voltage dip of 0.15 pu [13]

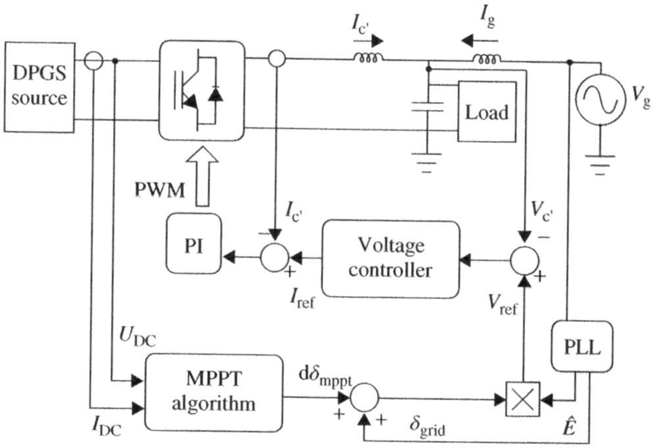

Figure 9.30 DPGS converter operating as shunt controller

Usually, the inverter output impedance is considered inductive owing to both the inductive component of the line impedance and the large inductor of the output filter. However, this is not always true because the closed-loop output impedance also depends on the control strategy. A possibility is to add an inductor in series with the inverter output in order to fix the output impedance. In the case where a large inductance is present in the grid side or where the grid impedance can be considered mainly inductive, the proposal of a DPGS control structure, improved with a small voltage sags compensation function, can be presented (Figure 9.30). In this, the voltage of the output filter capacitor V_c' is controlled directly while the current injection is controlled indirectly. The amplitude of the current depends on the difference between the grid voltage and the voltage on the AC capacitor V_c'. The phase displacement between these two voltages determines the injected active power (decided by the MPPT algorithm), and the voltage amplitude difference determines the reactive power exchange with the grid. The requested reactive power is limited by the fact that a voltage dip higher than 15% will force the system to disconnect (as requested by standards). The voltage error between V_{ref} and V_c' can be preprocessed by a repetitive or resonant controller, operating as a periodic signal generator of the fundamental component and of selected harmonics, in order to improve the system with a voltage harmonic compensation function. The use of a repetitive or resonant controller can ensure precise tracking of the selected harmonics and provide the reference for the inner loop [13].

The advantages provided by the shunt controller operational mode of the DPGS converter are highlighted by the case of distorting loads. In this situation, the voltage controller can compensate the harmonics introduced by the distorting loads, improving the quality of the load voltage profile. An example is presented in Figure 9.31 where a load voltage waveform affected by a THD of around 17% can lead to a voltage THD of around 2%. A detailed harmonic spectra comparison related to the example is shown in Figure 9.32.

9.3.2 *Reactive Power Capability*

DPGSs are becoming increasingly attractive structures for the proposal of an active low-voltage (LV) distribution network. Reactive power-handling capability is requested to DPGSs by the new national grid codes. Among them, the Italian standard CEI 0–21 establishes the DPGS participation to the voltage regulation through reactive power generation/absorption. It happens both to compensate voltage variations owing to the system active power injection at the PCC and to provide a service to the DSO. In particular,

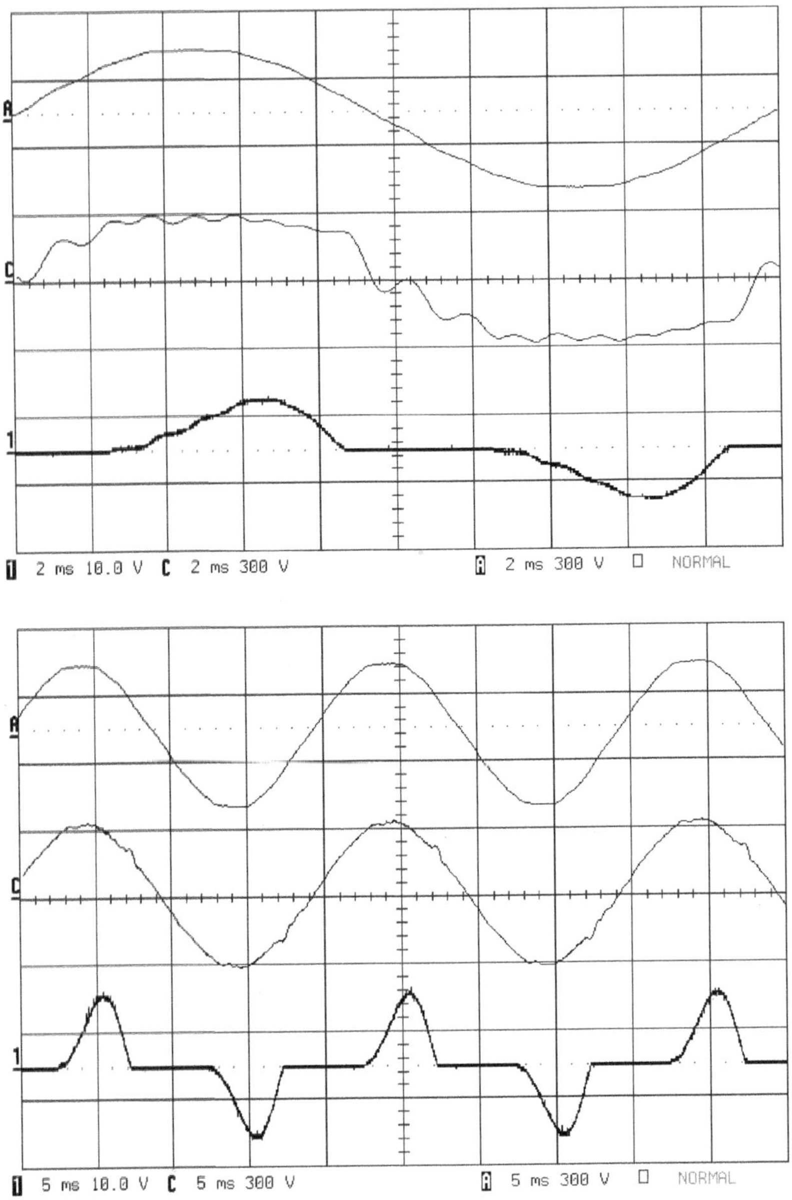

Figure 9.31 Experimental results obtained for the case of distorting load: (a) without shunt converter: A grid voltage (300 V/div), C load voltage (300 V/div), 1 load current (10 V/div); (b) with a shunt converter connected to the grid: A grid voltage (300 V/div), C load voltage (300 V/div), 1 load current (10 V/div) [13]

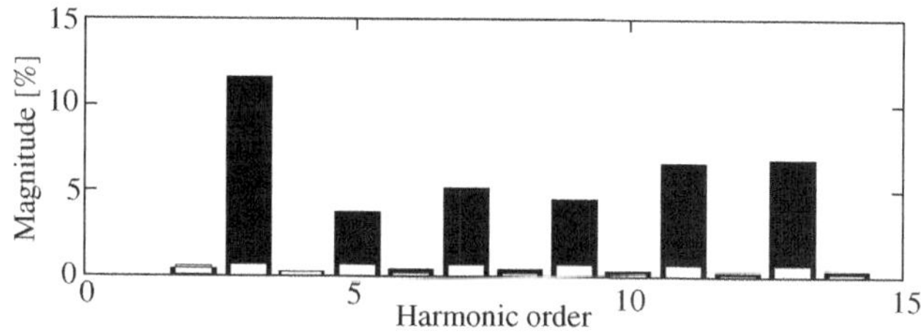

Figure 9.32 Load voltage harmonic spectrum in the case of distorting load: (black bars) without shunt converter; (white bars) with shunt converter connected to the grid [13]

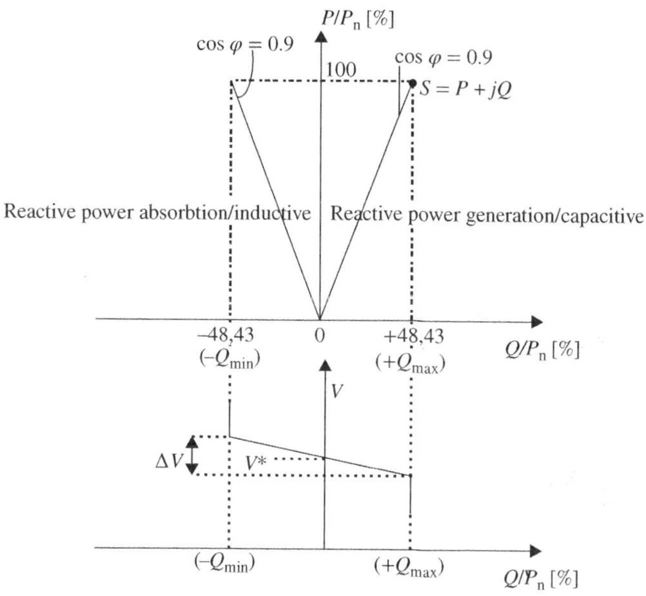

Figure 9.33 Triangular/rectangular reactive power capability curves for power converters over 6 kW (P_n is the rated active power) and a possible V/Q characteristic [37]

the reactive power exchange may be maintained inside the triangular reactive power capability curve, as defined by a power factor limit of 0.9 in the first case, or inside the rectangular reactive power capability curve (allowing for a power factor <0.9) in the second case (Figure 9.33). In order to allow this function, the DPGS may guarantee a continuous reactive power supply, regulated through a variable real-time power factor operation ($\cos\varphi = f(P)$), based on the local control system. If required by the DSO, the DPGS should provide reactive power following a variable reactive power curve ($Q = f(V)$) where the set points are defined through a remote control.

9.3.3 Voltage Support at Electric Power System Area

Recently, DPGSs based on renewable sources have represented one the most important worldwide research fields involving different matters. The main goals in this scenario have been to reduce costs and to improve efficiency. However, currently, a missing link remains between the performance improvement of a single DPGS unit because of the optimization of the power conversion stage, and the main EPS where the control and management is performed just for high-power plants integrated into the transmission network. Otherwise, the fulfillment of a real DPGS integration into the distribution network, based on a distribution management system (DMS) operation represents the last obstacle toward true exploitation of DPGS potential and the achievement of a low-voltage active network.

Twenty years ago, certain energy management system (EMS) functions, such as feeder voltage and loss optimization, began to be integrated within the DMS. However, in the past, all small- and medium-sized generations have been treated as "not-dispatchable," and among these were the wind turbine and PVSs. This means that all the power produced by these types of DPGS was simply injected into the grid with no possibility of control. However, this situation is now changing and the latest DMS will incorporate optimization functions for the short-term scheduling of the various energy and control resources available in the network.

A hierarchical control enabling real integration into the distribution network can be implemented based on a local control system (primary and secondary levels) performed on the grid-side DPGS converter, and a remote control system providing the set points for the local control (Figure 9.34). The idea is that every DPGS should have a distributed/local control system operating a coordinated active/reactive power control strategy carried out by means of three-level cascade loops. The local control system is responsible for operating the local voltage control and the active/reactive power exchange with the local loads and the main grid in universal operation (grid connected, stand-alone, or microgrids). The tertiary control is performed by the DMS, which can give set points to the inner control level in order that the DPGS provides a service to the DSO.

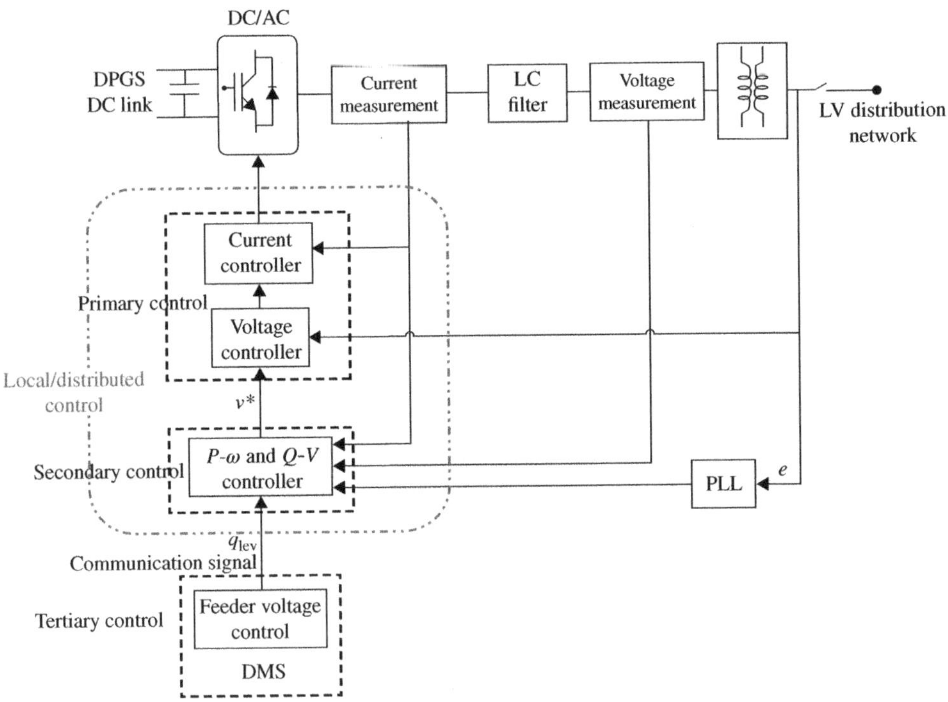

Figure 9.34 Hierarchical control of DPGS [37]

The primary control is in charge of the reliability and stability of the system, and thus, there is the need for a fast control strategy. The primary control structure can be designed based on Section 9.1.3.1. In contrast, the secondary control operates on a slower timeframe and deals with the power exchange between the LV distribution network and the DPGS, the voltage profile optimization and the frequency restoration. The primary and secondary controls form the local/distributed WTS control system and by using communication, the DPGS can receive signals from the DMS of DSO where the tertiary control is performed. The tertiary control completes the system with respect to what is presented in Section 9.2. As the inner loop is faster than the outer loop, a substantial dynamic decoupling is achieved between overlapped levels. In other words, the time-decomposition criterion requires that the dominant time constant of any external control loop be higher than the dominant time constants of all its internal loops [38]. Such a criterion is applied to the voltage control unit active/reactive power control unit and the set-point optimization unit.

Despite the traditional control theory of a DPGS converter, where the system is designed just to inject the maximum active power by means of an MPPT algorithm

$$Q_{\mathrm{REF}} = \sqrt{(S^2 - P^2_{\mathrm{MPPT}})} \tag{9.50}$$

in this new scenario, the active power reference can be changed dynamically based on the dispatched reactive power reference:

$$P^* = \sqrt{(S^2 - (Q^*)^2)} \tag{9.51}$$

Hence, in order to achieve a successful integration of the DPGS with high-performance power exchange, the secondary level control is designed based on (9.34) and (9.35) (considering the forcing of the voltage vector to be aligned with the d-axis of the reference frame). The local control system aims, at least, to limit the communication lines (Section 9.2.2.1), while it can be conceived that a central dispatcher periodically updates the controller's set points (tertiary control level) in order to exploit the capabilities of each DPGS for either present or foreseen grid-operating conditions.

In this hypothesis, the reference value Q^* is obtained from the product of the reactive power level q_{lev} by the unit capability limit Q_{LIM}, computed online according to the actual operating conditions.

$$Q^* = q_{\mathrm{LEV}} \cdot Q_{\mathrm{LIM}} \tag{9.52}$$

The output of the central dispatcher results in the reactive power level q_{lev} defined in closed loop and real time and was set in the interval [−1;+1] (corresponding, respectively, to the minimum value $Q_{\min} = -48.43\%P_n$ and the maximum value $Q_{\max} = +48.43\%P_n$) in order that the feeder voltage V_{P} expressed in pu corresponds to its reference value V_{PREF} expressed in pu.

$$q_{\mathrm{lev}} = \left[q_{\mathrm{rif}} - K_{\mathrm{PV}}\left(V_{\mathrm{PREF}} - V_{\mathrm{P}}\right) - K_{\mathrm{IV}} \int_0^t (V_{\mathrm{PREF}} - V_{\mathrm{P}})\mathrm{d}t \right]_{-1}^{+1} \tag{9.53}$$

In (9.53) q_{rif} is the reactive power level reference (assumed equal to zero in normal operation, K_{PV} and K_{IV} are the proportional and integral gains of the tertiary controller, respectively, tuned in such a way that the closed loop has a dominant time constant at least 10 times slower than that of the secondary controller.

9.4 Summary

Over recent years, DPGSs based on PV and small-wind turbines have been evolving at a fast pace. In order to fulfill a real integration of DPGSs with the main grid, a complex control system must be designed in order to achieve a universal operation (grid-connected mode, stand-alone mode and microgrid). In this chapter, starting from the analysis of the power stage and the control structure of a typical PV system and small WTS, the control units that contribute to performing a universal operation are reviewed in detail. These include the synchronization system and the islanding detection system, coordinated with both a supervision system and a cascade loops control structure, which modifies its references based on the supervision information. The universal operation relies on a flexible control structure designed with the goals of limiting the number of changes when moving from grid-connection mode to stand-alone operational mode, and of an optimum management of the transition among the different conditions. Flexibility is ensured by droop-control-based design. Reactive power processing is required in light of the new grid codes, and harmonic compensation strategy might be integrated inside the control structure in order to comply with power quality criteria.

References

1. Lasseter, R.H. and Piagi, P. (2004) Microgrid: a conceptual solution. 2004 Power Electronics Specialist Conference, PESC 2004, Aachen, Germany, pp. 4285–4290.
2. Hatziargyriou, N., Asano, H., Iravani, R., and Marnay, C. (2007) Microgrids. *IEEE Power and Energy Magazine*, **5** (4), 78–94.
3. Katiraei, F., Iravani, M.R., and Lehn, P.H. (2005) Micro-grid autonomous operation during and subsequent to islanding process. *IEEE Transactions on Power Delivery*, **20** (1), 248–257.
4. Villeneuve, P.L. (2004) Concerns generated by islanding. *IEEE Power and Energy Magazine*, **2** (3), 49–53.
5. IEEE Standard (2011) IEEE Std 1547.4. *IEEE Guide for Design, Operation and Integration of Distributed Resource Island Systems with Electric Power Systems*, IEEE.
6. IEEE Standard (2005) IEEE Std 1547.1. *IEEE Conformance Test Procedures for Equipment Interconnecting Distributed Resources with Electric Power Systems*, IEEE.
7. IEEE Standard (2003) IEEE Std 1547–2003. *IEEE Standard for Interconnecting Distributed Resources with Electric Power Systems*, IEEE.
8. Petrone, G., Spagnuolo, G., Teodorescu, R. *et al.* (2008) Reliability issues in photovoltaic power processing systems. *IEEE Transactions on Industrial Electronics*, **55** (7), 2569–2580.
9. IEC Standard (2004) 61727, Ed.2. *Photovoltaic PV Systems Characteristics of the Utility*, IEC.
10. Mastromauro, R.A., Liserre, M., and Dell'Aquila, A. (2008) Study of the effects of inductor nonlinear behavior on the performance of current controllers. *IEEE Transactions on Industrial Electronics*, **55** (5), 2043–2052.
11. Zmood, D.N. and Holmes, D.G. (2003) Stationary frame current regulation of PWM inverters with zero steady-state error. *IEEE Transactions on Power Electronics*, **18** (3), 814–822.
12. Mastromauro, R.A., Liserre, M., and Dell'Aquila, A. (2012) Control issues in single-stage photovoltaic systems: MPPT, current and voltage control. *IEEE Transactions on Industrial Informatics*, **8** (2), 241–254.
13. Mastromauro, R.A., Liserre, M., Kerekes, T., and Dell'Aquila, A. (2009) A single-phase voltage-controlled grid-connected photovoltaic system with power quality conditioner functionality. *IEEE Transactions on Industrial Electronics*, **56** (11), 4436–4444.
14. Vasquez, J.C., Mastromauro, R.A., Guerrero, J.M., and Liserre, M. (2009) Voltage support provided by a droop-controlled multifunctional inverter. *IEEE Transactions on Industrial Electronics*, **56** (11), 4510–4519.
15. Cagnano, A., De Tuglie, E., Liserre, M., and Mastromauro, R.A. (2011) Online optimal reactive power control strategy of PV inverters. *IEEE Transactions on Industrial Electronics*, **58** (10), 4549–4558.
16. Ropp, M., Newmiller, J., Whitaker, C., and Norris, B. (2008) Review of potential problems and utility concerns arising from high penetration levels of photovoltaics in distribution systems. 33rd IEEE Photovoltaic Specialists Conference, PVSC '08, May 11–16, 2008, pp. 1–6.
17. Teodorescu, R., Liserre, M., and Rodriguez, P. (2011) *Grid Converters for Photovoltaic and Wind Power Systems*, IEEE Press/John Wiley & Sons, Ltd. ISBN: 978–0470057513
18. Mirecki, A., Roboam, X., and Richardeau, F. (2007) Architecture complexity and energy efficiency of small wind turbines. *IEEE Transactions on Industrial Electronics*, **54** (1), 660–670.
19. Liserre, M., Nagliero, A., Mastromauro, R.A. *et al.* (2011) Universal operation of small/medium size renewable energy systems. Proceedings of PEIA 2011 (Power Electronics for Industrial Applications and Renewable Energy Conversion), Doha, Qatar, November 3–4, 2011, pp. 162–168.
20. Pogaku, N., Prodanovic, M., and Green, T.C. (2007) Modeling, analysis and testing of autonomous operation of an inverter- based microgrid. *IEEE Transactions on Power Electronics*, **22** (2), 613–625.

21. Nagliero, A., Mastromauro, R.A., Liserre, M., and Dell'Aquila, A. (2010) Monitoring and synchronization techniques for single-phase PV systems. IEEE 2010 International Symposium on Power Electronics Electrical Drives Automation and Motion (SPEEDAM), 14–16 June 2010, pp. 1404–1409.
22. Blaabjerg, F., Teodorescu, R., Liserre, M., and Timbus, A.V. (2006) Overview of control and grid synchronization for distributed power generation systems. *IEEE Transactions on Industrial Electronics*, **53**, 1398–1409.
23. Teodorescu, R. and Blaabjerg, F. (2004) Flexible control of small wind turbines with grid failure detection operating in stand-alone and grid failure detection operating in stand-alone and grid connected mode. *IEEE Transactions on Power Electronics*, **19** (5), 1323–1332.
24. Ghartemani, M.K., Khajehoddin, S.A., Jain, P.K., and Bakhshai, A. (2012) Problems of startup and phase jumps in PLL systems. *IEEE Transactions on Power Electronics*, **27** (4), 1830–1838.
25. De Mango, F., Liserre, M., Dell'Aquila, A., and Pigazo, A. (2006) Overview of anti islanding algorithms for PV systems. Part I: passive methods. 12th International Power Electronics and Motion Conference, PE-PEMC 2006, Portoroz, Slovenia, August 30 – September 1, 2006, pp. 1878–1883.
26. De Mango, F., Liserre, M., and Dell'Aquila, A. (2006) Overview of anti-islanding algorithms for PV systems. Part II: active methods. 12th International Power Electronics and Motion Conference, EPE_PEMC 2006, Portoroz, Slovenia, August 30 – September 1, 2006, pp. 1884–1889.
27. Timbus, A.V., Teodorescu, R., Blaabjerg, F., and Borup, U. (2004) Online grid measurement and ENS detection for PV inverter running on highly inductive grid. *IEEE Power Electronics Letters*, **2** (3), 77–82.
28. Pigazo, A., Liserre, M., Mastromauro, R.A. *et al.* (2009) Wavelet-based islanding detection in grid-connected PV systems. *IEEE Transactions on Industrial Electronics*, **56** (11), 4445–4455.
29. Orlando, N.A., Nagliero, A., Mastromauro, R.A. *et al.* Small wind turbines in grid-interactive microgrids. 38th Annual Conference on IEEE Industrial Electronics Society IECON 2012, October 25–28, 2012, pp. 1156–1161.
30. Nagliero, A., Mastromauro, R.A., Monopoli, V.G. *et al.* (2010) Analysis of a universal inverter working in grid-connected, stand-alone and micro-grid. 2010 IEEE International Symposium on Industrial Electronics (ISIE), July 4–7, 2010 pp. 650–657.
31. De Brabandere, K., Bolsens, B., Van den Keybus, J. *et al.* (2007) A voltage and frequency droop control method for parallel inverters. *IEEE Transactions on Power Electronics*, **22** (4), 1107–1115.
32. Guerrero, J.M., de Vicuna, L.G., Matas, J. *et al.* (2005) Output impedance design of parallel-connected UPS inverters with wireless load-sharing control. *IEEE Transactions on Industrial Electronics*, **52** (4), 1126–1135.
33. Nagliero, A., Mastromauro, R.A., Ricchiuto, D. *et al.* (2011) Gain-scheduling-based droop control for universal operation of small wind turbine systems. 2011 IEEE International Symposium on Industrial Electronics (ISIE), June 27–30, 2011, pp. 1459–1464.
34. Rocabert, J., Azevedo, G.M., Luna, A. *et al.* (2011) Intelligent connection agent for three-phase grid- connected microgrids. *IEEE Transactions on Power Electronics*, **26** (10), 2993–3005.
35. Nagliero, A., Mastromauro, R.A., Liserre, M., and Dell'Aquila, A. (2011) Harmonic control strategy for universal operation of wind turbine systems. 2011 International Conference on Power Engineering, Energy and Electrical Drives (POWERENG), May 11–13, 2011, pp. 1–5.
36. IEEE Standard (2008) 1547.2-2008 *IEEE Application Guide for IEEE Std 1547, IEEE Standard for Interconnecting Distributed Resources with Electric Power Systems*, IEEE.
37. Mastromauro, R.A., Orlando, N.A., Ricchiuto, D. *et al.* (2013) Hierarchical control of a small wind turbine system for active integration in LV distribution network. 2013 International Conference on Clean Electrical Power (ICCEP 2013), June 2013, pp. 461–468.
38. Corsi, S., Pozzi, M., Sabelli, C., and Serrani, A. (2004) The coordinated automatic voltage control of the italian transmission grid- Part I: reasons of the choice and overview of the consolidated hierarchical system. *IEEE Transactions on Power Systems*, **19** (4), 1723–1732.

10

Properties and Control of a Doubly Fed Induction Machine

Gonzalo Abad[1] and Grzegorz Iwanski[2]
[1] *Electronics and Computing Department, Mondragon University, Mondragon, Spain*
[2] *Institute of Control and Industrial Electronics, Warsaw University of Technology, Warszawa, Poland*

10.1 Introduction. Basic principles of DFIM

10.1.1 Structure of the Machine and Electric Configuration

The doubly fed induction machine (DFIM) or wound rotor induction machine (WRIM) are terms commonly used to describe an electrical machine, which has been used over many decades in various applications, often in the range of megawatts of power and also less commonly in the range of a few kilowatts. This concept of the machine is as an alternative to more common asynchronous and synchronous machines. It can be advantageous in applications that have a limited speed range, allowing a reduction in the size of the supplying power electronic converter as, for instance, in variable-speed generation, water pumping and so on.

The typical supply configuration of the DFIM is shown in Figure 10.1. The stator is supplied by three-phase voltages directly from the grid at constant amplitude and frequency, creating the stator magnetic field [1, 2]. The rotor is also supplied by three-phase voltages that take a different amplitude and frequency at steady state in order to reach different operating conditions of the machine (speed, torque, etc.). This is achieved by using a back-to-back three-phase converter, as represented in the simple schematic in the figure. This converter, together with the appropriate control strategy, is in charge of imposing the required rotor AC voltages to control the overall DFIM operating point and to perform the power exchange through the rotor to the grid. Although a voltage source converter is shown, different configurations or converter topologies could be utilized. Further details regarding the operation of the machine are described in subsequent sections.

Power Electronics for Renewable Energy Systems, Transportation and Industrial Applications, First Edition.
Edited by Haitham Abu-Rub, Mariusz Malinowski and Kamal Al-Haddad.

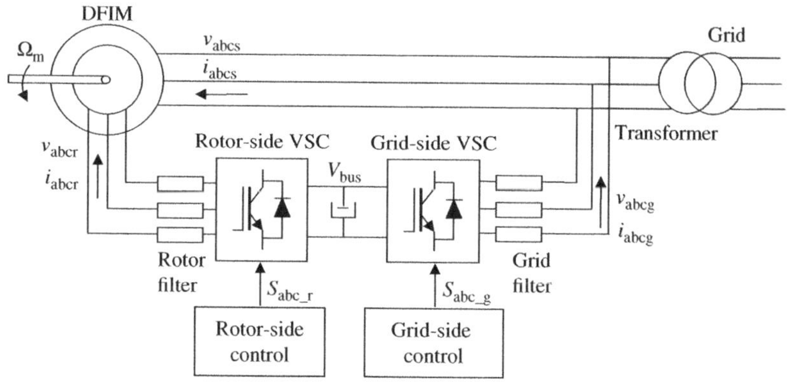

Figure 10.1 General supply configuration of DFIM

10.1.2 Steady-State Equivalent Circuit

10.1.2.1 Electric Equations

The relationships between the different frequencies of the machine are basics that must be known prior to the study of the electric equations of the DFIM. Thus, the equation that relates ω_s (frequency of stator voltages and currents), ω_r (frequency of rotor voltages and currents) and ω_m (rotor electrical speed) is [1] as follows:

$$\omega_s = \omega_r + \omega_m \tag{10.1}$$

The relation between the mechanical speed of the shaft Ω_m and the electrical speed depends on the pole pairs of the machine:

$$\omega_m = p\Omega_m \tag{10.2}$$

The units of these two equations are given in rad/s. The slip s of the machine is defined as follows:

$$s = \frac{\omega_s - \omega_m}{\omega_s} = \frac{\omega_r}{\omega_s} \tag{10.3}$$

The previous relations also hold if represented in different units, for example, in Hz or rev/min. In most applications, the stator windings are directly connected to the grid and thus, ω_s is constant. This frequency is also known as the synchronous frequency. However, ω_r obviously depends on the shaft's electrical speed ω_m, which leads to three operating modes of the machine dependent on the speed:

$$w_m < w_s \Rightarrow w_r > 0 \Rightarrow s > 0 \Rightarrow \text{Subsynchronous operation}$$

$$w_m > w_s \Rightarrow w_r < 0 \Rightarrow s < 0 \Rightarrow \text{Hypersynchronous operation}$$

$$w_m = w_s \Rightarrow w_r = 0 \Rightarrow s = 0 \Rightarrow \text{Synchronous operation}$$

Having reached this point, Figure 10.2 shows the equivalent steady-state circuit of a DFIM. It is an idealized model in which only one phase of the stator and rotor is represented. Owing to the symmetry in the machine, the other two phases are modeled as essentially equal. It has been assumed that the machine is symmetrical and balanced in structure and that the magnetization is linear. As stated before, it is supposed that the stator windings are supplied by three-phase voltages directly from the grid, while the rotor

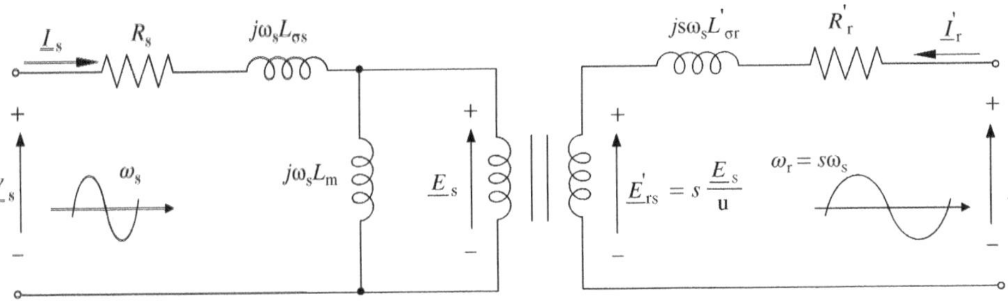

Figure 10.2 One-phase steady-state equivalent electric circuit of the DFIM

is also supplied by three-phase voltages but independent of the stator voltages. The most representative electric magnitude and parameters of the model are as follows:

Electric magnitudes of the stator and rotor (phasors):	Electric parameters of the stator and rotor:
$\underline{V}_s$: Supplied stator voltage	R_s: Stator resistance (Ω)
$\underline{V}'_r$: Supplied rotor voltage	R'_r: Rotor resistance (Ω)
$\underline{I}_s$: Stator current	L_m: Mutual inductance (H)
$\underline{I}'_r$: Rotor current	$L_{\sigma s}$: Stator leakage inductance (H)
$\underline{E}_s$: Induced emf in the stator	$L'_{\sigma r}$: Rotor leakage inductance (H)
$\underline{E}'_{rs}$: Induced emf in the rotor	N_s, N_r: Stator, rotor windings, number of turns per phase

The frequency of the stator circuit is ω_s, while the rotor frequency depends on the rotational speed, as revealed by expression (10.1). The leakage impedances of the stator and rotor are calculated according to the corresponding frequency of the stator and rotor. The relation between the stator and rotor turns per phase is provided by the u coefficient:

$$u = \frac{N_s}{N_r} \tag{10.4}$$

Some constructive variations can affect this coefficient, while the induced stator and rotor emfs are related by

$$\underline{E}'_{rs} = s\frac{\underline{E}_s}{u} \tag{10.5}$$

Note that the relation of the amplitudes depends on the slip (or rotational speed), whereas for the particular case when u=1 and the machine is at a standstill (s=1), both the induced stator and rotor emfs are equal.

In general, to facilitate the analysis, this circuit is modified to one that is a stator-reduced equivalent, referring the rotor magnitudes and parameters to the stator by means of the following equivalences:

$$R_r = R'_r u^2 \quad L_{\sigma r} = L'_{\sigma r} u^2 \quad \underline{I}_r = \frac{\underline{I}'_r}{u} \quad \underline{V}_r = \underline{V}'_r u \quad \underline{E}_{rs} = \underline{E}'_{rs} u \tag{10.6}$$

The notation adopted denotes real rotor magnitudes and parameters with the prime superscript ($'$), while the stator-reduced ones are without the prime superscript ($'$). This provides us with the equivalent phase circuit of Figure 10.3.

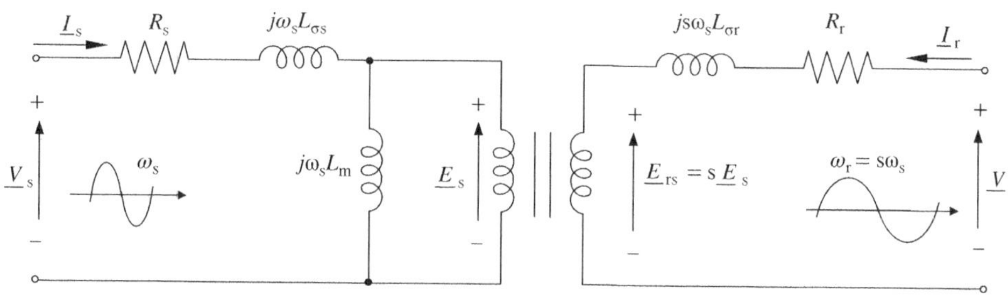

Figure 10.3 One-phase steady-state equivalent electric circuit of the DFIM with rotor parameters, current and voltages reduced to the stator

For practical analysis, it is more convenient if both the stator and rotor circuits present the same frequencies. Thus, the rotor circuit is "converted" to the stator frequency ω_s, simply by doing:

$$\underline{V}_r - s\underline{E}_s = (R_r + js\omega_s L_{\sigma r})\underline{I}_r \quad \Rightarrow \quad \frac{\underline{V}_r}{s} - \underline{E}_s = \left(\frac{R_r}{s} + j\omega_s L_{\sigma r}\right)\underline{I}_r \tag{10.7}$$

Therefore, the final equivalent circuit is obtained by referring the rotor to the stator, as shown in Figure 10.4, where all magnitudes are of the same frequency.

The electric equations of this equivalent steady-state circuit, including the stator and rotor fluxes, are summarized as follows:

Voltages:

$$\underline{V}_s = R_s\underline{I}_s + j\omega_s L_{\sigma s}\underline{I}_s + j\omega_s L_m(\underline{I}_s + \underline{I}_r) \tag{10.8}$$

$$\frac{\underline{V}_r}{s} = \frac{R_r}{s}\underline{I}_r + j\omega_s L_{\sigma r}\underline{I}_r + j\omega_s L_m(\underline{I}_s + \underline{I}_r) \tag{10.9}$$

Fluxes:

$$\underline{\Psi}_s = L_m(\underline{I}_s + \underline{I}_r) + L_{\sigma s}\underline{I}_s = L_s\underline{I}_s + L_m\underline{I}_r \tag{10.10}$$

$$\underline{\Psi}_r = L_m(\underline{I}_s + \underline{I}_r) + L_{\sigma r}\underline{I}_r = L_m\underline{I}_s + L_r\underline{I}_r \tag{10.11}$$

where $L_s = L_m + L_{\sigma s}$ being the stator inductance and $L_r = L_m + L_{\sigma s}$ being the rotor inductance. From all these electric equations, it is possible to derive the phasor diagram for any operating condition of the

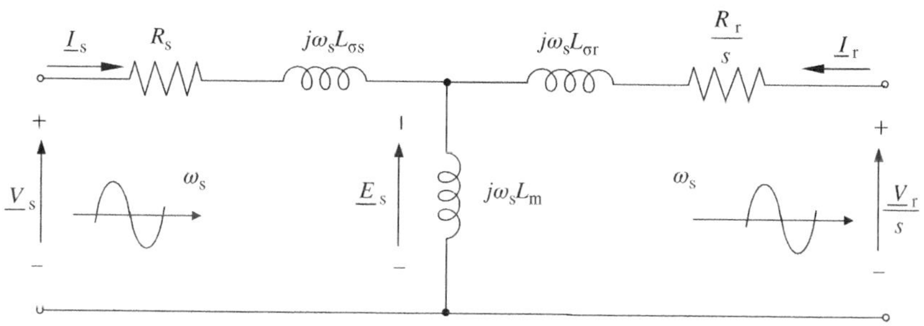

Figure 10.4 One-phase steady-state equivalent electric circuit of the DFIM referred to the stator

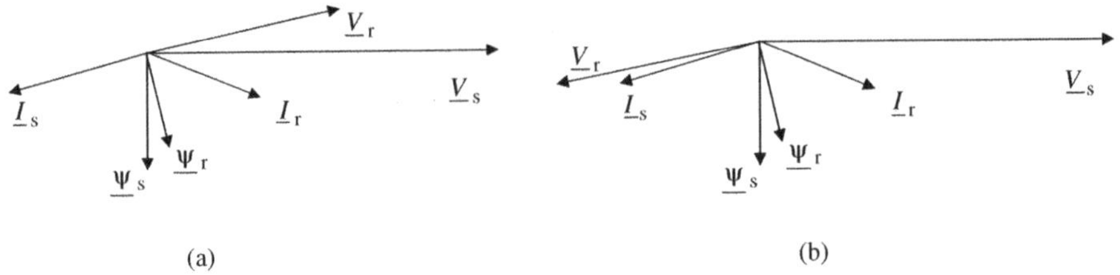

Figure 10.5 Phasor diagram in generator mode at $Q_s > 0$ of a multi-megawatt DFIM, (a) $s > 0$ and (b) $s < 0$ [1]

machine. Figure 10.5 illustrates an example of a multi-megawatt DFIM operating at subsynchronous and hypersynchronous speeds.

10.1.2.2 Power Flows and Modes of Operation

The basic active power balance of the DFIM reveals that adopting a motor convention, the addition of the stator active power P_s and the rotor active power P_r, is equal to the mechanical power in the shaft P_m minus the cupper losses in the stator and rotor ($P_{cu\_s} = 3R_s|\underline{I}_s|^2$ and $P_{cu\_r} = 3R_r|\underline{I}_r|^2$):

$$P_s + P_r = P_{cu\_s} + P_{cu\_r} + P_m \tag{10.12}$$

Positive values of P_s and P_r are interpreted as the power absorbed by the machine, whereas a positive value of P_m means mechanical power developed by the motor though the shaft. Therefore, the efficiency of the machine particular to motor or generator operation can be calculated according to the following formulae:

$$\eta = \frac{P_m}{P_s + P_r} \text{ if } P_m > 0 \qquad \eta = \frac{P_s + P_r}{P_m} \text{ if } P_m < 0 \tag{10.13}$$

On the other hand, the active stator and rotor powers, together with the reactive powers, can be calculated by the following classic formulae:

$$P_s = 3Re\{\underline{V}_s\underline{I}_s^*\} \qquad P_r = 3Re\{\underline{V}_r\underline{I}_r^*\} \tag{10.14}$$

$$Q_s = 3\mathrm{Im}\{\underline{V}_s\underline{I}_s^*\} \qquad Q_r = 3\mathrm{Im}\{\underline{V}_r\underline{I}_r^*\} \tag{10.15}$$

Hence, by substituting the stator and rotor voltage equations (10.8)–(10.9) into the power expressions (10.14)–(10.15), it is possible to develop power expressions showing their dependence on the machine's parameters, currents and slip:

$$P_s = 3R_s|\underline{I}_s|^2 + 3\omega_s L_m \mathrm{Im}\{\underline{I}_s\underline{I}_r^*\} \qquad P_r = 3R_r|\underline{I}_r|^2 - 3s\omega_s L_m \mathrm{Im}\{\underline{I}_s\underline{I}_r^*\} \tag{10.16}$$

$$Q_s = 3\omega_s L_s|\underline{I}_s|^2 + 3\omega_s L_m Re\{\underline{I}_r\underline{I}_s^*\} \qquad Q_r = 3s\omega_s L_r|\underline{I}_r|^2 + 3s\omega_s L_m Re\{\underline{I}_s\underline{I}_r^*\} \tag{10.17}$$

However, by definition, the electromagnetic torque developed in the shaft is equal to

$$P_{mec} = T_{em}\Omega_m = T_{em}\frac{\omega_m}{p} \tag{10.18}$$

Thus, by substituting in this expression, equivalent equations of power (10.16) and a torque expression that depends on the machine's parameters and currents can be found:

$$T_{em} = 3pL_m \mathrm{Im}\{\underline{I}_r^*\underline{I}_s\} \tag{10.19}$$

Further equivalent torque expressions can be derived, if in last expression, fluxes are included according to Equations 10.10 and 10.11:

$$T_{em} = 3p\frac{L_m}{L_s}\mathrm{Im}\{\underline{\Psi}_s\underline{I}_r^*\} = 3p\mathrm{Im}\{\underline{\Psi}_s^*\underline{I}_s\} = 3p\mathrm{Im}\{\underline{\Psi}_r\underline{I}_r^*\}$$
$$= 3\frac{L_m}{L_r}p\mathrm{Im}\{\underline{\Psi}_r^*\underline{I}_s\} = 3\frac{L_m}{\sigma L_r L_s}p\mathrm{Im}\{\underline{\Psi}_r^*\underline{\Psi}_s\} \tag{10.20}$$

where $\sigma = 1 - L_m{}^2/L_sL_r$. Furthermore, by neglecting the cupper losses in the active power equations (10.16), it is possible to develop some approximate power expressions, which can relate simply the power, torque and speed of the machine. These simple expressions are summarized in Table 10.1 and they can be useful in performing quick calculations. It must be highlighted that the relation between the stator and rotor voltage is also presented there, which is approximated by neglecting the voltage drop in the stator and rotor resistances and the leakage inductances of Equations (10.8) and (10.9). It is noticed that in assuming a constant stator voltage $\underline{V}_s$ owing to a direct grid connection, the necessary rotor voltage amplitude depends mainly on the slip or speed.

Therefore, it is possible to distinguish four possible combinations of torque (positive or negative) and speed (subsynchronism and hypersynchronism), which lead to the four quadrant modes of operation of the DFIM [3], as illustrated in Figure 10.6. Note that only at hypersynchronism, when the slip is negative, do both the stator and rotor active powers present an equivalent sign.

Table 10.1 List of simple and useful expressions

$P_s + P_r \cong P_m$	$P_r \cong -sP_s$	$P_m \cong (1-s)P_s$
$\lvert\underline{V}_r\rvert \approx \lvert s\underline{V}_s\rvert$		
$P_m = T_{em}\frac{\omega_m}{p}$	$P_s \cong T_{em}\frac{\omega_s}{p}$	$P_r \cong T_{em}\frac{\omega_r}{p}$

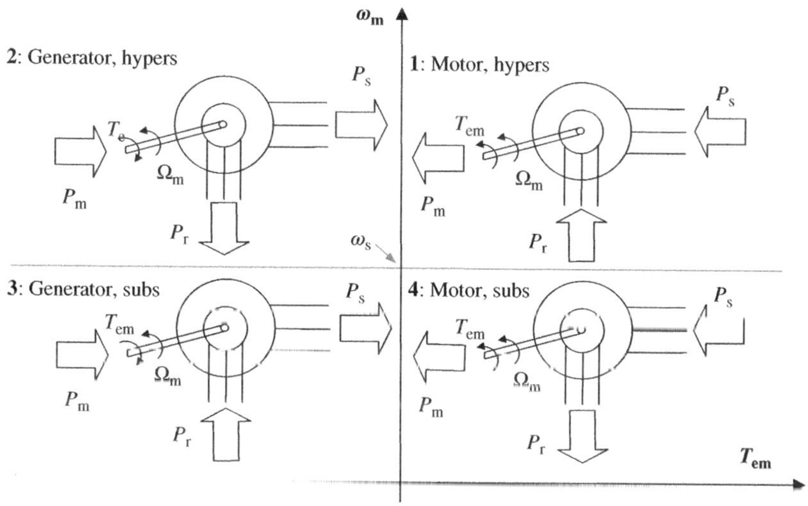

Figure 10.6 Four quadrant modes of operation of the DFIM attending to the active power

10.1.2.3 Steady-State Representative Torque Curves

It is interesting to develop further the circuit of Figure 10.4, deriving the equivalent Thevenin circuit at points *XY*, as illustrated in Figure 10.7. In this way, it is possible to calculate easily the rotor current and, therefore, to obtain a compact expression for the steady-state electromagnetic torque.

Firstly, knowing that the mechanical power P_m can be obtained by calculating the power in elements with slip dependence of the equivalent circuit, we have $|\underline{V}_r| = \frac{1-s}{s}|\underline{I}_r|\cos(\phi) + R_r\frac{1-s}{s}|\underline{I}_r|^2$ and then dividing by the speed provides the torque.

$$\left.\begin{aligned} |\underline{I}_r| &= \frac{|(\underline{V}_r/s) - (\underline{V}_{Th})|}{\sqrt{(R_{Th} + R_r/s)^2 + \omega_s^2(X_{Th} + X_{\sigma r})^2}} \\ \phi &= |\underline{V}_r - |\underline{I}_r \end{aligned}\right\} \Rightarrow T_{em} = \frac{3p}{s\omega_s}[R_r|\underline{I}_r|^2 - |\underline{V}_r||\underline{I}_r|\cos(\phi)] \tag{10.21}$$

Figure 10.8 illustrates the torque curves at different stator and rotor voltage relations. It can be seen that when the rotor voltage is set to zero, the typical curve of the squirrel-cage induction machine is obtained. By increasing the rotor voltage, the torque curves are shifted to the left or right depending on the phase shift imposed between the stator and rotor voltages. For instance, note that at relation $|V_r|/|V_s| = 0.025$, the torque is zero at slips ±0.025, while with phase shifts of 0 ° or 180 ° between the stator and rotor

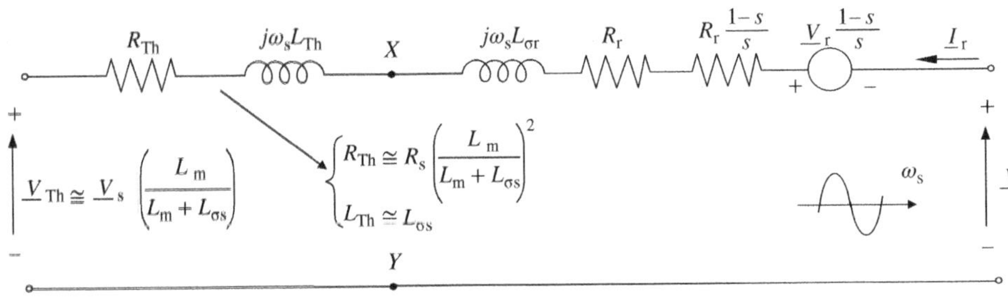

Figure 10.7 Equivalent Thevenin steady-state electric circuit of the DFIM

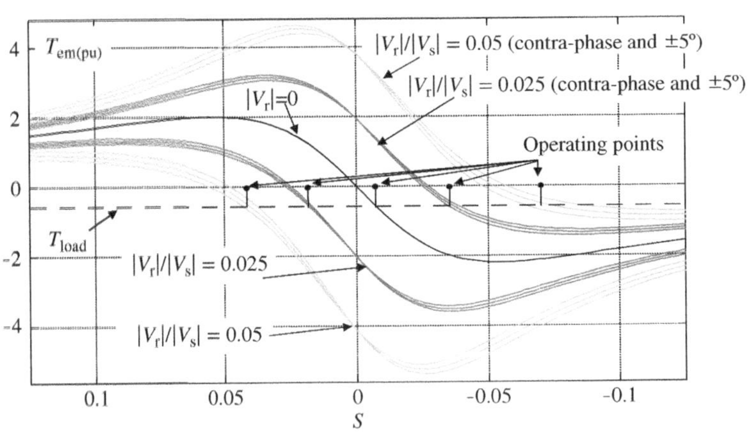

Figure 10.8 Torque curves versus slip at different rotor voltages

voltages, symmetric torque curves are achieved [4]. At a given constant torque load (generating), the steady-state operating points are obtained at the transitions of the T_{em} and T_{load} curves, also defining the slip. Therefore, by changing the rotor voltage, the speed (slip) is set and variable slip is obtained. Nevertheless, in general, the choice of the proper rotor voltage (amplitude and phase angle) is done by the closed-loop control strategy, as described in subsequent sections.

10.1.3 Dynamic Modeling

10.1.3.1 $\alpha\beta$ Model

In developing the dynamic $\alpha\beta$ model of the DFIM, space vector theory is applied to the basic electric equations of the machine and again, as in the steady-state model that was considered, the machine is assumed both ideal and linear. Figure 10.9 shows the three different rotating reference frames typically utilized to develop space vector-based models of the DFIM. The stator reference frame ($\alpha-\beta$) is a stationary reference frame, the rotor reference frame (DQ) rotates at ω_m and the synchronous reference frame (dq) rotates at ω_s. Subscripts "s", "r" and "a" are used to denote that one space vector is reference to the stator, rotor and synchronous reference frames, respectively. By using direct and inverse rotational transformation, a space vector can be represented in any of these frames [1].

Therefore, the three coils of the stator and rotor separately, by using space vector theory, can be represented by two stationary $\alpha\beta$ coils for the stator and two rotating coils DQ for the rotor, providing the following voltage equations:

$$\vec{v}_s^{\,s} = R_s \vec{i}_s^{\,s} + \frac{d\vec{\psi}_s^{\,s}}{dt} \qquad \vec{v}_r^{\,r} - R_r \vec{i}_r^{\,r} + \frac{d\vec{\psi}_r^{\,r}}{dt} \tag{10.22}$$

If both voltage equations are represented in stationary reference frame $\alpha\beta$, then the rotor equation must be multiplied by $e^{j\theta_m}$, which yields the following set of equations:

$$\vec{v}_s^{\,s} = R_s \vec{i}_s^{\,s} + \frac{d\vec{\psi}_s^{\,s}}{dt} \quad \Rightarrow \quad \begin{cases} v_{\alpha s} = R_s i_{\alpha s} + \dfrac{d\psi_{\alpha s}}{dt} \\ v_{\beta s} = R_s i_{\beta s} + \dfrac{d\psi_{\beta s}}{dt} \end{cases} \tag{10.23}$$

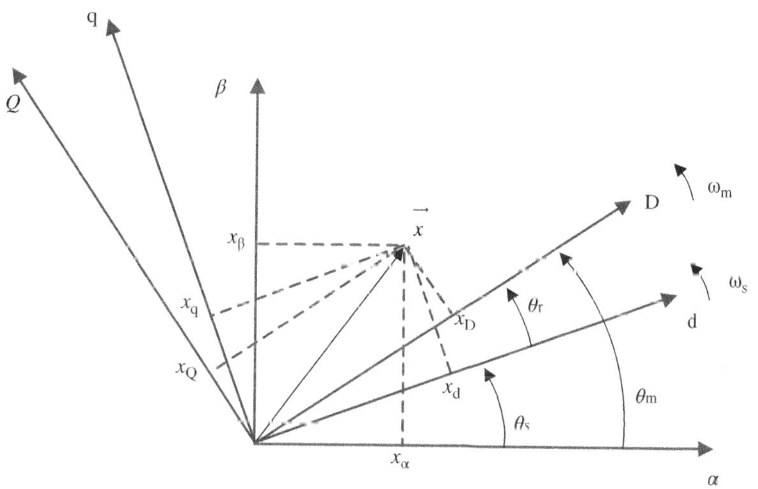

Figure 10.9 Different reference frames to represent space vectors of the DFIM

$$\vec{v}_r^s = R_r \vec{i}_r^s + \frac{d\vec{\psi}_r^s}{dt} - j\omega_m \vec{\psi}_r^s \quad \Rightarrow \quad \begin{cases} v_{\alpha r} = R_r i_{\alpha r} + \dfrac{d\psi_{\alpha r}}{dt} + \omega_m \psi_{\beta r} \\ v_{\beta r} = R_r i_{\beta r} + \dfrac{d\psi_{\beta r}}{dt} - \omega_m \psi_{\alpha r} \end{cases} \tag{10.24}$$

In a similar way, it is possible to derive the stator and rotor flux expressions in space vector form in a stationary reference frame:

$$\vec{\psi}_s^s = L_s \vec{i}_s^s + L_m \vec{i}_r^s \quad \Rightarrow \quad \begin{cases} \psi_{\alpha s} = L_s i_{\alpha s} + L_m i_{\alpha s} \\ \psi_{\beta s} = L_s i_{\beta s} + L_m i_{\beta s} \end{cases} \tag{10.25}$$

$$\vec{\psi}_r^s = L_m \vec{i}_s^s + L_r \vec{i}_r^s \quad \Rightarrow \quad \begin{cases} \psi_{\alpha r} = L_m i_{\alpha r} + L_r i_{\alpha r} \\ \psi_{\beta r} = L_m i_{\beta s} + L_r i_{\beta r} \end{cases} \tag{10.26}$$

Hence, from the set of equations derived, the $\alpha\beta$ equivalent circuit is developed as represented in Figure 10.10. There is one equivalent circuit for each $\alpha\beta$ coordinate, in which all the voltage, current and flux magnitudes are sinusoidal with a frequency of ω_s.

On the other hand, the active and reactive powers of the stator and rotor sides can be calculated according to the following equations:

$$P_s = \frac{3}{2}(v_{\alpha s} i_{\alpha s} + v_{\beta s} i_{\beta s}) \qquad P_r = \frac{3}{2}(v_{\alpha r} i_{\alpha r} + v_{\beta r} i_{\beta r}) \tag{10.27}$$

$$Q_s = \frac{3}{2}(v_{\beta s} i_{\alpha s} - v_{\alpha s} i_{\beta s}) \qquad Q_r = \frac{3}{2}(v_{\beta r} i_{\alpha r} - v_{\alpha r} i_{\beta r}) \tag{10.28}$$

While the electromagnetic torque, created by the DFIM, can be calculated by the following equivalent expressions:

$$T_{em} = \frac{3}{2} p \mathrm{Im}\{\vec{\psi}_r \vec{i}_r^{\,*}\} = \frac{3}{2} p(\psi_{\beta r} i_{\alpha r} - \psi_{\alpha r} i_{\beta r}) \tag{10.29}$$

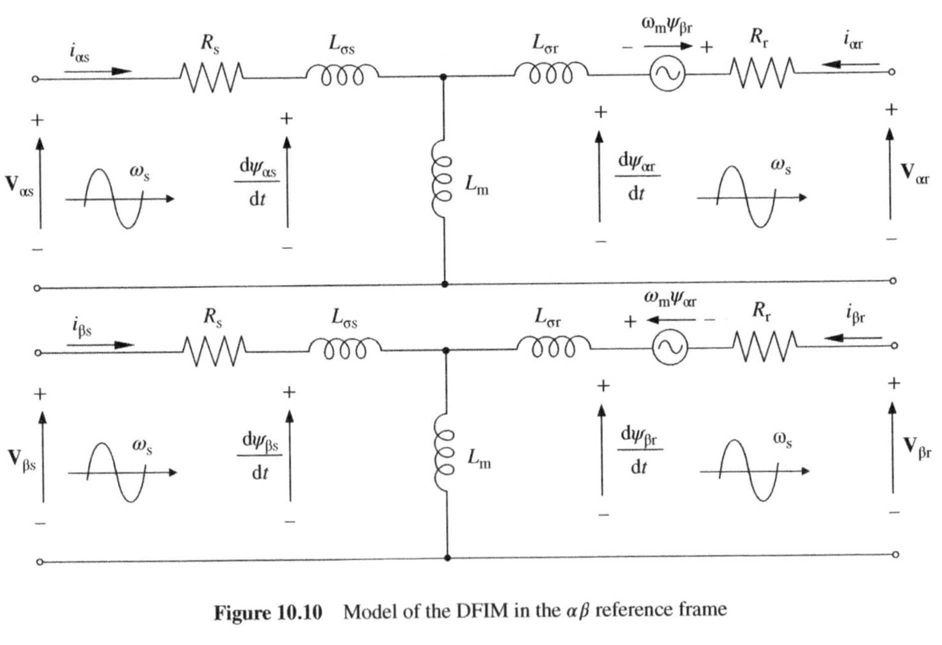

Figure 10.10 Model of the DFIM in the $\alpha\beta$ reference frame

$$T_{em} = \frac{3}{2}p\frac{L_m}{L_s}\mathrm{Im}\{\vec{\psi}_s \vec{i}_r^{\,*}\} = \frac{3}{2}p\mathrm{Im}\{\vec{\psi}_s^{\,*}\vec{i}_s\}$$
$$= \frac{3}{2}\frac{L_m}{L_r}p\mathrm{Im}\{\vec{\psi}_r^{\,*}\vec{i}_s\} = \frac{3}{2}\frac{L_m}{\sigma L_r L_s}p\mathrm{Im}\{\vec{\psi}_r^{\,*}\vec{\psi}_s\} = \frac{3}{2}L_m p\mathrm{Im}\{\vec{i}_s \vec{i}_r^{\,*}\} \quad (10.30)$$

where again $\sigma = 1 - L_m^{\,2}/L_s \cdot L_r$.

Thus, by rearranging expressions (10.23)–(10.26), we can develop several state-space representations of the DFIM, which are useful, for instance, for simulation or analysis purposes. The following expression shows one of them in which the state-space vector is composed of the stator and rotor fluxes:

$$\frac{d}{dt}\begin{bmatrix}\vec{\psi}_s^{\,s}\\ \vec{\psi}_r^{\,s}\end{bmatrix} = \begin{bmatrix}\frac{-R_s}{\sigma L_s} & \frac{R_s L_m}{\sigma L_s L_r}\\ \frac{R_r L_m}{\sigma L_s L_r} & \frac{-R_r}{\sigma L_r} + j\omega_m\end{bmatrix}\begin{bmatrix}\vec{\psi}_s^{\,s}\\ \vec{\psi}_r^{\,s}\end{bmatrix} + \begin{bmatrix}\vec{v}_s^{\,s}\\ \vec{v}_r^{\,s}\end{bmatrix} \quad (10.31)$$

Therefore, by adding the mechanical motion equation that describes the rotor speed behavior:

$$T_{em} - T_{load} = J\frac{d\Omega_m}{dt} \quad (10.32)$$

With J, the inertia of the rotor and T_{load}, the load torque applied to the shaft, we have constituted a model of the DFIM that can also be used for computer-based simulations.

10.1.3.2 dq model

The space vector model of the DFIM can be also represented in a synchronously rotating frame. For that purpose, by multiplying the voltage expressions (10.22) by $e^{-j\theta_s}$ and $e^{-j\theta_r}$, respectively, we obtain the dq voltage equations:

$$\vec{v}_s^{\,a} = R_s\vec{i}_s^{\,a} + \frac{d\vec{\psi}_s^{\,a}}{dt} + j\omega_s\vec{\psi}_s^{\,a} \quad \Rightarrow \quad \begin{cases} v_{ds} = R_s i_{ds} + \frac{d\psi_{ds}}{dt} - \omega_s\psi_{qs} \\ v_{qs} = R_s i_{qs} + \frac{d\psi_{qs}}{dt} + \omega_s\psi_{ds}\end{cases} \quad (10.33)$$

$$\vec{v}_r^{\,a} = R_r\vec{i}_r^{\,a} + \frac{d\vec{\psi}_r^{\,a}}{dt} + j\omega_r\vec{\psi}_r^{\,a} \quad \Rightarrow \quad \begin{cases} v_{dr} = R_r i_{dr} + \frac{d\psi_{dr}}{dt} - \omega_r\psi_{qr} \\ v_{qr} = R_r i_{qr} + \frac{d\psi_{qr}}{dt} + \omega_r\psi_{dr}\end{cases} \quad (10.34)$$

Similarly, the fluxes yield:

$$\vec{\psi}_s^{\,a} = L_s\vec{i}_s^{\,a} + L_m\vec{i}_r^{\,a} \quad \Rightarrow \quad \begin{cases} \psi_{ds} = L_s i_{ds} + L_m i_{dr} \\ \psi_{qs} = L_s i_{qs} + L_m i_{qr}\end{cases} \quad (10.35)$$

$$\vec{\psi}_r^{\,a} = L_m\vec{i}_s^{\,a} + L_r\vec{i}_r^{\,a} \quad \Rightarrow \quad \begin{cases} \psi_{dr} = L_m i_{ds} + L_r i_{dr} \\ \psi_{qr} = L_m i_{qs} + L_r i_{qr}\end{cases} \quad (10.36)$$

Hence, from the above equations, the equivalent electric circuit in dq is obtained, as depicted in Figure 10.11. However, this model representation with equivalent expressions of power and torque as in the αβ model, as well as the computer-based simulation model, can be developed further.

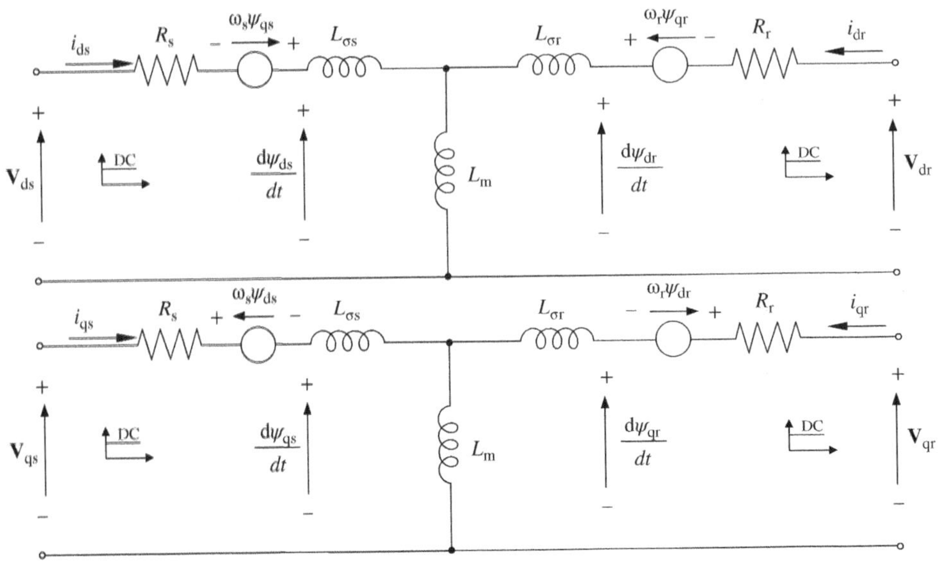

Figure 10.11 Model of the DFIM in *dq* reference frame

10.2 Vector Control of DFIM Using an AC/DC/AC Converter

10.2.1 Grid Connection Operation

This section describes the control of the DFIM. It is necessary to distinguish two different scenarios: the grid connection operation and the stand-alone operation. Both configurations have some significant control differences that will be described. However, only the rotor-side converter control is studied in this section, assuming that the DC bus voltage of the AC/DC/AC converter is set by the grid-side converter with any of the methods presented in Chapter 11.

10.2.1.1 Rotor Current Control Loops

Among the different alternative control methods that have been developed for the DFIM, only the vector control technique is studied in this section, which is probably the most extended and established one. For an easier understanding, the control is described in different steps: first, by studying the current control loops. After that, some interesting steady-state analyses are developed before finally describing the control at an unbalanced grid voltage.

In an equivalent way to the classic vector control techniques [3–5] of other different machines, the vector control of the DFIM is performed in a synchronously rotating *dq* frame, in which the *d*-axis is aligned, in this case, with the stator flux space vector, as illustrated in Figure 10.12 [1, 2]. Owing to this alignment choice, it will be shown later that the direct rotor current is proportional to the stator reactive power, and that the quadrature rotor current is proportional to the torque or active stator power. Therefore, from the model of the DFIM in a synchronous reference frame, by substituting Equations (10.35) and (10.36) into the rotor voltage equation (10.34), we obtain the rotor voltage as a function of the rotor currents and stator flux (note that $\psi_{qs} = 0$):

$$v_{dr} = R_r i_{dr} + \sigma L_r \frac{d}{dt} i_{dr} - \omega_r \sigma L_r i_{qr} + \frac{L_m}{L_s}\frac{d}{dt}|\vec{\psi}_s|$$

$$v_{qr} = R_r i_{qr} + \sigma L_r \frac{d}{dt} i_{qr} + \omega_r \sigma L_r i_{dr} + \omega_r \frac{L_m}{L_s}|\vec{\psi}_s| \tag{10.37}$$

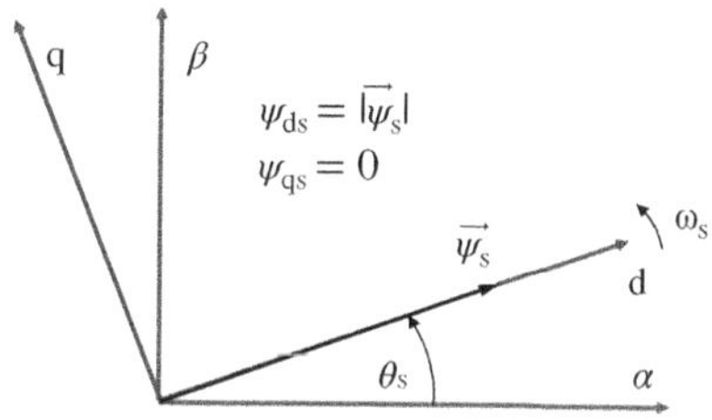

Figure 10.12 Synchronous rotating dq reference frame aligned with the stator flux space vector

From Equation (10.23), for instance, assuming that the voltage drop in the stator resistance is small, the stator flux is constant because the stator is connected directly to the grid at constant AC voltage; consequently, the term $d|\vec{\psi}_s|/dt$ is zero. These last two equations reveal that it is possible to perform dq rotor currents control, simply by using a regulator for each current component, as shown in Figure 10.13. Cross terms of Equation (10.37) can be included at the output of each regulator in order to assist the regulator. Note that the stator flux and ω_r must be estimated for that purpose; however, this is simple and does not add extra difficulties. For the reference frame transformation, the angle θ_r must be estimated. The control must be performed in dq coordinates, but then the rotor voltage and currents must be transformed into DQ coordinates. First, it is possible to obtain the angle of the stator voltage space vector, then subtract 90° from this estimated angle, and thus, obtain θ_s. A simple phase-locked loop (PLL) can be used to perform the stator voltage grid synchronization, providing robustness to the estimation and a rejection of small disturbances or harmonics. In subsequent sections, additional details are provided about how the current references are generated. Note that if the DFIM employed presents a different turn's ratio at the stator and rotor, it must be considered at the control stage. In the control block diagram presented in Figure 10.13, the current loops work with the rotor currents referred to the stator side, while the conversion to rotor-referred quantities is performed at the measurement stage for the currents and before the creation of the pulses for the converter for the voltages.

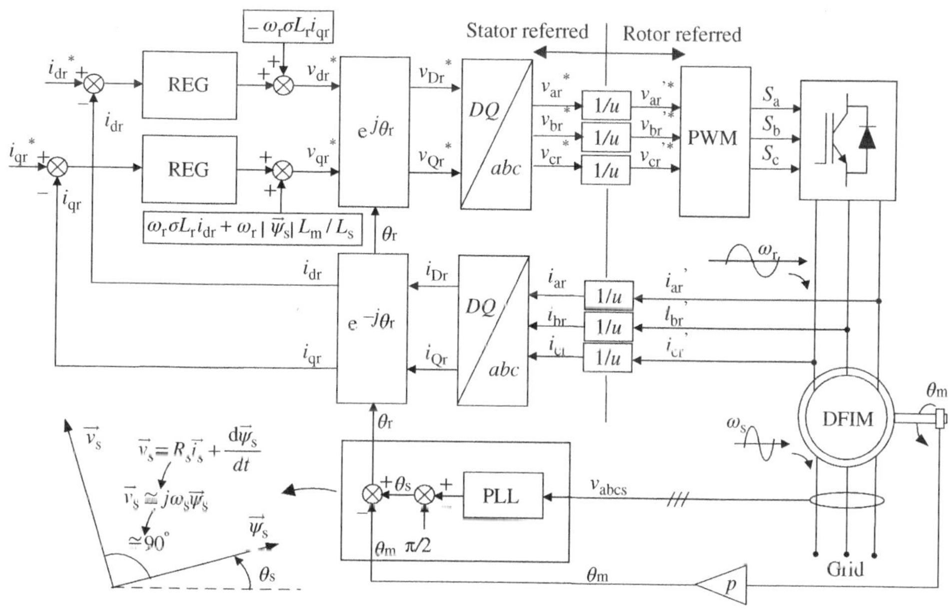

Figure 10.13 Current control loops of the DFIM

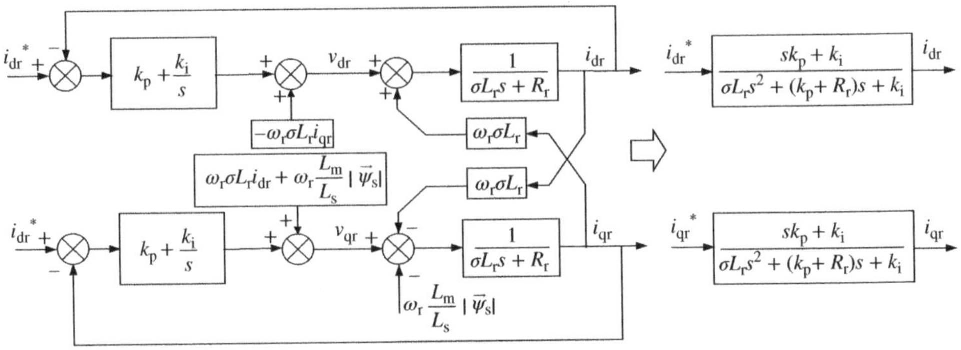

Figure 10.14 Equivalent second-order system of closed-loop current control with PI regulators

However, Figure 10.14 shows that when choosing equal proportional-integral (PI) regulators for both loops, employing compensation of the cross terms, and neglecting the effect of the voltage source converter and the possible delays in computation or measurements, the equivalent closed-loop systems of both current loops are equal to a second-order system with two poles and a zero that can be placed by classic control theory choosing the appropriate gains of the PI regulators.

10.2.1.2 Power and Speed Control Loops

Once the current control loops and the flux angle calculation have been studied, the complete control system can be introduced. As the *d*-axis of the reference frame is aligned with the stator flux space vector, the torque expression in the *dq* frame can be simplified as follows:

$$T_{em} = \frac{3}{2} p \frac{L_m}{L_s} (\psi_{qs} i_{dr} - \psi_{ds} i_{qr}) \quad \Rightarrow \quad T_{em} = -\frac{3}{2} p \frac{L_m}{L_s} |\vec{\psi}_s| i_{qr} \quad \Rightarrow \quad T_{em} = K_T i_{qr} \tag{10.38}$$

This means that the *q* rotor current component is proportional to the torque, that is, with i_{qr} it is possible to control the torque and, consequently, the speed of the machine if the application requires it. In a similar way, by developing the stator reactive power expression in the *dq* frame, we obtain a compact expression, which reveals that i_{dr} is responsible of Q_s.

$$Q_s = \frac{3}{2}(v_{qs} i_{ds} - v_{ds} i_{qs}) \Rightarrow Q_s = -\frac{3}{2}\omega_s \frac{L_m}{L_s} |\vec{\psi}_s| \left(i_{dr} - \frac{|\vec{\psi}_s|}{L_m} \right) \Rightarrow Q_s = K_Q \left(i_{dr} - \frac{|\vec{\psi}_s|}{L_m} \right) \tag{10.39}$$

Therefore, because of the orientation chosen, it can be seen that both rotor current components independently allow us to control the torque and reactive stator power. In this way, based on these expressions, Figure 10.15 illustrates the complete vector control of the DFIM. Maintaining the current loops studied in previous sections (Figure 10.13), a speed loop and a stator reactive power loop has been added. The necessity of the speed regulation depends on the application in which this machine is being used, and it could happen that the DFIM simply imposes an electromagnetic torque T_{em}, while the speed of the shaft is controlled by other elements.

However, with the Q_s loop, it is possible to control the magnetizing of the machine. As discussed before, because the stator of the machine is connected directly to the grid, the stator flux amplitude is

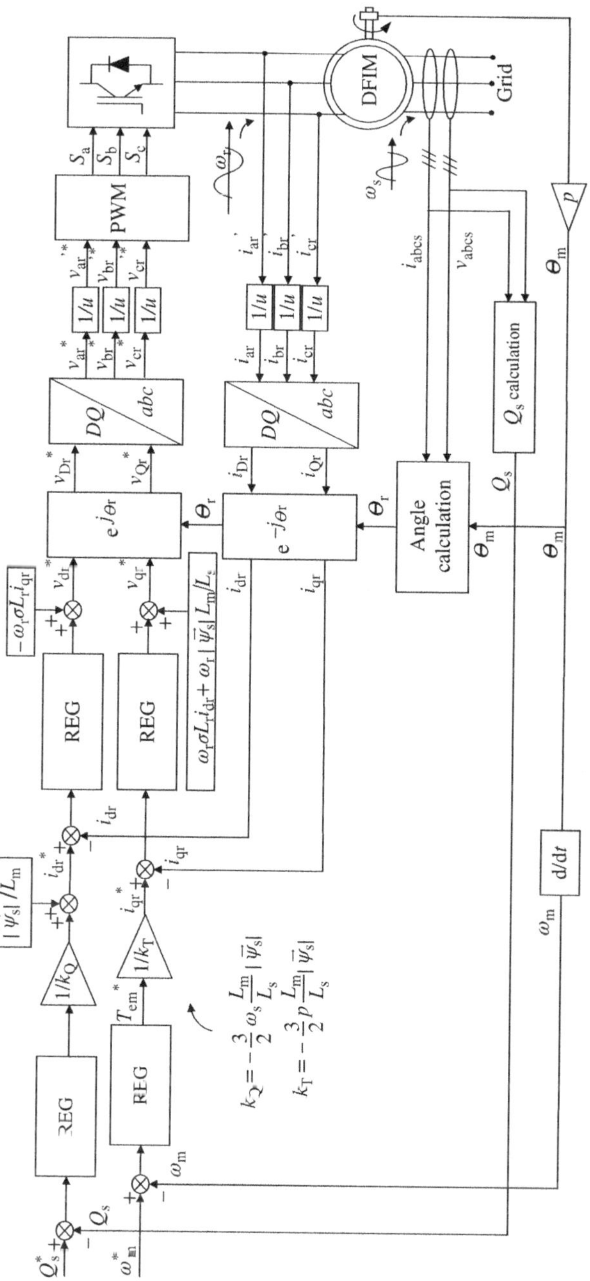

Figure 10.15 Complete vector control of the DFIM

constant and provided by the grid voltage: $|\vec{\psi}_s| \cong |\vec{v}_s|/\omega_s$, thus the stator flux equations reveal that

$$|\vec{\psi}_s| = \psi_{ds} = L_s i_{ds} + L_m i_{dr} \qquad \psi_{qs} = 0 = L_s i_{qs} + L_m i_{qr} \tag{10.40}$$

The stator flux level $|\vec{\psi}_s|$ must be created by choosing i_{ds} and i_{dr}, distributing thus the required amount of current between the rotor and the stator. Note that i_{qs} and i_{qr} are set by the torque and, therefore, there is no possibility to choose; however, depending on the Q_s set, it is possible to exchange bigger or lower stator and rotor currents. As an example, in applications of wind energy generation, different Q_s values might be demanded depending on the requirements of the grid codes and therefore, in such a case, the Q_s reference would be set directly by the grid system operator.

On the other hand, as done with the current loops, Figure 10.16 shows the equivalent closed-loop systems of the Q_s and ω_m loops, assuming that the current loops have been tuned much faster than the external loops and neglecting converter dynamics or measurement and computing delays. It can be seen that the simplified closed-loop systems yield into a first- and second-order system that can be tuned by choosing the appropriate gains of the PI regulators. Alternatively, if the application allows it, it is typical to set the rotor i_{dr} current to zero (eliminating the Q_s loop), minimizing the necessary rotor current and saving the dimensioning of the rotor windings design and the rotor-side converter. Obviously, this is achieved by the means of increasing the stator winding dimensioning.

Finally, Figure 10.17 shows the most representative magnitudes of a vector controlled DFIM, operating at constant torque in motor mode at variable speed. The stator voltage is kept constant owing to the direct grid connection, while the stator currents are also constant because T_{em} and Q_s are maintained constant. The speed ramp performed at the middle of the experiment provokes the variation of the rotor voltage and currents, which yields a variation of the rotor active and reactive powers.

10.2.1.3 Steady-State Analysis

In previous sections, we have seen that the model of the DFIM expressed in the *dq* frame is useful for developing the stator flux-oriented vector control strategy. In this section, the *dq* model of the DFIM is exploited further in order to calculate the most interesting magnitudes of the machine at a steady state [3]. Hence, assuming that by means of the control the DFIM is going to be required to operate at demands, namely T_{em}, ω_m and Q_s, the task consists of deducing the remaining electric magnitudes of the machine at steady-state operation such as stator and rotor currents, powers and slip. This analysis can be useful, for instance, for deducing the best magnetizing strategy in terms of efficiency and calculating the required rotor current at a given operation point for dimensioning the power electronic converter.

Therefore, firstly, it is necessary to calculate the stator flux amplitude of the machine. By assuming a steady state and a stator flux alignment with the *d*-axis, it is possible to obtain the following set of five

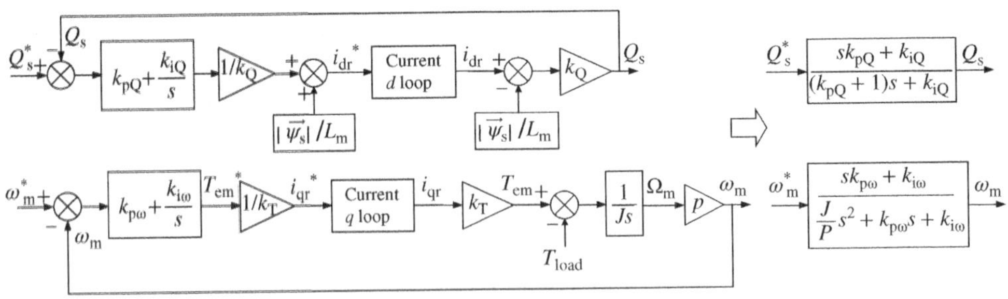

Figure 10.16 Equivalent closed-loop system of Q_s and ω_m loops (T_{load} considered equal to zero)

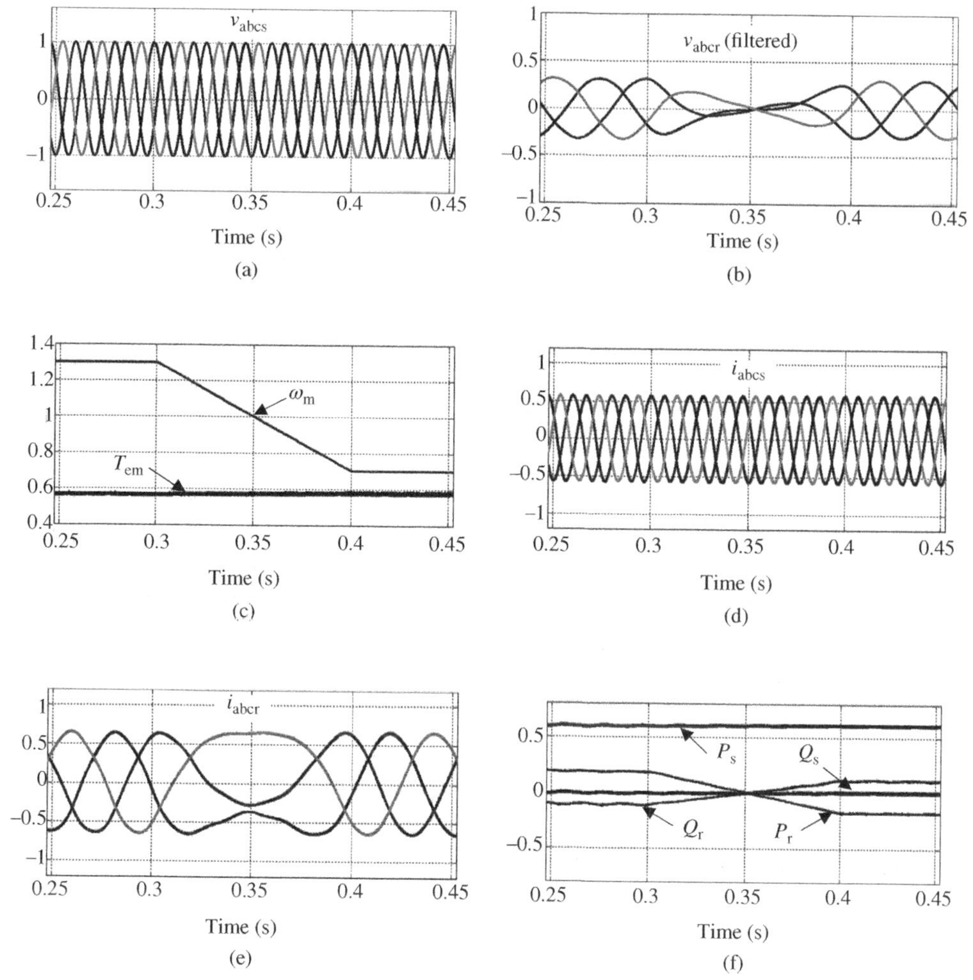

Figure 10.17 Most representative magnitudes of a multi-megawatt DFIM at constant torque in motor operation and variable speed

equations. From these five exact equations, it is easy to compute the stator flux:

$$\left.\begin{aligned} v_{ds} &= R_s i_{ds} \\ v_{qs} &= R_s i_{qs} + \omega_s \left|\vec{\psi}_s\right| \\ Q_s &= \frac{3}{2}\omega_s |\vec{\psi}_s| i_{ds} \\ T_{em} &= \frac{3}{2} p |\vec{\psi}_s| i_{qs} \\ |\vec{v}_s|^2 &= v_{ds}^2 + v_{qs}^2 \end{aligned}\right\} \Rightarrow |\vec{\psi}_s| = \sqrt{\frac{-B \pm \sqrt{B^2 - 4AC}}{2A}} \Rightarrow \begin{cases} A = \omega_s^{\,2} \\ B = \frac{4}{3} R_s T_{em} \omega_s - \left|\vec{v}_s\right|^2 \\ C = \left[\frac{2}{3}\frac{R_s}{L_m}\right]^2 \left[\left(\frac{Q_s}{\omega_s}\right)^2 + \left(\frac{T_{em}}{p}\right)^2\right] \end{cases} \tag{10.41}$$

Note that only the parameters of the machine, together with the stator voltage, torque and stator reactive power, are needed. Thus, once the flux is obtained, the remaining magnitudes can be calculated from the equations presented in previous sections. The procedure is detailed in Table 10.2.

In an alternative way, if instead of the Q_s reference $i_{dr} = 0$ is imposed, minimizing the rotor current, the steady-state computation procedure is slightly modified:

$$\left.\begin{array}{l} i_{ds} = \dfrac{|\vec{\psi}_s|}{L_s}, i_{qs} = -\dfrac{L_m}{L_s} i_{qr} \\ v_{ds} = R_s i_{ds} \\ v_{qs} = R_s i_{qs} + \omega_s |\vec{\psi}_s| \\ T_{em} = -\dfrac{3}{2} p \dfrac{L_m}{L_s} |\vec{\psi}_s| i_{qr} \\ |\vec{v}_s|^2 = v_{ds}^2 + v_{qs}^2 \end{array}\right\} \Rightarrow |\vec{\psi}_s| = \sqrt{\frac{-B \pm \sqrt{B^2 - 4AC}}{2A}} \Rightarrow \left\{\begin{array}{l} A = \left(\dfrac{R_s}{L_s}\right)^2 + \omega_s^2 \\ B = \dfrac{4}{3} \dfrac{R_s T_{em} \omega_s}{p} - |\vec{v}_s|^2 \\ C = \left(\dfrac{2}{3} \dfrac{R_s T_{em}}{p L_m}\right)^2 \end{array}\right. \quad (10.42)$$

Table 10.3 shows only the new required calculation steps, steps 4–10 of Table 10.2, are applied in the same way.

10.2.1.4 Current and Voltage Constraints

This section calculates the maximum capability curves of the machine considering the physical voltage and current limits that cannot be raised [4, 5]. Therefore, for the DFIM, the stator and rotor voltage and currents must be at every operating point, be smaller or equal to the established limits, which means

$$\begin{array}{ll} V_{s\_max}^2 \geq v_{ds}^2 + v_{qs}^2 & V_{r\_max}^2 \geq v_{dr}^2 + v_{qr}^2 \\ I_{s\_max}^2 \geq i_{ds}^2 + i_{qs}^2 & I_{r\_max}^2 \geq i_{dr}^2 + i_{qr}^2 \end{array} \quad (10.43)$$

Note that these limits can be established by the machine itself, by the converter, or by the grid connecting conditions. As the stator of the machine is connected directly to the grid, we cannot modify the stator voltage and, therefore, this variable is excluded from the analysis. For the rotor voltage, if we evaluate expression (10.37) at a steady state, neglecting the rotor resistance voltage drop:

$$\left.\begin{array}{l} v_{dr} = -\omega_r \sigma L_r i_{qr} \\ v_{qr} = \omega_r \left(\sigma L_r i_{dr} + \dfrac{L_m}{L_s} |\vec{\psi}_s|\right) \end{array}\right\} \Rightarrow V_{r\_max}^2 \geq (\omega_r \sigma L_r i_{qr})^2 + \omega_r^2 \left(\sigma L_r i_{dr} + \frac{L_m |\vec{\psi}_s|}{L_s}\right)^2$$

$$\Rightarrow \left(\frac{V_{r\_max}}{\omega_r \sigma L_r}\right)^2 \geq (i_{qr})^2 + \left(i_{dr} + \frac{L_m |\vec{\psi}_s|}{\sigma L_r L_s}\right)^2 \quad (10.44)$$

It is seen that the rotor voltage constraint is determined by circles in the plane i_{dr}, i_{qr}, with radius $V_{r\_max}/\omega_r \sigma L_r$ and centered at the point $-\frac{L_m |\vec{\psi}_s|}{\sigma L_r L_s}$. Note that depending on the operating point of the machine, more specifically depending on ω_r the radius of the voltage constraint will change. In a simpler way, considering the rotor current constraint, it is straightforward that it determines circles centered at the origin of the plane and with radius I_{r_max}. Finally, the for stator current constraints, it is necessary to consider expression (10.35), which yields

$$\left.\begin{array}{l} i_{ds} = \dfrac{|\vec{\psi}_s|}{L_s} - \dfrac{L_m}{L_s} i_{dr} \\ i_{qs} = -\dfrac{L_m}{L_s} i_{qr} \end{array}\right\} \Rightarrow I_{s\_max}^2 \geq \left(\frac{|\vec{\psi}_s|}{L_s} - \frac{L_m}{L_s} i_{dr}\right)^2 + \left(\frac{L_m}{L_s} i_{qr}\right)^2 \Rightarrow \left(I_{s\_max} \frac{L_s}{L_m}\right)^2 \geq \left(\frac{|\vec{\psi}_s|}{L_m} - i_{dr}\right)^2 + (i_{qr})^2 \quad (10.45)$$

Table 10.2 Procedure to derive the steady-state magnitudes with Q_s reference

	Given grid voltage: ω_s $\lvert\vec{v}_s\rvert$	**Given operating point**: Q_s ω_m T_{em}		
1. Stator flux	$\lvert\vec{\psi}_s\rvert = \sqrt{\frac{-B \pm \sqrt{B^2 - 4AC}}{2A}}$	$C = \left[\frac{2}{3}\frac{R_s}{L_m}\right]^2 \left[\left(\frac{Q_s}{\omega_s}\right)^2 + \left(\frac{T_{em}}{p}\right)^2\right]$	$B = \frac{4}{3}R_s T_{em}\omega_s - \lvert\vec{v}_s\rvert^2$	$A = \omega_s^2$
2. Stator currents	$i_{ds} = \frac{Q_s}{\frac{3}{3}\omega_s\lvert\vec{\psi}\rvert}$	$i_{qs} = \frac{T_{em}}{\frac{3}{2}p\lvert\vec{\psi}_s\rvert}$	$\lvert\vec{i}_s\rvert^2 = i_{ds}^2 + i_{qs}^2$	$\theta_{i_s} = a\tan\left(\frac{i_{qs}}{i_{ds}}\right)$
3. Rotor currents	$i_{dr} = \frac{\lvert\vec{\psi}_s\rvert - L_s i_{ds}}{L_m}$	$i_{qr} = -\frac{L_s}{L_m} i_{qs}$	$\lvert\vec{i}_r\rvert^2 = i_{dr}^2 + i_{qr}^2$	$\theta_{i_s} = a\tan\left(\frac{i_{qs}}{i_{ds}}\right)$
4. Stator voltages	$v_{ds} = R_s i_{ds}$	$v_{qs} = R_s i_{qs} + \omega_s\lvert\vec{\psi}_s\rvert$	$\lvert\vec{v}_s\rvert^2 = v_{ds}^2 + v_{qs}^2$	$\theta_{v_s} = a\tan\left(\frac{v_{qs}}{v_{ds}}\right)$
5. Slip	$\omega_r = \omega_s - \omega_m$		$s = \omega_r/\omega_s$	
6. Rotor voltages	$v_{dr} = R_r i_{dr} - \omega_r\sigma L_r i_{qr}$	$v_{qr} = R_r i_{qr} + \omega_r\sigma L_r i_{dr} + \omega_r\frac{L_m}{L_s}\lvert\vec{\psi}_s\rvert$	$\lvert\vec{v}_r\rvert^2 = v_{dr}^2 + v_{qr}^2$	$\theta_{v_r} = a\tan\left(\frac{v_{qr}}{v_{dr}}\right)$
7. Rotor fluxes	$\psi_{dr} = L_m i_{ds} + L_r i_{dr}$	$\psi_{qr} = L_m i_{qs} + L_r i_{qr}$	$\lvert\vec{\psi}_r\rvert^2 = \psi_{dr}^2 + \psi_{qr}^2$	$\theta_{\psi_r} = a\tan\left(\frac{\psi_{qr}}{\psi_{dr}}\right)$
8. Active powers	$P_m = T_{em}\frac{\omega_m}{p}$	$P_s = \frac{3}{2}(v_{ds}i_{ds} + v_{qs}i_{qs})$	$P_r = \frac{3}{2}(v_{dr}i_{dr} + v_{qr}i_{qr})$	
9. Reactive powers	$Q_s = \frac{3}{2}(v_{qs}i_{ds} - v_{ds}i_{qs})$	$PF_s = \cos(a\tan(Q_s/P_s))$	$Q_r = \frac{3}{2}(v_{qr}i_{dr} - v_{dr}i_{qr})$	$PF_r = \cos(a\tan(Q_r/P_r))$
10. Efficiency	$\eta = \frac{P_m}{P_s + P_r}$ if $P_m > 0$		$\eta = \frac{P_s + P_r}{P_m}$ if $P_m < 0$	

Table 10.3 Procedure to derive the steady-state magnitudes with $i_{dr} = 0$. Only the first three steps are shown, the rest are the same as in Table 10.2

Given grid voltage: ω_s $\lvert v_s \rvert$ **Given operating point**: $i_{dr} = 0$ ω_m T_{em}				
1. Stator flux	$\lvert\vec{\psi}_s\rvert = \sqrt{\dfrac{-B \pm \sqrt{B^2 - 4AC}}{2A}}$	$C = \left(\dfrac{2}{3}\dfrac{R_s T_{em}}{pL_m}\right)^2$	$B = \dfrac{4}{3}\dfrac{R_s T_{em}\omega_s}{p} - \lvert v_s\rvert^2$	$A = \left(\dfrac{R_s}{L_s}\right)^2 + \omega_s^2$
2. Rotor currents	$i_{dr} = 0$	$i_{qr} = \dfrac{T_{em}}{-\dfrac{3}{2}p\dfrac{L_m}{L_s}\lvert\vec{\psi}_s\rvert}$	$\lvert\vec{i}_r\rvert^2 = i_{dr}^2 + i_{qr}^2$	$\theta_{i_s} = a\tan\left(\dfrac{i_{qs}}{i_{ds}}\right)$
3. Stator currents	$i_{dr} = \dfrac{\lvert\vec{\psi}_s\rvert}{L_s}$	$i_{qs} = -\dfrac{L_m}{L_s}i_{qr}$	$\lvert\vec{i}_s\rvert^2 = i_{ds}^2 + i_{qs}^2$	$\theta_{i_s} = a\tan\left(\dfrac{i_{qs}}{i_{ds}}\right)$

This also determines circles with radius $I_{s\_max}\frac{L_s}{L_m}$ and centered at the point $\frac{|\vec{\psi}_s|}{Lm}$. On the other hand, by means of expressions (10.38) and (10.39), the torque is proportional to i_{qr}, while the stator reactive power is proportional to i_{dr}; this means that in the plane, both variables are determined by straight lines. This fact is graphically shown, together with the limit circle curves in Figure 10.18. By studying these curves, it is possible to determine whether the desired operating point of the machine is achievable.

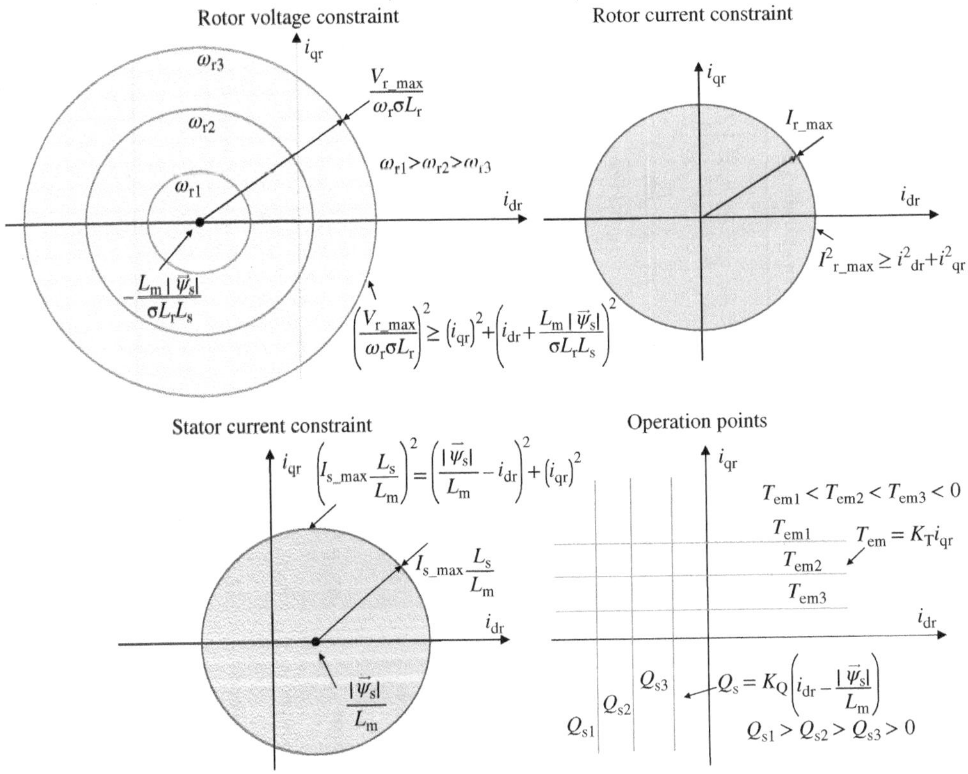

Figure 10.18 Voltage and current limit circles together with constant Q_s and T_{em} constant lines (only Q_s >0 and T_{em} <0 shown)

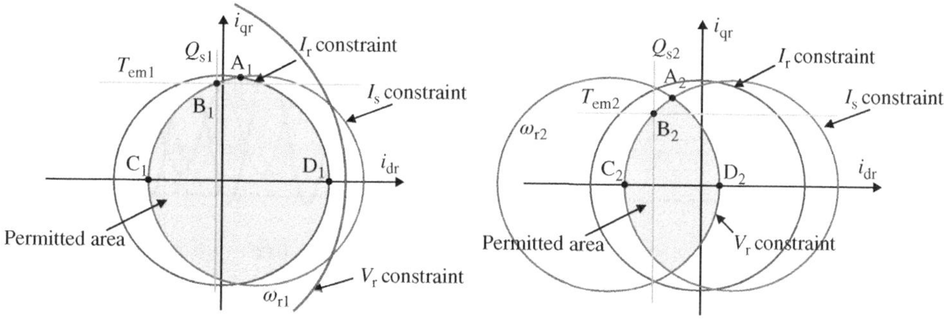

Figure 10.19 Two examples of operating points at low (1) and high (2) ω_r

Thus, Figure 10.19 shows two operational examples of a multi-megawatt DFIM in which the limits of stator and rotor currents (referred to the stator) are equal. Regarding situation 1, we can say that as the required operating point sets a small ω_r, the rotor voltage constraint is not limiting the operating point. At this speed, the achievable torques and reactive powers are within the shaded area. It can be seen that the maximum torque in generator mode is achieved at point A_1. Point B_1 shows one specific possible operating point. If we want to modify Q_s to more positive or more negative values, we need to move through curves A_1C_1 and A_1D_1, respectively, decreasing the torque capacity in generator mode. However, points C_1 and D_1 are the limits at which all the current is dedicated for reactive power, and thus, no torque can be created.

In situation 2, we note that with the higher ω_r required, the rotor voltage constraint now has an effect. At this speed, the achievable torques and reactive powers are within the shaded area (smaller than in the previous case). The maximum torque in generator mode is achieved at point A_2, which is smaller than A_1. Point B_2 shows one specific possible operating point. As before, by moving through curves A_2C_2 and A_2D_2, respectively, we can achieve different values of Q_s with the limits C_2, which is equal to C_1, and D_2, which is smaller than D_1.

10.2.1.5 Operation under Unbalanced Grid Voltage

In some scenarios and applications, the DFIM can be obliged to operate under unbalanced grid voltage [1]. This could be in a permanent way, for instance, in weak grids where nonlinear loads produce unbalances of grid voltage, or alternatively, for a limited period during grid faults, where the voltage seen by the DFIM could be unbalanced. In all cases, the unbalanced voltage affects directly the operation of the DFIM, which, if no further action is taken, degrades its performance, for example, through the appearance of torque oscillations and unbalanced exchange of stator currents.

In specialized literature, all possible voltage unbalances are classified, but Figure 10.20 only shows an example of a voltage unbalance caused by a phase-to-phase fault in the network. The unbalanced phenomenon can be studied by sequence decomposition, first enunciated by Fortescue. Thus, an unbalanced voltage implies the presence of a positive sequence and a negative sequence.

As demonstrated in the following, most problems caused by the unbalances can be overcome by introducing a precise amount of negative sequence in the rotor current references. In this way, the current references are the addition of two sequences: one synchronized with the positive sequence of the grid voltage, and the other synchronized with the negative sequence. In order to guarantee that the two sequences are well regulated, it is necessary to control each sequence independently. The original control loop is

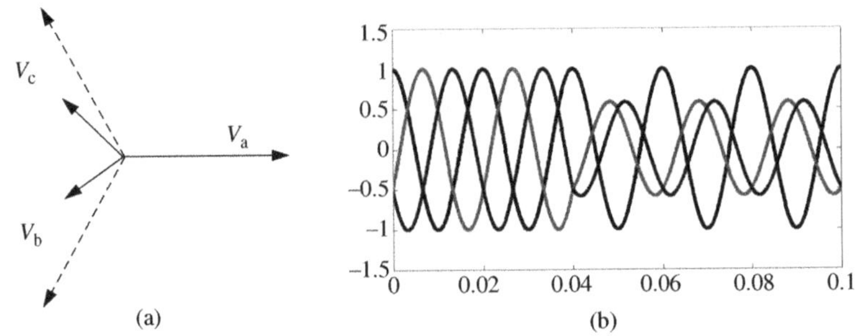

Figure 10.20 Unbalance caused by a dip type C phase-to-phase fault: (a) phasor diagram and (b) abc voltages

then substituted by two control loops: the first working in a positive rotating reference frame, and the other one working in an inversely rotating frame. This control philosophy is named the dual-vector control technique. Figure 10.21 illustrates an example of dual-vector control for a DFIM. The rotor current loops follow the same basics of classic vector control, as described in Figure 10.15.

Coupling terms are not included at control loops in this example, but could also be used (note that the presence of the negative sequence would require then to be recalculated). The measured rotor current must be split into its two sequences before being introduced to the current regulators. The sequence decomposition and the dual-vector control itself require the estimation of the θ_r angle. In general, this is calculated following the philosophy described in Figure 10.13, using the same PLL or, as preferred by some authors, using a more sophisticated synchronization method that handles positive and negative sequences of the stator voltage, synchronized with the positive sequence. The positive and negative rotor current references must be calculated from the stator active and reactive power references, as is described subsequently. Note that if the application requires it, the active power reference can come from the speed control loop, while the reactive power in this example does not include a closed-loop control, as utilized, for instance, in the control philosophy of Figure 10.15. The positive and negative rotor current references normally can be generated in order to meet different objectives of minimization: stator power oscillations, negative sequence of stator or rotor currents, rotor voltage and so on. In this section, only one of these strategies is studied in detail; the minimization of the negative sequence of the stator currents, which is a typical solution, adopted in DFIM-based wind turbines during voltage faults. Thus, in a general case, if no special action is taken, unbalanced stator voltages and currents cause oscillations in the stator active and reactive powers of double the grid frequency. The average powers are provided by the following matrix expression:

$$\begin{cases} P_s = P_{s0} + P_{scos}\cos(2\omega_s t) + P_{ssin}\sin(2\omega_s t) \\ Q_s = Q_{s0} + Q_{scos}\cos(2\omega_s t) + Q_{ssin}\sin(2\omega_s t) \end{cases} \qquad \begin{bmatrix} P_{s0} \\ Q_{s0} \end{bmatrix} = \frac{3}{2}\begin{bmatrix} v_{ds1} & v_{qs1} & v_{ds2} & v_{qs2} \\ v_{qs1} & -v_{ds1} & v_{qs2} & -v_{ds2} \end{bmatrix}\begin{bmatrix} i_{ds1} \\ i_{qs1} \\ i_{ds2} \\ i_{qs2} \end{bmatrix} \tag{10.46}$$

in which a zero negative sequence of stator current is desired. Thus, by inverting this matrix equation and making i_{ds2} and i_{qs2} zero, the stator positive currents are obtained. Once the stator currents are obtained, then by using the DFIM model expression (10.25) that relates stator and rotor currents and

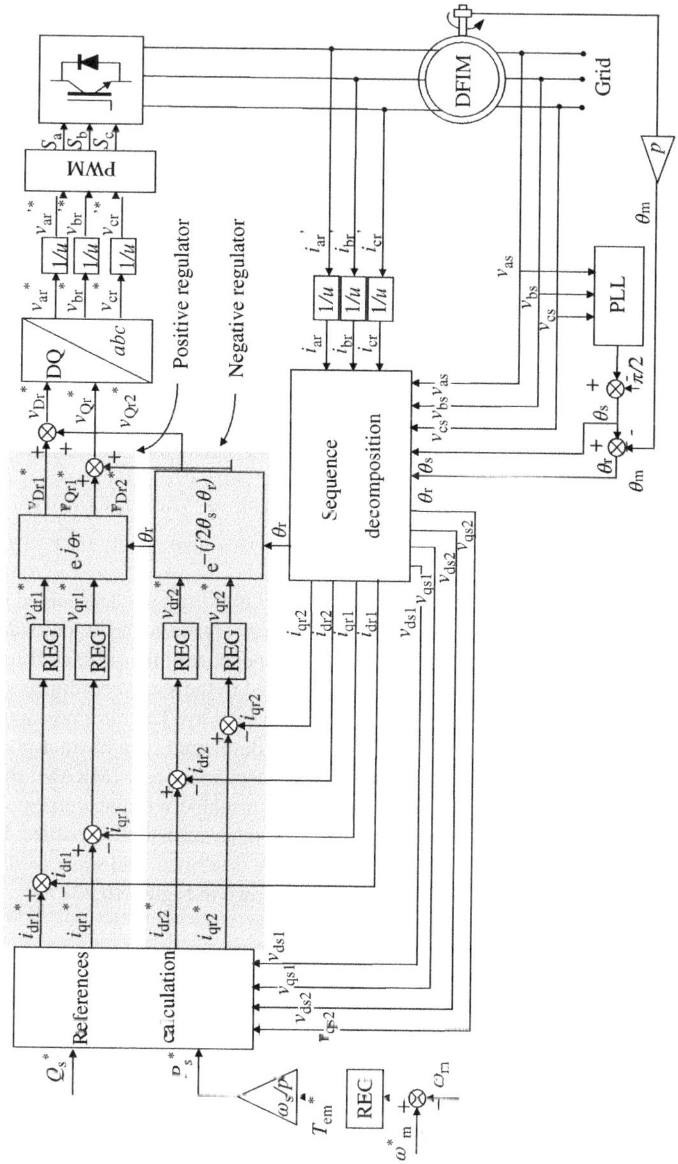

Figure 10.21 Dual-vector control block diagram of a DFIM

by assuming that the stator flux is in quadrature with the stator voltage:

$$\begin{bmatrix} i_{dr1} \\ i_{qr1} \\ i_{dr2} \\ i_{qr2} \end{bmatrix} = \frac{1}{\omega_s L_m} \begin{bmatrix} v_{qs1} \\ -v_{ds1} \\ -v_{qs2} \\ v_{ds2} \end{bmatrix} - \frac{L_s}{L_m} \begin{bmatrix} i_{ds1} \\ i_{qs1} \\ i_{ds2} \\ i_{qs2} \end{bmatrix} \tag{10.47}$$

Therefore, combining the last two expressions, the rotor current references, which achieve negative sequence elimination of stator currents, are obtained:

$$\begin{bmatrix} i_{dr1} \\ i_{qr1} \\ i_{dr2} \\ i_{qr2} \end{bmatrix} = -\frac{2}{3}\frac{L_s}{L_m}\frac{1}{v_{ds1}^2 + v_{qs1}^2} \begin{bmatrix} v_{ds1} & v_{qs1} \\ v_{qs1} & -v_{ds1} \\ 0 & 0 \\ 0 & 0 \end{bmatrix} \begin{bmatrix} P_{s0} \\ Q_{s0} \end{bmatrix} + \frac{1}{\omega_s L_m} \begin{bmatrix} v_{qs1} \\ -v_{ds1} \\ -v_{qs2} \\ v_{ds2} \end{bmatrix} \tag{10.48}$$

Thus, this expression is directly substituted in the block "reference calculation" of the general dual-control block diagram of Figure 10.21. In this, P_{s0} and Q_{s0} are the stator active and reactive power references $P_s{}^*$ and $Q_s{}^*$, which are at an average value and they oscillate as described by expression (10.49). Note that for the current loops, as well as for the voltage references generation, sequence decomposition must be performed. This task is graphically represented in Figure 10.22. When doing the decomposition and using the rotational transformation, oscillations at double the frequency appear that must be removed in order to obtain an acceptable performance of the loops. This can be achieved by using notch filters, low-pass filters, or a digital signal cancelation (DSC) method.

10.2.2 *Rotor Position Observers*

A rotor current and a rotor voltage fed by the power electronics converter are determined in a coordinate system rotating with a rotor, whereas the stator voltage and current are measured in stationary coordinates. Thus, the variable waveforms have different frequencies. Independent of the selected frame used for the control method realization, the variables should be transformed to the same coordinate system, and for this reason, information regarding the rotor position angle is necessary. The rotor position sensor can be replaced by the observer of rotor angular position. The most developed type of rotor position observer for a doubly fed induction generator is a model reference adaptive system (MRAS), the general idea of which is to compare one of the measured variables of the machine with the corresponding variable calculated (estimated) based on the machine parameters and other measured variables. In this section, some of already known rotor position MRAS observers will be described, and some modified structures will be shown. A general scheme of the MRAS observer is shown in Figure 10.23.

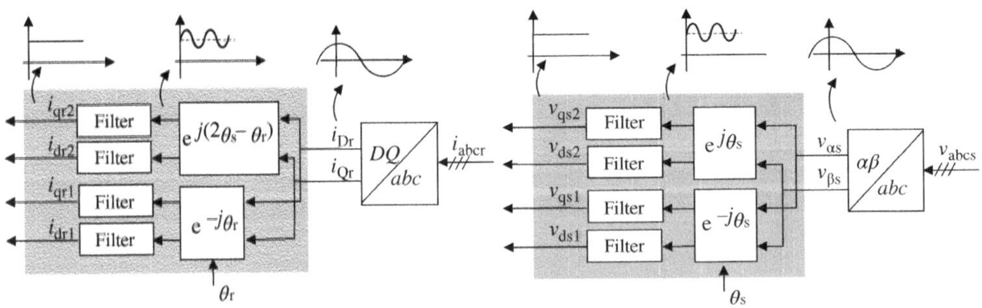

Figure 10.22 Sequence decomposition

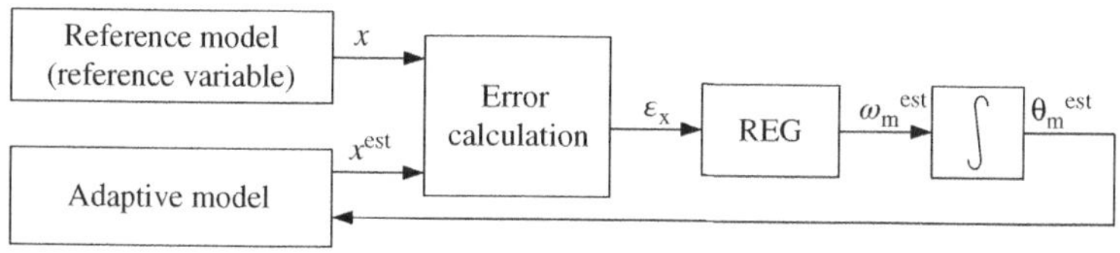

Figure 10.23 General scheme of the rotor speed and position MRAS observer for a doubly fed induction machine

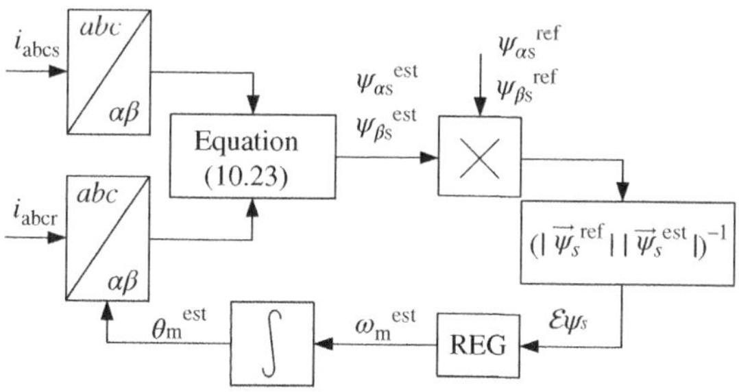

Figure 10.24 Rotor position MRAS observer based on the stator flux vectors synchronization

Different pairs of variables can be selected. In one of the first proposed MRAS observers for a DFIM, a stator flux vector has been used as a variable for error calculation [6]. As a reference model, the stator flux vector components $\psi_{\alpha s}{}^{ref}$, $\psi_{\beta s}{}^{ref}$ are calculated based on the integration of the stator voltage (according to Equation 10.23), whereas the variable of the adaptive model is an estimate of the stator flux vector components $\psi_{\alpha s}{}^{est}$, $\psi_{\beta s}{}^{est}$, calculated according to Equation 10.25, where machine parameters are used. Rotor current vector components in Equation 10.25 are represented in a stationary frame and in order to calculate properly the stator flux, the rotor current vector is transformed from the coordinate system rotated with rotor to the stationary coordinate system. Until the displacement between the reference and estimated stator flux is no longer equal to zero, the output signal of the controller responsible for the estimated rotor speed $\omega_m{}^{est}$, will be modified and, consequently, a proper transformation angle $\theta_m{}^{est}$ will be sought (Figure 10.24).

Finally, when the stator flux vectors are synchronized, the transformation angle obtained in the structure is equal to the actual rotor position angle $\theta_m{}^{est}$, such that it can be used further in the control algorithm, replacing the measured value of θ_m from Figure 10.13 and the subsequent figures with control structures. The error $\varepsilon_{\psi s}$ of the PI controller indicates a displacement between a reference and estimated flux, and it is calculated by a cross product of the vectors divided by the product of the vectors' magnitudes (Equation 10.49). Thus, the error $\varepsilon_{\psi s}$ is represented by a sine function of the angle between the estimated and actual stator flux vectors.

$$\varepsilon_{\psi s} = \frac{\psi_{\alpha s}^{ref}\psi_{\beta s}^{est} - \psi_{\beta s}^{ref}\psi_{\alpha s}^{est}}{|\vec{\psi}_s^{ref}||\vec{\psi}_s^{est}|} \tag{10.49}$$

A rotor position MRAS observer based on the stator flux vector's synchronization has a problem during operation in the case of a grid-connected DFIG and magnetization only from the stator side. Therefore, another structure based on the same concept has been proposed [6]. This method is based on the synchronization of the measured and estimated rotor currents, and the synchronization error ε_{ir} is calculated by Equation 10.50.

$$\varepsilon_{ir} = \frac{i_{\alpha r}^{est} i_{\beta r} - i_{\beta r}^{est} i_{\alpha r}}{|\vec{i}_r^{est}||\vec{i}_r|} \tag{10.50}$$

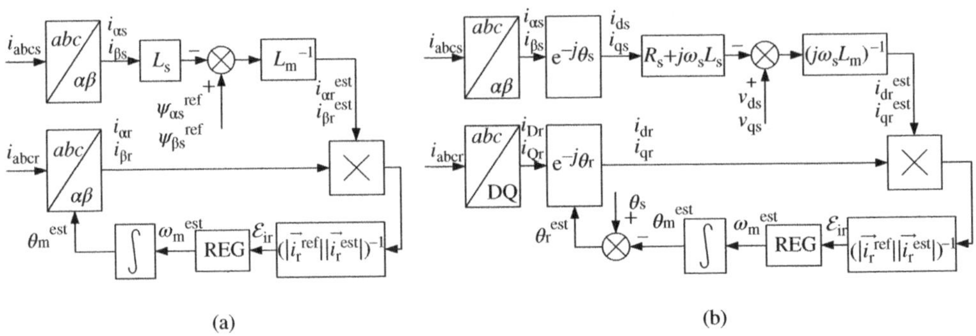

Figure 10.25 Rotor position MRAS observers based on (a) the flux–current model and (b) the voltage–current steady-state model of the doubly fed induction machine with a cross product used for error calculation

As the reference signal, estimated rotor current vector components $i_{\alpha r}{}^{est}$, $i_{\beta r}{}^{est}$ are used. The adaptive model is represented by the measured rotor current vector components $i_{\alpha r}$, $i_{\beta r}$ transformed to the $\alpha\beta$ stationary frame by the angle $\theta_m{}^{est}$ calculated from the estimated rotor speed $\omega_m{}^{est}$ as an output signal of PI synchronization controller (Figure 10.25a). The estimated rotor current components $i_{\alpha r}{}^{est}$ and $i_{\beta r}{}^{est}$ are calculated from the transformed Equation 10.25, whereas a stator flux is still calculated as an integration of the stator voltage. The rotor position angle θ_m, as well as the slip angle θ_r, can be estimated without the stator flux calculation. The method is based on the voltage–current equations of the machine, which are achieved by substitution of the stator flux in Equation 10.33 by Equation 10.35 to achieve Equation 10.51.

$$\vec{v}_s^{\,a} = R_s\vec{i}_s^{\,a} + L_s\frac{d\vec{i}_s^{\,a}}{dt} + L_m\frac{d\vec{i}_r^{\,a}}{dt} + j\omega_s L_s\vec{i}_s^{\,a} + j\omega_s L_m\vec{i}_r^{\,a}$$

$$\Rightarrow \begin{cases} v_{ds} = R_s i_{ds} + L_s\dfrac{di_{ds}}{dt} + L_m\dfrac{di_{dr}}{dt} - \omega_s L_s i_{qs} - \omega_s L_m i_{qr} \\ v_{qs} = R_s i_{qs} + L_s\dfrac{di_{qs}}{dt} + L_m\dfrac{di_{qr}}{dt} + \omega_s L_s i_{ds} + \omega_s L_m i_{dr} \end{cases} \quad (10.51)$$

The derivatives of the stator and rotor current vector components are neglected in an observer structure, because the synchronization process is much slower than the current changes in a *dq* frame, in which the observer is implemented. A voltage–current-model-based observer is presented in Figure 10.25b.

The operation of both presented MRAS observers in normal conditions is very similar. The error of the controller in both cases is represented by the sine function of the angle between the synchronized vectors. During the start-up of the system, when the vectors rotate with different frequencies prior to synchronization, this might mean that the error has an oscillating character until the system becomes synchronized. A modification of the observers from Figure 10.25a and b is possible by the use of both cross and dot products of the vectors. The structures of the modified observers are shown in Figure 10.26a and b, respectively. The angle between the synchronized vectors is determined by the tan^{-1} function, which in an extended form returns the angle in the range from $-\pi$ to π. This causes the error of the synchronizing controller to have still a periodic character, but considering that it is quite easy to find the discontinuity of the error waveform, the monotonicity of the error function can be obtained. The new error ε_{lin} is linear in the range from $-\pi$ to π (in a first period) and saturated when the tan^{-1} function overflows. It is possible to achieve full linearization of the ε_{lin} signal, but it is not necessary, especially when using the ε_{lin} function presented in Figure 10.27, because the error can be brought to zero in every period of the function and not just in the first period. This means, because the synchronization process is faster, that it is not necessary to wait one or more periods for synchronization of the variables.

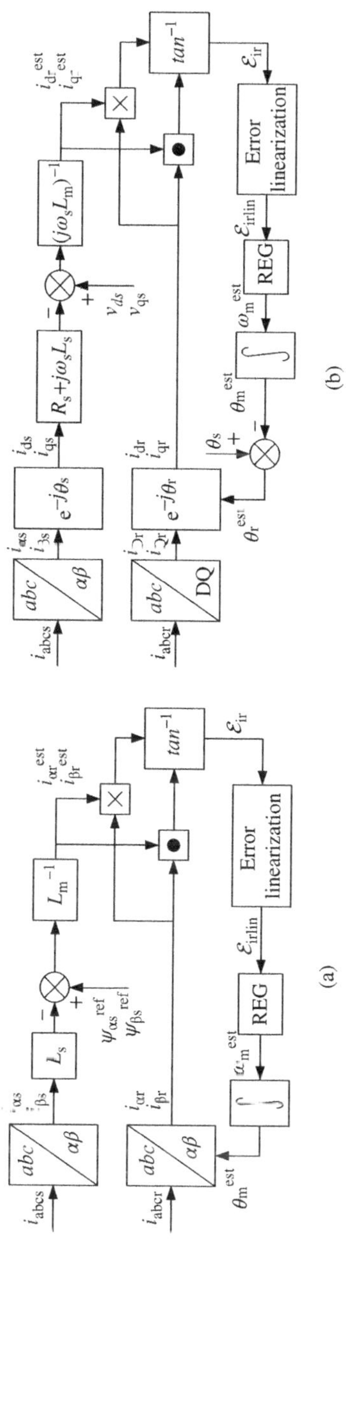

Figure 10.26 Modified rotor position MRAS observers based on (a) the flux–current model and (b) the voltage–current steady-state model of the DFIM with dot and cross products used for the error calculation

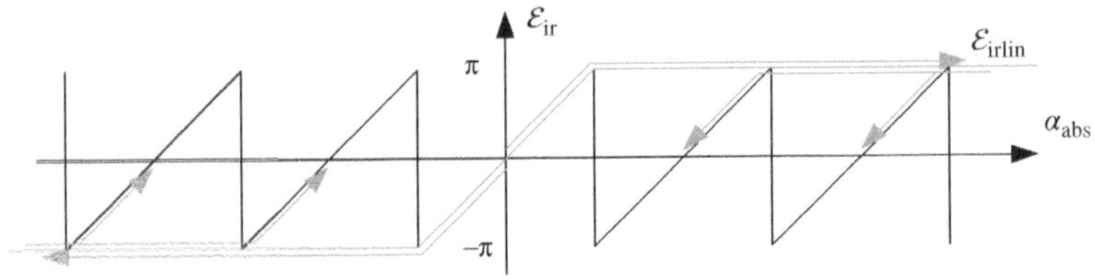

Figure 10.27 Method of elimination of periodic character of the controller error in a synchronization loop of the MRAS observer

Simulation results of the synchronization process using both the flux–current and voltage–current modified MRAS observers are shown in Figure 10.28. Figure 10.28a and b presents modified MRAS observers without the method of error linearization, whereas Figure 10.28c and d presents observers with error linearization. In both cases, with and without error linearization, there is not much difference between the responses of the observers. However, the error linearization, according to Figure 10.27, causes saturation of the error signal instead of its periodic shape and, consequently, it reduces significantly the time of the synchronization process.

10.2.3 Stand-alone Operation

10.2.3.1 Model of Stand-alone DFIM

A basic electrical circuit of a stand-alone DFIM is shown in Figure 10.29. The stator voltage is no longer set by the power network and results only on the operation of a back-to-back power converter. Considering that the rotor converter is responsible for the machine behavior, across a wider range than the grid converter, the current-controlled rotor-side converter has to be equipped with an outer control loop providing stability for the stator voltage. The main role of grid-side converter is the same as that in the grid-connected power system, which is the maintenance of the DC voltage to the reference level. In addition to the adequate control, voltage quality is obtained by stator-connected filtering capacitors, which compensate partially a reactive power of the machine or load. A separate issue of an autonomous generation unit is a four-wire system for the supply of single-phase loads. It can be obtained by the application of a delta/star matching transformer between the stator and load (recommended in cases where the stator voltage is higher than the voltage required by the load). Another way is by the application of a neutral wire connected to the star-connected stator and four-wire grid-side converter, as is shown as an option in Figure 10.29. The content of zero sequence components in a load current is typically not significant in high-power systems, and its generation through the grid-side converter should be enough to obtain satisfactory voltage quality. Moreover, the rotor-side converter cannot be in four-wire topology, because the machines are normally not four-slip rings designed.

The electrical circuit of a DFIM in a *dq* frame is described by Equations 10.33–10.36, presented in Section 10.1.3.2. The model of a three-wire stand-alone DFIM operating as a generator and equipped with stator connected filtering capacitors C_f is completed by Equation 10.52. This equation describes a stator circuit consisting of a filtering capacitor C_f and a load seen by the stator side. The stator current i_s equals a sum of the current of the filtering capacitor and the current i_o of the load seen by the stator, which is not the same as the physical load of the entire system. The stator load current i_o represents a sum of the physical load current i_{ld} and the current of the grid-side converter i_g, whose active component is positive or negative dependent on the rotor slip.

A rotor current is an algebraic difference of the magnetizing current i_m and a stator current i_s (Equation 10.53), whereas the magnetizing current depends on the stator voltage and magnetizing

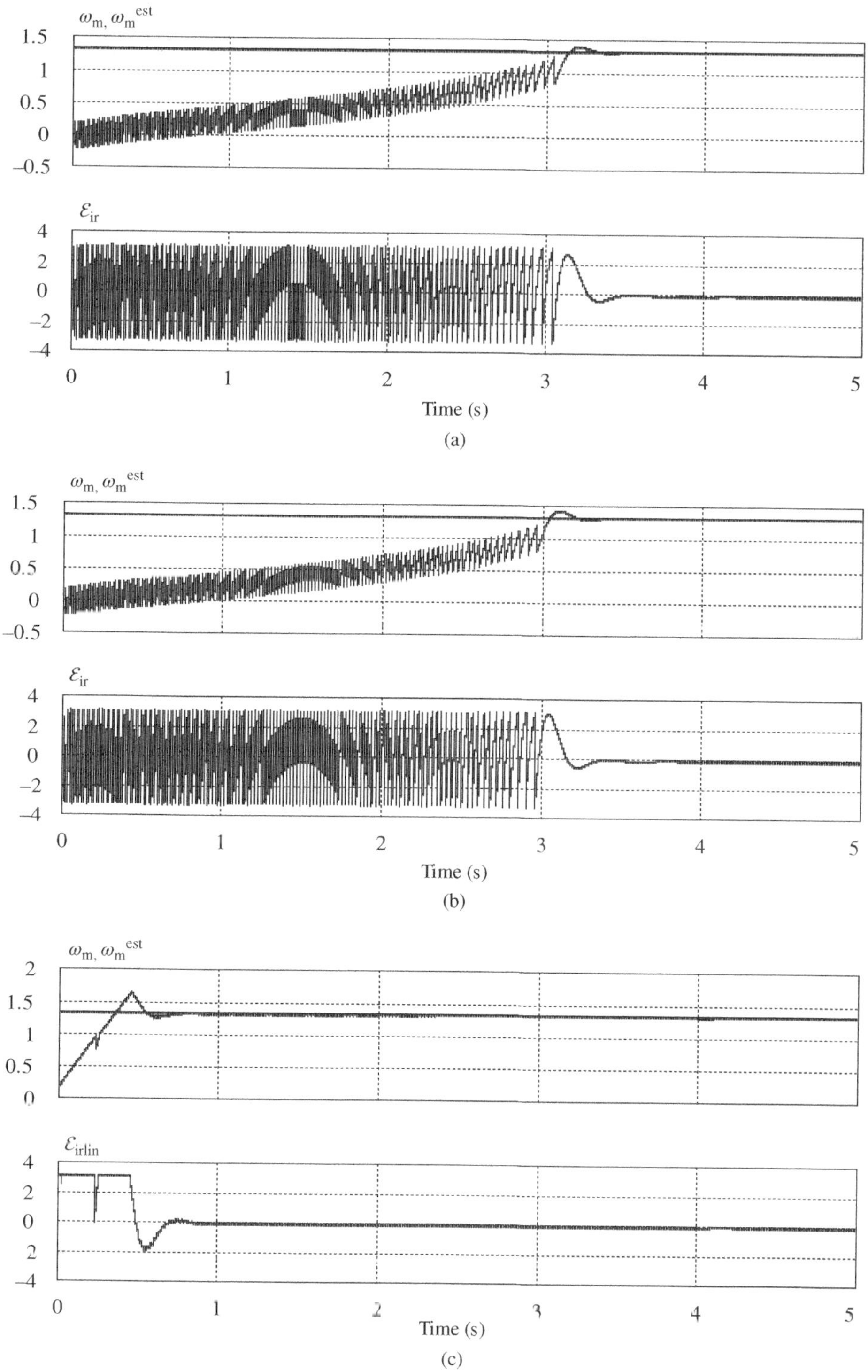

Figure 10.28 Simulation results of modified MRAS observers based on (a, c) the flux–current model and (b, d) the voltage–current model; (a, b) without and (c, d) with linearization of the controller error

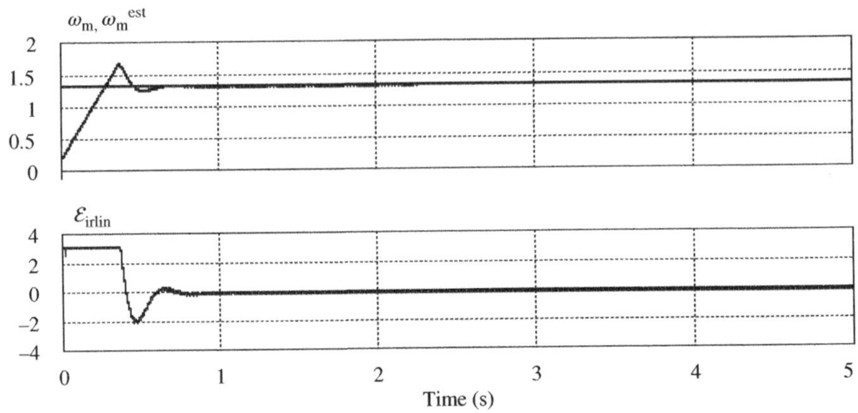

(d)

Figure 10.28 (*Continued*)

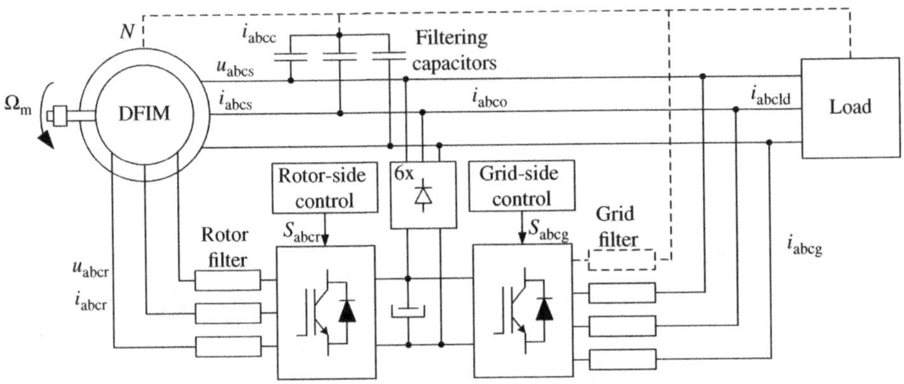

Figure 10.29 General scheme of the stand-alone doubly fed induction generator system with optional neutral wire

inductance (Equation 10.54).

$$\vec{i}_s^a = -C_f \frac{d\vec{v}_s^a}{dt} + \vec{i}_o^a - j\omega_s C_f \vec{v}_s^a \quad \Rightarrow \quad \begin{cases} i_{ds} = -C_f \dfrac{dv_{ds}}{dt} + i_{do} + \omega_s C_f v_{qs} \\ i_{qs} = -C_f \dfrac{dv_{qs}}{dt} + i_{qo} - \omega_s C_f v_{ds} \end{cases} \tag{10.52}$$

$$\vec{i}_r = \vec{i}_m - \vec{i}_s = \vec{i}_m - \vec{i}_o - \vec{i}_c \tag{10.53}$$

$$\vec{i}_m \approx -j\frac{|\vec{v}_s|}{\omega_s L_m} \tag{10.54}$$

10.2.3.2 Stator Voltage Control

A stand-alone operation of a power source requires an outer control loop to obtain high quality of the generated voltage. Considering that the power delivered by the stator is higher than the power delivered by the grid-side converter, it is better to control the generated voltage by the rotor-side converter responsible for excitation of the machine. Theoretically, it is possible to control the generated voltage by the grid-side converter using the same control of the rotor-side converter as used for the grid-connected system. However, it is very difficult to obtain a stable operation in the case where the fractional power grid converter stabilizes the AC voltage and, simultaneously, operates as a rectifier during subsynchronous rotor speed. A natural extension of the rotor current control structure for the rotor-side converter is an implementation of the stator voltage controllers. In addition, it can be applied as an additional loop as a feedforward of the stator current components i_{ds}, i_{qs} and the term responsible for the reference of the magnetizing current (Figure 10.30). Thus, the voltage controllers feed only a fractional part of the rotor current reference signals i_{dr}^*, i_{qr}^*. This means that the response of the control structure to the step change of the supplied load is faster.

The three basic structures of the outer control loop with feedforward are shown in Figure 10.31. In the first (Figure 10.31a), voltage controllers are implemented as in the typical structure used for a three-phase inverter equipped with an LC filter. This means that the output signal from the *d*-axis controller of the voltage vector is responsible for part of the reference rotor current *d* component signal, whereas the *q*-axis controller of the voltage vector returns part of the *q*-axis rotor current vector signal. With this control, the voltage vector components have visible oscillations (Figure 10.32a) because of coupling terms between the stator voltage and the stator current *dq* vector components (Equation 10.52), which are transferred to the couplings between the stator voltage and rotor current *dq* components because there is strong relation between the stator and rotor currents (Equation 10.53). Figure 10.32b presents the second structure of the voltage control, in which there is only one applied voltage controller responsible for the control of the voltage vector magnitude, the actual value of which is calculated based on *dq* or $\alpha\beta$ components of a stator voltage vector. This control utilizes Equation 10.54, which stays true, when the stator resistance and leakage inductance are neglected. The output signal is responsible for part of

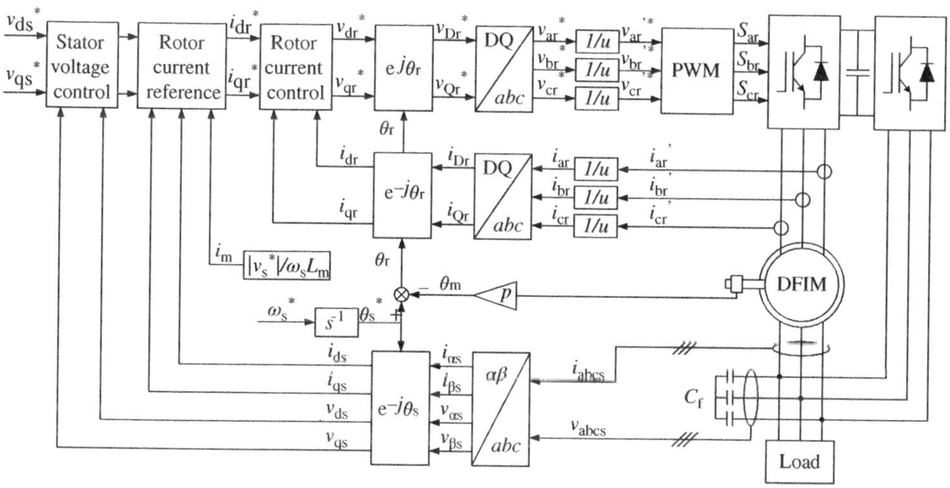

Figure 10.30 Block diagram of the stator voltage vector control of a stand-alone doubly fed induction generator

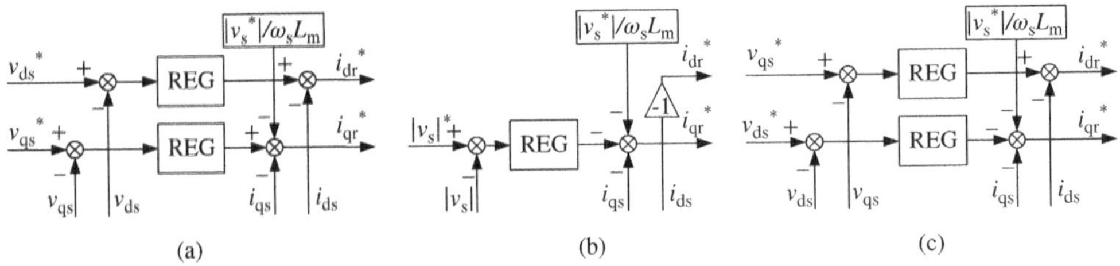

Figure 10.31 Three possible superior control structures of the stator voltage vector in a stand-alone DFIG

a reference *d* component of the rotor current in a frame oriented along the stator voltage vector. In the case of stator flux orientation, both the output signal of the controller and the reference magnetizing current signal should be added to the *d*-axis with a positive sign instead of the *q*-axis with negative sign. Control of the stator voltage magnitude eliminates oscillation during step-loading transients, but does not keep fixed the value of the voltage vector components, which vary depending on the load changes (Figure 10.32b). A third structure (Figure 10.31c) can be used to maintain the dynamic response with removed oscillations and simultaneous control of the stator voltage vector components. The results of the operation of these structures under the same conditions are shown in Figure 10.32c.

Figure 10.32 presents the results of a simulation of a 2 MW DFIM-based power system for the three different outer control loops shown in Figure 10.31. Simulations have been made under the same conditions in all cases. A rotor speed is changed from $0.66\omega_s$ to $1.33\,\omega_s$ in a time from zero to 0.5 s for an unloaded generator, and back from $1.33\,\omega_s$ to $0.66\,\omega_s$ in a time from 0.5 to 1 s for the generator loaded by a rated stator load. Next, the speed change cycle is repeated in a time from 1 to 2 s, but the load cycle is reversed, which means that from 1 to 1.5 s the stator of the machine is fully loaded, and from 1.5 to 2 s, it operates without load. The presented methods of control can be used not only for permanent stand-alone operation but also for synchronization of the power generation unit before connection to the grid [7].

During step unloading of the system at 1.5 s, when the entire load is disconnected, a significant voltage peak can occur; therefore, the power system and load should be adequately protected. The overvoltage is independent of the applied control and is caused by the energy stored in the machine's leakage field, which is released during step unloading. A simple and robust overvoltage protection can be configured by using a three-phase diode rectifier connected to the filtering capacitors on the AC side and to the DC link of a back-to-back converter on the DC side. Every voltage peak will be consumed by the DC link, whereas during normal operation, the diodes will be blocked by the DC-link voltage, which normally is higher than the rectified AC stator voltage.

10.2.3.3 Unbalanced Voltage Compensation

Stand-alone power systems rarely operate with a symmetrical load. Most of them are designed to be able to supply a single-phase load, which implies an unbalance in the load current. Using a matching transformer between the stator of the machine and the supplied load, the zero sequence component in a load current can be eliminated, but the negative sequence component will still influence the generated voltage. The negative sequence component in the stator voltage can be eliminated by adequate control of a back-to-back converter. It is obvious that an unbalanced stator current produces torque pulsations, and it is advisable to provide a negative sequence current through the grid-side converter. However, owing to the reduced power of the electronics converter in relation to the rated power of the machine, a grid-side converter might not be able to compensate fully a negative sequence, especially for significant load asymmetry at a rotor speed close to the limited speed range. In such case, in addition to the negative sequence current, a grid converter takes or delivers a high amount of slip power and might operate close to saturation.

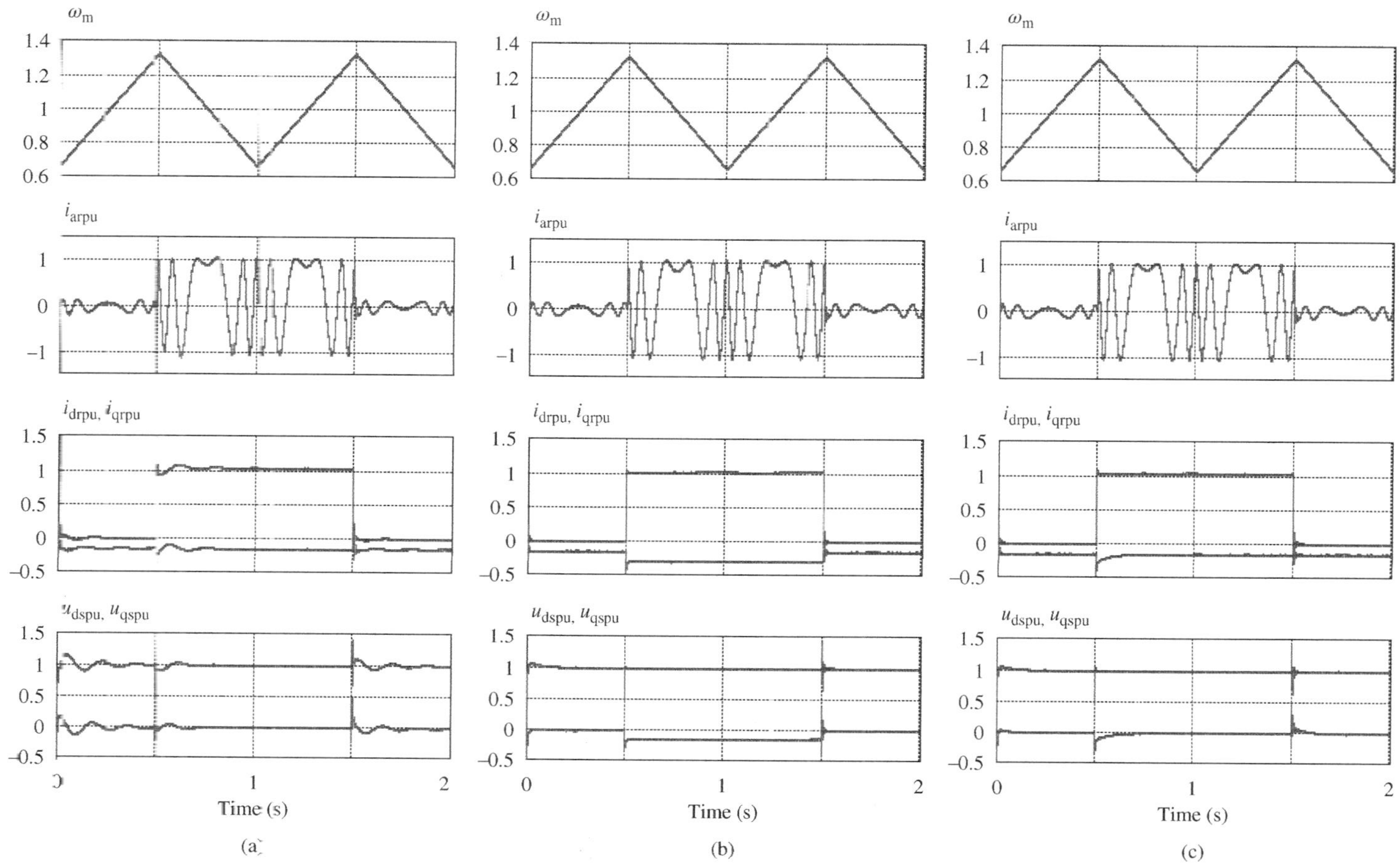

Figure 10.32 Simulation results of a general control from Figure 10.30 and for three different outer loops from Figure 10.31a–c, respectively, for the same speed and load conditions

Considering the above statements, it is necessary to introduce an adequate control method for both the rotor-side and grid-side converters, in order to obtain a sharing of the negative sequence component delivered by the DFIM system. Compensation of the negative sequence stator voltage by the rotor-side converter is possible by the generation of an adequate component in the rotor current. Owing to machine rotation, a stator-side unbalance influences the rotor current similarly in all three phases. A positive sequence component of the rotor current is represented by the slip frequency, whereas a negative sequence component of the rotor current is represented by a frequency equal to double synchronous minus slip frequency. A rotor converter control structure, including the part used for the elimination of a stator voltage negative sequence component, is presented in Figure 10.33.

A negative sequence component of the reference rotor current is partially obtained by the stator current feedforward. The rest of the rotor current negative sequence is referenced by controllers of negative sequence components of a stator voltage vector v_{ds2}, v_{qs2}. The other part of the control, responsible for the positive sequence, is taken from Figure 10.31c. Regarding an inner rotor current control loop, it is not necessary to implement sequence decomposition and to separate the control of the positive and negative sequences because, in an inner control loop, it is allowed a steady-state error. It is sufficient to eliminate a steady-state error in an outer control loop, which is obtained by separate controllers of the stator voltage positive and negative sequence components.

Figure 10.34 shows the results of the operation of an asymmetrically loaded stand-alone DFIM. Simulation results have been made for the same conditions as those set for the symmetrically loaded generator. An unbalanced load was built with the use of a Δ/Y transformer between the stator and load, wherein two phases of Y are fully loaded and one phase is not loaded. Torque pulsations produced under these conditions by the negative sequence components of the stator and rotor currents have a negative influence on the mechanical system.

Torque pulsations can be reduced by introducing an adequate control of a grid-side converter. A variant of the control method is shown in Figure 10.35. In addition to the positive sequence, a grid-side converter provides a negative sequence component of the current and the reference for the negative sequence is calculated by controllers of the stator current negative sequence components i_{ds2}, i_{qs2}. This improves the symmetry of the stator current and, consequently, reduces a negative sequence in the rotor current.

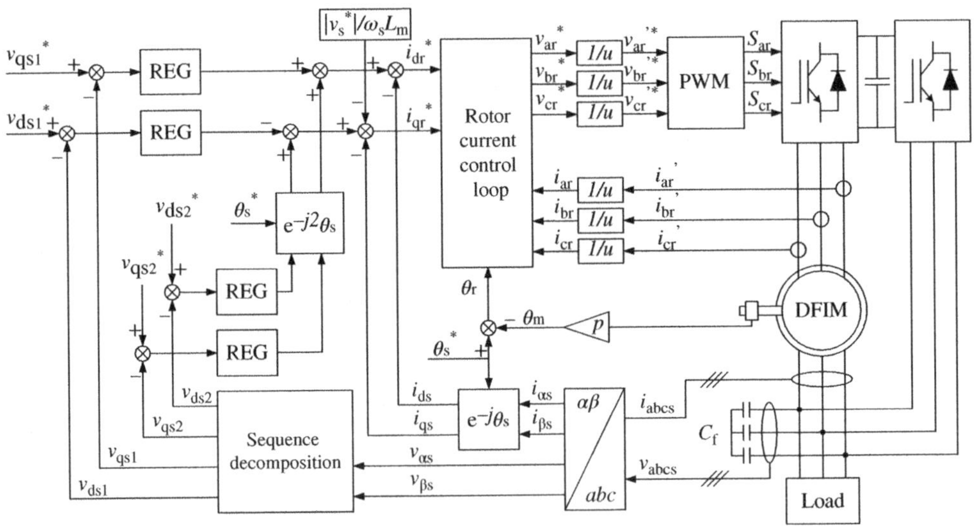

Figure 10.33 Block diagram of the stator voltage vector control with negative sequence compensation during unbalanced load supply

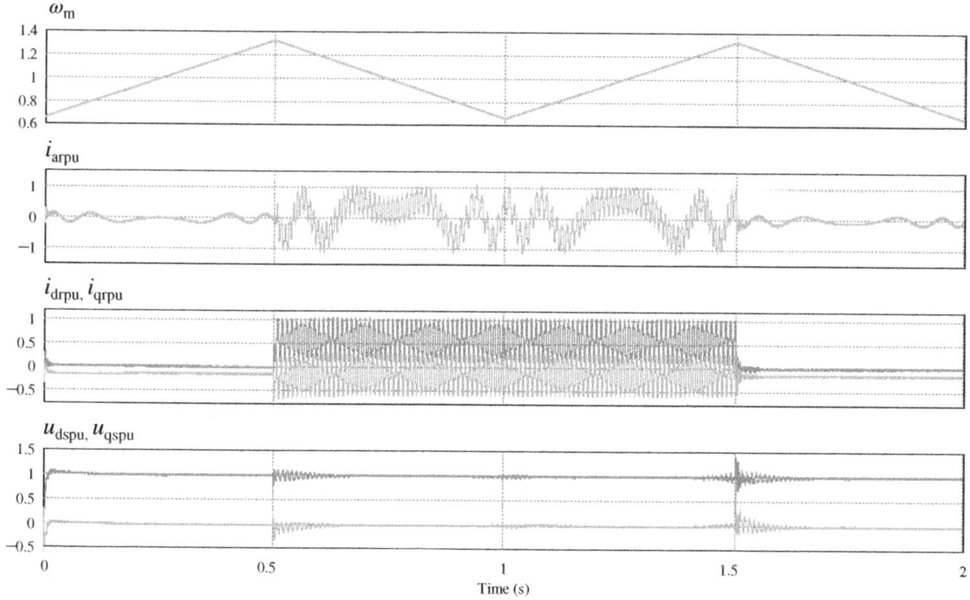

Figure 10.34 Simulation results of a stator voltage control with negative sequence compensation from Figure 10.33 during unbalanced load supply

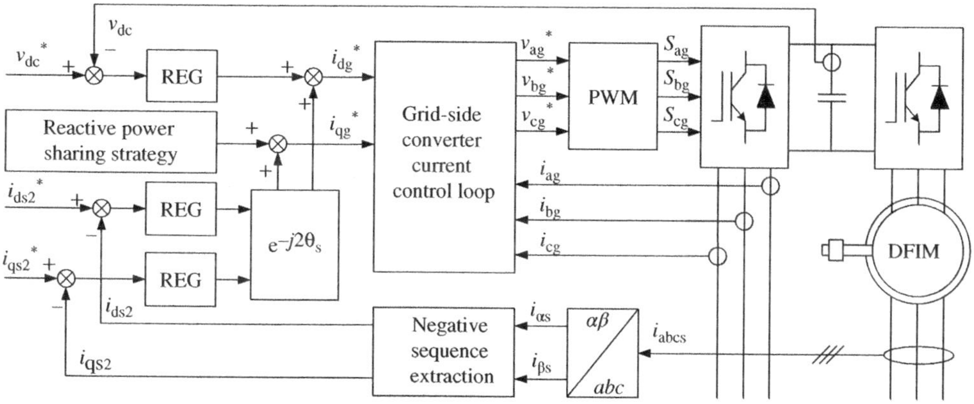

Figure 10.35 Block diagram of the grid-side converter control structure with a stator current negative sequence compensation during unbalanced load supply

The remaining part of the control, presented as a general block responsible for the grid converter current control, is the same as that presented for the typical grid-connected three-phase power inverter.

10.2.3.4 Nonlinear Load Supply

Similar to the case of the asymmetrically loaded stand-alone DFIM, a nonlinear load might produce electromechanical torque pulsations; it depends on the method of harmonics compensation. If all load current harmonics are delivered only from grid-side converter, which operates partially as an inverter/rectifier and partially as an active filter of the stator current, a load seen by the stator is linear and no torque pulsations occur. However, it might be necessary to compensate stator voltage harmonics by the rotor-side converter, when a grid converter operates with a current saturation. It is possible to implement independent controllers for each harmonic similar to the structure applied for the compensation of asymmetry. It requires an extraction of the component responsible for the representation of each harmonic and the transformation of this signal to the frame rotating with an angular speed corresponding to the controlled frequency.

Another possibility to obtain adequate rotor current reference signals is to implement a load current feedforward instead of the stator current feedforward, wherein a stator load current i_o (Figure 10.29) is taken into consideration. Signals i_{od} and i_{oq} of the stator load current include information that is more precise regarding the compensation necessary for the harmonics content, even if it is not measured directly but estimated. The estimation is based on the stator voltage and current measurements (Equation 10.52). The stator load current components are represented in a *dq* rotating frame and the harmonics are represented by oscillating components; therefore, something more sophisticated than a PI controller has to be used to obtain a reference rotor current, including harmonics content. A variant of the control structure can be based on a resonant controller in which a few elements that operate in parallel to the main PI controller are responsible for the regulation of selected harmonics [8]. In high-power systems, the most important harmonics are the fifth negative sequence and seventh positive sequence, typical for three-phase diode rectifiers and other three-phase loads. In a *dq* frame, both harmonics are represented by the common signal with a frequency of 300 Hz and both can be controlled by a common resonant controller designed for a 300 Hz resonant frequency. Part of the magnetizing current, uncompensated by stator filtering capacitors, is fed as a reference in the *q*-axis of the rotor current vector, as is shown in Figure 10.36.

Figure 10.37 presents the results of tests with the same conditions as those in Figures 10.32 and 10.34; however, instead of the rotor position sensor, a rotor speed and position observer from Figure 10.26b has been implemented.

A difference in relation to the grid connection mode is that the θ_s used in an estimator is replaced by a reference angle obtained by integration of the reference synchronous speed ω_s^*. During an initial

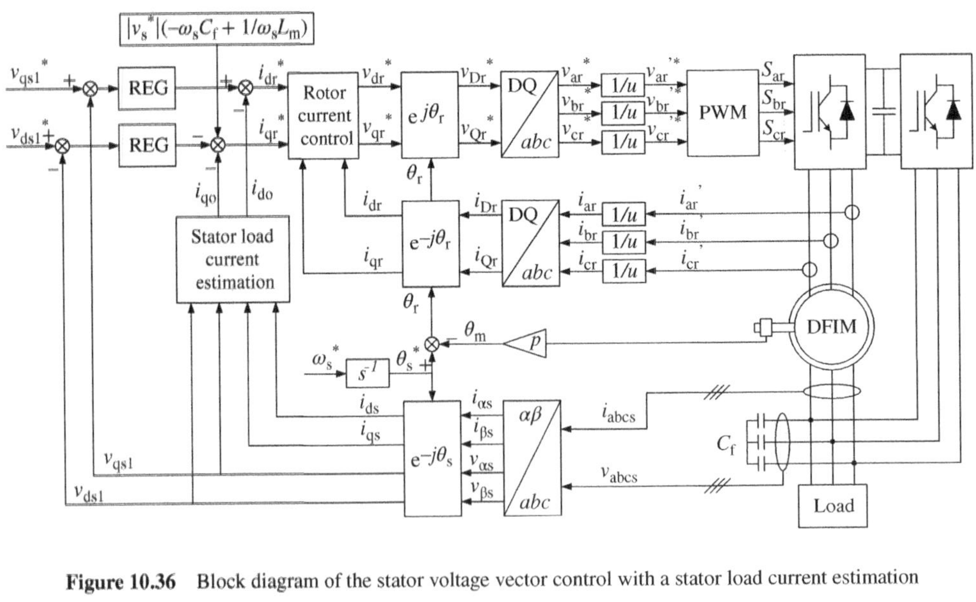

Figure 10.36 Block diagram of the stator voltage vector control with a stator load current estimation

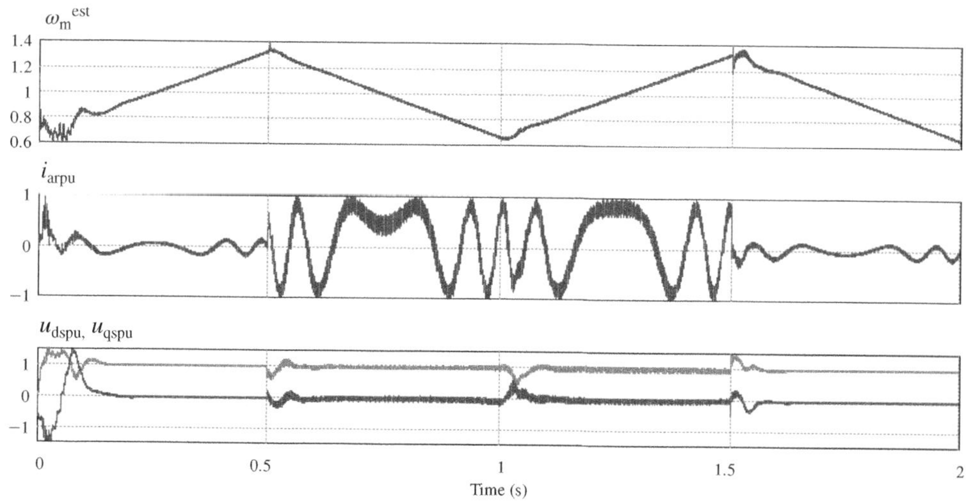

Figure 10.37 Simulation results of a stator voltage control with a stator load current estimation from Figure 10.36 during nonlinear load supply with implemented rotor position MRAS observer from Figure 10.26b

state, a stator-side overvoltage might occur, such that in a real system, a rotor current should be limited more than during normal operation in order to avoid the problem of overexcitation. An error of speed estimation implies an error in the estimated rotor position, and therefore, visible voltage dips and sags occur in the stator voltage waveform. However, it has to be noted that the acceleration and speed change characteristics near the speed range limits, presented in Figure 10.37, will never occur in a real system, even with an internal combustion engine used as the primary mover. For a nonlinear load, a three-phase diode rectifier loaded by a resistive load representing 75% of the stator power ($R_{load} = 0.5\,\Omega$) is applied, and between the stator and rectifier an inductor of value equal to 0.1 mH is placed.

Figure 10.38 presents stator-phase voltage, load-phase current and three-phase rotor current waveforms corresponding to Figure 10.37 registered in the period from 0.56 to 0.94 s, which is a period when the system supplies a nonlinear load of power equal to 75% of the rated stator power. A control method does not require the extraction of each separate harmonic and it is sufficient to compensate only the main harmonics, while less significant harmonics are canceled by the use of the stator-connected filtering capacitors. Similar to the case of the unbalanced load (Figure 10.35), adequate control of the grid-side converter can be introduced, which partially operates as an active filter for the reduction of the stator current harmonics and, therefore, for a reduction of the electromagnetic torque. This control might include selective compensation of the harmonics based on the harmonics component extraction or it might be realized based on the instantaneous values of the controlled variables.

10.3 DFIM-Based Wind Energy Conversion Systems

10.3.1 Wind Turbine Aerodynamic

This section analyzes the most important issues of DFIM-based pitch-regulated variable-speed wind turbines. In this type of modern wind turbine, the energy from the wind is captured mechanically by the blades, then it is converted to electric energy by the DFIM, and finally, this energy is delivered to the electricity network [1]. The theory of momentum is used to study the behavior of the wind turbine. Under certain ideal assumptions, the wind turbine can recover the power from wind given by

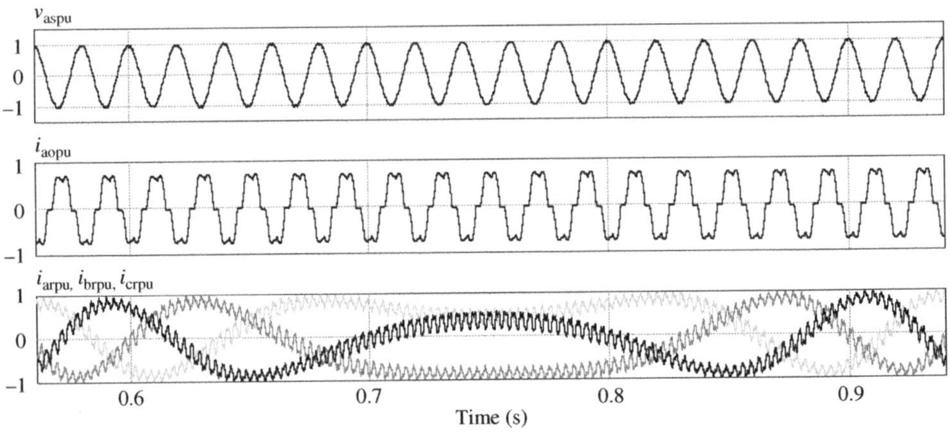

Figure 10.38 Simulation results presenting instantaneous phase variables from the same test as shown in Figure 10.37, but in an enlarged window

the expression:

$$P_t = \frac{1}{2}\rho\pi R^2 V_w^3 C_p \tag{10.55}$$

where ρ is the air density, R is the radius of the blades of the wind turbine, V_w is the wind speed and C_p is the power coefficient; a dimensionless parameter that expresses the effectiveness of the wind turbine in the transformation of the kinetic energy of the wind into mechanical energy. For a given wind turbine, this coefficient is a function of wind speed, the speed of rotation of the wind turbine Ω_t and the pitch angle β. C_p is often given as a function of the tip speed ratio λ, which is defined by

$$C_p = f(\lambda, \beta) \qquad \text{with} \qquad \lambda = \frac{R\Omega_t}{V_w} \tag{10.56}$$

The theoretical maximum value of C_p is given by the Betz limit: $C_{p\_theo\_max} = 0.593$. Figure 10.39 shows graphically an example of C_p curves.

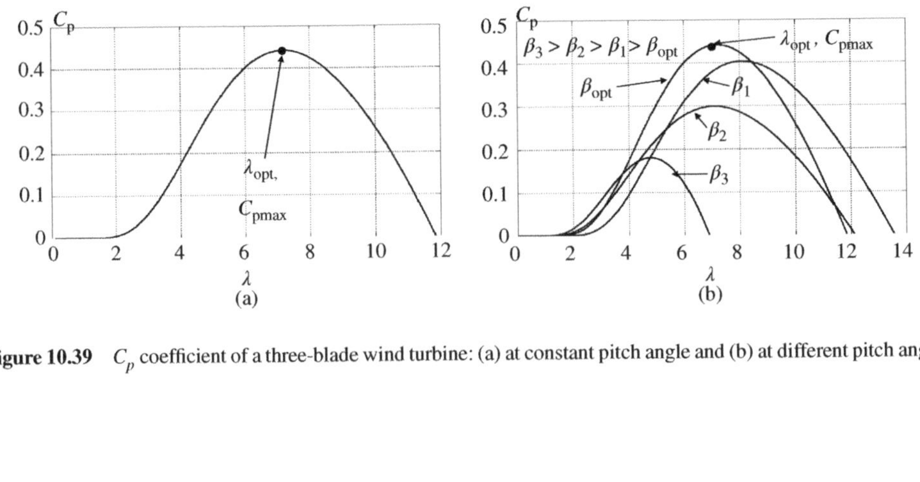

Figure 10.39 C_p coefficient of a three-blade wind turbine: (a) at constant pitch angle and (b) at different pitch angles

10.3.2 Turbine Control Zones

Under the assumption of an ideal mechanical coupling and neglecting possible losses of the gearbox, located between the turbine rotor and the DFIMs, the power generated by the wind turbine P_t is transmitted completely to the shaft of the DFIM; thus, $P_m = P_t$. Therefore, by evaluating expressions (10.55) and (10.56) for different wind speeds and different rotational speeds of the wind turbine (the pitch is maintained constant to its optimal position β_{opt}), the captured power from the wind can be represented in three-dimensional space, as illustrated in Figure 10.40a. This is a representative example of a multi-megawatt wind turbine for which a specific radius R has been selected. It can be seen that at each constant wind speed, the curve presents a maximum point of generated power at one specific rotational speed. This property of wind turbines is exploited by means of the variable speed control, which always tries to choose the appropriate rotational speed at any given wind speed, in order to capture the maximum available power from the wind. However, in Figure 10.40a, the points at which the wind turbine operates are also shown. It can be noted that it is not always possible to operate at the points of maximum power of the curves. This is because of the restriction of the maximum Ω_{t_nom} and minimum Ω_{t_min} rotational speeds of the wind turbine. Depending on the turbine's design, the speed is limited by efficiency and safety concerns. For most turbine designs, these speed limitations yield four zones of operation relating to the rotational speed. Projections of Figure 10.40a into the three planes illustrated in Figure 10.40b–d help to delineate these four operating zones (note that the gearbox relation holds: $N\Omega_t = \Omega_m$).

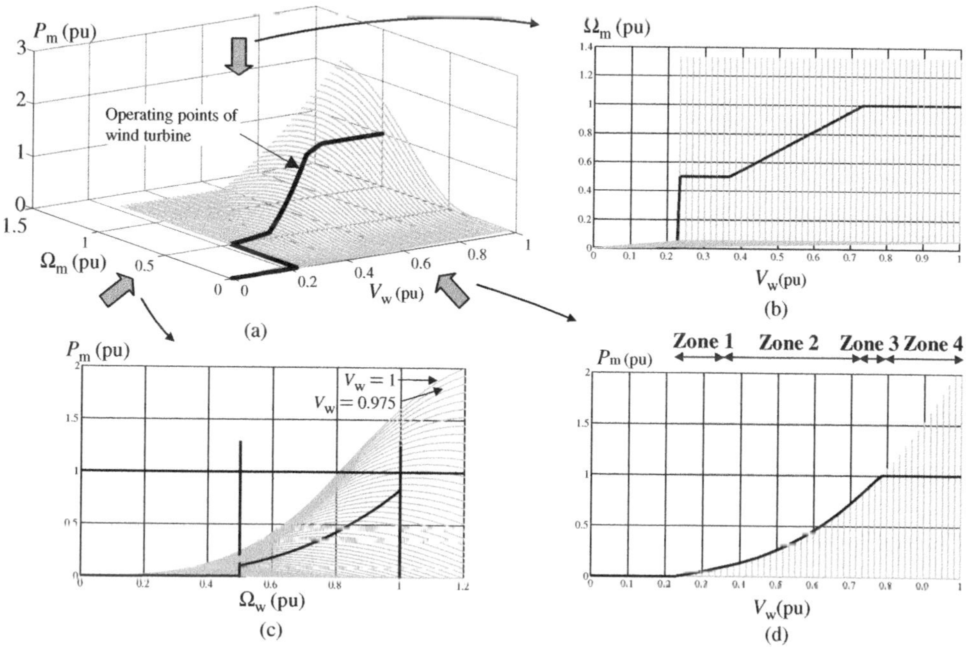

Figure 10.40 Power curves of a multi-megawatt pitch-regulated variable-speed wind turbine. (a) $P_m = f(V_w, \Omega_m)$, (b) projection into (V_w, Ω_m) plane, (c) projection into (P_m, Ω_m) plane, (d) projection into (P_m, V_w) plane, note that this curve is normally provided by manufacturers

ZONE 1: In this zone, the rotational speed is limited and maintained constant to its minimum value; therefore, it is not possible to extract the maximum power from the wind. In this example of a wind turbine, between wind speeds of 0.235 and 0.36 pu, the captured power is not maximum.

ZONE 2: In this zone, the rotational speed can be varied achieving the maximum point of the power curves. Within this zone, generally, a maximum power point tracking (MPPT) strategy is carried out. Therefore, between wind speeds of 0.36 and 0.73 pu, the rotational speed is modified, always seeking the maximum of the power curves.

ZONE 3: This zone appears in some turbine designs when the maximum rotational speed is reached but not that of the maximum generated power. For this turbine design, between wind speeds of 0.73 and 0.78 pu, the rotational speed must be maintained at the maximum value, although it is not possible to capture the maximum power from the wind and, therefore, it is not operating at the maximum of the power curves.

ZONE 4: This zone begins when the captured power is equal to that of the rated power, which for this turbine example is at 0.78 pu wind speed. In this zone, the generated power and speed are maintained constant at their maximum values by modifying the pitch angle. Thus, the blades are gradually pitched out of the wind, because it is possible to maintain constant captured power, even if the wind speed increases.

It must be highlighted that it is possible to find in the literature slightly alternative representations of the turbine control zones presented in this section.

10.3.3 Turbine Control

Having seen that the wind turbine can operate at four different zones according to the rotational speed, this section presents the general control strategy of the wind turbine (Figure 10.41a). The DFIM is vector controlled by the rotor-side converter, as described in Figure 10.15, for instance, but by omitting the speed control loop and translating to the wind turbine control block (if necessary), as is discussed later [1]. Thus, this new wind turbine control block generates the torque and pitch references in order to determine the four zones of operation. The grid-side converter is also vector controlled and is responsible for evacuating the generated power to the grid through the rotor of the machine and for controlling the DC bus voltage of the back-to-back converter. Q_s and Q_g references are generated by the reactive power generation strategy, assisting to the grid with the necessary reactive power, as demanded by the grid operator.

Figure 10.41b shows some possible wind turbine control block philosophies for each operating zone. As can be noted, zones 1 and 3 present the same control structure. The speed is regulated as constant with regard to the maximum or minimum value by means of the electromagnetic torque generated by the DFIM. Generally, in this situation, the pitch angle is maintained constant. On the other hand, in zone 4, the speed is also regulated as constant with regard to the maximum value, but in this case, by actuating the pitch angle and keeping the electromagnetic torque constant. Finally, the MPPT of zone 2 can be performed according to different control philosophies. For instance, it is possible to use an indirect speed control (ISC) by exploiting the fact that when the MPPT is achieved (operation at C_{p_max}), the generated power follows a cubic relation to the speed as follows:

$$V_w = \frac{R\Omega_t}{\lambda_{opt}} \Rightarrow P_t = \frac{1}{2}\rho\pi R^2\left(\frac{R\Omega_t}{\lambda_{opt}}\right)^3 C_{p\_max} \Rightarrow \begin{cases} P_t = k_{opt}\Omega_t^3 \\ k_{opt} = \dfrac{1}{2}\rho\pi R^5 \dfrac{C_{p\_max}}{\lambda_{opt}^3} \end{cases} \tag{10.57}$$

The newest grid codes for variable-speed wind turbines also impose some additional requirements related to grid support. On the one hand, the operator can demand a certain value of total reactive power exchange Q_{total} with the grid, within some predetermined limits. In general, this obliges an oversize of the design of the converter and/or machine of the wind turbine, in order to distribute the total reactive power demand between Q_s and Q_g and implement a special reactive power generation strategy, as depicted in

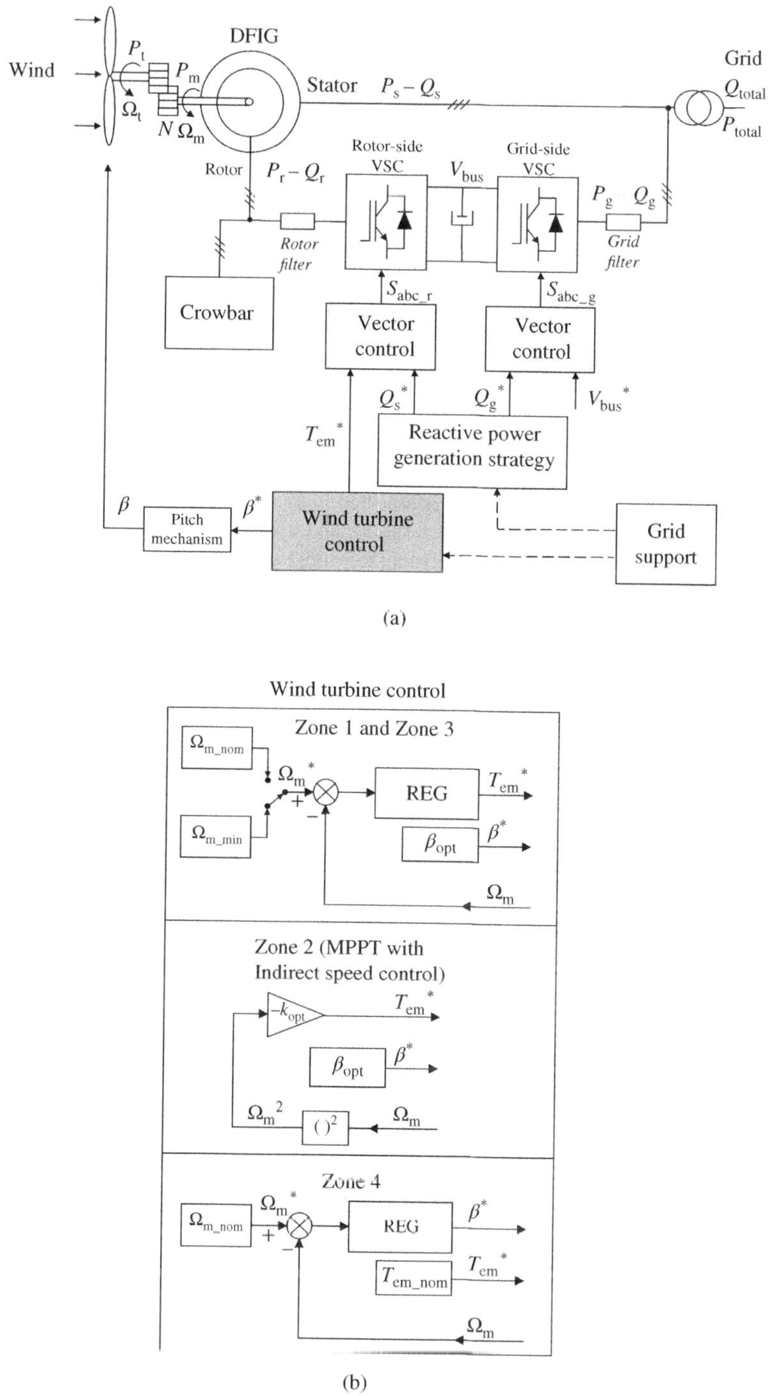

Figure 10.41 (a) Overall control of a DFIM-based wind turbine and (b) control loops at different zones of operation

Figure 10.41. On the other hand, the grid code can also impose the primary frequency control support of the grid by maintaining a certain reserve of power in the wind turbine. This can be achieved by avoiding the extraction of the maximum power from the wind, by working at speeds other than the optimum, or by moving further from the optimum pitch angle. In this way, the wind turbine will contribute to the balance of power generation in the grid, which is currently performed by conventional power plants. Finally, wind turbines might also be required to provide an inertial response to frequency changes, as done naturally by the synchronous generators of conventional power plants. Thus, as occurs with conventional generators, wind turbines at normal operation store kinetic energy in the rotating blades. If nothing special is done, this energy does not contribute to the inertia of the grid because the rotational speed is controlled by the converter control. Hence, it is also possible to emulate an inertial response in the wind turbines by adding the emulation block to the torque reference of the wind turbine [9], as shown in Figure 10.42. When a variation in the frequency above certain limits is detected, the proportional-derivative control law imposes a fast torque variation, which also varies the power delivered to the grid by the wind turbine. This is achieved by changing the rotational speed of the blades and, therefore, altering the stored kinetic energy of the wind turbine. This quick reaction, which is faster than the primary regulation of conventional power plants, contributes to the avoidance of quick frequency changes owing to sudden changes in power consumption, which can turn off power plants owing to the overshoot of the frequency deviation. Note that when the rotational speed reaches a pre-established maximum or minimum limit, the wind turbine is not able to contribute further to the inertial response and it will operate at a speed different from the optimum, not extracting the maximum power from the wind. However, it has contributed with additional power during the first critical seconds of the frequency variation. Then, in order to reach the optimum speed again, the wind turbine will operate oppositely to the frequency primary regulation. Note that a quick synchronization method, for instance, a PLL, is crucial in order to ensure quick and accurate information on the frequency variation.

10.3.4 *Typical Dimensioning of DFIM-Based Wind Turbines*

In general, DFIM-based variable-speed wind turbines are designed with some typical common characteristics. First, the number of pole pairs is typically two or three in order to achieve an efficient

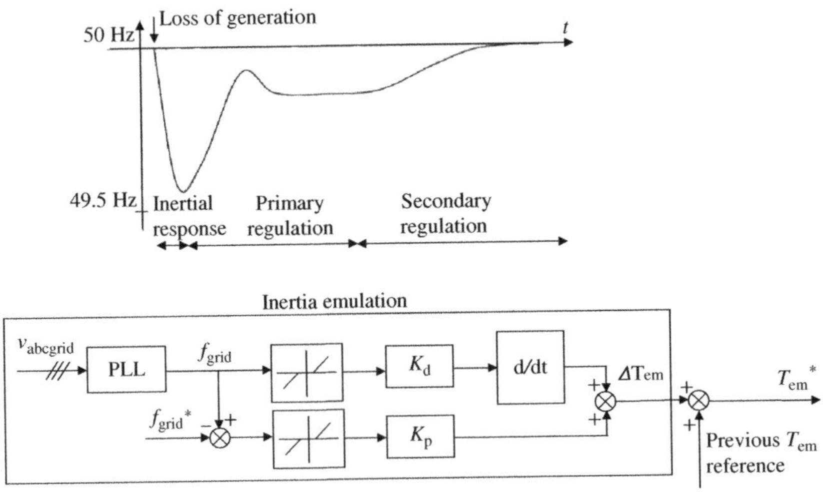

Figure 10.42 Grid frequency support by inertia emulation added to torque reference

machine design. The rated voltage provided at the rotor by the converter is normally chosen as low voltage, typically around 690 V, in an effort to reduce the size, price and losses of the converter. Then, the rated rotor voltage of the DFIM is chosen to be around three times larger than this voltage. This arrangement is deduced from the equations provided by Table 10.1, indicating that the machine can operate only under slips ±1/3, which means that the operating speed range can only be ±1/3 from synchronism. For this reason, it is not possible to achieve speeds near zero with this reduction of converter voltage; however, this is not really required by the application because the wind turbine can only provide significant power above a certain minimum speed of rotation, as seen in the earlier section. The start-up of the machine from zero speed is not possible in the conventional manner; however, this is not really a big drawback and is normally solved without any extra complexity. In addition, this limitation of the available slip brings one big advantage, that is, the power transmitted through the rotor and through the converter is 1/3 of the stator power (Table 10.1). This means that the converter is not only reduced in voltage but also reduced in power, leading to a very economical converter.

Regarding the stator voltage, it can be chosen as low voltage, similar to the converter voltage, that is, around 690 V, leading to a stator and rotor turn ratio of 1/3. For big multi-megawatt wind turbines, this stator voltage ratio can lead to operation with stator currents that are too high, leading to nonefficient machine designs. Thus, for the highest power scenarios, the manufacturers also tend to design medium voltage stators, reducing the required stator currents and reducing the size of the grid connection transformer. However, another transformer of reduced power is also needed for the grid-side converter if this solution is adopted. In this latter case, the stator and rotor turn ratios tend to be higher than 1. With regard to the rotor currents, if they are too high for the converter in big power wind turbines, a possible tendency is still to maintain the low-voltage converter, increasing the current capability by parallelizing converters, but not stepping up to the medium voltage. Finally, it must be highlighted that this reduced voltage and power converter does not allow the handling of problematic situations caused by severe grid faults. Therefore, as will be seen later, special hardware is normally incorporated to address this problem.

10.3.5 Steady-State Performance of the Wind Turbine Based on DFIM

Connecting the steady-state performance analysis of the DFIM described in Section 10.2 and the wind turbine operation studied in the earlier section, this section presents the most representative electric magnitudes of a multi-megawatt DFIM, operating in a pitch-regulated variable-speed wind turbine. Thus, the control of the wind turbine will impose the $P_m = f(\Omega_m)$ characteristic of Figure 10.30c. By considering that the maximum rotational speed corresponds to a minimum slip of $s = -0.2$, this yields the input power and torque curves as shown in Figure 10.43a. Thus, by evaluating the equations presented in Tables 10.2 and 10.3, it is possible to derive the steady-state behavior of the wind turbine at two different magnetizing levels, for instance, with $Q_s = 0$ and with $i_{dr} = 0$. Figure 10.43 presents the obtained steady-state results. It can be deduced that T_{em} and P_m are input to the shaft of the DFIM; therefore, both are the same regardless of the magnetizing strategy used. Note that the wind turbine application produces the highest torques and powers at the higher speeds (minimum slips). Similarly, Figure 10.43b shows how the input power is distributed between the stator and the rotor, depending mainly on the slip. Stator and rotor reactive powers are shown in Figure 10.43c and d.

By magnetizing the DFIM through the stator, it can be seen that this machine requires constant Q_s = 0.25 pu. With regard to the stator and rotor current modules of Figure 10.43e and f, it can be seen that they follow a very similar evolution to the input torque or power as shown in Figure 10.43a; higher input torques demand higher current modules. It can also be noticed that magnetizing the machine through the stator ($i_{dr} = 0$) requires higher stator currents and smaller rotor currents than magnetizing the machine through the rotor ($Q_s = 0$). Therefore, as seen in Figure 10.43h, for this specific machine, this implies better efficiency by magnetizing through the stator. Finally, the required rotor voltage amplitude is depicted in Figure 10.43g. Generally, it can be seen that the higher the slip in the module, the higher the required rotor voltage amplitude is, being almost zero at synchronism ($s \cong 0$). Neither magnetizing strategy

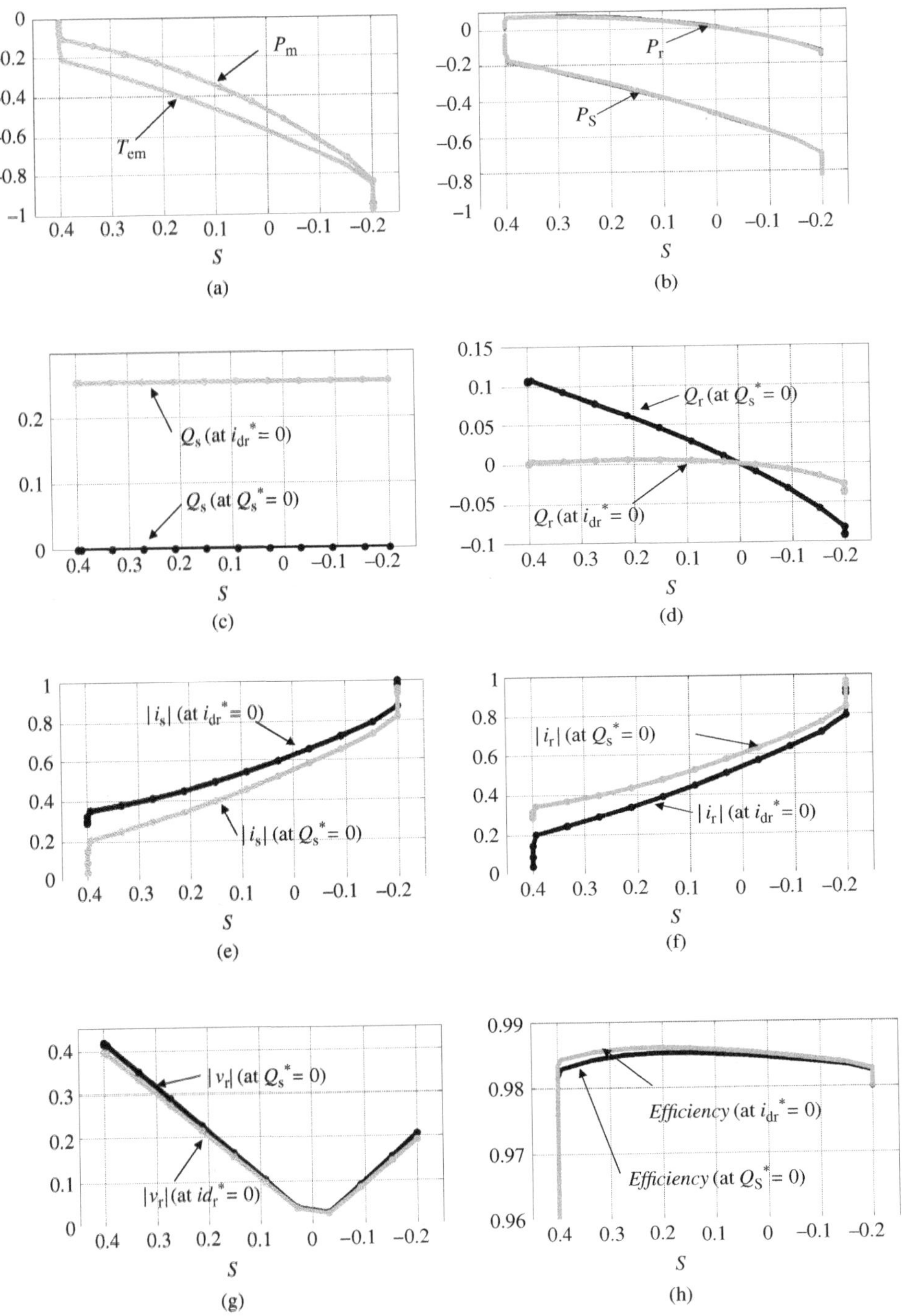

Figure 10.43 Steady-state most representative magnitudes in pu of a multi-megawatt pitch-regulated variable-speed wind turbine with $Q_s = 0$ and $i_{dr} = 0$ magnetizing strategies

produces a high difference of rotor voltage demand; however, for this specific example, the dimensioning of the converter in terms of the voltage is established by the operating point at slip $s = 0.4$.

10.3.6 Analysis of DFIM-Based Wind Turbines during Voltage Dips

10.3.6.1 Understanding the Loss of Control during Grid Voltage Dip

Voltage dips are known as sudden drops of grid voltage, caused by contingencies or faults occurring in the electricity grid. The study presented in this section focuses only on symmetric voltage dips and does not address asymmetric dips. When a voltage dip is seen directly by the stator of the DFIM, it is necessary to analyze the behavior of the stator flux in order to understand the difficulties derived from the perturbation caused by the dip. Thus, combining expressions (10.23) and (10.26) and removing the stator current, the following interesting expression is derived:

$$\frac{\mathrm{d}\vec{\psi}_s^{\,s}}{\mathrm{d}t} = \vec{v}_s^{\,s} - \frac{R_s}{L_s}\vec{\psi}_s^{\,s} + R_s\frac{L_m}{L_s}\vec{i}_r^{\,s} \tag{10.58}$$

It can be noticed that when the sudden voltage dip occurs, the stator flux cannot evolve to its final steady state as quickly as the stator voltage. Each phase of the stator flux evolves as the sum of a sinusoid (permanent and is not extinguished) and an exponential (also known as natural flux) with time constant L_s/R_s (which can be a few seconds). The term of the rotor current can make the flux decay more quickly (with rotor currents in contra-phase to the stator flux), as illustrated in Figure 10.44. Note that the rotor current is vector controlled by means of the rotor-side converter if control has not been lost.

On the other hand, it is useful to modify the equivalent electric circuit of the DFIM, as seen in Section 10.1.3, to a more compact equivalent circuit. For that purpose, from expressions (10.23)–(10.26), it is possible to achieve the following equation, which leads to the circuit as shown in Figure 10.45a:

$$\vec{v}_r^{\,r} = \frac{L_m}{L_s}\left(\vec{v}_s^{\,r} - j\omega_m\vec{\psi}_s^{\,r}\right) + \left[R_r + \left(\frac{L_m}{L_s}\right)^2 R_s\right]\vec{i}_r^{\,r} + \sigma L_r\frac{d}{dt}\vec{i}_r^{\,r} \tag{10.59}$$

It can be seen that the rotor current is established as a function of the stator and rotor voltages, the stator flux (is also rotor current dependent), and the equivalent resistances and inductances. Therefore, the space vector diagram at subsynchronous speed is as depicted as in Figure 10.45b. The sum of the dominant terms $\vec{v}_r^{\,r}$ and $j\omega_m\vec{\psi}_s^{\,r}$ must match approximately with $\vec{v}_s^{\,r}$.

Thus, when the machine operates at some point at steady state and a sudden stator voltage dip occurs, the sudden change in the stator voltage should be accompanied by a sudden rotor voltage change in order to prevent a high increase in the rotor current. Note that because the stator flux decreases slowly (several cycles depending on the machine), as seen in Figure 10.44, the required rotor voltage will be higher

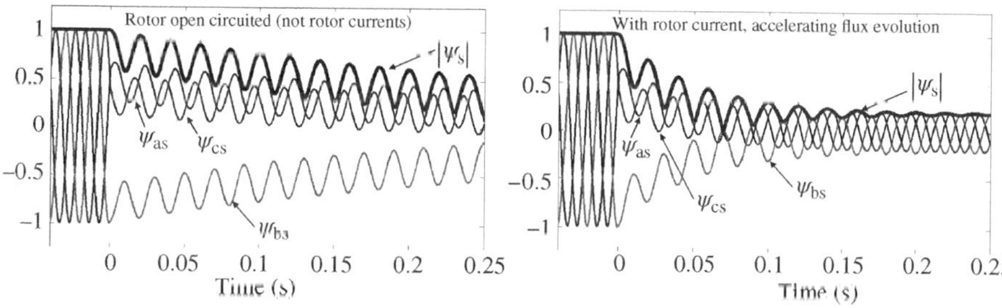

Figure 10.44 Stator flux evolution in pu during an 80% voltage dip

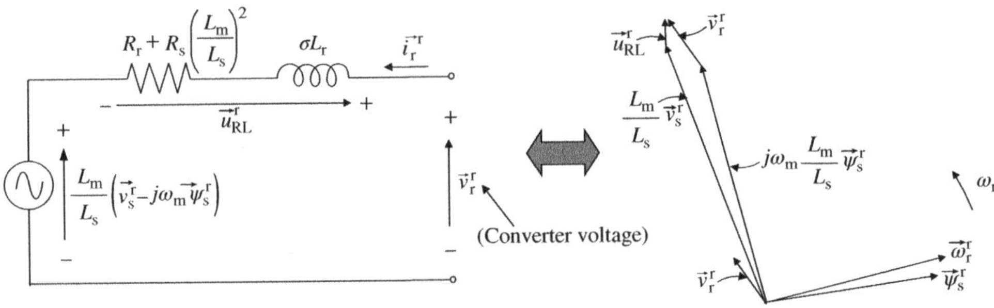

Figure 10.45 (a) Equivalent circuit of the DFIM for the analysis of voltage dips and (b) space vector diagram at subsynchronism in a generator mode

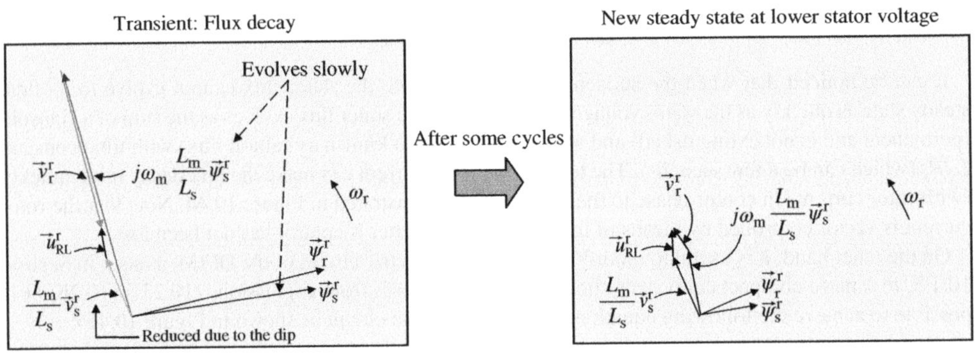

Figure 10.46 Evolution of the space vector magnitudes from the first instant when the stator voltage is reduced until the steady state reached at the dip

than at steady state, owing to the perturbation caused by the voltage dip. This is illustrated graphically in Figure 10.46. After some cycles, the new steady state can be similar to that of the moment prior to the dip, but with magnitudes of lower amplitude owing to the stator voltage decrease, which also leads to a lower T_{em} and Q_s. It should be noted that in order to not lose the control keeping the rotor currents within safe limits (no high increase), it is necessary to have an exceptionally higher rotor voltage amplitude, especially at the beginning of the voltage dip. The most unfavorable case would be when the stator voltage dip is 100%, requiring that the rotor voltage must replace totally the stator voltage that has disappeared. Unfortunately, as seen in earlier section, the typical dimensioning of the converter supplying the rotor can only provide a maximum of approximately 1/3 of the stator voltage. Therefore, owing to a rotor voltage limitation (converter limitation), these types of wind turbine cannot guarantee that during severe voltage dips the rotor current will be maintained within its safe limits, without losing the control. Generally, because of this voltage limitation, wind turbines based on DFIM incorporate an additional crowbar protection that solves these problematic severe voltage dips. It must be highlighted that very similar situations occur at hypersynchronous speeds and also with asymmetrical voltage dips. Therefore, it can be concluded that after a severe voltage dip, during the time the stator flux evolves, there is typically a short period when control is lost, which normally produces overcurrents at the stator and rotor of the machine, until the flux reaches one level at which the available converter voltage can guarantee the control of the machine. As said before, in order to protect the system against the overcurrents provoked by the loss of control, a crowbar protection is activated that simultaneously accelerates the flux evolution, while trying as quickly as possible, to recover control with the converter.

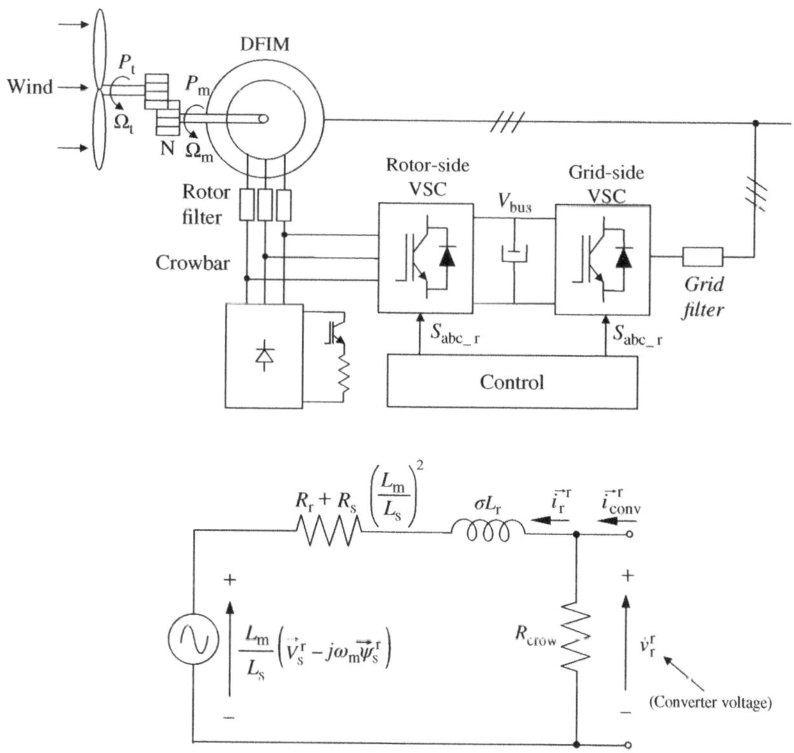

Figure 10.47 (a) System equipped with three-phase DC crowbar protection and (b) one phase equivalent circuit of the system when the crowbar is activated

10.3.6.2 Performance during Severe Grid Voltage Dips

As seen in the previous section, during significant voltage dips, the system needs a crowbar for protection from the overcurrents and overvoltages caused by the loss of control during the dip. In DFIM-based wind turbines, the crowbar is installed at the rotor terminals, as shown in Figure 10.47a, which prevents damage to the rotor converter. It is activated when one anomalous situation is detected. The rotor current is then diverted to the crowbar and the rotor converter is switched off. Figure 10.47b shows the equivalent circuit of the system using the rotor model deduced in the earlier section. As can be seen, when the crowbar is activated, the circuit becomes an impedance divider. In Figure 10.47a, the crowbar comprises a rectifier, a controlled switch and a resistance. There are several alternative crowbar topologies to the crowbar shown in this figure.

As demanded by the grid codes, in order to provide low voltage ride-through (LVRT) capability, the wind turbine must remain connected during the voltage dip; therefore, the crowbar must be activated and deactivated without disconnecting the DFIM from the grid. In this way, the sequence of events that typically occurs during severe voltage dips can be summarized as follows:

1. The DFIM is generating power at steady state at one specific operating point.
2. When the voltage dip occurs, there are a few milliseconds (typically 0.5–5) until the wind turbine's control detects the dip. Thus, during this period the system cannot guarantee control and, in general,

there is a high rise of currents through the rotor converter, which also provokes a DC bus voltage increase. The dip is detected by supervising the following anomalies:
   (a) overcurrent in the rotor,
   (b) overvoltage in the DC link,
   (c) grid voltage drop, detected by PLLs or synchronizing methods.
3. Once the dip has been detected, the crowbar is activated quickly, which demagnetizes the machine. The rotor converter is inhibited, keeping it safe and ensuring that the entire rotor current circulates through the crowbar. Depending on the machine's design, the time during which the crowbar is active can range for several cycles.
4. Once the flux has decayed and the available converter voltage can control the machine, the crowbar is disconnected and the rotor converter in activated again. In general, because at this moment the flux has not been totally damped, it is preferable to contribute to the total stabilization of the stator flux by injecting demagnetizing rotor currents by control. At the same time, as demanded by the grid codes, it is possible to provide progressively reactive power through the stator by increasing the corresponding *d* rotor current component. This situation will last until the grid voltage is progressively recovered, the fault is cleared, and normal operation is resumed.

This behavior is shown by means of a simulation-based example of a multi-megawatt wind turbine, which is affected by an 80% symmetric voltage dip. The control philosophy applied to the wind turbine is as studied in this chapter, that is, vector control, as shown in Figure 10.13. Hence, Figure 10.48 illustrates the most representative electric magnitudes of the DFIM during the dip. Once the dip occurs, the rotor-side converter is inhibited owing to the quick overcurrent detected in the rotor. At the same instant, the crowbar is activated, connecting the additional resistance path in the rotor and damping the big energy fluctuation of the machine. During the first instants of the dip, large torque and peak stator and rotor currents occur. After few milliseconds, the crowbar can be disconnected and at the same instant, the rotor converter is activated, which injects demagnetizing currents through the rotor, at the same time as injecting stator capacitive reactive power in ramp. Note that at approximately time instant 0.175 s, the stator and rotor current is at rated value, but Q_s is −0.2 owing to the voltage decrease in the grid. Finally, it must be highlighted that the DC bus voltage normally suffers a transient because the grid-side converter is also affected by the dip. If the connection time of the crowbar is low, this transient phenomenon could be more significant.

In addition, it could happen that the voltage dip is not sufficiently deep to activate the crowbar. Under that situation, the converter alone can control the system without the need of the crowbar protection.

The resistance of the crowbar R_{crow} must be chosen carefully. In general, it can be selected by a simulation-based analysis that attempts to find a compromise between the following aspects:

- If a very low value is chosen, the current during the dip will be very large. Thus, the crowbar elements should be oversized and the electromagnetic torque will present a big peak.
- The rotor current can be reduced by using a higher resistance. However, if the resistance is too big, the crowbar will not pull the rotor voltage low enough and the rotor current will circulate across the rotor converter via its freewheeling diodes, even if it is inactive, increasing the DC bus voltage. Therefore, it is very important that the resistance is sufficiently high that the diodes of the rotor converter do not operate, thus allowing the entire rotor current to circulate through the crowbar resistance.

It must be highlighted that the activation time of the crowbar is another critical parameter of the crowbar's operation. Although the crowbar is connected, the machine is not fully controlled, and it is impossible to generate the reactive power that most grid codes demand during voltage dips. On the other hand, a premature disconnection, when the natural flux is still too high, might cause saturation of the converter or the impossibility of controlling the machine, causing overcurrents in the rotor and DC bus

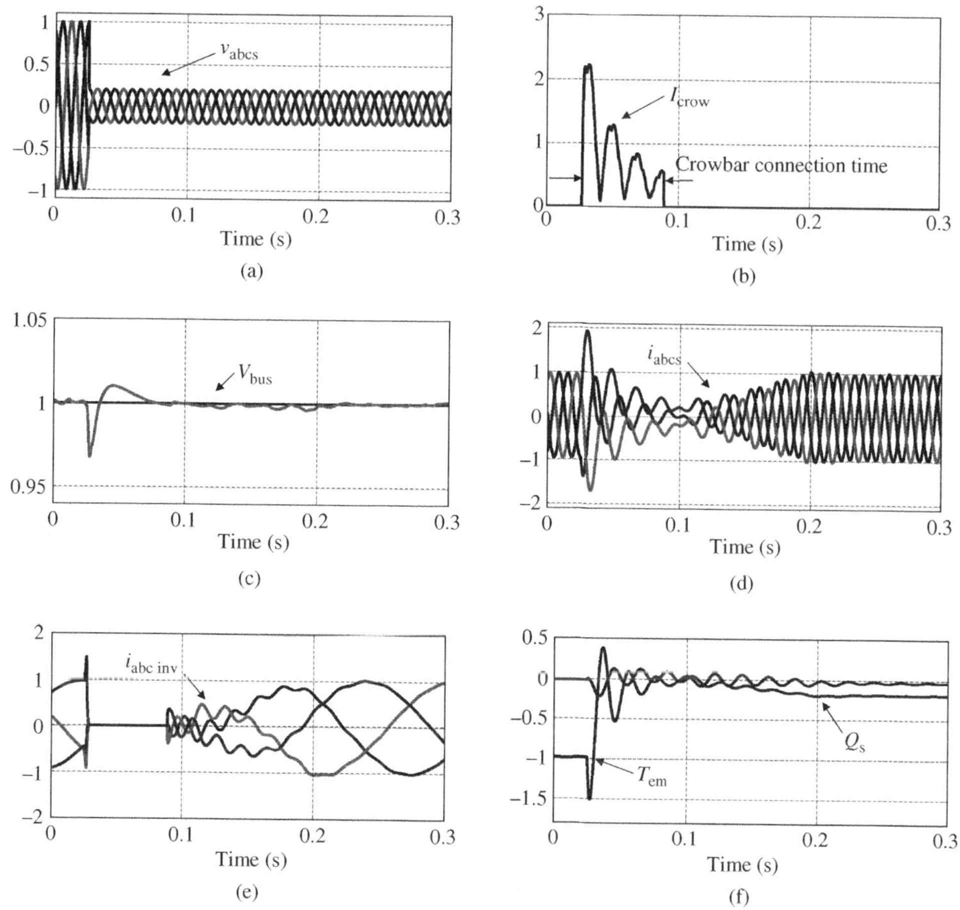

Figure 10.48 Most representative magnitudes of a multi-megawatt DFIM-based wind turbine affected by a symmetric 0.8 depth voltage dip

overvoltages. The connection time is then a compromise between safety and fulfillment of grid codes. Over the past years, the grid codes have toughened their requirements, demanding each year a quicker injection of reactive current.

References

1. Abad, G., López, J., Rodríguez, M.A. *et al.* (2011) *Doubly Fed Induction Machine: Modeling and Control for Wind Energy Generation*, John Wiley & Sons, Inc.
2. Pena, R., Clare, J.C., and Asher, G.M. (1996) Doubly fed induction generator using back-to-back PWM converters and its application to variable-speed wind energy generation. *IEE Proceedings - Electric Power Applications*, **143**, 231–241.
3. Wu, B., Lang, Y., Zargari, N., and Kouro, S. (2011) *Power Conversion and Control of Wind Energy Systems*, John Wiley & Sons, Inc.
4. Sul, K. (2011) *Control of Electric Machine Drive Systems*, John Wiley & Sons, Inc.
5. Bose, B.K. (2002) *Modern Power Electronics and AC Drives*, Prentice-Hall.

6. Cardenas, R., Pena, R., Clare, J. *et al.* (2008) MRAS observers for sensorless control of doubly-fed induction generators. *IEEE Transactions on Power Electronics*, **23** (3), 1075–1084.
7. Iwanski, G. and Koczara, W. (2008) DFIG based power generation system with UPS function for variable speed applications. *IEEE Transactions on Industrial Electronics*, **55** (8), 3047–3054.
8. Phan, V.-T. and Lee, H.-H. (2012) Performance enhancement of stand-alone DFIG systems with control of rotor and load side converters using resonant controllers. *IEEE Transactions on Industry Applications*, **48** (1), 199–210.
9. Morren, J., Haan, S.W., Kling, W.L., and Ferreira, J. (2006) Wind turbines emulating inertia and supporting primary frequency control. *IEEE Transactions on Power Systems*, **12** (1), 433–434.

11

AC–DC–AC Converters for Distributed Power Generation Systems

Marek Jasinski, Sebastian Stynski, Pawel Mlodzikowski and Mariusz Malinowski
Faculty of Electrical Engineering, Warsaw University of Technology, Warsaw, Poland

11.1 Introduction

Indirect AC–DC–AC converters consist of two connected *voltage source converters* (VSCs) as presented in Figure 11.1. The AC–DC–AC is a connection through a DC-link of two single-phase, three-phase, or multiphase AC circuits with different voltage amplitude u, frequency f, or phase angle ϕ. The major application of AC–DC–AC converters is in adjustable speed drives. However, recently, these converters have begun to play an increasingly important role in *distributed power generation systems* (DPGSs) and *sustainable AC and DC grids*.

There are several possibilities for an AC–DC–AC converter configuration [1–21]. In this chapter, only the most promising bidirectional topologies are presented and discussed. Recently, bidirectional AC–DC–AC converters are available on the market for different voltage levels [1, 2].

Both parts of the converter (i.e., AC–DC and DC–AC) can be controlled independently. However, in some cases, there is a need for improving the control accuracy and dynamics. Therefore, it is useful to use an additional link between both control algorithms, which operates as an *active power feedforward* (APFF). The APFF gives information about the active power on one side of the AC–DC–AC converter to the other side directly, and consequently, the stability of the DC voltage is improved significantly.

11.1.1 Bidirectional AC–DC–AC Topologies

There are several configurations possible for three-phase to three-phase AC–DC–AC full-bridge converters, which can connect two AC systems. The most popular is a two-level converter, as shown in Figure 11.2(a), which is used mostly in low voltage and low-power or medium-power applications, for example, adjustable speed drives. On the other hand, three-level *diode-clamped converters* (DCCs)

Power Electronics for Renewable Energy Systems, Transportation and Industrial Applications, First Edition.
Edited by Haitham Abu-Rub, Mariusz Malinowski and Kamal Al-Haddad.

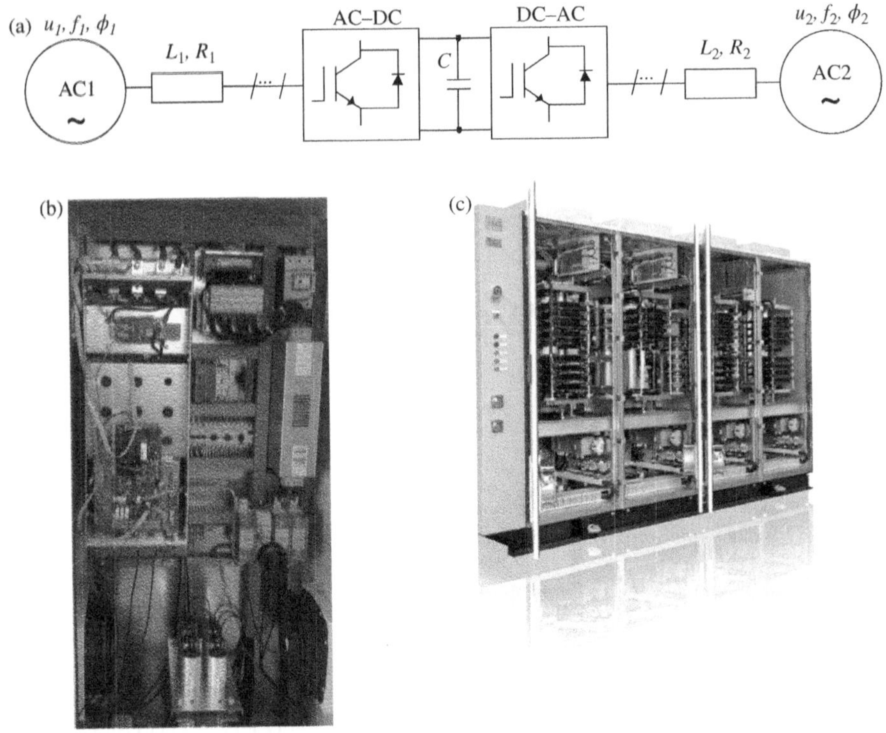

Figure 11.1 AC–DC–AC indirect converter (a) general block scheme (b) photo of commercially available AC–DC–AC converters manufactured by TWERD type: MFC810ACR, (c) photo of commercially available AC–DC–AC converters manufactured by ABB type: PCS6000 [1, 2]

[11–12] and *flying capacitor converters* (FCCs) [4, 6, 7] (Figure 11.2(b and c)) are becoming increasingly popular, but usually in the medium-voltage range for medium- and high-power applications [4], for example, marine propulsion, renewable energy conversion, rolling mills and railway traction. There are several advantages of multilevel converters, such as lower voltage stress of components, low current and voltage, total harmonic distortion factor and reduced volume of input passive filters. The main differences among the mentioned multilevel topologies are as follows [4]:

- DCC is the most popular topology and needs fewer capacitors. However, for higher voltage levels, it requires serially connected clamping diodes, which increases the losses and switching losses. In addition, for higher voltage levels, the DC capacitor voltage balancing cannot be achieved with classical modulations.
- FCC is less popular because it needs initialization of the FC voltage, and higher switching frequency is required (greater than 1.2 kHz, whereas for high-power applications the switching frequency is usually between 500 and 800 Hz) in high-power applications because of the FC limits, that is, capacitance versus volume.

Another group of AC–DC–AC converters are simplified topologies obtained by reducing the number of power electronic switches [8–10]. These attempts were based on the idea of replacing one of the semiconductor legs with a split capacitor bank and connecting a one-phase wire to its middle. In simplified topology, the lower number of switching devices, compared with that of a classical three-phase converter, corresponds to a reduced number of control channels and insulated-gate bipolar transistor

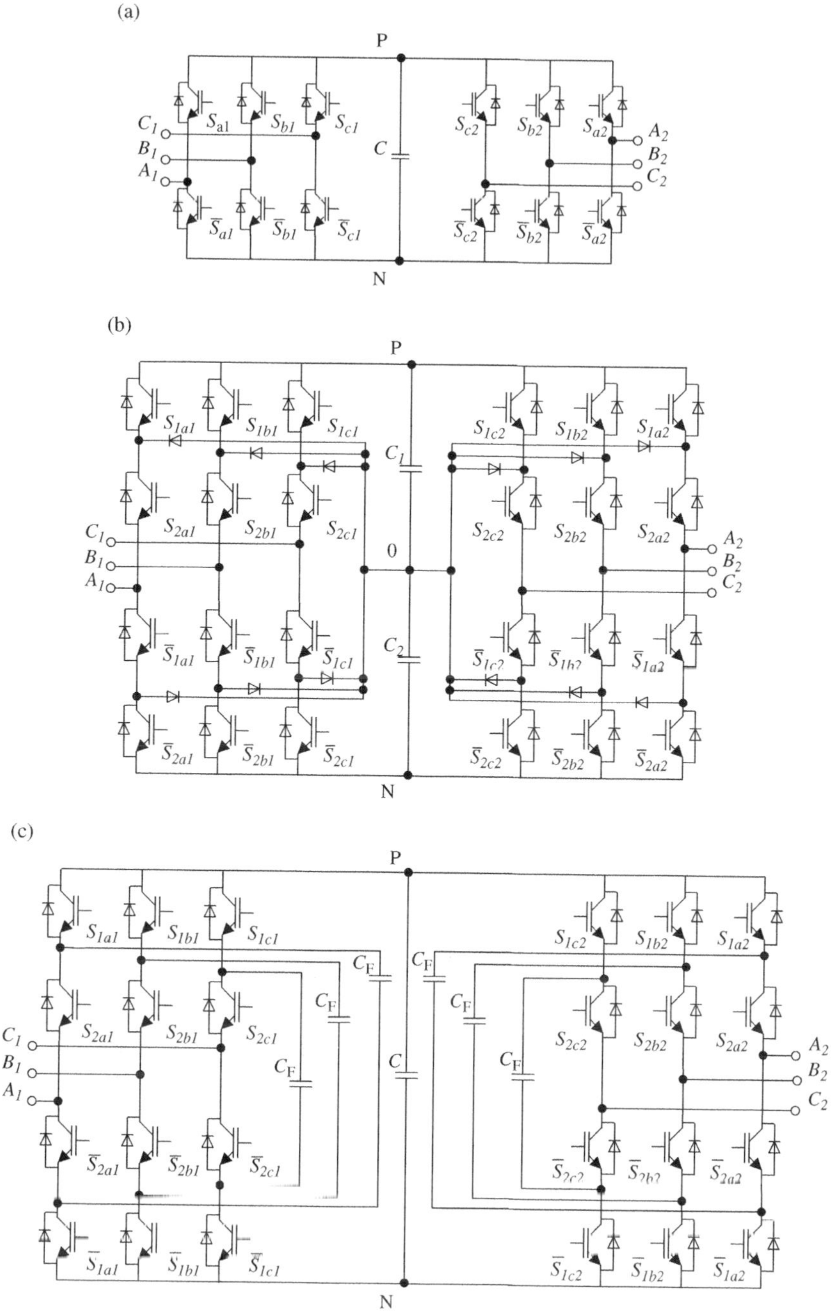

Figure 11.2 Fully controlled three-phase/three-phase, transistor-based AC–DC–AC converter: (a) two-level (2L-3/3), (b) three-level DCC (3L-DCC-3/3) and (c) three-level FCC (3L-FCC-3/3)

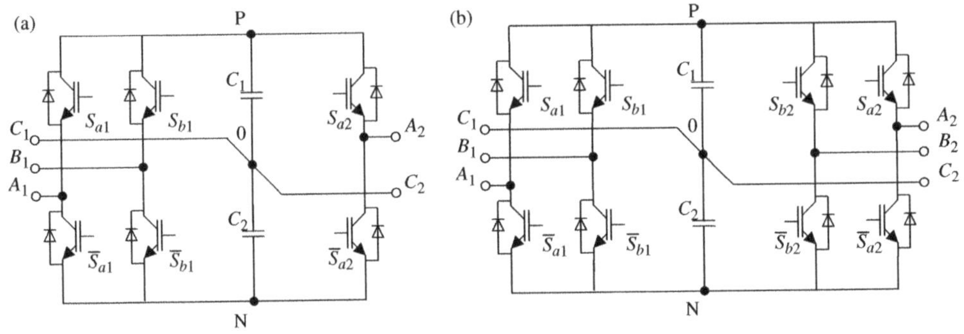

Figure 11.3 Simplified AC–DC–AC converters: (a) two-level three-phase/one-phase (S2L-3/1) and (b) two-level three-phase/three-phase (S2L-3/3)

(IGBT) driver circuits. Thus, connecting a three-phase AC system to a single-phase AC system is possible by using a single-standard three-leg integrated power module as the AC–DC–AC converter, as shown in Figure 11.3(a).

The same concept can be used to simplify an AC–DC–AC converter by connecting one three-phase system to another using only one additional leg, as is shown in Figure 11.3(b). Despite the advantages of these solutions, there is a necessity to develop new modulation techniques and to keep the DC voltage significantly high in order to maintain all nominal phase-to-phase converter voltages, which places higher voltage stress on the converter semiconductor devices.

This problem can be solved by application of a three-level DCC. When this technology was emerging in industry, not a single integrated power module product for three-level devices was available. Today, manufacturers are selling easy-to-use integrated half-bridge power modules with clamping diodes. New compact devices make it easier to improve the topology with a split capacitor in the DC-link. Thus, the improved topology of a simplified AC–DC–AC converter is shown in Figure 11.4, for applications of three-phase to single-phase, as well as for three-phase to three-phase systems. The first topology is dedicated only to low-power applications, whereas the second is devoted to low- and medium-power applications.

The simplified three-level DCC has several advantages compared with that of a simplified two-level topology [13]: reduction of machine torque pulsations (mechanical stress in cases where generator/motor application is decreased), additional zero vectors, and reduction in size of passive filters on the AC side.

11.1.2 Passive Components Design for an AC–DC–AC Converter

This section is devoted to the methods of passive components of an AC–DC–AC converter design. Among them, there are input filters (L or LCL), DC-link capacitors, and flying capacitors (FCs), which have a significant impact on the size, weight and final price of the AC–DC–AC converter.

11.1.3 DC-Link Capacitor Rating

In the literature, many design procedures [17–20] for DC-link capacitors are presented, where the minimum capacitance value is designed to limit the DC-link voltage ripple at a specified level. With the assumption of a balanced three-phase system and of ideal power electronics switches, the DC capacitor current can be expressed as:

$$C_{\mathrm{DC}}\frac{\mathrm{d}U_{\mathrm{DC}}}{\mathrm{d}t} = I_{\mathrm{DC}} - I_{\mathrm{load}} \approx \sum_{k=A}^{C} I_{\mathrm{L}k}S_k - \frac{P_{\mathrm{load}}}{U_{\mathrm{DC}}} \tag{11.1}$$

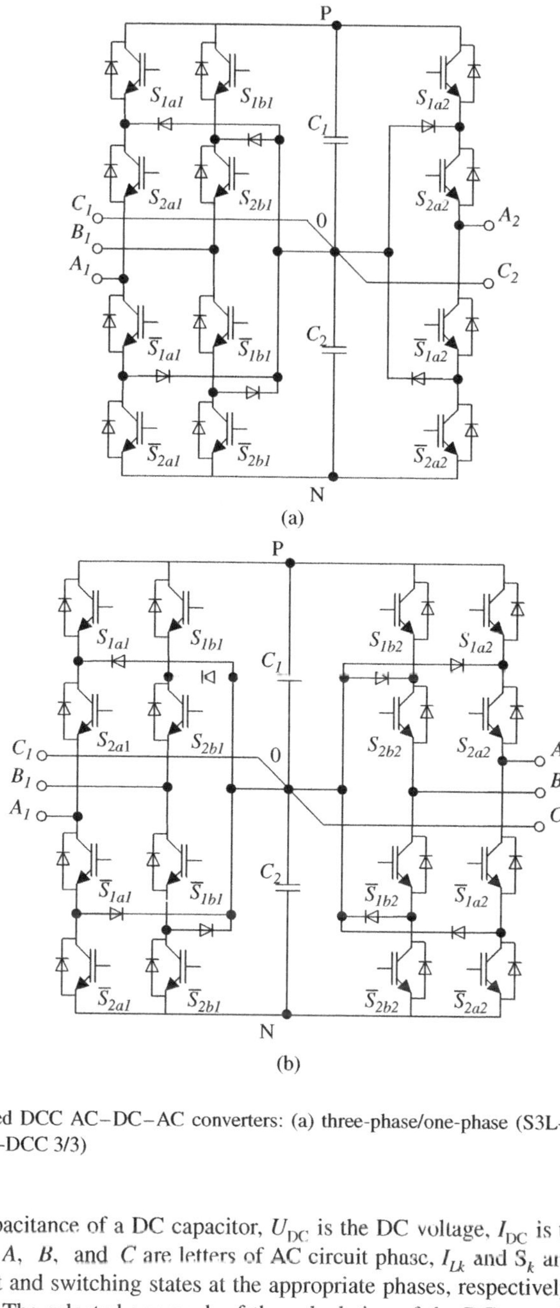

Figure 11.4 Simplified DCC AC–DC–AC converters: (a) three-phase/one-phase (S3L-DCC 3/1) and (b) three-phase/three-phase (S3L-DCC 3/3)

where C_{DC} is the capacitance of a DC capacitor, U_{DC} is the DC voltage, I_{DC} is the DC current, I_{load} is the load DC current, A, B, and C are letters of AC circuit phase, I_{Lk} and S_k are the AC circuit phase instantaneous current and switching states at the appropriate phases, respectively, and P_{load} is the load active power in DC. The selected approach of the calculation of the DC capacitors is focused on the following constraints:

- The voltage ripple, because of the high-frequency components of the modulated DC currents of both AC–DC and DC–AC converters, has to remain within desired limits,

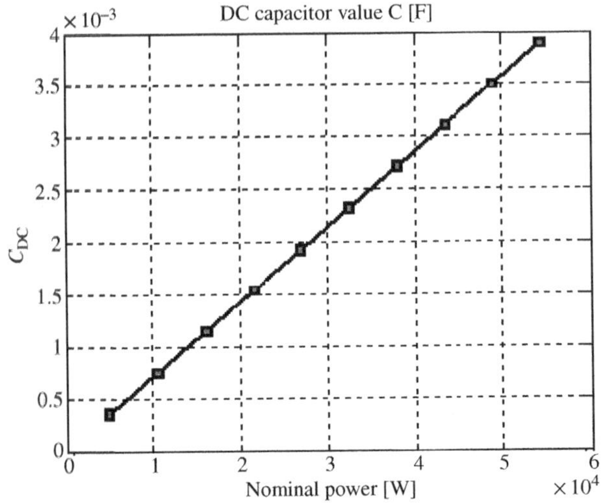

Figure 11.5 Value of the DC capacitor versus rated power up to 55 kW (Equation 11.4)

- The capacitor energy has to sustain the output power demand for a period equal to the time delay of the DC voltage control loop.

The final constraint determines the capacitor value. By assuming that T_{UT} is a time delay of the DC voltage control loop, and that ΔP_{load_max} is a variation of the maximal load power, the energy ΔW_{DC} exchanged by the DC capacitor can be estimated as:

$$\Delta W_{DC} = T_{UT} \Delta P_{load\_max} \tag{11.2}$$

Where T_{UT} is the sum of small time constant. From this equation, the maximal DC voltage variation during maximum power change is proportional to energy change:

$$\Delta U_{DC\_max} = \frac{\Delta W_{DC}}{U_{DC} C_{DC\_min}} \tag{11.3}$$

where ΔU_{DC_max} is the considered maximum DC voltage change within the transient load and C_{DC_min} is the minimum DC capacitance. Taking into account the maximal voltage variation ΔU_{DC_max} and by rearranging Equation 11.3, the minimal capacitance can be calculated as [20]:

$$C_{DC\_min} = \frac{T_{UT} \Delta P_{load\_max}}{U_{DC} \Delta U_{DC\_max}} \tag{11.4}$$

The minimum DC capacitance value in an AC–DC–AC converter with respect to different nominal powers is presented in Figure 11.5. Therefore, it can be assumed that for given nominal power, the DC capacitance depends mainly on the switching pattern of the AC–DC converters and on the quality (accuracy and dynamics) of the applied control method for the AC–DC–AC converter, that is, for a shorter sampling time T_s, the DC capacitance would be lower because the DC voltage regulation accuracy is improved. However, it should be noted that the switching frequency is limited by the switching losses of the power electronics devices applied in the AC–DC–AC converter.

11.1.4 Flying Capacitor Rating

FC capacitance, assuming sinusoidal converter output voltages and currents, can be calculated as [21]:

$$C_{FC} = \frac{I_{RMS}}{\Delta U_{C_{FC}} f_{Sw}} \tag{11.5}$$

where I_{RMS} is the RMS phase current flowing through the FC, $\Delta U_{C_{FC}}$ is the assumed maximum voltage ripple across the FC and f_{Sw} is the frequency of changes of the switching states (charging and discharging) applied to the FCC, which are used for FC balancing. It is important to note that f_{Sw} is not the same as the switching frequency, because the frequency of changes of the switching states applied to FCC depends on the type of modulation and, in the case of FC balancing, is typically a multiple of the switching frequency by a factor 0.5, 1 or 2.

11.1.5 L and LCL Filter Rating

An AC–DC–AC converter is connected to the grid through an AC filter (e.g., L or LCL). The inductance allows current control by the voltage drop on itself. In order to obtain better reduction of current ripples, the LCL T-type filter has been introduced.

For the LCL filter to fulfill the grid codes, or recommended practices for example, IEEE 519-1992 [21, 22] a variety of optimization criteria exist, such as minimum costs, losses, volume, weight, or design. However, it is a complex task because many parameters influence this process, for example, DC-link voltage level, phase current, modulation method, modulation index, switching frequency, fundamental frequency, and resonance frequency. Therefore, the design can be done using two different methods:

- simple trial-and-error method, which has to be supported by simulation to prove that the grid codes are fulfilled (simulation has to take into account all elements of the system),
- complicated iterative, which calculates directly the filter parameters fulfilling the grid code; however, detailed information is required regarding many parameters, for example, modulation method and modulation index [24, 25].

A general algorithm of a simple trial-and-error filter design is presented in Figure 11.6. The starting point for the design procedure is usually assumptions of the converter's rated power, grid, and switching frequencies. Those parameters allow the calculation of base values (base impedance Z_B, capacitance C_B and inductance L_B):

$$Z_B = \frac{U_{LL}^2}{S_N} \tag{11.6}$$

$$C_B = \frac{1}{\omega_g Z_B}; \quad L_B = \frac{Z_B}{\omega_g} \tag{11.7}$$

where U_{LL} is the line-to-line grid voltage and $\omega_g = 2\pi f_g$ is the grid voltage pulsation.

The first step is the design of the filter inductance. This should be designed carefully because a large inductance value may, by itself, attenuate higher frequencies and limit current ripples. On the other hand, it brings many disadvantages, such as greater inductor size, a decrease in the converter's dynamics and a smaller range of operation. The operation range is limited through the maximum reachable voltage drop across the inductance. It is dependent indirectly on the DC-link voltage. Thus, in order to achieve a large current through the inductance, a high DC-link voltage level or low input inductance is needed.

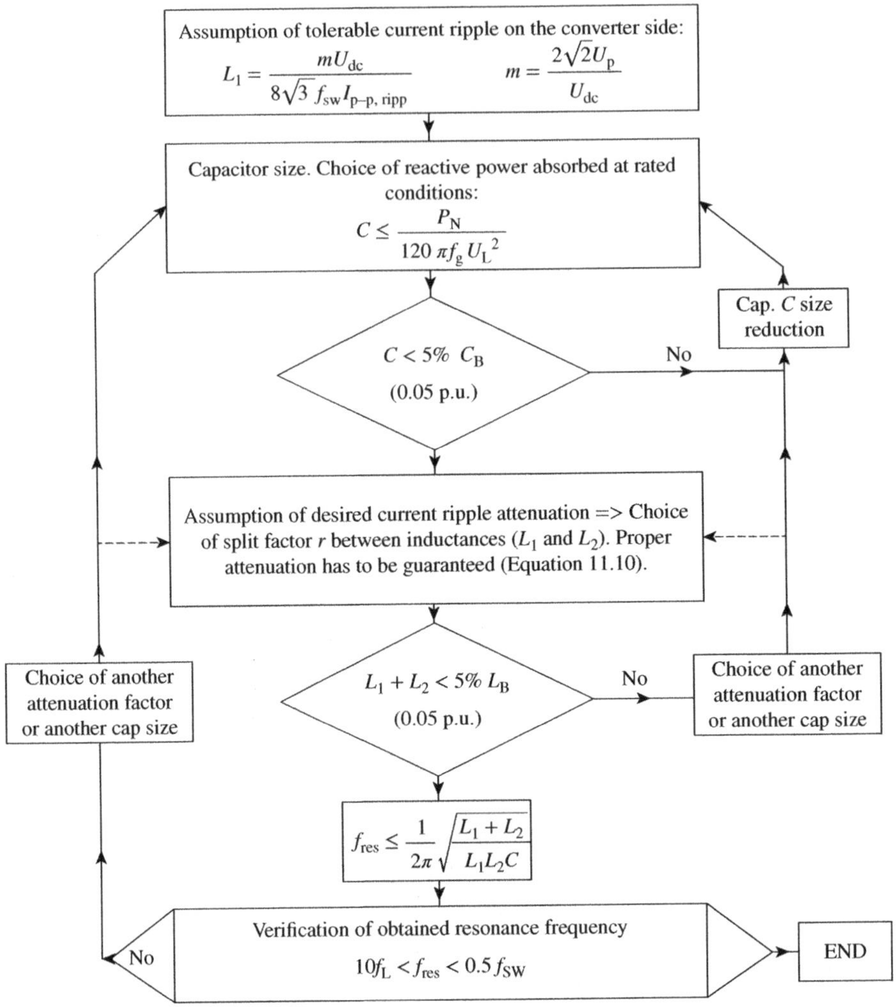

Figure 11.6 General design algorithm of low-pass LCL filter where L_1 is the converter-side inductance, L_2 is the grid-side inductance LCL filter, U_p is the converter RMS voltage, and $I_{p-p,ripp}$ is the peak-to-peak ripple current

The maximum total inductance of the input LCL filter may not be larger than [14]:

$$L_1 + L_2 < \frac{\sqrt{\frac{U_{DC}^2}{3} - U_m^2}}{\omega_g I_{Lm}} \tag{11.8}$$

where U_m is the grid maximum voltage value and I_{Lm} is the grid current maximum value.

In addition, the total value of the LCL filter inductance should be lower than 5% of the base impedance value, in order to limit the voltage drop during normal operation and to achieve excellent dynamics and reasonable cost [14, 26]. Based on the IEEE 519-1992 limitations and the assumption that peak ripples must not exceed 10% of the grid current, the converter-side inductance L_1 can be calculated, as shown in Figure 11.6 [14, 23]. Typically, the converter-side inductance L_1 is larger than the grid-side inductance L_2 in order to attenuate most of the current ripple. In this case, a split factor r, given by Liserre *et al.* [26]:

$$r = \frac{L_2}{L_1} \tag{11.9}$$

is set by the desirable ripple attenuation described as:

$$\frac{I_g(h_{sw})}{I_p(h_{sw})} = \frac{1}{|1 + r(1 - 4\pi^2)L_1 C f_{sw}^2|} \tag{11.10}$$

where $I_g(h_{sw})$ is the grid higher harmonics current and $I_p(h_{sw})$ is the converter higher harmonics current.

In order to limit the harmonic currents absorbed by the filter, the capacitor is limited to 5% of the reactive power absorbed at rated conditions (base capacitance) [26]. Therefore, the final capacitance of the LCL filter (Y-connected) cannot be larger than as presented in Figure 11.6. In the case of just an L filter, the value of the inductance can equal $L_1 + L_2 = L$.

Another constraint in the design process of the LCL filter is an obligation to maintain the resonance frequency f_{res} between 10 times the line frequency f_L and one-half of the switching frequency f_{sw} [14, 26]. These limits guarantee that there will be no resonance problems in both the lower and higher parts of the harmonic spectrum. The actual resonance frequency of the LCL filter may be calculated as in Figure 11.6.

11.1.6 Comparison

Table 11.1 presents a comparison of selected classical AC–DC–AC converter topologies: two-level three-phase/three-phase converter (2L-3/3); three-level three-phase/three-phase DCC converter (3L-DCC-3/3); three-level three-phase/three-phase FCC converter (3L-FCC-3/3), and simplified

Table 11.1 Comparison of selected AC–DC–AC converter topologies

	I	II	III	IV	V	VI	VII
2L-3/3	12	0/1	$8 \leftrightarrow 8$	1	1	$\frac{2}{3} \leftrightarrow \frac{2}{3}$	$3 \times L \leftrightarrow 3 \times L$
S2L-3/3	8	0/2	$4 \leftrightarrow 4$	$\sqrt{3}$	$\sqrt{3}$	$\frac{2\sqrt{3}}{3}^{a} \leftrightarrow \frac{2\sqrt{3}}{3}^{a}$	$2 \times \sqrt{3}L^{a} \leftrightarrow 2 \times \sqrt{3}L^{a}$
S2L-3/1	6	0/2	$4 \leftrightarrow 2$	$\sqrt{3}$	$\sqrt{3}$	$\frac{2\sqrt{3}}{3}^{a} \leftrightarrow \frac{\sqrt{3}}{2}$	$2 \times \sqrt{3}L^{a} \leftrightarrow \sqrt{3}L$
3L-DCC-3/3	24	12/2	$27 \leftrightarrow 27$	1	$\frac{1}{2}$	$\frac{1}{3} \leftrightarrow \frac{1}{3}$	$3 \times \frac{L}{3} \leftrightarrow 3 \times \frac{L}{2}$
S3L-DCC 3/3	16	8/2	$9 \leftrightarrow 9$	$\sqrt{3}$	$\frac{\sqrt{3}}{2}$	$\frac{\sqrt{3}}{3}^{a} \leftrightarrow \frac{\sqrt{3}}{3}$	$2 \times \frac{\sqrt{3}}{2}L^{a} \leftrightarrow 2 \times \frac{\sqrt{3}}{2}L^{a}$
S3L DCC 3/1	12	6/2	$9 \leftrightarrow 3$	$\sqrt{3}$	$\frac{\sqrt{3}}{2}$	$\frac{\sqrt{3}}{3}^{a} \leftrightarrow \frac{\sqrt{3}}{4}$	$2 \times \frac{\sqrt{3}}{2}L^{a} \leftrightarrow \frac{\sqrt{3}}{2}L$
3L-FCC-3/3	24	0/7[b]	$64 \leftrightarrow 64$	1	$\frac{1}{2}$	$\frac{1}{3} \leftrightarrow \frac{1}{3}$	$3 \times \frac{L}{2} \leftrightarrow 3 \times \frac{L}{2}$

[a]Simplified converter only between transistors legs.

[b]Including one standard electrolytic capacitor in DC link and six floating capacitors designed for rapid discharging.

I, Total number of IGBT transistors; II, Number of independent diodes/number of capacitors; III, Total number of converter states; IV, DC voltage level in p.u. in comparison to 2L-3/3; V, Blocking voltage for transistors in p.u. in comparison to 2L-3/3; VI, Minimum voltage level step change; VII, total L filter inductance for all phases.

topologies with reduced number of transistors: simplified AC–DC–AC two-level converter three-phase/one-phase (S2L-3/1), and three-phase/three-phase (S2L-3/3); simplified DCC AC–DC–AC converter three-phase/one-phase (S3L-DCC 3/1), and three-phase/three-phase (S3L-DCC 3/3).

11.2 Pulse-Width Modulation for AC–DC–AC Topologies

Pulse-width modulation (PWM) for AC–DC–AC converters can be divided into two groups, depending on the single-phase or three-phase topology of the converter. The *space vector modulation* (SVM) techniques for the different topologies of classical single-phase H-bridge AC–DC VSCs have been presented in Chapter 23. In this section, only the SVM techniques for the following are presented:

- classical three-phase two-level converter
- classical three-phase three-level converters: DCC and FCC
- simplified single-phase and three-phase two-level converter
- simplified single-phase and three-phase three-level DCC

Modulation techniques for VSCs are responsible for the generation of average output voltage (represented by different widths of short voltage pulses) equal to reference voltage with proper balancing of the additional DC voltage sources for multilevel converters. Depending on the VSC topologies and switching frequency, numerous PWM techniques can be listed [27, 28]. Among them, two types of PWM are used most often: *carrier-based* PWM (CB-PWM) and SVM. However, recent advances and the constant development of digital signal processing systems (allowing high-precision digital implementation of complex control algorithms) has meant that SVM has gained a superior position in research and industry. The digital implementation of SVM is characterized by its simplicity. Furthermore, classical SVM with symmetrical switching placement is equivalent to CB-PWM with an additional 1/4 peak amplitude triangular *zero sequence signal* (ZSS) of the third harmonic frequency [29].

SVM is based on a single- or three-phase circuit representation in a stationary rectangular coordinate system using *space vector* (SV) theory. The reference SV $\underline{U_{\text{ref}}} = U_\alpha + jU_\beta$ is described by its length and its angle:

$$U_{\text{ref}} = \sqrt{u_{\alpha,\text{ref}}^2 + u_{\beta,\text{ref}}^2}, \quad \varphi_{\text{m}} = \arctan\frac{u_{\alpha,\text{ref}}}{u_{\beta,\text{ref}}} \tag{11.11}$$

where $u_{\alpha,\text{ref}}$ and $u_{\beta,\text{ref}}$ are the α and β components of $\underline{U_{\text{ref}}}$ in a stationary rectangular coordinate system. To calculate the duration of the VSC switching states, a value of the modulation index is indispensable, which is proportional to the SV length with respect to the DC-link voltage. Usually, M is defined as:

$$M = \frac{\pi U_{\text{ref}}}{2U_{\text{DC}}}(n-1) \tag{11.12}$$

where n denotes the number of VSC output phase voltage levels. The allowable length of vector $\underline{U_{\text{ref}}}$ for each φ_{m} angle in the linear operation range is given by $U_{\text{ref}} = \frac{U_{\text{DC}}}{\sqrt{3}}$. Therefore, the linear modulation range of an n-level VSC is limited to:

$$M = 0.907(n-1) \tag{11.13}$$

which is the maximal converter linear range of operation [30].

11.2.1 *Space Vector Modulation for Classical Three-Phase Two-Level Converter*

Table 11.2 presents all possible switching states for a single leg of a two-level VSC (see Figure 11.2(a), generating output pole voltage u_{xp} between the phase terminal u_{x} and the DC-link "N" terminal $U_{\text{DC,N}}$, where x is the leg indicator.

Table 11.2 Switching states for single leg of the two-level VSC

States	S_x	$u_{xp} = u_x - U_{DC,N}$
0	OFF	0
1	ON	U_{DC}

Because the three-phase system is assumed symmetric, the Clarke transformation from natural *abc* into a stationary $\alpha\beta$ coordinate system can be used. Figure 11.7(a) shows a graphical representation of the SV $\alpha\beta$ voltage plane with possible output voltage vectors of the three-phase two-level converter. All voltage vectors are described by three numbers, which correspond to the switching states in leg *a*, leg *b*, and leg *c*, respectively. A three-phase two-level converter provides eight possible switching states, comprising six active (100, 110, 010, 011, 001, and 101) and two zero switching states (000 and 111). The active vectors divide the plane into six sectors, where reference vector U_{ref} is obtained by switching on (for proper time) two adjacent vectors. It can be seen that vector U_{ref} (see Figure 11.7(b)) can be implemented by the different ON/OFF switch sequences of V_1 and V_2, and that zero vectors V_0 and V_7 decrease the modulation index.

Reference vector U_{ref}, used to solve equations that describe times T_1, T_2, T_0, and T_7, is sampled with fixed sampling frequency $f_S = 1/T_S$. Digital implementation is described with the help of a simple trigonometric relationship for the first sector 11.14:

$$T_1 = \frac{2\sqrt{3}}{\pi} M T_S \sin\left(\frac{\pi}{3} - \varphi_m\right), \quad T_2 = \frac{2\sqrt{3}}{\pi} M T_S \sin\varphi_m \tag{11.14}$$

and is recalculated for the following sectors (from 2 to 6). After the T_1 and T_2 calculation, the residual sampling time is reserved for zero vectors V_0 and V_7 with condition $T_1 + T_2 < T_S$. Equation (11.14) is identical for all variants of SVM. The only difference is in the different placement of zero vectors $V_0(0\ 0\ 0)$ and $V_7(1\ 1\ 1)$. This gives different equations defining T_0 and T_7 for each method, but the total duration time of the zero vectors must fulfill the following conditions:

$$T_S - T_1 - T_2 = T_0 + T_7 \tag{11.15}$$

The most popular SVM method is modulation with symmetrical zero states' space vector PWM (SVPWM) where:

$$T_0 = T_7 = \frac{T_S - T_1 - T_2}{2} \tag{11.16}$$

Figure 11.8 shows gate pulses for and the correlation between time T_{ON}, T_{OFF}, and the duration of vectors V_1, V_2, V_0, and V_7. For the first sector, commutation delay can be computed as:

$$\begin{aligned} T_{a,ON} &= \frac{T_0}{2}, \quad T_{a,OFF} = \frac{T_0}{2} + T_1 + T_2 \\ T_{b,ON} &= \frac{T_0}{2} + T_1, \quad T_{b,OFF} = \frac{T_0}{2} + T_2 \\ T_{c,ON} &= \frac{T_0}{2} + T_1 + T_2, \quad T_{c,OFF} = \frac{T_0}{2} \end{aligned} \tag{11.17}$$

For classical SVM, times T_1, T_2, and T_0 are computed for one sector only. Commutation delay for the other sectors can be calculated with the help of a matrix:

$$\begin{bmatrix} T_{a,ON} \\ T_{b,ON} \\ T_{c,ON} \end{bmatrix} = \left[\begin{array}{c|c|c|c|c|c} \text{Sector 1} & \text{Sector 2} & \text{Sector 3} & \text{Sector 4} & \text{Sector 5} & \text{Sector 6} \\ 1\ 0\ 0 & 1\ 0\ 1 & 1\ 1\ 1 & 1\ 1\ 1 & 1\ 1\ 0 & 1\ 0\ 0 \\ 1\ 1\ 0 & 1\ 1\ 0 & 1\ 0\ 0 & 1\ 0\ 1 & 1\ 1\ 1 & 1\ 1\ 1 \\ 1\ 1\ 1 & 1\ 1\ 1 & 1\ 1\ 0 & 1\ 0\ 0 & 1\ 0\ 0 & 1\ 0\ 1 \end{array}\right] \begin{bmatrix} 0.5T_0 \\ T_1 \\ T_2 \end{bmatrix} \tag{11.18}$$

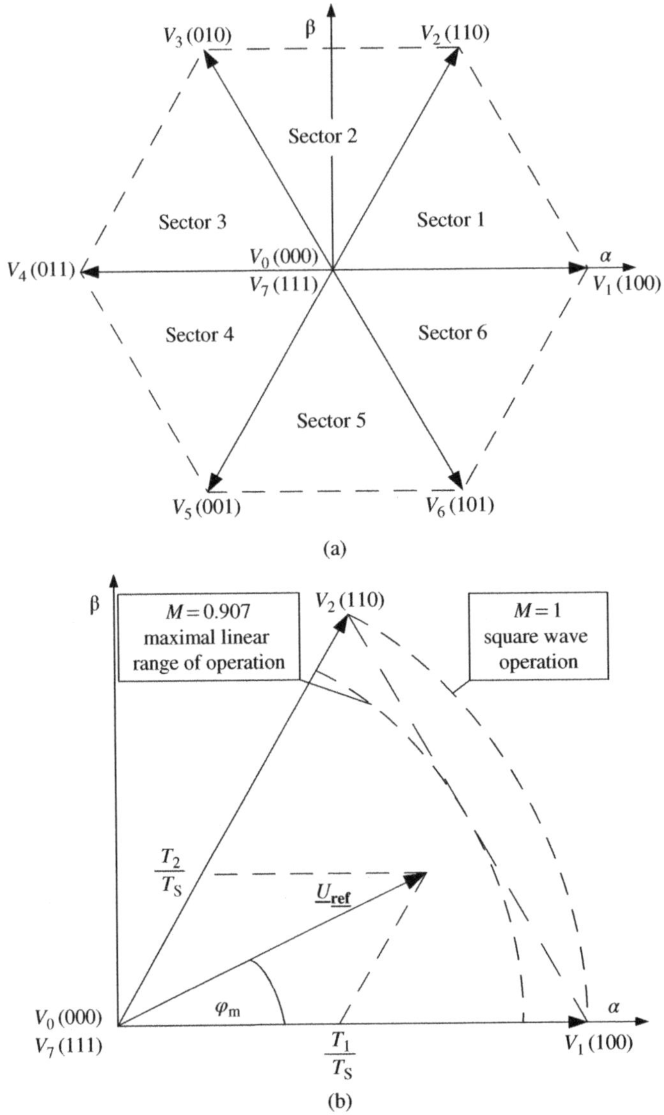

Figure 11.7 Space vector representation of three-phase two-level converter: (a) the $\alpha\beta$ voltage plane and (b) sector 1 with representation of $\underline{U}_{ref}$

$$\begin{bmatrix} T_{a,\text{OFF}} \\ T_{b,\text{OFF}} \\ T_{c,\text{OFF}} \end{bmatrix} = \begin{bmatrix} \text{Sector 1} & \text{Sector 2} & \text{Sector 3} & \text{Sector 4} & \text{Sector 5} & \text{Sector 6} \\ 1\ 1\ 1 & 1\ 1\ 0 & 1\ 0\ 0 & 1\ 0\ 0 & 1\ 0\ 1 & 1\ 1\ 1 \\ 1\ 0\ 1 & 1\ 1\ 1 & 1\ 1\ 1 & 1\ 1\ 0 & 1\ 0\ 0 & 1\ 0\ 0 \\ 1\ 0\ 0 & 1\ 0\ 0 & 1\ 0\ 1 & 1\ 1\ 1 & 1\ 1\ 1 & 1\ 1\ 0 \end{bmatrix} \begin{bmatrix} 0.5T_0 \\ T_1 \\ T_2 \end{bmatrix} \tag{11.19}$$

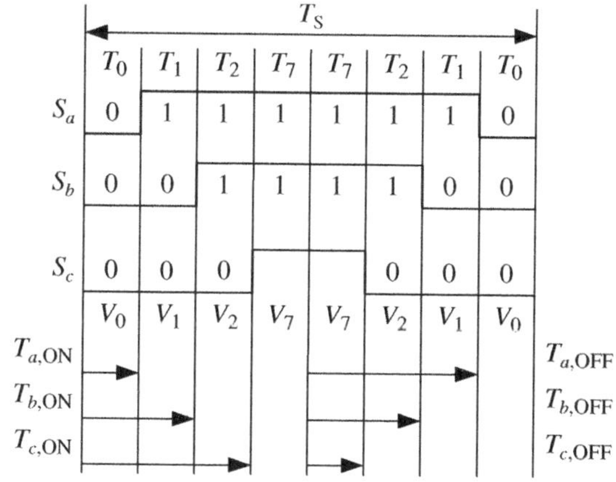

Figure 11.8 Vector placement in sampling time for three-phase SVM (SVPWM, $T_0 = T_7$) in sector 1

Table 11.3 Switching states for single leg of the three-level DCC

States	S_{1x}	S_{2x}	$u_{xp} - u_x - U_{DC,N}$
0	OFF	OFF	0
1	OFF	ON	$\frac{U_{DC}}{2}$
2	ON	ON	U_{DC}

Table 11.4 Switching states for single leg of the three-level FCC

States		S_{1x}	S_{2x}	$u_{xp} = u_x - U_{DC,N}$
0		OFF	OFF	0
1	A	ON	OFF	$U_{DC} - U_{FCx}$
	B	OFF	ON	U_{FCx}
2		ON	ON	U_{DC}

11.2.2 Space Vector Modulation for Classical Three-Phase Three-Level Converter

Tables 11.3 and 11.4 present all possible switching states for the single leg of a three-level DCC and FCC (see Figure 11.2(b and c)), respectively, generating output pole voltage u_{xp} between the phase terminal u_x and the DC-link "N" terminal $U_{DC,N}$, where x is the leg indicator.

Figure 11.9 shows a graphical representation of the SV $\alpha\beta$ voltage plane with possible output voltage vectors of the three phase three level converter, which is constructed from the main switching states: 0, 1, and 2.

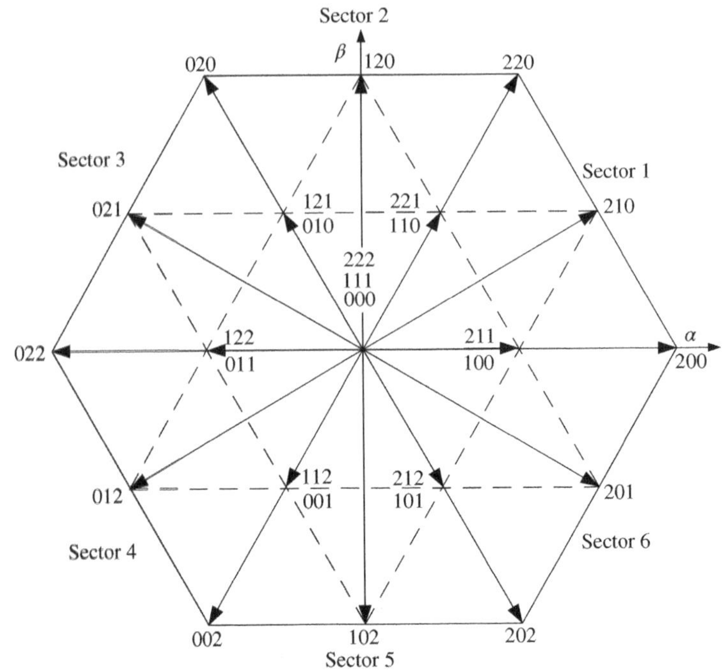

Figure 11.9 The $\alpha\beta$ voltage plane for three-phase three-level converter

For the three-phase three-level VSC, the 27 voltage vectors can be specified as follows: three zero vectors (000, 111, and 222); 12 internal small-amplitude vectors (100, 211, 110, 221, 010, 121, 011, 122, 001, 112, 101, and 212); six middle medium-amplitude vectors (210, 120, 021, 012, 102, and 201); six external large-amplitude vectors (200, 220, 020, 022, 002, and 202).

The external vectors divide the vector plane into six sectors (see Figure 11.9), which is similar to the two-level VSC. However, for the three-level VSC, each sector is divided into four triangular regions. Figure 11.10(a and b) shows that the first sector is divided into four regions with all possible switching states for the DCC and the FCC topologies, respectively. As FCC switching state 1 can be divided into two redundant switching states: A and B (see Table 11.4, highlighted in Figure 11.10(b)), for this topology, the number of possible switching states increases to 64.

Conventional SVM for multilevel VSC is based on the assumption that for the generation of reference vector U_{ref} all possible nearest vectors, including their redundant states, are used (V_1, V_2, and V_4 in the first; V_2, V_4, and V_5 in the second; V_2, V_3, and V_5 in the third; and finally, V_0, V_4 , and V_5 in the fourth region) with symmetrical placement of the zero and internal vectors.

The calculation of the region number of the location of the reference vector U_{ref} and switching times is based on two additional factors called the *small modulation indices* (m_1, m_2). Index m_1 and m_2 are the projection of the reference vector on the sector sides, limited by the external vectors (see Figure 11.10). According to the trigonometric dependence, the small modulation indices are calculated as follows:

$$m_1 = M\left(\cos\varphi - \frac{\sin\varphi_m}{\sqrt{3}}\right), \quad m_2 = 2M\frac{\sin\varphi_m}{\sqrt{3}} \tag{11.20}$$

In each sector, calculations are carried out to achieve vector switching times that are the same and where the difference is only in the power switch selection for the gating signal. Thus, the reference vector

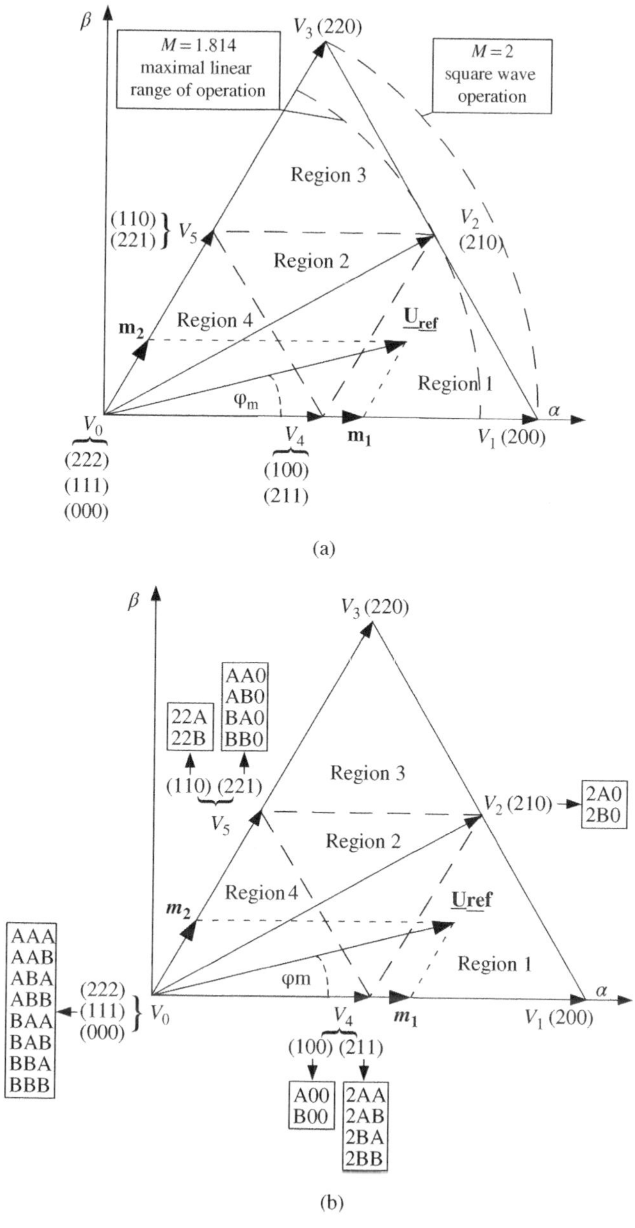

Figure 11.10 First sector with all possible switching states for (a) DCC and (b) FCC topologies

Table 11.5 Calculation of region number and switching times

Conditions	Region	Switching times
$m_1 > 1$	First	$T_1 = (m_1 - 1)T_S, T_2 = m_2 T_S, T_4 = (2 - m_1 - m_2)T_S, T_0 = T_3 = T_5 = 0$
$m_1 \leq 1, m_2 \leq 1, m_1 + m_2 > 1$	Second	$T_2 = (m_1 + m_2 - 1)T_S, T_3 = (m_2 - 1)T_S, T_5 = (1 - m_1)T_S, T_0 = T_1 = T_3 = 0$
$m_2 > 1$	Third	$T_2 = m_1 T_S, T_3 = (m_2 - 1)T_S, T_5 = (2 - m_1 - m_2)T_S, T_0 = T_1 = T_4 = 0$
$m_1 \leq 1, m_2 \leq 1, m_1 + m_2 \leq 1$	Fourth	$T_4 = m_1 T_S, T_5 = m_2 T_S, T_0 = (1 - m_1 - m_2)T_S, T_1 = T_2 = T_3 = 0$

is normalized to the first sector and after evaluation of the vector switching times, a proper transistor switching sequence is created for the reference position. Table 11.5 presents the calculation of the region number and switching times with respect to m_1 and m_2.

11.3 DC-Link Capacitors Voltage Balancing in Diode-Clamped Converter

Proper operation of the DCC requires that the voltage on each DC-link capacitor is stabilized on $\frac{1}{2}U_{DC}$. Unbalance in the DC-link gives inequality of redundant-internal vectors and distortions in middle vectors. In such cases, the generated voltage is different from the reference. In the DCC, the balancing of the DC-link capacitors depends on the switching states in all phases. When the phase is clamped to the neutral point of the DC-link (e.g., switching state 1 is chosen), it introduces additional current flowing from or to the neutral point. Consequently, inequalities occur in the charging and discharging of the DC-link capacitors. Other sources of imbalance in the capacitor voltages are asymmetries in the system:

- propagation of nonlinear IGBT gate signals (e.g., different IGBT ON/OFF times),
- equivalent series and parallel resistance of DC-link capacitors,
- differences in electrical parameters, such as each leg semiconductor devices and connection points.

It can be observed that selection of redundant-internal small-amplitude vectors always causes clamping of one or two phases to the neutral point of the DC-link. For example, when internal small-amplitude vector V_4 (redundant switching states 1 0 0 and 2 1 1) is chosen, this means that phase a is clamped to the neutral point of the DC-link for $T_4/2$, and for the same time, phases a, b and c are clamped to the neutral point of the DC-link. Therefore, an additional controller (proportional or proportional-integral (PI)) can be introduced, which depending on the imbalance in the capacitors voltages will change the time division between the redundant-internal small-amplitude vectors. In this simple way, the balancing of the capacitor voltages will be ensured without increasing losses.

For proper balancing of DC-link voltages based on the additional controller, the information on the difference between the capacitor voltages is insufficient [31, 32]. The second factor needed is the energy flow direction. The energy flow direction decides whether a selected redundant-internal small-amplitude vector will charge or discharge one of the DC-link capacitors. Therefore, the sign of the energy flow direction k should be used to determine the sign of the additional controller input error between the DC-link capacitor voltages. If we assume that internal small-amplitude vector V_4 is used to balance the DC-link voltages with a proportional controller, the ratio of T_4 charging time division of C_{DC1} (T_{4-DC1}) and C_{DC2} (T_{4-DC2}) is inversely proportional to the ratio of U_{DC1} and U_{DC2} voltages with the sign of the energy flow direction k:

$$T_{4-DC1} = \frac{U_{DC} - \frac{1+k}{2}U_{DC1} - \frac{1-k}{2}U_{DC2}}{U_{DC}}T_S, T_{4-DC2} = \frac{U_{DC} - \frac{1+k}{2}U_{DC2} - \frac{1-k}{2}U_{DC1}}{U_{DC}}T_S \quad (11.21)$$

where $T_4 = T_{4-DC1} + T_{4-DC2}$. Because of the sampling period, both of the DC-link capacitors during the calculated time T_4 will be charged or discharged. In the case of a PI controller in the aforementioned example, the ratio of T_4 charging time division will be the controller output signal.

Determination of the energy flow direction k in the case of *machine-side* VSC (MC) can be done by the determination of one of the following quantities [33]: instantaneous active power p_{MC}, electromagnetic torque m_e, torque angle σ, or current vector to voltage vector angle θ_{MC}. Choosing the appropriate quantity to use in the modulation algorithm depends on the purpose of the converter and the availability in the control. The active power consumed/produced by the MC can be estimated simply from the reference converter voltages U_{xc} and U_{yc} (reconstructed from switching states) and the actual stator currents I_{Sx}, I_{Sy}:

$$p_{MC} = \frac{3}{2}(I_{Sx}U_{xc} + I_{Sy}U_{yc}) \tag{11.22}$$

The sign of the power will define the operating mode of the converter: inverter or rectifier, that is, whether the energy is flowing from the capacitors to the motor, or vice versa. A similar procedure of calculation is used for electromagnetic torque m_e, which is the result of the vector product of stator flux Ψ_S and the stator current y component I_{Sy}:

$$m_e = p_b \frac{m_s}{2} \Psi_S I_{Sy} \tag{11.23}$$

where p_b is the number of pole pairs and m_e is the number of phases. In this case, the sign of the torque multiplied by mechanical speed Ω_m shows the direction of energy flow. In the case of the torque angle σ, no additional calculation is needed; only the sign of the σ is important. Therefore, current transformation to the dq stator flux rotating frame and the sign of I_d current determines the direction of energy flow. Table 11.6 presents the final determination of the direction of energy flow in MC.

The determination of energy flow direction k in *grid-side VSC* (GC) is similar and can be done by determination of one of the following quantities [33]: instantaneous active power p_{GC}, instantaneous active current I_d, or current vector to voltage vector angle θ_{GC}. The active power consumed/produced by the grid-side converter can be estimated simply from the reference converter voltages $U_{\alpha c}$ and $U_{\beta c}$ and the actual line currents:

$$p_{GC} = \frac{3}{2}(I_\alpha U_{\alpha c} + I_\beta U_{\beta c}) \tag{11.24}$$

Table 11.7 presents the final determination of the direction of energy flow in the line-side converter.

Table 11.6 Determination of direction of energy flow in MC

Parameter	Inverter (motoring) mode ($k = 1$)	Rectifier (regenerating) mode ($k = -1$)
p_{MC}	$p_{MC} > 0$	$p_{MC} < 0$
m_e, Ω_m	$m_e\Omega_m > 0$	$m_e\Omega_m < 0$
σ (I_d)	$\sigma > 0$ ($I_d > 0$)	$\sigma < 0$ ($I_d < 0$)
θ_{MC}	$\theta_{MC} < \frac{\pi}{2}$	$\theta_{MC} > \frac{\pi}{2}$

Table 11.7 Determination of direction of energy flow in GC

Parameter	Rectifier mode ($k = 1$)	Inverter mode ($k = -1$)
p_{GC}	$p_{GC} > 0$	$p_{GC} < 0$
I_d	$I_d > 0$	$I_d < 0$
θ_{GC}	$\theta_{GC} < \frac{\pi}{2}$	$\theta_{GC} > \frac{\pi}{2}$

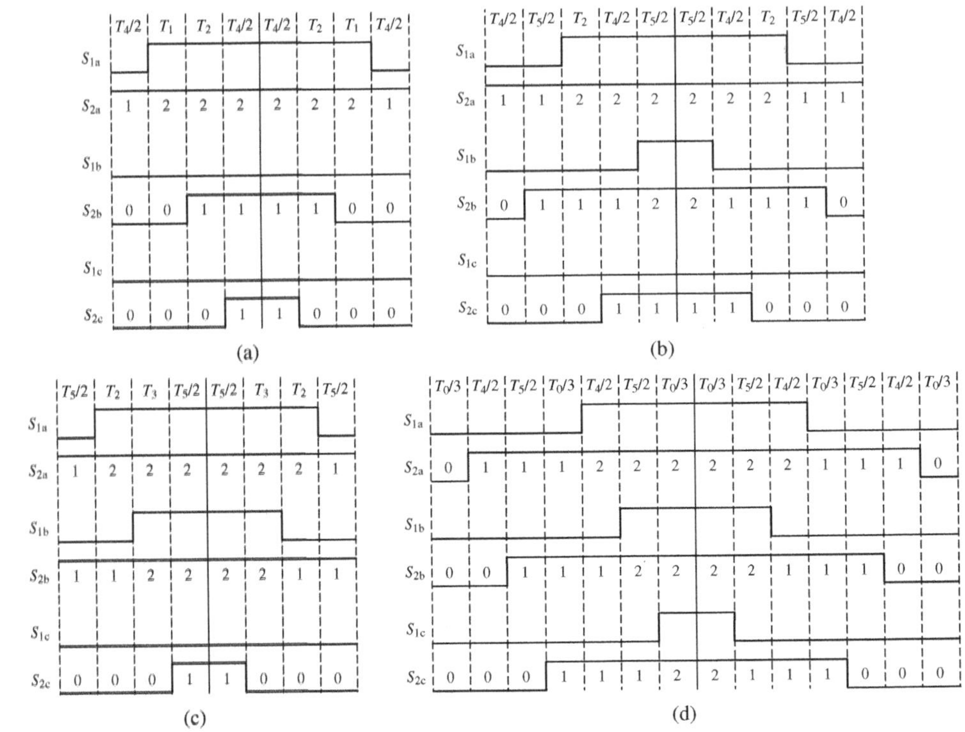

Figure 11.11 Switching sequence for DCC reference vector lying in (a) first, (b) second, (c) third, and (d) fourth regions in sector 1

Figure 11.11 presents examples of switching sequences for the DCC in sector 1, assuming that the ratio of the time division of the internal small-amplitude vector is $1/2$.

11.3.1.1 Flying Capacitors Voltage Balancing in Flying Capacitor Converter

Proper operation of the FCC requires that the FC voltage U_{FCx} should be half that of the DC-link voltage. With this condition, switching state 1 can be divided into two redundant states: 1A and 1B (highlighted in Table 11.4, Figure 11.10(b)), which generate the same output voltage $U_{\mathrm{FCx}} = U_{\mathrm{DC}}/2$. As the output voltage does not depend on the type of selected state, either 1A or 1B can be used for independent control of U_{FCx}. However, to decrease the number of commutations, only one state for each phase is chosen in the sampling period. For proper FC voltage balancing, only information about the sign of the phase current flowing through the FC is needed. Table 11.8 shows the selection between redundant states 1A or 1B based on the sign of phase current i_x.

Figures 11.12 and 11.13 present the switching sequence in all regions in sector 1 for different switching states 1A or 1B for all phases, respectively. When the selected redundant state 1A or 1B is changed, the next modulation period can contain additional switching between sampling periods: two in the second and four in the first, third, and fourth regions. These additional switching can damage the converter; all switches are changing their state, which can generate overvoltage. To eliminate this phenomenon and to provide better switching distribution between particular switches, a modification of the switching pattern should be introduced [6] (see Figure 11.14).

Table 11.8 Selection of redundant switching state for FC voltage balancing

Conditions	$i_x > 0$	$i_x < 0$
$U_{FCx} < \frac{U_{DC}}{2}$	1B	1A
$U_{FCx} > \frac{U_{DC}}{2}$	1A	1B

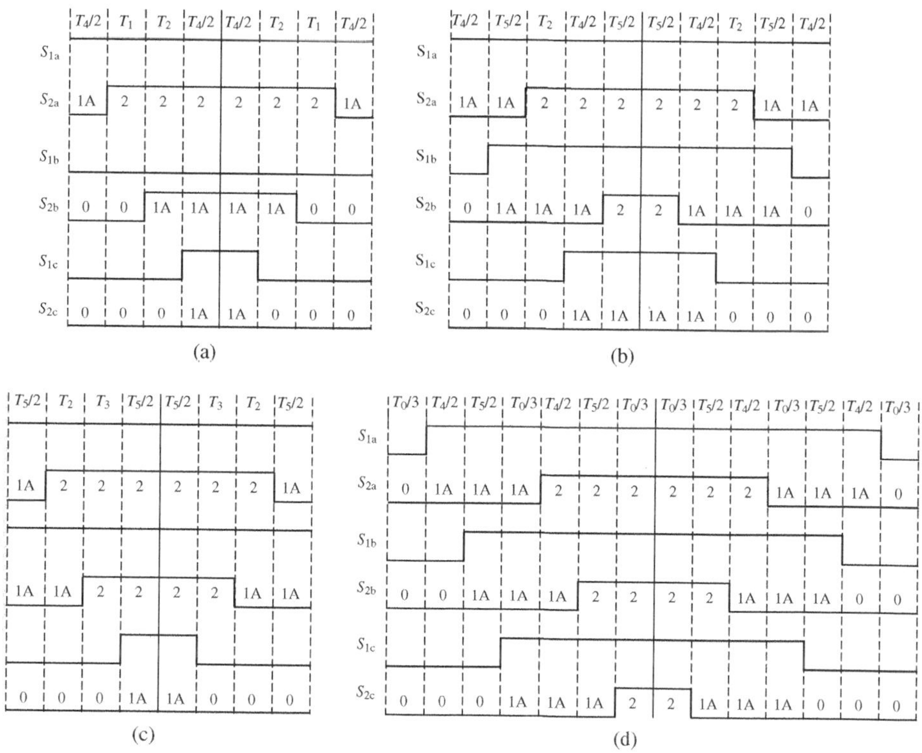

Figure 11.12 Switching sequence for FCC reference vector lying in (a) first, (b) second, (c) third, and (d) fourth regions in sector 1 for 1A state selection in each phase

11.3.2 Pulse-Width Modulation for Simplified AC–DC–AC Topologies

Modulation for simplified AC–DC–AC converters is also realized separately for the DC–AC and AC–DC stages. It can be considered in this case as a single-phase modulation for the DC–AC part, as shown in Figures 11.3(a) and 11.4(a), or as three-phase modulation for the DC–AC or AC–DC parts, as shown in Figures 11.3(b) and 11.4(b).

Single-phase modulation for two-level DC–AC converters is simple and is described by Equation (11). Slightly more complicated is the modulation for a single-phase three-level DC–AC converter [13] shown in Figure 11.4(b), because an additional zero state is included (Figure 11.15).

$$T_1 = \frac{0.5U_{DC} + \underline{U}_{ref}}{U_{DC}} T_S \qquad (11.25)$$
$$T_0 = T_S - T_1$$

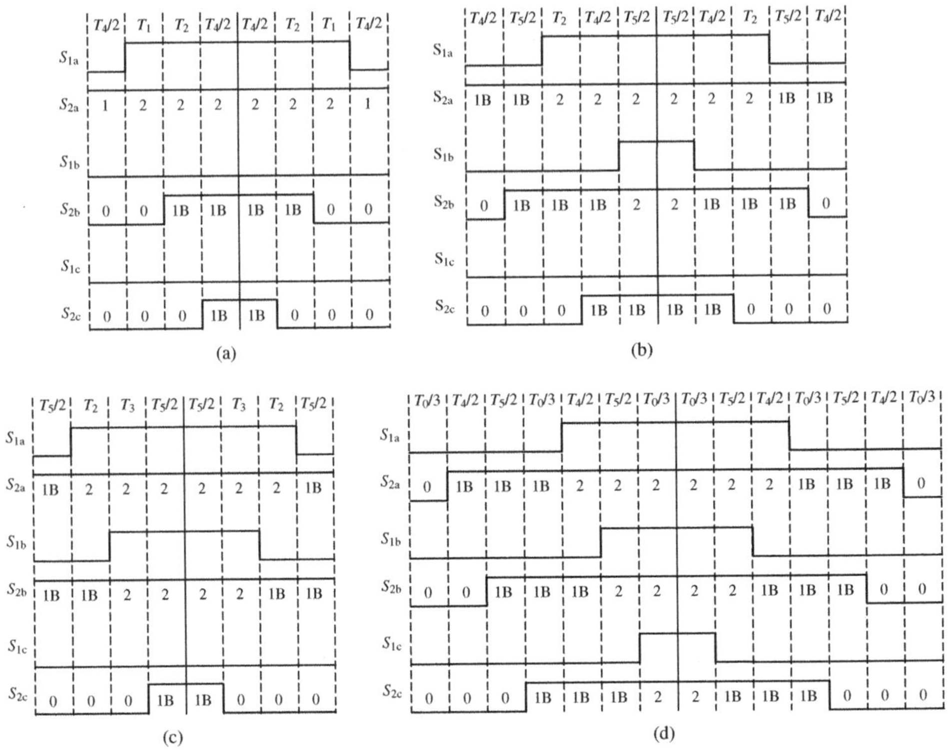

Figure 11.13 Switching sequence for FCC reference vector lying in (a) first, (b) second, (c) third, and (d) fourth region in sector 1 for 1B state selection in each phase

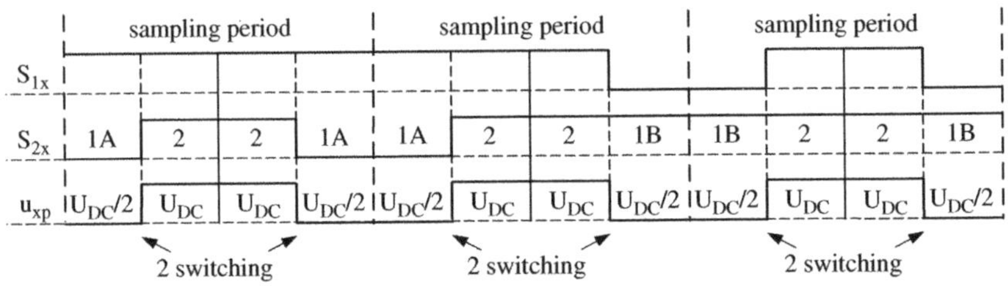

Figure 11.14 Modification of switching pattern in FCC

It can be utilized using a universal concept of time-domain duty-cycle computation technique (11.26) for single-phase multilevel converters, described as a *one-dimensional modulation* (1DM) technique and presented in Ref. [34].

$$
\begin{aligned}
T_2 &= \frac{|\underline{U}_{\text{ref}}|}{0.5\,U_{\text{DC}}} T_S; \quad T_1 = T_S - T_2; \quad \text{Re}(\underline{U}_{\text{ref}}) \geq 0 \\
T_0 &= \frac{|\underline{U}_{\text{ref}}|}{0.5\,U_{\text{DC}}} T_S; \quad T_1 = T_S - T_0; \quad \text{Re}(\underline{U}_{\text{ref}}) < 0
\end{aligned}
\tag{11.26}
$$

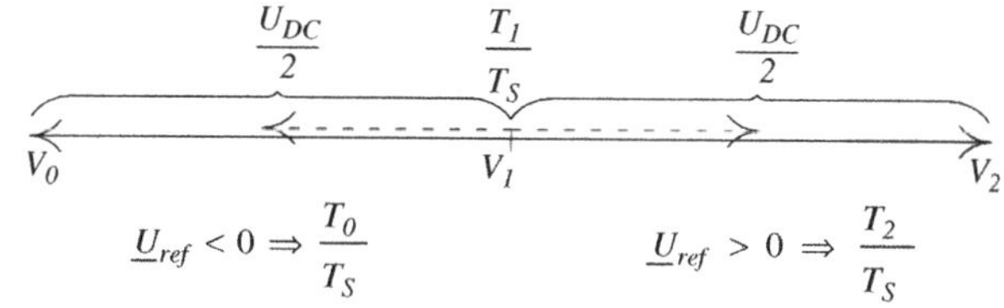

Figure 11.15 Vectors generated by 1DM technique DC/AC converter

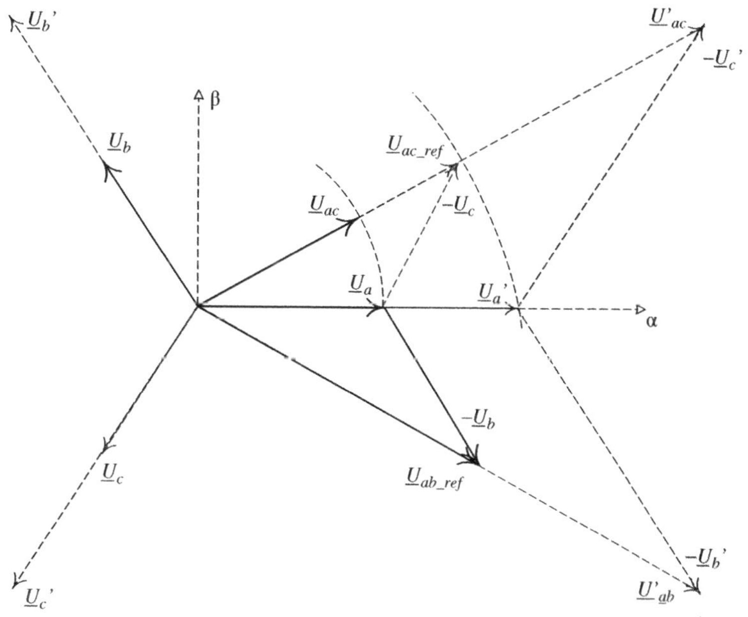

Figure 11.16 Generalized $\alpha\beta$ voltage plane for three-phase simplified converters: imaginary vectors (dotted line), $\underline{U}$, vectors achievable with standard DC-link voltage level, $\underline{U}'$, vectors achievable with high DC voltage, $\underline{U}_{ref}$, reference vectors

Modulation for a simplified three-phase converter requires a different approach than a classical topology, because only two reference signals in the *abc* coordinate system are needed to obtain a balanced three-phase voltage output, which implies that a higher voltage must be supplied to the DC-link to maintain the reference voltage vector in the linear modulation area (Figure 11.16).

SVM for a simplified converter is based on a modified transformation from $\alpha\beta$ to *ab* [35]:

$$U_{a\_ref} = \frac{3}{2} U_\alpha + \frac{\sqrt{3}}{2} U_\beta$$
$$U_{b\_ref} = \sqrt{3}\, U_\beta \tag{11.27}$$

Figure 11.17 shows two cases of a graphical representation of a SV $\alpha\beta$ voltage plane with four achievable voltage vectors of the three-phase two-level simplified converter shown in Figure 11.3(b). In the

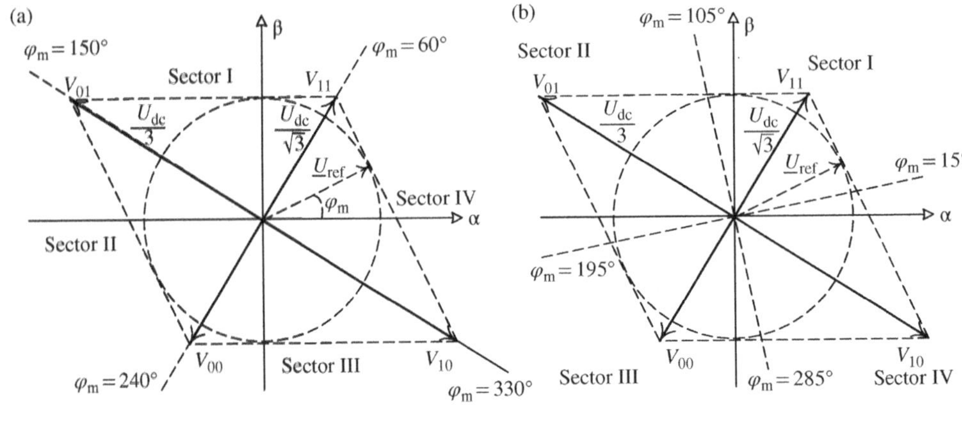

Figure 11.17 Space vector $V_{(Sa,\,Sb)}$ (Sa, Sb are defined in Table 11.2) representation of simplified three-phase two-level AC–DC–AC converter: (a) sector I ($60° \le \varphi_m < 150°$) and (b) sector I ($15° \le \varphi_m < 105°$)

first case (Figure 11.17(a)), all active vectors (00, 01, 11, 10) divide the $\alpha\beta$ plane in four sectors: sector I ($60° \le \varphi_m < 150°$), sector II ($150° \le \varphi_m < 240°$), sector III ($240° \le \varphi_m < 330°$), and finally, sector IV ($330° \le \varphi_m < 60°$). The reference voltage vector $\underline{U}_{ref}$, like in classical converters, could be obtained by switching two adjacent vectors, but this causes a voltage imbalance on DC-link capacitors. For the first sector, switching times can be computed as:

$$T_{10} = \frac{|\underline{U}_{ref}|}{0.5\,U_{DC}}\cos(30° + \varphi_m)T_S$$
$$T_{11} = T_S - T_{10} \tag{11.28}$$

The issue of voltage imbalance on DC-link capacitors can be solved by different divisions of the $\alpha\beta$ plane (Figure 11.17(b)). The number of sectors does not change, but the sectors are shifted 45° (e.g., sector I ($15° \le \varphi_m < 105°$)), which gives more vectors forming $\underline{U}_{ref}$ (e.g., V_{11}, V_{10}, V_{01} in sector I). For the first sector, switching state times can be computed as [10]:

$$T_{11} = T_S\frac{3|\underline{U}_{ref}|}{U_{DC1}+U_{DC2}}\cos(\varphi_m - 60°) - T_S\frac{U_{DC1}-U_{DC2}}{U_{DC1}+U_{DC2}}$$
$$T_{01} = T_S\frac{U_{DC1}}{U_{DC1}+U_{DC2}} - T_S\frac{\sqrt{3}|\underline{U}_{ref}|}{U_{DC1}+U_{DC2}}\cos(\varphi_m - 30°)$$
$$T_{10} = T_S\frac{U_{DC1}}{U_{DC1}+U_{DC2}} - T_S\frac{\sqrt{3}|\underline{U}_{ref}|}{U_{DC1}+U_{DC2}}\cos(\varphi_m - 90°)$$
$$T_s = T_{11} + T_{10} + T_{01} \tag{11.29}$$

Switching signals for a simplified three-phase, 2-level AC–DC or DC–AC converter (Fig. 11.3(b)) are shown in Fig. 11.18.

Another simplified three-phase converter is the three-level converter, as shown in Figure 11.4(b). A graphical representation of the nine achievable voltage vectors in the $\alpha\beta$ voltage plane for such a converter is shown in Figure 11.19.

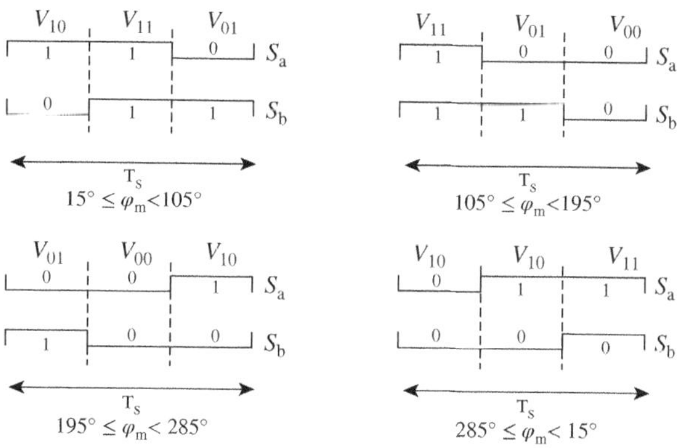

Figure 11.18 Switching sequence for simplified two-level AC–DC–AC converter in each of four sectors

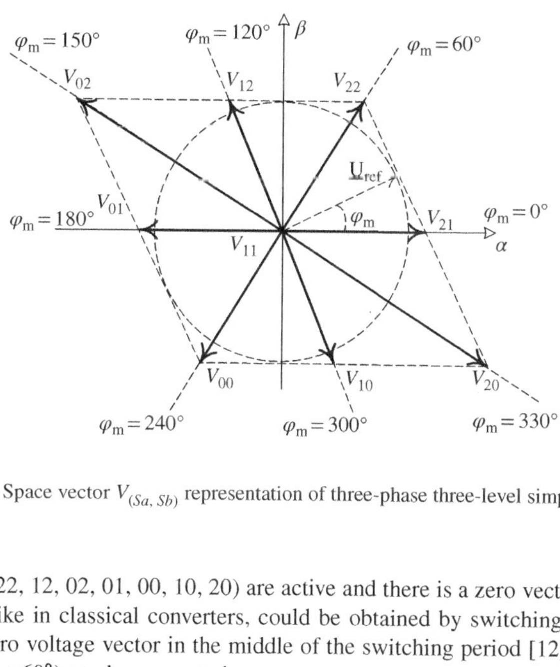

Figure 11.19 Space vector $V_{(Sa,\,Sb)}$ representation of three-phase three-level simplified converter

Eight of them (21, 22, 12, 02, 01, 00, 10, 20) are active and there is a zero vector (11). The reference voltage vector $\underline{U}_{ref}$, like in classical converters, could be obtained by switching two adjacent vectors with the use of the zero voltage vector in the middle of the switching period [12]. Switching times for the sector I ($0° \le \varphi_m < 60°$) can be computed as:

$$T_{21} = \frac{\sqrt{6}U_\alpha}{U_{DC}}T_S - \frac{\sqrt{2}U_\beta}{U_{DC}}T_S$$

$$T_{22} = \frac{2\sqrt{2}U_\beta}{U_{DC}}T_S$$

$$T_{11} = T_s - T_{21} - T_{22} \tag{11.30}$$

and switching sequences for each of the eight sectors are shown in Figure 11.20.

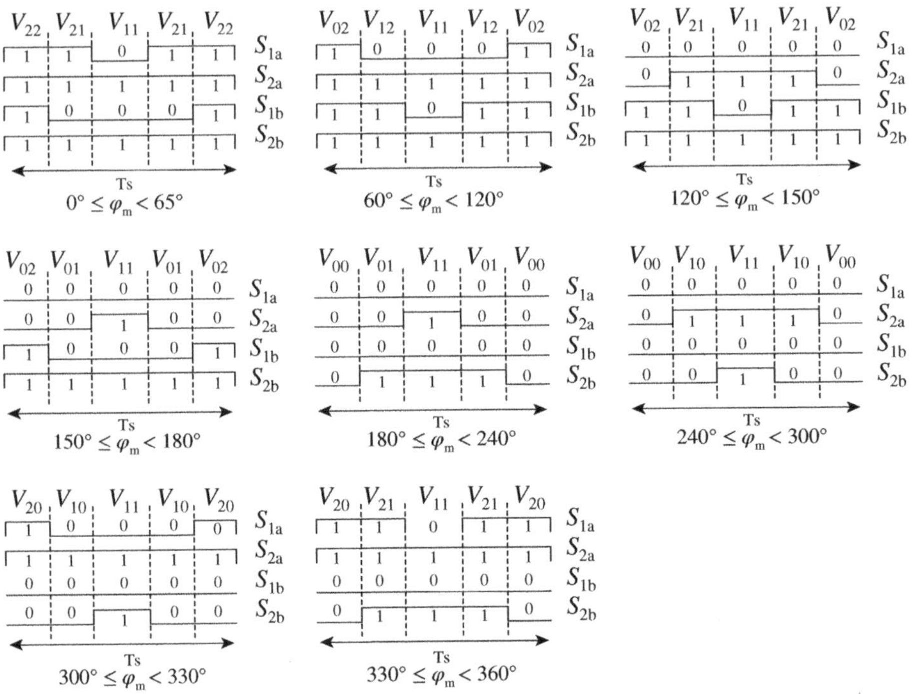

Figure 11.20 Switching sequences for three-level simplified converter in each of the eight sectors

11.3.3 Compensation of Semiconductor Voltage Drop and Dead-Time Effect

The voltage drop on semiconductor devices and the dead-time (between pairs of IGBT's) value distorts the phase current. In both cases, it is the result of increased or decreased amplitude (length) U_{ref} of the reference SV, depending on the sign of the phase current, respectively. To avoid that, the voltage drop and the dead-time effect compensation algorithms are used. The impact of voltage drop and dead-time effect of semiconductor devices on phase current distortion have been well investigated for the classical two-level and multilevel converters. These methods can be divided into two groups:

- modification of amplitude (length) U_{ref} of the reference SV [6, 36–38],
- modification of the duty cycles at the output of the modulator [6, 36, 38–40].

The solution is slightly more complicated in the case of simplified topologies. Because these converters use transistors in only two of three phases, the impact of voltage drop and dead-time effect of semiconductor devices is more significant, compared with that of classical topologies, owing to the different lengths of individual vectors. Moreover, these issues in reduced topologies cannot be considered as a mature. Therefore, in the following section, only reduced topologies will be considered.

Currents for ideal models of controlled switches and models including voltage drop on semiconductor devices are shown in Figure 11.21. In the other case, the converter currents are distorted significantly.

The effects of voltage drop on semiconductor devices are compensated by adding a correction to the reference signals on the input of the modulator [41]. In order to estimate the values of the voltage drops, the phase currents should be predicted for the next switching instant. Values of the predicted current in

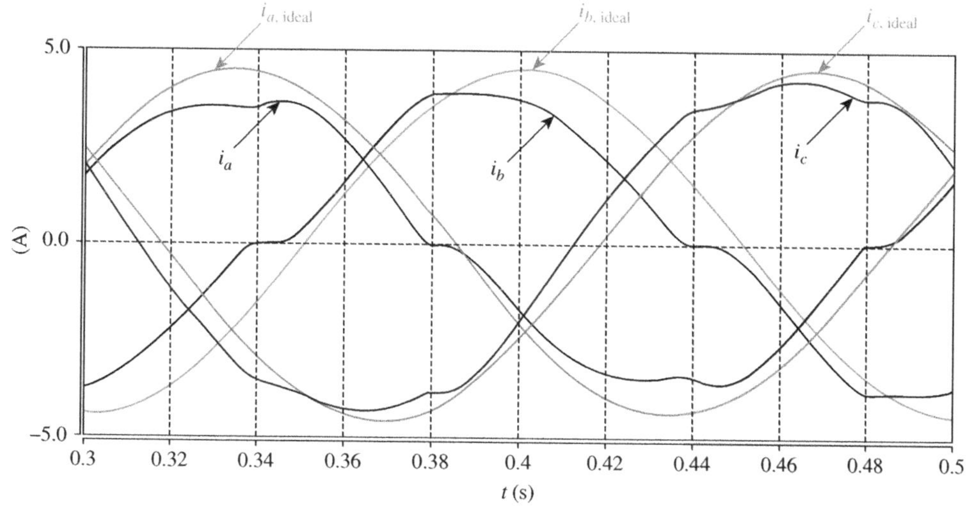

Figure 11.21 Ideal phase currents ($i_{a,\text{ideal}}, i_{b,\text{ideal}}, i_{c,\text{ideal}}$) and phase currents including voltage drop on semiconductor devices (i_a, i_b, i_c)

a and b phases are given by:

$$i_a^{\text{pred}} = i_a + \frac{i_a - i_{a\_\text{old}}}{2}; \quad i_b^{\text{pred}} = i_b + \frac{i_b - i_{b\_\text{old}}}{2} \tag{11.31}$$

where i is the value of the current in the present sample and i_{old} is the value of the current in the earlier sample. Predicted current values are compared with values calculated from transistor and diode data sheets to determine the state of elements in both phases and corresponding voltage drops:

$$\begin{aligned}
&i_{\text{on}}^{\text{tr}} = U_{\text{CEon}} G_{\text{off}}^{\text{tr}}; \quad i_{\text{on}}^{\text{di}} = U_{\text{FW}}^{\text{di}} G_{\text{off}}^{\text{di}} \\
&i_{a,b}^{\text{pred}} \leq i_{\text{on}}^{\text{tr}} \rightarrow u_{a,b}^{\text{tr}} = U_{\text{CEon}}; \; i_{a,b}^{\text{pred}} \leq i_{\text{on}}^{\text{di}} \rightarrow u_{a,b}^{\text{di}} = U_{\text{FW}} \\
&i_{a,b}^{\text{pred}} > i_{\text{on}}^{\text{tr}} \rightarrow u_{a,b}^{\text{tr}} = U_{\text{CEon}} + \frac{1}{G_{\text{on}}^{\text{tr}}} i_{a,b}^{\text{pred}} \\
&i_{a,b}^{\text{pred}} > i_{\text{on}}^{\text{di}} \rightarrow u_{a,b}^{\text{di}} = U_{\text{FW}} + \frac{1}{G_{\text{on}}^{\text{di}}} i_{a,b}^{\text{pred}}
\end{aligned} \tag{11.32}$$

where $i_{\text{on}}^{\text{tr}}$ and $i_{\text{on}}^{\text{di}}$ are the currents for the transistor and diode, respectively. U_{CEon} denotes the voltage drop on the transistor CE junction, U_{FW} denotes the voltage drop on the diode, $G_{\text{on}}^{\text{tr}}$ and $G_{\text{on}}^{\text{di}}$ are the conductances of the transistor and diode for conduction operation mode, respectively, $G_{\text{off}}^{\text{tr}}$ and $G_{\text{off}}^{\text{di}}$ are conductances of the transistor and diode for blocking operation mode, respectively, and finally, $u_{a,b}^{\text{tr}}$ and $u_{a,b}^{\text{di}}$ are the resulting voltage drop for the transistor and diode, respectively.

Using U_{α_ref} and U_{β_ref} transformed to the two-phase coordinate system, as in Equation 11.27, a mean value of voltage drop per sampling period is estimated. Assuming that at switching state 1, the phase voltage error is $u_{a,b}^{\text{di}} + u_{a,b}^{\text{tr}}$ (single transistor and a single DCC diode in a branch are conducting) and in the remaining states, it is $\pm 2u_{a,b}^{\text{di}}$ or $\pm 2u_{a,b}^{\text{tr}}$ we can write

$$D_{a'\pm} = \frac{|U_{\alpha\beta\text{ref}}^{*}|}{U_{\text{DC}}} = \frac{|U_{a'\text{ref}}|}{U_{\text{DC}}}; \quad D_{a'0} = 1 - D_{a'\pm} \tag{11.33}$$

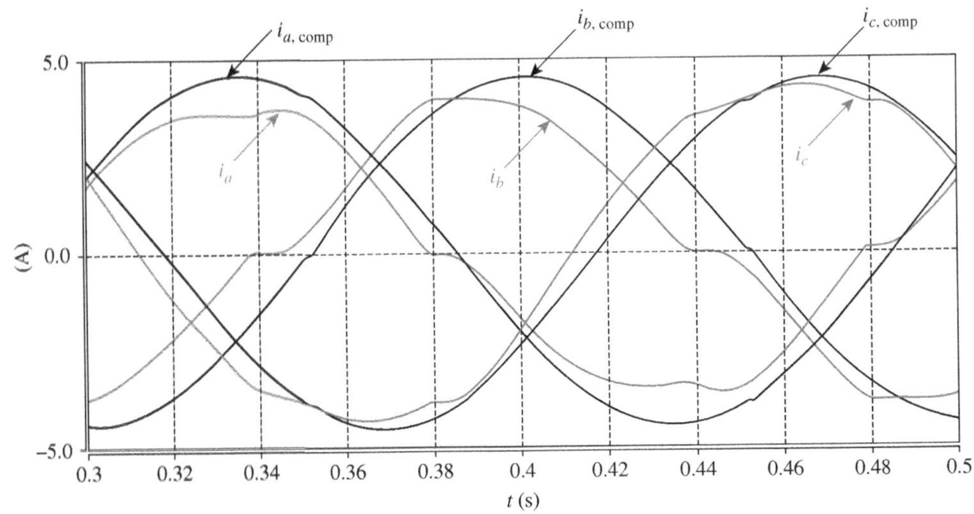

Figure 11.22 Phase currents without (i_a, i_b, i_c) and with ($i_{a,\text{comp}}$, $i_{b,\text{comp}}$, $i_{c,\text{comp}}$) voltage drop compensation

the duty cycles for zero states. Therefore, the correcting signals for voltage drop compensation for a leg $U_{a'}^{\text{cr}}$ (and similar b leg $U_{b'}^{\text{cr}}$) are:

$$\text{If } \operatorname{sgn}(i_a) < 0 \text{ then } \begin{cases} -U_{a'}^{\text{cr}} = D_{a'2}(2v_a^{\text{di}}) - D_{a'1}(u_a^{\text{di}} + u_a^{\text{tr}}) \\ -U_{a'}^{\text{cr}} = D_{a'_0}(2v_a^{\text{tr}}) - D_{a'1}(u_a^{\text{di}} + u_a^{\text{tr}}) \end{cases}$$

$$\text{If } \operatorname{sgn}(i_a) > 0 \text{ then } \begin{cases} U_{a'}^{\text{cr}} = D_{a'2}(2v_a^{\text{tr}}) - D_{a'1}(u_a^{\text{di}} + u_a^{\text{tr}}) \\ U_{a'}^{\text{cr}} = D_{a'_0}(2v_a^{\text{di}}) - D_{a'1}(u_a^{\text{di}} + u_a^{\text{tr}}) \end{cases} \tag{11.34}$$

where $\operatorname{sgn}(i_a)$ denotes the direction of phase current flow, $D_{a'^2_0}$ and $D_{b'^2_0}$ are the duty cycles for positive and negative voltage output states and $D_{a'1}$ denotes zero voltage. These voltage-correcting signals are transformed to the $\alpha\beta$ coordinate system by using the modified transform from Equation (11.27) and adding them to the reference voltages for the modulator. Currents with no compensation and with the applied compensation method for semiconductor voltage drop only are shown in Figure 11.22.

Another necessary compensation relates to the influence of the so-called dead time. Because IGBT devices turn off much more slowly than they turn on, there is a need for the introduction of a delay in control signals between complementary switching devices to prevent short-circuiting the DC-link. This effect is more visible for a low modulation index and low-frequency operation [41]. Figure 11.23 shows a comparison of phase currents without and with dead time included between complementary switching devices.

Correcting signals for limiting the effect of dead time dt are calculated based on current sign and saturation level $i_{dt,\text{SAT}}$; usually chosen experimentally to take into consideration the noise near the current zero-crossing. If phase current polarity, which depends on the measurement noise level as well as on the control delays, is not calculated properly, inadequate compensation can worsen the shape of the phase current. Therefore, in the range of $|i_{a,b,c}| < i_{dt,\text{SAT}}$ the dead-time correction value should be limited linearly. An example of a limiting function, modeled by two linear functions marked as 1 and 2, is shown in Figure 11.24.

Because a reduced converter generates voltages that are not equal in length in vector representation (Figure 11.19), after obtaining the above correction, it is necessary to choose which switching

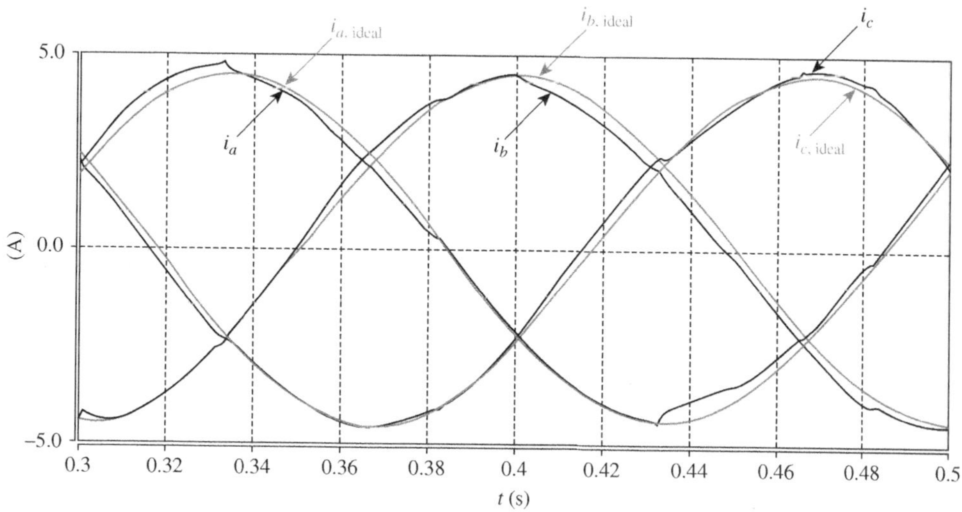

Figure 11.23 Phase currents without ($i_{a,\text{ideal}}$, $i_{b,\text{ideal}}$, $i_{c,\text{ideal}}$) and with (i_a, i_b, i_c) dead time

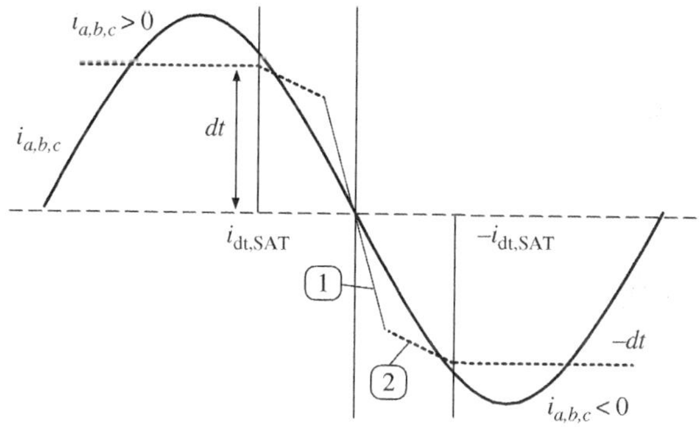

Figure 11.24 Phase current and correction function for dead-time compensation

combination should not only be increased or decreased, but also multiplied with a scaling constant. For long vectors (01 and 10 for the two-level converter and 02 and 20 for the three-level converter), the scaling constant is equal to 1/3, but for short vectors (00 and 11 for two-level converter 00, 01, 12, 22, 21 and 10 for three-level converter), the scaling constant is equal to $1/\sqrt{3}$. The effect of compensation for converters without and with dead-time compensation between the complementary switching devices is shown in Figure 11.25. As can be seen, the shape of phase currents i_a and i_b is improved mainly near the zero-crossing.

11.4 Control Algorithms for AC–DC–AC Converters

A typical application of an AC–DC–AC converter is the interconnection of an electrical machine (motor/generator) with the grid. In this case, advanced control methods should be used to obtain

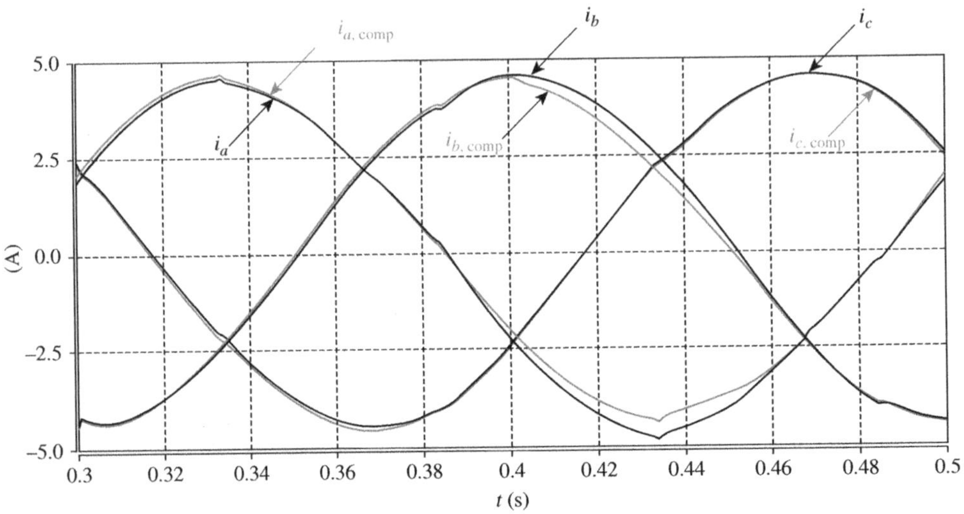

Figure 11.25 Phase currents without (i_a, i_b, i_c) and with ($i_{a,\text{comp}}$, $i_{b,\text{comp}}$, $i_{c,\text{comp}}$) dead-time compensation between complementary switching devices

high power quality for the grid-side converter (GC) and good energy utilization for the machine-side converter (MC). Control of both converters can be considered as a dual problem (Figure 11.26) [42–44]. We can distinguish among the most popular control methods *voltage-oriented control* (VOC) and *direct power control-space vector modulation* (DPC-SVM) for the GC, as well as *field-oriented control* (FOC) and *direct torque control-space vector modulation* (DTC-SVM) for the MC. All these methods are very well described in the literature [42–50]. This section presents the theoretical background of selected control methods and gives basic information regarding the design of the controllers.

11.4.1 Field-Oriented Control of an AC–DC Machine-Side Converter

The FOC can be divided into *direct field-oriented control* (DFOC) and *indirect field-oriented control* (IFOC). A simplified block diagram of the IFOC is presented in Figure 11.27. The IFOC seems to be more convenient than DFOC, especially for *permanent magnet synchronous machines* (PMSMs) because flux estimation is not necessary. The IFOC needs the coordinate system to rotate synchronously with the rotor flux angular speed Ω_{Ψ_r}. In this case, the system of coordinates is oriented with the d rotor flux linkage component, such that:

$$\underline{\psi}_r = \psi_r = \psi_{rd} \tag{11.35}$$

All variables are transformed from the natural abc into rotating the dq reference frame. Then, the referenced stator current values I_{Sdc} and I_{Sqc} are compared with the actual values of stator current component I_{Sd} and I_{Sq}, respectively. It should be stressed that (for steady state) I_{Sd} is equal to the magnetizing current, whereas the torque state, both dynamic and steady, is proportional to I_{Sq}. The current errors $e_{I_{Sd}}$ and $e_{I_{Sq}}$ are fed to two PI controllers, which generate referenced stator voltage components U_{Sdc} and U_{Sqc}, respectively. Furthermore, referenced voltages are transformed from rotating dq coordinates into the stationary $\alpha\beta$ coordinates system by using the rotor flux vector position angle γ_{Ψ_r}. Therefore, the obtained referenced voltage vector $\underline{U}_{Sc}$ is delivered to SVM, which generates the appropriate switching signals S.

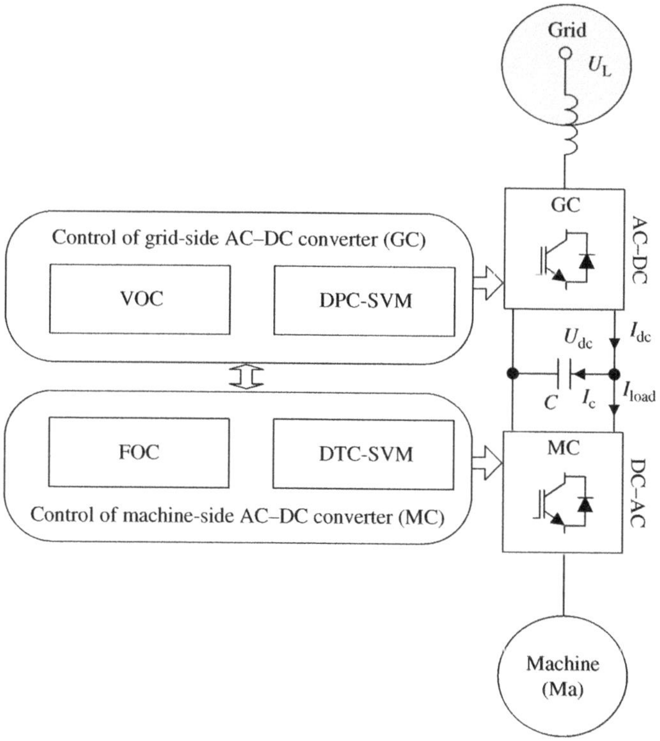

Figure 11.26 Space vector control methods for *grid-side converter* (GC) and *machine-side converter* (MC) in an AC–DC–AC indirect converter

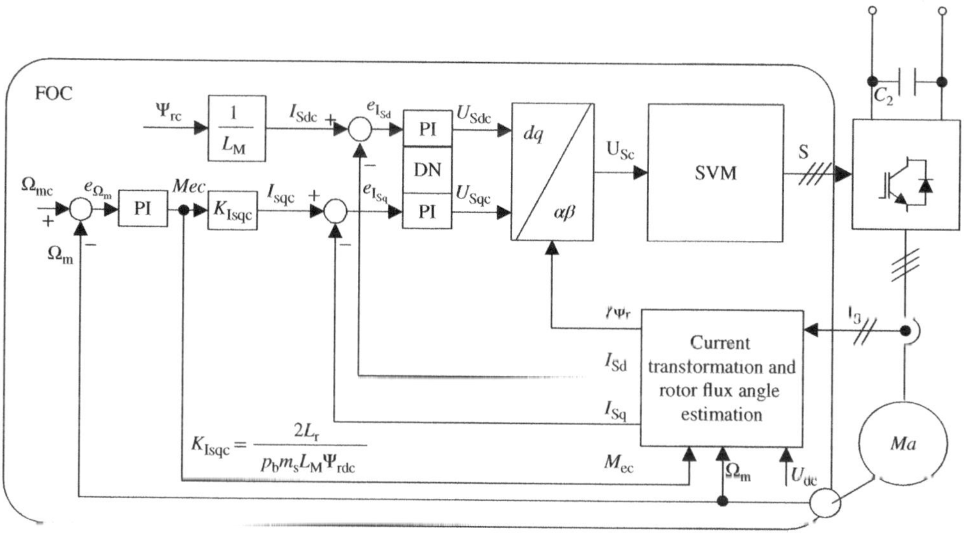

Figure 11.27 Indirect field-oriented control (IFOC), where Ψ_{rc} is the commanded rotor flux

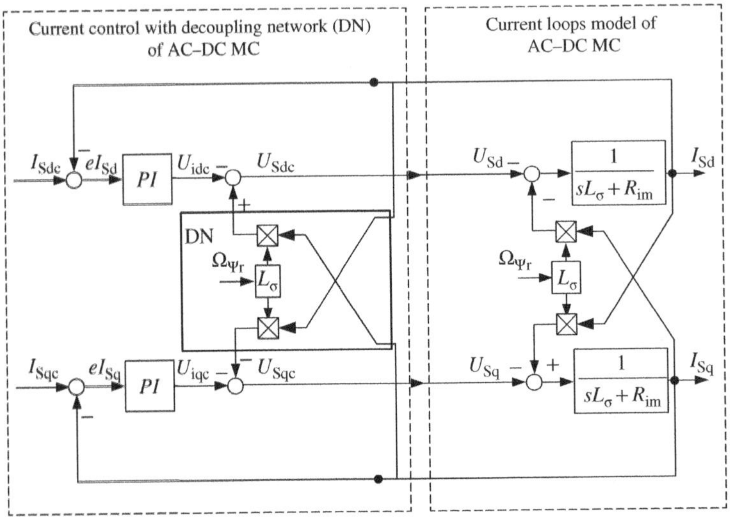

Figure 11.28 Current control with *decoupling network* (DN) for AC–DC converter, where Ω_{Ψ_r} is the angular frequency of the AC system

It should be taken into account that U_{Sdc} and U_{Sqc} are coupled with each other. Any change of d stator voltage component has an influence not only for d but also for the q current component (the same applies to the q component). Therefore, a *decoupling network* (DN) in the control loops is necessary. The solution for this phenomenon is presented in Figure 11.28. It should be noted that the decoupling between the d and q axes appears in all control methods, discussed in this chapter, for both GCs and MCs (VOC and FOC). In the case of direct torque control and direct power control, the SV modulated by the coupling phenomenon is reduced significantly and therefore, DN can be omitted.

11.4.2 Stator Current Controller Design

The model is very convenient to use for the synthesis and analysis of current controllers for MC. In the case of FOC, an asynchronous machine can be approximated and treated as a DC machine [48, 51]. However, it should be pointed out that presence of coupling requires an application of a DN, as presented in Figure 11.28.

Hence, it can be seen that referenced decoupled machine converter voltages should be generated (after simplification) as follows:

$$U_{\mathrm{Sdc}} = L_\sigma \frac{\mathrm{d}I_{\mathrm{Sd}}}{\mathrm{d}t} + R_{\mathrm{im}} I_{\mathrm{Sd}} + \Omega_{\Psi_r} L_\sigma I_{\mathrm{Sq}} \tag{11.36}$$

$$U_{\mathrm{Sqc}} = L_\sigma \frac{\mathrm{d}I_{\mathrm{Sq}}}{\mathrm{d}t} + R_{\mathrm{im}} I_{\mathrm{Sq}} - \Omega_{\Psi_r} L_\sigma I_{\mathrm{Sd}} \tag{11.37}$$

where $L_\sigma = \sigma L_{\mathrm{S}}$, and $R_{\mathrm{im}} = \frac{L_r^2 R_{\mathrm{S}} + L_{\mathrm{M}}^2 R_r}{L_r^2}$, $\sigma = 1 - \frac{L_{\mathrm{M}}^2}{L_{\mathrm{S}} L_r}$ is the total leakage factor, L_s is the stator windings self-inductance, and R_s, R_r are the stator and rotor windings resistances, respectively.

It simplifies the analysis and enables the derivation of analytical expressions for the parameters of stator current controllers. The control structure will operate in a discontinuous domain (digital implementation);

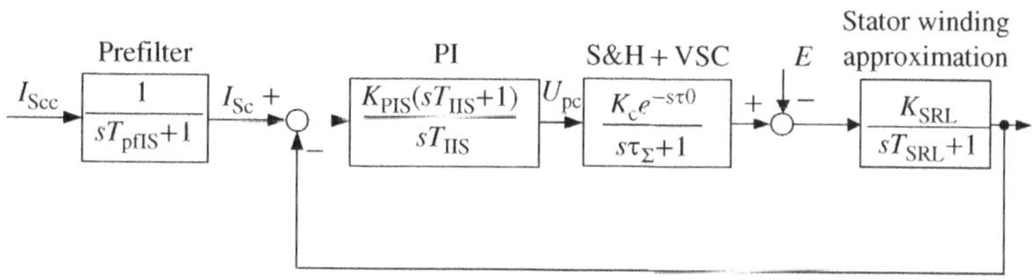

Figure 11.29 Stator current control loop approximation model where I_{Sc} is the referenced stator current before prefilter and T_{pfiS} is the prefilter time constant

therefore, it is necessary to take into account the sampling period T_s. This could be done by S&H block. Moreover, the statistical delay of the PWM in VSC $T_{PWM} = 0.5T_s$ should be taken into account. In the literature [6, 16, 47, 50, 51], the delay of the PWM is approximated from 0 to 2 sampling times T_s. Furthermore, $K_C = 1$ is the VSC gain, and τ_0 is the dead time of the VSC ($\tau_0 = 0$ for an ideal converter). The block S&H and VSC sum of their time constants is expressed by:

$$\tau_\Sigma = T_{PWM} + T_s \tag{11.38}$$

Therefore, the model of the stator current control loop can be approximated as in Figure 11.29. Please note that τ_Σ is the sum of small time constants, $T_{SRL} = L_o/R_{im}$ is a large time constant, and $K_{SRL} = 1/R_{im}$ is a gain approximation of the stator windings. Hence, T_{SRL} gives a dominant pole. Among several methods of analysis, there are two simple ways for the design of the controller parameters: modulus optimum (MO) and symmetry optimum (SO) [48]. With the assumption that the internal voltage induced in the machine winding (EMF) is $E =$ constant, for the circuit presented in Figure 11.29, the proportional gain and integral time constant of the PI current controller can be derived as [48, 51]

$$K_{PiS} = \frac{T_{SRL}}{2\tau_\Sigma K_{SRL}} \tag{11.39}$$

$$T_{IiS} = 4\tau_\Sigma \tag{11.40}$$

After simplifications, the closed-loop transfer function of the VSC can be approximated by the first-order transfer function as:

$$G_{CiS} \approx \frac{1}{1 + s4\tau_\Sigma} \tag{11.41}$$

11.4.3 Direct Torque Control with Space Vector Modulation

To avoid the drawbacks of the switching table in classical DTC, instead of hysteresis controllers and the switching table, PI controllers with the SVM block have been introduced, as in IFOC. Therefore, DTC with SVM (DTC SVM) joins DTC and IFOC features in one control structure, as shown in Figure 11.30.

DTC-SVM requires that the coordinate system rotates synchronously with the stator flux angular speed Ω_{Ψ_S}. In this case, the coordinate system is oriented with the x stator flux linkage component, such that

$$\underline{\psi}_S = \psi_S = \psi_{Sx} \tag{11.42}$$

Based on actual stator currents I_s, DC voltage U_{DC}, mechanical angular speed Ω_m, and switching signals S, the commanded electromagnetic torque M_{ec} is delivered from the outer PI speed controller (Figure 11.30). Then, M_{ec} and the commanded stator flux Ψ_{Sc} amplitudes are compared with the estimated

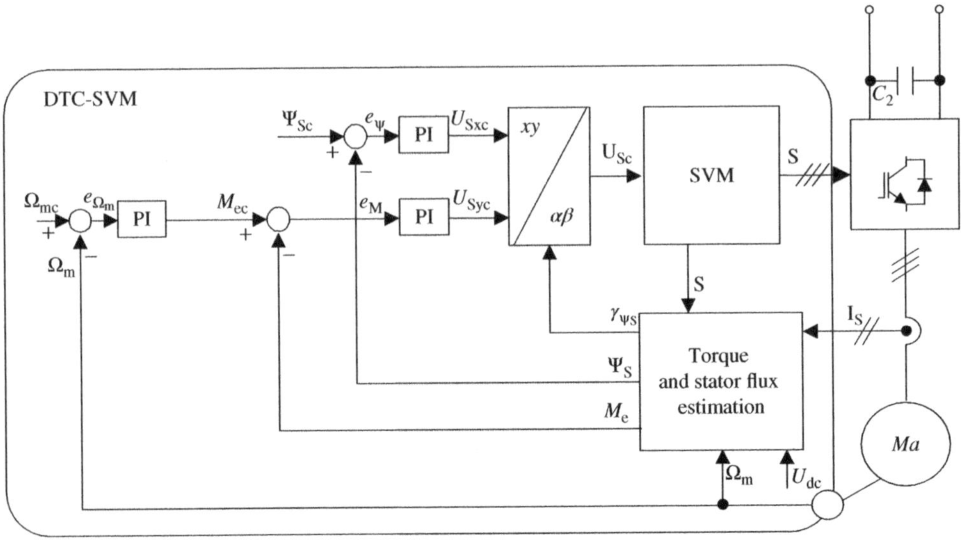

Figure 11.30 *Direct torque control with space vector modulation* (DTC-SVM) control for machine-side converter

values of electromagnetic torque M_e and stator flux Ψ_S. The torque e_M and stator flux e_Ψ errors are fed to the PI controllers. The output signals are the referenced stator voltage components U_{Sxc} and U_{Syc}, respectively.

Furthermore, voltage components in the rotating xy system of coordinates are transformed into $\alpha\beta$ stationary coordinates by using γ_{Ψ_S} stator flux position angle. The obtained referenced stator voltage SV $\underline{U}_{Sc}$ is delivered to the SVM, which generates the desired switching signals S for the MC.

11.4.4 Machine Stator Flux Controller Design

Figure 11.31 presents a block diagram of the PI-based stator flux magnitude control loop with neglected voltage drop on the stator resistance [47]. However, the τ_Σ delay introduced by the inverter is taken into consideration.

According to the SO criterion [48, 49], the plant transfer function can be written as:

$$G(s) = \frac{K_c e^{-s\tau_0}}{s(1 + s\tau_0)} \tag{11.43}$$

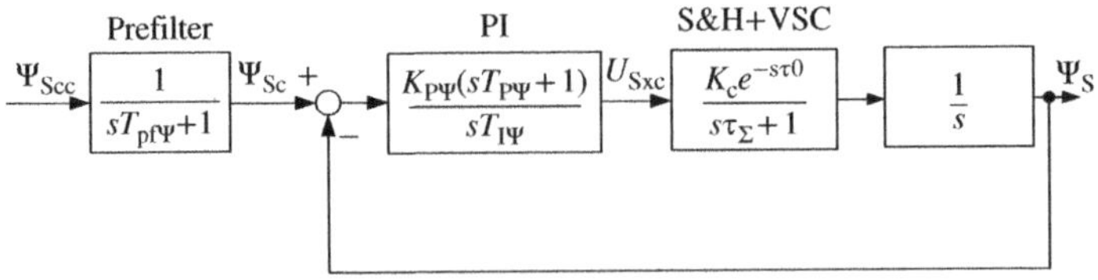

Figure 11.31 Block diagram of the stator flux magnitude control loop, where Ψ_{Scc} is the referenced stator flux before prefilter and $T_{pf\Psi}$ is the prefilter time constant, which equals the integral time constant of the flux PI controller

Therefore, according to the SO design technique, the stator flux PI controller parameters: proportional gain $K_{P\Psi}$ and the integral time constant $T_{I\Psi}$ can be calculated as [47, 48]

$$K_{P\Psi} = \frac{1}{2K_c(\tau_\Sigma + \tau_0)} = \frac{1}{3T_S} \tag{11.44}$$

$$T_{I\Psi} = 4(\tau_\Sigma + \tau_0) = 6T_S \tag{11.45}$$

11.4.5 Machine Electromagnetic Torque Controller Design

Figure 11.32 presents a block diagram of the simplified PI-based torque control loop with omitted coupling between the torque and stator flux. Consequently, the torque control loop is very simple Thus, for this model, not one criterion can be applied.

In this case, according to Ref. [47], the simple (practical) way to design the torque PI controller can be used. Starting from initial values, for example, proportional gain $K_{PM} = 1$ and an integral time constant $T_{IM} = 4T_S$, the value of K_{PM} increases cyclically. From these tests, the best value of K_{PM} for fast torque response without oscillation and defined overshoot can be selected.

11.4.6 Machine Angular Speed Controller Design

If the magnitude of the stator flux is constant, Ψ_S = const, the dynamics of the IM can be described as follows:

$$\frac{d\Omega_m}{dt} = \frac{1}{J}(M_e - M_{load}) \tag{11.46}$$

where J is the moment of inertia and M_{load} is the load torque.

Accordingly, Figure 11.33 shows a block diagram of the PI-based speed control loop, where $G'_{MC}(s)$ is the transfer function of the closed torque control loop with prefilter (to compensate the forcing element in the transfer function of the closed torque control loop).

Approximating the torque control loop by the first-order integral part [47], the simplified $G'_{MC}(s)$ transfer function can be written as:

$$G'_{MC}(s) = \frac{A'_m}{C'_m s + 1} \tag{11.47}$$

where:

$$A'_m = \frac{A_m K_{PM}}{C_m T_{IM} + A_m K_{PM}}; \quad C'_m = \frac{T_{IM}(A_m K_{PM} + B_m)}{C_m T_{IM} + A_m K_{PM}} \tag{11.48}$$

and:

$$A_m = \frac{p_b m_s \Psi_S}{2\sigma L_S}; \quad B_m = \frac{R_S L_r + R_r L_S}{\sigma L_S L_r}; \quad C_m = \frac{p_b^2 m_s \Psi_S^2}{2\sigma L_S} \tag{11.49}$$

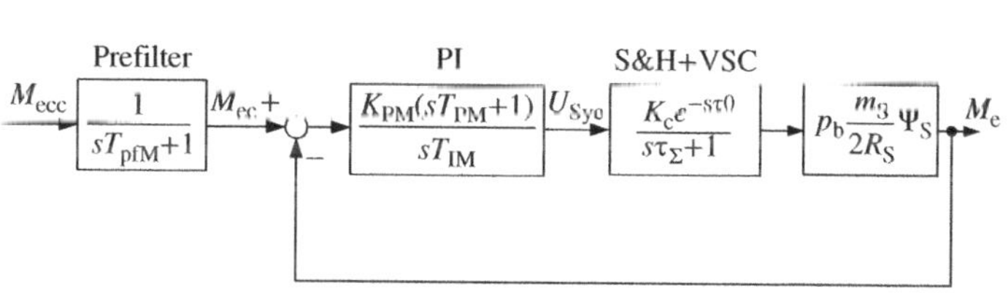

Figure 11.32 Block diagram of torque control loop, where M_{ecc} is the referenced electromagnetic torque before prefilter and T_{pfM} is the prefilter time constant, which equals the integral time constant of the torque PI controller

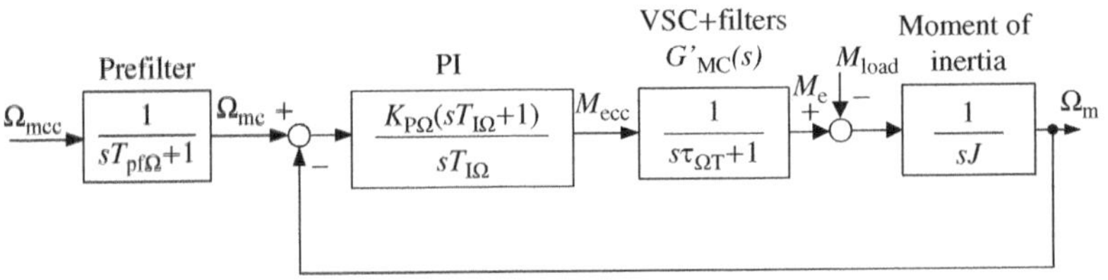

Figure 11.33 Block diagram of the speed control loop, where Ω_{mcc} is the referenced electromagnetic torque before prefilter and $T_{pf\Omega}$ is the prefilter time constant, which equals the integral time constant of the speed PI controller

where p_b is the machine pole pairs number, m_s is the phase number, σ is the total leakage factor, L_s is the stator windings self-inductance, L_M is the main magnetizing inductance, L_r is the rotor windings self-inductance, and R_s, R_r are the stator and rotor windings resistances, respectively.

According to the SO criterion [48, 49], the plant transfer function can be written as:

$$G(s) = \frac{K_C e^{-s\tau_0}}{sJ(sT_{\Omega T} + 1)} \tag{11.50}$$

where $K_C = A'_m$ is the gain of the plant and $T_{\Omega T}$ is the sum of small time constants and filters:

$$T_{\Omega T} = T_S + T_{PWM} + T_f = 1.5T_S + T_f \tag{11.51}$$

The speed PI controller parameters: proportional gain $K_{P\Omega}$, and the integral time constant $T_{I\Omega}$, can be calculated as [47, 48]

$$K_{P\Omega} = \frac{J}{2K_C(T_{\Omega T})} = \frac{J}{3(1.5T_S + T_f)} \tag{11.52}$$

$$T_{si} = 4(T_{\Omega T}) = 4(1.5T_S + T_f) \tag{11.53}$$

11.4.7 Voltage-Oriented Control of an AC–DC Grid-Side Converter

The VOC presented in Figure 11.34 guarantees high dynamic and static performance via an internal current control loop. It has become very popular and, consequently, it has been developed and improved [15, 53–55]. The goal of the control system is to maintain the DC-link voltage U_{DC}, at the required level, while currents drawn from the power system should be sinusoidal and in phase with the line voltage in order to satisfy the *unity power factor* (UPF) condition. The UPF condition is fulfilled when the line current vector $\underline{I}_L = I_{Lx} + jI_{Ly}$ is aligned with the phase voltage vector $\underline{U}_L = U_{Lx} + jU_{Ly}$ of the grid.

For the UPF condition, the commanded value of the reactive component grid current vector I_{Lyc} is set to zero. The reference value of I_{Lxc} is an active component of the grid current vector. After comparing commanded currents with actual values, the errors are delivered to the PI current controllers. Voltages generated by the controllers are transformed to $\alpha\beta$ coordinates by using the grid voltage vector position angle γ_{U_L}. Switching signals S for the GC are generated by an SVM.

11.4.8 Line Current Controllers of an AC–DC Grid-Side Converter

The model is very convenient to use for the synthesis and analysis of the current controllers for a grid-connected converter. However, the presence of coupling requires an application of a DN (as with FOC in Figure 11.28), as presented in Figure 11.35.

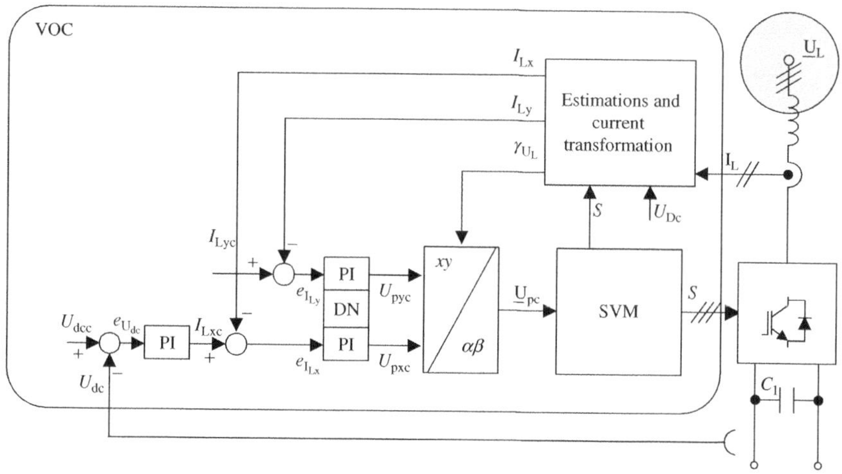

Figure 11.34 *Voltage-oriented control* (VOC)

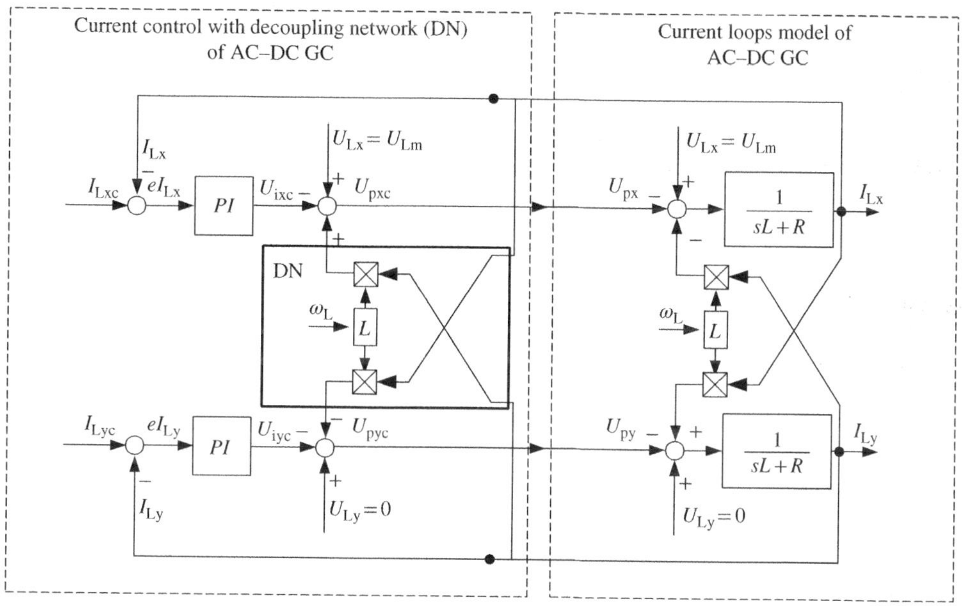

Figure 11.35 Current control with *decoupling network* (DN) for grid-side AC–DC converter

Hence, it can be seen clearly that referenced decoupled GC voltages should be generated as follows:

$$U_{pxc} = U_{Lx} - L\frac{dI_{Lx}}{dt} - RI_{Lx} + \omega_L LI_{Ly} \tag{11.54}$$

$$U_{pyc} = U_{Ly} - L\frac{dI_{Ly}}{dt} - RI_{Ly} + \omega_L LI_{Lx} \tag{11.55}$$

Therefore, the model of the current control loop can be presented as in Figure 11.36.

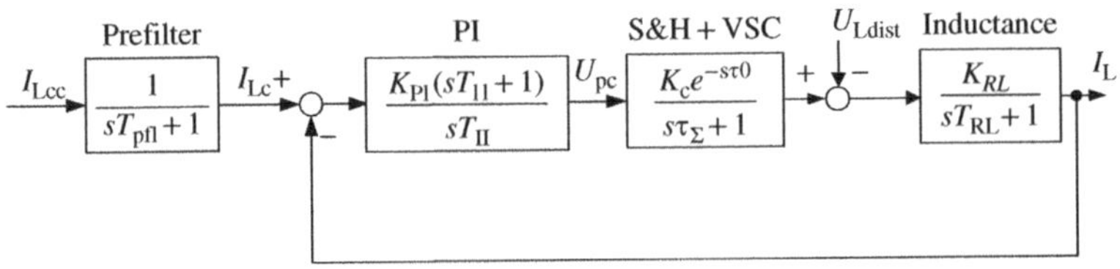

Figure 11.36 Current control loop model where I_{Lcc} is the referenced grid current before prefilter and T_{pfl} is the prefilter time constant, which equals the integral time constant of the current PI controller

Please note that τ_Σ is the sum of small time constants, $T_{RL} = L/R$ is a large time constant, and $K_{RL} = 1/R$ is a gain of input choke. Hence, T_{RL} gives a dominant pole. Between several methods of analysis, there are two simple ways for the design of the controller parameters: MO and SO [48]. With the assumption that disturbance $U_L = \text{const}$ for the circuit presented in Figure 11.5, the proportional gain and integral time constant of the PI current controller can be derived as [16]

$$K_{Pi1} = \frac{T_{RL}}{2\tau_\Sigma K_{RL}} \tag{11.56}$$

$$T_{Ii1} = 4\tau_\Sigma \tag{11.57}$$

After simplifications, the closed-loop transfer function of the VSC can be approximated by the first-order transfer function as:

$$G_{Ci1} \approx \frac{1}{1 + s4\tau_\Sigma} \tag{11.58}$$

For comparison, the parameters of the PI current controller calculated based on MO differ only in integral time constant:

$$T_{Ii1} = T_{RL} \tag{11.59}$$

The VSC with current controllers designed by MO can be approximated as:

$$G_{Ci1} \approx \frac{1}{1 + s2\tau_\Sigma} \tag{11.60}$$

For more information, please refer to Refs. [16, 53].

11.4.9 Direct Power Control with Space Vector Modulation of an AC–DC Grid-Side Converter

Direct power control with space vector modulation (DPC-SVM) [52] assures high dynamic and static performance via internal power control loops. Instead of line currents, the active and reactive powers are used as control variables (Figure 11.37). The command active power P_c is generated by the outer DC-link voltage controller, whereas command reactive power Q_c is set to zero for UPF operation. These values are compared with the estimated P and Q values, respectively [15, 16]. Calculated errors e_P and e_Q are delivered to the PI power controllers. Voltages generated by the power controllers are DC quantities, which eliminates steady-state error, as in VOC. Then, after transformation to stationary $\alpha\beta$ coordinates [16], the voltages are used for the generation of switching signals in the SVM block.

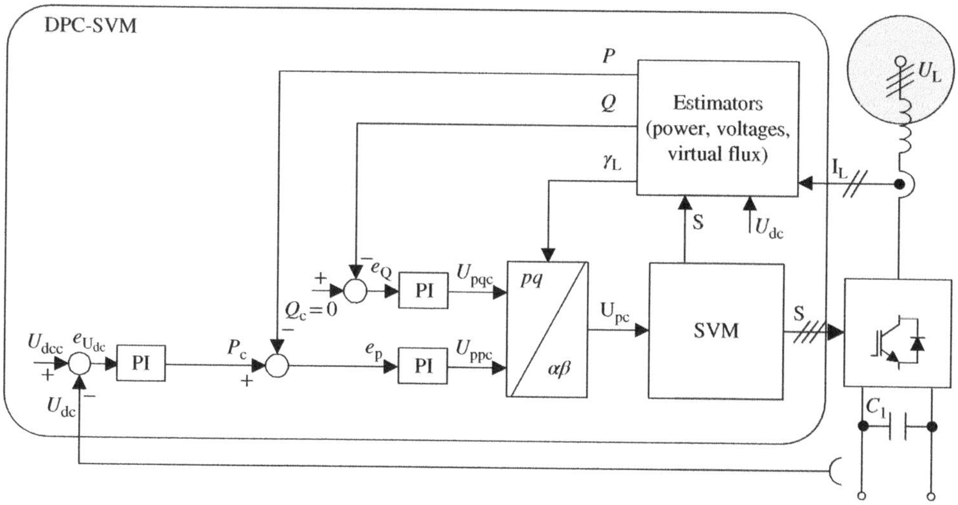

Figure 11.37 *Direct power control with space vector modulation* (DPC-SVM)

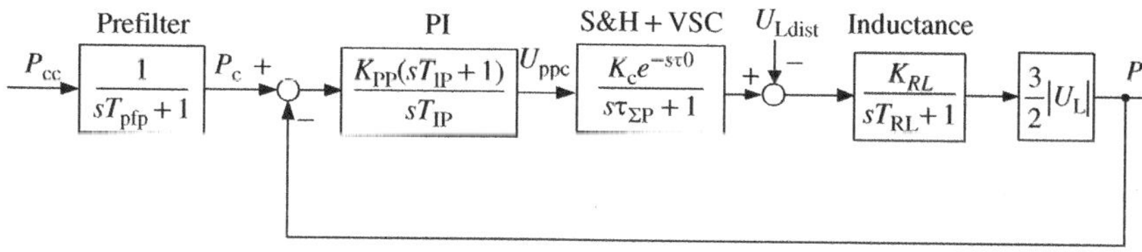

Figure 11.38 Power control loop with prefilter, where P_{cc} is the referenced active power before prefilter, T_{pfp} is the prefilter time constant, which equals the integral time constant of the power PI controller, and U_{Ldist} is the distortion of the grid voltage, for example, caused by higher harmonics

11.4.10 Line Power Controllers of an AC–DC Grid-Side Converter

The assumptions are the same as for the design of current controllers in VOC. Because the same block diagram can be applied to both P and Q power controllers, only the active power P control loop will be described briefly (Figure 11.38).

As in the case of VOC, SO is chosen because of its good response to a disturbance U_{Ldist} step change. With the assumption that U_L = const, an open- and closed-loop transfer function can be obtained easily [16]. The proportional gain and integral time constant of the PI current controller in DPC-SVM can be calculated as:

$$K_{PP1} = \frac{T_{RL}}{2\tau_{\Sigma P} K_{RL}} \frac{2}{3|U_L|} \tag{11.61}$$

$$T_{IP1} = 4\tau_{\Sigma P} \tag{11.62}$$

The PI controller with such parameters gives power-tracking performance with approximately 40% overshoot. Therefore, for decreasing the overshoot, a first-order prefilter can be used on the reference signal:

$$G_{Pf1} \approx \frac{1}{1 + sT_{pf1}} \tag{11.63}$$

where T_{pf1} usually equals a few $\tau_{\Sigma p}$ [48].

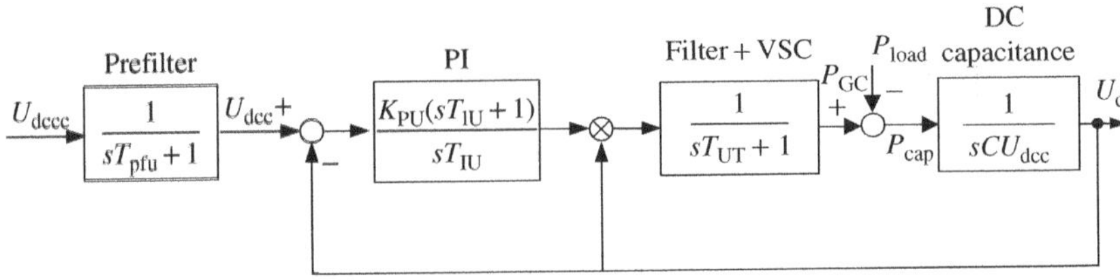

Figure 11.39 Block diagram for a simplified DC-link voltage control loop, where U_{dccc} is the referenced DC voltage before prefilter, T_{pfu} is the DC voltage prefilter time constant, and P_{GC} is the grid-side converter active power

In further investigation, a time delay of the prefilter is set to $4\tau_{\Sigma p}$. Finally, a closed control loop of the VSC can be approximated by a first-order transfer function as:

$$G_{Cp1} \approx \frac{1}{1 + s4\tau_{\Sigma p}} \tag{11.64}$$

11.4.11 DC-Link Voltage Controller for an AC–DC Converter

For calculation of the parameters of a DC-link voltage controller, the inner current or power control loop can be modeled with the first-order transfer function [16]. Therefore, the power control loop of a grid-connected converter can be approximated by further consideration of the first-order block with an equivalent time constant T_{IT} as:

$$G_{pz} \approx \frac{1}{1 + sT_{IT}} \tag{11.65}$$

where $T_{IT} = 2\tau_{\Sigma}$ for inner current or power controllers designed by MO or $T_{IT} = 4\tau_{\Sigma}$ for current/power controllers designed by SO criterion. Therefore, the DC-link voltage control loop can be modeled as in Figure 11.39.

For the sake of simplicity, it can be assumed that

$$T_{UT} = T_U + T_{IT} \tag{11.66}$$

where T_U is the DC voltage filter time constant, T_{UT} is a sum of time constants, and CU_{DCC} is an equivalent integration time constant. Hence, the proportional gain T_{PU} and integral time constant T_{IU} of the DC-link voltage controller can be derived from

$$K_{PU} = \frac{C}{2T_{UT}} U_{\mathrm{L\ or\ DCC}} \tag{11.67}$$

$$T_{IU} = 4T_{UT} \tag{11.68}$$

11.5 AC–DC–AC Converter with Active Power FeedForward

If the accuracy of power matching of an AC–DC GC and DC–AC MC is improved, then a reduction of the DC capacitance is possible, but still providing the same quality of DC voltage stabilization. Because of DC voltage control that is more accurate, the lifetime of the DC capacitor will also be improved. When the DC current I_{DC} of a grid converter is equal to the DC current I_{load} of the MC, no current will flow through the DC capacitor C. As a result, the DC voltage will be well stabilized on a desired constant level.

However, despite the very good dynamic behaviors of the presented control methods, the control of the DC voltage could be improved [55–60]. Hence, APFF of the MC side to GC side has to be introduced.

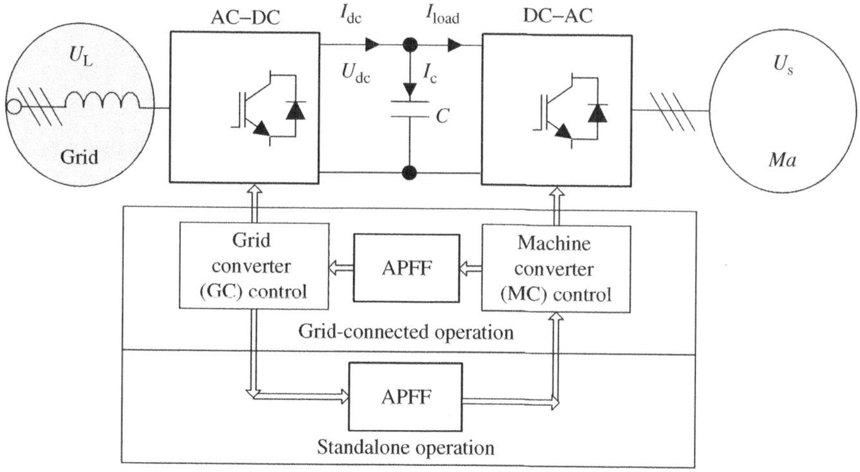

Figure 11.40 Simplified block diagram of the AC–DC–AC converter, which consists of an AC–DC and DC–AC converter, and APFF. Note: Direction of APFF changes in respect of grid-connected or standalone operation

The idea of APFF is to give instantaneous power value from the AC MC side directly to the GC active power control loop, faster than the DC voltage feedback control loop. Because control of the power flow between the GC and MC is more precise, the stability of the DC-link voltage would increase significantly. Therefore, the value of the DC capacitance can be reduced significantly, if this is necessary. On the other hand, if the DC capacitance is required to be high for energy storage needs, the APFF can increase the lifetime of the capacitor significantly [57, 58].

A simplified block diagram of the AC–DC–AC converter, which consists of a grid converter, machine converter, electrical machine, and APFF, is presented in Figure 11.40. It should be stressed here that in the case of operation with opposite direction power flow, when the MC control stabilizes the DC voltage and GC operates in inverting mode, the opposite direction of APFF is required, especially when standalone operation of GC is considered instead of stiff grid.

Hereafter, the AC–DC–AC converter with APFF in control loops will be described. I_{Lyc} is equal to 0 in the case of the UPF operation. Under this condition, the GC input active power is derived by:

$$P_{GC} = \frac{3}{2}(I_{Lx}\,U_{px} + I_{Ly}\,U_{py}) = \frac{3}{2}\,I_{Lx}U_{px} \tag{11.69}$$

Within a steady-state condition I_{Lx} = const and with the assumption that resistance of the input choke $R = 0$, from the machine side, the power consumed/produced by the MC is

$$P_{MC} = \frac{3}{2}(I_{Sx}\,U_{Sx} + I_{Sy}\,U_{Sy}) \tag{11.70}$$

If *virtual flux* (VF) [15, 16, 53, 54] is adopted, it can be noticed that the active power of the GC is proportional to the *virtual torque* (VT) [16, 45, 48]. Thanks to this, a convenient (as in the case of machines) equation for the active grid power calculation is derived:

$$P_{GC} = \frac{3}{2}\,\omega_L(I_{Ly}\Psi_{Lx} - I_{Lx}\Psi_{Ly}) = \omega_L I_{Ly}\Psi_{Lx} \tag{11.71}$$

where I_{Lx}, I_{Ly} are the line current components in a rotating reference frame and Ψ_{Lx}, Ψ_{Ly} are the grid VF components in a rotating reference frame.

On the MC, electromagnetic power can be approximated (neglecting power losses) as an active power delivered to the machine $P_e = P_{MC}$, so

$$P_{MC} = p_b \frac{m_S}{2} \Omega_m I_{Sy} \Psi_{Sx} \tag{11.72}$$

This is an insufficient assumption because of power losses P_{losses} in the real system; therefore, it should be written as

$$P_{MC} = p_b \frac{m_S}{2} \Omega_m I_{Sy} \Psi_{Sx} + P_{losses} \tag{11.73}$$

Let us take into account stand still $\Omega_m = 0$ when nominal torque is applied. In such a case, the electromagnetic power will be 0, but the MC power P_{MC} will have a nonzero value. Estimation of this power is problematic because the parameters of the machine and power electronics devices are required. Hence, for simplicity of the control structure, a power estimator based on the commanded stator voltage $\underline{U}_{Sc}$ and actual stator current $\underline{I}_S$ will be taken into consideration:

$$P_{MC} = \frac{3}{2}(I_{Sx} U_{Sxc} + I_{Sy} U_{Syc}) \tag{11.74}$$

11.5.1 Analysis of the Power Response Time Constant of an AC–DC–AC Converter

Based on Equation 11.58, the time constant of the GC response T_{IT} is determined. With the assumption that power losses of the converters can be neglected, the power-tracking performance can be expressed by

$$P_{GC} = \frac{1}{1 + s\,T_{IT}} P_{GCc} \tag{11.75}$$

Similarly, for the MC, it can be written as

$$P_{MC} = \frac{1}{1 + s\,T_{IF}} P_{MCc} \tag{11.76}$$

where T_{IF} is the equivalent time constant of the MC step response.

11.5.2 Energy of the DC-Link Capacitor

The DC-link voltage can be described as

$$\frac{dU_{DC}}{dt} = \frac{1}{C}(I_{DC} - I_{load}) \tag{11.77}$$

So,

$$U_{DC} = \frac{1}{C}\int (I_{DC} - I_{load})dt \tag{11.78}$$

Assuming the initial condition is in steady state, the actual DC-link voltage U_{DC} is equal to the commanded DC-link voltage U_{DCC}. Therefore, Equation 11.78 can be rewritten as

$$U_{DC} = \frac{1}{CU_{DCC}}\int (U_{DCC}I_{DC} - U_{DCC}I_{load})\,dt = \frac{1}{CU_{DCC}}\int (P_{DC} - P_{load})dt \tag{11.79}$$

where $P_{DC} - P_{load} = P_{cap}$. Therefore, we can obtain

$$U_{DC} = \frac{1}{CU_{DCC}}\int P_{cap}dt \tag{11.80}$$

If the power losses of the GC and MC are neglected (for simplicity), the energy storage variation of the DC-link capacitor will be the integral of the difference between the GC power P_{GC} and the MC power P_{MC}. Therefore, it can be written as

$$P_{GC} = P_{cap} + P_{MC} \tag{11.81}$$

From this equation, it can be concluded that for the most accurate control of the GC power P_{GC}, the commanded power P_{GCc} is

$$P_{GCc} = P_{capc} + P_{MCc} \tag{11.82}$$

where P_{capc} describes the power of the DC-link voltage feedback control loop and P_{MCc} is equal to the instantaneous APFF signal. After the necessary assumptions, we obtain [16]:

$$P_{GCc} = \frac{K_{PU}(1 + sT_{IU})}{s\,T_{IU}} eU_{dcf} U_{DCC} + \frac{1}{1 + s\,T_{IF}} P_{MCc} \tag{11.83}$$

From Equations (11.75) and (11.76), the open-loop transfer function of the grid and machine converter active powers can be expressed as

$$P_{GCo} = \frac{P_{GC}}{P_{GCc}} = \frac{1}{1 + s\,T_{IT}} \tag{11.84}$$

$$P_{MCo} = \frac{P_{MC}}{P_{MCc}} = \frac{1}{1 + s\,T_{IF}} \tag{11.85}$$

Based on these equations, the analytic model of the AC–DC–AC converter with APFF can be defined, as in Figure 11.41. Such a system can be described by an open-loop transfer function as

$$G_{Ao} = \frac{U_{DC}}{M_{ec}} \tag{11.86}$$

Assuming initial steady-state operation, $\omega_m = \omega_{mc}$ = const. and $U_{DC} = U_{DCC}$ = const., the transfer function of the AC–DC–AC converter drive can be derived.

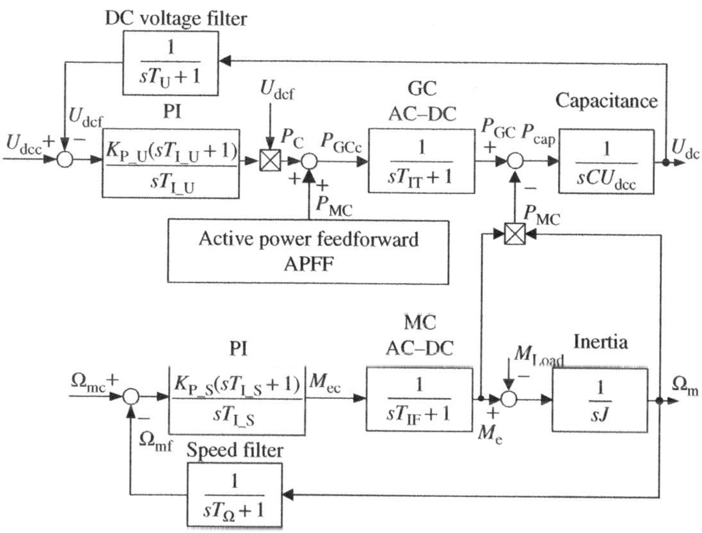

Figure 11.41 Simplified transmittances diagram of the AC–DC–AC converter with APFF

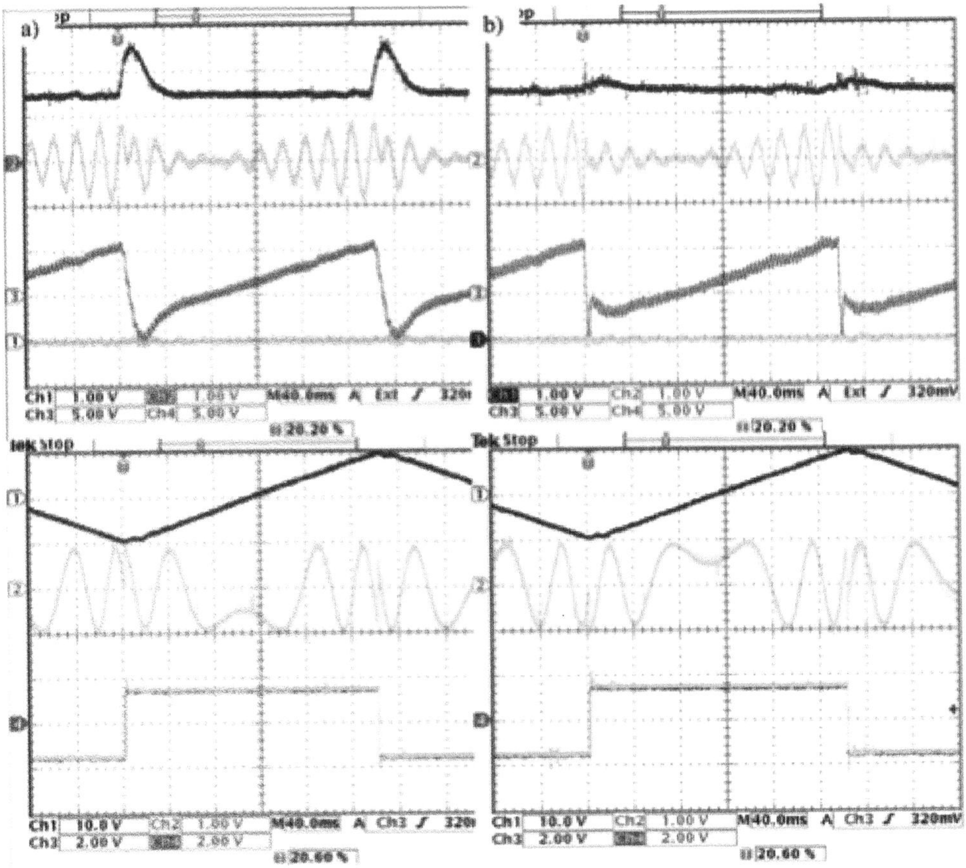

Figure 11.42 Torque reversal from −75% to 75% of nominal torque under DPTC-SVM: (a) only DC-link voltage feedback and (b) with APFF. From the top: DC-link voltage, grid current, active and reactive grid power, electrical machine mechanical speed, and stator current commanded and estimated electromagnetic torque

The active power of the machine converter can be calculated from the commanded values of the stator voltages and actual stator current. Assuming the case where $T_{IT} = T_{IF}$, theoretically, the DC-link voltage should not be affected by the change of the load P_{MC} power. However, in the real system, the unbalanced power difference that causes the fluctuations of the DC-link voltage occurs mainly because of the power estimation error, the moment of inertia of the rotor and low-pass filter in the speed control loop, and the saturation of the commanded control variable.

In Figure 11.42, waveforms illustrate the operation with a torque control loop. Torque reversal is applied when the mechanical speed reaches ±71% of the nominal speed. Very good stabilization of DC voltage can be seen with the use of APFF. The selected control method for the AC–DC–AC converter was *direct power and torque control with space vector modulations* (DPTC-SVM) [16]. The important advantage of this method is that there is no need to rescale the APFF signal.

Moreover, based on Figure 11.42, it can be concluded that for ideal control methods with sufficiently high sampling time, the DC voltage fluctuation during transient for reduced capacitors could be even smaller than for the higher value of capacitance without APFF. Moreover, for the circuit with APFF, it can be seen that the fluctuations of DC voltage during transient for $C = 1$ p.u. and for $C = 0.1$ p.u. are almost

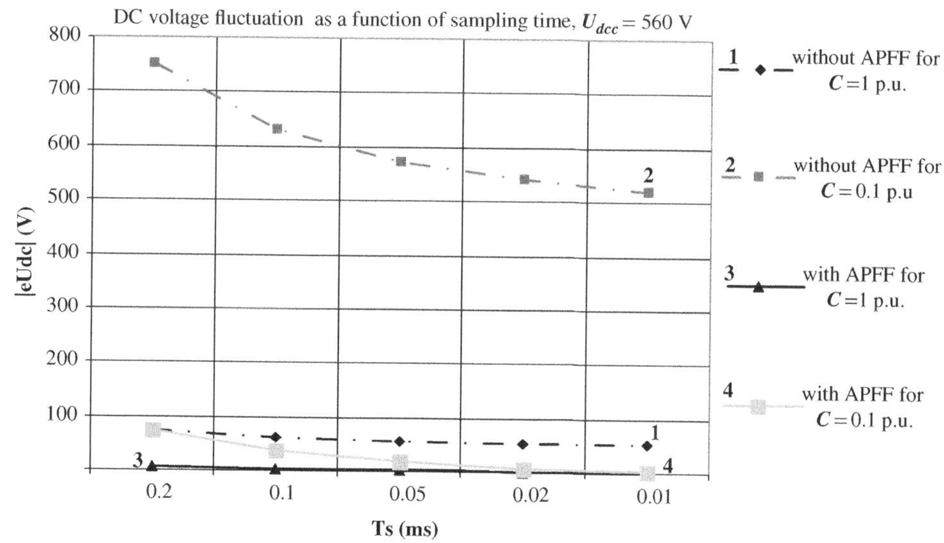

Figure 11.43 Module of the DC voltage fluctuation $|eU_{DC}|$ as a function of sampling time T_c, within transient in respect to load change from 0 to 1 p.u. with referenced DC voltage U_{DCC} = 560 V, and DPTC-SVM control method of the AC–DC–AC converter

equal for higher sampling times. Based on this, it can be stated that stabilization of the DC voltage in an AC–DC–AC converter with a controllable GC depends mainly on the quality (control loops accuracy and bandwidths) of the used DC voltage, the grid, and the MC power flow control methods. It can be expected that for high switching frequency, for example, ≥20 kHz, easily available for *silicon carbide devices* in wider power ranges than in the case of *silicon devices*, very accurate power flow control will be obtained. A simulation of DC voltage control accuracy in a simplified model (no switching losses) is presented as a function of sampling = switching frequency up to 100 kHz (Figure 11.43) with the DPTC-SVM control method applied [16].

11.6 Summary and Conclusions

This chapter has shown the different topologies of AC–DC–AC converters, which becoming increasingly popular in industrial applications, for example, wind turbines, smart grids, drives, and links connecting two electrical networks with different frequency or voltage levels. Among the presented configurations of AC–DC–AC converters, the simplified and classical two-level and multilevel converters are described and compared.

This chapter has also reviewed the modulation and control issues. Among the methods presented in this chapter, the FOC/VOC and DTC-SVM/DPC-SVM are described briefly. Selected issues relating to both the machine-side and grid-side, especially with APFF, have been discussed. Implemented APFFs have no negative impact on system performance in steady states, but do help to reduce DC capacitor and improve dynamic performance of the system. Selected laboratory and simulation tests with DPTC-SVM have been shown.

It is believed that owing to both continuing developments in power semiconductor components and to a reduction of their prices, as well as further elaboration of advanced control and modulation techniques, AC–DC–AC converters will have a significant impact on DPGSs.

References

1. Power Electronics Division TWERD (Polish: Zaklad Energoelektroniki TWERD) www.twerd.pl (accessed 18 December 2013).
2. ABB www.abb.com (accessed 18 December 2013).
3. Wu, B., Lang, Y. Zargari., *et al.* (2011) *Power Conversion and control of Wind Energy Systems*, John Wiley & Sons, Inc.
4. Kouro, S., Malinowski, M., Gopakumar, K. *et al.* (2010) Recent advances and industrial applications of multilevel converters. *IEEE Transactions on Industrial Electronics*, **57** (8), 2553–2580.
5. Friedli, T. (2010) Comparative evaluation of three-phase Si and SiC AC-AC converter systems comparative evaluation of three-phase Si and SiC AC-AC converter systems. PhD dissertation. ETH Zurich.
6. Styński, S. (2011) Analysis and control of multilevel AC–DC–AC flying capacitor converter fed from single–phase. PhD. thesis. Warsaw University of Technology, Warsaw, Poland.
7. Sedłak, M., Styński, S., Kaźmierkowski, M.P., and Malinowski, M. (2012) Three-level four-leg flying capacitor converter for renewable energy sources. *Electrical Review*, **R.88** (12a), 6–11.
8. Enjeti, P. and Rahman, A. (1990) A new single phase to three phase converter with active input current shaping for low cost AC motor drives. Industry Applications Society Annual Meeting, 1990, Conference Record of the 1990 IEEE, October 7–12, 1990, Vol. 2, No. 4, pp. 935–942.
9. Jacobina, C.B., de Rossiter Correa, M.B., da Silva, E.R.C., and Lima, A.M.N. (1999) Induction motor drive system for low-power applications. *IEEE Transactions on Industry Applications*, **35** (1), 52–61.
10. Blaabjerg, F., Neacsu, D., and Pedersen, J. (1999) Adaptive SVM to compensate DC-link voltage ripple for four-switch three-phase voltage-source inverters. *IEEE Transactions on Power Electronics*, **14**, 743–752.
11. Nabae, A., Takahashi, I., and Akagi, H. (1981) A new neutral-point-clamped PWM inverter. *IEEE Transactions on Industry Applications*, **IA-17** (5), 518–523.
12. Bor-Ren, L. and Ta-Chang, W. (2004) Space vector modulation strategy for an eight-switch three-phase NPC converter. *IEEE Transactions on Aerospace and Electronic Systems*, **40** (2), 553–566.
13. Teodorescu, R., Liserre, M., and Rodríguez, P. (2011) *Grid Converters for Photovoltaic and Wind Power Systems*, John Wiley & Sons, Ltd-IEEE Press, p. 416.
14. Teichmann, R., Malinowski, M., and Bernet, S. (2005) Evaluation of three-level rectifiers for low-voltage utility applications. *IEEE Transactions on Industrial Electronics*, **52** (2), 471–481.
15. Malinowski, M. (2001) Sensorless control strategies for three–phase PWM rectifiers. PhD thesis. Warsaw University of Technology, Warsaw, Poland.
16. Jasinski, M. (2005) Direct power and torque control of AC–DC–AC converter–fed induction motor drives. PhD thesis. Warsaw University of Technology, Warsaw, Poland.
17. Carlsson, A. (1998) The back-to-back converter control and design. PhD thesis. Lund Institute of Technology, Lund.
18. Moran, L., Ziogas, P.D., and Joos, G. (1992) Design aspects of synchronous PWM rectifier-inverter systems under unbalanced input voltage conditions. *IEEE Transactions on Industry Applications*, **28** (6), 1286–1293.
19. Winkelnkemper, M. and Bernet, S. (2003) Design and optimalization of the DC-link capacitor of PWM voltage source inverter with active frontend for low-voltage drives. Proceedings of the EPE 2003 Conference.
20. Malesani, L., Rossetto, L., Tenti, P., and Tomasin, P. (1993) AC/DC/AC PWM converter with minimum energy storage in the dc link. Applied Power Electronics Conference and Exposition, 1993. APEC'93. Conference Proceedings 1993, Eighth Annual, March 7–11, 1993, pp. 306–311.
21. Fazel, S.S., Bernet, S., Krug, D., and Jalili, K. (2007) Design and comparison of 4-kV neutral-point-clamped, flying-capacitor, and series-connected H-bridge multilevel converters. *IEEE Transactions on Industrial Electronics*, **43** (4), 1032–1040.
22. IEEE Std 519–1992. (1993) IEEE Recommended Practices and Requirements for Harmonic Control in Electrical Power Systems.
23. Bernet, S., Ponnaluri, S., and Teichmann, R. (2002) Design and loss comparison of matrix converters and voltage source converters for modern ac-drives. Transactions of Industrial Electronics Society, Special Edition Matrix Converters.
24. Jalili, K. (2009) Investigation of control concepts for high-speed induction machine drives and grid side PWM voltage source converters. PhD thesis. Dresden University of Technology.
25. Jalili, K. and Bernet, S. (2009) Design of LCL filters of active-front-end two-level voltage-source converter. *IEEE Transactions on Industrial Electronics*, **56** (5), 1674–1689.

26. Liserre, M., Blaabjerg, F., and Hansen, S. (2001) Design and control of an LCL-filter based three-phase active rectifier. Industry Applications Conference, 2001. Thirty-Sixth IAS Annual Meeting. Conference Record of the 2001 IEEE, September 30–October 4, 2001, Vol. 1, pp. 299–307.
27. Holmes, D.G. and Lipo, T.A. (2003) *Pulse Width Modulation for Power Converters, Principles and Practice*, John Wiley & Sons, New York.
28. Rodriguez, J. (2005) Tutorial on multilevel converters. Proceedings of International Conference PELINCEC 2005, October 2005.
29. Holmes, D.G. (1996) The significance of zero space vector placement for carrier–based pwm schemes. *IEEE Transactions on Industry Applications*, **32** (5), 1122–1129.
30. Jenni, F. and Wueest, D. (1993) The optimization parameters of space vector modulation. Proceedings of EPE 1993, September 1993, pp. 376–381.
31. Marchesoni, M. and Tenca, P. (2002) Diode-clamped multilevel converters: a practicable way to balance DC-link voltages. *IEEE Transactions on Industrial Electronics*, **49** (4), 752–765.
32. Pou, J., Zaragoza, J., Ceballos, S. *et al.* (2007) Optimal voltage-balancing compensator in the modulation of a neutral-point-clamped converter. Proceedings of IEEE International Symposium on Industrial Electronics, June 2007, pp. 719–724.
33. Kolomyjski, W. (2009) Modulation strategies for three-level pwm converter-fed induction machine drives. PhD dissertation. Warsaw University of Technology, Warsaw, Poland.
34. Leon, J., Portillo, R., Vazquez, S. *et al.* (2008) Simple unified approach to develop a time-domain modulation strategy for single-phase multilevel converters. *IEEE Transactions on Industrial Electronics*, **55**, 3239–3248.
35. Jacobina, C., da Silva, E., Lima, A., and Ribeiro, R. (1995) Vector and scalar control of a four switch three phase inverter. Industry Applications Conference, 1995. Thirtieth IAS Annual Meeting, IAS'95, Conference Record of the 1995 IEEE, Vol. 3, pp. 2422–2429.
36. Jeong, S.-G. and Park, M.–.H. (1991) The analysis and compensation of dead-time effects in pwm inverters. *IEEE Transactions on Industrial Electronics*, **38** (2), 108–114.
37. Ben-Brahim, L. (2004) On the compensation of dead time and zero-current crossing for a pwm–inverter–controlled ac servo drive. *IEEE Transactions on Industrial Electronics*, **51** (5), 1113–1118.
38. Dongsheng, Z. and Rouaud, D.G. (1999) Dead-time effect and compensations of three-level neutral point clamp inverters for high-performance drive applications. *IEEE Transactions on Power Electronics*, **14** (4), 782–788.
39. Leggate, D. and Kerkman, R.J. (1997) Pulse-based dead-time compensator for pwm voltage inverters. *IEEE Transactions on Industrial Electronics*, **44** (2), 191–197.
40. Oliveira, A.C., Jacobina, C.B., and Lima, A.M.N. (2007) Improved dead-time compensation for sinusoidal pwm inverters operating at high switching frequencies. *IEEE Transactions on Industrial Electronics*, **54** (4), 2295–2304.
41. Minshull, S.R., Bingham, C.M., Stone, D.A., and Foster, M.P. (2010) Compensation of nonlinearities in diode-clamped multilevel converters. *IEEE Transactions on Industrial Electronics*, **57**, 2651–2658.
42. Kazmierkowski, M.P., Krishnan, R., and Blaabjerg, F. (2002) *Control in Power Electronics*, Academic Press, p. 579.
43. Wu, R., Dewan, S.B., and Slemon, G.R. (1990) A PWM AC-to-DC converter with fixed switching frequency. *IEEE Transactions on Industry Applications*, **26** (5), 880–885.
44. Zhou, D. and Rouaud, D. (1999) Regulation and design issues of a PWM three-phase rectifier. Proceedings of the 25th Annual Conference IECON'99, Vol. 1, pp. 485–489.
45. Manninen, V. (1995) Application of direct torque control modulation technology to a line converter. Proceedings of the EPE 1995 Conference, pp. 1.292–1.296.
46. Blaschke, F. (1971) A new method for the structural decoupling of A.C. induction machines. Proceedings of the Conference Record IFAC, Duesseldorf, Germany, October 1971, pp. 1–15.
47. Zelechowski, M. (2005) Space vector modulated-direct torque controlled (dtc-svm) inverter-fed induction motor drive. PhD dissertation. Warsaw University of Technology, Warsaw, Poland.
48. Kazmierkowski, M.P. and Tunia, H. (1994) *Automatic Control of Converter–Fed Drives*, Elsevier, Amsterdam, London, New York, Tokyo, PWN Warszawa.
49. Levine, W.S. (2000) *Control System Fundamentals*, CRC Press.
50. Liserre, M. (2001) Innovative control techniques of power converters for industrial automation. PhD thesis. Politecnico di Bari, Italy.
51. Blasko, V. and Kaura, V. (1997) A new mathematical model and control of a tree-phase AC-DC voltage source converter. *IEEE Transactions on Power Electronics*, **12** (1), 116–123.

52. Malinowski, M., Jasinski, M. and Kaźmierkowski, M.P. (2004) Simple Direct Power Control of Three-Phase PWM Rectifier Using Space-Vector Modulation (DPC-SVM). *IEEE Transaction on Industrial Electronics*, **51** (2), 447–454.
53. Wilamowski, B.M. and Irwin, J.D. (2011) *The Industrial Electronics Handbook – Power Electronic and Motor Drives*, 2nd edn, CRC Press, Boca Raton, FL, London, New York.
54. Chien, W.S. and Tzou, Y.-Y. (1998) Analysis and design on the reduction of DC-link electrolytic capacitor for AC/DC/AC converter applied to AC motor drives. Power Electronics Specialists Conference, 1998. PESC 98 Record. 29th Annual IEEE Volume 1, May 17–22, 1998, pp. 275–279.
55. Gu, B.G. and Nam, K. (2002) A DC-link capacitor minimization method through direct capacitor current control. Industry Applications Conference, 2002. 37th IAS Annual Meeting. Conference Record of the Volume 2, October 13–18, 2002, pp. 811–817.
56. Liao, J.C. and Yen, S.N. (2000) A novel instantaneous power control strategy and analytic model for integrated rectifier/inverter systems. *IEEE Transactions on Power Electronics*, **15** (6), 996–1006.
57. EPCOS (0000) Aluminium Electrolic Capacitors – Applications, DataSheet, www.epcos.com (accessed 18 December 2013).
58. EPCOS (0000) Aluminium Electrolic Capacitors – General Technical Information, DataSheet, www.epcos.com (accessed 18 December 2013).
59. Sobczuk, D.L. (1999) Application of ANN for control of PWM inverter fed induction motor drives. PhD thesis. Warsaw University of Technology, Warszawa.
60. Jasinski, M., Rafal, K., Bobrowska-Rafal, M., and Piasecki, S. (2011) Grid interfacing of distributed energy sources by three-level BtB NPC converter under distorted grid voltage. 2011 Workshop on Predictive Control of Electrical Drives and Power Electronics (PRECEDE), October 14–15, 2011, pp. 30–35.

12

Power Electronics for More Electric Aircraft

Kaushik Rajashekara
Department of Electrical Engineering, University of Texas at Dallas, Texas, USA

12.1 Introduction

Transportation as a whole is estimated to be responsible for over 20% of the world's CO_2 discharges. According to the Intergovernmental Panel on Climate Change, global aviation contributes about 2% of the global CO_2 emissions caused by human activities [1]. This estimate includes emissions from all global aviation, including commercial and military. Global commercial aviation, including cargo, accounts for over 80% of this estimate. When other emissions such as nitrogen oxide and water vapor are considered, the estimated share of aviation's global emissions increases from 2 to 3%. In the United States, according to Environmental Protection Agency data, domestic aviation contributes about 3% of total CO_2 emissions. Although aviation is a relatively small source of the emissions contributing to global warming, it is of significance because it is probable that high-altitude emissions are disproportionately damaging to the environment.

The Advisory Council for Aeronautics Research in Europe has set several goals to be achieved by 2020 for air transportation [2]. These include a 50% reduction of CO_2 emissions through drastic reduction of fuel consumption; an 80% reduction of NO_X emissions; a 50% reduction of external noise; and a green product life cycle in terms of design, manufacturing, maintenance and disposal. The goals set by the International Civil Aviation Organization are to improve fuel efficiency by an average 2% per year until 2050 and to keep the global net carbon emissions from international aviation at the same level beyond 2020 [3]. Thus, the aerospace industry is facing challenges similar to those faced by the automotive industry in terms of improving emissions and fuel economy. Another similarity is the move toward replacing mechanical and pneumatic systems with electrical systems, thus transitioning toward "more electric" architectures.

To meet the above challenges in the automotive industry, significant work has been done by developing and commercializing the electric and hybrid vehicles. In the case of airplanes, more electric architecture is the emerging trend. The intention is to move as many aircraft loads as possible to electrical

Power Electronics for Renewable Energy Systems, Transportation and Industrial Applications, First Edition.
Edited by Haitham Abu-Rub, Mariusz Malinowski and Kamal Al-Haddad.

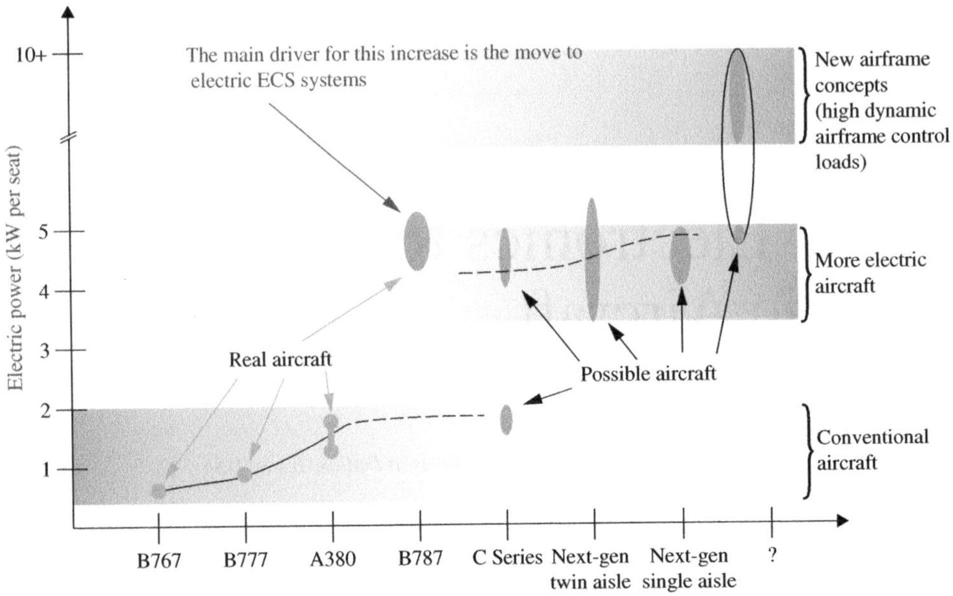

Figure 12.1 Increasing electrical power demand in civil aerospace market [4]

power resulting in simpler and efficient aircraft systems. This leads to lower fuel consumption, reduced emissions, reduced maintenance and possibly lower costs. The examples of aircraft loads under consideration are electric-powered environmental control systems (ECSs), electrical actuators and electric deicing. Electric starting of the engine and the conversion of all pneumatic and hydraulic units on the accessory gearbox (AGB) to an electrical system are also being investigated.

The electrical power being used by both civil and military aircraft is also growing [4–6]. As can be seen from Figure 12.1, the civil aerospace market is seeing rapidly increasing demands for electric power. Future trends in this market sector are not known exactly, but they could move toward ~10 kW per passenger seat, which represents a significant growth from today's level. In order to meet these requirements, a major re-evaluation of aircraft electrical generation and power distribution systems is being undertaken. The requirement for electrical power onboard aircraft is forecasted to rise dramatically in the future because of the following reasons:

- Additional electrical loads because of an increased use of electrical actuators and landing gear
- Increased cabin loads for better in-flight entertainment
- Information services and passenger comfort electrically operated ECSs
- Anti-icing of the wings
- Flight controls and other electrical loads

Passenger aircraft, such as the Boeing 787 and Airbus 380, employ a number of new electrical technologies including bleedless ECSs (in the Boeing 787). These loads are creating a substantial increase in the total electrical power drawn from the aircraft's engine-driven generators. For example, in the Boeing 787, each engine drives two generators each rated at 250 kVA producing a total power of 500 kVA. It is estimated that future electrical power requirements will exceed 500 kVA per engine. The available power generation capability of some aircraft is listed in Table 12.1.

Table 12.1 Power generation capability of some selected aircraft [6]

B717	2 × 40 kVA
B737NG	2 × 90 kVA
B767-400	2 × 120 kVA
B777	2 × 120 kVA and 2 × 20 kVA backup
A340	4 × 90 kVA
B747-X	4 × 120 kVA
A380	4 × 150 kVA and 2 × 120 kVA APU
B787	4 × 250 kVA and 2 × 225 kVA APU

12.2 More Electric Aircraft

In a traditional airplane, the jet engine is designed to produce thrust and to power the pneumatic, hydraulic and electrical systems, as shown in Figure 12.2. The pneumatic power is used for pressurization and cooling of the cabin, the starting of the main engines and for deicing the wings. The hydraulic power is used mainly for flight control actuators. The electrical power is used for supplying the power to all the electrical loads, including the computers and avionics systems. In addition, the engines drive the gearbox-mounted units, such as fuel, oil and hydraulic pumps. In a More Electric Aircraft (MEA) system, the jet engine is optimized to produce the thrust and the electric power, as shown in Figure 12.3. An electric machine is used for starting the engine and for generating electric power. Most of the loads are electrical, including the deicing and ECSs. The fuel, hydraulic and oil pumps are all driven by the electric motors.

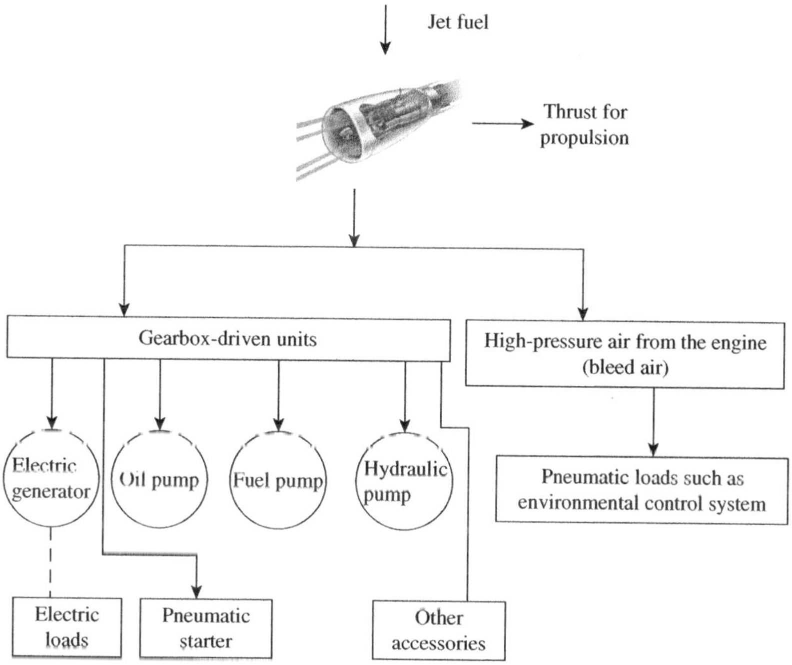

Figure 12.2 Traditional aircraft system

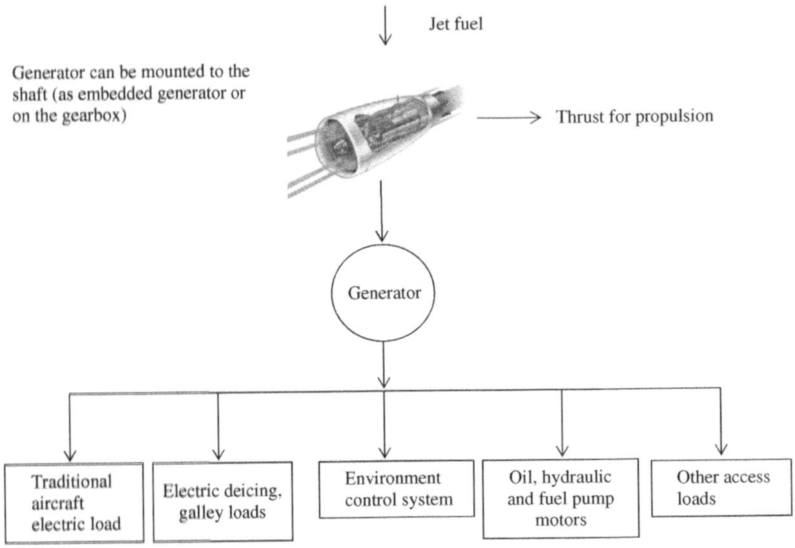

Figure 12.3 Typical more electric aircraft system

The benefits of the MEA are [5, 6] as follows:

- Better power availability throughout the flight envelope owing to the possible shift of power extraction from high spool to low spool, thus improving engine operability
- Availability of more electric power that enables the design of systems for sophisticated entertainment systems, seating comfort and so on
- Fewer constraints regarding certification of aircraft for commercial use
- Reliability and maintenance of aircraft is improved because electrical systems are easier to monitor and observe performance trends
- Enables faster diagnostics and better prediction of potential failures resulting in less downtime
- Reduction/elimination of bleed air to improve the overall performance of the engine
- Possibility of removal of engine accessory gearbox, thus reducing complexity and weight
- Increased overall performance and reduced fuel consumption and energy usage
- Eliminates high-temperature ducts and flammable fluids required in a traditional aircraft
- Reduced maintenance and ground support

More electric architecture and the associated components/subsystems are of significant interest to air framers, suppliers and the military. One of the primary motives behind the technology is the replacement of most (if not all) of the pneumatic and hydraulic systems in the aircraft by electrical systems. This has been reinforced further by the successful deployment of the Airbus 380 and Boeing 787 in the commercial arena. The MEA architecture offers significant overall system benefits in improving fuel efficiency, reducing emissions and enhancing reliability. On the other hand, the MEA concept imposes increasing demands on the generation, conversion and distribution of electrical power within the aircraft. MEA technologies are evolving continually, and there is great opportunity for improvement as systems continue to be refined and enhanced. The MEA concept is widely recognized as the future technology for the aerospace industry. The Airbus 380 and Boeing 787 systems are the two major aircraft programs that illustrate how the electric power generation and the increased use of power electronics in an aircraft are being achieved.

12.2.1 Airbus 380 Electrical System

The A380 was the first large civil aircraft to incorporate more electric architecture systems with variable-frequency (VF) power generation [6–8]. The A380 electrical system has the following:

- Four 150 kVA VF generators (370–770 Hz)
- Two constant frequency (CF) auxiliary power unit (APU) generators (nominal 400 Hz)
- One 70 kVA ram air turbine (RAT) for emergency purposes
- Four external power connections (400 Hz) for ground power

The major electrical system components of the A380 are shown in Figure 12.4. Each of the main 150 kVA AC generators is driven by the associated engine. The two APU generators are driven by their respective APUs. The main generator supplies power to the appropriate AC bus. The main AC buses cannot be paralleled because the output frequency of the generator depends on the speed of the engine driving the generator. Each output voltage of the generator is controlled by the respective generator

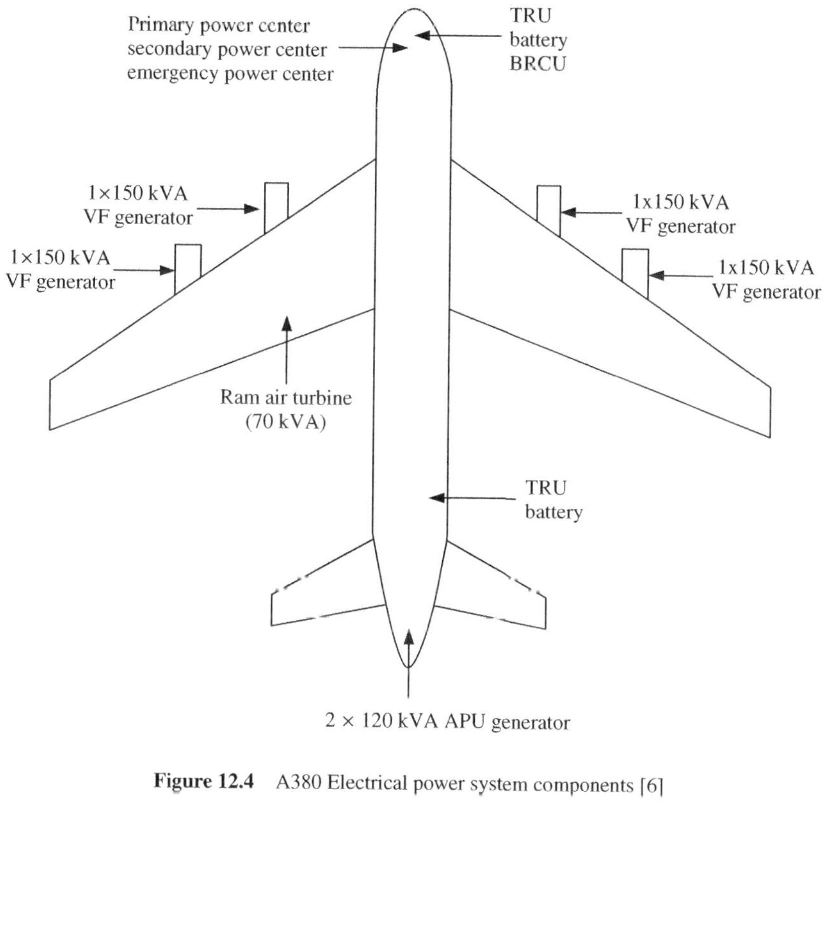

Figure 12.4 A380 Electrical power system components [6]

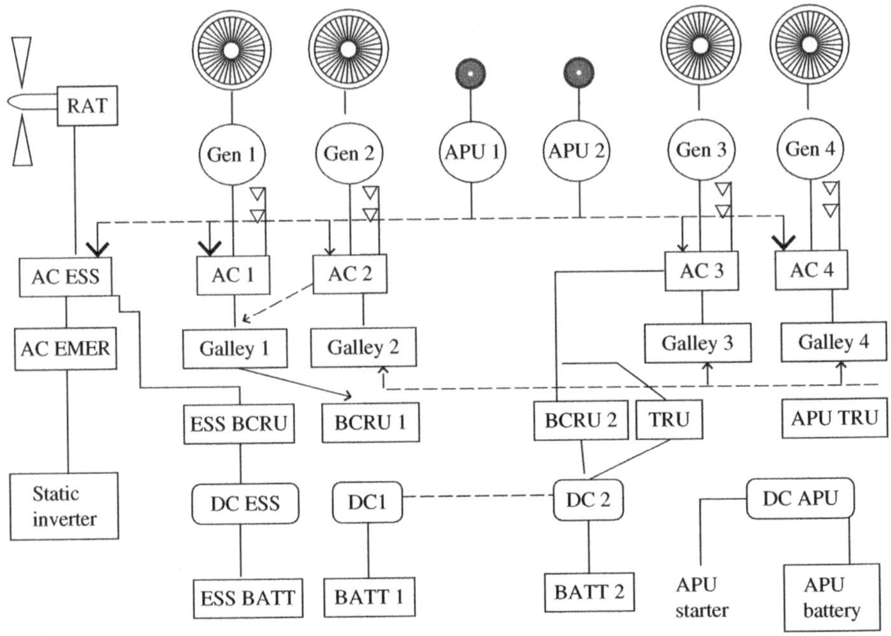

Figure 12.5 A380 DC electrical power architecture

excitation control unit (GCU). The main AC buses can also accept ground power input for servicing and support activities on the ground.

The main characteristics of the Airbus 380 power conversion and energy storage system are shown in Figure 12.5. AC1–AC4 are the AC buses at the output of the four generators, each driven by one of the four engines. There are three Battery Charge Regulator Units (BCRUs), which are based on regulated Transformer Rectifier Units (TRUs) connected to the AC bus. The AC bus also provides power to the galley loads. The DC system provides uninterrupted power capability to power the aircraft's electrical loads, including the control computers or IMA (Integrated Modular Avionics) cabinets, without power interruption during changes in system configuration.

The first A380 maiden flight took place on 27 April 2005. This plane, equipped with Rolls-Royce Trent 900 engines, flew from Toulouse Blagnac International Airport with a flight crew of six and landed successfully after 3 h and 54 min. Presently, the A380 is operating on scheduled flights by different airlines.

12.2.2 Boeing 787 Electrical Power System

The Boeing 787 has most of the features of an MEA system [6, 9, 10]. The electrical power system architecture is shown in Figure 12.6a and b. The electrical power generation system comprises the following:

- Two 250 kVA starter/generators driven by each engine. This results in 500 kVA of generated power per engine

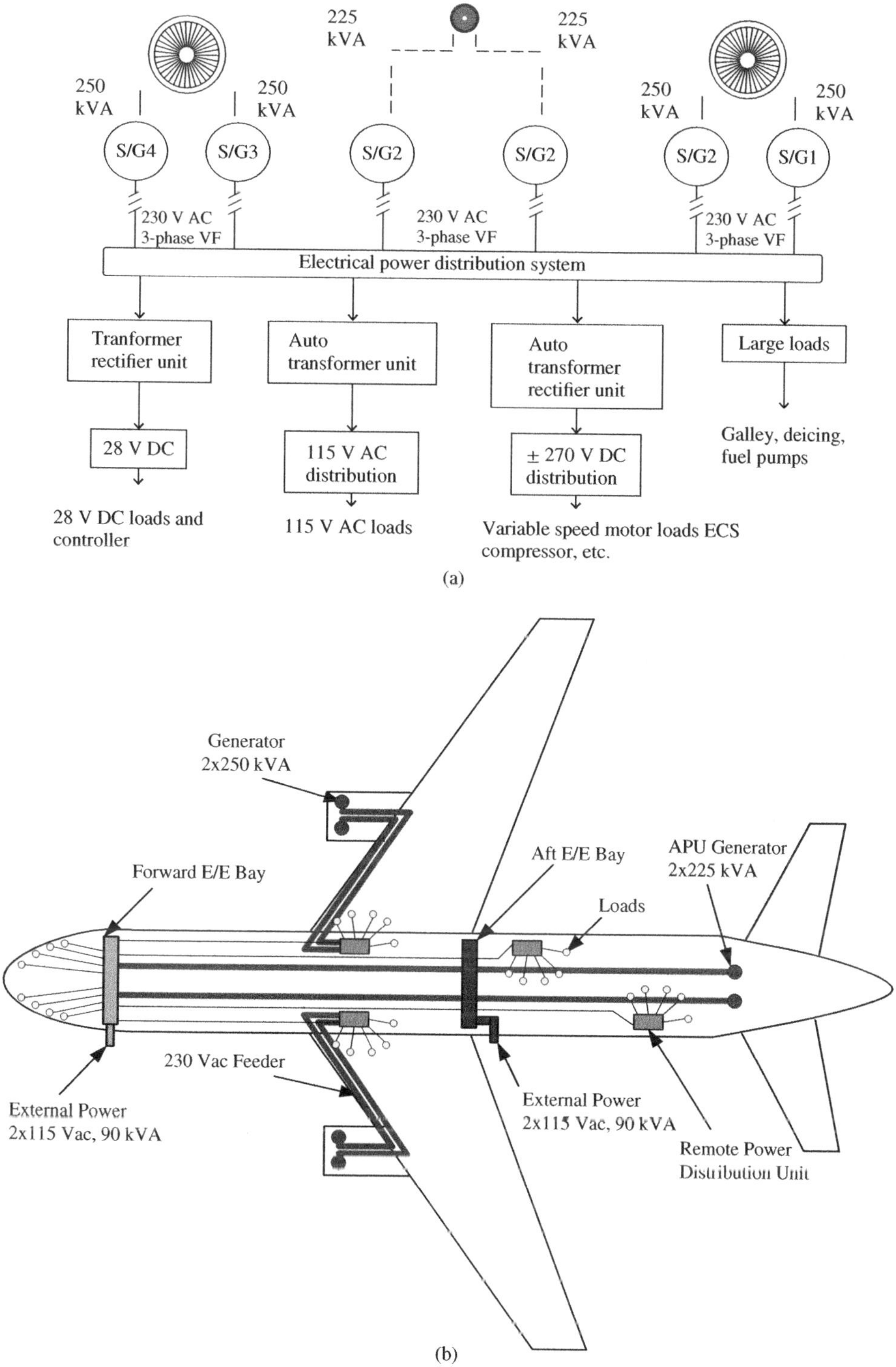

Figure 12.6 (a) Boeing 787 electrical distribution system. (b) Boeing 787 electrical power distribution system (physical locations) [10]

- Two 225 kVA APU starter/generators, each starter/generator driven by the APU engine
- Three-phase 230 V AC electric power generation with VF (380–760 Hz) compared with that of the conventional three-phase 115 V AC at 400 Hz arrangement. The increase in voltage from 115 to 230 V AC decreases losses in the cabled electrical distribution system

The maximum capability of combined power generation from the main engines and APUs is 1450 kVA. In addition to powering 230 V AC loads, a portion of the electrical power is converted into three-phase 115 V AC and 28 V DC power, in order to power many of the legacy subsystems that require conventional power supplies. As the bleed air is not used within the airframe, the ECS, cabin pressurization system, wing anti-icing system and other conventionally air-powered subsystems are all electrically powered. The only bleed air that is used from the engine is low-pressure fan air used for anti-icing the engine cowl. The main electrical loads are as follows:

- ECS and pressurization. Four electric compressors are used for cabin pressurization
- Electrically heated cargo bay
- Four electrical motor pumps, each of 100 kVA for the cooling loop of high-power motor controllers and galley refrigerators
- Wing deicing, which requires electrical power of the order of 100 kVA
- Flight controls
- Four electric motor pumps driven by 88 HP motor
- Electric brakes
- Landing gear, which is raised electrically

The maiden flight of the Boeing 787 took place on 15 December 2009, and flight testing was completed in mid-2011. The aircraft entered into commercial service on 26 October 2011.

12.3 More Electric Engine (MEE)

The embedded generation system, together with the use of electric pumps, will lead to a More Electric Engine (MEE) system [11–13]. The MEE replaces the current hydraulic, pneumatic and lubrication systems with electrical systems. This results in a lighter, more efficient, better performing, more reliable and less costly engine that can be more easily integrated into airframe systems.

A typical MEE architecture is shown in Figure 12.7. An electric machine with starter/generator capability is mounted on the high-pressure (HP) shaft of a two-spool engine and a generator is mounted on the low-pressure (LP) shaft of the engine. In addition, the oil, fuel and hydraulic pumps are driven electrically by their respective electric motors. The LP- and HP-shaft-mounted electric machines could be traditional wound-field synchronous machines or permanent magnet (PM) or switched reluctance machines. The motors driving the pumps could be PM brushless DC motors or induction machines. These motors are controlled using pulse-width modulation (PWM) inverters. The 28 V DC required for FADEC (Full Authority Digital Electronic Controller) and other flight controls are derived using a DC–DC converter operating from the 270 V DC supply. This 270 V DC power is obtained from the aircraft's APU or from a battery. For ground starting of the engine, the 270 V could also be derived from a ground-based APU.

12.3.1 Power Optimized Aircraft (POA)

The Power optimized aircraft (POA) is an MEE program that was demonstrated with the support of European funding [14–16]. POA was initiated in 2002 and was aimed to address and integrate technologies for a more efficient aircraft. The principal objective was to validate at an aircraft level, both qualitatively

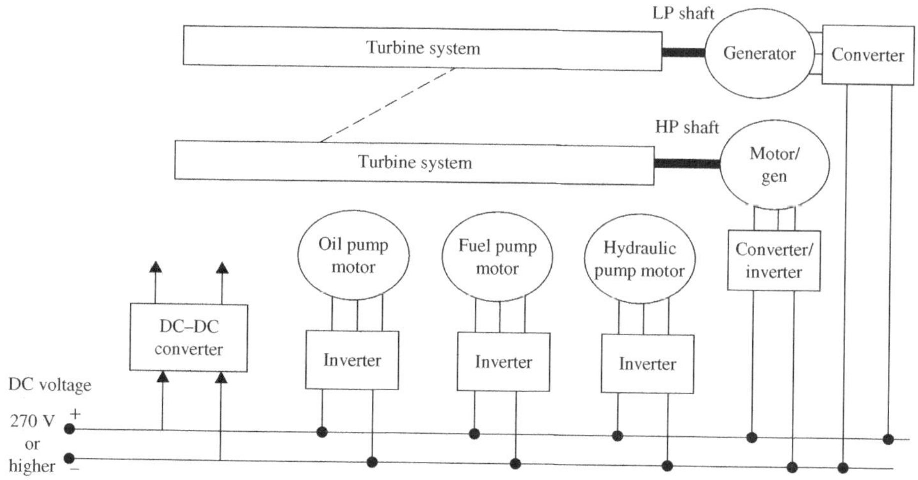

Figure 12.7 More electric engine system

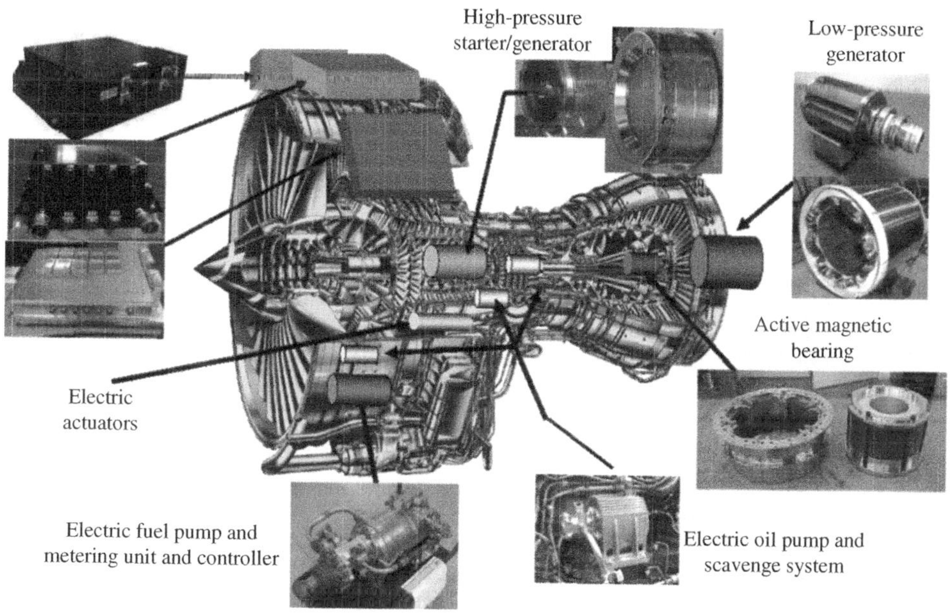

Figure 12.8 POA engine [16] (Reproduced by permission of Rolls Royce)

and quantitatively, the potential of the next-generation systems' equipment for effective reduction in consumption of non-propulsive power. The features of the POA system, as shown in Figure 12.8, are as follows:

- No external gearbox
- Electric-driven fuel pump

- Electric vane actuation
- Integral starter/generator on HP spool
- Fan shaft generator on LP spool
- Integrated power system
- Flight weight power electronics
- Oil-less LP turbine sump via active magnetic bearing

In 2008, the POA engine was tested successfully in both starting and generating modes, and a series of engine tests were conducted. Although most of the objectives were accomplished, but the reduction of aircraft total equipment weight and the reduction of aircraft fuel consumption by 5% appear too ambitious in hindsight. The embedded generation still needs significant development to ensure a viable whole engine solution. The POA project provided valuable insight into a very complex system and highlighted trends for future research and development.

12.4 Electric Power Generation Strategies

Since the beginning of the jet age, aircraft have become increasingly complex and they operate a vast array of electrical devices [17–19]. Modern military aircraft are equipped with powerful radars, sensors, weapon systems and sophisticated cockpit displays that require large amounts of electricity to operate. Commercial airliners need to provide power for environmental systems, galley equipment, cockpit displays, communication systems, weather radar, flight instruments and in-flight entertainment systems. Hence, the primary function of an aircraft's electrical system is to generate, regulate and distribute electrical power throughout the aircraft, and provide the required power to all the electrical loads. In the past, airplanes used small generators that supplied DC power only, typically at 28 V, to meet the electrical power requirements. The trend in the aircraft industry is to operate electrical components on many different voltages, both DC and VF AC, as in the Boeing 787. However, most present aircraft systems still use 115 V AC at 400 Hz and/or 28 V DC.

Aircraft are equipped with a number of power generation systems, including both primary and redundant backup systems, to supply power to critical equipment in an emergency. Primary power is usually provided by AC generators driven by the jet engines. Commercial and many military aircraft are equipped with an APU, which is essentially a mini jet engine being used as an additional power source. Many aircraft carry a RAT that can be deployed when needed to provide emergency power. If the main engine and the APU both fail, the RAT is generally deployed. The purpose of the RAT is to keep critical systems operating long enough for the aircraft to land safely. Different types of electrical power generation systems, currently being considered for aircraft, are shown in Figure 12.9 [6].

- CF 115 V AC, three phase, 400 Hz generation types integrated drive generator (IDG)
- Variable speed constant frequency (VSCF) cycloconverter
- VSCF with intermediate DC link
- VF 115 and 230 V, three-phase power generation (380–760 Hz)
- VF at 115 V and then converted to 270 V DC bus voltage
- Permanent magnet generators (PMGs) used for generating 28 V DC emergency electrical power for high integrity systems

The IDG is used for powering the majority of civil transport aircraft today. The constant speed drive (CSD) works like an automatic gearbox maintaining the shaft speed of the generator at a constant rpm to provide a CF output of 400 Hz, usually within 10 Hz or less. This has to cater for a 2:1 ratio in engine speed between maximum power and ground idle. The drawback of the hydromechanical CSD is that it needs to be maintained correctly in terms of oil cleanliness and level.

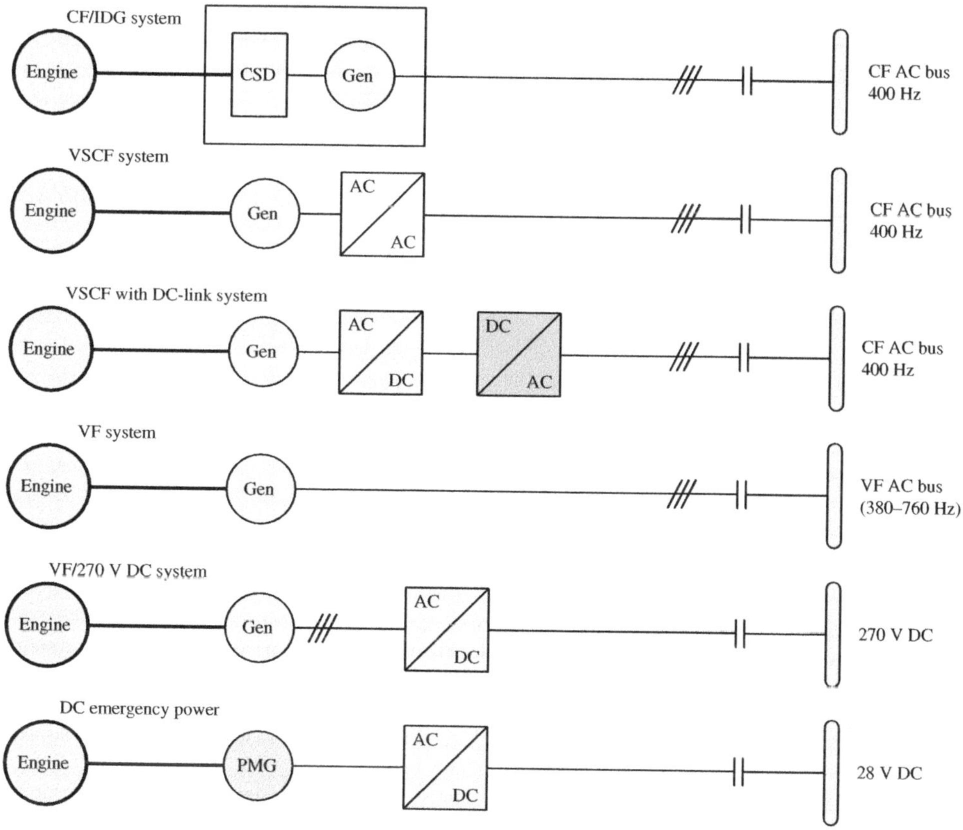

Figure 12.9 Electrical power generation strategies

In the case of variable speed generators, the VF power produced by the generator is converted electronically by the DC-link converters or cycloconverters to CF at 400 Hz with 115 V AC power. In the DC-link converters, the VF voltage of the generator is first converted to an intermediate DC power using AC–DC converters, before being converted (using inverters) to three-phase AC power at 400 Hz with 115 V. The AC–DC converters have been used on the B737, MD-90 and B777 airplanes. The cycloconverter converts directly the VF input voltage to a fixed frequency AC output. In this system, six phases are generated at relatively high frequencies in excess of 3000 Hz, and semiconductor power devices such as insulated-gate bipolar transistors switch between these multiple phases to electronically commutate the input and provide three phases of CF 400 Hz power. Cycloconverter systems have been used successfully deployed in military aircraft in the United States; the F-18, U-2 and the F-117 stealth fighter are a few examples. To date, no civil applications have used cycloconverters because the topology requires a large number of switches and a complex control system.

Due to limited availability of space in the aircraft and as weight is critical to aircraft engine thrust and fuel burn (and thus, the aircraft's range and engine horsepower per pound) three-phase 115 V, 400 Hz power has been the main system power in aircraft. It offers a distinct advantage over the usual 60 Hz used in utility power generation, notably in allowing smaller and lighter power supplies to be used. In MEAs, such as the A380 and B787, instead of 400 Hz, the output frequency of the generator is allowed to vary from about 380 to 760 Hz; thus, the engine speed is freely allowed to vary over a speed range of about 2:1. The wide variation of the frequency could have an effect on frequency-sensitive aircraft

Table 12.2 Recent civil and military aircraft power system developments

Generation type	Civil application		Military application
IDG/CF	B777	2 × 120 kVA	Eurofighter Typhoon
[115 V AC/400 Hz]	A340	4 × 90 kVA	
	B737NG	2 × 90 kVA	
	MD-12	4 × 120 kVA	
	B747-X	4 × 120 kVA	
	B717	2 × 40 kVA	
	B767-400	2 × 120 kVA	
VSCF (cycloconverter)			F-18C/D 2 × 40/45 kVA
[115 V AC/400 Hz]			F-18E/F 2 × 60/65 kVA
VSCF (DC link)	B777	2 × 20 kVA	
[115 V AC/400 Hz]	MD-90	2 × 75 kVA	
VF	Global Express	4 × 40 kVA	Boeing JSF 2 × 50 kVA
[115 V AC/380–760 Hz	Horizon	2 × 20/25 kVA	[X-32A/B/C]
typical]	A380	4 × 150 kVA	
VF	B787	4 × 250 kVA	F-22 Raptor 2 × 70 kVA
230 V AC			Lockheed Martin F-35
270 V DC			Under review

loads, the most obvious being the effect on the AC electric motors that are used in many aircraft systems. The VF voltage is converted to 270 V DC and then converted to VF AC to control the ECS compressor motors and fans, electrically driven hydraulic pumps, nitrogen generating systems (NGSs) and so on. The same converter could also be used for starting the engines electrically. VF is being widely adopted in the business jet community as the power requirements take them above the 28 V DC/12 kW limit of twin 28 V DC systems. Aircraft such as the Global Express has had VF designed in from the beginning.

A summary of the electrical power being generated in different airplanes is given in Table 12.2 [6, 7]. As can be seen, over the years the actual power generation in airplanes has increased gradually resulting in the incorporation of sophisticated safety, control and entertainment systems. However, the diversity of electrical power generation methods has introduced new aircraft systems' issues that need to be addressed. The power converters and the increased electrical loads generate extra heat inside the airplane, thus increasing the ECS requirements. In addition, the power switching devices create electromagnetic interference (EMI) to the other electronic systems that need to be addressed at the architecture level and by selection of systems' immune to EMI. Similarly, the adoption of VF can complicate motor loads, power conversion and protection requirements. As conventional circuit breakers cannot be used at high DC voltages, the US military has initiated the development of a family of 270 V DC protection devices.

Wound-field synchronous generators have been used in most civil and military airplanes as generators. Switched reluctance and PMGs have also been considered in a very few military applications. The advantages of wound-field synchronous machines are as follows:

- Known technology in aerospace applications
- Voltage can be controlled easily by controlling the field
- Fairly well-understood power electronics and control requirements
- Robust

The disadvantages of wound-field synchronous machines are as follows:

- Lower torque-to-inertia ratio compared with that of other types of machines, hence lower power density

- Lower efficiency compared with that of other AC machines
- Separate field and armature voltage control are required during motoring operation
- Need brushes if rotating field windings are used or slip rings if the armature is rotating

The PM machine allows operation without commutators or slip rings and the machine is much less susceptible to issues arising from leakage reactance and poor power factor. The PM machine is often found in a variety of configurations:

- conventional inner rotor configuration
- outer rotor configuration, where the rotor is located on the outer diameter of the stator
- axial gap configuration, where the air gap is not a cylinder
- a radial gap machine, but disk shaped with the rotor displaced axially from the stator

While the interior PM rotor design is predominately employed, the other configurations might have merit in applications that utilize their unique geometrical properties. In addition to the various types of rotor geometry, numerous magnetic materials are also available. PM machines also come in a wide variety of pole count and stator slot combinations. The most desirable features of the PM machine are its efficiency, size, weight and potential for quiet operation. However, these benefits are offset by cost and reliability/durability concerns. Reliability of the machine could be compromised by both the potential corrosion of the magnets and the fact that high temperatures and currents in the rotor can cause demagnetization and, therefore, loss of functionality.

A key requirement for many aircraft generation systems is a degree of fault tolerance, that is, an ability to continue operating at or near-rated power in the event of a single point fault in the generator or its associated converter. One way of achieving fault tolerance is to isolate the phase windings physically, magnetically and electrically. These can be three-phase or higher phase machines, where each phase of the machine is fed separately by a single-phase bridge inverter/converter. Consequently, a fault in one phase will not readily propagate to the adjacent phases. In addition, each phase winding is designed to have one per-unit self-reactance in order to limit the short-circuit current at fault conditions to its rated value. Furthermore, by employing a multiphase (usually >4) design, the machine can continue to provide a useful power output even with a fault in one of the phases. However, this must be balanced against the increased probability of a failure through having more phases. These design features, when combined with the high power density of PM machines, make a fault-tolerant PM machine an attractive option in a safety critical application. Several studies have reported on the possibility of using five-phase fault-tolerant PM machines for embedded aircraft generator applications [20].

The switched reluctance machine (SRM) has been investigated as an alternative to the wound-field synchronous machine as a starter/generator for aircraft engines and other applications [21]. The advantages are in reliability and fault tolerance. The magnetic and electric independence of the machine phases and absence of PMs improve reliability. The mechanical integrity of the rotor permits high-speed high-power density operation. The ability to operate in high-temperature environments and high-speed operation allows the possibility of direct drive and, hence, the elimination of a gearbox and hydraulic system accessories. The simplicity of the SRM translates into a very low cost and reliable machine. However, these machines are generally extremely noisy during operation and have higher torque pulsations, lower efficiency, larger size and weight (than PM machine); and the design has not been advanced to the same extent as the induction or PM machine.

The squirrel-cage induction machine being used as an induction generator in aircraft has not been popular because to obtain the power generation function, it needs to be at supersynchronous speed. Furthermore, a means of providing an excitation has to be incorporated. With the advancement of power converters and control topologies, both these functions can be achieved without much complexity. The induction machines have more fault-tolerant capability than PM machines. These machines have been used in many electric and hybrid vehicles for propulsion and as starter generators.

12.5 Power Electronics and Power Conversion

The power conversion depends on the type of generator being used and the nature of the required output voltage, that is, DC or fixed frequency AC or VF AC. It might also be required to convert power from one form to another within an aircraft's electrical system. Typical examples of power conversion are as follows:

- Conversion from DC to AC power; inverters are used to convert 28 V DC to 115 V AC single-phase or three-phase power.
- Conversion from 115 V AC to 28 V DC power. This function is achieved by using TRUs).
- Conversion from one AC voltage level to another.
- Battery charging. It is necessary to maintain the state of charge of the aircraft battery by converting 115 V AC to a 28 V DC battery charge voltage.
- Conversion to three-phase 115 V AC at 400 Hz from 270 V DC if the main bus voltage is 270 V DC. This conversion is required to power legacy equipment originally designed to operate using these voltages.

A typical power conversion system with various loads is shown in Figure 12.10.

Recent developments in power electronics have advanced active rectifier technologies to the point that they could replace the TRU [22, 23]. Furthermore, the active rectifier could facilitate the replacement of the synchronous machine by an induction machine, because the active rectifier can regulate the voltage at the AC bus of the aircraft. This has significant benefits for the engine by having lower weight and smaller volume of the machine. This enables engine nacelle lines to be redrawn with a reduced frontal area and,

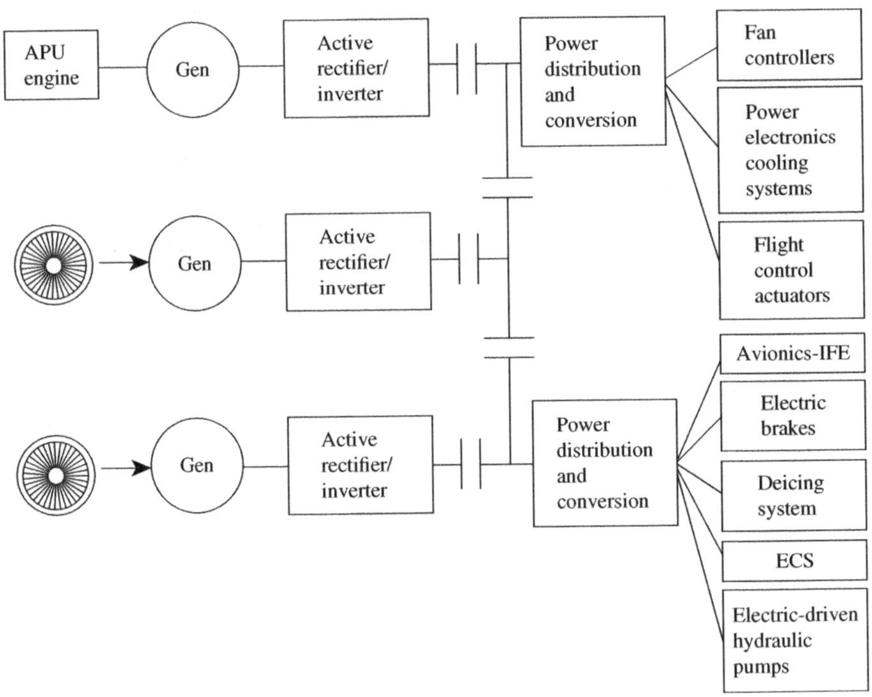

Figure 12.10 Typical power conversion system with various loads

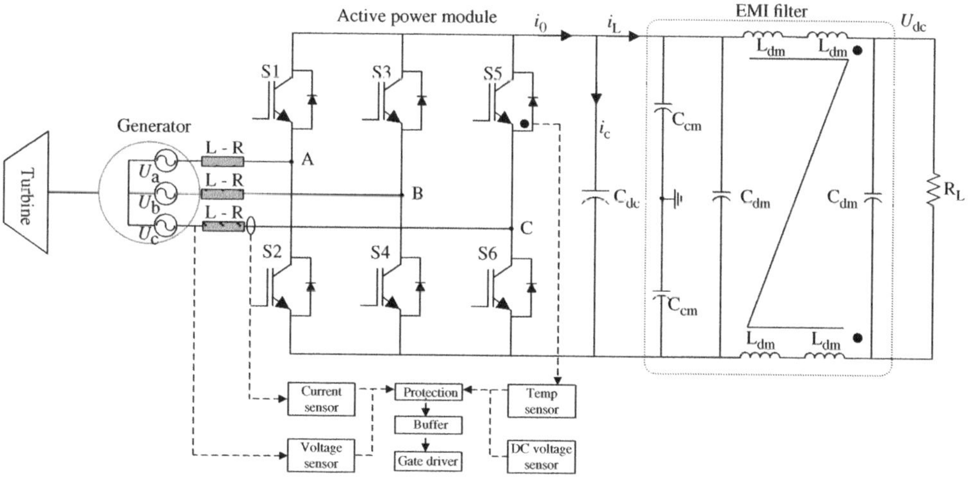

Figure 12.11 PWM active rectifier

therefore, a lower drag. In Figure 12.10, the power conversion from the VF AC output of the generator to controlled DC (±270 V DC) is achieved using an active PWM rectifier; a typical configuration is shown in Figure 12.11. In the rectifier mode, this converter acts as a three-phase boost converter to convert AC to DC. The advantages are that the current or voltage can be modulated with lower harmonics; the power factor can be controlled and can be made lagging or leading; and it can work as a voltage source or current source rectifier. The same power converter could be used as an inverter to convert the DC voltage to VF and variable voltage in order to run the same electric machine as a motor to start the engine. Hence, this topology is used in most starter/generator applications for starting the engine by converting the DC voltage (or battery voltage) to AC to run the electric machine as a motor to start the engine. Once the engine starts and overcomes the peak reactive torque, the electric machine works as a generator to produce the AC voltage at its output. This AC voltage is converted to DC by the active rectifier for powering the accessory motors and other electrical loads of the aircraft, as shown in Figure 12.12.

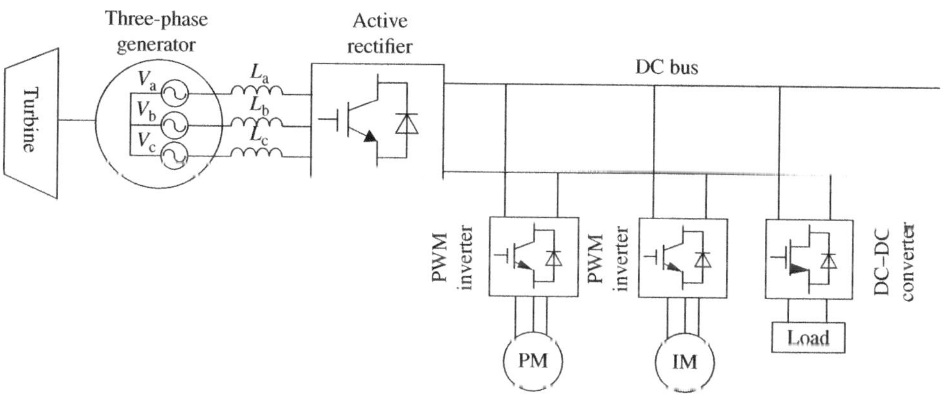

Figure 12.12 Typical DC distribution system with AC and DC loads

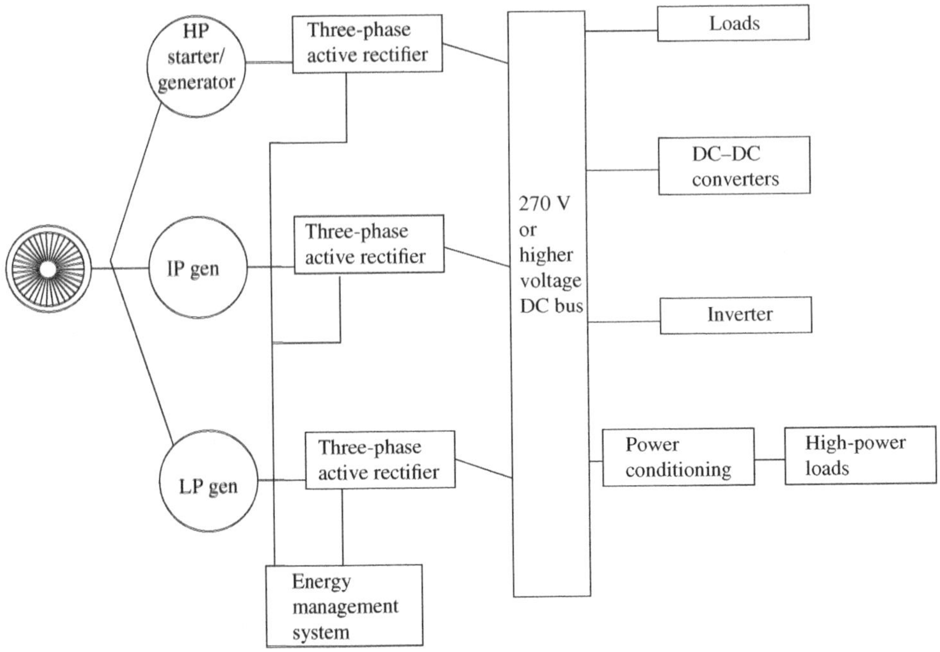

Figure 12.13 Power generation and distribution from a three-shaft engine

The future trend will be off-take of power from each shaft of the engine and load sharing between the buses. A three-shaft engine such as the Rolls-Royce Trent 1000 could be driving three generators with each output connected to a PWM rectifier, which are all paralleled on the DC side, as shown in Figure 12.13. This enables power to be combined from multiple generators operating at different frequency and voltage levels [17].

Although power electronics and electric machine technologies are well advanced, further work is needed in the following areas, particularly for MEAs [18].

- Mitigation of EMI/EMC and improving power quality
- Need for high-temperature materials and components, power devices and passive components
- Power device packaging required to withstand large temperature variations and high thermal cycle capability
- Alternative power conversion topologies with inherently reduced passive components
- Passive components with reduced weight and volume and with high-temperature capability with increased operating frequency
- High power density and high energy density components
- Availability of the components that meet aerospace requirements
- Fault-tolerant power conversion topologies
- Components/systems with long life with stringent vibration and thermal shock requirements

The main objectives are to obtain:

- Higher current density
- Higher power density: weight and volume must be significantly reduced
- High density interconnect

- High thermal conductivity
- Higher reliability and also redundancy
- Ability to withstand harsh environment

In addition to the above items, closer integration of the power electronics and electric machine might also be required. For the MEA, weight, volume including thermal management systems, reliability and redundancy take on a new importance. In addition to the weight and volume constraints, there are key differences in the operating environments for MEA applications. These include the following:

- High-temperature operation: Some applications require generators embedded within the engines making the power electronic systems operate at high temperatures in the range of 200–250 °C. This requires the use of high-temperature capability silicon carbide and gallium nitride power devices and also high-temperature passive devices with efficient packaging technologies
- High-altitude operation: High altitudes give lower ambient pressures that could result in corona discharge. Also, high-altitude flights are more vulnerable to cosmic rays that could cause damage or maloperation of power devices.

12.6 Power Distribution

The power distribution in an aircraft could be AC or DC, as shown in Figures 12.14 and 12.15 [22–26]. Each has its own advantages and limitations. There is an increasing interest in DC power distribution inside an aircraft. High-voltage DC (HVDC) power distribution of about 270 V DC provides a mass and volume advantage over its AC distribution owing to the reduction in the number of feeders from three to

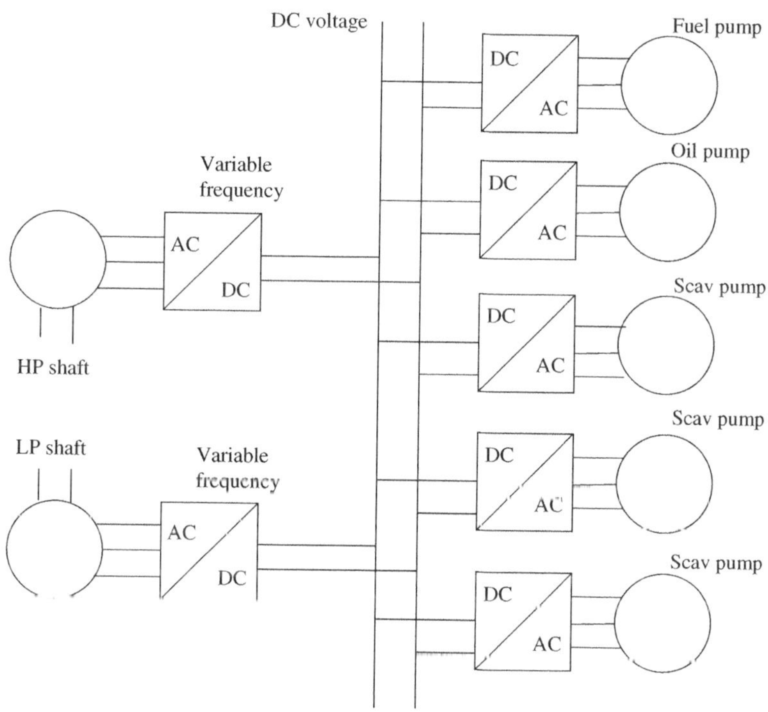

Figure 12.14 DC power distribution architecture

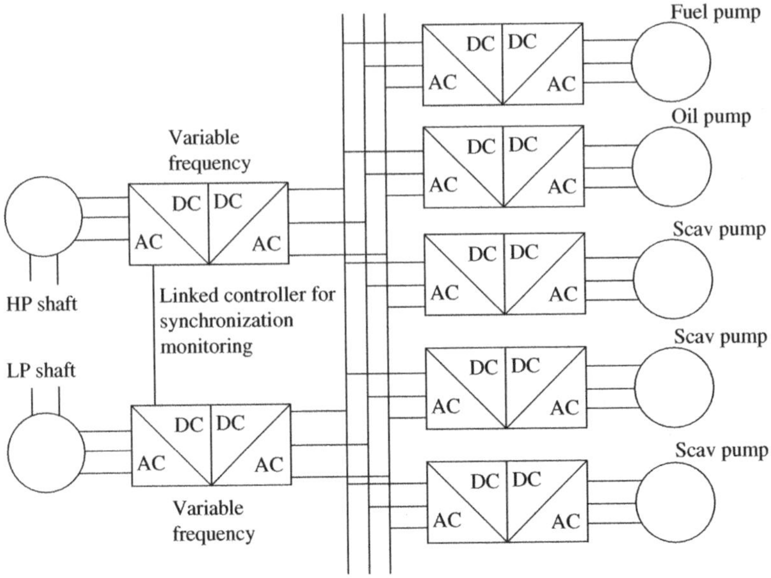

Figure 12.15 AC power distribution architecture

one. As the amount of power to be managed is of the order of MW, the cable size is a concern because of the increased amount of current. In order to avoid this problem, higher voltages have been considered for power distribution inside the aircraft. In the Boeing 787, this is achieved by using ±270 V DC (or 540 V total) rails centered on a zero volt return, but this needs another feeder cable. These HVDC systems enable more efficient use of generated power and aids in paralleling and load sharing between the generators. The 270 V DC is presently produced from TRUs or Autotransformer Rectifier Units (ATRUs) at the output of the generators driven by the engine. But these units are heavy to be deployed in MEA because the power to be handled is much higher than in a conventional aircraft. This is particularly applicable to VF generators where it is not possible to synchronize their outputs directly. A typical DC distribution system is shown in Figure 12.14. The DC distribution enables the use of high-efficiency DC–DC converters and inverters to provide power to the aircraft avionics and accessory motors. However, some issues still have to be addressed for full acceptance of DC distribution in aircrafts. In DC systems, the fault interruption needs specially designed DC contactors or circuit breakers that are bulky because a DC system offers no naturally occurring zero current at which fault can be interrupted.

For many years, AC distribution systems have been the standard for the primary power of aircraft at 115 V (phase voltage), three-phase, 400 Hz. All airports have this voltage for ground power equipment to connect to the aircraft. The use of AC distribution enables the use of a wide range of contactors, relays, and circuit breakers to switch the AC. As AC has a natural zero crossing point, the fault could easily be cleared compared with that of a DC distribution.

The limitations of AC distribution are as follows:

- The AC output from two generators cannot be easily paralleled because they have to be synchronized in magnitude, phase and frequency. Even if both generators are designed to the same specifications and run at the same speed, there could always be a phase difference in voltages of the generator output.
- The motors used for actuators, electric pumps and electric ECS are generally all AC motors. In order to control the speed and torque of these motors, the input frequency and voltage have to be varied over

a wide range. Hence, the distributed AC has to be first converted to DC using the power electronics and then converted to VF and variable-voltage AC using inverters.

- Depending on the frequency of the AC system, the reactive power also has to be managed in the distribution system.
- The feeders within the 115 V AC systems would be heavy, particularly for large loads. For example, the Wing Ice Protection System and the ECS will require both large currents and heavy feeders if supplied from a 115 V AC system.

12.6.1 High-voltage operation

Whether it is an AC or DC distribution system, the main advantage of a higher voltage system is the smaller cable size because of the reduced current for the same power [22, 27, 28]. Ground-based power systems always operate at a high voltage to minimize the I^2R losses and conductor size. Current aircraft electrical systems operate under well-established conditions, namely 115 V AC, altitudes of up to 60 000 ft and temperature cycles between −55 and 200 °C over a wide range of humidity. Moving to higher voltages (>270 V) could lead to several problems.

- The active and passive components capable of switching at these levels of voltage, with characteristics qualified for aerospace applications, have only recently come to the market and have limited availability.
- Although some aerospace-qualified DC contactors at 270 V DC are available, the technology of contactors operating at higher than 270 V DC and the protection systems are still not advanced for aerospace applications.
- The safety aspects related to the use of more than 540 V still needs to be investigated.
- Increased voltage can lead to undesirable effects from tracking and partial disruptive discharges, particularly at higher voltages because of the lower pressures at higher altitudes.

The breakdown voltage of an air gap in uniform field conditions is a function of the product of pressure and gap distance, as stated by Paschen's law, named after Friedrich Paschen in 1889. This law describes the breakdown voltage of parallel plates in a gas as a function of pressure and gap distance. According to his observation, the voltage necessary to arc across the gap decreased up to a point as the pressure was reduced. It then increased gradually, eventually exceeding its original value (Figure 12.16). He also found that decreasing the gap with normal pressure caused the same behavior in the voltage needed to cause an arc. Paschen observed that the minimum breakdown voltage between two conducting surfaces for a uniform electric field in air can be as low as 327 V. In practical terms, this means 5.8 kV that is required to breakdown a 1 mm gap at atmospheric pressure reduces to 1.1 kV at an altitude of 50 000 feet, where the atmospheric pressure is approximately one-tenth of that at sea level.

High-frequency switching in power conversion introduces new operating regimes for materials developed for more conventional DC and AC systems. The effect of microsecond level transients with high repetition rates on corona and insulating systems is not well understood and is of concern in electrical systems, particularly where PWM control is used. The use of DC reduces the peak voltage for the same current carried in a conductor and, therefore, increases the available voltage margin before corona begins. Furthermore, in DC, a system with a potential difference of X can be operated as a two-bus/wire system with voltage of $\pm X/2$ (as in the Boeing 787); the risk of corona onset can thereby be reduced still further. More investigation of corona is required; however, it can be clearly seen that DC has an advantage over AC with respect to levels of corona onset. The extent to which this advantage is reduced by transient effects emanating from power electronic converters requires further analysis.

The increased voltage rating does not always compensate the reduction in current rating because of the insulation thickness required for a given voltage. There should be a trade-off between the current reduction and increased insulation thickness. Moving to a higher voltage does not necessarily imply that

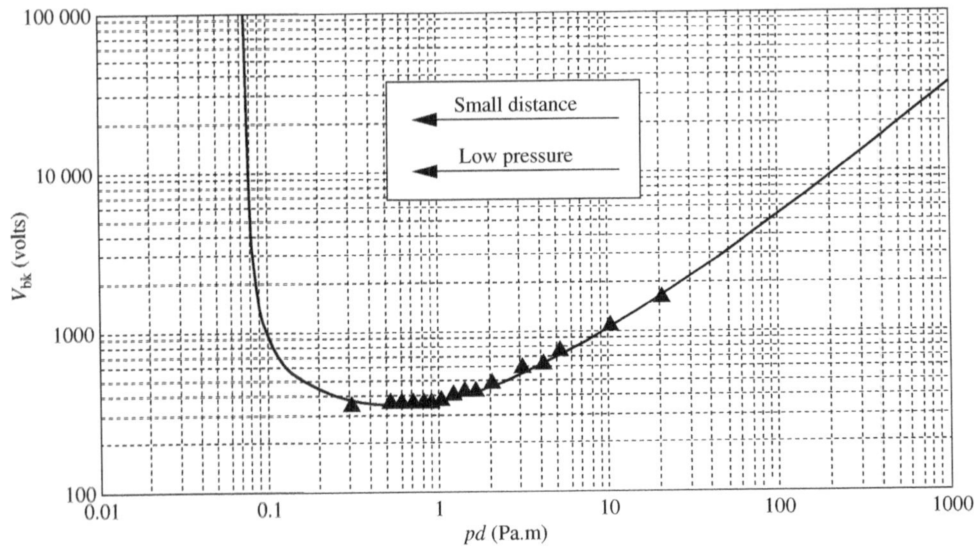

Figure 12.16 Illustration of Paschen's law (p is the pressure and d is the gap distance)

the system's weight will be reduced. By looking at the MEA as an entire integrated system and not as individual components, a decision related to the individual components and the voltage level has to be made. Hence, for a given aircraft, it is necessary to study the architecture of the entire system and understand the requirements of each individual component.

12.7 Conclusions

MEA technologies are evolving continually and there is much opportunity for improvement as systems continue to be refined and enhanced. As MEA technologies advance, smaller components will be used, continuing the reduction of cost of the components and improvements in operating efficiencies. The long-term goal is an "all-electric" aircraft with MEA being the evolutionary step. The transition to an all-electric aircraft, where the propulsors will be driven by the electric motors, is still many years in the future. Meanwhile, MEA will bridge two eras in aircraft technology as airplanes shed some of the traditional pneumatic and hydraulic systems for lighter, simpler, electric and electronic replacements.

Power electronics plays a significant role in the advancement of MEA technologies in terms of improving system efficiency, architecture, size and so on. In terms of power generation, in all likelihood, the aircraft's primary electrical power will remain as AC obtained from generators driven from the engines for the foreseeable future. Some interest has been shown in fuel cell technology, which could produce DC output for ground power, where its quiet operation would compare favorably with that of the APU. However, this is probably still some way off for both the military and commercial aircraft markets. The use of hybrid fuel cell APUs consisting of a solid oxide fuel cell and gas turbine has also been examined for powering electrical loads in an aircraft, but this technology is still in the feasibility study stage. The distribution of the primary power could be AC or DC and each approach has its merits and limitations. The move to a higher voltage generation and distribution would lead to a decrease in the mass of cables and loads and significant benefits from the use of higher voltages can be derived.

The main objectives are to obtain high power and volume density, high efficiency, reliability and the ability to withstand harsh environments. In addition, closer integration of power electronics and the electric machine might be required for operation in the hostile engine environment. In addition to the above requirements, achieving lower weight and volume is very critical in (MEA) systems.

References

1. Intergovernmental Panel on Climate Change (1999) IPCC Special Report Aviation and the Global Atmosphere, http://www.ipcc.ch/pdf/special-reports/spm/av-en.pdf (accessed 18 December 2013).
2. European Aeronautics: A Vision for 2020 (2001) Meeting Society's Needs and Winning Global Leadership, January 2001, http://www.acare4europe.org/sites/acare4europe.org/files/document/Vision%202020_0.pdf (accessed 18 December 2013).
3. ICAO (International Civil Aviation Organization) Aircraft Engine Emissions, http://www.icao.int/environmental-protection/Pages/aircraft-engine-emissions.aspx (accessed 18 December 2013).
4. McLoughlin, A. (2009) More electric – ready for take-off? European Power Electronics, EPE 2009, Barcelona, Spain.
5. Rosero, J.A., Ortega, J.A., Aldabas, E., and Romeral, L. (2007) Moving Towards a More Electric Aircraft. IEEE A&E Systems Magazine, pp. 3–9.
6. Moir, I. and Seabridge, A. (2008) *Aircraft Systems: Mechanical, Electrical and Avionics Subsystems Integration*, 3rd edn, John Wiley & Sons, Ltd.
7. Moir, I. (1999) More-electric aircraft – system considerations. IEE Colloquium on Electrical Machines and Systems for the More Electric Aircraft (Ref. No. 1999/180).
8. Avionics Today (2001) A380: More Electric Aircraft. Avionics Magazine (Oct. 1, 2001), http://www.aviationtoday.com/av/issue/feature/12874.html (accessed 18 December 2013).
9. Nozari, F. (2005) Boeing 787 no-bleed electrical systems architecture. Electric Platform Conference, April 19, 2005.
10. Sinnett, M. (2007) 787 no-bleed systems: saving fuel & enhancing operational efficiencies. Aero Quarterly, o4/2007.
11. Cloyd, J.S. (1997) A status of the United States Air Force's more electric aircraft initiative. IECEC Proceedings, pp. 681–686.
12. Jones, R.I. (1995) Consideration of all electric (Accessory) engine concept. *Proceedings of the IMechE*, **209**, 273–280.
13. Provost, M.J. (2002) The More Electric Aero-engine: a general overview from an engine manufacturer, *International Conference on Power Electronics, Machines and Drives*, (Conference Publication No. 487), pp. 246–251.
14. Faleiro, L. (2006) Summary of the European Power Optimised Aircraft (POA) project. 25th International Congress of the Aeronautical Sciences, ICAS 2006.
15. Joël, B. (2010) Key enablers for power optimized aircraft. 27th International Congress of the Aeronautical Sciences, ICAS 2010.
16. Hirst, M., McLoughlin, A., Norman, P.J., and Galloway, S.J. (2011) Demonstrating the more electric engine: a step towards the power optimised aircraft. *IET Electric Power Applications*, **5** (1), 3–13.
17. Avery, C.R., Burrow, S.G., and Mellor, P.H. (2007) Electrical generation and distribution for the more electric aircraft. 42nd International Universities Power Engineering Conference, UPEC 2007, September 4–6, 2007.
18. Rajashekara, K. (2010) Converging technologies for electric/hybrid vehicles and more electric aircraft systems. SAE Power Systems Conference, Paper No. 2010-01-1757, Fort Worth, TX, November 2–4, 2010.
19. Emadi, M. (2000) Ehsani, Aircraft Power Systems: Technology, State of the Art and Future Trends. IEEE AES Systems Magazine (Jan. 2000).
20. Sun, Z., Jason Ede, Wang, J. *et al.* (2007) *Experimental Testing of a 250kW fault-tolerant permanent magnet power generation system for large civil aero engine*, 5th International Energy Conversion Engineering Conference and Exhibition (IECEC), St.Louis, Missouri, AIAA 2007-4829.
21. da Silva, E.R. and Kankam, D. (1996) Potential Starter/Generator Technology for Future Aerospace Application. IEEE Aerospace and Electronic Systems Magazine (Oct. 1996), pp. 17–24.
22. Martin, A. (2009) A Review of Active rectification in aircraft ac systems. Proceedings of the More Open Electrical Technologies of the More Electric Aircraft Forum, Barcelona, Spain, September 2009.
23. Chang, J. and Wang, A. (2006) New VF power system architecture and evaluation for future aircraft. *IEEE Transactions on Aerospace and Electronic Systems*, **42** (2), 527–539.

24. Maldonado, M. (1999) Program Management and Distribution System for a More Electric Aircraft. IEEE AES Systems Magazine (Dec. 1999).
25. Buchheit, C., Bulin, G., Poujol, C. *et al.* (2009) More electric propulsion system. Proceedings of the More Open Electrical Technologies of the More Electric Aircraft Forum, Barcelona, Spain, September 2009.
26. Furmanczyk, K. and Stefanich, M. (2004) Demonstration of Very High Power Airborne AC to DC Converter, Paper Number: 2004-01-3210, Aerospace and Electronics 2004.
27. Cotton, I., Nelms, A., Husband, M. *et al.* (2006) High Voltage Aircraft Power Systems. IEEE Aerospace Systems Magazine.
28. McLoughlin, A. (2009) Engine powerplant electrical systems. Proceedings of the More Open Electrical Technologies of the More Electric Aircraft Forum, Barcelona, Spain, September 2009.

13

Electric and Plug-In Hybrid Electric Vehicles

Arash Shafiei, Giampaolo Carli and Sheldon S. Williamson
Department of Electrical and Computer Engineering, Concordia University, Montreal, Canada

13.1 Introduction

Conventional vehicles which use petroleum as the only source of energy, represent the majority of existing vehicles. However, the shortage of petroleum is considered one of the most critical worldwide issues, and increasingly costly fuel is becoming a major challenge for CV users. Moreover, CVs emit greenhouse gases (GHGs), which means that it is becoming harder for them to satisfy increasingly stringent environmental regulations.

One of the most attractive alternatives to CVs is electric vehicles (EVs) or zero-emission vehicles that consume electric energy only. However, owing to the limited energy densities of the commercial battery packs currently available, the performance of EVs are constrained to be being neighborhood vehicles with limitations of low speed, short autonomy and heavy battery packs. As a successful example, Canada-based ZENN's commercialized EV has an average speed of 25 mph and 30–40 miles driving range per charge.

Currently, hybrid electric vehicles (HEVs) are the most promising and practical solution. Their propulsion energy is usually from more than two types of energy storage devices or sources, one of which has to be electric. HEV drivetrains are divided into series and parallel hybrids. Series hybrids are electric-intensive vehicles, because the electric motor is the only traction source and the internal combustion engine (ICE) merely works at its maximum efficiency, as an onboard generator to charge the battery.

Keeping in mind the goals of creating an energy-wise, cost-effective and overall sustainable society, plug-in hybrid electric vehicles (PHEVs) have recently been widely touted as a viable alternative to both conventional and regular HEVs. PHEVs are equipped with sufficient onboard electric power to support daily driving (an average of 40 miles per day) in an all-electric mode, using only the energy stored in the batteries, that is, without consuming a drop of fuel. This means that the embedded ICE uses only a minimal amount of fossil fuel to support driving beyond 40 miles, which results in reduced GHG emissions.

Power Electronics for Renewable Energy Systems, Transportation and Industrial Applications, First Edition.
Edited by Haitham Abu-Rub, Mariusz Malinowski and Kamal Al-Haddad.

Fuel consumption of PHEVs can be reduced further by charging the battery from the grid. Thus, it is a valid assumption that moving into the future, a large number of PHEV users will most definitely exist and that the overall influence of charging the onboard energy storage system (ESS) cannot be neglected. Related literature firmly states that by 2020, the market share of PHEVs will increase by about 25%. Based on these data, the additional electric energy demanded from the distribution grid for five million PHEVs would be approximately 50 GWh per day. Furthermore, the typical charging time would be 7–8 h, which might make it hard to accommodate these additional loads in the load curve without increasing the peak load. In addition, the required additional charging energy would possibly have an impact on the utility system.

Expanding the electricity system in the conventional way with large generating plants located far from the load centers would require upgrading the transmission and distribution systems. Apart from the high costs, this could take many years before the right-of-way was obtained. Alternatively, smaller power plants based on renewable energy, such as wind energy, which is a cost-effective renewable energy in addition to many utilities could be incorporated, or solar energy could be installed in a fraction of that time on the distribution system; this is commonly referred to as "distributed generation (DG)." Photovoltaic (PV) presents a modular characteristic, which can be deployed easily on the rooftops and facades of buildings. Many corporations are adopting the green approach for distributed energy generation. For instance, *Google* has installed 9 MWh per day of PV on its headquarters, Googleplex, in Mountain View, California, which is connected to Mountain View's section of the electricity grid. Alternatively, it could be used for charging PHEVs during working hours, offering a great perk for environmentally concerned employees. The energy stored in the batteries could also be used for backup during faults. In Canada, the latest projections (2000) indicate that by 2010, renewable DG sources represented at least 5% of the total energy produced and 20% of cogeneration, an increase from actual figures of 1 and 4%, respectively. Therefore, from an environmental point of view, charging PHEVs with solar power is the most attractive solution.

This chapter aims primarily at addressing the practical issues for commercialization of current and future PHEVs and focuses on power electronics-based solutions for both current and future EV technologies. New PHEV power system architectures are discussed in detail. Key EV battery technologies are explained and some of the corresponding battery management issues are summarized. Advanced power electronics intensive charging infrastructures for EVs and PHEVs are also discussed in detail.

13.2 Electric, Hybrid Electric and Plug-In Hybrid Electric Vehicle Topologies

13.2.1 Electric Vehicles

Generally, EVs include electric motors for propulsion and ESSs, such as batteries, fuel cells, ultracapacitors, or combinations of these called hybrid energy storage systems (HESSs). EVs include not only electric cars but also other types of vehicles, such as electric bikes, electric boats, electric airplanes and so on. Nevertheless, the focus of this chapter is mainly on electric cars. Thus, unless otherwise stated, when talking about EVs we mean electric cars. EVs that are supplied from batteries, which are the most common type, are called battery electric vehicles (BEVs). The very first BEVs were made in the mid-nineteenth century because of the operational simplicity of electric motors compared with ICEs, that is, the inherent rotational movement of the shaft. However, over the decades, ICE cars replaced BEVs largely because of the low range of BEVs owing to low specific energy batteries.

The main reasons behind the BEV concept nowadays is reducing environmental pollutants and reducing fuel consumption. As batteries are the main source of energy in BEVs, they should be sized accordingly to provide a reasonable driving range and have low weight. To achieve this, many cells should be put in series and parallel in a battery pack to provide a suitable voltage and power rating for running electric motors. Various issues should be considered in designing battery packs. The charging method of

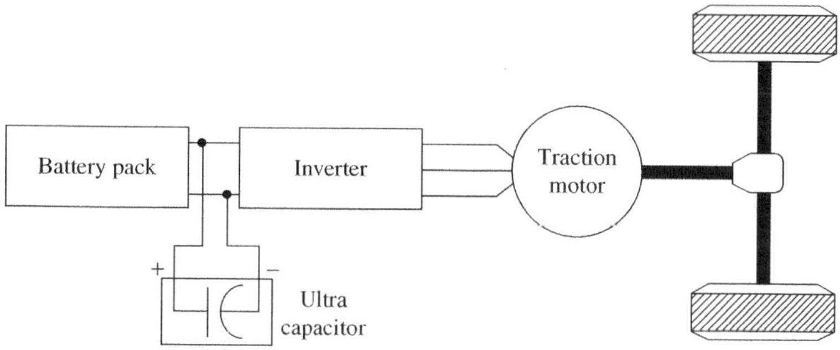

Figure 13.1 Typical drivetrain layout of an all-electric vehicle

battery packs is an important factor from the aspect of battery pack life cycle, which will be discussed later in this chapter. A typical BEV drivetrain topology is shown in Figure 13.1.

Generally, batteries have much lower specific energy compared with fossil fuels. Hence, to provide a reasonable mileage range, very heavy battery packs are needed. Thus, as shown in Figure 13.1, a high power density ultracapacitor bank is required, in order to form a hybridized ESS. This gives an EV an added power capability in addition to energy capacity. However, the above-mentioned issues and some other reasons, such as a lack of charging infrastructure, currently make the options of HEVs and PHEVs more attractive. Nevertheless, battery technology is improving steadily and EVs are the ultimate transportation goal.

13.2.2 Hybrid Electric Vehicles

An HEV utilizes two or more sources of energy for propulsion, for example, gasoline, natural gas, hydrogen, liquid nitrogen, compressed air, wind, solar, electricity and so on. If one of these sources is electricity, this vehicle is called an HEV. This electricity can be provided by a battery pack, fuel cell and so on. HEVs generally combine ICEs with electric motors to run the vehicle. As mentioned earlier, HEVs may include vehicles other than cars; however, here we mean hybrid electric cars, which are the most common type. The main purpose of using HEVs, the same as with EVs, is to reduce the amount of emissions and fuel consumption, which can be achieved in different ways. The simplest way can be just turning the ICE off during idle times, for example, waiting at stoplights; this is called stop–start control strategy. Another idea is to convert the kinetic energy of the car to electric energy during braking, instead of wasting this energy as heat in the brakes. This can improve the mile-per-gallon range by up to 15%. This figure is increasing as the efficiency of the wheel-to-battery path components is improving. Different configurations of ICE and electric motor are possible in an HEV. They will be described in the following sections.

13.2.2.1 Hybrid Electric Vehicle Topologies

Series Hybrid Electric Vehicle

In a series HEV, the ICE acts as a prime mover to run a generator. This generator charges a battery pack, which supplies power to the electric motors. In fact, the ICE is the primary source of power; however, the HEV runs by the direct mechanical coupling of the electric motor, not the ICE. The benefit arises because the ICE can be smaller compared with a car that relies solely on an ICE. Furthermore, compared with the ICE of a normal car that operates at different operating points, the ICE in a series hybrid will operate at the most efficient operating point most of the time, resulting in an improvement of the efficiency

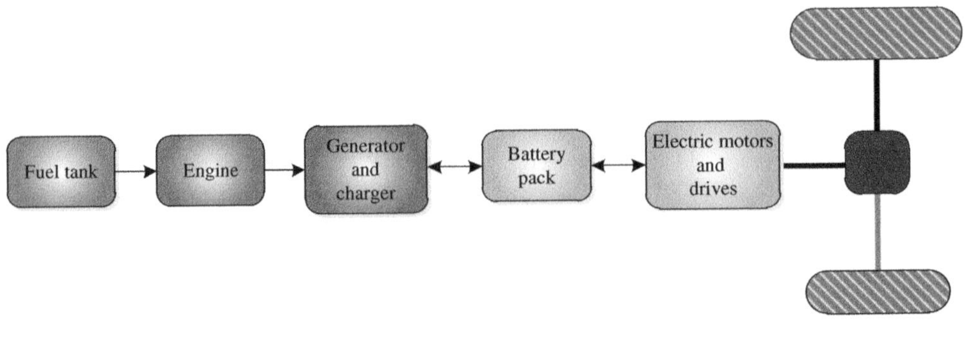

Figure 13.2 Typical layout of a series HEV

of the entire system. In terms of power flow analysis, the power from the generator could supply the electric motors directly. However, to smooth the power transients or variable power demand from the electric motors, it is necessary that the battery pack act as an energy buffer. Depending on the design and to minimize the stress on the battery pack leading to improved life cycle of the battery pack, banks of ultracapacitors might also be used to supply highly transient currents. However, for simplicity, they are not considered here. As mentioned earlier, during braking, the kinetic energy can be transformed into electrical energy using a generator; hence, the electric machine and motor drive in a series hybrid is designed to act as a motor generator. Generally, series hybrids are more efficient for low speeds and urban areas. A typical series hybrid power train is illustrated in Figure 13.2.

Parallel Hybrid Electric Vehicle

In a parallel HEV, in contrast to the series HEV, the ICE can run directly the HEV in parallel to the electric motor. In other words, the ICE or the electric motors or both of them can govern the transmission at the same time depending on the driving conditions and control strategy. A typical configuration of a parallel hybrid is shown in Figure 13.3. In general, a parallel hybrid is more efficient at high speeds, for example, highways. In today's parallel HEV market, the electric motors are of low power and usually less than 30 kW, supported with a relatively small-sized battery pack compared with EVs, because in parallel hybrids the electric motor is accompanied by the ICE and, therefore, does not need to supply all the power. Furthermore, regenerative braking can be supported in parallel HEVs.

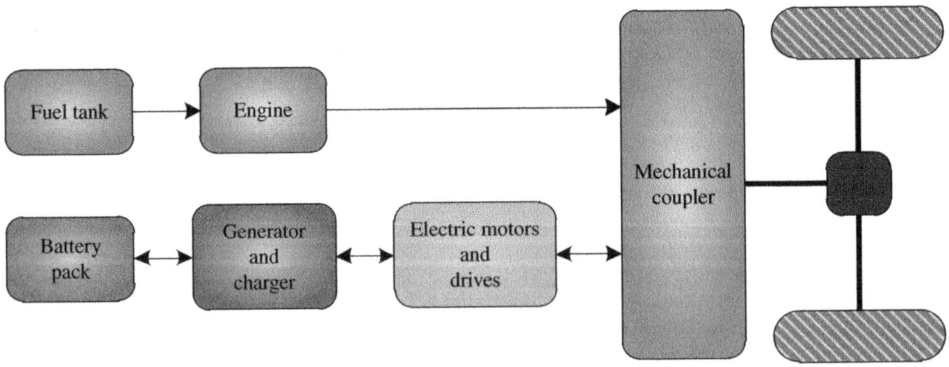

Figure 13.3 Typical layout of a parallel HEV

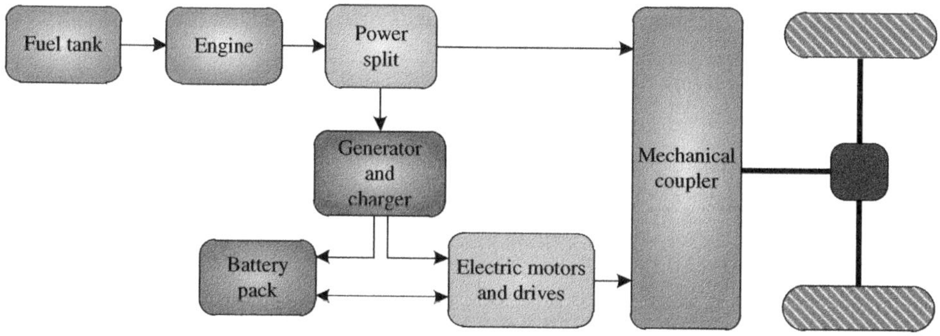

Figure 13.4 Typical layout of a series–parallel HEV

Series–Parallel (Combined or Complex) Hybrid Electric Vehicle
This configuration is also known as a power-split topology; it combines the features of both series and parallel HEVs. This configuration is shown in Figure 13.4. As shown in the figure, a power split is utilized to divide the output mechanical power from the ICE to the drive shaft or electric generator. This ensures that the battery pack is always charged to be able to run the electric motors when needed. A series–parallel hybrid can operate in either series hybrid mode or parallel hybrid mode depending on the driving conditions and supervisory control strategy. A series–parallel hybrid benefits from the advantages of both series and parallel hybrids, that is, it is efficient both in urban areas and on highways.

13.2.3 Plug-In Hybrid Electric Vehicles (PHEVs)

A PHEV is the combination of an HEV and an EV, which can be recharged using an electric plug. In fact, a PHEV benefits from both the hybrid nature of an HEV and the noticeable all-electric range (AER) of an EV. AER simply shows the distance that the PHEV or EV can cover using only the batteries. For instance, PHEV-30 means that the PHEV can cover 30 miles on electricity alone. In a simple HEV, the AER is relatively small because of the small capacity of the battery pack. However, in a PHEV, the vehicle can run for longer on batteries alone. A typical layout of a PHEV is shown in Figure 13.5.

The battery pack of a PHEV is much bigger and heavier than that in an HEV, in order that it can store the required amount of energy [1, 2]. The overall efficiency of PHEVs is much higher than ICE cars. The final usage price depends greatly on the price of electricity, because PHEVs require a relatively significant amount of energy to get charged. For example, charging a PHEV once per day approximately doubles the electrical energy consumption of a mid-sized home. Furthermore, the reduction of the amount of pollutants depends on the source of electricity used for charging, that is, natural gas, hydro, wind, solar and so on. PHEVs have three main topologies, which are the same as HEVs: series, parallel and series–parallel.

Generally, PHEVs can operate in three different modes: charge-depleting, charge-sustaining and blended mode. If the battery has enough charge, the PHEV can operate using only electricity until it reaches the end of state of the charge; this is called charge-depleting mode. The battery pack cannot provide enough energy and power for acceleration if its state of charge (SOC) is low. In contrast, the battery pack cannot absorb available energy from regenerative braking if it is fully charged. Thus, it is desirable that the SOC of the battery pack is kept within a range, for example, 60–80%. If the control strategy operates the ICE and other parts to achieve this, it is called the charge-sustaining mode. In some PHEVs, the control strategy works in such a way that for low speeds, for example, less than 60 km h^{-1}, the vehicle works in charge-depleting mode and for high speeds, it works in charge-sustaining mode. This is called the blended mode. In other cases, the PHEV might operate in different modes for different speed ranges depending on the driving conditions and control strategy. This mode is called the mixed mode of operation.

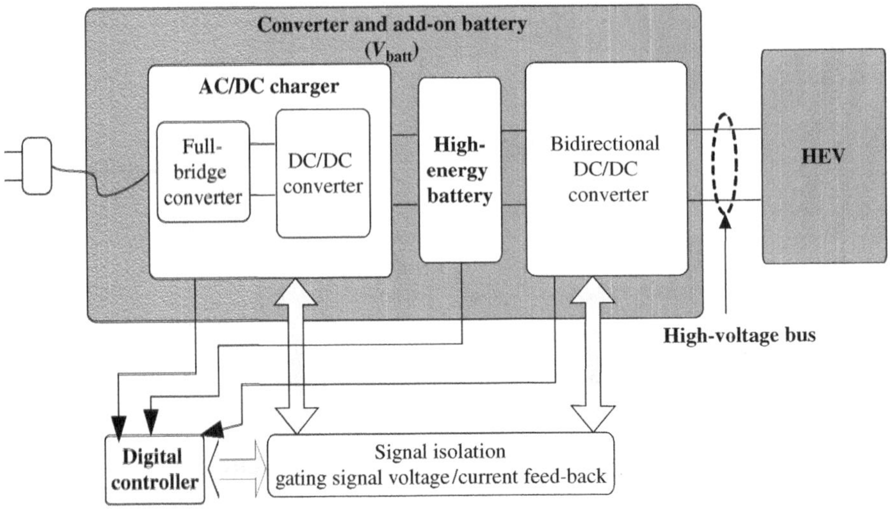

Figure 13.5 Typical layout of a PHEV power system

13.3 EV and PHEV Charging Infrastructures

13.3.1 EV/PHEV Batteries and Charging Regimes

Replacing conventional ICE vehicles with EVs and PHEVs on a large scale could result in tremendous benefits in saving our world from the dangerous and ever-increasing rate of emission of pollutants. The main benefits in moving toward EVs and PHEVs, such as pollution reduction and a decrease in oil consumption, are based on using batteries as a green source of energy. The chemical nature of batteries gives them a highly nonlinear behavior and makes them dependent on many factors, such as chemistry, temperature, aging, load profile, charging algorithm and so on. To have a specific amount of energy for a reasonable AER, tens or hundreds of cells should be connected in series and parallel to make the desirable voltage and current ratings of the battery pack. This causes the nonlinear behavior of the cells to be amplified in some aspects. Furthermore, some phenomena are observed only in battery packs and not in single cells, such as a thermal unbalance among the cells in the pack.

EV and PHEV battery packs are relatively expensive compared with the price of the entire car, owing to the high number of cells and chemistry types, such as lithium-based, protective circuits and so on. Accordingly, the life cycle of these battery packs is very important. Therefore, lower final cost for the customer can be achieved by increasing the battery pack life cycle, which results in a deferred need for the replacement of the entire pack. To get an idea about the price of battery packs, a real example from Honda Civic is mentioned here [3]. There was already news about the Honda Company regarding the battery packs of Honda Civics produced between 2006 and 2008. Apparently, some of the battery packs in second-generation Honda Civic hybrids, which went into production 5 years ago, are failing prematurely. According to regulations in California, there is a 10-year, 150 000-mile warranty requirement on the components of the hybrid system. The Honda Company has taken some action to solve the problem; however, some customers are not satisfied and prefer to change the battery packs themselves. The price of these battery packs is about $2000 excluding shipping and installation costs.

The above-mentioned case shows the importance of the price of battery packs in the commercialization of EVs and PHEVs on a large scale. A factor that affects greatly the life cycle of battery packs is the charging algorithm. However, other factors such as the charging time play an important role in the attraction of EVs and PHEVs. These topics and others related to this area should be handled mainly with a multilevel control and power system called the battery management system (BMS), which takes care of all or some of the aspects affecting batteries. The more accurate and comprehensive the BMS is, the more reliable, safer and faster the charging procedure can be performed. Designing a highly efficient BMS requires very good understanding of the behavior of single cells according to the variations of different parameters and the mutations of these behaviors in a packed state with a large number of cells.

In the following sections, at first we describe some basic definitions and aspects in the field of batteries, and then based on this, we provide some results such as appropriate charging algorithms, which improve the life cycle of batteries. In the following section, we do not want to describe very detailed mathematical definitions of different parameters of batteries that can be used for solving problems and design purposes; on the contrary, it is just intended to give basic definitions, which will help readers, who may be not familiar with these topics, understand the later sections.

13.3.1.1 Battery Parameters

Battery Capacity

This parameter can be assumed simply as the amount of charge that can be drawn from a fully charged battery until it becomes fully discharged. An important effect in batteries is that the higher amount of current drawn from a battery, the lower capacity the battery will have. Hence, theoretically, battery capacity is defined as the amount of current drawn from a battery that discharges it completely in exactly 1 h. For example, a battery capacity of 10 Ah means that if a constant current (CC) of 10 A is drawn from the battery, it will become discharged completely after 1 h. However, in practice, battery manufacturers might use other definitions. Usually, tables of different test results are provided showing the amount of time the battery runs under different CC loads and under different constant power loads. In practice, these tables provide much more practical information rather than standard definitions, because after production, different loads with different characteristics might be connected to the battery. Nevertheless, the amount of time that a battery runs is not predictable exactly, because not all the loads are CC or constant power loads. Even if the loads are one of these types, those tables are valid for new batteries and not for aged batteries. Therefore, in many design procedures, just rough estimates of battery runtime is calculated. The battery capacity is shown in the literature with letters such as "C" or "Q" or other notations. The main unit for battery capacity is the ampere hour (Ah); however, based on the size of the battery, alternative units might be used, such as milliampere-hour or even milliampere-second in the case of very small batteries.

C-Rate

This parameter is used to show the amount of current used for charging the battery, or that of a load drawn from the battery. We can consider the previous case of a 10 Ah battery as an example. When it is mentioned that the charging process is terminated when the charging current falls below $C/10$ rate (10 h rate), this means the charging should be stopped when the current becomes less than the amount of current with which the battery is discharged after 10 h, in other words: 10 Ah/10 h = 1 A.

State of Charge (SOC)

In its simplest form, SOC can be visualized as the percentage of the remaining water relative to the entire capacity of a water tank. In terms of charge, this means the percentage of charge available from a battery relative to the entire capacity of the battery. Representing the battery as a water tank gives a good approximation; however, it is not accurate because of some of the effects in the batteries such as relaxation effect will be described in the following sections. Furthermore, according to aging, the rated capacity of the battery reduces over time, and hence, for determining the SOC, the rated capacity should be measured or calculated regularly.

Depth of Discharge (DOD)
Again, using the water tank concept, the depth of discharge (DOD) can be assumed as the percentage of water that has been drawn from the water tank relative to the entire capacity of the tank. In terms of charge, the water can be replaced with electric charge. This parameter is usually used in discharge pattern recommendations. For example, the battery manufacturer might recommend the user not to exceed 30% DOD in relation to battery lifetime issues.

Energy Density
Energy density can be defined in two ways. One is "volumetric energy density," which is defined as the amount of available energy from a fully charged battery per unit volume ($\mathrm{Wh\,l^{-1}}$). The unit liter is used mainly for measuring the volume of liquids. Mostly, batteries have a liquid electrolyte and so in such cases, it easily makes sense; however, even for solid-state electrolytes such as lithium polymer batteries, the same unit is usually used. The other way of defining the energy density is "gravimetric energy density," which is usually referred to as "specific energy," and is defined as the available energy from a fully charged battery per unit weight ($\mathrm{Wh\,kg^{-1}}$). Based on application and the importance of the volume or weight, either definition can be used. In the case of EVs and PHEVs, the weight factor is usually more important than volume; hence, specific energy would usually be seen in the literature for this specific application.

Charging Efficiency
The chemical reactions inside the battery during charge and discharge are not ideal and there are always losses involved. In other words, not all the energy used to charge the battery is available during discharge. Some of this energy is wasted in other forms of energy dissipation such as heat energy dissipation. The charging efficiency can be defined as the ratio of available energy from the battery, because of a complete discharge, to the amount of energy needed to charge the battery completely. This parameter may be mentioned by other names such coulombic efficiency or charge acceptance. The types of losses that reduce the coulombic efficiency are mainly losses in the charging process owing to chemical reactions, such as electrolysis of water or other redox reactions in the battery. In general, the coulombic efficiency for a new battery is high; however, it reduces as the battery ages.

Hereafter, this chapter will discuss some aspects of batteries in the specific cases of EVs and PHEVs regarding charging battery packs. This will help greatly in the design of more efficient and flexible chargers based on battery behavior, which will finally lead to the improvement of the battery pack life cycle.

13.3.1.2 Important Characteristics of Common Battery Chemistries

There exist many types of batteries, which can be found in battery reference books such as [4]; however, a large number of them are just produced in laboratory conditions and are still under investigation. They are not commercialized because of many factors, such as non-maturity, low energy density, safety, high rate of toxic materials, price and so on. Hence, a small group of batteries is available commercially and within this, the most frequently used are Pb–acid, Ni–Cd, Ni–MH, Li–ion and Li–polymer. Batteries can be divided initially into two categories: primary and secondary. Primary batteries are simply those that can be used only once and after a full discharge cannot be used any more. This is because the chemical reactions happening inside them are irreversible. Secondary batteries, however, can be used many times through recharging. In the case of automotive and traction applications, secondary batteries are of most interest, because utilizing primary batteries in these applications seems unreasonable. Here, we will only consider secondary type batteries and when we mention batteries, we mean secondary batteries unless otherwise stated.

Lead–Acid (Pb–acid)

For over a century, lead–acid batteries have been utilized for various applications including traction. Their well-improved structure has led to valve-regulated lead–acid (VRLA) batteries, which can be considered as maintenance-free batteries, which is a desirable characteristic for PHEVs. In terms of efficiency, they have high efficiency in the range of 95–99%. The main disadvantage of lead–acid batteries is their weight, in other words, they have a low specific energy (30–40 Wh kg^{-1}) compared with their counterparts.

Nickel–Cadmium (Ni–Cd)

Considering low-power applications, nickel–cadmium (Ni–Cd) batteries also benefit from a mature technology, but considering traction applications, their specific energy is low. The typical specific energy for this type is 45–60 Wh kg^{-1}. They are used mainly where long life and price are of high importance. The main applications for this type are portable devices, but in cases that demand high instantaneous currents, there use is desirable. However, considering environmental issues, they do contain toxic metals [5].

Nickel–Metal Hydride (Ni–MH)

These have higher specific energy but lower cycle life compared with that of the previous types. In general, for the batteries of the same size, Ni–MH batteries can have up to two or three times the energy of a Ni–Cd type. The typical value for the specific energy of the present technology Ni–MH batteries is in the range of 75–100 Wh kg^{-1}. This type is widely used in EVs and PHEVs.

Lithium–Ion (Li–ion)

This type has noticeably high specific energy, specific power and great potential for technological improvements, providing EVs and PHEVs with perfect performance characteristics such as acceleration. Their specific energy is in the range of 100–250 Wh kg^{-1}. Because of their nature, Li–ion batteries can be charged and discharged at a faster rate than can Pb–acid and Ni–MH batteries, nominating them as good candidates for EV and PHEV applications. Furthermore, Li–ion batteries have outstanding potential for long life if managed in proper conditions; otherwise, their life cycle can be a disadvantage. One of the main reasons for this is the near absence of memory effect in Li-based batteries. However, safety issues are weak points of Li-based batteries. Overcharge of Li–ion batteries should be carefully prevented, as they have high potential for explosion owing to overheating caused by overcharging. They can easily absorb extra charge and thus could explode. Utilizing advanced BMSs can ensure a reliable range of operation of Li–ion batteries, even in cases of accidents. In addition, Li–ion batteries contain less environmentally harmful materials compared with that of nickel-based batteries.

Lithium–Polymer (Li–Po)

Li–Po batteries have the same energy density as Li–ion batteries but with lower cost. This specific chemistry is one with the greatest potential for EVs and PHEVs. Recently, there have been significant improvements in this technology. Formerly, the maximum discharge current of Li–Po batteries was limited to about 1C rate; however, recent enhancements have led to maximum discharge rates of almost 30 times the 1C rate. This improves and simplifies greatly the storage part of EVs and PHEVs in terms of power density, because in some cases, this can even eliminate the need for ultracapacitors. In addition, there have been outstanding improvements in charging times. Recent advances in this technology have led to some types that can reach over 90% SOC within a couple of minutes, which can increase significantly their attractiveness to EVs and PHEVs because of the noticeable reduction in charging time. Because this type is a solid-state battery, having solid electrolyte, the materials would not leak even in the case of an accident. One of the other advantages of this type is that it can be produced in any size or shape, which offers great flexibility to vehicle manufacturers.

13.3.1.3 Basic Requirements of EV/PHEV Batteries

The basic preferred characteristics of PHEV batteries can be summarized as follows [6]:

1. High specific energy that results in higher AER and fewer recharge cycles.
2. High specific power that results in high acceleration characteristics of PHEVs, owing to the high rates of current available from the battery without causing any permanent damage to the battery pack.
3. High number of charge/discharge cycles available and high safety mechanisms built into the battery because of high power ratings of battery packs.
4. Environmentally friendly aspect of the battery, that is, being recyclable and incorporating low amounts of toxic materials.

Cost is also an important concern for commercializing EVs and PHEVs in a large scale.

13.3.1.4 EV Battery Charging Methods

Generally, charging is the action of putting energy back into the battery in terms of charge or current. Different chemistries need different charging methods. Other factors that affect the choice of the charging method are capacity, required time and others. The following are the most common techniques used in this process.

Constant Voltage (CV)
As is clear from the name, "Constant Voltage" or CV is where a constant voltage is applied to the battery pack. This voltage is a value preset by the manufacturer. This method is usually accompanied with a current limiting circuit, especially for the beginning period of charging when the battery can easily accept high rates of current compared with that of its capacity. The current limitation value depends mainly on the capacity of the battery. Depending on the battery type to be charged, this preset voltage value is chosen. For example, for Li–ion cells, the value of $4.200 \pm 50\,\text{mV}$ is desirable. The accurate set point is necessary because overvoltage can damage the cell and undervoltage causes partial charge, which will reduce the battery life cycle over time. Therefore, the circuit used for charging, which can be a simple buck, boost, or buck/boost topology, depending on the voltage ratio of input and output, should be accompanied with a controller to compensate the source and load changes over time. When the cell reaches the preset voltage value, this causes the battery to enter standby mode, ready for later use. However, the duration of this idle time should not be very long and should be limited based on the manufacturer's recommendations. This method is usually used for lead–acid batteries and Li–ion batteries, using the current limiter to avoid overheating the battery, especially in the initial stages of the charging process [7].

Constant Current (CC)
CC charging means simply applying a CC to the battery with a low percentage of current ripples, regardless of the battery's SOC or temperature. This is achieved by varying the voltage applied to the battery by using control techniques such as current mode control to maintain the CC. The CC technique can be implemented using a "single rate current" or "split rate current." In single rate, only one preset current value is applied to the battery, which is useful in balancing the cells; however, backup circuits must be used to avoid overcharging. In split rate CC, different rates of current are applied based on the time of the charge, the voltage, or both during different stages of charging. This provides charging that is more accurate and balanced charging; however, backup circuits should still be used to avoid overvoltage of the cells. In some cases, the CC method with high rates and low duration can be utilized to extend the lifetime of batteries. However, this is a very sensitive procedure and it should be performed carefully. Ni–Cd and Ni–MH batteries are charged using this method. Ni–MH batteries can be damaged easily owing to overcharging; therefore, they should be monitored accurately during the charging process [8].

Taper Current

This can be used when the source is a non-regulated DC source. It is usually implemented with a transformer with a high output voltage compared with the battery voltage. A resistance should be used to limit the current flowing to the battery. A diode can also be used to ensure unidirectional power flow to the battery. In this method, the current starts at full rating and gradually decreases as the cell becomes charged. As an example, for a 24 V 12 A battery, the charging begins with 12 A when the battery voltage is 24 V. Then it drops to 6 A when the voltage reaches 25 V and then 3 A for 26 V, and finally, it is reduced to 0.5 A for 26.5 V. This is just a hypothetical example and the values are not necessarily valid. This technique is only applicable to sealed lead–acid (SLA) batteries. Taper charging has other disadvantages. As mentioned earlier, this technique uses transformers, which add to the weight of charger and generates heat.

Pulse Charge

This technique involves using short current pulses for charging. By changing the width of the pulses, the average of the current can be controlled. Furthermore, the charging provides two significant advantages. One is the noticeably reduced charging time and the other one is the conditioning effect of this technique, which improves greatly the battery life cycle. The intervals between pulses, called rest times, play an important role. They provide some time for the chemical reactions inside the battery to take place and stabilize. In addition, this method can reduce undesirable chemical reactions that might happen at the electrodes. These reactions can be gas formation and crystal growth, which are the most significant causes of life cycle reduction in batteries.

Reflex Charge

During the charging procedure, some gas bubbles appear on the electrodes and this is amplified during fast charging. This phenomenon is called "burping." Applying short discharge pulses or negative pulses, which can be achieved, for example, by short circuiting the battery for very small intervals in a current limited fashion, typically two to three times bigger than the charging pulses during the charging rest period, results in depolarizing the cell, which speeds up the stabilization process and, hence, the overall charging process. This technique is called, among others, "burp charging" or "negative pulse charging." Different control modes of charging along with waveforms and diagrams can be found in [9]. In addition, there are other charging methods such as current interrupt (CI), which will be explained in detail in Section 13.8.1.9.

Float Charge

For some applications, when the charging process is complete and the battery is fully charged, the batteries should be maintained at 100% SOC for a long time in order to be ready at the time of use. Uninterruptable power supplies (UPSs) are an example of one such application. The batteries should always remain fully charged. However, because of the self-discharge of batteries, they become discharged over time; for example, they may lose 20–30% of their charge per month. To compensate for self-discharge, a constant is applied, which is determined based on the battery chemistry and ambient temperature. This voltage is called the "float voltage." In general, the float voltage should be decreased with the increase in temperature. This causes a very low rate of current, for example, a C/300 to C/100 rate to the battery, which continuously compensates for the self-discharge rate, also prevents sulfate formation on the plates. This technique is not recommended for Li–ion and Li–Po batteries. Furthermore, this method is not necessary for EV/PHEVs, which are frequently used every day. In addition, float charging involves a protection circuit, which avoids overcharging. This circuit adjusts the float voltage automatically and interrupts charging at some intervals based on battery voltage and temperature.

Trickle Charge

Trickle charging is largely the same as float charging with just small differences. One difference is the usual absence of a protection circuit that avoids overcharging. Hence, it is very important to make sure in the design procedure that the charging current is less than the self-discharge rate. If so, they can be left connected to the battery pack for a long time.

13.3.1.5 Termination Methods

During the charging procedure, it is very important to know when to terminate the charging. This is because of two main reasons. One is to avoid undercharge, that is, to make sure that the battery is fully charged, not partially, in order to use the full capacity of the battery. The other is to avoid overcharging, which is very dangerous, especially in the case of high energy density lithium-based EV/PHEV battery packs. If not terminated on time, the overcharging of batteries can lead to over gassing of the cells, especially in liquid electrolyte cells, which results in an increase in the volume of individual cells that cannot be tolerated in a battery pack, which is rigidly packed. Another issue is the overheating of the cells, especially in lithium-based batteries, which can lead to an explosion and fire in the entire pack, because lithium is a very reactive material and combines easily with oxygen in the air. The one thing necessary to begin the combination is sufficient heat.

Choosing different termination criteria leads to different termination methods. Selecting the type of termination of charging process depends on different factors, such as the application and the environment in which the battery is used. The following are the different termination methods used in this process.

Time

Using time is one of the simplest methods, which is mainly used as a backup for fast charging or normally used for regular charging for specific types of batteries. This method can be implemented cheaply; however, because of diminishing battery capacity over time due to aging, the time should be set for a reduced capacity aged battery to avoid the overcharging of old batteries. Therefore, the charger would not work efficiently for new batteries and this leads to a reduction in battery lifetime.

Voltage

As mentioned earlier, voltage can be used as a termination factor. The charging process is stopped when the battery voltage reaches a specific value. However, this method has some inaccuracies, because real open-circuit voltage is obtained when the battery is left disconnected for some time after the charging. This is because chemical reactions happening inside the battery need some time to stabilize. Nevertheless, this method is widely used, usually with a CC technique in order to avoid overheating damage to the battery.

Voltage Drop (dV/dT)

In some chemistries like Ni–Cd, if charged using the CC method, the voltage increases up to the fully charged state point and then the voltage begins to decrease. This is because of oxygen build up inside the battery. This decrease is significant such that the negative derivative of the voltage versus time can be measured to be a sign of overcharge. When this parameter becomes positive, it shows that we are passing the fully charged state and the temperature begins to rise. After this point, the charging method can be switched to trickle or float charge, or terminated completely.

Current

In the last stages of charging, if the constant voltage method is used, the current begins to decrease as the battery reaches the fully charge state. A preset current value such as the C/10 rate can be defined and when the current goes below this value the charging would be terminated.

Temperature

In general, an increase in temperature is a sign of overvoltage. However, using temperature sensors adds greatly to the cost of the system. Nevertheless, for some chemistries such as Ni–MH, methods such as voltage drop are not recommended, because the voltage drop after the full charge state is not sufficiently significant to be relied on. In this case, the temperature increase is a good sign of overvoltage and can be used.

13.3.1.6 Cell Balancing

For applications demanding high power and energy such as EV/PHEVs, numerous cells should be connected in series to provide high voltages and connected in parallel to produce high currents. Hence, in general, high-power and high-energy rates for traction applications are achieved. This seems ideal; however, there are disadvantages involved. It is often claimed that single cells produced by different manufactures are rechargeable hundreds of times over; however, when connected in series their life cycle declines dramatically. This is because of cell imbalances. To get an idea about the significance of this effect, the results of a real experiment from [10] are mentioned here. In an experiment, 12 cells were connected in series. Despite the manufacturer claiming life cycles of 400 cycles, this was reduced to only 25–30 cycles when the cells were arranged in a string. This shows how devastating this effect can be. To deal with this, the reasons for cell imbalance should be understood and managed. Batteries are electrochemical devices. Even in the case of a simple resistor, there is a percentage of error. In the case of batteries, this is magnified. Two different cells produced in the same factory at the same time will have a slight difference in their parameters. One of these parameters is capacity difference. In the case of a battery pack, there are different reasons leading to cell imbalance. As mentioned in [11], there are four fundamental factors leading to cell imbalance: manufacturing variations, differences in self-discharge rate, differences in cell age and charge acceptance variance. Similarly, in [12], cell imbalance is classified as internal sources, which include "variations in charge storage volume" and "variations in internal battery impedance" and external sources resulting from "protection circuits" and "thermal differential across the battery pack."

To explain simply what is happening, again we refer to the water tank visualization of cells. Suppose different cells with different capacities are connected in series. It is like assuming that different water tanks with different volumes are connected using pipes at the bottom of tanks. If the first tank is supplied with water, the level of water in all the tanks rises evenly. After some time, those tanks with lower capacity become full of water whereas others are only partially filled with water. To fill completely the higher capacity tanks, there is no alternative other than overfilling the lower capacity tanks.

Returning to the real situation, it is easy to guess what happens in the case of battery strings. Fully charging the high-capacity cells involves overcharging lower capacity cells. This will lead to excessive gassing and premature drying out of lower capacity cells, and at the same time, sulfate formation in the partially charged cells will lead to a reduction in their life cycle. Overcoming this effect is the main task of cell equalization circuits and their control algorithms. It should be noted that in the case of EVs, the batteries are usually completely charged up to 100% SOC; hence, cell balancing is an important issue. However, in PHEVs, batteries are intended to be kept in the range of 40–80% so that they can provide sufficient energy, while being able to absorb regenerative power at the same time. Cell equalization techniques for series strings fall into three main groups: (i) charging, (ii) passive and (iii) active.

It is important to note that in cell balancing, the SOC is the key point and not voltage itself, although voltage is a good sign of the SOC. However, if other techniques that can determine the SOC are available more accurately, they may be used. As mentioned in [13], cell balancing in a series string really means equalizing the SOC of the cells, which is equivalent to voltage balancing. Voltage is a useful indicator of the SOC. Different SOC estimation techniques will be studied later.

1. Charging: Charging method is simply continuing charging the cells until they are all balanced to some extent. This implies overcharging the cells in a controlled manner, which leads to the full charge of the higher capacity cells. This method is applicable to lead–acid and nickel-based batteries as they can tolerate overcharge to some degree without significant damage. However, this should be implemented carefully because extra overcharge leads to overheating of the cells and, eventually, premature drying

of the electrolyte. Despite the simplicity and low cost of this method, there are disadvantages, such as low efficiency and long times required to obtain cell balance. Experimental results from [14] show that for actual cell equalization of 48 V batteries of a specific chemistry, weeks of time are required. Furthermore, results from [10] show that the extra time needed using this method increases with the square ratio of the number of cells added.

2. Passive: In this method, the extra energy in the lower capacity cells is dissipated in resistive elements connecting two terminals of the cells. This will provide enough time for the higher capacity cells to become fully charged. This method has also low efficiency because of the energy dissipation; nevertheless, it has a higher speed than the charging method. The passive technique is also cheap and easy to implement and the control algorithm can be easily designed.
3. Active: Active cell balancing involves using active electrical elements, such as transistors, op-amps and diodes to control the power flow between the different cells. This flow can be between groups of cells or single cells. Obviously, extra charge is removed from the lower capacity cells and transferred to the higher capacity cells. This speeds up the charging procedure significantly because no energy is dissipated. Just a small amount of energy is dissipated in the circuitry, which can be minimized using zero voltage or zero current switching techniques if possible.

Let us consider lithium–ion batteries, which are one of the most attractive candidates for EV/PHEVs. In this chemistry, the voltage should be carefully monitored and rigorously controlled in the typical range of 4.1–4.3 V per cell because the threshold voltage leading to the breaking down of the cell is very close to the fully charged cell voltage. As mentioned earlier, lithium batteries cannot tolerate being overcharged. Hence, the charging technique is not applicable to them. According to safety issues related to lithium-based batteries, active balancing is the only reliable cell equalization technique for them.

Various types of cell-balancing techniques can be found in the literature. Hence, there is a need to categorize them based on a certain criterion. Based on energy flow, they can be classified into four different groups: (i) dissipative, (ii) single cell to pack, (iii) pack to single cell and (iv) single cell to single cell. It is easy to imagine the operation of each category based on the name. There are advantages and disadvantages for each group. For instance, the dissipative shunting resistor technique is a low-cost technique. In addition, it is easy to control because of the simple structure leading to simple implementation [15].

In addition to the criterion of energy flow for categorization, cell-balancing techniques can be split into three main groups based on the circuit topology: (i) shunting, (ii) shuttling and (iii) energy converter. Non-dissipative techniques such as the pulse-width modulation (PWM)-controlled shunting technique have high efficiency, but need accurate voltage sensing and are somewhat complex to control [16]. In addition, the high number of elements leads to an expensive system. The use of resonant converters increases the efficiency significantly because of the very low switching losses; however, it increases the complexity of the control system [17].

Shuttling techniques work based on transferring the extra charge of a high-capacity cell or cells to an energy-storing component, such as a capacitor or a group of capacitors, and then transferring it to the low-capacity cell or cells [18]. The system would be cheaper if only one high-capacity capacitor was used; however, because of the existence of only one element for charge transfer, the speed of the equalization is lower compared with that of a group of capacitors used. Utilizing a group of low-capacity cells instead of one high-capacity cell is a good idea, although it increases the complexity of the control system.

Most techniques of energy converter cell equalization utilize transformers. The achieved isolation from transformers is an advantage; however, they suffer from more costly weight. A model and transfer function of the energy converter cell equalization system is derived in [19], which can be used for control design purposes.

The above-mentioned cell-balancing techniques are all summarized and explained, together with circuit topologies in [20]. The questions that arise are how much the cells should be balanced and whether the balance range should be in the range of volts or millivolts. As experiments from [13] show for lead–acid batteries, cell-to-cell voltage matching should be in the range of 10 mV, which corresponds to the SOC to provide reasonable improvements in life cycle. This is an important factor, because, for example, if the voltage matching should be in the range of 1 mV, this means that the sensors should be 10 times more accurate and also that the algorithm might need improvement for this case. This means more cost and might be complex. Therefore, there is a trade-off between expense and life cycle. This parameter should be verified experimentally for different chemistries, environments and applications.

As EV/PEHV battery packs do not possess a mature technology and because there are few experimental data available, contradictory claims may sometimes be seen in the literature, one of which is mentioned here. As mentioned earlier, battery packs used in HEVs are usually controlled to remain in the mid-range of the SOC. This is in order that the battery has the ability to absorb enough regenerative current, while still being able to provide sufficient power during acceleration. If the battery is in 100% SOC, absorbing regenerative current will lead to the overcharge of the battery. Cell overcharge is usually sensed through measuring the cell voltage. Some researchers believe that switched capacitor cell equalization techniques (shuttling method) are suitable candidates for applications with no end-of-charge state like HEVs, because there is no need for intelligent control and it can work in both charge and discharge modes [20]. On the other hand, some others believe that according to the nearly flat shape of the open-circuit terminal voltage of lithium–ion cells in the range of 40–80%, the suitability of charge shuttling methods for HEV applications is denied because of the negligible voltage deviation of the cells [15].

13.3.1.7 SOC Estimation

One important piece of information necessary for safe charging is the SOC. Charging algorithms are mainly based directly or indirectly on SOC. Hence, the knowledge of the SOC value is a key parameter in accurate charging. Unfortunately, measuring the SOC directly is impossible, or at least very difficult and expensive to implement. Therefore, generally the SOC is estimated based on other variables or states of the battery. This involves battery models based on which different estimation methods can be utilized or observers can be designed. Precise estimation of the SOC is not an easy task; however, in usual applications battery voltage, which is a sign of SOC, can be used. In the case of high-power/high-energy EV/PHEV battery packs, methods that are more accurate are advisable, despite being more expensive and complex to implement. The more accurate the SOC estimation, the better the charging algorithms can be implemented, which results in life cycle improvements.

As mentioned earlier, the SOC is mainly the ratio of available charge to the rated capacity of the cell. One of the important points in SOC estimation is the rated capacity change over time owing to the aging resulting from the degradation of the electrolyte, corrosion of plates and other factors. This issue is in the field of analyzing the state of health of the battery, and it is called "State of Health Estimation," which is a field of research not mentioned further here.

However, we will mention some SOC estimation techniques. One of the simplest methods is to *discharge* the battery completely and measure the SOC. Although simple, it is very time consuming and does not seem logical to discharge a battery completely just to measure the SOC. Knowledge of the SOC is useful for assessing the current situation of the battery; therefore, if the battery is discharged the state of the battery has changed and there is no more use for knowledge of the previous SOC. Furthermore, in the case EVs/PHEVs, this method is not applicable. Although this method is not used in battery packs, it may be used periodically after long intervals to calibrate other SOC methods.

Another method is *Ampere Hour Counting*, which measures and calculates the amount of charge entering the battery or leaving it through integrating the current over time. This is one of the most common methods used; however, there are some deficiencies. There are always inaccuracies in sensors and although very small, because of the integration over time, these errors can accumulate to a considerable value leading to significant errors. In addition, even supposing a very accurate current sensor, because this integration is implemented usually by digital circuits and numerical methods, there are always calculation errors involved, and again these can show up in larger errors over time. Even if assuming that both deficiencies could be solved in some way, there is another reason leading to inaccuracy. Even if the amount of charge entering the battery is calculated exactly, because of the coulomb efficiency mentioned earlier, less amount of charge is available, which is also dependent on the discharge rate when leaving the battery. One way to reduce these inaccuracies is to recalibrate the integration process each time a specific known set point, such as the fully discharged state, is reached.

Another method for SOC estimation is the *Measurement of Physical Characteristics of Electrolyte*. Obviously, this method is applicable mainly to liquid electrolyte batteries, not to solid electrolyte batteries like Li–Po. In this method, a chemical fact is used, which is the relation of change of some parameters in the electrolyte with the change of the SOC. One of these parameters is the density of the acid. There is an almost linear relation between change in acid density and SOC. This method is very well known, especially in lead–acid batteries. The density can be measured directly or indirectly using parameters, such as viscosity, conductivity, ion concentration, refractive index and ultrasonics.

As discussed earlier, the *Open-Circuit Voltage* of the batteries can be used as an indicator of the SOC. The uncertainty in this method is the fact that batteries under operation need some rest time for their open-circuit voltage to become stable. For some cases, this time can be up to a matter of hours. However, this method is also widely used. The key point in this method is the linear relation of open circuit voltage versus SOC in a specific range of SOC. This range and its slope are different in different chemistries, which should be taken into account.

There are other techniques categorized under soft computation techniques, such as fuzzy neural networks [21] or adaptive neuro-fuzzy modeling [22], which can also be utilized for SOC estimation. Other approaches that can be used include the heuristic interpretation of measurement curves mentioned in [23], such as coup de fouet, linear models, artificial neural networks, impedance spectroscopy, internal resistance and Kalman filters, which are more precise methods but more complicated to implement.

13.3.1.8 Charging Algorithms

Charging algorithms can be defined as the combination of what was mentioned previously and the control of all or part of the parameters affecting battery performance and life cycle in such a way that battery pack charging and the timely termination of charging can be achieved safely and efficiently. Managing the charging procedure of a high-power battery pack with hundreds of cells involves many issues, as mentioned earlier. The control of all these aspects needs efficient and accurate algorithms with reliable safety and backup circuits. The trend toward fast charging, in which huge amounts of current flow into the battery pack producing significant quantities of heat, needs accurate and reliable supervisory control algorithms to ensure safe charging. Managing this complicated task can be handled with some advanced control topics, such as fuzzy logic, supervisory control and decentralized control. In general, the chemistry of each battery needs its own charging algorithm. However, depending on the algorithm, it might be applicable to other types as well, although this should be done carefully according to life cycle issues.

For precise battery charging, the charge/discharge profile of the battery provided by the manufacturer may be used. However, the profile is valid for new batteries, and thus it is better to use other techniques such as data acquisition methods to acquire the charge/discharge profile of the battery. Novel techniques regarding this issue are regularly being introduced in the literature [24].

As mentioned earlier, lead–acid batteries have a mature technology and the infrastructure is already in place; however, they have poor life cycles of the order of just 300–400 cycles. Much effort has been put into research for increasing the life cycle of this chemistry because of its many advantages, such as cost and availability. This chemistry has a common algorithm, which includes four different stages, or perhaps just three based on the application. In the first stage, a predefined CC is applied to the battery pack, which charges the cells at high speed. In this stage, the cell voltages increase gradually because of the SOC increase. This is called the *Bulk Charge* stage. The process is continued until a predefined maximum voltage is reached. These values are recommended by the manufacturer in the datasheet. In the next stage, called the *Absorption Charge* stage, constant voltage is applied to the battery pack. At this stage, the current decreases gradually until it again reaches a predefined C-rate value. Now the cells

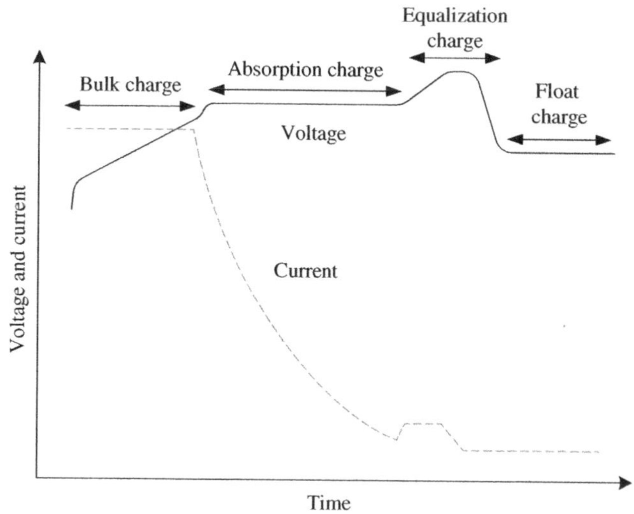

Figure 13.6 Charging algorithm of a typical lead–acid battery

are approximately charged, but not equalized because of cell imbalance. At this stage, a relatively higher voltage than the constant voltage stage can be applied to the pack to balance all the cells inside the pack. This stage is called the *Equalization Charge* stage. This can also be achieved with other techniques mentioned earlier, especially during the previous two stages. After some time, the charger switches to Float Charge mode to keep the battery in a ready state. Depending on the application, this stage can be omitted. This stage is called the *Float Charge* stage. This process is illustrated in Figure 13.6.

As the battery ages, its internal characteristics also change; hence, an adaptive charging algorithm should be used to take into account these changes. Experiments show that the value of the voltage of the third stage should be increased over time to obtain the same amount of energy as the battery ages [25–27]. The equalization stage is the key part of this algorithm and it has great influence on the life cycle of the battery. As mentioned, the voltage of this stage should be increased, but this increases the current and also the heat generated, which has a negative effect on the life cycle. One way to obtain the same amount of current with lower heat dissipation is by using pulses of current. This technique appears the same as pulse charging, but it is different, because the time intervals are significantly larger than the pulse charge periods, which are in the range of kHz. This method is called *Current Interrupt* or CI. This technique has shown significant life cycle improvements [28]. Using this algorithm, the battery can reach 50% of the initial capacity after 500 cycles, which is a significant improvement in life cycle. Although this algorithm is useful, it puts the battery under stress as it nears the end of its life because of its permanent increase in the overvoltage value. This algorithm can be implemented in an alternative way. Instead of using this method in each cycle, which puts high stress on the battery, it can be utilized every 10th cycle. This algorithm is called partial state-of-recharge cycling (PSOR) [28], which has approximately the same effect with the advantage of lower stress on the battery. This algorithm has been claimed to enable the battery to deliver up to 80% of the initial capacity after 780 cycles, which is a noticeable improvement in life cycle.

As can be seen, these complicated algorithms cannot be done using simple proportional-integral (PI) or proportional-integral-derivative (PID) controllers. They need digital signal processor (DSP)-based controllers to be programmable based on the chemistry and state of health of the battery and other factors. Different algorithms for improving the life cycle of batteries are being proposed and tested every day. This is a vast research area, which is developing and receiving more attention as EVs/PHEVs become increasingly popular.

13.4 Power Electronics for EV and PHEV Charging Infrastructure

In its simplest incarnation, a charging facility for EVs would consist merely of a unidirectional AC/DC converter–charger connected to the power grid. Power would simply flow on demand from the power grid, through a power conditioner, into the vehicle battery pack. Once the battery is fully charged, the connection to the grid no longer performs any useful work. This simple set up may have been appropriate for small private commercial vehicle fleets or where electric cars represented a very small fraction of the active road vehicles. However, as society's efforts to electrify our means of transportation intensify, it is clear that a smarter exploitation of the vehicle-to-grid (V2G) interaction is in order. To the power utility, the bulk of EVs connected to its grid appears as an energy storage agent that is too significant to be left untapped. This view is reinforced by the outcome of several statistical studies [29] that show that more than 90% of all vehicles are parked at all times, thus potentially connected to the grid. Assuming a 50% EV market penetration, simple calculations show that the total storage capacity available would be in the order of thousands of GWh. Therefore, the V2G connection should be bidirectional, giving the owner of each vehicle the ability to "sell" back a portion of this stored energy to the utility, presumably at an advantageous rate. The same requirement of bidirectionality also applies when the vehicle is connected to a microgrid powered by a distributed resource. In a grid-connected solar carport, for instance, many vehicles can be charged by PV panels (or by the grid or by both), depending on the load and insolation (time of day, meteorological conditions, time of year, etc.). In the case of overproduction, energy from the panels can be fed back to the grid for a profit, while the EV batteries function to buffer the characteristic solar intermittence. Similarly, the DC–DC converters that condition the power from the solar panels to each charging vehicle should also be bidirectional, in order to allow the owner of a plug-in EV (PEV) to exchange a portion of this energy with the operator of the microgrid (Figure 13.7).

These considerations demonstrate that bidirectionality is a highly desirable feature in any power conditioner utilized in vehicle charging–discharging applications, including interactions to and from the grid, microgrid, or residential loads and renewable energy generators. On this basis, the reader should note that the discussion that follows makes no distinction between V2G and grid-to-vehicle communication, both being classified by the acronym (V2G). Similarly, (V2H) will designate either the vehicle-to-home or the home-to-vehicle interface.

Other requirements for the optimal charging infrastructure are harder to identify. This is because of the pervasive lack of standardization involving battery technology and nominal voltage, safety strategy, connector configuration, communication protocols, location of charger (onboard or off-board) and more. In the following sections, these issues are treated with particular reference to their impact on local power generation and utilization.

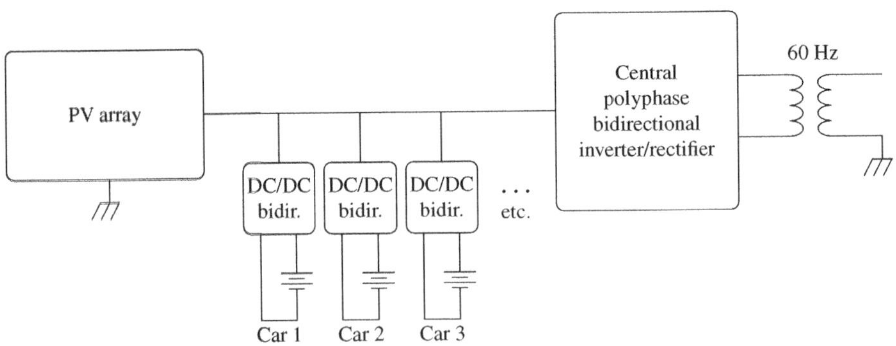

Figure 13.7 Typical PV-powered grid-tied carport architecture

13.4.1 Charging Hardware

Like any other means of transportation, EVs/PHEVs benefit markedly from minimizing their weight. These vehicles are even more sensitive to that issue, considering the unavoidable presence of heavy battery/ultracapacitor energy packs. The electronic power converters intended for the charging function can be bulky and heavy in their own right, and their deployment onboard seems to make little engineering sense. However, at the time of this writing, the great majority of PEVs in North America contains their own power rectifier and can connect directly to 120 or 240 V household plugs. This can be explained by two considerations. Firstly, although the household AC voltages are fully standardized, at least within a specific country, the DC nominal battery voltage for PEVs is definitely not standardized. Different manufacturers have adopted ad hoc energy storage technologies and safety strategies, resulting in strikingly different bus voltages and current requirements. An unsophisticated external converter could then be optimized for only one vehicle brand or model.

Secondly, some techniques have been developed that do not add significant weight to the vehicle. The critical idea is to utilize the power electronic circuitry that is already onboard in order to perform the rectifying function. This charging circuit is commonly referred to as an "integrated charger"; it makes use of the bidirectional inverter that drives the electric motor, as well as the windings of the motor itself. Figure 13.8 shows a well-known example of this concept.

With regard to Figure 13.8, it is important to realize that inductors LS1, LS2 and LS3 are not added magnetic devices, but the actual winding leakage inductances of the electric motor. Thus, the only added components are the two relays K1 and K2, which are activated in order to reconfigure the schematic from a three-phase motor driver, during normal vehicle propulsion operation, to a single-phase boost rectifier, during charging.

The above two considerations are consistent with relatively slow-charging strategies. In the first instance, it is because the amount of electric power available in a residential setting does not usually exceed 10 kW at a household plug; and in the second instance, it is because the electronics that drive a

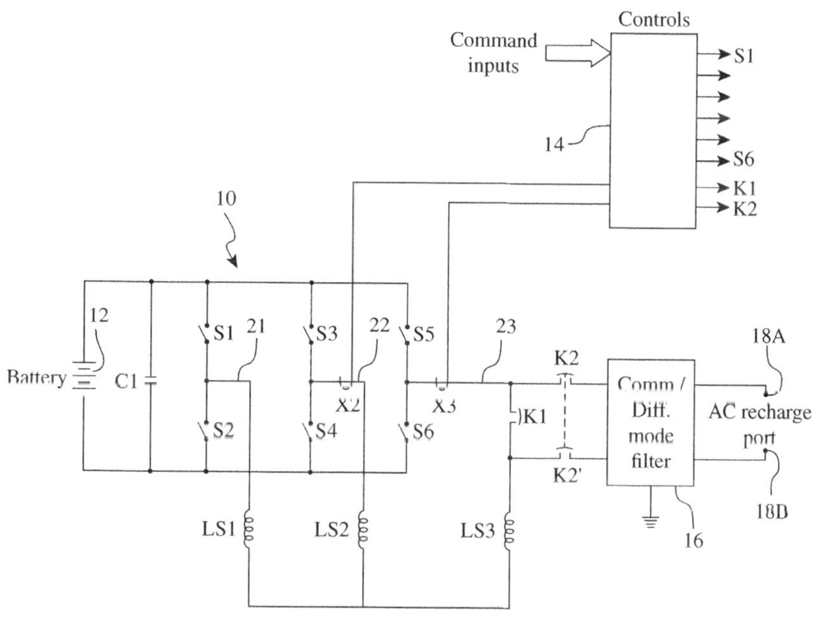

Figure 13.8 Integrated charger based on boost converter [30]

PEV electric machine is sized for its propulsion needs. Thus, the average charging power must be limited to a level comparable to that of the motor's rated power, which is of the order of 10–50 kW in smaller cars.

Slow-charging strategies are commonly referred to as Level 1 and Level 2. The former is associated with a connection to a regular AC household plug (120 V, 15 A), whereas the latter involves power that can be as high as 14.4 kW or 240 V at 60 A, which is also normally available in residential settings. Moreover, these power levels are compatible with the average generating capacity of microgrids and the corresponding distributed resources. Then, it would appear that whether a car is charged through a regular residential wall socket or a microgrid outlet, the available power levels justify the location of the rectifier onboard the vehicle.

However, EV manufacturers are quickly recognizing that long charging periods might be acceptable to consumers only if quick charging is available as well, albeit at higher cost. Two solutions are presently under consideration. The first solution is the so-called battery swapping, whereby a car owner simply drives to a service station and allows an automated system to replace safely a spent battery with a fully charged battery. Along the same lines, the battery could be of the redox flow type. In this case, the battery casing is not replaced; on the contrary, it is drained and then filled with fresh liquid electrolyte. In either forms, the obvious drawback is the need for the exact standardization of battery size, chemistry and capacity.

The second solution consists of allowing direct access to the battery DC terminals, so that a large off-board rectifier could be connected to re-energize the battery pack using powers of the order of up to several hundred kilowatts. This is known as Level 3 charging, allowing an electric "fill-up" service stop to last only a few minutes. In this instance, although the battery itself might not need a high level of standardization, it would be subject to extremely high currents at high voltages. This renders the practical implementation of this second solution strongly dependent on necessary improvements to battery and ultracapacitor technologies. Furthermore, a public charging station capable of servicing many cars simultaneously would represent a local load of several megawatts as seen by the grid.

Despite these difficulties, it is highly likely that either the battery swapping or the fast-charging strategy will eventually be available universally to complement, or even replace, the onboard charger.

13.4.2 Grid-Tied Infrastructure

Assuming that fast charging through direct DC connection becomes the method of choice, car owners will have two options. They may still prefer to slow charge their vehicles overnight by plugging it into an AC–DC charger (or electric vehicle supply equipment, EVSE), most probably in their homes. This converter will deliver relatively low power of the order of 5–10 kW because of the limitations of the residential connection, as mentioned earlier. However, as further explained in Section 15.5, this method might involve some financial returns. The alternative method will be to use a fast-charging public facility, corresponding to a familiar service gas station that is capable of multi-megawatt power transfers. Although the cost per kilowatt-hour will be high, the owner benefits from charge times in the order of minutes rather than hours.

In both cases, V2G capability enabled by smart grid technology will become a standard feature with all EVSEs, whether they are public, commercial, semi-public, or private. This will allow the subsistence of a very significant distributed storage resource at the disposal of electric utilities. More specifically, the PEV fleet will be optimally positioned to become a significant provider of some ancillary services and play a role in offering dispatchable peak power. These services to the electricity supplier will be analyzed separately.

13.4.2.1 PEVs as "Peakers"

A peaker is a small but nimble generating unit that can supply the grid with relatively fast response. Historically, natural gas turbines or small hydroelectric plants were the devices of choice for this task.

They are active for only a few hours every day, thus providing only limited energy. Thus, a substantial fleet of PEVs can carry out this task as a highly distributed resource without significantly depleting their batteries. Unfortunately, as long as peak power is not considered a "service," the utility operator will compensate the car owner solely for the energy sold, albeit at a higher peak-demand rate [31]. This might not constitute a strong incentive to the car owner who has to consider other factors, such as the additional battery and power electronics wear and tear for his vehicle. Nevertheless, future adjustments in energy market models are understood to address this among many other issues.

13.4.2.2 PEVs as Spinning and Non-spinning Reserve

One of the most lucrative ancillary services is the spinning and non-spinning reserve. The former consists of generators that are online, but normally run at very low capacity. In the case of a disruption, such as a failure in baseload generation or transmission, these generators are commanded to provide the missing power. They must be able to ramp up in less than 10 min and provide power for as long as 1 h or more. Non-spinning reserves are not online and are required to ramp up to full power within 30 min. Because this is a service, the utility company will pay for the availability of the power as well as its amount. In fact, this service is paid even when no power is ever delivered. A PEV owner can provide this service naturally and be reimbursed starting at the time he plugs his vehicle into the grid, even if the battery is never discharged. In addition, it must be noted that PHEVs have smaller battery capacity than AEVs, but contain an ICE that can be started on a V2G command to generate electricity and function as a spinning reserve as well.

13.4.2.3 PEVs as Voltage/Frequency Regulation Agents

An ancillary service that is even better tailored for PEVs is regulation. It consists of delivering or absorbing limited amounts of energy on demand and in real time. Normally, the request is automated in order to match exactly the instantaneous power generation with the instantaneous load. Failure to do so results in dangerous shifts in line frequency and voltage. The dispatched amount of energy has short duration, only of the order of a few minutes, but it is requested relatively frequently. Therefore, this is a continuous service. It is important to underline that the amount of energy involved is relatively small and changes direction quite rapidly and regularly, implying minimal PEV battery discharge for any reasonably short time interval. The near instantaneous response time and the distributed nature of the PEV fleet explains why regulation is probably the most competitive application for V2G from the point of view of the utility operators.

13.4.2.4 PEVs as Reactive Power Providers

Most electronic topologies used for the inverter/rectifier function in the interface of the PEV to the grid are fully capable of shaping the line current to have low distortion and varying amounts of phase shift with respect to the AC line voltage [32]. This implies that reactive power can be injected into the grid on demand and in real time. Furthermore, as reactive power translates in no net DC currents, this service can be provided without any added stress to the PEV battery.

13.5 Vehicle-to-Grid (V2G) and Vehicle-to-Home (V2H) Concepts

The advantages described in the preceding sections are not presently exploitable owing to a general lack of the required hardware infrastructure, as well as the thorny transition to new business models that include the V2G concept. The road map toward achieving this goal will probably consist of the following several milestones.

1. The first milestone is rather rudimentary because it does not yet require bidirectional converters. It will consist of a simple owner-selectable option afforded by the vehicle BMS user interface that allows the grid to schedule when to activate and deactivate charging. In return, the owner pays lower per-kilowatt-hour rates. Communication between the grid operator and the BMS can be done through existing cell phone technology, requiring no additional infrastructure or hardware.
2. The straightforward "grid-friendly" charging time-window strategy described earlier will evolve to include more sophisticated algorithms. For instance, the grid might broadcast any updates to the current per-kilowatt-hour cost and let the vehicle's BMS choose whether to activate charging. Some ancillary services, such as regulation "down" could become feasible, while regulation "up" will be limited by the lack of reverse power flow capability of the EVSE at this stage. The use of aggregators will also become widespread. Aggregators are intermediate communication and power distribution nodes between a group of vehicles, located in proximity to each other and to the grid. This allows the grid to macro-manage a single installment of several vehicles, corresponding to significant power-level blocks with somewhat predictable behavior, akin to other distributed energy resources. Furthermore, because the aggregator's consumption will be in the megawatt range, it will allow purchases of power on the wholesale market, reducing the cost for each participant vehicle.
3. Eventually, bidirectionality will become a standard feature for all EVSEs. However, this capability will not be harnessed immediately to achieve controlled reverse power flow to the grid. On the contrary, the PEV battery will, most likely, initially service the surrounding premises, probably the owner's home. This scenario, called V2H, will probably precede the full implementation of V2G [33] because it effectively bypasses several large infrastructure and technical issues needed for V2G, while achieving many of the same results. Through pricing incentives, a PEV parked at the residential premises and connected on the customer's side of the meter could be exploited to absorb energy from the grid during times of low demand and transfer it to the household appliances during times of high demand. Indirectly, this will shrink the power peaking for the grid while reducing the electricity bill for the user. It will also reduce overall transmission losses over the V2G strategy because line current will flow only in one direction, from grid to vehicle, and will then be consumed locally.
4. Moreover, if the household is geared with renewable source generators, the vehicle can immediately serve as storage and, during blackouts, as backup power. Although one can find some similarities between the concepts of V2H and V2G, there are important distinctions. In practical terms, these differences stem from the fact that V2H cannot take advantage of the high predictability deriving from statistical averages afforded by very high numbers of vehicles available for V2G operations. Simply stated, the real benefits of V2H are not easily estimated because they are dependent on many exceedingly uncertain variables, which include the number of available vehicles, commute schedule, time duration and distance; PEV energy storage capacity, presence and quantity of quasi-predictable local generation (e.g., solar panels); presence and quantity of unpredictable local generation (e.g., wind power); residence-specific energy consumption profile; and the presence of additional storage. Despite the fact that these issues will require complicated management algorithms in order to optimize the use of V2H, some benefits such as emergency backup are available immediately with relatively minor upgrades to the residential infrastructure. These upgrades consist mainly of the installation of a transfer switch to disconnect the residence from the grid during backup operation and to expand the design of the power converter to detect islanding conditions. Furthermore, the EVSE must be capable of controlling output current into the line when connected to the grid, but reverting to controlling output voltage when acting as a backup generator.
5. Full V2G implemented with automated options for V2H. The connection would be metered and it could include any locally generated renewable energy management.

13.5.1 Grid Upgrade

The electric transmission and distribution networks in most industrialized nations must consider changes and upgrades in order to fully benefit from the introduction of PEVs as distributed resources. First, we must consider the extent by which the current production capacity will have to be expanded. Various studies [34] have suggested that once the typical charging profile for a PEV is scrutinized and, hopefully, optimized – charging mostly at night – the installation of a new generation will be unnecessary or minimal at most. In fact, it will have the effect of diminishing reliance on more expensive load-following plants, as the overall 24 h demand curve will average closer to the base load. Therefore, the main effort should be in effectively introducing intelligence into the grid. The hardware and communication standards for implementing such intelligence are still under study. A wideband digital interface can take the form of PLC (power line communication) or utilize separate communication channels that have some market penetration already. In either case, the EV will most likely be treated as any other managed load by this smart grid, with the exception of a sophisticated onboard metering device that will have to be reconciled with the utility's pricing model. Presently the two major obstacles to the utilization of PEVs as distributed resources are the lack of bidirectionality in the power converters and the lack of recognized standards, both software protocols and hardware, for the smart grid function. Of the two, the former is by far the easiest to implement, given the well-established characterization of suitable power electronic topologies.

13.5.1.1 Renewable and Other Intermittent Resource Market Penetration

Owing to recent well-known trends, renewable resources are becoming increasingly prominent in the complex energy market mosaic. As long as their penetration level is low, they can be handled easily by the current infrastructure, but at present incremental rates, this will not be the case in the future. The intermittent nature of solar and wind generation will require a far more flexible compensation mechanism than is currently available. Because of this, large battery banks that act as buffers between the generator and the grid invariably accompany today's renewable energy installations. Wind power, in particular, is not only intermittent but it has no day-average predictability, as winds can differ hour-to-hour as easily at night as during the day, adding an extra amount of irregularity to an already varying load. This suggests that PEVs will be called on not only to perform the more manageable regulation tasks but also to aid in providing peak power. As noted earlier, this might not find approval with PEV owners unless the pricing model is modified. Nevertheless, it is reasonable to ask whether a large PEV contracted fleet could perform this task on a national (US) level. Studies have shown [35] that the answer is yes. With an overconfident 50% estimation for the market penetration of wind energy and 70 million PEVs available, peak power could be provided at the expense of approximately 7 kWh of battery energy per day or about 10–20% of an average PEV reserve.

13.5.1.2 Dedicated Charging Infrastructure from Renewable Resources

The traditional microgrid often relies on diesel generators as a single source of energy. Even in this case, any load fluctuations are quite difficult to negotiate, relying solely on the intrinsically slow ramp up speeds of the generator itself. The new trend toward integrating renewable resources into microgrids greatly amplifies this problem owing to their notorious intermittent nature. On the other hand, the dedicated generation from renewables for the explicit purpose of PEV charging is gaining more credibility as a means to eliminate transmission losses and greatly reduce the overall carbon footprint associated with EVs. Such installations would fall into two categories: (i) small installations with or without a grid tie and (ii) large installations with grid tie. Small installations can be somewhat arbitrarily defined at less than a total of 250 kW of peak production. This would be sufficient to slow charge about 20 vehicles and would certainly require local external storage in order to buffer the peaks and troughs in local energy production. This is more evident in the case of islanded installations; if any energy is produced in excess, it cannot be sent back to the grid, so it will need long-term storage capability. Large installations with a grid tie can inject or draw power to and from the grid as a means to equalize the grid during overproduction and draw from the line. However, depending on the number of vehicles connected, which can be accurately predicted with statistical methods, some of the PEV resource could be utilized to minimize the size of the external storage. Nonetheless, it appears that PEVs could alleviate the inherent issues associated with local renewable production for the dedicated purpose of PEV charging, but not eliminate them.

13.6 Power Electronics for PEV Charging

The PEV charging process will be enabled by the sophisticated power electronics circuits found in the EVSE. Such equipment will be optimally designed depending on the different possible sites and types of power connection. We will begin by looking at EVSE connected to the main power grid and then analyze dual-sourced systems such as grid-tied renewable energy installations dedicated to PEV charging. A short discussion on basic safety compliance strategy follows.

13.6.1 Safety Considerations

For off-board chargers, only a few important safety needs affect significantly the power converter design: (i) isolation of the battery pack with respect to the chassis and the grid terminals, (ii) ground fault interrupters (GFI) to detect any dangerous leakage current from either the grid or the battery circuit, (iii) connector interface and (iv) software. A typical EVSE and related connections are shown in Figure 13.9.

Two GFIs detect any breakdown or current leakage on either side of the isolation barrier in order to ensure complete protection to the user and to disconnect the high-power circuit immediately in case of fault. The battery pack is fully isolated from the chassis because it cannot be grounded properly during charging without heavily oversizing the connector cable. In fact, some existing safety recommendations require that an active breakdown test be performed on the battery pack prior to every charging cycle. At the time of writing, the de facto standard for Level 3 DC charging is the CHAdeMO standard developed by the Tokyo Electric Power Company. Although competing standards may eventually overtake it in popularity, the description of the CHAdeMO connector demonstrates the safety concerns involved. The connector itself will have the mechanical means to lock itself onto the car receptacle in order to prevent accidental removal when energized. It will carry the power leads and also the communication wires that include a controller area network (CAN) bus digital interface as well as several optically isolated analog lines for critical commands, such as on/off, start/stop and so on. Every analog signal sent by the PEV to the charger (or vice versa) is received and acknowledged through the analog lines. This analog interface is sturdier than a digital interface and less susceptible to electromagnetic interference. The CAN bus is activated only when more complex information is exchanged. Prior to the start charge command, the EVSE communicates its parameters to the PEV (maximum output voltage and currents, error flag convention, etc.) and the PEV communicates its parameters to the EVSE (target voltage, battery capacity, thermal limits, etc.), and thus a compatibility check is performed. During charging, the PEV continuously updates the EVSE with its instantaneous current request (every 100 ms or so) and all

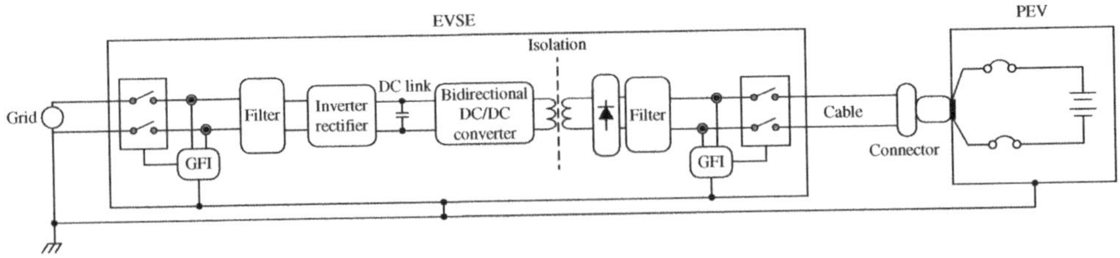

Figure 13.9 Typical EVSE safety configuration

accompanying status flags. Once charging is finished, the operator can safely unlock the connector and drive away.

As can be seen, the presence of safety devices, such as the GFIs as well as the sturdy method of analog and digital communication, renders the charging process extremely safe, leaving the power electronic designer of the EVSE with the relatively simple task of ensuring only the isolation barrier between the grid voltage and the PEV floating battery. In fact, the utilization of an isolation transformer can actually simplify some designs owing to the added voltage amplification capability afforded by the transformer's turns ratio. This could prove very beneficial if much higher battery voltages become necessary in order to increase storage capacity.

13.6.2 Grid-Tied Residential Systems

As noted earlier, only Levels 1 and 2 are feasible within the confines of a residential setting. This can be accomplished through integrated chargers when available or by an external EVSE. In the latter case, the most obvious circuit configuration is a single-phase bidirectional rectifier/inverter powered by a 240 V AC/60 A circuit that is readily available from the distribution transformer. The DC-link voltage is then processed by a bidirectional DC/DC converter that performs the isolation function. This simple topology, shown in Figure 13.10, can be called the canonical topology as it will be repeated, with minor changes, for most grid-tied systems irrespective of power rating.

In North America, the 240 V from the residential distribution transformer is in the form of a split 120 V supply, suggesting small modifications to the canonical topology. Figure 13.11 shows two possibilities. The two topologies in the figure are similar, but Figure 13.11b has better voltage utilization and is better equipped to counter unbalanced loads on the split supply [36]. For the DC/DC converter, many bidirectional isolated circuit topologies have been proposed [37]. Typical circuits are shown in Figure 13.12a and b.

When the two controlled bridges are driven independently in phase-shift modulation (PSM), these are generally referred to as dual-active bridge (DAB) topologies. In their simplest operation mode, when power needs to be transferred from the left-side circuit to the right-side circuit, for instance, the right-side insulated-gate isolated transistor (IGBT) switches are left undriven, leaving their antiparallel diodes in the form of a regular diode bridge. Under these circumstances, the topology becomes identical to a regular PSM converter, which is simple to operate, but not very flexible in terms of voltage gain. On the other hand, when both bridges are modulated, power transfer can be accomplished in both directions and with great variability ranges on the input and output voltages. In addition, zero voltage switching (ZVS) can be assured for all switches for reduced switching loss and generated electrical noise electro-magnetic interference (EMI). Other topologies [38, 39] based on the DAB have been proposed with purported additional benefits, such as better switch utilization, extended ZVS operating range and more flexible voltage amplification.

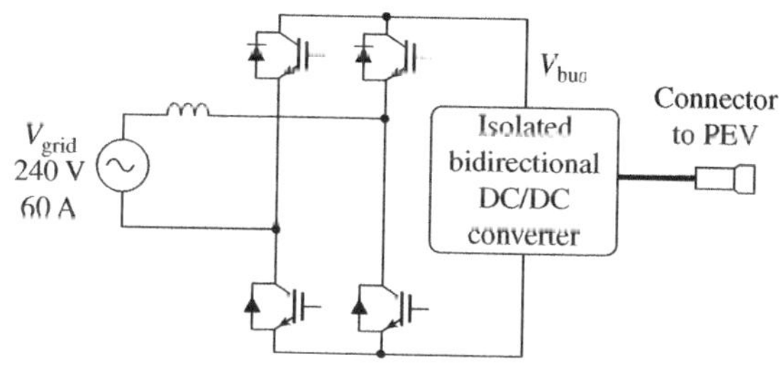

Figure 13.10 Canonical single-phase EVSE configuration

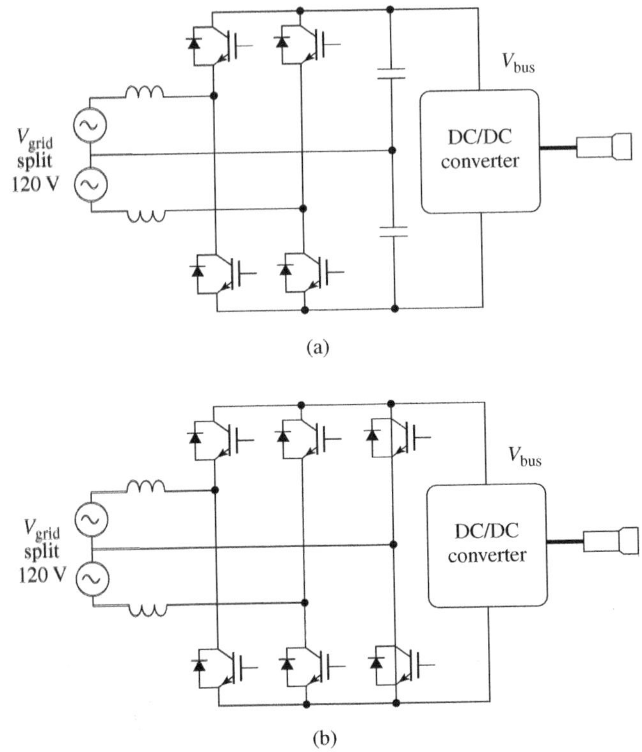

Figure 13.11 Split phase sourced EVSE configurations

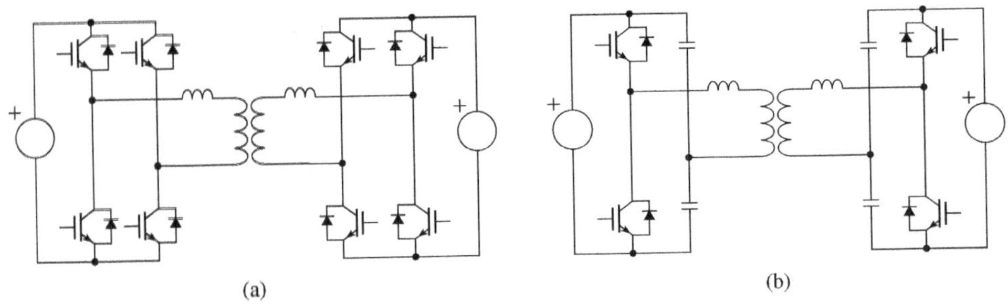

Figure 13.12 (a, b) Typical isolated bidirectional buck–boost DC/DC converter topologies

13.6.3 Grid-Tied Public Systems

A public parking/charging installation would deliver only Level 2 power, given the relatively long plug-in times. Because there are several parking locations in close proximity, the power configuration used for residential use might not optimal. On the contrary, a single transformer can be installed at the grid, delivering isolated power to all vehicles in the facility. By this process, cheaper and more efficient non-isolated DC/DC converters can be used without violating safety rules. Figure 13.13 illustrates this configuration

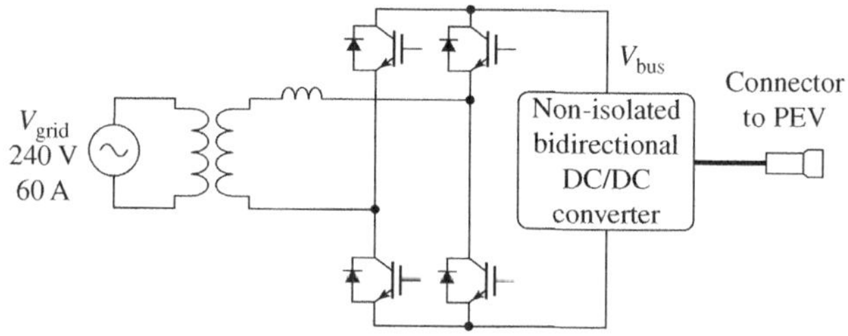

Figure 13.13 Configuration with isolation at the grid

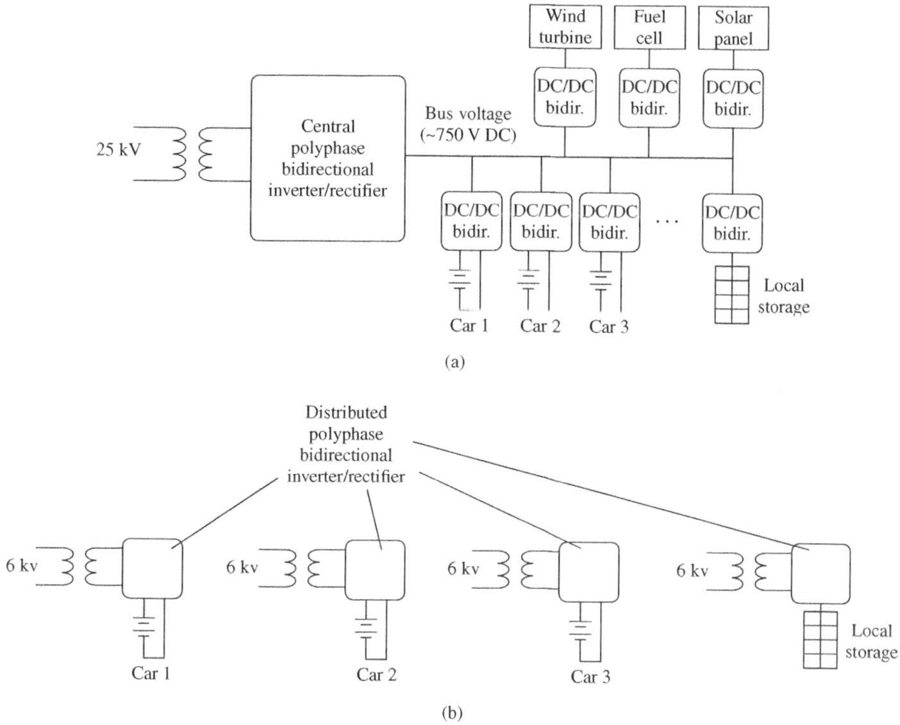

Figure 13.14 (a) Central architecture and (b) distributed architecture

for each charging station. For the entire installation, the architectures shown in Figure 13.14a and b are possible.

In the centralized architecture [40], a single large polyphase 50/60 Hz step-down transformer connects to the grid, providing isolation for the entire facility. A large bidirectional rectifier that produces a single high-voltage DC bus follows this. Each parking station uses inexpensive high-efficiency non-isolated DC/DC converters to process this bus voltage into the appropriate charging current for the individual PEVs. Because isolation is either desirable or required, especially on PV panels depending on local electrical codes, additional storage or generating resources, such as wind turbines and fuel cells, could also benefit from a simpler interface to the DC bus. Moreover, the single transformer connection guarantees that no DC current is injected into the grid, doing away with complicated active techniques to achieve the same purpose.

However, these advantages of the centralized configuration are somewhat offset by the following drawbacks: (i) the need for a bulky and usually inefficient line frequency transformer, (ii) an expensive high-power polyphase inverter/rectifier, (iii) single-fault vulnerability in the transformer and central inverter rectifier and (iv) lack of voltage amplification in each non-isolated DC/DC converter (otherwise afforded by the turns ratio of the high-frequency transformer in isolated topologies).

In a Level 3 (fast-charging) public facility, other technical challenges must be considered. For instance, with battery pack rated voltages in the range of 200–600 V, the overall currents required for fast charging will be of the order of thousands of amperes [43]. These currents must necessarily flow through cables and especially connectors, causing local thermal issues and loss of efficiency owing to ohmic losses. In addition, the charging stations will appear as concentrated loads to the grid, such that any power transients produced by the stations are very likely to cause local sags or surges.

The first issue can be countered partially by brute force methods such as the development of advanced sub-milliohm connectors and by minimizing cable lengths by placing the grid step-down transformer in physical proximity to the vehicle. It is obvious that any intervening power conditioning electronic circuitry should be added only when necessary. This suggests immediately that the architecture of the charging station should be distributed rather than central. As can be seen from Figure 13.15a and b, a distributed architecture could potentially reduce the number of processors from grid to battery from two to one. To be fair, this single stage may not be feasible when managing large input–output voltage ranges, especially if buck–boost operation is required (see discussion on the Z-converter later in this section). Nevertheless, if an additional DC/DC stage should prove necessary, it will be easily integrated locally with the inverter for improved efficiency. Furthermore, a central processor, in addition to constituting a single point of failure, as already noted, would have to be rated for the full service station power, which could be of the order of a megawatt. In contrast, a distributed architecture benefits from repeated circuitry (economies of scale), redundancy for higher reliability and the possibility of power conditioning in physical proximity to the vehicles, thereby reducing ohmic losses.

The issue of the deterioration of power line quality caused by the service station operating transients has only been studied for specific geographic locations [41], but possible voltage fluctuations of up to 10% have been reported, depending on the length of the feeding high-voltage transmission line. The obvious and perhaps sole approach to mitigate this problem is the integration of flywheel, battery, or ultracapacitor banks into the charging station. This storage will smooth out the load transients by delivering local power when needed and storing power during periods of lower demand. Moreover, it will average out the draw from the grid, such that the distribution equipment can be rated at much lower peak powers (by as much as 40%) [42].

The task of discriminating between the various available electronic topologies is made easier when considering the sheer power handled by fast chargers; to wit, up to 250 kW. Obviously, a good candidate

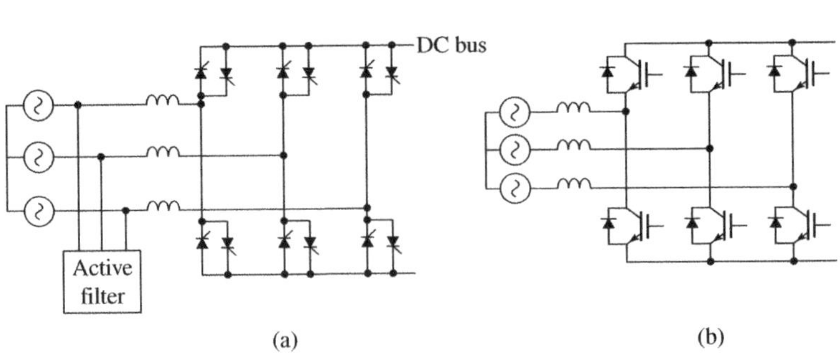

Figure 13.15 (a) Thyristor bridge and active line filter and (b) IGBT bridge

must be very efficient, inherently low-noise with low-component count and be capable of high-frequency operation in order to control physical size. For the inverter/rectifier section, we must also add the requirement that no significant harmonic content should be present in the line current. In order to obtain input currents that are sinusoidal and free of ripple noise, several methods of increasing complexity exist.

One method uses a three-phase thyristor bridge. The devices are very rugged and efficient in terms of conduction loss and have enough controllability to regulate approximately the DC bus [15]. In order to remove unwanted current harmonics, an active filter is added. This filter is based on IGBT devices, but it only processes a small proportion of the total power. The other method uses a fully controlled IGBT bridge in order to achieve excellent input current shaping for extremely low-input current distortion and well-regulated, ripple-free DC bus voltage.

Moreover, fewer components and much higher switching frequencies can be achieved resulting in smaller magnetic components. On the other hand, IGBTs have switching losses and more significant conduction losses than thyristors. However, other techniques, although less sophisticated, have the potential of realizing the required low current distortion limit without the addition of an active filter. The uncontrolled 12-pulse rectifier shown in Figure 13.16a and b can certainly do this, albeit with the addition of significant inductive filtering. As the output DC bus will not be regulated, the subsequent DC/DC converter design cannot be optimized. Using thyristors can achieve regulation of the bus and possibly still achieve the required input current shaping. It is important to note that of the four topologies mentioned here, only those in Figure 13.15a and b are bidirectional and, therefore, are the only choices if V2G is to be implemented. For the final DC/DC converter, all common basic topologies: boost, buck–boost, buck, Cuk, SEPIC and ZETA can be used, so long as they are rendered bidirectional by replacing the diode with a transistor device. In this case, these topologies function differently depending on the direction of the power flow (see Figure 13.17a–d).

Different design requirements might suggest different topologies [40], but some of these are more difficult to justify objectively. For instance, using the buck–boost/buck–boost (Figure 13.17c) produces a voltage inversion from positive to negative that might be undesirable. It also places higher electrical stress on the switches, it requires a more sophisticated design for the inductor and it draws pulsed current from the battery. Similarly, the ZETA/SEPIC topology has a higher part count, including a capacitive, rather than inductive energy-transferring element. On the other hand, as long as the DC bus is guaranteed to exceed the battery voltage – a requirement that is assured by the use of the controlled bridge discussed earlier – the buck/boost topology (Figure 13.17a), is quite attractive. Furthermore, this topology is readily modified in order to divide the task of handling a very large power flow among paralleled modules [41].

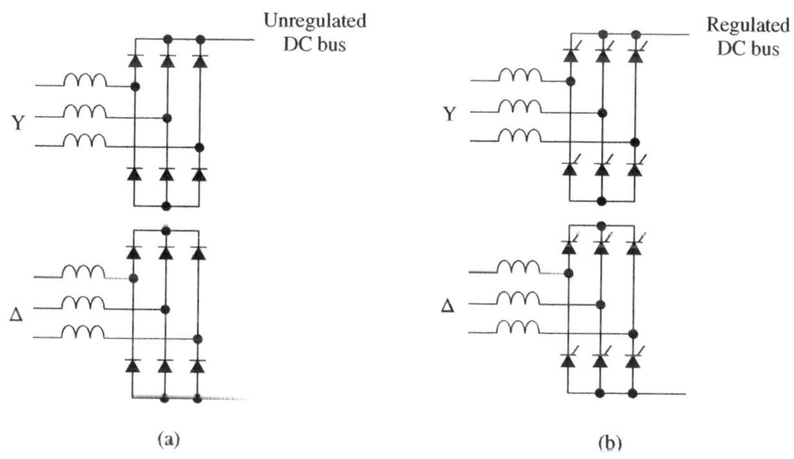

Figure 13.16 (a, b) Twelve-pulse rectifier circuits

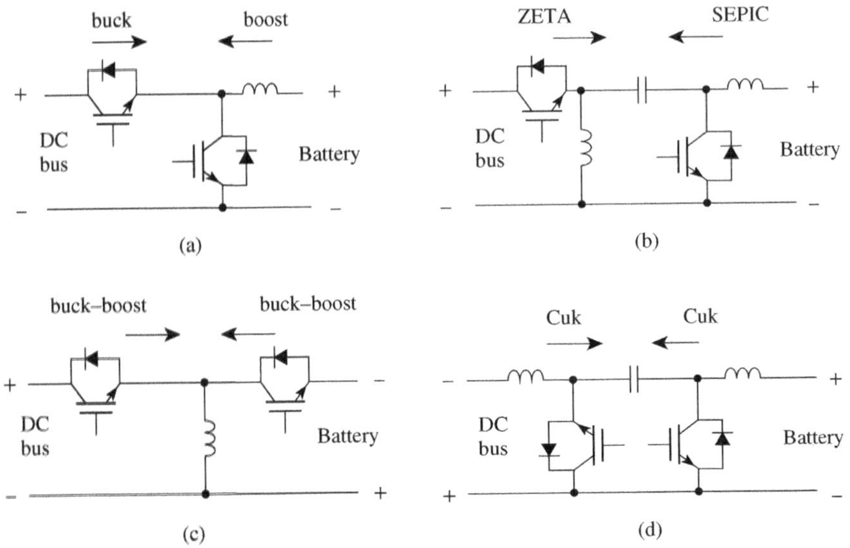

Figure 13.17 (a–d) Basic bidirectional non-isolated topologies

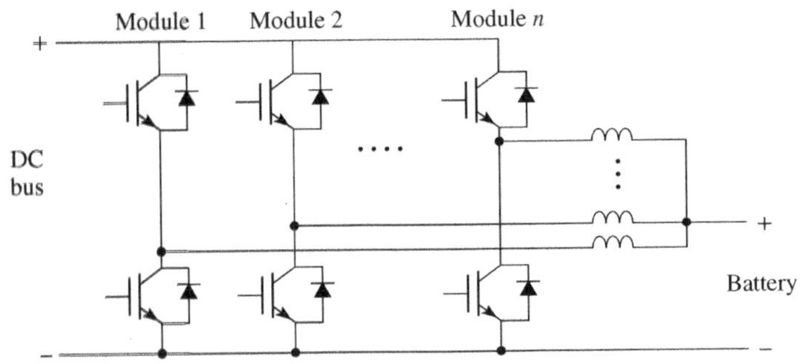

Figure 13.18 Interleaved modular approach for the DC/DC converter

This is shown in Figure 13.18. The amount of converted power can be split among n identical sections and the battery ripple current reduced greatly by the well-known technique of phase-shift interleaving. Using this circuit with $n = 3$ and a switching frequency of 2 kHz, for a typical 125 kW application, efficiencies as high as 98.5% have been reported.

13.6.4 Grid-Tied Systems with Local Renewable Energy Production

As noted earlier, when relatively large energy production from intermittent sources is to be tied into the grid, a statistically predictable PEV presence could serve the purpose of minimizing on-site dedicated storage. This would be the case for municipal carports powered by wind and/or solar generation, where the vehicles must be able to interact intelligently with both locally generated and grid-distributed power at the same time. The possible scenario described in Figure 13.14a may not be ideal when the renewable

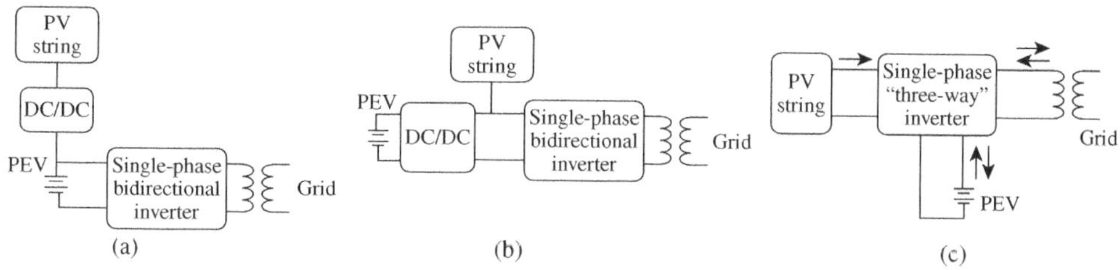

Figure 13.19 (a–c) Possible configurations for solar carport

resource is meant to generate the dominant share of PEV charging energy. On the contrary, by realizing the advantages of the distributed configuration, as in Figure 13.14b, one stage of conversion can be eliminated as long as a conversion topology with wide input–output voltage range capability can be found.

Figure 13.19a–c shows some possible configurations for one of the several charging stations in a solar carport. The architecture depicted in Figure 13.19a has the disadvantage of inserting a DC/DC converter into the main intended power flow from PV to battery. Moreover, the power drawn from a single-phase connection is pulsed at twice the line frequency. This pulsating power takes the form of an undesirably high ripple current into the battery. The configuration shown in Figure 13.19b removes the ripple issue, but adds an additional conversion stage between the grid and the battery. The configuration on Figure 13.19c requires a converter that is capable of bidirectional flow between the PEV and the grid, as well as the steering of PV power to either the PEV or the grid in a controlled fashion. Furthermore, ideally, this should be achieved by a single conversion stage for all power flow paths and with wide voltage range capability. A good candidate for this task is the Z-loaded inverter/rectifier topology shown in Figure 13.20a and b.

The operating characteristics of the Z-loaded converter have been described extensively in the literature [44–46]. The most salient feature of this conversion topology is its controllability through two distinct modulation modes within the same switching cycle, designated by duty cycle D and "shoot-through" duty cycle D_0. The gating patterns shown in Figure 13.20b describe the meaning of D and D_0. As can be seen, during period D_0, all four switches are closed simultaneously, causing the inductors to charge and ultimately boost the voltage across the capacitor, the battery and the grid terminals. Thus, D_0 can be understood as the duty cycle associated with operation, akin to that of a current-sourced inverter. During period D, on the other hand, the bridge operates in a manner similar to that of a voltage-sourced inverter, which is essentially a buck. Therefore, with the appropriate utilization of D and D_0, both buck and boost operations can be achieved, such that the battery voltage can be either higher or lower than the peak of the

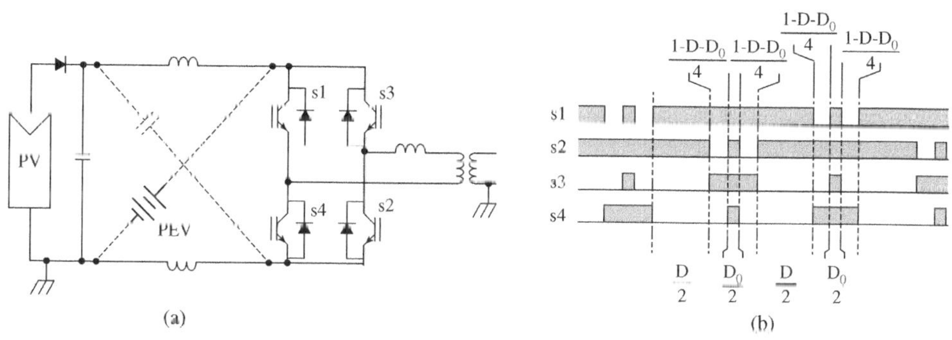

Figure 13.20 (a) Z-loaded rectifier and (b) gating pattern

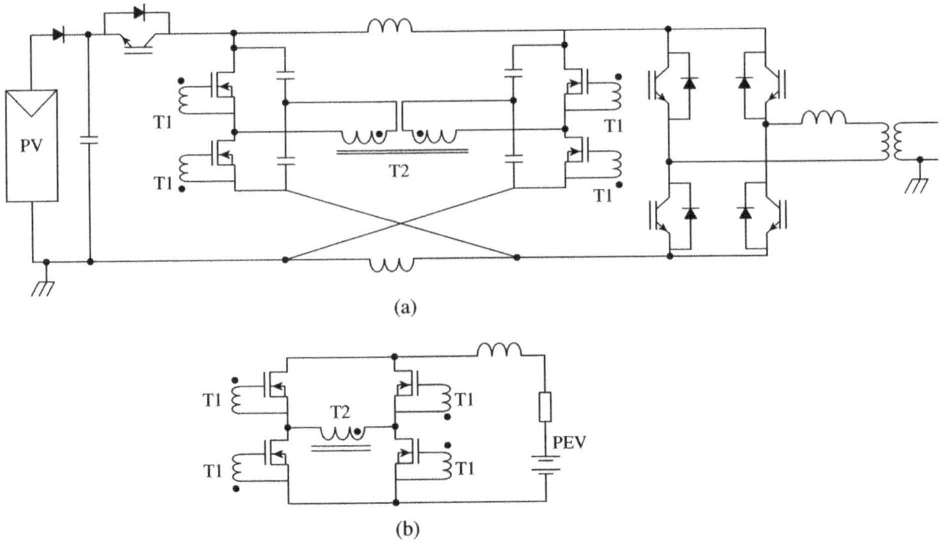

Figure 13.21 (a) Z-converter application to single-phase and (b) grid-tied PV charging station

line voltage. This allows a wide line and battery voltage range. Most significantly, owing to the double modulation, both the grid and the battery current can be controlled precisely in amplitude and shape (sinusoidal for the line current and ripple-less DC for the battery). The maximum power point tracking (MPPT) function for the PV string can then be achieved by managing the simple addition of these two power flows.

The topology shown in Figure 13.20a must be modified in order to achieve isolation of the battery pack. Therefore, the DAB converter shown in Figure 13.12b can be integrated, resulting in the detailed schematic of Figure 13.21a and b. The apparent complexity of the isolation stage is deceptive. In fact, it is a simple bidirectional converter using a small and inexpensive high-frequency transformer, which runs in open loop at full duty cycle, where all eight switches are driven by the same signal. In addition, as the duty cycle is always 100%, ZVS is assured, resulting in efficient operation executed by relatively small devices.

With the inclusion of the isolated DC/DC converter, the need for the 50/60 Hz isolation transformer might be called into question. In North America, the grounding of one side of the PV panel has traditionally been the required norm. Although the National Electric Code has allowed recent conditional exceptions to this safety regulation, utility companies have resisted this change mainly because a direct connection to the AC/DC bridge converter can inject dangerous levels of DC current into the distribution transformer. On the other hand, should this constraint become less binding in North America, as it is currently in Europe, other circuits could be proposed that could prove more reliable and efficient. Many so-called transformerless topologies have been proposed [47, 48] and Figure 13.22 depicts a simplified schematic for one such possibility.

In this case, the DC/DC conversion and the rectifier/inverter section are controlled separately, rendering the control strategy much simpler. On the other hand, the DC/DC converter is now governed by a feedback loop, meaning that it no longer takes advantage of the low switching loss normally associated with 100% duty cycle operation. With allowances from the regulatory safety agencies, the PV panels can be floating, as long as the circuit has additional protection afforded by GFIs and that it produces no leakage currents to ground during normal operation. The last requirement is attained only if the topology guarantees very little common-mode voltage on the PV panels during normal operation (note that this cannot be achieved

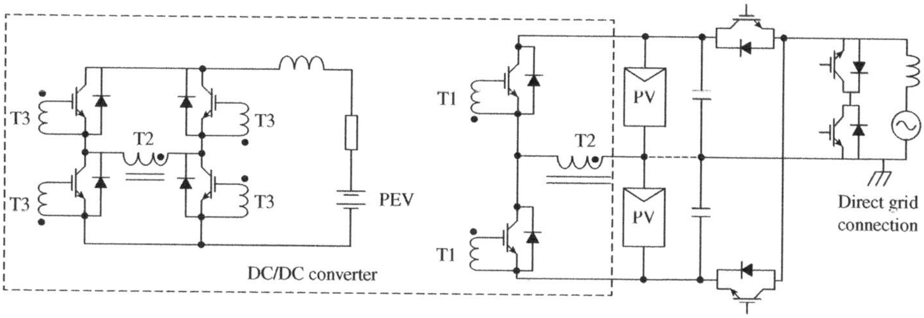

Figure 13.22 Transformerless topology

with the Z-converter). Nevertheless, the mid-point can still be grounded, as indicated by the dashed line in the figure, but at the expense of performance.

Whichever architecture is chosen, it is clear that the energy transfer cannot be controlled to satisfy fully any arbitrary current demands of the PV, the grid and the EV/PHEV battery, simultaneously. In fact, many renewable resources are themselves subject to MPPT control, such that the simple power balance in Equation (13.1) must be satisfied:

$$P_{\mathrm{MPPT}} = P_{\mathrm{PEV}} + P_{\mathrm{G}} \tag{13.1}$$

Here, MPPT is the power draw requested by the distributed resource. It has to equal the sum of the power absorbed by the grid and the PEV battery (PPEV and PG, respectively). As MPPT is determined by external factors, such as clouding in the case of PV, either PPEV or PG can be controlled independently, but not both. Which of these is controlled will depend heavily on how the PEV owner decides to utilize his vehicle storage resource. Thus, in installations where charging power comes primarily from intermittent sources, the need for a significant presence of additional storage on the premises will be diminished, but not eliminated.

References

1. Ehsani, M., Gao, Y., and Emadi, A. (2010) *Modern Electric, Hybrid Electric and Fuel Cell Vehicles, Fundamentals, Theory and Design*, 2nd edn, CRC Press, New York.
2. Husain, I. (2005) *Electric and Hybrid Vehicles, Design Fundamentals*, CRC Press, New York.
3. IEEE Spectrum Software Fix Extends Failing Batteries in 2006-2008 Honda Civic Hybrids: Is Cost Acceptable? http://spectrum.ieee.org/riskfactor/green-tech/advanced-cars/software-fix-extends-failing-batteries-in-20062008-honda-civic-hybrids-is-cost-acceptable (accessed 19 December 2013).
4. Crompton, T.R. (2000) *Battery Reference Book*, 3rd edn, Newnes.
5. Buchmann, I. (2001) *Batteries in a Portable World: A Handbook on Rechargeable Batteries for non-Engineers*, 2nd edn, Cadex Electronics Inc.
6. Dhameja, S. (2002) *Electric Vehicle Battery Systems*, Newnes.
7. Chen, L.R. (2008) Design of duty varied voltage pulse charger for improving li-ion battery-charging response. *IEEE Transaction on Industrial Electronics*, **56** (2), 480–487.
8. Park, S.Y., Miwa, H., Clark, B.T. *et al.* (2008) A universal battery charging algorithm for Ni-Cd, Ni-MH, SLA, and Li-Ion for wide range voltage in portable applications. Proceedings IEEE Power Electronics Specialists Conference, Rhodes, Greece, pp. 4689–4694.
9. Hua, C.C. and Lin, M.Y. (2000) A study of charging control of lead acid battery for electric vehicles. Proceedings IEEE International Symposium on Industrial Electronics, Vol. 1, pp. 135–140.
10. West, S. and Krein, P.T. (2000) Equalization of valve-regulated lead acid batteries: issues and life tests. Proceedings IEEE International Telecommunications Energy Conference, pp. 439–446.

11. Brost, R.D. (1998) Performance of valve-regulated lead acid batteries in EV1 extended series strings. Proceedings of IEEE Battery Conference on Applications and Advances, Long Beach, CA, pp. 25–29.
12. Bentley, W.F. (1997) Cell balancing considerations for lithium-ion battery systems. Proceedings IEEE Battery Conference on Applications and Advances, Long Beach, CA, pp. 223–226.
13. Krein, P.T. and Balog, R.S. (2002) Life extension through charge equalization of lead-acid batteries. Proceedings IEEE International Telecommunications Energy Conference, pp. 516–523.
14. Lohner, A., Karden, E., and DeDoncker, R.W. (1997) Charge equalizing and lifetime increasing with a new charging method for VRLA batteries. Proceedings IEEE International Telecommunications Energy Conference, pp. 407–411.
15. Moore, S.W. and Schneider, P.J. (2001) A review of cell equalization methods for lithium ion and lithium polymer battery systems. Proceedings SAE 2001 World Congress, Detroit, MI.
16. Nishijima, K., Sakamoto, H., and Harada, K. (2000) A PWM controlled simple and high performance battery balancing system. Proceedings IEEE 31st Annual Power Electronics Specialists Conference, Galway, Ireland, Vol. 1, pp. 517–520.
17. Isaacson, M.J., Hoolandsworth, R.P., and Giampaoli, P.J. (2000) Advanced lithium ion battery charger. Proceedings IEEE Battery Conference on Applications and Advances, pp. 193–198.
18. Pascual, C. and Krein, P.T. (1997) Switched capacitor system for automatic series battery equalization. Proceedings 12th Annual Applied Power Electronics Conference and Exposition, Atlanta, GA, Vol. 2, pp. 848–854.
19. Hung, S.T., Hopkins, D.C., and Mosling, C.R. (1993) Extension of battery life via charge equalization control. *IEEE Transactions On Industrial Electronics*, **40** (1), 96–104.
20. Cao, J., Schofield, N., and Emadi, A. (2008) Battery balancing methods: a comprehensive review. Proceedings IEEE Vehicle Power and Propulsion Conference, Harbin, China, pp. 1–6.
21. Lee, Y.S., Wang, W.Y., and Kuo, T.Y. (2008) Soft computing for battery state-of-charge (BSOC) estimation in battery string systems. *IEEE Transactions on Industrial Electronics*, **55** (1), 229–239.
22. Shen, W.X., Chan, C.C., Lo, E.W.C., and Chau, K.T. (2002) Adaptive neuro-fuzzy modeling of battery residual capacity for electric vehicles. *IEEE Transactions on Industrial Electronics*, **49** (3), 677–684.
23. Piller, S., Perrin, M., and Jossen, A. (2001) Methods for state-of-charge determination and their applications. *Journal of Power Sources*, **96** (1), 113–120.
24. Ullah, Z., Burford, B., and Dillip, S. (1996) Fast intelligent battery charging: neural-fuzzy approach. *IEEE Aerospace and Electronics Systems Magazine*, **11** (6), 26–34.
25. Atlung, S. and Zachau-Christiansen, B. (1994) Failure mode of the negative plate in recombinant lead/acid batteries. *Journal of Power Sources*, **52** (2), 201–209.
26. Feder, D.O., Jones, W.E.M. (1996) Gas evolution, dryout, and lifetime of VRLA cells an attempt to clarify fifteen years of confusion and misunderstanding. Proceedings IEEE International Telecommunications Energy Conference, pp. 184–192.
27. Jones, W.E.M. and Feder, D.O. (1996) Behavior of VRLA cells on long term float. II. The effects of temperature, voltage and catalysis on gas evolution and consequent water loss. Proceedings IEEE International Telecommunications Energy Conference, pp. 358–366.
28. Nelson, R.F., Sexton, E.D., Olson, J.B. *et al.* (2000) Search for an optimized cyclic charging algorithm for valve-regulated lead–acid batteries. *Journal of Power Sources*, **88** (1), 44–52.
29. Kempton, W., Tomic, J., Brooks, A., and Lipman, T. (2001) Vehicle-to-Grid Power: Battery, Hybrid, and Fuel Cell Vehicles as Resources for distributed Electric Power in California. UCD-ITS-RR-01-03.
30. Cocconi, A.G. (1994) Combined motor drive and battery recharge system. US Patent 5, 341, 075.
31. Kempton, W. and Kubo, T. (2000) Electric-drive vehicles for peak power in Japan. *Energy Policy*, **28**, 9–18.
32. Kisacikoglu, M.C., Ozpineci, B., and Tolbert, L.M. (2010) Examination of a PHEV bidirectional charger system for V2G reactive power compensation. IEEE Applied Power Electronics Conference, Palm Springs, CA.
33. Tuttle, D.P. and Baldick, R. (2012) The Evolution of Plug-in Electric Vehicle-Grid Interactions. The University of Texas at Austin Department of Electrical and Computer Engineering.
34. Jenkins, S.D., Rossmaier, J.R., and Ferdowsi, M. (2008) Utilization and effect of plug-in hybrid electric vehicles in the United States. Power Grid, Vehicle Power and Propulsion Conference.
35. Kempton, W. and Tomic, J. (2005) Vehicle-to-grid power implementation: from stabilizing the grid to supporting large-scale renewable energy. *Journal of Power Sources*, **144** (1), 280–294.
36. Wang, J., Peng, F.Z., Anderson, J. *et al.* (2004) Low cost fuel cell converter system for residential power generation. *IEEE Transactions on Power Electronics*, **19** (5), 1315–1322.
37. Han, S. and Divan, D. (2008) Bi-directional DC/DC converters for plug-in hybrid electric vehicle (PHEV). Applications, Applied Power Electronics Conference and Exposition, APEC.

38. Peng, F.Z., Li, H., Su, G.-J., and Lawler, J.S. (2004) A new ZVS bidirectional DC–DC converter for fuel cell and battery application. *IEEE Transactions on Power Electronics*, **19** (1), 54–65.
39. Xiao, H., Guo, L., and Xie, L. (2007) A new ZVS bidirectional DC-DC converter with phase-shift plus PWM control scheme. Applied Power Electronics Conference, APEC, pp. 943–948.
40. Du, Y., Zhou, X., Bai, S. *et al.* (2010) Review of non-isolated bi-directional DC-DC converters for plug-in hybrid electric vehicle charge station application at municipal parking decks. Applied Power Electronics Conference and Exposition (APEC), pp. 1145–1151.
41. Aggeler, D., Canales, F., Zelaya, H. *et al.* (2010) Ultra-fast DC-charge infrastructures for EV-mobility and future smart grids. Innovative Smart Grid Technologies Conference Europe (ISGT Europe), IEEE PES.
42. Bai, S., Du, Y., and Lukic, S. (2010) Optimum design of an EV/PHEV charging station with DC bus and storage system. Energy Conversion Congress and Exposition (ECCE), pp. 1178–1184.
43. Buso, S., Malesani, L., Mattavelli, P., and Veronese, R. (1998) Design and fully digital control of parallel active filters for thyristor rectifiers to comply with IEC-1000-3-2 standards. *IEEE Transactions on Industry Applications*, **34** (3), 508–517.
44. Peng, Z. (2003) Z-source inverter. *IEEE Transactions on Industry Applications*, **39** (2), 504–510.
45. Peng, F.Z., Shen, M., and Holland, K. (2007) Application of Z-source inverter for traction drive of fuel cell – battery hybrid electric vehicles. *IEEE Transactions on Power Electronics*, **22** (3), 1054–1061.
46. Carli, G. and Williamson, S. (2009) On the elimination of pulsed output current in Z-loaded chargers/rectifiers. Proceedings IEEE Applied Power Electronics Conference and Exposition, Washington, DC.
47. González, R., López, J., Sanchis, P., and Marroyo, L. (2007) Transformer-less inverter for single-phase photovoltaic systems. *IEEE Transactions on Power Electronics*, **22** (2), 693–697.
48. Kerekes, T., Teodorescu, R., and Borup, U. (2007) Transformer-less photovoltaic inverters connected to the grid. Proceedings IEEE Applied Power Electronics Conference and Exposition.

14

Multilevel Converter/Inverter Topologies and Applications

Baoming Ge[1], Fang Zheng Peng[1] and Yongdong Li[2]
[1]*Department of Electrical and Computer Engineering, Michigan State University, Michigan, USA*
[2]*Department of Electrical Engineering, Tsinghua University, Beijing, China*

14.1 Introduction

The demand for clean energy and energy saving has stimulated the rapid growth of sustainable energy generation and adjustable speed drives (ASDs). Today's wind and solar power unit ratings, as popular renewable energy sources, reach tens of megawatts. Many laminators, mills, conveyors, pumps, fans, blowers and compressors require large ASDs with medium voltage (several kV to tens of kV). In addition, flexible AC transmission system (FACTS) devices for the control of power flow, voltage and frequency are used to increase transmission capacity, reduce congestion, improve controllability and integrate more renewable energy into grids. FACTS devices are normally connected to medium-voltage distribution or to high-voltage transmission systems. For traditional two-level inverters used in the above applications, step-up transformers and/or zigzag transformers are used to reach the medium- and high-voltage levels and to synthesize output voltage to multipulse waveforms. However, transformers are bulky, lossy, high cost and present nonlinear/saturation problems that lead to difficulties in the system control.

Multilevel inverters provide a novel approach for reaching high voltage and reducing harmonics without the use of transformers. With no transformer and less switching frequency, multilevel inverters provide higher efficiency, smaller size and lower cost. Presently, there are many multilevel converter/inverter topologies, which include the typical cascaded, diode clamped and capacitor clamped multilevel structures [1–9]. The cascaded multilevel inverter (CMI) has the lowest component count and modularity, and it is most widely used for FACTS, medium-voltage ASDs and renewable energy power generation.

This chapter is dedicated to explaining the basic concept of multilevel converter/inverters, introducing the three typical topologies, and describing their pros and cons regarding their most suitable applications. A generalized multilevel inverter topology unifies the three typical multilevel inverters and has the ability to generate other multilevel topologies, such as modular multilevel converter (M^2C), magnetic-less

Power Electronics for Renewable Energy Systems, Transportation and Industrial Applications, First Edition.
Edited by Haitham Abu-Rub, Mariusz Malinowski and Kamal Al-Haddad.

multilevel DC/DC converters, and stacked multicell converters. As application examples in this chapter, Y-connection, Δ-connection and face-to-face-connected CMIs are applied to static var generation (SVG), static synchronous compensator (STATCOM) and FACTS. The Y-connection CMI is the most promising topology in application to ASDs and photovoltaic (PV) power systems.

Another emerging application of the multilevel concept is the multilevel DC/DC converter, which can achieve a high-voltage boost factor without magnetic components. The elimination of magnetic components not only reduces the converter's size and loss, but also is more suitable for high-temperature applications, because it avoids the saturation and instability factors owing to the dramatic decline of magnetic permeability with increasing temperature. This chapter introduces the magnetic-less multilevel DC/DC conversion of compact size, high efficiency and suitable for high-temperature application. Initially, a derivation of the generalized multilevel converter topology for the magnetic-less 3X DC/DC converter is presented. Subsequently, its 3X simplified version, multilevel modular capacitor clamped DC/DC converter (MMCCC) and the *n*X converter are discussed step by step. To reduce power loss further, the zero current switching (ZCS) method is introduced.

Multilevel converters involve a high fault possibility owing to the large number of components. This chapter also analyzes the fault tolerance and reliability of multilevel converters. The CMI is used as an example to illustrate this topic.

14.2 Fundamentals of Multilevel Converters/Inverters

14.2.1 *What Is a Multilevel Converter/Inverter?*

A standard three phase inverter is shown in Figure 14.1a, where the ideal switch represents the power semiconductor switch. Each phase leg outputs two values, the so-called two-level voltage. We use the phase leg A as an example to clarify this concept. As shown in Figure 14.1b, $u_A = V_{DC}$ when switch S_2 is on (S_1 is off to avoid shoot-through) and $u_A = 0$ when switch S_1 is on (S_2 is off to avoid shoot-through), and Figure 14.1c shows the two-level output voltage waveform. To generate three voltage values for three-level phase leg, one more voltage value has to be added, as shown in Figure 14.2a. In the similar manner, for the N-level phase leg, $(N-2)$ more voltage values should be added, as shown in Figure 14.2c. For example, a six-level voltage will be available in Figure 14.2d if $N=6$.

For multilevel inverters with more than three levels, Figure 14.2 just illustrates the multilevel concept, not a practical implementation, because some switch voltage stresses are high, for example, S_2 and S_1 should have the ability to support the entire DC-link voltage. Several practical topologies are shown in the following sections.

Multilevel inverters present many advantages: (i) lower dv/dt, (ii) the voltage waveform is less distorted with more voltage levels, (iii) it is possible to generate sinusoidal voltage, even with low switching frequency, and the resultant low loss is suitable for high-power systems.

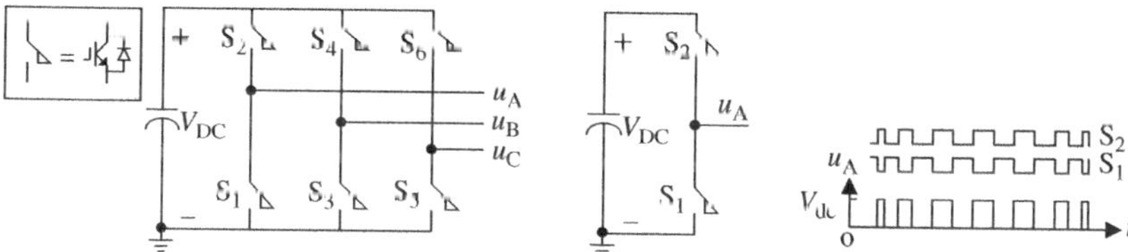

Figure 14.1 Standard three-phase two-level inverter. (a) Standard three-phase inverter, (b) single-phase two-level inverter and (c) two-level voltage

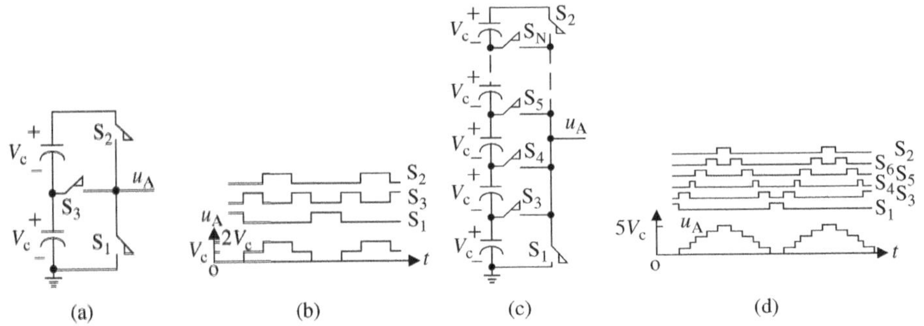

Figure 14.2 Single-phase multilevel inverters. (a) Three-level topology, (b) three-level voltage, (c) *N*-level topology and (d) six-level voltage

14.2.2 Three Typical Topologies to Achieve Multilevel Voltage

14.2.2.1 Diode Clamped Multilevel Inverter

The diode clamped (sometimes called neutral-point-clamped, NPC) multilevel inverter is the earliest one with the term multilevel [1]. Figure 14.3 shows one phase of the three-level inverter. It is the same as Figure 14.2a, where the two capacitors in series support the DC-link voltage, and the neutral point is located at the middle point of the two capacitors. The output voltage u_A can be of three values: $+V_{C1}$, 0 and $-V_{C2}$. The switches S_{a1} and S_{a2} are turned on for voltage $+V_{C1}$; the switches S'_{a1} and S'_{a2} are turned on for voltage $-V_{C2}$; and the switches S_{a2} and S'_{a1} are turned on for $u_A = 0$. The diodes D and D′ clamp the switch voltages to V_{C1} or V_{C2} when S_{a1} and S_{a2} are turned on or when S'_{a1} and S'_{a2} are turned on. For the high-level number *N*, each switch voltage stress is only one capacitor voltage, that is, V_{DC}/N, where V_{DC} is the DC-link voltage. Theoretically, this topology can be extended to any voltage level in high-voltage applications with low-voltage devices.

However, if all the devices, such as block diodes, active devices and capacitors have the same rating, each phase requires $(N-1)\times(N-2)$ additional clamping diodes to provide an *N*-level voltage. The diode number represents a quadratic increase. A large number of clamping diodes cause high cost and packaging problems for high-voltage level applications, and in addition, a special control is necessary to balance the capacitor voltages. Thus, the majority of practical applications for a diode clamped multilevel inverter are limited to below five levels.

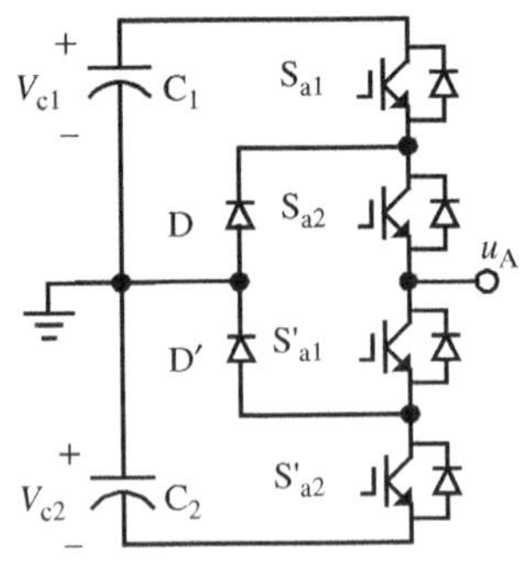

Figure 14.3 Diode clamped three-level inverter

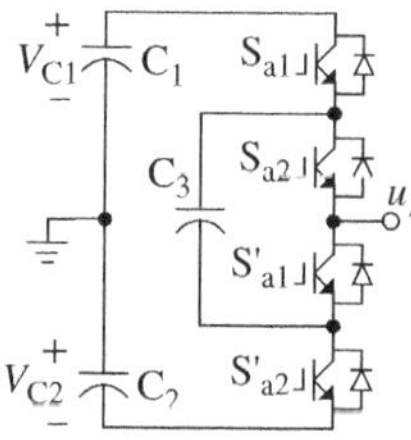

Figure 14.4 Capacitor clamped three-level inverter

14.2.2.2 Capacitor Clamped Multilevel Inverter

Another implementation of multilevel inverters is called the flying capacitor inverter or capacitor clamped multilevel inverter [2]. It uses capacitors to clamp the device voltage, rather than the diodes. Figure 14.4 presents a three-level topology. When S_{a1} and S_{a2} are on, $u_A = +V_{C1}$; when S'_{a1} and S'_{a2} are on, $u_A = -V_{C2}$; and when S_{a1} and S'_{a1} (or S_{a2} and S'_{a2}) are on, $u_A = 0$. This topology requires many capacitors. If all the devices, such as capacitors and power switches, have the same voltage rating, an N-level inverter will require $(N-1)\times(N-2)/2$ flying capacitors per phase. In addition, the complex control and high switching frequency are necessary to balance each capacitor charge. For utility applications, the capacitor number is high to reach high-voltage level.

14.2.2.3 Cascaded Multilevel Inverter

As shown in Figure 14.5a, to achieve an N-level output voltage, one phase inverter consists of $(N-1)/2$ full-bridge inverter modules connected in cascade [3–6]. Each full-bridge module generates three output voltage levels: $+V_C$, 0 and $-V_C$. Figure 14.5b and c uses a nine-level CMI to explain its operating principle. The four module output voltages contribute to the phase voltage u_{An}, that is, $u_{An} = V_{ca1} + V_{ca2} + V_{ca3} + V_{ca4}$, where V_{cax} is the output voltage of module x, and $x = 1, 2, 3$, and 4.

Fewer components are required in the CMI when compared with that of diode clamped and capacitor clamped multilevel inverters, which is well suited to high-voltage applications. For harmonic/reactive compensation purposes, there is no need for separate DC power supplies. If separate DC power supplies are available, for example, with PV panels and isolated transformers, the CMI can be applied to medium-voltage ASDs, PV power generation, and so on, and each full-bridge module can have its balanced DC voltage.

In addition to the step modulation in Figure 14.5c, many other modulation strategies have been reported, for example, the most popular carrier-based pulse-width modulation (PWM) and space vector modulation (SVM). Most carrier-based modulation schemes are from the carrier disposition strategy, for example, alternative phase opposition disposition, phase opposition disposition, and phase disposition (PD). The PD-based phase-shifted carrier PWM is widely applied in CMIs, because of its automatic balanced switching transitions between the modules.

14.2.3 Generalized Multilevel Converter/Inverter Topology and Its Derivations to Other Topologies

14.2.3.1 Generalized Multilevel Topology

Figure 14.6 shows the generalized multilevel inverter topology, and all switching devices, diodes, and capacitors support $1/(N-1)$ of the DC-link voltage [10]. As shown in Figure 14.6, we can obtain the cells.

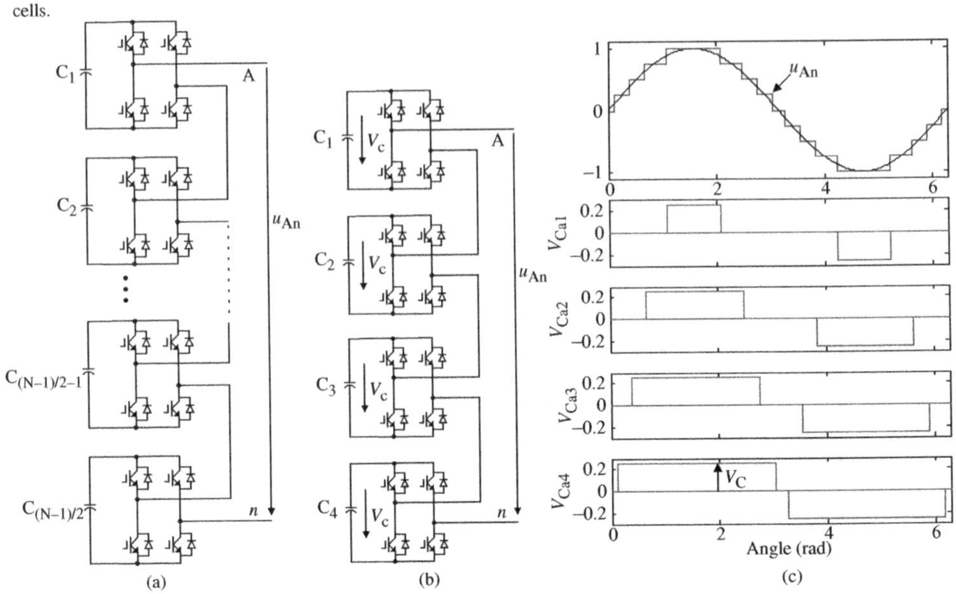

Figure 14.5 One phase of CMI. (a) General CMI, (b) nine-level inverter and (c) output voltage waveform of nine-level inverter

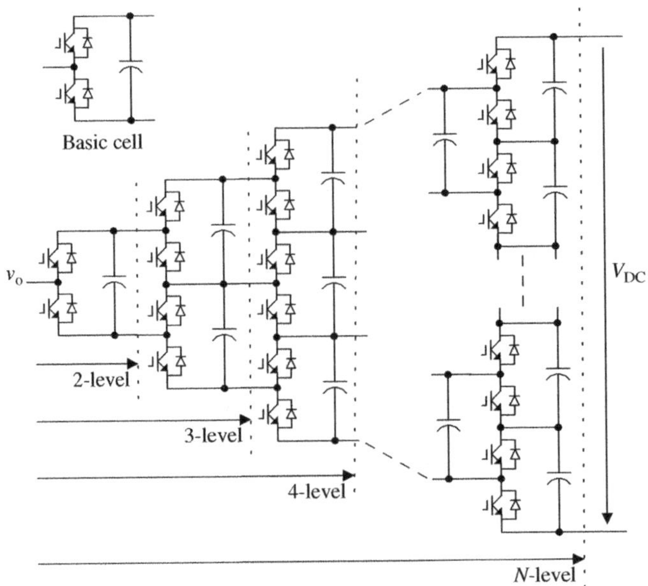

Figure 14.6 Generalized multilevel inverter topology

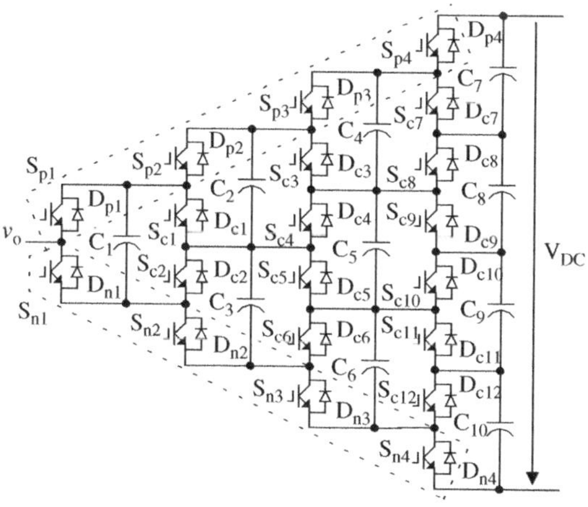

Figure 14.7 Five-level inverter phase leg

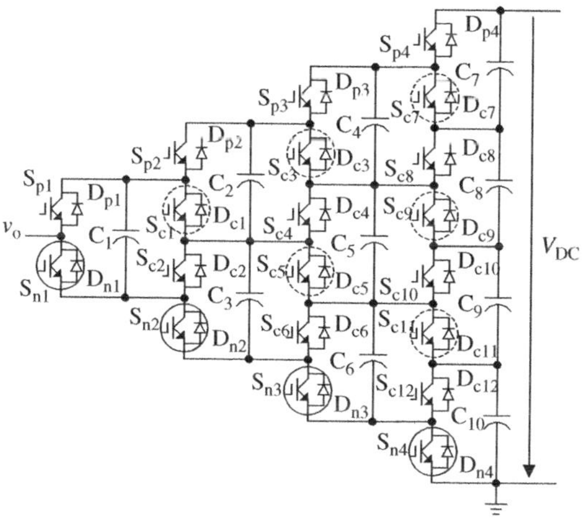

Figure 14.8 Switching states when $v_o = 0$

two-level inverter up to the "2-level" line, the three-level inverter will be there from the "3-level" line, and an *N*-level inverter is constructed up to the "*N*-level" line. All of them are built by using the "basic cell" and it looks like a horizontal pyramid of basic cells.

The operating principle can be explained by the five-level circuit shown in Figure 14.7. The outer switches (in the dashed-line block) are the main power switches to generate the desired voltage waveforms. The inner switches (outside the dashed-line block) are used to balance the capacitor voltages. This topology provides the self-balanced voltage with each component voltage stress of $V_{DC}/4$. Figure 14.8 shows the generation of zero voltage, Figures 14.9 and 14.10 present two ways to generate $V_{DC}/4$, where the circled devices in both solid and dashed lines are turned on and the uncircled devices are turned off. Furthermore, the solid-line circled devices are the on-state main power switches to produce the desired voltage level, and the dashed-line circled devices make the capacitor voltages balanced. As shown in Figure 14.8, the zero voltage is produced by the on-state switches S_{n1}–S_{n4}, and the capacitors C_1, C_3,

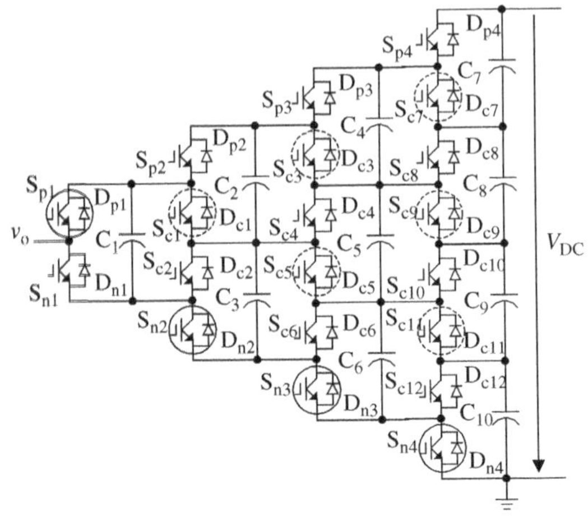

Figure 14.9 Switching states for $v_o = V_{DC}/4$

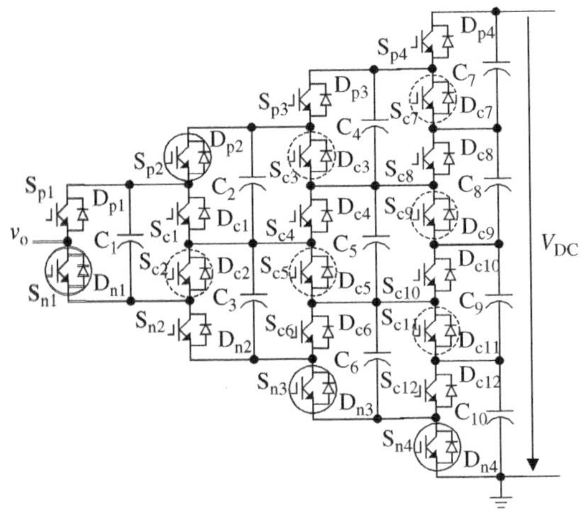

Figure 14.10 Alternative switching states for $v_o = V_{DC}/4$

C_6 and C_{10} are in parallel to balance their voltages through the on-state switches S_{c1}, S_{c5} and S_{c11}, that is, $V_{C1} = V_{C3} = V_{C6} = V_{C10}$. Similarly, $V_{C2} = V_{C5} = V_{C9}$ because of the on-state switches S_{c3} and S_{c9} and the on-state switch S_{c7} ensures the balanced voltages of the capacitors C_4 and C_8, that is, $V_{C4} = V_{C8}$. Figure 14.9 shows how to generate $v_o = V_{DC}/4$ and ensure that $V_{C1} = V_{C3} = V_{C6} = V_{C10}$, $V_{C2} = V_{C5} = V_{C9}$ and $V_{C4} = V_{C8}$. Other alternative switching states are shown in Table 14.1 to produce $v_o = V_{DC}/4$ and one of these is shown in Figure 14.10, where $V_{C3} = V_{C6} = V_{C10}$, $V_{C1} = V_{C2} = V_{C5} = V_{C9}$ and $V_{C4} = V_{C8}$. Table 14.1 lists the switching states necessary to produce $V_{DC}/2$, $3V_{DC}/4$ and V_{DC}, where the states of switches $S_{p1}-S_{p4}$ are independent and can represent fully the states of the entire circuit, and the states of the other switches are deduced by using complementary switching of adjacent switches.

For the reactive power and harmonic compensations, there is no need for separate DC power sources to feed the CMI and the diode clamped and capacitor clamped multilevel inverters, and the proper control

Table 14.1 Switching states to produce five voltage levels [10]

Output voltage	Capacitor* path	Switch states**			
		S_{p1}	S_{p2}	S_{p3}	S_{p4}
$0\ V_{DC}$	none	0	0	0	0
$V_{DC}/4$	$+C_1$	1	0	0	0
	$-C_1+C_2+C_3$	0	1	0	0
	$-C_3-C_2+C_4+C_5+C_6$	0	0	1	0
	$-C_6-C_5-C_4+C_7+C_8+C_9+C_{10}$	0	0	0	1
$V_{DC}/2$	$+C_2+C_3$	1	1	0	0
	$-C_1+C_4+C_5+C_6$	0	1	1	0
	$-C_3-C_2+C_7+C_8+C_9+C_{10}$	0	0	1	1
	$+C_1-C_3-C_2+C_4+C_5+C_6$	1	0	1	0
	$+C_1-C_4-C_5-C_6+C_7+C_8+C_9+C_{10}$	1	0	0	1
	$-C_1+C_2+C_3-C_6-C_5-C_4+C_7+C_8+C_9+C_{10}$	0	1	0	1
$3V_{DC}/4$	$+C_4+C_5+C_6$	1	1	1	0
	$-C_1+C_7+C_8+C_9$	0	1	1	1
	$+C_2+C_3-C_4-C_5-C_6+C_7+C_8+C_9+C_{10}$	1	1	0	1
	$+C_1-C_2-C_3+C_7+C_8+C_9+C_{10}$	1	0	1	1
V_{DC}	$+C_7+C_8+C_9+C_{10}$	1	1	1	1

*The capacitors are used to produce the desired output voltage level.
"+" denotes the capacitor connected positively to the output; "−" denotes the capacitor connected negatively.
**"1" denotes the on-state and "0" for the off-state.

can maintain the capacitor voltages, even if the voltage level number $N > 3$. These three multilevel converters have to employ isolated DC power sources or the complicated voltage balancing circuit when applied to motor drives or other related areas of real power transfer. However, the generalized multilevel inverter can achieve the self-balancing of each voltage level for both real power and reactive power conversions. Moreover, the diode clamped and capacitor clamped multilevel, CMIs and even traditional two-level inverters can be derived from this generalized topology. Figure 14.11 shows the traditional diode clamped multilevel inverter, deduced from Figure 14.6 by eliminating the clamping switches and

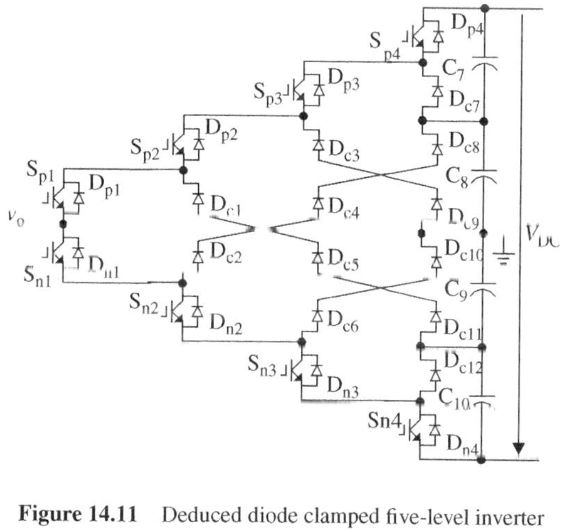

Figure 14.11 Deduced diode clamped five-level inverter

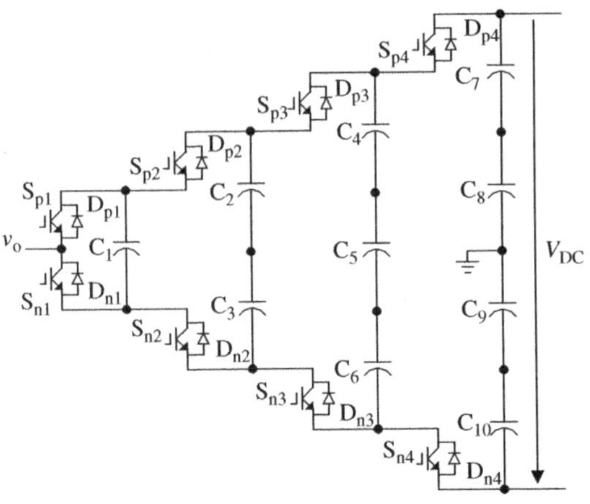

Figure 14.12 Deduced capacitor clamped five-level inverter

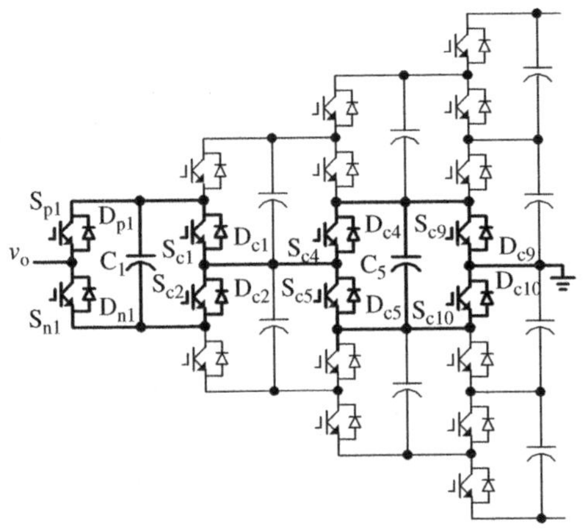

Figure 14.13 Deduced CMI through using only center part's cells (thick line)

capacitors, and swapping the diode clamped path. Figure 14.12 shows the capacitor clamped (or flying capacitor) multilevel inverter through eliminating the clamping switches and diodes of Figure 14.6. Figure 14.13 shows the CMI derived from Figure 14.6.

14.2.3.2 Derivations to Other Multilevel Topologies

New multilevel inverters are easily derived from the generalized multilevel inverter topology. For example, the diode and capacitor clamped multilevel inverter is obtained in Figure 14.14 when all clamping switches are eliminated from Figure 14.6. A new diode clamped multilevel inverter is shown in Figure 14.15 when the clamping switches and capacitors are eliminated from Figure 14.6. This figure presents the deduced zigzag multilevel [11] (Figure 14.16). The deduced M^2C is shown in Figure 14.17, and in practical use, two inductors should be connected in series to the upper and lower arms for protection.

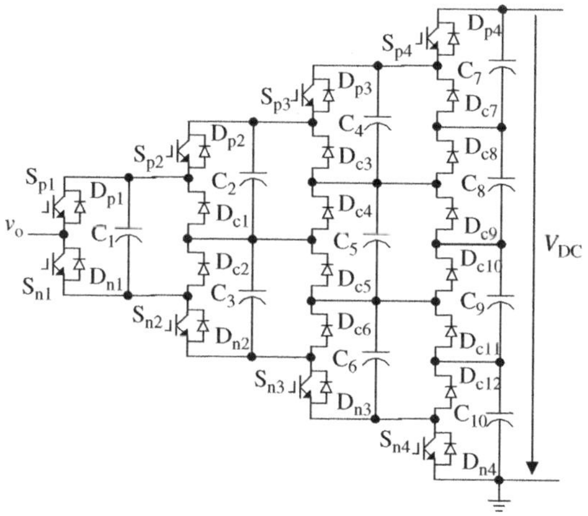

Figure 14.14 Diode and capacitor clamped five-level inverter

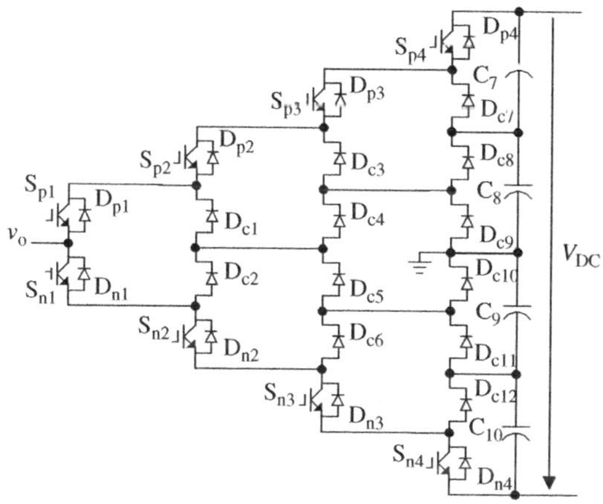

Figure 14.15 A new diode clamped five-level inverter

The M^2C was proposed in 2001, and currently is an attractive topology in application to medium- and high-voltage areas [12–14]. As shown in Figure 14.17, each module is a controlled voltage source. The terminal voltage V_{pi} or V_{ni} ($i = 1, 2$) can be either 0 V or V_C, regardless of the sign of the current $i_{a,i}$, assuming all modules have the same capacitor voltage V_C. The voltage V_{DC} and V_{AC} can be adjusted by controlling the modules of the upper and lower arms, but they have to meet the requirements of the following equation:

$$V_{DC} + |2V_{AC}| \leq 2 \times 2 \times V_C \tag{14.1}$$

If V_{DC} is constant and $V_{DC} = 2V_C$, the output voltage is limited to

$$|V_{AC}| \leq 2 \times V_C \tag{14.2}$$

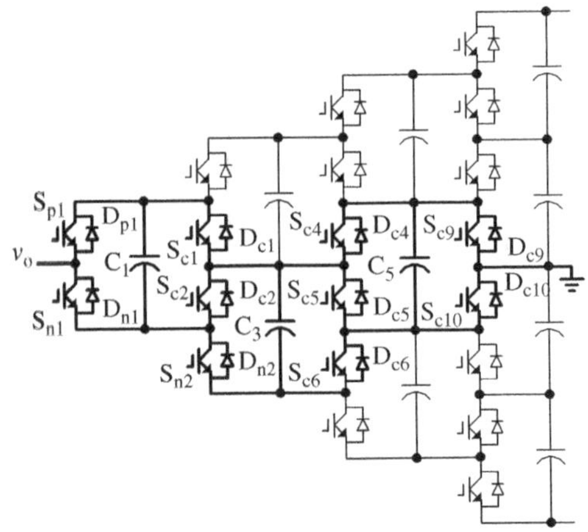

Figure 14.16 Zigzag multilevel inverter (thick line)

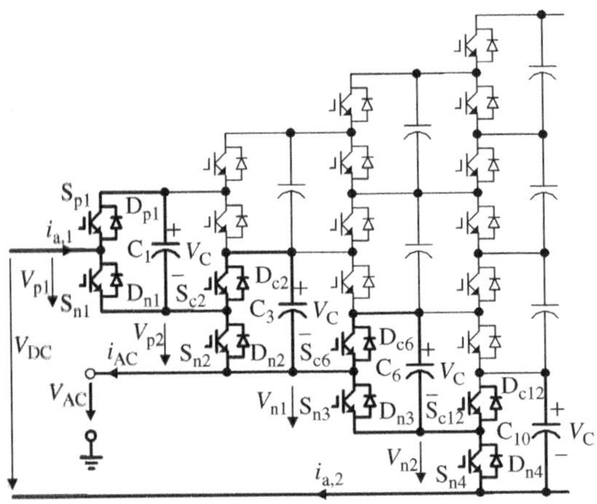

Figure 14.17 M^2C (thick line)

Figures 14.18 and 14.19 show the deduced Marx multilevel converter and the stacked multicell converter [15]. Figures 14.20–14.22 are the three deduced active NPC converters [15].

14.3 Cascaded Multilevel Inverters and Their Applications

14.3.1 Merits of Cascaded Multilevel Inverters Applied to Utility Level

With the growth of long-distance AC power transmission and load, the reactive power should be controlled to stabilize the power systems. For this purpose, the conventional 48-pulse inverter employs eight 6-pulse inverters and eight zigzag arrangement transformers for the SVG to reach high voltage. The transformers are expensive, lossy, bulky and present control difficulties owing to saturation, DC offset, and voltage surge.

The multilevel inverters can reach high voltage without the transformers of conventional SVG, but the diode clamped multilevel inverter requires $(N-1)\times(N-2)\times 3$ additional clamping diodes to provide an

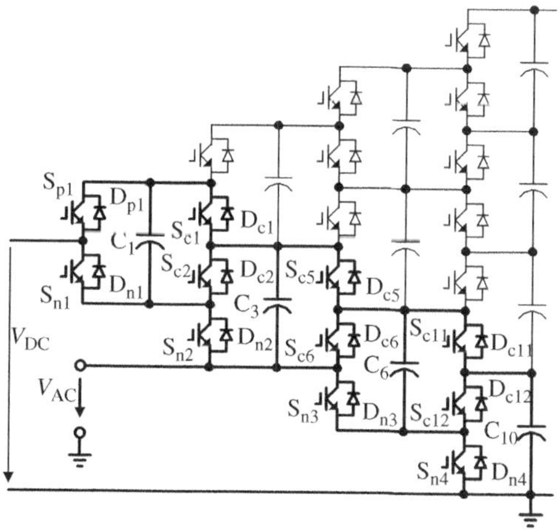

Figure 14.18 Marx multilevel converter (thick line)

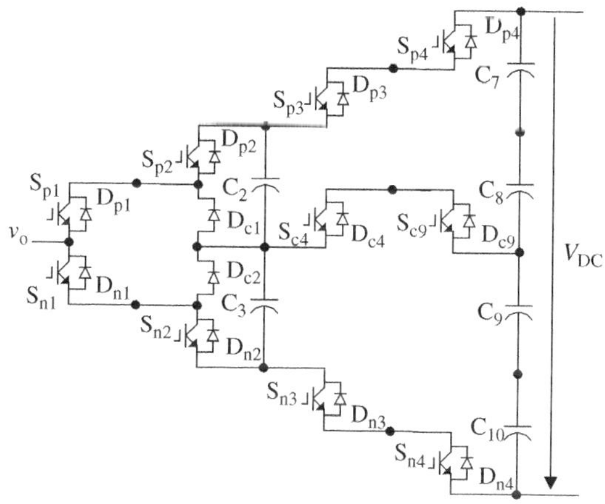

Figure 14.19 Stacked multicell converters

N-level voltage, and the capacitor clamped multilevel inverters require $(N-1)\times(N-2)\times 3/2+(N-1)$ capacitors. Moreover, the flying capacitor inverter's control is very complicated with high switching frequency to balance each capacitor voltage.

The CMI overcomes the above problems well [16, 17]. It needs the smallest component count with $(N-1)\times 3/2$ capacitors and $2(N-1)\times 3$ switches to perform the N-level three phase voltages.

14.3.2 Y-Connected Cascaded Multilevel Inverter and Its Applications

14.3.2.1 Y-Connected Cascaded Multilevel Inverter for Static var Generator

Figure 14.23 shows an example of the Y-connection structure of a three-phase 11-level CMI, which uses the switches to synthesize voltage waveforms instead of zigzag transformers [16–18]. Each phase consists of five cascaded H-bridge power modules, and an 11-level output voltage u_{Can} is produced in

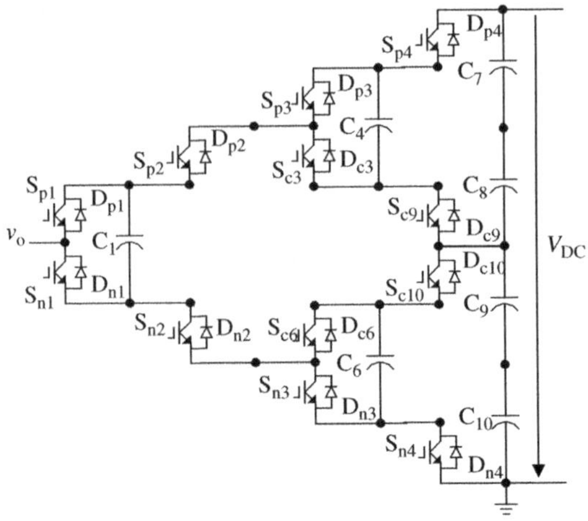

Figure 14.20 Active NPC converter I

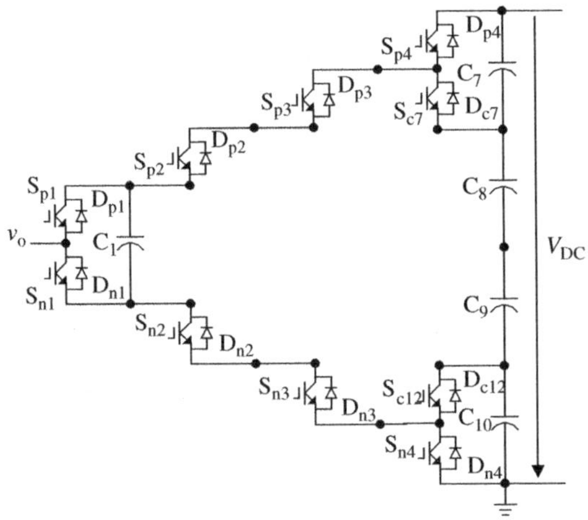

Figure 14.21 Active NPC converter II

Figure 14.24, where V_{Cai} $(i = 1, 2, \ldots, 5)$ denotes the module-i output voltage of phase a. The minimum harmonics can be achieved through designing the switching phase angles of each module.

14.3.2.2 Y-Connected Hybrid Cascaded Multilevel Inverter for Medium-Voltage Motor Drive

With additional isolated power sources, the CMI can be applied to control real power such as in applications to ASDs [4, 19] and PV power systems [20]. In fact, the CMI is the best-selling product in the medium-voltage ASDs market, even though it requires many transformer windings.

For medium-/high-voltage and high-power applications, the diode clamped or capacitor clamped multilevel leg can replace the conventional two-level leg in the H-bridge module of the CMI to reduce the

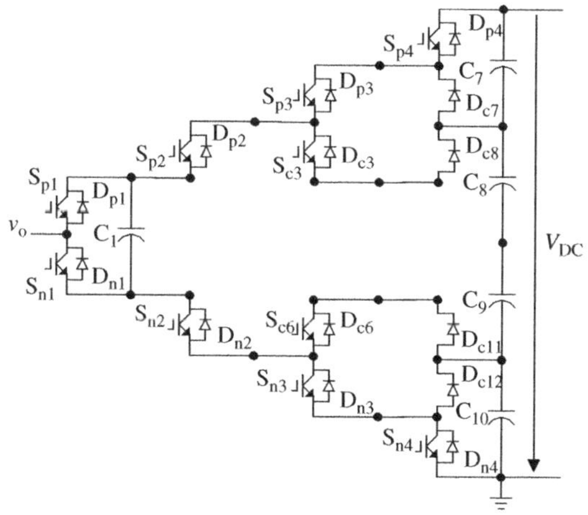

Figure 14.22 Active NPC converter III

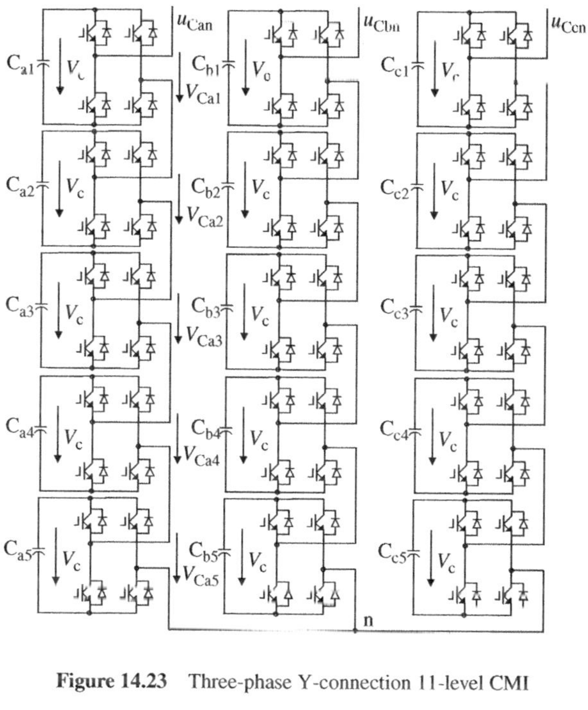

Figure 14.23 Three-phase Y-connection 11-level CMI

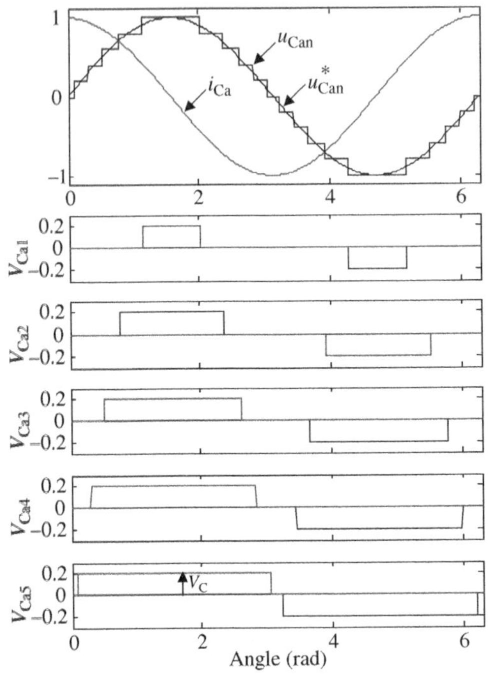

Figure 14.24 Multilevel output voltage

number of separate DC sources. Figure 14.25a shows an example, where the cascaded NPC-multilevel inverter feeds a three-phase induction motor [21]. Each phase consists of four NPC-based H-bridge modules connected in cascade, and each module can output five-level voltage with the configuration as in Figure 14.25b. Consequently, the 17-level phase-to-neutral voltage or 33-level line-to-line voltage will be produced, even though there are four modules per phase in Figure 14.25a. Four isolated DC sources per phase or 12 DC sources for the entire CMI are sufficient. However, for the traditional CMI, eight H-bridge modules and eight separate DC sources are necessary per phase, giving 24 DC sources for a three-phase CMI.

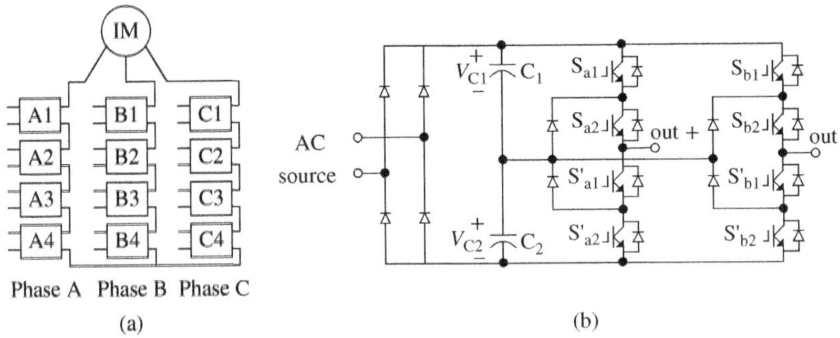

Figure 14.25 (a) Cascaded NPC-multilevel inverter-based induction motor drive and (b) NPC-based H-bridge module

14.3.2.3 Y-Connected Cascaded Multilevel Inverter for PV Power Generation

The CMI is most suited for application to PV power generation for a number of reasons:

1. The CMI requires separate DC voltage sources when applied to real power control and the separate PV panels suit this requirement.
2. The CMI can reach high voltage by using low-voltage devices, and therefore, the transformerless solution is possible.
3. A distributed maximum power point tracking (MPPT) can be achieved to maximize the solar power utilization, which can avoid/mitigate the power loss at mismatching conditions of series-connected PV panels.
4. It presents modularity, a simple layout, fewer components, and higher reliability when compared with that of the diode clamped and capacitor clamped multilevel inverters.

Figure 14.26 shows an example of a Y-connected CMI-based PV power system connected to the 4160 V grid, which requires a cascaded DC-link voltage larger than 3397 V [22]. Assuming that the PV panel's voltage changes in the range of 1–2, the minimum cascaded PV voltage could be 3397 V and the maximum voltage could be 6794 V. For the conventional CMI, the cascaded DC-link voltage is the same as the cascaded PV voltage. If we design each module's maximum DC-link voltage with 800 V, its minimum operating voltage will be 400 V and the nine modules are necessary to fulfill the CMI connected directly to the grid.

However, this kind of CMI does not have a boost function, which will lead to overrating the inverter by a factor of 2 in order to cope with wide PV panel's voltage changes. Furthermore, the partial shading

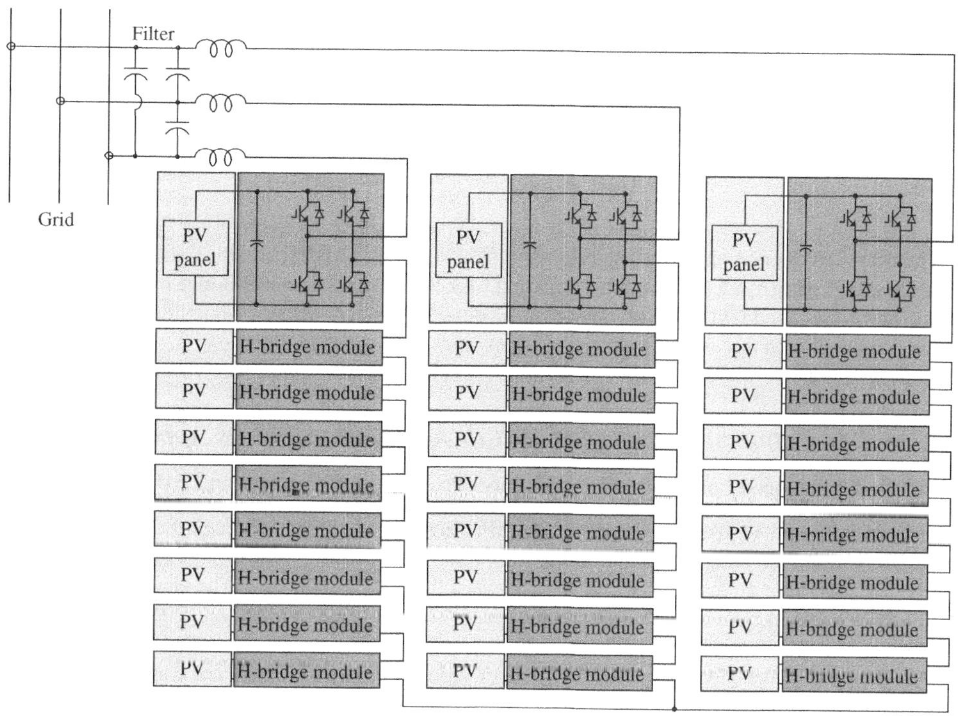

Figure 14.26 Traditional CMI for PV power system

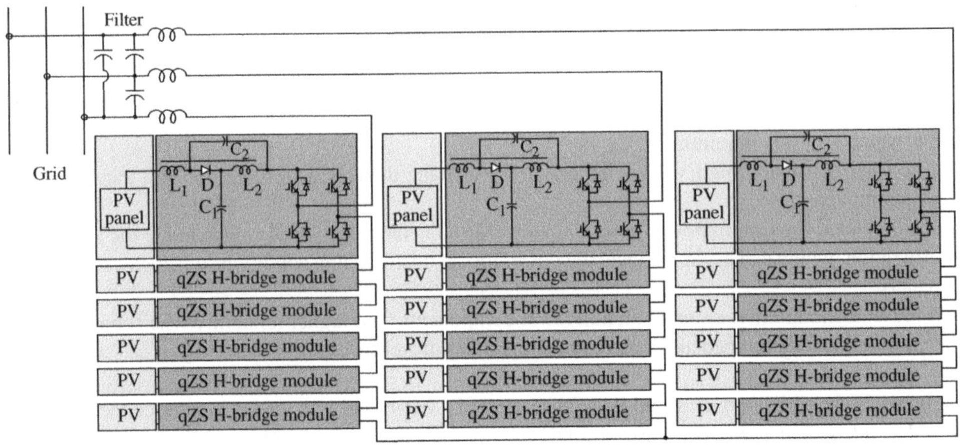

Figure 14.27 Quasi-Z-source CMI-based PV power generation system

might cause some modules' PV panels to have different maximum power points (MPP) from other PV panels, and the problem of unbalanced DC-link voltage will happen.

The quasi-Z-source inverter topology provides a suitable solution to boost the PV panel's low-voltage level to a high-voltage level, simultaneously achieving voltage inversion in a single-stage conversion [23, 24]. Figure 14.27 depicts a quasi-Z-source CMI topology, where each PV panel connects to the H-bridge through a quasi-Z-source network [22, 25, 26]. Each quasi-Z-source inverter module can achieve a constant DC-link peak voltage control regardless of the PV panel's voltage variation. This feature will ensure the constant DC-link peak voltage of each module. For the above case, we can design the DC-link peak voltage with 800 V, X modules in cascade, and the maximum modulation index M when the PV panel voltage is at its minimum value 400 V; thus, we obtain

$$800 \cdot X \cdot M = \frac{4160\sqrt{2}}{\sqrt{3}} = 3397\,\text{V} \tag{14.3}$$

The 400 V PV panel voltage can be boosted to 800 V when $D = 0.25$, and the maximum modulation index $M = 0.75$, then $X = 5.66$ if we solve Equation 14.3. Six quasi-Z-source H-bridge modules are chosen in the newly designed CMI. It is obvious that the quasi-Z-source CMI saves one-third power modules when compared with that of the traditional CMI as shown in Figure 14.26.

14.3.3 Δ-Connected Cascaded Multilevel Inverter and Its Applications

14.3.3.1 Unbalanced Load (Negative-Sequence) Problem of Y-Connection CMI

As shown in Figure 14.28a, the Y-connected CMI connected to the grid through the inductors is controlled to output the voltages u_{Ca}, u_{Cb} and u_{Cc} in phase with the grid voltages u_{Sa}, u_{Sb} and u_{Sc}, respectively. It should have no real power exchange between the CMI and the grid. The phase voltage and the grid-injected current have to be 90° out of phase, and the CMI behaves like a var generator.

However, the power system might have an unbalanced load. For this situation, the Y-connected CMI cannot compensate the unbalanced load to ensure the grid system with the balanced currents, even though it still can provide the reactive power compensation [27]. As shown in Figure 14.28a and b, the phase-c current i_{Lc} is zero owing to the open circuits between phase c and the other two phases a and b. If we want to get the balanced grid currents i_{Sa}, i_{Sb} and i_{Sc} as shown in Figure 14.28c, the three-phase Y-connected

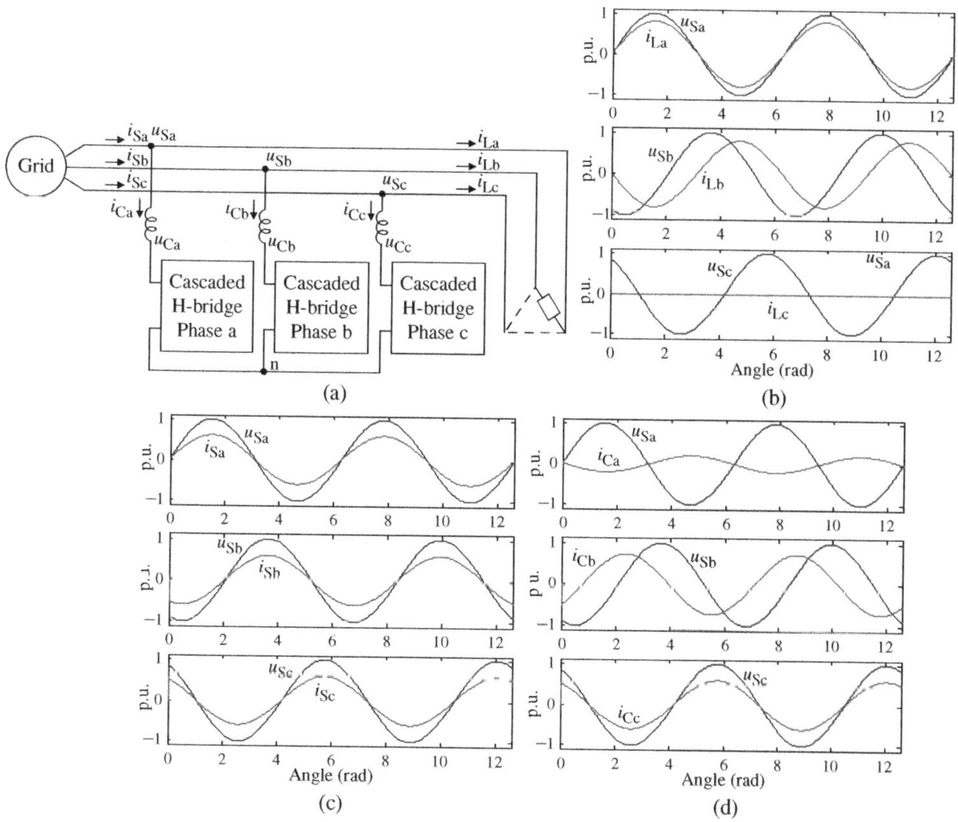

Figure 14.28 Imbalanced load problem of Y-connected STATCOM. (a) Whole system, (b) imbalanced load currents, (c) desired grid currents and (d) required STATCOM currents

CMI has to provide the currents shown in Figure 14.28d, which shows that there is real power exchange between the CMI and the grid. As a result, the CMI's DC capacitors will be overcharged/overdischarged. Therefore, the three-phase Y-connected CMI cannot compensate the unbalanced load.

14.3.3.2 Δ-Connected Cascaded Multilevel Inverter for STATCOM

Figure 14.29a shows the Δ-connected CMI-based STATCOM [27]. Each phase leg consists of the H-bridge modules and inductor in cascade. The three phase legs are connected in the Δ-shape. The line-to-line voltage presents the phase leg voltage waveform, as shown in Figure 14.29b.

The Δ-connected CMI-based STATCOM can compensate the unbalanced load, reactive power, and harmonics. The reactive power and harmonics compensations are the same as the Y-connected CMI. The compensation principle of unbalanced load currents can be explained in Figure 14.30. The balanced three-phase voltage source supplies the three-phase load in Figure 14.30a, and the load admittances are

$$Y^l_{ab} = G^l_{ab} + jB^l_{ab},\ Y^l_{bc} = G^l_{bc} + jB^l_{bc},\ Y^l_{ca} = G^l_{ca} + jB^l_{ca} \tag{14.4}$$

It is well known that the reactive power and system imbalance can be compensated by one Δ-connected reactive network. As shown in Figure 14.30b, the Δ-connected CMI-based STATCOM provides the pure

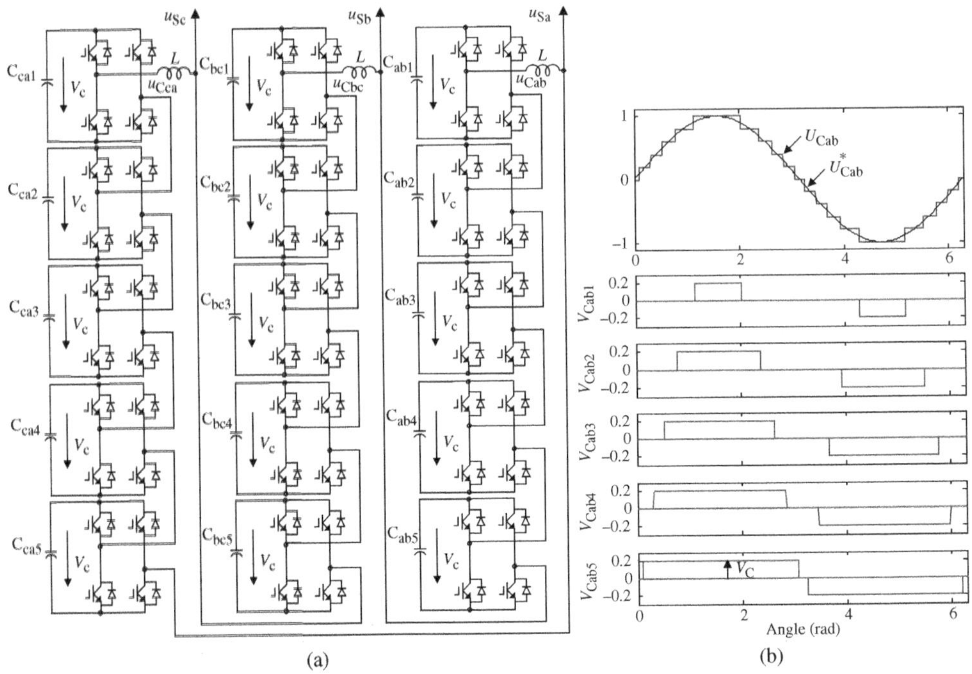

Figure 14.29 (a) Δ-connected CMI-based STATCOM and (b) single-phase voltage waveform

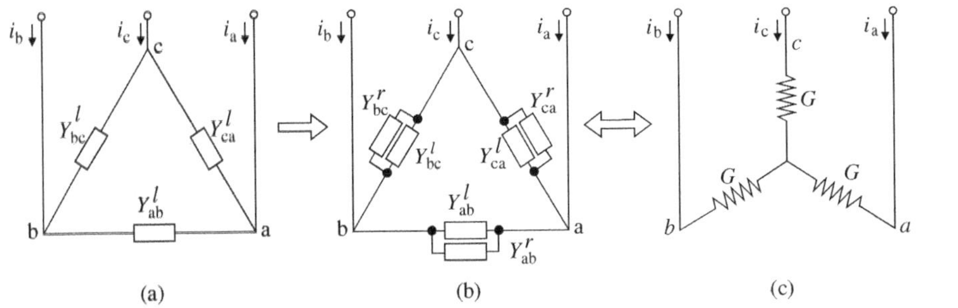

Figure 14.30 Compensation principle of the system with imbalanced load. (a) Imbalanced load, (b) compensation method and (c) after compensation

reactive compensation using

$$Y^r_{ab} = jB^r_{ab} = j\left[-B^l_{ab} + \frac{\left(G^l_{ca} - G^l_{bc}\right)}{\sqrt{3}}\right] \tag{14.5}$$

$$Y^r_{bc} = jB^r_{bc} = j\left[-B^l_{bc} + \frac{\left(G^l_{ab} - G^l_{ca}\right)}{\sqrt{3}}\right] \tag{14.6}$$

$$Y^r_{ca} = jB^r_{ca} = j\left[-B^l_{ca} + \frac{\left(G^l_{bc} - G^l_{ab}\right)}{\sqrt{3}}\right] \tag{14.7}$$

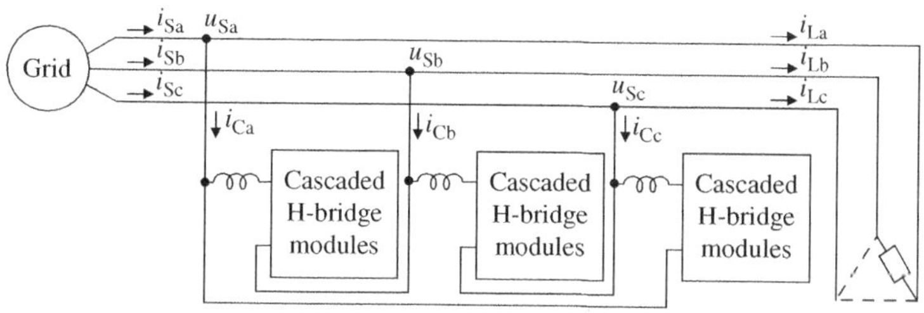

Figure 14.31 Δ-connected CMI-based STATCOM with imbalanced load

The resultant equivalent Y-connected system will be fully resistive and balanced in Figure 14.30c and the equivalent phase admittance meets $G = G^l_{ab} + G^l_{bc} + G^l_{ca}$ [27]. As a result, the compensation makes the system balanced. Figure 14.31 shows a case with an imbalanced load. The compensator can provide the currents shown in Figure 14.28d to create the system with the balanced currents shown in Figure 14.28c.

14.3.4 Face-to-Face-Connected Cascaded Multilevel Inverter for Unified Power Flow Control

14.3.4.1 Conventional UPFC Principle, Structure and Disadvantages

The conventional unified power flow control (UPFC) employs the back-to-back inverter topology, as shown in Figure 14.32, where inverter 1 is connected in parallel to the transmission line, and inverter 2 is connected in series to the transmission line. The two inverters have a common DC-link bus, and each inverter has a coupling transformer with the line. Because of the shared DC-link bus, both inverters can achieve real power exchange with each other. The injected series voltage of inverter 2 to the transmission line can have any phase angle difference from the line current, which provides the UPFC with flexible controllability to manage both real and reactive power flows through the line. Inverter 1 can provide or absorb the real power of inverter 2 and it can supply or absorb the reactive power for the transmission line. As a result, UPFC is the most versatile FACTS device. It can balance effectively the load of all the lines and allow the overall system to operate at its theoretical maximum capacity.

As shown in Figure 14.32, the voltage V_C makes the sending end voltage of the system become V_{SN} from the original voltage V_{SO}, and the angle between both voltages of sending and receiving ends changes from δ_0 to δ_N. As a result, the transmitted real and reactive powers from the sending end to the receiving end also change from

$$P = \frac{V_{SO} V_R \sin \delta_0}{X} \quad \text{and} \quad Q = -\frac{(V_R \cos \delta_0 - V_{SO}) V_{SO}}{X} \tag{14.8}$$

to

$$P' = \frac{V_{SN} V_R \sin \delta_N}{X} \quad \text{and} \quad Q' = -\frac{(V_R \cos \delta_N - V_{SN}) V_{SN}}{X} \tag{14.9}$$

The conventional UPFC has been put into practical applications with the following features: (i) both inverters have a common DC-link bus; (ii) there is real power exchange between both inverters and the transmission line; (iii) each inverter uses a transformer coupling to the transmission line. In addition, any utility-scale UPFC requires two high-voltage high-power (from several MVA to hundreds of MVA) inverters. Traditionally, these high-voltage high-power inverters have to use bulky and complicated zigzag transformers to reach their required VA and voltage ratings. Moreover, the zigzag transformer-based UPFCs are still too slow in dynamic response (up to minutes owing to the large time

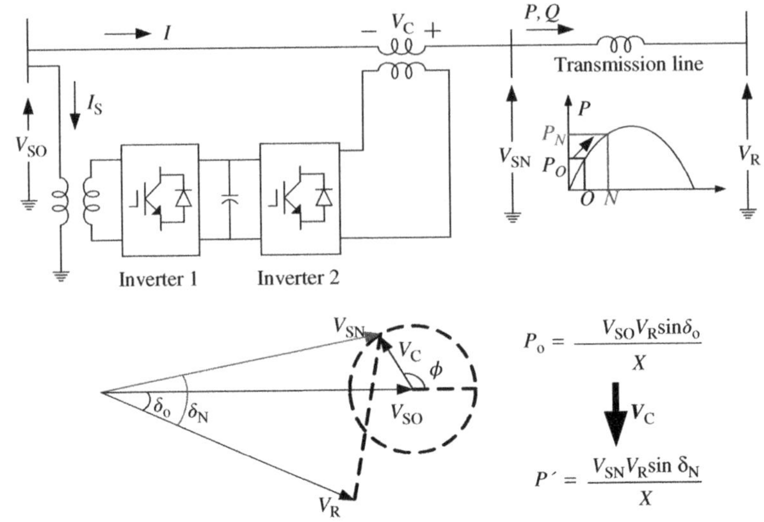

Figure 14.32 Conventional UPFC

constant of magnetizing inductance over resistance) and pose control challenges because of transformer saturation, magnetizing current, and voltage surge.

14.3.4.2 Face-to-Face-Connected Cascaded Multilevel Inverter for UPFC

There are several reasons why the CMI cannot be used directly with the conventional UPFC structure:

1. The CMI does not have a common DC-link bus, but the traditional UPFC requires a shared DC-link bus for two inverters.
2. The CMI does not allow real power flow from the AC side to the DC side, nor from the DC side to the AC side, because there are no sources for the CMI's modules.

The conventional UPFC structure is changed to fit the CMI. Figure 14.33 shows the face-to-face-connected CMI with the following features [28]:

1. Each module has only the DC capacitor for the reactive power, rather than any other sources.
2. Two CMIs do not have real power exchange with each other, nor any real power exchange with the transmission line.
3. Two CMIs are connected face-to-face on the AC side, rather than back-to-back on the DC-link bus.

In Figure 14.33, inverter 2 supplies the desired voltage V_C in the transmission line to control the transmitted real and reactive power flows over the line. Inverter 1 is connected in parallel to the transmission line and the injected current ($I_C - I$) has to be perpendicular to its own voltage $V_{SO} - V_C$. The current I_C of inverter 2 also has to present a difference of 90° angle from its own voltage V_C. This guarantees that there is no real power flowing into either CMI. For this new UPFC, inverter 1 no longer provides or absorbs the real power for inverter 2. Figures 14.34 and 14.35 show the equivalent circuit and phasor diagrams, respectively.

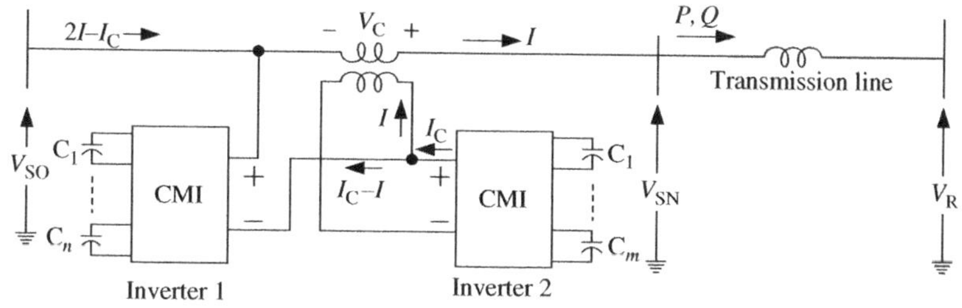

Figure 14.33 Face-to-face connected CMI for UPFC

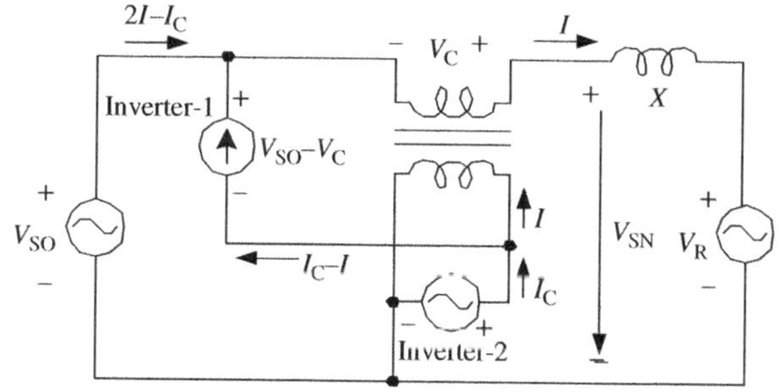

Figure 14.34 Equivalent circuit

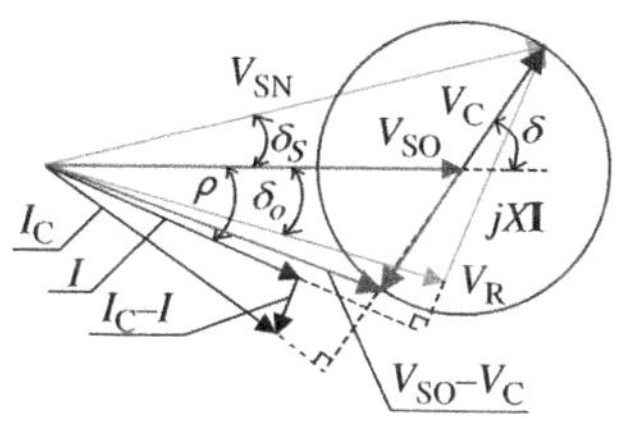

Figure 14.35 Phasor diagram

From Figures 14.34 and 14.35, inverter 1 and inverter 2 have the following respective real power equations:

$$P_1 = (V_{SO} - V_C) \cdot (I_C - I) = -V_{SO} I \cos\rho + V_{SO} I_C \cos(90^\circ - \delta) + V_C I \cos(\delta + \rho) = 0 \tag{14.10}$$

$$P_2 = V_C \cdot I_C = 0 \tag{14.11}$$

Solving the above two equations, one has the current I_C as

$$I_C = \frac{V_{SO} I \cos\rho - V_C I \cos(\delta + \rho)}{V_{SO} \sin\delta} \angle(\delta - 90^\circ) \tag{14.12}$$

The transmitted real power P and reactive power $-jQ$ can be expressed as

$$P - jQ = V_R \cdot \left(\frac{V_{SO} + V_C - V_R}{jX}\right)^* = \frac{V_{SO}V_R}{X}\sin\delta_0 + \frac{V_C V_R}{X}\sin(\delta_0 + \delta)$$
$$+j\left(\frac{V_{SO}V_R\cos\delta_0 - V_R^2}{X} + \frac{V_C V_R}{X}\cos\left(\delta_0 + \delta\right)\right) \quad (14.13)$$

The original transmitted real and reactive powers (i.e., without UPFC) are

$$P_0 = \frac{V_{SO}V_R}{X}\sin\delta_0, Q_0 = -\frac{V_{SO}V_R\cos\delta_0 - V_R^2}{X} \quad (14.14)$$

The net differences between the original (without UPFC) powers and the new (with UPFC) powers are the controllable real and reactive powers from the UPFC and can be expressed as

$$P_C = \frac{V_C V_R}{X}\sin(\delta_0 + \delta), Q_C = -\frac{V_C V_R}{X}\cos(\delta_0 + \delta) \quad (14.15)$$

Because the amplitude V_C and phase angle δ of the injected voltage phasor V_C can be of any values as commanded, the new UPFC provides a full controllable range of $-V_C V_R/X$ to $+V_C V_R/X$ for both real and reactive powers, P_C and Q_C, which are advantageously independent of the original sending end voltage and phase angle δ_0. Therefore, the theory proves that the new UPFC has the same functionality as the conventional UPFC.

The face-to-face configuration provides a way for achieving UPFC with the CMI. The advantages of the CMI provide the new features to the UPFC. For example, the CMI-based UPFC can reach any high-voltage level (from 15 kV to 35 kV, 69 kV and so on) by increasing the number of H-bridges without using complex step-up transformers. Each H-bridge can be modularized to achieve further low cost. Each phase leg consisting of n H-bridge modules can have $(n + 1)$ redundancy to improve reliability. Any faulty module can be bypassed to ensure the fail-safe operation of the entire system.

14.4 Emerging Applications and Discussions

14.4.1 Magnetic-less DC/DC Conversion

The DC/DC conversion is widely applied to PV power systems, hybrid electrical vehicles (HEVs), thermoelectric generators (TEGs), aerospace areas, and so on, and the ambient temperatures in these areas can be quite high. Conventional DC/DC converters employ an inductor and/or transformer, as shown in Figure 14.36. Consequently, the magnetic cores of the inductors and transformers present a bottleneck for high-temperature environments, because the permeability declines dramatically with increasing temperature. Thus, they are prone to saturation and instability as the temperature goes beyond a certain limit, which leads to the nonlinear relation of outputs to inputs, making effective control very difficult. Furthermore, inductors and/or transformers, as some of the hottest components in the system, are bulky, lossy, and are an obstacle to applying power converters in high temperatures and to reducing their size and cost.

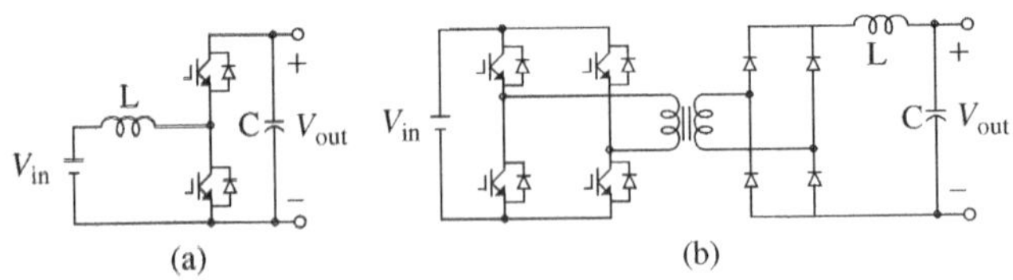

Figure 14.36 Conventional DC/DC converter. (a) Non-isolated type and (b) isolated type

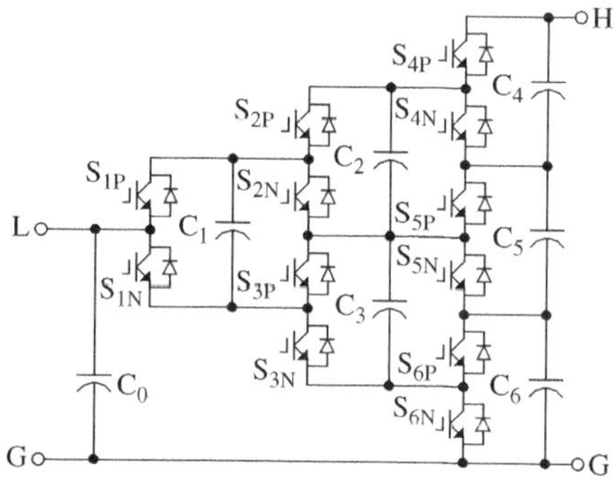

Figure 14.37 Four-level magnetic-less DC/DC converter

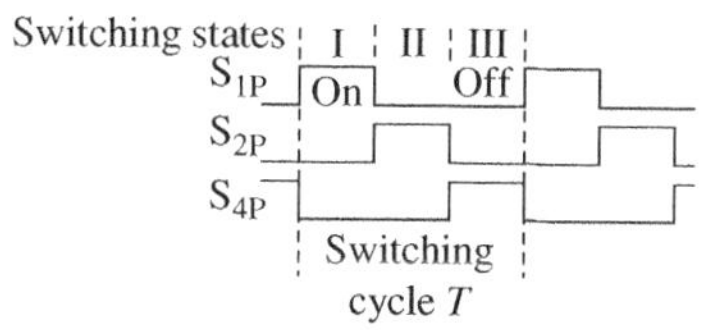

Figure 14.38 Converter's gating sequence

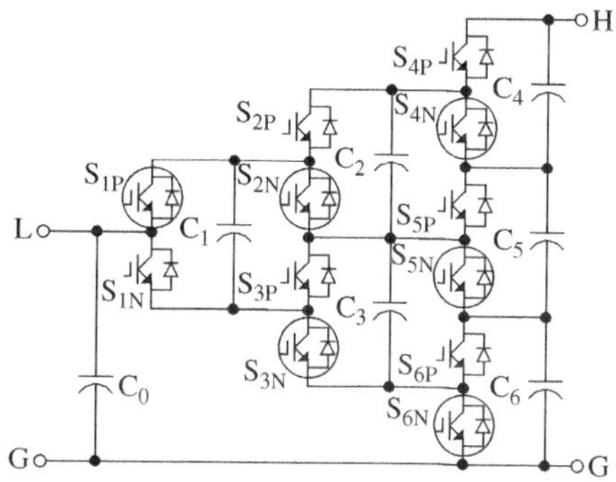

Figure 14.39 Converter's state I

Magnetic-less DC/DC converters are good candidates to overcome the above problems. Figure 14.37 shows an example derived from the generalized multilevel converter in Figure 14.6 [29]. The sample converter presents four-level and bidirectional power conversion and has no magnetic components. Figure 14.38 shows the gating sequence necessary to control the converter, where a fixed duty ratio of 1 : 3 is employed. The top switches S_{1P}, S_{2P}, and S_{4P} are independent and their states can describe the states of the entire converter. The states of other switches are derived by using the complementary switching of two adjacent switches. From the states I, II, and III of Figure 14.38, there are three respective switching states, as shown in Figures 14.39–14.41.

For the converter's state I in Figure 14.39, the capacitors C_1, C_3 and C_6 have the same voltage as the voltage between terminals L and G, that is, $V_{LG} = V_{C1} = V_{C3} = V_{C6}$, and the capacitors

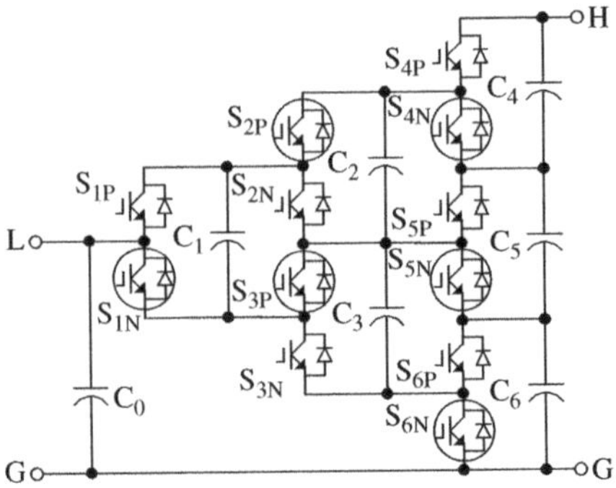

Figure 14.40 Converter's state II

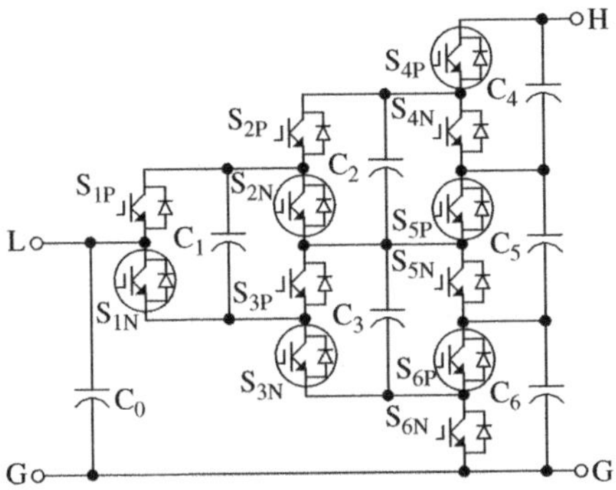

Figure 14.41 Converter's state III

C_2 and C_5 have the same voltage, that is, $V_{C2} = V_{C5}$. For the converter's state II in Figure 14.40, $V_{LG} = V_{C3} = V_{C6}$ and $V_{C1} = V_{C2} = V_{C5}$. For the converter's state III in Figure 14.41, $V_{LG} = V_{C6}$, $V_{C1} = V_{C3} = V_{C5}$ and $V_{C2} = V_{C4}$. After the three states of one control cycle, all the capacitor voltages are balanced, that is, $V_{LG} = V_{C1} = V_{C2} = V_{C3} = V_{C4} = V_{C5} = V_{C6}$. The voltage between terminals H and G is $V_{HG} = V_{C4} + V_{C5} + V_{C6}$ and $3V_{LG} = V_{HG}$. This four-level converter presents a voltage multiplier or divider with a factor of 3 or 1/3, respectively [29].

The circuit of Figure 14.37 can be simplified as the flying capacitor converter in Figure 14.42 [30]. The top switches S_{1P}, S_{2P}, and S_{3P} can be used to describe the converter's states, because the bottom switches

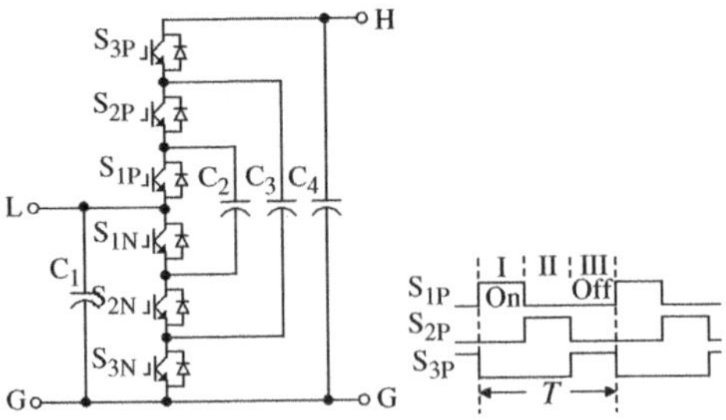

Figure 14.42 Flying capacitor DC/DC converter and gating sequence

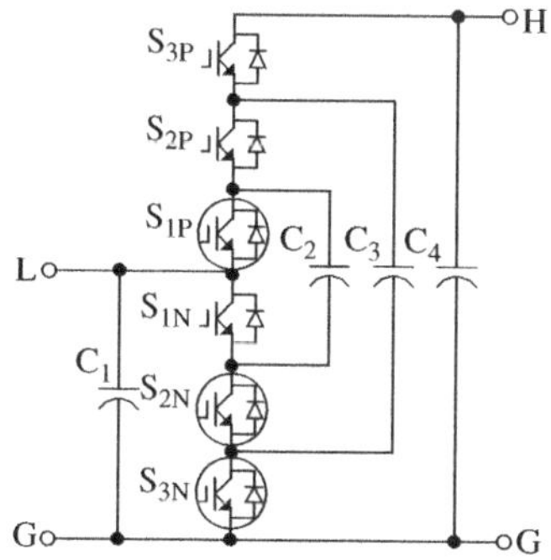

Figure 14.43 Converter's state I

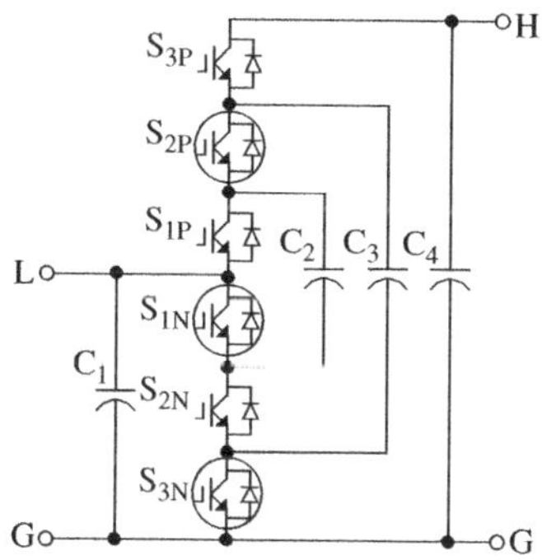

Figure 14.44 Converter's state II

have complementary states to the top switches if they are labeled with the same number, for example, switches S_{1P} and S_{1N} have the same number "1." From states I, II, and III of Figure 14.42, there are three respective converter's states, as shown in Figures 14.43–14.45.

In Figure 14.43, $V_{LG} = V_{C1} = V_{C2}$. In Figure 14.44, $V_{C3} = V_{C1} + V_{C2}$. Similarly, in Figure 14.45, $V_{HG} = V_{C1} + V_{C3}$. After one control cycle, the voltages of capacitors C_1, C_2, and C_3 meet $V_{C1} = V_{C2}$ and $V_{C3} = 2V_{C1}$. There is a fixed voltage ratio of 3 between V_{HG} and V_{LG}, that is, $V_{HG} = 3V_{LG}$ with a fixed duty ratio of 1 : 3 to control the bidirectional converter.

In addition, Figure 14.42 can be controlled to achieve $V_{HG} = V_{LG}$ and $V_{HG} = 2V_{LG}$. In total, there are three voltage relationships: $V_{HG} = V_{LG}$, $V_{HG} = 2V_{LG}$, and $V_{HG} = 3V_{LG}$, the so-called 1X, 2X, and 3X converters [31], respectively. Figure 14.46 shows the 1X converter's switching state, where the switches S_{1P}, S_{2P}, S_{3P}, S_{2N}, and S_{3N} are always on. As a result, all the capacitors connected in parallel have the same voltage and $V_{HG} = V_{LG}$. Figures 14.47 and 14.48 show the 2X converter's switching states, where the switches S_{3P} and S_{3N} are in the on-state and the capacitors C_3 and C_4, connected in parallel, have the same voltage, that is, $V_{HG} = V_{C4} = V_{C3}$. The converter's two states I and II alternate each for 50% duty ratio. As shown in Figure 14.47, switching state I of the 2X converter makes the switches S_{1P} and S_{2N} turn on and the capacitors C_2 and C_1, connected in parallel, have the same voltage, that is, $V_{LG} = V_{C1} = V_{C2}$. For switching state II of the 2X converter, as shown in Figure 14.48, two switches S_{1N} and S_{2P} are turned on, and both of the capacitors C_3 and C_4 are connected in parallel to the series of capacitors C_1 and C_2, so that there is the voltage relationship of $V_{HG} = V_{C4} = V_{C3} = V_{C1} + V_{C2} = 2V_{C1}$.

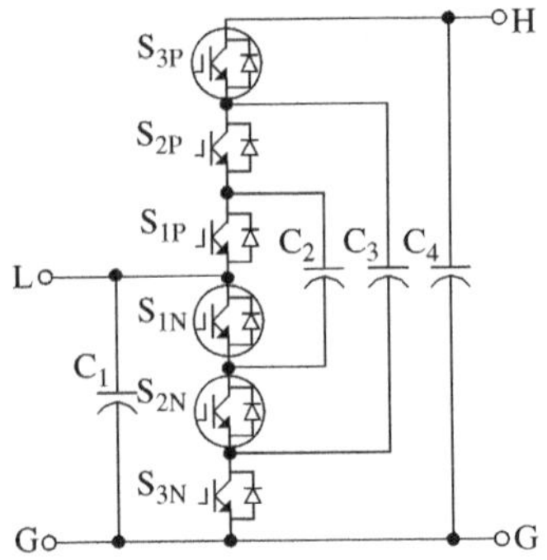

Figure 14.45 Converter's state III

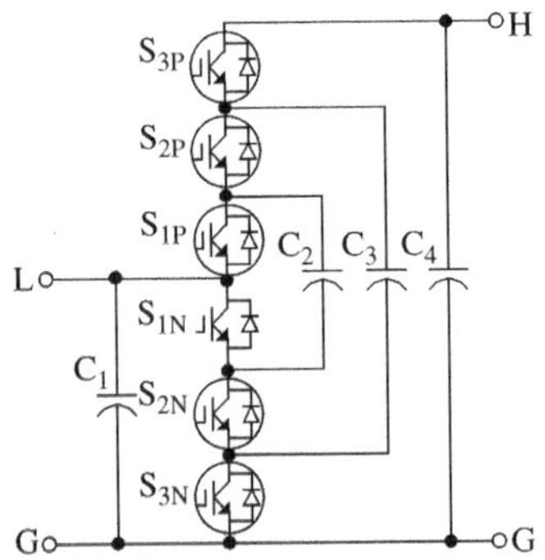

Figure 14.46 1X converter

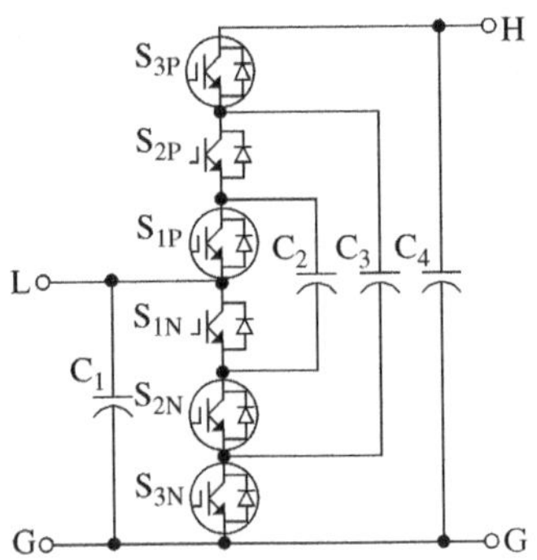

Figure 14.47 State I for 2X mode

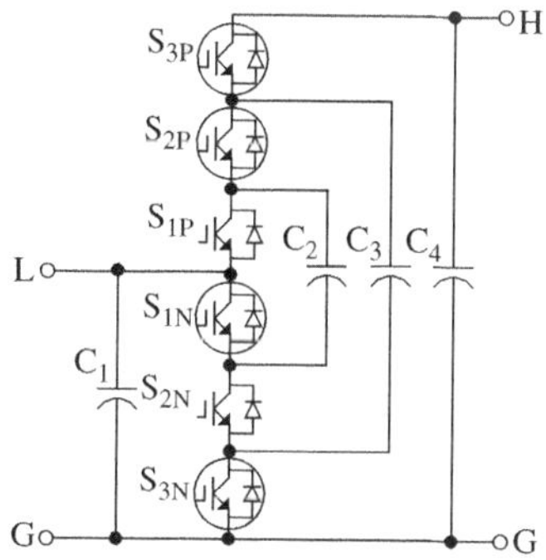

Figure 14.48 State II for 2X mode

In steady state, for separate 1X, 2X and 3X converters, all the capacitor voltages are well balanced. However, if we want to make the converter have a transition between 1X and 2X, or between 2X and 3X, the high transient current could flow through the devices and capacitors, owing to existing very large transient voltage differences. A PWM with gradually variable duty ratio and high switching frequency can limit this transient current. In addition, a small/stray inductance is required between capacitor C_1 and the switch leg [31].

In the 1X to 2X transient operation, the PWM duty ratio D will gradually decrease from 1 to 1/2 in order to limit the inrush current, and $D = 1$ for the 1X mode and $D = 1/2$ for the 2X mode. The same principle is applied to the transition from 2X to 3X, and the corresponding duty ratio D will gradually decrease from 1/2 to 1/3, and $D = 1/3$ for the 3X mode [31].

14.4.2 Multilevel Modular Capacitor Clamped DC/DC Converter (MMCCC)

The flying capacitor converter shown in Figure 14.42 has a smaller component count and lower voltage stress across the switching devices when compared with that of the generalized multilevel converter. However, the larger the required boost, the greater the losses incurred from the increased number of devices in the charge pump paths, because the input current always flows through n switching devices for the flying capacitor converter. The dominant voltage drop and power loss are inevitable. The control also becomes more complex with a higher boost ratio. Moreover, the flying capacitor structure is not modular, meaning that a fundamental, fully functioning unit of the circuit cannot be made. These issues offer the greatest argument against the use of the flying capacitor topology in situations involving high-voltage gain.

The MMCCC represents an attempt to surpass the flying capacitor converter's performance [32]. Figure 14.49 shows an example of the MMCCC, and it is a four-level converter with an output equal to four times the input. Figure 14.50 shows the modular block, which is used as the basic cell to extend the voltage boost ratio.

Figure 14.51 shows the gating sequence for Figure 14.49, and there are two switching states in one control cycle, as shown in Figure 14.52. For state I, $V_{C4} = V_{C3} + V_{in}$ and $V_{C2} = V_{C1} + V_{in}$; for state II, $V_{C1} = V_{in}$ and $V_{C3} = V_{C2} + V_{in}$. Then, after one cycle, $V_{C2} = 2V_{in}$, $V_{C3} = 3V_{in}$ and $V_{C4} = 4V_{in}$. Figure 14.53 redraws the MMCCC in the form of a flying capacitor circuit.

The MMCCC is expected to have high efficiency because (i) there are at most three switches used for charging a capacitor, even for achieving high voltage ratio; (ii) there are around $2/n$ times of currents flowing through the switches and capacitors when compared to that of the flying capacitor converter,

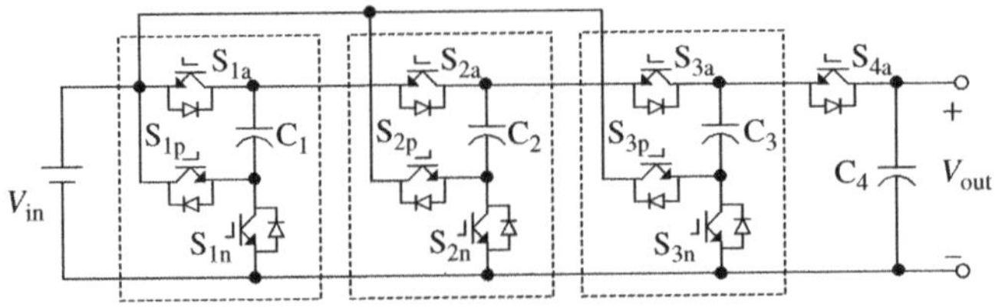

Figure 14.49 Four-level MMCCC with three modules

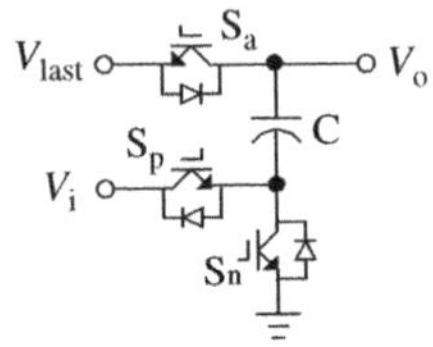

Figure 14.50 MMCCC module

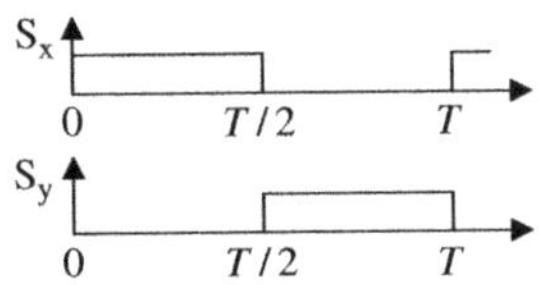

Figure 14.51 Gating sequence ($x = 4a$, 2a, 3p, 1p, and 2n; $y = 3a$, 1a, 2p, 3n and 1n)

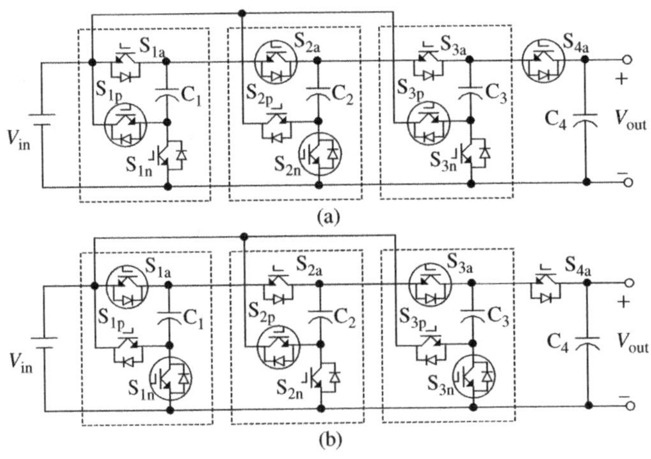

Figure 14.52 Two states of four-level MMCCC. (a) Switching state I and (b) switching state II

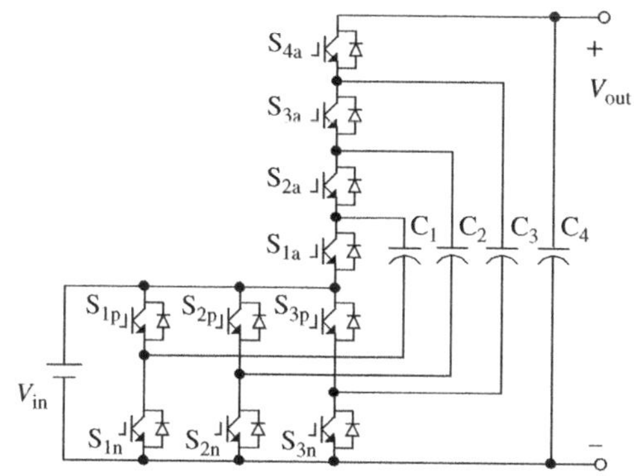

Figure 14.53 MMCCC in flying capacitor form

and total device power rating (TDPR) reduces. In addition, the control is very simple and the design is modular. However, the MMCCCs need the $(3n-2)$ switches and the voltage stresses of $(n-2)$ switches are twice the input voltage, whereas the flying capacitor converter just needs $2n$ switches and the voltage stress of each switch is equal to the input voltage.

14.4.3 nX DC/DC Converter

The *n*X DC/DC converter combines the modular structure and simple control of the MMCCC with the low switch count of the flying capacitor converter [33]. Figure 14.54 shows the 6X DC/DC converter. The power conversion is bidirectional in the respective buck and boost operations. The boost mode is used to clarify the converter's operating principle. Figure 14.55 redraws the 6X DC/DC converter in a modular structure, and the basic cell with 2X boost factor is shown in Figure 14.56. The 6X DC/DC converter is fulfilled through cascading three cells. The two switching states simplify the control required

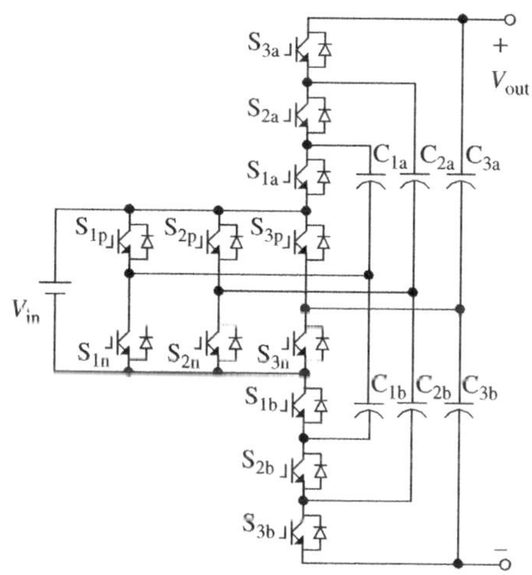

Figure 14.54 6X DC/DC converter

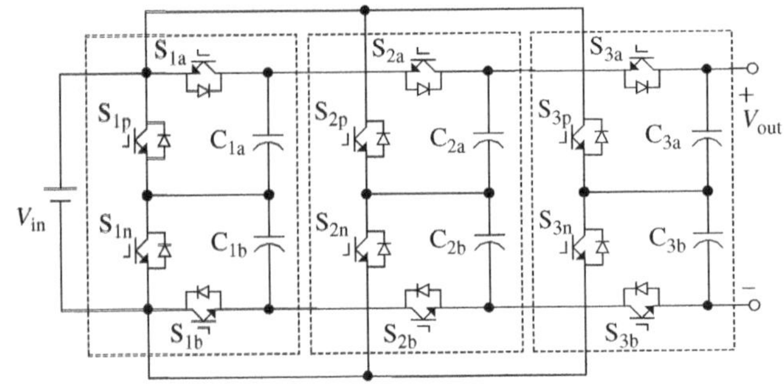

Figure 14.55 Modular structure of 6X DC/DC converter

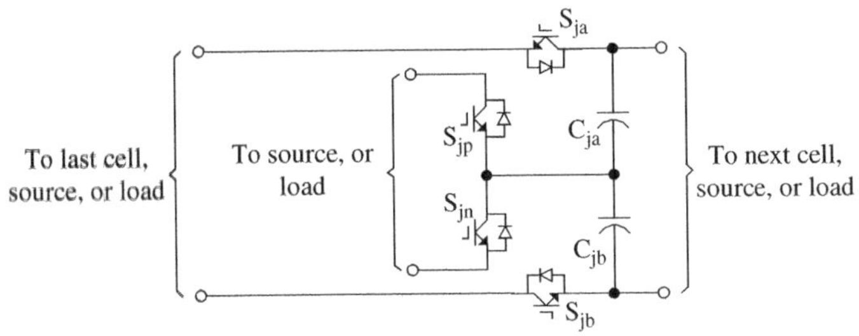

Figure 14.56 Basic module

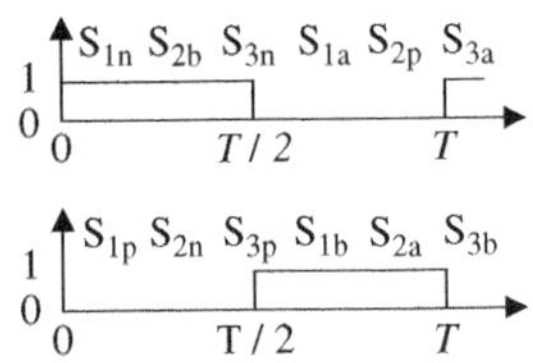

Figure 14.57 Gating sequence

to drive the converter. As shown in Figure 14.57, a complementary 50% duty cycle controls all of the semiconductors involved. Figure 14.58 shows the switching states of the 6X DC/DC converter, in which the circled switches are on. From Figure 14.58a, $V_{C2b} = V_{C1b} + V_{in}$, $V_{C1a} = V_{in}$ and $V_{C3a} = V_{C2a} + V_{in}$; from Figure 14.58b, $V_{C2a} = V_{C1a} + V_{in}$, $V_{C1b} = V_{in}$ and $V_{C3b} = V_{C2b} + V_{in}$. Finally, $V_{C1a} = V_{C1b} = V_{in}$, $V_{C2a} = V_{C2b} = 2V_{in}$, $V_{C3a} = V_{C3b} = 3V_{in}$ and $V_{out} = 6V_{in}$.

The nX converter's other unique features include the following:

1. There are two paths directly charging the load, which contributes to lower power loss during the energy transfer.
2. The maximum capacitor voltage is equal to $n \times V_{in}/2$.
3. Two output capacitors require lower capacitance because their interleaved complementary charge–discharge leads to low output voltage ripples, and they also have lower ripple currents because of their alternately receiving charge and continuously supplying the load.

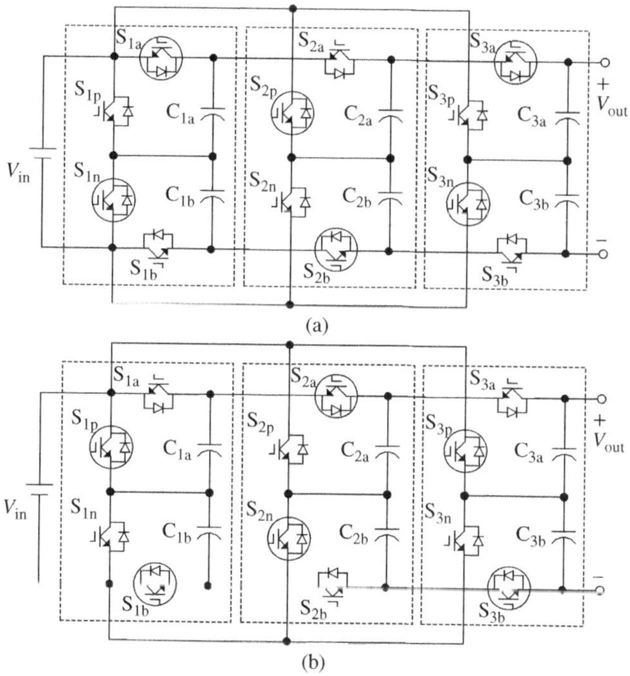

Figure 14.58 The nX converter's states. (a) Switching state I and (b) switching state II

4. The nX converter needs $2n$ switches, but the MMCCC requires $3n-2$ switches. Therefore, the nX converter will have higher efficiency, more compact packaging, and lighter weight in implementing a high-voltage boost converter when compared with that of the MMCCC and the flying capacitor converter.

14.4.4 Component Cost Comparison of Flying Capacitor DC/DC Converter, MMCCC and nX DC/DC Converter

The flying capacitor converter, the MMCCC, and the nX DC/DC converter have only capacitors and switches. The capacitor cost is dependent on the capacitor voltage stress, current rating, and capacitance requirement. The switch cost is related to the silicon area used. The TDPR provides a qualified indication of the total silicon area needed in a converter/inverter.

14.4.4.1 Total Device Power Rating

The TDPR of a converter/inverter is defined as

$$\mathrm{TDPR} = \sum_{m=1}^{N} V_m I_{m\text{-AV}} \tag{14.16}$$

where N is the number of switches, $I_{m-\mathrm{AV}}$ is the average current going through the mth switch and V_m is the peak voltage on the mth switch.

1. For an nX flying capacitor DC/DC converter, which is extended from the aforementioned 3X converter example in Figure 14.42, there are the $2n$ switches, and the maximum voltages and average currents of all the switches are equal to the input voltage and input current, respectively. Its TDPR will be

$$\text{TDPR} = 2n \cdot V_{in} \cdot I_{in} = 2n \cdot P_{in} \tag{14.17}$$

where V_{in} is the input voltage and I_{in} is the input current [33].

2. For the MMCCC shown in Figure 14.49, it is clear that output capacitor C_4, on average, supplies I_{out}, the load current, and that its charging current is I_{out}, giving an average current of zero through the capacitor over the entire period. The current flowing through S_{4a} must be $2I_{out}$ while it is on, because the duty ratio is 50%. The other switches also have the average current of $2I_{out}$ in one of the two switching states, because the charge is pumped from one capacitor to the next capacitor while the MMCCC is working and the charging current of one capacitor is the discharging current of the preceding capacitor. From Figure 14.52, S_{2a} and S_{3a} have to block the voltage $2V_{in}$, the other switches only need to block the voltage of V_{in}. Generalizing these results to the nX topology, it is clear that the n−2 switches must block twice the input voltage, while the remaining $2n$ switches need only block V_{in}. With all of this information clarified, the TDPR can now be calculated by Qian *et al.* [33]:

$$\text{TDPR} = 2n \cdot V_{in} \cdot 2I_{out} + (n-2) \cdot 2V_{in} \cdot 2I_{out} = \frac{8n-8}{n} P_{in}, n = 2, 3, 4, \ldots \tag{14.18}$$

3. For the nX DC/DC converter shown in Figure 14.54, the capacitors C_{3a} and C_{3b}, on average, supply I_{out}, load current. S_{3a} and S_{3b} have $2I_{out}$ current flowing through then because the duty ratio is 50%. Switches S_{1p}, S_{1n}, S_{2p} and S_{2n} have to transfer both currents of two charging paths, and thus, $4I_{out}$ current has to flow through them. The other switches carry $2I_{out}$ current. Switches S_{2a}, S_{3a}, S_{2b} and S_{3b} have to block the voltage $2V_{in}$, the other switches only need to block the voltage of V_{in}. In general, the (n−2) switches of complementary phase legs have to carry $4I_{out}$ current, even though the remaining switches only carry $2I_{out}$ current. The (n−2) switches have to block the voltage of $2V_{in}$, and the other switches just block the voltage of V_{in}. According to above analysis, we have [33]

$$\begin{aligned}\text{TDPR} &= (n-2) \cdot V_{in} \cdot 4I_{out} + (2+2) \cdot V_{in} \cdot 2I_{out} + (n-2) \cdot 2V_{in} \cdot 2I_{out} \\ &= \frac{8n-8}{n} P_{in}, n = 2, 4, 6, \ldots\end{aligned} \tag{14.19}$$

From Equations 14.18 and 14.19, the TDPRs of the nX converter and the MMCCC will approach $8P_{in}$ as the voltage gain n approaches infinity, but the flying capacitor converter's TDPR presents a continuous increase. It indicates that the nX converter and the MMCCC need less silicon area than the flying capacitor converter, and thus, they are more cost-effective in achieving high-voltage gain.

14.4.4.2 Capacitor Voltage, Current and Capacitance of Three Converters

The total capacitor voltage rating, the average charging–discharging current, and the capacitance requirement are listed in Table 14.2 for the comparison of the nX converter, the flying capacitor converter, and the MMCCC [33]. It shows that the flying capacitor converter and the MMCCC have the same total capacitor voltage rating to achieve the same voltage gain n if the input voltage is the same, but that the nX converter needs nearly half of that. The average charging–discharging currents of the output capacitor in the nX converter and the MMCCC are 1/(n-1) times that of the flying capacitor converter, and the other capacitors present 2/n times the charging–discharging currents when compared with that of the flying capacitor converter. In addition, for the nX converter and the MMCCC, the output capacitors have lower ripple current and lower capacitance requirement than the flying capacitor converter.

Table 14.2 Comparison of total capacitor voltage rating, average current, and capacitance

	Total capacitor voltage rating	Average charging–discharging current	Capacitance requirement
Flying capacitor	$\frac{(1+n)n}{2}V_{in}$	$I_j = \begin{cases} n \cdot I_{out}, & j = 1, 2, \ldots, n \\ (n-1) \cdot I_{out}, & j = n+1 \end{cases}$	$C_j = \begin{cases} \frac{I_{out}}{f_s \Delta V_j}, & j = 1, 2, \ldots, n \\ \frac{(n-1) \cdot I_{out}}{n f_s \Delta V_j}, & j = n+1 \end{cases}$
MMCCC	$\frac{(1+n)n}{2}V_{in}$	$I_j = \begin{cases} 2I_{out}, & j = 1, 2, \ldots, n-1 \\ I_{out}, & j = n \end{cases}$	$C_j = \begin{cases} \frac{I_{out}}{f_s \Delta V_j}, & j = 1, 2, \ldots, n-1 \\ \frac{I_{out}/2}{f_s \Delta V_j}, & j = n \end{cases}$
nX Converter	$\frac{(1+n/2)n}{2}V_{in}$ $(n = 2, 4, 6, \ldots)$	$I_{ja} = I_{jb} = \begin{cases} 2I_{out}, & j = 1, 2, \ldots, \frac{n}{2}-1 \\ I_{out}, & j = \frac{n}{2} \end{cases}$	$C_{ja} = C_{jb} = \begin{cases} \frac{I_{out}}{f_s \Delta V_j}, & j = 1, 2, \ldots, \frac{n}{2}-1 \\ \frac{I_{out}/2}{f_s \Delta V_j}, & j = \frac{n}{2} \end{cases}$

14.4.5 Zero Current Switching: MMCCC

ZCS can reduce effectively the switching loss of switches, and this is also true for the MMCCC, the nX DC/DC converter, the flying capacitor converter, and the generalized multilevel DC/DC converter. We can use the parasitic inductance to achieve the ZCS of these converters; this will ensure magnetic-less converter features. In addition, in theory, ZCS reduces the transient spikes of the converters and mitigates the electromagnetic noise. The MMCCC's ZCS is used to demonstrate the basic principle.

When the connection wire parasitic inductance L_{SW}, capacitor parasitic inductance L_{ESL} and the IGBT package parasitic inductance L_{Sp} are taken into account, the MMCCC has the modular block shown in Figure 14.59 [34]. Figure 14.60 shows its simplified circuit in order to demonstrate the ZCS principle, where L_s is the sum of L_{SW}, L_{ESL} and L_{Sp} because of these stray inductances in series when the current is flowing through. The required inductance for resonance is very small (nH-level) to achieve the ZCS. All modules should have equally distributed stray inductances; therefore, the modular design is important. A small air coil could be added in the connection wire if the stray inductance is not large enough.

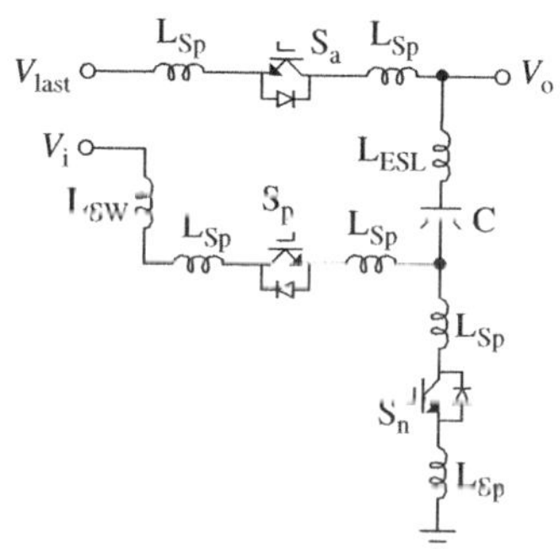

Figure 14.59 MMCCC's modular block with all the stray inductances

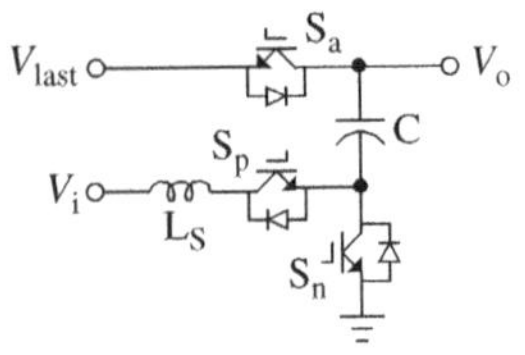

Figure 14.60 Simplified equivalent circuit

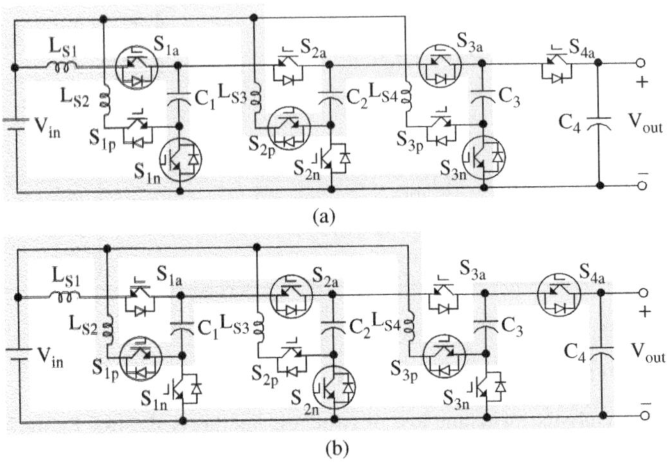

Figure 14.61 Two states of ZCS-MMCCC. (a) State I and (b) state II

Using three cells of Figure 14.60, we can build the four-level MMCCC and the two switching states I and II, as shown in Figure 14.61a and b, respectively. For the sake of simplification, we assume that the circuit uses: (i) ideal switches, (ii) ideal input voltage source and (iii) ideal capacitor, that is, no equivalent series resistance. The circuit's capacitances and stray inductances are designed to meet $C_1 = C_2 = C_3 = C_4$ and $L_{S2} = L_{S3} = L_{S4} = 2L_{S1}$, which guarantees the same resonant frequency. Figure 14.61a presents two resonant loops, and each of these is shown in Figure 14.62a and b [34]. The circuit shown in Figure 14.62a has the state equations as

$$V_{in} = L_{S1}\frac{di_{L_{S1}}}{dt} + v_{C1} \tag{14.20}$$

$$i_{L_{S1}} = C_1\frac{dv_{C1}}{dt} \tag{14.21}$$

The solutions are

$$i_{L_{S1}}(t) = \frac{\pi P_0}{4V_{in}}\sin\omega_r t \tag{14.22}$$

$$v_{C1}(t) = V_{in} - \frac{\pi P_0}{4V_{in}C_1\omega_r}\cos\omega_r t \tag{14.23}$$

where V_{in} is the input voltage, L_{S1} is the stray inductance, ω_r is the resonant frequency equal to $1/\sqrt{L_{S1}C_1}$ and P_o is the output power.

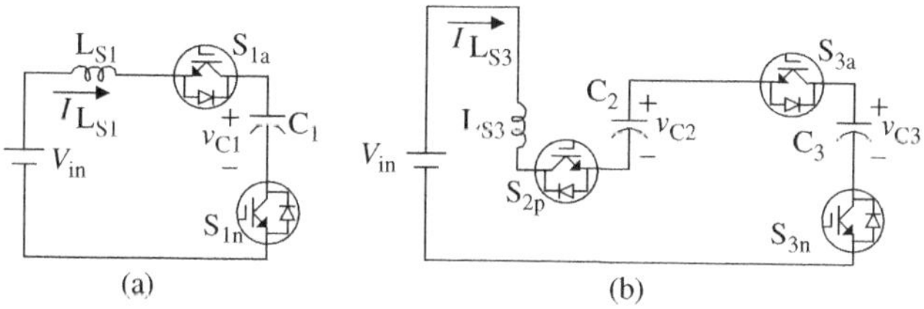

Figure 14.62 Two resonant loops of state I. (a) Loop I and (b) loop II

The circuit shown in Figure 14.62b has the state equations as

$$V_{in} = L_{S3}\frac{di_{L_{S3}}}{dt} + v_{C3} - v_{C2} \tag{14.24}$$

$$i_{L_{S3}} = C_3\frac{dv_{C3}}{dt} \tag{14.25}$$

$$i_{L_{S3}} = -C_2\frac{dv_{C2}}{dt} \tag{14.26}$$

The solutions are

$$i_{L_{S3}}(t) = \frac{\pi P_0}{4V_{in}}\sin\omega_r t \tag{14.27}$$

$$v_{C3}(t) = 3V_{in} - \frac{\pi P_0}{4V_{in}C_1\omega_r}\cos\omega_r t \tag{14.28}$$

Figure 14.61b also presents two resonant loops, as shown in Figure 14.63a and b. The circuit shown in Figure 14.63a has the state equations as

$$V_{in} = L_{S2}\frac{di_{L_{S2}}}{dt} + v_{C2} - v_{C1} \tag{14.29}$$

$$i_{L_{S2}} = C_2\frac{dv_{C2}}{dt} \tag{14.30}$$

$$i_{L_{S2}} = -C_1\frac{dv_{C1}}{dt} \tag{14.31}$$

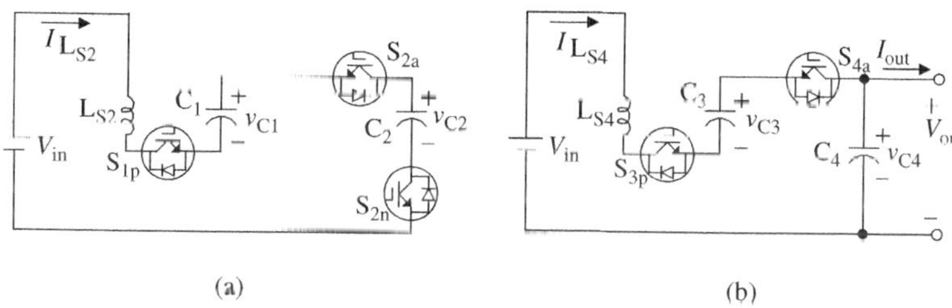

Figure 14.63 Two resonant loops of state II. (a) Loop I and (b) loop II

The solutions are

$$i_{L_{S2}}(t) = -\frac{\pi P_o}{4V_{in}} \sin \omega_r t \tag{14.32}$$

$$v_{C2}(t) = 2V_{in} + \frac{\pi P_o}{4V_{in} C_1 \omega_r} \cos \omega_r t \tag{14.33}$$

The circuit shown in Figure 14.63b has the state equations as

$$V_{in} = L_{S4}\frac{di_{L_{S4}}}{dt} + v_{C4} - v_{C3} \tag{14.34}$$

$$i_{L_{S4}} = C_4\frac{dv_{C4}}{dt} + I_{out} \tag{14.35}$$

$$i_{L_{S4}} = -C_3\frac{dv_{C3}}{dt} \tag{14.36}$$

where I_{out} is the output current.

The solutions are

$$i_{L_{S4}}(t) = -\frac{\pi P_0}{4V_{in}} \sin \omega_r t \tag{14.37}$$

$$v_{C4}(t) = 4V_{in} + \frac{\pi P_0}{4V_{in} C_1 \omega_r} \cos \omega_r t \tag{14.38}$$

The switching frequency has to equal the resonant frequency to achieve ZCS. The 50% duty ratio of each switch's on-state and the synchronous resonance ensure the switch turn on/off at zero current. The resonance makes the current going through L_S have two zero points per resonant cycle, and the zero current point's period is half the resonant cycle. ZCS is achieved through synchronizing the switching action to the zero current point. For example, in Figure 14.62a resonance happens between L_{S1} and C_1, every cycle when the current going through S_{1a} and S_{1n} increases from zero, and falls to zero after half a cycle; therefore, these two switches fulfill the ZCS. For Figure 14.62b, L_{S3} resonates with C_2 and C_3 in series and ZCS is achieved by the switches S_{2p}, S_{3a} and S_{3n}. For Figure 14.63a, switches S_{1p}, S_{2a} and S_{2n} will achieve ZCS owing to the resonance of L_{S2}, C_1 and C_2. For Figure 14.63b, switches S_{3p} and S_{4a} achieve the ZCS owing to the resonance of L_{S4}, C_3 and C_4 [34].

14.4.6 Fault Tolerance and Reliability of Multilevel Converters

Multilevel converters present high failure possibility because they employ many power semiconductors and capacitors. However, the redundant states and structures of multilevel converters definitely enhance their operational reliability, because the converters can still operate through using redundant modules or by modifying the modulation strategy in case of fault conditions. For the diode clamped and capacitor clamped multilevel inverters and CMIs, there are many different fault-tolerant operating approaches [35–38].

Fault tolerance and high reliability are demonstrated here through an example of CMI. As shown in Figure 14.64, the redundant modules are usually used and every module includes a bypass switch, as shown in Figure 14.65. The fault modules can be bypassed through placing switch S to the fault position *F*. The redundant cells are put in operation to re-establish the normal operation through recovering switch S to the position O.

Rather than using redundant modules, control and modulation techniques can also be applied to sustain the CMI's fault-tolerant operation. When a fault occurs, the three phases might have different module

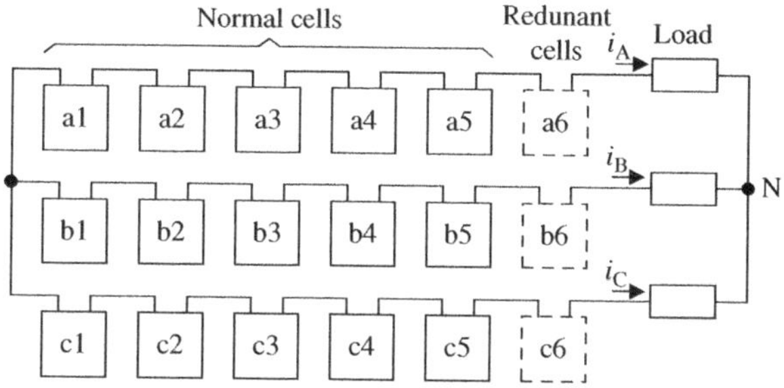

Figure 14.64 CMI with redundant modules

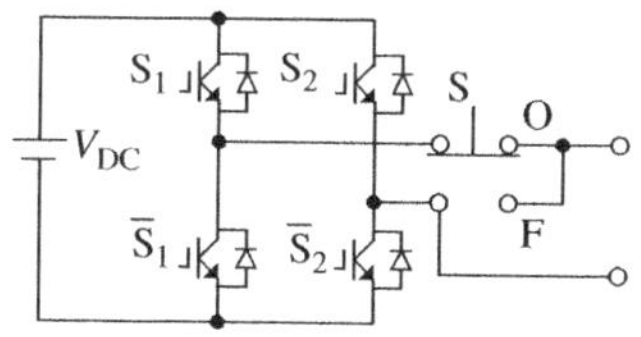

Figure 14.65 Module with bypass switch

numbers, and the unbalanced output voltages produce an unbalanced load current. We can bypass as many modules as necessary to recover balanced operation. Figure 14.66a and b shows the phasor diagrams of an 11-level CMI operating in normal and fault conditions, respectively, in which two modules are at fault in phase c and one module in phase b. When each phase bypasses two modules, that is, two normal modules of phase a, and one fault module and one normal module of phase b, and two fault modules of phase c, the system is restored to balanced operation with a lower voltage, as shown in Figure 14.66c. If we keep all normal modules operational and just bypass the faulty modules, the line-to-line voltage of three phases can still be balanced in amplitude, even though the phase angle is not balanced. As shown in Figure 14.66d, the balanced line-to-line voltage magnitude is provided by the inverter after the phase shift of phases b and c [35].

Another well-known fault-tolerant operation is to utilize the redundant states. In a fault condition, the modulator just uses as many of the healthy modules as it can in the possible switching states in order to produce the maximum line-to-line voltage possible. The switching states involved in the faulty modules are removed.

However, no matter whatever method is used, the maximum output voltage possible will be lower than the inverter's rating and its specific new value is related to the fault type.

14.5 Summary

This chapter presented the fundamentals of multilevel converters/inverters and compared three popular topologies, that is, diode clamped multilevel inverter, flying capacitor multilevel inverter and CMI. The generalized multilevel inverter unified the above three topologies and provided an approach to deduce the new multilevel converter/inverter. For example, the M^2C, magnetic-less multilevel DC/DC converter, flying capacitor DC/DC converter and several new multilevel inverters were deduced from the generalized multilevel inverter topology. The utility applications of CMIs were presented, and the Y-connection and Δ-connection CMIs were analyzed when applied to the reactive power compensation and unbalanced

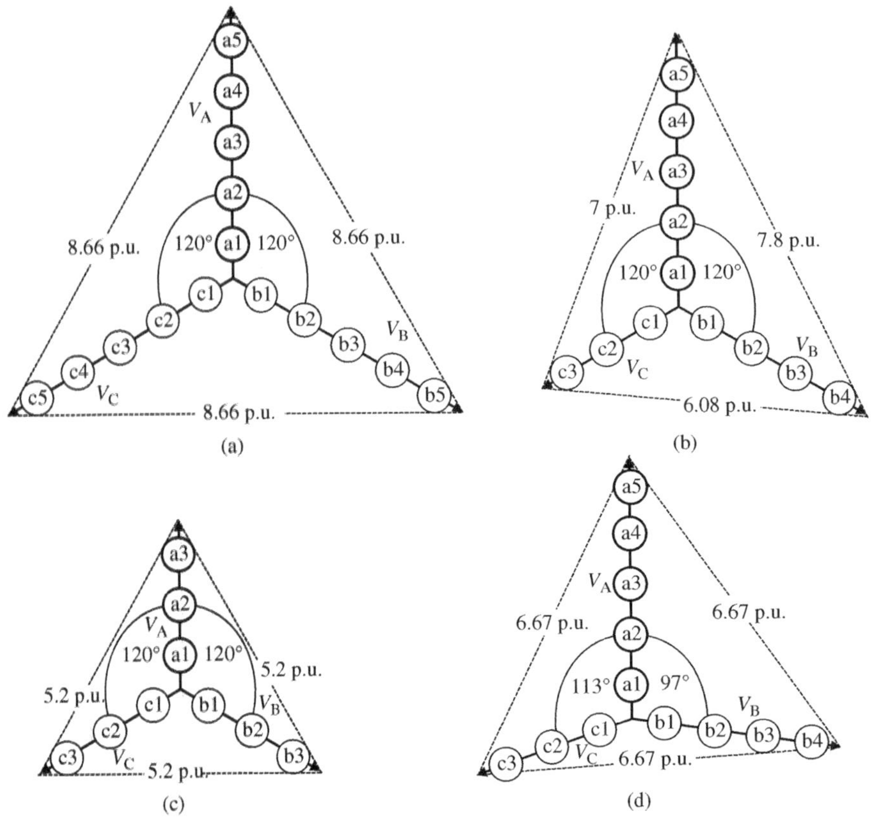

Figure 14.66 Fault-tolerant operation of CMI. (a) Normal operation, (b) fault operation, (c) new balanced system by bypassing modules, and (d) phase-shift operation

current compensation. The Δ-connection CMI was good at both reactive power and unbalanced current compensation, but the Y-connection CMI only could compensate the reactive power. A CMI-based UPFC scheme, that is, face-to-face UPFC, was presented to eliminate the complicated transformer of conventional UPFCs. In addition, the CMI demonstrated many advantages when applied to motor drives and PV power generation, and example applications were introduced. The quasi-Z-source CMI provided more advantages than the conventional CMI in applications to PV power systems, because it could balance the DC-link voltages of all modules and boost the PV panel's low-voltage level to a higher voltage level. The multilevel converter/inverter presented high possibility of failure because of the large number of components used. However, the tolerant structure of the multilevel converter/inverter provided fault-tolerant operation, which could enhance system reliability. This chapter used the CMI as an example to discuss the fault tolerance and reliability of multilevel inverters.

Four magnetic-less multilevel DC/DC converters were introduced. First, the generalized multilevel inverter was used to obtain a magnetic-less DC/DC converter, and the voltage multiplier/divider was achieved by using the fixed duty ratio. The flying capacitor DC/DC converter, as a simplified multilevel DC/DC converter, was introduced to save power switches. Furthermore, the MMCCC provided a modular structure, simple control and lower losses than the flying capacitor DC/DC converter, but it required many more switches. The nX converter combined the advantages of the flying capacitor DC/DC converter and the MMCCC and presented a modular structure, simple control and low switch count. The ZCS was demonstrated and applied to the MMCCC to reduce the loss and current spike. The component costs of the flying capacitor DC/DC converter, the MMCCC and the nX converter were compared, and the nX converter presented more advantages compared with that of the other two topologies.

Acknowledgment

This chapter was made possible by NPRP-EP grant No. X-033-2-007 from the Qatar National Research Fund (a member of Qatar Foundation). The statements made herein are solely the responsibility of the authors.

References

1. Nabae, A., Takahashi, I., and Akagi, H. (1981) A new neutral-point-clamped PWM inverter. *IEEE Transactions on Industrial Applications*, **IA-17**, 518–523.
2. Meynard, T.A. and Foch, H. (1992) Multi-level conversion: high voltage choppers and voltage-source inverters. 23rd Annual IEEE Power Electronics Specialists Conference, PESC'92 Record, 29 June 29– July 3, 1992, Vol. 1, pp. 397–403.
3. Peng, F.Z. and Lai, J. (1994) A static var generator using a staircase waveform multilevel voltage-source converter. Proceedings PCIM/Power Quality, Dallas/Ft. Worth, TX, pp. 58–66.
4. Hammond, P.W. (1997) A new approach to enhance power quality for medium voltage AC drives. *IEEE Transactions on Industrial Applications*, **33**, 202–208.
5. Rodriguez, J., Lai, J.-S., and Peng, F.Z. (2002) Multilevel inverters: a survey of topologies, controls, and applications. *IEEE Transactions on Industrial Electronics*, **49**, 724–738.
6. Peng, F.Z., Lai, J.-S., and McKeever, J. (1995) A multilevel voltage-source converter system with balanced DC voltages. 26th Annual IEEE Power Electronics Specialists Conference, PESC'95, pp. 1144–1150.
7. Abu-Rub, H., Holtz, J., Rodriguez, J., and Ge, B. (2010) Medium-voltage multilevel converters – state of the art, challenges, and requirements in industrial applications. *IEEE Transactions on Industrial Electronics*, **57** (8), 2581–2596.
8. Lai, J.-S. and Peng, F.Z. (1995) Multilevel converters-a new breed of power converters. Thirtieth IAS Annual Meeting, IAS'95, pp. 2348–2356.
9. Sinha, G. and Lipo, T.A. (2000) A four-level inverter based drive with a passive front end. *IEEE Transactions on Power Electronics*, **15**, 285–294.
10. Peng, F.Z. (2001) A generalized multilevel inverter topology with self voltage balancing. *IEEE Transactions on Industrial Applications*, **37**, 611–618.
11. Zhang, F., Yang, S., Peng, F. Z., and Qian, Z. (2008) A zigzag cascaded multilevel inverter topology with self voltage balancing. Twenty-Third Annual IEEE Applied Power Electronics Conference and Exposition, APEC 2008, pp. 1632–1635.
12. Lesnicar, A. and Marquardt, R. (2003) An innovative modular multilevel converter topology suitable for a wide power range. Proceedings of 2003 IEEE Bologna Power Tech Conference, pp. 1–6.
13. Marquardt, R. (2010) Modular multilevel converter: an universal concept for HVDC-networks and extended dc-bus-applications. 2010 International Power Electronics Conference, pp. 502–507.
14. Hagiwara, M. and Akagi, H. (2009) Control and experiment of pulse width-modulated modular multilevel converters. *IEEE Transactions on Power Electronics*, **24**, 1737–1746.
15. Wang, C. and Li, Y. (2009) A survey on topologies of multilevel converters and study of two novel topologies. IEEE 6th International Power Electronics and Motion Control Conference, IPEMC'09, May 17–20, 2009, pp. 860–865.
16. Peng, F.Z., Lai, J.-S., McKeever, J., and VanCOevering, J. (1996) A multilevel voltage-source inverter with separate dc sources for static var generation. *IEEE Transactions on Industry Applications*, **32**, 1130–1138.
17. Peng, F.Z. and Lai, J.-S. (1996) Dynamic performance and control of a static var generator using cascade multilevel inverters. Conference Record IEEE-IAS Annual Meeting, pp. 1009–1015.
18. Peng, F.Z., McKeever, J.W., and Adams, D.J. (1998) A power line conditioner using cascade multilevel inverters for distribution systems. *IEEE Transactions on Industry Applications*, **34** (6), 1293–1298.

19. Tolbert, L.M., Peng, F.Z., and Habetler, T.G. (1999) Multilevel converters for large electric drives. *IEEE Transactions on Industry Applications*, **35**, 36–44.
20. Villanueva, E., Correa, P., Rodriguez, J., and Pacas, M. (2009) Control of a single-phase cascaded H-bridge multilevel inverter for grid-connected photovoltaic systems. *IEEE Transactions on Industrial Electronics*, **56** (11), 4399–4406.
21. Ge, B., Peng, F.Z., de Almeida, A.T., and Abu-Rub, H. (2010) An effective control technique for medium-voltage high-power induction motor fed by cascaded neutral-point-clamped inverter. *IEEE Transactions on Industrial Electronics*, **57** (8), 2659–2668.
22. Xue, Y., Ge, B., and Peng, F.Z. (2012) Reliability, efficiency, and cost comparisons of MW-scale photovoltaic inverters. 2012 IEEE Energy Conversion Congress and Exposition (ECCE), September 15–20, 2012, pp. 1627–1634.
23. Anderson, J. and Peng, F. Z. (2008) Four quasi-Z-source inverters. IEEE PESC 2008, June 15–19, 2008, pp. 2743–2749.
24. Ge, B., Abu-Rub, H., Peng, F.Z. *et al.* (2013) An energy-stored quasi-Z-source inverter for application to photovoltaic power system. *IEEE Transactions on Industrial Electronics*, **60** (10), 4468–4481.
25. Sun, D., Ge, B., Peng, F.Z. *et al.* (2012) A new grid-connected PV system based on cascaded H-bridge quasi-Z source inverter. 2012 IEEE International Symposium on Industrial Electronics (ISIE), May 28–31, 2012, pp. 951–956.
26. Liu, Y., Ge, B., Abu-Rub, H., and Peng, F. Z. (2013) A modular multilevel space vector modulation for Photovoltaic quasi-Z-source cascade multilevel inverter. APEC2013, March 17–21, 2013, pp. 714–718.
27. Peng, F.Z. and Jin, W. (2004) A universal STATCOM with delta-connected cascade multilevel inverter. IEEE 35th Annual Power Electronics Specialists Conference, PESC 04, pp. 3529–3533.
28. Wang, J. and Peng, F.Z. (2004) Unified power flow controller using the cascade multilevel inverter. *IEEE Transactions on Power Electronics*, **19** (4), 1077–1084.
29. Peng, F.Z., Zhang, F., and Qian, Z. (2003) A magnetic-less dc-dc converter for dual-voltage automotive systems. *IEEE Transactions on Industrial Applications*, **39** (2), 511–518.
30. Zhang, F., Du, L., Peng, F.Z., and Qian, Z. (2008) A new design method for high-power high-efficiency switched-capacitor dc-dc converterse. *IEEE Transactions on Power Electronics*, **23** (2), 832–840.
31. Qian, W., Cha, H., Peng, F.Z., and Tolbert, L.M. (2012) 55-kW variable 3X dc-dc converter for plug-in hybrid electric vehicles. *IEEE Transactions on Power Electronics*, **27** (4), 1668–1678.
32. Khan, F.H. and Tolbert, L.M. (2007) A multilevel modular capacitor-clamped dc-dc converter. *IEEE Transactions on Industrial Applications*, **43** (6), 1628–1638.
33. Qian, W., Cao, D., Cintron-Rivera, J.G. *et al.* (2012) A switched-capacitor dc–dc converter with high voltage gain and reduced component rating and count. *IEEE Transactions on Industrial Applications*, **48** (4), 1397–1406.
34. Cao, D. and Peng, F.Z. (2010) Zero-current-switching multilevel modular switched-capacitor dc–dc converter. *IEEE Transactions on Industrial Applications*, **46** (6), 2536–2544.
35. Lezana, P., Pou, J., Meynard, T.A. *et al.* (2010) Survey on fault operation on multilevel inverters. *IEEE Transactions on Industrial Electronics*, **57** (7), 2207–2218.
36. Song, W. and Huang, A.Q. (2010) Fault-tolerant design and control strategy for cascaded H-bridge multilevel converter-based STATCOM. *IEEE Transactions on Industrial Electronics*, **57** (8), 2700–2708.
37. Ma, M., Hu, L., Chen, A., and He, X. (2007) Reconfiguration of carrier-based modulation strategy for fault tolerant multilevel inverters. *IEEE Transactions on Power Electronics*, **22** (5), 2050–2060.
38. Khomfoi, S. and Tolbert, L.M. (2007) Fault diagnosis and reconfiguration for multilevel inverter drive using AI-based techniques. *IEEE Transactions on Industrial Electronics*, **54** (6), 2954–2968.

15

Multiphase Matrix Converter Topologies and Control

SK. Moin Ahmed[1,2], Haitham Abu-Rub[1] and Atif Iqbal[3,4]

[1]*Department of Electrical and Computer Engineering, Texas A&M University at Qatar, Doha, Qatar*
[2]*Department of Electrical Engineering, Universiti Teknologi Malaysia, Johor Bahru, Malaysia*
[3]*Department of Electrical Engineering, Qatar University, Doha, Qatar*
[4]*Department of Electrical Engineering, Aligarh Muslim University, Aligarh, India*

15.1 Introduction

Power electronic converters are currently employed in numerous industrial and household applications such as motor drives and power system operation and control (FACTS (flexible AC transmission systems), HVDC (high-voltage DC), static var compensation, power quality improvement, active filtering, the linking of two different frequency power systems such as 50 Hz and 60 Hz, etc.). The main function of a power electronic converter is to convert uncontrolled power to controlled power. Broadly classified power electronic converters are AC/DC, DC/AC, DC/DC, and AC/AC. The classical approach to AC/AC conversion is the use of thyristor devices called cycloconverters. The major shortcoming of such topology is its limited range of output frequency (only one-fourth of the input frequency value). Another topology is based on bidirectional power switches that are arranged in the form of array or matrix called matrix converters [1–3]. Matrix converters transform uncontrolled AC (fixed voltage amplitude, fixed frequency) into controlled AC (variable voltage amplitude, variable frequency) without any intermediate conversion stage. The major advantages of a matrix converter are the sinusoidal source-side current, controlled source-side power factor, lack of a bulky DC-link capacitor, and no limitation on output frequency range. The major disadvantage is its lower output voltages: 86.6% in a three-phase input and three-phase output configuration. The output voltage reduces further to 78.86% in three-phase input and five-phase output matrix converters and 76.94% in a three-phase input and seven-phase output configuration. Broadly classified, matrix converters are of two types: direct and indirect. In an indirect topology, it is treated as a combination of a controlled rectifier and an inverter with a fictitious DC link. In the direct topology, all the switches are considered as a single unit. The control approaches are different in the two topologies. This chapter elaborates on multiphase [4–6] (three-phase input and five-phase output and five-phase input and three-phase output) AC/AC power electronic converters, encompassing existing and new and emerging

Power Electronics for Renewable Energy Systems, Transportation and Industrial Applications, First Edition.
Edited by Haitham Abu-Rub, Mariusz Malinowski and Kamal Al-Haddad.

topologies and controls. Three control approaches are principally discussed: carrier-based pulse-width modulation (PWM), direct duty ratio-based PWM (DPWM) and space vector PWM (SVPWM). The theoretical background along with analytical detail and simulation models is presented, followed by the experimental results.

15.2 Three-Phase Input with Five-Phase Output Matrix Converter

15.2.1 Topology

The general power circuit topology of a three-to-five-phase matrix converter is illustrated in Figure 15.1. There are five legs, where each leg has three bidirectional power switches connected in series. Each power switch is bidirectional in nature with antiparallel connected insulated-gate bipolar transistors (IGBTs) and diodes. The input source is identical to a three-to-three-phase matrix converter developed in [7–9]. A small LC filter is connected at the source side to eliminate ripple, and the output is five phases with 72° phase displacement between each phases.

The switching function is defined as $S_{jp} = \{1 \text{ for a closed switch, } 0 \text{ for an open switch}\}$, j = {a, b, c} (input), p = {A, B, C, D, E} (output). The switching constraint is $S_{ap} + S_{bp} + S_{cp} = 1$.

15.2.2 Control Algorithms

The control of a matrix converter depends on whether it is viewed as an indirect type or a direct type. In indirect matrix converters it is undertaken as two units, a rectifier at the source side and inverter at the load side and in addition a fictitious DC link is assumed. The control techniques in this perspective is an extension of controlled AC/DC and PWM DC/AC converters. Hence, similar control is employed as for conventional AC/DC and DC/AC converters. In the case of direct types, the methodology is different and customized techniques need to be used. In the following, the techniques applied

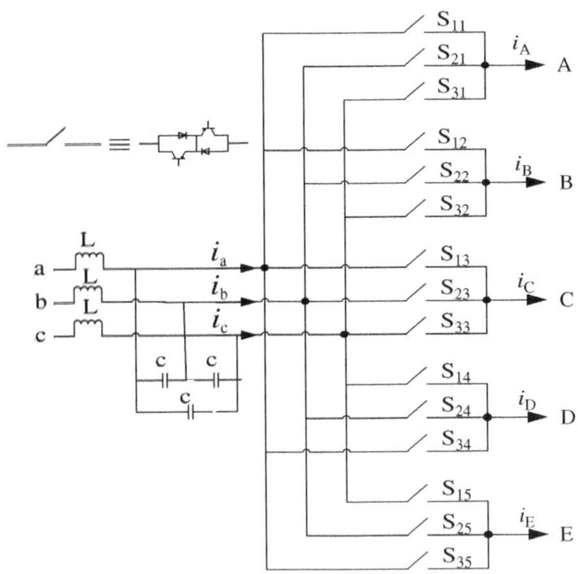

Figure 15.1 Block schematic of a three-phase to five-phase matrix converter [10] (Reproduced by permission of IEEE)

for direct matrix converters are discussed. PWM techniques for a direct matrix converter are realized by sinusoidal carrier-based PWM schemes [11, 12], Direct Duty Ratio based PWM schemes and space vector PWM. These techniques as applied to non-square direct matrix converters, in which three-phase input and multiphase (five and seven) outputs are discussed [13–16].

15.2.2.1 Sinusoidal Carrier-Based PWM Technique

A. General Description

In this section, a balanced three-phase system is considered at the input side. The input voltages and output voltages are as follows:

$$\begin{aligned} u_a &= U\sin(\omega t)\\ u_b &= U\sin(\omega t - 2\pi/3)\\ u_c &= U\sin(\omega t - 4\pi/3) \end{aligned} \tag{15.1}$$

$$\begin{aligned} v_A &= V\sin(\omega_o t - \varphi)\\ v_B &= V\sin\left(\omega_o t - 2\frac{\pi}{5} - \varphi\right)\\ v_C &= V\sin\left(\omega_o t - 4\frac{\pi}{5} - \varphi\right)\\ v_D &= V\sin\left(\omega_o t - 6\frac{\pi}{5} - \varphi\right)\\ v_E &= V\sin\left(\omega_o t - 8\frac{\pi}{5} - \varphi\right) \end{aligned} \tag{15.2}$$

The suffix with small letters indicates the input voltage, and the suffix with capital letters represents the output voltages. When implementing the PWM, the duty ratios of the bidirectional switches have to be calculated. However, in order to decouple the frequency of the output voltage from the input voltage frequency, the outputs are assumed in a synchronously rotating reference frame and the input remains in the stationary reference frame. By doing this, no input frequency term will appear in the output voltage equations. The duty ratios for output phase "A" are assumed as [1, 2]:

$$\begin{aligned} d_{aA} &= k_A\cos(\omega t - \varphi)\\ d_{bA} &= k_A\cos(\omega t - 2\pi/3 - \varphi)\\ d_{cA} &= k_A\cos(\omega t - 4\pi/3 - \varphi) \end{aligned} \tag{15.3}$$

Therefore, the output five-phase voltages can be obtained using the above duty ratios as follows:

$$\begin{aligned} v_A &= u_a d_{aA} + u_b d_{bA} + u_c d_{cA}\\ v_B &= u_a d_{aB} + u_b d_{bB} + u_c d_{cB}\\ v_C &= u_a d_{aC} + u_b d_{bC} + u_c d_{cC}\\ v_D &= u_a d_{aD} + u_b d_{bD} + u_c d_{cD}\\ v_E &= u_a d_{aE} + u_b d_{bE} + u_c d_{cE} \end{aligned} \tag{15.4}$$

For phase "A":

$$\begin{aligned} v_A = k_A U[&\cos(\omega t)\bullet\cos(\omega t - \varphi) + \cos(\omega t - 2\pi/3)\bullet\cos(\omega t - 2\pi/3 - \varphi) + \cos(\omega t - 4\pi/3)\\ &\bullet\cos(\omega t - 4\pi/3 - \varphi)] \end{aligned} \tag{15.5}$$

Using geometrical manipulation by expanding the cosine terms, one obtains the following for phase "p":

$$v_p = \frac{3}{2} k_p U \cos(\varphi) \qquad p \in A, B, C, D, E \tag{15.6}$$

In Equation (15.6), the cos(φ) term indicates that the output voltage is dependent on the phase shift φ. Thus, the output voltages are independent of the input frequency and are only governed by the amplitude U of the input voltage and k_p is the reference output voltage time-varying modulating signal of the desired output frequency ω_o. The five-phase reference output voltages can be represented as follows:

$$\begin{aligned} k_A &= m\cos(\omega_o t) \\ k_B &= m\cos(\omega_o t - 2\pi/5) \\ k_C &= m\cos(\omega_o t - 4\pi/5) \\ k_D &= m\cos(\omega_o t - 6\pi/5) \\ k_E &= m\cos(\omega_o t - 8\pi/5) \end{aligned} \tag{15.7}$$

Therefore, from Equation (15.6), the output voltages can be written as

$$v_P = \left[\frac{3}{2} mV \cos(\varphi)\right] \cos\left(\omega_o t - \frac{2n\pi}{5}\right) \qquad n = 0, 1, 2, 3, 4 \tag{15.8}$$

where m is the modulation index.

B. Application of Offset Duty Ratio

In the above discussion, the duty ratios given by Equation (15.3) are sinusoidal and, hence, will achieve negative values during the half-cycle, which does not carry any physical meaning [13–15]. This is due to the fact that the duty ratios signify the duration of the switching "on" of the semiconductor switches, which can never attain negative values. The duty ratio must satisfy the constraint $0 \le d_{ap}, d_{bp}, d_{cp} \le 1$, where p refers to one of the output phases. Therefore, an offset in the duty ratios should be injected (Equation 15.3), so that the net resultant duty ratios of individual transistors remain positive. Also the offset duty ratios should be added equally to all the output phases to guarantee that the effect of output voltage vector produced by the offset duty ratios is not reflected in the load. This means that the offset duty ratios can only add the common-mode voltages in the output. In general, the summation of the duty ratios are zero:

$$d_{ap} + d_{bp} + d_{cp} = k_p \cos(\omega t - \varphi) + k_p \cos(\omega t - 2\pi/3 - \varphi) + k_p \cos(\omega t - 4\pi/3 - \varphi) = 0 \tag{15.9}$$

Absolute values (positive values) of the duty ratios are added to eliminate negative components from individual duty ratios, in other words it can be said that the negative values are clipped because of the shifting of the duty ratios upward. Hence, the lowest values of individual offset duty ratios should be

$$\begin{aligned} D_{ap}(t) &= |k_p \cos(\omega t - \varphi)|, \\ D_{bp}(t) &= |k_p \cos(\omega t - 2\pi/3 - \varphi)| \\ D_{cp}(t) &= |k_p \cos(\omega t - 4\pi/3 - \varphi)| \end{aligned} \tag{15.10}$$

Effective duty ratios are $d_{ap} + D_{ap}(t)$, $d_{bp} + D_{bp}(t)$, $d_{cp} + D_{cp}(t)$. The net duty ratio $d_{ap} + D_{ap}(t)$ should be within the range of 0 to 1. Therefore,

$$0 \le d_{ap} + D_{ap}(t) \le 1 \tag{15.11a}$$

can be written as

$$0 \leq k_p \cos(\omega t - \rho) + |k_p \cos(\omega t - \rho)| \leq 1 \tag{15.11b}$$

For the extreme case, $0 \leq 2.|k_p| \leq 1$. The maximum value of k_p or, in other words, m in Equation (15.7) is equal to 0.5. Hence, the offset duty ratios corresponding to the three input phases are

$$D_{ap}(t) = |0.5\cos(\omega t - \varphi)|, D_{bp}(t) = |0.5\cos(\omega t - 2\pi/3 - \varphi)|, D_{cp}(t) = |0.5\cos(\omega t - 4\pi/3 - \varphi)| \tag{15.12}$$

Now if the value of "m" is placed in Equation (15.8), the maximum value of output voltage that can be achieved is $0.75 \times V$, where V is the input voltage magnitude. In this way, the modulation index can be enhanced by modifying the duty ratio.

The original duty ratio is modified by injecting the offset, and hence the resulting duty ratios are obtained as [15]

$$\begin{aligned} d_{aA} &= D_{aA}(t) + k_A \cos(\omega t - \rho) \\ d_{bA} &= D_{bA}(t) + k_A \cos(\omega t - 2\pi/3 - \rho) \\ d_{cA} &= D_{cA}(t) + k_A \cos(\omega t - 4\pi/3 - \rho) \end{aligned} \tag{15.13}$$

In one switching period, the output phase has to be connected to any of the input phases. This implies that the sum of the duty ratios of Equation (15.13) must be equal to unity. However, it is seen that the summation $D_{ap}(t) + D_{bp}(t) + D_{cp}(t)$ does not reach unity. Hence, there is further scope for modifying the duty ratios. Thus, another offset duty ratio $[1 - \{D_{ap}(t) + D_{bp}(t) + D_{cp}(t)\}]/3$ is injected into $D_{ap}(t), D_{bp}(t), D_{cp}(t)$ in Equation (15.13). It is to be noted that injecting this new offset duty ratio into all the switches will not change the output voltages and input currents. Similarly, the duty ratios are modified for the other output phases in Equation (15.14). The finally modified duty ratios for all five phases are shown in Figure 15.2.

$$\begin{aligned} d_{ap} &= D_{ap}(t) + k_p \cos(\omega t - \rho) \\ d_{bp} &= D_{bp}(t) + k_p \cos(\omega t - 2\pi/3 - \rho) \\ d_{cp} &= D_{cp}(t) + k_p \cos(\omega t - 4\pi/3 - \rho) \qquad \text{p} = \text{A, B, C, D and E} \end{aligned} \tag{15.14}$$

While the modulating signals k_A, k_B, k_C, k_D, k_E are assumed to be five-phase sinusoidal references as given in Equation (15.7), the input voltage capability is not fully utilized for output voltage generation. This is because there is still some unused space in the magnitude of the duty ratios. To utilize this, an additional common-mode term equal to $-0.5\max\{(k_A, k_B, k_C, k_D, k_E) + \min(k_A, k_B, k_C, k_D, k_E)\}$

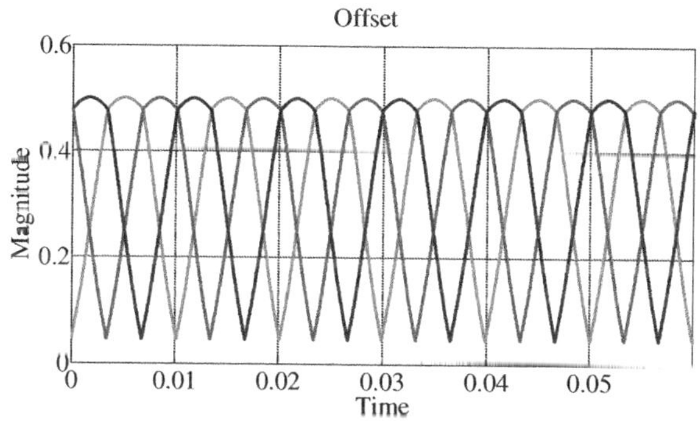

Figure 15.2 Modified offset duty ratios for all input phases [14]

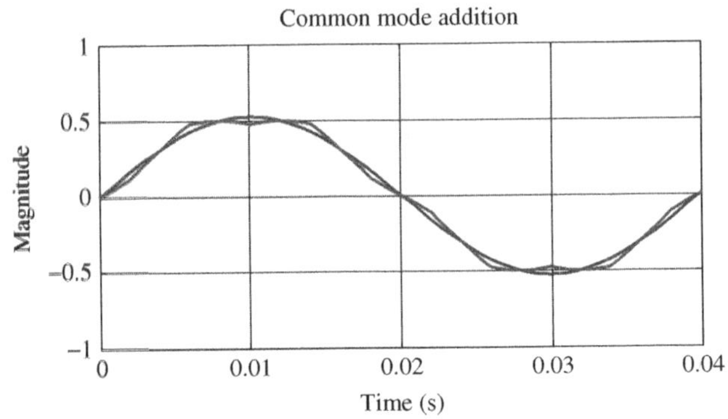

Figure 15.3 With and without common-mode added reference for an output phase [14]

is injected, which can further enhance the modulation index. This technique is most readily used in voltage source inverters [17]. The increase in the output voltages in the case of five phases is 5.15%. Thus, the amplitude of k_A, k_B, k_C, k_D, k_E can be enhanced from 0.5 to 0.5257, which is about a 5.15% increase. This is shown in Figure 15.3.

C. Without Common-Mode Voltage Addition

After adding the offsets and other constants, the overall duty ratios are obtained for output phase "A" as [15]

$$\begin{aligned} d_{aA} &= D_{aA}(t) + (1 - \{D_{aA}(t) + D_{bA}(t) + D_{cA}(t)\})/3 + k_A \times \cos(\omega t - \rho) \\ d_{bA} &= D_{aA}(t) + (1 - \{D_{aA}(t) + D_{bA}(t) + D_{cA}(t)\})/3 + k_A \times \cos(\omega t - 2\pi/3 - \rho) \\ d_{cA} &= D_{aA}(t) + (1 - \{D_{aA}(t) + D_{bA}(t) + D_{cA}(t)\})/3 + k_A \times \cos(\omega t - 4\pi/3 - \rho) \end{aligned} \quad (15.15)$$

D. With Common-Mode Voltage Addition

The duty ratio for output phase A can be written as [15]

$$\begin{aligned} d_{aA} &= D_{aA}(t) + (1 - \{D_{aA}(t) + D_{bA}(t) + D_{cA}(t)\})/3 \\ &\quad + [k_A - \{\max(k_A, k_B, k_C, k_D, k_E) + \min(k_A, k_B, k_C, k_D, k_E)\}/2] \times \cos(\omega t - \rho) \\ d_{bA} &= D_{aA}(t) + (1 - \{D_{aA}(t) + D_{bA}(t) + D_{cA}(t)\})/3 \\ &\quad + [k_A - \{\max(k_A, k_B, k_C, k_D, k_E) + \min(k_A, k_B, k_C, k_D, k_E)\}/2] \times \cos(\omega t - 2\pi/3 - \rho) \\ d_{cA} &= D_{aA}(t) + (1 - \{D_{aA}(t) + D_{bA}(t) + D_{cA}(t)\})/3 \\ &\quad + [k_A - \{\max(k_A, k_B, k_C, k_D, k_E) + \min(k_A, k_B, k_C, k_D, k_E)\}/2] \times \cos(\omega t - 4\pi/3 - \rho) \end{aligned}$$

where

$$\begin{aligned} D_{aA}(t) &= |0.5\cos(\omega t - \rho)| \\ D_{bA}(t) &= |0.5\cos(\omega t - 2\pi/3 - \rho)| \\ D_{bA}(t) &= |0.5\cos(\omega t - 4\pi/3 - \rho)| \end{aligned} \quad (15.16)$$

The five-phase output voltages can be written as

$$\begin{aligned}
k_A &= m\cos(\omega_o t)\\
k_B &= m\cos(\omega_o t - 2\pi/5)\\
k_C &= m\cos(\omega_o t - 4\pi/5)\\
k_D &= m\cos(\omega_o t - 6\pi/5)\\
k_E &= m\cos(\omega_o t - 8\pi/5)
\end{aligned} \tag{15.17}$$

where ω is the input frequency in rad/s, ω_o is the output frequency in rad/s and m is the modulation index. For unity, power factor ρ has to be chosen as zero. The power factor control is governed by the choice of phase shift angle ρ.

15.2.2.2 Direct Duty Ratio-Based PWM Technique

In this section, the PWM technique based on the direct calculation of duty ratios in conjunction with the generalized three- to k-phase topology of a matrix converter has been described [18–21]. The DPWM uses the concept of per-phase output averaged over one switching period. This PWM scheme is modular and flexible in nature, and can therefore be employed to generalized converter circuit topology with arbitrary output phase numbers [22].

One switching period with a switching time span of T_s can be divided into two subperiods. These subperiods correspond to the rising slope of the triangular carrier signal T_1 and the falling slope of the triangular carrier wave T_2. The input three-phase sinusoidal signals can have different values at different instances of time. The maximum among the three input signals is termed Max, the medium amplitude among three input signals is termed Mid, and the smallest magnitude is represented as Min. During interval T_1 (positive slope of the carrier), the line-to-line voltage between Max and Min ($\text{Max}\{v_A, v_B, v_C\} - \text{Min}\{v_A, v_B, v_C\}$) phases is utilized to directly compute the duty ratio. In this computation, the medium amplitude of the input signal is not considered. The output voltage should initially follow the Max signal of the input and should then follow the Min signal of the input. During interval T_2, the two line voltages between Max and Mid ($\text{Max}\{v_A, v_B, v_C\} - \text{Mid}\{v_A, v_B, v_C\}$) and Mid and Min ($\text{Mid}\{v_A, v_B, v_C\} - \text{Min}\{v_A, v_B, v_C\}$) are first computed. The larger of the two is used for the calculation of the duty ratio in order to obtain the high modulation index and also to satisfy the volt-second principle. Two different cases can arise in time interval T_2 depending on the relative magnitude of the input voltages. If Max−Mid is greater than Mid−Min, the output should follow Max for a specific time period and then follow Mid for a specific time period. This situation is denoted case I, which is further explained in the following section. Similarly, if Max−Mid is less than Mid−Min, the output should first follow Mid of the input signal and then Min of the input signal, which is named case II. Therefore, the DPWM approach uses two line input voltages out of the three to synthesize output voltages. All three input phases are utilized to conduct current during each switching period. Cases I and II and the generation of the gating signals are further explained in the following section.

A. Case I

For the condition when Max−Mid ≥ Mid−Min, the generation of the gating signal for the kth output phase is illustrated in Figure 15.4 for one switching period. To obtain the switching pattern, at first the duty ratio $D_{k1}, k \in a, b, c, \ldots$ is calculated and then compared with the high-frequency triangular carrier signal to generate the kth output phase switching pattern. The gating pattern for the kth leg of the matrix converter is directly derived from the output switching pattern. The switching pattern is obtained assuming that Max is phase "A" of the input, Mid is phase "B," and Min is phase "C." The switching pattern changes

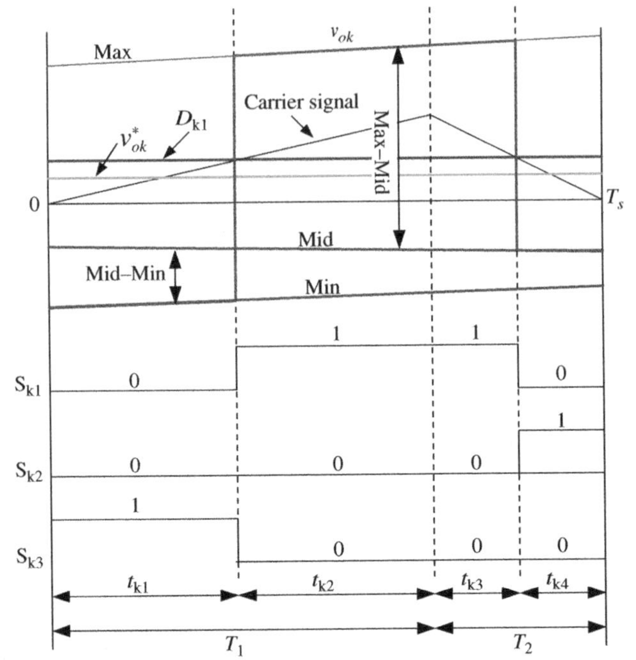

Figure 15.4 Output and switching pattern for kth phase in case I [21] (Reproduced by permission of IEEE)

according to the variation in the relative magnitude of the input phases. The output follows Min of the input signal if the magnitude of the duty ratio is greater than the magnitude of the carrier and the slope of the carrier is positive. The output follows Max of the input signal if the magnitude of the carrier is greater than the magnitude of the duty ratio, regardless of the slope of the carrier. Finally, the output tracks Mid if the magnitude of the carrier signal is less than the magnitude of the duty ratio and the slope of the carrier is negative. Thus, the resulting output phase voltage changes accordingly Min→Max→Max→Mid. These transition periods are termed t_{k1}, t_{k2}, t_{k3} and t_{k4}; these four subintervals can be expressed as

$$\begin{aligned} t_{k1} &= D_{k1}\delta T_s \\ t_{k2} &= (1 - D_{k1})\delta T_s \\ t_{k3} &= (1 - D_{k1})(1 - \delta)T_s \\ t_{k4} &= D_{k1}(1 - \delta)T_s \\ T_s &= t_{k1} + t_{k2} + t_{k3} + t_{k4} \end{aligned} \tag{15.18}$$

where D_{k1} is the kth phase duty ratio value, when case I is under consideration and δ is defined by $\delta = T_1/T_s$, which refers to the fraction of the slope of the carrier. Now, by using the volt-second principle of the PWM control, the following equation can be obtained [22]:

$$v_{ok}^* T_s = \int_0^{T_s} v_{ok} dt = \text{Min}\{v_A, v_B, v_C\}.t_{k1} + \text{Max}\{v_A, v_B, v_C\}.(t_{k2} + t_{k3}) + \text{Mid}\{v_A, v_B, v_C\}.t_{k4} \tag{15.19}$$

Substituting the time interval expressions from Equation (15.18) into Equation (15.19) yields

$$v_{ok}^{*} = \frac{1}{T_s}\int_0^{T_s} v_{ok}dt - D_{k1}\begin{pmatrix}\delta.\text{Min}\{v_A, v_B, v_C\} - \delta.\text{Mid}\{v_A, v_B, v_C\}+ \\ \text{Mid}\{v_A, v_B, v_C\} - \text{Max}\{v_A, v_B, v_C\}\end{pmatrix} + \text{Max}\{v_A, v_B, v_C\} \quad (15.20)$$

where T_s is the sampling period, v_{ok}^{*}, v_{ok} are the reference and actual average output voltages of phase "k," respectively and v_A, v_B, v_C are the input side three-phase voltages. Max, Mid and Min refer to the maximum, medium and minimum values and D_k represents the duty ratio of the power switch.

The duty ratio is obtained from Equation (15.20) as

$$D_{k1} = \frac{\text{Max}\{v_A, v_B, v_C\} - v_{ok}^{*}}{\Delta + \delta(\text{Mid}\{v_A, v_B, v_C\} - \text{Min}\{v_A, v_B, v_C\})} \quad (15.21)$$

where $\Delta = (\text{Max}\{v_A, v_B, v_C\} - \text{Mid}\{v_A, v_B, v_C\})$

Similarly, the duty ratios of other output phases can be obtained and can subsequently be used for the implementation of the PWM scheme.

B. Case II

Now considering the condition when Max−Mid < Mid−Min. The output voltage and the required switching sequence can once again be derived following the same methodology discussed in the earlier section. The output voltage signal and the switching pattern are shown in Figure 15.5. Similar to the earlier section, here a high-frequency triangular carrier signal is also compared with that of the duty ratio value D_{k2} to generate the switching pattern. The only difference in this case is that at the interval when the magnitude of the carrier signal is greater than the magnitude of the duty ratio and the slope is negative,

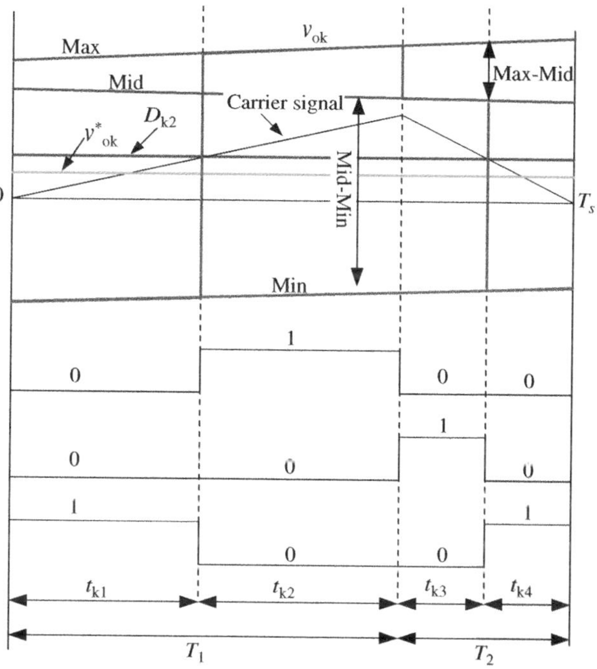

Figure 15.5 Output and switching pattern for kth phase in case II [21] (Reproduced by permission of IEEE)

then the output should follow Mid instead of Max. Contrary to case I, for this situation, the output must follow Max of the input. The time intervals t_{k1}, t_{k2}, t_{k3} and t_{k4} are the same as in Equation (15.18), and now the output phase voltage changes with the sequence Min→Max→Mid→Min. The volt-second principle is now applied to derive the equation for the duty ratio. The volt-second principle equation can be written as

$$v_{ok}^{*}T_s = \int_0^{T_s} v_{ok}\mathrm{d}t = \mathrm{Min}\{v_A, v_B, v_C\}.(t_{k1} + t_{k4}) + \mathrm{Max}\{v_A, v_B, v_C\}.t_{k2} + \mathrm{Mid}\{v_A, v_B, v_C\}.t_{k3} \quad (15.22)$$

Now, once again substituting the time expression from Equation (15.18) into Equation (15.22), one obtains

$$\begin{aligned} v_{ok}^{*} = \frac{1}{T_s}\int_0^{T_s} v_{ok}\mathrm{d}t = D_{k2}(&\mathrm{Min}\{v_A, v_B, v_C\} - \delta.\mathrm{Max}\{v_A, v_B, v_C\} - \mathrm{Mid}\{v_A, v_B, v_C\} + \delta.\mathrm{Mid}\{v_A, v_B, v_C\}) \\ &+ \delta.\mathrm{Max}\{v_A, v_B, v_C\} - \delta.\mathrm{Mid}\{v_A, v_B, v_C\} + \mathrm{Mid}\{v_A, v_B, v_C\} \end{aligned} \quad (15.23)$$

The duty ratio can now be written as [22]

$$D_{k2} = \frac{\delta.\Delta + (\mathrm{Mid}\{v_A, v_B, v_C\} - v_{ok}^{*})}{\delta.\Delta + (\mathrm{Mid}\{v_A, v_B, v_C\} - \mathrm{Min}\{v_A, v_B, v_C\})} \quad (15.24)$$

The switching pattern for the bidirectional power switching devices can be generated by considering the switching states of Figures 15.4 and 15.5. Depending on the output pattern, the gating signals can be obtained. If the output pattern of phase "*k*" is Max (or Mid, Min), then the output phase "*k*" is connected to the input phase whose voltage is Max (or Mid, Min). The PWM algorithm can be understood by the block diagram given in Figure 15.6.

The maximum, minimum and medium values of the input voltages are first computed. The information about their relative magnitudes is given to the next computation block along with the commanded output phase voltages. The computation block either uses Equation (15.21) or Equation (15.24) to generate the duty ratios, depending on the relative magnitude of the input voltages. The duty ratio obtained goes to the PWM block. The PWM block calculates the time subinterval using Equation (15.18). The gating pattern is then derived accordingly and given to the bidirectional power semiconductor switches of the matrix converter.

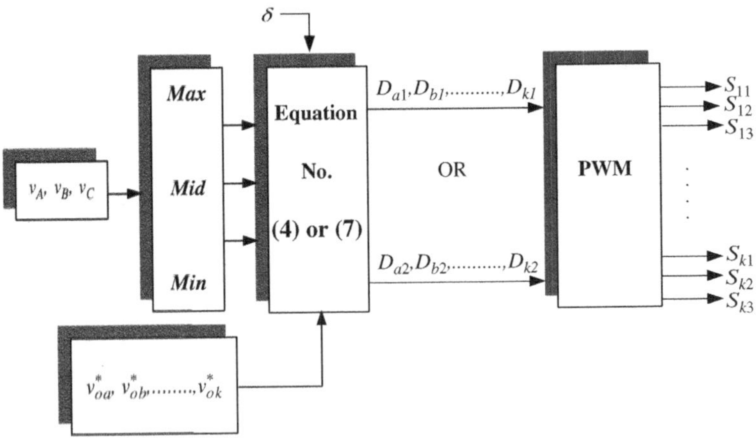

Figure 15.6 Block diagram of gate signal generation for a three-phase to *k*-phase matrix converter [21] (Reproduced by permission of IEEE)

C. DDPWM for Three-to-Five-Phase Matrix Converter

The principle of the working of direct DPWM is illustrated with the help of a matrix converter topology with three-phase input and five-phase output [22]. The output pattern for phases a, b, c, d and e is shown for a particular switching period in Figures 15.7–15.11, considering only case I. Since case II is similar to case I with only minor modification, the output patterns and subsequently the switching pattern can also be derived in a similar fashion and hence these are not elaborated further.

The direct DPWM has the major advantage of modularity; hence each phase of the outputs can be modulated separately to follow their references. Depending on how the references or target output voltages are created, two methods are developed. One method is termed "without harmonic injection" and the other is called "with harmonic injection." The maximum output voltage reaches half the input voltage if simple sinusoidal reference voltages are assumed. The magnitude of the output voltages or ratio of output to input voltage can be increased by subtracting the common mode (which is the third harmonic) of the input from the input voltage itself. The magnitude of the common-mode voltage that is to be subtracted can be varied. Sometime it is one-sixth of the input magnitude. The optimum value

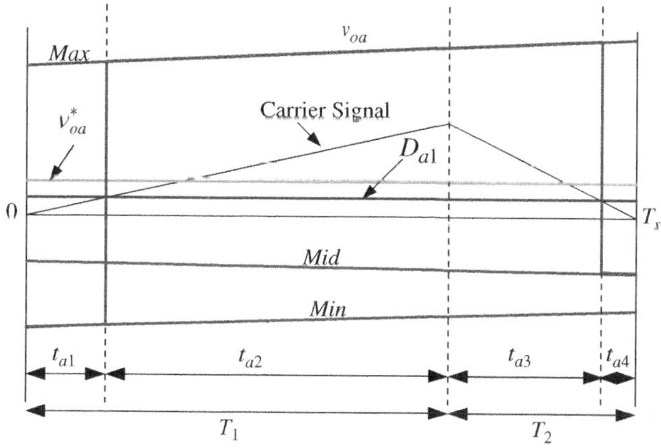

Figure 15.7 Output pattern of phase "a" [21] (Reproduced by permission of IEEE)

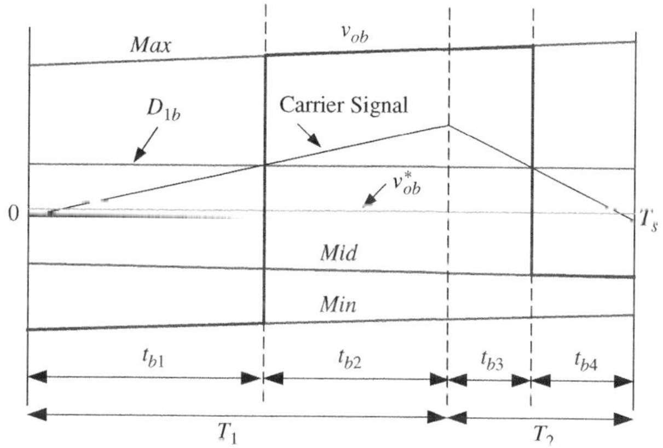

Figure 15.8 Output pattern of phase "b" [21] (Reproduced by permission of IEEE)

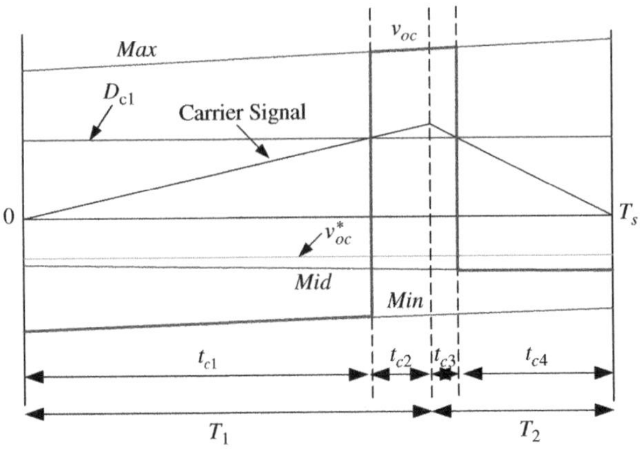

Figure 15.9 Output pattern of phase "c" [21] (Reproduced by permission of IEEE)

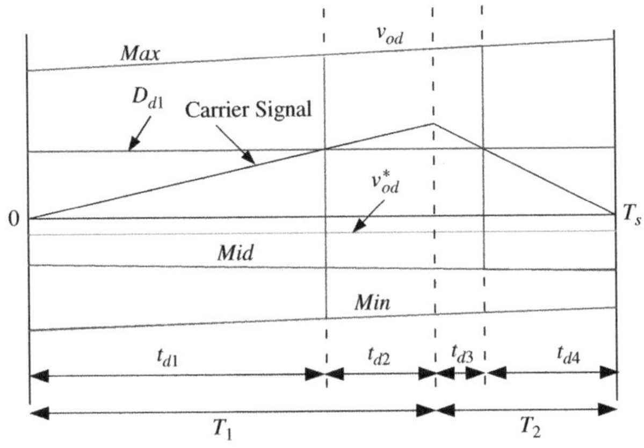

Figure 15.10 Output pattern of phase "d" [21] (Reproduced by permission of IEEE)

of the injected harmonic is seen as 25% of the input maximum magnitude. Hence, by employing the common-mode voltage addition scheme, the output voltage magnitude can reach up to 0.75 of the input voltage value, which is an almost a 50% increase.

The output voltage magnitude can be further enhanced by injecting the third harmonic of the output frequency in the reference output voltage or modulating signal. Thus, by injecting one-sixth of the magnitude of the third harmonic of output, the voltage transfer ratio goes up to 0.866 in the case of a three-to-three-phase matrix converter. This is a 15.5% increase compared to that of the harmonic injection in the input side voltage alone. Note that here the harmonic injection is done at both the input side and the output side. It is also important to note that the value of 15.5% is same as the amount of enhancement of the modulation index in the case of a three-phase voltage source inverter, which is achieved by harmonic injection when compared to that of a simple sinusoidal carrier-based scheme.

In the case of a multiphase voltage source inverter, a similar concept of the nth harmonic can be used for the enhancement of the modulation index. By injecting the nth harmonic of magnitude $M_n = -(M_1 \sin(\pi/2n))/n$, where n is the number of phases, the output voltage can be increased by

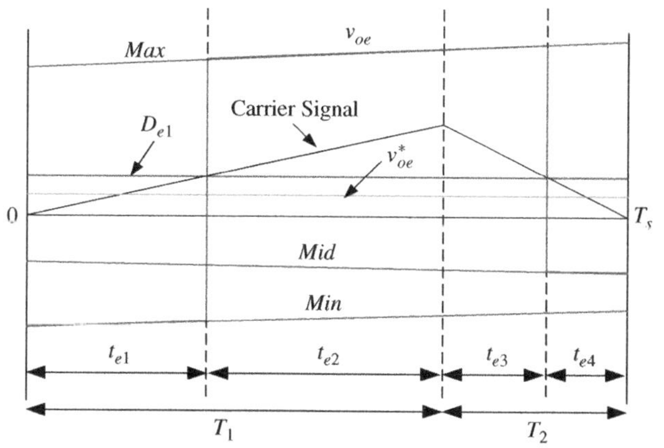

Figure 15.11 Output pattern of phase "e" [21] (Reproduced by permission of IEEE)

$1/\cos(\pi/2n)$. The same approach can be employed to enhance the output voltage magnitude of the matrix converter. The output voltage becomes 75% of the input voltage by only injecting the third harmonic. This increase is the same as that achieved in the three-to-three-phase matrix converter. Now in the case of three-to-five-phase matrix converter, the third harmonic of output cannot be injected, hence the fifth harmonic of the output frequency is injected. The maximum output voltage magnitude thus achieves 78.86% of the input voltage magnitude by injecting both the third (at the input side) and fifth (at the output side) harmonics in the linear modulation region. Hence, the overall gain in the output is 5.15%. It is to be noted here that the same amount of enhancement is achieved by the fifth harmonic injection in a five-phase voltage source inverter [23]. The output voltage references are now given as [22]

$$
\begin{aligned}
v^*_{\text{oa}} &= \sqrt{\frac{2}{5}} q V_{\text{in-rms}} \cos(\omega_{\text{o}} t) - \varphi(t) \\
v^*_{\text{ob}} &= \sqrt{\frac{2}{5}} q V_{\text{in-rms}} \cos\left(\omega_{\text{o}} t - 2\frac{\pi}{5}\right) - \varphi(t) \\
v^*_{\text{oc}} &= \sqrt{\frac{2}{5}} q V_{\text{in-rms}} \cos\left(\omega_{\text{o}} t - 4\frac{\pi}{5}\right) - \varphi(t) \\
v^*_{\text{od}} &= \sqrt{\frac{2}{5}} q V_{\text{in-rms}} \cos\left(\omega_{\text{o}} t + 4\frac{\pi}{5}\right) - \varphi(t) \\
v^*_{\text{oe}} &= \sqrt{\frac{2}{5}} q V_{\text{in-rms}} \cos\left(\omega_{\text{o}} t + 2\frac{\pi}{5}\right) - \varphi(t)
\end{aligned} \tag{15.25}
$$

where

$$
\varphi(t) = \frac{\sqrt{6}}{12} V_{\text{in-rms}} \cos(3\omega_{\text{i}} t) + \frac{\sqrt{6}}{48.5} q V_{\text{in-rms}} \cos(5\omega_{\text{o}} t) \tag{15.26}
$$

The output reference is the sum of the fundamental and the third and fifth harmonic components, where the sinusoidal output references ride on a common-mode voltage $\varphi(t)$, which is a combination of the third and fifth harmonic components. The arguments $\omega_{\text{i}}, \omega_{\text{o}}$ are the input and output frequencies, respectively. The input and output references, along with their common-mode voltages and the relative magnitude of the input voltages, are fed to the "duty ratio calculation" block. The implementation block diagram of the proposed method of the PWM is shown in Figure 15.12.

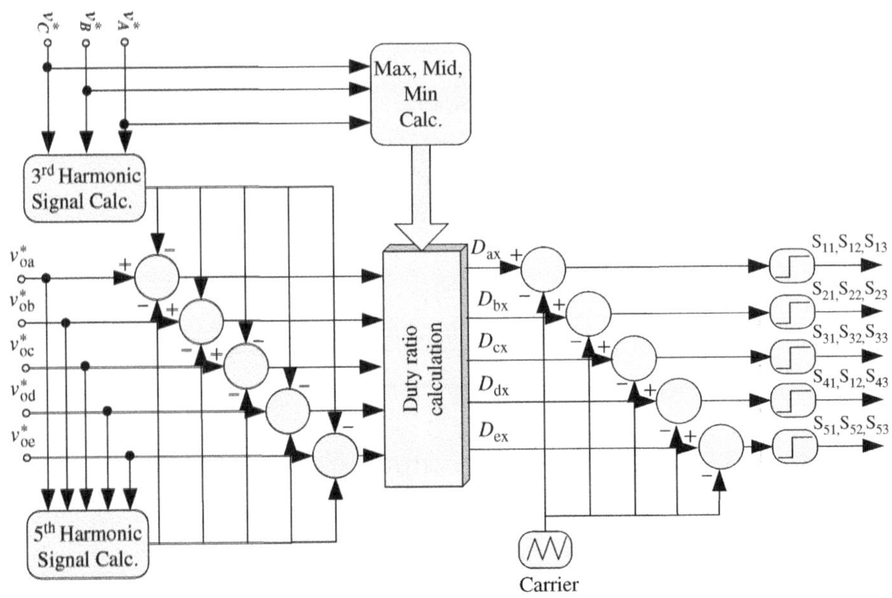

Figure 15.12 Modulation implementation block using harmonic injection [21] (Reproduced by permission of IEEE)

The outputs ($D_{ax}, D_{bx}, D_{cx}, D_{dx}, D_{ex}, x = 1$ or 2) are compared with that of a high-frequency carrier signal in order to generate the output voltage pattern. The gate signals are then derived directly from the generated output voltage patterns. The simulation and experimental results are presented in the later sections.

15.2.2.3 Space Vector PWM Technique

A. General Description

The space vector PWM technique is highly popular in conjunction with voltage source inverters because of higher DC bus utilization and its easier digital realization. The concept of space vectors can be extended to the matrix converter. In the matrix converter, the space vector algorithm [24–30] is based on the representation of the input current and output line voltages on the space vector planes. In matrix converters, each output phase is connected to each input phase depending on the state of the switches. For a three-to-five-phase matrix converter [10, 31], the total number of bidirectional power semiconductor switches is 15.

With this number of switches the total combination of switching is 2^{15}; however, for the protection of the switches the following conditions are considered:

- Input phases should never be short circuited during operation (to protect the source from short circuiting).
- Output phases should never be open circuited during operation (to protect the inductive load).

Considering the above two constraints, the switching combinations are reduced to 3^5, that is, 243 for connecting output phases to input phases. These switching combinations can be put into five groups.

The switching combinations are represented as [31]

$$\{p, q, r\}$$

where p, q and r represent the number of output phases connected to input phase A, phase B and phase C, respectively.

1. $p, q, r \in 0, 0, 5 | p \neq q \neq r$: All output phases are connected to the same input phase. This group consists of three possible switching combinations, that is, all output phases are connected to either the input phase A, or input phase B, or input phase C. {5, 0, 0} represents the switching condition when all the output phases connect to input phase A. {0, 5, 0} represents the switching condition when all output phases connect to input phase B. {0, 0, 5} represents the switching condition when all output phases connect to input phase C. These vectors have zero magnitude and frequency. These are called zero vectors and they are used for space vector PWM implementation [31].
2. $p, q, r \in 0, 1, 4 | p \neq q \neq r$: Four of the output phases are connected to the same input phase and the fifth output phase is connected to any of the other two remaining input phases. Here, 4 means that four different output phases are connected to input phase A. The number 1 means that one output phase other than the previous four is connected to input phase B and input phase C is not connected to any output phase. As such six different switching states exist ({4, 1, 0}, {1, 4, 0}, {1, 0, 4}, {0, 1, 4}, {0, 4, 1}, {4, 0, 1}). Out of these each switching state can have a further five different combinations, that is, every switching state has ${}^5C_4 \times {}^1C_1 = 5$ combinations. Hence, this group consists of $6 \times 5 = 30$ switching combinations in total. These vectors have variable amplitudes at a constant frequency in space. This means that the amplitude of the output voltages depends on the selected input line voltages. In this case, the phase angle of the output voltage space vector does not depend on the phase angle of the input voltage space vector. The 30 combinations in this group determine the 10 prefixed positions of the output voltage space vectors which are not dependent on α_i. A similar condition is also valid for the current vectors. The 30 combinations in this group determine the six prefixed positions of the input current space vectors, which are not dependent on α_o.
3. $p, q, r \in 0, 2, 3 | p \neq q \neq r$: Three of the output phases are connected to the same input phase and the two other output phases are connected to any of the other two input phases. As such six different switching states exist ({3, 2, 0}, {2, 3, 0}, {2, 0, 3}, {0, 2, 3}, {0, 3, 2} and {3, 0, 2}). Out of these, each switching permutation can have a further 10 different combinations, that is, every switching permutation has ${}^5C_3 \times {}^2C_2 = 10$ combinations. This group, therefore, consists of $6 \times 10 = 60$ switching combinations. These vectors also have variable amplitudes at a constant frequency in space. The 60 combinations in this group determine the 10 prefixed positions of the output voltage space vectors, which are not dependent on α_i. A similar condition is also valid for current vectors. The 60 combinations in this group determine the six prefixed positions of the input current space vectors, which are not dependent on α_o.
4. $p, q, r \in 1, 1, 3 | p \neq q \neq r$: Three of the output phases are connected to the same input phase and the two other output phases are connected to the other two input phases, respectively. As such three different switching states exist ({3, 1, 1}, {1, 3, 1}, {1, 1, 3}). Out of these, each switching state can have a further 20 different combinations, that is, every switching permutation has ${}^5C_3 \times {}^2C_1 \times {}^1C_1 = 20$ combinations. This group is, therefore, made up of $3 \times 20 = 60$ switching combinations. These vectors have fluctuating amplitudes with variable frequency in space. It makes sense that the amplitude of the output voltages depends on the particular input line voltages. As a result, the phase angle of the output voltage space vector depends on the phase angle of the input voltage space vector. The 60 combinations in this group do not determine any prefixed positions of the output voltage space vector. The locus of the output voltage space vectors forms ellipses in different orientations in space as α_i is varied. A similar condition can be obtained for the current vectors. For the space vector modulation technique, these switching states are not used in the matrix converter since the phase angle of both the input and output vectors cannot be independently controlled.
5. $p, q, r \in 1, 2, 2 | p \neq q \neq r$: Two of the output phases are connected to the same input phase, the two other output phases are connected to another input phase and the fifth output phase is connected to the third input phase. As such three different switching states exist ({1, 2, 2}, {2, 1, 2} and {2, 2, 1}). Each switching state can have a further 30 different combinations, that is, every switching permutation has ${}^5C_2 \times {}^3C_2 \times {}^1C_1 = 30$ combinations. This group thus consists of $3 \times 30 = 90$ switching combinations. These vectors also have variable amplitude and variable frequency in space. That is, the amplitude of the output voltages depends on the selected input line voltages. In this case, the phase angle of

the output voltage space vector depends on the phase angle of the input voltage space vector. The 90 combinations in this group do not determine any prefixed positions of the output voltage space vector. The locus of the output voltage space vectors form ellipses in different orientations in space as α_i is varied. A similar condition is also valid for current vectors. For the space vector modulation technique, these switching states are also not used in the matrix converter since the phase angle of both the input and output vectors cannot be independently controlled.

The active switching vectors that can be used for the space vector PWM of a three-to-five-phase matrix converter are as follows:

Group 1:{5, 0, 0} consists of three vectors
Group 2:{4, 1, 0} consists of 30 vectors
Group 3:{3, 2, 0} consists of 60 vectors

B. Space Vector Control Strategy

There are 243 permissible switching combinations in a three-to-five-phase matrix converter that are eligible for the selection of a space vector PWM. However, only 93 active switching vectors are used for the matrix converter space vector modulation technique. These active vectors are divided into four groups [31]:

Group 1: {5, 0, 0} consists of three vectors, these are called "zero vectors."
Group 2: {4, 1, 0} consists of 30 vectors, these are called "medium vectors."
Group 3: {3, 2, 0} consists of 30 vectors in which the two adjacent output phases are connected to the same input phase, these are called "large vectors."
Group 4: {3, 2, 0} consists of 30 vectors in which the two alternate output phases are connected to the same input phase, these are called "small vectors."

The vectors are named according to their length or magnitude.

In the space vector PWM strategy for a three-to-five-phase matrix converter, only the switching states of Groups 1, 2 and 3 can be used. The switching states in Group 4 cannot be used since the corresponding switching space vectors (SSVs) are rotating with time and would not be suitable. The input current SSVs and output voltage SSVs of each switching state in Groups 2 and 3 are illustrated in Figures 15.13 and 15.14, respectively.

The large vectors and medium vectors are represented as "L" and "M", respectively. The small vectors are not considered for space vector PWM implementation. The letters "L" and "M" refer to the large and medium vectors, respectively and the numbers in front of the letters are the vector numbers.

For each combination, the input and output line voltages can be expressed in terms of space vectors as

$$\vec{V_i} = \frac{2}{3}\left(V_{ab} + V_{bc}.e^{j\frac{2\pi}{3}} + V_{ca}.e^{j\frac{4\pi}{3}}\right) = V_i.e^{j\alpha_i} \tag{15.27}$$

$$\vec{V_o} = \frac{2}{5}\left(V_{AB} + V_{BC}.e^{j\frac{2\pi}{5}} + V_{CD}.e^{j\frac{4\pi}{5}} + V_{DE}.e^{j\frac{6\pi}{5}} + V_{EA}.e^{j\frac{8\pi}{5}}\right) = V_o.e^{j\alpha_O} \tag{15.28}$$

In a similar fashion, the input and output line currents space vectors are defined as [31]

$$\vec{I_i} = \frac{2}{3}\left(I_a + I_b.e^{j\frac{2\pi}{3}} + I_c.e^{j\frac{4\pi}{3}}\right) = I_i.e^{j\beta_i} \tag{15.29}$$

$$\vec{I_o} = \frac{2}{5}\left(I_A + I_B.e^{j\frac{2\pi}{5}} + I_C.e^{j\frac{4\pi}{5}} + I_D.e^{j\frac{6\pi}{5}} + I_E.e^{j\frac{8\pi}{5}}\right) = I_o e^{j\beta_O} \tag{15.30}$$

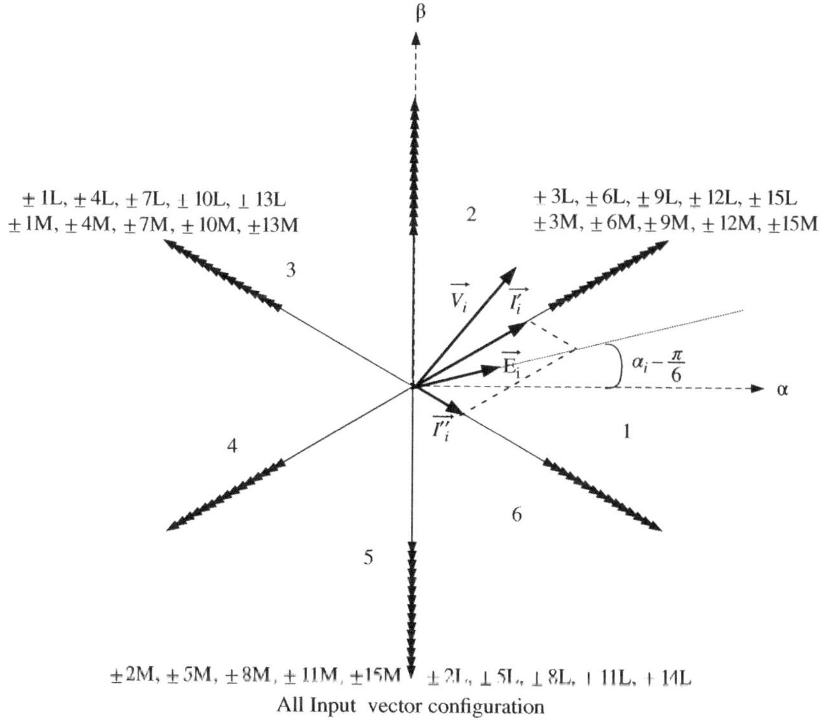

Figure 15.13 Input current space vectors corresponding to the permitted switching combinations for group 3: {3, 2, 0} (all vectors) [10] (Reproduced by permission of IEEE)

where α_i and α_o are the input and the output voltage vector phase angles, respectively, while β_i and β_o are the input and output current vector phase angles, respectively.

The SVM algorithm can be written as

1. select the appropriate switching states
2. calculate the duty cycle for each switching state

During one switching period T_s, the switching states whose switching state vectors are adjacent to the desired output voltage (and input current) vectors should be selected, and the zero switching states are applied to complete the switching period to provide the maximum output to input voltage transfer ratio.

The aim of the space vector control strategy is to generate the desired output voltage vector with the constraint of a unity input power factor, nevertheless the power factor can also be varied. For this purpose, let $\vec{V}_o$ be the desired output line voltage space vector and $\vec{V}_i$ be the input line voltage space vector at a given time. The input line to neutral voltage vector $\vec{E}_i$ is defined by

$$\vec{E}_i = \frac{1}{\sqrt{3}}\vec{V}_i . e^{-j\frac{\pi}{6}} \tag{15.31}$$

In order to obtain a unity input power factor, the direction of the input current space vector $\vec{I}_i$ has to be same as that of $\vec{E}_i$. Assume that $\vec{V}_o$ and $\vec{I}_i$ are in sector 1 (there are 6 sectors in input side and 10 sectors in output side as the input side is three phase and output side is five phase). In Figure 15.14, for large and

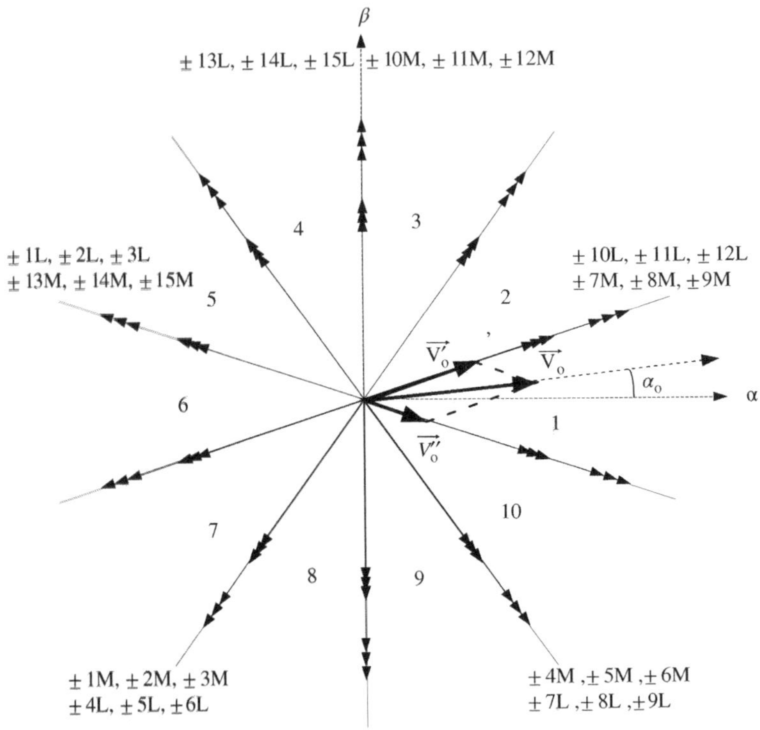

Figure 15.14 Output voltage space vectors corresponding to the permitted switching combinations for group 3: {3, 2, 0} (large and medium vectors) [10] (Reproduced by permission of IEEE)

medium vector configurations, $\vec{V}_o'$ and $\vec{V}_o''$ represent the components of $\vec{V}_o$ along the two adjacent vector directions. Similarly $\vec{I}_i$ is resolved into components $\vec{I}_i'$ and $\vec{I}_i''$ along the two adjacent vector directions. Possible switching states that can be utilized to synthesize the resolved voltage and current components (assuming both input and output vectors are in sector 1) are [31]

$$\vec{V}_o' \ : \pm 10L, \pm 11L, \pm 12L \text{ and } \pm 7M, \pm 8M, \pm 9M$$

$$\vec{V}_o'' \ : \pm 1L, \pm 2L, \pm 3L \text{ and } \pm 13M, \pm 14M, \pm 15M$$

$$\vec{I}_i' \ : \pm 3L, \pm 6L, \pm 9L, \pm 12L, \pm 15L \text{ and } \pm 3M, \pm 6M, \pm 9M, \pm 12M, \pm 15M$$

$$\vec{I}_i'' \ : \pm 1L, \pm 4L, \pm 7L, \pm 10L, \pm 13L \text{ and } \pm 1M, \pm 4M, \pm 7M, \pm 10M, \pm 13M$$

The output voltage and input current vectors can be synthesized simultaneously by selecting the common switching states of the output voltage components and input current components. Looking at sector 1, the common switching states between voltage and current are $\pm 10L, \pm 12L, \pm 7M, \pm 9M$ and $\pm 1L, \pm 3L, \pm 13M, \pm 15M$. From two switching states with the same number but opposite signs, only one should be used since the corresponding voltage or current space vectors are in opposite directions. Switching states with positive signs are used to calculate the duty cycle of the switching state. The selection of switching states for implementing space vector PWM can also be explained as follows [31].

Considering small variation in input voltage during the switching cycle period, the desired $\overrightarrow{V}_{o}'$ can be approximated by utilizing four (two medium and two large) switching configurations corresponding to four space vectors in the same direction of $\overrightarrow{V}_{o}'$ and one zero voltage configuration.

Among the six possible switching configurations, the two giving the higher output voltage values corresponding to large vectors and the two giving medium voltage values corresponding to medium vectors with the same direction of $\overrightarrow{V}_{o}'$ are chosen. In the same way, four different switching configurations and one zero voltage configuration are used to define $\overrightarrow{V}_{o}''$. With reference to the example shown in Figures 15.13 and 15.14, the input voltage $\overrightarrow{V}_{i}$ has a phase angle $0 \leq \alpha_i \leq \frac{\pi}{3}$. In this case, the line voltages V_{AB} and $-V_{CA}$ assume the higher values. Then according to the switching table of large and medium vectors, the configurations used to obtain $\overrightarrow{V}_{o}'$ are $+10L$ and $-12L$ for large vectors and $+7M$ and $-9M$ for medium vectors, while for $\overrightarrow{V}_{o}''$ are $+1L$, $-3L$ and $+13M$, $-15M$. These eight space vector combinations can be utilized to determine the input current vector direction as also shown in Figure 15.13. These configurations are associated with vector directions adjacent to the input current vector position.

There are 60 switching combinations for different sector combinations. These combinations for large and medium vectors are shown in Table 15.1.

By applying the space vector modulation technique, the on-time ratio δ of each configuration can be obtained by solving two systems of algebraic equations. In particular, utilizing configurations $+10L$, $-12L$ and $+7M$, $-9M$ to generate $\overrightarrow{V}_{o}'$ and to set the input current vector direction, one can write

$$\delta_{+10L}.|L|.V_{ab} - \delta_{-12L}.|L|.V_{ca} + \delta_{+7M}.|M|.V_{ab} - \delta_{-9M}.|M|.V_{ca} = V_o' = \frac{5}{3}|\overrightarrow{V_o}|.|L+M|.\sin\left(\frac{\pi}{10}+\alpha_o\right) \tag{15.32}$$

$$\begin{aligned}
\delta_{+10L}\frac{2}{\sqrt{3}}i_D &= I_i' = |\overrightarrow{I_i'}|\frac{2}{\sqrt{3}}\sin\left[\frac{\pi}{6}-\left(\alpha_i-\frac{\pi}{6}\right)\right]\\
\delta_{-12L}\frac{2}{\sqrt{3}}i_D &= I_i'' = |\overrightarrow{I_i'}|\frac{2}{\sqrt{3}}\sin\left[\frac{\pi}{6}+\left(\alpha_i-\frac{\pi}{6}\right)\right]\\
\delta_{+7M}\frac{2}{\sqrt{3}}i_C &= I_i' = |\overrightarrow{I_i'}|\frac{2}{\sqrt{3}}\sin\left[\frac{\pi}{6}-\left(\alpha_i-\frac{\pi}{6}\right)\right]\\
\delta_{-9M}\frac{2}{\sqrt{3}}i_C &= I_i'' = |\overrightarrow{I_i'}|\frac{2}{\sqrt{3}}\sin\left[\frac{\pi}{6}+\left(\alpha_i-\frac{\pi}{6}\right)\right]
\end{aligned} \tag{15.33}$$

Table 15.1 Space vector choice for space vector PWM in different sectors [31]

Sector # of V_o	Sector # of I_i		
	1 or 4	2 or 5	3 or 6
1 or 6	$\pm10L, \pm12L, \pm1L, \pm3L, \pm7M, \pm9M, \pm13M, \pm15M$	$\pm12L, \pm11L, \pm3L, \pm2L, \pm15M, \pm14M, \pm9M, \pm8M$	$\pm11L, \pm10L, \pm2L, \pm1L, \pm14M, +13M, +8M, +7M$
2 or 7	$\pm10L, \pm12L, \pm4L, \pm6L, \pm7M, +9M, +1M, +3M$	$\pm12L, \pm11L, \pm6L, \pm5L, \pm3M, \pm2M, \pm9M, \pm8M$	$\pm11L, +10L, +5L, +4L, \pm2M, \pm1M, \pm8M, \pm7M$
3 or 8	$\pm13L, \pm15L, \pm4L, \pm6L, \pm1M, \pm3M, \pm10M, \pm12M$	$\pm15L, \pm14L, \pm6L, \pm5L, \pm12M, \pm11M, \pm3M, +2M$	$\pm14L, \pm13L, \pm5L, \pm4L, +11M, +10M, +2M, \pm1M,$
4 or 9	$\pm13L, \pm15L, \pm7L, \pm9L, \pm10M, \pm12M, \pm4M, \pm6M$	$\pm15L, \pm14L, \pm9L, \pm8L, \pm12M, \pm11M, \pm6M, \pm5M$	$\pm14L, \pm13L, \pm8L, \pm7L, \pm11M, +10M, +5M, +4M$
5 or 10	$\pm1L, \pm3L, \pm7L, \pm9L, \pm4M, \pm6M, \pm13M, \pm15M$	$\pm3L, \pm2L, \pm9L, \pm8L, \pm15L, +14L, +6L, +5L$	$\pm2L, \pm1L, \pm8L, \pm7L, \pm14M, \pm13M, \pm5M, \pm4M$

Considering a balanced system of sinusoidal supply voltages expressed as

$$\begin{aligned} V_{ab} &= |\vec{V}_i| \cos(\alpha_i) \\ V_{bc} &= |\vec{V}_i| \cos\left(\alpha_i - \frac{2\pi}{3}\right) \\ V_{ca} &= |\vec{V}_i| \cos\left(\alpha_i - \frac{4\pi}{3}\right) \end{aligned} \tag{15.34}$$

the solution of the system of Equations (15.32) and (15.33) gives

$$\begin{aligned} \delta_{+10L} &= q.|L|.\frac{10}{3\sqrt{3}} \sin\left(\frac{\pi}{10} + \alpha_o\right).\sin\left(\frac{\pi}{3} - \alpha_i\right) \\ \delta_{-12L} &= q.|L|.\frac{10}{3\sqrt{3}} \sin\left(\frac{\pi}{10} + \alpha_o\right).\sin(\alpha_i) \\ \delta_{+7M} &= q.|M|.\frac{10}{3\sqrt{3}} \sin\left(\frac{\pi}{10} + \alpha_o\right).\sin\left(\frac{\pi}{3} - \alpha_i\right) \\ \delta_{-9M} &= q.|M|.\frac{10}{3\sqrt{3}} \sin\left(\frac{\pi}{10} + \alpha_o\right).\sin(\alpha_i) \end{aligned} \tag{15.35}$$

where $q = \frac{|\vec{V}_o|}{|\vec{V}_i|}$ is the voltage transfer ratio between input source and output load. L and M correspond to large and medium vectors, respectively. With the same procedure, utilizing configurations $+1L$, $-3L$ and $+13M$, $-15M$ to generate $\vec{V}_o''$ and to set the input current vector direction yields [31]:

$$\begin{aligned} \delta_{+1L} &= q.|L|.\frac{10}{3\sqrt{3}} \sin\left(\frac{\pi}{10} - \alpha_o\right).\sin\left(\frac{\pi}{3} - \alpha_i\right) \\ \delta_{-3L} &= q.|L|.\frac{10}{3\sqrt{3}} \sin\left(\frac{\pi}{10} - \alpha_o\right).\sin(\alpha_i) \\ \delta_{+13M} &= q.|M|.\frac{10}{3\sqrt{3}} \sin\left(\frac{\pi}{10} - \alpha_o\right).\sin\left(\frac{\pi}{3} - \alpha_i\right) \\ \delta_{-15M} &= q.|M|.\frac{10}{3\sqrt{3}} \sin\left(\frac{\pi}{10} - \alpha_o\right).\sin(\alpha_i) \end{aligned} \tag{15.36}$$

The results obtained are valid for $-\pi/10 \le \alpha_o \le \pi/10$ and for $0 \le \alpha_i \le \pi/3$.

By applying a similar procedure for the other possible pairs of angular sectors, the required switching configurations and the on-time ratio of each configuration can be determined.

Note that the values of the on-time ratios (or duty cycle) should be positive. Furthermore, the sum of the ratios must be lower than or equal to unity. By adding Equations (15.35) and (15.36) with the above constraints, one can write

$$\delta_{+10L} + \delta_{-12L} + \delta_{+1L} + \delta_{-3L} + \delta_{+7M} + \delta_{-9M} + \delta_{+13M} + \delta_{-15M} \le 1 \tag{15.37}$$

The maximum value of the voltage transfer ratio can be determined as $q = 0.7886$ for a three-phase input to five-phase output matrix converter.

C. Maximum Output in n-Phase by m-Phase Matrix Converter

One can relate the maximum output voltage in an n-phase to m-phase matrix converter with the maximum output voltage achievable in an equivalent m-phase voltage source inverter and the length of the largest space vector of an n-phase voltage source inverter. A general relationship is given as [31]

$$\text{Maximum possible output in } n \text{ by } m \text{ matrix converter} = \left(\frac{\text{maximum output in } m\text{-phase inverter in linear range}}{\text{length of the largest space vector of } n\text{-phase inverter}}\right).$$

The maximum output expression for n-phase by m-phase matrix converters is correlated with n-phase and m-phase inverters. In an n-phase by m-phase matrix converter, the input is n-phase and the output is m-phase. It can be reimaged as two inverters (one is of n-phase output and the other is of m-phase output) are connected back to back. In the case of the m-phase inverter, the maximum output in the linear range can be written as [23]

$$\frac{1}{\{2.\cos(\pi/(2.m))\}} \tag{15.38}$$

The above term will be divided by the maximum vector length of n-phase inverter to obtain the maximum output value for an "n" by "m" phase matrix converter in the linear modulation range. The V_{DC} is written in the formula of Table 15.2 to show the exact vector length equation for an inverter. In this case, V_{DC} is equal to unity.

D. Commutation Requirements

Once the phase angles of the input current and output line voltage are known, the eight space vectors are required to synthesize the space vector PWM (since the input is three phase). These eight space vectors are utilized until α_i or α_o changes the angular sector. One of the zero space voltage vectors should be employed in each switching cycle to obtain a symmetrical switching waveform. The sequence of switching of the resulting nine (eight active and one zero) space vectors should be defined in order to minimize the number of switch commutations [31].

Table 15.2 Maximum modulation index formulation [10]

Matrix converter configuration (n-input × m-output)	Maximum output to input voltage formula	Maximum modulation index (%)
3×3	$\frac{1/\{2.\cos(\pi/6)\}}{2/3V_{DC}}$	86.66
3×5	$\frac{1/\{2.\cos(\pi/10)\}}{2/3V_{DC}}$	78.86
3×6	$\frac{1/\{2.\cos(\pi/12)\}}{2/3V_{DC}}$	77.65
3×7	$\frac{1/\{2.\cos(\pi/14)\}}{2/3V_{DC}}$	76.93
3×9	$\frac{1/\{2.\cos(\pi/18)\}}{2/3V_{DC}}$	76.15
5×3	$\frac{1/\{2.\cos(\pi/6)\}}{0.6472V_{DC}}$	89.21
5×5	$\frac{1/\{2.\cos(\pi/10)\}}{0.6472V_{DC}}$	81.23

Table 15.3 Space vector switching sequence [10]

Space vector	A	B	C	D	E	Vector no.
0/2	b	b	**b**	b	b	0
$+7M/2$	b	b	**a**	b	**b**	I
$+13M/2$	**b**	b	**b**	b	**a**	II
$-10L/2$	**a**	b	b	**b**	a	III
$+1L/2$	a	**b**	**b**	**a**	a	IV
$-3L/2$	a	**c**	**c**	**a**	a	V
$+12L/2$	**a**	c	c	**c**	a	VI
$-15M/2$	**c**	c	**c**	c	**a**	VII
$-9M/2$	c	c	**a**	c	**c**	VIII
0	c	c	**c**	c	c	IX
$-9M/2$	c	c	**a**	c	**c**	VIII
$-15M/2$	**c**	c	**c**	c	**a**	VII
$+12L/2$	**a**	c	c	**c**	a	VI
$-3L/2$	a	**c**	**c**	**a**	a	V
$+1L/2$	a	**b**	**b**	**a**	a	IV
$-10L/2$	**a**	b	b	**b**	a	III
$+13M/2$	**b**	b	**b**	b	**a**	II
$+7M/2$	b	b	**a**	b	**b**	I
0/2	b	b	**b**	b	b	0

Reproduced by permission of IEEE

With reference to the α_i and α_o values considered in Figures 15.13 and 15.14, the available space vectors and their sequence of switching is listed in Table 15.3 assuming both the input and output reference vectors in sector I. The first column lists the different space vectors that will be used for the space vector PWM. The second to sixth columns list the input and output phases that will be connected during the switching period. The capital letter denotes the output phases (five phase) and the small letter indicates the input phases (three phase). The sequence of application of space vectors can be defined such that the number of switchings in one sampling period is the minimum. The switching sequence in one sample period in sector I (both input and output reference vectors) is listed in Table 15.3.

To obtain symmetrical switching, at first a zero vector is applied followed by eight active vectors in a half-sampling period. The mirror image of the switching sequence is employed in the second half of the sampling period. The time of applications of active and zero vectors are divided into two; hence, the total time of application is also halved. It is observed that when applying vector $+7M$ after zero vector, only one state is changed, the input phase "a" is now connected to the output phase "C." In the next transition from $+7M$ to $+13M$, two states are changed. Each change in the switching is shown by a bold and underlined alphabet.

It should be noted that in this way only 12 commutations are required in each half-sampling period. Once the configurations are selected and sequenced, the on-time ratio of each configuration is calculated using Equations (15.35) and (15.36) given for the appropriate sector.

15.3 Simulation and Experimental Results

A Matlab/Simulink model can be used to verify the PWM algorithm presented above. A simulation example is presented here. The input voltage is kept at 100 V peak to show the exact voltage transfer ratio at the output side, and the switching frequency of the devices is kept at 6 kHz. (Different parameters can be set in the Matlab/Simulink model.) The load connected to the Matrix converter is R–L with parameter values $R = 15\,\Omega$ and $L = 10\,\text{mH}$. The above parameters are utilized for the carrier-based modulation scheme. Similarly, the DDPWM parameters are $R = 15\,\Omega$, $L = 12\,\text{mH}$, output frequency = 40 Hz and switching frequency = 10 kHz. In the case of the space vector modulation technique, the parameters are $R = 12\,\Omega$, $L = 40\,\text{mH}$, output frequency = 40 Hz and switching frequency = 6 kHz. The operation of the topology of the matrix converter is tested for a wide range of frequencies, from as low as 1 Hz to higher frequencies, considering deep-flux weakening operation. The simulation results are shown for different modulation techniques (Figures 15.15 to 15.23).

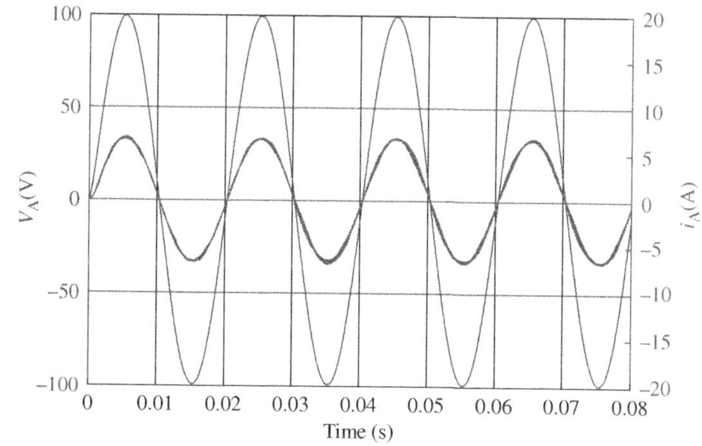

Figure 15.15 Input voltage and current at 50 Hz (carrier-based modulation)

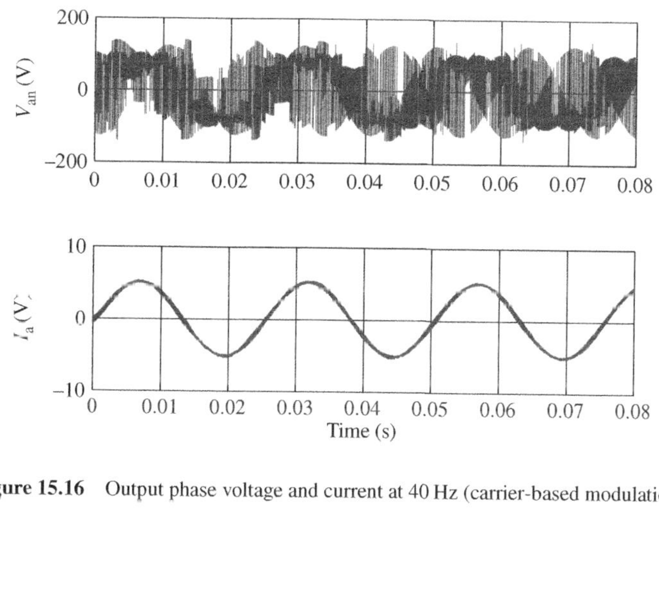

Figure 15.16 Output phase voltage and current at 40 Hz (carrier-based modulation)

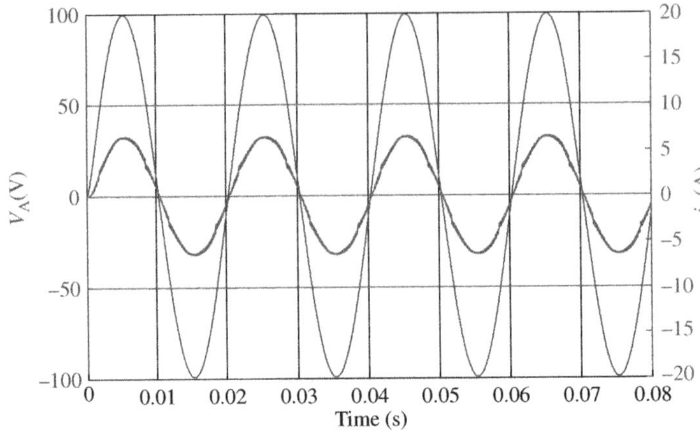

Figure 15.17 Input voltage and current at 50 Hz (DDPWM)

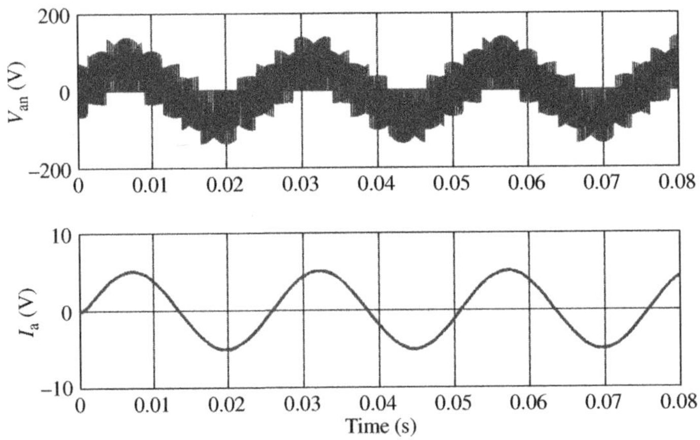

Figure 15.18 Output phase voltage and current at 40 Hz (DDPWM)

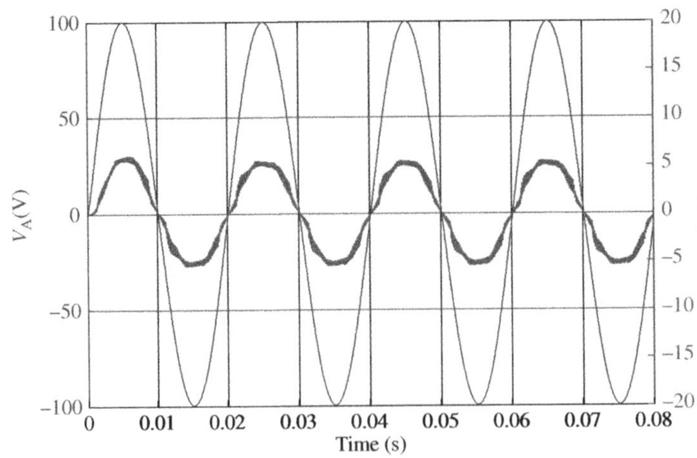

Figure 15.19 Input voltage and current at 50 Hz (SVPWM)

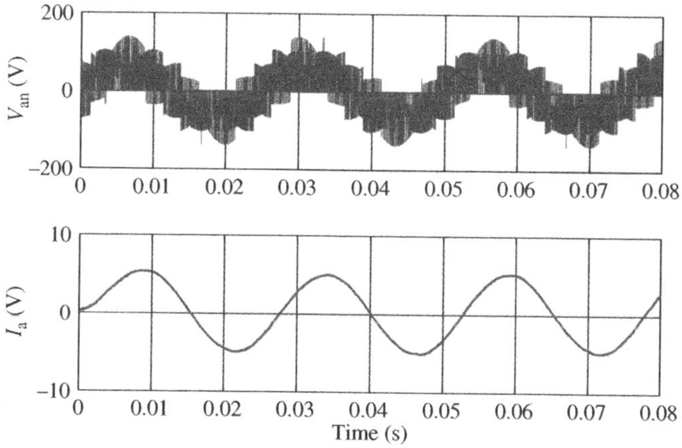

Figure 15.20 Output phase voltage and current at 40 Hz (SVPWM)

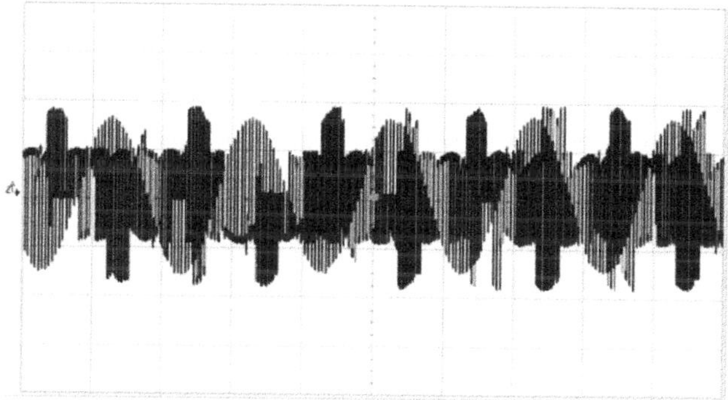

Figure 15.21 Experimental output line voltage for carrier-based PWM technique (200 V/div, 10 ms/div)

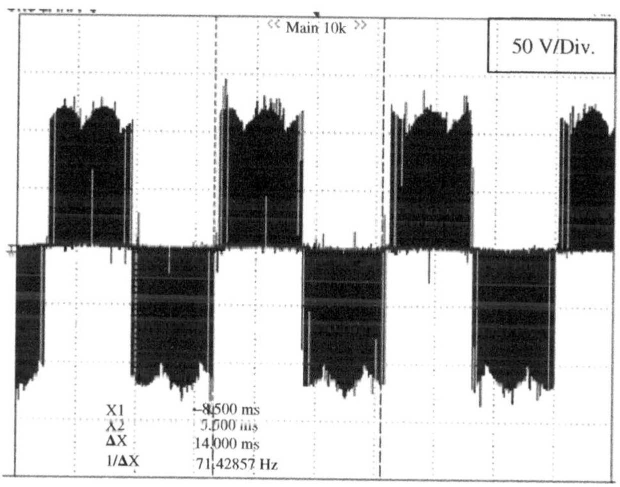

Figure 15.22 Experimental output phase voltage for DDPWM technique (50 V/div, 5ms/div) [21] (Reproduced by permission of IEEE)

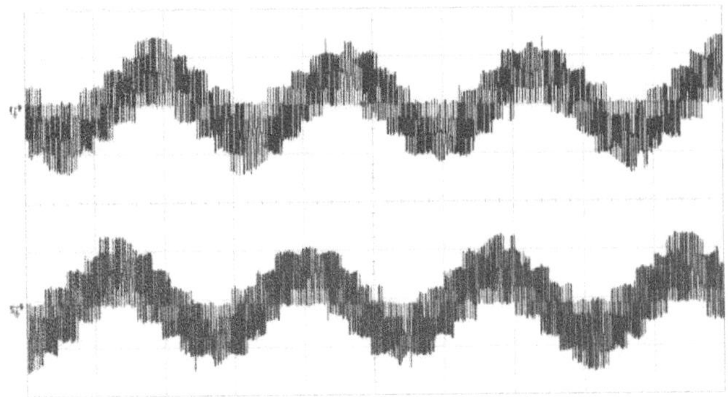

Figure 15.23 Experimental output phase voltages for SVPWM technique (100 V/div, 5ms/div) [10]

15.4 Matrix Converter with Five-Phase Input and Three-Phase Output

15.4.1 Topology

The power circuit topology of a five-to-three phase matrix converter [32] is shown in Figure 15.24. There are three legs with each leg having five bidirectional power switches connected in parallel. Each power switch can conduct in both directions and have antiparallel connected IGBTs and diodes. The output is similar to the three-to-three-phase matrix converter, and the input is five phases with 72° phase displacement between each with small LC filters. Such a topology can be used in conjunction with a five-phase wind energy generator. The switching function is defined as S_{kj} = {1 for a closed switch, 0 for

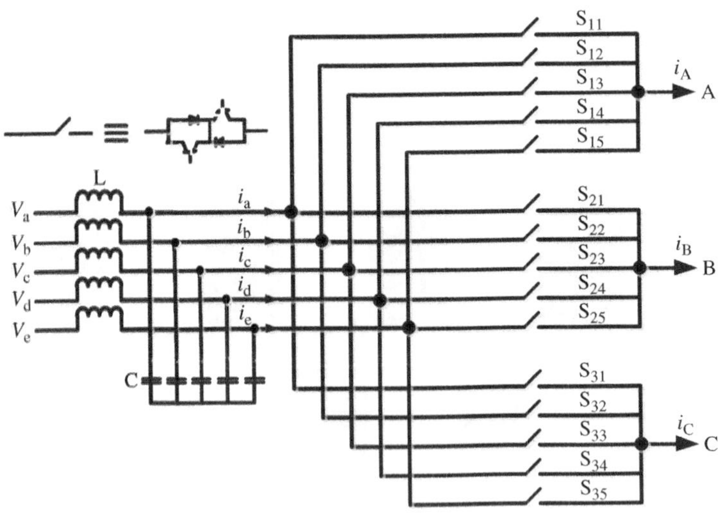

Figure 15.24 Five-to-three-phase direct matrix converter [33]

an open switch}, k = {a, b, c, d, e} (input), j = {A, B, C} (output). The switching constraint is $S_{aj} + S_{bj} + S_{cj} + S_{dj} + S_{ej} = 1$. The load to the matrix converter can be a motor drive or utility grid system. The requirements on the matrix converter control depend on the type of load. The modulation of this converter topology (input phase > output phase) is similar to that of the converter with output phase > input phase. Carrier-based PWM and space vector PWM can be developed for this topology and is further elaborated in the following section.

15.4.2 Control Techniques

15.4.2.1 Carrier-Based PWM Modeling

A. General Description

The input five-phase system is assumed as

$$\begin{aligned} v_a &= |V|\cos(\omega t), v_b = |V|\cos(\omega t - 2\pi/5), v_c = |V|\cos(\omega t - 4\pi/5) \\ v_d &= |V|\cos(\omega t + 4\pi/5), v_e = |V|\cos(\omega t + 2\pi/5) \end{aligned} \tag{15.39}$$

The three-phase output voltage duty ratios should be calculated in such a way that the frequency of the output voltages remains independent of the input frequency [31]. In other words, the three-phase output voltages can be considered a synchronous reference frame and the five-phase input voltages can be considered to be in a stationary reference frame, so that the input frequency term will be absent in output voltages. Considering the above, the duty ratios of output phase j are chosen as

$$\begin{aligned} \delta_{aj} &= k_j\cos(\omega t - \rho), \delta_{bj} = k_j\cos(\omega t - 2\pi/5 - \rho) \\ \delta_{cj} &= k_j\cos(\omega t - 4\pi/5 - \rho), \delta_{dj} = k_j\cos(\omega t + 4\pi/5 - \rho) \\ \delta_{ej} &= k_j\cos(\omega t + 2\pi/5 - \rho) \end{aligned} \tag{15.40}$$

The input and output voltages are related as

$$\begin{bmatrix} V_A \\ V_B \\ V_C \end{bmatrix} = \begin{bmatrix} \delta_{aA} & \delta_{bA} & \delta_{cA} & \delta_{dA} & \delta_{eA} \\ \delta_{aB} & \delta_{bB} & \delta_{cB} & \delta_{dB} & \delta_{eB} \\ \delta_{aC} & \delta_{bC} & \delta_{cC} & \delta_{dC} & \delta_{eC} \end{bmatrix} \begin{bmatrix} v_a \\ v_b \\ v_c \\ v_d \\ v_e \end{bmatrix} \tag{15.41}$$

Therefore, the phase j output voltage can be obtained by using the above duty ratios:

$$\begin{aligned} V_j = k_j|V|\,[&\cos(\omega t) \bullet \cos(\omega t - \rho) + \cos(\omega t - 2\pi/5) \bullet \cos(\omega t - 2\pi/5 - \rho) + \cos(\omega t - 4\pi/5) \\ &\bullet \cos(\omega t - 4\pi/5 - \rho) + \cos(\omega t + 4\pi/5) \bullet \cos(\omega t + 4\pi/5 - \rho) + \cos(\omega t + 2\pi/5) \\ &\bullet \cos(\omega t + 2\pi/5 - \rho)] \end{aligned} \tag{15.42}$$

In general, Equation (15.42) can be written as

$$V_j = \frac{5}{2}k_j|V|\cos(\rho) \tag{15.43}$$

In Equation (15.43), the $\cos(\rho)$ term indicates that the output voltage is influenced by ρ. Thus, the output voltage V_j is independent of the input frequency and only depends on the amplitude $|V|$ of the input voltage and k_j is a reference output voltage time-varying modulating signal for the output phase j with the desired output frequency ω_o. The three-phase reference output voltages can be represented as

$$k_A = m\cos(\omega_o t),\ k_B = m\cos(\omega_o t - 2\pi/3),\ k_C = m\cos(\omega_o t - 4\pi/3) \tag{15.44}$$

Therefore, from Equation (15.43), the output voltages are obtained as

$$V_A = \left[\frac{5}{2}m\,|V|\cos(\rho)\right]\cos(\omega_o t),\; V_B = \left[\frac{5}{2}m\,|V|\cos(\rho)\right]\cos\left(\omega_o t - 2\frac{\pi}{3}\right)$$
$$V_C = \left[\frac{5}{2}m\,|V|\cos(\rho)\right]\cos\left(\omega_o t - 4\frac{\pi}{3}\right) \tag{15.45}$$

B. Application of Offset Duty Ratio

In the above discussion [32], duty ratios become negative (see Equation (15.44)), which is not practically realizable. For the switches connected to output phase j, at any instant the conditions $0 \le d_{aj}, d_{bj}, d_{cj}, d_{dj}, d_{ej} \le 1$ and $d_{aj} + d_{bj} + d_{cj} + d_{dj} + d_{ej} = 1$ should be valid. Therefore, offset duty ratios should to be added to the existing duty ratios, so that the net resultant duty ratios of individual switches are always positive. Furthermore, the offset duty ratios should be added equally to all the output phases to ensure that the effect of the resultant output voltage vector produced by the offset duty ratios is null in the load. That is, the offset duty ratios can only add the common-mode voltages in the output. Considering the case of output phase j:

$$d_{aj} + d_{bj} + d_{cj} + d_{dj} + d_{ej} = k_j\cos(\omega t - \rho) + k_j\cos(\omega t - 2\pi/5 - \rho) + k_j\cos(\omega t - 4\pi/5 - \rho)$$
$$+ k_j\cos(\omega t + 4\pi/5 - \rho) + k_j\cos(\omega t + 2\pi/5 - \rho) = 0 \tag{15.46}$$

The sum of all the duty ratios is zero because the duty ratios contain equal amounts of positive and negative components. Absolute values of the duty ratios are added to cancel the negative components from individual duty ratios. Thus, the minimum individual offset duty ratios should be

$$\begin{aligned}
D_a(t) &= |d_{aj}| = |k_j\cos(\omega t - \rho)| \\
D_b(t) &= |d_{bj}| = |k_j\cos(\omega t - 2\pi/5 - \rho)| \\
D_c(t) &= |d_{cj}| = |k_j\cos(\omega t - 4\pi/5 - \rho)| \\
D_d(t) &= |d_{dj}| = |k_j\cos(\omega t + 4\pi/5 - \rho)| \\
D_e(t) &= |d_{dj}| = |k_j\cos(\omega t + 2\pi/5 - \rho)|
\end{aligned} \tag{15.47}$$

The effective duty ratios are

$$\delta'_{aj} = d_{aj} + D_a(t), \delta'_{bj} = d_{bj} + D_b(t), \delta'_{cj} = d_{cj} + D_c(t), \delta'_{dj} = d_{dj} + D_d(t), \delta'_{ej} = d_{ej} + D_e(t) \tag{15.48}$$

The net duty ratios of $0 \le \delta'_{aj}, \delta'_{bj}, \delta'_{cj}, \delta'_{dj}, \delta'_{ej} \le 1$ should be within the range 0–1. For the worst case with respect to a five-phase input:

$$0 \le 2.|k_j| \times 2\cos(\pi/5) \le 1 \tag{15.49}$$

The maximum value of k_j is equal to 0.309 or $\sin(\pi/10)$. In any switching cycle, the output phase should not be open circuited. Thus, the sum of the duty ratios in Equation (15.47) must equal unity. But the summation $D_a(t) + D_b(t) + D_c(t) + D_d(t) + D_e(t)$ is less than or equal to unity. Hence, another offset duty ratio $[1 - \{D_a(t) + D_b(t) + D_c(t) + D_d(t) + D_e(t)\}]/5$ is added to $D_a(t), D_b(t), D_c(t), D_d(t)$ and $D_e(t)$ in Equation (15.49). The addition of this offset duty ratio in all switches will maintain the output voltages and input currents unaffected. The above explanation finally derives the maximum modulation index for five-phase input with three-phase output from Equation (15.45) as $\frac{5}{2}k_j = \frac{5}{2} \times \sin\left(\frac{\pi}{10}\right) = 0.7725$ or 77.25%. If k_A, k_B, k_C are chosen to be three-phase sinusoidal references as given in Equation (15.44), the input voltage capability is not fully utilized for output voltage generation and the output magnitude remains only 77.25% of the input magnitude. To overcome this, an additional common-mode term equal to $[\{\max(k_A, k_B, k_C) + \min(k_A, k_B, k_C)\}/2]$ is added, as in the carrier-based PWM principle for two-level inverters. Thus, the amplitude of (k_A, k_B, k_C) can be enhanced from 0.309 to 0.3568.

C. Without Common-Mode Voltage Addition

In the earlier section, two offsets are added to the original duty ratios to form the effective duty ratio which can be compared to that of the triangular carrier wave to generate the gating signals for the bidirectional power switches. The output phase voltage magnitude will reach 77.25% of the input voltage magnitude with this method. To further enhance the output voltage magnitude, common-mode voltages of the output reference signals are added to formulate the new duty ratios, as discussed in the following section.

D. With Common-Mode Voltage Addition

The duty ratios can further be modified by the injection of common-mode voltage of the output voltage references to improve the output voltage magnitude. The output voltage magnitude increases and reaches its limiting value of 89.2% of the input magnitude. The common-mode voltage that is added to obtain new duty ratios is

$$V_{cm} = -\frac{V_{Max} - V_{Min}}{2} \tag{15.50}$$

where

$$V_{MAX} - \max\{k_A, k_B, k_C\}, V_{MIN} = \min\{k_A, k_B, k_C\} \tag{15.51}$$

The duty ratio for output phase p can be written as [32]

$$\begin{aligned}
\delta_{aj} &= D_a(t) + (1 - \{D_a(t) + D_b(t) + D_c(t) + D_d(t) + D_e(t)\})/3 + [k_j + V_{cm}] \times \cos(\omega t - \rho) \\
\delta_{bj} &= D_b(t) + (1 - \{D_a(t) + D_b(t) + D_c(t) + D_d(t) + D_e(t)\})/3 + [k_j + V_{cm}] \times \cos(\omega t - 2\pi/5 - \rho) \\
\delta_{cj} &= D_c(t) + (1 - \{D_a(t) + D_b(t) + D_c(t) + D_d(t) + D_e(t)\})/3 + [k_j + V_{cm}] \times \cos(\omega t - 4\pi/5 - \rho) \\
\delta_{dj} &= D_d(t) + (1 - \{D_a(t) + D_b(t) + D_c(t) + D_d(t) + D_e(t)\})/3 + [k_j + V_{cm}] \times \cos(\omega t + 4\pi/5 - \rho) \\
\delta_{ej} &= D_e(t) + (1 - \{D_a(t) + D_b(t) + D_c(t) + D_d(t) + D_e(t)\})/3 + [k_j + V_{cm}] \times \cos(\omega t + 2\pi/5 - \rho)
\end{aligned} \tag{15.52}$$

where $j \in A, B, C$

15.4.2.2 Space Vector Model of Five-to-Three-Phase Matrix Converter

A. General Description

The space vector algorithm is based on the representation of the five-phase input current and three-phase output line voltages on the space vector plane [33]. In matrix converters, each output phase is connected to each input phase depending on the state of the switches. For a five-to-three-phase matrix converter, there are 15 switches with a total combination of switching possibilities of $2^{15} = 32768$. However, for the safe switching of the matrix converter, the following constraints must be met (as for three-phase input and multiphase output):

- Input phases should never be short circuited to protect the source
- Output phases should never be open circuited at any switching time due to the inductive nature of the load

Considering the above two constraints, the switching combinations are reduced to 5^3 or 125 different switching combinations connecting output phases to input phases. These switching combinations can be analyzed in three different categories.

The switching combinations are represented as $\{p, q, r, s, t\}$ where p, q, r, s and t represent the count of output phases that may be 0, 1, 2, or 3 connected to input phases A, B, C, D and E.

1. $\{p, q, r, s, t\} \in \{0, 0, 0, 0, 3\} | p \neq q \neq r \neq s \neq t$: All the output phases are connected to the same input phase. This group consists of five possible switching combinations, that is, all the output phases connect to the same input phases A, B, C, D, or E. {3, 0, 0, 0, 0} represents the switching condition when all output phases connect to input phase A. {0, 3, 0, 0, 0} represents the switching condition when all output phases connect to input phase B. {0, 0, 3, 0, 0} represents the switching condition when all output phases connect to input phase C. {0, 0, 0, 3, 0} represents the switching condition when all the output phases connect to input phase D. {0, 0, 0, 0, 3} represents the switching condition when all output phases connect to input phase E. These so-called zero vectors have zero magnitude and frequency.
2. $\{p, q, r, s, t\} \in \{0, 0, 0, 1, 2\} | p \neq q \neq r \neq s \neq t$: Two of the output phases are connected to the same input phase and the third output phase is connected to any of the other four input phases. Here the element "2" denotes that two different output phases are connected to input phase E and element "1" denotes that the remaining output phase is connected to input phase D. Input phases A, B and C are not connected to any output phases. As such ${}^5P_3 = 20$ different permutations exist. Out of these, the "1" switching state can have a further three different combinations. That is, every switching permutation has ${}^3C_2 \times {}^1C_1 = 3$ combinations. Therefore, there are $20 \times 3 = 60$ switching combinations in all. These vectors have variable amplitude, but constant frequency in space implying output voltage amplitudes depend on the selected input line voltages. Moreover, the phase angle of the output voltage space vector is not dependent on the phase angle of the input voltage space vector. The 60 combinations in this group determine the six prefixed positions of the output voltage space vectors that are not dependent on α_i. A similar condition is also valid for current vectors in which the same 60 combinations determine the 10 prefixed positions of the input current space vectors that are not dependent on α_o.
3. $\{p, q, r, s, t\} \in \{0, 0, 1, 1, 1\} | p \neq q \neq r \neq s \neq t$: All output phases are connected to three out of five different input phases and the two other input phases are in open condition. As such ${}^5C_3 = 10$ different combinations exist. Next, the "1" element switching combination can have further six different permutations for every switching combination with ${}^3P_3 = 6$ permutations, yielding $10 \times 6 = 60$ total switching combinations. These vectors are variable in amplitude and frequency in space, which implies that the output voltage amplitudes are dependent on the selected input line voltages. In this case, the phase angle of the output voltage space vector depends on the phase angle of the input voltage space vector. Also of note, the 60 combinations in this category do not determine any prefixed positions of the output voltage space vector. The locus of the output voltage space vectors form ellipses in different orientations in space as α_i varies. A similar condition is also valid for current vectors as α_0 varies. For the space vector modulation technique proposed in the paper, these switching states are not used since the phase angle of both the input and output vectors cannot be independently controlled.

In general, the active switching vectors used for space vector PWM of a five-to-three-phase matrix converter are as follows:

Category 1: {3, 0, 0, 0, 0} consisting of five vectors
Category 2: {2, 1, 0, 0, 0} consisting of 60 vectors
Category 3: {1, 1, 1, 0, 0} consisting of 60 vectors

B. SVPWM Control Strategy

The general topology of a five-to-three-phase matrix converter is shown in Figure 15.24. It consists of 15 bidirectional switches that allow any output phase to be connected to any input phase. As the converter is supplied by a voltage source, the input phases should never be short circuited, and because of the presence of inductive loads the output phases should not be interrupted. With these constraints, there are 125 permitted switching combinations as described in the earlier section. However, only 65 of the active switching vectors can be considered for this matrix converter modulation technique, which are divided in three groups:

Group 1: {3, 0, 0, 0, 0} consists of five vectors called zero vectors in which all the output phases are connected to the same input phase.

Group 2: {2, 0, 1, 0, 0} consists of 30 vectors in which two output phases are connected to the same input phase and the other output phase is connected to a non-adjacent input phase. This can be seen in group 2 configuration in Figure 15.25.

Group 3: {2, 1, 0, 0, 0} consists of 30 vectors in which, like group 2, two output phases are connected to the same input phase, but differs in that the remaining output phase is connected to an adjacent input phase. This can be seen in group 3 configuration in Figure 15.25. These vectors are called medium vectors.

In the SVPWM control strategy for a five-to-three-phase matrix converter, only the switching states of categories 1 and 2 described in the earlier section are utilized. The switching states in category 3 are not used since the corresponding SSVs are rotating with time. Category 2 consists of medium and large vectors as defined earlier.

The proposed space vector control strategy will use group 2 configuration of switching states along with the zero vectors of group 1, consisting of a total number of space vectors of 30 + 5 = 35. The input current SSVs and output voltage SSVs of each switching state are shown in Figures 15.26 and 15.27, respectively.

For each combination, the input and output line voltages can be expressed in terms of space vectors as

$$\begin{aligned}\overrightarrow{V_{\text{i-LL}}} &= \frac{2}{5}\left(V_{\text{ac}} + V_{\text{bd}}.e^{j\frac{2\pi}{5}} + V_{\text{ce}}.e^{j\frac{4\pi}{5}} + V_{\text{da}}.e^{j\frac{6\pi}{5}} + V_{\text{eb}}.e^{j\frac{8\pi}{5}}\right)\\ &= \frac{2}{5}\times 2\cos\left(\frac{\pi}{10}\right)\ V_{\text{i}}.\ e^{j\alpha_{\text{i}}} = \frac{1}{5}\sqrt{10+2\sqrt{5}}V_{\text{i}}.\ e^{j\alpha_{\text{i}}}\\ &= 0.7608\ V_{\text{i}}.\ e^{j\alpha_{\text{i}}}\end{aligned} \tag{15.53}$$

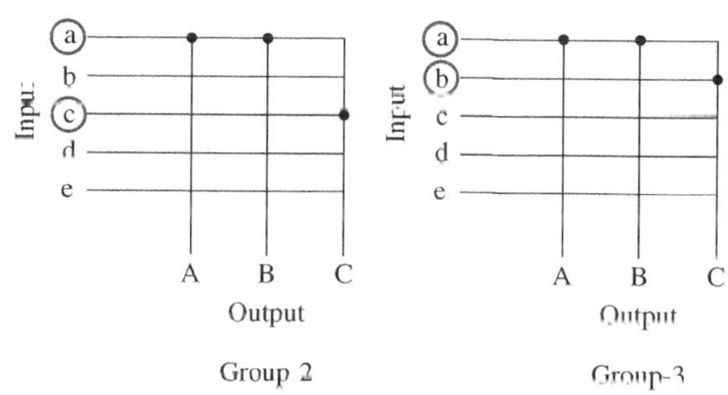

Figure 15.25 Sample configuration of groups 2 and 3 [33]

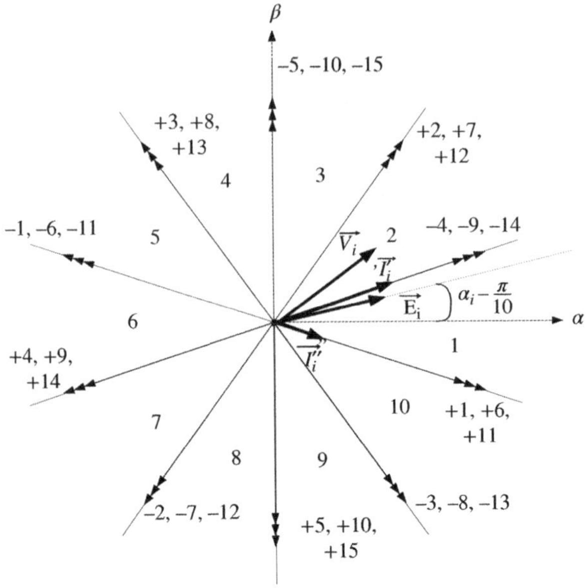

Figure 15.26 Input current space vectors corresponding to the permitted switching combinations for group 2: {2, 0, 1, 0, 0} [33]

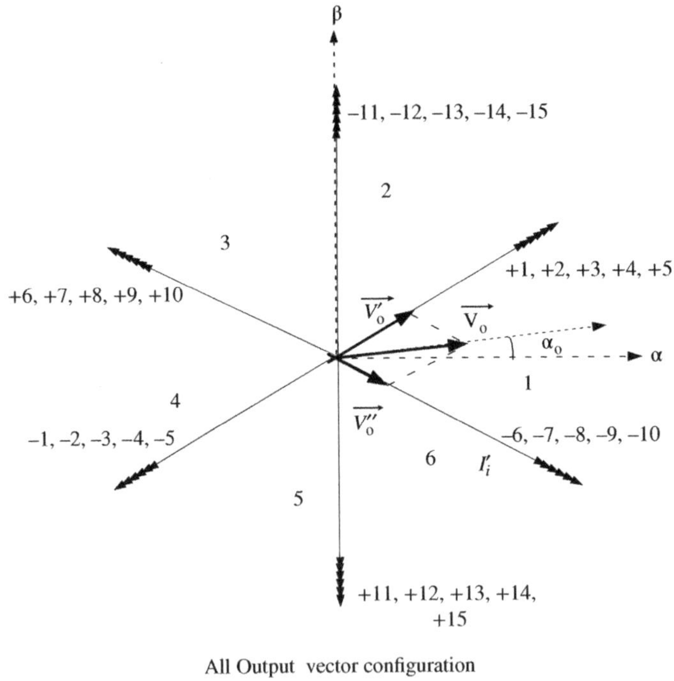

Figure 15.27 Output voltage space vectors corresponding to the permitted switching combinations for group 2: {2, 0, 1, 0, 0} [33]

$$\overrightarrow{V_{\text{o-LL}}} = \frac{2}{3}\left(V_{\text{AB}} + V_{\text{BC}}.e^{j\frac{2\pi}{3}} + V_{\text{CA}}.e^{j\frac{4\pi}{3}}\right)$$
$$= \frac{2}{3} \times 2\cos\left(\frac{\pi}{6}\right) V_{\text{o}}.e^{j\alpha_{\text{o}}} = \frac{2}{\sqrt{3}} V_{\text{o}}.e^{j\alpha_{\text{o}}} \tag{15.54}$$

where $V_{\text{i-LL}}$ and $V_{\text{o-LL}}$ are the input and output line-to-line voltages, respectively. V_{i} and V_{o} are input and output lines to neutral voltages, respectively.

In the same way, space vectors of the input and output line currents are defined as

$$\vec{I_{\text{i}}} = \frac{2}{5}\begin{pmatrix} I_{\text{a}} + I_{\text{b}}.e^{j\frac{2\pi}{5}} + I_{\text{c}}.e^{j\frac{4\pi}{5}} \\ +I_{\text{d}}.e^{j\frac{6\pi}{5}} + I_{\text{e}}.e^{j\frac{8\pi}{5}} \end{pmatrix} = I_{\text{i}}.e^{j\beta_{\text{i}}} \tag{15.55}$$

$$\vec{I_{\text{o}}} = \frac{2}{3}\left(I_{\text{A}} + I_{\text{B}}.e^{j\frac{2\pi}{3}} + I_{\text{C}}.e^{j\frac{4\pi}{3}}\right) = I_{\text{o}}e^{j\beta_{\text{O}}} \tag{15.56}$$

α_{i} and α_{o} are the phase angles of the input and the output voltage vector, respectively, whereas β_{i} and β_{o} are phase angles of the input and output current vector, respectively.

The SVPWM algorithm firstly selects the appropriate switching states and secondly calculates the duty cycle for each selected switching state. During one switching period T_{s}, the switching states whose SSVs are adjacent to the desired output voltage (input current) vector should be selected, and the zero switching states are then applied to complete the switching period to provide the maximum output to input voltage transfer ratio.

The aim of the proposed space vector control strategy is to generate the desired output voltage vector with the constraint of a unity input power factor. For this purpose, let $\vec{V_{\text{o}}}$ be the desired output line voltage space vector and $\vec{V_{\text{i}}}$ be the input line voltage space vector at a given time. The input line to neutral voltage vector $\vec{E_{\text{i}}}$ is defined by

$$\vec{E_{\text{i}}} = \frac{1}{2\cos\left(\frac{\pi}{10}\right)} \vec{V_{\text{i}}}.e^{-j\frac{\pi}{10}} \tag{15.57}$$

In order to obtain a unity input power factor, the direction of the input current space vector $\vec{I_{\text{i}}}$ has to be the same as that of $\vec{E_{\text{i}}}$. Assume that $\vec{V_{\text{o}}}$ and $\vec{I_{\text{i}}}$ are in sector 1 of the 6 possible sectors in output side and 10 possible sectors in the input side. In Figure 15.27, the vector configurations $\vec{V'_{\text{o}}}$ and $\vec{V''_{\text{o}}}$ represent the components of $\vec{V_{\text{o}}}$ along the two adjacent vector directions. Similarly, $\vec{I_{\text{i}}}$ is resolved into components $\vec{I'_{\text{i}}}$ and $\vec{I''_{\text{i}}}$ along the two adjacent vector directions. The possible switching states that can be utilized to synthesize the resolved voltage and current components (assuming both input and output vectors are in sector 1) are as follows:

$$\vec{V'_{\text{o}}} : +1, +2, +3, +4, +5$$

$$\vec{V''_{\text{o}}} : -6, -7, -8, -9, -10$$

$$\vec{I'_{\text{i}}} : -4, -9, -14$$

$$\vec{I''_{\text{i}}} : +1, +6, +11$$

The output voltage and input current vectors can be synthesized simultaneously by selecting the common switching states of the output voltage components and input current components, those being +1, −4, 6 and +9. Two switching states with the same number but opposite signs have the corresponding voltage or current space vectors in opposite directions, and hence only one SSV should be used. SSVs with the positive signs are used to calculate the duty cycle of the switching state. If the duty cycle is

Table 15.4 Space vector choice for SVPWM in different sectors

Sector no. of I_i	Sector no. of V_o		
	1 or 4	2 or 5	3 or 6
1 or 6	±1, ±4, ±6, ±9	±1, ±4, ±11, ±14	±6, ±9, ±11, ±14
2 or 7	±2, ±4, ±7, ±9	±2, ±4, ±12, ±14	±7, ±9, ±12, ±14
3 or 8	±2, ±5, ±7, ±10	±2, ±5, ±12, ±15	±7, ±10, ±12, ±15
4 or 9	±3, ±5, ±8, ±10	±3, ±5, ±13, ±15	±8, ±10, ±13, ±15
5 or 10	±1, ±3, ±6, ±8	±1, ±3, ±11, ±13	±6, ±8, ±11, ±13

positive, the switching state with a positive sign is selected, otherwise the switching state with a negative sign is selected. The selection for switching states for implementing SVPWM can also be explained as follows:

Due to the small variation in input voltage during the switching cycle period, the desired $\vec{V}'_o$ can be approximated by utilizing two switching configurations corresponding to five space vectors in the same direction of $\vec{V}'_o$ and one zero voltage configuration.

Among the five possible switching configurations, the two SSVs with the highest voltage values corresponding to large vectors with the same sense of $\vec{V}'_o$ are chosen. In the same way, two different switching configurations and one zero voltage configuration are used to define $\vec{V}''_o$. With reference to the example shown in Figures 15.26 and 15.27, the input voltage $\vec{V}_i$ has a phase angle of $0 \le \alpha_i \le \frac{\pi}{5}$. In this case, the line voltages V_{ac} and $-V_{da}$ assume the highest values. Then, according to the switching table of large vectors, the configuration used to obtain $\vec{V}'_o$ are +1 and −4 and for $\vec{V}''_o$ are −6 and +9. These four configurations can also be utilized to determine the input current vector direction, as shown in Figure 15.26. These configurations are associated with the vector directions adjacent to the input current vector position. There are 60 switching combinations for the different sector combinations. These vector combinations are shown in Table 15.4.

A balanced system of sinusoidal non-adjacent line voltages can be expressed as

$$
\begin{aligned}
V_{ac} &= 0.7608 \times |\vec{V}_i| \cos(\alpha_i) \\
V_{bd} &= 0.7608 \times |\vec{V}_i| \cos\left(\alpha_i - \frac{2\pi}{5}\right) \\
V_{ce} &= 0.7608 \times |\vec{V}_i| \cos\left(\alpha_i - \frac{4\pi}{5}\right) \\
V_{da} &= 0.7608 \times |\vec{V}_i| \cos\left(\alpha_i + \frac{4\pi}{5}\right) \\
V_{eb} &= 0.7608 \times |\vec{V}_i| \cos\left(\alpha_i + \frac{2\pi}{5}\right)
\end{aligned}
\tag{15.58}
$$

Applying the space vector modulation technique, the on-time ratio δ of each configuration can be obtained by solving two systems of algebraic equations.

In particular, utilizing configurations +1, −4 to generate $\vec{V}'_o$ and to set the input current vector direction, one can write

$$\delta_1^+ \times V_{ac} - \delta_4^- \times V_{da} = V'_o = \frac{3}{5}|\overrightarrow{V_o}|.\left(\frac{2}{\sqrt{3}}\right).\sin\left(\frac{\pi}{6} + \alpha_o\right) \tag{15.59}$$

$$\delta_1^+ P.i_A = I'_i = |\vec{I'_i}|P.\sin\left[\frac{\pi}{10} - \left(\alpha_i - \frac{\pi}{10}\right)\right]$$
$$\delta_4^- P.i_A = I''_i = |\vec{I'_i}|P.\sin\left[\frac{\pi}{10} + \left(\alpha_i - \frac{\pi}{10}\right)\right] \quad (15.60)$$

where $P = \frac{2}{5} \times 2\cos\left(\frac{\pi}{10}\right) = \frac{2}{5}\left[\frac{\sqrt{2(5+\sqrt{5})}}{2}\right] = 0.7608$

Equation (15.59) can be derived from Equation (15.58) as follows:

$$\delta_1^+ \times 0.7608.\cos(\alpha_i) - \delta_4^- \times 0.7608.\cos\left(\alpha_i + \frac{4\pi}{5}\right) = \frac{3}{5}\frac{|\vec{V_o}|}{|\vec{V_i}|}.\left(\frac{2}{\sqrt{3}}\right).\sin\left(\frac{\pi}{6} + \alpha_o\right)$$

$$Or,\ \delta_1^+ \cos(\alpha_i) - \delta_4^- \cos\left(\alpha_i + \frac{4\pi}{5}\right) = q \times 0.9106 \times \sin\left(\frac{\pi}{6} + \alpha_o\right) \quad (15.61)$$

where $q = \frac{|\vec{V_o}|}{|\vec{V_i}|}$ = modulation index or voltage transfer ratio.

Combining Equations (15.60) and (15.61) gives

$$\delta_1^+ \cos\left(\alpha_i - \frac{\pi}{2}\right) - \delta_4^- \cos\left(\alpha_i + \frac{3\pi}{10}\right) = 0 \quad (15.62)$$

The solution of the transcendental Equations (15.61) and (15.62) gives

$$\delta_1^+ = q \times 1.549 \times \sin\left(\frac{\pi}{6} + \alpha_o\right).\cos\left(\frac{3\pi}{10} + \alpha_i\right) \quad (15.63)$$

$$\delta_4^- = q \times 1.549 \times \sin\left(\frac{\pi}{6} + \alpha_o\right).\cos\left(\alpha_i - \frac{\pi}{2}\right) \quad (15.64)$$

With the same procedure, utilizing the configurations −6, +9 to generate $\vec{V''_o}$ and to set the input current vector direction yields

$$\delta_6^- = q \times 1.549 \times \sin\left(\frac{\pi}{6} - \alpha_o\right).\cos\left(\frac{3\pi}{10} + \alpha_i\right) \quad (15.65)$$

$$\delta_9^+ = q \times 1.549 \times \sin\left(\frac{\pi}{6} - \alpha_o\right).\cos\left(\alpha_i - \frac{\pi}{2}\right) \quad (15.66)$$

The results obtained are valid for $-\pi/6 \leq \alpha_o \leq \pi/6$ and $0 \leq \alpha_i \leq \pi/5$.

Applying a similar procedure for the other possible pairs of angular sectors, the required switching configurations and the on-time ratio of each configuration can be determined.

Note that the values of the on-time ratios should be greater than zero, as required for the feasibility of the control strategy. Furthermore, the sum of the ratios must be lower than unity. By adding Equation (15.63) to Equation (15.66), with the constraint

$$\delta_1^+ + \delta_4^- + \delta_6^- + \delta_9^+ \leq 1$$

the maximum value of the voltage transfer ratio can be determined as $q = 1.0444$ or 104.44%.

C. Generalized Maximum Output of Three-to-n-Phase Matrix Converter

The generalization for the maximum possible output of the space vector PWM algorithm for an n-phase input to three-phase output matrix converter can be calculated. The input contain $2n$ sectors and the output consists of six sectors for the n-phase input and three-phase output configuration of a matrix converter when represented using a space vector diagram. After detailed trigonometric analysis, the on-time duty ratios are obtained for the period $-\pi/6 \le \alpha_o \le \pi/6$ and $0 \le \alpha_i \le \pi/n$ as

$$\delta_{\mathrm{I}}^{+} = q \times k \times \sin\left(\frac{\pi}{6} + \alpha_o\right) . \cos\left(\frac{(n-2)\pi}{2n} + \alpha_i\right) \tag{15.67}$$

$$\delta_{\mathrm{II}}^{-} = q \times k \times \sin\left(\frac{\pi}{6} + \alpha_o\right) . \cos\left(\alpha_i - \frac{\pi}{2}\right) \tag{15.68}$$

$$\delta_{\mathrm{III}}^{-} = q \times k \times \sin\left(\frac{\pi}{6} - \alpha_o\right) . \cos\left(\frac{(n-2)\pi}{2n} + \alpha_i\right) \tag{15.69}$$

$$\delta_{\mathrm{IV}}^{+} = q \times k \times \sin\left(\frac{\pi}{6} - \alpha_o\right) . \cos\left(\alpha_i - \frac{\pi}{2}\right) \tag{15.70}$$

where $k = \frac{\sqrt{3}}{2\times\sin\frac{\pi}{n}\times\cos\frac{\pi}{2n}}$.

Applying a similar procedure for the other possible pairs of angular sectors, the required switching configurations and the on-time ratio of different sector configurations can be determined. By adding Equation (15.67) to Equation (15.70), with the constraints as in the earlier section, the sum of the on-time ratios for the general case becomes

$$\delta_{\mathrm{I}}^{+} + \delta_{\mathrm{II}}^{-} + \delta_{\mathrm{III}}^{-} + \delta_{\mathrm{IV}}^{+} \le 1 \tag{15.71}$$

The maximum value of the voltage transfer ratio for n-to-three-phase matrix converter can be determined as

$$q = \frac{2 * \cos^2\left(\frac{\pi}{2n}\right)}{\sqrt{3}} \tag{15.72}$$

The maximum modulation indexes for different configurations of an n-to-three-phase matrix converter are displayed in Table 15.5.

Table 15.5 Maximum modulation index formulation

Matrix converter configuration ($n \times 3$)	Maximum output to input voltage formula	Maximum modulation index (%)
3×3	$\frac{2 * \cos^2\left(\frac{\pi}{6}\right)}{\sqrt{3}}$	86.66
5×3	$\frac{2 * \cos^2\left(\frac{\pi}{10}\right)}{\sqrt{3}}$	104.44
6×3	$\frac{2 * \cos^2\left(\frac{\pi}{12}\right)}{\sqrt{3}}$	107.74
7×3	$\frac{2 * \cos^2\left(\frac{\pi}{14}\right)}{\sqrt{3}}$	109.76
9×3	$\frac{2 * \cos^2\left(\frac{\pi}{18}\right)}{\sqrt{3}}$	111.99
11×3	$\frac{2 * \cos^2\left(\frac{\pi}{22}\right)}{\sqrt{3}}$	113.13

It is observed that the gain in the output voltage increases with an increase in the number of input phases.

15.5 Sample Results

A Matlab/Simulink model can be used for the matrix converter using both carrier-based and space vector control to verify the operation [31, 33]. The operation of a five-to-three-phase matrix converter is shown for different input voltages and frequencies (considering a variable speed generator) for a fixed output voltage and frequency applicable for grid application. The output duty ratios and the output voltage spectrum for the carrier-based PWM technique are shown in Figures 15.28 and 15.29, respectively.

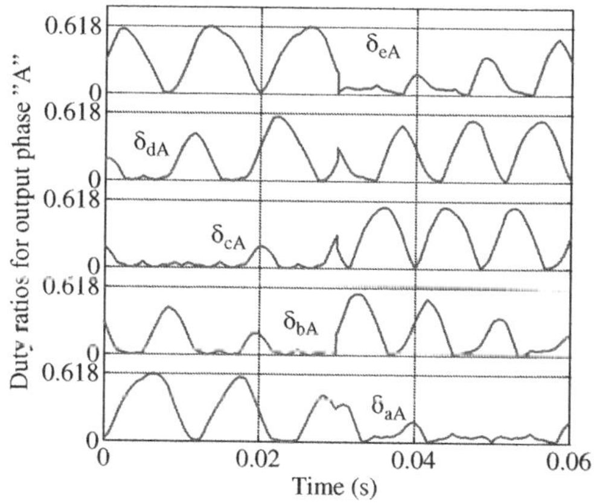

Figure 15.28 Output duty ratios [31]

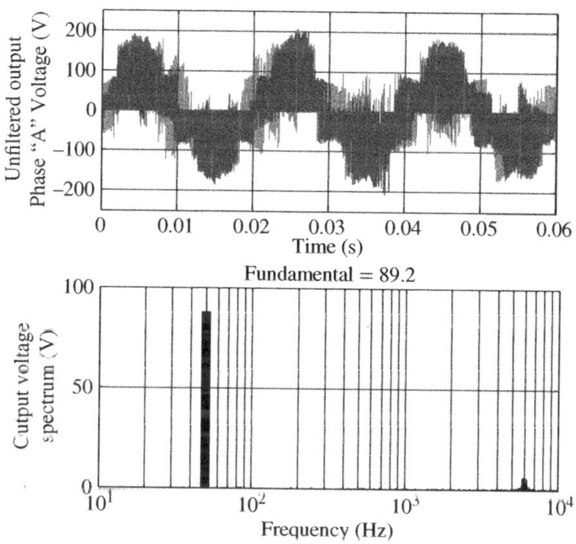

Figure 15.29 Spectrum of output voltages [31]

The view of the input and output voltages is shown in Figure 15.30 for this scheme. Initially, the input voltage is kept at 100 V peak with a frequency of 50 Hz to show the exact gain at the output side in the case of the SVPWM technique. The matrix converter with input phase number greater than output phase number can be operated in buck or boost mode, depending on the source-side voltage magnitude. The switching frequency of the devices is held at 6 kHz for simulation purposes. The maximum output

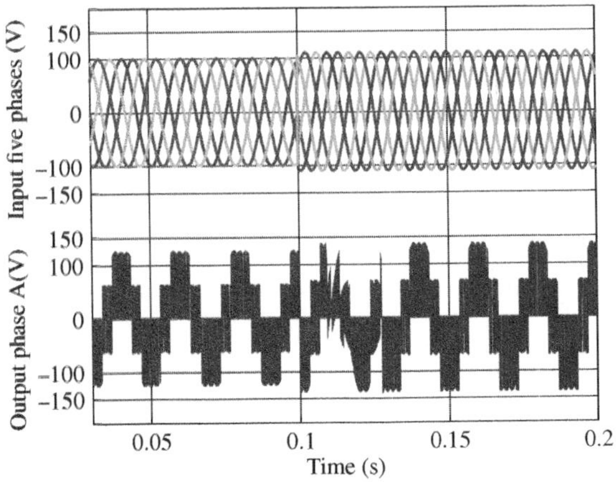

Figure 15.30 Input five-phase voltages with output phase "A" voltage [33]

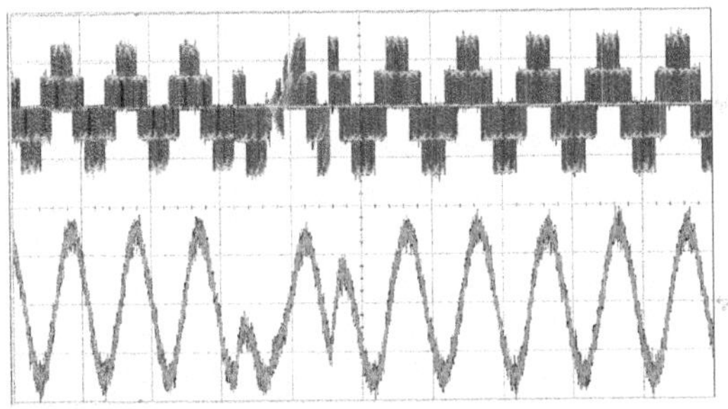

Figure 15.31 Output phase voltage (100 V, 20 ms/div) and phase current: (5 A, 20 ms/div) [33]

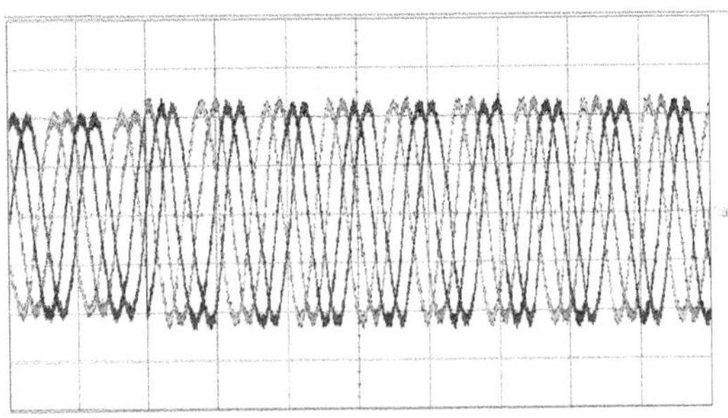

Figure 15.32 Input side five-phase voltages (50 V, 20 ms/div) (four voltages shown due to oscilloscope channel restriction)

voltage for a 100 V input is 104.4 V. The input voltage is increased by 110 V with 55 Hz frequency at 0.1 s, keeping the output voltage and frequency constant. The RL load connected to the matrix converter has values $R = 10\,\Omega$ and $L = 30$ mH. The input voltages along with the output "A" phase voltage are depicted in this figure. The experimental results are shown in Figures 15.31 and 15.32.

Acknowledgment

This chapter was made possible by an NPRP Grant No. 04-152-2-053 from the Qatar National Research Fund (a member of Qatar Foundation). The statements made herein are solely the responsibility of the authors.

References

1. Empringham, L., Kolar, J.W., Rodriguez, J. *et al.* (2013) Technological issues and industrial application of matrix converters: a review. *IEEE Transaction Industrial Electronics*, **60** (10), 4260–4271.
2. Clare, J. and Wheeler, P. (2005) New technology: matrix converters. *IEEE IES Newsletter*, **52** (1), 10–12.
3. Szczepanik, J. and Sienko, T. (2007) A new concept of application of multiphase matrix converter in power systems. Proceedings EUROCON'07, September 9–12, 2007, pp. 1535–1539.
4. Szczepanik, J. and Sienko, T. (2008) New control strategy for multiphase matrix converter. Proceedings International Conference on System Engineering, Las Vegas, NV, August 19–21, 2008, pp. 121–1267.
5. Szczepanik, J. and Sienko, T. (2009) Control scheme for a multiphase matrix converter. Proceedings EUROCON'09, Saint Petersburg, Russia, May 18–23, 2009, pp. 545–551.
6. Helle, L., Larsen, K.B., Jorgensen, A.H. *et al.* (2004) Evaluation of modulation schemes for three-phase to three-phase matrix converters. *IEEE Transactions on Industrial Electronics*, **51** (1), 158–171.
7. Apap, M., Clare, J.C., Wheeler, P.W., and Bradley, K.J. (2003) Analysis and comparison of AC-AC matrix converter control strategies. Proceedings IEEE Power Electronics Specialists Conference, Vol. 3, pp. 1287–1292.
8. Pena, R., Cardenas, R., Reyes, E. *et al.* (2009) A topology for multiple generation system with doubly fed induction machines and indirect matrix converter. *IEEE Transactions on Industrial Electronics*, **56** (10), 4181–4193.
9. Wang, B. and Venkataramanan, G. (2006) A carrier-based PWM algorithm for indirect matrix converters. Proceedings IEEE-PESC 2006, pp. 2780–2787.
10. Iqbal, A., Ahmed, S.K., and Abu-Rub, H. (2012) Space vector PWM technique for a three to five phase matrix converter. *IEEE Transactions on Industry Applications*, **48** (2), 697–707.
11. Satish, T., Mohapatra, K.K., and Mohan, N. (2007) Carrier-based control of matrix converter in linear and over-modulation modes. Proceedings 2007 Summer Computer Simulation Conference, San Diago, CA, pp. 98–105.
12. Ahmed, S.K.M., Iqbal, A., Abu-Rub, H., and Khan, M.R. (2010) Carrier-based PWM technique of a novel three-to-seven-phase matrix converter. International Conference on Electrical Machiones (ICEM'10), Rome, Italy, September 3–6, 2010, CD-ROM Paper No. RF-004944.
13. Ahmed, S.M., Iqbal, A., Abu-Rub, H. *et al.* (2011) Simple carrier-based PWM technique for a three to nine phase matrix converter. *IEEE Transactions on Industrial Electronics*, **58** (11), 5014–5023.
14. Iqbal, A., Ahmed, S.K.M., Abu-Rub, H., and Khan, M.R. (2010) Carrier based PWM technique for a novel three-to-five phase matrix converter. Proceedings European PCIM, Nuremberg, Germany, May 2–6, 2010, CD ROM Paper No. 173.
15. Lee, K. B. and Blaabjerg, F. (2008) Sensorless DTC SVM for induction motor driven by a matrix converter using parameter estimation strategy. *IEEE Transactions on Industrial Electronics*, **55** (2), 512–521.
16. Abu-Rub, H., Iqbal, A., and Guzinski, J. (2012) *High Performance Control of AC Drives with Matlab/Simulink Models*, John Wiley & Sons, Ltd, ISBN: 978-0-470-97829-0.
17. Yoon, Y.-D. and Sul, S.-K. (2006) Carrier-based modulation technique for matrix converter. *IEEE Transactions on Power Electronics*, **21** (6), 1691–1703.
18. Loh, P.C., Rong, R., Blaabjerg, F., and Wang, P. (2009) Digital carrier modulation and sampling issues of matrix converter. *IEEE Transactions on Power Electronics*, **24** (7), 1690–1700.
19. Li, Y. and Choi, N.-S. (2009) Carrier based pulse width modulation for matrix converter. Proceedings Applied Power Electronics Conference Exposition, pp. 1709–1715.

20. Li, Y., Choi, N.-S., Han, B.-M. *et al.* (2008) Direct duty ratio pulse width modulation method for matrix converter. *International Journal of Control Automation and System*, **6** (5), 660–669.
21. Ahmed, S.K.M., Iqbal, A., and Abu-Rub, H. (2011) Generalized duty ratio based pulse width modulation technique for a three-to-*k* phase matrix converter. *IEEE Transactions on Industrial Electronics*, **58** (9), 3925–3937.
22. Iqbal, A., Levi, E., Jones, M., and Vukosavic, S.N. (2006) Generalised sinusoidal PWM with harmonic injection for multi-phase VSIs. IEEE 37th Power Electronics Specialists Conference (PESC), Jeju, Korea, CD-ROM Paper No. ThB2-3.0, pp. 2871–2877.
23. Dujic, D., Jones, M., and Levi, E. (2009) Generalised space vector PWM for sinusoidal output voltage generation with multiphase voltage source inverter. *International Journal of Industrial Electronics and Drives*, **1** (1), 1–13.
24. Huber, L. and Borojevic, D. (1991) Space vector modulator for forced commutated cycloconverters. Proceedings IEEE IAS Annual Meeting, pp. 1032–1041.
25. Huber, L. and Borojevic, D. (1995) Space vector modulated three-phase to three-phase matrix converter with input power factor correction. *IEEE Transactions on Industry Applications*, **31** (6), 1234–1246.
26. Igney, J. and Braun, M. (2005) Space vector modulation strategy for conventional and indirect matrix converters. Proceedings EPE Conference Dresden, Germany, CD-ROM Paper.
27. Casadei, D., Serra, G., Tani, A., and Zarri, L. (2002) Matrix converter modulation strategies: a new general approach based on space vector representation of the switch state. *IEEE Transactions on Industrial Electronics*, **49** (2), 370–381.
28. Bradaschia, F., Cavalcanti, M.C., Neves, F.A.S., and de Souza, H.E.P. (2009) A modulation technique to reduce switching losses in matrix converter. *IEEE Transactions on Industrial Electronics*, **56** (4), 1186–1195.
29. Casadei, D., Serra, G., Tani, A., and Zarri, L. (2009) Optimal use of zero vectors for minimizing the output current distortion in matrix converters. *IEEE Transactions on Industrial Electronics*, **56** (2), 326–336.
30. Ahmed, S.K., Iqbal, A., Abu-Rub, H., and Khan, M.R. (2010) Space vector PWM technique for a novel 3 to 5 phase matrix converter. IEEE Energy Conversion Congress and Exposition (ECCE), Atlanta, GA, pp. 1875–1880.
31. Ahmed, S.M., Abu-Rub, H., and Iqbal, A. (2012) Pulse width modulation control of a direct AC-AC power converter with five-phase input and three-phase output. *International Journal of Automation and Power Engineering*, **1** (8), 186–192.
32. Saleh, M., Iqbal, M.A., Ahmed, S.K.M. *et al.* (2011) Carrier based PWM technique for a three-to-six phase matrix converter for supplying six-phase two-motor drives. IEEE IECON'11, Melbourne, Australia, November 7–10, 2011, pp. 3470–3475.
33. Ahmed, S.K.M., Abu-Rub, H., and Salam, Z. (2013) Space vector PWM technique for a direct five-to-three-phase matrix converter. IEEE IECON'13, Vienna, Austria, November 10–13, 2013.

16

Boost Preregulators for Power Factor Correction in Single-Phase Rectifiers

Hadi Y. Kanaan[1] and Kamal Al-Haddad[2]
[1]*Department of Electrical and Mechanical Engineering, Saint-Joseph University – ESIB, Mar Roukoz, Lebanon*
[2]*Department of Electrical Engineering, Ecole de Technologie Supérieure, Montreal, Canada*

16.1 Introduction

In most industrial and domestic applications, the electrical source is a single-phase alternative application. In this perspective, and more especially in the power supply and telecommunications fields, single-phase rectifiers represent nowadays one of the major equipments used for electrical energy conversion. Due to their inherent nonlinearity, these electronic interfaces generally cause power quality degradation, which is materialized by the creation of undesirable current and, consequently, voltage harmonics into the mains.

The presence of such harmonics may increase the losses in the power flow network and cause over-voltages in some parts of it. Furthermore, high-order harmonics may induce the electro-magnetic interference (EMI) phenomenon that could disturb the operation of electronic equipment located nearby. In order to avoid these problems and thus increase the reliability and efficiency of the power network, these power electronics devices should comply with international standards [1–8]. This can be achieved by integrating power factor correction (PFC) circuits or preregulators into the basic rectifier topology, in order to emulate a purely resistive operation and to ensure a near-unity power factor at the front end [9–11]. Single-stage high-power factor rectifiers are discussed in [12–14]. Although they ensure good performance in terms of source current total harmonic distortion (THD), DC output voltage regulation and robustness with respect to a load or mains voltage disturbance, these converters are characterized by a high number of high-switching-frequency power semiconductors, leading to a reduced efficiency, a relatively high control complexity, implementation of which generally requires gate isolation and blanking-time programming, and low robustness to control defects [15]. These drawbacks are avoided in

Power Electronics for Renewable Energy Systems, Transportation and Industrial Applications, First Edition.
Edited by Haitham Abu-Rub, Mariusz Malinowski and Kamal Al-Haddad.

the two-stage topologies where a classical four-diode bridge is connected in series with a switch-mode DC/DC high-switching-frequency converter. The diode bridge ensures the rectifying process, whereas the role of the DC/DC converter is to modulate the DC current on the bridge output so as to retrieve a practically sine wave current from the grid. Three types of two-stage rectifiers may exist. Buck [16] and some of the buck–boost [17] rectifiers generally operate in a discontinuous mode. They are therefore characterized by a high harmonic level, and thus need a high filtering effort on the AC side. Other buck–boost preregulators, such as the cascaded boost–buck [18], the Cùk [19–22], the SEPIC [22–37] and the Sheppard–Taylor topologies [38–52], have an inductive input side that ensures smooth variation of the source current, but this benefit comes at the expense of double the number of inductors and capacitors.

Boost-type PFCs are the most widely used topologies in single-phase high-power factor rectifiers. This is mainly due to the reduced number of high-frequency switches and the smooth behavior of the rectifier input current. The basic boost converter, which is discussed in Section 16.2, is the most popular in such applications [53–59]; however, its main drawback remains the control detuning phenomenon that it presents at the zero-crossings of the source voltage, which increases the source current distortion and limits the performance of this type of PFC. To overcome this problem, a two-switch asymmetric half-bridge converter [60–62] is introduced and studied in Section 16.3. In addition, in order to reduce the power losses when operating at high-switching frequencies, the interleaved dual-boost converter could be employed [63–77]. This highly efficient topology is presented in Section 16.4.

On the other hand, reliable numerical models for power converters are increasingly required for both academic and industrial purposes. The aim is to elaborate highly precise mathematical representations of widely used converters. The usefulness of such virtual models could be emphasized in several aspects. Specifically, they allow, first, a systematic design of well-tuned control systems that improve the time response of the converter; second, a pre-evaluation of the operating regime, as well as the analysis of the static and dynamic performances of the converter; third, a better selection of the system parameters and components; fourth, fast simulations, which makes these models suitable for real-time applications such as the hardware-in-the-loop (HIL) technique widely used in the industry to test hardware controllers before being integrated into the real plants [78]; and fifth, the avoidance of the elaboration of a real laboratory prototype, which can be costly and consuming in terms of time and effort. Several modeling techniques for switch-mode converters exist in the literature [79–95]; however, the state-space averaging method appears as the simplest and the most adapted method to the control design process.

In this chapter, a single-switch basic boost, a two-switch asymmetric half-bridge boost and an interleaved dual-boost topology are studied, respectively. The averaged model for the first two topologies has been developed, and the same methodology can be systematically applied for the third topology. On the basis of the obtained models, control algorithms were then designed and implemented. The reported simulation or experimental results show the effectiveness of these converters connected to their respective control systems and validate the theoretical issues made out through the chapter.

16.2 Basic Boost PFC

16.2.1 Converter's Topology and Averaged Model

The converter's topology is depicted in Figure 16.1. A single-phase diode bridge is connected in cascade with a typical boost DC/DC converter. The converter operates in a continuous current mode (CCM) during almost the entire mains cycle (except in very limited local regions surrounding the zero-crossings of the source voltage), and therefore slow-recovery diodes could be used in the AC-side stage of the rectifier. The controlled switch Q is generally a MOSFET and the boost-diode D should be a fast-recovery diode. Switch Q is controlled using the pulse-width modulation (PWM) technique, ensuring a fixed-switching frequency imposed by the modulation carrier. The duty cycle of Q is adjusted so that the power factor becomes unity and the DC output voltage is stabilized around a desired value with a negligible ripple.

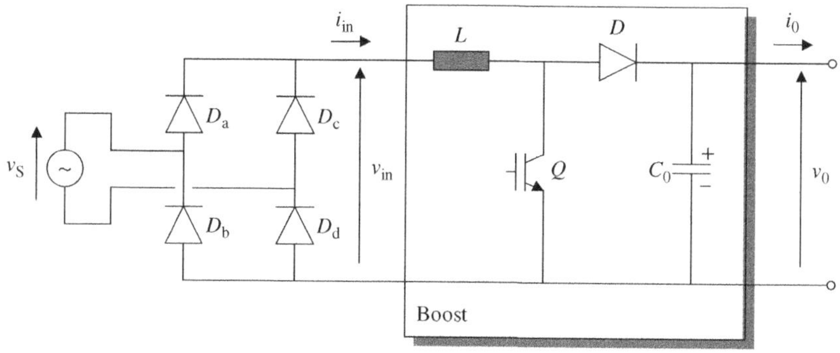

Figure 16.1 Single-phase boost PFC rectifier

In CCM, the basic boost converter has two configurations, as illustrated in Figure 16.2. The electrical equations of the converter can thus be stated as follows:

$$v_{in} = L\frac{di_{in}}{dt} + (1 - s_Q)\, v_0 \tag{16.1a}$$

$$(1 - s_Q)\, i_{in} = \frac{v_0}{R_0} + C_0\frac{dv_0}{dt} \tag{16.1b}$$

where v_{in} denotes the rectified source voltage at the DC side of the diode bridge, v_0 is the output voltage, i_{in} is the current in inductor L, R_0 is the DC load and s_Q is the switching function of Q, defined as

$$s_Q = \begin{cases} 0 & \text{when Q is turned off} \\ 1 & \text{when Q is turned on} \end{cases} . \tag{16.2}$$

The current waveform over one switching cycle, as illustrated in Figure 16.3a, shows its ability to track any reference i^*_{in}, assuming that the following two assumptions are satisfied over the entire mains cycle:

$$v_0 > v_{in} \tag{16.3}$$

and

$$\left|\frac{di_{in}}{dt}\right| > \left|\frac{di^*_{in}}{dt}\right| \tag{16.4}$$

Unfortunately, for this topology, assumption Equation (16.4) is not totally satisfied. In fact, near the zero-crossings of the AC source voltage where v_{in} has very small values, the current i_{in} grows slowly when Q is on and will be able to reach its reference only after a relatively large period of time, as described in Figure 16.3b. The tracking disability, which occurs only twice in an operation cycle, is called the control detuning phenomenon. It is characterized by the detuning angle that can be expressed as

$$\delta = 2\tan^{-1}\left(\frac{L\omega_0\,\hat{i}_s}{\hat{v}_s}\right) \tag{16.5}$$

where $\hat{v}_s$ and $\hat{i}_s$ denote the peak values of the source voltage and current, respectively and ω_0 is the main angular frequency.

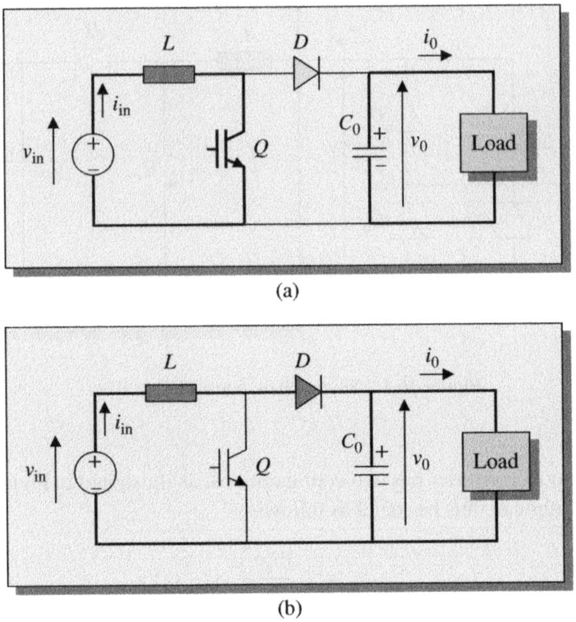

Figure 16.2 Basic boost configurations: (a) when Q is turned on and (b) when Q is switched off

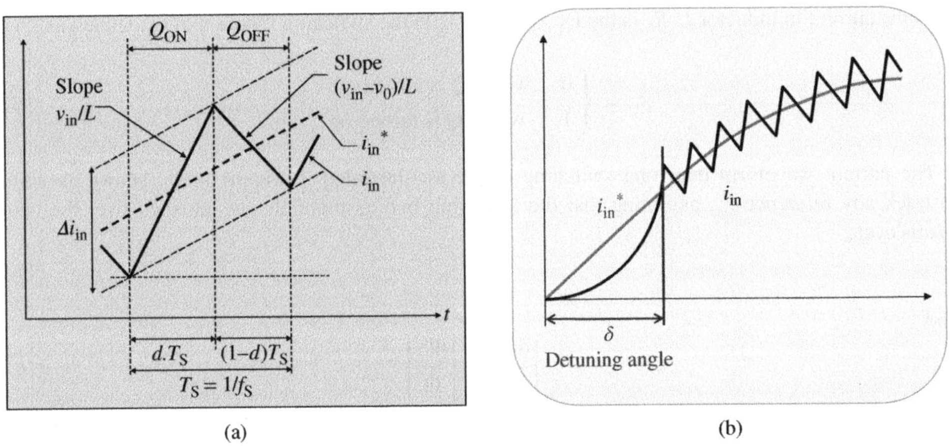

Figure 16.3 (a) Current waveform during one switching cycle and (b) the control detuning phenomenon

The application of the averaging technique to system Equation (16.1) yields the following nonlinear state representation of the converter [56, 57]:

$$v_{in} = L\frac{di_{in}}{dt} + (1-d)\,v_0 \tag{16.6a}$$

$$(1-d)\,i_{in} = \frac{v_0}{R_0} + C_0\frac{dv_0}{dt} \tag{16.6b}$$

where d represents the duty cycle of the switch Q. Equation (16.6) represents, thus, a single-input two-output system, having d as the control input and i_{in} and v_0 as the output variables. The input voltage v_{in} is considered, in a control point of view, as a disturbance signal.

16.2.2 Steady-State Analysis

In the steady-state desired regime, the input current i_{in} has a rectified sine wave shape (similar to v_{in}), whereas the output voltage v_0 is fixed at a desired value v_0^*. Replacing these expressions into Equation (16.6a) yields theoretically:

$$d^*(t) = \begin{cases} 1 - \dfrac{\hat{v}_s}{v_0^* \cos\varphi} \sin\left(\omega_0 t - \varphi\right) & \text{if } 0 \le \omega_0 t \le \pi \\ 1 + \dfrac{\hat{v}_s}{v_0^* \cos\varphi} \sin(\omega_0 t - \varphi) & \text{if } \pi \le \omega_0 t \le 2\pi \end{cases} \tag{16.7}$$

where d^* is the desired instantaneous value of the duty cycle and

$$\varphi = \tan^{-1}\left(\frac{L\omega_0 \hat{i}_s}{\hat{v}_s}\right) = \frac{\delta}{2} \tag{16.8}$$

Figure 16.4 shows clearly the control saturation phenomenon. Its angular duration is equal to ϕ. In order to reduce this time interval, the value of the inductor L must be limited. On the other hand, by using Equation (16.7) into Equation (16.6b), we obtain the following approximated expression of the output voltage ripple:

$$\left(\frac{\Delta v_0}{v_0}\right)^* \cong \frac{\hat{v}_s \hat{i}_s}{2C_0 \omega_0 v_0^{*2} \cos\varphi} \tag{16.9}$$

which is useful for the design of the output capacitor, given a maximum value for the voltage ripple.

16.2.3 Control Circuit

The control circuit associated with the converter is presented in Figure 16.5. The date signal of switch Q is delivered by a PWM generator comparing the current control signal v_{ci}, given by the current regulator, with a saw-tooth carrier. The frequency of the carrier is chosen with respect to the operating limits. A multiplier/divider block is used for the generation of the current reference waveform. It executes the following relationship [55]:

$$i_{in,ref} = \frac{R_{MD} K_{iAC} v_{in} (v_{cv} - 1.5)}{K_{rms}^2 \, V_{in}^2} \tag{16.10}$$

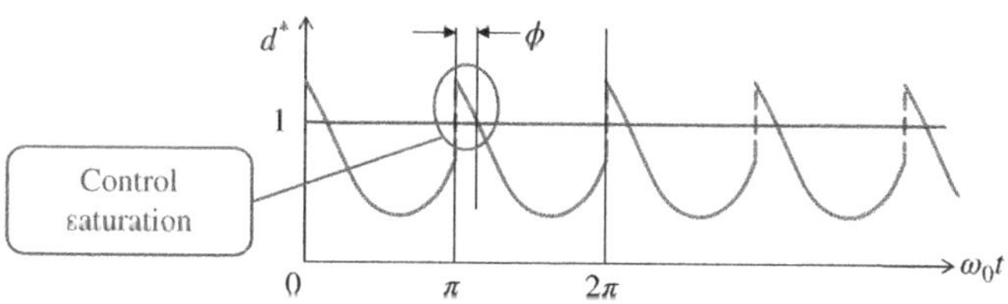

Figure 16.4 Control saturation

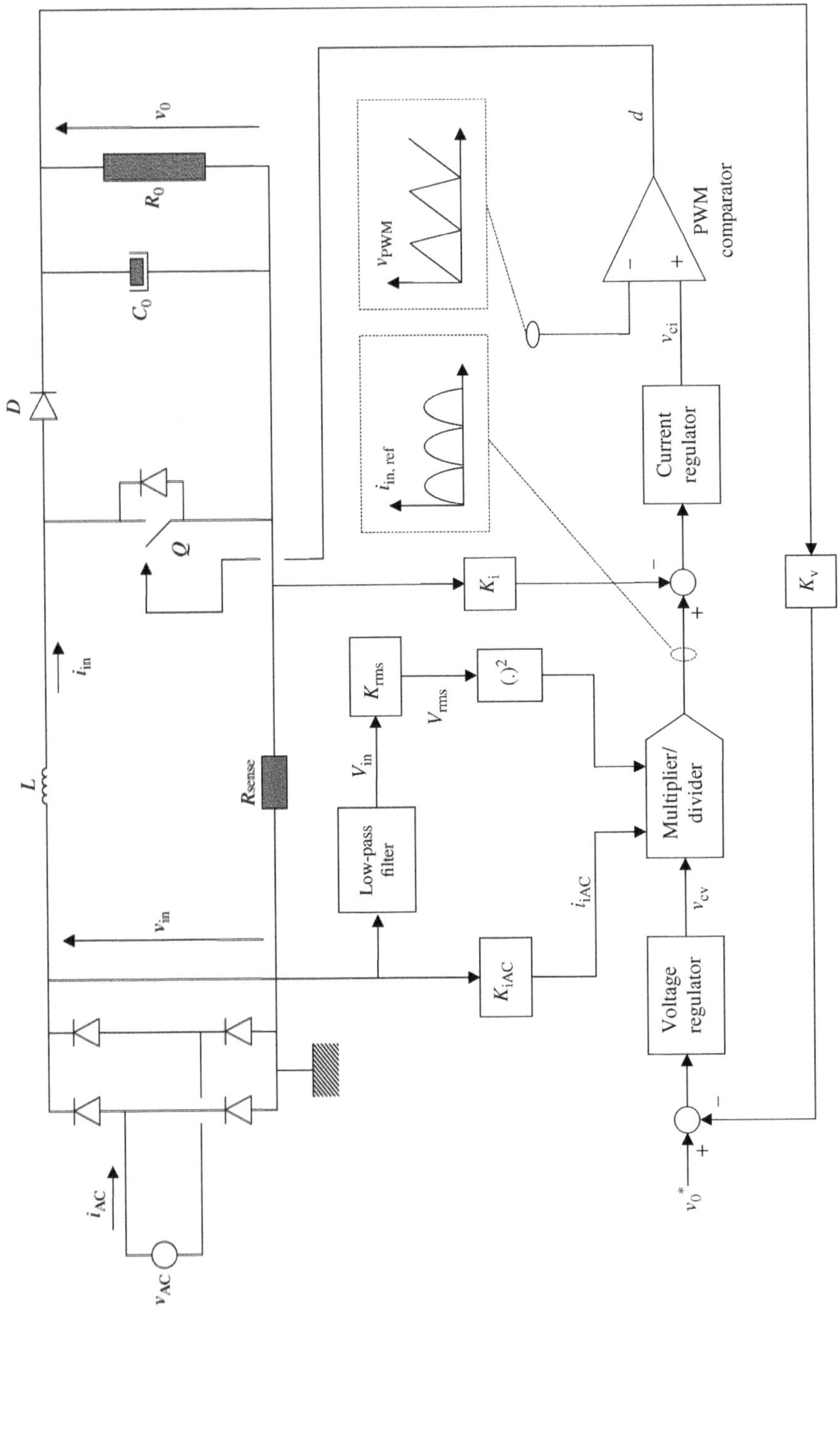

Figure 16.5 Control circuit for the basic boost preregulator [56]

v_{cv} denotes the voltage control signal and V_{in} is the mean value of the rectified input voltage v_{in}, which is obtained through a second-order low-pass filter characterized by the following transfer function:

$$F(s) = \frac{1}{\left(1 + \frac{s}{2\pi.18}\right)^2} \tag{16.11}$$

Finally, K_i and K_v are the feedback gains of the inner current loop and outer voltage loop, respectively, and R_{sense} represents the equivalent impedance of the current sensor.

16.2.4 Linear Control Design

16.2.4.1 Small-Signal Modeling

In this section, linear regulators for both the current inner loop and the outer voltage loop are designed on the basis of a small-signal model of the converter, expressed in the frequency domain. According to [56], such a model is represented by the block diagram of Figure 16.6. The corresponding transfer functions are expressed as follows:

$$G_{di}(s) = G_{di0} \cdot \frac{1 + \frac{s}{\omega_{zdi}}}{1 + \frac{2\xi_{pdi}s}{\omega_{pdi}} + \frac{s^2}{\omega_{pdi}^2}} \tag{16.12}$$

$$G_{iv}(s) = G_{iv0} \cdot \frac{1 - \frac{s}{\omega_{ziv}}}{1 + \frac{s}{\omega_{piv}}} \tag{16.13}$$

$$G_{pi}(s) = G_{pi0} \cdot \frac{1 + \frac{s}{\omega_{zpi}}}{1 + \frac{2\xi_{ppi}s}{\omega_{ppi}} + \frac{s^2}{\omega_{ppi}^2}} \tag{16.14}$$

$$G_{pv}(s) = G_{pv0} \cdot \frac{1}{1 + \frac{s}{\omega_{ppv}}} \tag{16.15}$$

where

$$G_{di0} = \frac{2\,V_0}{R_0(1-D)^2}, G_{iv0} = \frac{R_0(1-D)}{2}, G_{pi0} = \frac{1}{R_0(1-D)^2}, G_{pv0} = \frac{1}{2(1-D)}$$

$$\omega_{zdi} = 2\omega_{zpi} = \omega_{piv} = \omega_{ppv} = \frac{2}{R_0C_0}, \omega_{ziv} = \frac{R_0(1-D)^2}{L}$$

$$\omega_{pdi} = \omega_{ppi} = \frac{1-D}{\sqrt{LC_0}}, \xi_{pdi} = \xi_{ppi} = \frac{1}{2R_0(1-D)}\sqrt{\frac{L}{C_0}}$$

Note that $v_{in\sim}$, $i_{in\sim}$, $v_{0\sim}$ and $d_\sim$ represent, respectively, the small-signal components of v_{in}, i_{in}, v_0 and d, whereas V_0 and D are the respective static values of v_0 and d.

16.2.4.2 Design of the Linear Regulators

By neglecting the presence of the perturbation signal $v_{in\sim}$, the block diagram of the control system is conformed to Figure 16.7, where the PWM comparator and the multiplier/divider block are replaced by their respective small-signal gains K_{PWM} and K_{MD}. The current regulator is designed in order to (i)

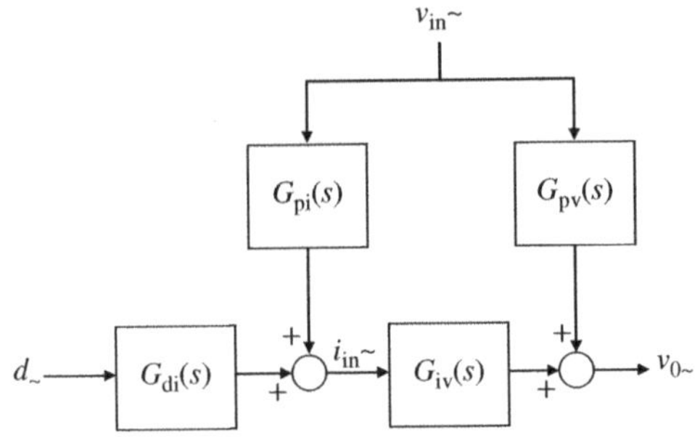

Figure 16.6 Small-signal equivalent block diagram of the converter

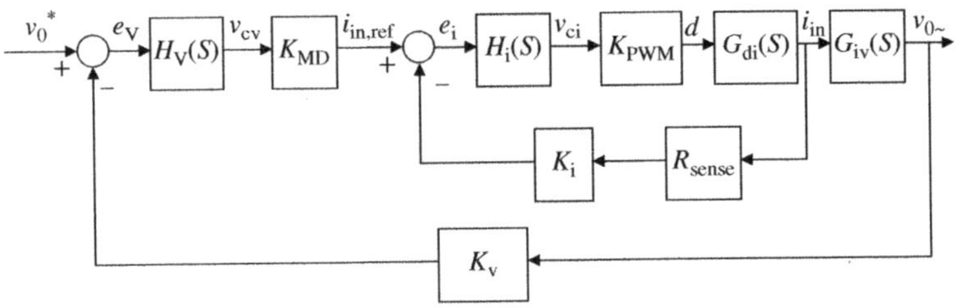

Figure 16.7 Linear control scheme for the basic boost preregulator

ensure closed-loop stability, (ii) compensate the high-frequency components in the controlled current and (iii) reduce significantly the dynamic error in the low-frequency region. Meanwhile, the voltage regulator is elaborated on the basis of an optimizing criterion: the closed-loop transfer function is set to be a second-order low-pass filter with a damping factor equal to 0.707. Under these considerations, the regulators are chosen as follows:

$$H_i(s) = K_{Hi} \cdot \frac{1 + \frac{s}{\omega_{zHi}}}{s\left(1 + \frac{s}{\omega_{pHi}}\right)} \tag{16.16}$$

$$H_v(s) = K_{Hv} \cdot \frac{1 + \frac{s}{\omega_{zHv}}}{s\left(1 + \frac{s}{\omega_{pHv}}\right)} \tag{16.17}$$

The numerical values of all the control system's parameters that were adopted for the simulations are as follows:

Rated load power $P_0 = 500\,W$
Mains voltage RMS–value $V_{AC} = 120\,V$
Desired output voltage $v_0^* = 410\,V$
Mains frequency $f_0 = 60\,Hz$
Switching frequency $f_S = 250\,kHz$
DC inductor $L = 200\,\mu H$
DC output capacitor $C_0 = 440\,\mu F$
Line equivalent impedance $L_s = 30\,\mu H$

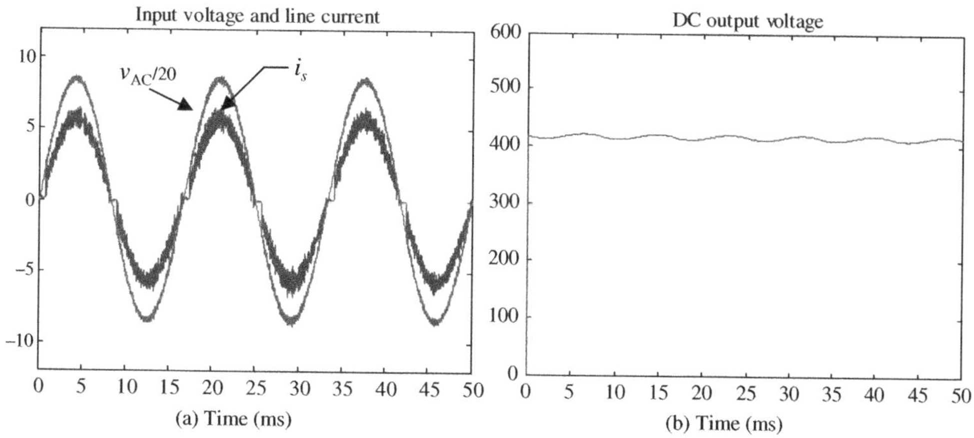

Figure 16.8 (a, b) Steady-state waveforms of the line current and the output voltage

AC input capacitor $C_s = 1\ \mu F$
PWM comparator dynamic gain $K_{PWM} = 0.181\ V^{-1}$
Multiplier/divider parameters $R_{MD} = 3300\ \Omega$, $K_{iAC} = 1.28\ \mu S$, $K_{rms} = 1/51$, $K_{MD} = 0.1$
Current sensor equivalent impedance $R_{sense} = 0.1\ \Omega$
Current loop feedback gain $K_i = 1$
Voltage loop feedback gain $K_v = 3/410$
Inner regulator parameters $K_{Hi} = 1880$ krd/s, $\omega_{zHi} = 100\pi$ krd/s, $\omega_{pHi} = 250\pi$ krd/s
Outer regulator parameters $K_{Hv} = 190$ rd/s, $\omega_{zHv} = 13.52$ rd/s, $\omega_{pHv} = 40\pi$ rd/s

16.2.5 Simulation Results

In order to evaluate the performance of the considered control scheme, numerical simulations are carried out using a digital version of the converter implemented in Matlab/Simulink. The performance is analyzed in both steady-state and transient regimes. A first-order low-pass filter is added in order to reduce the switching-frequency components in the input voltage v_{in}. The maximum value of the duty cycle d is limited at 0.94 through a saturation block. Figures 16.8–16.10 illustrate the simulated behavior of the converter using the linear control scheme. A current THD of 9.38% is obtained in the nominal steady-state regime. The output voltage tends to stabilize around its desired value, with a 2% ripple, even when a set-point offset or a load disturbance is applied.

16.3 Half-Bridge Asymmetric Boost PFC

In this section, a two-switch asymmetric half-bridge boost type PFC is used to improve the power factor at the input side of a single-phase diode rectifier. The topology of the converter is presented in Figure 16.11. The pair of synchronized switches Q_1 and Q_2 is controlled in order to achieve perfect current tracking at the AC source and voltage regulation across the DC load. In this perspective, three PWM control strategies for the two-switch asymmetric half-bridge boost-type PFC are proposed and compared. The first control scheme employs a simple hysteretic based current controller, whereas the two other carrier-based schemes use, respectively, a conventional proportional-integral (PI) current regulator and a model nonlinearity compensating controller. These two last control schemes are designed

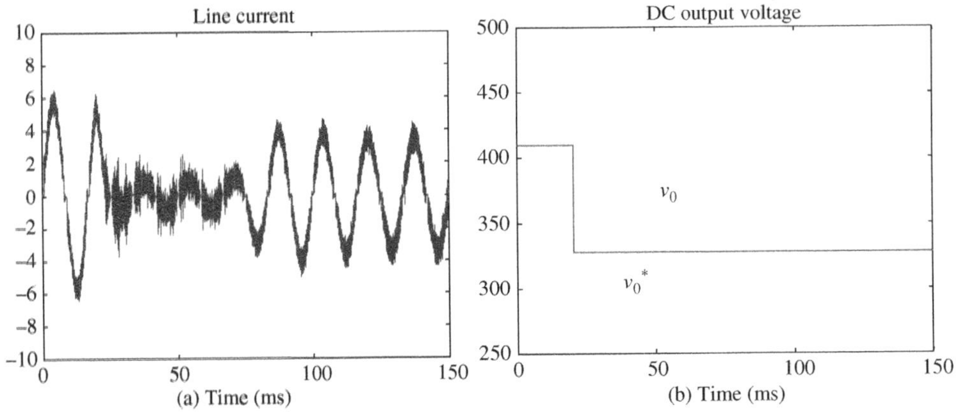

Figure 16.9 (a, b) Tracking performance

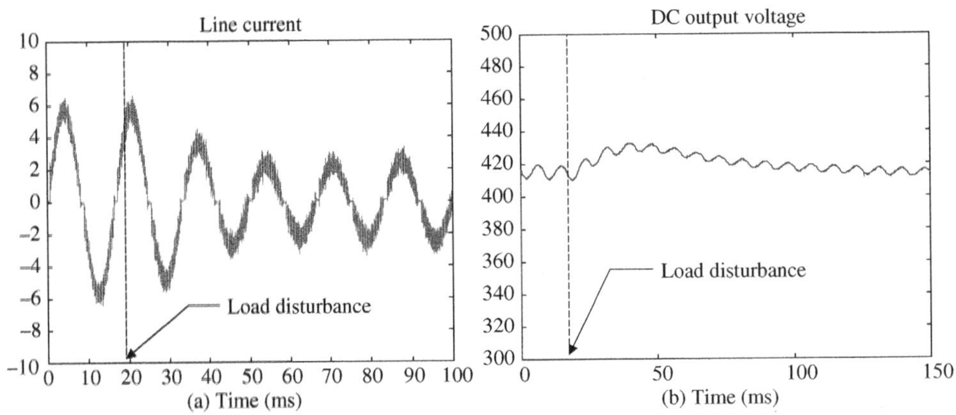

Figure 16.10 (a, b) Regulation performance

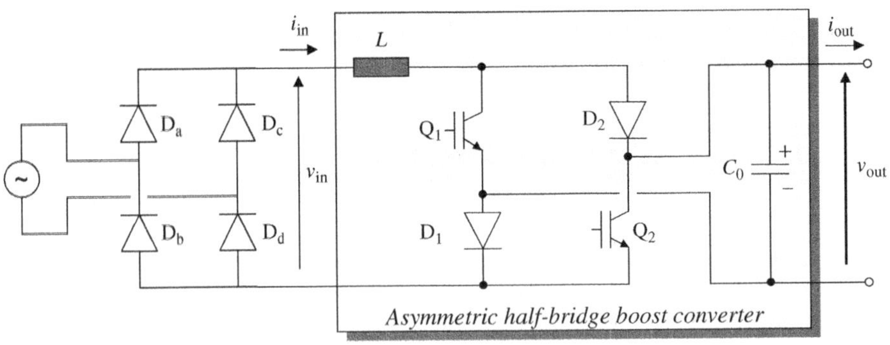

Figure 16.11 Two-switch asymmetric half-bridge boost PFC

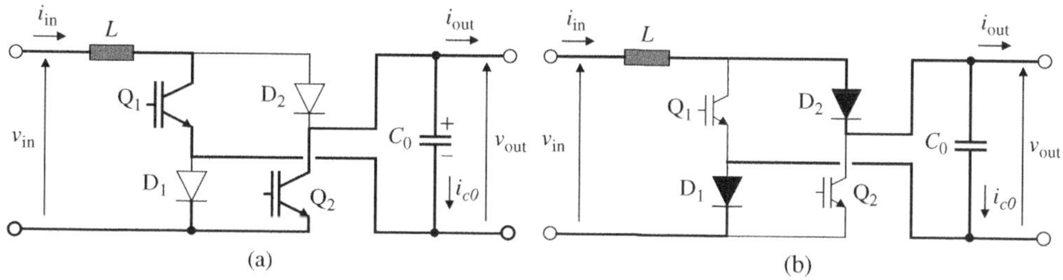

Figure 16.12 Operating sequences in CCM/CVM: electrical configuration when Q_1 and Q_2 are turned on (a) or turned off (b)

on the basis of a state-space averaged model of the converter. The performance of the proposed control schemes is then analyzed through numerical simulations on Matlab/Simulink, and a comparative evaluation is presented in terms of source current THD, input power factor and DC-voltage regulation following start-up.

16.3.1 CCM/CVM Operation and Average Modeling of the Converter

The operating sequences of the converter in the most common case of a CCM in the input inductor L and a continuous voltage mode (CVM) in the output capacitor C_0 are illustrated in Figure 16.12. The circuit has two configurations depending on the state of the main switches Q_1 and Q_2. The waveforms of the system's variables over a switching cycle are illustrated in Figure 16.13, where s_Q represents the gate signal common to Q_1 and Q_2. To ensure periodicity, the output voltage should always be set to a value higher than the input voltage. It is worth noting, at this stage, that at the zero-crossings of the input voltage and the input current slopes at the switch-on or switch-off of Q_1 and Q_2 are non-zero, and therefore the detuning problem from which the conventional boost PFC suffers is avoided in this case. This is considered as the main benefit of the proposed topology compared to that of the conventionally used topology.

Moreover, the switching frequency of switches Q_1 and Q_2 is either time-varying (if a hysteretic flip-flop controller is used for the line current shaping) or fixed (if a saw-tooth carrier-based pulse-width modulator is used along with a continuous current controller). However, it will be assumed in both cases, especially as far as the calculation of the reactive components L and C_0 is concerned, that the lowest switching frequency is far beyond the bandwidth of the current and voltage loops. Consequently, a simplified averaged model of the converter may be used in order to analyze the low-frequency operation of the converter in the steady-state regime and thus to choose the required reactive components.

In average mode, the state representation of the converter in the CCM/CVM case is expressed as follows:

$$\begin{cases} L\dfrac{di_{in}}{dt} = v_{in} - (1-2d)\cdot v_{out} \\ C_0\dfrac{dv_{out}}{dt} = (1-2d)\cdot i_{in} - i_{out} \end{cases} \tag{16.18}$$

where $d(t)$ denotes the common duty cycle of switches Q_1 and Q_2.

In the desired regime where v_{in} and i_{in} follow rectified sine waves and v_{out} is practically constant, the duty cycle would be

$$d^*(t) = \begin{cases} \dfrac{1}{2} - \dfrac{\hat{v}_{in}}{2v_{out}\cos\varphi}\sin(\omega_0 t - \varphi) & \text{if } 0 \le \omega_0 t \le \pi \\ \dfrac{1}{2} + \dfrac{\hat{v}_{in}}{2v_{out}\cos\varphi}\sin(\omega_0 t - \varphi) & \text{if } \pi \le \omega_0 t \le 2\pi \end{cases} \tag{16.19}$$

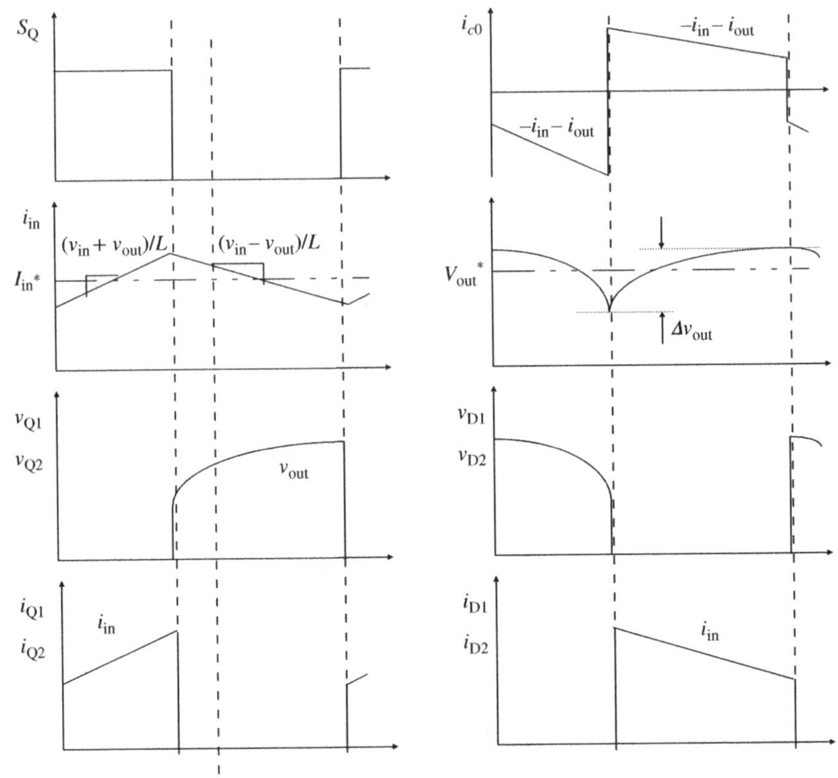

Figure 16.13 Waveforms over a switching period

and the control saturation problem, which was inevitable for the basic boost topology, can be avoided in this case if v_{out} is maintained greater than $\hat{v}_{in}/\cos\varphi$, $\hat{v}_{in}$ being the peak value of v_{in}.

16.3.2 Small-Signal Averaged Model and Transfer Functions

For the design of a linear control system based on PI regulators, a small-signal model of the converter is required. This is obtained by a linearizing system (Equation 16.18) around the static operating point defined by

$$V_{out}^* = \frac{V_{in}}{1-2D^*} = \frac{1}{1-2D^*} \cdot \frac{2\hat{v}_{in}}{\pi}$$
$$I_{in}^* = \frac{I_{out}^*}{1-2D^*} = \frac{V_{out}^*}{R_0(1-2D^*)} \quad (16.20)$$

Note that in a static operation, the duty cycle is always limited between 0 and 0.5, whereas in a dynamic regime, it can vary theoretically between 0 and 1.

The linearization process yields

$$\frac{\mathrm{d}}{\mathrm{d}t}\begin{bmatrix}\Delta i_{in}\\ \Delta v_{out}\end{bmatrix} = \mathbf{A}\cdot\begin{bmatrix}\Delta i_{in}\\ \Delta v_{out}\end{bmatrix} + \mathbf{B}\cdot\Delta d + \mathbf{P}\cdot\Delta v_{in} \quad (16.21)$$

with

$$\mathbf{A} = \begin{bmatrix} 0 & -\dfrac{1-2D^*}{L} \\ \dfrac{1-2D^*}{C_0} & -\dfrac{1}{R_0C_0} \end{bmatrix}, \mathbf{B} = \begin{bmatrix} \dfrac{2\,V^*_{\text{out}}}{L} \\ -\dfrac{2I^*_{\text{in}}}{C_0} \end{bmatrix}, \mathbf{P} = \begin{bmatrix} \dfrac{1}{L} \\ 0 \end{bmatrix}$$

where $R_0 = v_{\text{out}}/i_{\text{out}}$ designates the DC load. In Equation (16.21), **A** defines the state matrix, **B** the input matrix and **P** the disturbance matrix. In addition, for any state variable $z \in \{i_{\text{in}}, v_{\text{out}}\}$, Δz denotes its assumed small variation around its static value Z^*, that is, $\Delta z = z - Z^*$. The frequency-domain representation of the converter is obtained by applying the Laplace transform to the state Equations (16.21). It yields

$$\mathbf{X}(s) = (s\mathbf{I}_2 - \mathbf{A})^{-1}\mathbf{B} \cdot D(s) + (s\mathbf{I}_2 - \mathbf{A})^{-1}\mathbf{P} \cdot V_{\text{in}}(s) \tag{16.22}$$

where $\mathbf{I}_2$ denotes the 2-by-2 identity matrix, s is the Laplace operator, $\mathbf{X}(s) = [I_{\text{in}}(s), V_{\text{out}}(s)]^T$, $D(s)$ and $V_{\text{in}}(s)$ are the Laplace transforms of the state vector $[\Delta i_{\text{in}}, \Delta v_{\text{out}}]^T$, the control input Δd and the disturbance input Δv_{in}, respectively. The development of expression (16.22) leads to the following input–output transfer functions:

$$G_{i_{\text{in}},d}(s) \equiv \left.\frac{I_{\text{in}}(s)}{D(s)}\right|_{V_{\text{in}}=0} = \frac{2\,V^*_{\text{out}}}{L} \cdot \frac{s + \omega_{z1}}{s^2 + \omega_{p1}s + \omega^2_{p2}} \tag{16.23.a}$$

$$G_{v_{\text{out}},d}(s) \equiv \left.\frac{V_{\text{out}}(s)}{D(s)}\right|_{V_{\text{in}}=0} = -\frac{2\,V^*_{\text{out}}}{R_0C_0(1-2D^*)} \cdot \frac{s - \omega_{z2}}{s^2 + \omega_{p1}s + \omega^2_{p2}} \tag{16.23.b}$$

with

$$\omega_{z1} = \frac{2}{R_0C_0}, \omega_{z2} = \frac{R_0(1-2D^*)^2}{L}, \omega_{p1} = \frac{1}{R_0C_0}, \omega_{p2} = \frac{1-2D^*}{\sqrt{LC_0}}$$

It is on the basis of these transfer functions that the linear control scheme, which ensures unity power factor at the source, as well as voltage stabilization at the DC bus, will be designed.

Similarly, the disturbance transfer functions can also be derived from Equation (16.22) and are obtained as follows:

$$F_{i_{\text{in}},v_{\text{in}}}(s) \equiv \left.\frac{I_{\text{in}}(s)}{V_{\text{in}}(s)}\right|_{D=0} = \frac{1}{L} \cdot \frac{s + \omega'_{z1}}{s^2 + \omega_{p1}s + \omega^2_{p2}} \tag{16.24a}$$

$$F_{v_{\text{out}},v_{\text{in}}}(s) \equiv \left.\frac{V_{\text{out}}(s)}{V_{\text{in}}(s)}\right|_{D=0} = \frac{1-2D^*}{LC_0} \cdot \frac{1}{s^2 + \omega_{p1}s + \omega^2_{p2}} \tag{16.24b}$$

where

$$\omega'_{z1} = \omega_{p1} = \frac{1}{R_0C_0}$$

16.3.3 Control System Design

The control circuit is depicted in Figure 16.14a. It consists of two successive loops: the inner or current loop is designed to ensure the wave shaping of the DC input current i_{in} and, consequently, the improvement of the input power factor, while the outer or voltage loop is intended to regulate the DC load voltage v_{out} and to stabilize it around a desired set point.

The inner controller would be either a hysteretic flip-flop, a carrier-based linear regulator, or a carrier-based nonlinearity compensation controller designed on the basis of the input–output feedback linearization technique. The outer controller is a linear PI controller represented by the transfer function $H_v(s)$. K_i, K_{v0} and K_{v1} are scaling factors for the current and voltage loops, respectively. To ensure high stability of the control system, the outer loop is designed to be sufficiently slower than the inner loop.

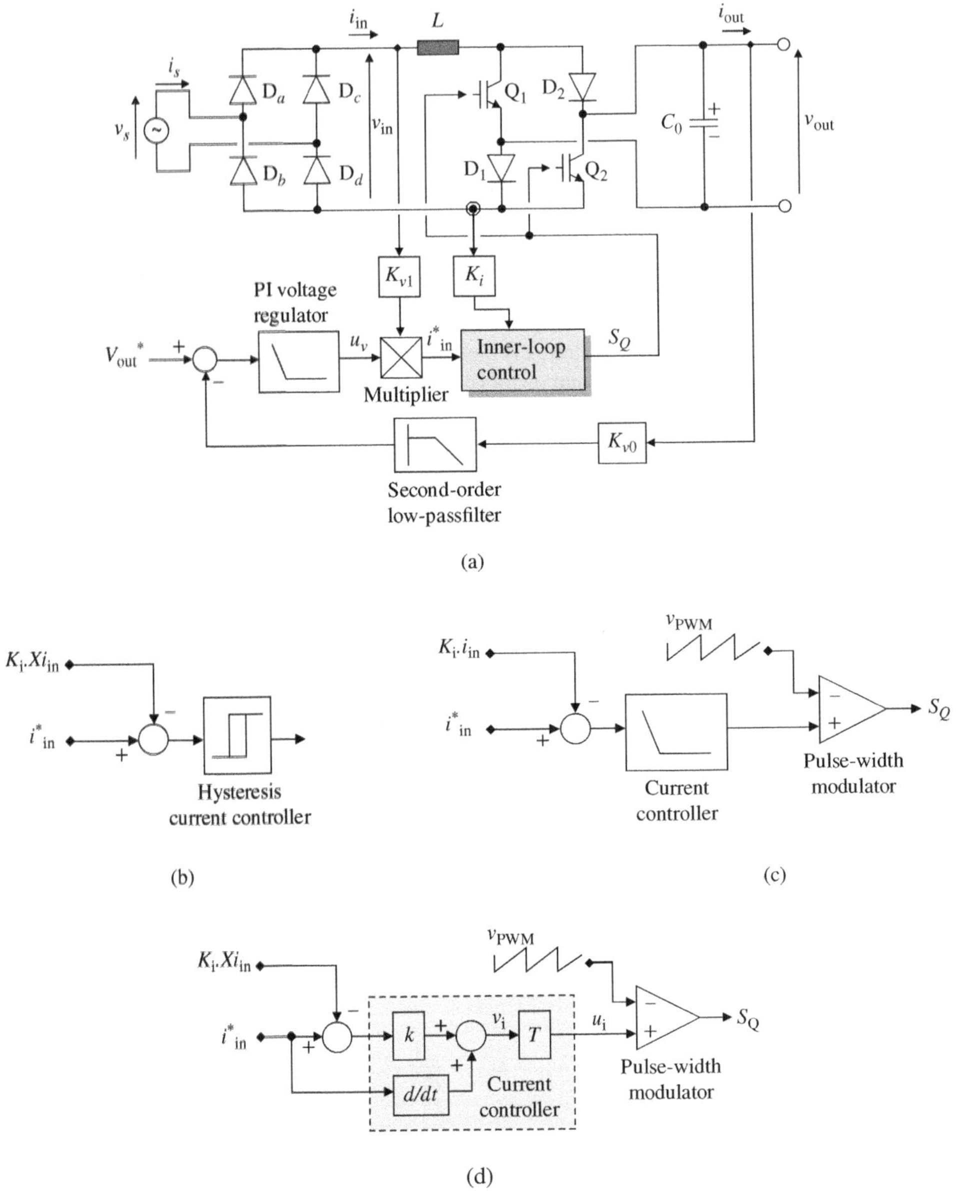

Figure 16.14 (a) Control system with different control strategies for the inner loop, (b) hysteretic control, (c) carrier-based linear control and (d) carrier-based nonlinearity compensating control [97]

In addition, in order to emulate a pure resistor behavior, the current reference must have the same shape as the rectified source voltage v_{in}, with an adjustable magnitude. An analog multiplier is used for this purpose. Furthermore, in order to avoid the distortion of the current reference, the voltage control signal u_c should be harmonic-free. This can be achieved adequately either by tuning the outer regulator, which will have the additional task of filtering the 120 Hz component (at twice the mains frequency) in the output DC voltage v_{out}, or by inserting a suitable second-order low-pass filter, represented by the transfer function $F_v(s)$, in the voltage feedback path.

16.3.3.1 Hysteretic Control

In this first case, the inner controller is a hysteretic flip-flop with a 1.5 A width, as illustrated in Figure 16.14b. The current width is calculated in order to get a maximum switching frequency near 50 kHz all over a main cycle. This would allow a convenient comparison with the other carrier-based control strategies, where the switching frequency is fixed at 50 kHz.

Despite its simplicity of implementation and robustness to disturbances, this control strategy suffers mainly from a time-varying switching frequency, rendering the whole system less reliable in the event of parameters' variations and increasing the filtering effort because of a more distributed harmonics in the current spectrum.

16.3.3.2 Carrier-Based Linear Control

In this second case, the inner controller is a linear PI regulator, represented by the transfer function $H_i(s)$, as illustrated in Figure 16.14c. The parameters of both inner and outer regulators are calculated using the lead-lag compensation method on the basis of the transfer functions established in the previous section. They are chosen in order to ensure high dynamic performance in the system bandwidth. Having in mind that the current loop is in fact embedded into the voltage loop, the design of the outer regulator will thus make use of the following open-loop transfer function that takes account of the presence of the inner loop:

$$G_{v_{out},i_{in}^*}(s) \equiv \left.\frac{V_{out}(s)}{I_{in}^*(s)}\right|_{V_{in}=0} = -\frac{L}{K_i R_0 C_0 (1-2D^*)} \cdot \frac{s-\omega_{z2}}{s+\omega_{z1}} \tag{16.25}$$

The inner control signal u_i given by the current regulator is compared to a saw-tooth carrier in order to generate a fixed-frequency PWM gate signal for the pair of switches.

Despite its relatively good performance in terms of regulation and reliability, this control approach disregards the nonlinear aspect of the converter and, hence, cannot match the required dynamics in the case of large set-point offsets.

16.3.3.3 Carrier-Based Nonlinearity Compensating Control

In the third case, the inner controller is a nonlinear controller. It is designed on the basis of the single-input single-output (SISO) input/output feedback linearization principles in order to achieve nonlinearity compensation of the inner subsystem. A nonlinear block, represented in Figure 16.14d by $\boldsymbol{T}$, is thus used in order to transform the inner subsystem model (which is originally nonlinear) into a linear canonical model. This would simplify considerably the design of the inner regulator. The control signal u_i generated by the nonlinear block is then compared with a saw-tooth carrier in order to generate a fixed-frequency PWM gate signal for the pair of switches. The outer controller is still a linear PI controller.

System (Equation 16.18) has a relative degree of 1. Thus, the derivation of the inner control law is based on the first equation in (Equation 16.18). It yields

$$d \stackrel{\Delta}{=} \frac{1}{2}\left(1+\frac{u_i}{\hat{v}_{PWM}}\right) = \frac{Lv_i - v_{in} + v_{out}}{2v_{out}} \tag{16.26}$$

where u_i denotes the output of the inner controller, v_i is the new control input of the linearized inner subsystem and $\hat{v}_{PWM}$ is the peak value of the saw-tooth modulation signal. Replacing Equation (16.26) into model Equation (16.18) gives the following canonical form of the inner subsystem:

$$\frac{di_{in}}{dt} = v_i \tag{16.27}$$

For the tracking problem at hand, we may choose the following control law:

$$v_i = \frac{di^*_{in}}{dt} - k(i_{in} - i^*_{in}) \quad \text{with} \quad k > 0 \tag{16.28}$$

The internal dynamics, which are characterized by a first-order subsystem:

$$C_0\frac{d(v^2_{out})}{dt} = 2v_{in}i^*_{in} - 2\frac{v^2_{out}}{R_0} \tag{16.29}$$

are stable, having a bounded solution.

16.3.4 Numerical Implementation and Simulation Results

In order to highlight the performance of the two-switch asymmetric half-bridge boost converter in PFC applications, a virtual version of the control system of Figure 16.14a has been implemented using Matlab/Simulink. The numerical values of the structural parameters and operating conditions are listed, which are as follows:

Mains voltage RMS−value $V_S = 120$ V
Rated load power $P_{out} = 1$ kW
Voltage reference $V_{out}{}^* = 250$ V
Mains frequency $f_0 = 60$ Hz
Carrier frequency $f_S = 50$ kHz
Carrier peak value $\hat{v}_{PWM} = 1$ V
DC inductor $L = 1$ mH
Series inductor resistance $R_L = 0.1\ \Omega$
DC capacitor $C_0 = 1$ mF
Feedback scaling factors $K_i = 1\ \Omega$, $K_{v0} = 0.05$
Feedforward scaling factor $K_{v1} = 0.05$
Hysteretic width $h = 1.25$ A
Inner current PI controller $H_i(s) = 15\dfrac{1+\frac{s}{15}}{s}$
Inner nonlinear control parameter $k = 30000$
Outer voltage PI controller $H_v(s) = 20\dfrac{1+\frac{s}{30}}{s}$
Voltage filter $F_v(s) = \dfrac{1}{1+\frac{\sqrt{2}}{300}s+\left(\frac{s}{300}\right)^2}$

The converter is implemented according to its switching-function-based model [62]. The simulations are carried out using a fixed-step ode5 (Dormand-Prince) solver. The step size is 1 μs. The results obtained for a 1 kW load are given in Figures 16.15–16.17 for the hysteretic, carried-based linear and carrier-based nonlinearity compensating control schemes, respectively.

Table 16.1 compares the major characteristics of the proposed control algorithms, namely the line current THD, the input power factor, the maximum switching frequency, the high-frequency current ripple near the positive and negative peaks of the source current, and the DC output voltage ripple at 120 Hz. It is found that the carrier-based control schemes exhibit slightly better steady-state performance than

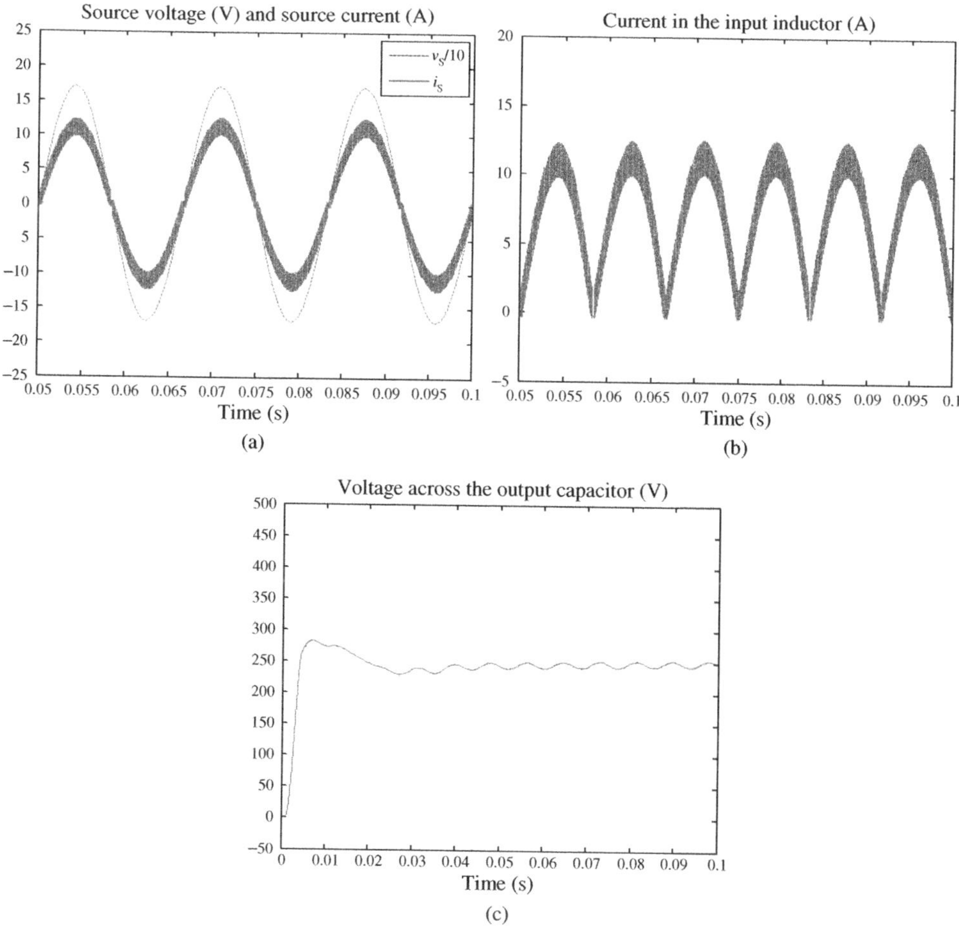

Figure 16.15 Simulation results given by the hysteretic-based control system at a rated load: steady-state waveforms of (a) the source voltage (v_S) and the source current (i_S), (b) the current in the input inductor (i_{in}) and (c) transient waveform of the voltage across the output capacitor (v_{out}) following a start-up [97]

the hysteretic-based performance in terms of source current THD and input power factor. As for the carrier-based control systems, the control input generated by the nonlinearity compensating controller (see Figure 16.17c) lies within the saturation limits and presents relatively small high-frequency fluctuations, unlike the fluctuations given by the conventional PI regulator (see Figure 16.16c). In addition, these figures show that, in all three cases, i_{in} and v_{out} are always positive, which justifies the assumption of a CCM/CVM operation made previously. Note finally that in all three control cases, the control detuning problem, from which the conventional Boost PFC suffers, does not exist.

16.4 Interleaved Dual-Boost PFC

In order to increase the efficiency and power density of boost PFCs, and to help reduce the EMI level injected into the utility line, interleaved technology has been proposed [63–67] as a serious candidate

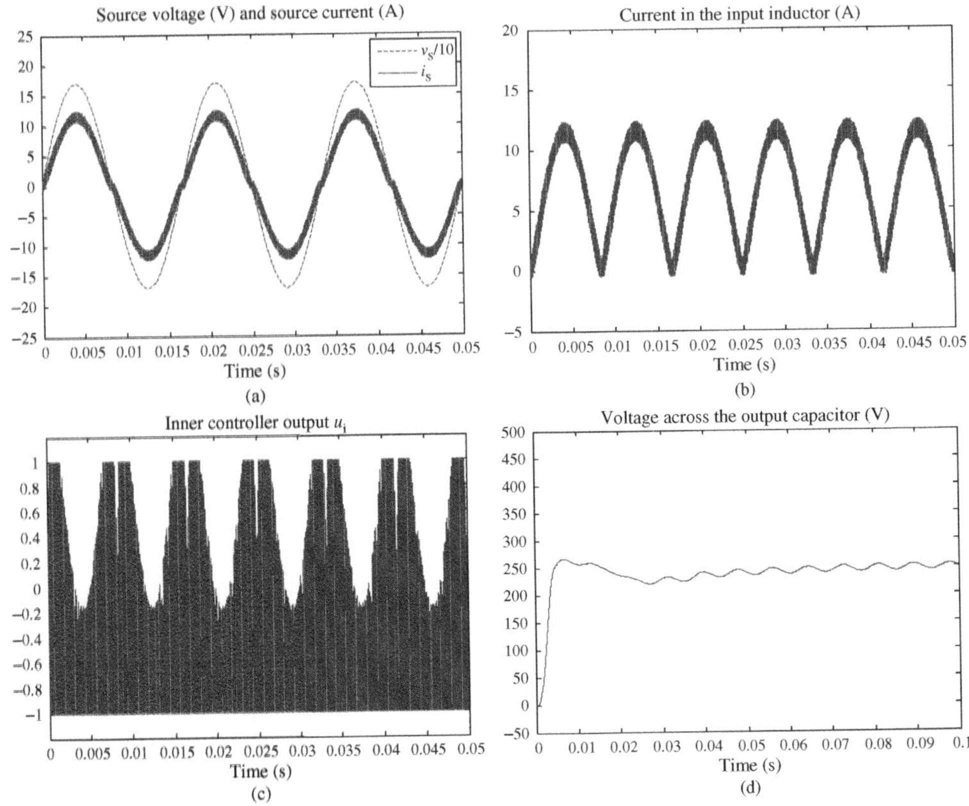

Figure 16.16 Simulation results given by the linear control system at rated load: steady-state waveforms of (a) the source voltage (v_s) and the source current (i_s), (b) the current in the input inductor (i_{in}), (c) the inner controller output (u_i) and (d) transient waveform of the voltage across the output capacitor (v_{out}) following a start-up [97]

Table 16.1 Comparative performance evaluation of the control systems applied to the two-switch asymmetric half-bridge boost PFC

Performance characteristics	Hysteretic control	Linear control	Nonlinear control
Line current THD (%)	9.2	9	8.6
Input power factor	0.9958	0.9959	0.9962
Maximum switching frequency (kHz)	50	50	50
Current ripple (%)	10.6	10.9	10.4
Output voltage ripple (%)	4	4	4

to overcome all the drawbacks found in a single device AC/DC boost preregulator topology. These drawbacks are the emission level of harmonics at the switching frequency of the device, and the level of EMI that propagate into the line side; furthermore, efficiency increase in the conversion stage is also one of the key factors. To help achieve these goals, different techniques have been proposed. Two-switch

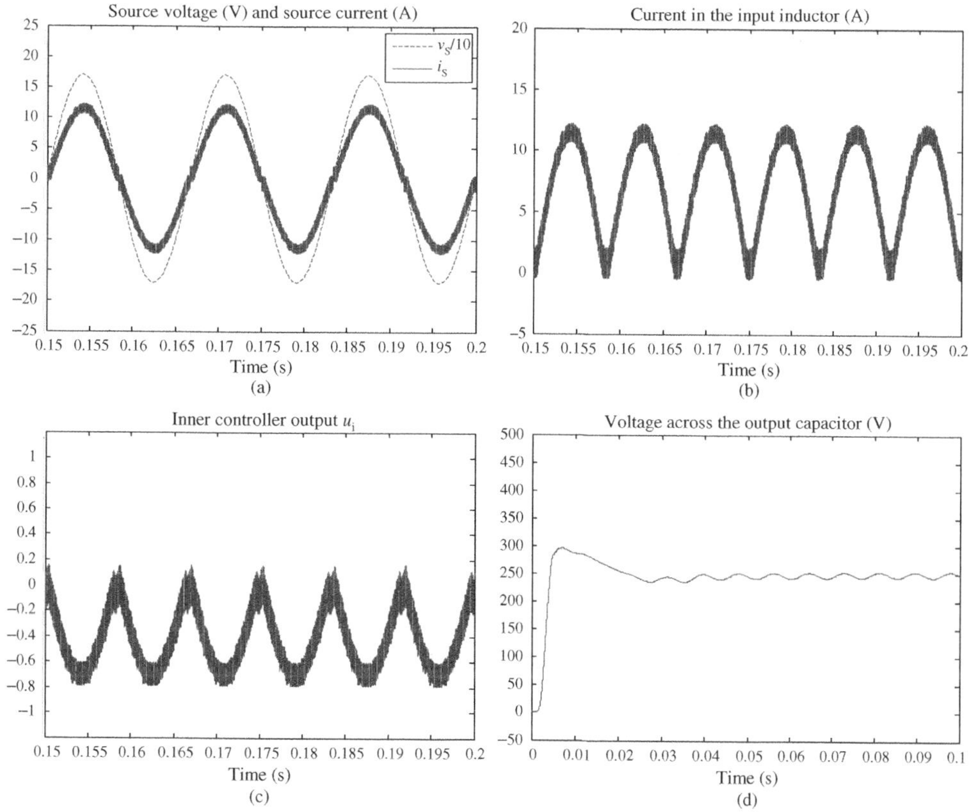

Figure 16.17 Simulation results given by the nonlinearity compensating control system at rated load: steady-state waveforms of (a) the source voltage (v_s) and the source current (i_s), (b) the current in the input inductor (i_{in}), (c) the inner controller output (u_i) and (d) transient waveform of the voltage across the output capacitor (v_{out}) following a start-up [97]

passive and active shaping techniques have been studied over the years [68–74], with the accent on topology and on the use of a very simple controller. The initial results yielded an increase in the power density, reducing the level of emission at low frequencies while increasing slightly the efficiency of the power conversion unit. Moreover, the size of the energy storage bulky inductor that is used for the boost stage can be reduced by using a high-frequency stage, assuming the shaping of the current created by the low-frequency switch. This active technique exhibits the same advantages obtained using passive/active harmonic filtering using active devices instead of passive devices, technology largely used for power network depollution [75]. Despite their acceptable cost, passive components are bulky, sensitive to line impedance changes and inflict more difficult control on the converter. On the contrary, the use of active devices permits us to obtain good tracking and regulation, which result in a system that is more robust toward line perturbations. PFC interleaved converters are connected to the AC main; they generate two types of harmonics. Low-frequency harmonics in the range of a kilohertz are generated at the multiple frequencies of the low-frequency switch Q_{LF} operation. High-frequency harmonics in the range of hundreds of kilohertz are generated at the multiples of high-frequency switch Q_{HF} operation. The low frequencies fall within the IEEE-519 regulation that defines the shape of the current

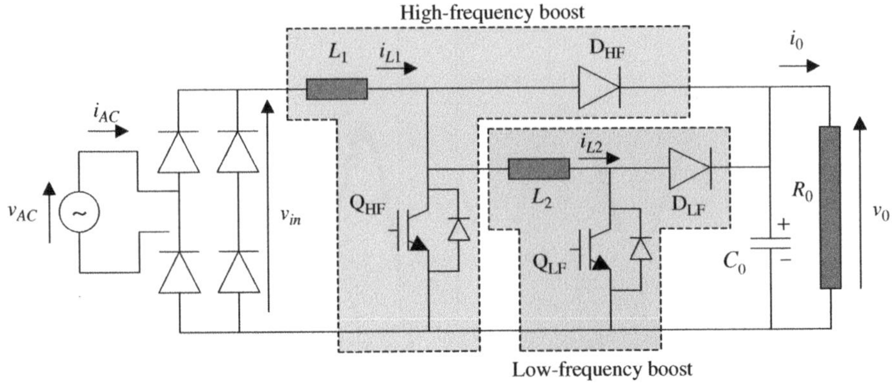

Figure 16.18 Topology of the dual-boost PFC-based single-phase rectifier [77]

injected into the network, whereas the high frequencies fall under the conducted EMI regulation IEC 61000–2, which defines the spikes and edges of the line current and the permitted level of emission. Both continuous and discontinuous conduction modes of operation for single phase as well as three-phase topology have been evaluated [71, 72], but the modeling and control technique for the interleaved converters are still being worked on. Simple control using flip-flops has been extensively used for these topologies [68, 70, 74]. In this section, the development of a model-based fixed-switching-frequency control system and its validation on a 1 kW prototype of interleaved single-phase boost converters are presented.

The topology under study [68] has two main switches Q_{HF} and Q_{LF} operating at different frequencies. The main switch assumes the control of up to 90% of the energy transferred to the load and operating at 5 kHz. The low-rated switch operates at a high frequency of 50 kHz, contributes up to 10% of the rated power, and permits the cancelation of the harmonics generated by the low-frequency switch Q_{LF}. The low-frequency switch operates in continuous conduction mode, whereas the high-frequency switch operates in discontinuous mode.

16.4.1 Converter Topology

The single-phase interleaved dual-boost PFC rectifier is presented in Figure 16.18. It consists of

- a diode bridge that ensures the AC-to-DC conversion
- a high-rated low-frequency typical boost topology that ensures mainly line current wave shaping by modulating the current i_{L2} in inductor L_2 and control of the power transmitted to the load
- a low-rated high-frequency boost converter, which is intended to compensate the high-order harmonics in the input current i_{L1}, by modulating the low-frequency ripple of i_{L2}.

A pure resistive load R_0 is connected across the DC-bus capacitor filter C_0. Furthermore, the low-frequency Q_{LF} and high-frequency Q_{HF} switches are controlled synchronously at multiple frequencies and the switch-on of Q_{LF} coincides with a switch-on of Q_{HF}. Denoting by f_{HF} and f_{LF} the switching frequencies of Q_{HF} and Q_{LF}, respectively, we thus have

$$f_{HF} = m \cdot f_{LF} \quad \text{with} \quad m \in N \quad \text{and} \quad m >> 1 \tag{16.30}$$

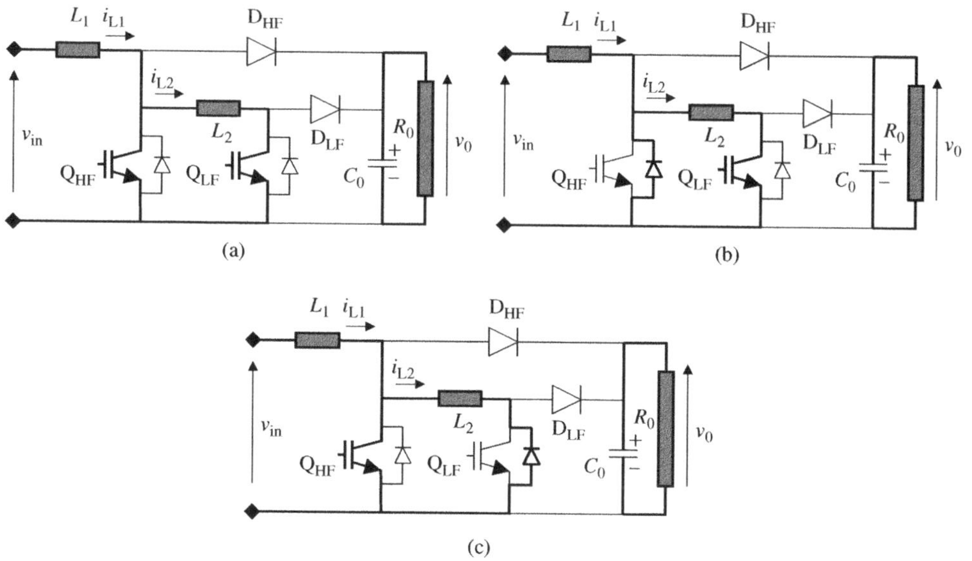

Figure 16.19 Configurations of the PFC when Q_{HF} and Q_{LF} are both on and (a) $0 < i_{L2} < i_{L1}$, (b) $0 < i_{L1} < i_{L2}$ and (c) $i_{L2} < 0 < i_{L1}$

16.4.2 Operation Sequences

As the PFC has two switches, four different configurations are possible:

16.4.2.1 Case 1: Q_{HF} Is on and Q_{LF} Is on

When both switches Q_{HF} and Q_{LF} are turned on, the configuration of the PFC becomes as described in Figure 16.19. If the input inductor current i_{L1} (which is always positive) is greater than i_{L2}, and i_{L2} is greater than zero, the following equations can be written (Figure 16.19a):

$$
\begin{aligned}
L_1 \frac{di_{L1}}{dt} &= v_{in} > 0 \\
L_2 \frac{di_{L2}}{dt} &= 0 \\
C_0 \frac{dv_0}{dt} + \frac{v_0}{R_0} &= 0
\end{aligned}
\tag{16.31}
$$

When i_{L1} is lower than i_{L2}, the parasitic diode of switch Q_{HF} conducts and the configuration of the PFC is as given in Figure 16.19b, where Equations (16.31) are still valid. Similarly, the same equations are applicable if current i_{L2} becomes negative (Figure 16.19c). In this case, it is the parasitic diode of switch Q_{LF} that conducts.

16.4.2.2 Case 2: Q_{HF} Is off and Q_{LF} Is on

When only switch Q_{LF} is turned on, the configuration of the PFC becomes as described in Figure 16.20. If the input inductor current i_{L1} is greater than i_{L2} and i_{L2} is greater than zero, the high-frequency boost-diode

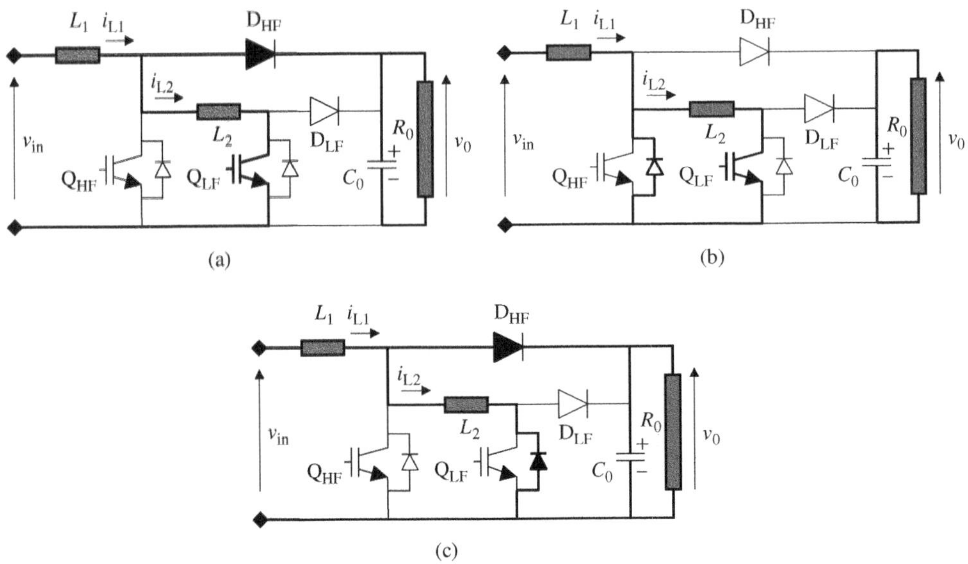

Figure 16.20 Configurations of the PFC when Q_{HF} is off, Q_{LF} is on and (a) $0 < i_{L2} < i_{L1}$, (b) $0 < i_{L1} < i_{L2}$ and (c) $i_{L2} < 0 < i_{L1}$

D_{HF} conducts, and the following equations can be written (Figure 16.20a) as

$$L_1 \frac{di_{L1}}{dt} = v_{in} - v_0$$
$$L_2 \frac{di_{L2}}{dt} = v_0 > 0$$
$$C_0 \frac{dv_0}{dt} + \frac{v_0}{R_0} = i_{L1} - i_{L2} \tag{16.32}$$

A proper modulation of current i_{L1} requires an output voltage v_0 greater than the peak value of the input voltage v_{in}. On the other hand, when i_{L1} is lower than i_{L2}, it is the parasitic diode of switch Q_{HF} that conducts the current $(i_{L2} - i_{L1})$ instead of the boost-diode D_{HF}. In this case, the configuration of the PFC is as given in Figure 16.20b, where the system's equations become as given by Equation (16.31). This configuration is identical to the configuration given in Figure 16.19b.

In the case where i_{L2} is negative (Figure 16.20c), the parasitic diode of switch Q_{LF} conducts and Equations (16.32) are still applicable.

16.4.2.3 *Case 3:* Q_{HF} *Is on and* Q_{LF} *Is off*

When only switch Q_{HF} is turned on, the configuration of the PFC becomes as described in Figure 16.21. If the input inductor current i_{L1} is greater than i_{L2} and i_{L2} is greater than zero, the low-frequency boost-diode D_{LF} conducts, and the following equations can be written (Figure 16.21a) as

$$L_1 \frac{di_{L1}}{dt} = v_{in} > 0$$
$$L_2 \frac{di_{L2}}{dt} = -v_0 < 0$$

$$C_0\frac{dv_0}{dt}+\frac{v_0}{R_0}=i_{L2} \tag{16.33}$$

On the other hand, when i_{L1} is lower than i_{L2}, the parasitic diode of switch Q_{HF} conducts and the configuration of the PFC is as given in Figure 16.21b, where the system's equations remain the same as Equation (16.33).

In the case where i_{L2} is negative (Figure 16.21c), it is the parasitic diode of switch Q_{LF} that conducts the current i_{L2} instead of D_{LF}. The PFC configuration thus becomes identical to the configuration given in Figure 16.19c, and Equations (16.31) are the ones that apply.

16.4.2.4 Case 4: Q_{HF} Is off and Q_{LF} Is off

When both switches Q_{HF} and Q_{LF} are turned off, the configuration of the PFC becomes as described in Figure 16.22. If the input inductor current i_{L1} is greater than i_{L2} and i_{L2} is greater than zero, the low-frequency boost-diode D_{LF} and the high-frequency boost-diode D_{HF} conduct simultaneously. The system's equations are thus as follows (Figure 16.22a):

$$\begin{aligned} L_1\frac{di_{L1}}{dt}&=v_{in}-v_0\\ L_2\frac{di_{L2}}{dt}&=0\\ C_0\frac{dv_0}{dt}+\frac{v_0}{R_0}&=i_{L1} \end{aligned} \tag{16.34}$$

On the other hand, when i_{L1} is lower than i_{L2}, the parasitic diode of switch Q_{HF} conducts the current $(i_{L2}-i_{L1})$ instead of D_{HF} and the configuration of the PFC becomes as given in Figure 16.22b. This case is identical to the case depicted in Figure 16.21b, and the system's equations are given by Equation (16.33).

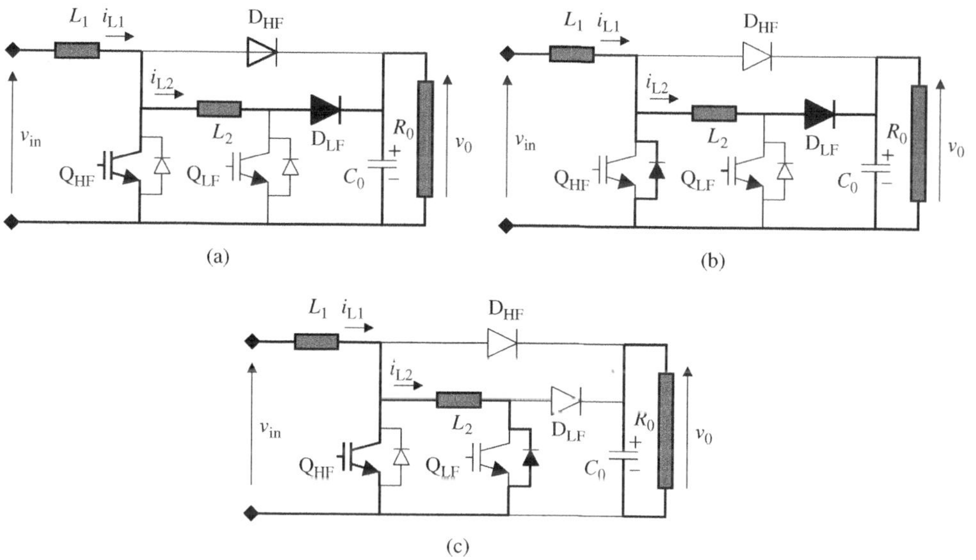

Figure 16.21 Configurations of the PFC when Q_{HF} is on, Q_{LF} is off and (a) $0<i_{L2}<i_{L1}$, (b) $0<i_{L1}<i_{L2}$ and (c) $i_{L2}<0<i_{L1}$

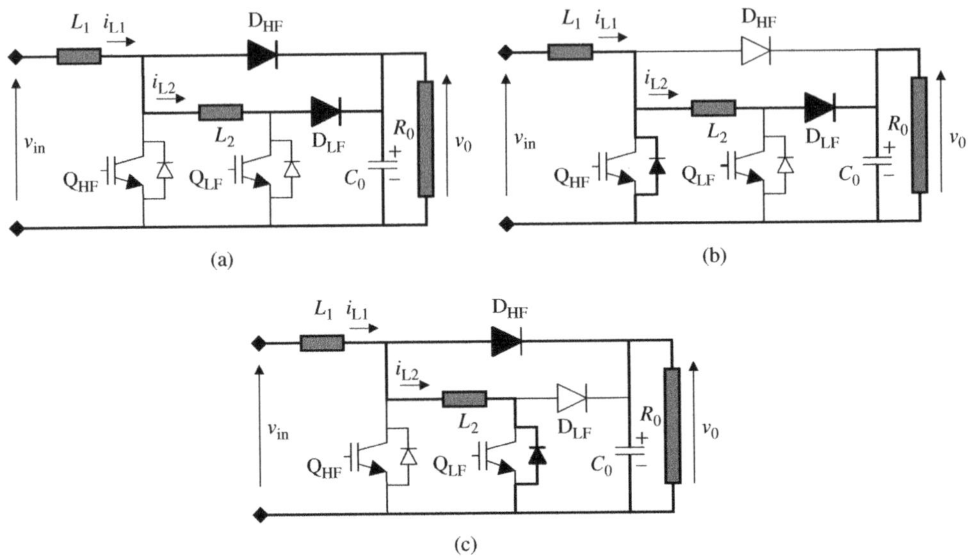

Figure 16.22 Configurations of the PFC when Q_{HF} and Q_{LF} are both off, and (a) $0 < i_{L2} < i_{L1}$, (b) $0 < i_{L1} < i_{i2}$, and (c) $i_{L2} < 0 < i_{L1}$

In the case where i_{L2} is negative (Figure 16.22c), the parasitic diode of switch Q_{LF} conducts the current i_{L2} instead of D_{LF}. The PFC configuration thus becomes identical to the configuration given in Figure 16.20c and Equations (16.32) are the ones that apply.

16.4.3 Linear Control Design and Experimental Results

The control system is described in Figure 16.23. The gate signals of switches Q_{HF} and Q_{LF} are generated by comparing the current loops control signals (at the output of the inner regulators) with the modulation signals $v_{PWM,HF}$ and $v_{PWM,LF}$, respectively. Both the modulation signals, at fixed frequency, are saw-tooth type and are synchronized at the turning-on of switch Q_{LF}. K_{iL1}, K_{iL2} and K_v are the feedback scaling gains of the currents and voltage loops. The feedforward scaling gain $K_{v,in}$ is inserted in order to reduce the level of v_{in} at the input of the control circuit.

The adopted design of the inner regulators $H_{iL1}(s)$ and $H_{iL2}(s)$ neglects the cross-coupling between the duty cycle d_{HF} and current i_{L2}, on the one hand and between the duty cycle d_{LF} and current i_{L1}, on the other [76, 77]. The structure and parameters of these regulators are determined so that the inner closed-loop transfer functions correspond to a second-order low-pass filter. Furthermore, regarding the outer voltage loop, the voltage regulator $H_{v0}(s)$ is similarly designed in order to ensure that the outer loop has the same dynamic characteristics as an optimized linear second-order low-pass filter.

The converter with the control system shown in Figure 16.23 was realized using the following numerical values:

Load power $P_0 = 1\,\text{kW}$
RMS source voltage $V_{AC} = 110\,\text{V}$
Mains frequency $f_0 = 60\,\text{Hz}$
Low switching frequency $f_{LF} = 5\,\text{kHz}$

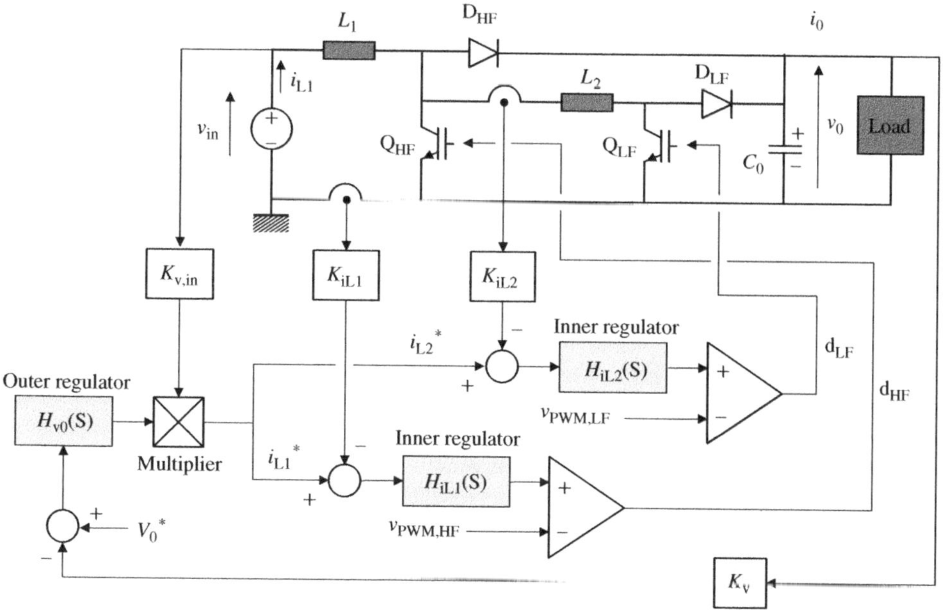

Figure 16.23 Control system [77]

High-switching frequency $f_{HF} = 50\,\text{kHz}$
Input DC inductor $L_1 = 1\,\text{mH}$, $R_{L1} = 10\,\text{m}\Omega$
Low-frequency inductor $L_2 = 2\,\text{mH}$, $R_{L2} = 10\,\text{m}\Omega$
Output DC capacitor $C_0 = 2400\,\mu\text{F}$
DC output voltage $V_0 = 250\,\text{V}$

Inner current regulators $H_{iL1}(s) = \dfrac{5000 + \frac{5}{\pi}s}{s \cdot \left(1 + \frac{s}{50000\pi}\right)}$

$$H_{iL2}(s) = \frac{500 + \frac{s}{2\pi}}{s \cdot \left(1 + \frac{s}{5000\pi}\right)}$$

DC-voltage regulator $H_{v0}(s) = 20 \cdot \dfrac{s+5}{s \cdot (s+70)}$
Feedback scaling gains $K_{iL1} = K_{iL2} = K_v = 1$
Feedforward scaling gain $K_{v,in} = 1/24$
PWM comparators gains $K_{PWM,HF} = K_{PWM,LF} = 1$

Figure 16.24 shows both line current and line voltage when the converter delivers 800 W to the load. One can note that the converter is operating at a unity displacement factor.

Figure 16.25 shows the efficiency of the power converter drawn as a function of the load output power for three types of operation. The first type is defined when Q_{LF} operates alone; the maximum efficiency obtained is 90% at 940 W. The second type is defined when both Q_{HF} and Q_{LF} are functioning simultaneously; the maximum efficiency measured is 92%. Finally, the third type of measurement is performed when only Q_{HF} is operating; the maximum efficiency measured is 87%. The use of the dual-boost type of converter ensures a trade-off between power quality and efficiency.

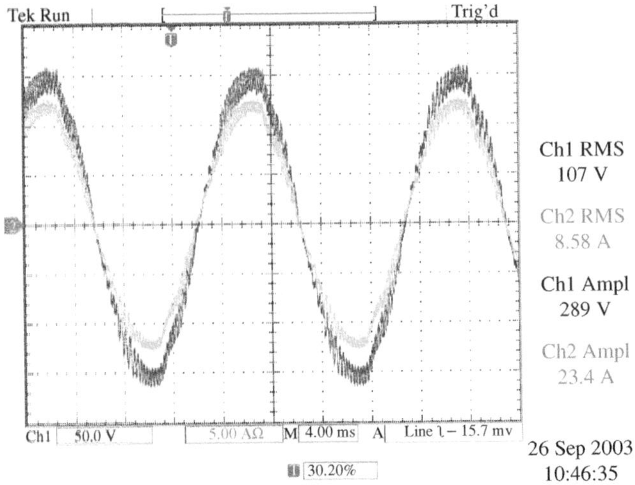

Figure 16.24 Line current and voltage for $P_0 = 800\,\text{W}$ [50 V/div, 5 A/div] [77]

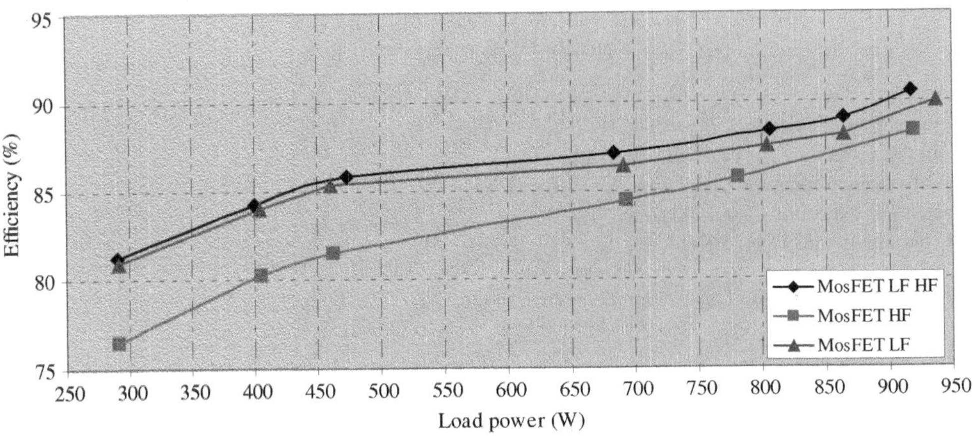

Figure 16.25 Converter efficiency with respect to the load power for three types of operations [77]

16.5 Conclusion

Boost-type preregulators are the most used topologies in various industrial applications at low and medium power levels. Among the reasons for their attractiveness are robustness with regard to supply fluctuations, spikes, voltage transition, lightning and easy to design, their control circuit is generally current mode type that is simple to develop, they make good use of power semiconductor devices and finally their cost is very competitive.

The basic boost converter topology is the most popular in such applications; however, its main drawbacks remain, first, the control detuning phenomenon that is present at the zero-crossing of the source voltage, which increases the source current distortion and limits the performance of this type of PFC; and second, the inherent power losses, which increase drastically when the converter operates at higher switching frequencies. To overcome the first problem, a boost structure consisting of a two-switch asymmetric half-bridge converter can be considered. As for the second problem, an interleaved boost preregulator that operates at different switching frequencies can be employed.

References

1. De Keulenaer, H. (2003) *The Hidden Cost of Poor Power Quality*, European Copper Institute, October 2003.
2. Billinton, R. (2001) Methods to Consider Customer Interruption Costs in Power System Analysis. Technical Report Task Force 38.06.01, CIGRE, Paris.
3. Gutierrez Iglesias, J. and Bartak, G. (2002) Power Quality in European Electricity Supply Networks – 1st Edition. Technical Report, Eurelectric, Brussels.
4. Chapman, D. (2002) The cost of poor power quality. *Power Quality Application Guide*, **2** (1).
5. IEEE (1992) IEEE Std 519-1992. *IEEE Recommended Practices and Requirements for Harmonic Control in Electric Power Systems*, June 1992, IEEE.
6. IEC (International Electrotechnical Commission) Subcommittee 77A (1992) IEC 555–2 (EN 60555–2). *Disturbance in Supply Systems Caused by Household Appliance and Similar Electrical Equipment, Part 2: Harmonics*, September 1992.
7. IEC (International Electrotechnical Commission) 61000-3-2 (2009), Electromagnetic compatibility (EMC) – Part 3-2: *Limits – Limits for harmonic current emissions*, Edition 3.2.
8. Key, T.S. and Lai, J.-S. (1993) Comparison of standards and power supply design options for limiting harmonic distortion in power systems. *IEEE Transactions on Industry Applications*, **29** (4), 688–695.
9. Singh, B., Singh, B.N., Chandra, A. *et al.* (2003) A review of single-phase improved power quality AC-DC converters. *IEEE Transactions on Industrial Electronics*, **50** (5), 962–981.
10. Kanaan, H.Y. and Al-Haddad, K. (2011) A unified approach for the analysis of single-phase power factor correction converters. Proceedings 37th Annual Conference of the IEEE Industrial Electronics Society (IECON'11), Melbourne, Australia, November 7–10, 2011.
11. Kanaan, H.Y. and Al-Haddad, K. (2012) A comparative study of single-phase power factor correction converters: Modeling, steady-state characteristics, current tracking ability and design settings. Proceedings 21st International Symposium on Power Electronics, Electrical Drives, Automation and Motion (SPEEDAM'12), Sorrento, Italy, June 20–22, 2012.
12. Hui, S.Y., Chung, H.S.–.H., and Yip, S.–.C. (2000) A bidirectional AC-DC power converter with power factor correction. *IEEE Transactions on Power Electronics*, **15** (5), 942–949.
13. Qiao, C. and Smedley, K.M. (2000) A topology survey of single-stage power factor corrector with a boost type input-current-shaper. APEC 2000, 15th IEEE Annual Meeting, Vol. 1, pp. 460–467.
14. Salmon, J.C. (1993) Techniques for minimizing the input current distortion of current-controlled single-phase boost rectifiers. *IEEE Transactions on Power Electronics*, **8** (4), 509–520.
15. Mohan, N., Undeland, T.M., and Robbins, W.P. (1995) *Power Electronics: Converters, Applications, and Design*, 2nd edn, John Wiley & Sons, Inc.
16. Kanaan, H., Al-Haddad, K., Chaffaï, R., and Duguay, L. (2001) Averaged modeling and linear control of a new buck-type single-phase single-switch PWM rectifier. Proceedings IETA'01, Cairo, Egypt, December 19–21, 2001.
17. Ghanem, M.C., Al-Haddad, K., and Roy, G. (1996) A new control strategy to achieve sinusoidal line current in a cascade buck-boost converter. *IEEE Transactions on Industrial Electronics*, **43** (3), 441–449.
18. Redl, R., Balogh, L., and Sokal, N.O. (1994) A new family of single-stage isolated power-factor-correctors with fast regulation of the output voltage. IEEE PESC'94 Record, pp. 1137–1144.
19. Tseng, C.-J. and Chen, C.-L. (1999) A novel ZVT PWM Cuk power-factor corrector. *IEEE Transactions on Industrial Electronics*, **46**, 780–787.
20. Costa, D.B. and Duarte, C.M.C. (2004) The ZVS PWM active clamping Cuk converter. *IEEE Transactions on Industrial Electronics*, **51** (1), 54–60.
21. Fuad, Y., de Koning, W.L., and van der Woude, J.W. (2004) On the stability of the pulse-width-modulated Cuk converter. *IEEE Transactions on Circuits and Systems II: Express Briefs*, **51** (8), 412–420.
22. Sabzali, A.J., Ismail, E.H., Al Saffar, M.A., and Fardoun, A.A. (2009) A new bridgeless PFC SEPIC and Cuk rectifiers with low conduction and switching losses. Proceedings International Conference on Power Electronics and Drive Systems (PEDS'09), pp. 550–556.

23. Ismail, E.H. (2009) Bridgeless SEPIC rectifier with unity power factor and reduced conduction losses. *IEEE Transactions on Industrial Electronics*, **56** (4), 1147–1157.
24. Petersen, L. (2001) Input-current-shaper based on a modified SEPIC converter with low voltage stress. Proceedings IEEE PESC'01, pp. 666–671.
25. Al-Saffar, M.A., Ismail, E.H., Sabzali, A.J., and Fardoun, A.A. (2008) An improved topology of SEPIC converter with reduced output voltage ripple. *IEEE Transactions on Power Electronics*, **23** (5), 2377–2386.
26. Chung, H.S.-H., Tse, K.K., Hui, S.Y.R. *et al.* (2003) A novel maximum power point tracking technique for solar panels using a SEPIC or Cuk converter. *IEEE Transactions on Power Electronics*, **18** (3), 717–724.
27. de Melo, P.F., Gules, R., Romaneli, E.F.R., and Annunziato, R.C. (2010) A modified SEPIC converter for high-power-factor rectifier and universal input voltage applications. *IEEE Transactions on Power Electronics*, **25** (2), 310–321.
28. Ye, Z., Greenfeld, F., and Liang, Z. (2009) Single-stage offline SEPIC converter with power factor correction to drive high brightness LEDs. Proceeings 24th Annual IEEE Applied Power Electronics Conference and Exposition (APEC'09), pp. 546–553.
29. Lam, J., Jain, P.K., and Agarwal, V. (2008) A novel SEPIC type single-stage single switch electronic ballast with very high power factor and high efficiency. Proceedings IEEE Power Electronics Specialists Conference (PESC'08), pp. 2861–2866.
30. Shen, C.-L., Wu, Y.-E. and Chen, M.-H. (2008) A modified SEPIC converter with soft-switching feature for power factor correction. Proceedings International Conference on Industrial Technology (ICIT'08), pp. 1–6.
31. Kanaan, H.Y., Al-Haddad, K., Sauriole, G., and Chaffaï, R. (2012) Modeling, design and control of a SEPIC power factor corrector for single-phase rectifiers: Experimental validation. *International Journal of Power Electronics (IJPELEC)*, **4** (3), 221–239.
32. Kanaan, H.Y., Al-Haddad, K., and Fnaiech, F. (2004) Switching-function-based modeling and control of a SEPIC power factor correction circuit operating in continuous and discontinuous current modes. Proceedings IEEE ICIT'04, Hammamet, Tunisia, December 8–10, 2004, Vol. 1, pp. 431–437.
33. Kanaan, H.Y. and Al-Haddad, K. (2005) A novel averaged-model-based control of a SEPIC power factor corrector using the input/output feedback linearization technique. Proceedings of IEEE PESC'05, Recife, Brazil, June 12–16, 2005, pp. 565–571.
34. Kanaan, H.Y. and Al-Haddad, K. (2005) A comparative analysis of nonlinear current control schemes applied to a SEPIC power factor corrector. Proceedings 31st Annual Conference of the IEEE Industrial Electronics Society (IECON'05), Raleigh, NC, November 6–10, 2005, pp. 1104–1109.
35. Kanaan, H.Y. and Al-Haddad, K. (2008) Small-signal averaged model and carrier-based linear control of a SEPIC-type power factor correction circuit. Proceedings INTELEC'08, San Diego, CA, September 14–18, 2008.
36. Kanaan, H.Y. and Al-Haddad, K. (2011) Design, study, modeling and control of a SEPIC power factor corrector in single-phase rectifiers. Proceedings ELECTRIMACS'11, Paris, France, June 6–8, 2011.
37. Kanaan, H.Y., Al-Haddad, K., Sauriole, G., and Chaffaï, R. Practical design of a single-phase SEPIC power factor corrector with DC-voltage regulation. Proceedings IEEE ISIE'06, Montreal, Quebec, July 09–13, 2006.
38. Zhao, L., Zhang, B., Ma, H., and Liu, X. (2005) Research and experiment of Sheppard-Taylor topology. Proceedings IECON'05, November 6–10, 2005.
39. Tse, C.K. and Chow, M.H.L. (1998) New single-stage PFC regulator using the Sheppard-Taylor topology. *IEEE Transactions on Power Electronics*, **13** (5), 842–851.
40. Tse, C.K. and Chow, M.H.L. (1996) Single stage high power factor converter using the Sheppard-Taylor topology. 27th Annual IEEE Power Electronics Specialists Conference Records (PESC'96), June 23–27, 1996, Vol. 2, pp. 1191–1197.
41. Kanaan, H.Y., Al-Haddad, K., and Salloum, G. (2009) Study, modelling and control of a single-phase power factor corrector based on the Sheppard-Taylor topology. *International Journal of Power Electronics (IJPELEC)*, **1** (4), 434–459.
42. Kanaan, H.Y., Hayek, A., and Al-Haddad, K. (2006) Modeling and control of a single-phase Sheppard-Taylor based power factor corrector. Proceedings IEEE International Conference on Industrial Technology (ICIT'06), Mumbai, India, December 15–17, 2006, pp. 2785–2790.
43. Kanaan, H.Y., Hayek, A., and Al-Haddad, K. (2007) Averaged model based control of a Sheppard-Taylor PFC with nonlinearity compensation. Proceedings 20th IEEE Canadian Conference on Electrical and Computer Engineering (CCECE'07), Vancouver, BC, April 22–26, 2007.
44. Kanaan, H.Y., Hayek, A., and Al-Haddad, K. (2007) Small-signal averaged model and carrier-based linear control of a Sheppard-Taylor PFC. Proceedings IEEE International Symposium on Industrial Electronics (ISIE'07), Vigo, Spain, June 4-7, 2007, pp. 527–532.

45. Kanaan, H.Y., Hayek, A., and Al-Haddad, K. (2007) Comparative study of two average-model-based PWM control schemes for a Sheppard-Taylor PFC. Proceedings 38th IEEE Power Electronics Specialists Conference (PESC'07), Orlando, FL, June 17–21, 2007.
46. Kanaan, H.Y., Hayek, A., and Al-Haddad, K. (2008) Study, modeling and control of a single-phase power factor corrector based on the Sheppard-Taylor topology. Proceedings ELECTRIMACS'08, Quebec, Canada, June 8–11, 2008.
47. Kanaan, H.Y., Al-Haddad, K., Hayek, A., and Mougharbel, I. (2009) Design, study, modelling and control of a new single-phase high power factor rectifier based on the single-ended primary inductance converter and the Sheppard–Taylor topology. *IET Proceedings – Power Electronics*, **2** (2), 163–177.
48. Kanaan, H.Y., Hayek, A., and Al-Haddad, K. (2007) A new single-phase power factor corrector based on the SEPIC and Sheppard-Taylor topologies. Proceedings 29th International Telecommunications Energy Conference (INTELEC'07), Rome, Italy, September 30–October 4, 2007.
49. Kanaan, H.Y., Hayek, A., and Al-Haddad, K. (2008) Multi-loops control design for a new Sheppard-Taylor based power factor corrector with model-nonlinearity compensation. Proceedings IEEE International Symposium on Industrial Electronics (ISIE'08), Cambridge, UK, June 30–July 2, 2008.
50. Kanaan, H.Y., Hayek, A., and Al-Haddad, K. (2008) Small-signal averaged model and carrier-based linear control of a new Sheppard-Taylor-based PFC. Proceedings 34th Annual Conference of the IEEE Industrial Electronics Society (IECON'08), Orlando, FL, November 10–13, 2008.
51. Kanaan, H.Y. and Al-Haddad, K. (2009) Design, study, modeling and control of a modified Sheppard-Taylor PFC. Proceedings 35th Annual Conference of the IEEE Industrial Electronics Society (IECON'09), Porto, Portugal, November 3–5, 2009.
52. Kanaan, H.Y. and Al-Haddad, K. (2011) A modified Sheppard-Taylor power factor corrector operating in discontinuous capacitor voltage mode. Proceedings 20th IEEE International Symposium on Industrial Electronics (ISIE'11), Gdansk, Poland, June 27–30, 2011.
53. Fallaha, C., Kanaan, H.Y., and Al-Haddad, K. (2004) Implementation of a dSPACE-based digital controller for a single-phase UPF two stage boost rectifier. Proceedings IEEE PESC'04, Aachen, Germany, June 20–25, 2004.
54. Zhou, C., Ridley, R.B., and Lee, F.C. (1990) Design and analysis of a hysteretic boost power factor correction circuit. Proceedings IEEE PESC'90, pp. 800–807.
55. Noon, J.P. (1998) UC3855A/B high performance power factor pre-regulator. Unitrode Application Notes, Section U-153, pp. 3.460-3.479.
56. Kanaan, H., Al-Haddad, K., Chaffaï, R., and Duguay, L. (2000) Susceptibility and input impedance evaluation of a single phase unity power factor rectifier. 7th IEEE ICECS'2 K Conference, Beirut, Lebanon, December 17–20, 2000.
57. Kanaan, H., Al-Haddad, K., Chaffaï, R., and Duguay, L. (2003) Impédance d'entrée et susceptibilité d'un redresseur monophasé non polluant. *Revue Internationale de Génie Electrique, Hermès*, **6** (1–2), 187–224.
58. Kanaan, H. and Al-Haddad, K. (2002) A comparative evaluation of averaged model based linear and non-linear control laws applied to a single-phase two-stage boost rectifier. Proceedings RTST'02, Beirut & Byblos, Lebanon, March 4–6, 2002.
59. Fallaha, C., Kanaan, H.Y., and Al-Haddad, K. (2004) Design of a digital linear controller for a single-phase UPF two-stage boost rectifier using the D-Space Tool of Matlab/Simulink. Proceedings IEEE ISCCSP'04, Hammamet, Tunisia, March 21–24, 2004, pp. 11–14.
60. Salmon, J.C. (1993) Circuit topologies for single-phase voltage-doubler boost rectifiers. *IEEE Transactions on Power Electronics*, **8** (4), 521–529.
61. Kanaan, H.Y., Al-Haddad, K. and Fadel, M. (2011) Modeling and control of a two-switch asymmetrical half-bridge Boost power factor corrector for single-phase rectifiers. Proceedings Workshop on Power Electronics in Industrial Applications and Renewable Energy (PEIA'11), Doha, Qatar, November 3–4, 2011.
62. Kanaan, H.Y. and Al Haddad, K. (2009) Modeling and simulation of a DC-DC dual-Boost converter in continuous and discontinuous current modes using the switching functions approach. Proceedings 3rd International Conference on Modeling, Simulation, and Applied Optimization (ICMSAO'09), Sharjah, United Arab Emirates, January 20–22, 2009.
63. Balog, R. and Redl, R. (1993) Power factor correction with interleaved Boost converter in continuous inductor current mode. APEC 93, pp. 167–174.
64. Miwa, B.A., Otten, D.M., and Schlect, M. (1992) High efficiency power factor correction using interleaving techniques. Proceedings IEEE Applied Power Electronics Conference 1992, pp. 557–568.
65. Pinto, J.A.C. *et al.* (1997) A power factor correction pre-regulator AC-DC interleaved Boost with soft-commutation. INTELEC 97, pp. 121–125.

66. Chen, C.H. and Pong, M.H.(1998) Input current analysis of interleaved Boost converters operating in discontinuous inductor current mode. IEEE PESC 1998, pp. 905–910.
67. Braga, H.A.C. and Barbi, I. (1999) A 3 KW unity power factor rectifier based on a two cell Boost converter using a new parallel-connection technique. *IEEE Transactions on Power Electronics*, **14**, 209–217.
68. Yoshida, T., Shiizuka, O., Miyachita, O., and Ohniwa, K. (2000) An improved technique for the efficiency of high-frequency switch-mode rectifiers. *IEEE Transactions on Power Electronics*, **15** (6).
69. Barbosa, P. *et al* (2001) Interleaved three-phase Boost rectifiers operated in the discontinuous conduction mode: Analysis, design considerations and experimentation. *IEEE Transactions on Power Electronics*, **16** (5).
70. Zumel, P., Gracia, O., Cobos, J. A., and Uceda, J. (2002) Exploring interleaved converters as an EMI reduction technique in power converters. IEEE- PESC 2002, pp. 1219–1224.
71. Saito, T., Yashizawa, M., Torikai, H., and Tazaki, S. (2002) Analysis of interleaved converters with WTA based switching. INTELEC 2002, Vol. 3, pp. 405–415.
72. Gallo, C.A. *et al.* (2002) Soft-switched PWM high frequency with PFC converter using Boost-Flyback converter interleaved. PESC 2002, pp. 356–360.
73. Singh, B.N., Chandra, A., Rastgoufard, P., and Al-Haddad, K. (2002) Single-phase switch mode boost rectifier: an improved design/control applied to three phase AC-DC converters to power up telecommunication system. 24th Annual International Telecommunications Energy Conference (INTELEC'02), September 29–October 3, 2002.
74. Veerachary, M., Senjyu, T., and Uezato, K. (2003) Maximum power point tracking of coupled inductor interleaved Boost converter supplied PV system. *IEEE Proceedings Electric Power Application*, **150** (1), 71–80.
75. Zhang, J., Lee, F.C., and Jovanhovic, M. (2003) An improved CCM single stage PFC converter with a low frequency auxiliary switch. *IEEE Transactions on Power Electronics*, **18** (1).
76. Kanaan, H.Y., Sauriole, G., and Al-Haddad, K. (2009) Small-signal modelling and linear control of a high efficiency dual boost single-phase power factor correction circuit. *IET Proceedings – Power Electronics*, **2** (6), 665–674.
77. Kanaan, H.Y., Marquis, A., and Al-Haddadvz, K. (2004) Small-signal modeling and linear control of a dual boost power factor correction circuit. Proceedings IEEE PESC'04, Aachen, Germany, June 20–25, 2004, Vol. 4, pp. 3127–3133.
78. Abourida, S. and Murere, G. (2001) *Simulateur temps-réel de systèmes électriques*, RT-LAB & Artémis, RT-Opal, UQTR, April 2001.
79. Wester, G.W. and Middlebrook, R.D. (1972) Low-frequency characterization of switched DC-to-DC converters. Proceedings IEEE Power Processing and Electronics Specialists Conference, Atlantic City, NJ, May 22–23, 1972.
80. Musavi, F., Al-Haddad, K., and Kanaan, H. (2004) A novel large signal modeling and dynamic analysis of paralleled DC/DC converters with automatic load sharing control. Proceedigs IEEE ICIT'04, Hammamet, Tunisia, December 8–10, 2004, Vol. 1, pp. 536–541.
81. Musavi, F., Al-Haddad, K., and Kanaan, H. (2005) A large signal averaged modelling and control of paralleled DC/DC converters with automatic load sharing. Proceedings IEEE APEC'05, Austin, TX, March 6–10, 2005, Vol. 2, pp. 1353–1358.
82. Middlebrook, R.D. and Cuk,S. (1976) A general unified approach to modeling switching-converter power stages. Proceedings IEEE PESC'76, Cleveland, OH, June 8–10, 1976.
83. Erickson, R.W., Cuk, S., and Middlebrook, R.D. (1982) Large-scale modeling and analysis of switching regulators. *IEEE PESC Record*, 240–250.
84. Sanders, S.R., Noworolski, J.M., Liu, X.Z., and Verghese, G.C. (1991) Generalized averaging method for power conversion circuits. *IEEE Transactions on Power Electronics*, **6** (2), 251–259.
85. Vorperian, V. (1990) Simplified analysis of PWM converters using the model of the PWM switch: Parts I and II. *IEEE Transactions on Aerospace and Electronic System*, **26**, 490–505.
86. Jatskevich, J., Wasynczuk, O., Walters, E.A., and Lucas, C.E. (2000) A globally continuous state-space representation of switched networks. CCECE'00, Vol. 1, pp. 559–563.
87. Krein, P.T., Bentsman, J., Bass, R.M., and Lesieutre, B.L. (1990) On the use of averaging for the analysis of power electronic systems. *IEEE Transactions on Power Electronics*, **5** (2), 182–190.
88. Sun, J. and Grotstollen, H. (1997) Symbolic analysis methods for averaged modeling of switching power converters. *IEEE Transactions on Power Electronics*, **12** (3), 537–546.
89. Aoun, M., El-Maalouf, M., Rouhana, N., Kanaan, H.Y., and Al-Haddad, K. (2012) Average modeling and linear control of a Buck-Boost KY converter. Proceedings ISCCSP'12, Rome, Italy, May 2–4, 2012.

90. Kanaan, H.Y. and Al-Haddad, K. (2005) A comparison between three modeling approaches for computer implementation of high-fixed-switching-frequency power converters operating in a continuous mode. Proceedings CCECE'02, Winnipeg, Canada, May 12–15, 2002, Vol. 1, pp. 274–279.
91. Kanaan, H.Y. and Al-Haddad, K. (2005) Modeling and simulation of DC-DC power converters in continuous and discontinuous modes using the switching functions approach. Proceedings ICMSAO'05, Sharjah, United Arab Emirates, February 1–3, 2005.
92. Kanaan, H.Y. and Al-Haddad, K. (2005) Modeling and simulation of DC-DC power converters in CCM and DCM using the switching functions approach: application to the Buck and Cùk converters. Proceedings 6th IEEE PEDS'05, Kuala Lumpur, Malaysia, November 28–December 1, 2005, Vol. 1, pp. 468–473.
93. Kanaan, H.Y., Al-Haddad, K., and Hayek, A. (2008) Modeling techniques applied to switch-mode power converters: Application to the Boost-type single-phase full-bridge rectifier. Proceedings HSI'08, Krakow, Poland, May 25–27, 2008.
94. Wu, R., Dewan, S.B., and Slemon, G.R. (1991) Analysis of an AC-to-DC voltage source converter using PWM with phase and amplitude control. *IEEE Transactions Industry Applications*, **27** (2), 355–364.
95. Lee, J.W., Nowicki, E., and Alfred, C. (2000) A computational small-signal modeling technique for switch mode converters, CCECE'00, Vol. 1, pp. 587–591.
96. Slotine, J.-J.E. and Li, W. (1991) *Applied Nonlinear Control*, Prentice-Hall.
97. Kanaan, H.Y., Al-Haddad, K., and Fadel, M. (2013) Modeling and control of a two-switch asymmetrical half-bridge boost power factor corrector for single-phase rectifiers. Proceedings 22nd IEEE International Symposium on Industrial Electronics (ISIE'13), Taipei, Taiwan, May 28–31, 2013.

17

Active Power Filter

Ahmed M. Massoud[1,3], Shehab Ahmed[2] and Ayman S. Abdel-Khalik[3]

[1]*Electrical Engineering Department, Qatar University, Doha, Qatar*

[2]*Electrical and Computer Engineering Department, Texas A&M at Qatar, Doha, Qatar*

[3]*Electrical Engineering Department, Alexandria University, Egypt*

17.1 Introduction

The proliferation of power electronics-based devices and equipment has significantly affected the power quality of the grid, which changes its sinusoidal nature through adding harmonic distortion. Power electronics applications have penetrated several venues in our life. In residential applications, power electronics-based devices can be found not only in devices such as personal computers but also in refrigerators, freezers, air conditioning systems, cookers and lighting. Extending up to commercial applications, power electronics applications embrace air conditioning systems, heaters, central refrigeration, uninterruptible power supply, elevators and lighting. In industrial applications, power electronics exist in pumps, compressors, fans, blowers, arc furnaces, induction heating and arc welding. In addition, interfacing renewable energy sources to the grid necessitates converters to comply with the grid voltage and frequency requirements. This power electronics-based equipment increases the share of nonlinear loads compared with that of linear loads, and hence degrades the power quality of the grid. Simultaneously, power electronics-based loads are sensitive to harmonic distortion. This chapter starts with the definition of harmonics, effects and negative consequences of harmonics, international standards for harmonics, types of harmonics, harmonic current sources and harmonic voltage sources. Then, the different methods for harmonics mitigation are introduced, including passive filters and active filters. Definitions of loading power and power factor in the 3D space current coordinate system are introduced. The voltage source inverter (VSI) and current source inverter (CSI) such as an active power filter (APF) are discussed. The series and shunt APF are then introduced, presenting the different configurations for hybrid filters. APFs in high-power applications are discussed. In addition, the APF switching frequency choice methodology is presented. Different harmonic extraction techniques such as P–Q theory, cross-vector theory, the instantaneous power theory using the rotating P–Q–R reference frame, synchronous reference frame, adaptive interference canceling technique, capacitor voltage control, time-domain correlation function technique and identification by Fourier series are presented and discussed. Then the shunt APF is modeled, presenting simulation and practical results of a low-power prototype. The series active power principle of operation is then presented. Finally, the unified power quality conditioner (UPQC) is presented.

Power Electronics for Renewable Energy Systems, Transportation and Industrial Applications, First Edition.
Edited by Haitham Abu-Rub, Mariusz Malinowski and Kamal Al-Haddad.

17.2 Harmonics

Harmonics are components of a periodic wave having frequencies that are integral multiples of the fundamental power network frequency, which can be presented through Fourier series. Harmonics are not transient phenomena and therefore should not be confused with transient phenomena such as spikes, dips and oscillations. Harmonics are typically the by-product of power electronics-based loads, which are proliferating tremendously in power systems. They occur frequently because of the large number of personal computers and electronic devices (single-phase loads), variable frequency drives (AC and DC), switch-mode power supplies, low-power consumption lamps, electrical arc furnaces, arc welding machines, transformer and induction motor saturation and any equipment that has nonlinear voltage/current characteristics (i.e., nonlinear loads). Most of these loads distort the load current. Because of the equivalent impedance of the grid, the supply voltage gets distorted. As the short-circuit level increases (stiff power system network), the grid impedance is low. Therefore, the effect of distorted current on the supply voltage quality is less than that of the low short-circuit level (weak power system network). For example, the IEEE-519 standard objective is to limit the harmonic injection from individual customers so as to limit the overall harmonic distortion in the grid voltage. The process is mutual between the customer and the utility; that is, the customer should act according to the standard to limit the distorted current injection and the utility should act to limit the voltage distortion for the supplied electric power. The harmonic level is measured through a term called *total harmonic distortion* (THD) of either voltage or current. The THD is used as a measure in the international standards for determining the allowable level of THD at certain voltage level and short-circuit capability. The THD is defined as the ratio between the rms of the harmonic components and the rms of the fundamental component. Measurement location in a power system network is important for assessing the harmonics, and therefore the term *point-of-common coupling* (PCC) arises. The PCC is the location where the harmonic voltage and current distortions are measured or calculated. The PCC can be on either the primary or the secondary winding of utility transformers or at the service entrance of the considered facility.

17.3 Effects and Negative Consequences of Harmonics

High levels of harmonic distortion can lead to distribution system problems and may cause a shutdown of other equipment, especially if their operation is sensitive to zero crossing [1]. Distortion problems can be severe in small/weak power networks where the equivalent power network impedance is significant, resulting in increased voltage distortion at the PCC [2]. Harmonics may adversely affect most of the power system components such as conductors, capacitors, transformers, motors and generators through overheating these components. In transformers and reactors, the eddy current and hysteresis losses increase and a resonance may occur between the transformer inductance and line capacitance. In capacitors, as the reactive power increases, the dielectric losses increase, which may cause capacitor failure, lifetime decreases and the possibility of overvoltage and resonance. In cables, skin and proximity effects increase heat, and so dielectric breakdown of cable insulation may occur. Harmonics affect switchgear through changing the rate of rise of the transient recovery voltage and the operation of the blow-out coil, and the current-carrying parts will be affected by skin and proximity effects. The time-delay characteristics of the relay may be affected by harmonics. Relays may be affected by harmonics also through interference, especially microprocessor based. Also, false tripping may occur. In generators, the rotor heating increases and torque ripples are produced, which adds to maintenance requirements and may eventually lead to failure. Motor losses increase due to harmonics and eddy currents and skin effects increase. In addition, it may cause misoperation of electronic devices (which depends on voltage zero crossing detection or sensitivity to the voltage waveform). It also affects protective devices, depending on the design features and principles of operation and causes interference with telephones. It may introduce either series or parallel resonances in the power system network.

17.4 International Standards for Harmonics

In 1981, the IEEE introduced the IEEE standard 519 "IEEE Recommended Practices and Requirements for Harmonic Control in Electrical Power Systems" to provide guidelines for dealing with harmonics presented by nonlinear loads so that power quality problems could be mitigated. The IEEE 519-1992 limits for voltage and current harmonics are shown in Tables 17.1–17.4 for the current distortion limits in distribution systems, current distortion limits in subtransmission systems, current distortion limits in transmission systems and voltage distortion limits, respectively, [3].

The THD_V is defined as the ratio of the rms of the voltage harmonics to the rms of the fundamental component, as in Equation (17.1).

$$THD_V = \frac{\sqrt{V_2^2 + V_3^2 + V_4^2 + \ldots + V_h^2}}{V_1} \tag{17.1}$$

Table 17.1 Current distortion limits (in % of I_L) for general distribution systems (120–69 000 V) [3]

I_{SC}/I_L	$h<11$	$11 \leq h < 17$	$17 \leq h < 23$	$23 \leq h < 35$	$35 \geq h$	TDD
<20	4	2	1.5	0.6	0.3	5
20–50	7	3.5	2.5	1	0.5	8
50–100	10	4.5	4	1.5	0.7	12
100–1000	12	5.5	5	2	1	15
>1000	15	7	6	2.5	1.4	20

where I_{sc} is the short-circuit current, I_L is the maximum fundamental demand current, TDD is the total demand distortion and h is the harmonic order.

Table 17.2 Current distortion limits (in % of I_L) for general subtransmission systems (69 001–161 000 V) [3]

I_{SC}/I_L	$h<11$	$11 \leq h < 17$	$17 \leq h < 23$	$23 \leq h < 35$	$35 \geq h$	TDD
<20	2	1	0.75	0.3	0.15	2.5
20–50	3.5	1.75	1.25	0.5	0.25	4
50–100	5	2.25	2	0.75	0.35	6
100–1000	6	2.75	2.5	1	0.5	7.5
>1000	7.5	3.5	3	1.25	0.7	10

where I_{sc} is the short-circuit current, I_L is the maximum fundamental demand current, TDD is the total demand distortion and h is the harmonic order.

Table 17.3 Current distortion limits (in % of I_L) for general transmission systems (>161 000 V) [3]

I_{SC}/I_L	$h<11$	$11 \leq h < 17$	$17 \leq h < 23$	$23 \leq h < 35$	$35 \geq h$	TDD
<50	2	1	0.75	0.3	0.15	2.5
≥50	3	1.5	1.15	0.45	0.22	3.75

where I_{sc} is the short-circuit current, I_L is the maximum fundamental demand current, TDD is the total demand distortion and h is the harmonic order.

Table 17.4 Voltage distortion limits (in % of V_L) [3]

PCC voltage	Individual harmonic magnitude (%)	THD_V (%)
≤69 kV	3	5
69–161 kV	1.5	2.5
≥ 161 kV	1	1.5

where THD_V is the voltage THD.

The THD_I is defined as in Equation (17.1) but for current components. The TDD has the same definition as THD_I except that it is referred to as *the maximum fundamental demand current.*

The EN 61000 [4] (EN stands for Euro Norm which is identical to the International Electrotechnical Commission (IEC)) has been divided into six main parts. These are classified into, first, a general part dealing with general considerations (IEC 61000-1-x); second, an environmental part defining the characteristics of the environment where equipment is applied (IEC 61000-2-x); a third part defining the permissible levels of emissions that can be generated (IEC 61000-3-x); a fourth part providing detailed guidelines for measurement equipment and test procedures (IEC 61000-4-x); a fifth part providing guidelines in the application of equipment for ensuring electromagnetic compatibility among electrical and electronic apparatus or systems (IEC 61000-5-x); and a miscellaneous part with generic standards defining the immunity and emission levels required for equipment in general categories or for specific types of equipment (IEC 61000-6-x). The EN 61000-3-x is of high interest in the permissible harmonic level provision to the network. The EN 61000-3-2 controls the level of distortion of the current drawn by all equipment rated up to 16 A, and EN61000-3-3 controls the level of voltage changes that equipment rated up to 16 A will impose on the AC supply. Previously, the harmonic standard was called *IEC 555 Part 2* (*1987*), and the flicker standard was called *IEC 555 Part 3* (*1987*); these are now referred to as *EN61000-3-2* and *EN61000-3-3 Standards.*

17.5 Types of Harmonics

17.5.1 Harmonic Current Sources

In load (naturally)-commutated converters with inductive loads, the harmonic current content and characteristics are less dependent on the AC side. Therefore, this type of harmonic source behaves as a harmonic current source [5]. Thus, it is called *a harmonic current source* and is represented as a current source. Figure 17.1(a) presents a three-phase uncontrolled rectifier feeding an inductive load, which is considered a common example of the harmonic current sources. Figure 17.1(b)–(e) represents the supply current and its spectrum, and the voltage at the PCC and its spectrum, respectively, where the supply voltage is 220 V at 50 Hz and the source impedance is 0.1 mH and 0.1 Ω. The nonstiff supply (supply with high internal impedance) significantly affects the voltage at the PCC and, therefore, propagates the harmonics through the power system. This is clear when the same system is simulated but for a source impedance of 1 mH and 0.1 Ω. Figure 17.2(a)–(d) represents the supply current and its spectrum and the voltage at the PCC and its spectrum, respectively.

17.5.2 Harmonic Voltage Sources

Another typical harmonic source is that caused by diode rectifiers with DC capacitors. As the impedance of capacitors decreases at higher frequencies [5], connecting large capacitance to the DC side of the

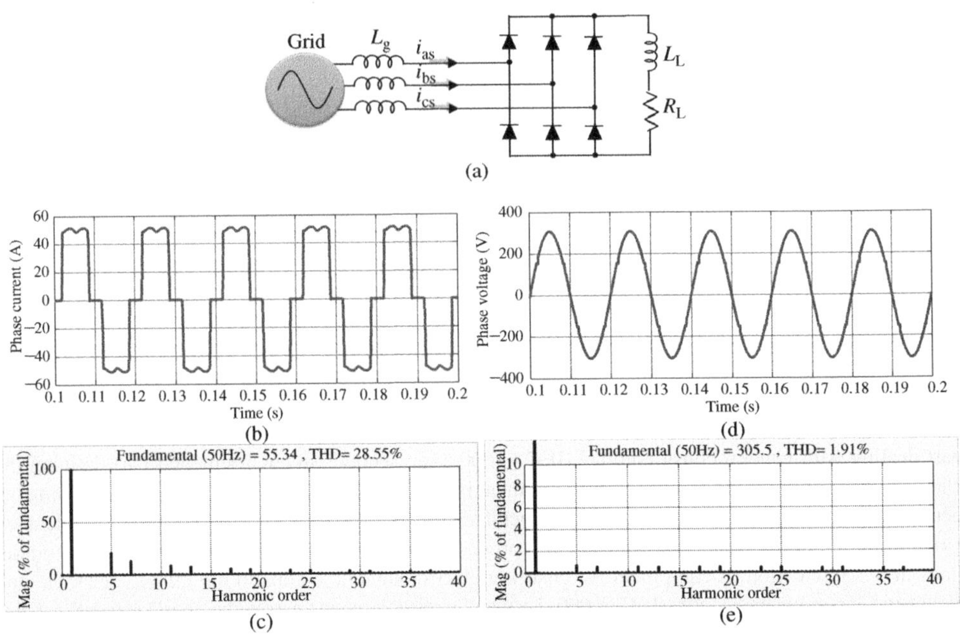

Figure 17.1 Harmonic current source: (a) three-phase uncontrolled rectifier, (b) and (c) source current and its spectrum and (d) and (e) supply voltage and its spectrum (supply impedance 0.1 mH and 0.1 Ω)

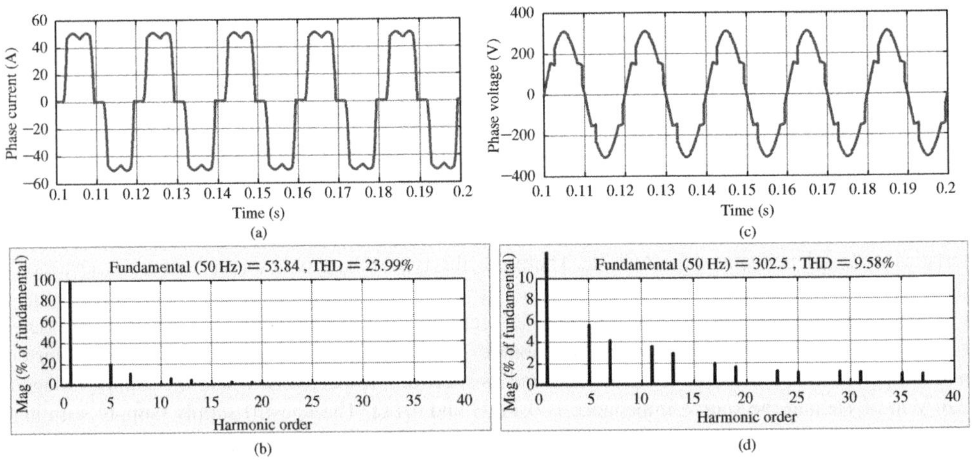

Figure 17.2 Harmonic current source: (a) and (b) source current and its spectrum and (c) and (d) supply voltage and its spectrum (supply impedance 1 mH and 0.1 Ω)

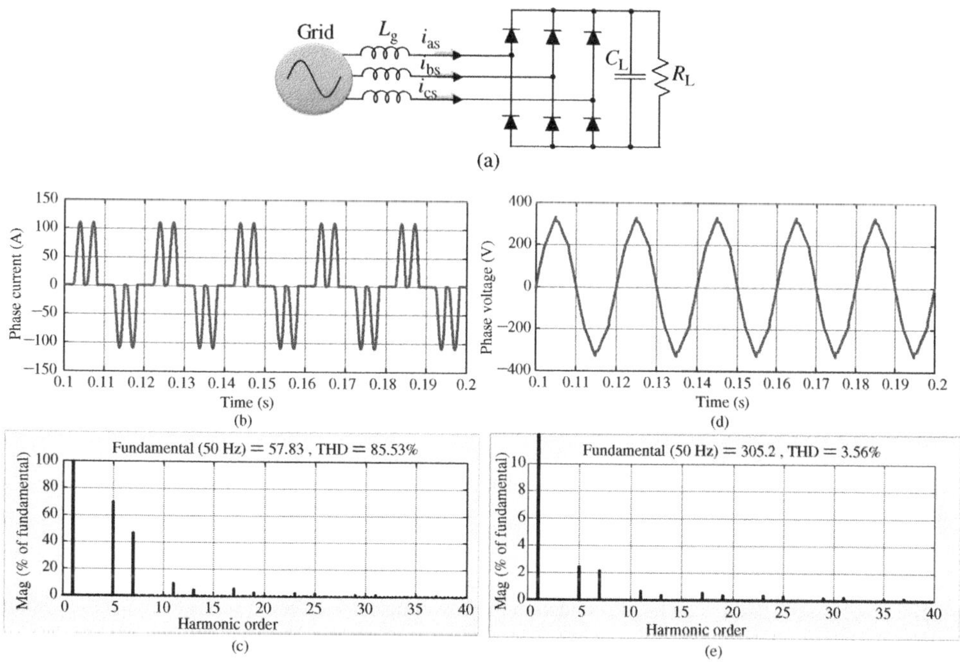

Figure 17.3 Harmonic voltage source: (a) three-phase uncontrolled rectifier, (b) and (c) source current and its spectrum and (d) and (e) supply voltage and its spectrum (supply impedance 0.1 mH and 0.1 Ω)

diode rectifier causes lower impedance for harmonics. Figure 17.3(a) presents a three-phase uncontrolled rectifier feeding a capacitive load. Figure 17.3(b)–(e) represents the supply current and its spectrum, and the voltage at the PCC and its spectrum, respectively, where the supply voltage is 220 V at 50 Hz and the source impedance is 0.1 mH and 0.1 Ω. The harmonic current amplitude on the AC side is affected by the impedance of the AC side; therefore, a diode rectifier behaves like a voltage source (rather than a current source). This is clear when the same system is simulated but for a source impedance of 1 mH and 0.1 Ω. Figure 17.4(a)–(d) represents the supply current and its spectrum, and the voltage at the PCC and its spectrum, respectively.

17.6 Passive Filters

In order to cancel out or mitigate harmonics, two main approaches have been considered in the last decades. In the first approach, the AC input current for the power electronics-based load is considered to be as near as possible to a sinusoidal waveform and in phase with the supply AC voltage. Therefore, the load behaves virtually as a resistive load, provided that the power electronics is of sufficient controllability (bandwidth) to achieve the targeted objective. This can be seen in active front-end rectifiers (force-commutated converters). However, this approach fails in load-commutated converters.

The second approach is to use an additional system component for harmonic mitigation, namely power filters. Power filters were first introduced for harmonic compensation as passive power filters. Passive filters consist of inductors, capacitors and damping resistors connected either in series or in parallel. Various types of passive filters with different filtering characteristics and for different application requirements

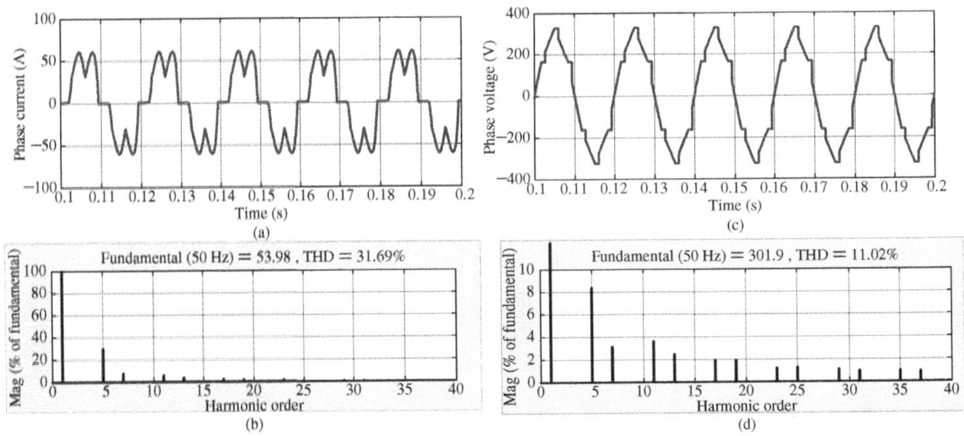

Figure 17.4 Harmonic voltage source: (a) and (b) source current and its spectrum and (c) and (d) supply voltage and its spectrum (supply impedance 1 mH and 0.1 Ω)

are available. Passive filters can be introduced in the power network either in series or in parallel. Shunt passive filters are filters tuned at the required harmonic component to be filtered. The filter order can be of first, second, or third order, based on the filtering requirements. The shunt passive filter is greatly affected by the source impedance, which in turn is very difficult to measure. The series passive filter is connected in series with the network to provide a high impedance path of the harmonic current, preventing its flow. Traditionally, the passive filter with thyristor-controlled reactor (TCR) is used to filter harmonics, in addition to compensating reactive current components [6]. Advantages of passive filters are easy maintenance, lower costs and less complexity. However, in practice, these filters have many disadvantages [7–11], which can be enumerated as follows:

- When both harmonic and fundamental current components flow into the filter, the filter should be rated considering both components.
- The filter may be overloaded with harmonic current increase.
- Parallel resonance between the AC source and passive filter causes overvoltage.
- Detuning of the harmonic frequency with aging of passive components.
- Filtering characteristics are dependent on the source impedance that is not accurately known and may change with network configuration.
- The passive filter may cause series resonance, with the AC source leading to large currents flowing through the filter.
- The operating frequency of the AC system varies around its nominal value as the loading condition of the system changes, and should therefore be considered in filter design.
- Passive filters mitigate only the frequencies they are tuned to.
- There are constraints on the choice of filter component in order to reach a compromise between the requirements of both the AC current harmonic and the reactive power supplied.

17.7 Power Definitions

To avoid the demerits of the passive filter, the APF has been proposed. Before detailing the APF, the authors will shed light on definitions that will ease understanding of the concept and principle of operation of the APF.

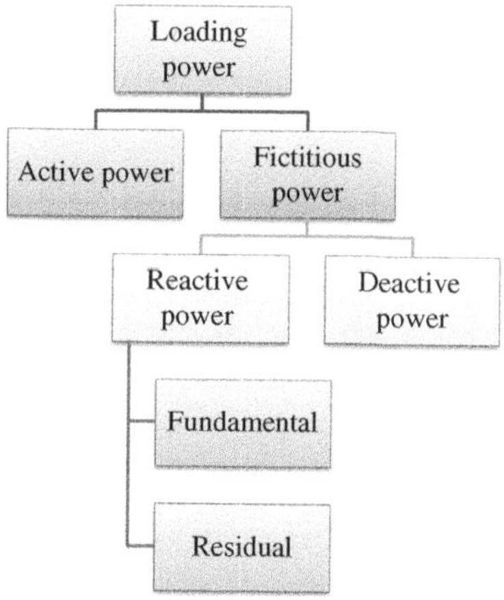

Figure 17.5 Loading power components

17.7.1 *Loading Power and Power Factor*

Traditionally, and considering only sinusoidal waveforms, active power, defined as the average of the product of voltage and current, has been recognized to cause real electric energy flow between two subsystems [12]. Conversely, reactive power, defined as the orthogonal term to active power, has been considered to cause no real energy flow, but increases the steady-state current in sinusoidal systems. The power factor is therefore defined as the ratio between active power and apparent power (the resultant of the active power and reactive power components). However, considering the harmonics, the definition of the power factor will differ with the introduction of new terms that are clarified in the following sections.

17.7.2 *Loading Power Definition*

Loading power [13] can be divided, as shown in Figure 17.5, into the following:

- *Active power*, which is defined as the average rate of the energy transfer from the source to the load over an interval of time. Active current [14] is defined as the minimum effective current associated with the transfer of the average power to the load over a predefined period.
- *Fictitious power* includes all components of power that cause the loading power to be higher than the active power.
- *Reactive power* can be divided into fundamental power and residual power. The *fundamental reactive power* is the cross-product of the fundamental component of voltage and current, while the *residual reactive power* component is the difference between the reactive power and fundamental reactive power. For a single-phase circuit, reactive power is the power that circulates between the source and the load, while in polyphase circuits, reactive power is the power that circulates between the source and the phases and the power that circulates among phases.
- *Deactive power* is associated with uncorrelated voltage and current waveforms. For this reason, the deactive power can be called *uncorrelated fictitious power*.

17.7.3 *Power Factor Definition in 3D Space Current Coordinate System*

The conventional 2D current coordinate system decomposes the load current into active and reactive components where this is effective only in the case of linear loads. But, in the case of nonlinear loads,

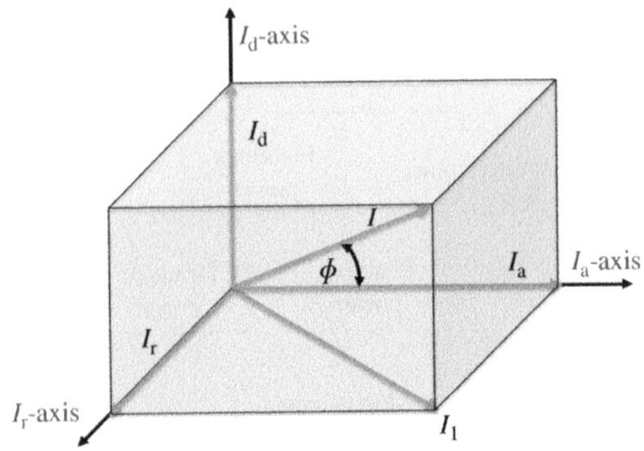

Figure 17.6 3D space current coordinate system

the nonfundamental current cannot be represented in 2D, and therefore a new current coordinate system to present the directly distorted portion of system current is required in the case of nonlinear loads (3D coordinate). In this definition, the current is expressed in terms of three mutually orthogonal components [13]: active, reactive and distorted components, as shown in Figure 17.6. Therefore, their rms fulfill the following relationships

$$I^2 = I_a^2 + I_r^2 + I_d^2 \tag{17.2}$$

$$I^2 = I_1^2 + I_d^2 \tag{17.3}$$

$$I_1^2 = I_a^2 + I_r^2 \tag{17.4}$$

where I is the system rms current, I_a is the fundamental active rms component of I, I_r is the fundamental reactive rms component of I, I_d is the distorted rms component of I and I_1 is the fundamental rms component of I.

If the supply voltage is a nondistorted sinusoidal waveform, then only the fundamental component of the current contributes to average power flow. So the power factor angle shown in Figure 17.4 can be expressed as

$$\cos\varphi = \frac{I_a}{I} = \frac{\text{Fundamental active rms component of } I}{\text{System rms current}} \tag{17.5}$$

If the supply voltage is a distorted sinusoidal waveform, then only the fundamental component of the current and the harmonic current components that correspond to the harmonics in the supply voltage contribute to the average power. But the contribution of the harmonic components to the active power is small and can be neglected with respect to the fundamental component. Therefore, the power factor can be defined generally as

$$\cos\ \varphi = \frac{\text{Fundamental active power}}{\text{Apparent power}} \tag{17.6}$$

This helps in evaluating the performance of the APF. If only I_r is compensated, this power electronic device is named as a STATCOM. If I_d or both (I_d and I_r) are compensated, then it is named as an APF. However, the power factor cannot be brought to unity till totally compensating I_d and I_r. The conventional power factor definition (displacement factor), which is defined in the 2D coordinate system, is misleading especially in industries that highly source harmonics into power systems, for example, the steel and aluminum industries, where reducing the distorted current components is the key factor for increasing the power factor.

17.8 Active Power Filters

APFs, as in passive filters, can be divided into AC and DC power filters [15]. DC APFs have been designed to compensate for current and/or voltage harmonics on the DC side of thyristor converters for high-power applications such as high-voltage DC (HVDC) systems and high power drives [16–20]. The DC APF can be considered as a special case of the AC APF. AC APFs are generally known as *APFs*, and are also called *active power line conditioners*, *active filters* and *active power quality conditioners* [21]. APFs can be employed in power system for providing compensation for harmonics, reactive power, and/or neutral current. APFs can also be used to eliminate voltage harmonics, to regulate terminal voltage, to suppress voltage flicker, and to improve voltage balance in three-phase systems [21]. APFs, compared to that of passive filters, have better harmonic compensation characteristics with impedance variation of the AC power line and the frequency variation of harmonic currents. A voltage source pulse-width modulation (PWM) inverter is commonly used as APF and var compensator. A voltage source APF has a capacitor on the DC side with a regulated DC voltage, whereas a current source APF has an inductor with a regulated DC current. Although the voltage source type is better in terms of losses and filter capacity to eliminate PWM carrier harmonics, the current source type is better in terms of the dynamics of compensating current as well as reliability and protection [8]. However, generally the APF experiences few disadvantages [11]:

- The initial and running costs, losses and complexity of APFs are higher than for passive filters.
- The trade-off in the design of APFs with high-power rating and fast dynamic current response.

Depending on the topology, the APF can be classified as series, shunt, or combined shunt series (UPQC), as shown in Figure 17.7. Combinations of APF and passive filters are known as *hybrid filters* [21].

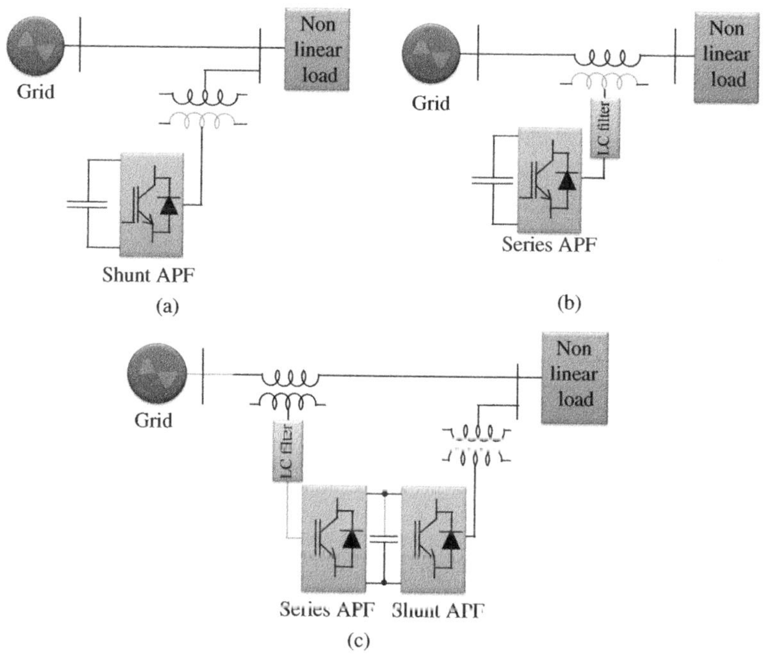

Figure 17.7 (a) Shunt APF, (b) series APF and (c) UPQC

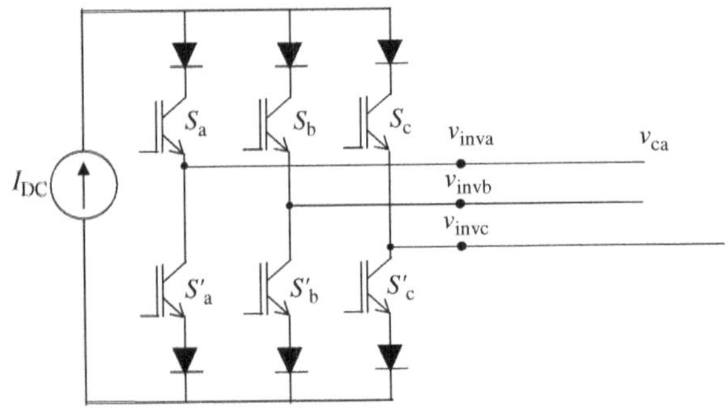

Figure 17.8 Current source inverter

The maximum harmonic order to be suppressed by the APF has no theoretical limit and is governed by the switching frequency employed for the APF. Theoretically, injecting PWM current enables the harmonic components of orders not greater than the pulse number per half-cycle to be mitigated [22]. If the harmonic components change in magnitude and frequency, the APF continues functioning without introducing any change in the system hardware, which is considered an advantage compared with that of the passive filter.

17.8.1 Current Source Inverter APF

The CSI behaves as a nonsinusoidal current source, providing the harmonic current requirements of the nonlinear load. A diode is used in series with the self-commutating device (insulated-gate bipolar transistor (IGBT)) for reverse voltage blocking (as shown in Figure 17.8), which can be replaced with IGBT of bidirectional voltage blocking capability or integrated gate-commutated thyristor (IGCT) [21]. Also, GTO-based configurations can be used but have a restricted switching frequency. The CSIs are considered sufficiently reliable, but have higher losses and require high inductance. Moreover, they are difficult to use in multilevel applications compared with that of VSI.

17.8.2 Voltage Source Inverter APF

The VSI APF is based on a VSI (shown in Figure 17.9) and has a self-supporting DC voltage bus with a large DC capacitor [21]. It has become more dominant since it is lighter, cheaper and easily expandable to multilevel applications as the cascaded type, diode clamped and flying capacitor multilevel inverter.

17.8.3 Shunt Active Power Filter

Shunt APFs have the advantage of carrying only the compensation current in addition to a small active fundamental current component to compensate for system losses and sustain a constant voltage across the DC capacitor [23]. Nevertheless, the shunt APF can be employed for reactive power compensation as well. It is common to attempt to use shunt APFs for harmonic compensation of harmonic voltage sources. Nevertheless, the impedance of the diode rectifier to harmonics is not much higher than that of the source side. Therefore, the compensating current injected by the shunt active filter flows into both the source and diode rectifier. Therefore, shunt active filters cannot cancel the harmonics completely but may cause problems such as enlarging the DC ripple and AC peak current of the diode rectifier. To avoid

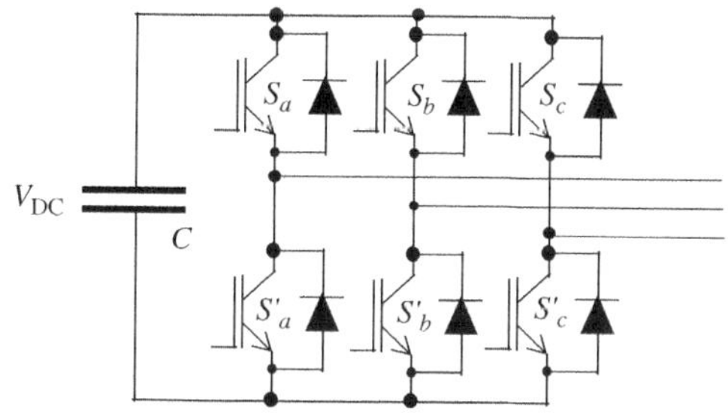

Figure 17.9 Voltage source inverter

these problems, Peng. showed that added series inductance is required [5] if a shunt APF is used. In Ref. [5], the use of a series active filter is proposed to overcome these problems.

17.8.4 Series Active Power Filter

The series APF does not compensate for load current harmonics but acts as a high impedance to current harmonics from the source side. Series APFs are less common industrially than their counterpart shunt APFs. This is because they have to handle the full load current, which increases their current rating considerably compared with that of the parallel filters. This effect may be worse on the secondary side of the coupling transformer, increasing the I^2-R losses and the physical size of the filter. The main advantage of series filters over the parallel version is that they can eliminate voltage-waveform harmonics and balance the three-phase voltages at the PCC of interest [23]. The approaches available for detecting harmonic voltage are similar to that for detecting harmonic current in the shunt APF. Table 17.5 summarizes the salient features of both shunt and series APFs.

17.8.5 Hybrid Filters

The series APF, which constitutes high impedance for high-frequency harmonics [23], can be accompanied by a parallel passive filter to provide a path for the harmonic currents of the load as shown in Figure 17.10(c). Alternatively, a shunt APF is designed to eliminate only part of the low-order current harmonics while the passive filter is designed to eliminate the bulk of the load–current harmonics [23], as shown in Figure 17.10(a). The main drawback is that it contains many power components, especially the passive filter. Since filters are permanently connected to the system, this approach is only suitable for a single load with a predefined harmonic source.

A shunt APF in series with passive filters is shown in Figure 17.10(b) and suits medium- and high-voltage applications where the passive filter reduces the voltage stress on the APF switches [23]. To reduce the complexity of the active and passive filter, the active filter is connected in series while the passive filter is connected in parallel [23], as shown in Figure 17.10(c). This configuration overcomes the problems of source impedance interaction with the passive filter.

The UPQC for the distribution system shown in Figure 17.10(d) is different in operation, purpose and control strategy from the unified power flow controller for transmission systems [15]. The functions being performed by the series active filter are harmonic isolation between the subtransmission system and the distribution system, voltage regulation and voltage flicker/imbalance compensation at the PCC. The functions performed by the shunt active filter are harmonic current and/or negative sequence current compensation and DC-link voltage regulation between both active filters.

Table 17.5 Comparison of shunt and series APFs

	Shunt APF	Series APF
Connection	It is connected in distribution networks in shunt	It is connected in distribution networks in series
APF acts as	The shunt APF acts as a current source	The series APF acts as a voltage source
Rating	It is rated at the full load voltage but the current rating is partially the harmonics and reactive current components	It is rated at the full load current but the voltage rating is partially the compensated voltage components
Function	Current harmonics filtering Reactive current compensation Current unbalance mitigation	Voltage harmonics mitigation Voltage sag or swell mitigation Voltage unbalance mitigation Current harmonics filtering Reactive current compensation Current unbalance mitigation
Compensation characteristics	The compensation characteristics of the shunt APF are independent of the source impedance for current source loads	The compensation characteristics of the series APF are independent of the source impedance and the load impedance for voltage source loads
Application considerations	Injected current flows into the load side and may cause overcurrent when applied to a voltage source load	A low-impedance parallel branch (parallel passive filter or power factor improvement capacitor bank) is needed when applied to an inductive or current source load
Considered loads	Inductive or current source loads or harmonic current sources	Capacitive or voltage source loads or harmonic voltage sources

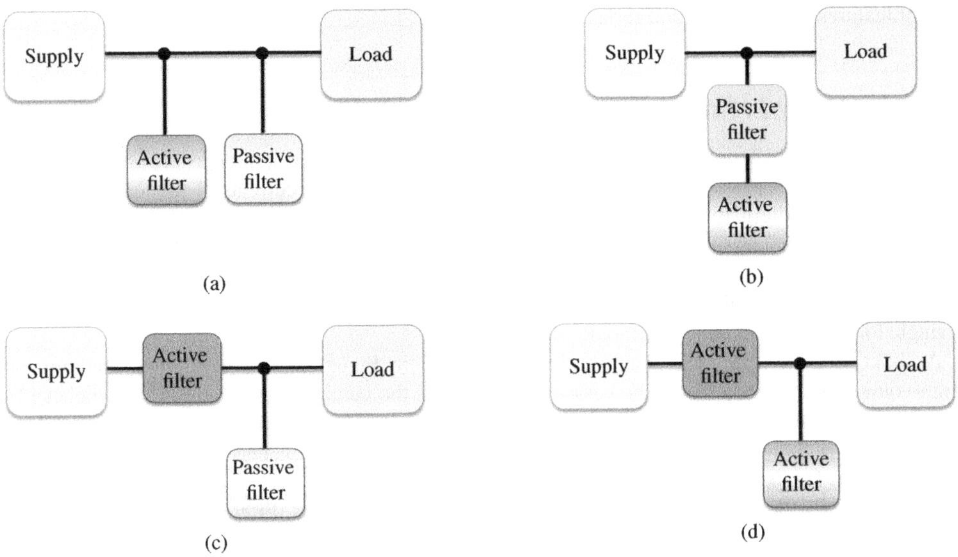

Figure 17.10 Filter combinations: (a) parallel active and passive filters, (b) active filter in series with the parallel passive filter, (c) series active and parallel passive filter and (d) unified power quality conditioner

17.8.6 High-Power Applications

The implementation of high-power APF filters with the conventional two-level VSI may not be cost-effective, because of the lack of high-switching frequency high-power single devices that can control the current flow at high-power ratings [23, 25]. Harmonic pollution in the high-power ranges is not the major problem, as in low-power ranges. High-power applications include power subtransmission grids and high-power AC and DC drives, as well as DC transmission systems. The effect of harmonics generated in the low-power range is minimized, either naturally or by the installation of several medium- and low-power APFs downstream, which contribute to the compensation. Static var compensation is the major concern and is usually catered for by using traditional static power conditioners/filters, such as several sets of synchronous condensers connected in parallel or in multilevel inverter var compensators. One of the applications of APFs in high-power systems is the Japanese bullet train (Shinkansen), which uses a parallel combination of several APFs.

17.9 APF Switching Frequency Choice Methodology

An important parameter in APF design is the order "m_h" of the highest harmonic to be reduced [26]. The required filtering capability of the APF can be defined as follows:

$$f_{af} = m_h \cdot f_s \tag{17.7}$$

where f_s is the fundamental frequency of the power network. The frequency f_{af} is chosen based on the maximum switching frequency of the APF (f_{maxAPF}). If the required f_{af} is higher than the APF maximum switching frequency capability f_{maxAPF}, it is impossible to control the APF line currents, hence the harmonics are not fully compensated. The APF maximum switching frequency f_{maxAPF} determines the filtering capability of the active filter through the following relationship, $2f_{af} < f_{maxAPF} < 10f_{af}$, as shown in Figure 17.11. The lowest factor results when semiconductors are the limiting aspect, whereas the highest factor is applicable to low-power APFs using semiconductor devices with the high-switching frequency capability such as metal-oxide semiconductor field-effect transistors (MOSFETs) and IGBTs. A high-pass passive filter can be used to reduce the harmonics beyond the APF capabilities. Another design step is based on the control system as it is implemented in a digital structure. The control algorithms are usually implemented in DSP, where most of the available DSPs have a high-processing capability that mostly will not limit the capability of the APF. For VSIs, the switching frequency f_{maxAPF} is dependent on the inverter power requirement. Generally, this power clearly limits the maximum value of f_{VSI}. To circumvent this limit for high-power VSIs, the two-level VSI topology can be changed to a multilevel VSI (number of levels greater than 2).

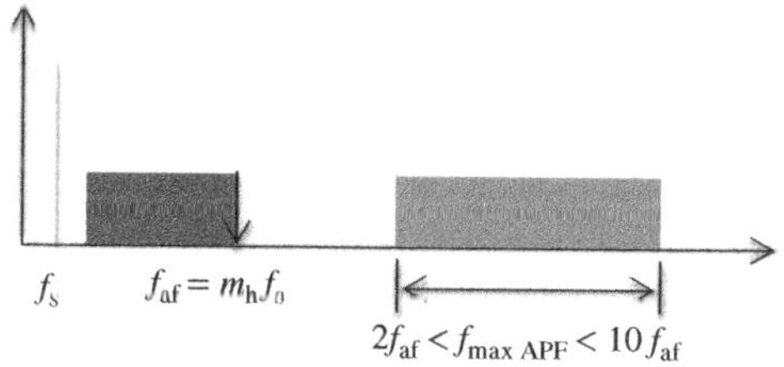

Figure 17.11 APF switching frequency choice methodology

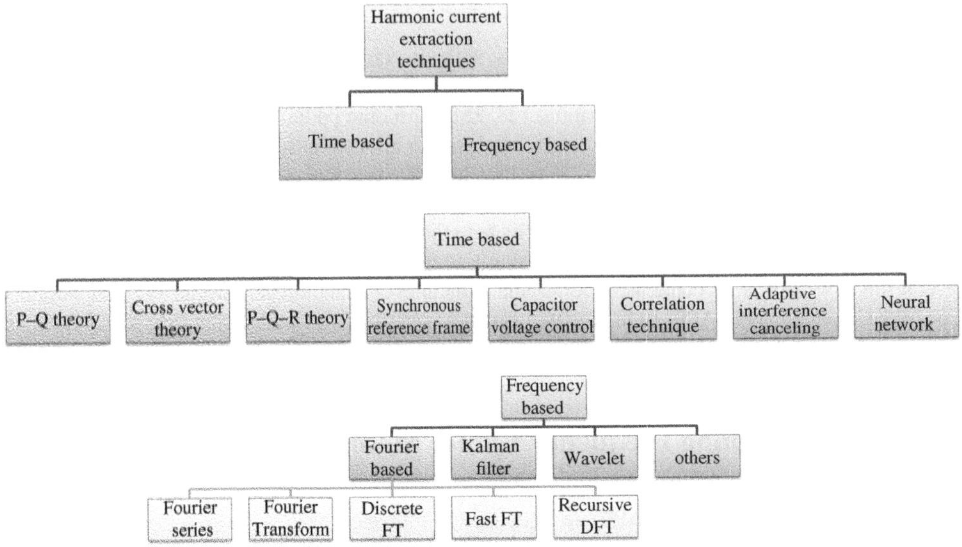

Figure 17.12 Classification of harmonic current extraction techniques

17.10 Harmonic Current Extraction Techniques (HCET)

In this section, even though the attention is given to the shunt APF and its harmonic current extraction techniques, nevertheless and analogously this can be reflected on the series APF and its voltage harmonic extract techniques. Various power theories and techniques have been developed for harmonic current extraction technique (HCET) where the authors tried to shed light on the most important and common techniques. HCET can be classified into time-based and frequency-based techniques, as shown in Figure 17.12.

17.10.1 P–Q Theory

The *P–Q* theory (the instantaneous power theory or the instantaneous reactive power theory) based on the instantaneous value concept was proposed in 1983 [24, 27]. In the instantaneous power theory and using Clarke transformation, the three-phase voltages and currents can be expressed as follows:

$$\begin{pmatrix} v_\alpha \\ v_\beta \\ v_0 \end{pmatrix} = \sqrt{\frac{2}{3}} \begin{pmatrix} 1 & \frac{-1}{2} & \frac{-1}{2} \\ 0 & \frac{\sqrt{3}}{2} & \frac{-\sqrt{3}}{2} \\ \frac{1}{\sqrt{2}} & \frac{1}{\sqrt{2}} & \frac{1}{\sqrt{2}} \end{pmatrix} \begin{pmatrix} v_a \\ v_b \\ v_c \end{pmatrix} \quad \text{and} \quad \begin{pmatrix} i_\alpha \\ i_\beta \\ i_0 \end{pmatrix} = \sqrt{\frac{2}{3}} \begin{pmatrix} 1 & \frac{-1}{2} & \frac{-1}{2} \\ 0 & \frac{\sqrt{3}}{2} & \frac{-\sqrt{3}}{2} \\ \frac{1}{\sqrt{2}} & \frac{1}{\sqrt{2}} & \frac{1}{\sqrt{2}} \end{pmatrix} \begin{pmatrix} i_a \\ i_b \\ i_c \end{pmatrix} \tag{17.8}$$

If the three-phase system has three wires, the zero-sequence current components will not exist, and therefore the contribution will be due to the $\alpha-\beta$ components only.

The instantaneous active power for a three-phase circuit can be defined in $\alpha-\beta$ coordinates as

$$p = v_\alpha i_\alpha + v_\beta i_\beta = \widetilde{p} + \overline{p} \tag{17.9}$$

where $\overline{p}$ is the average instantaneous active power. It is related to the active power transferred from the source to the load through the three phases (this is the main required component in the power system). $\widetilde{p}$ is alternating instantaneous active power (the active power exchanged between the source and the load).

The instantaneous reactive power is defined as

$$q = v_\beta i_\alpha - v_\alpha i_\beta = \widetilde{q} + \overline{q} \tag{17.10}$$

where $\overline{q}$ is the average instantaneous reactive power. $\widetilde{q}$ is the alternating instantaneous reactive power.

The instantaneous zero-sequence power is defined as

$$p_0 = v_0 i_0 = \widetilde{p}_0 + \overline{p}_0 \tag{17.11}$$

where $\overline{p}_0$ is the mean value of the instantaneous zero-sequence active power. This is related to the active power transferred from the source to the load through the zero-sequence voltage and current components. $\widetilde{p}_0$ is the alternating instantaneous zero-sequence active power, which is the power exchanged between the source and the load (as seen in Figure 17.13).

The oscillating active and reactive power are undesirable as they are a result of the harmonics. In some cases, the average reactive power is not desirable (in reactive power compensation for unity power factor operation). The oscillating active and reactive power can be extracted through filtering the total active and reactive power.

For a three-phase nonlinear load, if the reactive and harmonic producing currents are to be compensated, the compensating current can be expressed as

$$\begin{pmatrix} i_{c\alpha} \\ i_{c\beta} \end{pmatrix} = \begin{pmatrix} v_\alpha & v_\beta \\ -v_\beta & v_\alpha \end{pmatrix}^{-1} \begin{pmatrix} \widetilde{p} \\ q \text{ or } \widetilde{q} \end{pmatrix} \tag{17.12}$$

where $\widetilde{p}$ is the AC component of the instantaneous active power. Then using the Clarke inverse transformation, the three-phase currents contributed by the active filter can be calculated. If the voltage at the PCC is not balanced and/or nonsinusoidal, the compensated currents are no longer sinusoidal. The harmonic components in the voltages at the PCC can be eliminated by filtering. This approach is preferred when the superimposed harmonic voltage components are of high frequencies. The phase-locked loop can be employed for extracting the positive sequence component of the voltage at the point of the common coupling. In a three-phase four-wire system, the compensating current can be expressed as follows:

$$\begin{pmatrix} i_{c\alpha} \\ i_{c\beta} \\ i_{c0} \end{pmatrix} = \frac{1}{v_{\alpha\beta}^2 v_0} \begin{pmatrix} v_0 v_\alpha & -v_0 v_\beta & 0 \\ v_0 v_\beta & v_0 v_\alpha & 0 \\ 0 & 0 & v_{\alpha\beta}^2 \end{pmatrix} \begin{pmatrix} \widetilde{p} \\ q \text{ or } \widetilde{q} \\ p_0 \end{pmatrix} \tag{17.13}$$

where $v_{\alpha\beta}^2 = v_\alpha^2 + v_\beta^2$

If the voltage of the PCC is unbalanced and sinusoidal, the source current after compensation is also not balanced and sinusoidal [28]. If the voltage applied is balanced and nonsinusoidal, the source current is distorted. Therefore, in the case of unbalanced and/or nonsinusoidal voltage, the compensating current

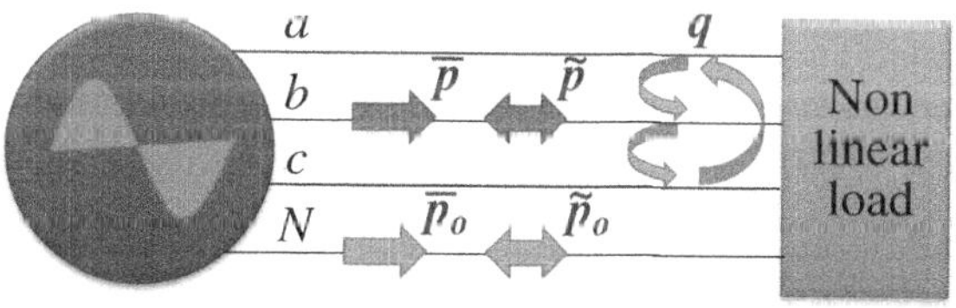

Figure 17.13 The power transferred to the load/oscillating between source and load

should be proportional to the fundamental positive sequence voltage component at the PCC. Therefore, the source current can be expressed as follows:

$$\vec{i}_{s} = \frac{P}{V_{f}^{+2}} \vec{v}_{f}^{+} \tag{17.14}$$

where V_f^+ is the rms of the fundamental positive sequence of the voltage vector and can be defined as follows:

$$V_{f}^{+2} = \frac{1}{T}\int_{T} \left(v_{1f}^{+2} + v_{2f}^{+2} + v_{3f}^{+2}\right) dt \tag{17.15}$$

where v_{1f}^{+2}, v_{2f}^{+2} and v_{3f}^{+2} are the components of the fundamental positive sequence of voltage vector.

17.10.2 Cross-Vector Theory

The cross-vector theory or modified P–Q theory [29, 30] defines an instantaneous real power p and three instantaneous imaginary powers. The instantaneous reactive power theory can be defined in the a–b–c reference frame as

$$q = v \times i = \begin{pmatrix} q_a \\ q_b \\ q_c \end{pmatrix} \quad \text{and} \quad p = v \cdot i \tag{17.16}$$

It can be defined in the α–β–0 reference frame as

$$\begin{pmatrix} p \\ q_0 \\ q_\alpha \\ q_\beta \end{pmatrix} = \begin{pmatrix} v_0 & v_\alpha & v_\beta \\ 0 & -v_\beta & v_\alpha \\ v_\beta & 0 & -v_0 \\ v_\alpha & v_0 & 0 \end{pmatrix} \begin{pmatrix} i_0 \\ i_\alpha \\ i_\beta \end{pmatrix} \tag{17.17}$$

where the instantaneous imaginary powers (q_0, q_α and q_β) are not independent. They are related as follows:

$$q_\alpha v_\alpha + q_\beta v_\beta + q_0 v_0 = 0 \tag{17.18}$$

The currents can be expressed as follows:

$$\begin{pmatrix} i_0 \\ i_\alpha \\ i_\beta \end{pmatrix} = \frac{1}{v_{0\alpha\beta}^2} \begin{pmatrix} v_0 & 0 & v_\beta & -v_\alpha \\ v_\alpha & -v_\beta & 0 & v_0 \\ v_\beta & v_\alpha & -v_0 & 0 \end{pmatrix} \begin{pmatrix} p \\ q_0 \\ q_\alpha \\ q_\beta \end{pmatrix} \tag{17.19}$$

where

$$v_{0\alpha\beta}^2 = v_0^2 + v_\alpha^2 + v_\beta^2 \tag{17.20}$$

Similarities and differences between the three-phase four-wire part in P–Q theory and the cross-vector theory can be summarized as follows. The P–Q theory considers the zero-sequence circuit as a single-phase circuit independent of the α-phase and β-phase circuits and deals with the zero-sequence current as an instantaneous active current only [31]. This means that no instantaneous reactive current exists in the zero-sequence circuit, while in the cross-vector theory, in the zero-sequence, α-phase and β-phase circuits are considered. This implies that the zero-sequence current can be divided into zero-sequence instantaneous active and reactive current components. The compensating current in the cross-vector theory can be expressed as follows:

$$\begin{pmatrix} i_{c0} \\ i_{c\alpha} \\ i_{c\beta} \end{pmatrix} = \frac{1}{v_{0\alpha\beta}^2} \begin{pmatrix} v_0 & 0 & v_\beta & -v_\alpha \\ v_\alpha & -v_\beta & 0 & v_0 \\ v_\beta & v_\alpha & -v_0 & 0 \end{pmatrix} \begin{pmatrix} \tilde{p} \\ q_0 \\ q_\alpha \\ q_\beta \end{pmatrix} \tag{17.21}$$

The cross-vector theory does not separate the instantaneous active power into its $\alpha\beta$ and zero-sequence components.

Therefore, the elimination of the neutral current is not possible in this theory [31].

17.10.3 *The Instantaneous Power Theory Using the Rotating* P–Q–R *Reference Frame*

Three power components are defined as linearly independent in the $P-Q-R$ reference frames [32]. Thus, the three current components can be controlled independently by compensating for the three instantaneous power components in the $P-Q-R$ reference frames. As shown in Figure 17.14(a), a new $\alpha'-\beta'-0$ reference frame is established by rotating the 0-axis of the $\alpha-\beta-0$ reference frame by θ_1, aligning the α-axis with the projected voltage space vector on the $\alpha-\beta$ plane. The current space vector in the $\alpha'-\beta'-0$ reference frame can be described as

$$\begin{pmatrix} i'_\alpha \\ i'_\beta \\ i_0 \end{pmatrix} = \begin{pmatrix} \cos\theta_1 & \sin\theta_1 & 0 \\ -\sin\theta_1 & \cos\theta_1 & 0 \\ 0 & 0 & 1 \end{pmatrix} \begin{pmatrix} i_\alpha \\ i_\beta \\ i_0 \end{pmatrix} \tag{17.22}$$

where $\theta_1 = \tan^{-1}\left(\frac{e_\beta}{e_\alpha}\right)$

Next, as shown in Figure 17.14(b), the $P-Q-R$ reference frame can be formed by rotating the β'-axis of the $\alpha'-\beta'-0$ reference frame by θ_2, aligning the α'-axis with the voltage space vector. The current space vector on the $P-Q-R$ reference frame is described as

$$\begin{pmatrix} i_p \\ i_q \\ i_r \end{pmatrix} = \begin{pmatrix} \cos\theta_2 & 0 & \sin\theta_2 \\ 0 & 1 & 0 \\ -\sin\theta_2 & 0 & \cos\theta_2 \end{pmatrix} \begin{pmatrix} i'_\alpha \\ i'_\beta \\ i_0 \end{pmatrix} \tag{17.23}$$

where $\theta_2 = \tan^{-1}\left(\frac{e_0}{e_{\alpha\beta}}\right)$ and $e_{\alpha\beta} = \sqrt{e_\alpha^2 + e_\beta^2}$

The instantaneous active/reactive powers can be defined, respectively, by the scalar/vector products of the voltage and the current space vectors.

$$p = \vec{e}_{pqr} \cdot \vec{i}_{pqr} = e_p i_p \tag{17.24}$$

$$q = \vec{e}_{pqrr} \times \vec{i}_{pqr} = [0, -e_p i_r, e_p i_q]^t \tag{17.25}$$

Therefore, the instantaneous active and reactive powers can be expressed as follows:

$$\begin{pmatrix} p \\ q_r \\ q_q \end{pmatrix} = e_\varphi \begin{pmatrix} 1 & 0 & 0 \\ 0 & 1 & 0 \\ 0 & 0 & -1 \end{pmatrix} \begin{pmatrix} i_\varphi \\ i_q \\ i_r \end{pmatrix} \tag{17.26}$$

So the current space vector can be expressed as follows:

$$\begin{pmatrix} i_\varphi \\ i_q \\ i_r \end{pmatrix} = \frac{1}{e_\varphi} \begin{pmatrix} 1 & 0 & 0 \\ 0 & 1 & 0 \\ 0 & 0 & -1 \end{pmatrix} \begin{pmatrix} p \\ q_r \\ q_q \end{pmatrix} \tag{17.27}$$

As can be noted, the instantaneous active and reactive powers are defined in three main current components that are linearly independent. Compensating for the instantaneous reactive power (q_r and q_q), the two current components can be controlled independently. To eliminate the neutral current, the current space vector is located on the $\alpha-\beta$ plane. Therefore, controlling i'_r as in the following equation by

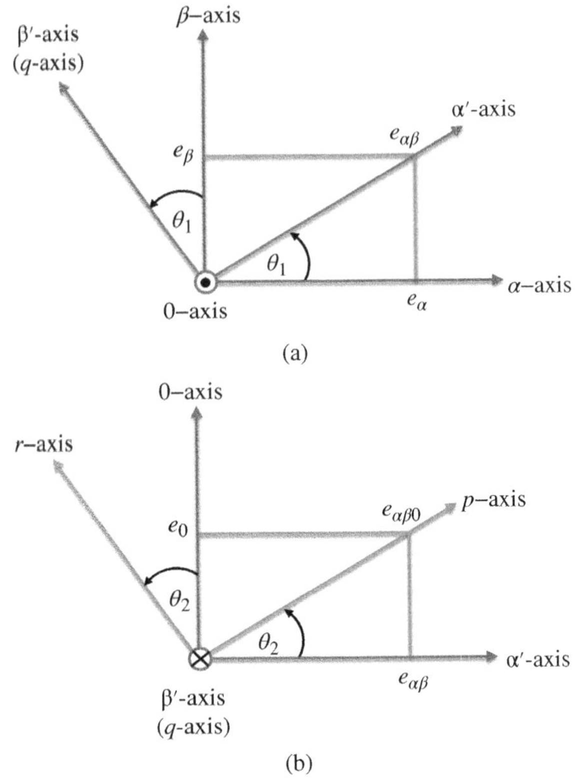

Figure 17.14 Physical meaning of the P–Q–R reference frames: (a) relation between $\alpha-\beta-0$ reference frame and $\alpha'-\beta'-0$ reference frames (viewed from the top of the 0-axis) and (b) relation between $\alpha'-\beta'-0$ reference frames and P–Q–R reference frames (viewed from the bottom of the q-axis)

compensating for the q-axis instantaneous reactive power properly, the neutral current of the system can be eliminated.

$$i'_r = -i_p \tan\theta_2 = -i_p \frac{e_0}{e_{\alpha\beta}} \tag{17.28}$$

If i_q is controlled to zero by compensating the r-axis instantaneous reactive power q_r, the current space vector is $\vec{i}_{rp}$ which is located on the projection line of the voltage space vector to the $\alpha-\beta$ plane. In this case, the current space vector is minimized as the neutral current is removed.

Note that P–Q theory does not observe power conservation since the instantaneous zero-sequence reactive power is not defined by the theory. Cross-vector theory defines one instantaneous real power and three instantaneous imaginary powers that observe power conservation, where the three instantaneous imaginary powers are linearly dependent, which means they cannot be separately compensated. The P–Q–R theory [33, 34] takes advantage of both P–Q theory and cross-vector theory. The defined instantaneous power observes power conservation. Both instantaneous real and imaginary powers can be defined in the zero-sequence circuit in three-phase four-wire systems. The three power components are linearly independent.

17.10.4 Synchronous Reference Frame

A synchronous reference frame [2, 7] is based on defining the instantaneous power in the synchronous reference frame. The load currents are transformed from the a–b–c stationary reference frame to the d–q synchronous rotating reference frame as

$$\begin{pmatrix} i_d \\ i_q \end{pmatrix} = \frac{2}{3} \begin{pmatrix} \sin\theta & \sin\left(\theta - \frac{2\pi}{3}\right) & \sin\left(\theta + \frac{2\pi}{3}\right) \\ \cos\theta & \cos\left(\theta - \frac{2\pi}{3}\right) & \sin\left(\theta + \frac{2\pi}{3}\right) \end{pmatrix} \begin{pmatrix} i_a \\ i_b \\ i_c \end{pmatrix} \tag{17.29}$$

where θ is the angle of rotation of the d–q coordinates and is equal to ωt, where ω is the power supply angular frequency. The d and q current components represent the active and reactive power components of the current, respectively. The currents can be decomposed into:

$$i_d = \bar{i}_d + \tilde{i}_d \tag{17.30}$$

$$i_q = \bar{i}_q + \tilde{i}_q \tag{17.31}$$

where $\bar{i}_d$ and $\bar{i}_q$ are the fundamental active and reactive current components, respectively, while $\tilde{i}_d$ and $\tilde{i}_q$ are the harmonics active and reactive current components, respectively. A low-pass filter is used to extract the DC components. The moving-average process is another filtering method [35].

17.10.5 Adaptive Interference Canceling Technique

The adaptive interference canceling technique [27] maintains the system in the best operating state by continuously self-tuning and self-adjusting. The noise-canceling technique is illustrated in Figure 17.15. A signal is transmitted over a channel to a sensor that receives the signal s plus an uncorrelated noise n_q. The combined signal and noise, $s + n_q$, forms the "primary input" to the canceler [27]. A second sensor receives a noise n_1 that is uncorrelated with the signal but is correlated in some unknown way with the noise n_q. This sensor provides the "reference input" to the canceler [27]. The noise n_1 is filtered to produce an output y, which is an approximate replica of n_q. This output is subtracted from the primary input $s + n_q$ to form the system output, $s + n_q - y$ [27] (Figure 17.15).

The fundamental components in the load current and the AC source voltage are mutually correlated. In the detecting system in Figure 17.15, the AC source voltage is used as the reference input and the load current is used as the primary input, that is, the fundamental component acts as the noise while harmonics act as the signal.

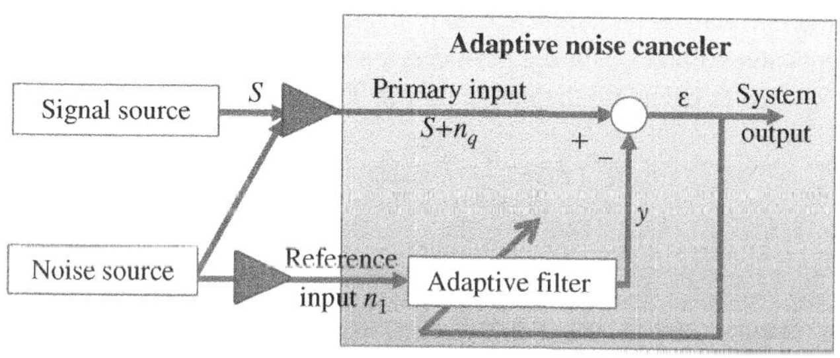

Figure 17.15 Adaptive noise canceling concept [27]

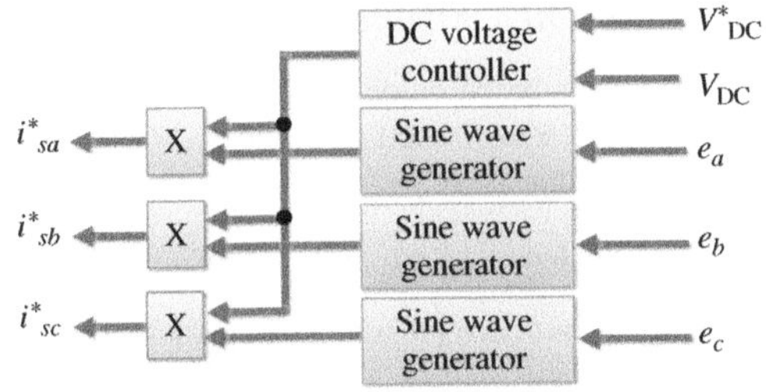

Figure 17.16 Capacitor voltage control technique

17.10.6 Capacitor Voltage Control

This technique depends on regulating the DC-bus voltage of the power inverter [34]. Harmonic extraction is performed using the capacitor voltage control as in Refs. [28, 36–39]. This technique is based on power balance, where the supply real power should be equal to the load real power plus the inverter losses. To maintain power balance, the capacitor should instantaneously compensate for the difference between the supply and the load power (Figure 17.16(a)). Controlling the capacitor voltage using a PI controller results in an output that is proportional to instantaneous power balance changes. Multiplying this output by the per-unit voltages of the PCC results in the reference supply current. With this concept, the control circuit can be significantly simplified. For obtaining a balanced current under nonideal mains voltages, only the fundamental positive sequence of the grid voltage (v_{sm}) is used as the phase reference to calculate the desired mains current.

e_a, e_b and e_c are the sensed three-phase voltages. Therefore, u_a, u_b and u_c can be defined as follows:

$$u_a = \frac{e_a}{v_{sm}}, \quad u_b = \frac{e_b}{v_{sm}} \quad \text{and} \quad u_c = \frac{e_c}{v_{sm}} \tag{17.32}$$

The reference supply current can expressed as

$$i^*_{sa} = I_{sm}u_a, \ i^*_{sb} = I_{sm}u_b \ \text{and} \ i^*_{sc} = I_{sm}u_c \tag{17.33}$$

17.10.7 Time-Domain Correlation Function Technique

Enslin [11] proposed division of the apparent power into two orthogonal components, namely real power and fictitious power. Fictitious power is also subdivided into two orthogonal components: reactive and deactive power using correlation techniques. Enslin uses autocorrelation to calculate the rms values of the measured currents and voltages over the period T. Cross-correlation between the current and voltage over the period dT is used to calculate the active power. The autocorrelation is defined as

$$R_{vv}(\tau) = \frac{1}{\mathrm{d}T}\int_0^{\mathrm{d}T} v(t)\cdot v(t-\tau)\,\mathrm{d}t \tag{17.34}$$

From Equation (17.34), the voltage rms is defined as

$$V = (R_{vv}(0))^{\frac{1}{2}} \tag{17.35}$$

The cross-correlation is

$$R_{vi}(\tau) = \frac{1}{T}\int_0^{\mathrm{d}T} v(t)\cdot i(t-\tau)\cdot \mathrm{d}t \tag{17.36}$$

From Equation (17.36), the active power can be defined as

$$P = R_{vi}(0) \tag{17.37}$$

So the active current component is

$$i_a(t) = \frac{P}{V^2} \cdot v(t) \tag{17.38}$$

17.10.8 Identification by Fourier Series

A periodic current $i(t)$ with a zero DC value can be represented [40] by

$$i(t) = \sum_{n=1}^{\infty}[a_n \cos(n\omega_f t) + b_n \sin(n\omega_f t)]$$

where ω_f is the fundamental angular frequency. With a numerical implementation of a moving Fourier series, the coefficients become

$$a_n(k) = \frac{2}{N} \sum_{j=k-(N-1)}^{k} i(jT_s)\cos(n\omega_f jT_s) \tag{17.40}$$

$$b_n(k) = \frac{2}{N} \sum_{j=k-(N-1)}^{k} i(jT_s)\sin(n\omega_f jT_s) \tag{17.41}$$

where T_s is the sampling period and N is an integer. Or recursively,

$$a_n(k) = a_n(k-1) + \frac{2}{N}[i(kT_s) + i((k-N)T_s)]\cos(n\omega_f kT_s) \tag{17.42}$$

$$b_n(k) = b_n(k-1) + \frac{2}{N}[i(kT_s) + i((k-N)T_s)]\sin(n\omega_f kT_s) \tag{17.43}$$

with $N = T_f/(2T_s)$. The identification response time corresponds to half the fundamental period. In this way, the individual current harmonics can be identified.

17.10.9 Other Methods

A method to generate the current reference for shunt APFs is presented in Ref. [41], which uses a neural network to extract the fundamental sinusoid from a distorted load current waveform. In Ref. [42], an artificial neural network (ANN) with a large number of inputs and neurons is used, where the harmonic estimation uses training to determine the weights for different neurons. In Ref. [43], an adaptive neural network is used to determine adaptively the fundamental and harmonic components, instead of training the neurons. The application of DFTs and FFTs for harmonic current extraction can be found in Ref. [44]. However, the recursive discrete Fourier transform (RDFT), presented in Ref. [44], can renew the spectrum values immediately after input of a new data set. Thus, the RDFT is more suitable for real-time implementation than DFT and FFT. The Kalman filter uses a mathematical model of the states to be estimated and a Kalman filter recursive estimation algorithm is presented in Ref. [45]. Wavelet bases are based on the definition of the active and reactive powers in the time frequency domain using the complex wavelet transform [46]. The sub-band of interest is the one that covers the fundamental.

17.11 Shunt Active Power Filter

The shunt APF [47–61] is the most widely used type of filter. It acts as a harmonic current source that injects same magnitude antiphase current to eliminate the load harmonic and reactive components of the current. Figure 17.17 shows the shunt APF single-line diagram.

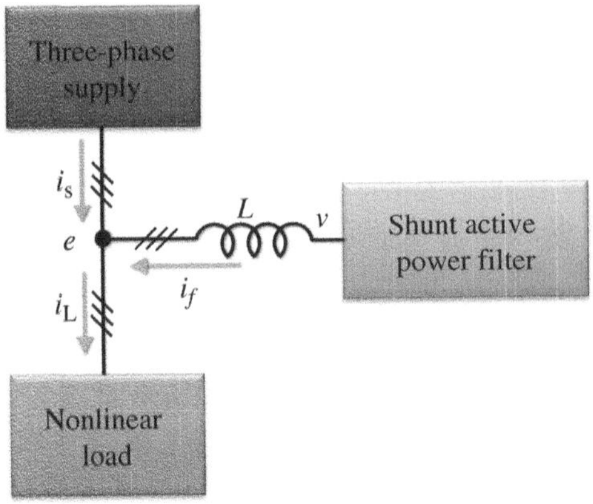

Figure 17.17 Shunt APF single-line diagram [62]

There are two important controlling parts in the shunt APF design [62–67]. The first is the harmonic extraction technique that has been discussed earlier in this chapter, and the second is the current control technique. The current control technique can be a predictive current control technique [43, 63], ramp comparison current control technique [43], or hysteresis current control technique [43]. The reference currents can be the harmonic current to be extracted (active filter current) or the supply current (grid current), to be as near as possible to a sinusoidal waveform and if needed with unity power factor. Figure 17.18 shows a block diagram for the shunt APF using instantaneous reactive power theory as

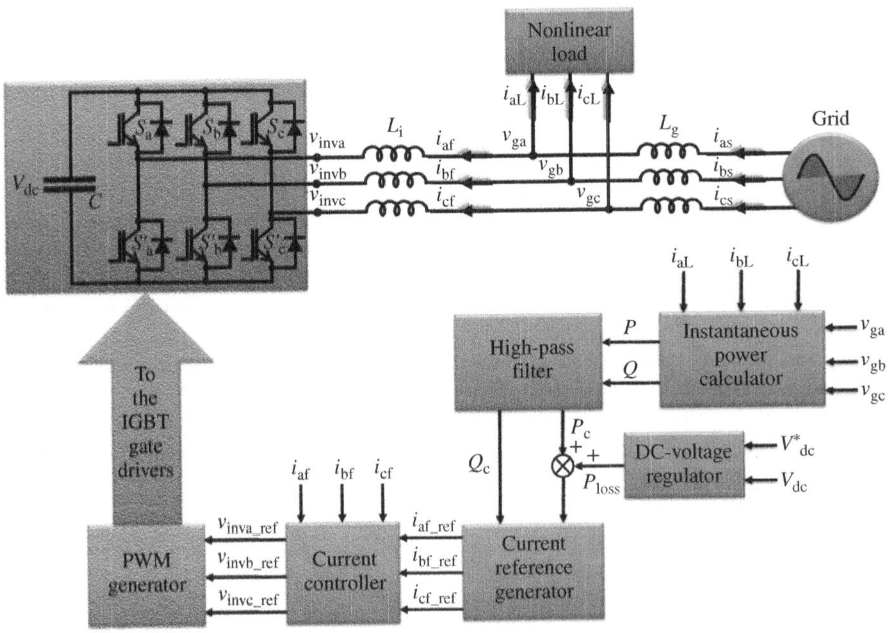

Figure 17.18 Block diagram of shunt APF using instantaneous reactive power theory

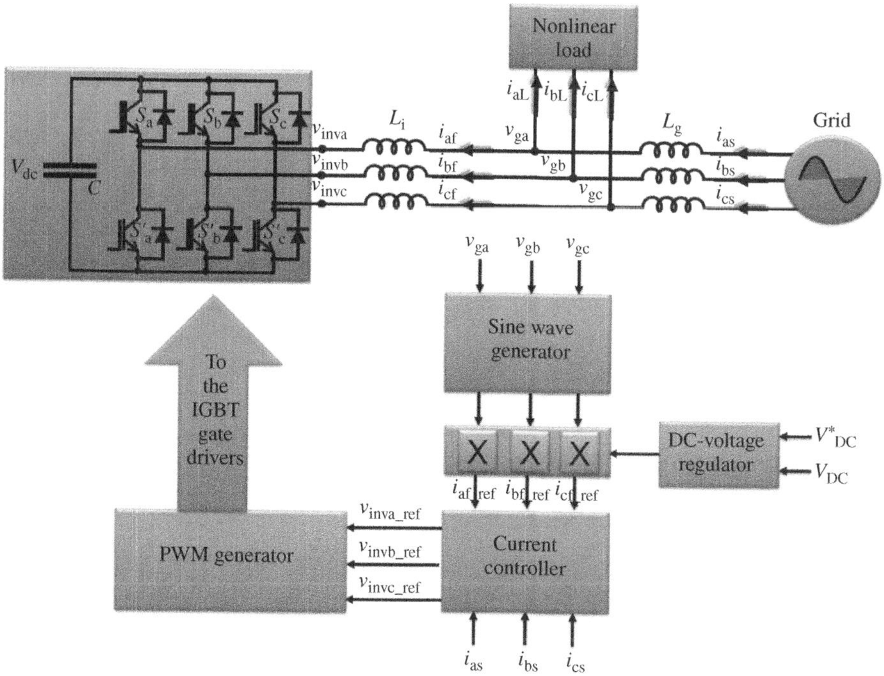

Figure 17.19 Block diagram of shunt APF using capacitor voltage control

the HCET. Figure 17.19 shows a block diagram for the shunt APF using the capacitor voltage control as the HCET. Figure 17.20 shows a block diagram for the shunt APF using the synchronous reference frame as the HCET.

17.11.1 Shunt APF Modeling

As shown in Figure 17.18, the inverter output line voltages are expressed as follows:

$$\begin{aligned} v_{ab}(t) &= v_{DC}(t) \cdot [s_a - s_b] \\ v_{bc}(t) &= v_{DC}(t) \cdot [s_b - s_c] \\ v_{ca}(t) &= v_{DC}(t) \cdot [s_c - s_a] \end{aligned} \tag{17.44}$$

where s_a, s_b and s_c are the switching states of the upper inverter switches, which are "1" for the on-state and "0" for the off-state. $v_{DC}(t)$ is the instantaneous capacitor voltage. For balanced supply voltages and equal interfacing inductances, the APF voltages with respect to the neutral point of the supply are defined as

$$\begin{aligned} v_{an}(t) &= \frac{1}{3}[v_{ab}(t) - v_{ca}(t)] \\ v_{bn}(t) &= \frac{1}{3}[v_{bc}(t) - v_{ab}(t)] \\ v_{cn}(t) &= \frac{1}{3}[v_{ca}(t) - v_{bc}(t)] \end{aligned} \tag{17.45}$$

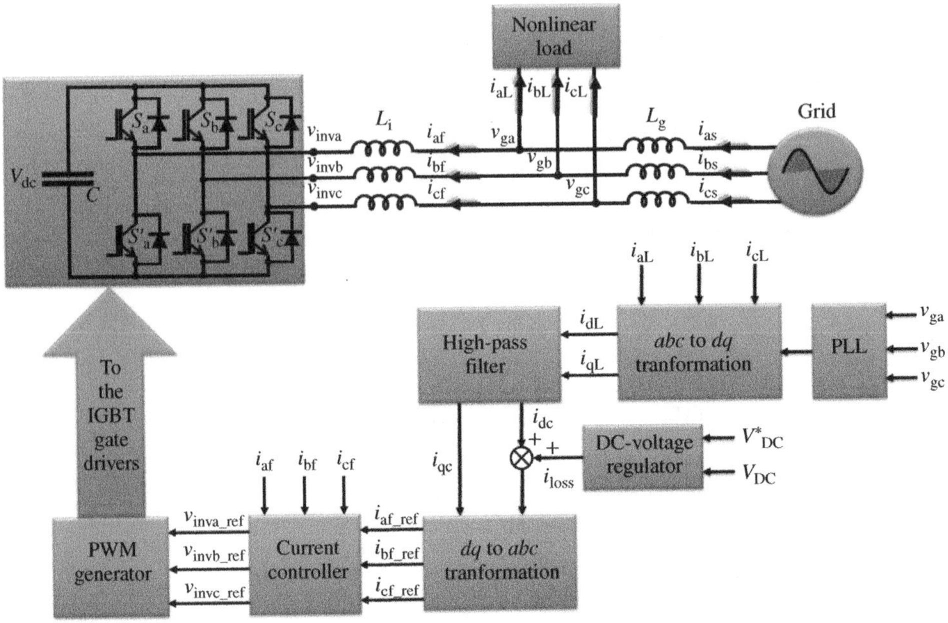

Figure 17.20 Block diagram of shunt APF using synchronous reference frame

From Equations (17.44) and (17.45), the phase-to-neutral voltages are as follows:

$$v_{an}(t) = \frac{1}{3}v_{\mathrm{DC}}(t)\cdot[2\cdot s_a - s_b - s_c]$$
$$v_{bn}(t) = \frac{1}{3}v_{\mathrm{DC}}(t)\cdot[2\cdot s_b - s_a - s_c]$$
$$v_{cn}(t) = \frac{1}{3}v_{\mathrm{DC}}(t)\cdot[2\cdot s_c - s_a - s_b] \tag{17.46}$$

The three active filter currents i_a, i_b and i_c can be expressed as

$$L\frac{\mathrm{d}i_a(t)}{\mathrm{d}t} = v_{an}(t) - e_a(t)$$
$$L\frac{\mathrm{d}i_b(t)}{\mathrm{d}t} = v_{bn}(t) - e_b(t)$$
$$L\frac{\mathrm{d}i_c(t)}{\mathrm{d}t} = v_{cn}(t) - e_c(t) \tag{17.47}$$

Substituting Equation (17.45) into Equation (17.45) yields

$$L\frac{\mathrm{d}i_a(t)}{\mathrm{d}t} = \frac{1}{3}v_{\mathrm{DC}}(t)\cdot[2\cdot s_a - s_b - s_c] - e_a(t)$$
$$L\frac{\mathrm{d}i_b(t)}{\mathrm{d}t} = \frac{1}{3}v_{\mathrm{DC}}(t)\cdot[2\cdot s_b - s_a - s_c] - e_b(t)$$
$$L\frac{\mathrm{d}i_c(t)}{\mathrm{d}t} = \frac{1}{3}v_{\mathrm{DC}}(t)\cdot[2\cdot s_c - s_a - s_b] - e_c(t) \tag{17.48}$$

Equation (17.48) shows three independent currents. In simulation, the three currents must be checked to sum to zero. Equation (17.48) cannot be used for either unbalanced supply voltages or unequal interfacing inductances. The capacitor current can be expressed in terms of the filter currents i_a, i_b and i_c as follows:

$$i_{DC}(t) = [2s_a - 1] \cdot i_a(t) + [2s_b - 1] \cdot i_b(t) + [2s_c - 1] \cdot i_c(t) \tag{17.49}$$

The DC voltage is

$$i_{DC}(t) = -C \cdot \frac{dv_{DC}(t)}{dt} \tag{17.50}$$

Substituting Equation (17.48) into Equation (17.49) yields

$$-C \cdot \frac{dv_{DC}(t)}{dt} = [2s_a - 1] \cdot i_a(t) + [2s_b - 1] \cdot i_b(t) + [2s_c - 1] \cdot i_c(t) \tag{17.51}$$

Equations (17.48) and (17.50) represent the state-space model in the form

$$\dot{X} = A \cdot C + B \cdot U \tag{17.52}$$

where A, B and U are defined as follows:

$$A = \begin{bmatrix} 0 & 0 & 0 & \frac{2 \cdot s_a - s_b - s_c}{3 \cdot L} \\ 0 & 0 & 0 & \frac{2 \cdot s_b - s_a - s_c}{3 \cdot L} \\ 0 & 0 & 0 & \frac{2 \cdot s_c - s_b - s_a}{3 \cdot L} \\ \frac{1 - 2 \cdot s_a}{C} & \frac{1 - 2 \cdot s_b}{C} & \frac{1 - 2 \cdot s_c}{C} & 0 \end{bmatrix}$$

$$C = \begin{bmatrix} i_a(t) \\ i_b(t) \\ i_c(t) \\ v_{dc}(t) \end{bmatrix}, \quad B = \begin{bmatrix} \frac{-1}{L} & 0 & 0 \\ 0 & \frac{-1}{L} & 0 \\ 0 & 0 & \frac{-1}{L} \\ 0 & 0 & 0 \end{bmatrix} \quad \text{and } U = \begin{bmatrix} e_a(t) \\ e_b(t) \\ e_c(t) \end{bmatrix}$$

Better shunt APF performance can be obtained by increasing the switching frequency, sampling frequency and capacitance, but there are limitations.

(i) Increasing the switching frequency increases the semiconductor switching losses.
(ii) The sampling frequency depends on the execution time of the control algorithm.
(iii) Increasing capacitance increases the size and cost of the capacitors.

Therefore, the lowest capacitance and switching frequency of acceptable filtering operation, to meet the IEEE-519 standards, is the main goal. In symmetrical PWM, the sampling frequency is the same as the switching frequency, while it is twice the switching frequency in asymmetrical PWM.

The rapid development in power electronics is expected to provide a solution to the limitations/problems faced with APF design and proliferation, especially in the medium voltage range. In APF design, several factors should be considered, for example, buffering element, semiconductor devices, bandwidth, interfacing inductance, control technique employed and harmonic current extraction technique. In VSI, the buffering element is a capacitor while it is an inductor in CSI. In APF, the grid voltage level, connected nonlinear load and switching frequency of APF determine the size and cost of the required capacitor or inductor for either VSI or CSI. Also, the limitation of ripples should be considered to avoid thermal overloading of the capacitors or inductors. The semiconductor device type and its rating justify its employment in APF. In VSI-based APF, a semiconductor device with its freewheeling diode is needed, while in CSI based APF, a semiconductor device with a blocking voltage capability

Table 17.6 The system parameters (simulation and practical implementation)

Parameter	Value
V^{*}_{DC}	400 V
Supply voltage	110 V
C	2200 μF
L	10 Mh
Sampling frequency	24.42 kHz
Switching frequency	12.21 kHz
Diode bridge load	70 Ω

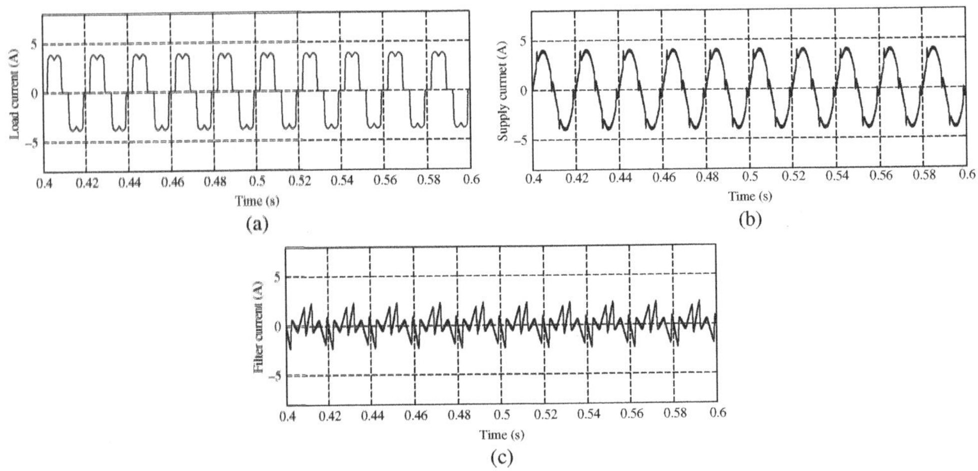

Figure 17.21 Simulation results: (a) load current, (b) supply current and (c) active filter current

is needed. As the rating of the semiconductor device increases, its switching frequency capability decreases. Extending the range of the semiconductor devices can be achieved through a series or parallel connection of semiconductor devices. Also, multilevel configuration is a route for extending the range. Hybrid configuration can be achieved by using passive filters combined with APF, as presented earlier in this chapter. The state-space model of the active filter is simulated using Matlab/Simulink. The parameters used in the simulation (and also the practical) are shown in Table 17.6. The nonlinear load is a three-phase uncontrolled bridge rectifier feeding a 70 Ω resistive load. Figure 17.21a–c shows the load current, supply current, active filter current and inverter output line voltage and its spectrum, respectively.

Figure 17.22 presents the load current, supply current, active filter current and DC voltage for a sudden reduction of the loading by 50% at 0.6 s. Figure 17.23 shows the load and supply currents after compensating the harmonic current by the shunt APF. Figure 17.24a–c shows the practical results for the load current, supply current, active filter current and inverter output line voltage with their spectra.

17.11.2 Shunt APF for Three-Phase Four-Wire System

The main source of harmonic currents produced by single-phase nonlinear loads is the switch-mode power supplies in computers. These switch-mode power supplies are connected between phases to neutral

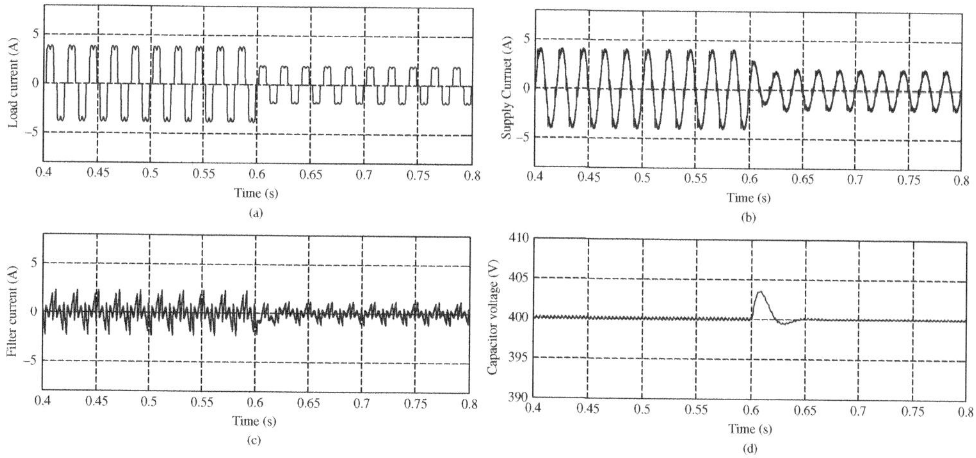

Figure 17.22 Simulation results for decreasing the loading by 50%: (a) load current, (b) supply current, (c) active filter current and (d) DC voltage

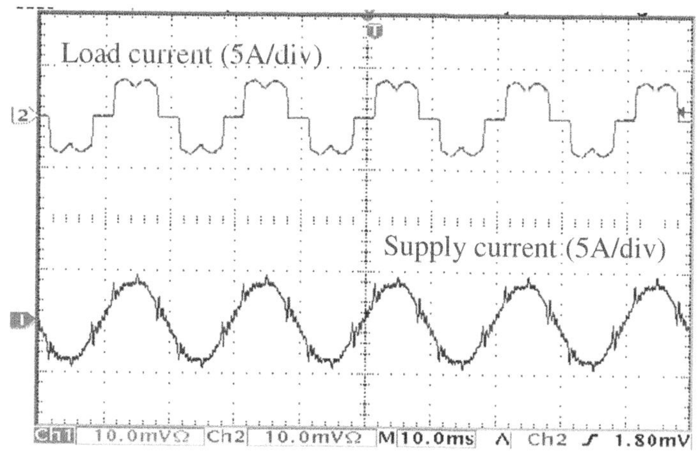

Figure 17.23 Practical results for the load and supply currents (5 A/div)

in a three-phase four-wire system introducing zero-sequence harmonics (triple harmonics), which results in neutral current increase. In order to mitigate this sort of harmonics, another configuration for the APF dedicated for four-wire applications can be employed. In three-phase four wire shunt APF applications, two main topologies for VSI-based APF are commonly used, namely the four-leg three-phase APF and the three-leg split capacitor APF. These two configurations were presented in the early 1990s. The four wire shunt APF offers better output voltage control, but at the expense of having greater complexity (e.g., 3D SVM instead of conventional SVM) for proper current controllability.

The three-leg split capacitor APF has only three legs, which reduces the complexity in current control; however, the zero-sequence current path is the DC capacitors, which adds a degree of unbalance to the capacitor voltage. Figure 17.25 shows a block diagram for the four-wire shunt APF, while Figure 17.26 shows the block diagram for the split capacitor shunt APF.

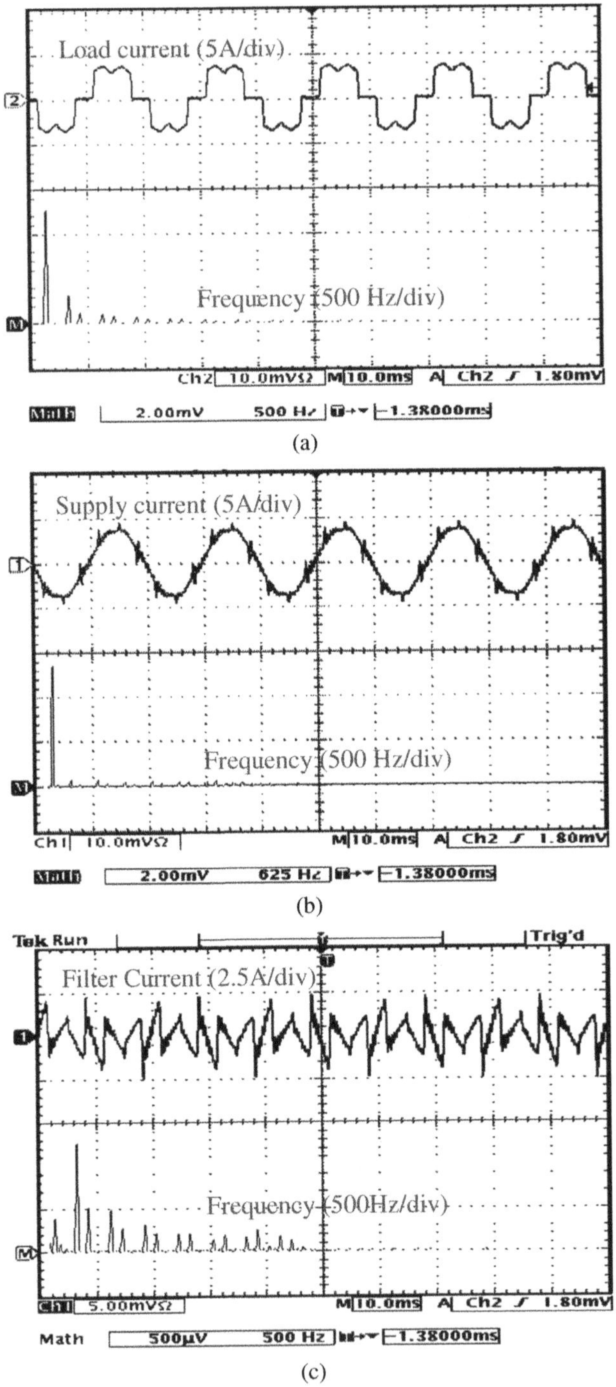

Figure 17.24 Practical results: (a) the load current and its spectrum (5 A/div), (b) the supply current and its spectrum (5 A/div) and (c) the active filter current and its spectrum (2.5 A/div) [62]

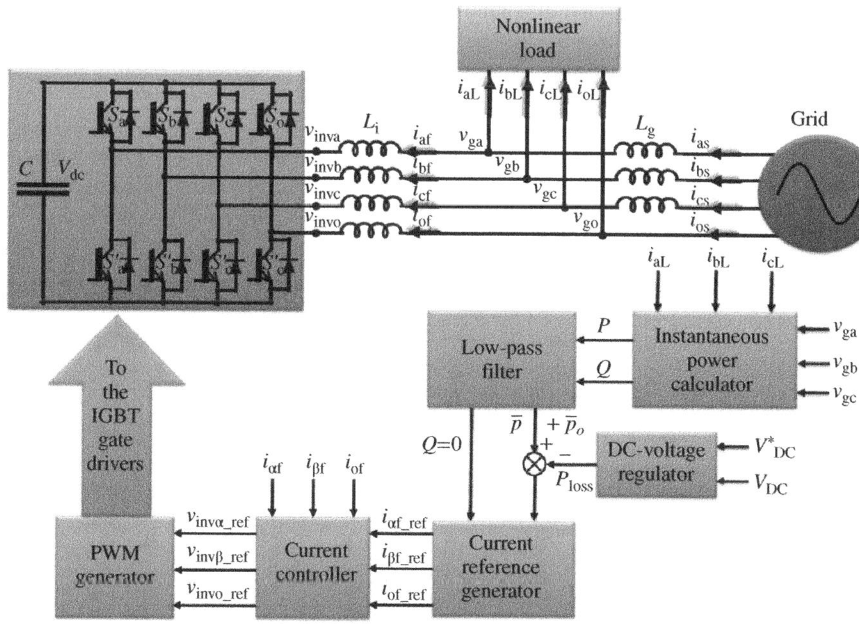

Figure 17.25 Four-wire three-phase shunt APF block diagram

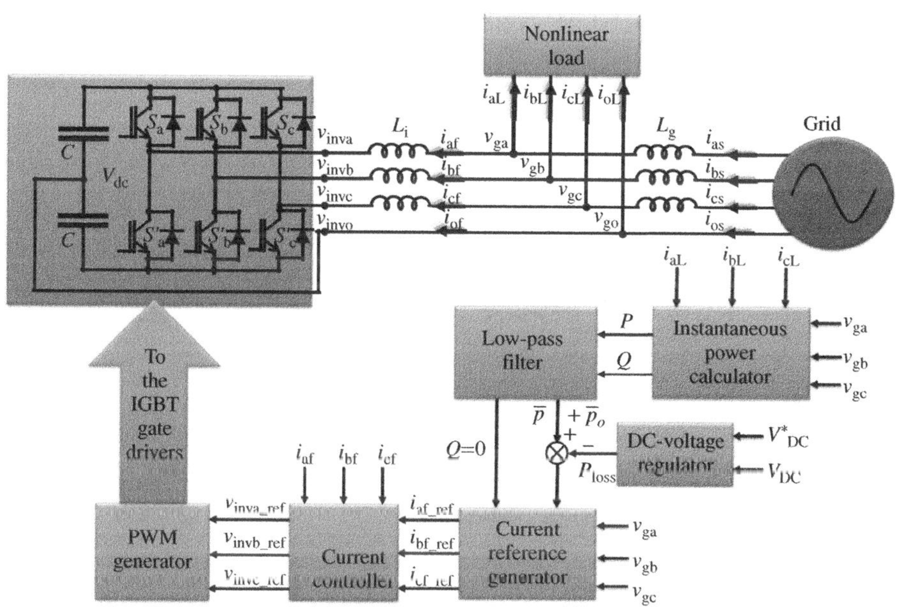

Figure 17.26 Split capacitor three-phase shunt APF block diagram

In balanced and sinusoidal voltage conditions, harmonic and reactive power mitigation can be obtained using all the harmonic extraction techniques mentioned earlier [52]. Nevertheless, in distorted/unbalanced grid voltage, the harmonic extraction techniques differ in their ability to extract the compensating currents.

The main objective of the harmonic extraction technique is to get the supply to deliver only the average active power required by the load. Moreover, the source should not deliver the zero-sequence active power, that is, the voltage at the PCC will not contribute to the source power. Therefore, in cross-vector theory, the required supply currents (after being compensated) can be expressed as follows

$$\begin{pmatrix} i_0 \\ i_\alpha \\ i_\beta \end{pmatrix} = \frac{1}{v_{0\alpha\beta}^2} \begin{pmatrix} 0 & 0 & v_\beta & -v_\alpha \\ v_\alpha & -v_\beta & 0 & v_0 \\ v_\beta & v_\alpha & -v_0 & 0 \end{pmatrix} \begin{pmatrix} \overline{p} + \overline{p_0} \\ 0 \\ 0 \\ 0 \end{pmatrix} \tag{17.53}$$

where

$$v_{0\alpha\beta}^2 = v_\alpha^2 + v_\beta^2 \tag{17.54}$$

while in capacitor voltage control, the targeted source current can be expressed as follows [52]

$$\begin{pmatrix} i_0 \\ i_\alpha \\ i_\beta \end{pmatrix} = \frac{\overline{p} + \overline{p}_0}{(v_{0\alpha\beta}^2)_{\mathrm{DC}}} \begin{pmatrix} v_0 \\ v_0 \\ v_0 \end{pmatrix} \tag{17.55}$$

where $(v_{0\alpha\beta}^2)_{\mathrm{DC}}$ is the DC component of $v_{0\alpha\beta}^2$

while in a synchronous reference frame, the targeted source current can be expressed as follows [52]

$$\begin{pmatrix} i_0 \\ i_\alpha \\ i_\beta \end{pmatrix} = \left(\frac{p}{\sqrt{v_\alpha^2 + v_\beta^2}} \right)_{\mathrm{DC}} \frac{1}{\sqrt{v_\alpha^2 + v_\beta^2}} \begin{pmatrix} 0 \\ v_\alpha \\ v_\beta \end{pmatrix} \tag{17.55}$$

An alternative is to consider the fundamental positive sequence of the voltage at the PCC. The supply current (after being compensated) should be in phase with this voltage. Therefore, the supply reference current can be expressed as follows [52]:

$$\begin{pmatrix} i_0 \\ i_\alpha \\ i_\beta \end{pmatrix} = \frac{\overline{p} + \overline{p}_0}{v_{1\alpha}^{+2} + v_{1\beta}^{+2}} \begin{pmatrix} 0 \\ v_{1\alpha}^+ \\ v_{1\beta}^+ \end{pmatrix} \tag{17.56}$$

17.12 Series Active Power Filter

A series APF [68–80] is connected in series in the distribution system to compensate for current and voltage harmonics, and voltage unbalance through acting as a series-connected voltage source or a series-connected impedance. The series-connected impedance will act as a harmonic isolator, preventing the harmonic from flowing by introducing an infinite impedance at the tuned frequency of the harmonic component to be compensated and presenting a zero impedance at the fundamental frequency, permitting the fundamental current flow. The series-connected voltage source operation will compensate the disturbance (harmonics unbalance, etc.) in the grid voltage by injecting the required voltage vector(s). The principle of operation is the dual of that of the shunt APF. It can compensate voltage unbalance, voltage distortion, current distortion and voltage sag. Based on the nature of the disturbance and the required action, the series APF injects a voltage vector to be added to the grid voltage to achieve the assigned task. Practically, the series APF is connected in the distribution system with parallel passive

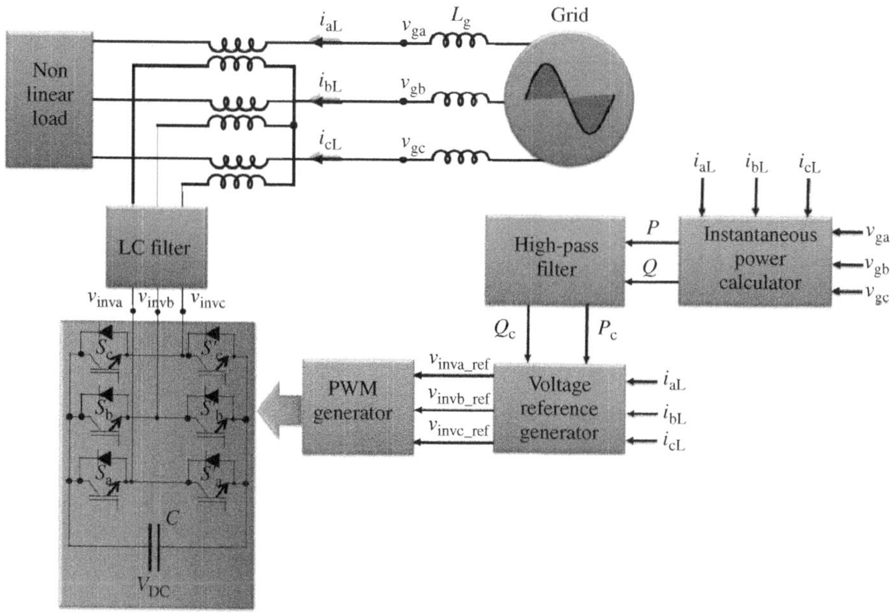

Figure 17.27 Block diagram of series APF

filters in order to enhance the passive filter performance and mitigate the problems they introduce into the distribution system. In this case, the series APF compensates for current system distortion caused by nonlinear loads by inserting a high impedance path for the harmonic current, which forces the harmonic current to pass through the passive filter connected in parallel with the load. The high impedance introduced by the series APF is generated through a voltage of the same frequency as the current harmonic component is to be compensated. Voltage imbalance is compensated by compensating the negative- and zero-sequence components of the system fundamental frequency. Figure 17.27 presents a block diagram for the series APF. The series APF is effective in compensating voltage harmonic sources. The main disadvantage of the series APF is that the inverter has to be rated at the full load current. The series APF is mainly a voltage-controlled VSI. Nevertheless, another type of inverter such as a CSI can be employed. Instantaneous power theory is employed as the harmonic voltage extraction technique. There are several configurations for the so-called *hybrid filters* (series APF and passive filters). The rating of the series active filter depends mainly on the voltage vector to be injected, as the current that the series APF carries is the line current. Figure 17.28 shows a series APF with parallel passive filter. The passive filter consists of three main stages: third-order harmonic, fifth-order harmonic and high-pass filter.

17.13 Unified Power Quality Conditioner

The main objective of the UPQC [81–85] shown in Figure 17.29 is to compensate for voltage harmonic/unbalance/flicker, reactive power, negative sequence current and harmonic currents, therefore enhancing the power quality at the PCC in the distribution system. A UPQC is a device that is similar in its configuration to the unified power flow conditioner (UPFC). The UPQC is typically a combined shunt APF and series APF connected back-to-back from the DC side. The capabilities of the two APFs (shunt and series) are inherited into the UPQC, performing both the shunt and series compensation. The main drawback of the UPQC is its complexity and high cost.

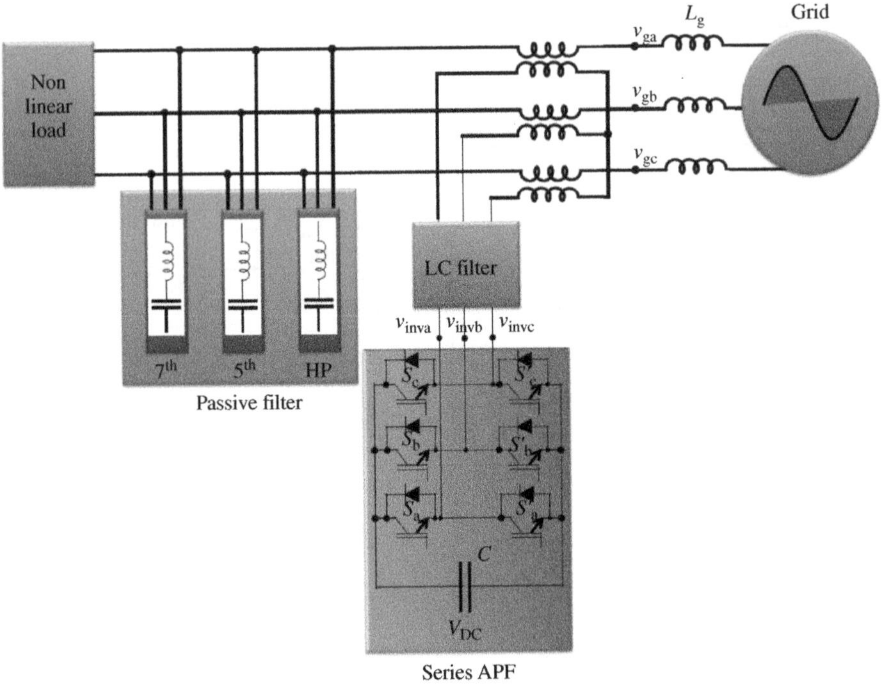

Figure 17.28 Hybrid filter (combined series APF and parallel passive filter)

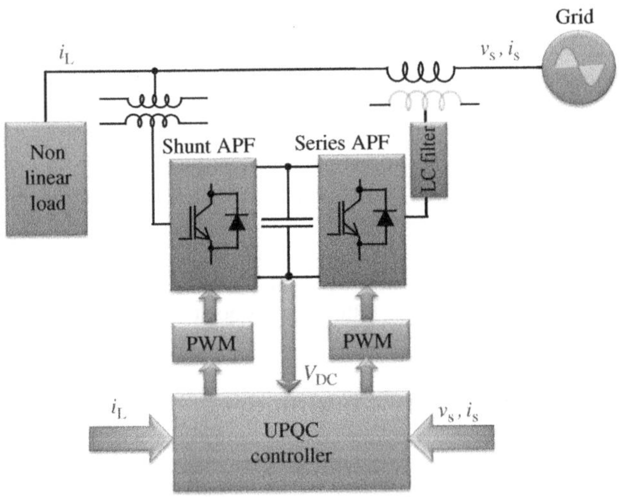

Figure 17.29 UPQC

The UPQC can be classified based on configuration (single-phase, three-phase three-wire, three-phase four-wire, etc.). It can also be classified based on converter type (voltage source-based, current source-based, etc.)

As explained earlier in this chapter, the shunt APF is employed for mitigating the current-related power quality problems, whereas the series APF is employed for mitigating voltage-related power quality problems.

If a distribution system is experiencing a lower quality supply voltage and current, installing a UPQC is a more cost-effective device than using two separate devices (shunt APF and series APF) to compensate for low-power quality. The main function of the UPQC is to compensate for supply voltage quality issues such as sags, swells, unbalance, flicker, harmonics and for load current quality problems such as harmonics, unbalance, reactive current and neutral current.

Figure 17.29 shows a single-line diagram of the UPQC system configuration. The UPQC consists of the following system components:

- Two back-to-back inverters where one is connected across the load and acts as a shunt APF and the second is connected in series and acts as a series APF.
- A shunt interfacing inductor which is used to interface the shunt APF to the grid. It acts as a first-order filter to smooth inverter output current. An isolation transformer may be used for electrical isolation of the inverter.
- A common DC link (either a capacitor for a VSI or an inductor for a CSI).
- An LC filter that acts as a low-pass filter to eliminate the high-frequency switching ripples.
- The series-connected transformer used to interface the series APF to the grid.

In the UPQC, the shunt APF is controlled in current control mode to track the reference current generated by the shunt APF control algorithm. For example, in order to mitigate the current harmonic components, the shunt APF should inject a current based on the following equation:

$$i_{\text{shf}}(t) = i_{\text{s}}^{*}(t) - i_{\text{L}}(t) \tag{17.57}$$

where $i_{\text{shf}}(t)$, $i_{\text{s}}^{*}(t)$ and $i_{\text{L}}(t)$ represent the shunt APF current, reference source current and load current, respectively.

The series APF is controlled in voltage control mode to generate a voltage in series with the supply to achieve the referenced voltage at the PCC. The principle of operation of the series APF can be summarized as in the following equation:

$$v_{\text{ser}}(t) = v_{\text{L}}^{*}(t) - v_{\text{s}}(t) \tag{17.58}$$

where $v_{\text{ser}}(t)$, $v_{\text{L}}^{*}(t)$ and $v_{\text{s}}(t)$ represent the series APF voltage, reference load voltage and supply voltage, respectively.

The UPQC can be classified based on the number of phases/wires, such as three-phase three-wire, three-phase four-wire, and single-phase two-wire. In a single-phase system, the reactive and harmonic currents are the components of interest (to be compensated). In a three-phase three-wire system, the reactive, harmonic and unbalance current components are of interest (to be compensated). The three-phase four-wire system has an additional function for neutral current mitigation. The UPQC can be classified based on configuration into a left-shunt UPQC and a right-shunt UPQC. This is based on the location of the shunt APF with respect to the series APF. The right-shunt UPQC has the shunt APF connected at the PCC, then the series APF between the PCC and the supply, while the left-shunt UPQC has the series APF connected between the PCC and the supply and the shunt APF is connected at the point between the series APF and the supply. The most commonly used topology is the right-shunt APF, as the current flowing through the series transformer is sinusoidal, and thus a better UPQC performance is introduced.

The UPQC is simulated using Matlab/Simulink. The load is a three-phase uncontrolled rectifier bridge feeding load of 70 Ω. The supply is a three-phase balanced supply of 110 Vrms at 50 Hz but distorted

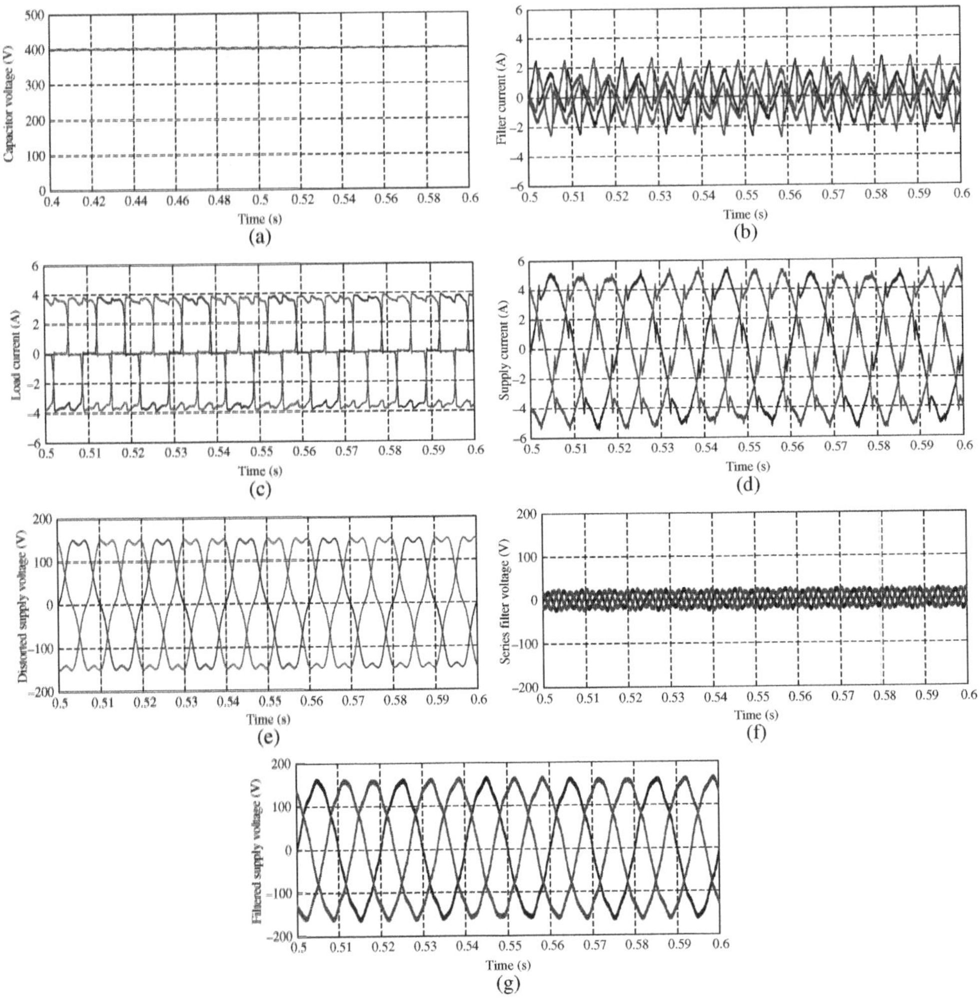

Figure 17.30 UPQC simulation results: (a) the DC voltage, (b) the three-phase shunt APF currents, (c) the three-phase load currents, (d) the three-phase supply currents, (e) the three-phase supply voltage, (f) the three-phase series APF voltages and (g) the three-phase supply voltages after being filtered

with 10% fifth-order harmonics. The supply impedance is 0.1 mH. The shunt APF acts to control the DC capacitor voltage and to cancel the harmonic currents injected by the nonlinear load. The capacitor voltage harmonic extraction technique is employed for extracting the harmonic currents. The shunt APF is interfaced to the grid through a 1 : 1 transformer. The switching frequency of the shunt APF is 12 kHz, as also for the series APF. The series APF acts to cancel the harmonic voltages to make the voltage at the PCC pure sinusoidal. The series APF is connected through an LC filter of 20 μF and 2 mH and a transformer of 1 : 10 turns ratio. The synchronous reference frame harmonic extraction technique is employed for harmonic voltage extraction. Figure 17.30 presents the simulation results of the UPQC. Figure 17.30(a)–(g) presents the DC voltage, the three-phase shunt APF currents, the three-phase load currents, the three-phase supply currents, the three-phase supply voltage, the three-phase series APF voltages and the three-phase supply voltages after being filtered, respectively.

Acknowledgment

This chapter was made possible by an NPRP Grant No. 04-250-2-080 from the Qatar National Research Fund (a member of Qatar Foundation). The statements made herein are solely the responsibility of the authors.

References

1. Rukonuzzaman, M. and Nakaoka, M. (2002) Single-phase shunt active power filter with knowledge-based harmonic detection algorithm. Power Conversion Conference (PCC), Osaka, Japan, 2002, Vol. 2, pp. 778–783.
2. Soares, V., Verdelho, P., and Marques, G.D. (2000) An instantaneous active and reactive current component method for active filters. *IEEE Transactions on Power Electronics*, **15** (4), 660–669.
3. IEEE Std 519-1992. (1993) *IEEE Recommended Practices and Requirements for Harmonic Control in Electrical Power Systems. IEEE Std.* 519–1992 pp. 1–112
4. IEC/TR EN 61000-1-1. *Electromagnetic Compatibility (EMC).*
5. Peng, F.Z. (1998) Application issues of active power filters. *IEEE Industry Applications Magazine*, **4** (5), 21–30.
6. Rahmani, S., Hamadi, A., and Al-Haddad, K. (2009) A new combination of shunt hybrid power filter and thyristor controlled reactor for harmonics and reactive power compensation. Electrical Power and Energy Conference (EPEC), IEEE, October 22–23, 2009, pp. 1–6.
7. Takeda, M., Ikeda, K., Teramoto, A., and Aritsuka, T. (1988) Harmonic current and reactive power compensation with an active filter. 19th Annual IEEE Power Electronics Specialists Conference 1988. PESC'88 Record, April 11–14, 1988, Vol. 2, pp. 1174–1179.
8. Moran, L.A., Dixon, J.W., and Wallace, R.R. (1995) A three phase active power filter operating with fixed switching frequency for reactive power and current harmonic compensation. *IEEE Transactions on Industrial Electronics*, **42** (4), 402–408.
9. Cheng, P.T., Bhattacharya, S., and Divan, D.M. (1998) Control of square-wave inverters in high-power hybrid active filter systems. *IEEE Transactions on Industry Applications*, **34** (3), 458–472.
10. Plaisant, A. and Reeve, J. (1999) An active filter for AC harmonics from HVDC converters. Basic concepts and design principles. Power Engineering Society Summer Meeting, 1999. IEEE, July 18–22, 1999, Vol. 1, pp. 395–400.
11. Pittorino, L.A., Horn, A., and Enslin, J.H.R. (1996) Power theory evaluation for the control of an active power filter. IEEE 4th AFRICON, 1996, Vol. 2, pp. 676–681.
12. Kim, H. and Akagi, H. (1997) The instantaneous power theory based on mapping matrices in three-phase four-wire systems. Power Conversion Conference, Vol. 1, pp. 361–366.
13. Lim, Y.C., Park, J.K., Jung, Y.G. *et al.* (1995) Development of a simulator for compensation performance evaluation of hybrid active power filter using three-dimensional current co-ordinate. International Conference on Power Electronics and Drive Systems, 1995, Vol. 1, pp. 427–432.
14. Marshall, D.A. and Van Wyk, J.D. (1991) An evaluation of the real-time compensation of fictitious power in electric energy networks. *IEEE Transactions on Power Delivery*, **6**, 1774–1780.
15. Akagi, H. (1996) New trends in active filters for power conditioning. *IEEE Transactions on Industrial Applications*, **32** (6), 1312–1322.
16. Xiao, G. Liu, J., Yang, J., and Wang, Z. (2000) The experimental studies on active power filter for HVDC system. PIEMC 2000 Proceedings: Third International Power Electronics and Motion Control Conference, August 15–18, 2000, Vol. 3, pp. 1376–1379.
17. Verdelho, P. and Marques, G.D. (1993) Multiple applications control system for the PWM voltage converter connected to the AC mains. Fifth European Conference on Power Electronics and Applications, September 13–16, 1993, Vol. 8, pp. 42–46.
18. Pereira, M., Wild, G., Huang, H., Sadek, K. (2002) Active filters in HVDC systems: actual concepts and application experience. 2002 International Conference on Power System Technology Proceedings. PowerCon 2002, October 13–17, 2002, Vol. 2, pp. 989–993.
19. Gole, A.M. and Meisingset, M. (2001) An AC active filter for use at capacitor commutated HVDC converters. *IEEE Transactions on Power Delivery*, **16** (2), 335–341.
20. Akagi, H. (2001) Large static converters for industry and utility applications. *Proceedings of the IEEE*, **89** (6), 976–983.

21. Singh, B., Al-Haddad, K., and Chandra, A. (1999) A review of active filters for power quality improvement. *IEEE Transactions on Industrial Electronics*, **46** (5), 960–971.
22. Choe, G.H., Wallace, A.K., and Park, M.H. (1988) Control technique of active power filter for harmonic elimination and reactive power control. IEEE Industry Applications Society Annual Meeting, 1988, Vol. 1, pp. 859–866.
23. El-Habrouk, M., Darwish, M.K., and Mehta, P. (2000) Active power filters: a review. *IEEE Proceedings Electric Power Applications*, **147** (5), 403–413.
24. Akagi, H., Kanazawa, Y., and Nabae, A. (1984) Instantaneous reactive power compensators comprising switching devices without energy storage components. *IEEE Transactions on Industry Applications*, **IA-20** (3), 625–630.
25. Pahmer, C., Capolino, G.A., and Henao, H. (1994) Computer-aided design for control of shunt active filter. 20th International Conference on Industrial Electronics, Control and Instrumentation, IECON'94, Vol. 1, pp. 669–674.
26. Akagi, H., Kanazawa, Y., and Nabae, A. (1983) Generalized theory of the instantaneous reactive power in three-phase circuits. IPEC'83, International Power Electronics Conference, Tokyo, Japan, pp. 1375–1386.
27. Luo, S. and Hou, Z. (1995) An adaptive detecting method for harmonic and reactive currents. *IEEE Transactions on Industrial Electronics*, **42** (1), 85–89.
28. Herrera, R.S. and Salmeron, P. (2009) Instantaneous reactive power theory: a reference in the nonlinear loads compensation. *IEEE Transactions on Industrial Electronics*, **56** (6), 2015–2022.
29. Peng, F.Z., Ott, G.W., and Adams, D.J. (1998) Harmonic and reactive power compensation based on the generalized instantaneous reactive power theory for three-phase four-wire systems. *IEEE Transactions on Power Electronics*, **13** (6), 1174–1181.
30. Peng, F.Z. and Lai, J.S. (1996) Generalized instantaneous reactive power for three-phase power systems. *IEEE Transactions on Instrumentation and Measurement*, **45** (1), 293–297.
31. Herrera, R.S., Salmeron, P., and Kim, H. (2008) Instantaneous reactive power theory applied to active power filter compensation: different approaches, assessment, and experimental results. *IEEE Transactions on Industrial Electronics*, **55** (1), 184–196.
32. Kim, H. and Akagi, H. (1999) The instantaneous power theory on the rotating p-q-r reference frames. IEEE 1999 International Conference on Power Electronics and Drive Systems, PED'99, July, Hong Kong, pp. 422–427.
33. Kim, H., Blaabjerg, F., Bak-Jensen, B., and Jaeho, C. (2002) Instantaneous power compensation in three-phase systems by using p–q–r theory. *IEEE Transactions on Power Electronics*, **17** (5), 701–710.
34. Kim, H., Blaabjerg, F., and Jensen, B.B. (2002) Spectral analysis of instantaneous powers in single-phase and three-phase systems with use of p–q–r theory. *IEEE Transactions on Power Electronics*, **17** (5), 711–720.
35. De Oliveira, L.E.L., da Silva, L.E.B., da Silva, V.F. *et al.* (2002) Improving the dynamic response of active power filters based on the synchronous reference frame method. Seventeenth Annual IEEE Applied Power Electronics Conference and Exposition, 2002, Vol. 2, pp. 742–748.
36. Huang, H.J. and Wu, J.C. (1999) A control algorithm for three-phase three-wired active power filters under non-ideal mains voltages. *IEEE Transactions on Power Electronics*, **14**, 753–760.
37. Singh, B. and Kothari, A.D.P. (1998) Analysis of a novel active filter for balancing and reactive power compensation. Power Electronics and Variable Speed Drives, 1998, Conference Publication No. 456, IEEE, pp. 57–62.
38. Furuhashi, T., Okuma, S., and Uchikawa, Y. (1990) A study on the theory of instantaneous reactive power. *IEEE Transactions on Industrial Electronics*, **37**, 86–90.
39. Bruyant, N. and Machmoum, M. (1998) Simplified digital-analogical control for shunt active power filters under unbalanced conditions. Seventh International Conference on Power Electronics and Variable Speed Drives, 1998. (IEE Conference Publication No. 456), September 21–23, pp. 11–16.
40. Abrahamsen, F. and David, A. (1995) Adjustable speed drive with active filtering capability for harmonic current compensation. 26th Annual IEEE Power Electronics Specialists Conference, June 18–22, 1995, Vol. 2, pp. 1137–1143.
41. Gao, D., Lu, Q., and Sun, X. (2002) Design and performance of an active power filter for unbalanced loads. International Conference Power System Technology, 2002, Vol. 4, pp. 2496–2500.
42. Round, S.D. and Mohan, N. (1993) Comparison of frequency and time domain neural network controllers for an active power filter. Proceedings of IEEE Industrial Electronics, Control and Instrumentation, Vol. 2, pp. 1099–1104.
43. Massoud, A.M., Finney, S.J., Grant, D.M., and Williams, B.W. (2006) Predictive current controlled shunt active power filter using three-level cascaded type inverter. The 3rd IET International Conference on Power Electronics, Machines and Drives, 2006, pp. 388–393.
44. Dolen, M. and Lorenz, R.D. (2000) An industrially useful means for decomposition and differentiation of harmonic components of periodic waveforms. IEEE Industry Applications Society (IAS) Conference, Vol. 2, pp. 1016–1023.

45. Girgis, A.A., Chang, W.B., and Makram, E.B. (1991) A digital recursive measurement scheme for online tracking of power system harmonics. *IEEE Transactions on Power Delivery*, **6**, 1153–1160.
46. Driesen, J. and Belmans, R. (2002) Active power filter control algorithms using wavelet-based power definitions. 10th International Conference Harmonics and Quality of Power, October 6–9, 2002, Vol. 2, pp. 466–471.
47. Aredes, M. and Watanabe, E.H. (1995) New control algorithms for series and shunt three-phase four-wire active power filters. *IEEE Transactions on Power Delivery*, **10** (3), 1649–1656.
48. Akagi, H. (1997) Control strategy and site selection of a shunt active filter for damping of harmonic propagation in power distribution systems. *IEEE Power Engineering Review*, **17** (1), 58.
49. Hafner, J., Aredes, M., and Heumann, K. (1997) A shunt active power filter applied to high voltage distribution lines. *IEEE Power Engineering Review*, **17** (1), 51–52.
50. Hafner, J., Aredes, M., and Heumann, K. (1997) A shunt active power filter applied to high voltage distribution lines. *IEEE Transactions on Power Delivery*, **12** (1), 266–272.
51. Akagi, H. (1997) Control strategy and site selection of a shunt active filter for damping of harmonic propagation in power distribution systems. *IEEE Transactions on Power Delivery*, **12** (1), 354–363.
52. Montero, M.I.M., Cadaval, E.R., and Gonzalez, F.B. (2007) Comparison of control strategies for shunt active power filters in three-phase four-wire systems. *IEEE Transactions on Power Electronics*, **22** (1), 229–236.
53. Akagi, H., Fujita, H., and Wada, K. (1999) A shunt active filter based on voltage detection for harmonic termination of a radial power distribution line. *IEEE Transactions on Industry Applications*, **35** (3), 638–645.
54. Mishra, M.K., Joshi, A., and Ghosh, A. (2000) A new algorithm for active shunt filters using instantaneous reactive power theory. *IEEE Power Engineering Review*, **20** (12), 56–58.
55. Elmitwally, A., Abdelkader, S., and El-Kateb, M. (2000) Neural network controlled three-phase four-wire shunt active power filter. *IEEE Proceedings on Generation, Transmission and Distribution*, **147** (2), 87–92.
56. Chandra, A., Singh, B., Singh, B.N., and Al-Haddad, K. (2000) An improved control algorithm of shunt active filter for voltage regulation, harmonic elimination, power-factor correction, and balancing of nonlinear loads. *IEEE Transactions on Power Electronics*, **15** (3), 495–507.
57. Chang, G.W. (2001) A new method for determining reference compensating currents of the three-phase shunt active power filter. *IEEE Power Engineering Review*, **21** (3), 63–65.
58. Al-Zamil, A.M. and Torrey, D.A. (2001) A passive series, active shunt filter for high power applications. *IEEE Transactions on Power Electronics*, **16** (1), 101–109.
59. Jintakosonwit, P., Fujita, H., and Akagi, H. (2002) Control and performance of a fully-digital-controlled shunt active filter for installation on a power distribution system. *IEEE Transactions on Power Electronics*, **17** (1), 132–140.
60. Jintakosonwit, P., Akagi, H., Fujita, H., and Ogasawara, S. (2002) Implementation and performance of automatic gain adjustment in a shunt-active filter for harmonic damping throughout a power distribution system. *IEEE Transactions on Power Electronics*, **17** (3), 438–447.
61. Mendalek, N., Al-Haddad, K., Fnaiech, F., and Dessaint, L.A. (2003) Nonlinear control technique to enhance dynamic performance of a shunt active power filter. *IEEE Proceedings on Electric Power Applications*, **150** (4), 373–379.
62. Massoud, A.M., Finney, S.J., and Williams, B.W. (2004) Predictive current control of a shunt active power filter. 2004 IEEE 35th Annual Power Electronics Specialists Conference. PESC 04, June 20–25 2004, Vol. 5, pp. 3567–3572.
63. Massoud, A.M., Finney, S.J., Cruden, A.J., and Williams, B.W. (2007) Three-phase, three-wire, five-level cascaded shunt active filter for power conditioning, using two different space vector modulation techniques. *IEEE Transactions on Power Delivery*, **22** (4), 2349–2361.
64. Massoud, A.M., Finney, S.J., and Williams, B.W. (2004) Practical issues of three-phase, three-wire, voltage source inverter-based shunt active power filters. 11th International Conference on Harmonics and Quality of Power, September 12–15, 2004, pp. 436–441.
65. Zhang, H., Finney, S.J., Massoud, A., and Williams, B.W. (2008) An SVM algorithm to balance the capacitor voltages of the three-level NPC active power filter. *IEEE Transactions on Power Electronics*, **23** (6), 2694–2702.
66. Massoud, A.M., Finney, S.J., and Williams, B.W. (2004) Review of harmonic current extraction techniques for an active power filter. 11th International Conference on Harmonics and Quality of Power, 2004, September 12–15, pp. 154–159.
67. Peng, F.Z., Akagi, H., and Nabae, A. (1990) A new approach to harmonic compensation in power systems-a combined system of shunt passive and series active filters. *IEEE Transactions on Industry Applications*, **26** (6), 983–990.

68. Peng, F.Z., Akagi, H., and Nabae, A. (1990) A new approach to harmonic compensation in power systems-a combined system of shunt passive and series active filters. *IEEE Transactions on Industry Applications*, **26** (6), 983–990.
69. Peng, F.Z., Akagi, H., and Nabae, A. (1990) A study of active power filters using quad-series voltage-source PWM converters for harmonic compensation. *IEEE Transactions on Power Electronics*, **5** (1), 9–15.
70. Akagi, H., Tsukamoto, Y., and Nabae, A. (1990) Analysis and design of an active power filter using quad-series voltage source PWM converters. *IEEE Transactions on Industry Applications*, **26** (1), 93–98.
71. Fujita, H. and Akagi, H. (1991) A practical approach to harmonic compensation in power systems-series connection of passive and active filters. *IEEE Transactions on Industry Applications*, **27** (6), 1020–1025.
72. Aredes, M. and Watanabe, E.H. (1995) New control algorithms for series and shunt three-phase four-wire active power filters. *IEEE Transactions on Power Delivery*, **10** (3), 1649–1656.
73. Fujita, H. and Akagi, H. (1997) An approach to harmonic current-free AC/DC power conversion for large industrial loads: the integration of a series active filter with a double-series diode rectifier. *IEEE Transactions on Industry Applications*, **33** (5), 1233–1240.
74. Barbosa, P.G., Santisteban, J.A., and Watanabe, E.H. (1998) Shunt-series active power filter for rectifiers AC and DC sides. *IEEE Proceedings on Electric Power Applications*, **145** (6), 577–584.
75. Fujita, H. and Akagi, H. (1998) The unified power quality conditioner: the integration of series and shunt-active filters. *IEEE Transactions on Power Electronics*, **13** (2), 315–322.
76. Moran, L.A., Pastorini, I., Dixon, J., and Wallace, R. (1999) A fault protection scheme for series active power filters. *IEEE Transactions on Power Electronics*, **14** (5), 928–938.
77. Moran, L., Pastorini, I., Dixon, J., and Wallace, R. (2000) Series active power filter compensates current harmonics and voltage unbalance simultaneously. *IEEE Proceedings on Generation, Transmission and Distribution*, **147** (1), 31–36.
78. Al-Zamil, A.M. and Torrey, D.A. (2001) A passive series, active shunt filter for high power applications. *IEEE Transactions on Power Electronics*, **16** (1), 101–109.
79. Salmeron, P. and Litran, S.P. (2010) Improvement of the electric power quality using series active and shunt passive filters. *IEEE Transactions on Power Delivery*, **25** (2), 1058–1067.
80. Tian, J., Chen, Q., and Xie, B. (2012) Series hybrid active power filter based on controllable harmonic impedance. *IET Power Electronics*, **5** (1), 142–148.
81. Fujita, H. and Akagi, H. (1998) The unified power quality conditioner: the integration of series and shunt-active filters. *IEEE Transactions on Power Electronics*, **13** (2), 315–322.
82. Khadkikar, V. and Chandra, A. (2008) A new control philosophy for a unified power quality conditioner (UPQC) to coordinate load-reactive power demand between shunt and series inverters. *IEEE Transactions on Power Delivery*, **23** (4), 2522–2534.
83. Han, B., Bae, B., Kim, H., and Baek, S. (2006) Combined operation of unified power-quality conditioner with distributed generation. *IEEE Transactions on Power Delivery*, **21** (1), 330–338.
84. Kwan, K.H., So, P.L., and Chu, Y.-C. (2012) An output regulation-based unified power quality conditioner with kalman filters. *IEEE Transactions on Industrial Electronics*, **59** (11), 4248–4262.
85. Teke, A., Saribulut, L., and Tumay, M. (2011) A novel reference signal generation method for power-quality improvement of unified power-quality conditioner. *IEEE Transactions on Power Delivery*, **26** (4), 2205–2214.

18A

Hardware-in-the-Loop Systems with Power Electronics: A Powerful Simulation Tool

Ralph M. Kennel[1], Till Boller[2] and Joachim Holtz[2]
[1]*Institute for Electrical Drive Systems and Power Electronics, Technische Universitaet Muenchen, Munich, Germany*
[2]*Institute for Electrical Machines and Drives, University of Wuppertal, Wuppertal, Germany*

18A.1 Background

The basic ideas of hardware-in-the-loop (HiL) systems with power electronics (PHiL systems) are different from the usual HiL systems (HiL systems), because PHiL systems include real energy flow in the simulation. This provides some special features, which are presented in an example of a three-phase electric load for industrial voltage source inverters (VSIs) for AC-drive applications. For this purpose, the inverter under test (IUT) is not connected to a real machine, but to a second inverter or inverters instead, which behaves like an electrical machine. The overall modulation frequency (f_{pwm}) of the so-called Virtual Machine is increased by sequential switching of parallel-connected standard inverters. The parallel-connected inverters can be of the same type as the IUT; hence, there exists no power limit for the drive inverter testing with respect to the product range of the manufacturer.

18A.1.1 Hardware-in-the-Loop Systems in General

The application of hardware-in-the-loop (HiL) systems has become "hyped" during the last decade. In particular, HiL systems with power electronics (so-called power hardware-in-the-loop (PHiL) systems) have become the focus of power engineers. The difference from HiL systems used some 20–30 years ago is that PHiL systems integrate real energy motion in the simulation. Originally, HiL systems were used to test products such as Programmable Logic Controls (PLCs) before delivery to market. The product under test was connected to a HiL system, whose main purpose was to provide digital and analog communication signals in the same way that the real world would provide it once the product was in the field.

Power Electronics for Renewable Energy Systems, Transportation and Industrial Applications, First Edition.
Edited by Haitham Abu-Rub, Mariusz Malinowski and Kamal Al-Haddad.

Published 2014 by John Wiley & Sons, Ltd.

During this test, the proper (signal processing) performance of the hardware included in the simulation system had to be assured. However, modern PHiL systems have a different background purpose. It not only simulates the signals (analog or digital) of the hardware under test, but also has the possibility to include real energy in the simulation process.

Usual simulation programs deal with numbers and respective mathematical equations only. Any physical restriction or limitation has to be considered by a respective model. Any mistake in the model results in failures during simulation. The physical behavior of energy has some specific characteristics. Including real energy in the simulation process would consider these natural characteristics without the need for designing a respective simulation model.

Power electronics provide the possibility of including real energy in the simulation without "loosing" it during the process. Shifting energy from one storage device (inductance, capacitance, etc.) to another and feeding back the respective physical quantities to the simulation program results in a PHiL system with real energy as a part of the simulation.

The following sections consist a report on experiences with a PHiL system used for the simulation of electrical machines for testing power electronic products.

18A.1.2 "Virtual Machine" Application

Power inverter testing (e.g., a burn-in test) before the product is delivered to the customer is an important issue for the avoidance of early failures in the customers' application. Generally, the manufacturer of power inverters for drive applications has to use a multitude of different electrical machine units (see Figure 18A.1) to emulate the industrial application as close as possible to reality. However, in most cases, the inverter under test (IUT) is connected to a machine under no load condition. Thus, the inverter operates just for a short time during acceleration and deceleration at maximum power, and sometimes not even that. Therefore, with respect to the very different time constants of electrical machines and power electronics, the inverters are usually oversized. Hence, the IUT is not really tested at real power levels and, in fact, this type of testing is more stressful for the electrical machine than for the IUT.

An alternative method is to connect the IUT to an electrical test bench – a so-called virtual machine (VM) – instead of a real machine [2–7]. In fact, this is a hardware-in-the-loop system with power electronics (PHiL). As a basic idea, the power stage of the VM can be equal to that of the IUT, for example, a two-level voltage source inverter (VSI) in the product power range. For energy saving, an active front end might be connected to the inverter of the VM.

A specific drawback of the concept, according to [5], is the need for a complex filter between the IUT and the VM in order to avoid the proportional-integral (PI)-current controllers of the IUT and VM

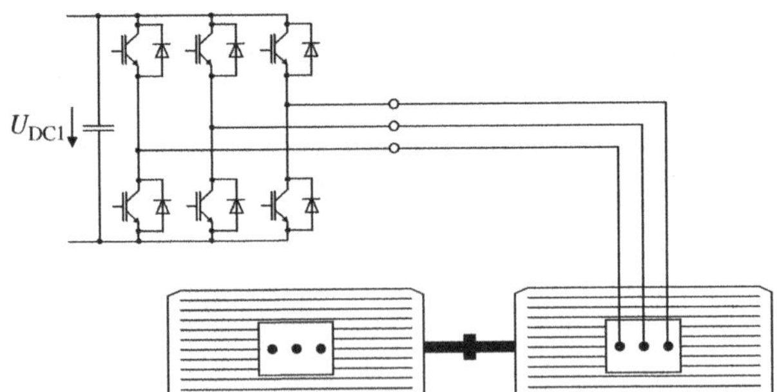

Figure 18A.1 Final test of power electronics products (drive inverter) [1]

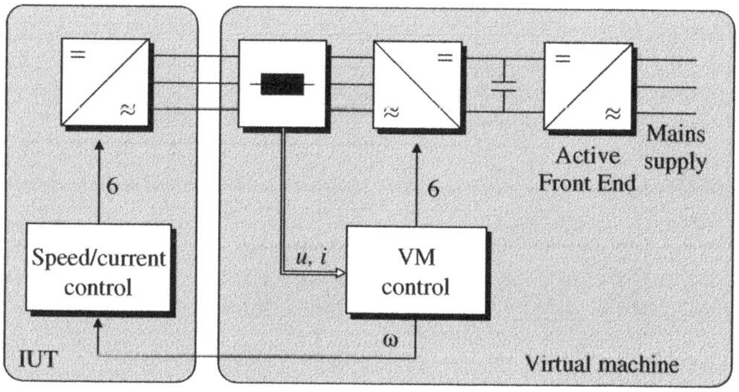

Figure 18A.2 Schematic of the test bench: (a) IUT with the superimposed controller (b) VM comprising a second inverter, an active front end and the control structure [1]

operating against each other. Another approach is described in [7], in which the requirement for a special filter is avoided; however, the IUT can only operate under open-loop control, for example, *v/f* control because the current is already controlled by the VM.

However, inverter manufacturers want to test the complete drive inverter, that is, the inverter in interaction with the control unit, which in most cases is a field-oriented control. To overcome the drawbacks mentioned earlier, a different approach is presented here. The proposed setup consists of another power inverter, an active front end and appropriate coupling inductances, as illustrated in Figure 18A.2. The VM has to behave at the output terminals like a real machine in order to emulate the respective electrical machine. For this purpose, a digital signal processor (DSP)-based control system calculates the machine states depending on the outputs of the IUT and modulates the respective response of an electrical machine. This approach in the behavior of the VM is identical to the characteristics of a real machine.

18A.2 Increasing the Performance of the Power Stage

As the VM has to enforce a specific behavior of voltages and/or currents to the IUT, the performance of the VM must be slightly superior to that of the IUT. There has to be slightly higher DC-link voltage to impress any current, as defined by the machine model, and there has to be higher dynamics to overrule the dynamics of the IUT. A slightly higher DC-link voltage is not really a problem, because an increase by some 20–50 V is more than sufficient. The tolerances and reserves of most inverter products are sufficient to allow this modification. Thus, all that is needed is just an adjustment of the control parameters.

However, higher dynamics require higher switching frequency. A doubling or tripling of the switching frequency in an industrial inverter is not simple, because the thermal design of the inverter product is already done with limited reserves; therefore, it cannot allow steady-state operation at significantly higher switching frequencies. To overcome this problem, the concept of "sequential switching" is used within the VM.

18A.2.1 Sequential Switching

Sequential switching is a well-known concept for increasing the switching frequency [8–16], especially in boost converters. The basic idea is to use parallel power devices and to switch them sequentially; the first pulse will switch the first device, the second pulse will switch the second device and so on (see Figure 18A.3). This concept distributes the switching losses to all the devices in parallel.

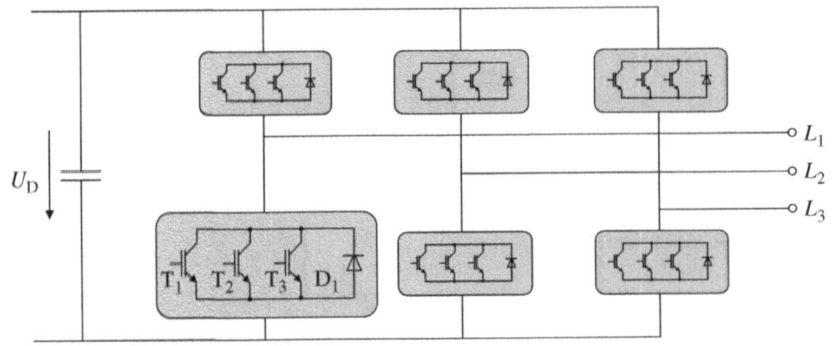

Figure 18A.3 Sequential switching of parallel power devices in a voltage source inverter [1]

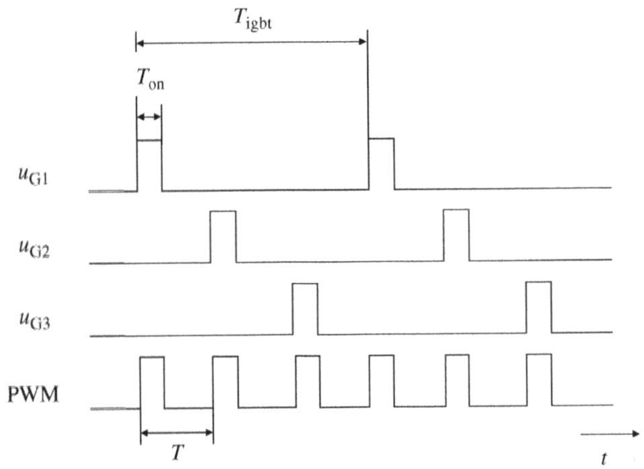

Figure 18A.4 Example of pulse sequence for sequential switching of paralleled power devices [1]

In this concept, the switching losses of each semiconductor device can be reduced without decreasing the pulse-width modulation (PWM) frequency of the complete inverter. On the contrary, the modulation frequency of the parallel-connected devices can even be increased, because the PWM frequency of the complete inverter, consisting of three power stages in parallel, is three times the switching frequency of each individual power device (Figures 18A.3 and 18A.4).

Regarding the increased material cost of the solution with paralleled power devices, this concept is mainly interesting for projects with low piece numbers and high development effort. With this background, a test bench for drive inverters – a "Virtual Machine" – is an excellent application.

The expression "interleaved switching" is used in the literature [8–16] in two different ways. In some papers, "interleaved switching" is used to describe a concept where a power device is switched on while the corresponding device is still in a conductive condition. The gate signals are really "interleaving." In most cases, however, especially with respect to boost converters, "interleaved switching" is used to describe the situation where parallel power devices are never operated in parallel. Any particular power device is switched on only as long as all other corresponding devices are turned off. In fact, the expression "interleaved switching" does not describe this concept properly; a more correct description is "sequential switching."

Applying this concept to a structure of three parallel insulated-gate bipolar transistors (IGBTs) (Figure 18A.3), the switching frequency of each power device is only a third of the total switching frequency (Figure 18A.4). The PWM pulses are not transferred to the IGBT gates simultaneously, but sequentially. While pulse no. 1 is activating IGBT T_1, pulse no. 2 activates IGBT T_2 and pulse no. 3 activates IGBT T_3. In this example, sequential switching reduces the switching frequency of each power device to a third of the switching frequency of the complete inverter.

In Figure 18A.4, the maximum switching-on period T_{on} of each semiconductor is limited to one-third of the switching period T_{igbt}. Even with a modulation index of 100%, the conduction period of each power device only encompasses a third of the complete conduction period. This fact provides some additional thermal reserve in the power stages, which could be used to increase the switching frequency further.

In contrast to the active devices (e.g., the IGBTs), the diode in Figure 18A.3 operates at full load and PWM frequency all the time. This is the reason why "interleaved switching" is so far only used in DC/DC applications, where there is no paralleling of the freewheeling diodes. However, in the case of a three-phase VSI, it is not possible to bring the reverse diodes to a sequential switching mode without additional provisions. The concept of "magnetic freewheeling control" is a proper solution for that.

18A.2.2 Magnetic Freewheeling Control

Using standard IGBTs in parallel, it is possible to realize inverters for higher switching frequencies than usual. Current load and thermal losses are split between the paralleled devices. Parallel connecting of complete inverters is even more advantageous for small- and medium-sized enterprises (SMEs), because this avoids the design effort for special inverters for high switching frequencies. The concept of sequential switching, however, causes some problems with respect to the freewheeling diodes integrated in each IGBT device. These diodes are connected in parallel as well, but they cannot be switched actively. With respect to statistical tolerances, the diodes do not have identical characteristics. Consequently, the diode with lowest internal resistance will accept a major part of the load current and heat up more than the others do. Owing to the negative temperature coefficient, the electrical conductivity of this specific device will "improve" even more, resulting in an even greater part of the load current. This effect will continue to increase until overload and possible destruction of the specific diode.

Of course, it would be possible to use controllable devices like thyristors instead of diodes, which would also guarantee sequential switching during the "freewheeling" periods. In this case, however, standard IGBT modules could no longer be used and a special design of the power stage would be necessary.

To make standard IGBT modules with integrated freewheeling diodes usable for sequential switching, there must be an apparatus providing the freewheeling current to flow through the parallel diodes in the same sequential scheme as in the paralleled IGBTs. This can be obtained by a rather simple magnetic concept called "magnetic freewheeling control" [1, 17–20].

To achieve the behavior of magnetic freewheeling, the windings of the coupling inductor are split to several parallel coils mounted on the same magnetic core. Each IGBT/diode branch is connected separately to one of the coils; all coils of a half-bridge use the same magnetic core (see Figure 18A.5). The coupling inductor is necessary and used between the IUT and the VM (Figure 18A.2) and the realization of magnetic coupling of the inverter output phases is only a moderate modification of the decoupling inductance.

The basic idea is presented in Figure 18A.5. The internal freewheeling diodes (marked as D_1, D_2, D_3), which are integrated in the IGBT modules and therefore connected in parallel as well, are coupled by an inductor with separate coils. The inductance for the magnetic freewheeling control (shown here as the series connection of an ideally coupled inductance L_m and a stray inductance L_σ) must handle the high switching frequency of the current pulses. The main (mutual) inductance L_m guarantees the load current commutating from one power device to the next, as well as the decoupling of the IUT and the VM (note: in the latter application, point "A" is connected to the IUT). The leakage inductance L_σ of each output

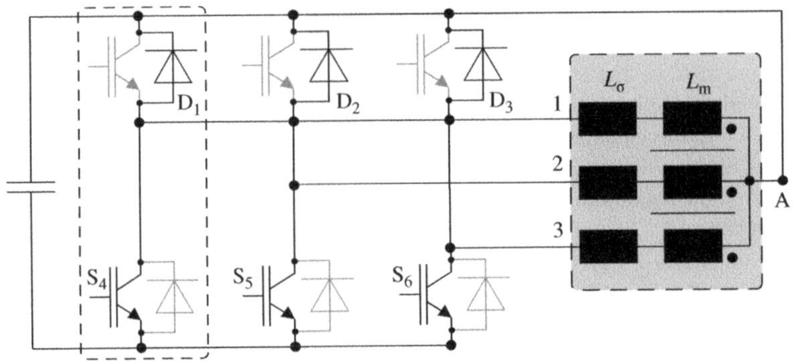

Figure 18A.5 Basic circuit for "magnetic freewheeling control" [1]

phase is a significant part of the magnetic freewheeling control concept. The following aspects have to be considered for designing the inductance:

- A significant leakage inductance guarantees that the load current will not commutate to an improper freewheeling path during turn-off of the IGBT.
- The commutation time for the load current from one power device to the next is proportional to the values of the leakage inductances involved.
- During turn-off, the leakage inductance interacts with internal capacitances of the diode.

In addition, any parasitic leakage inductance in each phase, which is inevitable when connecting complete inverters instead of single power devices or inverter branches/legs, must not disturb the magnetic freewheeling control, but it can be considered when calculating the complete leakage inductance per phase. Of course, a well-balanced value of mutual and leakage inductances must be chosen in order to guarantee the commutation of the load current from the previous to the next inverter, as well as the sequential switching of the antiparallel diodes by magnetic freewheeling control. In this manner, the magnetic freewheeling control (according to [17–20]) forces the load current to flow during the freewheeling periods only through the diode corresponding to the IGBT that was switched on before.

Figure 18A.6 demonstrates the experimental results of the circuit shown in Figure 18A.5 and shows clearly that the freewheeling current is flowing through each diode during each third pulse only. There is only a very small current in the diode when one of the parallel diodes is in duty. During the conductive pulses (each third only), the major part of the load current has to be carried by the respective freewheeling diode. Nevertheless, a small part of the load current commutates to the other two freewheeling diodes, which should not be active. However, this current does not contribute significantly to the switching losses or to the conduction losses of the diode. and hence it can be neglected.

The "magnetic freewheeling control" concept allows sequential switching of the freewheeling diodes by a magnetic coupling between the individual parallel-connected IGBTs. This can be used by specially designed power stages with paralleled IGBT modules mounted rather close together on a heat sink (see Figure 18A.7). The concept, however, is assigned to the sequential switching of complete inverters just by connecting them by cables and it is applicable to standard three-phase VSIs as produced by many manufacturers [17]. The respective structure is presented in Figure 18A.8.

Each output phase of the several inverters is connected separately to one coil. Owing to the leakage inductance between the coils, it is possible that the load current does not commutate to all diodes of the parallel inverters, but to the freewheeling diode in the same arm with another conducting IGBT. If the IGBT of the next inverter is turned on, the load current commutates from the previous to the next inverter supported by the mutual inductance of the common core. The size of the inductance does not

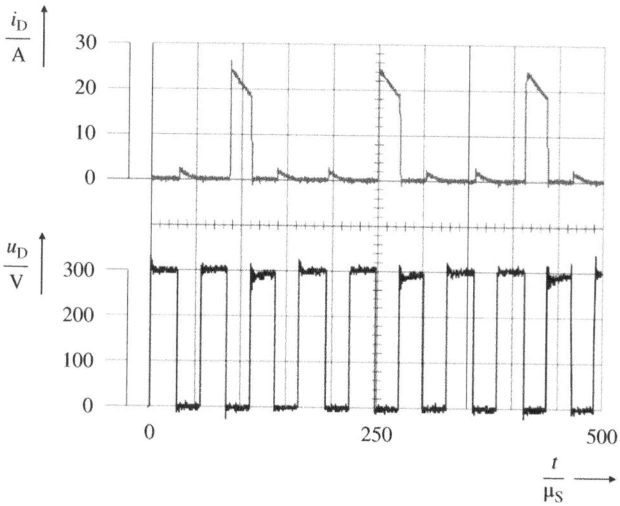

Figure 18A.6 Experimental freewheeling current in a setup of three parallel IGBT modules with magnetic freewheeling control [1]

Figure 18A.7 Inverter branch/leg with five paralleled IGBT modules

increase significantly in comparison with the usual choke, because the magnetic circuit is identical to an inductance with a single coil.

As an additional benefit, the concept of magnetic freewheeling control enables power electronics manufacturers to increase the power range of their products without using specially designed power devices. In particular, small- and medium-sized companies can use it as a tool to enhance the range of their products based on the same power electronic components (e.g., standard IGBT modules) that they use anyway.

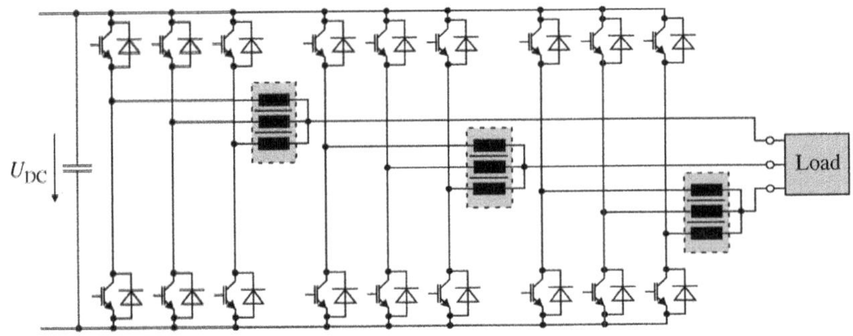

Figure 18A.8 Sequential switching in a three-phase VSI [1]

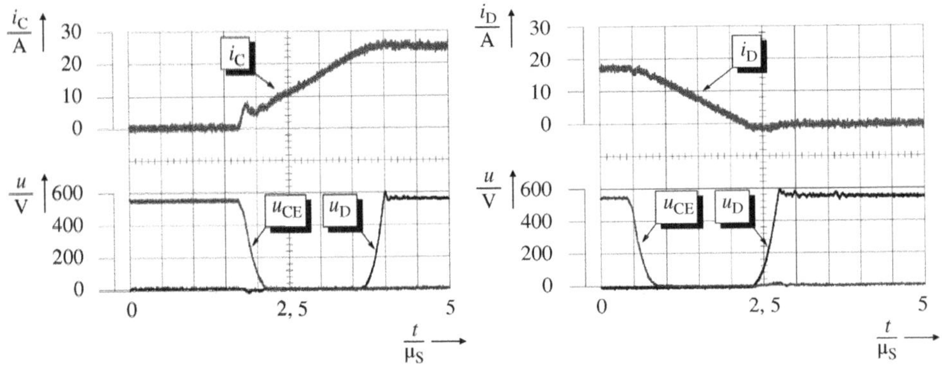

Figure 18A.9 Soft switching during "magnetic freewheeling control" (a) IGBT current and (b) diode current [1]

Because of sequential switching, the turn-off process of the diode and the turn-on process of the IGBT are decoupled. In fact, this is a soft switching behavior instead of the hard switching during load current commutation from one inverter branch to the next. The IGBT switches at zero current (see Figure 18A.9), and the diode has a reduced reverse recovery current owing to the limited di/dt at turn-off (see Figure 18A.9). During the commutation, the voltage of the diode and IGBT are zero and thus, the voltage is shared in the same proportion between the two involved leakage inductances. Therefore, this leads to a further decrease in the switching losses.

18A.2.3 Increase in Switching Frequency

In the case of five paralleled inverters, the switching frequency of each IGBT is only a fifth of the PWM frequency. Furthermore, the maximum turn-on time of the IGBTs is limited to a fifth of the switching period of the IGBTs, and thus the conduction losses are additionally reduced. With this reduction of the switching and conduction losses, it is possible to increase the power capability and ongoing, the overall switching frequency can be increased without overcoming the maximum switching losses of an individual inverter.

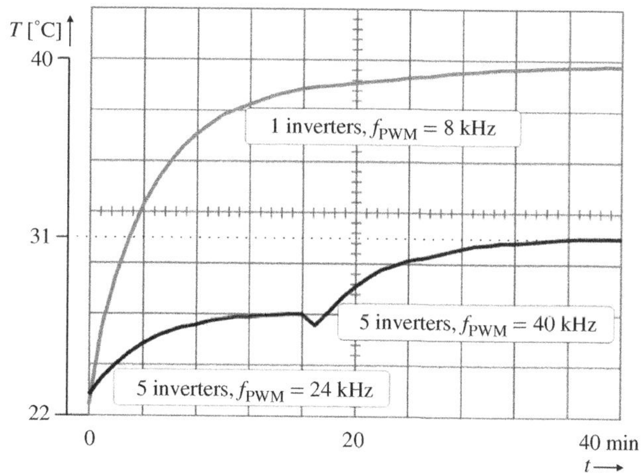

Figure 18A.10 Measured temperature of cooling fins in normal and sequential switching modes. In sequential operation, the frequency was changed at $t = 16$ min [1]

Figure 18A.10 compares the thermal behavior of a single inverter with that of the sequential switching of five parallel-connected inverters with magnetic freewheeling control. Even after increasing the switching frequency of the complete setup with five paralleled inverters by 40 kHz, the temperature of the heat sink is lower than that for the single inverter (with 8 kHz switching frequency). With respect to the effects mentioned earlier (limited duration of conduction losses, soft switching by magnetic freewheeling control), there are additional thermal reserves that could be used to operate with even higher switching frequencies.

18A.3 Machine Model of an Asynchronous Machine

18A.3.1 Control Problem

As the IUT usually comprises a current control as a part of a field-oriented control, another current controller within the VM would cause problems. Both current controllers would attempt to control the same current, as there is no chance for the current from the IUT to go a different way other than into the VM (see Figure 18A.2). The approach according to [5] overcomes this problem by implementing a T-filter instead of using a simple inductor between the IUT and the VM. Here, the current has a chance to deviate between the IUT and VM, and thus neither current controller affects the other. In this case, however, the current in the VM is no longer similar to the current in the IUT; the "Virtual Machine" has a significant "leakage current."

Another solution is to use a state space controller within the VM, which can "overrule" a PI current controller of the IUT. To set up a state space controller, however, special skills are necessary, which are not available in many small- and medium-sized companies.

Another approach is to use a specially designed power stage for the VM [6], but this is also not an economical solution for small- and medium-sized companies.

A very successful solution was the use of an "inverted" machine model, which is described in the following section.

18A.3.2 "Inverted" Machine Model

The test bench has to act at its output terminals like a real machine in order to emulate the respective electrical machine. As an example, an induction motor has been emulated in this project. To avoid the need for a current controller, the model of the electrical machine should not calculate the machine current depending on the machine voltage, but vice versa. An "inverted" machine model, calculating the machine voltage as a response to the machine current (which is controlled by the IUT), does not need a current controller on the VM side. Therefore, the equivalent circuit in Figure 18A.11 has been used to derive the rotor-based machine model of an induction motor [21]. According to the voltage response, calculated by the "inverted" machine model inside the VM (see Figure 18A.12), the respective induced voltage of the induction motor is modulated by the power stage of the VM. This approach avoids any need for an additional current controller.

Nevertheless, this approach is able to represent the characteristics of a real machine. The current depends on the stator leakage inductance $L_{s\sigma}$. In our case, this is the coupling inductance between the two inverters, which also guarantees the magnetic freewheeling control for sequential switching of the diodes, as mentioned in the preceding chapters. The stator current $\boldsymbol{i}_s$ yields to a rotor flux Ψ_r (see Figure 18A.12). The machine responds with an induced voltage at its terminals, which depends on the angular velocity ω and the rotor flux Ψ_r. The angular velocity is the integral of the electrical torque minus the load torque

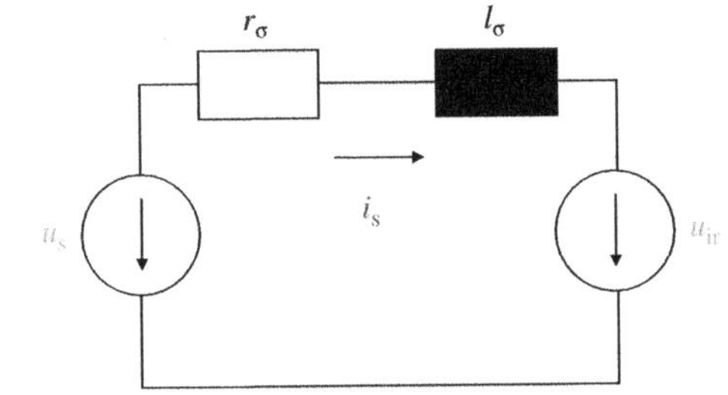

Figure 18A.11 Equivalent circuit of an induction machine [1]

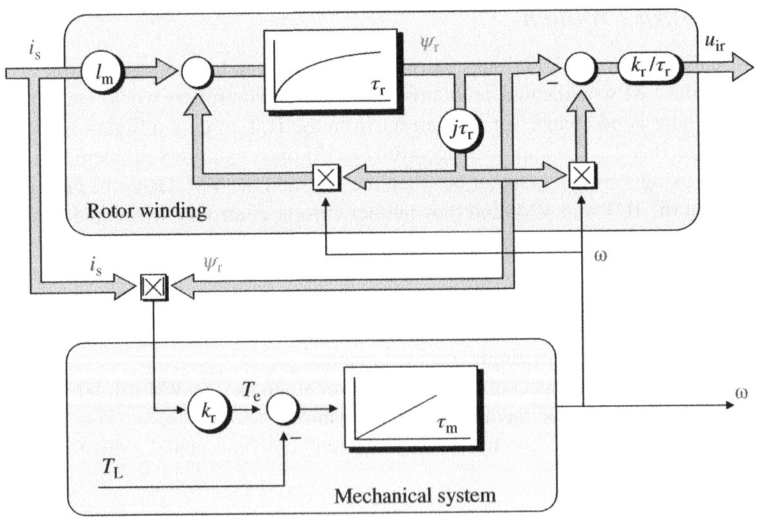

Figure 18A.12 Current-based machine model for evaluating the induced voltage and the angular velocity of the virtual machine [1]

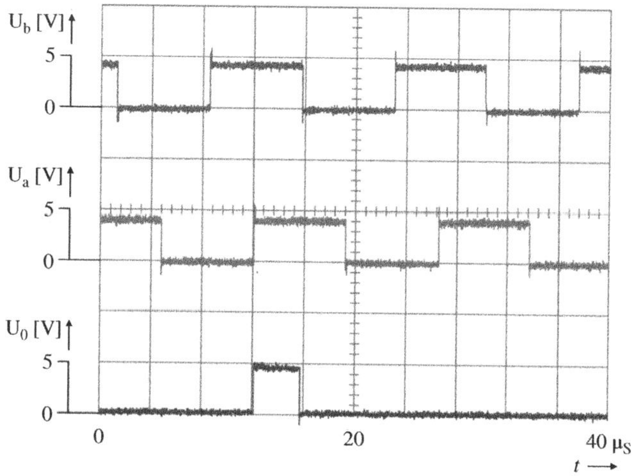

Figure 18A.13 Output signals of encoder emulation [1]

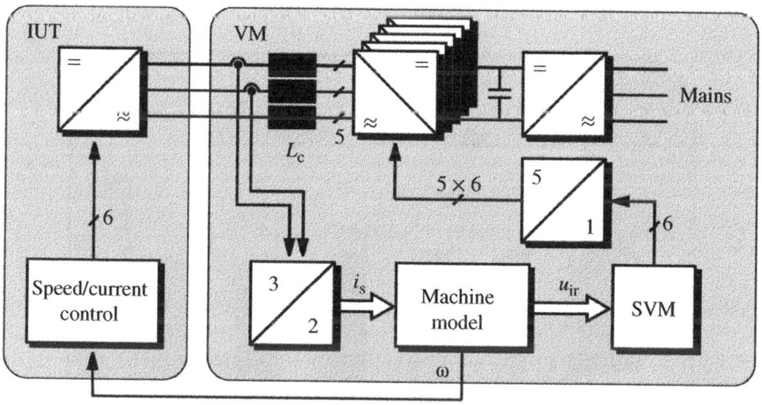

Figure 18A.14 Setup of the test bench: (a) the five parallel-connected inverters, the active front end, the coupling inductances and the control structure are displayed; and (b) the IUT with the superimposed controller is shown [1]

(see Figure 18A.12), which is an important quantity, because it simulates different load conditions for the VM and serves as input for the superimposed speed controller of the IUT. To emulate this velocity for the speed variable drive, an encoder emulator based on an FPGA is implemented. This emulator generates the traces of an optical encoder (A, B and zero) depending on the angular velocity calculated by the machine model. The output of the emulator is compatible with a standard incremental encoder interface (see Figure 18A.13), and hence can be connected to any inverter, which provides an incremental encoder input.

18A.4 Results and Conclusions

18A.4.1 Results

Figure 18A.14 shows the block diagram of the complete VM setup used for obtaining the experimental results. Figure 18A.15 presents the respective laboratory setup.

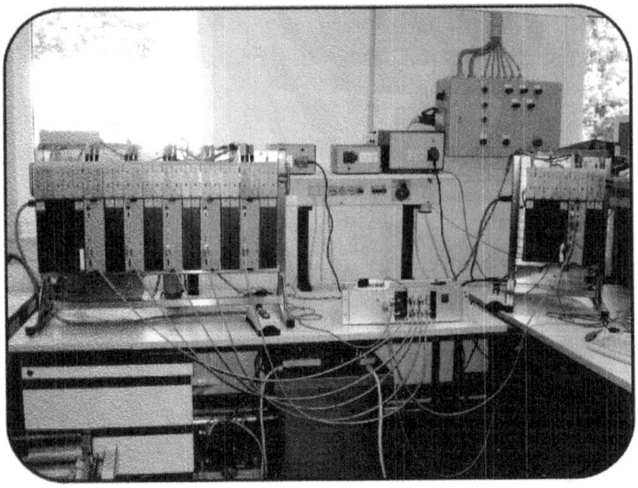

Figure 18A.15 Laboratory setup: (a) the five parallel-connected inverters together with the active front end are located and the coupling inductances are mounted on the back; and (b) there is the IUT [1]

Figure 18A.16 Virtual machine – five inverters in parallel [1]

The test setup consists of five industrial standard 14 kVA two-level VSIs (see Figure 18A.16) that are connected in parallel and switched sequentially. The rated switching frequency is 8 kHz. The calculation of the machine model and the generation of the PWM pulses are done by a DSP controller (see Figure 18A.17). The PWM pulses are distributed to the various inverters by a field-programmable gate array (FGPA).

The VM can be parameterized via a PC user interface (Figure 18A.18). During the test, the load torque can be changed and thus, load steps can be emulated. The parameters of the induction machine emulated by the VM are listed in Table 18A.1. The IUT is of the same type as the parallel-connected inverters of the test bench and is controlled by a field-oriented control.

Figure 18A.17 Real-time system for the VM control [1]

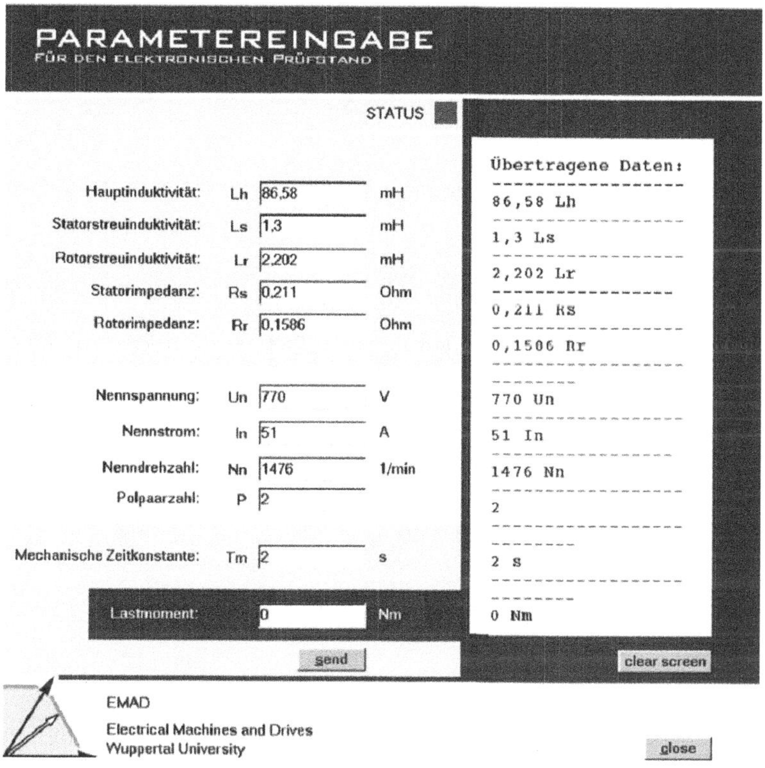

Figure 18A.18 Control screen of the virtual machine [1]

Table 18A.1 Parameters of the virtual machine

$U_N = 380\,V$	$m = 81.49\,mH$
$I_N = 22\,A$	$L_s = 84.18\,mH$
$f_N = 50\,Hz$	$L_r = 86.36\,mH$
$p = 2$	$R_s = 292\,m\Omega$
$J = 0.01\,kg\,m^2$	$R_r = 232\,m\Omega$

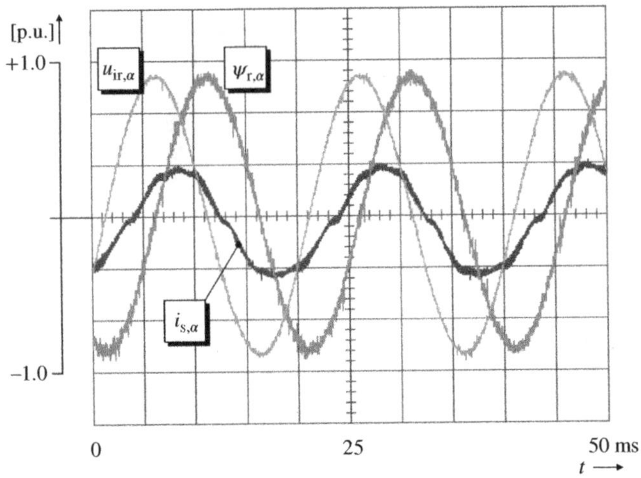

Figure 18A.19 α-Components of the machine model input current, the rotor flux and the induced voltage and rated speed and slight load torque [1]

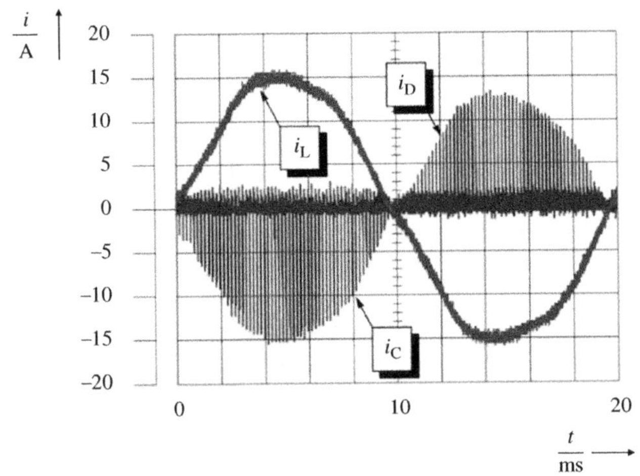

Figure 18A.20 "Load" current and inverter branch current [1]

The states of the VM at rated speed and slight torque load are illustrated in Figure 18A.19. These are direct results from the computation of the machine model (Figure 18A.12). The current shows the marks typical of a machine current with VSI supply – nonlinearities of the IUT (voltage drop) cannot be compensated completely by the field-oriented control of the IUT. Voltage and flux, however, are exactly sinusoidal, as these results are from the calculations of the machine model of the VM show.

Figure 18A.20 shows the "load current" of the complete VM in comparison with the current in one of the five paralleled inverter branches/legs. The current pulses in the power devices can be seen clearly.

The diagram in Figure 18A.21 shows the stator currents as well as the speed during a speed reversal.

The oscillogram in Figure 18A.22 shows a speed step from standstill to rated speed. The IUT operates in speed control mode. It is obvious that it is only during the acceleration process that the inverter delivers

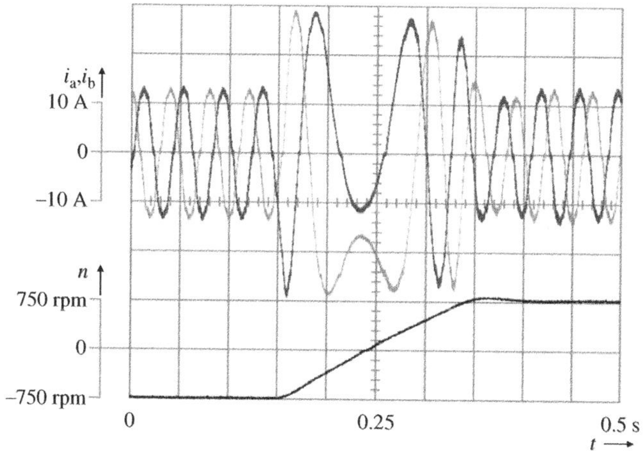

Figure 18A.21 Speed reversal – stator current in stator coordinates [1]

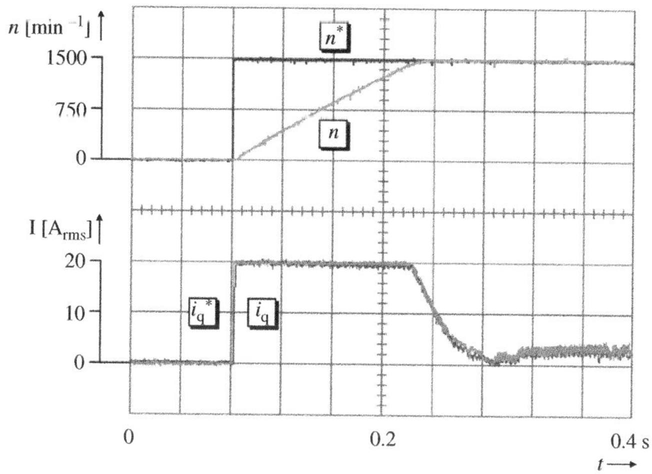

Figure 18A.22 Speed step from standstill to rated speed – q-component of the field-oriented current controller of the IUT [1]

maximum torque. The characteristics of the VM are equal to those of a real induction motor. Currents and speeds are presented in field orientation. Figure 18A.23 shows the same situation with currents, voltages, and fluxes in stator coordinates. The reference speed is increased slowly and the superimposed field-oriented controller of the IUT controls the speed and the output current. The amplitude of the rotor flux is kept constant owing to the field-oriented control. As the actual speed increases up until the nominal speed, the induced voltage also rises until the nominal value at nominal rotor flux. This is exactly the representative of the behavior of a real induction motor. As in Figure 18A.21, all values are calculated by the VM.

Figure 18A.24 displays a load step at rated speed. At time $t = 0.08$ s, a load change is simulated by the VM. The superimposed speed controller of the IUT reacts and the torque current i_q makes sure that the

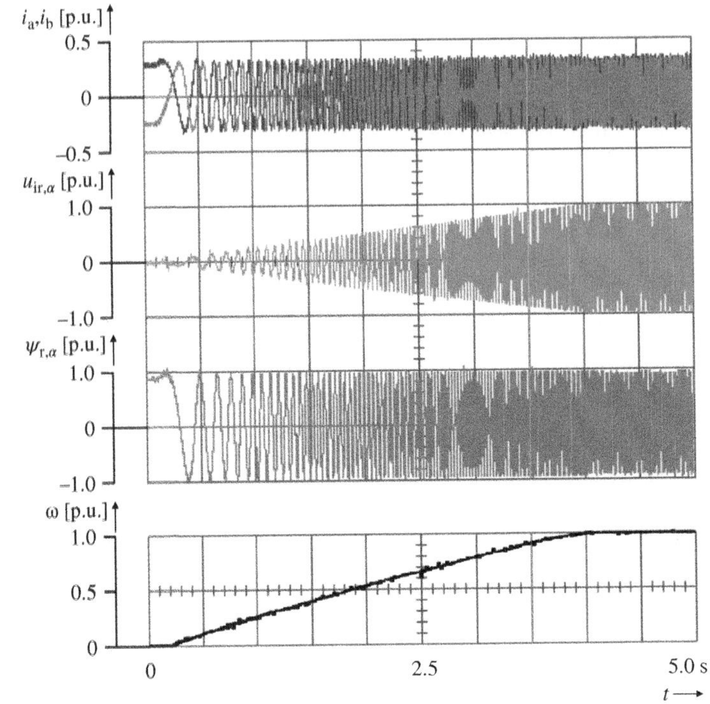

Figure 18A.23 Acceleration from standstill to nominal speed [1]

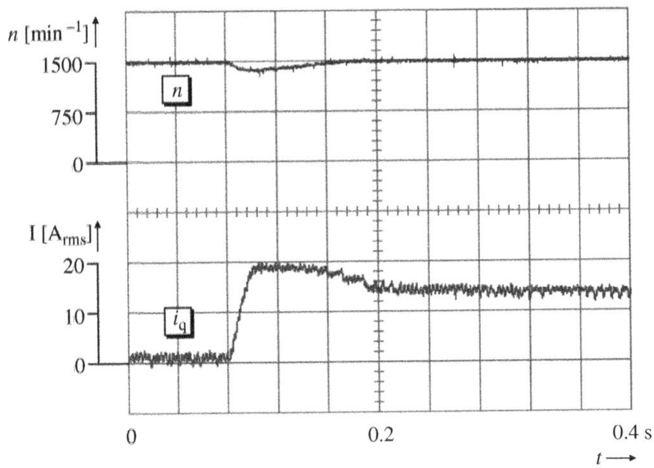

Figure 18A.24 Load step from 0% to 75% rated load at rated speed rated speed of the VM can be kept constant. Consequently, the IUT can be tested under real power levels solely by adjusting the load torque of the VM.

18A.4.2 Conclusions

The report has pointed out that sequential switching is not only applicable to IGBT modules with integrated diodes mounted closely on a heat sink, but also to complete industrial standard two-level VSIs. Thus, the power capability and the overall switching frequency of these products can be increased by using several of them in parallel with sequential switching. Using this approach, a novel electrical test bench for drive inverters can be set up with a minimum effort on design and development. This provides an alternative for manufacturers of drive inverters to test their complete product range under real power levels without the requirement for a multitude of real machines.

The results prove that the presented VM is a HiL system allowing an inverter to be tested at real power levels without the need for installing and operating real machines. The VM has the same characteristics as a real induction motor. Different machines and their respective load conditions can be emulated by software, which means that the drive IUT can operate in its normal mode (as usual). No modification has to be done to the inverter or to the control unit.

References

1. Kennel, R.M., Boller, T., and Holtz, J. (2011) Hardware-in-the-loop systems with power electronics – a powerful simulation tool. Workshop on Power Electronics in Industrial Applications and Renewable Energies (PEIA 2011), Doha, Qatar, December 2011.
2. Boller, T., Holtz, J., and Kennel, R. (2010) *Sequentiell schaltende Umrichter als elektronische Last für Antriebsumrichter*, SPS/IPC/Drives, Nürnberg (in German).
3. Ralph, K. (2009) Virtual machine – replacement of electrical (Load) machines by a hardware-in-the-loop system. 13th European Conference on Power Electronics and Applications, EPE 2009, Barcelona, Spain, September 2009.
4. Boller, T. and Kennel, R. (2009) Virtual machine – a hardware in the loop test for drive inverters. 13th European Conference on Power Electronics and Applications, EPE 2009, Barcelona, Spain, September 2009.
5. Monti, A., D'Arco, S., and Deshmukh, A. (2008) A new architecture for low cost power hardware in the loop testing for power electronics equipment. IEEE International Symposium on Industrial Electronics, ISIE 2008, June 30 2008-July 2 2008, pp. 2183–2188.
6. Wenzel, A. (2008) *Gröling, Schumacher: "Hoch-dynamische, elektronische HiL-Echtzeit-Lastsimu-lation zur Umrichterprüfung", 3*, VDE/VDI-Tagung, Böblingen (in German).
7. Slater, H.J., Atkinson, D.J., and Jack, A.G. (1998) Real-time emulation for power equipment development. Part 2: the virtual machine. *IEE Proceedings - Electric Power Applications*, **145** (3), 153–158.
8. Kleveland, F., Undeland, T.M., and Langelid, J.K. (2000) Increase of output from IGBTs in high power high frequency resonant load inverters. IEEE Annual Meeting of Industry Applications Society, Rome, 2000, CD-ROM.
9. Abo Zied, H., Makky, A.-R.A.M., Mutschler, P., and Stier, S. (2002) A modular IGBT converter system for high frequency induction heating applications. 23rd International Conference and Exhibition on Power Electronics, Intelligent Motion, Power Quality PCIM 2002 Europe, Nürnberg, Germany, 2002, CD-ROM.
10. Stier, S.H. and Mutschler, P. (2004) A modular IGBT converter system for high frequency induction heating applications. German-Korean Symposium 2004 on Power Electronics and Electrical Drives KOSEF 2004, Aachen, Germany, 2004, S. 164–171.
11. Pérez-Pinal, F.J. and Cervantes, I. (2006) Simple almost zero switching losses for interleaved boost converter. 3rd International Conference on Power Electronics Machines and Drives PEMD 2006, Dublin, Republic of Ireland, 2006.
12. Chang, H.-H., Tseng, S.-Y., and Huang, J.G. (2006) Interleaving boost convertsers with a single-capacitor Turn-off Snubber. 37th Annual IEEE Power Electronics Specialists Conference PESC, Jeju, 2006, CD-ROM.
13. Hu, Y., Xie, Y., Tian, H., and Mei, B. (2006) Characterisitics analysis of two channel interleaved boost converter with intergrated coupling inductors. 37th Annual IEEE Power Electronics Specialists Conference PESC, Jeju, 2006, CD-ROM.

14. Singh, R.P., Khambadkone, A.M., Samudra, G.S., and Liang, Y.C. (2006) An FPGA based digital control design for high-frequency DC-DC converters. 37th Annual IEEE Power Electronics Specialists Conference PESC, Jeju, 2006, CD-ROM.
15. Zhang, X. and Huang, A. (2006) MVRC and its tolerance analysis for microprocessor power management. 37th Annual IEEE Power Electronics Specialists Conference PESC, Jeju, 2006, CD-ROM.
16. Asiminoaei, L., Aeloiza, E., Kim, J.H. *et al.* (2006) Parallel interleaved inverters for reactive power and harmonic compensation. 37th Annual IEEE Power Electronics Specialists Conference PESC, Jeju, 2006, CD-ROM.
17. Boller, T., Kennel, R., and Holtz, J. (2010) Increased power capability of standard drive inverters by sequential switching. IEEE-ICIT 2010, Valparaiso, Chile, March 2010.
18. Kennel, R. and Holtz, J. (2007) Industrial servo drives - interleaved or sequential switching for increasing the switching frequency. ICPE07 (International Conference on Power Electronics), Daegu, Korea, October 2007.
19. Ferreira, A.C. (2007) Improved output power of multiphase inverters by sequential switching. PhD thesis. Wuppertal University, Germany, June 2007 (in German).
20. Kennel, R.M., Boller, T., and Holtz, J. (2011) Replacement of electrical (Load) drives by a hardware-in-the-loop system. International Aegean Conference on Electric Machines and Power Electronics and Electromotion (ACEMP 2011), Istanbul, Turkey, September 2011.
21. Holtz, J. (1995) The representation of AC machine dynamics by complex signal flow graphs. *Transactions on Industrial Electronics*, **42** (3), 263–271.
22. Ferreira, A.C. and Kennel, R. (2007) Emissionsfreier elektronischer Prüfstand. *Antriebstechnik*, **5**, 44–45 (in German).
23. Holtz, J. (1994) Pulsewidth modulation for electronic power conversion. *Proceedings of the IEEE*, **82** (8), 1194–1214.

18B

Real-Time Simulation of Modular Multilevel Converters (MMCs)

Luc A. Grégoire[1], Jean Bélanger[2], Christian Dufour[2], Handy. F. Blanchette[1] and Kamal Al-Haddad[1]
[1]*Electrical engineering department, Ecole de Technologie Supérieure, Montréal, Canada*
[2]*OPAL-RT TECHNOLOGIES INC, Montréal, Canada*

18B.1 Introduction

Real-time simulation offers several advantages to speed up the development of new product. One of these advantages being the possibility to test and develop controllers when the hardware is not yet available. This is a serious advantage in the case of high-order multilevel converter, like modular multilevel converter (MMC) topology. Its physical size could raises serious issues for most laboratories, without even mentioning the cost to build such a complex structure. It can also be useful to analyze the interaction between several MMC and conventional high-voltage DC (HVDC) systems installed on the same power grid. Furthermore, it can perform factory acceptance test of the control system before the installation in the field. Nowadays, real-time simulators are often used simply to accelerate simulations, which could take several hours for simulation runs of a few seconds with a power grid having two or three converter stations using conventional single-processor simulation software. This chapter introduces bases of real-time simulation: its advantages and its constraints. Using these bases, real-time simulation of an MMC will be undertaken. This topology was first introduced in [1], it is made of many identical cells connected in series. Its modularity makes it suitable for various applications from medium voltage in a drive system using only a few cells [2] to a large HVDC transmission system containing a wide range of cells [3, 4]. Connecting many of these cells in series reduces the voltage level that each sustains, decreasing the price of each component, reducing the switching losses and smaller dV/dt at its AC bus, while producing a sinusoidal waveform with a very low total harmonic distortion (THD) eliminating the use of bulky reactive component filter.

18B.1.1 Industrial Applications of MMCs

This topology was first tested by ABB in 1997. It consists of a 10 km overhead transmission line with a 3 MW capability at ±10 kV between Hällsjön and Grängesberg in Sweden. It was used as a proof of

---

Power Electronics for Renewable Energy Systems, Transportation and Industrial Applications, First Edition.
Edited by Haitham Abu-Rub, Mariusz Malinowski and Kamal Al-Haddad.

concept and to establish the capability of this new topology. The MMC was named HVDC light by ABB and was first used in a commercial project in Australia between Mullumbimby and Bungalora. Its voltage rating was ±80 kV with a power rating of 180 MW commissioned in 2000. Not long after, Siemens commercialized a similar topology as HVDC PLUS. Its first commercial project was a submarine HVDC link connecting San Francisco city center to a substation in the Pittsburgh area, and it was commissioned in 2010 [5].

As of today, MMC projects being built are point-to-point converters only. Though actual HVDC networks have been discussed theoretically, protection system for such network still needs to be developed. ABB announced in November 2012 that they achieved an HVDC breaker called hybrid HVDC breaker [6]. Now that it has been used in a point-to-point setup, it will be tested in HVDC grid and should soon be commercialized. These new developments could change the future of power transportation.

18B.1.2 Constraint Introduced by Real-Time Simulation of Power Electronics Converter in General

Until now, big differences still exist between what can be achieved with standard or offline simulation software and real-time simulator. The major constraint is in the time available to solve the differential equations. This describes the operation sequences of power electronic circuit. Offline simulation usually uses a precise variable-step solver. They will use two different orders of discretization, one higher than the other, and will iterate reducing the time step of every iteration until the difference between the two solutions coincides within a pre-set tolerance [7]. This process is very efficient for typical simulation of system with few disturbances. However, it can be very slow especially in power electronic where stiff system with repetitive switching of fast semiconductor needs to be solved. On the other hand, real-time simulation uses several processors, operating in parallel and fixed-step solver and has a fix period of time to solve the differential equations. If a time step of 50 μs is chosen to discretize a system, the real-time simulator has to solve the differential equation within that period. Larger model, with more state-space equations, will naturally take more time to be solved; in this case, there are very few solutions to obtain acceptable results. One can increase the chosen time step, risking instability or inaccuracy. The computing power of real-time simulator has increased over the past decade following Moore's law [8] and is suited to simulate relatively small model. However, the real-time simulation of very large power system requires to decouple the system in smaller subsystems that can be solved in parallel [9, 10]. Nowadays, most real-time simulators will achieve a time step between 10 and 50 μs when using several general-purpose processors or digital signal processors (DSPs) and between 100 ns and 1 μs when using field gate programmable array (FPGA).

In the case of power electronic or circuit that contains fast switching devices, the chosen time step is very important since it will determine the accuracy that can be achieved by the pulse-width modulation (PWM) circuit to generate the gating signals. A switching frequency of 10 kHz has a period of 100 μs. If one chooses the time step of the real-time simulator equal to 50 μs, then there is a maximum of 50% error on the time of occurrence of the switching event. Such inaccuracy may produce unrealistic transients and harmonics that could be confused with faulty controllers. This has motivated the usage of super-fast computer subsystem, where the time step can be reduced further, or an interpolation scheme in order to achieve accurate switching frequency [11]. Moreover, this is one of the reasons to justify the use of FPGA to solve such a problem and consequently increase the popularity of the technology. FPGA chips operate at a 100–400 MHz clock frequency much lower than 2–4 GHz used for the general-purpose processors. However, processors are composed of hundreds of DSP blocks operating in parallel to achieve very small time step between 100 ns and 1 ms. This goal is still not possible with the most powerful commercial computer using general-purpose processors because of the processor communication latency. Furthermore, firing signal of power devices generated from actual power electronic controller can be connected directly to FPGA digital input pins, which are connected to the model of the converter simulated in the

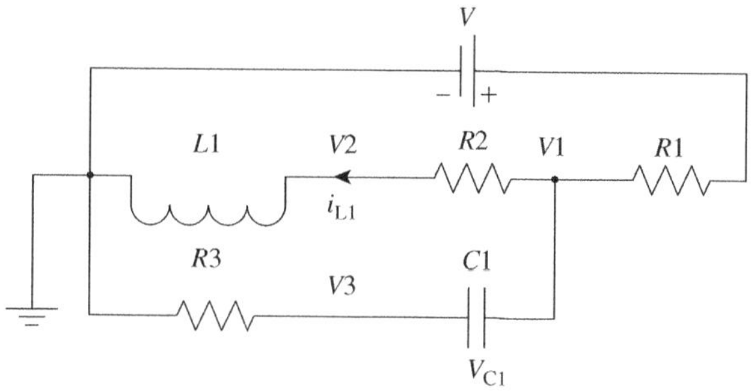

Figure 18B.1 Circuit illustrating state-space versus nodal approach

same FPGA. The result is that the total latency between the firing order, measured at the controller output and the resulting currents computed by the FPGA model could be less than 200 ns to 1 μs. Such small time delay is acceptable for most controllers working with sampling of 10–100 μs and with PWM carrier frequency up to about 50 kHz. In such a case, the accuracy will be as good as variable-step solver. Such low latency and accuracy cannot yet be achieved using general-purpose computer because of the typical latency of the PCI communication system of 2–3 μs.

Moreover, another important parameter to be considered is the type of modeling technique chosen to discretize the circuit. The two main well-known techniques are the nodal and the state space methods. Depending of the circuit topology, one can be more advantageous than the other. Taking shortcut and making this simpler than it actually is, time of execution in real-time simulation comes down to the size of the matrix and its sparsity; since the latter needs to be inverted each time there is a change in the circuit topology, caused by a switching event. In the case of the circuit illustrated in Figure 18B.1, a state-space approach would generate a two-by-two matrix to discretize or inverse, as shown in Equation (18B.1), since there are only two state variables. The nodal approach would yield a four-by-four matrix to solve, as shown in Equation (18B.2). This simple example illustrates the same concept that can be applied to large circuit.

$$\begin{bmatrix} \dot{v}_{C1} \\ \dot{i}_{L1} \end{bmatrix} = \begin{bmatrix} -\dfrac{1}{(R1+R3)\,C1} & -\dfrac{R1}{(R1+R3)C1} \\ \dfrac{R1}{(R1+R3)L1} & -\dfrac{R2+(R1//R3)}{L1} \end{bmatrix} \times \begin{bmatrix} v_{C1} \\ i_{L1} \end{bmatrix} + \begin{bmatrix} -\dfrac{1}{(R1+R3)\,C1} \\ -\dfrac{R3}{(R1+R3)L1} \end{bmatrix} \times V \qquad (18B.1)$$

$$\begin{bmatrix} i1 \\ i2 \\ i_{L1} \\ i3 \end{bmatrix} = \begin{bmatrix} \dfrac{1}{R1} & -\dfrac{1}{R1} & 0 & 0 \\ -\dfrac{1}{R1} & \dfrac{1}{R1}+\dfrac{1}{R2}+C1s & -\dfrac{1}{R2} & -C1s \\ 0 & \dfrac{1}{R2} & \dfrac{1}{L1s}+\dfrac{1}{R2} & 0 \\ 0 & -C1s & 0 & C1s+\dfrac{1}{R3} \end{bmatrix} \times \begin{bmatrix} V \\ V1 \\ V2 \\ V3 \end{bmatrix} \qquad (18B.2)$$

Most simulation software uses one or the other method without giving the choice to the user. It is only brought up here to stay as broad and general as possible. Also when it comes to FPGA implementation, very few off-the-shelf tools are now available [12, 13]. Many users still have to develop their own FPGA model despite of its complexity and researchers are still trying to develop general-purpose electrical solvers, which would eliminate this complex task of implementing models and solvers in FPGA chip. Furthermore, one must keep in mind that one of the most difficult operations to be accomplished on FPGA is the division; therefore, it renders that inverting a matrix is an important research topic to complete it in a timely fashion suitable for a real-time simulation.

18B.1.3 MMC Topology Presentation

An MMC topology is constituted by an equal number of cells, presented in Figure 18B.2a, that are distributed in the upper and lower limb, shown in Figure 18.2b. The cell includes switches S1, S2 and the DC-bus capacitor. When a cell is ON, the capacitor voltage is applied to its output using the upper switch S1 of the cell. When a cell is OFF, it is bypassed using the lower switch S2.

The voltage obtained at the mid-point of the converter arm is given by the number of conducting cells in each limb. At any given time, only half of the cells from one arm are conducting. For instance, if a converter contains 100 cells, only 50 cells distributed between the upper and lower limbs of an arm are conducting at any giving time. Figure 18B.3 shows the voltage seen at mid-point in the following cases:

- 1 cell in the upper limb and 49 in the lower limb are conducting, mid-point is near HVDC+.
- 25 cells in the upper limb and 25 in the lower limb are conducting, mid-point is zero.
- 49 cells in the upper limb and 1 in the lower limb are conducting, mid-point is near HVDC.

When a cell is conducting, its voltage will vary according to the limb current. The voltage is then regulated with rigorous algorithm to choose which cell to turn ON or turn OFF. Though this topology was proposed a few years back, only the increase in the computation power of controller made it possible to accurately control it. Since then, many methods have been proposed to control this converter topology requiring individual control of each cell capacitor voltage [14–16], to cite only a few.

The number of cells plays two roles in this topology. The number of cell is linked to the quality of produced voltage. When more than 12 levels are used, it demonstrates that the gain on the THD becomes almost negligible, as shown in Figure 18B.4 [17].

In the case of MMC used for HVDC application, the extremely high number of cells used reduces the stress on each component and also offers a redundancy improving reliability. Furthermore,

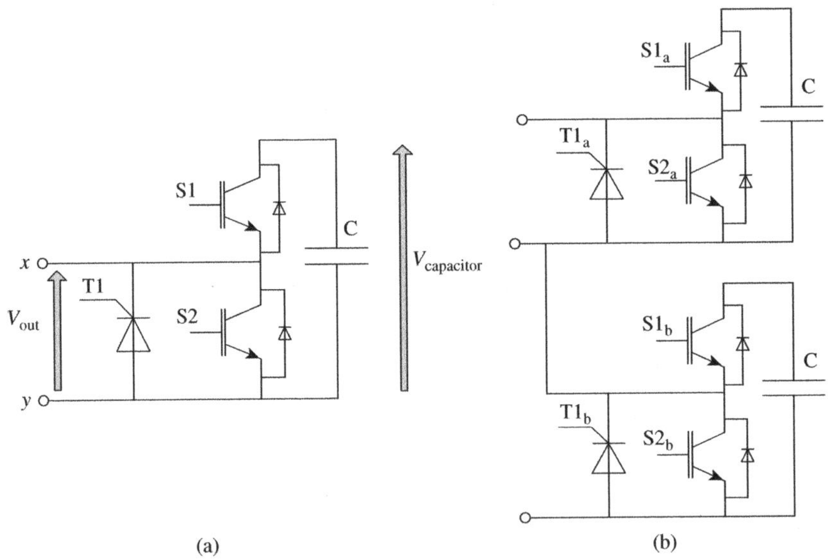

Figure 18B.2 (a) Single MMC cell with its protection and (b) MMC limb

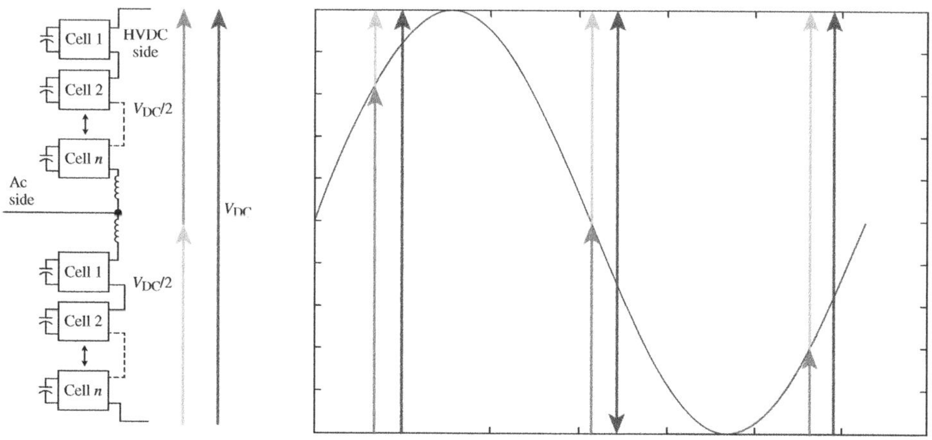

Figure 18B.3 Standard exemple of one MMC arm for HVDC link

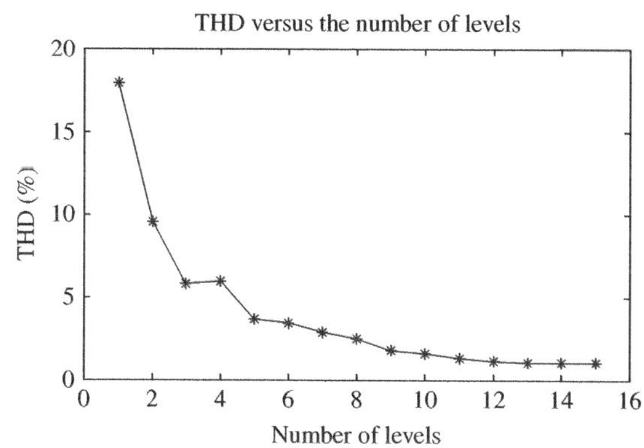

Figure 18B.4 THD versus the number of levels

increasing the number of cells reduces the switching frequency of each individual cell up to once per cycle [4].

Since this topology is a voltage source converter (VSC), it has complete control on the power flow, both active and reactive. Also unlike most HVDC classic topology, it does not require a network to synchronize itself, allowing to make a black start, since the voltage is imposed by the converter. But all of these advantages come with the price of complex control law that needs to be optimized and tested.

18B.1.4 Constraints of Simulating MMCs

When it comes to real-time simulation of MMCs, there are two major problems to resolve. The first being the considerable size of the model, whether a state-space or a nodal approach is taken, and the second being the tremendous amount of input/output required to control the converter. Bear in mind that the main purpose to real-time simulation is to be interfaced with a real controller. When it comes to thousands of cells to be controlled, it can only be assumed that even more signals are required for control of the converter. Most of the processing time was due to the *I/O* management and data transfer between the external controller and the MMC model. It is therefore obvious that MMC simulators with 1000 cells per limb would require an *I/O* processing time much larger than 25 μs, which is unacceptable.

18B.1.4.1 Solving Large State-Space System

One way to surmount the first problem is to exploit certain advantages of MMC topologies. Having a rather large inductance in each limb, this generates a very "strong state" on the AC side; where the current variation is rather slow. The DC side being often connected to a DC cable, the capacitance of the cable also generates a "strong state" on the DC side, where the HVDC voltage variation is slow. There two "strong state" can be used to decouple each limb. Once decoupled, it is possible to spread the computation burden over multiple computing units achieving parallel processing. Furthermore, one limb can be divided into smaller subcircuit without any extra efforts on computation time.

18B.1.4.2 Solving I/O Management Problem

Having met the requirement to decouple large system, the only problem remaining is the one concerning the amount of *I/O* to be dealt with. Architecture of real-time simulator will be discussed later, but for now what need to be understood is that most real controller and simulator platforms use custom-made card with a communication link to its computation unit. More *I/O* implies more data to be sent over the communication link and therefore requires more time. If more time is required for *I/O* communication, this leaves less time for computation of the model. What have been done in the first part will actually help resolve the second issue. Spreading the model across more computation unit reduces the amount of data that each must exchange with the *I/O* solving the second problem. Furthermore, simulating MMC cells directly on FPGA chips, which are managing *I/O* channels, also minimizes data transfer between external controllers and simulator main processors. Such technique is now used by most advanced real-time simulators.

Separating the large state-space systems formed by the MMCs coupled to the AC network in order to achieve parallel processing can be achieved in several ways but might involve the use of artificial delays [18] or multirate simulation [19]. As of now, there are no formal and easy methods to achieve parallel processing of complex power electronic circuits coupled to large AC circuits.

18B.1.4.3 System-Wide Simulation Simulated Faster Than Real Time

Several studies target the behavior of several converters in a large network or the development of controllers before the manufacturing of controller prototype boards. In these cases, the controller algorithm can also be simulated in the same simulator simulating the MMC system and the grid. Consequently, no external *I/O* are required and it is then possible to simulate faster than real time; that is, a typical simulation run of 60 s could take 60 s with a real-time simulator or only a few seconds with a simulator running faster than the real time. In real-time simulation, the acceleration factor is 1, where the time step used and the time required to execute the model has a ratio of 1. For faster than real-time simulation, the acceleration factor is greater than 1 since the time required to solve the equation of the model is smaller than the time step used. Like before, if the model is decoupled and spread over many computation units, its acceleration factor is increased above 1 while the use of conventional signal-processor simulation software may have an acceleration factor much lower than 1, that is, the simulation time of a 10 s case could take several minutes or event hours depending on the network and MMC size. As the control development requires to analyze hundreds of contingencies and to optimize several parameters, it is obvious that fast simulation tools exploiting multicore processors and FPGAs will become essential as model complexity increases.

18B.2 Choice of Modeling for MMC and Its Limitations

As mention earlier, time is a very important constrain in real-time simulation; choosing the appropriate level of modeling for a specific application helps to reduce the required computation time. The level of modeling can be classified into three main categories namely: detailed model, switching function model and average model. Each of them will give accurate results but they all have different limitations.

18B.2.1.1 Detailed Model

There exists different level of modeling in the so-called detailed model. Most of them offer too much detail which is not useful for real-time simulation used to design, optimize and test control systems. The highest level of details could be qualified as "SPICE" modeling, where all the parasitic capacitors of the power switch and strain inductance of the PCB are taken into consideration. This type of modeling is used to calculate losses that will occur during switching. Even though this is a very important part of a real design, real-time simulation should not, but also could not, be used to evaluate switching losses and electromagnetic interference (EMI). Taking a numerical approach, the time constant of components such as picofarad and nanohenry is around nanoseconds; these types of time step cannot be achieved today in real time even with FPGA. Processor technology advancements may however make this dream possible during the next decade.

The model where the switch and diode are considered as linear components can also be considered as a detailed model. Every semiconductor is represented by an impedance: small when conducting and high when blocking. Whether a state-space approach or a nodal approach is used, a new set of matrices need to be computed and inverted each time there is a change in switch status. This approach has been demonstrated using Hypersim [20] or the State-Space Nodal (SSN) solver [21, 22] with 100 cells/arms MMC at time step in the 30 μs range.

18B.2.1.2 Switching Function

Switching functions or event-based dynamic system [23] can be interpreted as a switch case; for a certain input, certain behaviors are expected. In the case of Figure 18B.2, the switching function can take this form:

- If S1 = 1 and S2 = 0, $V_{out} = V_{capacitor}$
- If S1 = 0 and S2 = 1, $V_{out} = 0$
- Else $V_{out} = 0$

This implies that the switching is complementary and that there are no conducting losses, that is, ideal switch. To introduce the switching losses, the current must be taken into account. Therefore, the switching function becomes

- If S1 = 1 and S2 = 0, $V_{out} = V_{capacitor} - I_s * R_{on}$
- If S1 = 0 and S2 = 1, $V_{out} = 0$ $I_s * R_{on}$
- Else $V_{out} = 0$

The flexibility of switching function makes it a very powerful tool, but it requires a very good understanding of the circuit in order to predict and have a contingency for every possible case. Unlike detailed model, this can result into an unnatural behavior and discontinuity that is not present in real life. The gain is in the rapidity of execution, which makes it a very good candidate for real-time simulation. Also when the limitations are known, it does not prevent the use of this model in all other supported mode. In this model, the number of states is not reduced, meaning that an integrator is required for every simulated capacitor cell. A detailed example is given in Section 18.4.2.

18B.2.1.3 Average Model

The term average model here is not only intended as in the classical way. In classic average model, the duty cycle is given as input instead of PWM, but here the overall voltage of every cell's capacitor is also averaged out across all the cells, making an ideal regulation of voltage of all the cells. This type of modeling is the easiest to implement, but it is also the one offering the most limitation. The main interest of this implementation is to study the behavior of the converter in a larger network where the regulation of each cell is of little interest. Similar to the switching function, the cell output is given by a simple equation decoupling it from the large system. The rest of the system will see the converter as a variable impedance as it would be with a detailed model. Again, a detailed example of the implementation is given in Section 18.4.1.

One drawback of this modeling is that it needs a special implementation to support the high-impedance mode occurring when no pulses are applied to the converter. In this mode, the output of the converter, a voltage source, is only controlled by its current when no pulses are applied to the switches. Normally, if the voltage applied to the limb is higher than the sum of all the capacitor voltages of this limb, current should circulate through the antiparallel diode of the switches of the cells, charging the cells' capacitor to voltage applied to the input of the limb. Once all the cells are charged, the current should become zero, since the antiparallel diode is not polarized anymore, and stay at zero until either of the pulses is applied again to the converter or the antiparallel diodes are polarized. Different schemes can be used to achieve this behavior, a voltage source is controlled by a voltage, but if it is not well implemented the response of the model can become erratic.

All three different types of modeling presented here serve a specific purpose, understanding the limitation of each model helps one to determine whether or not this implementation is suited for his application.

18B.3 Hardware Technology for Real-Time Simulation

In the mid-1960s, real-time simulation was achieved using analog simulator, where real linear and nonlinear components were used to model and simulate a circuit [24]. Not long after hybrid simulators, part analog part digital, were introduced and then with the evolution in the microprocessor speed, fully digital simulators were achieved. Even though the first digital simulators were limited, their smaller size and versatility made them more attractive and their popularity was powered by the increase in computer power capability over the last 15 years. For these reasons, analog and hybrid simulators are hardly used nowadays and would not be explained further here. As for digital simulators, two main technologies divide them; the first one uses sequential programming embedded on DSP or microprocessor and the second type makes use of parallel programming on FPGA. Because of their differences and their complementarity, it is not rare to see both technologies in one simulator, taking advantage of each one of them. Their respective features are discussed in the following section.

18B.3.1 Simulation Using Sequential Programming with DSP Devices

DSP is digital signal processor optimized for certain applications. It receives sequential programming: a series of instructions which are executed subsequently and repeated in a loop. These instructions need to be understood by the processor, what can be called low-level language. But it has to be entered by a user high-level language. The gap between those two levels is the different programming language, such as C, C++ and java. Every manufacturer has a different machine code that can only be understood by their hardware. Using a common language by the user, like C, manufacturers make compilers that are compatible with their hardware. Nowadays, high-end processor can execute billions of instructions per second. In order to achieve further more computation power, as mentioned earlier, it is possible to execute a different set of instruction in parallel using multiple processors sharing a high-speed communication link.

In real-time simulation, the most sophisticated processors are used in specially designed hardware. The code required is generated using software like Matlab/Simulink, so users do not require to bother about writing code. When multiple processors are available in parallel, users also rely on software to easily distribute the computational burden among them.

One of the greatest examples of real-time simulation in parallel is the Hypersim simulator, developed at IREQ, the engineering department of Hydro-Quebec [25]. It can simulate a large network, thousands of nodes, cluster of hundreds of CPU while the allocation of the processor unit to simulate each network subsystems is fully automated [26]. Other real-time simulator would normally require the intervention of advanced user in order to distribute the computation load over multiple computing unit [27, 28].

18B.3.2 Simulation Using Parallel Programming with FPGA Devices

FPGA offers much more flexibility when it comes to executing instructions; it actually allows user to develop its own instruction set. Logic operators like NAND-gate or XOR-gate, basic arithmetic like sum, multiplication are some of the components available. By using these functions, users can make an optimized set of instruction for a specific application. On older FPGA generation, only fixed-point representation was available, but since 2009, built-in operators supporting floating point are now available.

In FPGA implementation, signals travel as fast as their propagation allows. The number of operations that can be done will depend on the design; the route that signals need to take for a desired logic. On FPGA, clock signals are used to ensure that the expected result has reached its destination and is synchronized with other signals. Figure 18B.5 shows the concept of propagation to a simple circuit and the synchronization of its output.

From Figure 18B.5, since only one level of logic gate is required to obtain A, it can be supposed that it will be ready before B that needs two levels of logic. C might change when A is ready and change again when B is ready. To avoid uncertain value at the output, a register is added and synchronized with a clock signal. By doing so, D will be synchronized with the clock and its value will be accurate as long as the period of the clock is long enough for the inputs 1, 2 and 3 to pass through the logic resulting into C. In Figure 18B.5, the results of A and B are being processed simultaneously and independently; this is the major advantage of FPGA referred to as parallel processing. From simple logic to large matrix multiplication can be performed in parallel and the result for the global solution is found in the end where all the different solutions are joined and synchronized.

Understanding that the process can be synchronized to a specific clock, it is also possible to use time multiplexing or pipelining, allowing the same logic to be used for different processes. If the clock of the process is slower than the clock of the FPGA, it becomes possible to execute the same process using the same logic. For example, if the process shown in Figure 18B.5 can be obtained one FPGA clock period, but its inputs are only ready be executed, then every five FPGA clock periods. By multiplexing the input, using selector and demultiplexing the output, the same logic could be used up to five times to calculate the same process. The chronogram of Figure 18B.6 shows the time multiplexing with only two different

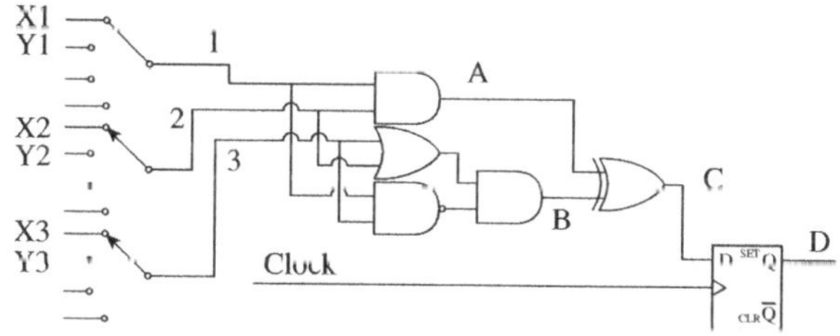

Figure 18B.5 Example of logic propagation

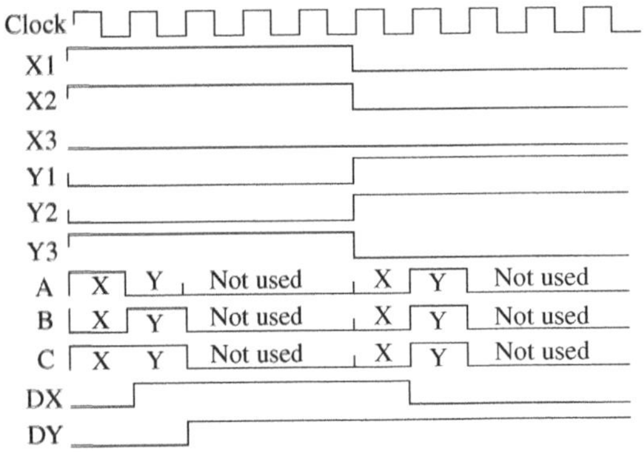

Figure 18B.6 Chronogram of time-multiplexed process

processes. Process X has inputs X1, X2 and X3; process Y has inputs Y1, Y2 and Y3; and each process, X and Y, yields the results of A, B and C in time. At every clock, the values from the different processes X and Y are applied to the logic and their results are shown on the chronogram. After the two processes, the logic is not used and its result is not important. The results of DX and DY are updated when available and stay there until the next clock of the slow process.

Such design can ensure that none of the resources are left idling during the different processes, but it requires very accurate synchronization and design.

The next example is more related to simulation: the implementation of a forward Euler integrator. The FGPA has a clock of 5 ns and the integrator time step is 20 ns; it is then possible to use pipelining. Figure 18B.7 shows the block schematic used in this example. The input A receives the multiplexed in time values to be integrated. B is the result of the values multiplied by the integration time step, 20 ns. D is actually the output of the sum C with a four-step delay, making the forward Euler integrator. The result in D can then be demultiplexed to send the integrated values to the right process. Here, the integration time step was chosen to facilitate the representation in a chronogram of the system in Figure 18B.8. Such a small time step is unlikely to be chosen since it would require a very high level of precision if one chooses to use a fixed-point or a floating-point representation.

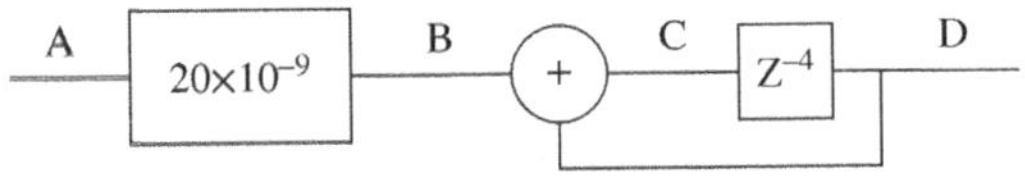

Figure 18B.7 FPGA integrator using pipelining

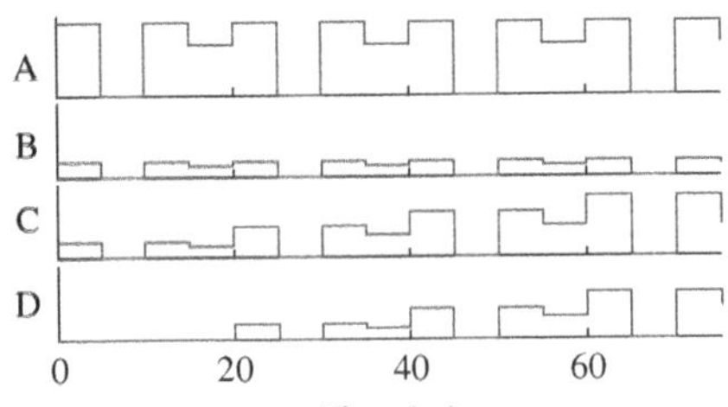

Figure 18B.8 Chronogram of pipelined integrator

The great versatility of the FPGA also creates its main drawbacks: complexity to implement models and excessive time to generate the bitstream. The example given earlier clearly demonstrates that the programming complexity is much larger than using high-level language like C++ or very high-level language like Simulink and code generators (RTW). Such complexity limits the number of specialists who can develop and maintain models. The debugging is also very difficult and long.

Because it is an FPGA, each individual gate needs to be programmed and interconnected when generating the code. With the size of FPGA and models getting larger and larger, the required time to compile the code, or bitstream, also increases. Meaning that if the configuration of your model changes, you need to recompile a new version of your bitstream, which may take several hours.

One option to avoid these two drawbacks is the use of embedded solver on the FPGA [29]. This allows testing many different circuit configurations and if needed it is also possible to make some changes and recompile a new bitstream.

18B.4 Implementation for Real-Time Simulator Using Different Approach

These are simple examples to give the reader the bases allowing him to implement its design. Matlab/Simulink was used to implement and test these implementations, but similar results could be achieved with any other simulation software. For both examples, all cells from a limb are represented by an equivalent voltage source. The only difference is in how the voltage is computed. The equivalent circuit is shown in Figure 18B.9. V_{UA}, V_{UB} and V_{UC} represent the upper limb equivalent voltage, whereas V_{LA}, V_{LB} and V_{LC} represent the lower limb equivalent voltage.

This method of decoupling is adequate since there are two very large states in the model; the large arm inductance ensures a slow variation of the current and the large cell capacitors ensure a slow varying voltage. Measuring the current from the arm inductance, the equivalent voltage from all the conducting cells is computed. In order to break an algebraic loop, a forward Euler integration method is used; this would not affect much the stability of the circuit since it is introduced at a point where there are dominating poles.

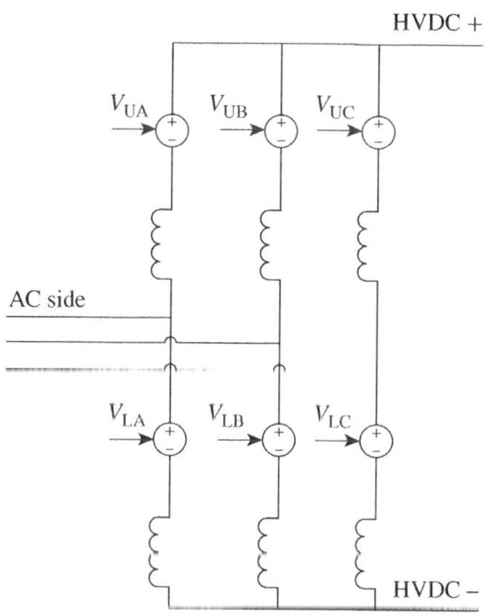

Figure 18B.9 Equivalent decoupled circuit

18B.4.1 Sequential Programming for Average Model Algorithm

This type of modeling can be used to test the inner and outer controls for converters that would be connected to a larger network. This allows estimating the load flow, verifying contingency test or general behavior of the overall network without having to bother regulating each individual capacitor cell.

For this model, the following assumptions are made:

- The current in one limb is the same for all the cells forming that limb, naturally because all are connected in series.
- All the capacitors have the same value; the integration of the current will result in the same voltage variation for all the capacitors of conducting cells.
- Only the number of conducting cells is required as input to the model, it is assumed that the choice of which cell is turned ON within a limb is made by a local and independent controller, who is not part of the model.

Figure 18B.10 shows a block diagram for one limb. It has the limb current and the number of cells ON as input, and the sum of all conducting cells' voltage as output.

The limb current is multiplied by the time step and it is divided by the capacitor value, giving the voltage variation of any conducting cells. Then this voltage variation is multiplied by the number of conducting cells and the result is added to the previous voltage value of all cells. The total voltage value is divided by the total number of cells obtaining the capacitor voltage of a single cell; this is how the regulation of all the cells is made to the same voltage. Finally, the voltage of a single cell is multiplied by the number of conducting cells generating the equivalent voltage for a single arm.

This technique is simple and could even be implemented in a variable-step solver with small modification. The only drawback is that natural rectification using antiparallel diodes from the cells' switches is not considered in this implementation as well as the possibility to simulate faults inside the limb and to test the individual cell voltage regulator.

Using this implementation, the HVDC grid of Figure 18B.11 was simulated faster than real-time simulation. This configuration is the DC-grid benchmark proposed by the CIGRE work group B4-57. The converter A1 is connected to a larger network, modeled by two voltage sources. Converters B1, B2 and B3 are connected to a different network but also have an AC link between one and the other. Converters C1, C2, D1 and F1 are offshore wind farm and E1 is an isolated offshore load. All the offshore converters are connected through underground cables for their HVDC link. Converters on land use overhead transmission of power lines to interconnect among them.

Figure 18B.12 shows the response at the converters C1 and C2. Only the phase A is monitored in this case, but all phases are available. In this test, there is a three-phase fault on the AC side between C1 and C2. When the fault occurs, line between C1 and C2 is opened at each end for two cycles and then it is reclosed. At this point, the fault has been cleared. Figure 18B.12 shows the voltage at each converter. On C2 side, at reclosing an overvoltage is seen. This overvoltage can vary according to the angle at which

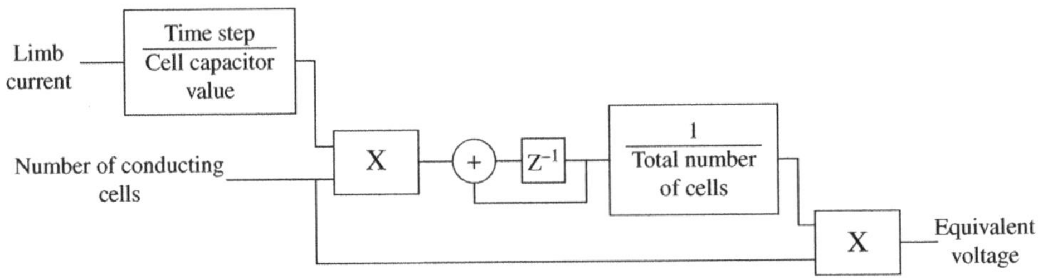

Figure 18B.10 Block diagram for average MMC model

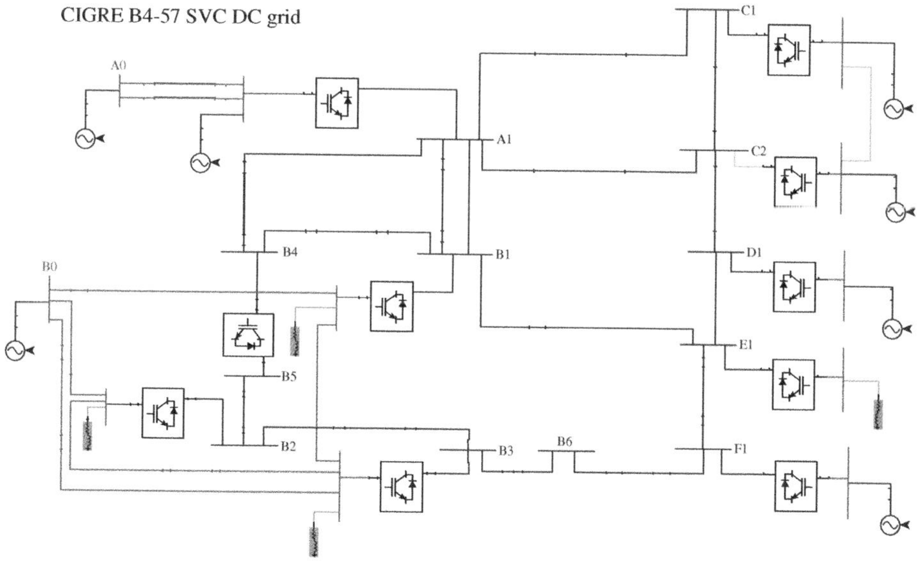

Figure 18B.11 CIGRE B4-57 HVDC grid

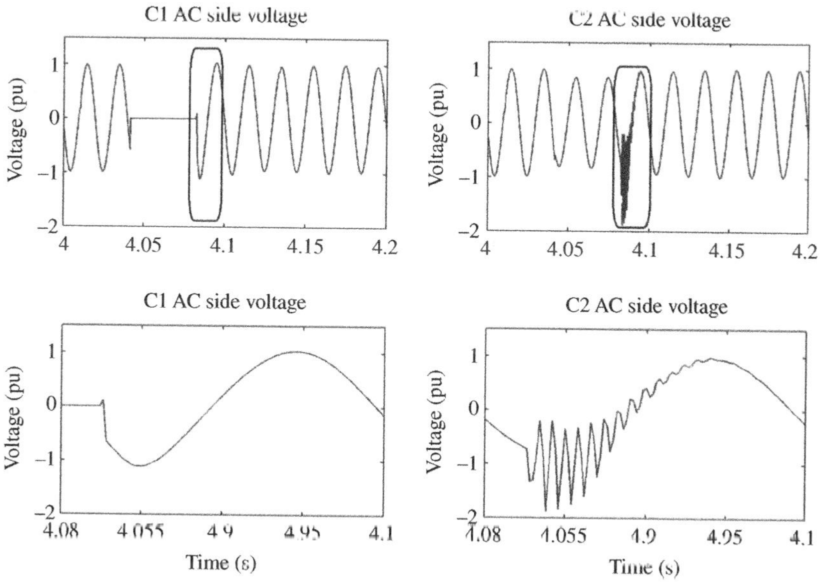

Figure 18B.12 CIGRE benchmark AC fault

the breaker is reclosed. Using this model and a sequencer, a series of tests can be generated to make a Monte Carlo study to identify the V2% [30].

As no *I/O*s were used in this model, it was possible to simulate it faster than the real time. Using 11 processors of an eMEGAsim simulator, an acceleration factor of 4 was achieved. In the case of Monte Carlo study, where thousands of simulations are required, this acceleration factor is very significant.

18B.4.2 Parallel Programming for Switching Function Algorithm

As it has been previously discussed, parallel programming can be implemented on FPGA. Taking advantage of both parallel processing and time multiplexing, a very large MMC can be simulated on FPGA with a very small time step of 250 ns. The choice of the time step of 250 ns is not based on stability of the circuit but rather to have very accurate firing instant for each cell.

Table 18B.1 gives the switching function for Figure 18B.2 that will be implemented on FPGA.

One can note that the mathematics behind this model is still relatively simple; the challenge comes in the implementation to achieve the small computation step. The arm current is obtained from the model running on DSP, where the complete network can easily be implemented using standard simulation software. The gate signals, S1 and S2, come from digital input connected to the FPGA. The simulation time step on the DSP is 100 times slower than the one of the FPGA; therefore instead of sending the instantaneous voltage output of all the cells, only the average over the DSP time step is send in a similar way that only the duty cycle of PWM can be applied when the simulation step is slower than the PWM period.

There are two distinct processes that need to be implemented: the switching function and the integrator. The integrator uses the same method as the one used in Figure 18B.7, in this case 10 signals are pipelined over 250 ns or 25 FPGA steps. During the demultiplexing of the integrator results, the capacitor voltages of the conducting cells are summed to achieve the equivalent voltage for the limb. Another important part of the logic is the implementation of the switching function that determines which cell is conducting and which capacitor is charging. Figure 18B.13 shows the block diagram of the process and the number of FPGA step each process requires.

Table 18B.1 Switching function of MMC cell

Cases	Arm current	S1	S2	Cell's voltage $V_{out}(T)$	Capacitor voltage $V_c(T)$
1	X	0	1	0	$V_c(T-T_s)$
2	X	1	0	$V_c(T)$	$V_c(T-T_s)+1/C*I(T-T_s)*dt$
3	X	1	1	Not considered	
4	>0	0	0	$V_c(T)$	$V_c(T-Ts)+1/C*I(T-T_s)*dt$
5	<0	0	0	0	$V_c(T-T_s)$
6	=0	0	0	High impedance	$V_c(T-T_s)$

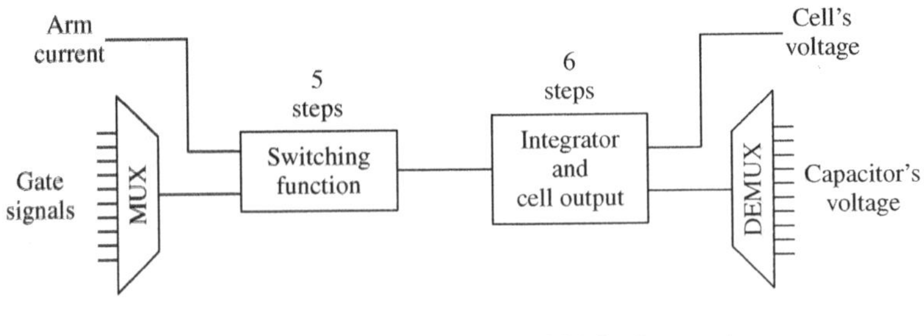

Figure 18B.13 Block diagram of FPGA implementation

Figure 18B.14 Real-time simulator with *I/Os* chassis

Note that the overall process takes 11 time steps, using an internal clock of 10 ns for the FPGA, which means that the first capacitor value will be available after 110 ns. As all of them are time multiplexed by a group of 10, the last capacitor voltage is available after 21 time steps or 210 ns. Here, the advantage of pipelining is very clear, by adding more capacitors in the pipeline, only one more time step is required to obtain the value. In this case, there are still four steps available to add more logic if required, allowing more flexibility as it has been demonstrated in [19].

The implementation from Figure 18B.13 was used to simulate a converter with 500 cells per half-limb, for a total of 3000 cells. It can either use an internal controller, embedded on the FPGA, or an external controller, via optical fiber. In this example, every cell is using two optical fibers for communication: one for receiving data and the other for sending data. Figure 18B.14 shows simulator used to simulate the converter. In the center of the picture is the main simulator where the model is computed with a 250 ns time step. The other racks on each side are only used to manage all the optical fibers that are needed to control the simulation.

Figure 18B.15 shows results obtained when changing the power reference. Reactive power is stable at −0.3 pu and the active power changes from 0 to 0.5 pu. Looking at the voltage and current, one can see the phase shift of the current as the active power increases.

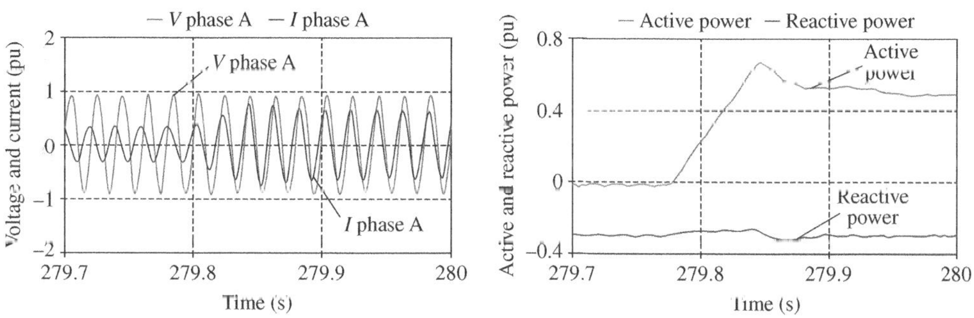

Figure 18B.15 Results of a step on power reference

This is only one of many tests that can be applied to such a system. Using the FPGA implementation allows a very low latency between the *I/O*s and the model. In this case, only the MMC is simulated on FPGA and the remaining of the network is simulated using processors. In 2011, Nari-Relays Electric Co. in China used the HiL results for the Nanhui MMC demonstration project, 20 MVA/60 kV two-terminal MMC HVDC project.

18B.5 Conclusion

This chapter presented an overview of real-time simulation with a practical application of the different technology. As discussed, the digital simulators are widely used but different technologies are available. Nowadays, understanding the application before acquiring a real-time simulator can help identify the best suited type for the application.

Standard single-processor offline simulation tool does not offers adequate solution to achieve real-time simulation, but it is possible to implement its own design using the different methods proposed in this chapter. Multicore processors, DSP and FPGA, are evolving very fast and therefore so do real-time simulation.

General-purpose electrical solvers are available and being developed to facilitate the use of FPGA technologies by abstracting the inner construction of FPGA chips, as this is done with general-purpose processors. Such FPGA-based solvers should evolve very fast over the next years. This chapter mainly focuses on EMTP simulation, which is the best suited for power electronic simulation, but some softwares are now offering a mix simulation EMTP/phasor; slow components like transmission networks are simulated with phasor algorithm and this simulation is coupled with an EMTP simulation where fast systems, like power electronics, are simulated. Looking to the past 10 years, one can expect that the use of real simulation will keep growing and it seems like it is only limited by the need of the industries.

References

1. Lesnicar, A. and Marquardt, R. (2003) An innovative modular multilevel converter topology suitable for a wide power range. 2003 IEEE Power Tech Conference Proceedings, Bologna, Italy, p. 6.
2. Hiller, M., Krug, D., Sommer, R., and Rohner, S. (2009) A new highly modular medium voltage converter topology for industrial drive applications. 2009 13th European Conference on Power Electronics and Applications (EPE'09), pp. 1–10.
3. Rajasekar, S. and Gupta, R. (2012) Solar photovoltaic power conversion using modular multilevel converter. 2012 Students Conference on Engineering and Systems (SCES), pp. 1–6.
4. Peralta, J., Saad, H., Dennetiere, S., and Mahseredjian, J. (2012) Dynamic performance of average-value models for multi-terminal VSC-HVDC systems. 2012 IEEE Power and Energy Society General Meeting, pp. 1–8.
5. Zhang, X.Y., Wu, Z.J., Hao, S.P., and Xu, K. (2012) A study on the new grid integration solutions of offshore wind farms. *Advanced Materials Research*, **383**, 3610–3616.
6. Callavk, M., Blomberg, A., Häfner, J., and Jacobson, B. (2012) The Hybrid HVDC Breaker an Innovation Breakthrough Enabling Reliable HVDC Grids, http://new.abb.com/docs/default-source/default-document-library/hybrid-hvdc-breaker – an-innovation-breakthrough-for-reliable-hvdc-gridsnov2012finmc20121210_clean.pdf?sfvrsn=2 (accessed 27 December 2013).
7. Hartley, T.T., Beale, G.O., and Chicatelli, S.P. (1994) *Digital Simulation of Dynamic Systems: A Control Theory Approach*, Prentice Hall, Englewood Cliffs, NJ.
8. Schaller, R.R. (1997) Moore's law: past, present and future. *Spectrum IEEE*, **34**, 52–59.
9. Baracos, P., Murere, G., Rabbath, C., and Jin, W. (2001) Enabling PC-based HIL simulation for automotive applications. 2001 IEEE International Electric Machines and Drives Conference (IEMDC 2001), pp. 721–729.
10. Abourida, S., Dufour, C., Bélanger, J. *et al.* (2002) Real-time PC-based simulator of electric systems and drives. Seventeenth Annual IEEE Applied Power Electronics Conference and Exposition, 2002. APEC 2002, pp. 433–438.
11. Dufour, C., Bélanger, J., Ishikawa, T., and Uemura, K. (2005) Advances in real-time simulation of fuel cell hybrid electric vehicles. Proceedings of 21st Electric Vehicle Symposium (EVS-21), Monte Carlo, Monaco.

12. Typhoon-HIL http://www.typhoon-hil.com/ (accessed 27 December 2013).
13. OPAL-RT Technologies Inc. eFPGAsim Power Electronic Real-Time Simulator. http://www.opal-rt.com /new-product/efpgasim-power-electronic-real-time-simulator(accessed 27 December 2013).
14. Hagiwara, M. and Akagi, H. (2009) Control and experiment of pulsewidth-modulated modular multilevel converters. *IEEE Transactions on Power Electronics*, **24**, 1737–1746.
15. Antonopoulos, A., Angquist, L., and Nee, H.P. (2009) On dynamics and voltage control of the modular multilevel converter. 13th European Conference on Power Electronics and Applications, 2009. EPE'09, pp. 1–10.
16. Zhao, Y., Hu, X.-H., Tang, G.-F., and He, Z.-Y. (2010) A study on MMC model and its current control strategies. 2nd IEEE International Symposium on Power Electronics for Distributed Generation Systems (PEDG), 2010, pp. 259–264.
17. Arrillaga, J., Liu, Y.H., and Watson, N.R. (2007) *Flexible Power Transmission: The HVDC Options*, John Wiley & Sons, Inc..
18. Hui, S. and Fung, K. (1997) Fast decoupled simulation of large power electronic systems using new two-port companion link models. *IEEE Transactions on Power Electronics*, **12**, 462–473.
19. Gregoire, L.-A., Belanger, J., and Li, W. (2012) FPGA-based real-time simulation of modular multilevel converter HVDC systems. *Journal of Energy and Power Engineering*, **6**, 1119–1125.
20. Le-Huy, P., Giroux, P., and Soumagne, J. (2011) Real-time simulation of modular multilevel converters for network integration studies. Proceedings of IPST, p. 6.
21. Dufour, C., Mahseredjian, J., and Bélanger, J. (2011) A combined state-space nodal method for the simulation of power system transients. *IEEE Transactions on Power Delivery*, **26**, 928–935.
22. Saad, C.D.H., Mahseredjian, J., Dennetière, S., and Nguefeu, S. (2013) Real time simulation of MMCs using the state-space nodal approach. Presented at the International Conference on Power Systems Transients, Vancouver, Canada.
23. Zeigler, B.P., Praehofer, H., and Kim, T.G. (2000) *Theory of Modeling and Simulation: Integrating Discrete Event and Continuous Complex Dynamic Systems*, Academic Press, San Diego, CA.
24. Hudson, J.E., Hunter, E.M., and Wilson, D.D. (1966) EHV-DC simulator. *IEEE Transactions on Power Apparatus and Systems*, **PAS-85**, 1101–1107.
25. OPAL-RT Technologies Inc. HYPERSIM Power System Real-Time Simulator, http://www.opal-rt.com /new-product/hypersim-power-system-real-time-simulator (accessed 27 December 2013).
26. Gagnon, R., Fecteau, M., Prud'Homme, P. *et al.* (2012) Hydro-Québec strategy to evaluate electrical transients following wind power plant integration in the Gaspésie transmission system. *IEEE Transactions on Sustainable Energy*, **3**, 880.
27. OPAL-RT Technologies Inc. ePOWERgrid Product Family Overview, http://www.opal-rt.com/epowergrid -product-family-overview (accessed 27 December 2013).
28. RTDS Technologies High Speed Power System Studies, http://www.rtds.com/applications/high-speed-power -system-studies/high-speed-power-system-studies.html (accessed 27 December 2013).
29. Dufour, C., Cense, S., Ould-Bachir, T. *et al.* (2012) General-purpose reconfigurable low-latency electric circuit and motor drive solver on FPGA. IECON 2012-38th Annual Conference on IEEE Industrial Electronics Society, pp. 3073–3081.
30. Paquin, J.-N., Bélanger, J., Snider, L.A. *et al.* (2009) Monte-carlo study on a large-scale power system model in real-time using emegasim. 2009 IEEE Energy Conversion Congress and Exposition (ECCE 2009), San Jose, CA, pp. 3194–3202.

19

Model Predictive Speed Control of Electrical Machines

José Rodríguez[1], Marcelo A. Pérez[1], Héctor Young[1] and Haitham Abu-Rub[2]
[1]*Department of Electronics, Universidad Técnica Federico Santa María, Valparaíso, Chile*
[2]*Department of Electrical and Computer Engineering, Texas A&M University Qatar, Doha, Qatar*

19.1 Introduction

Speed control improves the efficiency of electrical machines under different operating conditions. This is particularly important when the operating range is wide, such as in pumps and fans. Speed control must also be used when a constant speed is required, for example, in paper mills and in robotic equipment, even if the load changes. Electrical machines are usually controlled using a piece of power electronics equipment called an inverter. Inverters only work with semiconductors in on-state and off-state, hence they are very efficient and can synthesize a fully controllable AC waveform with adjustable frequency and amplitude.

During the last decades, only two speed control strategies for electrical drives have reached widespread industrial application: field-oriented control (FOC) and direct torque control (DTC). However, huge advances in microprocessor technology have made it possible to develop more intelligent and sophisticated control strategies for electrical drives. One of these strategies is model predictive control [1, 2], which has recently been recognized as one of the most successful techniques that originated from modern control theory [3]. The term predictive control refers to a wide range of control methods that share a common operating principle: to calculate, via optimization of the system model, the values of the actuating variables so that over a given interval of time, the system output follows a desired trajectory into the future [4]. The application of model predictive control is straightforward in power electronics because of the availability of precise models and a finite number of control inputs given by the converter switching states.

Predictive control was initially used to replace the inner current controllers in FOC [5, 6] and then to control torque and flux [7, 8]. More recently, predictive schemes have been proposed to control the speed directly [9]. The main advantages of the predictive algorithm are its simple implementation and its flexibility; it is easily adaptable to different AC machines such as permanent magnet synchronous motors [10] and brushless [11] or doubly fed inductor generators [12]. The algorithm has been

Power Electronics for Renewable Energy Systems, Transportation and Industrial Applications, First Edition.
Edited by Haitham Abu-Rub, Mariusz Malinowski and Kamal Al-Haddad.

successfully applied to different converters, for example, cascaded H-bridge [13, 14], flying capacitors [15] and matrix converters (MCs) [16, 17] and to different configurations such as multiphase drives [18] and inverters with LC output filters [19].

In this chapter, an overview of classical speed control of machine drives is presented, followed by an introduction to the model predictive control strategy, along with a description of the algorithm and control scheme for current control. The predictive torque control (PTC) is developed and applied to an MC to demonstrate the simplicity of applying this method to different converters and control objectives. Finally, the technique is adapted to directly control the speed of the machine, including electrical and mechanical dynamics in the same model and cost function.

19.2 Review of Classical Speed Control Schemes for Electrical Machines

Speed control of AC machines using power electronics inverters has been the subject of intense research during the last decades. However, since its introduction in the 1970s, only a few machine speed controllers have been widely used in industrial drives. The first method adopted for industrial application was FOC [20–22]. Although it was initially developed for induction motors [23, 24], the method was later successfully adapted to synchronous motors [25, 26] and other electrical machines [27, 28]. Based on linear control principles, FOC uses pulse-width modulation (PWM) [29] to linearize the discrete behavior of the converter while employing linear models for current and flux. DTC is another well-established speed control strategy [30]. This method was introduced in the 1980s and its main difference from FOC is that it considers the discrete nature of the converter in the controller strategy. It is based on nonlinear hysteresis controllers and a predefined switching table [31]. Like FOC, this method was developed for induction motors [32, 33] and was later applied to permanent magnet synchronous machines (PMSMs) [34, 35] and brushless DC machines [36], among others. Each of the aforementioned AC machine control techniques have their own advantages and drawbacks, but they have become the industry standards for control strategies [37].

In this section, the model of an induction machine is developed, and the two standard speed control machines are depicted.

19.2.1 Electrical Machine Model

Any rotating electrical machine can be modeled by the equations of stator and rotor circuits. Considering a squirrel-cage induction motor, the equations are

$$\boldsymbol{v}_{\mathrm{s}} = r_{\mathrm{s}}\boldsymbol{i}_{\mathrm{s}} + \frac{\mathrm{d}\boldsymbol{\psi}_{\mathrm{s}}}{\mathrm{d}t} \tag{19.1}$$

$$0 = r_{\mathrm{r}}\boldsymbol{i}_{\mathrm{r}} + \frac{\mathrm{d}\boldsymbol{\psi}_{\mathrm{r}}}{\mathrm{d}t} - j\omega\boldsymbol{\psi}_{\mathrm{r}} \tag{19.2}$$

where $\boldsymbol{v}_{\mathrm{s}}$ is the voltage applied to the machine terminals, $\boldsymbol{i}_{\mathrm{s}}$ is the stator current, $\boldsymbol{\psi}_{\mathrm{s}}$ is the stator flux, r_{s} is the stator resistance, $\boldsymbol{i}_{\mathrm{r}}$ is the rotor current, $\boldsymbol{\psi}_{\mathrm{r}}$ is the rotor flux, r_{r} is the stator resistance and ω is the rotational speed of the machine.

Although these equations define the dynamical behavior of currents and fluxes, the flux linkage equations must be included to complete the model. These equations are

$$\begin{aligned} \boldsymbol{\psi}_{\mathrm{s}} &= l_{\mathrm{s}}\boldsymbol{i}_{\mathrm{s}} + l_{\mathrm{m}}\boldsymbol{i}_{\mathrm{r}} \\ \boldsymbol{\psi}_{\mathrm{r}} &= l_{\mathrm{m}}\boldsymbol{i}_{\mathrm{s}} + l_{\mathrm{r}}\boldsymbol{i}_{\mathrm{r}} \end{aligned} \tag{19.3}$$

where l_{s} is the stator self-inductance, l_{r} is the rotor self-inductance and l_{m} is the mutual inductance.

The electrical torque T_e can be obtained as the amplitude of the cross-product of rotor and stator flux:

$$T_e = \frac{p}{2}\frac{l_m}{l_m^2 - l_s l_r}|\boldsymbol{\psi}_r \times \boldsymbol{\psi}_s| \quad (19.4)$$

where p is the number of poles of the machine.

19.2.2 Field-Oriented Control

The FOC method takes advantage of the decoupled dynamics between electromagnetic torque and the real (or direct) component of the rotor flux when the machine is represented in a rotating frame [38]. Considering the rotating transformation in phase with the direct component of the rotor flux, that is $\psi_{rd} = 0$, the dynamics of the direct component of the rotor flux is as follows:

$$\psi_{rd} = \frac{l_m}{\tau_r s + 1} i_{sd} \quad (19.5)$$

where $\tau_r = l_r/r_r$. Because the quadrature component is zero, the amplitude of the rotor flux is

$$\Psi_r = \psi_{rd} \quad (19.6)$$

On the other hand, the torque can be expressed as

$$T_e = \frac{p}{2}\frac{l_m}{l_r}|\boldsymbol{\psi}_r \times \boldsymbol{i}_s| = \frac{p}{2}\frac{l_m}{l_r}|\boldsymbol{\psi}_r i_{sq}| \quad (19.7)$$

Considering that the flux dynamics are slower than the current, the torque can be controlled using the imaginary component of the stator current. Therefore, according to the previous expressions 19.5 and 19.7, the rotor flux can be directly controlled by the real component of the stator current. And since the flux reference is defined as constant, the torque can be controlled by the imaginary component of the same current.

The block diagram in Figure 19.1 shows a standard implementation of FOC. Both the real and imaginary components of the stator current are controlled using linear proportional-integral (PI) controllers, which give the required stator voltages. After a coordinate transformation from the rotating frame. To the stationary three-phase frame, a PWM is used to generate the switching pulses that synthesize the fundamental stator voltage proportional to the reference value given by the linear current controllers. The speed is also controlled using a linear PI controller, which gives the reference of the imaginary stator current. As can be seen in the controller diagram, Figure 19.1, a number of coordinate transformations must be made in order to properly calculate the controller variables.

Among the PWM techniques, the carrier-based and space vector techniques are the two most implemented in industrial drives. The carrier-based PWM shown in Figure 19.2a generates the gating pulses comparing the sinusoidal reference with a high-frequency triangular waveform. This carrier signal defines the switching frequency. This type of modulation is very simple to implement using either analog or digital circuits, but its voltage utilization factor is low. The space vector PWM, shown in Figure 19.2b, synthesizes the reference voltage using the adjacent vectors produced by the converter and calculates how much time the vectors must be applied for. This technique requires digital implementation, but its voltage utilization factor is better than carrier-based PWM. Recently, a carrier-based PWM with a modified modulation signal that improves the utilization factor was proposed. This approach is called min–max modulation, because the modified reference is calculated by adding a common mode given by

$$\boldsymbol{m}_s^* = \boldsymbol{m}_s + \frac{\max(\boldsymbol{m}_s) - \min(\boldsymbol{m}_s)}{2} \quad (19.8)$$

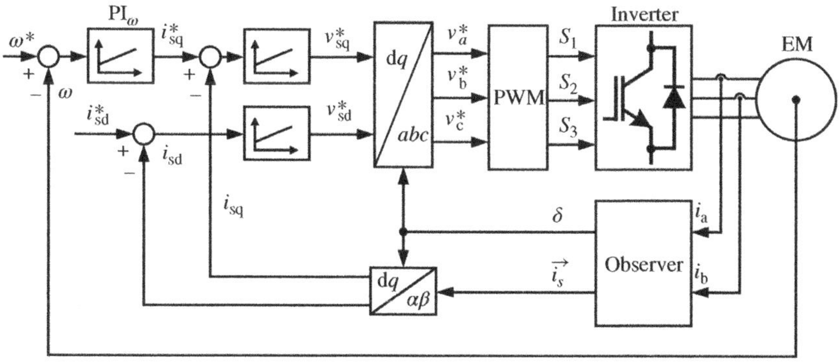

Figure 19.1 Field-oriented control (FOC) block diagram

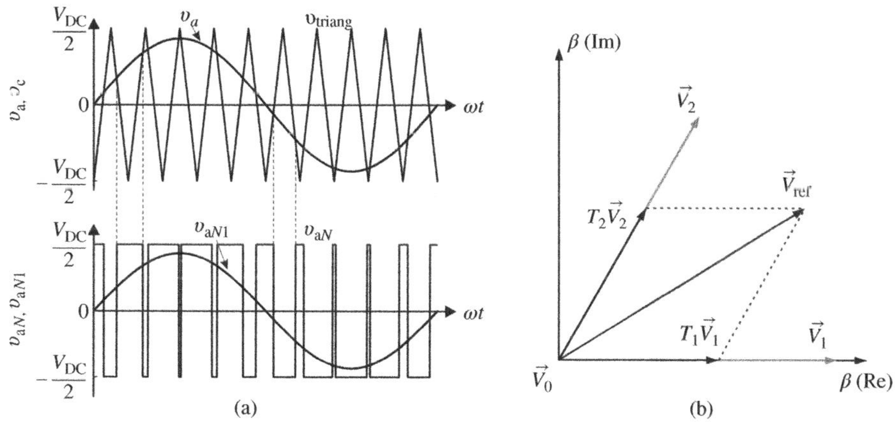

Figure 19.2 Working principle of (a) carrier-based PWM and (b) space vector modulation

where $\boldsymbol{m}_s$ is the original modulation signal and $\boldsymbol{m}_s^*$ is the modified modulation signal. This modulation has a utilization factor similar to space vector modulation (SVM) under certain conditions.

FOC strategy has a very high performance, as seen in Figure 19.3, which shows a speed reversal of an induction machine. The machine speed reverses smoothly with very fast torque dynamics. A low ripple in torque, flux and currents is also observed.

19.2.3 Direct Torque Control

The DTC strategy does not use a modulation method such as FOC, because it considers the discrete nature of the converter and the effect of each switching state on the torque and flux [30]. If the resistive losses are considered negligible, the voltage generated by the inverter can change the behavior of the stator flux very rapidly, as shown in Figure 19.4, according to the following equation:

$$\Delta\boldsymbol{\psi}_s \approx \boldsymbol{v}_s T_s \qquad (19.9)$$

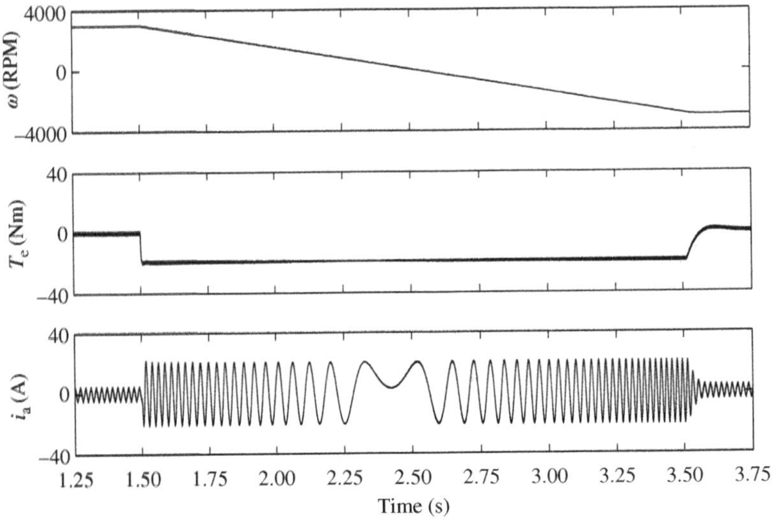

Figure 19.3 Speed reversal of an induction machine using field-oriented control

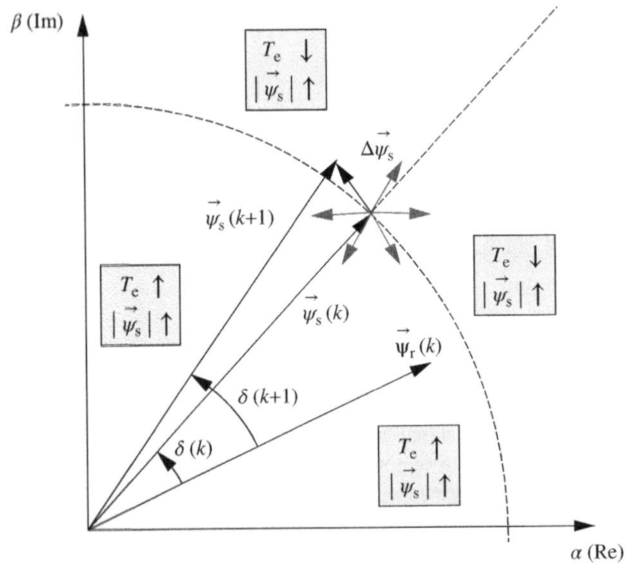

Figure 19.4 Direct torque control working principle

where $\Delta\boldsymbol{\psi}_s$ is the variation of the stator flux and T_s is the sampling time. The working principle of DTC is based on the change in stator flux produced by the stator voltage. The relationship between rotor and stator flux is given by

$$\boldsymbol{\psi}_r = \frac{k_s}{\tau_r'\, s + 1}\boldsymbol{\psi}_s \tag{19.10}$$

where k_s and τ_r' are machine parameters.

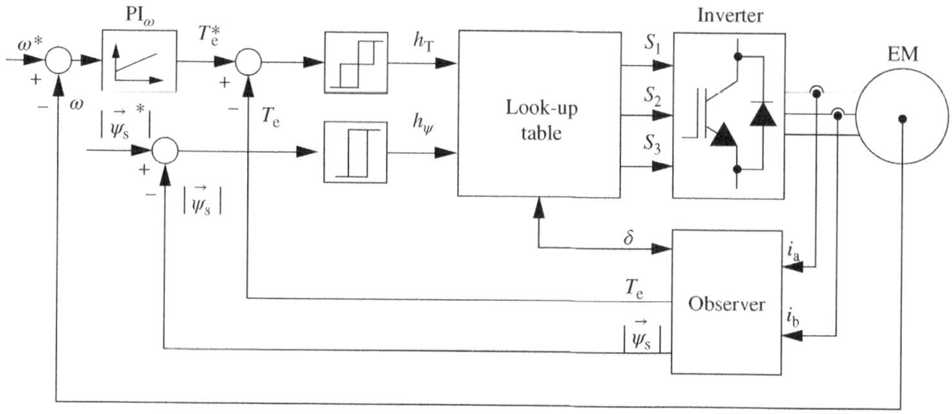

Figure 19.5 Block diagram of direct torque control

The torque can be expressed in terms of rotor and stator flux [39]:

$$T_e = \frac{p}{2}\frac{l_m}{l_m^2 - l_s l_r}\boldsymbol{\psi}_r \times \boldsymbol{\psi}_s = \frac{p}{2}\frac{l_m}{l_m^2 - l_s l_r}\psi_r \psi_s \sin(\delta) \tag{19.11}$$

where δ is the angle between both fluxes, and ψ_r, ψ_s, are the rotor and stator flux amplitudes, respectively. Therefore, the input voltage can simultaneously control the amplitude and angle of the stator flux. Given that the rotor flux dynamics are slower than the stator flux, the input voltage can control the torque and speed.

The key concept behind DTC is that the converter can produce a finite number of voltage combinations, and, depending on the flux angle, it is possible to select four of them to produce positive and negative changes in torque and flux. These selected combinations are predefined in a look-up table, which contains the required changes in torque and flux and also the angles of the stator flux.

The block diagram of DTC is shown in Figure 19.5. The stator currents are measured to estimate the values of the torque and flux vectors. The speed is controlled using a PI controller that generates the torque reference. The torque and stator flux references are compared with the actual values, and the errors are sent to hysteresis comparators, which in turn generate the required change signals in each variable. These variables, in combination with the flux angles, are included in a look-up table. This table also contains the gate signals needed to generate the voltage combinations that will follow the references.

The dynamic performance of DTC is slightly faster than FOC, as shown in Figure 19.6, where a speed reversal of the machine is produced using this controller. The ripple is larger than FOC while the switching frequency is not fixed [40–42]. However, it is possible to control this switching frequency using a variable hysteresis band.

As can be seen, DTC is a completely different approach than FOC, as it is based on nonlinear control and predefined control actions given by the discrete nature of the converter. Implementing DTC is simpler than FOC, because DTC has no modulations and does not require any angle-dependent coordinate transformations.

19.3 Predictive Current Control

Model predictive control has become a very promising control technique because of its intuitive working principle, easy implementation and the advances in digital processor technology, which make its real time implementation possible.

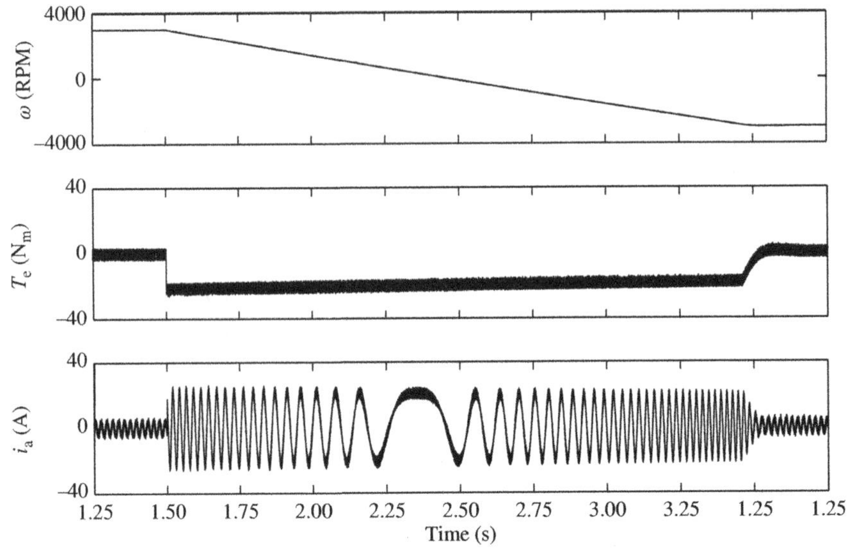

Figure 19.6 Speed reversal of an induction machine using direct torque control

The working principle of model predictive control is to predict the behavior of the system based on a model for each one of the switching states, and then select the state that optimizes the system's behavior, based on a defined cost function.

When the control objective is the stator current, predictive control can be included in the FOC scheme, replacing the PI controllers. This approach, namely predictive current control (PCC), is the simplest approach and it will be used to explain the key concepts behind the method, including predictive model, cost function, predictive algorithm and control scheme.

From the machine model, it is possible to replace the stator flux in the stator equation, obtaining

$$\boldsymbol{v}_\mathrm{s} = r_\mathrm{s}\boldsymbol{i}_\mathrm{s} + l_\mathrm{s}\frac{\mathrm{d}\boldsymbol{i}_\mathrm{s}}{\mathrm{d}t} + l_\mathrm{m}\frac{\mathrm{d}\boldsymbol{i}_\mathrm{r}}{\mathrm{d}t} \tag{19.12}$$

This is the only equation required to implement the PCC, as shown in the following.

19.3.1 Predictive Model

One of the most important elements in predictive control is the model. In the case of AC machines, this refers to the discrete form, usually first-order forward Euler, of current and/or flux equations. In this case, neglecting the derivative of the rotor current, the stator current at the next sample time can be calculated as follows:

$$\boldsymbol{i}_\mathrm{s}(k+1) = \left(1 - \frac{r_\mathrm{s}}{l_\mathrm{s}}T_\mathrm{s}\right)\boldsymbol{i}_\mathrm{s}(k) + \frac{1}{l_\mathrm{s}}T_\mathrm{s}\boldsymbol{v}_\mathrm{s}(k) \tag{19.13}$$

It is worth noting that the stator current corresponds to a first-order discrete model that often appears in the models throughout this chapter. The stator voltage $\boldsymbol{v}_\mathrm{s}$ is synthesized by the inverter and can be expressed as

$$\boldsymbol{v}_\mathrm{s}(k) = \boldsymbol{s}(k)V_\mathrm{DC}, \tag{19.14}$$

where V_DC is the constant DC voltage of the inverter and $\boldsymbol{s}(k)$ is the vector of the switching states. This model must be evaluated for each one of the possible switching states the inverter can generate. The

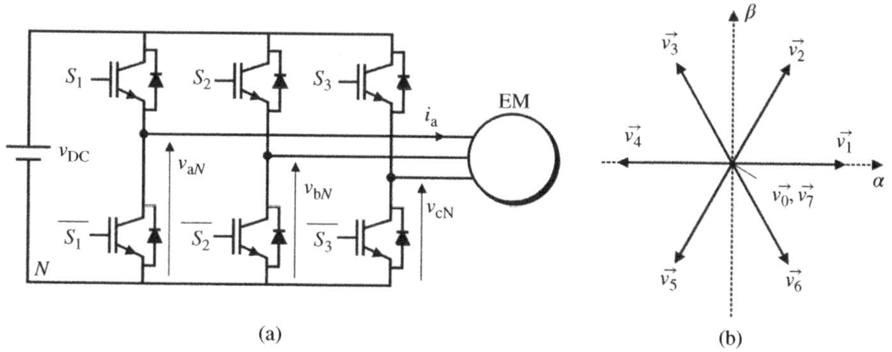

Figure 19.7 (a) Two-level voltage source inverter power circuit and (b) voltage vectors generated by the inverter

Table 19.1 Switching states and voltage vectors

S_1	S_2	S_3	Voltage vector $\vec{v}$
0	0	0	$\vec{v}_0 = 0$
1	0	0	$\vec{v}_1 = \frac{2}{3} v_{DC}$
1	1	0	$\vec{v}_2 = \frac{1}{3} v_{DC} + j\frac{\sqrt{3}}{3} v_{DC}$
0	1	0	$\vec{v}_3 = -\frac{1}{3} v_{DC} + j\frac{\sqrt{3}}{3} v_{DC}$
0	1	1	$\vec{v}_4 = -\frac{2}{3} v_{DC}$
0	0	1	$\vec{v}_5 = -\frac{1}{3} v_{DC} - j\frac{\sqrt{3}}{3} v_{DC}$
1	0	1	$\vec{v}_6 = \frac{1}{3} v_{DC} - j\frac{\sqrt{3}}{3} v_{DC}$
1	1	1	$\vec{v}_7 = 0$

converter traditionally used in low-voltage drives is the two-level voltage source inverter (2L-VSI) shown in Figure 19.7a. This converter has six active switching states and two zero switching states, as shown in Figure 19.7b. The output voltage vectors, as a function of the switching states, are presented in Table 19.1.

19.3.2 Cost Function

The cost function is usually related to the error in the variable of interest, in this case the stator current. To perform the optimization, absolute values or a convex function such as a quadratic can be used. Throughout this chapter, quadratic error cost functions are used. In the case of PCC, the cost function is

$$g = (\boldsymbol{i}_s^* - \boldsymbol{i}_s(k+1))^2 \tag{19.15}$$

where $\boldsymbol{i}_s^*$ is the value of the reference current. It is important to note that the current variables are in three phase, and therefore the cost function is composed of the sum of three quadratic errors. This cost function requires the reference at the next sample time. The reference can be obtained by extrapolating from the previous values or by using the actual reference if its dynamics is slower than the sample time.

19.3.3 Predictive Algorithm

One of the main advantages of the predictive method is that the control algorithm is always the same. Other components such as the model, the cost function and the switching states may change depending on the load, the control objectives and the converter topology, respectively.

As shown in Figure 19.8, the algorithm starts with the measurements of voltages and currents. Then, if necessary, an observer is employed to estimate control variables and, if a low sample frequency is used, for delay compensation [43]. Later, in a loop – or in parallel if the hardware allows it – the next sample time variables for each switching state are predicted using the discrete model, and the cost function is evaluated and minimized. Finally, the optimal switching state is selected and applied to the converter.

19.3.4 Control Scheme

In this case, the current predictive control is inserted into an FOC scheme, replacing the PI controllers, as shown in Figure 19.9. It is worth noting that no modulation is needed, and therefore the number of coordinate transformations required is lower than in the traditional FOC scheme.

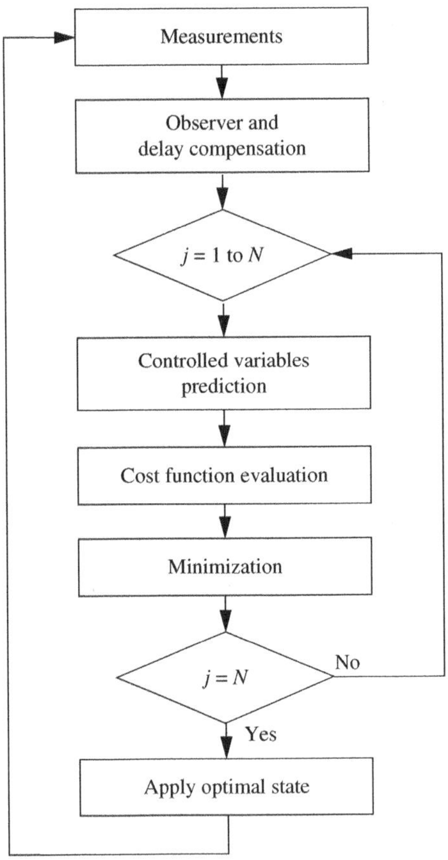

Figure 19.8 Predictive control algorithm flowchart

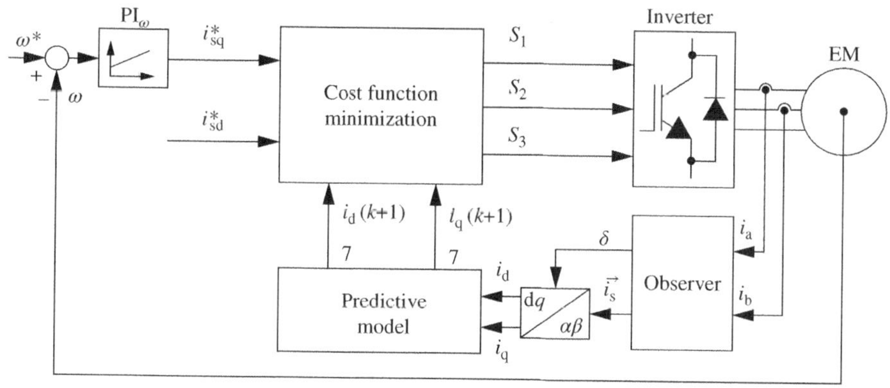

Figure 19.9 Block diagram of FOC with a predictive current control loop

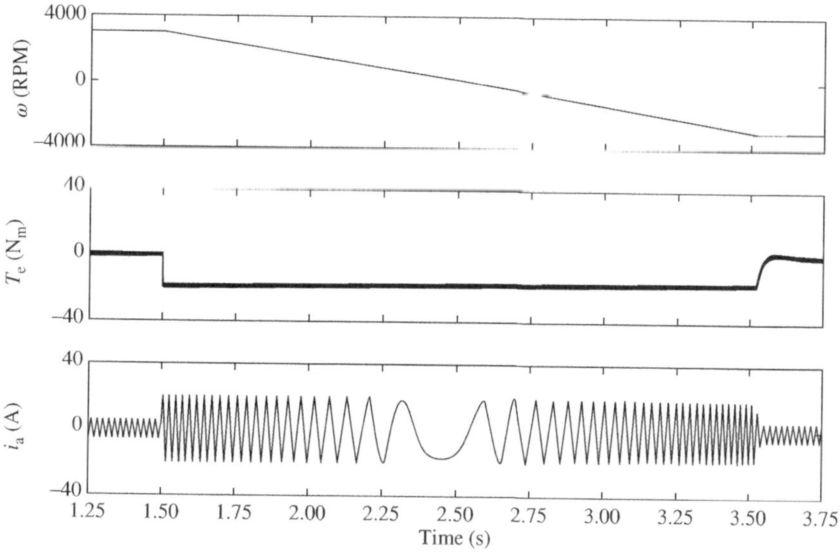

Figure 19.10 Speed reversal of an induction machine using FOC with predictive current control

The speed reversal in Figure 19.10 shows a slightly faster dynamic of currents, torque and flux and also a higher ripple than the original FOC.

19.4 Predictive Torque Control

In addition to controlling the current, predictive control can efficiently control other variables such as flux and torque and even speed. The control scheme shown in this section is called PTC, and it can be considered as an alternative to FOC and DTC.

19.4.1 Predictive Model

The model of the machine is obtained using the stator dynamic (Equation 19.1) and torque (Equation 19.4). Using the first-order Euler discrete model, the predicted flux is

$$\boldsymbol{\psi}_s(k+1) = \boldsymbol{\psi}_s(k) + T_s(\boldsymbol{v}_s(k) - r_s\boldsymbol{i}_s(k)) \tag{19.16}$$

Considering that the rotor flux dynamics is slower than the stator flux, the next step is to approximate the value of the rotor flux by using the actual rotor flux $\boldsymbol{\psi}_r(k+1) \approx \boldsymbol{\psi}_r(k)$. The torque prediction is given by

$$T_e(k+1) = \frac{p}{2}\frac{l_m}{l_m^2 - l_s l_r}\boldsymbol{\psi}_r^{\mathrm{T}}(k)\boldsymbol{J}\boldsymbol{\psi}_s(k+1) \tag{19.17}$$

where l_s, l_r and l_m are the stator, rotor and mutual inductances as defined in Section 19.2.1. The matrix $\boldsymbol{J}$ is used to calculate the cross-product amplitude, and it can be defined in three-phase *abc*, two-phase stationary $\alpha\beta$ or rotating *dq* frames [44].

19.4.2 Cost Function

The cost function used to control the torque and flux is

$$g = (T_e(k+1) - T_e^*)^2 + \lambda_\Psi(|\boldsymbol{\psi}_s(k+1)| - |\boldsymbol{\psi}_s^*|)^2 \tag{19.18}$$

where T_e^* is the reference torque, $\boldsymbol{\psi}_s^*$ is the reference flux and λ_Ψ is a weighting factor that establishes a relative importance between both terms. This factor corresponds to a design parameter of PTC [45].

19.4.3 Predictive Algorithm

The predictive algorithm is the same as the PCC, and it must be evaluated for the same switching states. The only difference is that the torque and flux are evaluated instead of the currents and cost function.

19.4.4 Control Scheme

The control scheme for PTC is shown in Figure 19.11. The observer uses the load currents to generate the rotor fluxes, stator fluxes and a delay-compensated version of the sampled currents. These variables are used to generate the torque and stator flux predictions for each converter state.

As shown in the earlier section, the predictions are evaluated in the cost function using the flux and torque references, and the optimal state is selected and applied directly to the converter. In this control scheme, the torque reference is generated by an external PI controller that controls the speed. Figure 19.12 shows the experimental waveforms of a speed reversal using PTC. The results show a dynamic similarity to DTC in both torque and current. As with the case of DTC, the control of the switching frequency has also been addressed in the literature [46, 47].

One of the main advantages of PTC over DTC is the simplicity of control implementation when the circuit topology changes [48, 49]. For example, a three-level neutral-point-clamped inverter requires the use of several switching tables and an algorithm to select one of these tables for each sample time. Another advantage of PTC is its flexibility to incorporate different control objectives, such as minimization of the switching frequency [50], reduction of the common-mode voltage [16], or control of the input reactive power [51, 52]. The next section will address both concepts: topology changes and additional control objectives.

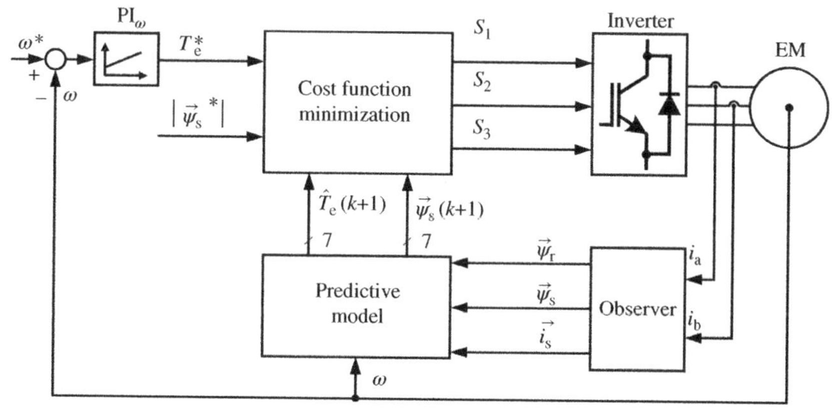

Figure 19.11 Predictive torque control block diagram

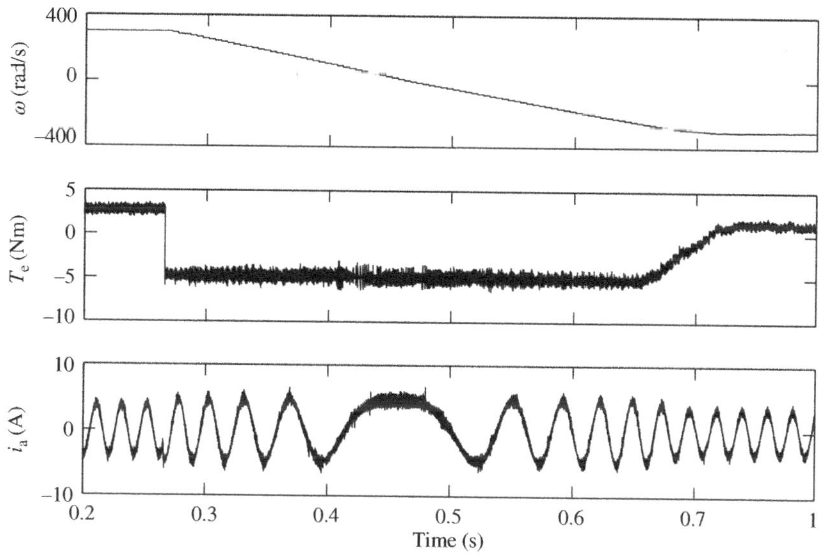

Figure 19.12 Speed reversal of an induction machine using predictive torque control

19.5 Predictive Torque Control Using a Direct Matrix Converter

The MC is a direct converter where the output is directly related to the input, without energy storage elements. The circuit topology of the MC is shown in Figure 19.13. It is composed of an arrangement of nine bidirectional power switches that can connect each output phase to any of the input phases. The output voltage is generated by connecting the input voltage to the load by using a proper combination of switches. The converter input current has a PWM waveform, which requires an LC filter to reduce the switching harmonics and to generate sinusoidal input currents. In this section, this converter is used to demonstrate the flexibility of predictive control when a different converter is used and different function objectives are evaluated.

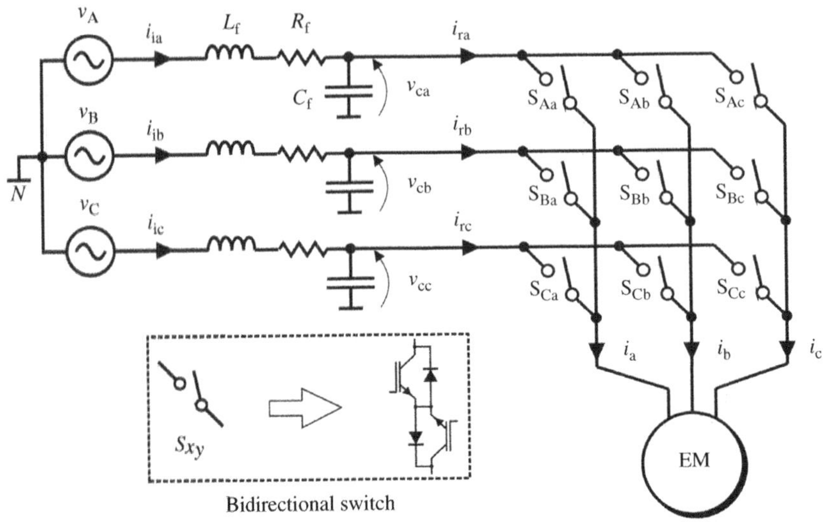

Figure 19.13 Power circuit of an electric machine fed by a matrix converter

19.5.1 *Predictive Model*

The predictive model is very similar to the one shown in the earlier section. The main difference between a VSI and an MC are the switching states. The MC has 27 possible combinations of output voltage instead of eight switching states as with 2-L inverters. Furthermore, these output voltages depend on the input voltage, producing moving switching states. Therefore, the only change of the model is in the way the output voltages are calculated.

$$\mathbf{v}_s(k) = \mathbf{S}(k)\mathbf{v}_i \tag{19.19}$$

This behavior makes this converter extremely difficult to control with classic techniques; however, predictive control can be applied directly as in VSI. This is one of the major advantages of predictive control, because no matter how complex a converter may be, predictive control only needs to evaluate the possible switching states.

19.5.2 *Cost Function*

As a first step, the cost function used to control the machine is the same as in Equation (19.18), controlling the torque and the flux with a weighting factor for this last term.

19.5.3 *Predictive Algorithm*

The only difference of the predictive algorithm is the number of evaluations of the model, because the MC has 27 possible combinations. Therefore, the algorithm is the same as the 2L-VSI.

19.5.4 *Control Scheme*

The PTC control scheme of an electric machine fed by an MC is shown in Figure 19.14. This scheme has a very similar structure to that of PTC using a VSI, but the input voltage is now required to calculate

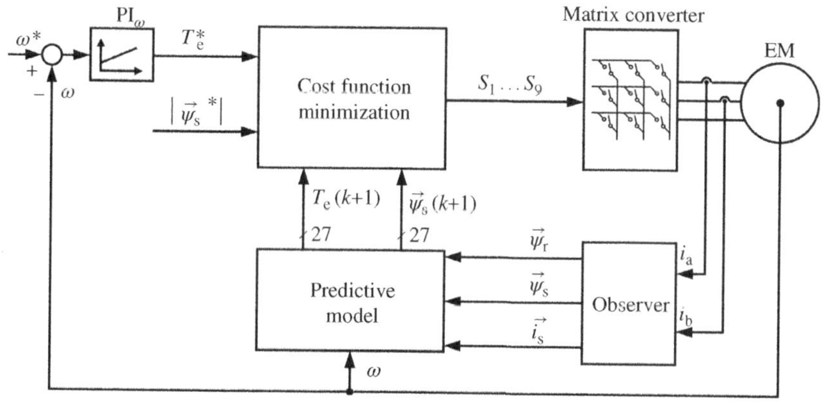

Figure 19.14 Block diagram of predictive torque control of an electric machine fed by a matrix converter

the output voltage combinations. Predictive control is used to control torque and flux, and an external PI is used to control speed.

Experimental results of a speed reversal using a matrix converter are shown in Figure 19.15. The torque, flux and output currents show a very good dynamics, similar to when a 2L-VSI is used. However, it is worth noting that the input currents are very noisy, and the reactive power is not controlled. This behavior occurs because these variables are not modeled and not included in the cost function. These issues are addressed in the following section.

19.5.5 Control of Reactive Input Power

In order to control the input reactive power and reduce the noise in the currents, a model of the input filter and the reactive power prediction is required.

The model of the input filter is given by

$$\boldsymbol{v}_i = R_f \boldsymbol{i}_i + L_f \frac{d\boldsymbol{i}_i}{dt} + \boldsymbol{v}_c$$

$$\boldsymbol{i}_i = \boldsymbol{i}_r + C_f \frac{d\boldsymbol{v}_c}{dt} \tag{19.20}$$

where $\boldsymbol{v}_i$ and $\boldsymbol{v}_c$ are the grid and capacitor voltages; $\boldsymbol{i}_i$ and $\boldsymbol{i}_r$ are the input and converter currents; and R_f, L_f and C_f are the filter parameters.

A discrete approximation of this equation is used to calculate the prediction of the input current. In this case, a second-order discrete approximation [53] or an exact discrete model [54] must be used to avoid resonances.

$$\boldsymbol{i}_i(k+1) = a_{11}\,\boldsymbol{i}_i(k) + a_{12}\,\boldsymbol{v}_c(k) + b_{11}\,\boldsymbol{i}_r(k) + b_{12}\boldsymbol{v}_i(k) \tag{19.21}$$

where $a_{11}, a_{12}, b_{11}, b_{12}$ are parameters of the exact discrete model. If the input voltage is considered constant between two consecutive samples, the input reactive power is then calculated as

$$Q(k+1) = \boldsymbol{v}_i^T(k)\boldsymbol{J}\boldsymbol{i}_i(k+1) \tag{19.22}$$

where the matrix $\boldsymbol{J}$ is the same matrix used to calculate torque in Section 19.4.1. To control the input reactive power, the cost function must include a third term related to this variable:

$$g = (T_e(k+1) - T_e^*)^2 + \lambda_\psi(\boldsymbol{\psi}_s(k+1) - \boldsymbol{\psi}_s^*)^2 + \lambda_Q(Q(k+1) - Q^*)^2 \tag{19.23}$$

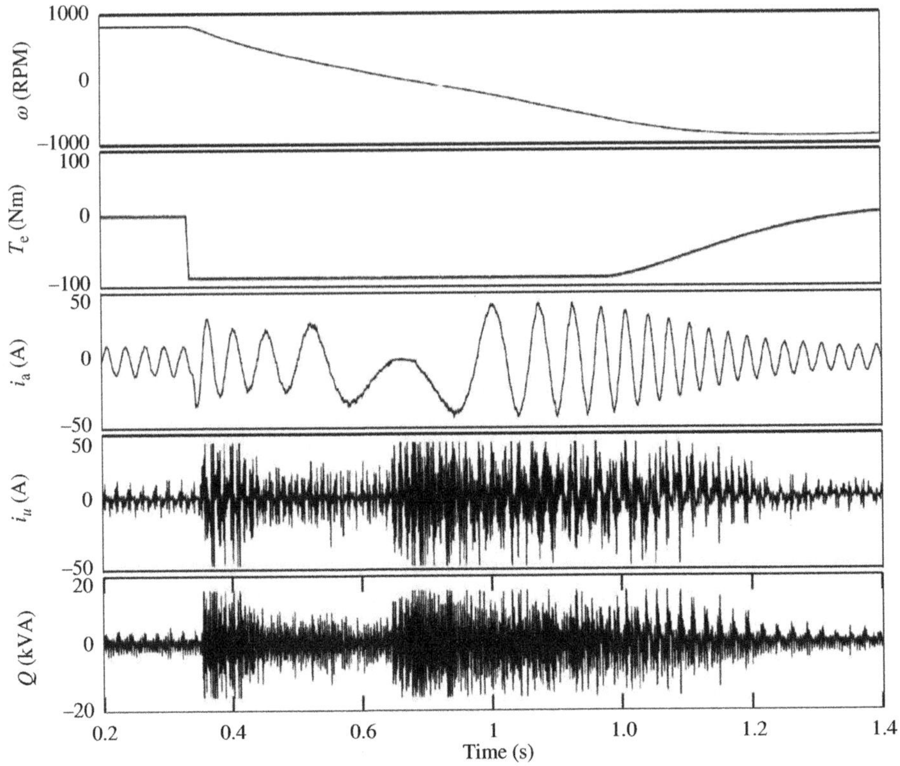

Figure 19.15 Speed reversal of an induction machine fed by a matrix converter using PTC. No reactive input power control

where Q^* is the reference for the reactive power and λ_Ψ and λ_Q are weighting factors to give relative importance to each term.

Therefore, the algorithm includes the model of the input current and reactive power and also, in the evaluation of the cost function, the term related to reactive power. The control scheme has the same structure as the previous one. The main differences are the filter variables (voltage and current), which are now used in the predictive model. Experimental results in Figure 19.16 show that the speed, torque and output current are similar to the previous approach, but the input current becomes sinusoidal and reactive power is zero with a very small ripple.

This section has demonstrated the flexibility of predictive control when applied to a very complex power topology, such as a matrix converter, to include the input reactive power control as an additional control objective.

19.6 Predictive Speed Control

All the approaches presented in the earlier sections use a linear PI controller to generate the reference torque and to control the speed. Predictive control, however, has the capability to control the speed directly by including the mechanical dynamics of the machine in the model.

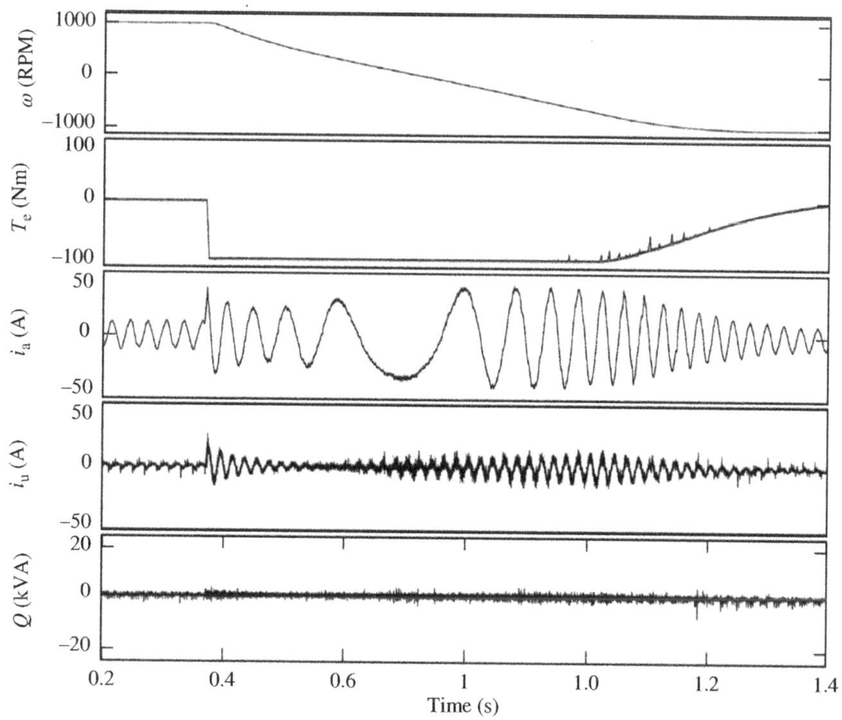

Figure 19.16 Speed reversal of an induction machine fed by a matrix converter using PTC. With reactive input power control

This section presents the predictive speed control of a permanent magnet synchronous motor (PMSM). The stator model of a PMSM in a rotating *dq* frame is [9]

$$\boldsymbol{v}_s = R_s\boldsymbol{i}_s + L_s\frac{d\boldsymbol{i}_s}{dt} + \omega_m\frac{\partial L_s}{\partial\theta_m}\boldsymbol{i}_s + j\omega_m\boldsymbol{\psi}_{pm} \tag{19.24}$$

where ω_m is the mechanical speed, θ_m is the mechanical angle, $\boldsymbol{\psi}_{pm}$ is the permanent magnet flux vector, and R_s and L_s are the stator windings parameters and the j in the last term stands for a quadrature component. The electrical torque in a PMSM is given by

$$T_e = \frac{3}{2}\cdot p\cdot(\Psi_{pm}i_{sq} + (L_{sd} - L_{sq})i_{sd}i_{sq}) \tag{19.25}$$

where L_{sd} and L_{sq} are the stator inductances in the direction of the magnet flux and in quadrature, respectively, and Ψ_{pm} is the amplitude of the permanent magnet flux. To control the speed machine with the predictive approach, the mechanical model of the machine is required. This model is given by

$$J_m\frac{d\omega_m}{dt} = p\cdot(T_e - T_l) \tag{19.26}$$

where J_m is the inertia of the rotor and T_l is the load torque.

The machine speed is modeled using a second-order discrete model in order to have a direct feed-through from the input $\boldsymbol{v}_s$. The load torque is obtained using an estimator.

19.6.1 Predictive Model

The observer uses the load current and speed to estimate a time-delay compensated version of the sampled currents, the mechanical speed and the load. The predictions of the load current, electrical torque, and speed in the next step time are as follows:

$$\boldsymbol{i}_s(k+1) = \boldsymbol{i}_s(k) - \frac{T_s}{L_s}\left(R_s + \omega_m(k)\frac{\partial L_s}{\partial \theta_m}\right)\boldsymbol{i}_s(k) + \frac{T_s}{L_s}(\boldsymbol{v}_s(k) - \omega_m(k)\boldsymbol{J}\psi_{pm}(k)) \quad (19.27)$$

$$T_e(k+1) = \frac{3}{2}\cdot p \cdot \Psi_{pm}\boldsymbol{i}_{sq}(k+1) + \frac{3}{2}\cdot p \cdot (L_{sd} - L_{sq})\boldsymbol{i}_{sd}(k+1)\boldsymbol{i}_{sq}(k+1) \quad (19.28)$$

$$\omega_m(k+1) = \omega_m(k) + \frac{T_s p}{J_m}\cdot(T_e(k) - T_l(k)) \quad (19.29)$$

where T_s is the sample time and $\boldsymbol{J}$ is the matrix defined to calculate the torque in Section 19.4.1; here it is used to calculate the quadrature component. It is worth noting that Equation (19.27) is expressed in the three-phase *abc* frame and, therefore, the stator current must be rotated to the synchronous *dq* frame in order to be used in Equation (19.28).

19.6.2 Cost Function

The cost function used in predictive speed control is composed of three terms: a steady-state F_{ss}, a speed transient F_{st}, and a set of constraints-related components F_{cn}:

$$g = F_{ss} + \lambda_{st}\delta(\Delta\omega_m)F_{st} + \lambda_{cn}F_{cn} \quad (19.30)$$

The weighting factors λ_{st} and λ_{cn} are used to define the relative importance of these terms compared with that of the speed error.

The cost function for the steady-state operation is defined as

$$F_{ss} = (\omega_m^* - \omega_m)^2 + \lambda_T(T_l - T_e)^2 + \lambda_d i_d^2 \quad (19.31)$$

The first term uses the speed error to generate the reference tracking. The second term stabilizes the torque at its steady-state value as the load torque, and the last term introduces an optimum value for the torque-per-ampere ratio, thus minimizing the direct component of the current.

The function $\delta(\Delta\omega_m)$ establishes an operating range for the speed transient term. This function depends on the speed error and is given by

$$\delta(\Delta\omega_m) = \begin{cases} \Delta\omega_m^2 & \text{if } \Delta\omega_m^2 < \hat{\delta} \\ \hat{\delta} & \text{if } \Delta\omega_m^2 \geq \hat{\delta} \end{cases} \quad (19.32)$$

where $\hat{\delta}$ is a design parameter related to the maximum speed change allowed. This term assigns a quadratic weight to F_{st} when the value of the speed change is smaller than $\hat{\delta}$ and a constant weight of F_{st} when the speed change is larger than $\hat{\delta}$.

The cost function for the speed transient condition F_{st} is given by

$$F_{st} = \lambda_T'(\text{sign}(\Delta\omega_l)\,\hat{T}_e - T_e)^2 + \lambda_d' i_d^2 \quad (19.33)$$

The first term depends on the error between the actual torque and its maximum allowed value $\hat{T}$, controlling its dynamical change; the second term reduces the dynamic change of the torque-per-ampere ratio.

Finally, the term F_{cn} contains all the constraints required to protect the drive and is given by

$$F_{cn} = \sum_{k=1}^{N} f_k \tag{19.34}$$

As seen in the previous equation, this term can be composed of several terms, each one of them defined by a nonlinear function that takes high values when the magnitude of the corresponding constrained variable grows beyond the limits imposed by the drive specifications. For example, to implement a restriction for the stator current, f_{is} is given by

$$f_{is} = \begin{cases} |i_s|^2 - \hat{i}_s^2 & \text{if } |i_s| > \hat{i}_s \\ 0 & \text{if } |i_s| \leq \hat{i}_s \end{cases} \tag{19.35}$$

This function is zero if the stator current amplitude is smaller than a predefined maximum limit $\hat{i}_s$ and it grows with a quadratic ratio if the stator current lies beyond this limit.

19.6.3 Predictive Algorithm

The predictive algorithm is the same as in the previous sections, but the model prediction and cost function evaluation are different.

19.6.4 Control Scheme

Figure 19.17 shows the control scheme of the predictive speed control. In this case, the control is done entirely inside the predictive algorithm. Therefore, no external current or speed controllers are required.

The weighting factors are adjusted heuristically through simulations. In both functional components F_{ss} and F_{st}, the λ_T and λ_d factors are adjusted to obtain effective torque tracking and acceptable current ripple, respectively. In contrast, λ_{st} is adjusted considering the trade-off between precise speed control F_{ss} and aggressive control action F_{st}; in other words, high dynamics and rejection of high-frequency effects. Further details about weighting factor design are given in the references [55, 56].

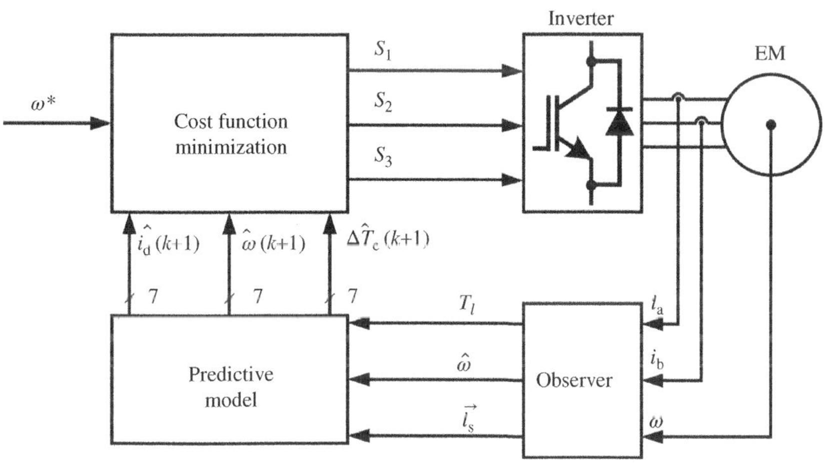

Figure 19.17 Block diagram of predictive speed control

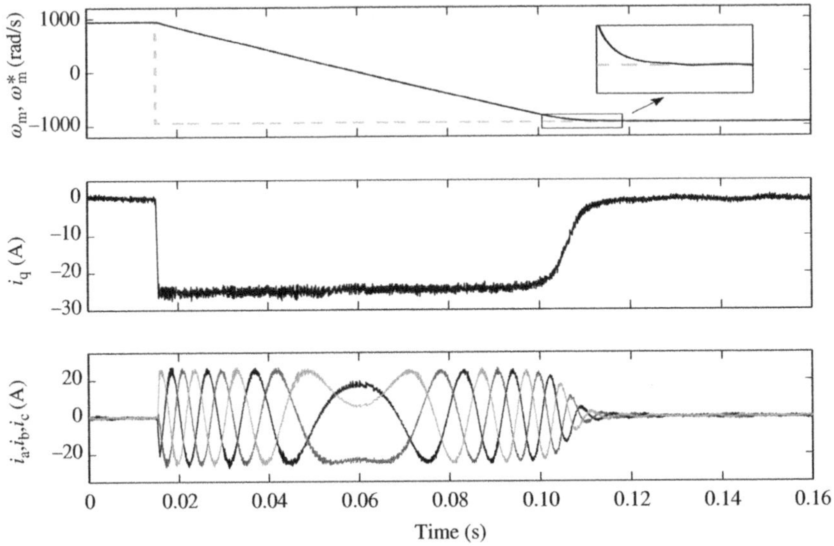

Figure 19.18 Speed reversal of a permanent magnet synchronous machine using predictive speed control

The experimental results in Figure 19.18 show a very good control of speed without overshooting. The torque and currents also exhibit a fast dynamic behavior. It is remarkable that this control strategy generates sinusoidal stator currents without sinusoidal references, modulation, or linear controllers.

19.7 Conclusions

Over the course of more than five decades, the electrical community learned that in order to inject controlled sinusoidal currents in the stator of an AC machine, sinusoidal references, a linear controller and a pulse-width modulator were necessary, according to the classical linear control theory. This chapter shows that this is not strictly true.

This chapter has demonstrated that MPC is conceptually different and simpler than classical techniques like FOC and DTC. MPC offers a high flexibility to control different circuit topologies and to manage several control objectives, without adding significant complexity.

The advances in modern control theory and in microprocessors have made it possible to apply MPC in power electronics and drives in a natural and simple way. The results shown demonstrate that, in principle, MPC allows for high-quality speed control of AC machines. This method shows a behavior similar to the well-established drive controllers. However, there are a number of aspects that must be clarified in the near future in order to bring MPC to industrial application in drives. First, a rigorous comparison with the existing techniques must be done in order to assess the possible advantages of MPC in terms of performance and simplicity. In addition, the design procedure for MPC must be improved to a more simple and systematic form, as other standard control strategies typically are. In particular, a simpler and more systematic procedure must be found to calculate the weighting factors used in the quality functions.

Finally, the authors consider that MPC can dramatically change the way electrical energy is controlled using power semiconductors, and it offers challenging opportunities for research in power electronics and electrical drives.

Acknowledgment

This chapter was made possible by an NPRP Grant No. 04-077-2-028 from the Qatar National Research Fund (a member of Qatar Foundation) and the chilean Comision Nacional de Investigacion Cientifica y Tecnologica (CONICYT) scholarship. The statements made herein are solely the responsibility of the authors.

References

1. Garcia, C.E., Prett, D.M., and Morari, M. (1989) Model predictive control: theory and practice – a survey. *Automatica*, **11**, 335–348.
2. Kouro, S., Cortes, P., Vargas, R. *et al.* (2009) Model predictive control – a simple and powerful method to control power converters. *IEEE Transactions on Industrial Electronics*, **56** (6), 1826–1838.
3. Goodwin, G.C. and Quevedo, D.E. (2003) Finite alphabet control and estimation. *International Journal of Control*, **1**, 412–430.
4. Qin, S.J. and Badgwell, T.A. (2003) A survey of industrial model predictive control technology. *Control Engineering Practice*, **11**, 733–764.
5. Hua, C.-C., Wu, C.-W., and Chuang, C.-W. (2009) A digital predictive current control with improved sampled inductor current for cascaded inverters. *IEEE Transactions on Industrial Electronics*, **56** (5), 1718–1726.
6. Rodriguez, J., Pontt, J., Silva, C.A. *et al.* (2007) Predictive current control of a voltage source inverter. *IEEE Transactions on Industrial Electronics*, **54** (1), 495–503.
7. Correa, P., Pacas, M., and Rodriguez, J. (2007) Predictive torque control for inverter-fed induction machines. *IEEE Transactions on Industrial Electronics*, **54** (2), 1073–1079.
8. Miranda, H., Cortes, P., Yuz, J.I., and Rodriguez, J. (2009) Predictive torque control of induction machines based on state-space models. *IEEE Transactions on Industrial Electronics*, **56** (6), 1916–1924.
9. Fuentes, E.J., Rodriguez, J., Silva, C. *et al.* (2009) Speed control of a permanent magnet synchronous motor using predictive current control. Proceedings of the. IEEE 6th International Power Electronics and Motion Control Conference IPEMC'09, May 17–20, 2009, pp. 390–395.
10. Morel, F., Lin-Shi, X., Retif, J.-M. *et al.* (2009) A comparative study of predictive current control schemes for a permanent-magnet synchronous machine drive. *IEEE Transactions on Industrial Electronics*, **56** (7), 2715–2728.
11. Wipasuramonton, P., Zhu, Z., and Howe, D. (2006) Predictive current control with current-error correction for PM brushless AC drives. *IEEE Transactions on Industry Applications*, **42** (4), 1071–1079.
12. Xu, L., Zhi, D., and Williams, B. (2009) Predictive current control of doubly fed induction generators. *IEEE Transactions on Industrial Electronics*, **56** (10), 4143–4153.
13. Cortes, P., Wilson, A., Kouro, S. *et al.* (2010) Model predictive control of multilevel cascaded H-bridge inverters. *IEEE Transactions on Industrial Electronics*, **57** (8), 2691–2699.
14. Perez, M., Cortes, P., and Rodriguez, J. (2008) Predictive control algorithm technique for multilevel asymmetric cascaded H-bridge inverters. *IEEE Transactions on Industrial Electronics*, **55** (12), 4354–4361.
15. Lezana, P., Aguilera, R., and Quevedo, D. (2009) Model predictive control of an asymmetric flying capacitor converter. *IEEE Transactions on Industrial Electronics*, **56** (6), 1839–1846.
16. Vargas, R., Ammann, U., Rodriguez, J., and Pontt, J. (2008) Predictive strategy to control common-mode voltage in loads fed by matrix converters. *IEEE Transactions on Industrial Electronics*, **55** (12), 4372–4380.
17. Correa, P., Rodriguez, J., Rivera, M. *et al.* (2009) Predictive control of an indirect matrix converter. *IEEE Transactions on Industrial Electronics*, **56** (6), 1847–1853.
18. Barrero, F., Prieto, J., Levi, E. *et al.* (2011) An enhanced predictive current control method for asymmetrical six-phase motor drives. *IEEE Transactions on Industrial Electronics*, **58** (8), 3242–3252.
19. Laczynski, T. and Mertens, A. (2009) Predictive stator current control for medium voltage drives with LC filters. *IEEE Transactions on Power Electronics*, **24** (11), 2427–2435.
20. Blaschke, F. (1972) The principle of field orientation as applied to the new TRANSVECTOR closed-loop control system for rotating field machines. *Siemens Reviews*, **39** (5), 217–220.
21. Leonhard, W. (2001) *Control of Electrical Drives*, Springer.
22. Novotny, D.W. and Lipo, T.A. (eds) (1996) *Vector Control and Dynamics of AC Drives*, Oxford University Press.
23. Zhen, L. and Xu, L. (1998) Sensorless field orientation control of induction machines based on a mutual MRAS scheme. *IEEE Transactions on Industrial Electronics*, **45** (5), 824–831.

24. Kerkman, R., Rowan, T., and Leggate, D. (1992) Indirect field-oriented control of an induction motor in the field-weakening region. *IEEE Transactions on Industry Applications*, **28** (4), 850–857.
25. Hofmann, H., Sanders, S., and EL-Antably, A. (2004) Stator-flux-oriented vector control of synchronous reluctance machines with maximized efficiency. *IEEE Transactions on Industrial Electronics*, **51** (5), 1066–1072.
26. Jain, A. and Ranganathan, V. (2011) Modeling and field oriented control of salient pole wound field synchronous machine in stator flux coordinates. *IEEE Transactions on Industrial Electronics*, **58** (3), 960–970.
27. Xu, L., Zhen, L., and Kim, E.-H. (1998) Field-orientation control of a doubly excited brushless reluctance machine. *IEEE Transactions on Industry Applications*, **34** (1), 148–155.
28. Singh, G., Nam, K., and Lim, S. (2005) A simple indirect field-oriented control scheme for multiphase induction machine. *IEEE Transactions on Industrial Electronics*, **52** (4), 1177–1184.
29. Holtz, J. (1994) Pulsewidth modulation for electronic power conversion. *Proceedings of the IEEE*, **82** (8), 1194–1214.
30. Takahashi, I. and Noguchi, T. (1986) A new quick-response and high-efficiency control strategy of an induction motor. *IEEE Transactions on Industry Applications*, **22** (5), 820–827.
31. Buja, G. and Kazmierkowski, M. (2004) Direct torque control of PWM inverter-fed ac motors – a survey. *IEEE Transactions on Industrial Electronics*, **51** (4), 744–757.
32. Lascu, C., Boldea, I., and Blaabjerg, F. (2000) A modified direct torque control for induction motor sensorless drive. *IEEE Transactions on Industry Applications*, **36** (1), 122–130.
33. Bertoluzzo, M., Buja, G., and Menis, R. (2006) Direct torque control of an induction motor using a single current sensor. *IEEE Transactions on Industrial Electronics*, **53** (3), 778–784.
34. Faiz, J. and Mohseni-Zonoozi, S. (2003) A novel technique for estimation and control of stator flux of a salient-pole pmsm in dtc method based on mtpf. *IEEE Transactions on Industrial Electronics*, **50** (2), 262–271.
35. Tang, L., Zhong, L., Rahman, M.F., and Hu, Y. (2003) A novel direct torque control for interior permanent-magnet synchronous machine drive with low ripple in torque and flux-a speed-sensorless approach. *IEEE Transactions on Industry Applications*, **39** (6), 1748–1756.
36. Liu, Y., Zhu, Z., and Howe, D. (2005) Direct torque control of brushless dc drives with reduced torque ripple. *IEEE Transactions on Industry Applications*, **41** (2), 599–608.
37. Casadei, D., Profumo, F., Serra, G., and Tani, A. (2002) FOC and DTC: two viable schemes for induction motors torque control. *IEEE Transactions on Power Electronics*, **17** (5), 779–787.
38. Holtz, J. (1995) The representation of ac machine dynamics by complex signal flow graphs. *IEEE Transactions on Industrial Electronics*, **42** (3), 263–271.
39. Sorchini, Z. and Krein, P. (2006) Formal derivation of direct torque control for induction machines. *IEEE Transactions on Power Electronics*, **21** (5), 1428–1436.
40. Tang, L., Zhong, L., Rahman, M., and Hu, Y. (2004) A novel direct torque controlled interior permanent magnet synchronous machine drive with low ripple in flux and torque and fixed switching frequency. *IEEE Transactions on Power Electronics*, **19** (2), 346–354.
41. Casadei, D., Serra, G., Tani, A. *et al.* (2003) Performance analysis of a speed-sensorless induction motor drive based on a constant-switching-frequency DTC scheme. *IEEE Transactions on Industry Applications*, **39** (2), 476–484.
42. Idris, N. and Yatim, A. (2004) Direct torque control of induction machines with constant switching frequency and reduced torque ripple. *IEEE Transactions on Industrial Electronics*, **51** (4), 758–767.
43. Cortes, P., Rodriguez, J., Silva, C., and Flores, A. (2012) Delay compensation in model predictive current control of a three-phase inverter. *IEEE Transactions on Industrial Electronics*, **59** (2), 1323–1325.
44. Fuentes, E. and Kennel, R. (2011) Sensorless-predictive torque control of the PMSM using a reduced order extended Kalman filter. 2011 Symposium on Sensorless Control for Electrical Drives (SLED), September 2011, pp. 123–128.
45. Cortes, P., Kazmierkowski, M., Kennel, R. *et al.* (2008) Predictive control in power electronics and drives. *IEEE Transactions on Industrial Electronics*, **55** (12), 4312–4324.
46. Preindl, M., Schaltz, E., and Thogersen, P. (2011) Switching frequency reduction using model predictive direct current control for high-power voltage source inverters. *IEEE Transactions on Industrial Electronics*, **58** (7), 2826–2835.
47. Cortes, P., Rodriguez, J., Quevedo, D., and Silva, C. (2008) Predictive current control strategy with imposed load current spectrum. *IEEE Transactions on Power Electronics*, **23** (2), 612–618.
48. Escalante, M., Vannier, J.-C., and Arzande, A. (2002) Flying capacitor multilevel inverters and DTC motor drive applications. *IEEE Transactions on Industrial Electronics*, **49** (4), 809–815.

49. Casadei, D., Serra, G., and Tani, A. (2001) The use of matrix converters in direct torque control of induction machines. *IEEE Transactions on Industrial Electronics*, **48** (6), 1057–1064.
50. Lee, K.-B., Song, J.-H., Choy, I., and Yoo, J.-Y. (2002) Torque ripple reduction in dtc of induction motor driven by three-level inverter with low switching frequency. *IEEE Transactions on Power Electronics*, **17** (2), 255–264.
51. Vargas, R., Rodriguez, J., Ammann, U., and Wheeler, P. (2008) Predictive current control of an induction machine fed by a matrix converter with reactive power control. *IEEE Transactions on Industrial Electronics*, **55** (12), 4362–4371.
52. Vargas, R., Ammann, U., Hudoffsky, B. *et al.* (2010) Predictive torque control of an induction machine fed by a matrix converter with reactive input power control. *IEEE Transactions on Power Electronics*, **25** (6), 1426–1438.
53. Fuentes, E.J., Silva, C.A. and Yuz, J.I. (2012) Predictive Speed Control of a Two-Mass System Driven by a Permanent Magnet Synchronous Motor. *IEEE Transactions on Industrial Electronics*, **59** (7), 2840–2848, doi: 10.1109/TIE.2011.2158767.
54. Cortes, P., Ortiz, G., Yuz, J. *et al.* (2009) Model predictive control of an inverter with output LC filter for ups applications. *IEEE Transactions on Industrial Electronics*, **56** (6), 1875–1883.
55. Cortes, P., Kouro, S., La Rocca, B. *et al.* (2009) Guidelines for weighting factors design in model predictive control of power converters and drives. IEEE International Conference on Industrial Technology, 2009. ICIT 2009, February 2009, pp. 1–7.
56. Chang, Y.-T. and Lai, Y.-S. (2009) Parameter tuning method for digital power converter with predictive current-mode control. *IEEE Transactions on Power Electronics*, **24** (12), 2910–2919.

20

The Electrical Drive Systems with the Current Source Converter

Marcin Morawiec and Zbigniew Krzeminski
Department of Automatic Control of Electrical Drives, Gdansk University of Technology, Gdansk, Poland

20.1 Introduction

Digital-circuit engineering allows the use of high-efficiency microprocessors or signal processor applications in power converters. These high-performance processors have high-frequency clock timing and cache memory. The arithmetic-logic operations with floating point processors are not the cause, and the operation time is shorter than a few years ago. As a result, it is possible to implement more complicated applications with advanced control systems. The circuits used for power conversion applied in drives with induction motors (IMs) are classified into two groups: voltage source inverters (VSIs) and current source inverters (CSIs). The VSIs are used more often than the CSIs because of their better properties. Nowadays, the development of power electronic devices has enormous influence on applications of systems based on the CSI and creates new possibilities.

In the 1980s, CSIs were the main commonly used electric machine feeding devices. Characteristic features of those drives were the motor electromagnetic torque pulsations, the voltage and the current with higher harmonic content. The CSI comprised a thyristor bridge and large inductance and commutation capacitors. Serious problems in such drive systems were unavoidable in overvoltage cases during the thyristor commutation, as the CSI current is supplied in a cycle from a DC-link circuit to the machine-phase winding. The thyristor CSI has been replaced by the transistor reverse-blocking IGBT (RBIGBT) devices, where the diode is series connected and placed in one casing with the transistor. Power transistors, like RBIGBT or silicon carbon (SiC), used in the modern CSIs guarantee superior static and dynamic drive characteristics.

The electric drive development trends are focused on the high-quality system. The use of current sources for the electric machine control ensures better drive properties than in the case of voltage sources, where it may be necessary to use an additional passive filter at the inverter output. Pulse-width modulation (PWM) with properly chosen DC-link inductor and input–output capacitors results in sinusoidal inverter output currents and voltages. Methods of calculating proper inductance in the DC link were proposed in Refs. [1–6]. Properties of the DC-link circuit of the current source converter (CSC) force the utilization

Power Electronics for Renewable Energy Systems, Transportation and Industrial Applications, First Edition.
Edited by Haitham Abu-Rub, Mariusz Malinowski and Kamal Al-Haddad.

of two fully controlled inverters to supply the system with an electric motor. The first – CSI – generates the current output vector to supply the IM. The second – current source rectifier (CSR) – generates a DC voltage to supply a DC-link circuit. The strategy for controlling the output current vector of CSI can be realized in two ways [1–8]. The first is based on changes in the modulation index while the value of the current in the DC-link circuit remains constant [5, 6]. The second method is based on changes in the DC-link current. In this case, the CSI is working with a constant maximum value of the PWM modulation index. Control of the modulation index in CSI is used in drive systems, where a high dynamic of the electromagnetic torque should be maintained [5, 6]. A high current in the DC-link circuit is a reason for high power losses in CSI. The simplified control method is the scalar control: current to slip (*I/s*). This method is very simple to implement, but the drive system has average performance (only one controller is necessary, the current in the DC link is kept at a constant value by a proportional-integral (PI) controller).

The drive system quality is closely connected with the machine control algorithm. The space vector concept, introduced in 1959 by Kovacs and Racz, opened a new path in the electric machine mathematical modeling field. The international literature on the subject presents drive systems with the CSI feeding an IM with the control system based on the coordinate system orientation in relation to the rotor flux vector (FOC – field-oriented control). Such control consisted of the DC-link circuit current stabilization [5, 6, 8]. In such control systems, the control variables are the inverter output current components. This control method is presented in Ref. [9], where the authors analyze the control system based on direct torque control. The control process where the control variable is the inverter output current may be called the *current control* of an IM supplied by the CSI.

Another control method for a CSI-fed IM uses the DC-link circuit voltage and the motor slip as control variables. That type of control may be called the *voltage control* of a CSI-fed IM, as the DC-link circuit voltage and angular frequency of the current vector are the control variables. The proposed control strategy is based on nonlinear multi-scalar control [1–4]. Nonlinear control may result in better properties in cases where the IM is fed by the CSI. To achieve independent control of flux and rotor speed, a new nonlinear control scheme is proposed. In this control method, the inverter output currents are not controlled variables. The voltage in the DC link and the pulsation of the output current vector are the controlled variables that can be obtained by nonlinear transformations and are proposed by the authors in Refs. [1–4, 10]. The multi-scalar model is called the *extended model* because the mathematical model contained the DC-link current and output capacitor equations. This full mathematical model of an induction machine with the CSI is used to derive a new multi-scalar model. In the proposed method, the output current vector coefficients are not controlled variables. The output current vector and the flux vector are used to achieve new multi-scalar variables and a new multi-scalar model. The control system structure may be supported on PI controllers and nonlinear decoupling or different controllers, for example, sliding-mode controllers, the backstepping control method, or fuzzy neural controllers.

20.2 The Drive System Structure

The configuration of the drive system with the CSI is presented in Figure 20.1. The integral parts of the system are the inductor in the DC link and the output capacitors. In Figure 20.1, the structure with the chopper as an adjustable voltage source is presented.

The chopper with the small inductor L (a few mH) forms the large dynamic impedance of the current source. In the proposed system, the transistors form a commutator that transforms the DC current into an AC current. The current is controlled by the voltage source e_d in the DC link. In this way, the system with CSI remains voltage controlled and the differential equation for the DC link may be integrated with the differential equation for the stator. The inductor limits current ripples during commutations of the transistors. The transistors used in this structure are called *RBIGBT transistors*. This structure has diode series connected with an IGBT in the same transistor module. The double six transistor bridge CSC, in which bidirectional energy transmission is possible, is shown in Figure 20.2.

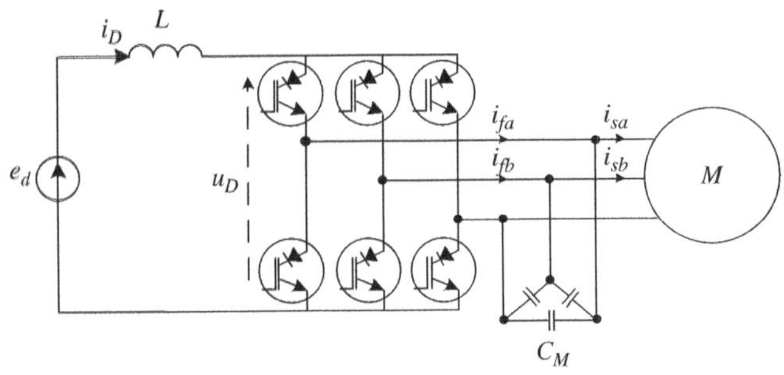

Figure 20.1 The CSI with the chopper

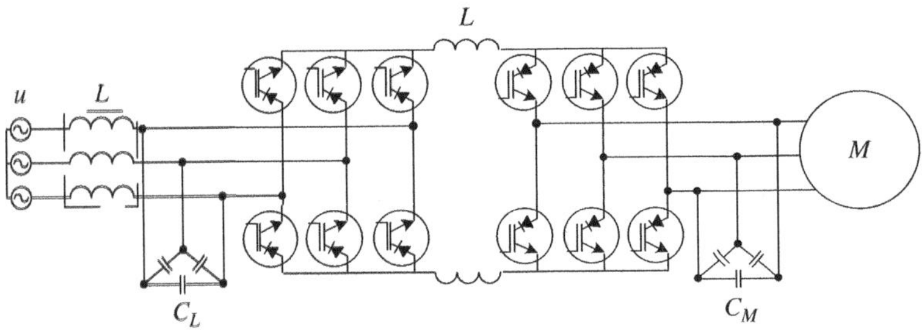

Figure 20.2 The current source converter

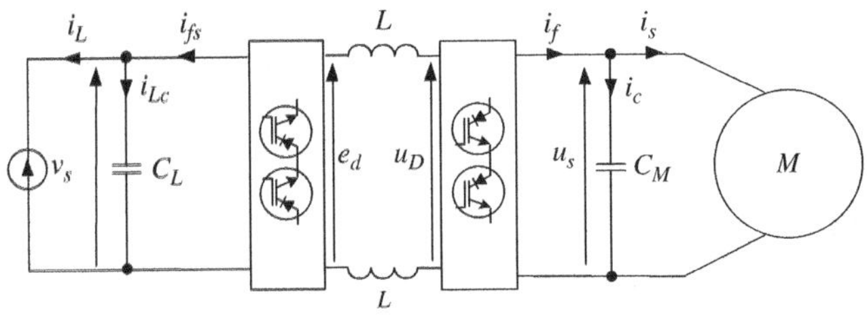

Figure 20.3 The simplified scheme of the CSC

In order to avoid resonance problems, the CSI or CSC structure parameters (input–output capacitors and inductor) ought to be properly chosen. The transistor CSI or CSC structures should guarantee sinusoidal stator current and voltage of the IM if the parameters are selected by an iteration algorithm [11].

It is possible to simplify the scheme of CSC in the way shown in Figure 20.3. The RBIGBT transistors are considered as unidirectional switches connecting the supply to the DC link and the DC link to the load. The models of the capacitors and DC link are introduced below.

The mathematical model of the CSC, determined by a commutation function, was presented in Ref. [4]. In this chapter, the differential equations of capacitors C_L and C_M and the DC-link model are presented below. The input capacitors' model in the $\alpha\beta$ stationary coordinate system is as follows:

$$\frac{dv_{s\alpha}}{d\tau} = \frac{1}{C_L}(i_{fs\alpha} - i_{L\alpha}) \tag{20.1}$$

$$\frac{dv_{s\beta}}{d\tau} = \frac{1}{C_L}(i_{fs\beta} - i_{L\beta}) \tag{20.2}$$

$$\frac{du_{s\alpha}}{d\tau} = \frac{1}{C_M}(i_{f\alpha} - i_{s\alpha}) \tag{20.3}$$

$$\frac{du_{s\beta}}{d\tau} = \frac{1}{C_M}(i_{f\beta} - i_{s\beta}) \tag{20.4}$$

$$\frac{di_D}{d\tau} = \frac{1}{L}(e_d - R_d i_D - u_D) \tag{20.5}$$

where $\alpha\beta$ is the stationary coordinate system; $i_{f\alpha}$, $i_{f\beta}$ are the output six transistors bridge current; $i_{fs\alpha}$, $i_{fs\beta}$ are the input six transistors bridge current; $i_{s\alpha}$, $i_{s\beta}$ are the stator current vector coefficients; $u_{s\alpha}$, $u_{s\beta}$ are the stator voltage vector coefficients; $v_{s\alpha}$, $v_{s\beta}$ are the supply voltage vector coefficients; L is the DC-link inductance; R_d is theinterior inductor inductance; C_M is the output capacity of capacitor; C_L is the input capacity of capacitor; and i_D is the DC-link current.

The voltage u_D in the DC link is the input voltage to the machine side CSI. This voltage is determined by the equation:

$$u_D \approx \frac{i_{f\alpha}u_{s\alpha} + i_{f\beta}u_{s\beta}}{i_D} \tag{20.6}$$

The output vector current $\vec{i}_f$ is determined:

$$i_{fx} = K \cdot i_D \tag{20.7}$$

When $K \approx 1$ the output current $|\vec{i}_f| = i_{fx} = i_D$. If $i_{fx} = i_D$ in an xy coordinate system (Figure 20.20), in which the x axis is overlapped with output current vector $\vec{i}_f$, one obtains:

$$\frac{di_{fx}}{d\tau} = \frac{1}{L}(e_d - R_d i_{fx} - u_{sx}) \tag{20.8}$$

20.3 The PWM in CSCs

The methods provided to form the output current's i_f primary harmonic were called *modulations in the literature*. The PWM deserves particular attention. The method is based on the rotation of the reference space vector. This method was shown in Refs. [1, 2, 12–14]. The average values of output current vector components in the sampling time are received by turning on switch states in cycles. In accordance with space vector theory, the projection of three-phase currents by space vectors in the $\alpha\beta$ stationary coordinate system is possible. The three-phase currents (i_{sa}, i_{sb}, i_{sc}) can be specified by reference to the space vector module marked I_0 and rotated with angular speed ω_0 in the $\alpha\beta$ orthogonal system (Figure 20.4).

The PWM CSI uses six unidirectional switches in two bridge legs: the first in the upper part (switches 1, 3, 5) and the second in the lower part of the bridge (switches 4, 6, 2). The switches from Figure 20.5 should be chosen in such a way that continuous current is assured in the DC link. This condition will be fulfilled if two switches suitably conduct. Active vectors are denoted as I_1 to I_6. Three passive vectors

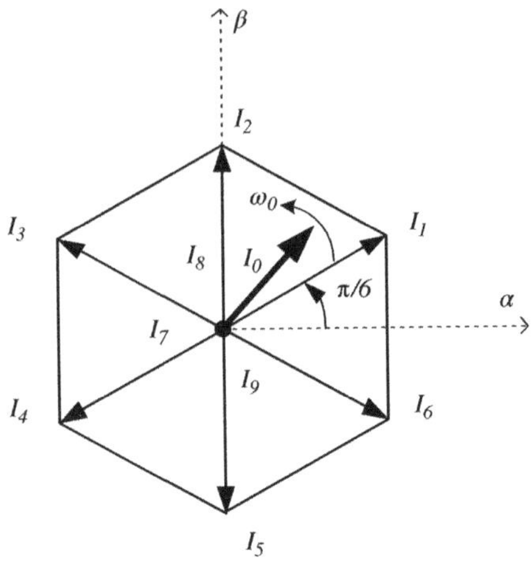

Figure 20.4 The current space vector projections

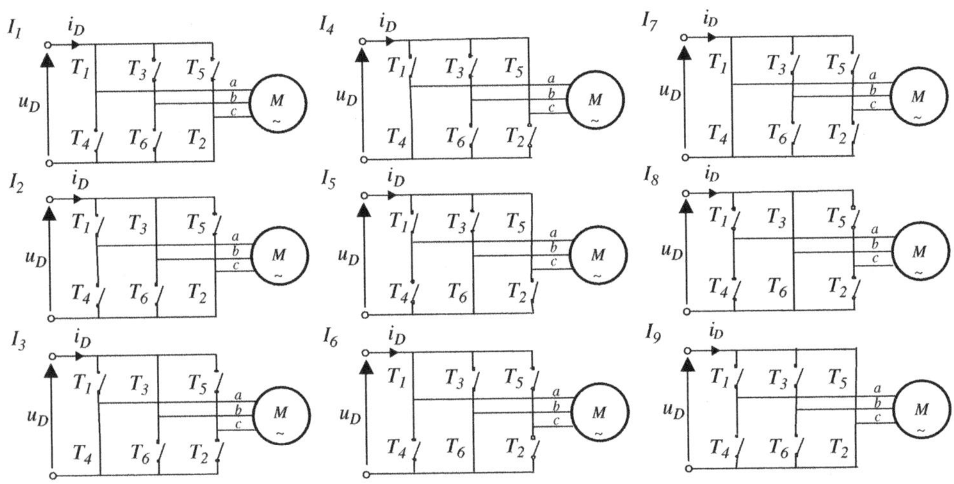

Figure 20.5 Configuration of the CSI switches

(zero vectors) are denoted as I_7, I_8, I_9 in Figure 20.5. The inverter output current vectors from I_1 to I_9 are presented in Table 20.1.

The reference vector I_0, composed with adjacent active space vector coefficients, is shown in Figure 20.6. In the orthogonal $\alpha\beta$ stationary system, the value of module I_0 follows from the transform of the three-phase values to a two-phase system, where the invariant power will be taken into account:

$$|I_0| = \sqrt{2} \cdot i_D \tag{20.9}$$

where $|I_0|$ is the module of output current vector and i_D is the DC-link current.

Table 20.1 The space vector states

Switches	I_1	I_2	I_3	I_4	I_5	I_6	I_7	I_8	I_9
1	1	0	0	0	0	1	1	0	0
2	1	1	0	0	0	0	0	0	1
3	0	1	1	0	0	0	0	1	0
4	0	0	1	1	0	0	1	0	0
5	0	0	0	1	1	0	0	0	1
6	0	0	0	0	1	1	0	1	0

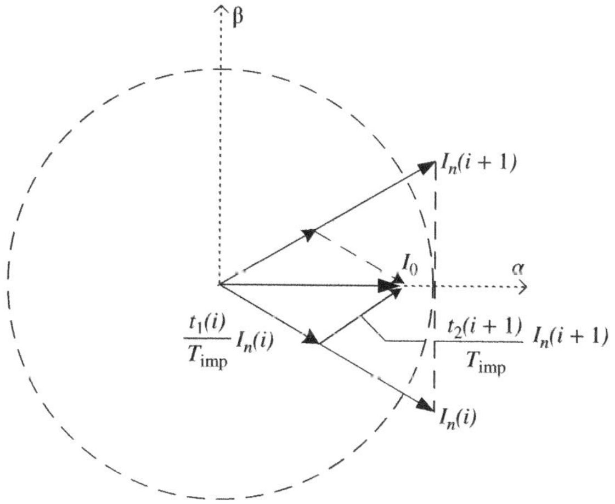

Figure 20.6 The method to compute average space vector values in each cycle time

The values of the output current vectors when the passive vectors are turned on are equal to zero. Therefore,

$$I_{n\alpha} = I_{n\alpha}(i) + I_{n\alpha}(i+1) \tag{20.10}$$

$$I_{n\beta} = I_{n\beta}(i) + I_{n\beta}(i+1) \tag{20.11}$$

$$I_{n\alpha} = \frac{t_1(i)}{T_{\text{imp}}} I_{n\alpha}(i) + \frac{t_2(i+1)}{T_{\text{imp}}} I_{n\alpha}(i+1) \tag{20.12}$$

$$I_{n\beta} = \frac{t_1(i)}{T_{\text{imp}}} I_{n\beta}(i) + \frac{t_2(i+1)}{T_{\text{imp}}} I_{n\beta}(i+1) \tag{20.13}$$

where $t_{1,2}$ is the active vector times; t_0 is the passive vector time; T_{imp} is the cycle time; $I_{\alpha,\beta}$ is the space vector coefficient in $(\alpha\beta)$ (index i is the number of the active vector, n is the sector number).

Using Equations (20.10)–(20.13) the active and passive vector times are obtained

$$t_1 = \frac{I_{n\alpha} \cdot I_{n\beta}(i+1) - I_{n\beta} \cdot I_{n\alpha}(i+1)}{I_{n\alpha}(i) \cdot I_{n\beta}(i+1) + I_{n\beta}(i) \cdot I_{n\alpha}(i+1)} \tag{20.14}$$

$$t_2 = \frac{-I_{n\alpha} \cdot I_{n\beta}(i) + I_{n\beta} \cdot I_{n\alpha}(i)}{I_{n\alpha}(i) \cdot I_{n\beta}(i+1) + I_{n\beta}(i) \cdot I_{n\alpha}(i+1)(i+1)} \tag{20.15}$$

$$t_0 = T_{\text{imp}} - (t_1 + t_2) \tag{20.16}$$

The inverter output current coefficients are shown in Figure 20.7.
The FFT of stator current and stator voltage are shown in Figures 20.8 and 20.9.

20.4 The Generalized Control of a CSR

The CSR in the CSC can be treated as a voltage-controlled source. The transistor modulation of the CSR can be similar to that of the CSI (Section 20.3) The current control of CSR was presented in Refs. [15, 16] and voltage control was presented in Refs. [14].

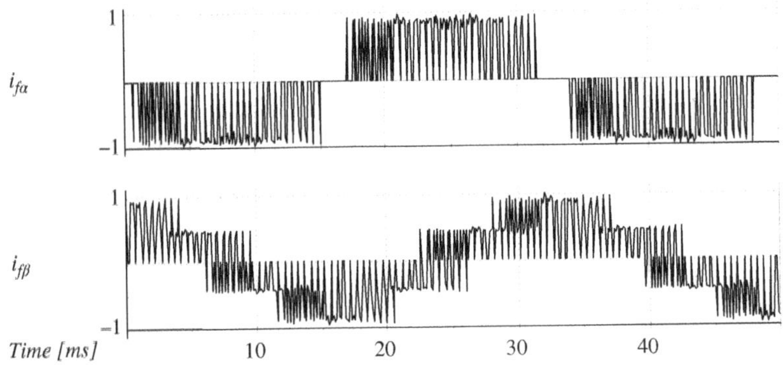

Figure 20.7 The output current vector coefficients in per unit values

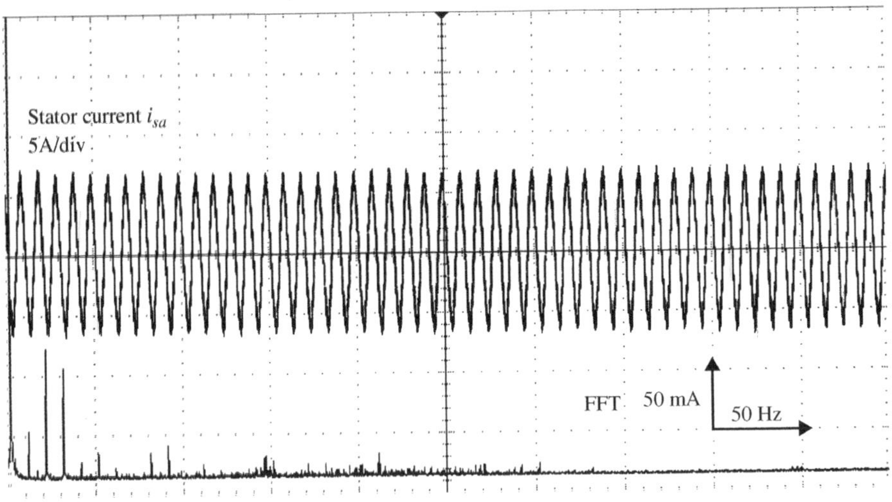

Figure 20.8 The FFT of stator current (5.5 kW induction motor)

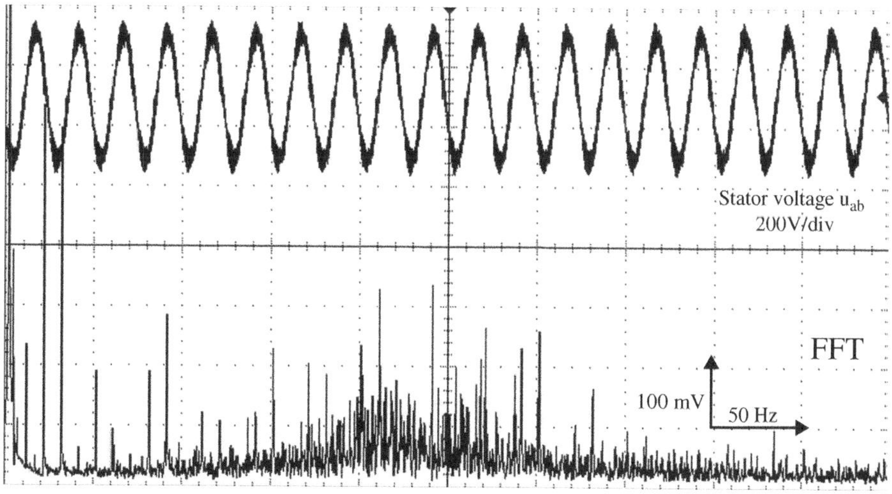

Figure 20.9 The FFT of stator voltage (5.5 kW induction motor)

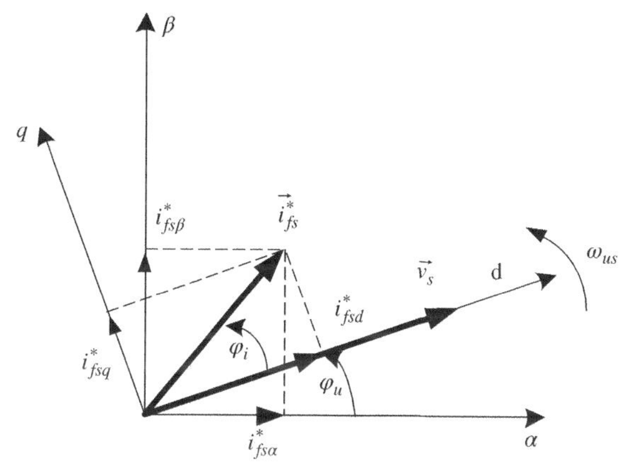

Figure 20.10 The supply current vector and voltage in the *dq* system

On the assumption that the CSR is an ideal converter of energy, the dependence can be written

$$i_{fsd}|v_s| \approx e_d i_D \tag{20.17}$$

where dq is the coordinate system connected with the supply vector voltage (Figure 20.10).

Taking into account the current control of the CSR [15, 16] in which the control variables of a machine are stator current vector components, one obtains

$$i^*_{fsd} \approx \frac{e^*_d i^*_f}{|v_s|} \tag{20.18}$$

where the reference values are indexed by "*," i^*_f is the output current vector module from the machine control system and e^*_d is from, for example, the current PI controllers (Figure 20.11).

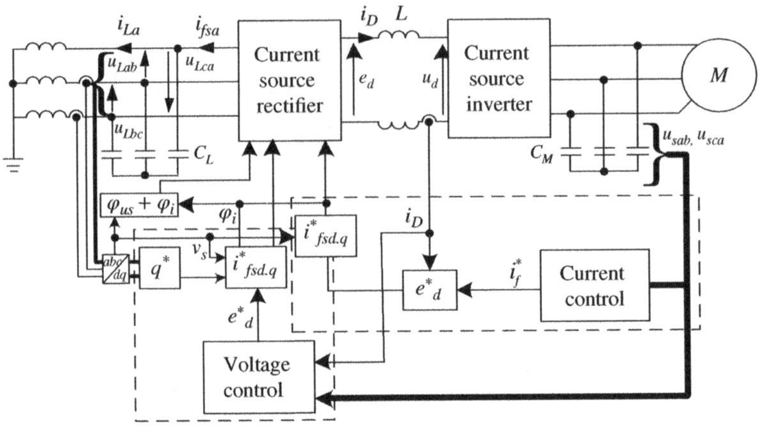

Figure 20.11 The control scheme of the CSC in which the machine can be controlled by voltage or the current control system

The reference current module is as follows:

$$i_{fs}^* = \sqrt{i_{fsd}^{*2} + i_{fsq}^{*2}} \tag{20.19}$$

where the i_{fsq}^* is determined by

$$i_{fsq}^* = \frac{v_{s\alpha} i_{fs\beta} - v_{s\beta} i_{fs\alpha}}{|v_s|} = \frac{e_q^* i_{fs}^*}{|v_s|} \tag{20.20}$$

and e_q^*:

$$e_q^* = \frac{q^*}{i_D}$$

where

$$q^* = v_{s\alpha} i_{L\beta} - v_{s\beta} i_{L\alpha} \tag{20.21}$$

The reference current angel $\vec{i}_{fs}^{\,*}$ is determined as

$$\varphi_i = \text{arctg}\left(\frac{i_{fsq}^*}{i_{fsd}^*}\right) \tag{20.22}$$

When the voltage control of a machine is taken into account, the above equations must be modified:

$$i_{fsd}^* \approx \frac{e_d^* i_D}{|v_s|} \tag{20.23}$$

$$i_{fsq}^* \approx \frac{e_q^* i_D}{|v_s|} \tag{20.24}$$

$$i_{fs}^* = \frac{i_D}{|v_s|}\sqrt{e_d^{*2} + e_d^{*2}} \tag{20.25}$$

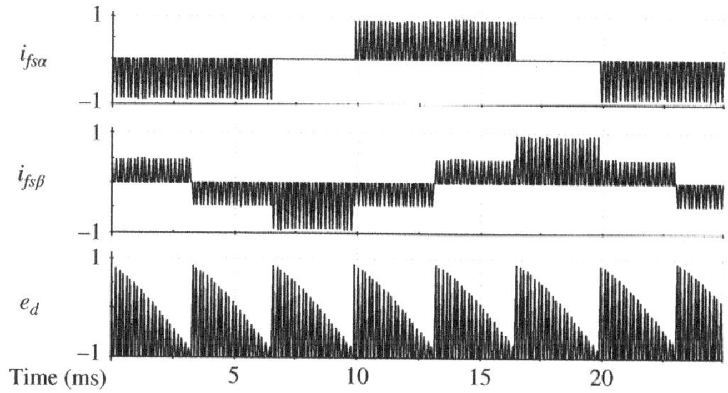

Figure 20.12 The input current vector components and measurement voltage e_d

The input current reference to the PWM is

$$i^*_{fs\alpha} = M_i i^*_{fs} \cos(\varphi_{us} + \varphi_i) \tag{20.26}$$

$$i^*_{fs\beta} = M_i i^*_{fs} \cos(\varphi_{us} + \varphi_i) \tag{20.27}$$

where $M_i < 1.0$ pu is the modulation index.

The times of the active and passive vectors are similar to Equations (20.14)–(20.16):

$$t_{1s} = \frac{i^*_{fs\alpha} \cdot I_{n\beta}(i+1) - i^*_{fs\beta} \cdot I_{n\alpha}(i+1)}{(I_{n\alpha}(i) \cdot I_{n\beta}(i+1) + I_{n\beta}(i) \cdot I_{n\alpha}(i+1))} \tag{20.28}$$

$$t_{2s} = \frac{i^*_{fs\alpha} \cdot I_{n\beta}(i) - i^*_{fs\beta} \cdot I_{n\alpha}(i)}{(I_{n\alpha}(i) \cdot I_{n\beta}(i+1) + I_{n\beta}(i) \cdot I_{n\alpha}(i+1))} \tag{20.29}$$

$$t_{0s} = T_{\mathrm{imp}} - (t_{1s} + t_{2s}) \tag{20.30}$$

where t_{1s}, t_{2s} are the active vector times; t_{0s} is the passive vector time; i is the number of active vectors and n is the number of sectors.

In Figure 20.12, the currents $i_{fs\alpha,\beta}$ and voltage e_d are shown. In Figure 20.13, the line current $i_{L\alpha}$ and supply voltage u_{Lab} are presented. In $i_{L\alpha}$, the current subharmonics are visible, because in the experiment the CSC was without the input inductors L.

20.5 The Mathematical Model of an Asynchronous and a Permanent Magnet Synchronous Motor

The mathematical model of an asynchronous and a synchronous machine are obtained under the assumptions presented in Refs. [1–4, 10]. The differential equations of an asynchronous machine for vectors $(\vec{i}_s, \vec{\psi}_r)$ are as follows:

$$\frac{di_{sx}}{d\tau} = -\frac{R_s L_r^2 + R_r L_m^2}{L_r w_\sigma} i_{sx} + \frac{R_r L_m}{L_r w_\sigma} \psi_{rx} + \omega_0 i_{sy} + \omega_r \frac{L_m}{w_\sigma} \psi_{ry} + \frac{L_m}{w_\sigma} u_{sx} \tag{20.31}$$

$$\frac{di_{sy}}{d\tau} = -\frac{R_s L_r^2 + R_r L_m^2}{L_r w_\sigma} i_{sy} + \frac{R_r L_m}{L_r w_\sigma} \psi_{ry} - \omega_0 i_{sx} - \omega_r \frac{L_m}{w_\sigma} \psi_{rx} + \frac{L_m}{w_\sigma} u_{sy} \tag{20.32}$$

$$\frac{d\psi_{rx}}{d\tau} = -\frac{R_r}{L_r} \psi_{rx} + (\omega_0 - \omega_r)\psi_{ry} + \frac{R_r L_m}{L_r} i_{sx} + u_{rx} \tag{20.33}$$

$$\frac{d\psi_{ry}}{d\tau} = -\frac{R_r}{L_r} \psi_{ry} - (\omega_0 - \omega_r)\psi_{rx} + \frac{R_r L_m}{L_r} i_{sy} + u_{ry} \tag{20.34}$$

$$\frac{d\omega_r}{d\tau} = \frac{L_m}{JL_r}(\psi_{rx} i_{sy} - \psi_{ry} i_{sx}) - \frac{1}{J} T_L \tag{20.35}$$

where ω_0, ω_r are the proper angular speeds of the (*xy*) coordinate system and rotor speed; J is the machine inertia; T_L is the load torque; R_s is the stator resistance; R_r is the rotor resistance; L_s is the stator inductance; L_r is the rotor inductance; and L_m is the mutual inductance.

$$w_\sigma = \sigma L_r L_s$$

and dissipation coefficient σ is defined as

$$\sigma = 1 - \frac{L_m^2}{L_s L_r}$$

Vector coefficients are indicated by x and y. The xy coordinate system is rotating with ω_0.

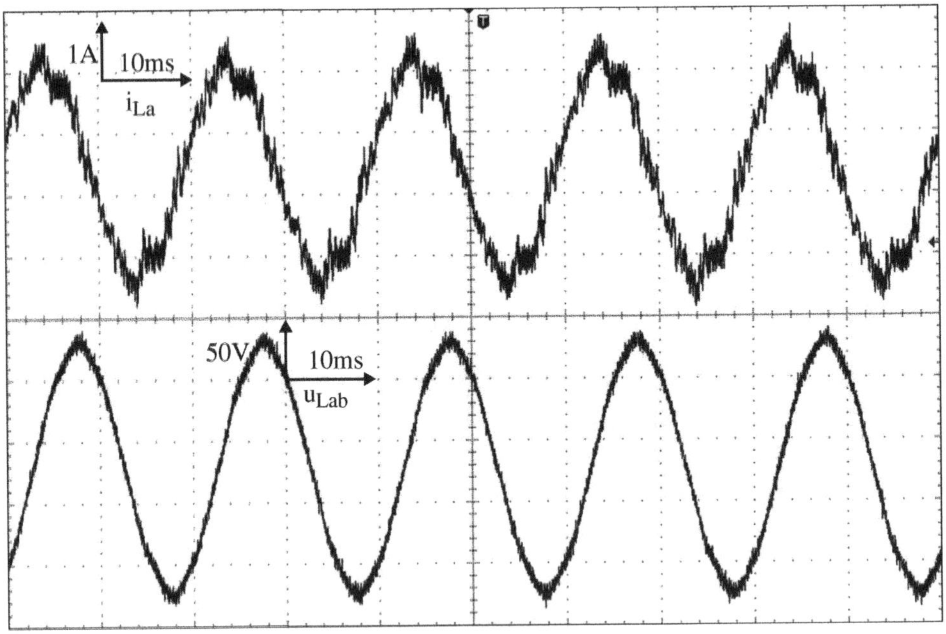

Figure 20.13 The line current i_{La} and supply voltage u_{Lab}

The differential equations of the permanent magnet synchronous machine (PMSM) in the $\alpha\beta$ stationary system are as follows [17–19]:

$$\frac{di_\alpha}{d\tau} = -\frac{R_s}{L_s} i_\alpha + \frac{1}{L_s}\omega_r \psi_{f\beta} + \frac{1}{L_s} u_{s\alpha} \tag{20.36}$$

$$\frac{di_\beta}{d\tau} = -\frac{R_s}{L_s} i_\beta - \frac{1}{L_s}\omega_r \psi_{f\alpha} + \frac{1}{L_s} u_{s\beta} \tag{20.37}$$

$$\frac{d\omega_r}{d\tau} = \frac{1}{J}(\psi_\alpha i_\beta - \psi_\beta i_\alpha) - \frac{1}{J}T_L \tag{20.38}$$

and

$$\psi_\alpha = L_s i_\alpha + \psi_{f\alpha} \tag{20.39}$$

$$\psi_\beta = L_s i_\beta + \psi_{f\beta} \tag{20.40}$$

$$\psi_{f\alpha} = \psi_f \cos\theta_r \tag{20.41}$$

$$\psi_{f\beta} = \psi_f \sin\theta_r \tag{20.42}$$

$$e_\alpha = \frac{d\psi_{f\alpha}}{d\tau} = -\psi_f \omega_r \sin\theta_r \tag{20.43}$$

$$e_\beta = \frac{d\psi_{f\beta}}{d\tau} = \psi_f \omega_r \cos\theta_r \tag{20.44}$$

where $i_{\alpha,\beta}$ is the stator current vector coefficients, $\psi_{f\alpha,\beta}$ is the permanent magnet flux vector coefficients; $u_{s\alpha,\beta}$ is the stator voltage vector coefficients; ω_r is the rotor angular speed; θ_r is the rotor position; R_s is the stator resistance; L_s is the stator inductance; T_L is the load torque; and J is the motor inertia.

The interior permanent magnet motor (*IPMSM*) mathematical model is as follows [18, 19]:

$$\frac{di_d}{d\tau} = -\frac{R_s}{L_d} i_d + \frac{L_q}{L_d}\omega_r i_q + \frac{1}{L_d} u_d \tag{20.45}$$

$$\frac{di_q}{d\tau} = -\frac{R_s}{L_q} i_q - \frac{L_d}{L_q}\omega_r i_d - \frac{1}{L_q}\omega_r \psi_f + \frac{1}{L_q} u_q \tag{20.46}$$

$$\frac{d\omega_r}{d\tau} = \frac{1}{J}[\psi_f i_q + (L_d - L_q) i_d i_q - T_L] \tag{20.47}$$

$$\frac{d\theta_r}{d\tau} = \omega_r \tag{20.48}$$

where ω_r is the rotor angular speed; θ_r is the rotor position; i_d, i_q are the stator current vector coefficients in dq; L_d, L_q are the stator inductances; R_s is the stator resistance; J is the motor inertia; T_L is the load torque; ψ_f is the module of the flux vector; and u_d, u_q are the stator voltage vector coefficients in dq.

20.6 The Current and Voltage Control of an Induction Machine

20.6.1 Field-Oriented Control

The FOC is the one of the current control methods. FOC is the most popular control method used in electrical drive systems and has been presented in many papers, for example, in Refs. [5–7, 20]. Control without decouplings is the simplest version of control. The mathematical model is connected with the (dq) coordinate system in which the d-axis is aligned to the rotor flux vector. The control variables are

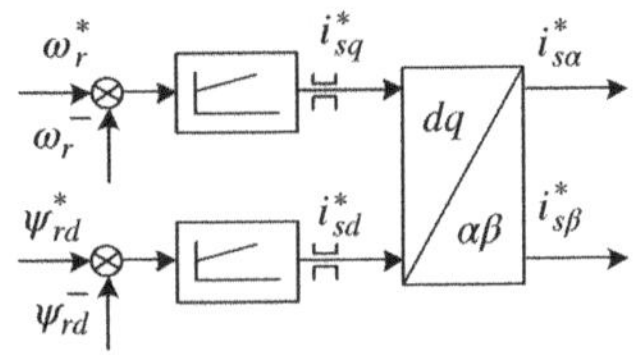

Figure 20.14 The FOC control system structure with PI controllers without decoupling variables

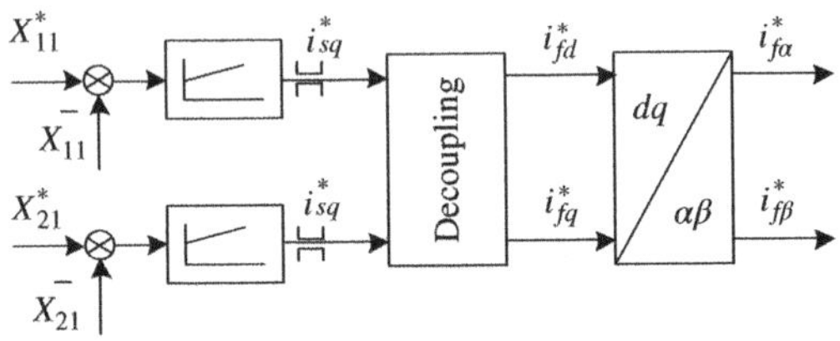

Figure 20.15 The FOC control system structure with PI controllers and decoupling variables

the PI controller outputs and the reference values of the PWM. The PI-controller outputs are the stator current components i_{sq} and i_{sd}. The voltage e_d is determined with the help of an additional PI controller. The scheme of the FOC control system is shown in Figures 20.14 and 20.15.

The mathematical model of an induction machine in the (dq) coordinate system with the rotor flux vector is determined:

$$\frac{\mathrm{d}i_{sd}}{\mathrm{d}\tau} = -\frac{R_s L_r^2 + R_r L_m^2}{L_r w_\sigma} i_{sd} + \frac{R_r L_m}{L_r w_\sigma} \psi_{rd} + \omega_r i_{sq} + \frac{L_r}{w_\sigma} u_{sd} \tag{20.49}$$

$$\frac{\mathrm{d}i_{sq}}{\mathrm{d}\tau} = -\frac{R_s L_r^2 + R_r L_m^2}{L_r w_\sigma} i_{sq} - \omega_r \frac{L_m}{w_\sigma} \psi_{rd} - \omega_r i_{sd} + \frac{L_r}{w_\sigma} u_{sq} \tag{20.50}$$

$$\frac{\mathrm{d}\psi_{rd}}{\mathrm{d}\tau} = -\frac{R_r}{L_r} \psi_{rd} + \frac{R_r L_m}{L_r} i_{sd} \tag{20.51}$$

$$\frac{\mathrm{d}\psi_{rq}}{\mathrm{d}\tau} = -(\omega_r - \omega_r)\psi_{rd} + \frac{R_r L_m}{L_r} i_{sq} \tag{20.52}$$

$$\frac{\mathrm{d}\omega_r}{\mathrm{d}\tau} = \frac{L_m}{J L_r} \psi_{rd} i_{sq} - \frac{1}{J} T_L \tag{20.53}$$

For the capacitors the model can be obtained:

$$\frac{\mathrm{d}u_{sd}}{\mathrm{d}\tau} = \frac{1}{C_M}(i_{fd} - i_{sd}) + \omega_{\psi r} u_{sq} \tag{20.54}$$

$$\frac{\mathrm{d}u_{sq}}{\mathrm{d}\tau} = \frac{1}{C_M}(i_{fq} - i_{sq}) - \omega_{\psi r} u_{sd} \tag{20.55}$$

where: $\omega_{\psi r} = \omega_r + \frac{R_r L_m}{L_r} \frac{i_{sq}}{\psi_{rd}}$ is from (20.52).

In Equations (20.49)–(20.53), representing the cage IM mathematical model, the stator current vector components appear but the direct control variables do not. Therefore, control variables must be introduced into the mathematical model (20.49)–(20.53). Subtracting from and adding to Equation (20.49) the component i_{sd} and to Equation (20.50) the component i_{sq}, one obtains

$$\frac{di_{sd}}{d\tau} = -\frac{R_s L_r^2 + R_s L_m^2 + L_r w_\sigma}{L_r w_\sigma} i_{sd} + \frac{R_r L_m}{L_r w_\sigma} \psi_{rd} + \frac{L_r}{w_\sigma} u_{sd} + v_1 \tag{20.56}$$

$$\frac{di_{sq}}{d\tau} = -\frac{R_s L_r^2 + R_r L_m^2 + L_r w_\sigma}{L_r w_\sigma} i_{sq} - \omega_r \frac{L_m}{w_\sigma} \psi_{rd} + \frac{L_r}{w_\sigma} u_{sq} + v_2 \tag{20.57}$$

where

$$v_1 = \omega_{\psi r} i_{sq}^* + i_{sd}^*$$

$$v_2 = i_{sq}^* - \omega_{\psi r} i_{sd}^*$$

the control variables are indicated: i_{sd}^*, i_{sq}^*.

Using the linearization method described in Ref. [10], the following relations are obtained, where m_1 is the subordinated regulator output in the speed control line and m_2 is the subordinated regulator output in the flux control line:

$$v_1 = \frac{1}{T_1} m_1 - \frac{R_r L_m}{L_r w_\sigma} \psi_{rd} - \frac{L_r}{w_\sigma} u_{sd} \tag{20.58}$$

$$v_2 = \frac{1}{T_1} m_2 + \omega_r \frac{L_m}{w_\sigma} \psi_{rd} - \frac{L_r}{w_\sigma} u_{sd} \tag{20.59}$$

where

$$\frac{1}{T_i} = \frac{R_s L_r^2 + R_r L_m^2 + L_r w_\sigma}{L_r w_\sigma}$$

The control variables have the form:

$$i_{sd}^* = \frac{v_1 - \omega_{\psi r} v_2}{\omega_{\psi r}^2 + 1}$$

$$i_{sq}^* = \frac{v_2 + \omega_{\psi r} v_1}{\omega_{\psi r}^2 + 1}$$

The above equations are obtained under the assumption that $C_M \approx 0$.

The transients of state variables for the FOC are shown in Figure 20.16 (motor start-up) and in Figure 20.17 (motor reverse to –0.8 pu).

20.6.2 The Current Multi-Scalar Control

The author in Ref. [10] showed the mathematical model of an induction machine which is called the *multi-scalar model*. The following new state of the variables was chosen:

$$x_{11} = \omega_r \tag{20.60}$$

$$x_{12} = \psi_{r\alpha} i_{s\beta} - \psi_{r\beta} i_{s\alpha} \tag{20.61}$$

$$x_{21} = \psi_{r\alpha}^2 + \psi_{r\beta}^2 \tag{20.62}$$

$$x_{22} = \psi_{r\alpha} i_{s\alpha} + \psi_{r\beta} i_{s\beta} \tag{20.63}$$

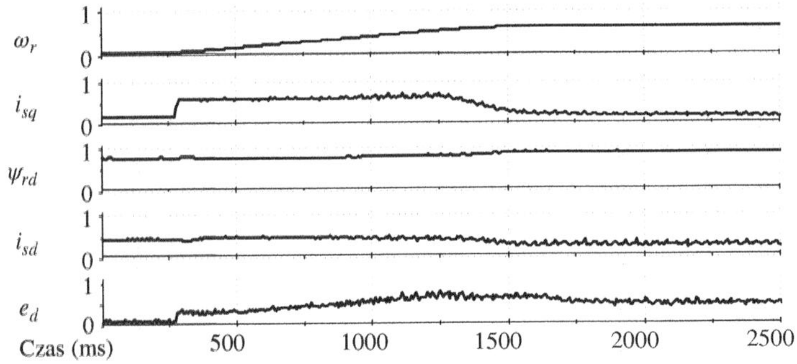

Figure 20.16 The transients of state variables when the motor is starting up, up to 0.7 pu, 5.5 kW induction motor (experimental test)

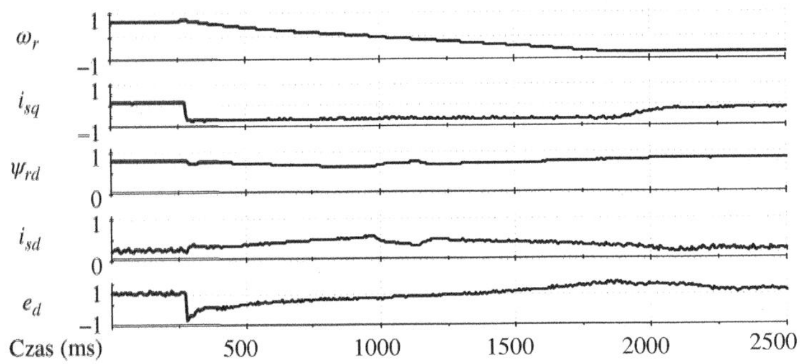

Figure 20.17 The transients of state variables when the motor is reversing, to −0.8 pu, 5.5 kW induction motor (experimental test)

where x_{11} is the rotor angular speed; x_{12} is the electromagnetic torque; x_{21} is the square of rotor flux; x_{22} is the additional variables; and $\alpha\beta$ is the stationary system.

Using Equations (20.31)–(20.35) and differentiating Equations (20.60)–(20.63), the multi-scalar model of an induction machine supplied by a VSI is obtained [10]:

$$\frac{\mathrm{d}x_{11}}{\mathrm{d}\tau} = \frac{L_m}{L_r J}x_{12} - \frac{T_L}{J} \tag{20.64}$$

$$\frac{\mathrm{d}x_{12}}{\mathrm{d}\tau} = -T_i x_{12} - x_{11}\left(x_{22} + \frac{L_m}{w_\sigma}x_{21}\right) + \frac{L_r}{w_\sigma}u_{11} \tag{20.65}$$

$$\frac{\mathrm{d}x_{21}}{\mathrm{d}\tau} = -2\frac{R_r}{L_r}x_{21} + 2\frac{R_r L_m}{L_r}x_{22} \tag{20.66}$$

$$\frac{\mathrm{d}x_{22}}{\mathrm{d}\tau} = -T_i x_{22} + x_{11}x_{22}\frac{R_r L_m}{w_\sigma L_r}x_{21} + \frac{R_r L_m}{L_r}i_s^2 + \frac{L_r}{w_\sigma}u_{22} \tag{20.67}$$

where

$$T_i = \frac{R_s L_r + R_r L_s}{w_\sigma}$$

$$i_s^2 = i_{s\alpha}^2 + i_{s\beta}^2$$

$$u_{11} = \psi_{r\alpha} u_{s\beta} - \psi_{r\beta} u_{s\alpha} \tag{20.68}$$

$$u_{22} = \psi_{r\alpha} u_{s\alpha} + \psi_{r\beta} u_{s\beta} \tag{20.69}$$

The decoupling controls of an induction machine supplied by a VSI are as follows:

$$u_{11} = \frac{w_\sigma}{L_r}\left[x_{11}\left(x_{22} + \frac{L_m}{w_\sigma}x_{21}\right) + \frac{1}{T_i}m_1\right] \tag{20.70}$$

$$u_{22} = \frac{w_\sigma}{L_r}\left[-x_{11}x_{12} - \frac{R_r L_m}{L_r}i_s^2 - \frac{R_r L_m}{L_r w_\sigma}x_{21} + \frac{1}{T_i}m_2\right] \tag{20.71}$$

An induction machine supplied by a CSI may be controlled in the same way as the VSI. The generalized control is provided by an induction machine multi-scalar model formulated for the voltage inverter machine control Equations (20.64)–(20.67). The variables u_{11}, u_{22} can be treated as extended multi-scalar variables. Differentiating Equations (20.68) and (20.69), one obtains

$$\frac{du_{11}}{d\tau} = -\frac{R_r}{L_r}u_{11} - x_{11}u_{22} + \frac{R_r L_m}{L_r}q_s - \frac{1}{C_M}x_{12} + v_{11} \tag{20.72}$$

$$\frac{du_{22}}{d\tau} = -\frac{R_r}{L_r}u_{22} + x_{11}u_{11} + \frac{R_r L_m}{L_r}p_s + \frac{1}{C_M}x_{22} + v_{22} \tag{20.73}$$

where

$$p_s = u_{s\alpha} i_{s\alpha} + u_{s\beta} i_{s\beta} \tag{20.74}$$

$$q_s = i_{s\alpha} i_{s\beta} - i_{s\beta} u_{s\alpha} \tag{20.75}$$

The variables in Equations (20.70) and (20.71) can be treated as input variables to the internal control system and denoted by index "*." The tracking errors of extended variables are

$$e_{u1} = u_{11}^* - u_{11} \tag{20.76}$$

$$e_{u2} = u_{22}^* - u_{22} \tag{20.77}$$

Nonlinear decoupling controls by feedback linearization are obtained from Equations (20.72) and (20.73):

$$v_{11} = -\frac{R_r}{L_r}v_{p1} - \frac{R_r L_m}{L_r}q_s + \frac{1}{C_M}x_{12} + x_{11}u_{22} \tag{20.78}$$

$$v_{22} = -\frac{R_r}{L_r}v_{p2} - \frac{R_r L_m}{L_r}p_s + \frac{1}{C_M}x_{22} - x_{11}u_{11} \tag{20.79}$$

The control variables of an induction machine supplied by a CSI are as follows:

$$i_{f\alpha} = -C_M \frac{v_{11}\Psi_{r\beta} - v_{22}\Psi_{r\alpha}}{x_{21}} \tag{20.80}$$

$$i_{f\beta} = C_M \frac{v_{11}\Psi_{r\alpha} + v_{22}\Psi_{r\beta}}{x_{21}} \tag{20.81}$$

The control system of extended multi-scalar variables with PI controllers is shown in Figure 20.18.

The transients of state variables for current multi-scalar control are shown in Figure 20.19. Two decoupled subsystems are obtained. The current i_D is proportional to x_{12} because the additional PI controller of i_D is applied.

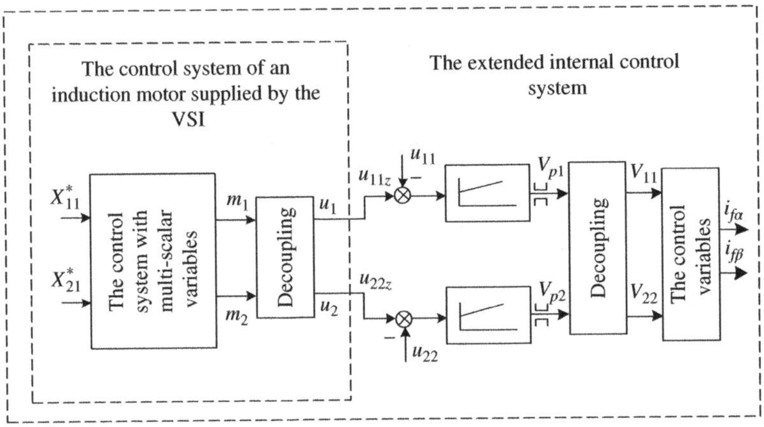

Figure 20.18 The FOC control system structure with PI controllers and decoupling variables

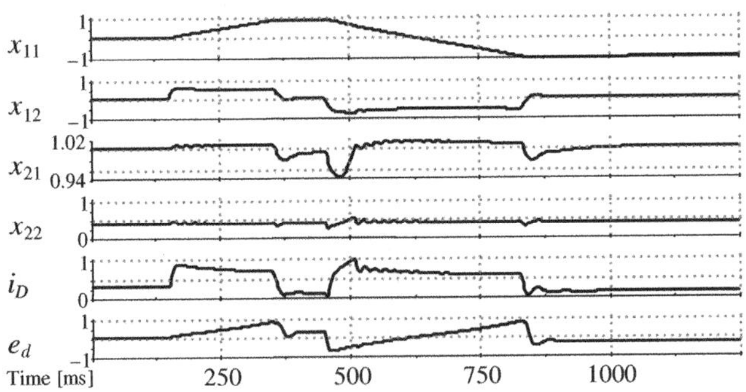

Figure 20.19 The transients of the variables for the current multi-scalar control system when the motor is starting up and reversing 5.5 kW induction machine (experimental test)

20.6.3 The Voltage Multi-Scalar Control

Another control method of a CSI-fed IM was called *voltage control*. The DC-link circuit voltage e_d and angular frequency ω_{if} of the current vector are the control variables. Proposed control strategy is based on nonlinear multi-scalar control [1–4, 10]. To achieve independent control of the flux and rotor speed, a new nonlinear control scheme is proposed. In this control method, the inverter output currents are not controlled variables. The voltage in DC-link e_d and pulsation of output current vector ω_{if} are the controlled variables, which can be obtained by nonlinear transformations and are proposed by the authors in Refs. [1–4]. The multi-scalar model is called the *extended model* because the mathematical model contained the DC-link current and the output capacitor equations. This full mathematical model of an induction machine with the CSI is used to derive a new multi-scalar model. In the proposed method, the output current vector coefficients are not controlled variables. The output current vector and the flux vector are used to achieve the multi-scalar variables and the multi-scalar model. The control system structure may be supported on PI controllers and nonlinear decouplings or different controllers, for example, sliding-mode controllers, the backstepping control method, or fuzzy neural controllers. The author in Ref. [4] revealed proof of stability of the simplified multi-scalar control while the parameters of the CSI are optimally selected. When the capacitance C_M is neglected, the stator current vector $\vec{i}_s$ is about 5% out of phase to $\vec{i}_f$, while the nominal torque is set. Then, the control variables and decoupling are not precisely obtained. The error is smaller than 2% because the PI controllers improved it.

In order to compensate for these errors, the capacity C_M to mathematical model is applied.

The variables in the rotating (xy) frame of reference system are presented in Figure 20.20.

For xy system in stationary state, Equations (20.3) and (20.4) lead to dependencies:

$$i_{sx} = i_{fx} + \omega_{if} C_M u_{sy} - C_M \frac{du_{sx}}{d\tau} \tag{20.82}$$

$$i_{sy} = -\omega_{if} C_M u_{sx} - C_M \frac{du_{sy}}{d\tau} \tag{20.83}$$

In xy – system $(\frac{du_{sx}}{d\tau}, \frac{du_{sy}}{d\tau} \approx 0)$ or in the stationary state:

$$i_{vx} = i_{sx} \approx i_{fx} + \omega_{if} C_M u_{sy} \tag{20.84}$$

$$i_{vy} = i_{sy} \approx -\omega_{if} C_M u_{sx} \tag{20.85}$$

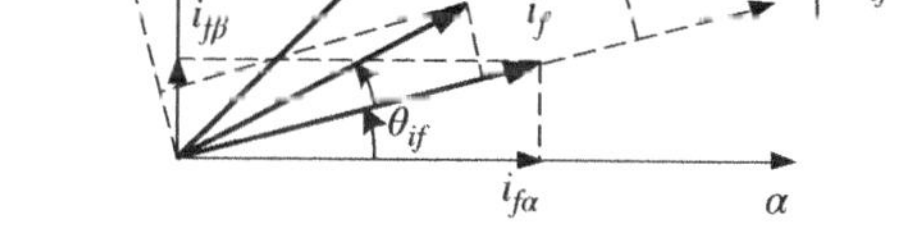

Figure 20.20 The variables in the rotating frame of reference

The new mathematical model of the drive system is obtained from Equations (20.84)–(20.85), through differentiation and using (20.3)–(20.4) and (20.8) in the *xy* coordinate system:

$$\frac{\mathrm{d}i_{vx}}{\mathrm{d}\tau} = -\frac{R_d}{L}i_{fx} + \frac{1}{L}e_d - \frac{1}{L}u_{sx} - \omega_{if}C_M i_{vy} - \omega_{if}^2 C_M^2 u_{sx} \tag{20.86}$$

$$\frac{\mathrm{d}i_{vy}}{\mathrm{d}\tau} = -\omega_{if}C_M i_{fx} + \omega_{if}C_M i_{vx} - \omega_{if}^2 C_M^2 u_{sy} \tag{20.87}$$

$$\frac{\mathrm{d}\psi_{rx}}{\mathrm{d}\tau} = -\frac{R_r}{L_r}\psi_{rx} + (\omega_{if} - \omega_r)\psi_{ry} + \frac{R_r L_m}{L_r}i_{sx} \tag{20.88}$$

$$\frac{\mathrm{d}\psi_{ry}}{\mathrm{d}\tau} = -\frac{R_r}{L_r}\psi_{ry} - (\omega_{if} - \omega_r)\psi_{rx} + \frac{R_r L_m}{L_r}i_{sy} \tag{20.89}$$

$$\frac{\mathrm{d}u_{sx}}{\mathrm{d}\tau} = \frac{1}{C_M}(i_{fx} - i_{sx}) + \omega_{if}u_{sy} \tag{20.90}$$

$$\frac{\mathrm{d}u_{sy}}{\mathrm{d}\tau} = -\frac{1}{C_M}i_{sy} - \omega_{if}u_{sx} \tag{20.91}$$

Substituting Equations (20.84)–(20.85) for the multi-scalar variables (20.60)–(20.63) in the *xy* system, one obtains:

$$x_{11} = \omega_r \tag{20.92}$$

$$x_{12} = \psi_{rx}i_{sy} - \psi_{ry}i_{sx} \approx \psi_{rx}i_{vy} - \psi_{ry}i_{vx} = -i_{fx}\psi_{ry} - \omega_{if}C_M x_{32} \tag{20.93}$$

$$x_{21} = \psi_{rx}^2 + \psi_{ry}^2 \tag{20.94}$$

$$x_{22} = i_{sx}\psi_{rx} + i_{sy}\psi_{rx} \approx i_{vx}\psi_{rx} + i_{vy}\psi_{rx} = i_{fx}\psi_{rx} + \omega_{if}C_M x_{31} \tag{20.95}$$

and

$$d_{31} = \psi_{rx}u_{sy} - \psi_{ry}u_{sx} \tag{20.96}$$

$$d_{32} = \psi_{rx}u_{sx} + \psi_{ry}u_{sy} \tag{20.97}$$

For the closed-loop system and $(\omega_{if} - \omega_r) \approx 0$: $d_{31} \approx \omega_r$ and $d_{32} \approx 0$, therefore do not affect on the control system stability.

The multi-scalar model for the multi-scalar variables has the form:

$$\frac{\mathrm{d}x_{11}}{\mathrm{d}\tau} = \frac{L_m}{JL_r}x_{12} - \frac{1}{J}T_L \tag{20.98}$$

$$\frac{\mathrm{d}x_{12}}{\mathrm{d}\tau} = -\left(\frac{R_d}{L} + \frac{R_r}{L_r}\right)x_{12} + \frac{1}{L}u_{sx}\psi_{ry} - x_{11}x_{22} - \frac{R_r L_m}{L_r}i_{sy}i_D + v_1 \tag{20.99}$$

$$\frac{\mathrm{d}x_{21}}{\mathrm{d}\tau} = -\frac{R_r}{L_r}x_{21} + R_r\frac{L_m}{L_r}x_{22} \tag{20.100}$$

$$\frac{\mathrm{d}x_{22}}{\mathrm{d}\tau} = -\left(\frac{R_d}{L} + \frac{R_r}{L_r}\right)x_{22} - \frac{1}{L}u_{sx}\psi_{rx} + \frac{R_r L_m}{L_r \;{}_{sx}}i_D + x_{11}x_{12} + v_2 \tag{20.101}$$

where

$$v_1 = -\frac{1}{L} e_d \psi_{ry} + \omega_{if} A_1 \tag{20.102}$$

$$v_2 = \frac{1}{L} e_d \psi_{rx} + \omega_{if} A_2$$

$$A_1 = x_{22} - \frac{R_d}{L} C_M d_{32} - C_M \frac{R_r L_m}{L_r} p_s \tag{20.103}$$

$$A_2 = -x_{12} + \frac{R_d}{L} C_M d_{31} + C_M \frac{R_r L_m}{L_r} q_s$$

The compensation of nonlinearities in differential equations leads to the following expressions for the control variables v_1 and v_2 appearing in differential Equations (20.99) and (20.101):

$$v_1 = \frac{1}{T_i} m_1 - \frac{1}{L} u_{sx} \psi_{ry} + x_{11} x_{22} + \frac{R_r L_m}{L_r} i_{sy} i_D \tag{20.104}$$

$$v_2 = \frac{1}{T_i} m_2 + \frac{1}{L} u_{sx} \psi_{ry} - x_{11} x_{22} - \frac{R_r L_m}{L_r} i_{sx} i_D \tag{20.105}$$

and the control variables:

$$e_d = L \frac{v_2 A_1 - v_1 A_2}{\psi_{rx} A_1 + \psi_{ry} A_2} \tag{20.106}$$

$$\omega_{if} = \frac{v_1 \psi_{rx} + v_2 \psi_{ry}}{\psi_{rx} A_1 + \psi_{ry} A_2} \tag{20.107}$$

where

$$\frac{1}{T_i} = \frac{R_d}{L} + \frac{R_r}{L_r}$$

The decoupled two subsystems are obtained:

- the electromechanical subsystem

$$\frac{dx_{11}}{d\tau} = \frac{L_m}{J L_r} x_{12} - \frac{1}{J} T_L \tag{20.108}$$

$$\frac{dx_{12}}{d\tau} = \frac{1}{T_i}(-x_{12} + m_1) \tag{20.109}$$

- the electromagnetic subsystem

$$\frac{dx_{21}}{d\tau} = -\frac{R_r}{L_r} x_{21} + \frac{R_r L_m}{L_r} x_{22} \tag{20.110}$$

$$\frac{dx_{22}}{d\tau} = \frac{1}{T_i}(-x_{22} + m_2) \tag{20.111}$$

The voltage multi-scalar control system is shown in Figure 20.21.

The transients of the multi-scalar variables and i_D when the IM is starting up to 1.0 pu are presented in Figure 20.22. In Figure 20.23, the motor reverse is shown from 0.7 to −0.7 pu. In Figure 20.24, the load torque has been changed after 900 ms.

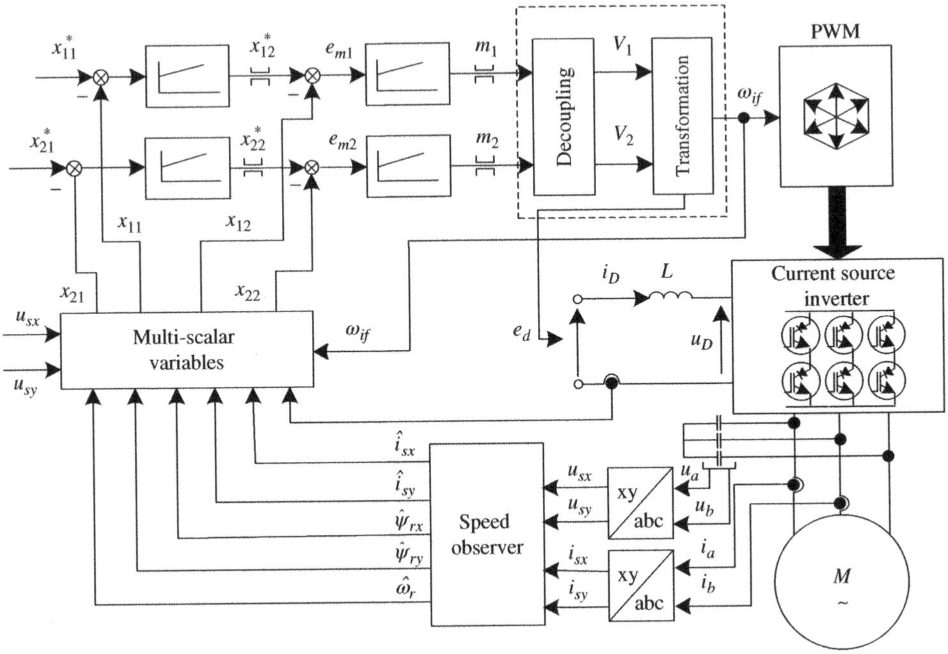

Figure 20.21 The voltage multi-scalar control system structure

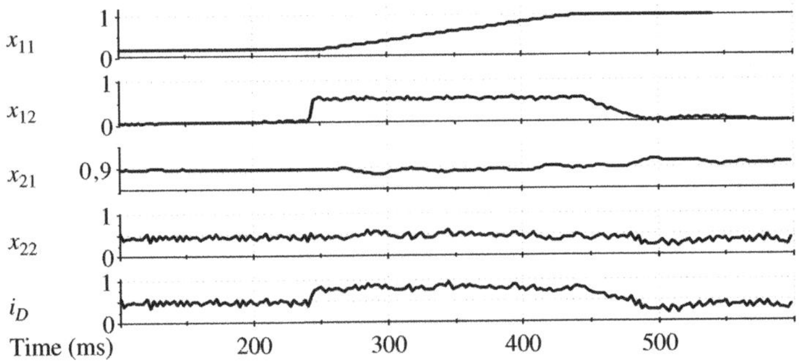

Figure 20.22 The multi-scalar variables and i_D transients when the motor is starting up, 5.5 kW induction machine (experimental test)

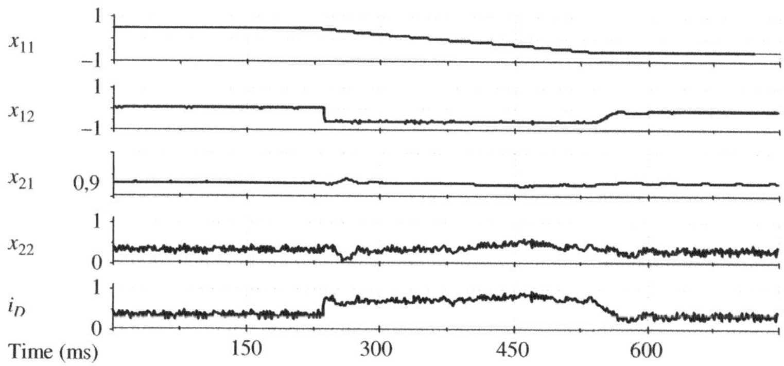

Figure 20.23 The multi-scalar variables and i_D transients when the motor is reversing, 5.5 kW induction machine (experimental test)

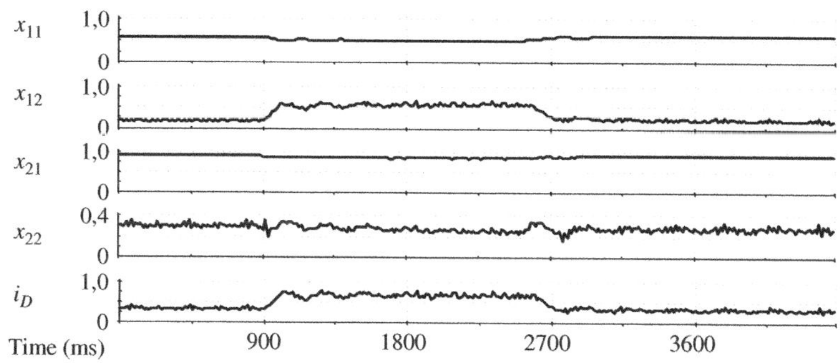

Figure 20.24 The multi-scalar variables and i_D transients when the load torque is being turned on, 5.5 kW motor (experimental test)

20.7 The Current and Voltage Control of Permanent Magnet Synchronous Motor

20.7.1 The Voltage Multi-scalar Control of a PMSM

The control method strategies for the PMSM were presented in Refs. [17–19]. These strategies are based on a constant power angle (the stator current component $\delta - \pi/2$). The same strategy can be implemented in the PMSM control system supplied by the CSI (to simplify the development of the controller). This assumption leads to nonoptimal results for the motor at all operating conditions [22]. In order to obtain the optimal working of the PMSM, the maximum torque per ampere MTPA of the stator current injection was proposed [22].

The voltage control of a PMSM supplied by the CSI can be similar in approach to an IM control system. The mathematical model of the drive system in the xy coordinate system is determined:

$$\frac{di_x}{d\tau} = -\frac{R_s}{L_s} i_x + \frac{1}{L_s} \psi_{fy} \omega_r + \frac{1}{L_s} u_{sx} + \omega_{if} i_y \tag{20.112}$$

$$\frac{di_y}{d\tau} = -\frac{R_s}{L_s} i_y - \frac{1}{L_s} \omega_r \psi_{fx} + \frac{1}{L_s} u_{sy} - \omega_{if} i_x \tag{20.113}$$

$$\frac{d\psi_x}{d\tau} = -R_s i_x + u_{sx} + \omega_{if} \psi_y \tag{20.114}$$

$$\frac{d\psi_y}{d\tau} = -R_s i_y + u_{sy} - \omega_{if} \psi_x \tag{20.115}$$

and Equations (20.8), (20.90) and (20.91), ψ_x, ψ_y are the stator flux vector coefficients.

Taking into account Equations (20.82), (20.83) and Equations (20.60), (20.61) and (20.63) in the (xy) coordinate system, the multi-scalar variables have the form [23]:

$$x_{11} = \omega_r \tag{20.116}$$

$$x_{12} = -i_{fx} \psi_y - C_M \omega_{if} g_{32} \tag{20.117}$$

$$x_{22} = \psi_{fx} i_{fx} + L_s (i_x^2 + i_y^2) + C_M \omega_{if} g_{31} \tag{20.118}$$

where are introduced:

$$g_{31} = \psi_{fx} u_{sy} - \psi_{fy} u_{sx} \tag{20.119}$$

$$g_{32} = \psi_{fx} u_{sx} + \psi_{fy} u_{sy} \tag{20.120}$$

For the closed-loop system and $(\omega_{if} - \omega_r) \approx 0$: $g_{31} \approx \omega_r$ and $g_{32} \approx 0$, therefore do not affect on the control system stability.

Using Equations (20.116)–(20.118) and (20.36)–(20.38), the multi-scalar model has the form:

$$\frac{dx_{11}}{d\tau} = \frac{1}{J} x_{12} - \frac{1}{J} T_L \tag{20.121}$$

$$\frac{dx_{12}}{d\tau} = -\frac{R_d}{L} x_{12} + \frac{1}{L} u_{sx} \psi_{fy} + L_s x_{11} i_s^2 - x_{11} x_{22} + v_1 \tag{20.122}$$

$$\frac{dx_{22}}{d\tau} = -\frac{R_d}{L} x_{22} + \left(\frac{R_d L_s}{L} - 2R_s\right) i_s^2 + 2p_s - \frac{1}{L} u_{sx} \psi_{fx} + v_2 \tag{20.123}$$

where

$$v_1 = \omega_{if} \left(x_{22} - \frac{R_d C_M}{L} g_{32} - L_s i_s^2 - \frac{1}{L} e_d \psi_{fy} \right)$$

$$v_2 = \frac{1}{L} e_d \psi_{fx} + \omega_{if} \left(\frac{R_d C_M}{L} g_{31} - x_{12} \right)$$

The compensation of nonlinearities in differential equations leads to the following expressions for the control variables, v_1 and v_2, appearing in the differential Equations (20.122) and (20.123):

$$v_1 = \frac{1}{T_i} m_1 - \frac{1}{L} u_{sx} \psi_{fy} - L_s x_{11} i_s^2 + x_{11} x_{22} \tag{20.124}$$

$$v_2 = \frac{1}{T_i} m_2 - \left(\frac{R_d L_s}{L} - 2R_s\right) i_s^2 - 2p_s + \frac{1}{L} u_{sx} \psi_{fx} \tag{20.125}$$

The control variables are

$$\omega_{if} = \frac{\psi_{fx} v_1 + \psi_{fy} v_2}{A_1 \psi_{fx} + A_2 \psi_{fy}} \tag{20.126}$$

$$e_d = L \frac{A_1 v_2 - A_2 v_1}{A_1 \psi_{fx} + A_2 \psi_{fy}} \tag{20.127}$$

where

$$A_1 = C_M R_s - \frac{R_d}{L} C_M g_{32} - C_M u_s^2 + x_{22}$$

$$A_2 = \frac{R_d}{L} C_M g_{31} - x_{12} - C_M R_s q_s$$

$$u_s^2 = u_{sx}^2 + u_{sy}^2$$

$$i_s^2 = i_x^2 + i_y^2$$

$$\frac{1}{T_i} = \frac{R_d}{L}$$

p_s, q_s are the active and reactive stator power Equations (20.74) and (20.75).

The xy coordinate system is shown in Figure 20.25. The control scheme of a PMSM is shown in Figure 20.26.

The transients of the multi-scalar variables and e_d, i_D when the PMSM is reversing to −1.0 pu are presented in Figure 20.27. The $x_{22}^* \approx 0$. The i_D and u_a, the stator voltage transients in stationary state, are presented in Figure 20.28. The i_D oscillations appear because the inductor L was very small. These DC-link current oscillations had no influence on the transient stator current and voltage.

20.7.2 The Current Control of an Interior Permanent Magnet Motor

The control systems based on the stator current components in dq, connected with rotor, system were called the current control of the PMSM/IPMSM supplied by the CSI. These control systems were the focus in Refs. [18, 19, 22]. In Refs. [21, 24], the authors used the output capacitor model to obtain

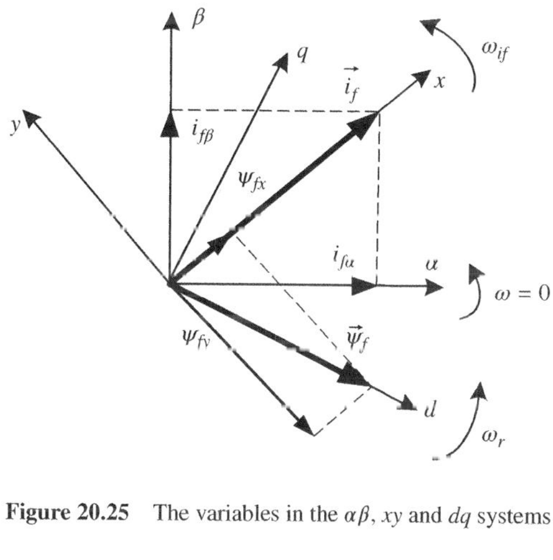

Figure 20.25 The variables in the $\alpha\beta$, xy and dq systems

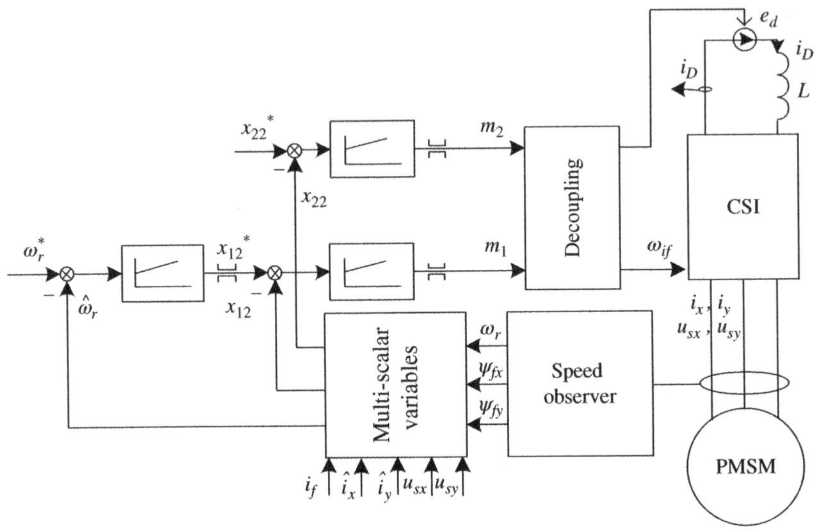

Figure 20.26 The PMSM multi-scalar control system

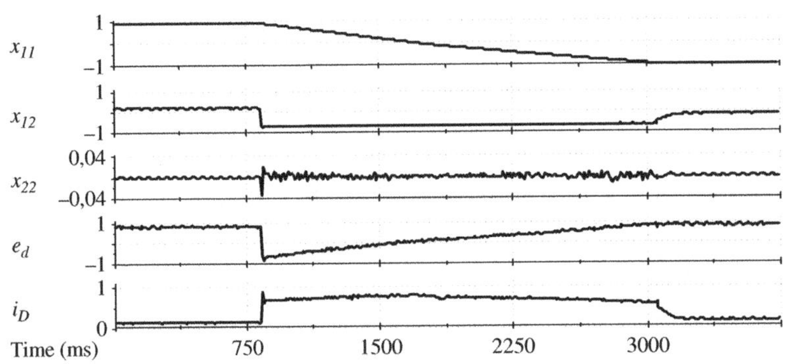

Figure 20.27 The transients when the PMSM is reversing from 1.0 to −1.0 pu (experimental test)

the extended current control variables. The output capacitor current variable control is more precisely determined than not taken into consideration. The current control allows control of the supply's reactive power and a unit power factor [21, 24].

The simplified control of a PMSM generates the output current vector, which is shifted in relation to the permanent flux vector at about $\gamma = \pi/2$. Such a control system is shown in Figure 20.29.

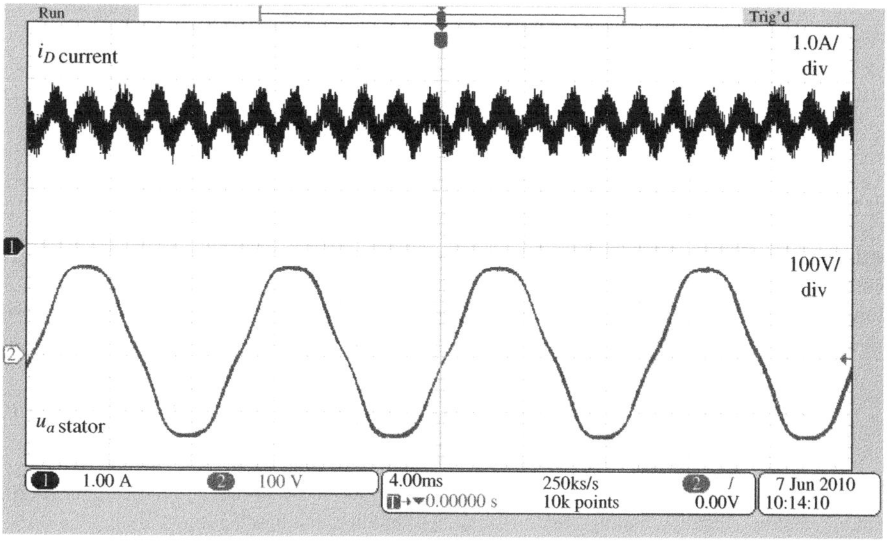

Figure 20.28 The dc-link current i_D and stator voltage u_a transients (experimental test)

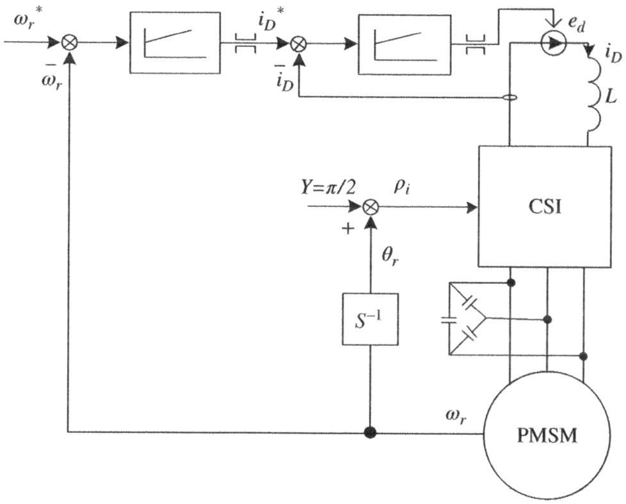

Figure 20.29 The current control of a PMSM in which the output current vector is shifted in relation to the permanent flux vector

For a short-impulse time T_{imp}, the capacitor model dependencies (20.3) and (20.4) in the dq coordinate system can be approximated by the difference equations. The voltage vector coefficient has the form

$$u_d = -\frac{\omega_r T_{\text{imp}} u_{\text{qprev}}}{a_1} + \frac{u_{\text{dprev}}}{a_1} - \frac{\omega_r T_{\text{imp}}^2}{C_M a_1}(u_{fq} - i_q) + \frac{T_{\text{imp}}}{C_M a_1}(i_{fd} - i_d) \tag{20.128}$$

$$u_q = \frac{\omega_r T_{\text{imp}} u_{\text{dprev}}}{a_1} + \frac{u_{\text{qprev}}}{a_1} + \frac{\omega_r T_{\text{imp}}^2}{C_M a_1}(u_{fd} - i_d) + \frac{T_{\text{imp}}}{C_M a_1}(i_{fq} - i_q) \tag{20.129}$$

where the index prev means the previous value of the variables and $a_1 = \omega_r^2 T_{\text{imp}}^2 + 1$, i_{fd}, i_{fq} are the output vector components.

Using Equations (20.128) and (20.129) in Equations (20.45) and (20.46), the mathematical model of the drive system takes the form:

$$\frac{di_d}{d\tau} = -\frac{1}{T_{i1}} i_d + \frac{L_q}{L_d}\omega_r i_q - \frac{\omega_r T_{\text{imp}} u_{\text{qprev}}}{L_d a_1} + \frac{\omega_r T_{\text{imp}}^2}{C_M L_d a_1} i_q + \frac{u_{\text{dprev}}}{L_d a_1} + v_1 \tag{20.130}$$

$$\frac{di_q}{d\tau} = -\frac{1}{T_{i2}} i_q - \frac{L_d}{L_q}\omega_r i_d - \frac{1}{L_q}\omega_r \psi_f + \frac{\omega_r T_{\text{imp}} u_{\text{dprev}}}{L_q a_1} - \frac{\omega_r T_{\text{imp}}^2}{C_M L_q a_1} i_d + \frac{u_{\text{qprev}}}{L_q a_1} + v_2 \tag{20.131}$$

where

$$\frac{1}{T_{i1}} = \frac{R_s}{L_d} + \frac{T_{\text{imp}}}{C_M L_d a_1}$$

$$\frac{1}{T_{i2}} = \frac{R_s}{L_q} + \frac{T_{\text{imp}}}{C_M L_d a_1}$$

$$v_1 = \frac{T_{\text{imp}}}{C_M L_d a_1} i_{fd} - \frac{\omega_r T_{\text{imp}}^2}{C_M L_d a_1} i_{fd}$$

$$v_2 = \frac{\omega_r T_{\text{imp}}^2}{C_M L_q a_1} i_{fd} + \frac{T_{\text{imp}}}{C_M L_q a_1} i_{fq}$$

The compensation of nonlinearities in the differential equations leads to the following expressions for control variables v_1 and v_2 appearing in differential Equations (20.130) and (20.131):

$$v_1 = \frac{1}{T_i} m_1 - \frac{L_q}{L_d}\omega_r i_q + \frac{\omega_r T_{\text{imp}} u_{\text{qprev}}}{L_d a_1} - \frac{\omega_r T_{\text{imp}}^2}{C_M L_d a_1} i_q - \frac{u_{\text{dprev}}}{L_d a_1} \tag{20.132}$$

$$v_2 = \frac{1}{T_i} m_2 + \frac{L_d}{L_q}\omega_r i_d + \frac{1}{L_q}\omega_r \psi_f - \frac{\omega_r T_{\text{imp}} u_{\text{dprev}}}{L_q a_1} + \frac{\omega_r T_{\text{imp}}^2}{C_M L_q a_1} i_d - \frac{u_{\text{qprev}}}{L_q a_1} \tag{20.133}$$

The control variables of the IPMSM have the following form:

$$i_{fd} = C_M a_1 \frac{L_q T_{\text{imp}} \omega_r v_2 + v_1 L_d}{T_{\text{imp}} a_1} \tag{20.134}$$

$$i_{fq} = -C_M a_1 \frac{\omega_r T_{\text{imp}} L_d v_1 - v_2 L_q}{T_{\text{imp}} a_1} \tag{20.135}$$

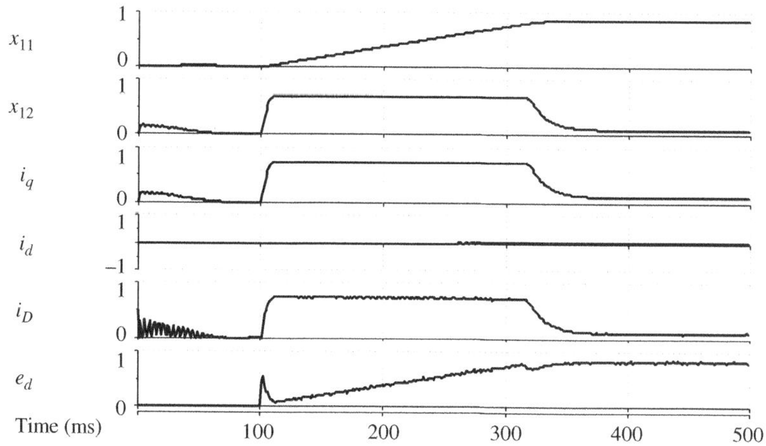

Figure 20.30 The transients for current control of an IPMSM when the motor is starting up, up to 1.0 pu (simulation test)

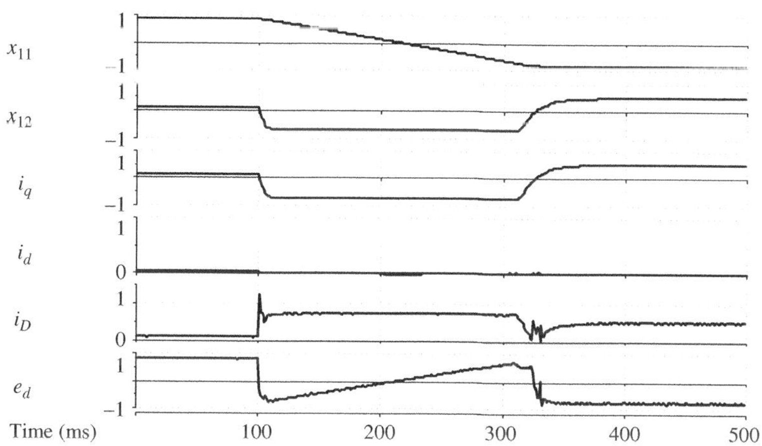

Figure 20.31 The transients for current control of an IPMSM when the motor is reversing to −1.0 pu (simulation test)

In Figures 20.30 and 20.31, the transients for the current control of the IPMSM are presented. The stator current coefficient i_d is stabilized near zero $i_d \approx 0$. After 100 ms the rotor speed is changed to 0.8 pu (Figure 20.30) and from 0.9 to −0.9 pu (Figure 20.31).

20.8 The Control System of a Doubly Fed Motor Supplied by a CSC

The ring shaped motor can be supplied on the rotor side by the CSC, whereas on the stator side by the supply network. Such the drive systems were used in the 19th century and called doubly fed induction motors. They were thyristor generator systems with the inductor in the DC link with high values of inductance and commutation capacitors. Currently, it is possible to use the RBIGBT transistors or SiCs in the CSC. In the modern CSC, experienced control systems can be used, for example, the backstepping control, the sliding-mode control, or the multi-scalar control, with PI controllers of the stator active and reactive power. The control system of a doubly fed machine can be the current and the voltage (similar to the squire-cage induction machine). In this section, the voltage control will only be presented.

If the rotor is supplied by the CSC, then the following equations can be written as

$$\frac{di_{rfx}}{d\tau} = \frac{1}{L}(e_d - R_d i_{rfx} - u_{rx}) \tag{20.136}$$

$$\frac{du_{rx}}{d\tau} = \frac{1}{C_M}(i_{rfx} - i_{rx}) + \omega_{irf} u_{ry} \tag{20.137}$$

$$\frac{du_{ry}}{d\tau} = -\frac{1}{C_M} i_{ry} - \omega_{irf} u_{rx} \tag{20.138}$$

where $i_{rfx} = Ki_D \approx i_D, K = 1$

Vectors in $\alpha\beta_{\text{rotor}}$ and $\alpha\beta_{\text{stator}}$ are shown in Figure 20.32.

When the xy coordinate system is chosen, the ω_0 in Equations (20.31)–(20.34) is equal to $\omega_0 = \omega_{irf} + \omega_r$. The mathematical model of a doubly fed motor for $(\vec{i}_r, \vec{\psi}_s)$ in the xy coordinate system is as follows [20]:

$$\frac{di_{rx}}{d\tau} = -\frac{L_s^2 R_r + L_m^2 R_s}{L_s w_\sigma} i_{rx} + \frac{R_s L_m}{L_s w_\sigma} \psi_{sx} + \omega_{irf} i_{ry} - \frac{L_m}{w_\sigma} \omega_r \psi_{sy} + \frac{L_s}{w_\sigma} u_{rx} - \frac{L_m}{w_\sigma} u_{sx} \tag{20.139}$$

$$\frac{di_{ry}}{d\tau} = -\frac{L_s^2 R_r + L_m^2 R_s}{L_s w_\sigma} i_{ry} + \frac{R_s L_m}{L_s w_\sigma} \psi_{sy} - \omega_{irf} i_{rx} - \frac{L_m}{w_\sigma} \omega_r \psi_{sx} + \frac{L_s}{w_\sigma} u_{ry} - \frac{L_m}{w_\sigma} u_{sy} \tag{20.140}$$

$$\frac{d\psi_{sx}}{d\tau} = -\frac{R_s}{L_s} \psi_{sx} + \frac{R_s L_m}{L_s} i_{rx} + (\omega_{irf} + \omega_r)\psi_{sy} + u_{sx} \tag{20.141}$$

$$\frac{d\psi_{sy}}{d\tau} = -\frac{R_s}{L_s} \psi_{sy} - (\omega_{irf} + \omega_r)\psi_{sx} + u_{sy} \tag{20.142}$$

$$\frac{d\omega_r}{d\tau} = \frac{L_m}{JL_s}(\psi_{sx} i_{ry} - \psi_{sy} i_{rx}) - \frac{1}{J} T_L \tag{20.143}$$

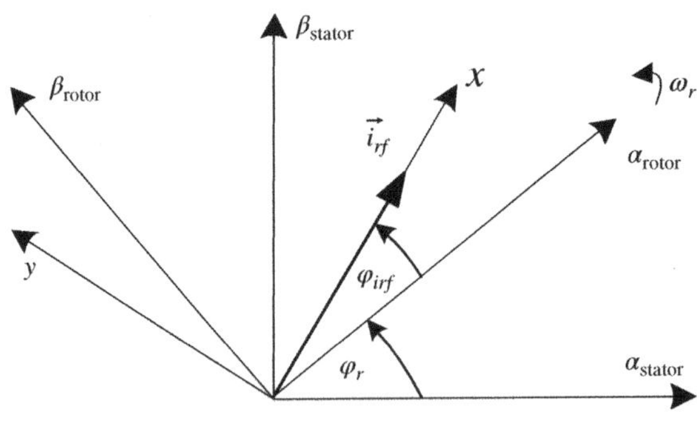

Figure 20.32 The vectors in $\alpha\beta_{\text{rotor}}$ (coordinate system connected with the rotor), $\alpha\beta_{\text{stator}}$ (coordinate system connected with the stator) and xy (coordinate system connected with output current vector $\vec{i}_{rf}$)

where i_{rx}, i_{ry} are the rotor current vector components; ψ_{sx}, ψ_{sy} are the stator flux vector components;and the remaining variables are defined in Section 20.5.

The multi-scalar variables in the $\alpha\beta$ system were chosen [20]:

$$z_{11} = \omega_r \tag{20.144}$$

$$z_{12} = \psi_{s\alpha} i_{r\beta} - \psi_{s\beta} i_{r\alpha} \tag{20.145}$$

$$z_{21} = \psi_{s\alpha}^2 + \psi_{s\beta}^2 \tag{20.146}$$

$$z_{22} = \psi_{s\alpha} i_{r\alpha} + \psi_{s\beta} i_{r\beta} \tag{20.147}$$

where x_{11} is the rotor angular speed; x_{12} is the electromagnetic torque; x_{21} is the square of rotor flux; x_{22} is the additional variables; and $\alpha\beta$ is the stationary system.

Taking into account the xy coordinate system and (20.136)–(20.147) under assumption ($\frac{du_{rx}}{d\tau}$, $\frac{du_{ry}}{d\tau} \approx 0$), the multi-scalar variables (20.144), (20.146) have the form:

$$z_{12} = -i_{rfx}\psi_{sy} - C_M \omega_{ifr} e_{32} \tag{20.148}$$

$$z_{22} = i_{rfx}\psi_{sx} - C_M \omega_{ifr} e_{31} \tag{20.149}$$

where

ω_{irf} is the output current vector angular speed, $i_{rfx} = K i_D \approx i_D$, $(K = 1)$ and e_{31}, e_{32} are defined:

$$e_{31} = u_{rx}\psi_{sy} - u_{ry}\psi_{sx} \tag{20.150}$$

$$e_{32} = u_{rx}\psi_{sx} + u_{ry}\psi_{sy} \tag{20.151}$$

In (20.148)–(20.149) $C_M \omega_{ifr} e_{32}$ and $C_M \omega_{ifr} e_{31}$ are near to zero therefore do not effect on closed-loop system stability.

The multi-scalar model for (20.150)–(20.151) is as follows:

$$\frac{dz_{12}}{d\tau} = -\left(\frac{R_s}{L_s} + \frac{R_d}{L}\right) z_{12} + \frac{1}{L} u_{rx}\psi_{sy} - \frac{R_s L_m}{L_s} i_{ry} i_D + z_{11} z_{22} - u_{sy} i_D + v_1 \tag{20.162}$$

$$\frac{dz_{22}}{d\tau} = -\left(\frac{R_s}{L_s} + \frac{R_d}{L}\right) z_{22} - \frac{1}{L} u_{rx}\psi_{sx} + \frac{R_s L_m}{L_s} i_{rx} i_D - z_{11} z_{12} + u_{sx} i_D + v_2 \tag{20.163}$$

where

$$v_1 = -\frac{1}{L} e_d \psi_{sy} + \omega_{irf} A_1 \tag{20.164}$$

$$v_2 = \frac{1}{L} e_d \psi_{sx} + \omega_{irf} A_2 \tag{20.165}$$

$$A_1 = \left(i_{rx}\psi_{sx} + i_{ry}\psi_{sy} - \frac{R_s L_m}{L_s} C_M \left(i_{rx} u_{rx} + i_{ry} u_{ry}\right) - C_M (u_{sx} u_{rx} + u_{sy} u_{ry}) - \frac{R_d}{L_d} C_M e_{32}\right) \tag{20.166}$$

$$A_2 = \left(\left(i_{rx}\psi_{sy} - i_{ry}\psi_{sx}\right) - \frac{R_s L_m}{L_s} C_M (i_{ry} u_{rx} - i_{rx} u_{ry}) - C_M (u_{sy} u_{rx} - u_{sx} u_{ry}) - \frac{R_d}{L_d} C_M e_{31}\right) \tag{20.167}$$

The control variables of a doubly fed machine supplied by the CSI are

$$e_d = L \frac{v_2 A_1 - v_1 A_2}{\psi_{sx} A_1 + \psi_{sy} A_2} \tag{20.168}$$

$$\omega_{ifr} = \frac{v_1 \psi_{sx} + v_2 \psi_{sy}}{\psi_{sx} A_1 + \psi_{sy} A_2} \tag{20.169}$$

In stationary state, the p_s and q_s can be expressed as [20]

$$p_s = \frac{1}{L_s}u_{sf2} + \frac{L_m}{L_s}\frac{u_{sf1}z_{12} - u_{sf2}z_{22}}{z_{21}} \tag{20.170}$$

$$q_s = -\frac{1}{L_s}u_{sf1} + \frac{L_m}{L_s}\frac{u_{sf1}z_{22} + u_{sf2}z_{12}}{z_{21}} \tag{20.171}$$

where

$$u_{sf1} = u_{s\alpha}\Psi_{s\beta} - u_{s\beta}\Psi_{s\alpha} \approx 1 \tag{20.172}$$

$$u_{sf2} = u_{s\alpha}\Psi_{s\alpha} + u_{s\beta}\Psi_{s\beta} \approx 0 \tag{20.173}$$

Equations (20.170) and (20.171) are more complicated for direct using in the control system. To obtain more convenient relationships, the machine stationary state has been taken into account in which $u_{sf1} \approx 1$, $u_{sf2} \approx 0$. In stationary state, the p_s and q_s can be expressed [20, 25]:

$$p_s \sim \frac{L_m}{L_s}z_{12} \tag{20.174}$$

$$q_s \sim -\frac{1}{L_s} + \frac{L_m}{L_s}z_{22} \tag{20.175}$$

All the variables must be determined in the same reference system.

Figure 20.33 presents the sensorless control system of a doubly fed machine. Two controllers of active and reactive power and two controllers of z_{12} and z_{22} are introduced in control system.

Figures 20.34 and 20.35 present the transients when the reference active and reactive power is changing. The set value for active power has been changed from −0.8 to 0.8 pu. When $p_s, q_s > 0$ then the active or reactive power is generated to the supply. When $p_s, q_s < 0$ then the machine is fed from the supply.

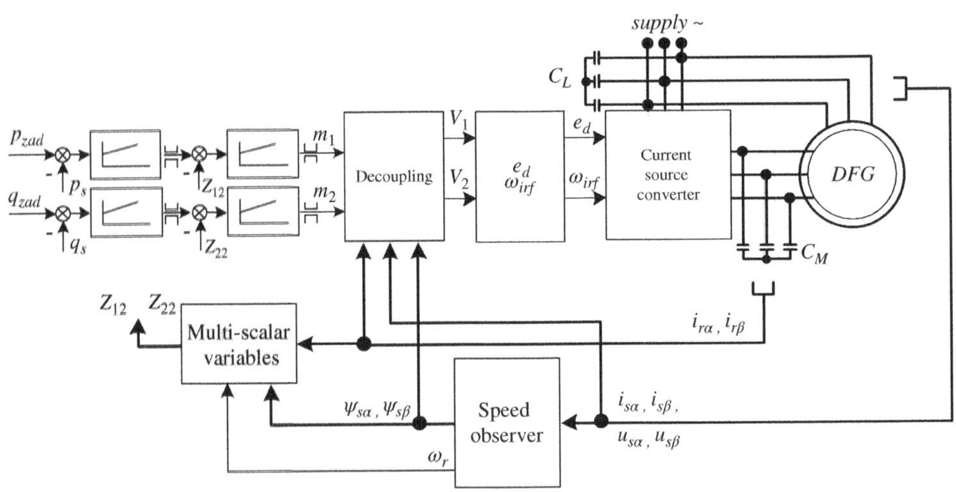

Figure 20.33 The sensorless multi-scalar control system of a doubly fed machine supplied by a CSC

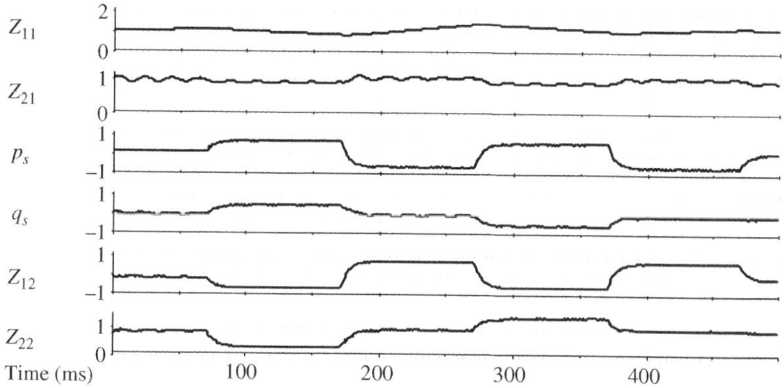

Figure 20.34 The transients of multi-scalar variables and the active and reactive power of a doubly fed machine supplied by CSC (simulation test)

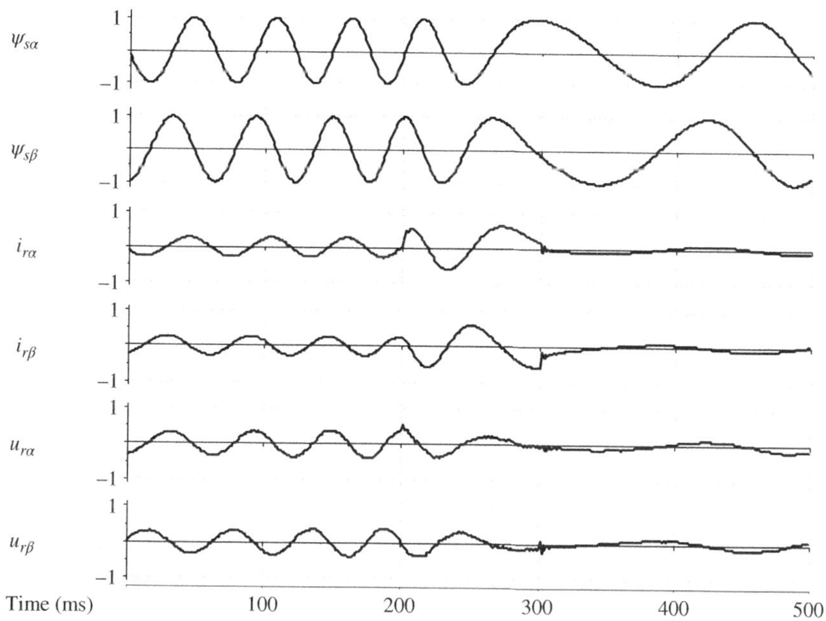

Figure 20.35 The transients of variables in the $\alpha\beta$ coordinate system (simulation test)

20.9 Conclusion

In this chapter two approaches to the control of a machine supplied by CSC were presented. The first is a voltage multi-scalar control based on the PI-controllers. The voltage approach seems to be a better solution than the second current control, although the control system structure is more complicated than the current control structure. The voltage in the dc-link is the control variable obtained directly from the decouplings. The current in the dc-link is not kept at a constant value but its value depends on the machine's working point. The current control gives higher losses in the dc-link inductor and higher transistor power losses than the voltage control. The power losses can be minimized by the modulation index control method but the control system is more complicated. Both control systems lead to control path decoupling and sinusoidal stator current and voltage when the space vector modulation of transistors is applied.

References

1. Glab (Morawiec), M., Krzeminski, Z., and Włas, M. (2005) The PWM current source inverter with IGBT transistors and multiscalar model control system. 11th European Conference on Power Electronics and Applications, IEEE 2005.
2. Glab (Morawiec), M., Krzeminski, Z., and Lewicki, A. (2007) Multiscalar control of induction machine supplied by current source inverter. PCIM 2007, Nuremberg 2007.
3. Morawiec, M., Krzeminski, Z., and Lewicki, A. (2010) Voltage multiscalar control of induction machine supplied by current source converter. IEEE International Symposium on Industrial Electronics (ISIE) 2010.
4. Morawiec, M. (2007) Sensorless control of induction machine supplied by current source inverter. PhD thesis. Gdansk University of Technology.
5. Fuschs, F. and Kloenne, A. (2004) dc-link and dynamic performance features of PWM IGBT current source converter induction machine drives with respect to industrial requirements. IPEMC 2004, August 14–16, 2004, Vol. 3.
6. Klonne, A. and Fuchs, W.F. (2003) High dynamic performance of a PWM current source converter induction machine drive. EPE 2003, Toulouse, France.
7. Kwak, S. and Toliyat, H.A. (2006) A current source inverter with advanced external circuit and control method. *IEEE Transactions on Industry Applications*, **42** (6), 1496–1507.
8. Salo, M. and Tuusa, H. (2005) Vector-controlled PWM current-source-inverter-fed induction motor drive with a new stator current control method. *IEEE Transactions on Industrial Electronics*, **52** (2), 523–531.
9. Nikolic, A.B. and Jeftenic, B.I. (2006) Improvements in direct torque control of induction motor supplied by CSI. IEEE Industrial Electronics - 32nd Annual Conference on Industrial Electronics, IECON 2006.
10. Krzeminski, Z. (1987) Nonlinear control of induction motor. Proceedings of the 10th IFAC World Congress, Munich, Germany, 1987.
11. Morawiec, M. Sensorless control of induction motor supplied by current source inverter, *Induction Motors, Modelling and Control* R. Araujo, InTech, Open Science Open Minds, (2005), **52** (2), pp. 523–531.
12. Ledwich, G. (1991) Current source inverter modulation. *IEEE Transactions on Industrial Electronics*, **6** (4), 618–623.
13. Lee, D., Kim, D., and Chung, D. (1996) Control of PWM current source converter and inverter system for high performance induction motor drives. *IEEE Transactions on Power Electronics*, **2**, pp: 1100–1105.
14. Lewicki, A., Krzemiński, Z., and Morawiec, M. (2008) Sterowanie moca bierna w falowniku pradu sterowanym napieciowo. *Polish Electrical Review*, **R84** (11), 344–347.
15. Zhang, C., Li, Y., Wang, Q., and Li, C. (2009) The optimal control policies of three-phase current source rectifier based on over-modulation technique. International Conference on Sustainable Power Generation and Supply, SUPERGEN '09.
16. Wiseman, J.C. and Wu, B. (2005) Active damping control of a high-power PWM current-source rectifier for line-current THD reduction. *IEEE Transactions on Industrial Electronics*, **52** (3), 758–764.
17. Zhou, J. and Wang, Y. (2002) Adaptive backstepping speed controller design for a permanent magnet synchronous motor. *IEE Proceedings-Electric Power Applications*, **149** (2), 165–172.
18. Tang, L. and Su, G.-J. (2011) A novel current angle control scheme in a current source inverter fed surface-mounted PMSM drive. Energy Conversion Congress and Exposition 2011.
19. Lee, H.-J., Jung, S., and Sul, S.-K. (2011) A current controller design for current source inverter-fed PMSM drive system. 8th International Conference on Power Electronics 2011.
20. Krzeminski, Z. (2002) Sensorless multiscalar control of double fed machine for wind power generators. Power Conversion Conference, PCC, Osaka, Japan, 2002.
21. Wu, B. (2006) *High-Power Converters and AC Drives*, Wiley-IEEE Press.
22. Uddin, M.N. and Chy, M.M.I. (2007) Development and implementation of a nonlinear controller incorporating flux control for IPMSM. The 33rd Annual Conference of the IEEE Industrial Electronics Society (IECON), Taipei, Taiwan, November 5–8, 2007.

23. Morawiec, M. (2012) The adaptive backstepping voltage control of permanent magnet synchronous motor supplied by current source inverter. *IEEE Transactions on Industrial Informatics*, TII, **9**, (2).
24. Li, Y.W., Pande, M., Zargari, N.R., and Wu, B. (2009) DC-Link current minimization for high-power current-source motor drives. *IEEE Transactions on Power Electronics*, **24**, 232–240.
25. Blecharz, K., Krzeminski, Z., and Bogalecka, E. (2009) Control of a doubly-fed induction generator in wind park during and after line-voltage distortion. ELECTROMOTION 2009. 8th International Symposium on Advanced Electromechanical Motion Systems and Electric Drives Joint Symposium 2009.

21

Common-Mode Voltage and Bearing Currents in PWM Inverters: Causes, Effects and Prevention

Jaroslaw Guzinski
Faculty of Electrical and Control Engineering, Gdansk University of Technology, Gdansk, Poland

21.1 Introduction

Accelerated degradation of bearings of induction motors (IMs) operating with voltage inverters is an effect of parasitic current flow, defined as a bearing current. The first reports on bearing currents were published nearly 100 years ago [1]. The observed phenomena were reported only for high-power machines as a result of magnetic asymmetry [2]. Its value is negligible compared with the bearing currents that occur in machines with inverter-type supplies [3]. Bearing failure is now the most common failure of AC machines operating in adjustable speed drives (ASDs). Owing to the high number of ASDs, this type of failure requires special attention. The bearing current in modern ASDs is closely connected with the appearance of the common-mode (CM) voltage resulting from the operation of a voltage inverter with pulse-width modulation (PWM). For that operation, long-term current flow through motor bearings, with a density exceeding the ratings of rolling elements can completely destroy them. It is reported that a current density of $J_b \geq 0.1$ A mm^{-2} has no noticeable impact on the life of the bearing, but a current density of $J_b \geq 0.7$ A mm^{-2} can significantly shorten its lifetime [3].

CM voltage refers to the three-phase natural system and is defined as the voltage between the neutral point of the three-phase star-connected load and the potential of protective earth (PE). Figure 21.1 shows the equivalent circuit for a three-phase voltage converter with a motor, cable and classical grid supply circuit [4].

In Figure 21.1, CM voltage is denoted as u_N. The parts of circuits with dotted lines denote the parasitic capacitances between the circuit elements and the PE.

An analysis of CM voltage for the complete circuit is difficult because of the lack of complete information on the individual parasitic components present on the supply side of the converter. However, a

Power Electronics for Renewable Energy Systems, Transportation and Industrial Applications, First Edition.
Edited by Haitham Abu-Rub, Mariusz Malinowski and Kamal Al-Haddad.

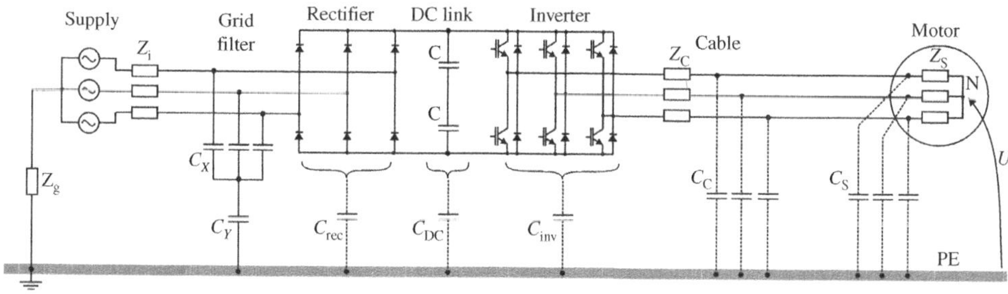

Figure 21.1 The structure of the IM drive for common-mode current flow analysis [4]

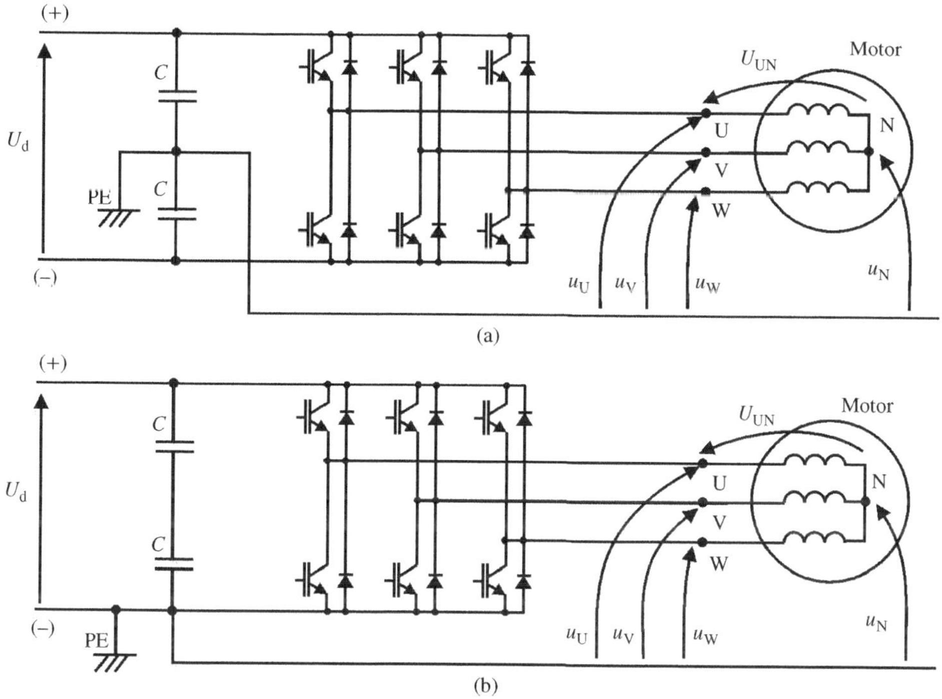

Figure 21.2 Simplified structures of three-phase voltage inverter and notation of output voltages for CM analysis: (a) midpoint reference and (b) negative point reference

small serial impedance of the CM circuit elements allows it to be accepted for the analysis. It is usual to take the midpoint of the DC link as a reference point connected to the PE (Figure 21.2a). However, in a real converter, access to the midpoint of the DC link is troublesome, so observation of the CM voltage according to Figure 21.2a is difficult. So, for practical reasons, it is convenient to take into account the positive or negative potential of the DC link. For this purpose, in this chapter, the negative potential has been assumed for the purpose of analysis (Figure 21.2b).

According to Figure 21.2, the expressions for the voltages u_N, u_V and u_W are as follows.

$$u_U = u_{UN} + u_N \tag{21.1}$$

$$u_V = u_{VN} + u_N \tag{21.2}$$

$$u_W = u_{WN} + u_N \tag{21.3}$$

which leads to

$$u_U + u_V + u_W = u_{UN} + u_{VN} + u_{WN} + 3u_N \tag{21.4}$$

because in a three-phase system

$$u_{UN} + u_{VN} + u_{WN} = 0 \tag{21.5}$$

so the CM voltage is

$$u_N = \frac{u_U + u_V + u_W}{3} \tag{21.6}$$

In the voltage inverters operating with PWM used nowadays, the most popular modulation algorithm is vector modulation. It is widely known as space vector modulation (SVM). For a three-phase inverter during SVM operations, the inverter output voltage is adequate for the state of inverter switches. For six switches, there are $2^6 = 64$ combinations, but only eight have a technical sense. Between them, six states are noted as active and two as passive. During the active state, the output voltage is −zero and for the passive state the output voltage is zero and the load terminals are shortened. The inverter output voltages corresponding with each of eight vectors are listed in Table 21.1.

The voltage u_0 appearing in Table 21.1 is the counterpart of u_N in the orthogonal $\alpha\beta$ system when the next power-invariant transformation ABC → $\alpha\beta$ is used:

$$\begin{bmatrix} x_0 \\ x_\alpha \\ x_\beta \end{bmatrix} = \begin{bmatrix} \frac{1}{\sqrt{3}} & \frac{1}{\sqrt{3}} & \frac{1}{\sqrt{3}} \\ \frac{\sqrt{2}}{\sqrt{3}} & -\frac{1}{\sqrt{6}} & -\frac{1}{\sqrt{6}} \\ 0 & \frac{1}{\sqrt{2}} & -\frac{1}{\sqrt{2}} \end{bmatrix} \begin{bmatrix} x_A \\ x_B \\ x_C \end{bmatrix} \tag{21.7}$$

where x denotes an arbitrary variable, for example, current or voltage.

Table 21.1 The components of the inverter output voltage vectors[a]

	Binary notation of the inverter switches states and corresponding voltage values[b]							
	100	110	010	011	001	101	000	111
u_U	U_d	U_d	0	0	0	U_d	0	U_d
u_V	0	U_d	U_d	U_d	0	0	0	U_d
u_W	0	0	0	U_d	U_d	U_d	0	U_d
u_0		$\frac{2U_d}{\sqrt{3}}$	$\frac{U_d}{\sqrt{3}}$	$\frac{2U_d}{\sqrt{3}}$	$\frac{U_d}{\sqrt{3}}$	$\frac{2U_d}{\sqrt{3}}$	0	$\sqrt{3U_d}$
u_α	$\frac{\sqrt{2U_d}}{\sqrt{3}}$	$\frac{U_d}{\sqrt{6}}$	$-\frac{U_d}{\sqrt{6}}$	$-\frac{\sqrt{2U_d}}{\sqrt{3}}$	$-\frac{U_d}{\sqrt{6}}$	$\frac{U_d}{\sqrt{6}}$	0	0
u_β	0	$\frac{U_d}{\sqrt{2}}$	$\frac{U_d}{\sqrt{2}}$	0	$-\frac{U_d}{\sqrt{2}}$	$-\frac{U_d}{\sqrt{2}}$	0	0

[a]Voltages are referenced to the negative potential of the inverter input circuit (Figure 21.1).
[b]The binary value means that for adequate state, the upper switch is 1, on; 0, off. The bottom switch is in the opposite state.

The analysis of Table 21.1 shows for that the voltage generated by the three-phase voltage inverter, the zero voltage component u_0 is nonzero. Variation of u_0 is characterized by significant transient changes. The difference between the minimum and maximum values in the orthogonal coordinate $\alpha\beta$ system is equal to $\sqrt{3}U_d$.

In a real system, the zero voltage component can be observed by measuring the voltage at the neutral point of the motor stator windings (Figure 21.3).

The occurrence of CM output voltage of the inverter and the presence of parasitic capacitance in the motor (Figure 21.3) and other ASD components causes the flow of zero sequence current (Figure 21.4).

Voltage drop across the serial impedance of the cable for CM is small compared with the value of u_0. Therefore, it is possible to provide an equivalent circuit for a CM current as shown in Figure 21.5 – voltage u_0 is given directly on the cable input; that is, the inverter is represented as the source of u_0. Figure 21.5 shows the equivalent circuit for the CM current for the motor and feeder cable.

The elements noted in the structure shown in Figure 21.5 are as follows:

Cable parameters:

- R_c, cable resistance;
- L_c, cable inductance;
- C_c, cable capacitance.

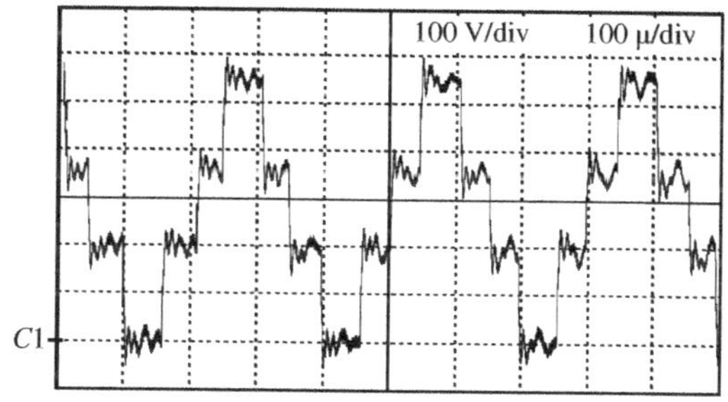

Figure 21.3 The voltage u_N waveform in the case of a voltage inverter-type supply

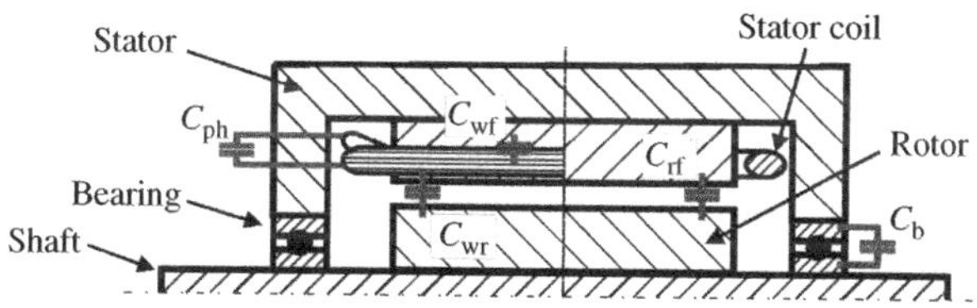

Figure 21.4 Parasitic capacitances in induction motor

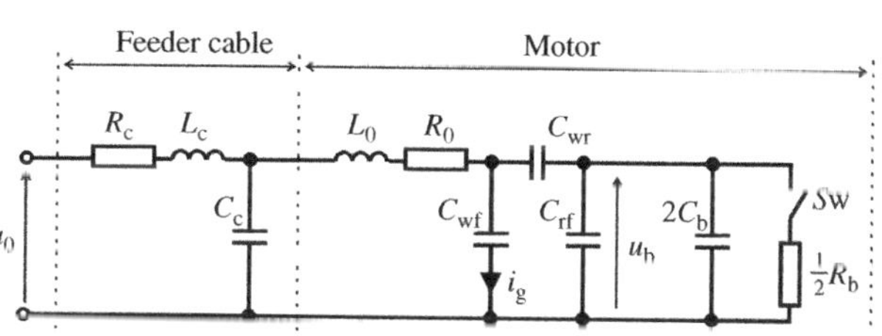

Figure 21.5 Common-mode equivalent circuit for induction motor and motor cable [5]

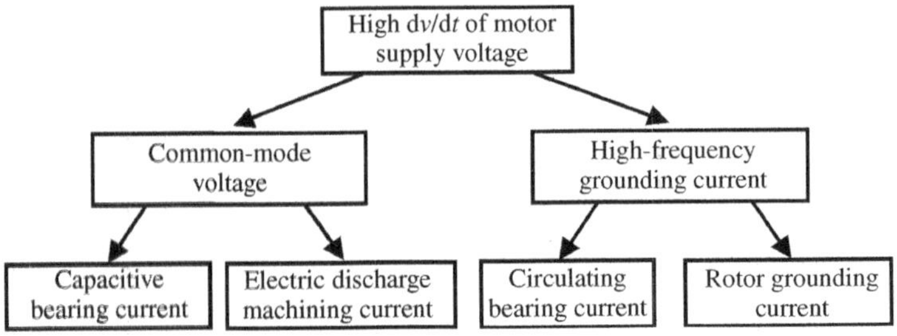

Figure 21.6 Diagram of bearing currents [5]

Motor parameters:

- C_{wf}, motor winding-to-frame capacitance;
- C_{wr}, motor winding to rotor capacitance;
- C_{rf}, motor rotor to frame capacitance;
- C_b, motor bearings equivalent capacitance;
- R_b, motor bearings equivalent resistance;
- S_w, switch modeling breakdown of the bearing lubrication film.

CM voltage with high values of dv/dt causes a current bearing and shaft voltage in the motor. As shown in [5] in the inverter supply drives, there are several types of bearing currents, such as

capacitive bearing current,
electric discharge bearing current,
circulating bearing current or shaft current, related to the shaft voltage effect and
rotor ground current.

The primary reason for each of these types of currents is high dv/dt values at the motor terminals. But each type of current is related to a different motor physical phenomenon (Figure 21.6).

21.1.1 Capacitive Bearing Current

Capacitive bearing current i_{bcap} is flowing in electrical circuits where bearings equivalent capacitance C_b exists. According to [5], the maximal value of this current is 5–10 mA in the case of a bearing temperature of $T_b \approx 25\,°C$ and a motor mechanical speed of $n \geq 100$ rpm. An increase in either temperature T_b or motor speed or both will result in an increase in i_{bcap} as well. However, the value of i_{bcap} is relatively small compared with that of the other components of the CM current, so usually i_{bcap} is assumed to be harmless in terms of the motor bearing life.

21.1.2 Electrical Discharge Machining Current

Electrical discharge machining (EDM) current i_{bEDM} is a result of the breakdown of bearing oil film. The oil film is a thin insulating layer of the lubricant with a dielectric strength of the order of 15 kV mm^{-1}. The oil film thickness depends on the bearing type and size [6] and for a typical motor bearing is close to 0.5 μm, which corresponds to a voltage breakdown of approximately 7.5 V [5]. The bearing voltage u_b corresponds to the CM voltage u_0 according to the voltage distribution in a capacitive voltage divider with C_{wr} and C_{rf}. If the voltage u_b exceeds the oil film breakdown stress, then an impulse of machine discharging current appears. According to the circuit structure shown in Figure 21.5, it corresponds to S_W switch on state. In accordance with the data given in [3, 7], maximal values of i_{bEDM} take values in the range 0.5–3 A. The bearing voltage u_b is independent of the motor size [8], which causes the i_{bEDM} current to be more dangerous especially for small-power motors. This is because of the lower elastic contact surface between bearing balls and races, which increases the current density.

21.1.3 Circulating Bearing Current

Circulating bearing current and shaft voltage are related with current flow through the motor stator winding and frame capacitance. It is a high-frequency grounding current i_g. The i_g flow excites the magnetic flux ψ_{circ} which circulates through the motor shaft. The flux ψ_{cir} induces the shaft voltage u_{sh}. If u_{sh} is large enough, then the oil film in the bearings breaks down and the circulating bearing current, i_{bcir}, appears. The circuit for i_{bcir} flow contains the motor frame, shaft and both bearings (Figure 21.7).

According to the data given in [3], the maximal value of i_{bcir} is in the range of 0.5–20 A depending on motor size; the largest values are observed for high-power motors. The measurement of i_{bcir} requires the use of a Rogowski coil on the motor shaft. The coil must be placed as far as possible from the stator coil out-hang [10] (Figure 21.8).

Measurement of i_{bcir} is complex, requiring special equipment and access inside the motor, so for practical reasons i_{bcir} could also be estimated on the basis of knowledge of the grounding current i_g measurement and type of bearings used in the motor. In the case of a standard motor with a pair of conventional bearings, the circulating bearing current is [3]

$$i_{bcir(max)} \leq 0,4 \cdot i_{g(max)} \tag{21.8}$$

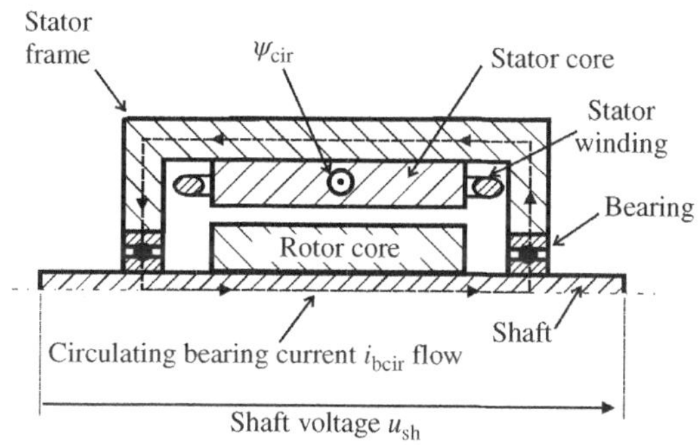

Figure 21.7 The flow path of the circulating bearing current i_{bcir} [9]

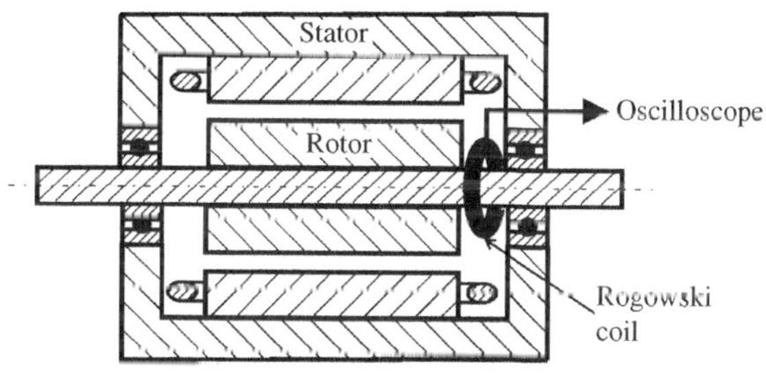

Figure 21.8 The use of a Rogowski coil for circulating current i_{bcir} measurement [10]

The estimation of shaft voltage in a simple way is also possible, as was presented in [3]. It was observed empirically, based on a series of tests, that the shaft voltage u_{sh} is proportional to the grounding current and length of the stator core l_{Fe}:

$$u_{sh} \sim i_g l_{Fe} \tag{21.9}$$

Obviously, the length l_{Fe} is proportional to the motor frame size H:

$$l_{Fe} \sim H \tag{21.10}$$

The grounding current i_g is proportional to stator winding-to-frame parasitic capacitance C_{wf}, which is proportional to the square of the motor frame size:

$$i_g \sim C_{wf} \sim H^2 \tag{21.11}$$

which finally leads to the empirical relation [3]:

$$u_{sh} = i_g l_{Fe} \sim H^2 \cdot H = H^3 \tag{21.12}$$

The measurement of shaft voltage is complex. It requires the use of high-quality brushes on both ends of the motor shaft to assure a good contact area. Because of the very small value of u_{sh}, it has to be amplified (Figure 21.9).

The voltage shaft values measured in laboratory conditions range from fractions of a volt to several volts [2, 11] (Figure 21.10). In the case of high u_{sh}, it is possible to protect the motor by the use of slip rings on the shaft and brushes connected with the earth potential. However, durability and mechanical problems limit the application of such protection.

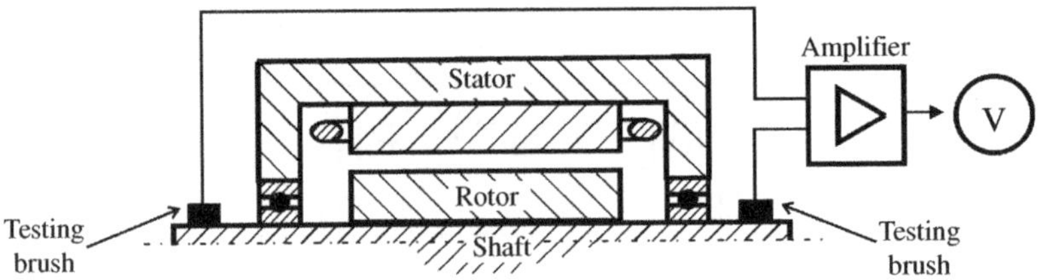

Figure 21.9 Shaft voltage measurement

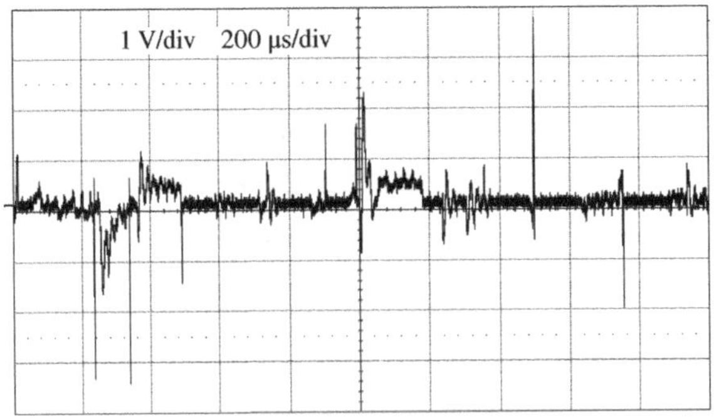

Figure 21.10 Example of shaft voltage waveform (motor: 1.5 kW)

21.1.4 Rotor Grounding Current

The appearance of rotor grounding current i_{rg} is possible only when the motor rotor has a galvanic connection with the earth potential through the driven load. If the impedance of the stator–rotor electric circuit is significantly lower than stator–frame impedance, then part of the total grounding current i_g flows as total grounding current i_{rg}. The amplitude of i_{rg} can reach very large values and quickly destroy the motor bearings [3].

21.1.5 Dominant Bearing Current

The dominant bearing current is dependent on the motor mechanical size H. The proper relation has been formulated in [3] and the dominant components are machine discharging current i_{bEDM} if $H < 100$ mm, both machine discharging current i_{bEDM} and circulating bearing current i_{bcir} if $100\,\text{mm} < H < 280\,\text{mm}$ and circulating bearing current i_{bcir} if $H > 280$ mm.

When the motor size increases, the motor circulating bearing current increases simultaneously.

21.2 Determination of the Induction Motor Common-Mode Parameters

Knowledge of the motor's CM parameters is indispensable for modeling of the inverter-type drive. As outlined in Section 21.1, the detailed CM circuit of the motor is complex (Figure 21.5). An additional difficulty is the variability of some parameters of the circuit. In the literature [3, 5, 12–14], the analytical dependencies used for calculating the parameters of the circuit shown in Figure 21.5 are presented. However, the calculations are complex and require a lot of motor data that are difficult to access from datasheets and simple measurements, although some of the results are presented in the literature (Figure 21.11) [1].

For practical reasons in most cases, the simpler motor CM circuit is considered, as shown in Figure 21.12 [8, 9, 14].

Inductance L_0 and resistance R_0 are the leakage inductance and substitute resistance, respectively, of the motor stator windings. These parameters are easy to measure in the configuration shown in Figure 21.13.

The measurement of parasitic capacitance C_0 needs deeper analysis. As shown in Figure 21.5, the equivalent capacitance C_0 which is used in the simplified model of CM circuit is a combination of a

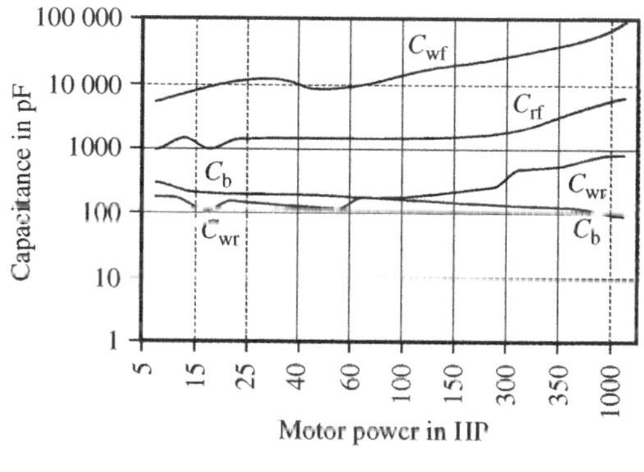

Figure 21.11 Simplified structure of the motor CM circuit [1]

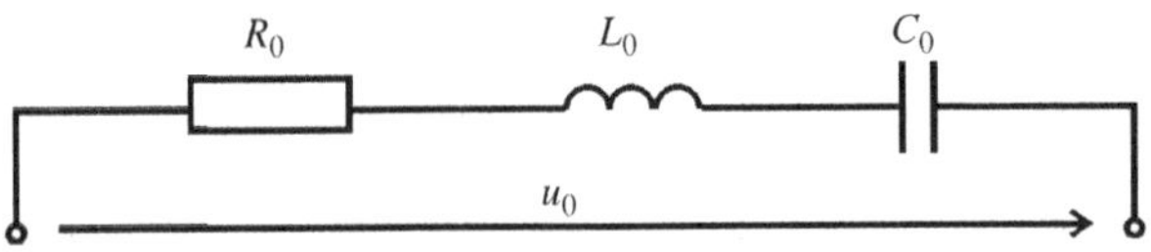

Figure 21.12 Simplified structure of the motor CM circuit

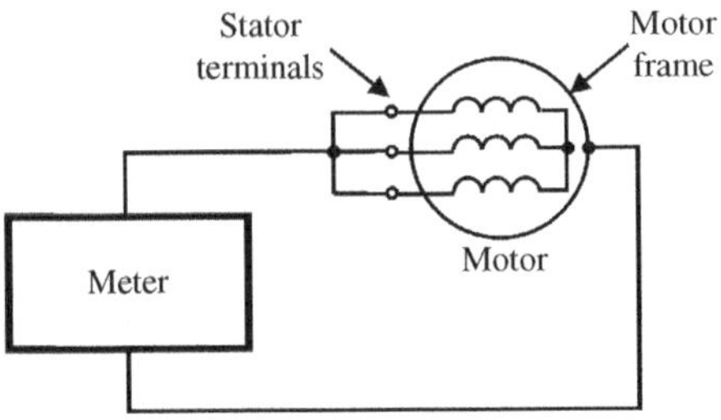

Figure 21.13 Measurement circuit for determining the parameters L_0 and R_0 of the motor

few parasitic elements: C_{wr}, C_{wf}, C_{rf} and C_b. The topology of the circuit can also change in the case of oil film breakdown. In addition, the bearing capacitance is nonlinear, depending mainly on the motor speed. These phenomena can lead to measurement errors when typical measuring equipment such as an electronic RLC bridge meter is used. This equipment operates at low-voltage power and frequency, which may differ significantly from the value and frequency of CM voltage. Therefore, the C_0 capacitance measurement should be performed with a frequency and voltage similar to those observed during normal operation of the inverter.

The solution to the measurement of C_0 could be the creation of a serial resonance circuit (Figure 21.14). The supply source of the circuit could be an inverter, which supplies the motor during normal operation. In this way, it is possible to measure C_0 at the same voltage and frequency as those which appear during normal operation of the drive system. This approach requires access to a PWM algorithm to generate a square wave voltage waveform with a constant modulation factor of 0.5 and the possibility of changing the modulation frequency.

The inverter is connected to a resonant circuit L_{res}, C_{res} including dumping resistance R_{res}. The measurement is done at a frequency f_{res} corresponding to the frequency component of the common voltage, that is, the inverter switching frequency f_{imp}:

$$f_{imp} = f_{res} = \frac{1}{2\pi\sqrt{L_{res}C_{res}}} \tag{21.13}$$

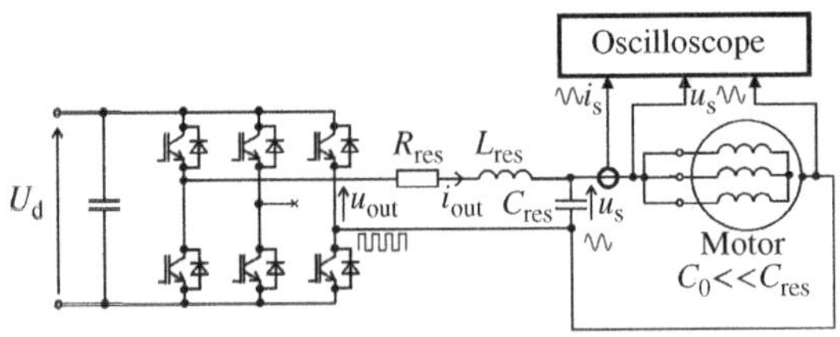

Figure 21.14 The system for CM capacitance of the motor

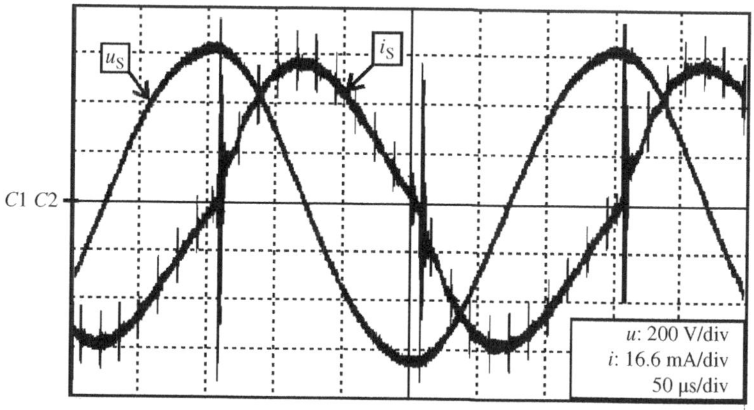

Figure 21.15 Examples of the u_S and i_S waveforms measured by the test bench presented in Figure 21.10 (motor: 1.5 kW, $U_{S(RMS)} = 430$ V, $I_{0(max)} = 46$ mA, $f_{res} \approx f_{imp} = 3.3$ kHz, $C_0 = 2.7$ nF)

The value of C_{res} must be significantly larger than the expected motor CM capacitance C_0. If this condition is fulfilled, the impact of C_0 on the resonant frequency is negligible. Because the value of C_0 is in the range of nanofarads, the capacitor C_{res} should have a value of microfarads. The resistance R_{res} should be chosen to limit the inverter output current to a safe level and to provide a good quality factor of the resonant circuit.

The sample waveforms of current and voltages collected by the measurement test bench shown in Figure 21.14 are presented in Figure 21.15.

With the measured RMS values of motor voltage and current, the motor CM impedance is

$$Z_{s0} = \frac{U_{S(RMS)}}{I_{S(RMS)}} \tag{21.14}$$

and the capacitive reactance is

$$X_{C0} = \sqrt{Z_{s0}^2 - R_0^2} - X_{L0} \tag{21.15}$$

where

$$X_{C0} = 2\pi f_{imp0} \tag{21.16}$$

and finally the motor CM capacitance sought is

$$C_0 = \frac{1}{2\pi f_{imp} X_{C0}}. \tag{21.17}$$

In Table 21.2, the experimentally measured values of C_0 for some typical industrial IMs are given.

Table 21.2 Induction motor capacitances C_{s0} measured using the test bench presented in Figure 21.14 and the serial resonance method

H (mm)	P_n (kW)	p (–)	U_n Y (V)	n_n (rpm)	C_0 (nF)
90	1.5	2	300	1420	2.9
90	1.5	2	400	1410	3.53
112	1.5	8	400	720	4.57
132	10	2	173	1455	6.4

The motor capacitances C_0 given in Table 21.2 were measured for a rotor at standstill. In the exact model of the CM circuit, the bearing capacitance is a function of motor speed [12, 13]. So for measurement purposes, the tested motor should be driven at different speeds using an external machine. With that test, the characteristic $C_0 = f(\omega_r)$ should be created. However, C_b is very small in comparison to other parasitic capacitances, and its influence on the total CM capacitance of the motor can be negligible. A typical value of the bearing capacity is close to 190 pF and is much smaller than the dominant capacitances between the stator windings and the motor frame for the motor sizes from 80 to 315. For that motor, that value is in the range of a few nanofarads up to tens of nanofarads [1, 13].

21.3 Prevention of Common-Mode Current: Passive Methods

The first and most important thing that must be implemented to reduce the influence of CM voltage and current on the system is a proper cabling and earthing system. Manufacturers of converters recommend the use of symmetrical multicore motor cables, which prevents the CM at fundamental frequency. Also a short and low-impedance path for the return of CM current to the inverter must be provided. The best way to do that is to use shielded cables where shield connections have to be made with 360° termination on both sides. Also, a high-frequency bonding connection must be made between the motor and load machine frame and the earth. It is recommended that flat-braided strips of copper wire should be used and the strip should be at least 50 mm wide [15].

If the above-mentioned conditions are fulfilled, the elimination or reduction of the high-frequency CM current can be done by increasing the impedance or by using a specially designed motor. Some solutions are shown in Table 21.3.

21.3.1 Decreasing the Inverter Switching Frequency

Decreasing the inverter switching frequency is the simplest way of reducing the CM current. Most industrial inverters offer the possibility of changing that value within a wide range. With a decrease in f_{imp}, the

Table 21.3 Methods for CM current reduction and elimination [5]

Reduction of electric discharge current i_{bEDM}
Ceramic bearings
Common-mode passive filters
Systems for active compensation of CM voltage
Decreasing the inverter switching frequency
Motor shaft grounding by using brushes
Conductive grease in the bearings
Reduction of circulating bearing current i_{bcir}
Common-mode choke
Systems for active compensation of CM voltage
Decreasing inverter switching frequency
Use of one or two insulated bearings
Use of one or two ceramic bearings
dv/dt filter
Reduction of rotor grounding current i_{gr}
Common-mode choke
Systems for active compensation of CM voltage
Decreasing inverter switching frequency
Use of one or two insulated bearings
Use of one or two ceramic bearings
Shielded cable for motor supply

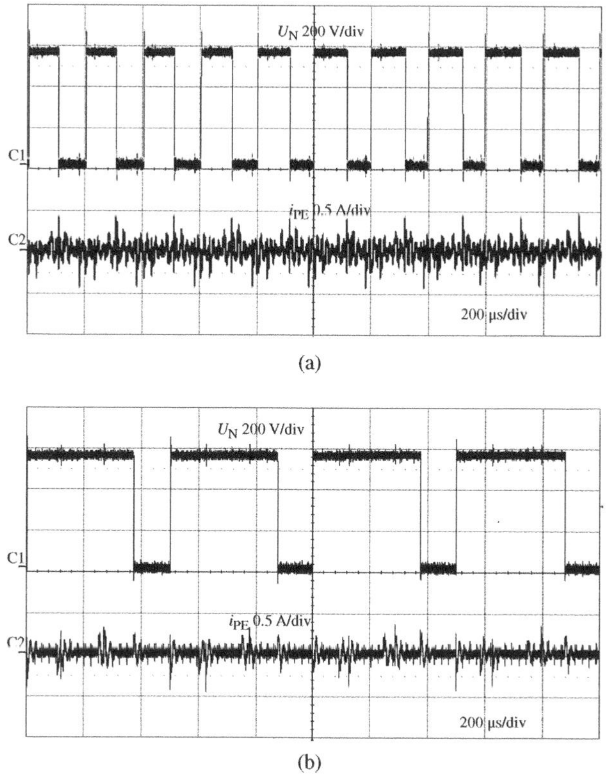

Figure 21.16 Reduction of CM current by decreasing the inverter switching frequency: (a) f_{imp} = 5 kHz and (b) f_{imp} = 2 kHz

dv/dt is not changed and the CM current peaks will not decrease. However, their frequency is reduced and the value of total RMS CM current is decreased (Figure 21.16).

The disadvantage of decreasing f_{imp} is that the THD of the motor supply current is decreased as well.

21.3.2 Common-Mode Choke

The most commonly used component for limiting the CM current is the CM choke (Figure 21.17).

CM choke is constructed with three symmetrical windings on a toroidal core. The mutual inductance M between the windings is identical. The choke inductance is negligible for differential modem (DM) current because the total flux in the core is zero for three-phase symmetrical currents. However, the inductance of CM choke L_{CM} is significant for a CM current circuit. The equivalent electrical circuit for a system with CM choke is presented in Figure 21.18.

The design process for CM choke selection requires previous measurement of the CM voltage u_0, which allows the choke core flux to be specified [16]:

$$\psi_0 = \frac{1}{N_{CM}} \int u_0 \mathrm{d}t \tag{21.18}$$

where N_{CM} is the number of turns of one CM winding.

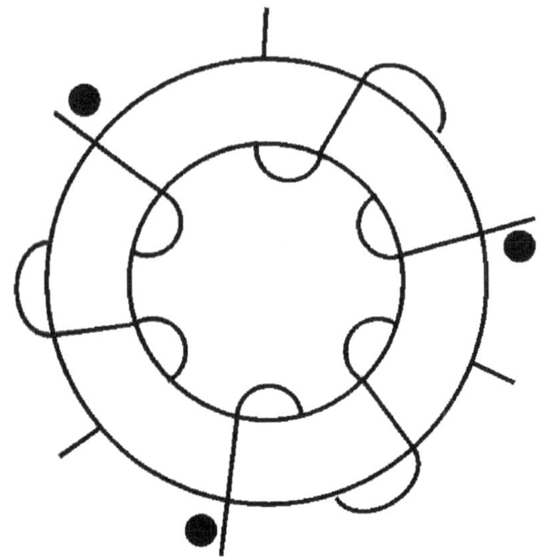

Figure 21.17 Common-mode choke for a three-phase system

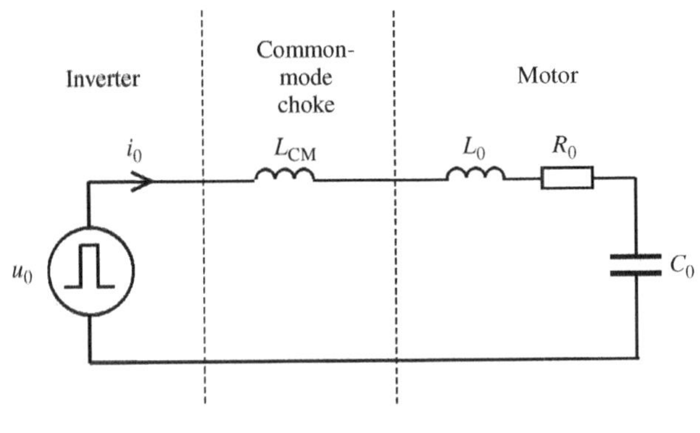

Figure 21.18 The equivalent electrical circuit with CM choke use

The magnetic flux density of the CM choke is

$$B_0 = \frac{\psi_0}{S_{CM}} = \frac{1}{S_{CM}N_{CM}} \int u_0 \mathrm{d}t \tag{21.19}$$

where S_{CM} is the CM choke core cross-sectional area.

The value of B_{CM} must be less than the saturation value B_{sat} of the core material. Nowadays, the manufacturers offer materials dedicated to CM cores with $B_{sat} = 1-1.2\,\mathrm{T}$.

Inductance of a CM choke with known core dimensions is determined by the relationship:

$$L_{CM} = \frac{\mu S_{CM} N_{CM}^2}{l_{Fe0}} \tag{21.20}$$

where l_{Fe0} is the average path of the flux in the CM choke core.

The maximum value of the zero sequence current $I_{0\,max}$ is proportional to the ratio of l_{Fe0}/N_{CM} [16]. A decrease in the core size reduces the flux path of l_{Fe0}, increases the number of turns N_{CM}, and finally decreases $I_{0\,max}$. At the same time, the dimensions of the maximal windings should be taken into account.

The search for the optimal CM choke is a complex iterative process, requiring proper equipment for measurement of u_0 and $I_{0\,max}$. Therefore, the manufacturers of magnetic materials suggest specific solutions and complete CM reactor cores fitted for specific motor power. Particularly good properties are characterized by a toroidal core made of nanocrystalline materials.

For high-power motors, the use of several simple CM cores is a good solution. In that case, the windings of CM choke is simply the motor cable ($N_{CM} = 1$). This construction is easy to use in industry because CM cores are installed directly on motor cable and no terminals have to be used. It is worth to see that only the shield have to be cut off in the place where the cores are installed.

Figures 21.19 and 21.20 present waveforms of CM current that shows the effects of CM choke use. In Figure 21.19, an electric drive with a 1.5 kW IM is operating without a CM choke. One can see current pulses of up to 1.4 A, which have been strongly limited in the case of CM choke use with inductance $L_{CM} = 14$ mH (Figure 21.20).

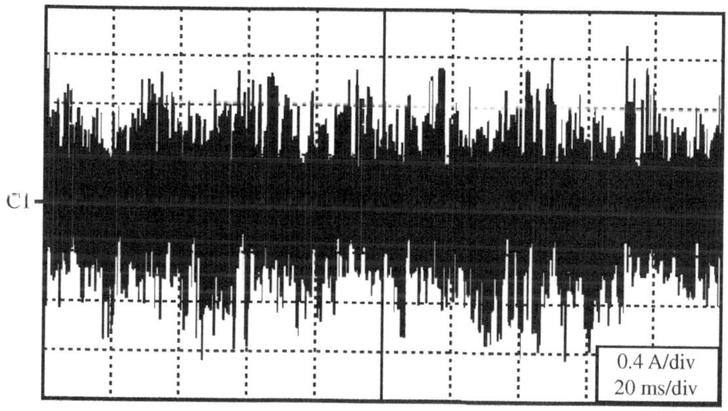

Figure 21.19 CM current measured in PE wire for a 1.5 kW induction motor operating with an inverter but *without* a CM choke (switching frequency: 3.3 kHz)

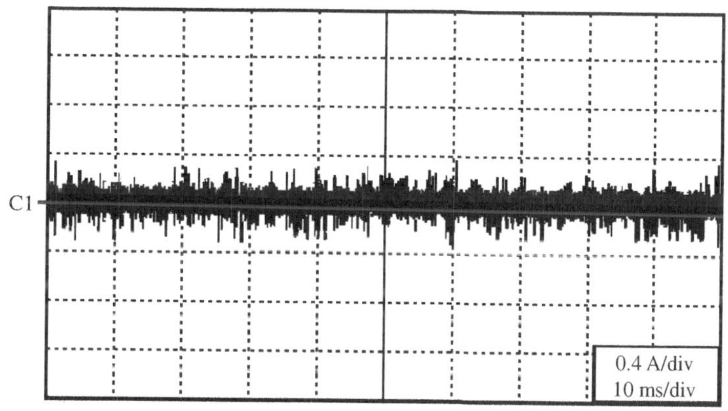

Figure 21.20 CM current measured in PE wire for a 1.5 kW induction motor operating with an inverter and *with* a CM choke (switching frequency: 3.3 kHz)

21.3.3 Common-Mode Passive Filter

A CM passive filter is a combination of elements that give higher impedance for a CM circuit and creation of an alternative path for CM current flow with a motor bypass. The classical structure is presented in Figure 21.21 [16, 17].

The inverter output filter presented in Figure 21.21 is a combination of two filters: a differential and a CM filter. The differential filter is an LC low-pass filter used to smooth the motor supply voltage. With a DM filter, the motor supply voltage is close to sinusoidal in shape.

Elements L_1, C_1 and R_1 are parts of the DM filter, while M_1, R_0 and C_0 are parts of the CM filter. The CM choke M_1 increases the reactance of the CM circuit. The resistance R_0 and capacitance C_0 create a low-impedance electrical circuit parallel to the high-impedance motor CM circuit. In this way, an alternative path for CM current flow is created and most of the CM current bypasses the motor.

Design processes for a DM filter can be found in the literature, for example, in [18]. For a CM filter, it is necessary to select CM choke inductance $M_1 \gg L_1$. The capacitance C_0 is determined using the known values of inductances M_1 and M_2 and the resonance frequency f_{res}:

$$C_0 = \frac{1}{4\pi^2 f_{res}^2 M_1} \tag{21.21}$$

while the resistance R_0 is

$$R_0 = \frac{\sqrt{M_1/C_0}}{Q_{CM}} \tag{21.22}$$

and Q_{CM} is the quality factor of the CM filter ($Q_{CM} = 5{-}8$).

The equivalent circuit of the inverter output filter is presented in Figure 21.22.

For CM current, the DM filter parameters are connected in parallel. It is obvious that according to the condition $L_1 \ll M_1$, the influence of L_1 on the CM circuit structure can be omitted.

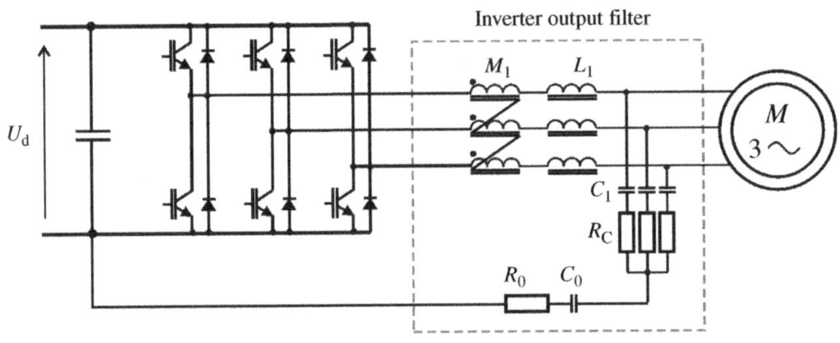

Figure 21.21 Inverter output filter for differential and common modes [16, 17]

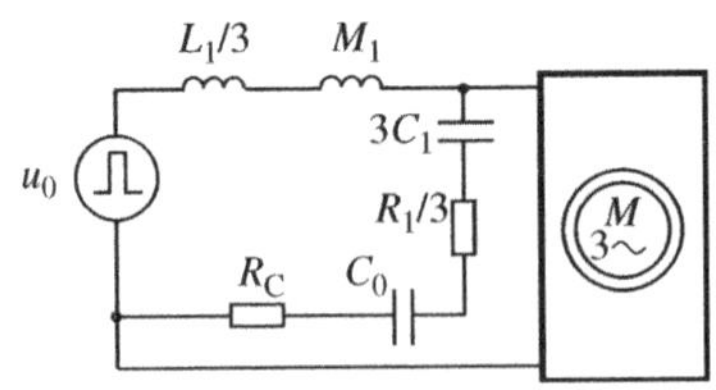

Figure 21.22 Equivalent circuit of the inverter output filter [16, 17]

21.3.4 Common-Mode Transformer

The advantages of using a CM choke could be improved through the use of a CM transformer [14]. The construction of a CM transformer is nearly the same as that of a CM choke, but a fourth additional winding is added (Figure 21.23).

The additional winding has the same number of turns as the other windings. With CM transformer, the CM current could be reduced up to 25% [14].

When the CM transformer is used, an additional inductance L_t and resistance R_t appear, as shown in Figure 21.24.

Inductances $l_{\sigma t}$ are leakage inductances of the primary and secondary windings of the CM transformer. Magnetizing inductance is identified as L_t. In the equivalent circuit of the CM transformer, the primary winding comprises three windings connected in parallel, $U-U'$, $V-V'$ and $W-W'$, as indicated in Figure 21.23.

The structure of the equivalent circuit of the inverter, CM transformer and motor is presented in Figure 21.25.

The resistance R_t operates as damping resistance in the resonance circuit. If leakage inductances $l_{\sigma t}$ are omitted, the CM current relation is as follows [14]:

$$I_0(s) = \frac{sL_tC_0 + R_tU_d}{s^3L_tL_0C_0 + s^2(L_t + L_0)C_0R_t + sL_t + R_t} \tag{21.23}$$

Analysis of the relationship (21.23) allows the selection of R_t, which can reduce both the peak and RMS value of the CM current. According to [14], resistance R_t satisfying this condition should be within the acceptable range:

$$2Z_{00} \leq R_t \leq \frac{1}{2}Z_{0\infty} \tag{21.24}$$

where Z_{00} and $Z_{0\infty}$ are characteristic impedances of the electrical circuit presented in Figure 21.25 for $R_t = 0$ and $R_t = \infty$, respectively.

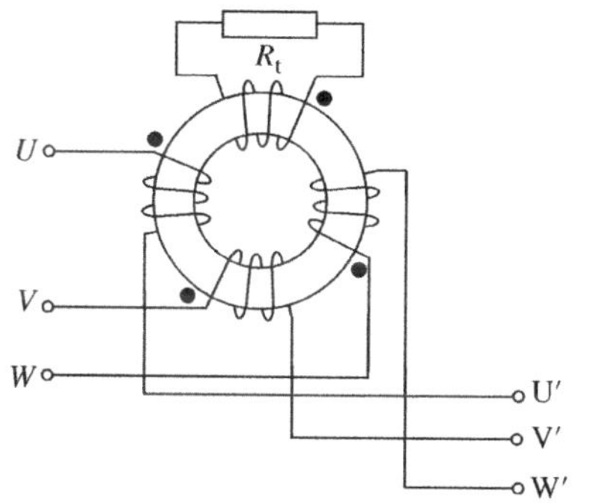

Figure 21.23 Common-mode transformer

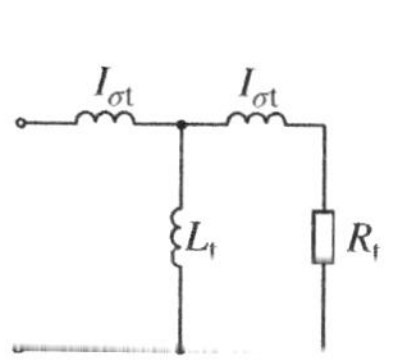

Figure 21.24 The equivalent circuit of the common-mode transformer

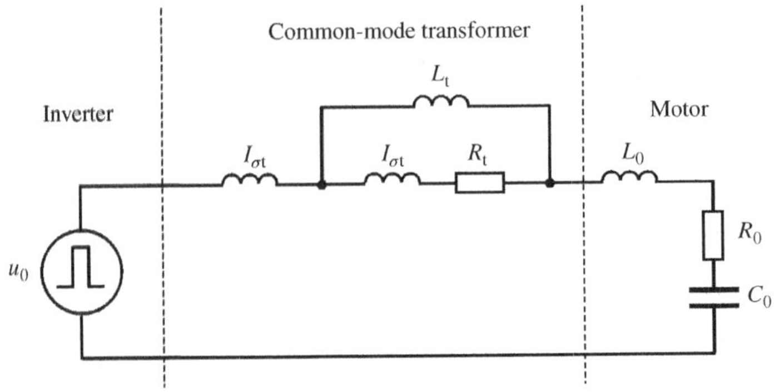

Figure 21.25 The structure of the equivalent circuit of the inverter, CM transformer and motor

21.3.5 Semiactive CM Current Reduction with Filter Application

The semiactive CM current reduction method has been presented in [19]. The idea is to use additional windings on DM inductors which are connected in series. That windings are installed between the filter's neutral point N and the converter DC link midpoint N' (Figure 21.26).

In the filter structure presented in Figure 21.26, inductor windings L_1, capacitances C_1 and resistors R_C are elements of the DM LC filter. In addition, the inductors are detectors of CM voltage and operate as CM transformers. CM current flowing between N–N' points through L_2 windings forces the appropriate voltage in L_1 windings, which compensates the CM voltage generated by the inverter.

The semiactive solution has also been presented in [20], where a classical CM transformer has been used (Figure 21.27).

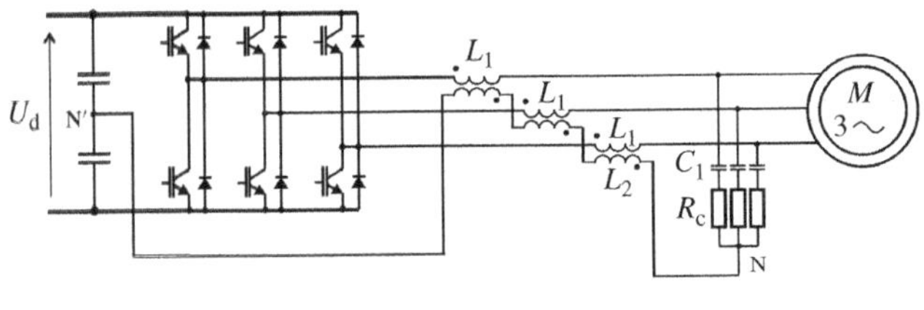

Figure 21.26 Semiactive CM current reduction system with coupled inductors [19]

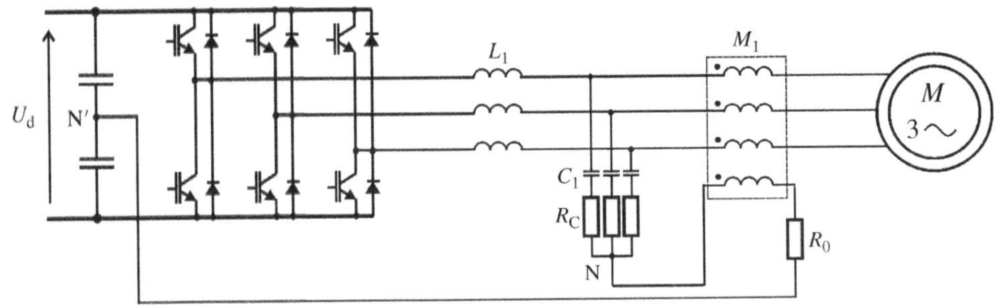

Figure 21.27 Semiactive CM current reduction system with CM transformer [20]

In Figure 21.27, elements L_1, C_1 and R_C are used for DM operation and additionally for CM voltage detection. The connection between the points N and N' assures the CM current flow and produces the primary current in the CM transformer M_1. In this way, the CM voltage generated by the inverter is reduced as well.

21.3.6 Integrated Common-Mode and Differential-Mode Choke

In some inverter applications, it is possible to use an integrated CM and DM choke [21]. The aim of the integration of the two chokes is to minimize the geometrical dimensions of the reactor and reduce the cost of materials by reducing the amount of copper used. In Figures 21.28 21.30, the choke structure, equivalent circuit and implementation in the converter are presented.

The integrated choke is made in the form of two coils wound on two toroidal cores. The dimensions of the toroidal cores are such that the outer diameter of core 2 is slightly smaller than the inner diameter of core 1. Core 1 is made of a magnetic material with high permeability, whereas core 2 has low magnetic permeability. A suitable difference in permeability is achieved of the ferrite for core 1 and iron powders for core 2. Winding AB is wound on both cores simultaneously, while winding CD is wound on both

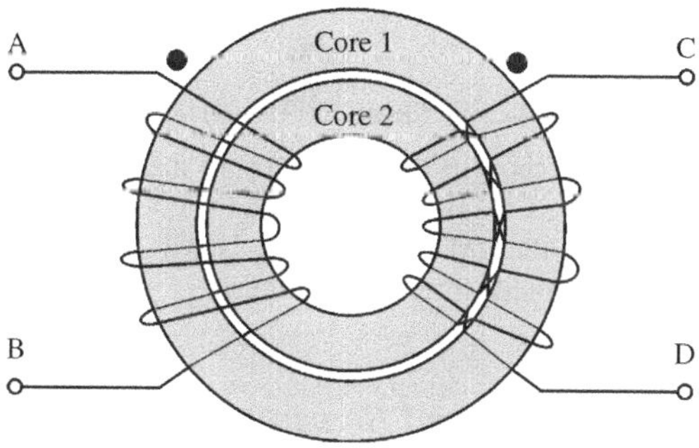

Figure 21.28 The integrated CM and DM choke – structure [21]

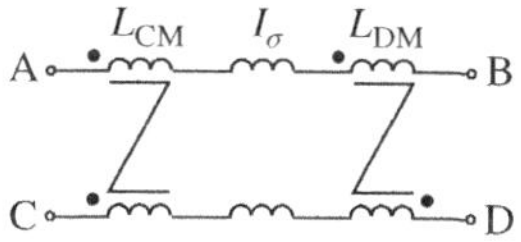

Figure 21.29 The integrated CM and DM choke – equivalent electrical circuit

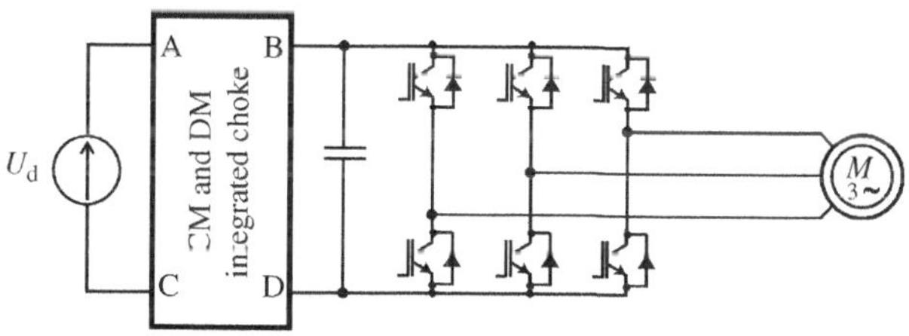

Figure 21.30 The integrated CM and DM choke – application for the converter

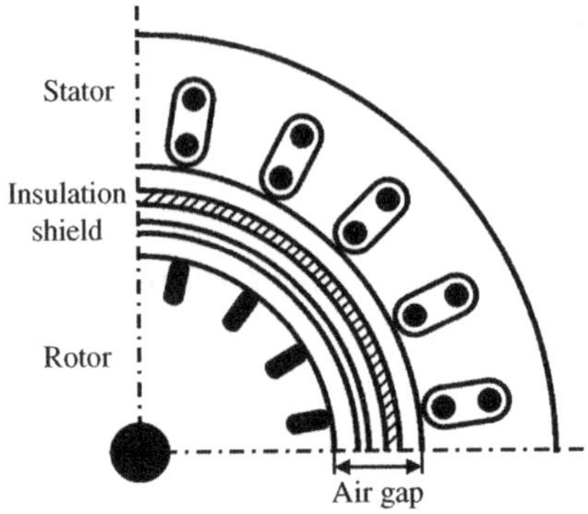

Figure 21.31 Structure of the electrostatically shielded induction motor [12]

cores in the form of an interlaced figure of eight shape. The number of turns of windings AB and CD is the same, that is, $N_{AB} = N_{CD} = N$. If leakage inductance is negligible, then the ratio of L_{CM} and L_{DM} is dependent on the magnetic reluctance $\Re_1$, $\Re_2$ of the cores:

$$\frac{L_{CM}}{L_{DM}} = \frac{\Re_2}{4\Re_1} \tag{21.25}$$

The disadvantage of an integrated choke is the higher cost of manufacture of the windings because of the complex interleaving of winding CD. The limitation of the choke use is that it can only be applied in the DC circuit of the inverter. So the CM part of the choke can only limit the CM voltage forced by external source, for example, by grid-side inverter operation.

21.3.7 Machine Construction and Bearing Protection Rings

Suitable design of electrical machines could also reduce or eliminate the bearing currents. Such construction solutions could be as follows [3]:

- shielding the rotor by placing a metal shield between the stator and the rotor and connecting the shield to earth potential – this solution is known as an electrostatically shielded IM [12] (Figure 21.31),
- using insulated bearings in which the outer surface of the outer race is covered with a nonconductive oxide layer, and
- using hybrid bearings in which ceramic rolling elements are used.

Among the above solutions, only the insulated bearings are commercially available. Other solutions are still under development.

Another market solution for bearing protection is the installation of bearing protection rings. Round-shaped rings with brushes inside are connected to PE and installed on one or both sides of the motor shaft. This installation creates a low-impedance path between the motor shaft and earth. So the shaft current flows to earth directly without bearing reactance.

21.4 Active Systems for Reducing the CM Current

Besides the passive methods for the reduction of CM current, active systems are also used. These are active serial filters [21, 22] (Figure 21.32) or low-power auxiliary inverters [23, 24] (Figure 21.33).

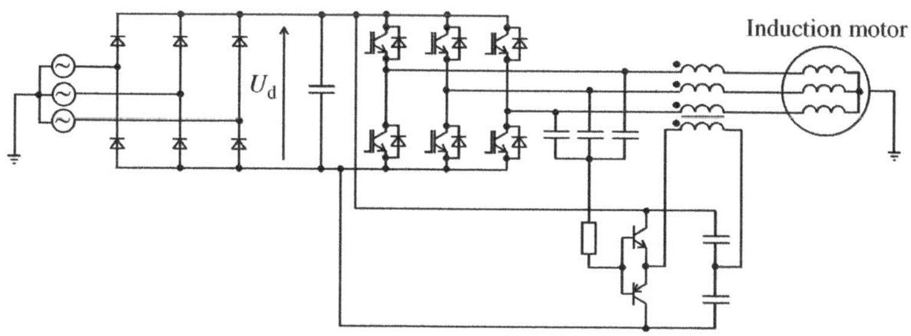

Figure 21.32 Active filter for CM current reduction [22, 25]

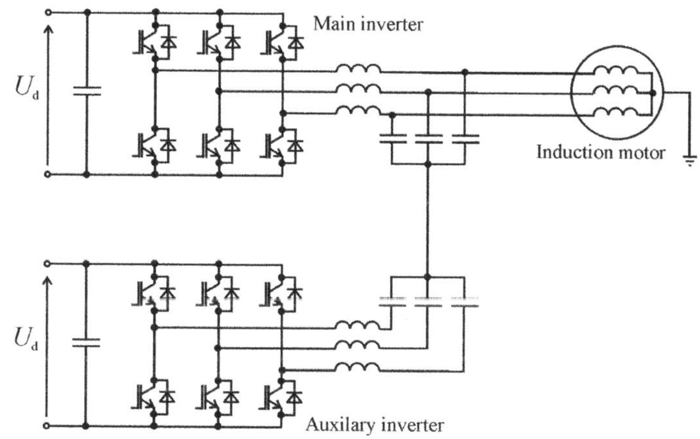

Figure 21.33 The use of an auxiliary inverter for CM current reduction [23, 24]

The practical solution presented in Figure 21.32 requires a pair of complementary bipolar transistors. Control of the transistors is done without the use of an inverter digital control system. The transistors are directly triggered by a CM voltage signal. Unfortunately due to the lack of suitable high-speed high-voltage bipolar transistors, the solution is limited to low-power electrical systems.

This disadvantage is eliminated in the system presented in Figure 21.33. The auxiliary inverter uses the same kind of transistors as the main inverter. Only the power range is smaller because of the value of the CM current. To eliminate the CM current, the auxiliary inverter generates an opposite CM voltage. Such auxiliary inverter operation almost completely eliminates the CM current. However, a small CM voltage may still be present in the system as a result of differences in the transistor switching parameters of both inverters and irregularities in the dead-time compensation algorithm. Correction by the dead-time compensation algorithm allows complete elimination of the CM voltage [24].

21.5 Common-Mode Current Reduction by PWM Algorithm Modifications

Both passive and converter-based classical methods to reduce CM current are relatively expensive and require the installation of additional elements and/or modifications of the converter topology.

However, the reduction of CM current in the drive system is possible in a simpler way that does not require complex changes in the inverter topology. It is possible to reduce CM current by changing the PWM algorithms. The PWM has to be changed to eliminate or reduce the CM voltage generated by the inverter. Various proposals for such modification of the PWM are presented in the literature [26, 27].

Nowadays, the space vector PWM method is widely used in voltage inverters. With SVM, the CM voltage could be reduced by

- removal of zero/passive vectors,
- use of a nonzero/active vectors that have the same value of the CM voltage.

The CM voltage reduction possibilities can be explained using data given in Table 21.4.

In Table 21.4, it can be seen that the voltage u_0 for each of the parity-numbered and non-parity-numbered vectors is consistent. The graphical representation of the vectors is presented in Figure 21.34.

In the case of an inverter operating with a classical SVM, the waveform voltage at the star point of the load related to negative potential of the DC link is as presented in Figure 21.35.

Table 21.4 Common-mode voltage for voltage inverter output vectors

Types of vectors	Nonzero/active						Zero/passive	
Notation	U_{w4}	U_{w6}	U_{w2}	U_{w3}	U_{w1}	U_{w5}	U_{w0}	U_{w7}
Binary code	100	110	010	011	001	101	000	111
Decimal code	4	6	2	3	1	5	0	7
u_N	$\frac{1}{3}U_d$	$\frac{2}{3}U_d$	$\frac{1}{3}U_d$	$\frac{2}{3}U_d$	$\frac{1}{3}U_d$	$\frac{2}{3}U_d$	0	U_d
u_0 ($\alpha\beta$ coordinates)	$\frac{1}{\sqrt{3}}U_d$	$\frac{2}{\sqrt{3}}U_d$	$\frac{1}{\sqrt{3}}U_d$	$\frac{2}{\sqrt{3}}U_d$	$\frac{1}{\sqrt{3}}U_d$	$\frac{2}{\sqrt{3}}U_d$	0	$\sqrt{3U_d}$
Notation[a]	NP	P	NP	P	NP	P	Z	Z

[a]Vector notation: NP, non-parity; P, parity and Z, zero.

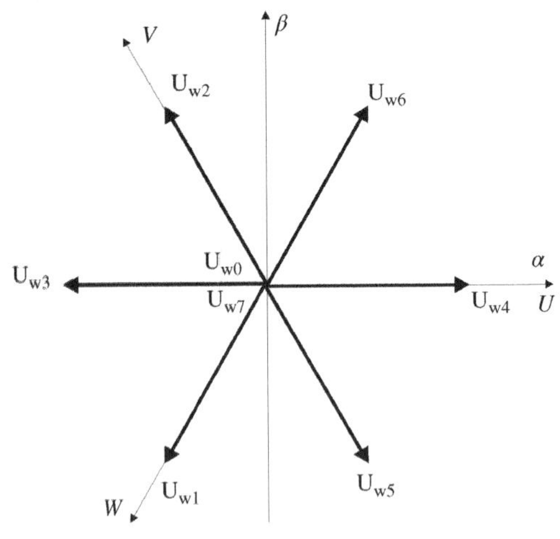

Figure 21.34 Graphic representation of the inverter output vectors

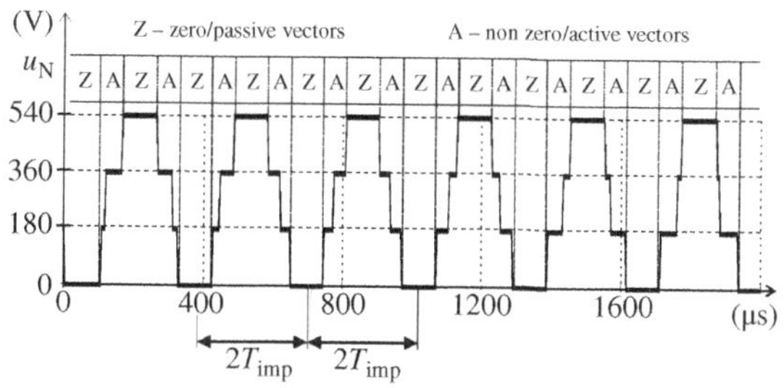

Figure 21.35 The common-mode voltage for the inverter operating with classical SVM (DC link voltage $U_d = 540\,V$)

With regard to the waveform shown in Figure 21.35, one can see that the highest variations of u_N are between zero vectors, being from 0 up to 540 V. Therefore, the elimination of these vectors can significantly reduce CM voltage and accordingly the CM current. Simultaneously, it can be observed that if the consecutive active vectors have the same value of u_0, the CM voltage will be of direct current voltage and the current will not flow through the motor parasitic capacitances.

On the basis of the analysis of the relationship between voltage vectors and the corresponding values of u_0, it is possible to propose modifications of PWM which can lead to the reduction of bearing currents. This solution was proposed in [28, 29], where modification of the SVM algorithm was done by elimination of zero vectors and use of parity or non-parity active vectors only.

The problem that occurs when using only parity active vectors or non parity active vectors is that the inverter maximal output voltage is limited. The problem is underlined when the inverter operation is limited to the modulation region only. This is due to the fact that in the classical SVM algorithm, the output voltage vector u_{out}^{com} is generated for each modulation period as a combination of the two active and two passive vectors (Figure 21.36).

The use of active vectors only in the SVM modulator makes it possible to generate the maximum amplitude of the inverter output voltage as shown in Figure 21.37.

21.5.1 *Three Non-parity Active Vectors (3NPAVs)*

If the zero vectors are eliminated, the output inverter voltage can be generated using, for example, only non-parity active vectors, as presented in Figure 21.38. This method is named three non-parity active vectors (3NPAVs).

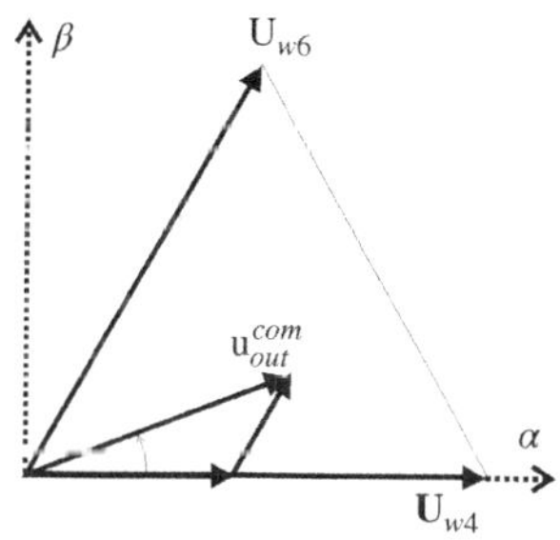

Figure 21.36 Generation of the inverter output voltage by the classical SVM algorithm

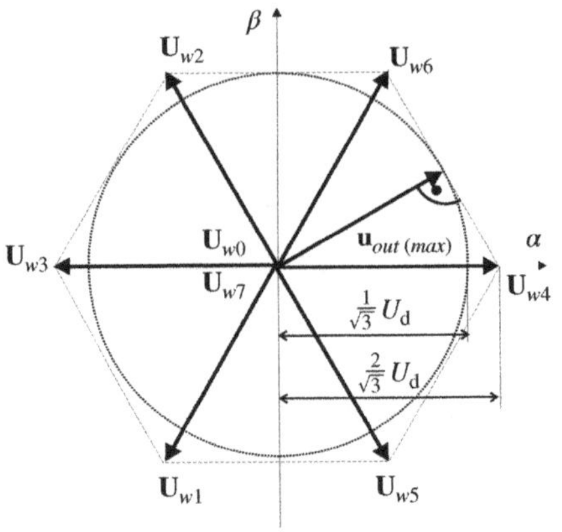

Figure 21.37 The range of inverter output voltage with classical SVM

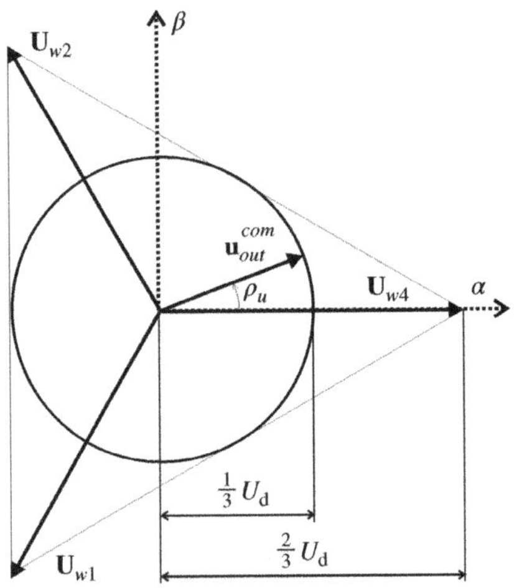

Figure 21.38 The creation of inverter output voltage in the 3NPAV algorithm

In Figure 21.38, the inverter output voltage vector $\boldsymbol{u}_{\text{out}}^{\text{com}}$ is generated using vectors $\boldsymbol{U}_{w4}$, $\boldsymbol{U}_{w2}$ and $\boldsymbol{U}_{w1}$, which are indicated in Table 21.4 as non-parity vectors. The switching times of particular vectors denoted as $\boldsymbol{t}_4$, $\boldsymbol{t}_1$ and $\boldsymbol{t}_2$ are defined as

$$u_{\text{out}}^{\text{com}} \cdot T_{\text{imp}} = U_{w4}t_4 + U_{w2}t_2 + U_{w1}t_1 \tag{21.26}$$

$$T_{\text{imp}} = t_4 + t_2 + t_1 \tag{21.27}$$

The vector switching times can be determined from the relations [30]:

$$\boldsymbol{t}_4 = \frac{1}{3}\left(1 + \frac{2\boldsymbol{U}_{\text{out}}^{\text{com}}}{U_{\text{d}}}\left(\cos\left(\frac{\pi}{3} + \rho_{\text{u}}\right) + \sin\left(\frac{\pi}{6} + \rho_{\text{u}}\right)\right)\right) \tag{21.28}$$

$$t_2 = \frac{1}{3}\left(1 - \frac{2U_{\text{out}}^{\text{com}}}{U_{\text{d}}}\cos\left(\frac{\pi}{3} + \rho_{\text{u}}\right)\right) \tag{21.29}$$

$$t_1 = \frac{1}{3}\left(1 - \frac{2U_{\text{out}}^{\text{com}}}{U_{\text{d}}}\sin\left(\frac{\pi}{6} + \rho_{\text{u}}\right)\right) \tag{21.30}$$

where $U_{\text{out}}^{\text{com}}$ and ρ_{u} are the magnitude and angle position of the reference inverter output voltage vector $\boldsymbol{U}_{\text{out}}^{\text{com}}$.

The use of non-parity active vectors leads to a constant value of the CM voltage:

$$u_0 = \frac{U_{\text{d}}}{\sqrt{3}} \tag{21.31}$$

and obviously does not allow CM current flow.

The main drawback of the modified PWM is that the maximum value of the inverter output phase voltage is limited to the value $U_{\text{d}}/3$, which leads to deformation of the motor current at higher ranges of motor speed.

21.5.2 Three Active Vector Modulation (3AVM)

The inverter output voltage can be increased simultaneously with CV voltage reduction using a method similar to the standard SVM. The modification of SVM uses all active vectors – no zero vectors are used [28, 29]. In [30], the modified modulation method is called three active vector modulation and is abbreviated as 3AVM. In 3AVM, a voltage vector plane is divided into six sectors (Figure 21.39).

If both parity and non-parity active vectors are used, the CM voltage value has a value $U_{\text{d}}/3$ when the output vector goes from one sector to the adjacent sector. Therefore, the CM voltage frequency is equal to six times the frequency of the first harmonic of the inverter output voltage. This is a much lower frequency of u_0 than that appears in classical SVM. Therefore, despite variation in u_0, the CM current

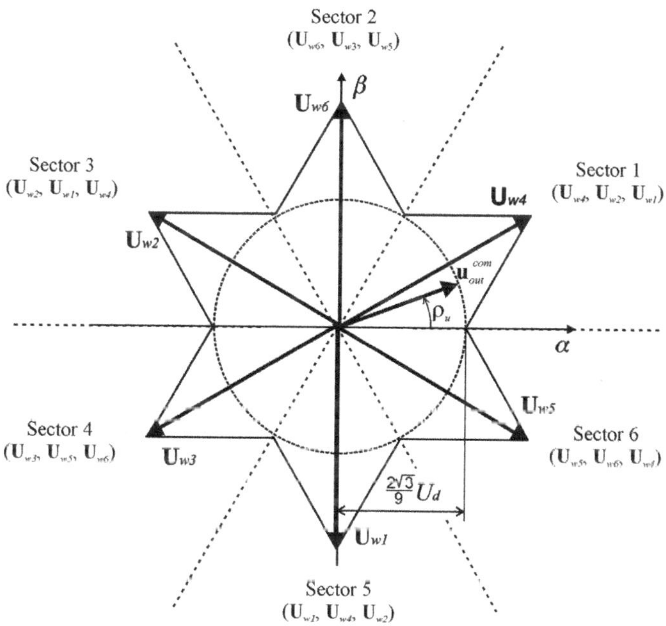

Figure 21.39 Vectors and sectors in the three active vector modulation algorithm

is significantly reduced. Simultaneously, the amplitude of the inverter output voltage is 15.5% higher in comparison with the 3NPAV algorithm.

21.5.3 Active Zero Voltage Control (AZVC)

Another PWM method for decreasing CM current is active zero voltage control (AZVC). The method presented in [30, 31] is characterized by the replacement of zero vectors by two reverse active vectors (AZVC-2) or one active vector (AZVC-1).

The principle of modulation by the AZVC-2 method is presented in Figure 21.40.

For the case shown in Figure 21.40, the output voltage vector is the result of the particular active vectors U_{w4} and U_{w6}. The angular position of the output voltage vector is determined by times t_4 and t_6, respectively. Both vectors correspond to active vectors with standard SVM. In comparison with SVM, instead of a zero vector, the reverse active vectors U_{w5} and U_{w2} are generated. Because the values of t_5 and t_2 are the same, the reverse vectors do not change the angle position of the output voltage vector – only the output voltage vector length is reduced. Thus, in method AZVC-2, the switching time calculations are as follows:

$$u_{out}^{com} \cdot T_{imp} = \mathbf{U}_{w4}t_4 + \mathbf{U}_{w6}t_6 + \mathbf{U}_{w2}t_2 + \mathbf{U}_{w5}t_5 \tag{21.32}$$

$$t_4 = T_{imp} \cdot \frac{u_{wy\alpha} \cdot \mathbf{U}_{w6\beta} - u_{wy\beta} \cdot \mathbf{U}_{w6\alpha}}{\mathbf{U}_d \cdot w_t} \tag{21.33}$$

$$t_6 = T_{imp} \cdot \frac{-u_{wy\alpha} \cdot \mathbf{U}_{w4\beta} + u_{wy\beta} \cdot \mathbf{U}_{w4\alpha}}{\mathbf{U}_d \cdot w_t} \tag{21.34}$$

$$t_2 = t_5 = \frac{1}{2}(T_{imp} - t_4 - t_6) \tag{21.35}$$

where

$$w_t = \mathbf{U}_{w4\alpha} \cdot \mathbf{U}_{w6\beta} - \mathbf{U}_{w4\beta} \cdot \mathbf{U}_{w6\alpha} \tag{21.36}$$

The principle of modulation by AVC1, presented in [27], is illustrated in Figure 21.41.

For the case shown in Figure 21.41, the vectors U_{w4} and U_{w3} have been properly chosen to assure the decrease in the magnitude of the output vector U_{out}. These vectors are, respectively, attached to the times

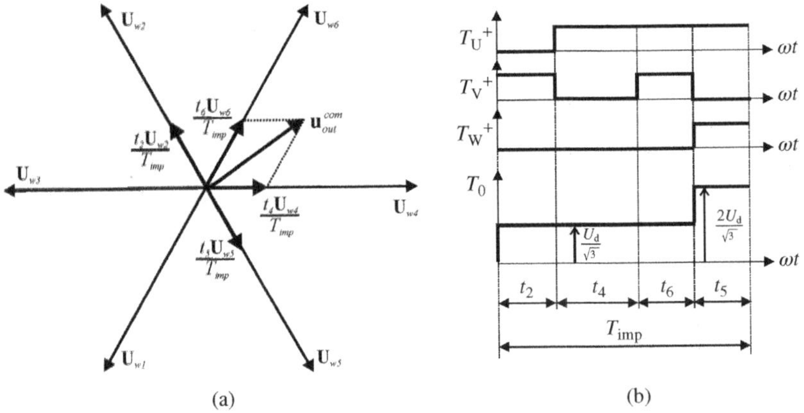

Figure 21.40 The principle of AZVC-2 modulation: (a) vectors and (b) control signals (T_U^+, T_V^+, T_W^+) and output voltage waveform u_{out}

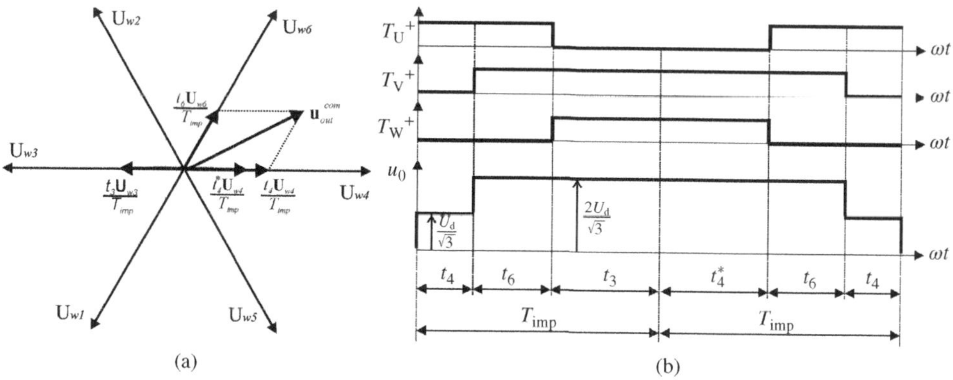

Figure 21.41 The principle of AZVC-2 modulation: (a) vectors and (b) control signals $T_U{}^+$, $T_V{}^+$ and $T_W{}^+$ and output voltage waveform u_{out}

t_4 and t_3^*, where $t_4^* = t_3$. The sum of times t_4^* and t_3 is equivalent to t_0 in classical SVM. In AZVC-1, the reverse vectors are chosen such that one of them must be the same as one of the active vectors determining the angle position of u_{out}^{com}; that is, it is U_{w4} for the combination presented in Figure 21.41. In the method AZVC-1, the relation between vectors and switching times is

$$u_{out}^{com} \cdot T_{imp} = U_{w4}t_4 + U_{w6}t_6 + U_{w3}t_3 + U_{w4}t_4^* \tag{21.37}$$

The switching times t_4 and t_6 are determined by (21.33) and (21.34), whereas t_4^* and t_3 take the values:

$$t_3 = t_4^* = \frac{1}{2}(T_{imp} - t_4 - t_6) \tag{21.38}$$

Both AZVC methods make it possible to obtain the same maximum output voltage of the inverter as the SVM method. In comparison with the classical SVM, both the amplitude and frequency of the CM voltage are reduced and finally the CM current is also limited.

21.5.4 Space Vector Modulation with One-Zero Vector (SVM1Z)

The previously presented methods operate without zero vectors. This is a disadvantage, because it leads to problems with inverter output current measurement. In the voltage inverter operating with classical SVM, the output current has to be sampled synchronously with SVM. Simultaneously the sampling instant must be in the middle of the zero vectors period. If the samples are taken in that instant, then the instantaneous value of the current is nearly equal to the fundamental harmonic for the inverter output current [32–34] (Figure 21.42).

The reduction of CM current while leaving the synchronous measurement unchanged is possible if only one of the zero vectors is eliminated, $\boldsymbol{U}_{w0}$ or $\boldsymbol{U}_{w7}$. Thus, the modified SVM algorithm is called the *S*pace *V*ector *M*odulation with *One*-*Z*ero vector (SVM1Z) modulator. An example of the control signals and output voltage waveforms for SVM1Z is presented in Figure 21.43.

The example shown in Figure 21.43 consists of a sequence including one-zero vector, $\boldsymbol{U}_{w0}$ and two active vectors, $\boldsymbol{U}_{w4}$ and $\boldsymbol{U}_{w6}$. When $\boldsymbol{U}_{w6}$ is switched off, the same zero vector, $\boldsymbol{U}_{w0}$, is applied. The number of transistors switching for one PWM cycle increases but will not influence synchronous sampling of the inverter output current.

If one-zero vector is eliminated, the amplitude of the CM voltage is reduced by 33%. Unfortunately, dv/dt is increased for the instant of transition from active vector to zero vector, that is, from $\boldsymbol{U}_{w6}$ to $\boldsymbol{U}_{w0}$. The comparison can be observed in Figure 21.44.

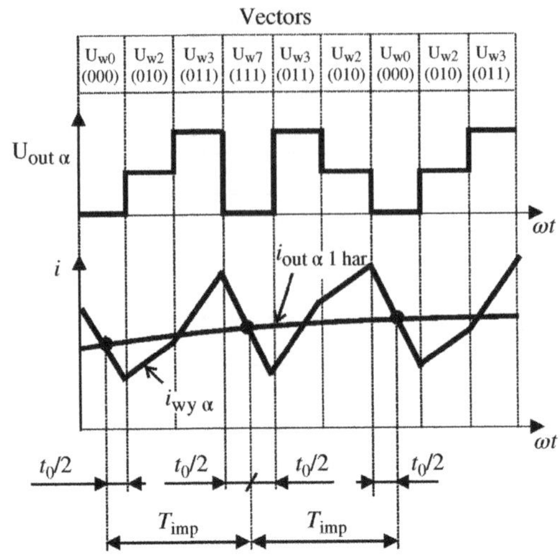

Figure 21.42 The relationship between current sampling instant and inverter output vectors in classical SVM

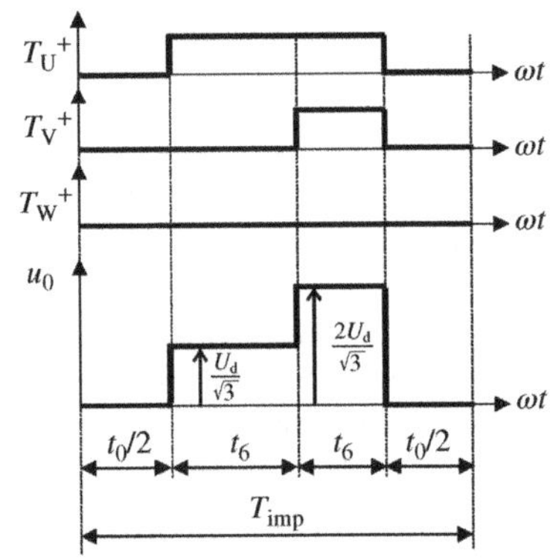

Figure 21.43 An example of control signals and inverter output voltage waveforms in the SVM1Z method

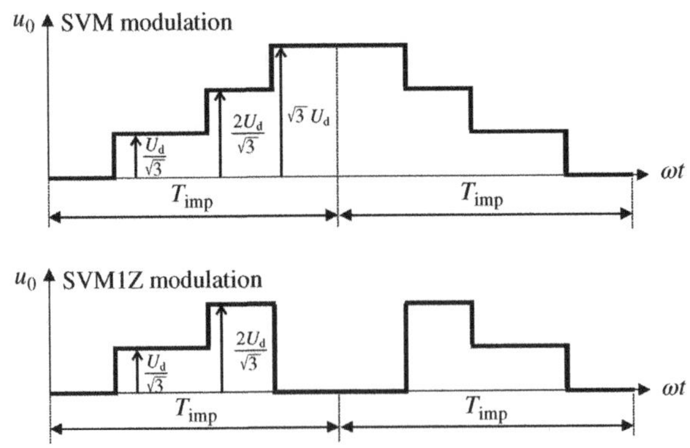

Figure 21.44 The comparison of CM voltage waveforms in the SVM and SVM1Z algorithms

The quantitative effect of the SVM1Z algorithm on CM current reduction is difficult to evaluate in an analytical manner. This is due to the fact that a model of the electrical machine CM is nonlinear and not perfectly defined. For these reasons, the usefulness of the SVM1Z method needs to be tested in practical applications.

The use of SVM1Z methods can bring significant benefits in the case of CM filter use. With increasing frequency of u_0, the core of the CM choke will have a smaller diameter, or if the diameter is kept constant the CM current will decrease.

A comparison of the SVM, AZVC-2 and SVM1Z methods is presented in Figures 21.45–21.47.

As can be seen in the waveforms presented in Figures 21.35–21.37, the shape of the motor current is the same with SVM and SVM1Z. In the case of SVM1Z, the amplitude of the CM voltage decreases. The smallest CM voltage occurs in the case of AZVC-2 control, but the current shape is the worst.

The practical implementation of SVM1Z is simpler than that of NPAV, AVM and AZVC. At the same time, the inverter output voltage is not reduced. So SVM1Z is implemented in some industrial inverters nowadays (Figure 21.48).

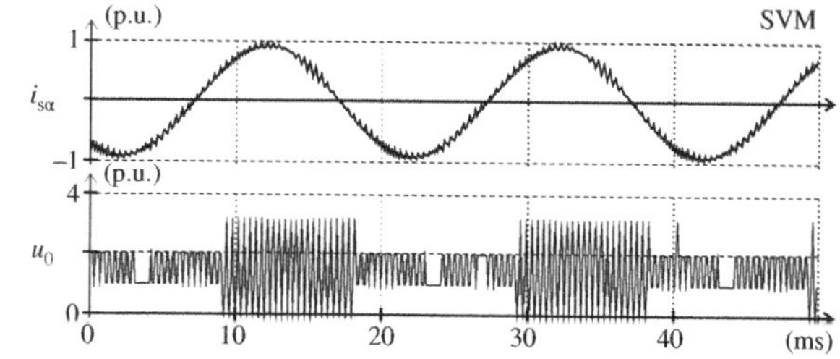

Figure 21.45 The motor current and CM voltage waveforms in the case of classical SVM control

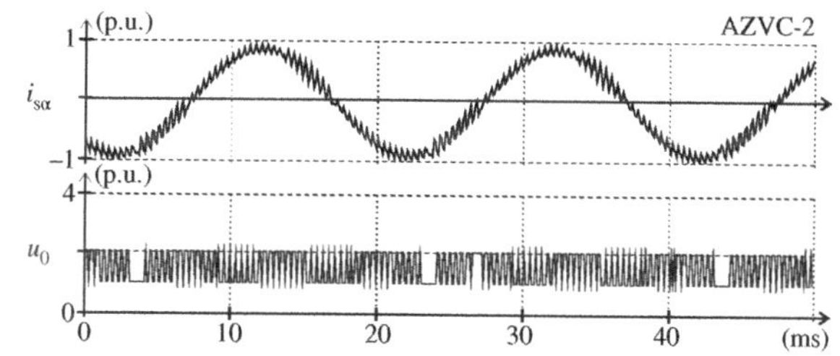

Figure 21.46 The motor current and CM voltage waveforms in the case of AZVC-2 control

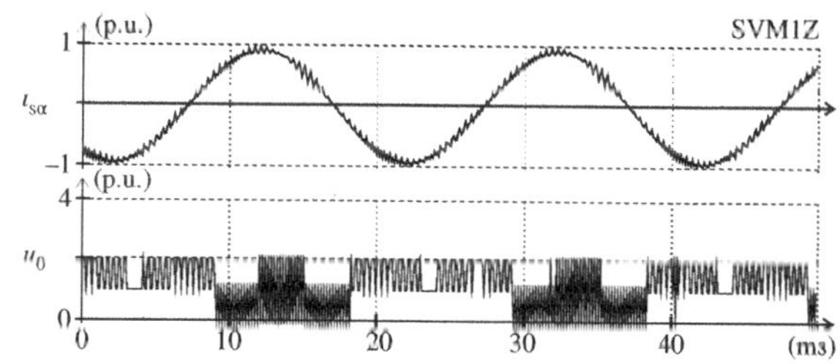

Figure 21.47 The motor current and CM voltage waveforms in the case of SVM1Z control

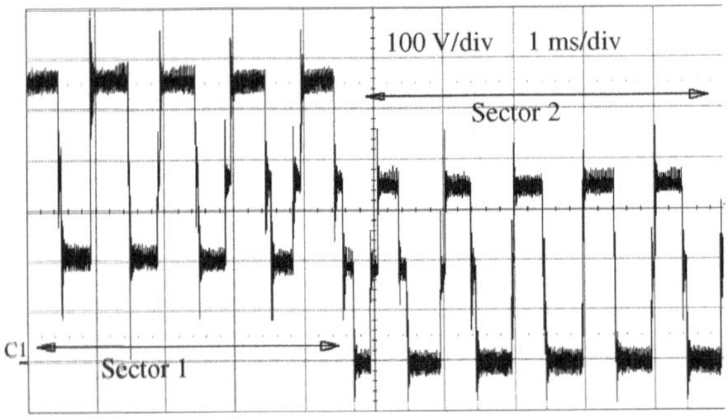

Figure 21.48 The motor current and CM voltage waveforms in the case of SVM1Z control with different zero vectors for output voltage position sectors

In the case of SVM1Z, in each of the position sectors of the inverter output voltage vector, different zero vectors could be used. This minimizes the switching of transistors when the output vector crosses the border between the sectors.

21.6 Summary

Voltage inverters operating with PWM are a source of CM voltage. This voltage forces the CM current to flow through the parasitic capacitances of the motor. Part of the current flows through the motor bearings, causing their accelerated degradation.

Nowadays, the most common type of failure of IMs is bearing failures. Therefore, the most reasonable course of action is to seek and implement solutions that will help prevent this damage.

CM current reduction can be done either by increasing the impedance of the proper electrical circuit or by reducing the CM voltage. The impedance could be increased by using the appropriate passive filter, the CM filter. Reduction of the CM voltage is done by active solutions – mainly by methods for modification of the PWM.

The solutions that use PWM modification are cheaper compared to the solutions involving the use of filters. With PWM modifications, no additional elements are required; only the appropriate changes and extensions in the control software are required. However, these PWM changes will result in increases in the number of transistor switching, which will cause the efficiency of the inverter to decrease, and a more efficient inverter cooling system could be required.

Owing to the complexity of the motor CM electrical circuit and the nonlinearities of some elements, the usefulness of different PWM controls needs to be tested in real conditions. Nowadays, the analytical or simulation models presented are insufficient for estimation of the CM current.

References

1. Busse, D., Erdman, J., Kerkman, R.J. *et al.* (1995) Bearing currents and their relationship to PWM drives. IEEE 21st International Conference on Industrial Electronics, Control, and Instrumentation, IECON 1995, November 6–10, 1995, Vol. 1, pp. 698–705.
2. Conraths H.J., Giebler F., and Heining H.D. (1999) Shaft-voltages and bearing currents – new phenomena in inverter driven induction machines. 8th European Conference on Power Electronics and Applications EPE'99, Lausanne, France, September 7–9, 1999, pp. 1–9.
3. Muetze, A. and Binder, A. (2007) Practical rules for assessment of inverter-induced bearing currents in inverter-fed AC motors up to 500 kW. *IEEE Transactions on Industrial Electronics*, **54** (3).

4. Pairodamonchai, P. and Sangwongwanich, S. (2011) Exact common-mode and differential-mode equivalent circuits of inverters in motor drive systems taking into account input rectifiers. IEEE 11th International Conference on Power Electronics and Drive Systems (PEDS), Singapore, December 5–8, 2011.
5. Muetze, A. and Binder, A. (2003) High frequency stator ground currents of inverter-fed squirrel-cage induction motors up to 500 kW. 10th European Conference on Power Electronics and Applications, EPE'03, Toulouse, France, September 2–4, 2003.
6. Bhushan, B. (2002) *Introduction to Tribology*, John Wiley &Sons, Inc., New York.
7. Muetze, A. and Binder, A. (2006) Calculation of influence of insulated bearings and insulated inner bearing seats on circulating bearing currents in machines of inverter-based drive systems. *IEEE Transactions on Industry Applications*, **42** (4), 965–972.
8. Binder, A. and Muetze, A. (2008) Scaling effects of inverter-induced bearing currents in AC machines. *IEEE Transactions on Industry Applications*, **44** (3), 769–776.
9. Muetze, A. and Binder, A. (2007) Calculation of circulating bearing currents in machines of inverter-based drive systems. *IEEE Transactions on Industrial Electronics*, **54** (2), 932–938.
10. Dymond, J.H. and Findlay, R.D. (1997) Comparison of techniques for measurement of shaft currents in rotating machines. *IEEE Transactions on Energy Conversion*, **12** (4), 363–367.
11. Strom, J.P., Koski, M., Muittari, H., and Silventoinen, P. (2007) Analysis and filtering of common mode and shaft voltages in adjustable speed AC drives, 12th European Conference on Power Electronics and Applications, EPE'07, Aalborg, Denmark, September 2–5, 2007.
12. Busse, D.F., Erdman, J., Kerkman, R. *et al.* (1996) The effects of PWM voltage source inverters on the mechanical performance of rolling bearings. Eleventh Annual Applied Power Electronics Conference, APEC'96, San Jose, CA, March 3–7, 1996.
13. Busse, D.F., Erdman, J.M., Kerkman, R.J., Schlegel, D.W., and Skibinski, G.L. (1997) An evaluation of the electrostatic shielded induction motor: a solution for rotor shaft voltage buildup and bearing current. *IEEE Transactions on Industry Applications*, **33** (6), 1563–1570.
14. Ogasawara, S. and Akagi, H. (1996) Modeling and damping of high-frequency leakage currents in PWM inverter-fed AC motor drive systems. *IEEE Transactions on Industry Applications*, **22** (5).
15. ABB Drives (2011) Bearing Currents in Modern AC Drive Systems. Technical Guide No. 5. ABB.
16. Akagi, H., Hasegawa, H., and Doumoto, T. (2004) Design and performance of a passive EMI filter for use with a voltage-source PWM inverter having sinusoidal output voltage and zero common-mode voltage. *IEEE Transactions on Power Electronics*, **19** (4), 1069–1076.
17. Akagi, H. (2002) Prospects and expectations of power electronics in the 21st century. Power Conversion Conference, PCC'02, Osaka, Japan, April 2–5, 2002.
18. Abu-Rub, H., Iqbal, A., and Guzinski, J. (2012) *High Performance Control of AC Drives with Matlab/Simulink Models*, John Wiley & Sons, Inc., New York.
19. Wang, X., Huang, K., and Xu, B. (2007) A new inverter output passive filter topology for PWM motor drives, 8th International Conference on Electronic Measurement and Instruments ICEMI'2007, Xi'An, China, August 16–18, 2007.
20. X. Chen, D. Xu, F. Liu, J. Zhang: A novel inverter-output passive filter for reducing both differential – and common-mode dv/dt at the motor terminals in PWM drive systems, *IEEE Transactions on Industrial Electronics*, **54** (1), 419–426.
21. Lai, R., Maillet, Y., Wang, F., Wang, S., Burgos, R., and Boroyevich, D. (2010) An integrated EMI choke for differential-mode and common-mode noise suppression. *IEEE Transactions on Power Electronics*, **25** (3).
22. Sun, Y., Esmaeli, A., and Sun, L. (2006) A new method to mitigate the adverse effects of PWM inverter. 1st IEEE Conference on Industrial Electronic and Applications, ICIEA'06, Singapore, May 24–26, 2006.
23. Kikuchi, M. and Kubota, H. (2009) A novel approach to eliminating common mode voltages of PWM inverter with a small capacity auxiliary inverter. 13th European Conference on Power Electronics and Applications EPE'09, Barcelona, Spain, September 8–10, 2009.
24. Kikuchi, M., Naoyuki, A.Z., Kubota, H. *et al.* (2009) Investigation of common-mode voltages of PWM inverter with a small capacity auxiliary inverter. 12th International Conference on Electrical Machines and Systems, ICEMS'09, Tokyo, Japan, November 15–18, 2009.
25. Kempski, A., Smolenski, R., Kot, E., and Fedyczak, Z. (2004) Active and passive series compensation of common mode voltage in adjustable speed drive system. 39th IEEE Industry Applications Conference, IAS 2004, Seattle, WA, October 3–7, 2004.

26. Zitselsberger, J. and Hofmann, W. (2003) Reduction of bearing currents by using asymmetric space-vector-based switching patterns. European Conference on Power Electronics and Applications, EPE'03, Toulouse, France, September 2–4, 2003.
27. Zitzelsberger, J. and Hofmann, W. (2004) Reduction of bearing currents in inverter fed drive applications by using sequentially positioned pulse modulation. *EPE Journal*, **14** (4), 19–25.
28. Cacciato, M., Consoli, A., Scarcella, G., and Testa, A. (1999) Reduction of common-mode currents in PWM inverter motor drives. *IEEE Transactions on Industry Applications*, **35** (2), 469–476.
29. Cacciato, M., Consoli, A., Scarcella, G. *et al.* (2009) Modified space-vector-modulation technique for common mode currents reduction and full utilization of the DC bus. Twenty-Fourth Annual IEEE Applied Power Electronics Conference and Exposition, APEC 2009, Singapore, November 14–15, 2009.
30. Hofmann, W. and Zitzelsberger, J. (2006) PWM-control methods for common mode voltage minimization – a survey. International Symposium on Power Electronics, Electrical Drives, Automation and Motion, SPEEDAM 2006, Taormina (Sicily), Italy, May 23–26, 2006.
31. Lai, Y.S. and Shyu, F.-S. (2004) Optimal common-mode voltage reduction PWM technique for inverter control with consideration of the dead-time effects – part I: basic development. *IEEE Transactions on Industry Applications*, **40** (6), 1605–1612.
32. Blasko, V., Kaura, V., and Niewiadomski, W. (1998) Sampling of discontinuous voltage and current signals in electrical drives – a system approach. *IEEE Transactions on Industry Applications*, **34** (5), 1123–1130.
33. Briz, F., Díaz-Reigosa, D, Degner, M.W. *et al.* (2010) Current sampling and measurement in PWM operated AC drives and power converters. The 2010 International Power Electronics Conference IPEC, Sapporo, Japan, June 21–24, 2010.
34. Holtz, J. and Oikonomou, N. (2008) Estimation of the fundamental current in low-switching-frequency high dynamic medium-voltage drives. *IEEE Transactions on Industrial Applications*, **44** (5), 1597–1605.

22

High-Power Drive Systems for Industrial Applications: Practical Examples

Lazhar Ben-Brahim[1] and Teruo Yoshino[2]
[1]*Department of Electrical Engineering, Qatar University, Doha, Qatar*
[2]*Power Electronics Systems Division of Toshiba Mitsubishi-Electric Industrial Systems Corporation, Tokyo, Japan*

22.1 Introduction

Most natural gas (NG) reserves are located in areas (Middle East or Africa) far from large consumers, such as the United States, China and Japan. Therefore, the gas needs to be transported over long distances. However, NG is voluminous and the reduction of its volume is necessary prior to its transportation. NG volume can be reduced by around 600 times by using a liquefaction process; thus, making it more economical for storage and transportation [1]. The obtained liquefied natural gas (LNG) is a natural gas that has been cooled to −160 °C changing it to a liquid. To carry out this task, several large refrigeration compressors are used for each LNG plant. The operation of these high-powered compressors requires a large drive system. Traditionally, these compressors, whose power exceeds 100 MW, are driven by large gas turbines (GTs) [2]. To start a GT, it is often combined with a 10–20 MW motor drive called a GT starter or helper [3]. Recent advances in power electronics technology have offered hope for the development of multi-megawatt electric drives to drive large-capacity compressors in modern LNG plants, instead of GT drives. These electric drives, also called variable frequency drives (VFDs), provide operational flexibility for driving compressors in LNG plants. VFDs are characterized by better efficiency, lower downtime, lower maintenance costs and lower emissions when compared with that of GTs. In general, VFDs are more suitable than GTs in liquefaction applications and are more environmentally friendly drives. Megawatt VFDs in LNG plants are connected to the grid-through power converters/inverters. Today, it is hard to connect a single power converter directly to medium-voltage grids for several reasons. Therefore, various architectures of multilevel inverters have emerged as a solution to this problem. VFDs are classified according to these converter topologies. The suitability of these converters for use

Power Electronics for Renewable Energy Systems, Transportation and Industrial Applications, First Edition.
Edited by Haitham Abu-Rub, Mariusz Malinowski and Kamal Al-Haddad.

in LNG plants and the techniques used to control the megawatt VFDs are discussed. This chapter starts with an overview of LNG plants, outlines conventional GT, drives technoeconomic and environmental implications and introduces various electric drive technologies used for LNG plants, highlighting their limitations, technological problems and their impact on future LNG plants.

22.2 LNG Plants

An LNG plant is a facility that receives an NG feed from gas wells, processes it and liquefies it for shipping to distant customers. Usually, the plant comprises several processing "trains." Many plants start with a single train and additional trains are installed later to cope with additional gas reserves. Figure 22.1(a) shows an example of a simplified train of an LNG plant overall flow scheme. In a typical scheme, the feed gas is first pretreated to remove any impurities that might hinder the liquefaction process or affect the quality of the final product. Acid gases, sulfur compounds, water and mercury are among the impurities removed from the feed gas. The obtained product, called dry sweet gas, is then cooled by refrigerant streams to separate the heavier hydrocarbons. The remaining gas is cooled further to approximately −160 °C to make it fully LNG. The resulting LNG is properly stored and made ready for transportation by custom-made ships. The liquefaction process represents 30–40% of the overall cost of an LNG plant [4]. Many liquefaction processes are available in the market; Figure 22.1(b) shows that Philips optimized cascade LNG process, which is widely used. The liquefaction process is usually carried out in various stages using three pure component refrigerants: propane, ethylene and methane at various pressure levels. The refrigerants are circulated using centrifugal compressors driven by large GTs or motors. Traditionally, large GTs with power exceeding 100 MW drive the liquefaction compressors. In order to start a GT, it is often combined with a 10–20 MW motor called a GT starter or helper.

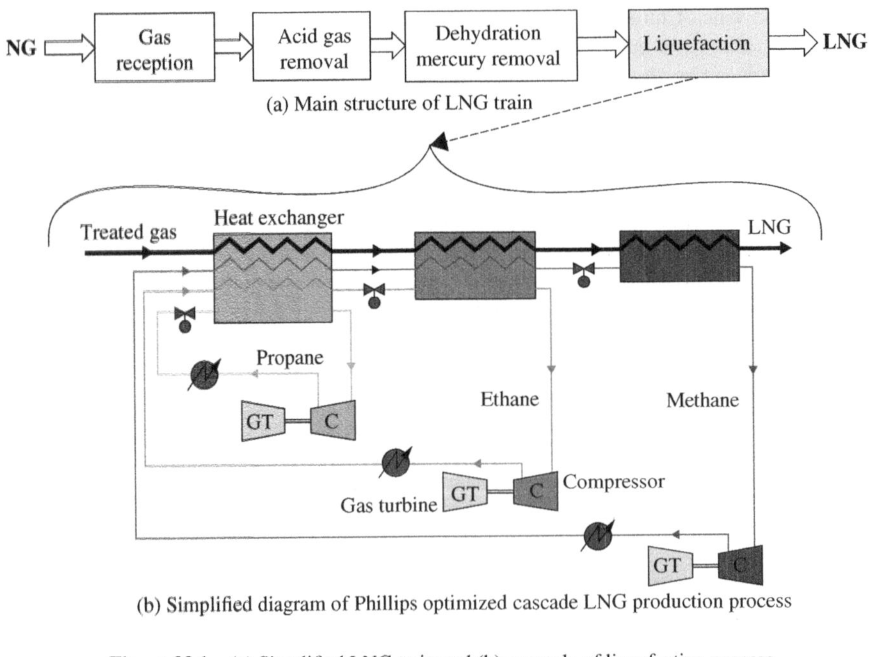

Figure 22.1 (a) Simplified LNG train and (b) example of liquefaction process

22.3 Gas Turbines (GTs): the Conventional Compressor Drives

Traditionally, large GTs are used for power generation applications. GT's reliability and performance have improved through these applications. This successful use of GTs in generation helped in the adopting of larger ones for particular mechanical-drive applications.

Several major LNG plants are embracing GTs as drivers for liquefaction compressors. However, the change from generator-drive applications to mechanical-drive applications brings new considerations, such as variable speed operation, starting method, torsional and lateral acceptability, continuous service reliability and so on [5].

22.3.1 Unit Starting Requirements

When a single-shaft generator-drive GT starts driving an unloaded generator, a modestly sized starting motor (or other starting means) can bring the GT to self-sustaining speed and torque. With a single-shaft GT mechanical drive, the net available torque from the standard GT starting system (starter motor/converter) is inadequate to start a typical load compressor. To overcome this shortfall, in most GT compressor drive applications, an adjustable speed electric motor drive (ASD) is used, as shown in Figure 22.2. This starting ASD can also be used as a helper drive on the train. The ASD provides the entire starting torque, the magnitude of which will vary according to the particular characteristics of the process compressors and the method and degree of unloading achieved [5]. The need for starting is only one issue regarding the use of GT drives. GTs have limited speed/power characteristics, as outlined in Figure 22.3. As the power increases, the speed decreases and LNG compressors need high-power high-speed operation.

22.3.2 Temperature Effect on GT Output

Because the GT is an air-breathing engine, its performance is changed by ambient weather conditions, such as humidity and ambient temperature. The effect of humidity is thought to be less important than

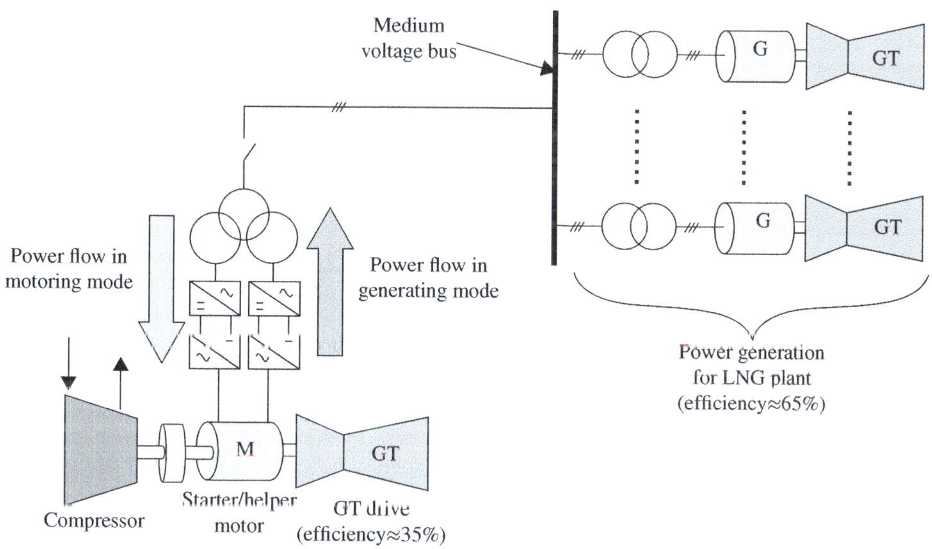

Figure 22.2 Conventional compressor drive system

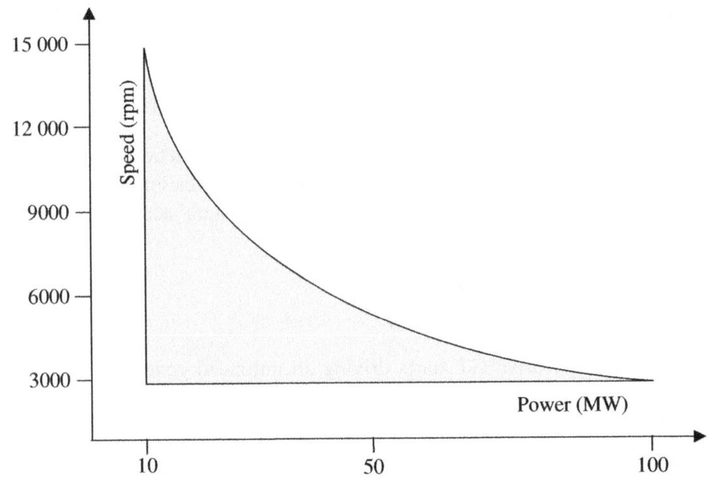

Figure 22.3 GT speed versus power characteristics

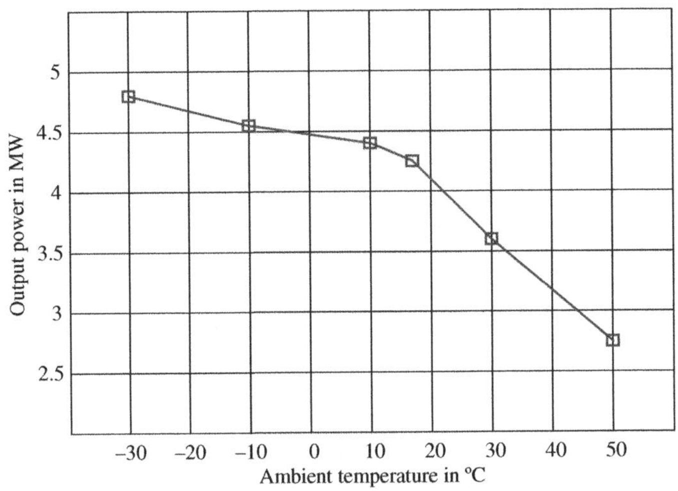

Figure 22.4 Effect of ambient temperature on GT output power

temperature change. Figure 22.4 shows how the output of a GT is significantly affected by the ambient temperature. This is one of the most critical issues in GT drives and to overcome this, a helper motor is needed to produce the lost power.

22.3.3 Reliability and Durability

Conventionally, the GT's control system has full flexibility to coordinate with the sequencing and control requirements of the load equipment/process, including the precise speed control of the train. The GT's control system uses redundant controllers and redundant sensors to enhance the reliability.

As shown in the above, all GTs, by the nature of their physics and design, have inherent limitations:

(a) Limited speed range especially for high-power GTs
(b) Not self-starting, slow starting and cannot accelerate a load compressor
(c) As a mechanical drive, a GT has lower efficiency than the combined cycle type GT used for generation
(d) Lower lifetime owing to high thermal and mechanical stresses caused by high-ambient temperatures
(e) Output power depends on ambient temperatures and efficiency degrades with the rise of temperature
(f) Complexity and sensitivity of the machine
(g) Periodic maintenance is required
(h) Limited vendor competition

22.4 Technical and Economic Impact of VFDs

In LNG plants, the GT-driven compressor power rating is approximately 100–120 MW. To replace GTs of this size, electric motors should be capable of delivering such power at the LNG compressor speed. These types of motors are widely used as generators with capacity exceeding 250 MW. In LNG applications, two-pole motors of this size can be used. VFDs using two-pole motors are technically and economically viable solutions for the required LNG power ranges. In such cases, VFDs can be used to drive a compressor with higher efficiency, as illustrated in Figure 22.5. Suppliers of electric motors and VFDs are proposing various drives to control these high-power motors. The following sections will highlight the latest technological achievements in these fields.

Table 22.1 provides a comparison of the GT and VFD characteristics. Using VFDs has a number of significant advantages over the conventional GT in terms of maintenance, reliability, efficiency and air and noise pollution. Conventional GTs require extensive and costly periodic maintenance. On the other hand, VFDs require less maintenance and, thus, offer greater availability and efficiency for LNG production. Moreover, LNG production using VFDs is not affected by ambient temperature, as in the

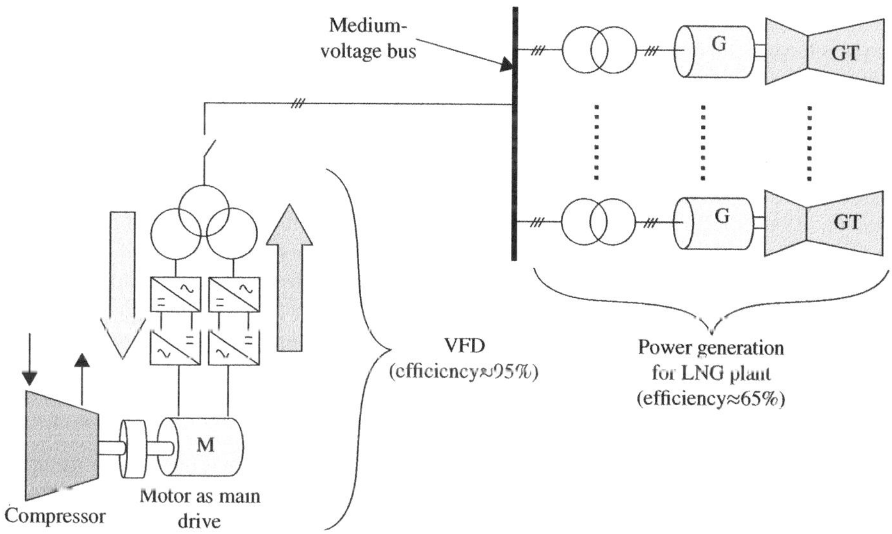

Figure 22.5 VFD-based compressor drive system with around 55% overall efficiency

Table 22.1 Comparison of the characteristics of GT-drive and VFD for LNG compressor

	GT (Gas turbine) drives	VFDs
Initial investment cost	Low	High (including power generation)
Running and maintenance cost	High	Low
Efficiency	<35%	>95% for VFD only ~55% (including combined cycle (CC) GT unit generating electric power for VFD)
Availability	95% (shutdown 15 days/year)	Almost 100% (no major shutdown is needed)
Maintenance	Periodic maintenance (15 days/year)	More less maintenance free
Shaft length	Long	Short
Noise level	Very high	Moderate
Speed control range	Limited	Wide
Greenhouse gasses production	High	Low (including the CC power generation unit)

case of using GTs. Ambient temperature changes affect the performance of GTs, as shown in Figure 22.4, which is why in the case of a GT drive, a starter motor is used as a helper motor in order to overcome the shortage of power experienced at higher ambient temperatures.

22.5 High-Power Electric Motors

For large-capacity applications such as LNG plants, two types of electric motors are mainly used: induction motors (IMs) and/or synchronous motors (SMs). Each type of motor has its own specific features and it is selected according to the application type, the electric power network and environmental conditions [6]. In practice, depending on the application case, a detailed comparison between the various motors is necessary in order to select the most suitable type. The following general facts are observed during the selection of motor type. Although SMs are more complex owing to the need for an exciter and controller to operate, they are characterized by higher efficiency, controllable power factor and constant rotating speed regardless of load variation. On the other hand, IMs are simpler and are of lower cost. For high-speed rotation applications, IMs are preferred because the rotor structure is simple and can be made stiff. SMs are most likely selected for large-capacity applications because it is easier to realize a machine with large torque output. Figure 22.6 shows the concepts of the trade-off trends between motor capacity and rotating speed. Note that the numbers given in the figure are approximate ranges of motor power and speed. Figure 22.6 also shows the physical limits of manufactured motors. The higher the rotating speed, the higher is the centrifugal force applied to the rotor, which increases as the rotor radius increases. A larger motor should withstand the larger centrifugal force of its rotor compared with a smaller motor rotating at the same speed. The centrifugal force F is proportional to the square of rotor speed and the radius of the rotor

$$F \propto r\omega^2 = r\left(2\pi\frac{n}{60}\right)^2 \tag{22.1}$$

where r is the radius of the rotor, ω is the angular rotor speed (in rad/s) and n is the rotating speed in rpm.

The rotor can be manufactured from several types of materials, such as steel with conductors of aluminum or copper. The mechanical strengths of these materials determine the stiffness of the rotor. Therefore, the rotor has finite stiffness and a limit to the centrifugal force it can withstand. As a result, when the rotating speed is higher, the motor capacity is most likely smaller.

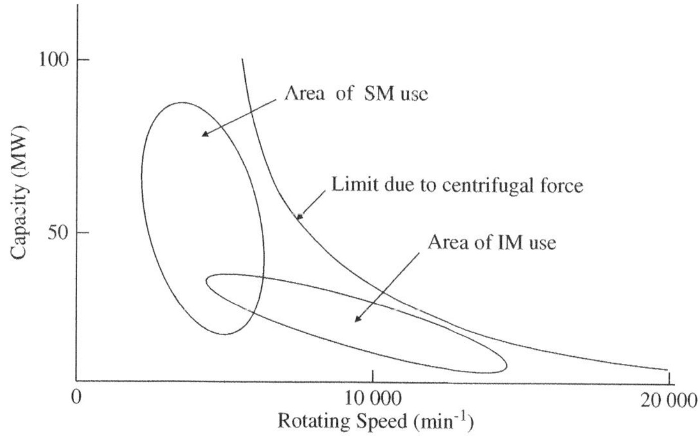

Figure 22.6 Areas of use of induction and synchronous motors

22.5.1 State-of-the-Art High-Power Motors

Technologies for very high-speed and large-capacity motors in the range of several tens of megawatts are introduced by showing actual motor examples. For very high-speed applications, special considerations are taken regarding the mechanical structure of the rotor and bearing. Although of higher efficiency, for large-capacity applications, cooling considerations is taken into account to reduce further the losses in the motors. Furthermore, an explosion proof construction is generally required for use in hazardous locations such as LNG plants [7].

22.5.1.1 Induction Motor

An example of a very high-speed and large-capacity IM rated at 3.3 MW and 11 430 rpm is shown in Figure 22.7. The motor can be used with the drive inverter for adjustable speed operation. The IM illustrated is a squirrel cage type with a totally enclosed water-to-air-cooled (TEWAC) enclosure. The air

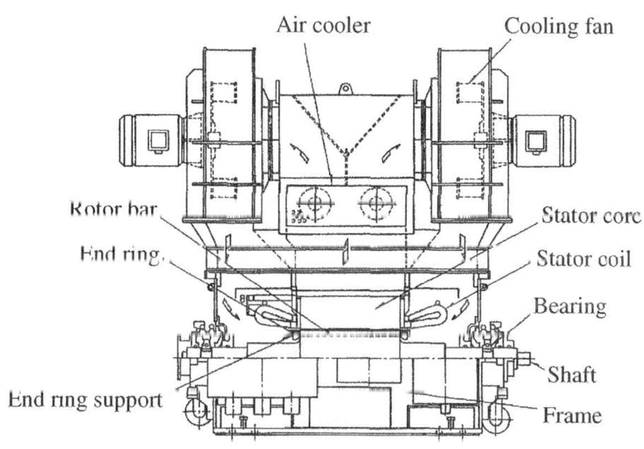

Figure 22.7 3.3 MW, 11 430 rpm induction motor construction

Figure 22.8 8 MW, 10 000 rpm induction motor with magnetic bearing (Reproduced by permission of TMEIC)

cooler and cooling fans are mounted on the top of the motor's main structure and the cooling air is circulated within the motor enclosure. The cooling air is forced to circulate by the auxiliary fans, as shown in Figure 22.7. Note that the cooling system is closed, and that no air is introduced from the outside atmosphere. The air is cooled by exchanging the heat using water. The motor is a two-pole machine and the rotor type is solid for very high-speed applications. For the optimal rotor design, a mechanical strength analysis was performed beforehand with the finite element method. The bearing is another of the important items of a very high-speed motor. In Figure 22.7, the tilting pad bearing is a sleeve type, which is selected to guarantee a stable operation over a wide range of speed.

Figure 22.8 shows an external view of another very high-speed motor example rated at 8 MW and 10 000 rpm. This is a similar motor to the motor shown in Figure 22.7, but a magnetic bearing is used to levitate the rotor shaft. The absence of mechanical contact offers good maintainability and small loss in rotation. However, the system needs a controller to regulate the shaft position control.

22.5.1.2 Synchronous Motor

The structure of a synchronous motor rated at 25 MW and 3600 rpm is illustrated in Figure 22.9(a). The motor comprises the rotor, the stator and a TEWAC enclosure to comply with the safety requirements of LNG and similar plants. The cooling system does not introduce air from outside, and the airflow is forced to circulate within the enclosure, as shown in Figure 22.9(b). In this case, the air is driven by the fans installed at the rotor shaft. The heated air is cooled by heat exchange with the external secondary cooling media, which in this case is water. The cylindrical rotor has two poles and it is cut from a bulk of forged alloy. The shaft is flexible and has structure with good balance for adjustable speed from 70–105% of the rated rotating speed. The motor is suitable for adjustable speed operation with the high-voltage and

(a)

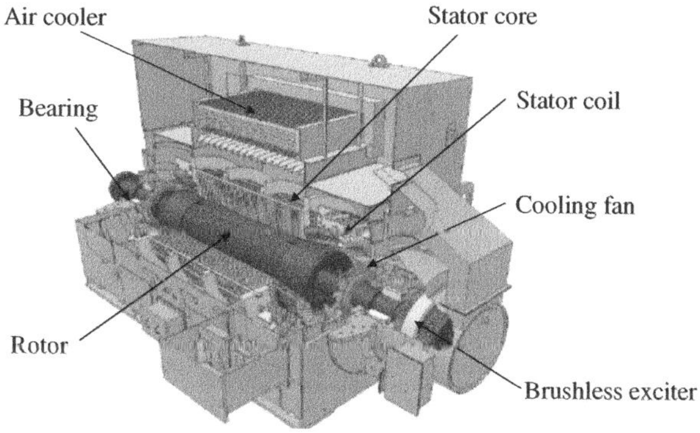

(b)

Figure 22.9 External view and construction of 25 MW, 3600 rpm synchronous motor: (a) external view and (b) construction (Reproduced by permission of TMEIC)

large-capacity drive inverters, which are introduced in the following section. The stator is supported by a spring structure, which reduces mechanical vibration.

By applying the technologies described above, much larger capacity two-pole SMs can be realized. Figure 22.10 shows a two-pole synchronous motor rated at 53 MW and 3000 rpm.

22.5.2 Brushless Excitation for SM

The synchronous motor needs an excitation to flow current in the rotor winding. Conventionally, the DC current is supplied by an external excitation circuit using a thyristor rectifier. The excitation current flows through the sliding contacts between the brush and the slip ring on the rotor shaft. The conceptual construction of the excitation circuit is shown in Figure 22.11. However, the brushed excitation needs periodic maintenance. In addition, the difficulty increases with much higher speeds or larger capacities because of the larger current flowing through the brushes and the higher rotating speed of the slip rings.

Recently, a brushless excitation has been applied by installing an auxiliary excitation generator and a rectifier on the rotor shaft. A brushless excitation is shown in Figure 22.12, illustrating the main difference to brushed excitation (shown in Figure 22.11), which is where the excitation is done by the stationary winding, that is, the stator. The auxiliary excitation generator outputs an AC current and the rotating rectifier rectifies the AC current to a DC current to flow in the rotor of the SM. In fact, two methods are

Figure 22.10 53 MW, 3000 rpm synchronous motor (Reproduced by permission of TMEIC)

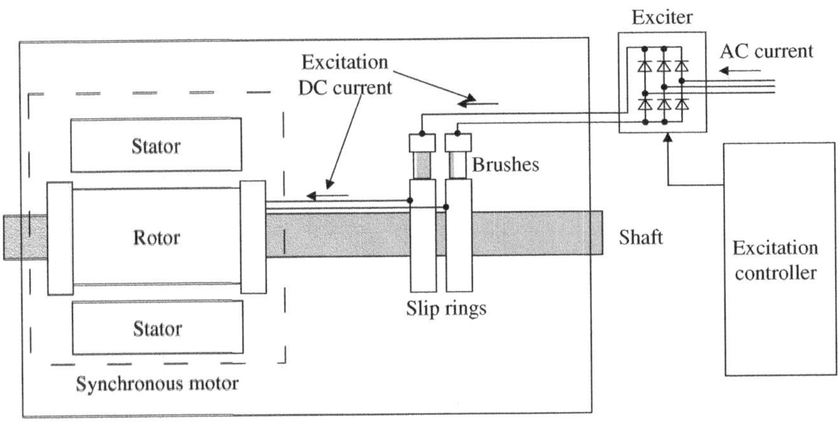

Figure 22.11 Brushed excitation for SM using brushes and slip rings

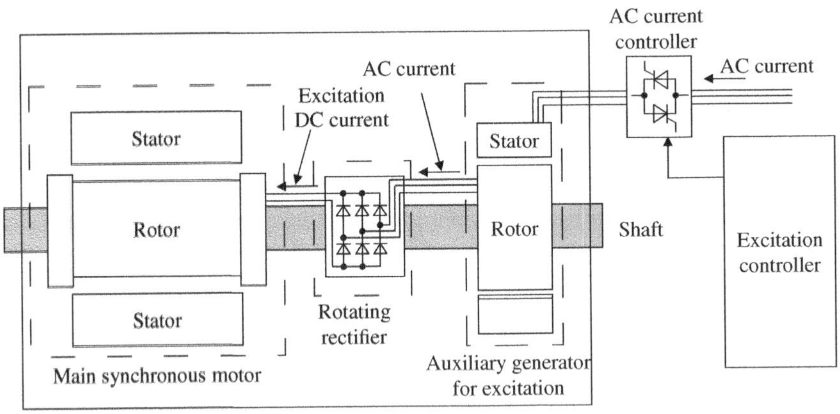

Figure 22.12 Brushless excitation of SM

applied to excite the stator of the auxiliary excitation generator. One is to apply a DC current and the other is to apply an AC current, as shown in Figure 22.12.

In the case of an AC current excitation, the auxiliary excitation generator operates as it does in the wound-rotor induction generator. This type of circuit is used in large-capacity adjustable speed SMs, because it can generate steady power while the rotating speed of the main motor varies. By regulating the AC current supplied from the external power source using the thyristor switches or equivalent circuits, the output AC current from the auxiliary excitation generator can be controlled and then the field current can be controlled.

In the case of a DC current excitation, the auxiliary excitation generator operates as it does in the synchronous generator. In this case, the auxiliary excitation generator cannot produce power when the main motor is at a standstill. Therefore, this circuit is only applied to the synchronous motor operating at constant speed.

22.6 High-Power Electric Drives

Large-capacity motors rated at several tens of megawatts are realized by the technologies described earlier. For adjustable speed operation of such large-capacity motors, to match up with the capacity of the converters, combinations of multiple inverters are used to drive the motor. To synthesize the outputs from multiple inverters, the following methods of connections are used in practice:

- Parallel connection of inverters through balance reactors at the output
- Synthesis of inverter outputs through a multiwinding transformer

Two types of connections with dual converters are shown in Figure 22.13. For these systems, installation space is required for the balance reactors or for the transformers, in addition to the motor and the inverters.

Multiwinding motors are driven by more than one converter. In this case, the armature windings are separated into groups and each group's input terminals are connected to one of the converter/inverter set. Figure 22.14 shows a two windings motor driven by dual converters.

22.7 Switching Devices

Together with the development of semiconductor switching devices for higher voltage and larger capacity, the drive inverter is also developed for higher voltage and for larger capacity [8]. This section introduces briefly the development history of the semiconductor switching devices used for high-power VFDs. Thyristors, insulated-gate bipolar transistors (IGBTs), injection-enhanced insulated-gate transistors (IEGTs) and gate-commutated thyristors (GCTs) are the main switching devices used in high-voltage large-capacity drive inverters.

The thyristor was the first semiconductor switching device developed for large-capacity use. The thyristor was commercialized in 1952 and developed for higher voltage and larger current ratings. In the 1990s, thyristors rated at several kilovolts and several kiloamperes were commercialized, as shown in Figure 22.15.

Thyristors are only on-controlled devices, gate turn-off thyristors (GTOs) were developed as on–off controlled thyristors. However, GTOs require anode reactors, snubber circuits and bulky gate-drive circuits. Owing to semiconductor technologies, GCTs were developed to replace GTOs. The GCT shows better gate controllability and can realize large-capacity drive inverters with simpler circuit configurations.

Transistors able to handle large power were commercialized in the 1980s. The IGBT was developed to deal with higher voltages than conventional transistors. The first generation of the IGBT was developed

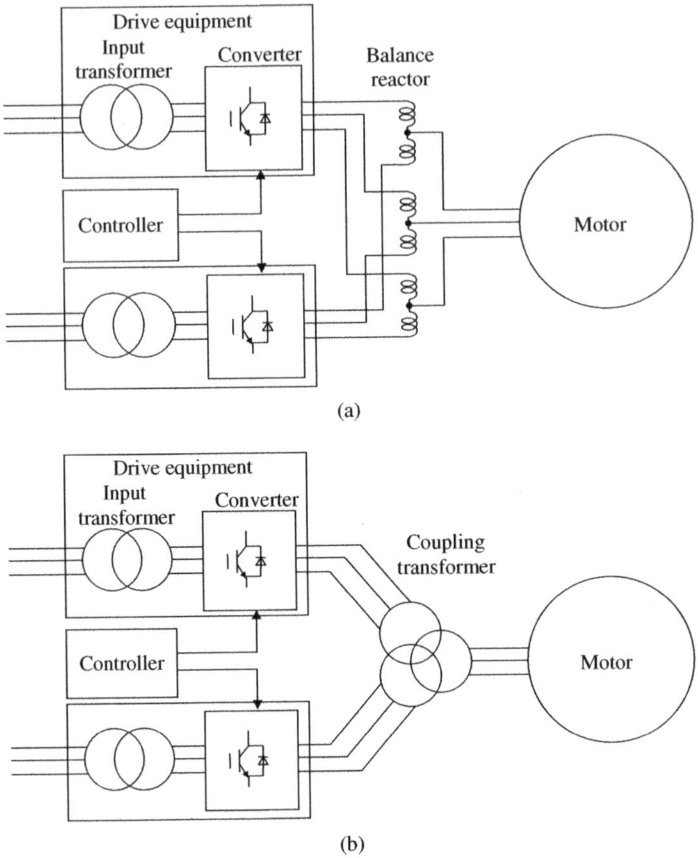

Figure 22.13 Two possible connections of dual converters for driving a motor: (a) parallel connection using reactor and (b) synthesis with transformer

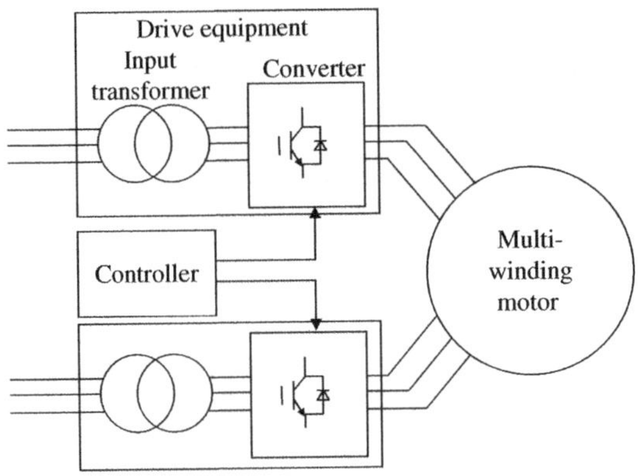

Figure 22.14 Multiwinding motor drive using dual converters

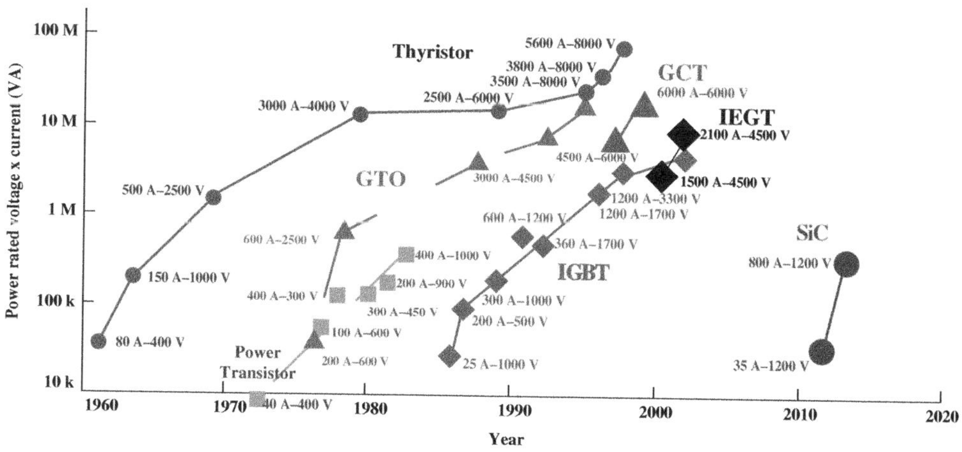

Figure 22.15 Development of semiconductor switching devices

in the 1980s. The ongoing development of IGBTs continues to improve performance on a yearly basis. At present, the sixth generation of IGBTs is commercialized. Furthermore, the IEGT was developed to handle large currents in the range of kiloamperes. The voltage drop in the IEGT is kept low while large current conducts by the injection-enhanced effect.

The semiconductor switching devices introduced earlier are all made of the silicon wafer. The next generation of semiconductor switching devices, actively under development, will be made of silicon carbide (SiC) wafer, which is a better material than silicon and more suited to high-voltage and large-current applications. However, at present, the development is only at the stage for small-capacity and low-voltage applications. It seems to take time before the technology is sufficiently advanced for application to high-voltage and large-capacity drive inverters rated in the megawatt range.

22.7.1 *High-Power Semiconductor Devices*

This section explains the operation and features of the high-voltage large-current semiconductor switching devices applied in large-capacity inverters.

22.7.1.1 Thyristor

The symbol representing a thyristor and a sample photo of a high-power thyristor are shown in Figure 22.16 [9, 10]. Thyristor are turned off mainly by the induced AC voltage of the motor, namely the load of the thyristor bridge. Thus, such equipment is called a load-commutated inverter (LCI). In the past, self-commutated voltage source inverters (VSIs) were developed using thyristors with auxiliary circuits to force the thyristors to turn off. These days such types of inverters are rarely used because the development of self-commutated semiconductor switching devices can achieve better results.

22.7.1.2 IGBT

The symbol for the IGBT and a picture of an example are shown in Figure 22.17 [11]. IGBTs are easily turned on and off simply by applying a gate voltage signal. IGBTs are widely used in voltage source

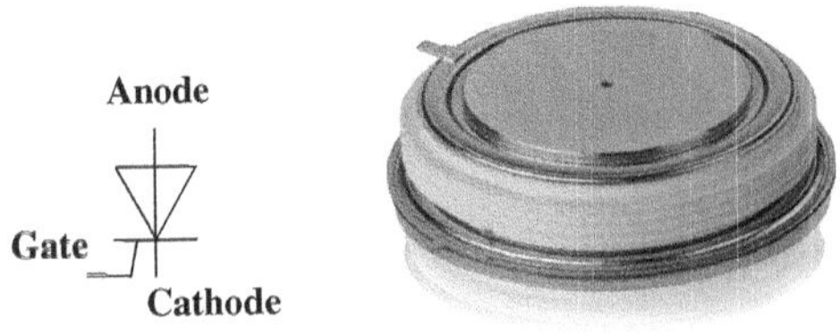

Figure 22.16 Thyristor symbol and a typical device example (Courtesy of ABB)

Figure 22.17 IGBT symbol and a typical device example (Courtesy of Mitsubishi Electric)

self-commutated inverters, especially for low-power drives. Typically, in the IGBT package, antiparallel diodes are also installed, as illustrated by the symbol.

22.7.1.3 IEGT

The IEGT is also a semiconductor switching device that can be turned on and off by the gate voltage control. The IEGT has two points of improvement and can handle high voltage and large current. An example of an IEGT is shown in Figure 22.18 [12]. Unlike the IGBT, the high-voltage IEGT applies an internal structure to increase the carrier density in the wafer in order to keep the on-state voltage low, and thus, the loss is also kept low. Usually, the IGBT package is of modular type where only the bottom of the package can be cooled by the heat sink. In contrast, the IEGT package is of presspack type, as

Figure 22.18 IEGT device (Courtesy of Toshiba)

Figure 22.19 GCT device (Courtesy of Mitsubishi Electric)

shown in Figure 22.18, and both sides of the package can be cooled using two heat sinks. This structure of package is suitable for large-current applications.

22.7.1.4 GCT

The GCT is a semiconductor switching device that has similar internal structure in the silicon wafer to the thyristor. However, it is apparently different from the thyristor, because it is a self-turn-off device. In order to deal with the large current of the power circuit, the gate terminal has a ring shape, as shown in Figure 22.19 [13]. The GCT is a semiconductor switching device suitable for VSIs and it has the following features:

- Small conduction loss even rated at high voltage
- Two-side cooling possible because of presspack structure

The GCT device and its gate-drive unit (GDU) are often housed together to form an integrated GCT (IGCT).

22.8 High-Power Converter Topologies

VFDs are categorized according to the configuration of the used power electronics inverter. Two main configurations are used: (1) current source inverter (CSI) and (2) VSI. In each configuration there are various topologies used. For CSI drives, the topology most often used is the LCI, which is described in the following section.

22.8.1 LCI

The LCI, as its name indicates, relies on commutation by the connected motor (or load). LCI covers a large variety of industrial applications in a power range up to 100 MW [14]. Using thyristors as the main commutation devices, LCI-based large-capacity drives are possible. Figure 22.20(a) shows an LCI circuit comprising a thyristor converter bridge and a thyristor inverter bridge. Series-connected thyristors can make the LCI capable of generating high-voltage output. The thyristor inverter is commutated with the help of motor back-emf voltage. As a result, LCI drives are limited to just SMs. LCIs have been used widely in high-power applications owing to the robustness and low cost of thyristors. Although this

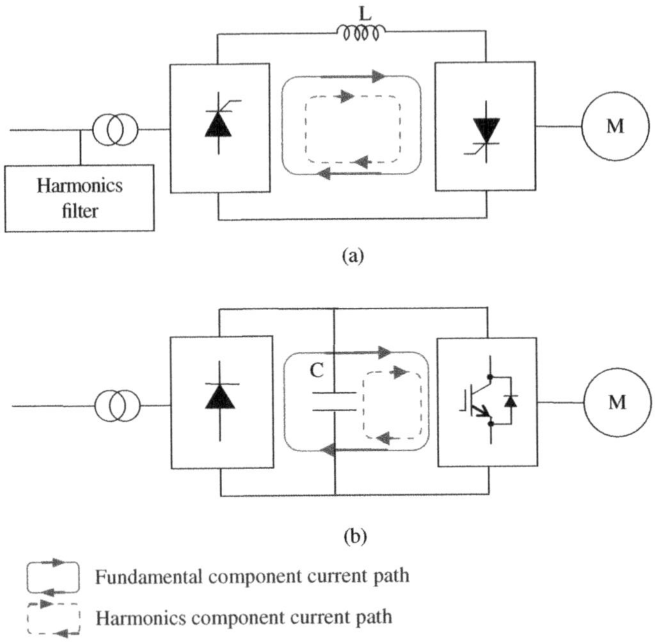

Figure 22.20 Inverter types and harmonics current flows in LCI and VSI: (a) LCI and (b) VSI

technology is known for its simplicity, LCIs suffer from many drawbacks, such as high-grid harmonics, poor power factor and larger motor torque ripple [3]. Therefore, an LC filter is used to mitigate the undesired harmonics injected into the grid by LCI drives. The capacitor of the tuned LC filter might cause some undesirable phenomena in the power network such as resonance.

22.8.2 VSI

Currently, GCTs, IGBTs and IEGTs are available for high-power applications and constitute the main switching devices for VSIs. Electric drives based on VSI are characterized by lower harmonics content, higher power factor and lower torque pulsations than LCI drives. Figure 22.20(b) shows a low-cost VSI-based drive system. The front-end AC/DC converter can be a rectifier (case of Figure 22.20(b)) or a PWM converter to allow four-quadrant operation including regeneration of power.

As VSIs exhibit lower current harmonics than the LCIs, subsynchronous torsional interactions and interharmonics are of lesser concern [15].

To simplify the understanding of the harmonics effects on the drive based on LCI (Figure 22.20(a)) and the drive based on VSI (Figure 22.20(b)), the harmonics flow is shown in Figure 22.20. These harmonics and interharmonics in LCI depend on the speed of the motor and might cause torsional interactions [15]. Note that the harmonics current flows shown in the figure do not take into account the mechanism of AC/DC conversion. The interaction between inverter and rectifier harmonics in LCIs are an important concern [16]. The frequency of these grid harmonics produced by the LCI are speed dependent and, therefore, might decrease in order to match the torsional resonance frequency of the drive system [15]. With a typical design of LCIs, the bulky intermediate reactor is not enough to lower the current harmonics in the DC circuit, which affects the harmonics in the grid, as shown in Figure 22.20. Unlike LCIs, VSIs dominant harmonic components are located at a relatively high-switching frequency and can be absorbed

Table 22.2 Comparison of LCI and VSI drives

Comparison parameter	LCI drives	VSI drives
Grid-side harmonics	For 12-pulse LCI <11% harmonics filter required power grid resonance concerns	<5% harmonics filter not required (PWM rectifier is used)
Ride-through capability	Limited	Robust
Torque ripples	<7% (12-pulse) large torque ripple. Need for special compressor design. Torsional vibration issues	<0.5% use of standard compressor and less concerns for torsional issues
Efficiency	High	High
Reliability	Well-established know-how, long history	Field-proven technology, newer technology for high-power applications with good applications references

by the large DC capacitance, as shown in Figure 22.20(b). Therefore, the harmonics of the rectifier's AC line currents are reduced significantly. Moreover, these harmonics are lowered further by the line impedance, which makes the VSI drive system preferable to the LCI drive system.

22.8.3 Summary

Table 22.2 compares the characteristics of LCI and VSI drives. The VSI outperforms the LCI for LNG plants. VSIs are more suitable for high-power VFD applications. Although the VSI devices' (IEGT, IGCT) power ratings have improved recently, they cannot handle the megawatt ratings of VFDs that are required for LNG plants. New topologies of inverters called multilevel inverters (MLIs) have emerged that provide better performance when working at higher power demands [17–20].

22.9 Multilevel VSI Topologies

The main circuit of the drive equipment that supplies the motor with the necessary variable frequency power is the inverter, which is a DC-to-AC converter. The inverter can control the rotating speed and the torque of the motor by varying its output voltage and frequency.

This section introduces briefly the main inverter circuit structure and the principle of operation of multilevel inverters, which are used for high-power applications such as in LNG plants. In general, the inverter for which the number of voltage steps in the output voltage waveform is larger than or equal to three is called the MLI [21]. For better understanding of MLIs, the description starts from the simplest inverter, which is the two-level inverter.

22.9.1 Two-Level Inverters

The circuit topology of the two-level inverter is shown in Figure 22.21(a). The inverter is a three-phase bridge consisting of a DC voltage source E and six semiconductor switching devices. The three-phase bridge consists of three similar legs. In the explanation, it is assumed that the midpoint of the DC voltage source is at the ground potential. The leg structure of phase A is used for the description of the principle of operation. The leg has two switching devices U and X. U connects the positive terminal of the DC voltage and the AC terminal. X connects the negative terminal of the DC voltage source and the AC terminal. When the device U is on, the phase A outputs the voltage step of $+Ed/2$. When, the device X

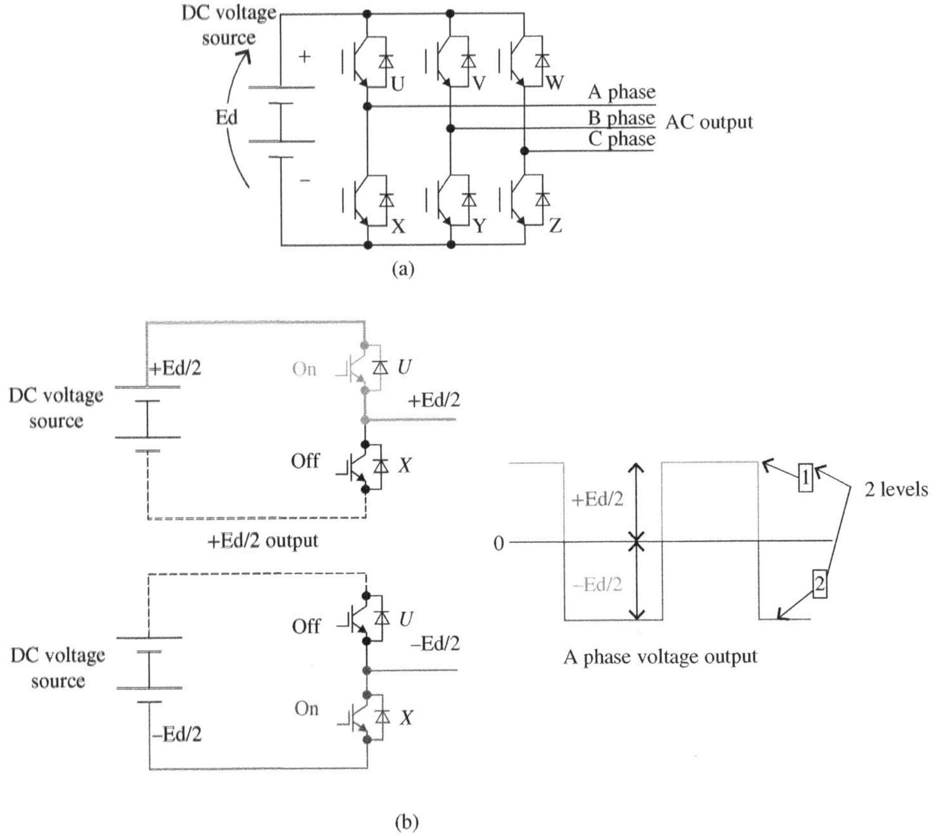

Figure 22.21 Circuit configuration and operation principle of two-level inverter. (a) Three-phase VSI bridge and (b) operation of one leg and output voltage levels

is on, it outputs voltage step of $-Ed/2$. Therefore, the phase A outputs two voltage levels, $+Ed/2$ and $-Ed/2$, as shown in Figure 22.21(b).

By appropriately controlling the width ratio between $+E/2$ and $-E/2$ using PWM control, the voltage waveform can be obtained with the desired voltage amplitude and frequency. By applying the voltage waveform to the motor, owing to the impedance of the motor, the current of high-frequency components around the PWM carrier frequency is blocked and mostly the current component of the desired frequency flows in the motor. The output voltage waveforms of the two-level PWM inverter are shown in Figure 22.22, both for phase-to-neutral and for phase-to-phase.

22.9.2 *Multilevel Inverters*

As MLIs can handle high voltage/current using IGBTs, IEGTs, or IGCTs, they dominate high-power applications. Using the same switching frequency, the output waveforms are of better quality than the classical two-level inverter topologies. MLIs are characterized by smaller filtering elements, high modularity, high flexibility and multilevel voltage outputs with less harmonics components. Thus, MLIs topologies are more appropriate for high-power drive systems.

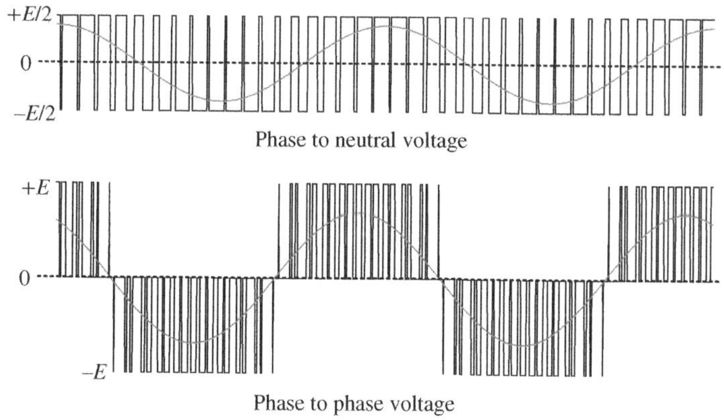

Figure 22.22 Voltage waveforms of two-level PWM inverter

22.9.2.1 Three-Level Inverters

Among the circuit topologies of MLIs is the three-level inverter; a circuit topology where the neutral point (NP) is connected to the AC output through a diode. The circuit topology is shown in Figure 22.23. This type of three-level inverter is called the neutral-point-clamped (NPC) inverter [22].

The DC voltage source consists of two voltage sources: the positive voltage source and the negative voltage source. The connection point of the two voltage sources is the NP 0. Twelve semiconductor switching devices are used in the circuit. The leg of phase A consists of switches A1, A2, A3 and A4. When A1 and A2 are on, the output of phase A is connected to the positive terminal of the DC voltage source, P. When A3 and A4 are on, the output of phase A is connected to the negative terminal

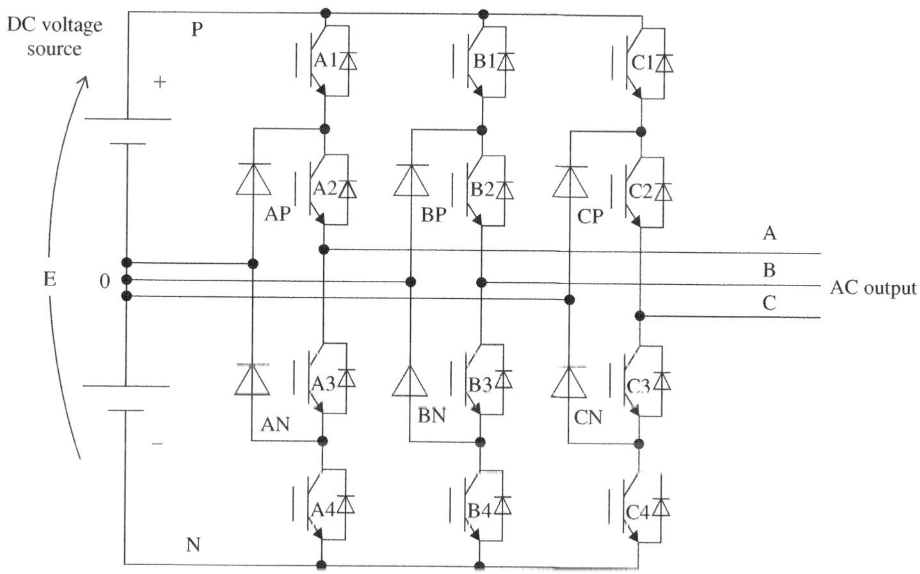

Figure 22.23 Circuit configuration of NPC three-level inverter

of the DC voltage source N. The diodes AP and AN are connected to the NP. The output voltage is as follows:

- A1 and A2 are on, then $v_A = +E/2$
- A3 and A4 are on, then $v_A = -E/2$
- A2 and A3 are on, then $v_A = 0$

The phase voltage output has three levels: $+E/2$, $-E/2$, or 0, as shown in Figure 22.24.

By appropriately controlling the ratio of the widths of $+E/2$, 0 and $-E/2$ using PWM control, the output voltage amplitude and frequency are controllable. The high-frequency components of the output current are suppressed by the impedance of the motor. Therefore, only the fundamental frequency component of the current flows in the motor. The phase-to-neutral and phase-to-phase of the output voltage waveforms for the NPC (three-level) inverter are shown in Figure 22.25.

When similar semiconductor switching devices (same voltage rating) are used, the three-level inverter can handle voltage levels twice those of the two-level inverter. As a result, the circuit topology of the three-level inverter is suitable for high-voltage applications. Furthermore, the three-level inverter outputs more voltage levels, and its voltage waveform is more sinusoidal than the two-level inverter.

Other topologies of three-level inverters exist; however, the NPC type is widely used in industry, especially for LNG plants. Other three-level inverter topologies are rarely used for large-capacity applications rated at several tens of megawatts.

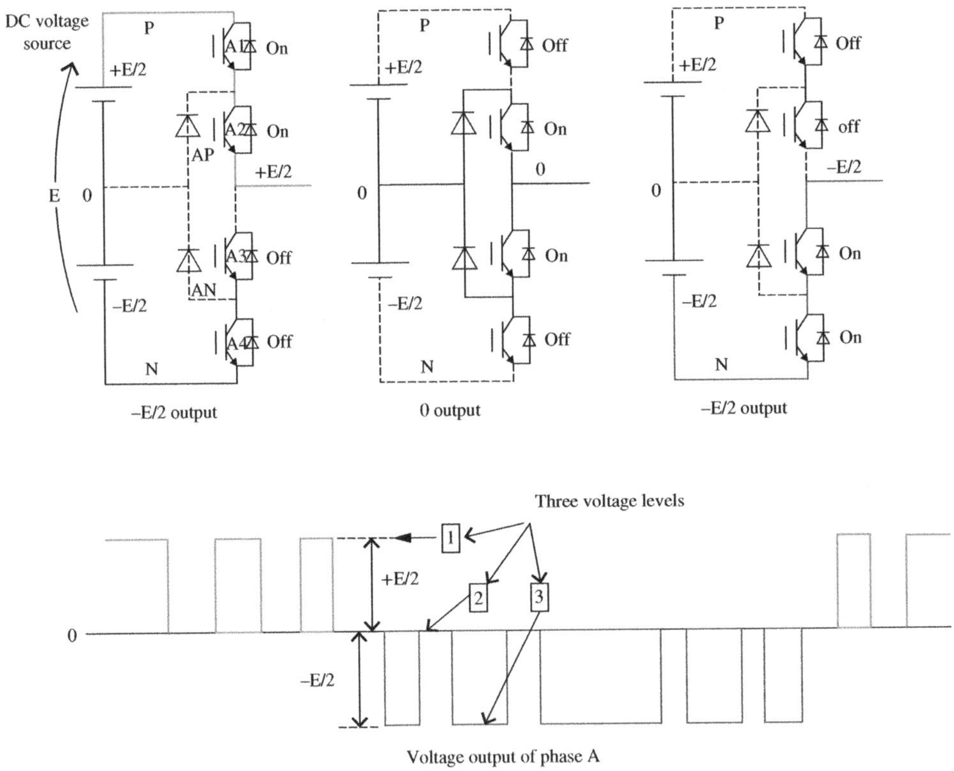

Figure 22.24 Operation principle of three-level NPC inverter

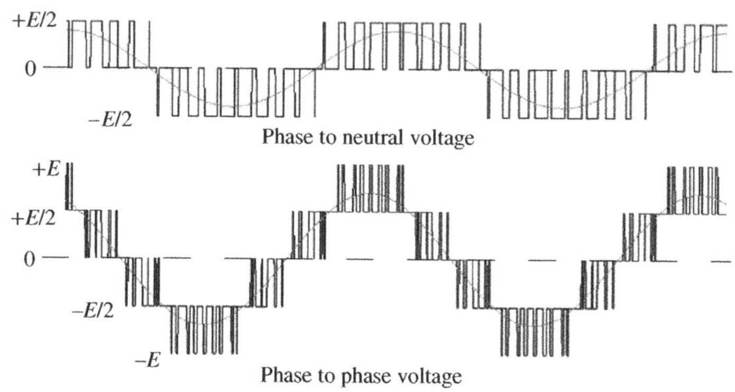

Figure 22.25 Voltage waveforms of NPC three-level inverter

22.9.2.2 Five-Level Inverters (Star Connection of two NPC Inverters)

Using NPC as a cell, a five-level inverter is introduced for high-power applications. The circuit topology is constructed from a combination of three-level inverters introduced above. The circuit topology is shown in Figure 22.26 [23].

A single-phase inverter consists of two legs of a three-level inverter. One output terminal of a single-phase inverter is connected to the neutral, while the other output is one phase output of the five-level inverter, which is connected to the motor terminal. It is worth noting that the middle point 0 of the DC voltage sources of the single-phase inverter has different potential from the neutral bus of the star-connected total system. This circuit topology can output the following five voltage levels or

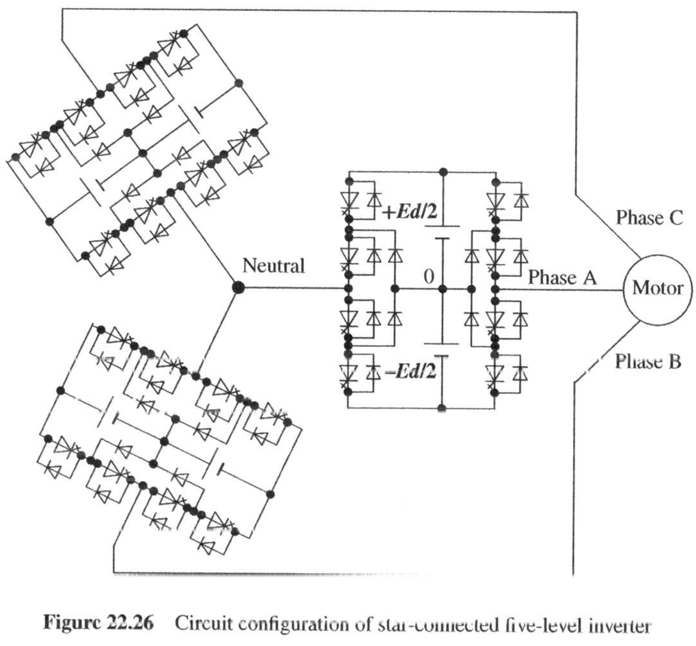

Figure 22.26 Circuit configuration of star-connected five-level inverter

Table 22.3 Combination of voltage levels of star-connected circuit

Phase A voltage to neutral V_{AN}	Leg A1 output voltage V_1 0 to neutral	Leg A2 output voltage V_2 0 to phase A
$-E$	$-E/2$	$+E/2$
$-E/2$	$-E/2$	0
	0	$+E/2$
	$-E/2$	$-E/2$
0	0	0
	$+E/2$	$+E/2$
$+E/2$	0	$-E/2$
	$+E/2$	0
$+E$	$+E/2$	$-E/2$

steps: $+E$, $+E/2$, 0, $-E/2$ and $-E$. The reference point of these levels is the neutral bus and is obtained by combination of the output voltages from the two legs of the single-phase inverter, as shown in Table 22.3. In addition, the voltage amplitude is doubled compared with that of the three-level inverter. The voltage waveforms are shown in Figure 22.27. As the number of voltage levels is larger, the voltage waveform approaches the sinusoidal waveform more closely than does the three-level inverter. As the phase-to-phase voltage has nine levels of voltage, the motor current ripples are smaller and the currents are almost sinusoidal waveforms.

A developed 30 MW, five-level drive inverter using an NPC cell is shown in Figure 22.28 and its output voltage and current waveforms are shown in Figure 22.29. In the example, or in similar drive inverters, large-capacity semiconductor switching devices, such as IEGTs or GCTs, introduced in an earlier section, are used. The number of semiconductor switching devices is 24, as shown in the circuit drawing of Figure 22.26. By using large-capacity semiconductor switching devices, large-capacity drive inverters rated around 20–30 MW can be realized with a small number of semiconductor switching devices. The rated output voltage of the drive inverter shown in Figure 22.28 is 7.2 kV. The drive inverter can be considered to have high reliability because it realizes high voltage and large capacity with the simple circuit topology and with a small number of devices.

Furthermore, the low-frequency harmonics are canceled out with the special PWM pulse pattern in the five-level inverter shown in Figure 22.28. This pulse pattern suppresses the low-frequency torque ripples,

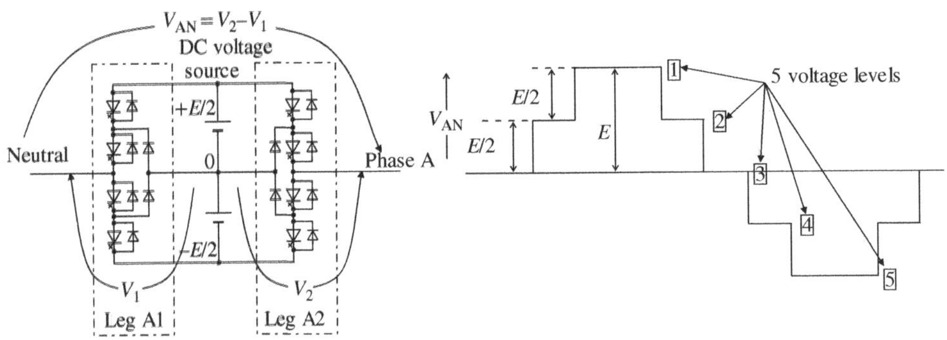

Figure 22.27 Operation principle of star-connected five-level inverter

Figure 22.28 30 MW NPC-cell-based five-level inverter set (Reproduced by permission of TMEIC)

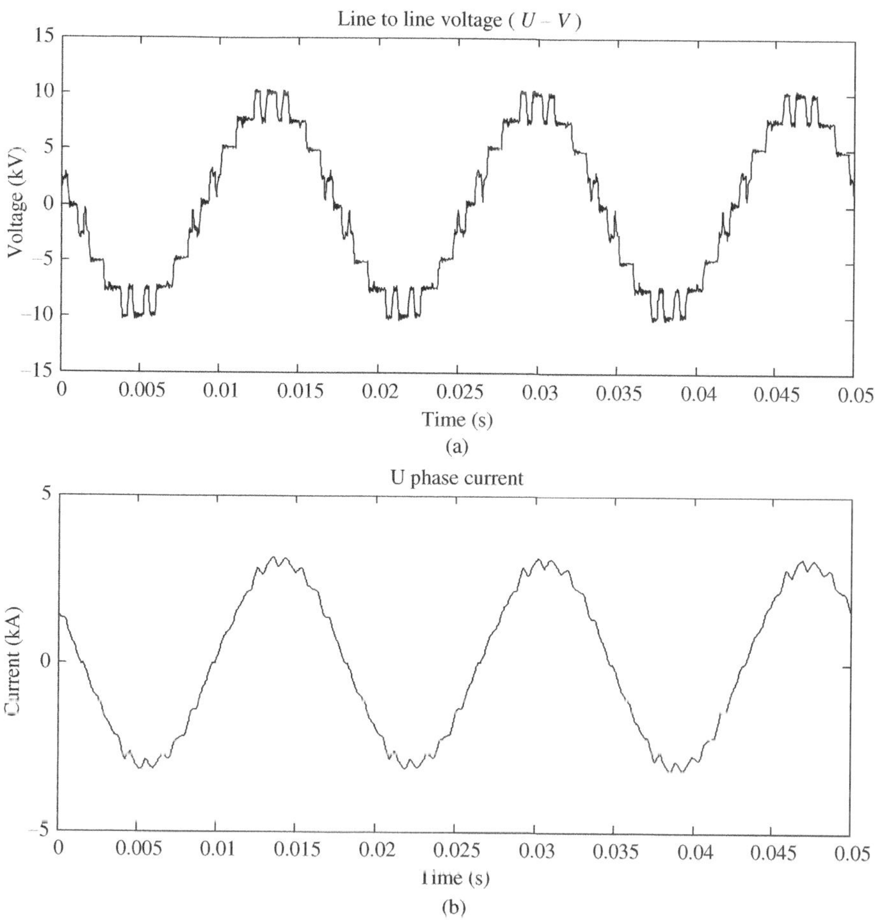

Figure 22.29 Output voltage and current waveforms of a five-level inverter

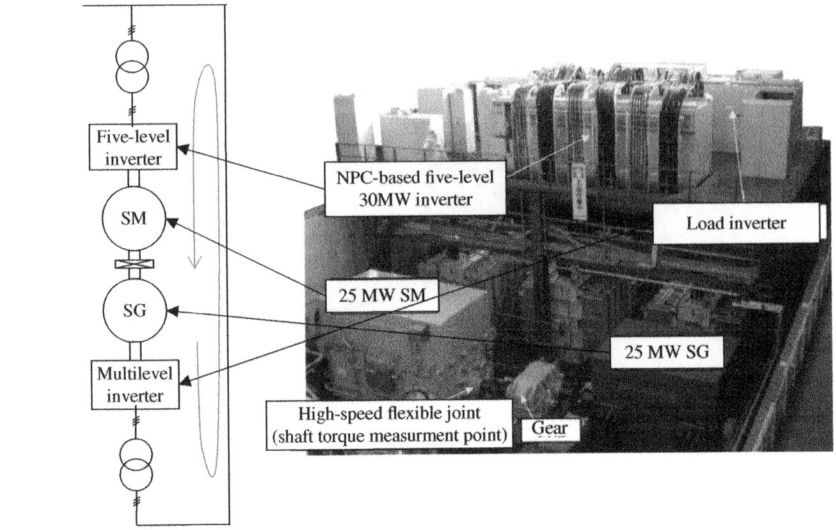

Figure 22.30 25 MW drive system test bench

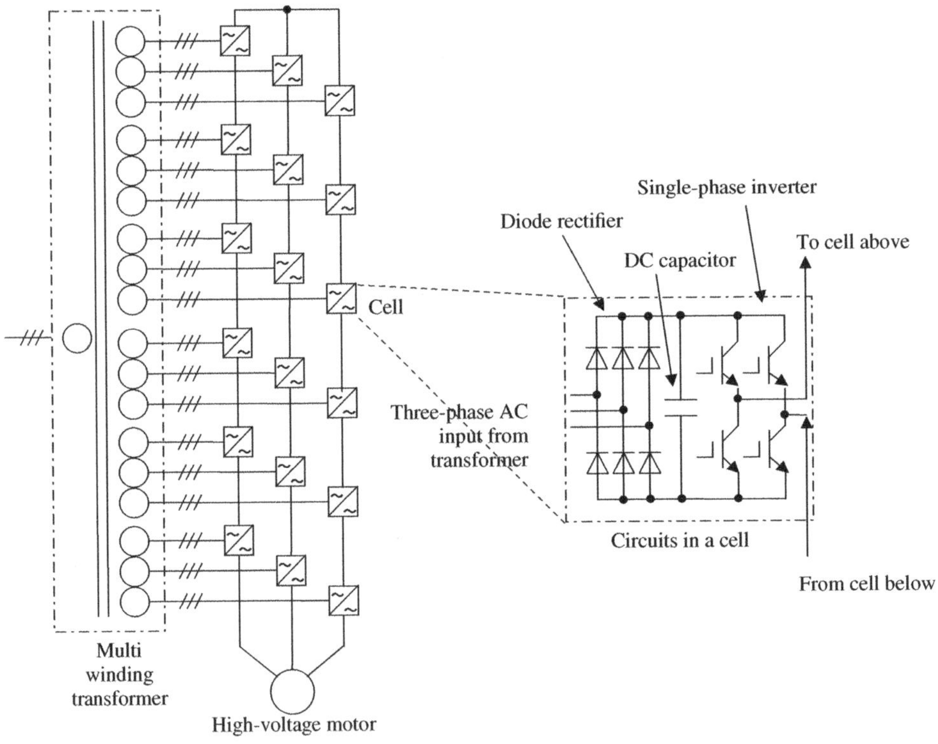

Figure 22.31 Typical multilevel inverter using full-bridge single-phase multicells connected in series

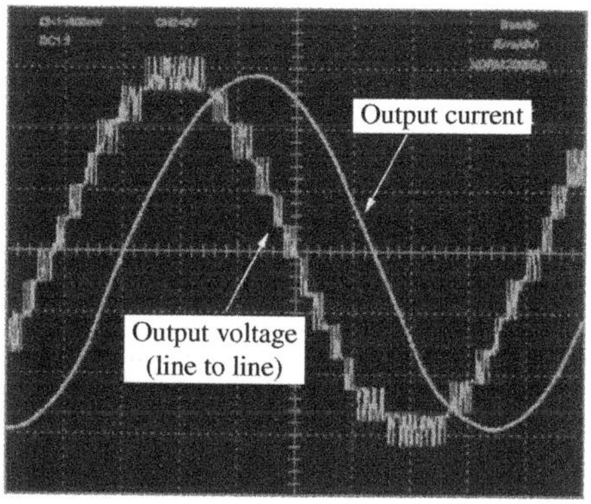

Figure 22.32 Line-to-line output voltage waveforms of multicell six-series cells

which could coincide with the low mechanical resonance of the large-capacity compressor of an LNG plant [24].

In order to test the 30 MW, NPC-based electric drive system, and assess its performance, a proper test bench was developed. A 25 MW test bench facility based on the recuperation of the mechanical energy in electrical energy to be fed back to the electric drive is realized, as shown in Figure 22.30. Thus, tests are carried out with minimum power drawn from the power line; it is only the losses that are needed from the power grid.

22.9.2.3 Multilevel Inverters (Cells in Series Connection)

One of the most popular circuit topologies of the MLI is the series connection of several cells. The circuit topology and the detailed circuit in the cell are shown in Figure 22.31. The output section of the cell consists of a single-phase inverter. In practical application, the single-phase inverter is made of an IGBT with a modular type package and its AC output voltage is around 1000 Vrms. Then, to obtain the 6 kV output, six cells are connected in series, and a large number of IGBTs is used in the inverter-based drive system.

Figure 22.32 shows the output voltage waveform of the drive inverter with six cells in series. The number of voltage levels is large and near sinusoidal waveforms can be obtained.

For supplying the DC power to each cell, a multiwinding transformer is used. The multiwinding transformer provides insulation among cells and reduces low-order harmonic current components to the AC power system by shifting the phases of the winding from each other.

22.10 Control of High-Power Electric Drives

A general block diagram of the VFD is shown in Figure 22.33 where some practical issues are highlighted. A drive system should overcome these technical issues if it is to operate in an industrial environment such as LNG plants. For LNG plants, electric drive efficiency, reliability and robustness to power network disturbances are the main concerns of the plant operators. These issues are still challenging tasks for the designers, researchers and drive manufacturers.

The following three main techniques are available to manufacturers for the control of electric drives:

1. *V*/*f* control
2. Field-oriented control (FOC) or vector control (VC)
3. Direct torque control (DTC)

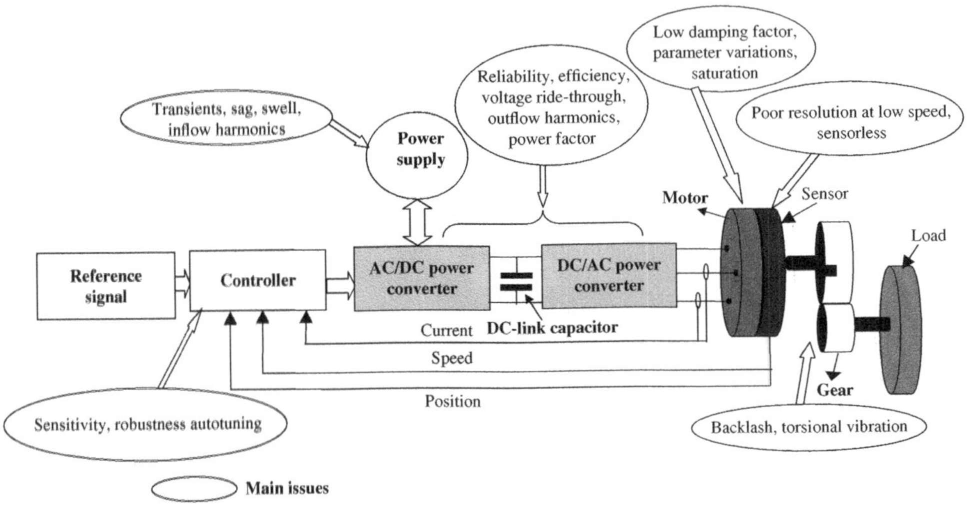

Figure 22.33 Practical issues in a drive system

The *V/f* method keeps the ratio *V/f* constant when controlling an IM. It consists of proportionally controlling, in an open-loop manner, the voltage and frequency in order to keep the flux constant in an IM, as shown in Figure 22.34(a). This scalar open-loop frequency control method operates without a speed sensor. Although simple, this control technique suffers from many drawbacks, such as the steady-state inaccuracy and poor dynamic performance. Decoupling between flux and torque levels is not possible in this control technique, and thus, independent control of flux and torque is not possible. Poor performance at low speed is observed in the *V/f* method. Transient operation is uncontrollable in the *V/f* controlled drive and the system might become unstable. Techniques have been developed to improve the stability of the *V/f* control [25]. The FOC or VC methods (Figure 22.34(b)) control independently the torque and flux components. Therefore, implementation of direct flux control is possible. In general, a speed/position

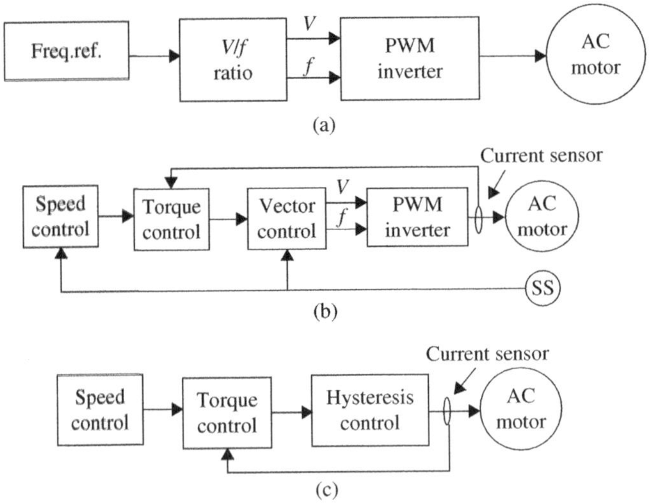

Figure 22.34 Evolution of AC motor drive control techniques: (a) *V/f*, (b) FOC or VC and (c) DTC

sensor is required for the implementation of the VC, which includes two current controllers: one for the flux component and the other for the torque component. These controllers are realized in the dq frame rotating with the rotor flux. A transformation of the voltage references (output of current controllers) from the rotating frame to the stationary frame is needed. A PWM inverter will output voltages proportional to voltage references. The DTC was proposed in 1985 [26, 27]. In these preliminary proposals, two hysteresis controllers were used for the control of stator flux and torque of the motor. The simplicity of this DTC technique is made possible by avoiding coordinate transformation and by using a look-up table to select the appropriate voltage vector of the inverter. The DTC can achieve the decoupling between flux and torque without using a speed sensor and the PWM technique. Owing to the simultaneous control of the flux and the torque, the DTC scheme exhibits a higher dynamic performance. However, hysteresis controls with their variable switching frequencies and limited number of voltage vectors in the inverter control are known to cause torque ripples in the motor and poor steady-state performance at low speed.

Unlike small motors, large motors have low stator/rotor resistance values; thus, a low damping factor is obtained. The low damping factor makes the control of high-power motors difficult and prone to instability. Small motors (i.e., IMs) have higher damping factors, are easier to control and are unlikely to be unstable. To illustrate this, Figure 22.35 shows the pole positions for small and large IMs as a function of operating frequency. These low damping factors of large IMs make their control vulnerable and might easily cause oscillations [25]. A new approach to improve the damping factor of high-power motors using current feedback is proposed in [25]. A virtual resistance based on current feedback is obtained to realize active damping; as a result, controllers developed for small motors can also be used for control of high-power motors.

22.10.1 PWM Methods

In many high-power applications, energy saving is critical; therefore, the minimization of energy losses is necessary. VFD losses might come from the motor as well as from the power converter. For reduced environmental pollution, the motor, converter and controller should all be selected such that energy saving and high efficiency are obtained. Losses in the PWM power converter are mainly due to the devices' switching losses. In large-capacity applications, the switching losses might represent an important percentage of the total drive energy; therefore, the minimization of switching losses is a crucial task.

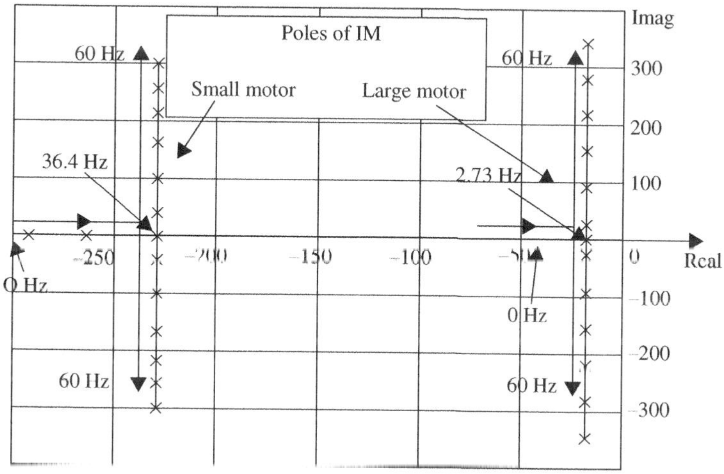

Figure 22.35 Example of motor poles versus operating frequency for small and large IMs

Many PWM techniques were developed to deal with the control of inverters, such as the sinusoidal pulse-width modulation (SPWM) and space vector pulse-width modulation (SVPWM) strategies. Various discontinuous pulse-width modulation (DPWM) strategies were developed to reduce the switching losses of the inverter [28–38]. DPWM techniques are based on the principle that power devices in each bridge leg do not switch during a predetermined time during a cycle of output voltage. Thus, the DPWM strategy reduces the switching losses in the inverter.

MLIs based on the NPC topology are used extensively for high-power VFD applications, especially in LNG plants. NPC inverter output voltages can be clamped to the positive, negative, or NP) (point O in Figure 22.37) voltage. The NP voltage is assumed to be at zero level. In reality, NP voltage fluctuates, which could instigate PWM inverter instability even with balanced load [39]. Many PWM techniques to decrease the variations of NPC NP voltage have been proposed [40–48].

In most VFD applications, the three-phase Y-connected load is used where the neutral current path is not connected, which offers more choices in the inverter voltage selection. In fact, the load in such an application will only act in response to the input line-to-line voltages. A promising discontinuous PWM technique is based on the principle of applying non-sinusoidal phase voltages while keeping the line-to-line voltages input sinusoidal [46–49]. Figure 22.36 illustrates the conventional PWM and the DPWM input reference voltages. Note that the motor operating under the DPWM method, shown in Figure 22.36(b), is not affected by the phase voltages. The implementation of this DPWM method is shown in Figure 22.37 and can be easily plugged in for any NPC inverter in order to achieve three goals

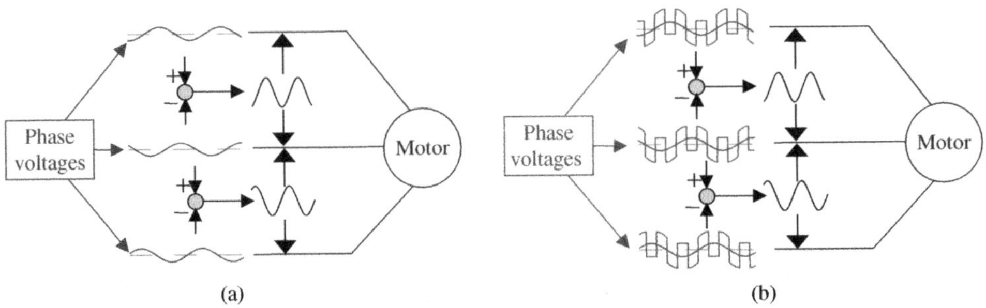

Figure 22.36 PWM modulation techniques for multilevel NPC inverters: (a) conventional PWM method and (b) discontinuous PWM method

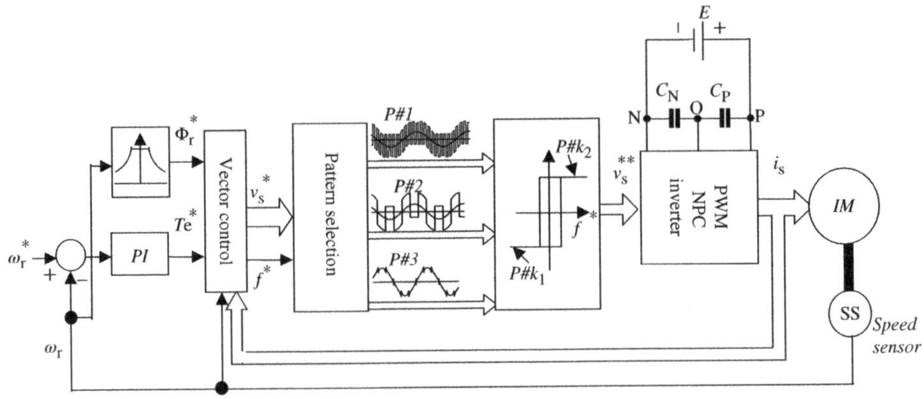

Figure 22.37 Implementation of DPWM in [49]

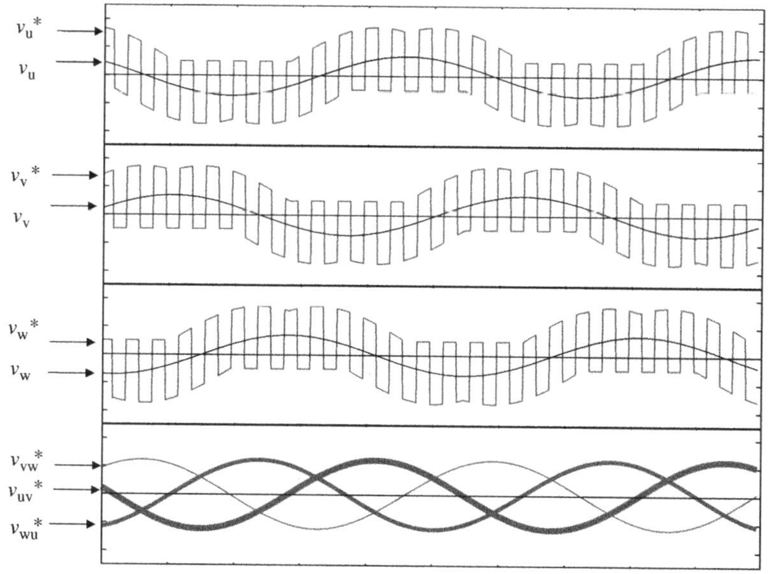

Figure 22.38 DPWM method for multilevel NPC inverters presented in [49]

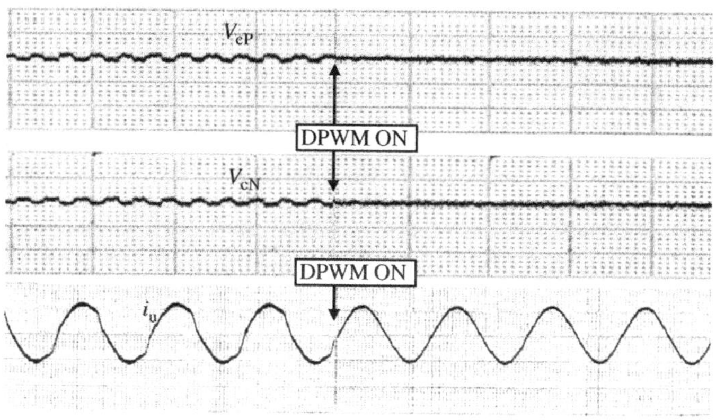

Figure 22.39 Experimental results for NPC inverter with and without DPWM technique, as proposed in [49] (upper: neutral point capacitor's voltages; bottom: phase *u* current waveforms)

simultaneously: (1) prevention of self-commutated semiconductor switching device minimum on-pulse, (2) reduction of switching losses and (3) reduction of NP fluctuation in NPC inverters. Figure 22.38 shows an example of the pattern used in the DPWM method for multilevel NPC inverters in order to avoid minimum on-pulse of the switching device, reduce the switching losses and decrease the NP voltage fluctuations, as shown in Figure 22.39 [49].

22.11 Conclusion

The impacts of megawatt VFDs for LNG plants were discussed. LNG operators are looking for higher plant availability and improved plant operational flexibility, independent of the effect of ambient temperature. Using electric VFDs to drive LNG plant compressors offers better performance than when using conventional GT drives. Unlike GTs, VFDs reduce maintenance costs and downtime, reduce plant air emissions and enhance safety. Following recent technological developments, in principle, a VFD-driven LNG plant is possible. Motor and VFD manufacturers are developing large electric drives with ratings of 100–120 MW using the recently developed technology of power electronics converters. This chapter presented a few examples of actual high-power VFDs that can realize up to 100 MW using four sets of 30 MW NPC-based multilevel VSIs. This NPC-cell-based VFD improves reliability because it uses large-capacity semiconductor devices. For megawatt drive systems, the losses represent a significant amount of power. Therefore, some PWM techniques to reduce the losses were presented in this chapter. Other techniques were included in the VFDs to make it sufficiently robust to ride-through voltage sags in power networks.

Acknowledgment

This work was supported by NPRP Grants No. 08-548-2-223 and 09-426-2-160 from the Qatar National Research Fund (a member of Qatar Foundation). The statements made herein are solely the responsibility of the authors. L. Ben-Brahim would like also to thank RasGas Company for the funding of Industrial Electronics chair.

References

1. Sannino, A., Liljestrand, L., Rothman, B. *et al.* (2007) All-electric LNG liquefaction plants: technical challenges and possible concept solutions. IEEE Industry Applications Conference, 42nd IAS Annual Meeting. Conference Record 2007.
2. Tessarolo, A., Zocco, G., and Tonello, C. (2011) Design and testing of a 45-MW 100-Hz quadruple-star synchronous motor for a liquefied natural gas turbo-compressor drive. *IEEE Transactions on Industry Applications*, **47** (3), 1210–1219.
3. Ben-Brahim, L., Benammar, M., Yoshino, T. *et al.* (2012) Electric drives for LNG plants. Proceedings of the 3rd Gas Processing Symposium, Doha, Qatar, March 2012.
4. Shukri, T. (2004) LNG selection technology. *Hydrocarbon Engineering*, **9** (2), 71–76.
5. Ekstrom, T.E. and Garrison, P.E. Gas Turbines for Mechanical Drive Application. (1994) Report GER-3701B, GE Industrial and Power Systems, Schenectady, NY, http://site.ge-energy.com/prod_serv/products/tech_docs/en/downloads/ger3701b.pdf (accessed 23 December 2013).
6. Dabbousi, R., Savinovic, D., and Anundsson, Y. (2008) A comparison between IMs and SMs for applications in the oil and gas industry. Petroleum and Chemical Industry Technical Conference, 2008 PCIC 2008, September 22–25, 2008.
7. Dabbousi, R., Balfaqih, H., Anundsson, Y., and Savinovic, D. (2008) A comparison of totally enclosed motor coolers commonly used in the COG industry. Proceedings of the 5th Petroleum and Chemical Industry Conference Europe – Electrical and Instrumentation Applications, 2008. PCIC Europe 2008, pp. 1–5.
8. Yoshino, T., Yoshimura, M., Tobita, M., Ota, S. *et al.* (2008) Power electronics conserves the environment for better future. Proceedings of PESC 2008, pp. XCIX–CIV.
9. IEC 60617 *Graphical Symbols for Diagrams*, http://std.iec.ch/iec60617 (accessed: 9 May 2013).
10. ABB Phase Control and Bi-directionally Controlled Thyristors, ABB Web sitehttp://www05.abb.com/global/scot/scot256.nsf/veritydisplay/f71d894539bb0d22c125787e002b5b88/$file/ABB_Flyer_PCT-BCT_2012_Web.pdf (accessed 23 December 2013).
11. Mitsubishi Electric HVIGBT Module (High Voltage Insulated Gate Bipolar Transistor), Mitsubishi Electric Web site, http://www.MitsubishiElectric.com/semiconductors/ (accessed 9 May 2013).
12. Kitagawa, M., Omura, I., Hasegawa, S. *et al.* (1993) A 4500 V injection enhanced insulated gate bipolar transistor (IEGT) operating in a mode similar to a thyristor. Electron Devices Meeting, 1993. IEDM '93.
13. Yamamoto, M., Satoh, K., Nakagawa, T., and Kawakami, A. (1998) GCT (gate commutated turn-off) thyristor and gate drive circuit. Power Electronics Specialists Conference, 1998. PESC 98 Record, May 17–22, 1998, Vol. 2, pp. 1711–1715.

14. Beuermann, M., Fischer, W., Kalbfleisch, P. *et al.* (2008) Modular load commutated inverters: a proven concept for high power applications. Industry Applications Society Annual Meeting, 2008. IAS '08. IEEE.
15. Fujii, T., Masuda, H., Ogashi, Y. *et al.* (2011) Study of subsynchronous torsional interaction with voltage source inverter drive for LNG plant. IEEE PSEC Conference Proceedings 2011.
16. Huetten, V., Zurowski, R.M., and Hilscher, M. (2008) Torsional interharmonic interactions study of 75MW direct-driven VSDS motor compressor trains for LNG duty. Proceedings of the 37th Turbomachinery Symposium, 2008, pp. 57–66.
17. Hosoda, H. and Peak, S. (2010) Multi-level converters for large capacity motor drive. The 2010 International Power Electronics Conference, IPEC 2010, June 2010.
18. Baccani, R., Zhang, R., Toma, T. *et al.* (2007) Electric systems for high power compressor trains in oil and gas applications-system design, validation approach, and performance. Proceedings of the 36th Turbomachinery Symposium, 2007.
19. Rodrigues, J., Lai, J., and Peng, F. (2002) Multilevel Inverters: a survey of topologies, controls, and applications. *IEEE Transactions on Industrial Electronics*, **49** (4), 724–738.
20. Abu-Rub, H., Holtz, J., Rodriguez, J., and Baoming, G. (2010) Medium-voltage multilevel converters-state of the art, challenges, and requirements in industrial applications. *IEEE Transactions on Industrial Electronics*, **57** (8), 2581–2596.
21. Kouro, S., Malinowski, M., Gopakumar, K., *et al.* (2010) Recent Advances and Industrial Applications of Multilevel Converters. *IEEE Transactions on Industrial Electronics*, **57** (8), 2553–2580.
22. Nabae, A., Takahashi, I., and Akagi, H. (1981) A new neutral-point clamped PWM inverter. *IEEE Transactions on Industry Applications*, **IA-17**, 518–523.
23. Tsukakoshi, M., Al Mamun, M., Hashimura, K., and Hosoda, H. (2009) Study of large VSI drive system for oil and gas industry. Proceedings of the 38th Turbomachinery Symposium, 2008, pp. 261–265.
24. Tsukakoshi, M., Al Mamun, M., Hashimura, K. *et al.* (2010) Novel torque ripple minimization control for 25 Mw variable speed drive system fed by multilevel voltage source inverter. Proceedings of 29th Turbomachinery Symposium, 2010, pp. 193–200.
25. Ben-Brahim, L. (1998) Improvement of the stability of a V/f controlled Induction Motor drive system. Proceeding IEEE IECON'98, Aachen, Germany, August/September 1998, Vol. 2, pp. 859–864.
26. Takahashi, I. and Noguchi, T. (1986) A new quick response and high efficiency control strategy of an induction motor. *IEEE Transactions on Industry Applications*, **22**, 820–827.
27. Depenbrock, M. (1988) Direct Self-Control (DSC) of inverter-fed induction machine. *IEEE Transactions on Power Electronics*, **3**, 420–429.
28. Ahmet, M. and Onur, N. (2011) A generalized scalar PWM approach with easy implementation features for three-phase, three-wire voltage-source inverters. *IEEE Transactions on Power Electronics*, **24** (5), 1385–1395.
29. Depenbrock, M. (1977) Pulse width control of a 3-phase inverter with non-sinusoidal phase voltages. 1977 IEEE International Semiconductor Power Conversion Conference, pp. 399–403.
30. Kolar, J.W., Ertl, H., and Zach, F.C. (1991) Influence of the modulation method on the conduction and switching losses of a PWM converter system. *IEEE Transactions on Industry Applications*, **27**, 1063–1075.
31. Alexander, D.R. and Williams, S.M. (1993) An optimal PWM algorithm implementation in a high performance 125 kVA inverter. 1993 IEEE Applied Power Electronics Conference and Exposition Conference, pp. 771–777.
32. Zhao, D. and Pavan Kumar Hari, V.S.S. (2010) Space-vector-based hybrid pulse width modulation techniques for reduced harmonic distortion and switching loss. *IEEE Transactions on Power Electronics*, **25** (3), 760–774.
33. Nguyen, N.V., Nguyen, B.X., and Lee, H.H. (2011) An optimized discontinuous PWM method to minimize switching loss for multilevel inverters. *IEEE Transactions on Industrial Electronics*, **58** (9), 3958–3966.
34. Khan, H., Miliani, E.H., and Drissi, K.E.K. (2012) Discontinuous random space vector modulation for electric drives: a digital approach. *IEEE Transactions on Power Electronics*, **27** (12), 4944–4951.
35. Sun, P., Liu, C., and Lai, J.S. (2012) Three phase dual buck inverter with unified pulsewidth modulation. *IEEE Transactions on Power Electronics*, **27** (3), 1159–1167.
36. Ojo, O. (2004) The generalized discontinuous PWM scheme for three-phase voltage source inverters. *IEEE Transactions on Industrial Electronics*, **51** (6), 1280–1289.
37. Hava, A.M., Kerkman, R.J., and Lipo, T.A. (1998) A high-performance generalized discontinuous PWM algorithm. *IEEE Transactions on Industry Applications*, **34** (5), 1059–1071.
38. Younis, M.A.A., Rahim, N.A., and Mekhilef, S. (2010) High efficiency THIPWM three phase inverter for grid connected system. 2010 IEEE Symposium on Industrial Electronics and Applications Conference, pp. 88–93.

39. Liu, H., Choi, N., and Cho, G. (1991) DSP based space vector PWM for three-level inverter with DC-link voltage balancing. IEEE IECON'91, pp. 197–203.
40. Kim, H.-J., Lee, H.-D., and Sul, S.-K. (2001) A new PWM strategy for common-mode voltage reduction in neutral-point-clamped inverter-fed AC motor drives. *IEEE Transactions on Industry Applications*, **37** (6), 1840–1845.
41. Newton, C. and Sumner, M. (1997) Neutral Point Control for Multi-Level Inverters: theory, design and operational limitations. IEEE Industry Applications Society Conference Record, pp. 1336–1343.
42. Liu, H.L. and Cho, G.H. (1994) Three-level space vector PWM in low index modulation region avoiding narrow pulse problem. *IEEE Transactions on Power Electronics*, **9** (5), 481–486.
43. Lee, Y.H., Suh, B.S., and Hyun, D.S. (1996) A novel PWM scheme for a three-level voltage source inverter with GTO thyristors. *IEEE Transactions on Industry Applications*, **32** (2), 260–268.
44. Tanaka, S., Miura, K., Watanabe, T., Tadakuma, S., and Ikeda, H. (1992) Consideration on PWM control for Neutral-Point-Clamped inverter. *Transactions IEE of Japan*, **112-D** (6), 553–560 (in Japanese).
45. Miyazaki, S., Kurosawa, R., and Ben-Brahim, L. (1993) Characteristics improvement of a PWM controlled three level GTO inverter. National Convention Record I.E.E. Japan, March 1993 (in Japanese).
46. Ben-Brahim, L. (1999) Improvement of a PWM controlled NPC GTO inverter for AC motor drives. ISIE99 Proceedings, Slovenia, July 1999.
47. Ben-Brahim, L. and Tadakuma, S. (2001) A new PWM control for GTO minimum on-pulse compensation. Industry Applications Conference, IAS Annual Meeting, 2001, Vol. 2, pp. 1015–1022.
48. Ben-Brahim, L. and Tadakuma, S. (2006) A novel multilevel carrier-based PWM control method for GTO inverter in low index modulation region. *IEEE Transactions on Industry Applications*, **42** (1), 121–127.
49. Ben-Brahim, L. (2008) A discontinuous PWM method for balancing the neutral point voltage in three-level inverter-fed variable frequency drives (VFDs). *IEEE Transactions on Energy Conversion*, **23** (4), 1057–1063.

23

Modulation and Control of Single-Phase Grid-Side Converters

Sebastian Styński and Mariusz Malinowski
Faculty of Electrical Engineering, Warsaw University of Technology, Warsaw, Poland

23.1 Introduction

Recent advances in the field of energy conversion (e.g., distributed power generation systems, traction and adjustable speed drives (ASD)) show a focus on voltage source converters (VSCs) [1–5]. VSCs are designed to meet the demands of high efficiency, robustness and low harmonics injection into power systems or low torque pulsation (corresponding to an increased motor lifetime due to decreased shaft vibrations). Expectations related to energy-saving and power quality solutions cause more frequent replacement of input diode rectifiers with active front end (AFE) because of the following main features [6]:

- Bidirectional power flow (allows regenerative breaking),
- Nearly sinusoidal input current with low harmonics distortion,
- High (including unity)-input power factor operation,
- Sufficient dynamics to follow load variation,
- Adjustment and stabilization of the DC-link voltage U_{DC} (distorted 100 Hz oscillation),
- Operation under line voltage distortion (harmonics, sags, etc.).

Nowadays, AFE in AC–DC–AC energy conversion systems has started to be the standard solution. One of the important applications of medium-power AC–DC–AC fed ASD is high speed rail (HSR). Several different types of railway traction power systems (RTPSs) are used throughout the world [7, 8]; however, in modern HSR, there is a trend toward using medium-voltage AC single-phase RTPSs. Therefore, issues related to the energy conversion for ASD used in the HSR become extremely important. In such systems, the VSCs connected to the grid through a step-down transformer are widely used as the AC–DC AFE.

Initially, as the AC–DC VSC in single-phase RTPS, an H-bridge converter (H-BC) fed from single low-voltage winding of the step-down transformer was used [8, 9]. The H-BC provides three levels of

Power Electronics for Renewable Energy Systems, Transportation and Industrial Applications, First Edition.
Edited by Haitham Abu-Rub, Mariusz Malinowski and Kamal Al-Haddad.

converter output voltage u_{CONV}: 0 and $\pm U_{DC}$. However, owing to the low converter-switching frequency for high power, the harmonics injected into the RTPS can be very high [9]. The current harmonics of the transformer high-voltage side generated by the VSC may have detrimental effects on the RTPS components and other loads, and is of great concern. Therefore, instead of a single H-BC, a parallel connected H-BC fed from two symmetrical low-voltage windings of the step-down transformer was introduced [10–14]. The main advantage of such a solution, due to the interleaved pulse-width modulation (PWM) between parallel connected H-BC modules, is the reduction in the current harmonics distortion of the transformer high-voltage side (five levels of converter output voltage on transformer high-voltage side is obtained: 0, $\pm U_{DC}$ and $\pm 2U_{DC}$).

Recently, other multilevel VSC topologies have become attractive to medium-power conversion [15, 16]. The idea of multilevel converters is based on a series connection of semiconductor devices with more than one DC voltage source. VSC operation above typical semiconductor voltage limits, with reduced voltage stress, lower common mode voltages, reduced harmonics distortion and lower filter requirements are some of the well-known advantages that have made this topology popular in both research and the industry. In particular, two multilevel topologies are widely used in ASD industrial applications: the diode clamped converter (DCC) and the flying capacitor converter (FCC). H-DCC was introduced into single-phase RTPS by Hitachi [17] in 1995 (Japan), and has received further attention over the years [18–21]. Despite many publications, the H-FCC has not been commercially applied to single-phase RTPS. Other multilevel cascade and hybrid topologies, which can be compared with classically used parallel-connected H-BC, have also been proposed [15], but they are not fully accepted by industry in transportation systems. A comparison between the multilevel H-DCC, H-FCC and parallel-connected H-BC – if the same voltage and current rating semiconductor devices are used – can be summarized as follows:

- the H-DCC and H-FCC input voltage and output DC-link voltage may be doubled, retaining the magnitude of input current constant,
- the H-DCC and H-FCC are fed from a single low-voltage winding with the same input current, as for the parallel connected H-BC but with doubled input voltage,
- the H-DCC uses additional semiconductor devices (clamping diodes),
- the H-FCC uses additional DC voltage sources (flying capacitors (FCs),
- a more complex PWM strategy is required for the H-DCC and H-FCC.

The choice between these topologies: H-DCC, H-FCC and parallel-connected H-BC, as well as choice of value of input voltage u_{line} (especially for the H-DCC and H-FCC), depends not only on the technical requirements of the power circuit, such as the DC-link voltage. Equally important – if not more – are the economic requirements: the VSC and transformer volume, size and weight, component prices (semiconductor devices, especially) and so on. Therefore, it is not possible to clearly indicate the best VSC solution for HSR, and only a comparison for similar conditions can be made.

This chapter – divided into two main parts – is devoted to the modulation and control of single-phase grid-side converters in RTPS applications. The first part presents the analysis and comparative studies of PWM techniques with unipolar switching for the aforementioned converter topologies: the parallel-connected H-BC, the H-DCC and the H-FCC. Particular emphasis is placed on the impact of individual modulation techniques and topologies on the quality of the grid current and the harmonic content generated by the converter. The second part is devoted to the current control of single-phase VSC, where the basic structures of the *dq* synchronous reference frame are presented – proportional integral current control (PI-CC) and the *abc* natural reference frame, proportional resonant current control (PR-CC). Moreover, the production of current and DC-link voltage controllers with tuning methods for PI and PR regulators is discussed. At the end, the active power feed forward (APFF) algorithm required to improve the dynamics of the DC-link stabilization (reduction of overvoltage in transient states) is provided.

23.2 Modulation Techniques in Single-Phase Voltage Source Converters

The current harmonics generated by the VSC – as a result of the PWM strategy applied to VSC – are particularly important for single-phase RTPSs. The current harmonics distortion depends on the number of output voltage pattern-forming states visible from the transformer high-voltage side. The number of output voltage pattern-forming states is defined as follows: when one active and one zero switching state is applied symmetrically to the VSC, the output waveform will have three voltage-forming states arranged as follows: zero–active–zero. Therefore, a higher number of switching states applied to the VSC gives a lower current harmonics distortion. An increased number of output voltage pattern-forming states results in a decreasing input current i_{line} distortion.

Each VSC topology is characterized by a different method of output voltage pattern generation, dependent on the modulation technique applied. Moreover, each modulation technique can provide different numbers of output voltage pattern-forming states. Therefore, the comparison between modulation techniques for different topologies for single-phase RTPSs should be performed under strictly defined conditions. In this section, the following PWM techniques in the single-phase VSC are presented:

- hybrid pulse-width modulation (HPWM) and unipolar pulse-width modulation (UPWM) for parallel-connected H-BCs,
- one-dimensional nearest two vectors (1D-N2V) and one-dimensional nearest three vectors (1D-N3V) modulations for H-DCCs,
- 1D-N2V, 1D-N3V and one-dimensional nearest three with two redundant vectors (1D-N(3 + 2R)V) modulations for H-FCCs.

Taking into account the large number of PWM strategies and their possible modifications, the presentation will be based on the following assumptions.

- for each topology switching period, T_S is the same,
- each semiconductor device can only be switched two times (in ON–OFF–ON or OFF–ON–OFF sequences, where the switching is defined as an ON–OFF or OFF–ON state change) per period,
- for each modulation type, the minimum number of switching states is used in order to fulfill the modulation objectives.

Next, the comparison will be carried out for all VSC topologies and PWM techniques mentioned earlier. In order to perform the comparative study under similar conditions, for the control system the following assumptions have been made:

- transistors are ideal switches (switching losses are omitted),
- time delay between calculation and realization of transistor duty cycle is neglected,
- there is no dead time between pairs of transistors,
- grid voltage is u_{grid} is ideal sinusoidal,
- transformer low-voltage side short circuit impedance is calculated as 25% of nominal impedance (typically between 20% and 30%),
- the charge of the DC-link capacitors C_{DC} and the FC is constant, regardless of grid voltage u_{grid} and DC-link voltage U_{DC} amplitude changes.

23.2.1 *Parallel-Connected H-Bridge Converter (H-BC)*

Figure 23.1 shows a single-phase parallel-connected H-BC with a common DC-link fed from the transformer with two low-voltage windings [10–14]. Among the main advantages of this topology are:

- low harmonics distortion of current i_{grid} for transformer high-voltage side at the hybrid and the unipolar modulations,

- high reliability – in the case of one module failure, the H-BC still works with the same amplitude of U_{DC} but at twice reduced power and with an increased THD factor of i_{grid}.

Despite the obvious advantages introduced by parallel connection of the VSC, this is also a major disadvantage because of:

- the high cost, weight and size of a step-down transformer with two secondary windings,
- the increased current and voltage rating of the semiconductor devices.

Each module of the converter shown in Figure 23.1 has two legs, with insulated gate bipolar transistors (IGBTs) denoted S_x and $\overline{S_x}$, where $\overline{S_x}$ is the negation of S_x and x is the leg indication: a or b. The IGBT is switched ON when the gate signal is 1 and switched OFF when the gate signal is 0. All switching states for one module are shown in Table 23.1.

To calculate the duration of the switching states, the modulation index M is indispensable, which is the proportion of the control algorithm reference $u_{CONV,ref}$ with respect to the DC-link voltage U_{DC}:

$$M = \frac{u_{CONV,ref}}{U_{DC}} \tag{23.1}$$

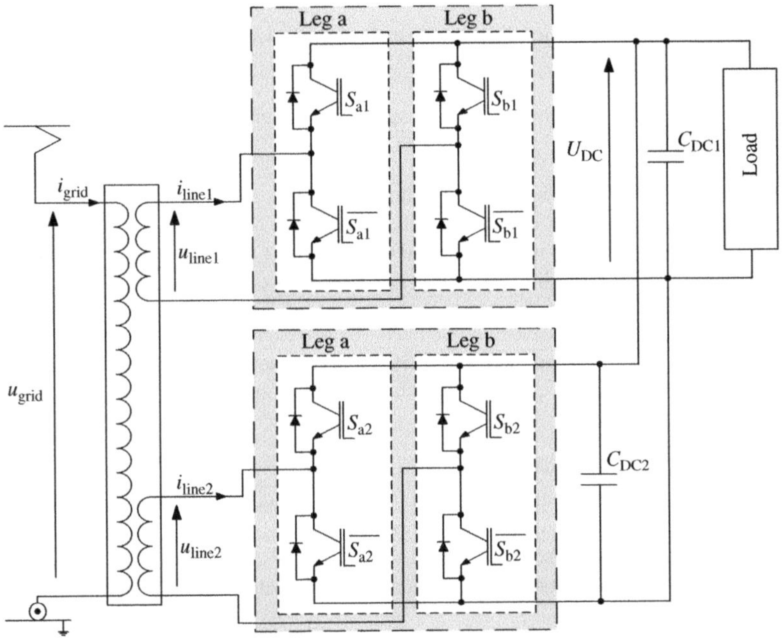

Figure 23.1 Single-phase five-level parallel-connected H-bridge converter

Table 23.1 Switching states for single H-BC module

Switch number	Levels of output t voltage u_{CONV}			
	$-U_{DC}$	0		U_{DC}
S_a	0	0	1	1
S_b	1	0	1	0

where $u_{CONV,ref}$ is the reference amplitude of the converter output voltage. Note that $|u_{CONV,ref}|$ cannot be greater than U_{DC}; thus, $\boldsymbol{M} \in \langle -1, 1 \rangle$.

The decision as to which output voltage level is applied (positive or negative) depends on $\boldsymbol{M}$ (Equation 23.1):

$$\begin{aligned} &\text{positive} \quad 0 \leftrightarrow U_{DC} \leftrightarrow 0 \quad \text{for } \boldsymbol{M} \in (0, 1\rangle \\ &\text{negative} \quad 0 \leftrightarrow -U_{DC} \leftrightarrow 0 \quad \text{for } \boldsymbol{M} \in \langle -1, 0) \end{aligned} \tag{23.2}$$

The switching state time durations for every u_{CONV} level are assigned as follows: T_0 for $u_{CONV} = 0$ and T_1 for $u_{CONV} = \pm U_{DC}$, and are calculated as:

$$\text{for } \boldsymbol{M} \in (0, 1\rangle \begin{cases} T_1 = MT_s \\ T_0 = T_s - T_1 \end{cases} \tag{23.3}$$

$$\text{for } \boldsymbol{M} \in \langle -1, 0) \begin{cases} T_1 = -MT_s \\ T_0 = T_s - T_1 \end{cases} \tag{23.4}$$

The IGBT state and its time duration depends on the PWM technique applied: the HPWM or UPWM.

The HPWM [22–24] for the single H-BC module is based on the assumption that only two among four transistors are pulse-width modulated in each sampling period (Figure 23.2). This means that in each period, only one leg is modulated and – as a consequence – only one of two possible redundant

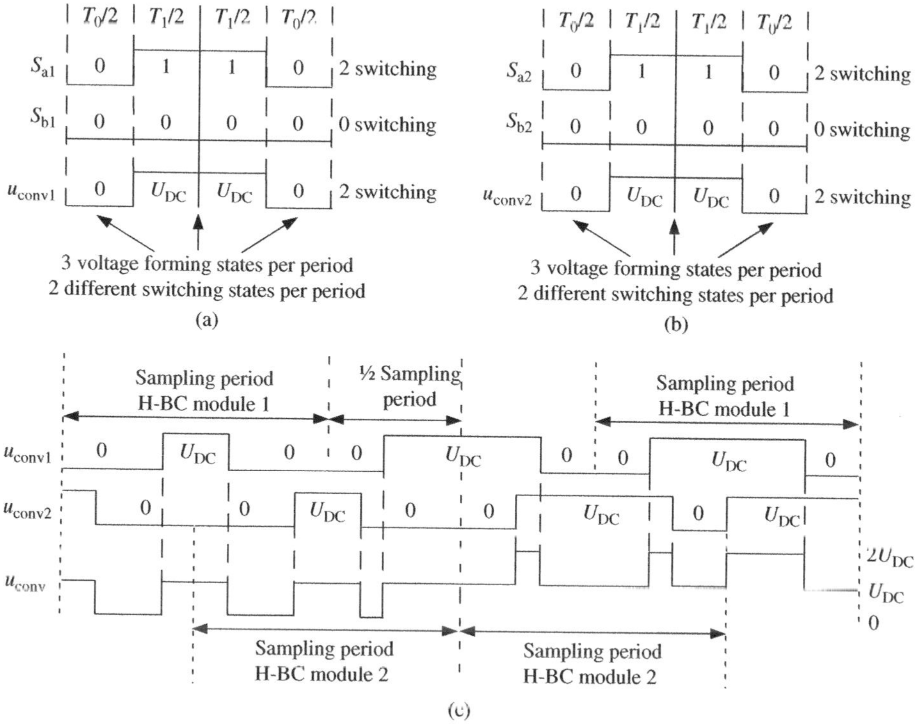

Figure 23.2 Duty cycles for parallel-connected H-BC with hybrid modulation: (a) module 1, (b) module 2 and (c) parallel-connected H-BC u_{CONV} voltages with interleaved sampling periods $\frac{1}{2}T_S$ between modules

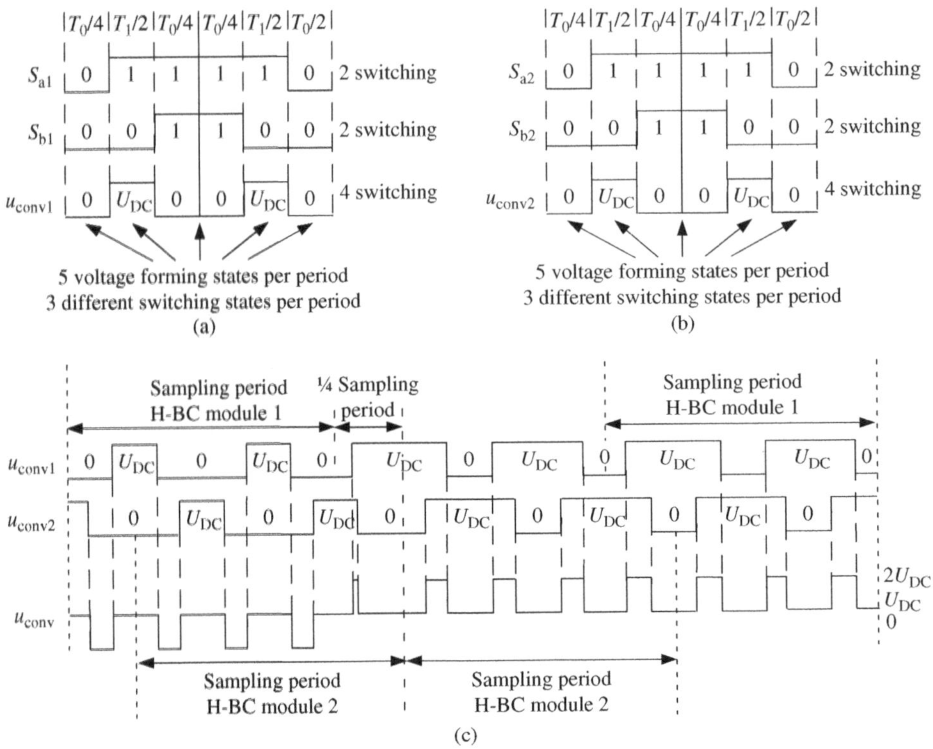

Figure 23.3 Duty cycles for parallel-connected H-BC with unipolar modulation: (a) module 1, (b) module 2 and (c) parallel-connected H-BC u_{CONV} voltages with interleaved sampling periods $^1/_4 T_S$ between modules

states for the $u_{CONV} = 0$ voltage level is used. Switching states are applied symmetrically with respect to the middle of the sampling period. Which leg is modulated depends on the sign of u_{CONV}:

$$\text{leg} \begin{cases} a \text{ for positive } u_{CONV} \text{ sign} \\ b \text{ for negative } u_{CONV} \text{ sign} \end{cases} \tag{23.5}$$

The HPWM applies two switching states in each sampling period for a single H-BC, which gives three output voltage u_{CONV} forming states for module 1 (Figure 23.2(a)) and module 2 (Figure 23.2(b)). For two parallel-connected H-BC modules, the interleaved modulation is applied, which means that for each the same switching pattern and duty cycles of the HPWM can be used, but the sampling periods are shifted $^1/_2 T_S$ between modules. Interleave modulation gives five levels on the transformer high-voltage side: 0, $\pm U_{DC}$, $\pm 2U_{DC}$ and five output voltage u_{CONV} forming states (Figure 23.2(c)).

The other technique, UPWM [22, 24], for the single H-BC module is based on the assumption that all switches are modulated in every sampling period (Figure 23.3). As a result, two redundant states for zero voltage level $u_{CONV} = 0$ are applied. Thus, the UPWM applies three switching states in each sampling period for the single H-BC, which gives five output voltage u_{CONV} forming states for module 1 (Figure 23.3(a)) and module 2 (Figure 23.3(b)). For the parallel-connected H-BC, interleaved modulation is applied, which means the same switching pattern and duty cycles of the UPWM can be used, but the sampling periods are shifted $^1/_4 T_S$ between modules. Interleave modulation gives five levels on the transformer high-voltage side: 0, $\pm U_{DC}$, $\pm 2U_{DC}$, and nine output voltage u_{CONV} forming states (Figure 23.3(c)).

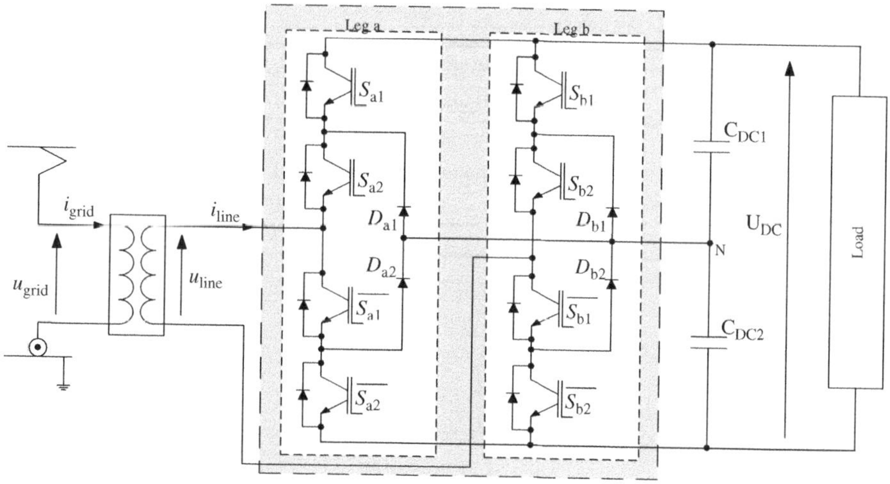

Figure 23.4 Single-phase five-level H-diode clamped converter

23.2.2 H-Diode Clamped Converter (H-DCC)

The DCC was proposed in 1981 by Nabae *et al.* [25]. Figure 23.4 presents the single-phase H-DCC [18–21], which has two legs, with IGBTs denoted S_{x1}, S_{x2}, $\overline{S_{x1}}$ and $\overline{S_{x2}}$, where $\overline{S_{x1}}$ and $\overline{S_{x2}}$ are the negation of S_{x1} and S_{x2}, respectively and $_x$ is the leg indication: *a* or *b*. Two clamping diodes D_{x1} and D_{x2} are parallel connected to S_{x2} and $\overline{S_{x1}}$ for each leg, and the clamping point *N* is connected to the center of the series capacitors C_{DC1} and C_{DC2} in the DC-link. Clamping diodes conduct the current during the generation of the intermediate H-DCC voltage levels.

Three different switching states generating three different output pole voltages can be distinguished for the single leg:

- 0, when S_{x1} and S_{x2} are turned OFF (output pole voltage equals 0),
- 1, when S_{x1} is turned OFF and S_{x2} is turned ON (output pole voltage equals $\frac{1}{2}U_{DC}$, leg is clamped by D_{x1} and D_{x2} diodes),
- 2, when S_{x1} and S_{x2} are turned ON (output pole voltage equals U_{DC}).

The single-phase H-DCC gives nine possible switching states, allowing five levels of output voltage to be obtained: three redundant for $u_{CONV} = 0$, two redundant for whichever $u_{CONV} = \pm\frac{1}{2}U_{DC}$ and one for whichever $u_{CONV} = \pm U_{DC}$. The output voltage level depends on *M* (Equation 23.1):

$$\begin{aligned}
&\text{upper positive} \quad \tfrac{1}{2}U_{DC} \leftrightarrow U_{DC} \leftrightarrow \tfrac{1}{2}U_{DC} \quad \text{for } M \in \langle 0.5, 1\rangle \\
&\text{lower positive} \quad 0 \leftrightarrow \tfrac{1}{2}U_{DC} \leftrightarrow 0 \quad \text{for } M \in \langle 0, 0.5\rangle \\
&\text{lower negative} \quad 0 \leftrightarrow -\tfrac{1}{2}U_{DC} \leftrightarrow 0 \quad \text{for } M \in \langle -0.5, 0\rangle \\
&\text{upper negative} - \tfrac{1}{2}U_{DC} \leftrightarrow -U_{DC} \leftrightarrow -\tfrac{1}{2}U_{DC} \quad \text{for } M \in \langle -1, -0.5\rangle
\end{aligned} \tag{23.6}$$

All possible switching states for the H-DCC are shown in Table 23.2. For every u_{CONV} level, the switching state duration times are assigned as follows:

- T_0 for $u_{CONV} = 0$,
- T_1 for $u_{CONV} = \pm\frac{1}{2}U_{DC}$,
- T_2 for $u_{CONV} = \pm U_{DC}$,

Table 23.2 Switching states for H-DCC

Switch number	Levels of output voltage u_{CONV}								
	$-U_{DC}$	$-\frac{1}{2}U_{DC}$		0			$\frac{1}{2}U_{DC}$		U_{DC}
S_{a1}	0	0	0	0	0	1	1	0	1
S_{a2}	0	1	0	0	1	1	1	1	1
S_{b1}	1	1	0	0	0	1	0	0	0
S_{b2}	1	1	1	0	1	1	1	0	0

and are calculated as:

$$\text{for } M \in (0.5,\ 1\rangle \begin{cases} T_2 = 2(M - 0.5)T_s \\ T_0 = T_s - T_2 \end{cases} \tag{23.7}$$

$$\text{for } M \in (0,\ 0.5\ \rangle \begin{cases} T_1 = 2MT_s \\ T_0 = T_s - T_1 \end{cases} \tag{23.8}$$

$$\text{for } M \in (-0.5,\ 0\rangle \begin{cases} T_1 = -2MT_s \\ T_0 = T_s - T_1 \end{cases} \tag{23.9}$$

$$\text{for } M \in \langle -1, -0.5\rangle \begin{cases} T_2 = -2(M + 0.5)T_s \\ T_1 = T_s - T_2 \end{cases} \tag{23.10}$$

In order to provide the correct H-DCC operation, the voltage on each DC-link capacitor should be stabilized on $\frac{1}{2}U_{DC}$. Switching states used for the balancing of capacitor voltages U_{DC1} and U_{DC2} – depending on output voltage level and sign of line current – are shown in Table 23.3. For each sampling period, one (over the entire calculated time T_1) or two (each at the appropriate parts of calculated time T_1) redundant states affecting the DC-link capacitor voltages can be used. Switching state selection and duration time depend on the PWM technique applied, which are described subsequently: 1D-N2V or 1D-N3V modulations.

The 1D-N2V modulation for the H-DCC is based on the assumption that only two switching states are applied in each sampling period [20, 26]. To reduce the number of switching, between output voltage transition from $u_{CONV} = 0$ to $u_{CONV} = \pm 1/2U_{DC}$, only $S_{x1}S_{x2}$ (01) is selected from the three redundant $u_{CONV} = 0$ (Table 23.2). Figure 23.5 presents the H-DCC positive output voltage-switching patterns for

Table 23.3 H-DCC switching states used for DC-link capacitors voltage balancing

Output voltage u_{CONV} level	DC-link voltages	Line current	Switching state S_{a1}	S_{a2}	S_{b1}	S_{b2}
$\frac{1}{2}U_{DC}$	$U_{DC1} > U_{DC2}$	$i_{line} > 0$	0	1	0	0
		$i_{line} < 0$	1	1	0	1
	$U_{DC1} < U_{DC2}$	$i_{line} > 0$	1	1	0	1
		$i_{line} < 0$	0	1	0	0
$-\frac{1}{2}U_{DC}$	$U_{DC1} > U_{DC2}$	$i_{line} > 0$	0	1	1	1
		$i_{line} < 0$	0	0	0	1
	$U_{DC1} < U_{DC2}$	$i_{line} > 0$	0	0	0	1
		$i_{line} < 0$	0	1	1	1

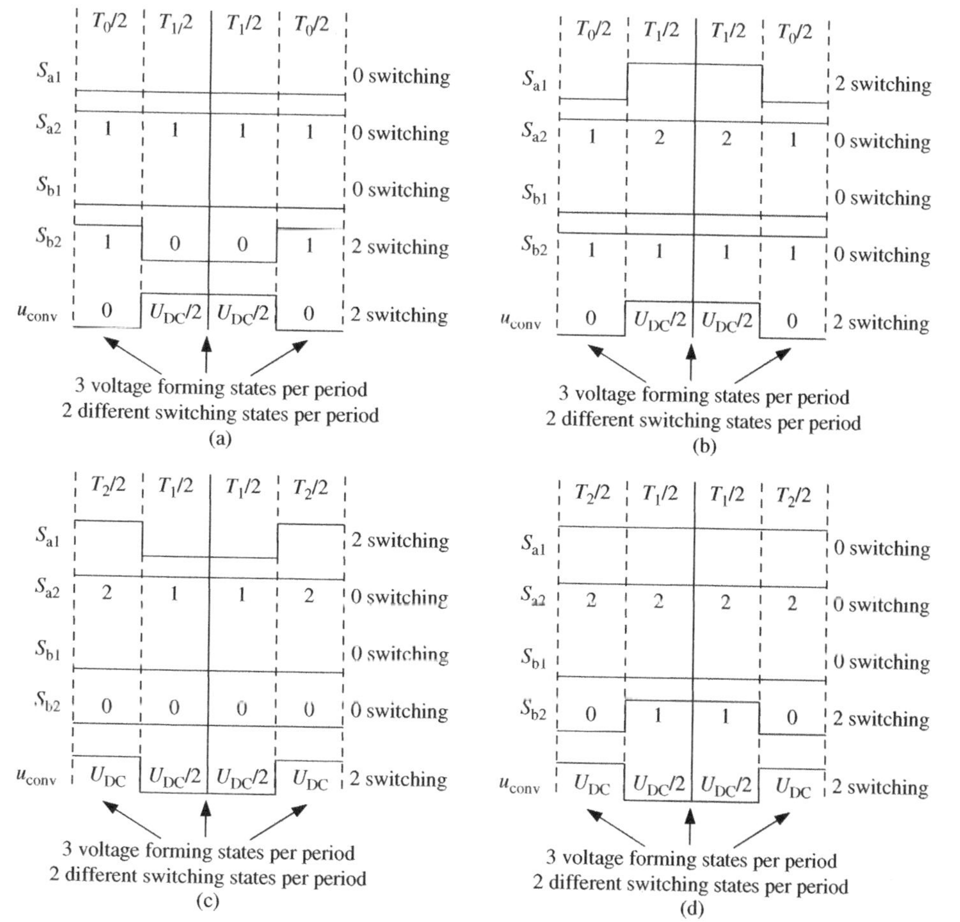

Figure 23.5 Duty cycles for H-DCC with 1D-N2V modulation generating u_{CONV} output voltage level: (a) and (b) positive lower $0 \leftrightarrow \frac{1}{2}U_{DC} \leftrightarrow 0$, (c) and (d) positive upper $\frac{1}{2}U_{DC} \leftrightarrow U_{DC} \leftrightarrow \frac{1}{2}U_{DC}$ for 10 and 21 redundant switching state selections, respectively

the 1D-N2V modulation. The 1D-N2V modulation allows obtaining the symmetrical duty cycle placement of three u_{CONV} voltage-forming states for all output voltage levels.

Another modulation technique for H-DCC is the 1D-N3V. Figure 23.6 presents the positive output voltage-switching patterns for 1D-N3V modulation, which is based on the assumption that both redundant switching states, for whichever output voltage level $u_{CONV} = \pm\frac{1}{2}U_{DC}$ are applied in each sampling period [20, 26, 27]. The 1D-N3V modulation allows symmetrical duty cycle placement to be obtained for the seven (see Figure 23.6(a)) and five (see Figure 23.6(b)) output voltage u_{CONV} forming states for lower and upper output voltage levels, respectively. Redundant switching states are applied for the appropriate part of the calculated time T_1, which corresponds to the voltage ratio between DC-link capacitors C_{DC1} and C_{DC2}. It means that if U_{DC1} is larger than U_{DC2}, C_{DC1} will be charged less than C_{DC2}. The ratio of T_1 charging time division of C_{DC1} (T_{1-DC1}) and C_{DC2} (T_{1-DC2}) is inversely proportional to the ratio of the U_{DC1} and U_{DC2} voltages:

$$T_{1-DC1} = \frac{U_{DC1}}{U_{DC2}}T_1, \quad T_{1-DC2} = \frac{U_{DC2}}{U_{DC1}}T_1 \tag{23.11}$$

where $T_1 = T_{1-DC1} + T_{1-DC2}$. As a result, during the sampling period, both DC-link capacitors over the calculated time T_1 will be charged or discharged.

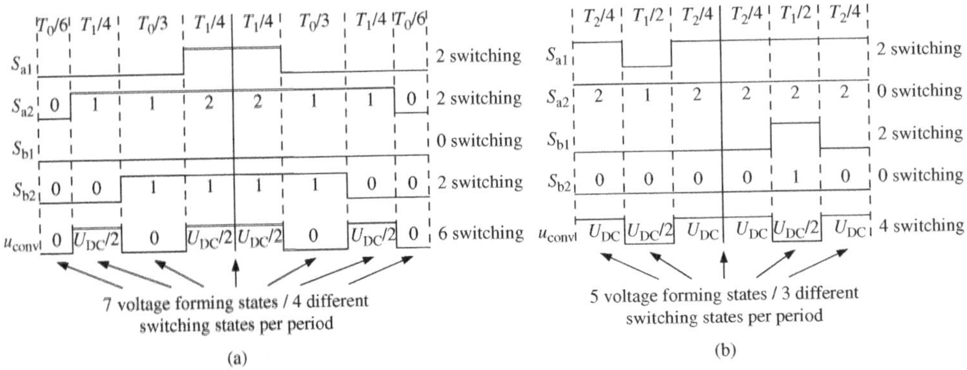

Figure 23.6 Duty cycles for the H-DCC with 1D-N3V modulation generating u_{CONV} output voltage level: (a) positive lower $0 \leftrightarrow \frac{1}{2}U_{DC} \leftrightarrow 0$ and (b) positive upper $\frac{1}{2}U_{DC} \leftrightarrow U_{DC} \leftrightarrow \frac{1}{2}U_{DC}$.

It should be noted that the self-balancing of the DC-link capacitors with a symmetrical distribution of T_1, even for balanced U_{DC1} and U_{DC2} voltages, is not possible in practice because of:

- nonlinear IGBT gate signal propagation,
- different IGBT ON/OFF times,
- DC-link capacitor equivalent series and parallel resistances,
- differences in electrical parameters, such as each leg's semiconductor devices and connection points.

Therefore, an additional controller (proportional or proportional-integral (PI)) should be used in order to determine the distribution of T_1, providing the equalization of the U_{DC1} and U_{DC2} voltages.

23.2.3 *H-Flying Capacitor Converter (H-FCC)*

Figure 23.7 presents the single-phase H-FCC [28–30], which has two legs, with IGBTs denoted S_{x1}, S_{x2}, $\overline{S_{x1}}$ and $\overline{S_{x2}}$, where $\overline{S_{x1}}$ and $\overline{S_{x2}}$ are the negation of S_{x1} and S_{x2}, respectively and x is the leg indication: a or b. The so-called *FC* C_{FCx} are parallel connected to S_{x2} and $\overline{S_{x2}}$ for each leg. The FCs are used to generate the intermediate H-FCC voltage levels, which can be obtained – in contrast to the H-DCC – separately for each leg from the other leg-switching states. Balancing of the FC voltages uses two redundant switching states for each H-FCC leg and is independent of the switching states in the other leg.

Four different switching states (including two redundant states for FC voltage balancing), generating three different output pole voltages can be distinguished for a single leg of the converter:

- 0, when S_{x1} and S_{x2} are turned OFF (output pole voltage equals 0),
- 1, when:
  - S_{x1} is tuned OFF and S_{x2} is turned ON,
  - S_{x1} is tuned ON and S_{x2} is turned OFF,

  where both generate the same $\frac{1}{2}U_{DC}$ output pole voltage (depending on the selected state, $u_{CONV} = U_{FCx}$ or $u_{CONV} = U_{DC} - U_{FCx}$),
- 2, when S_{x1} and S_{x1} are turned ON (output pole voltage equals U_{DC}).

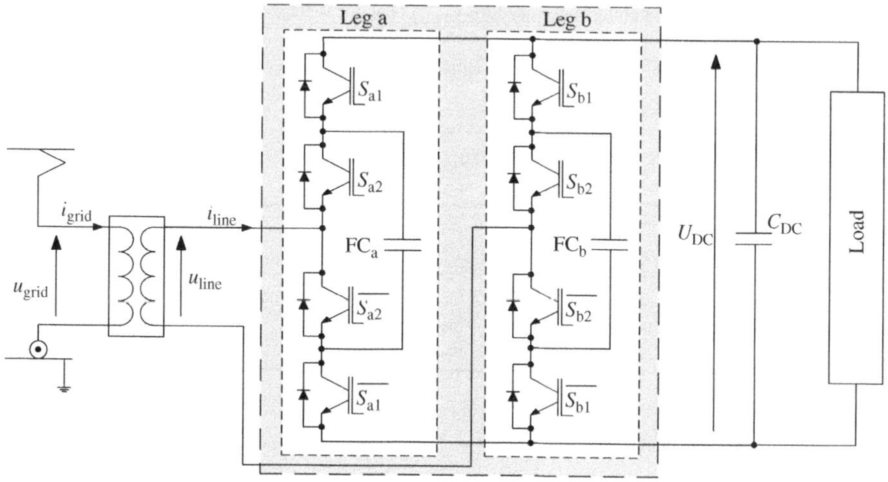

Figure 23.7 Single-phase five-level H-flying capacitor converter

Table 23.4 Switching states for H-FCC

Switch number	Levels of output voltage u_{CONV}															
	$-U_{DC}$	$-\frac{1}{2}U_{DC}$				0						$\frac{1}{2}U_{DC}$				U_{DC}
S_{a1}	0	0	0	0	1	0	0	1	1	0	1	0	1	1	1	1
S_{a2}	0	0	0	1	0	0	1	0	0	1	1	1	0	1	1	1
S_{b1}	1	0	1	1	1	0	0	0	1	1	1	0	0	0	1	0
S_{b2}	1	1	0	1	1	0	1	1	0	0	1	0	0	1	0	0

In the case of the single-phase H-FCC, it gives 16 possible switching state combinations. Thus, the H-FCC allows five levels of module u_{CONV} output voltage to be obtained (Table 23.4): six redundant for $u_{CONV} = 0$, four redundant for whichever $u_{CONV} = \pm\frac{1}{2}U_{DC}$, and one for whichever $u_{CONV} = \pm U_{DC}$. Which output voltage level is applied is dependent on M (Equations 23.1 and 23.6). For every u_{CONV} level, the switching state duration times are assigned as follows:

- T_0 for $u_{CONV} = 0$,
- T_1 for $u_{CONV} = \pm\frac{1}{2}U_{DC}$, which can be divided – with respect to the modulation technique used – into two times: T_A and T_B (where $T_1 = T_A + T_B$) assigned for leg a and b, respectively (in each sampling period only one or two FC capacitors can be used in order to fulfill the modulation objectives),
- T_2 for $u_{CONV} = \pm U_{DC}$,

and are calculated in the same way as for the H-DCC (Equations 23.7–23.9, 23.10).

As mentioned earlier, in order to provide proper H-FCC operation, the voltage on each FC capacitor should be stabilized on $\frac{1}{2}U_{DC}$. The balancing of single FC voltage C_{FCa} or C_{FCb} is independent of the switching state in the other leg. Therefore, switching states depending on the output voltage level and sign of the line current – used only for FC voltage balancing in leg a (leg b is similar) are shown in Table 23.5.

Table 23.5 H-FCC switching states used for C_{FCa} capacitor voltage balancing

Output voltage u_{CONV} level	DC-link voltages	Line current	Switching state			
			S_{a1}	S_{a2}	S_{b1}	S_{b2}
$\frac{1}{2}U_{DC}$	$U_{FCa} > \frac{1}{2}U_{DC}$	$i_{line} > 0$	1	0	0	0
		$i_{line} < 0$	0	1	0	0
	$U_{FCa} < \frac{1}{2}U_{DC}$	$i_{line} > 0$	0	1	0	0
		$i_{line} < 0$	1	0	0	0
$-\frac{1}{2}U_{DC}$	$U_{FCa} > \frac{1}{2}U_{DC}$	$i_{line} > 0$	0	1	1	1
		$i_{line} < 0$	1	0	1	1
	$U_{FCa} < \frac{1}{2}U_{DC}$	$i_{line} > 0$	1	0	1	1
		$i_{line} < 0$	0	1	1	1

For each sampling period:

- one (over the entire calculated time T_1, when only one FC is used),
- two (each at time T_A and T_B, when both FCs are used),
- all four (each at the appropriate part of the calculated times T_A and T_B)

redundant states affecting the FC capacitor voltages can be used. Switching state selection and duration time depend on the applied PWM technique: 1D-N2V, 1D-N3V, or 1D-N(3 + 2R)V modulations, respectively.

1D-N2V modulation is based on the assumption that only two switching states are applied in each sampling period, when only one leg is pulse-width modulated [20, 26]. Which leg is switched depends on the sign of u_{CONV}, similar to the parallel-connected H-BC (Equation 23.5). As a consequence, for a positive sign of u_{CONV}, only one of two, whichever $u_{CONV} = \pm\frac{1}{2}U_{DC}$, redundant states for each lower or upper voltage level is selected. The duration of redundant state T_A in leg a is equal to T_1 with proper $u_{CONV} = 0$ and $u_{CONV} = \pm U_{DC}$ output voltage switching states. Leg b is permanently in the 0 state. The selection of one of two possible redundant states for leg a depends on the conditions given in Table 23.5. If u_{CONV} has a negative sign, the opposite situation occurs – leg b is modulated and leg a is permanently in the 0 state. To reduce the number of switching between $u_{CONV} = 0$ and redundant $u_{CONV} = \pm\frac{1}{2}U_{DC}$ output voltage state, the $S_{x1}S_{x2}$ (00) state is chosen from six redundant $u_{CONV} = 0$ output voltage states (Tables 23.4 and 23.6). The reduced switching states of the 1D-N2V modulation are presented in Table 23.6. Figure 23.8 presents the positive output voltage-switching patterns for the H-FCC with the 1D-N2V modulation. It allows symmetrical duty cycle placement to be obtained for the three forming states of the output voltage u_{CONV} for all levels, with two different switching states used in each sampling period.

The next type, the 1D-N3V modulation, is based on the assumption that both legs, a and b, are working in each sampling period to generate the output voltage level $u_{CONV} = \frac{1}{2}U_{DC}$ [20, 26, 27]. However,

Table 23.6 Switching states for H-FCC with 1D-N2V modulation

Switch number	Levels of output voltage u_{CONV}						
	$-U_{DC}$	$-\frac{1}{2}U_{DC}$		0	$\frac{1}{2}U_{DC}$		U_{DC}
S_{a1}	0	0	0	0	1	0	1
S_{a2}	0	0	0	0	0	1	1
S_{b1}	1	1	0	0	0	0	0
S_{b2}	1	0	1	0	0	0	0

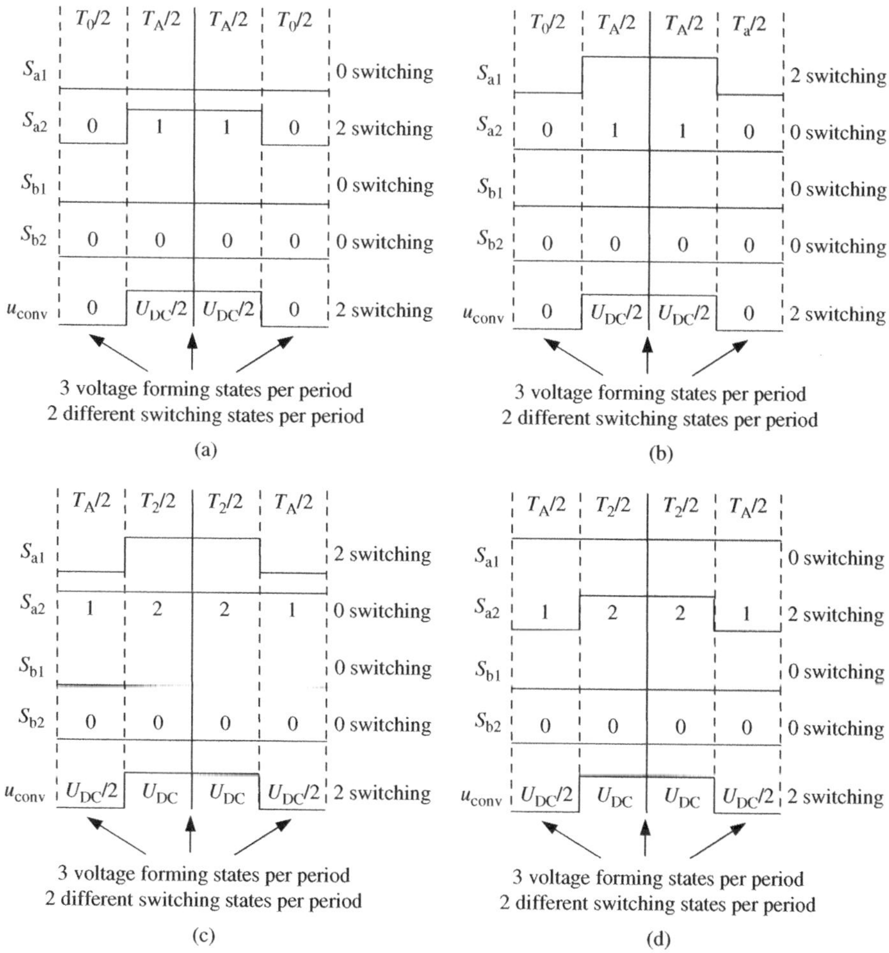

Figure 23.8 Duty cycles for the H-FCC with 1D-N2V modulation generating u_{CONV} output voltage level: (a) positive lower $0 \leftrightarrow \frac{1}{2}U_{DC} \leftrightarrow 0$ and (b) positive upper $\frac{1}{2}U_{DC} \leftrightarrow U_{DC} \leftrightarrow \frac{1}{2}U_{DC}$

for each leg only one of two, whichever $u_{CONV} = \pm\frac{1}{2}U_{DC}$, redundant states for the lower or upper voltage level is selected over the calculated time $T_A = T_B$ equal to $\frac{1}{2}T_1$. Switching state selection for leg b depends – similar to 1D-N2V modulation – on the conditions given in Table 23.5. Such a selection results in four possible combinations of switching states generating output voltage $u_{CONV} = +\frac{1}{2}U_{DC}$. To reduce the number of switching between output voltage $u_{CONV} = 0$ and $u_{CONV} = \pm\frac{1}{2}U_{DC}$, for each aforementioned combination, only two (from six redundant $u_{CONV} = 0$ output voltage states) are chosen:

- one permanently: $S_{x1}S_{x2}$ (00) or $S_{x1}S_{x2}$ (11) (for positive and negative voltage levels, respectively),
- one depending on the selected combination: (0101) for (0100) and (1101), (0110) for (0100) and (1110), (1001) for (1000) and (1101) and finally (0110) for (1000) and (1110).

All lower $0 \leftrightarrow \frac{1}{2}U_{DC} \leftrightarrow 0$ and upper $\frac{1}{2}U_{DC} \leftrightarrow U_{DC} \leftrightarrow \frac{1}{2}U_{DC}$ positive output voltage level switching state combinations are shown in Figures 23.9 and 23.10, respectively.

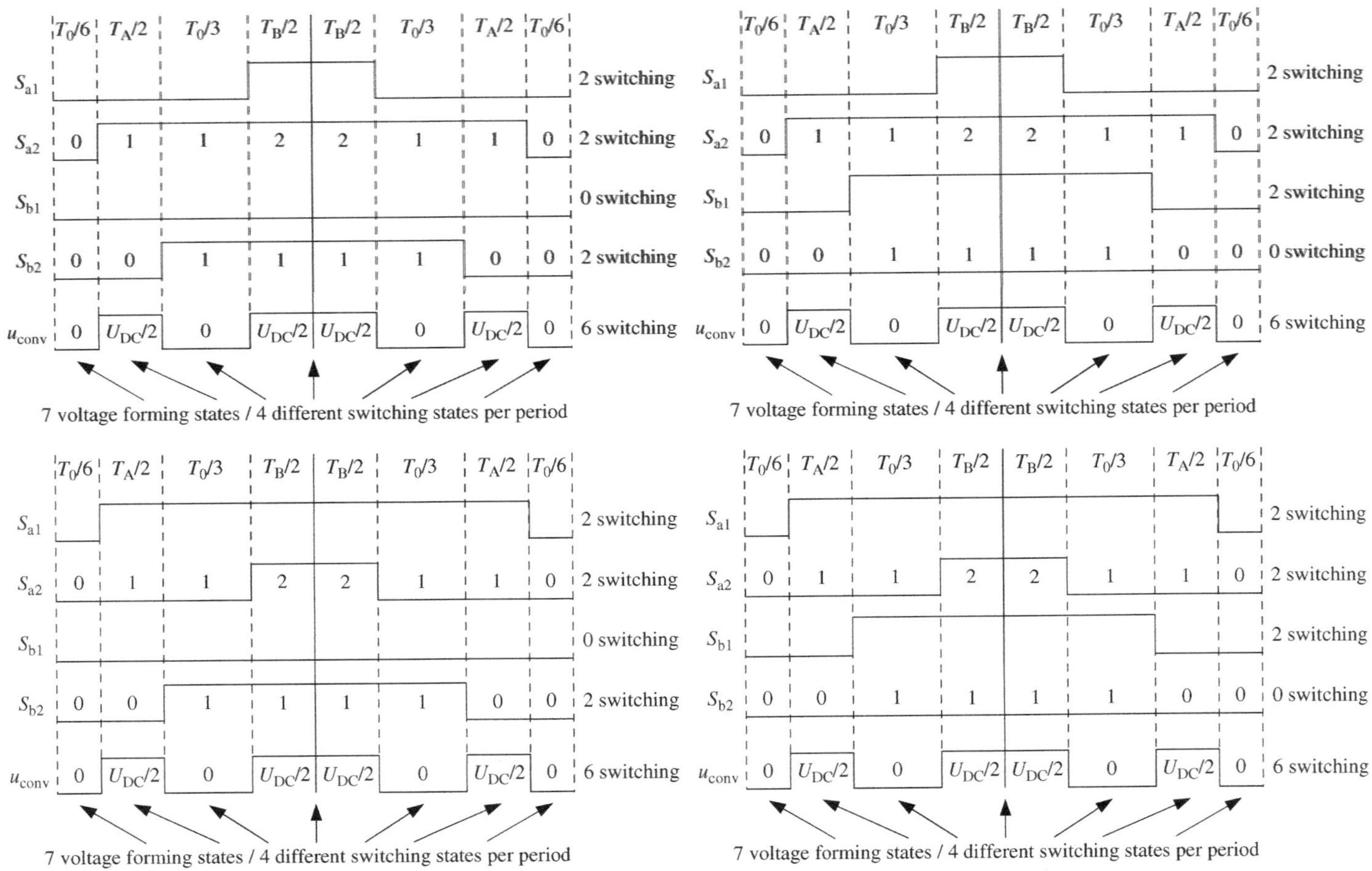

Figure 23.9 Duty cycles for the H-FCC with 1D-N3V modulation generating u_{CONV} lower positive output voltage level $0 \leftrightarrow \frac{1}{2}U_{DC} \leftrightarrow 0$ with different redundant switching states

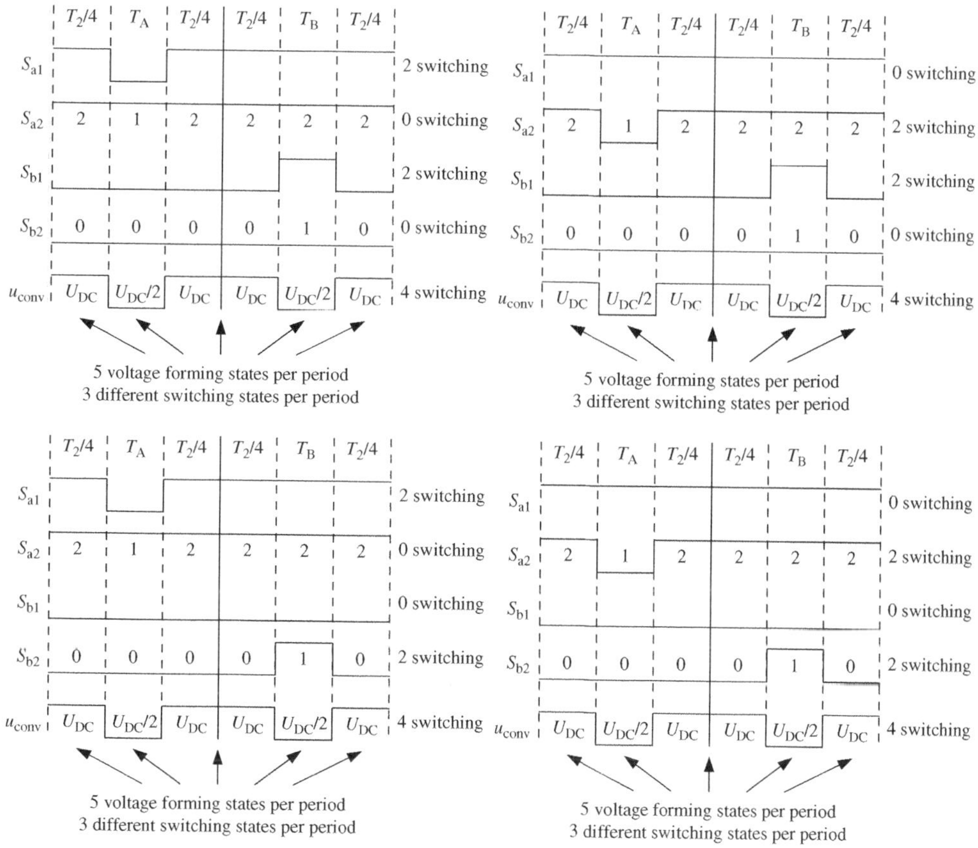

Figure 23.10 Duty cycles for the H-FCC with 1D-N3V modulation generating u_{CONV} positive upper output voltage level $\frac{1}{2}U_{DC} \leftrightarrow U_{DC} \leftrightarrow \frac{1}{2}U_{DC}$ with different redundant switching states

The symmetrical duty cycle placement of the four different switching states used for the lower output voltage level gives seven forming states of output voltage u_{CONV} (Figure 23.9), but the upper positive voltage level redundant states are placed nonsymmetrically with respect to the middle of the sampling period; however, the output pulses of u_{CONV} are generated symmetrically (Figure 23.10). Three different switching states used for the generation of the upper u_{CONV} levels (combination of redundant intermediate with the proper $S_{x1}S_{x2}$ (20) or $S_{x1}S_{x2}$(02) state) gives five forming states of the u_{CONV} voltage.

It is possible to increase the H-FCC output pulse frequency over the frequency generated by the 1D-N3V modulation, which is not possible for the H-DCC. This can be achieved by the method proposed in Ref. [31] the 1D-N(3+2R)V modulation. The 1D-N(3+2R)V modulation is based on the assumption that in every sampling period all four redundant $u_{CONV} = \pm\frac{1}{2}U_{DC}$ output voltage switching states are used (Figure 23.11). As a result, both FCs are charged and discharged in each period, where the redundant switching state charging and discharging of the FCs are applied for the appropriate part of the calculated time $T_A = T_B$ (for FC_a and FC_b, respectively) equal to $\frac{1}{2}T_1$. It means, for example, that if U_{FCa} is bigger than $\frac{1}{2}U_{DC}$, C_{FCa} will be more charged than discharged. The ratio of T_A and T_B between charging (T_{A+}, T_{B+}) and discharging (T_{A-}, T_{B-}) state time division is a proportional function of U_{DC}

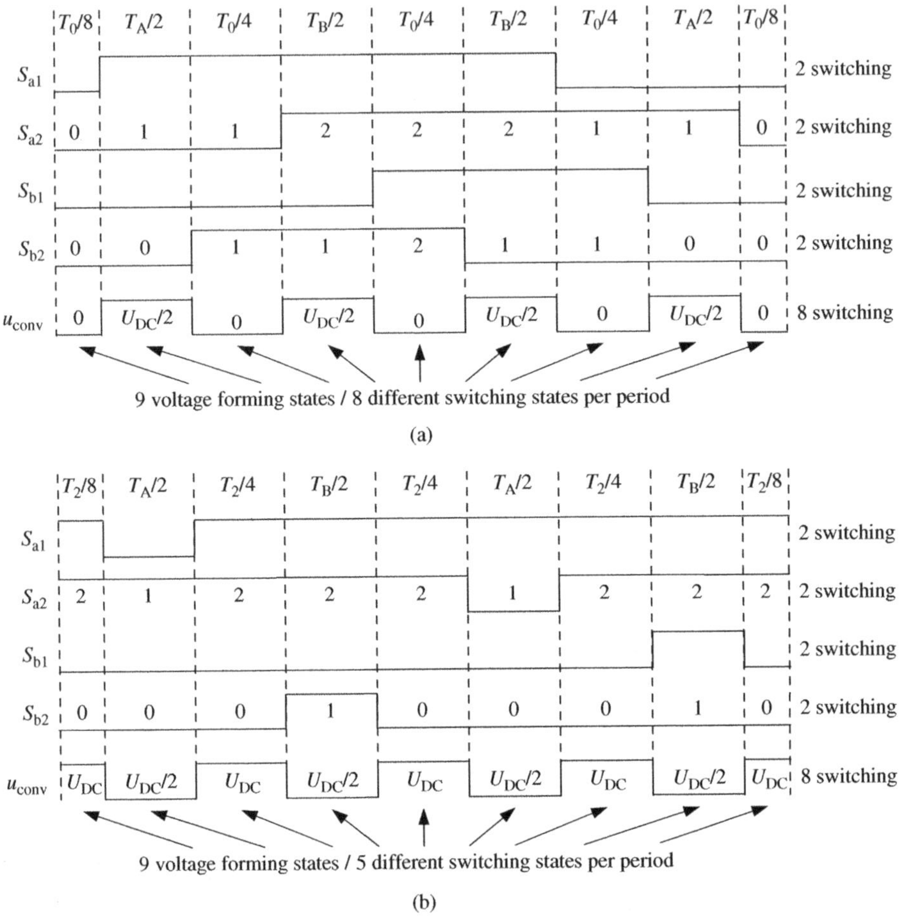

Figure 23.11 Duty cycles for the H-FCC with 1D-N(3 + 2R)V modulation generating u_{CONV} output voltage level: (a) positive lower $0 \leftrightarrow \frac{1}{2}U_{DC} \leftrightarrow 0$ and (b) positive upper $\frac{1}{2}U_{DC} \leftrightarrow U_{DC} \leftrightarrow \frac{1}{2}U_{DC}$

and FC voltages U_{FCa} and U_{FCb}:

$$T_{A+} = \frac{U_{DC} - U_{FCa}}{U_{DC}}\frac{T_1}{4}, \quad T_{A-} = \frac{U_{FCa}}{U_{DC}}\frac{T_1}{4}, \quad T_{B+} = \frac{U_{DC} - U_{FCb}}{U_{DC}}\frac{T_1}{4}, \quad T_{B-} = \frac{U_{FCb}}{U_{DC}}\frac{T_1}{4}. \tag{23.12}$$

where $\frac{1}{2}T_1 = T_{A+} + T_{A-} = T_{B+} + T_{B-}$. Such an approach to the nonsymmetrical switching of each IGBTs allows the same output voltage waveform to be obtained as for the parallel-connected H-BC with UPWM – nine forming states of output voltage u_{CONV} for all output voltage levels with:

- eight different switching states used per period for the lower u_{CONV} output voltage,
- five different switching states used per period for the upper u_{CONV} output voltage.

The self-balancing of the FC voltages (assuming $T_{A+} = T_{A-}$ and $T_{B+} = T_{B-}$) is not possible in practice, for reasons similar to the 1D-N3V modulation for H-DCC (in addition to the FC's equivalent series and parallel resistances).

Table 23.7 Number of output voltage u_{CONV} forming states/switching per period

PC-H-BC		H-DCC		H-FCC		
HPWM	UPWM	1D-N2V	1D-N3V	1D-N2V	1D-N3V	1D-N(3 + 2R)V
Lower levels of output voltage u_{CONV}: $0 \leftrightarrow \pm{}^{1}/_{2}U_{DC} \leftrightarrow 0$						
5/4	9/8	3/2	7/6	3/2	7/6	9/8
Upper levels of output voltage u_{CONV}: $\pm{}^{1}/_{2}U_{DC} \leftrightarrow \pm U_{DC} \leftrightarrow \pm{}^{1}/_{2}U_{DC}$						
5/4	9/8	3/2	5/4	3/2	5/4	9/8

23.2.4 Comparison

The reference for the comparison study of the single-phase VSC topologies and PWM techniques presented is the single-phase parallel-connected H-BC with unipolar modulation used in RTPSs. Such a converter, owing to interleaved modulation, provides an increased number of output voltage u_{CONV} forming states and very low current harmonics distortion on the transformer high-voltage side.

Table 23.7 presents the number of output voltage u_{CONV} forming states visible from the transformer high-voltage side and the number of total switching states used in the sampling period to generate the u_{CONV} voltage waveform. The comparison of PWM techniques shows the clear advantage of the parallel-connected H-BC with the UPWM and the H-FCC with the 1D-N(3 + 2R)V modulation with respect to the highest number of forming states.

Figures 23.13–23.15 present the steady-state operation of the 1 MW parallel-connected H-BC, H-DCC and H-FCC with the UPWM, 1D-N3V and the 1D-N(3 + 2R)V modulations, respectively. The simplified simulation model for the aforementioned cases is shown in Figure 23.12, with the main electrical parameters of the power circuits and control data given in Table 23.8. For all cases, the converter voltage harmonics distortion $THD(u_{CONV})$ factor is similar. However, with respect to 1 kHz sampling frequency bands related to the output pulse's frequency:

- quadrupled for the parallel-connected H-BC (Figure 23.13(e-f)),
- doubled for the lower (Figure 23.14(e)) and tripled for the upper (Figure 23.14(f)) output voltage levels for the H-DCC,
- quadrupled for the H-FCC (Figure 23.15(e-f)),

shifting in the direction of higher harmonics can be observed:

- 4 kHz for the parallel-connected H-BC (Figure 23.13(g)),
- 2 kHz and 3 kHz for the H-DCC (Figure 23.14(g)),
- 4 kHz for the H-FCC (Figure 23.15(g)).

Thus, a similar voltage distortion THD(u_{CONV}) factor for higher frequencies provides a lower harmonics distortion of the grid current.

Table 23.9 presents a comparison of the grid current distortion THD(i_{grid}) for all topologies and modulation techniques presented, with different load values. It can be observed that the parallel-connected H-BC with the UPWM, as well as the H-FCC with 1D-N(3 + 2R)V PWM modulation – thanks to the increased number of output voltage u_{CONV} forming states – provides the lowest grid current distortion THD(i_{grid}). However, if the H-FCC with 1D-N(3 + 2R)V modulation and parallel-connected H-BC with UPWM;

- provide similar harmonics injection into the power system,
- uses all switches in the sampling period to generate output voltage,
- uses the same voltage and current class semiconductors (which guarantees similar switching losses),

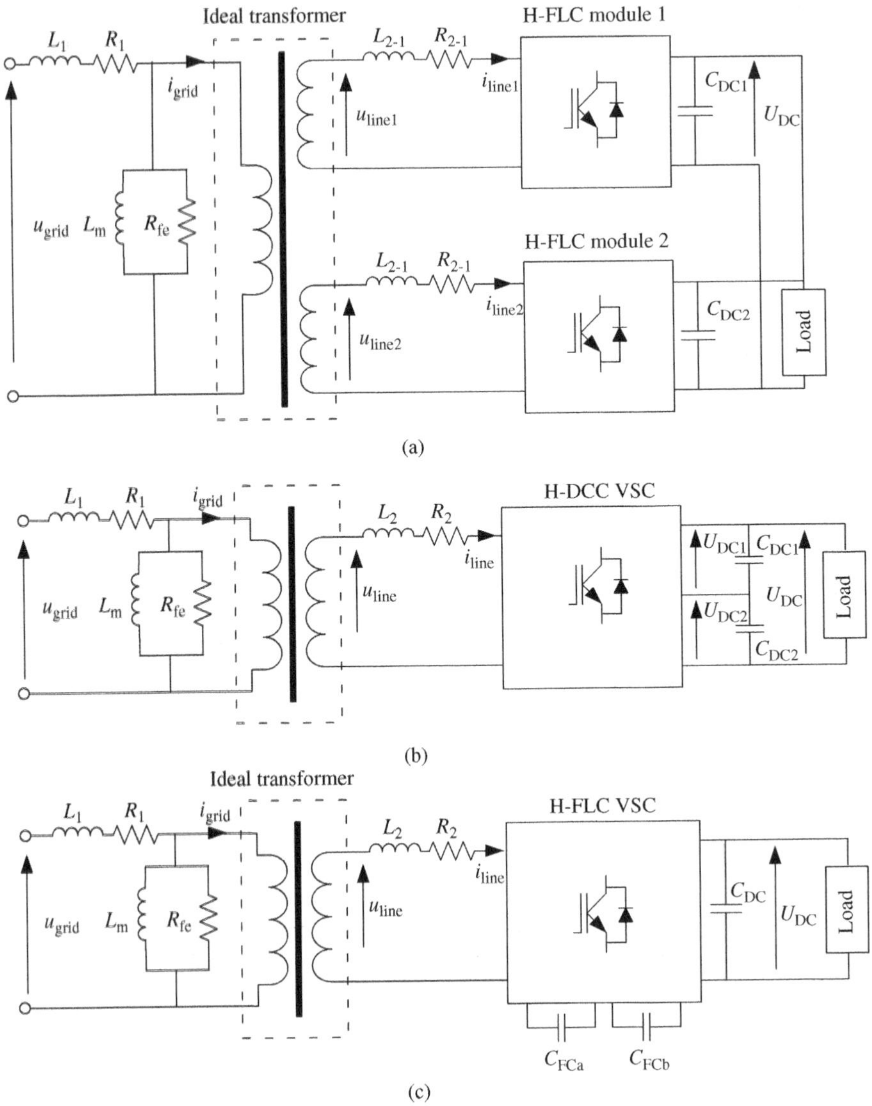

Figure 23.12 Simplified model of 1 MW single-phase converters: (a) parallel-connected H-BC, (b) H-DCC and (c) H-FCC

the advantages of the H-FCC solution are:

- a reduced number of transformer low-voltage side windings,
- a reduced transformer volume, size and weight,
- component price reductions due to the decreased voltage class semiconductors.

On the other hand, the parallel-connected H-BC has a high reliability in the case of one module failure – the VSC works with the same amplitude of U_{DC} but with twice reduced power and with an

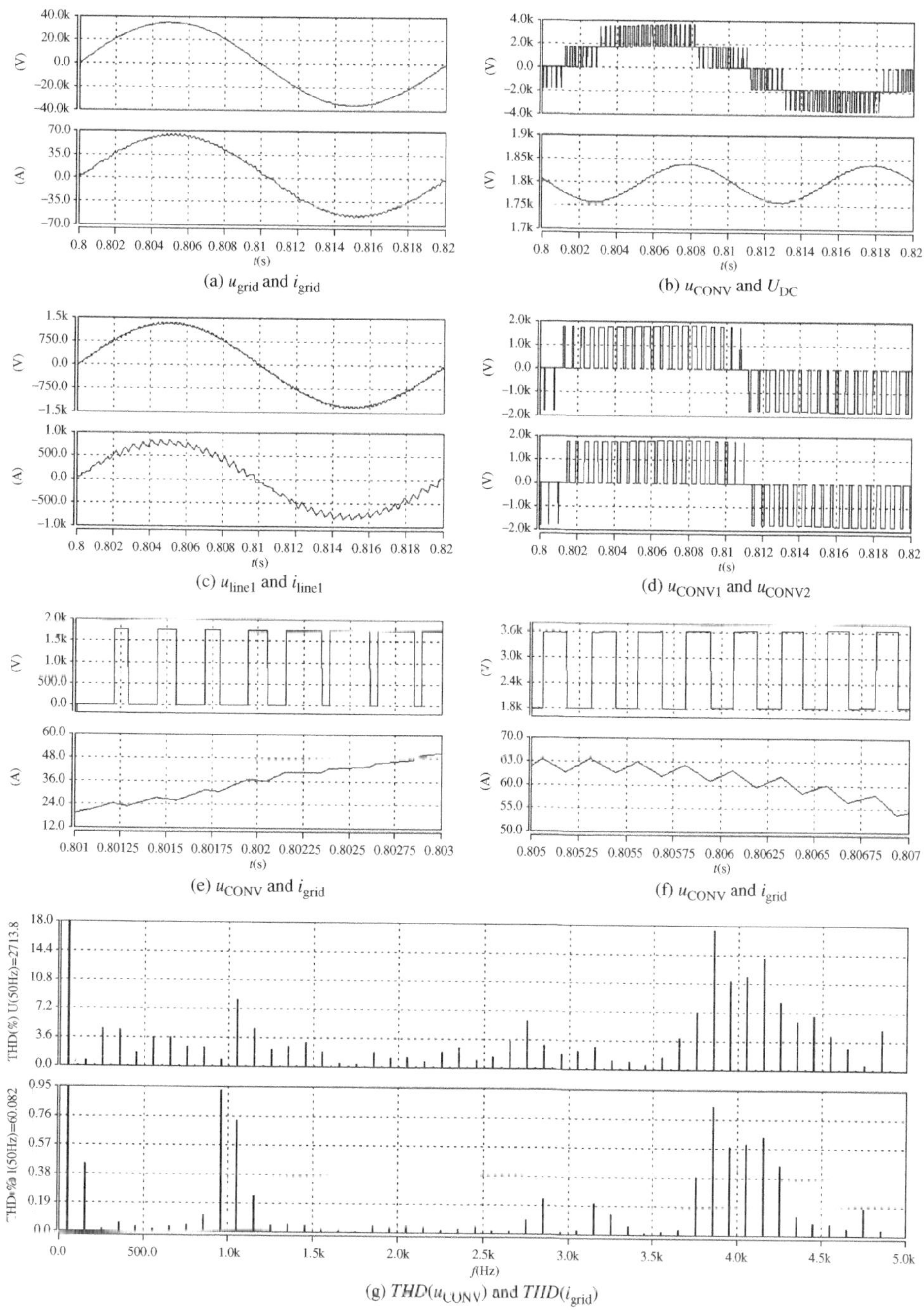

Figure 23.13 Operation of parallel-connected H-BC with UPWM: (a) grid voltage u_{grid} and current i_{grid}, (b) converter voltage u_{CONV} and DC-link voltage U_{DC}, (c) module 1 line voltage u_{line1} and current i_{line1}, (d) module 1 and 2 voltage u_{CONV1} and u_{CONV2} (e) and (f) two periods zoom for lower and upper u_{CONV} converter voltage level and grid current i_{grid} and (g) grid current and converter voltage harmonics

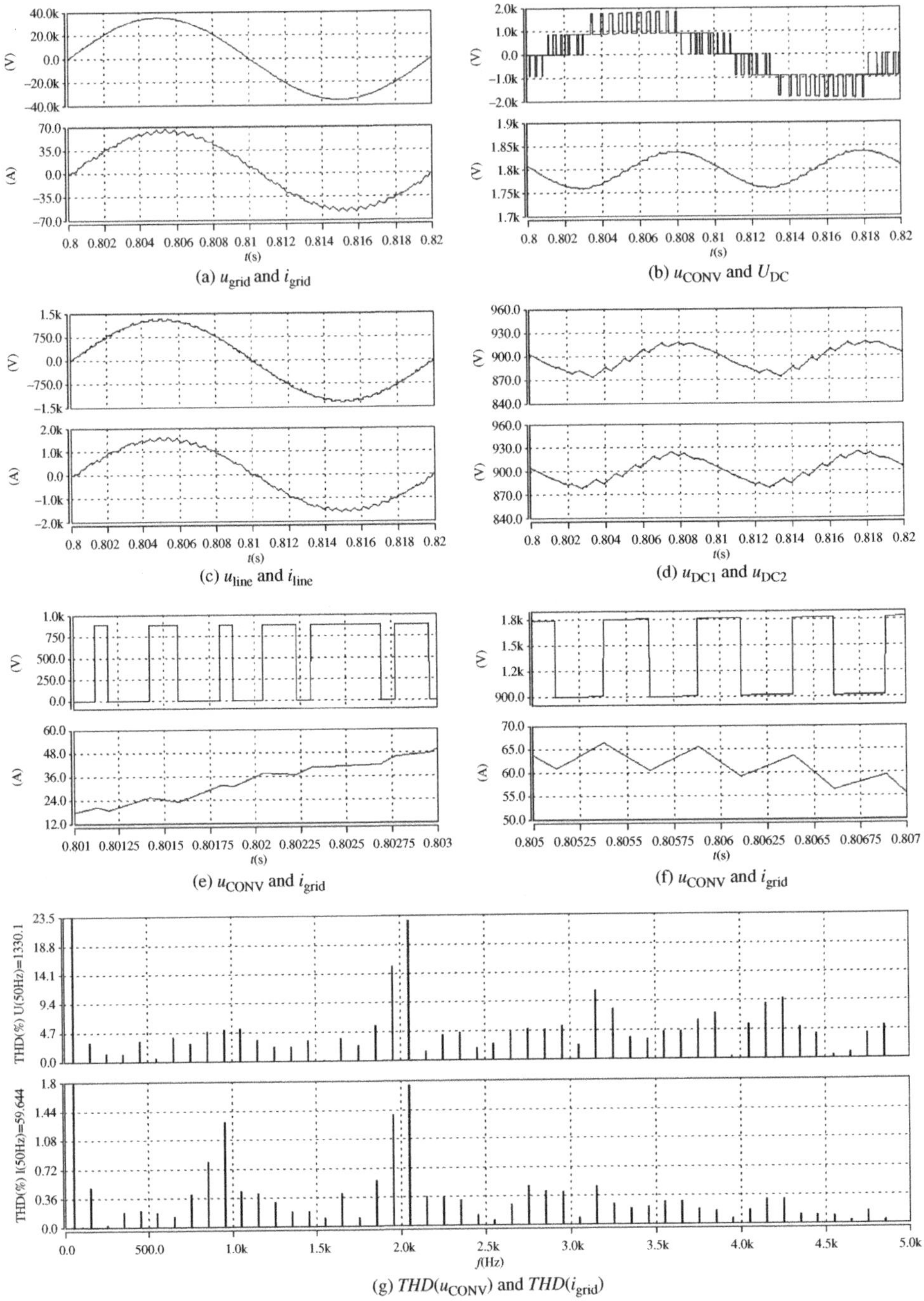

Figure 23.14 Operation of H-DCC with 1D-N3V: (a) grid voltage u_{grid} and current i_{grid}, (b) converter voltage u_{CONV} and DC-link voltage U_{DC}, (c) line voltage u_{line} and current i_{line}, (d) upper and lower DC-link capacitor voltage U_{DC1} and U_{DC2}, (e) and (f) two periods zoom for lower and upper u_{CONV} converter voltage level and grid current i_{grid} (g) grid current and converter voltage harmonics spectrum $THD(u_{CONV})$ and $THD(i_{grid})$.

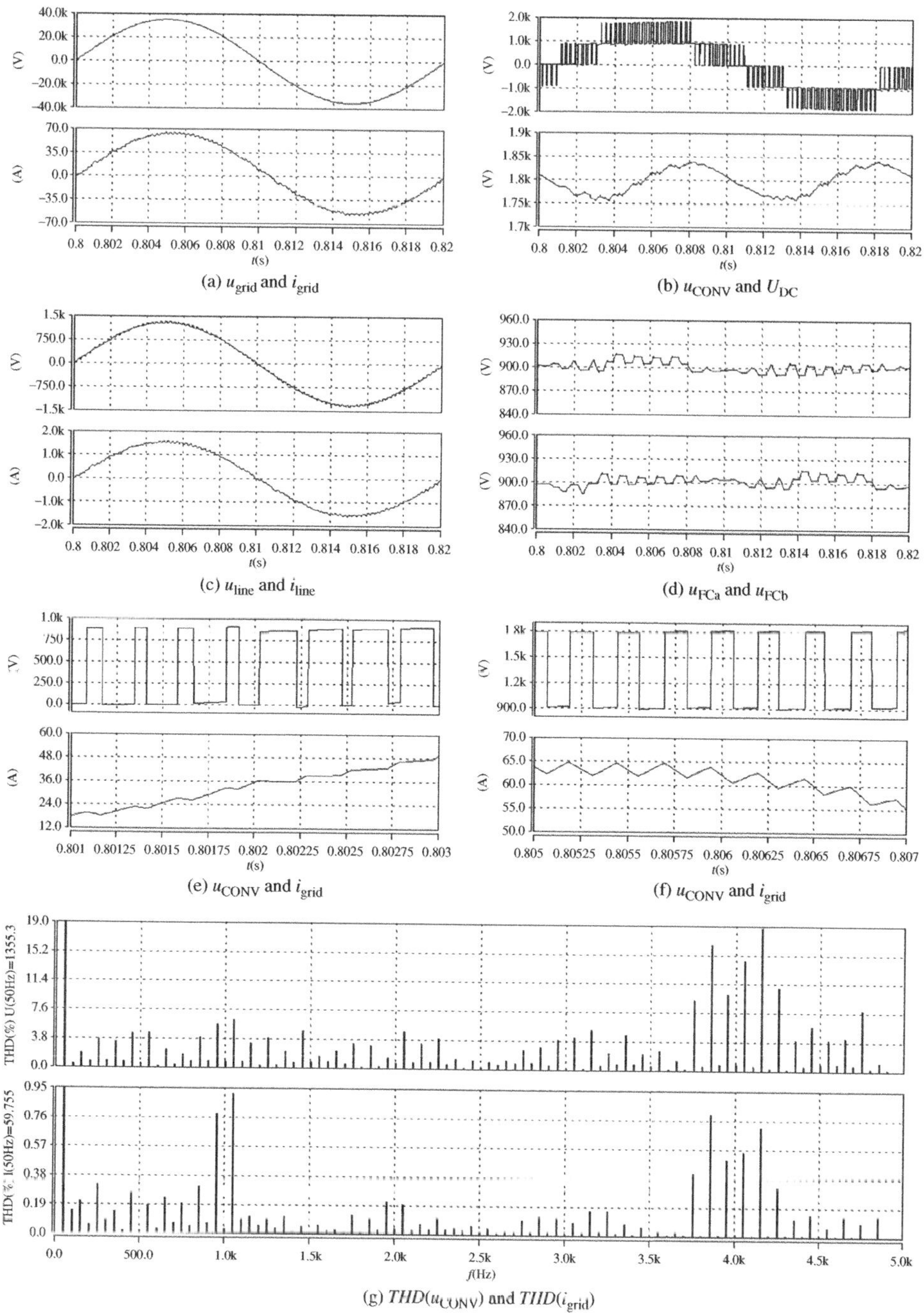

Figure 23.15 Operation of H-FCC with 1D-N(3 + 2R)V: (a) grid voltage u_{grid} and current i_{grid}, (b) converter voltage u_{CONV} and DC-link voltage U_{DC}, (c) line voltage u_{line} and current i_{line}, (d) flying capacitor voltages U_{FCa} and U_{FCb}, (e) and (f) two periods zoom for lower and upper u_{CONV} converter voltage level and grid current i_{grid} and (g) grid current and converter voltage harmonics spectrum $THD(u_{CONV})$ and $THD(i_{grid})$.

Table 23.8 Parameters of 1 MW single-phase converters: PC-H-BC, H-DCC and H-FCC

Simulation parameter	Symbol	Value		
Topology		PC-H-BC	H-DCC	H-FCC
Transformer power/primary side	S_{n1}	1 MVA		
Grid voltage/primary side	$u_{\text{grid,RMS}}$	25 kV/50 Hz		
Grid current/primary side	$i_{\text{grid,RMS}}$	40 A		
Transformer power/secondary sides	$S_{n2-1} = S_{n2-2}, S_{n2}$	500 kVA	1 MVA	
Line voltage/secondary sides	$u_{\text{line1,RMS}} = u_{\text{line2,RMS}}, u_{\text{line,RMS}}$	950 V/50 Hz	1900 V/50 Hz	
Line current/secondary sides	$i_{\text{line1,RMS}} = i_{\text{line2,RMS}}, i_{\text{line,RMS}}$	526 A	1052 A	
Primary leakage inductance	L_1	24 mH		
Primary winding resistance	R_1	3 Ω		
Magnetizing inductance	L_{m}	24 H		
Core loss resistance	R_{fe}	60 kΩ		
Secondary leakage inductance	$L_{2-1} = L_{2-2}, L_2$	1.44 mH	0.72 mH	
Secondary winding resistance	$R_{2-1} = R_{2-2}, R_2$	15 mΩ		
Sampling frequency	f_s	1 kHz		
Capacitance of DC-link capacitors	$C_{\text{DC1}} = C_{\text{DC2}}, C_{\text{DC}}$	23.4 mF	46.8 mF	23.4mF
DC-link voltage	U_{DC}	1800 V		
Capacitance of flying capacitors	$C_{\text{FC}a} = C_{\text{FC}b}$			11.7 mF

Table 23.9 Comparison of grid current distortion $THD(i_{\text{grid}})$

PC-H-BC		H-DCC		H-FCC		
HPWM	UPWM	1D-N2V	1D-N3V	1D-N2V	1D-N3V	1D-N(3 + 2R)V
1 MW output power, 100% load, $R_{\text{Load}} = 3.25\ \Omega$						
3.11%	2.03%	6.93%	3.61%	6.47%	3.46%	2.11%
0.33 MW output power, 33% load, $R_{\text{Load}} = 9.75\ \Omega$						
8.39%	5.30%	21.13%	9.78%	17.66%	9.56%	6.46%

increased THD factor of i_{grid}. For the H-FCC in the case of one leg failure (as well as for the H-DCC), there is the possibility of switching off the broken leg and connecting the phase from the broken leg to the midpoint of the connection of the DC-link capacitors. The advantage of such a solution is that full power is delivered to the load, but of course at the cost of an increased THD factor of the i_{grid} However, it is necessary to modify the PWM strategy, which is not required with the parallel-connected H-BC.

23.3 Control of AC–DC Single-Phase Voltage Source Converters

Figure 23.16 presents the single-phase equivalent circuit of the VSC. According to Figure 23.16, the VSC can be described as:

$$u_{\text{line}} = u_{\text{CONV}} + u_{\text{i}} \tag{23.13}$$

where the voltage drop in inductor u_{i} is defined as:

$$u_{\text{i}} = L\frac{\mathrm{d}i_{\text{line}}}{\mathrm{d}t} + Ri_{\text{line}} \tag{23.14}$$

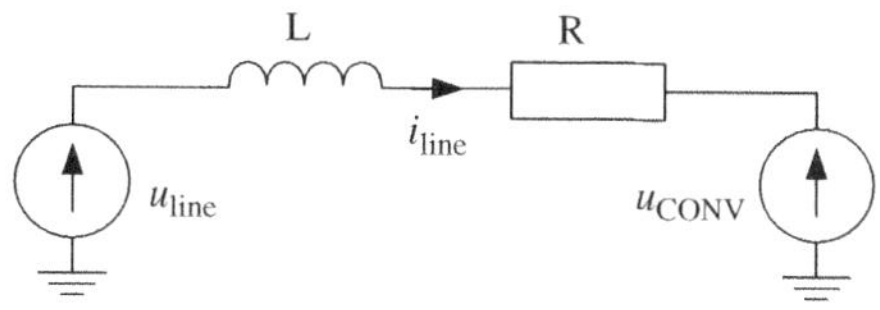

Figure 23.16 Single-phase equivalent circuit of the voltage source converter

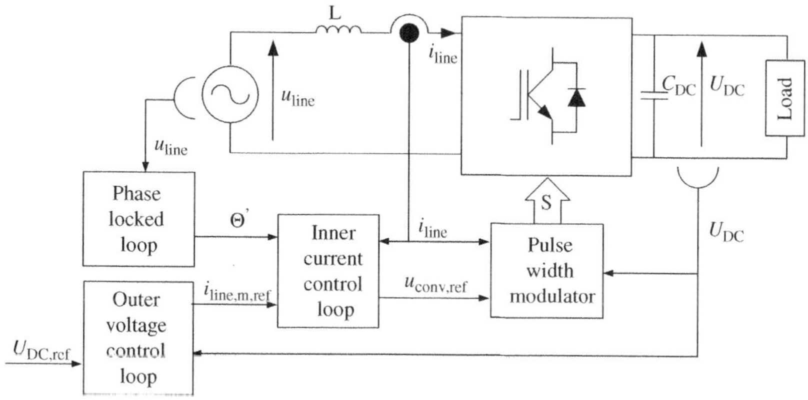

Figure 23.17 Simplified current control structure of single-phase VSC

and L and R denote, respectively, the inductance and resistance of the inductor. The converter output voltage u_{CONV} is controllable and depends on the applied switching states' S table (e.g., Tables 23.1, 23.2 and 23.4), constructed by the individual switching states and DC-link voltage level (Figure 23.17):

$$u_{CONV} = U_{DC}S \tag{23.15}$$

Owing to changes in the u_{CONV} magnitude and phase, the voltage drop on the inductor u_{l} can be controlled directly and thus the line current i_{line} can be controlled indirectly. Figure 23.17 presents a simplified current control structure for the single-phase VSC, which consists of inner current control loop and the outer voltage control loop. Synchronization with grid voltage for high (including unity) input power factor operation is provided by a phase-locked loop (PLL) algorithm.

Figure 23.18 presents the general phasor diagrams of the VSC for the rectifier and inverter modes of operation, with and without unity power factor conditions, according to Equation (23.13). The goal of the outer voltage control loop is to regulate the DC-link voltage U_{DC} to follow reference value $U_{DC,ref}$, while the inner current control loop is designed to keep line current i_{line} sinusoidal and in phase with line voltage u_{line}. This means that the current control should provide VSC unity power factor for rectification and inverting modes of operation.

23.3.1 Single-Phase Control Algorithm Classification

The current control should ensure a constant switching frequency with a specified switching pattern from a filter design point of view. This requirement can only be satisfied with PWM-based current control, allowing the implementation of modern PWM techniques [32]. Unless high dynamic requirements

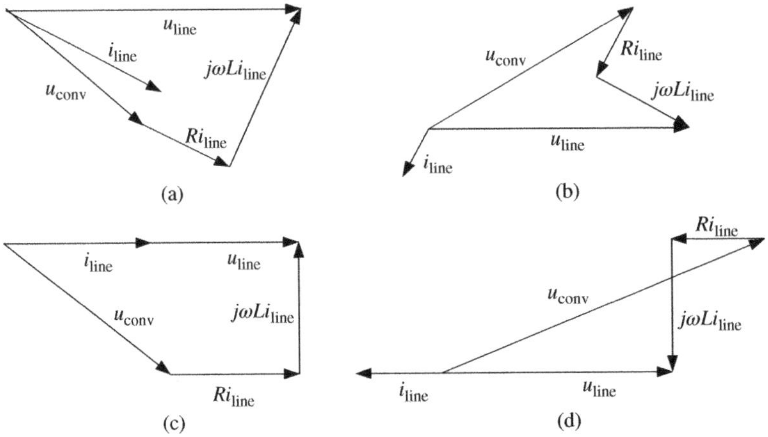

Figure 23.18 Phasor diagrams of the VSC for: (a), (c) rectification mode, (b), (d) invertering mode, (a), (b) nonunity power factor, (c), (d) unity power factor

are given, a linear controller is the most suitable for PWM-based current control. Linear controllers are designed on an average model of the converter based on PWM, which is responsible for the transformation of continuous switching functions into discrete switching pattern functions. Two such systems, PI- and proportional-resonant (PR) based current control (PI-CC and PR-CC, respectively) schemes, have gained superior position [33–36].

PI-based current control is mainly used in *dq* synchronous reference frame rotating with the grid voltage, in which control variables become DC values. The PI controller – composed of proportional gain and an integrator – tracking DC reference is able to eliminate steady-state error. However, in *abc* natural reference frame, the PI controller exhibits two well-known drawbacks: the inability to track the AC reference without steady-state error and poor disturbance rejection capability.

The tracking of periodical signals and rejection of periodical disturbances problem in the *abc* natural reference frame can be solved by using PR-based current control. A PR controller – composed of proportional gain and a resonant integrator – achieves a very high gain around the resonance frequency and almost no gain outside this frequency. Therefore, the PR controller is capable of not only eliminating steady-state amplitude but also phase error. The implementation of a PR controller in *abc* natural reference frame is straightforward, and in comparison to PI-based control the complexity of the control algorithm is considerably reduced.

As far as PR-CC can be used directly, for PI-CC a transformation from a natural to synchronous reference frame is needed. In three-phase systems, coordinate transformation is straightforward. However, in single-phase, it is necessary to create a second quantity in-quadrature with the real one. There are a few mathematical methods available to achieve a set of in-quadrature signals based on the measured single-phase grid voltage [37, 38]:

- $T/4$ transport delay technique, where T is the fundamental grid frequency period,
- Hilbert transform,
- inverse Park transform.

These methods, while allowing for correct quadrature signal generation (QSG), are complex, nonlinear and significantly dependent on fundamental grid frequency changes. Algorithms that are able to eliminate these drawbacks are based on adaptive filtering, allowing in-quadrature generation based both on the phase and frequency synchronization: the second-order adaptive filter and the second-order generalized integrator (SOGI), of which SOGI is most suitable [39].

Synchronization with grid voltage for high (including unity) input power factor operation for both CC methods is also needed. The typical hardware solution with grid voltage zero crossing detection (discontinuous and dependent on the occurrence of the event) is based on line voltage measurement. Filtering of the measured signal in order to determine the zero crossing generates errors, because of the low speed of synchronization and possible distortions arising from the zero crossing detection under distorted voltage. Solutions based on mathematical algorithms – PLLs – allow for fast and accurate synchronization while eliminating the influence of interferences. PLL needs in-quadrature signals, which can be obtained for a single-phase system with SOGI. The SOGI-PLL detects the input phase-angle faster than conventional PLL without steady-state oscillations [39].

As mentioned, unlike PR-CC, a PI-CC transformation from natural to synchronous reference frame is needed. As for both solutions PLL is necessary, the SOGI does not increase the complexity of the control algorithm. However, coordinate transformation from *abc* to *dq* and then inverse transformation is necessary for PI-CC, which makes it more complicated.

Another important issue for current control is the ability to compensate grid harmonics in order to improve power quality. Power quality – beyond the control of amplitude and phase – is one of the responsibilities of current controllers. In *dq*, synchronous coordinates harmonics compensation is based on low-pass and high-pass filtering with PI regulators. For specific harmonic coordinates, frame rotating with harmonic frequency should be implemented. Filtered signals are controlled by two PI regulators in *dq* coordinates and again transformed to rotating coordinates with fundamental frequency. The obtained signals are added to signals referenced for the PWM modulator. In *abc* natural reference frame harmonics, compensation is based on a resonant controller. For specific harmonics, there is no need for additional coordinate transformation and filtering. For single harmonics, an additional resonant controller is parallel connected to a fundamental frequency PR controller. Thus, in *dq* coordinates, two PI controllers are needed, whereas in *abc* only one resonant part is used.

In this section, the basic structures of the *dq* synchronous reference frame current control, PI-CC, and the *abc* natural reference frame current control, PR-CC, are presented. Moreover, the development of current and DC-link voltage controllers with tuning methods for PI, PR, and PMR is discussed. Ultimately, the APFF algorithm used to improve the dynamics of the DC-link stabilization is presented.

23.3.2 DQ Synchronous Reference Frame Current Control – PI-CC

A characteristic feature for *dq* synchronous reference frame current control is coordinate transformation from stationary to synchronous rotating $\alpha\beta/dq$ and opposite from $dq/\alpha\beta$, which is very natural for a three-phase system:

$$\begin{bmatrix} k_d \\ k_q \end{bmatrix} = \begin{bmatrix} \cos(\gamma_{u_{\text{line}}}) & \sin(\gamma_{u_{\text{line}}}) \\ -\sin(\gamma_{u_{\text{line}}}) & \cos(\gamma_{u_{\text{line}}}) \end{bmatrix} \begin{bmatrix} k_\alpha \\ k_\beta \end{bmatrix} \tag{23.16}$$

$$\begin{bmatrix} k_\alpha \\ k_\beta \end{bmatrix} = \begin{bmatrix} \cos(\gamma_{u_{\text{line}}}) & -\sin(\gamma_{u_{\text{line}}}) \\ \sin(\gamma_{u_{\text{line}}}) & \cos(\gamma_{u_{\text{line}}}) \end{bmatrix} \begin{bmatrix} k_d \\ k_q \end{bmatrix} \tag{23.17}$$

where an angle of the voltage vector $\gamma_{u_{\text{line}}}$ is defined as

$$\begin{aligned} \sin(\gamma_{u_{\text{line}}}) &= \frac{u_{\beta,\text{line}}}{\sqrt{u_{\alpha,\text{line}}^2 + u_{\beta,\text{line}}^2}} \\ \cos(\gamma_{u_{\text{line}}}) &= \frac{u_{\alpha,\text{line}}}{\sqrt{u_{\alpha,\text{line}}^2 + u_{\beta,\text{line}}^2}} \end{aligned} \tag{23.18}$$

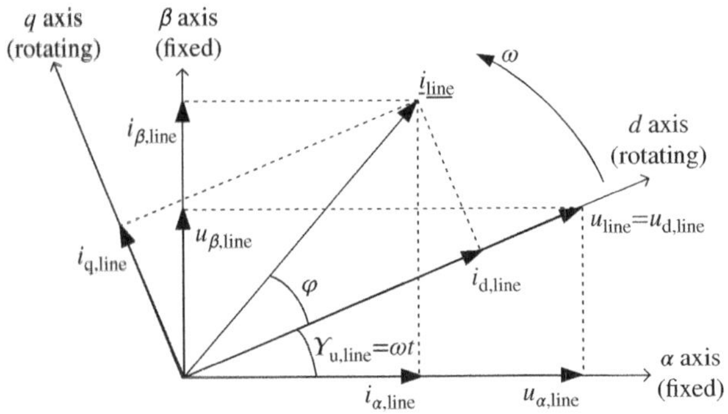

Figure 23.19 Coordinate transformation of line current vector $\underline{i}_{\text{line}}$ and line voltage vector $\underline{u}_{\text{line}}$ from stationary $\alpha\beta$ to synchronous rotating dq reference frame coordinates

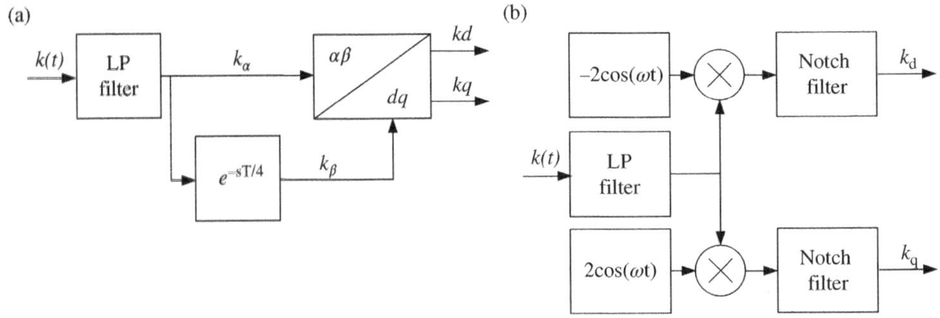

Figure 23.20 Methods to obtain a virtual dq synchronous rotating coordinate system in single-phase converter: (a) based on $^1/_4$ delay of α-axis and (b) based on two notch filters tuned at twice the line frequency

The line current vector $\underline{i}_{\text{line}}$ is split into two rectangular components $\underline{i}_{\text{line}} = [i_{d,\text{line}}, i_{q,\text{line}}]$ in voltage-oriented dq coordinates (Figure 23.19). The component $i_{q,\text{line}}$ determinates the reactive power, whereas $i_{d,\text{line}}$ determines the active power flow. Thus, the reactive and the active power can be controlled independently. Unity power factor operation can be achieved when the line current vector $\underline{i}_{\text{line}}$ is aligned with the line voltage vector $\underline{u}_{\text{line}}$.

The situation is slightly more complicated in the case of a single-phase system, because the coordinates are virtual and need special algorithms to obtain the $\alpha\beta$ or dq system. Among the most attractive solutions is a method based on delaying the α-axis by $^1/_4$ of a line voltage period to obtain the β-axis (Figure 23.20(a)) or the use of notch filters with a narrow stop-band (e.g., second-order Butterworth) tuned at twice the line frequency [40, 41] (Figure 23.20(b)).

The voltage equations (Equation 23.13) in the dq synchronous reference frame in accordance with the presented transformation (Equations 23.16 and 23.17) are as follows:

$$u_{d,\text{line}} = Ri_{d,\text{line}} + L\frac{\mathrm{d}i_{d,\text{line}}}{\mathrm{d}t} + u_{d,\text{CONV}} - \omega Li_{q,\text{line}}$$

$$u_{q,\text{line}} = Ri_{q,\text{line}} + L\frac{\mathrm{d}i_{q,\text{line}}}{\mathrm{d}t} + u_{q,\text{CONV}} + \omega Li_{d,\text{line}} \tag{23.19}$$

The block diagram of the control for the single-phase converter in the dq synchronous reference frame is shown in Figure 23.21. Measured $i_{d,\text{line}}$ and $i_{q,\text{line}}$ currents are compared with reference values $i_{d,\text{line,ref}}$

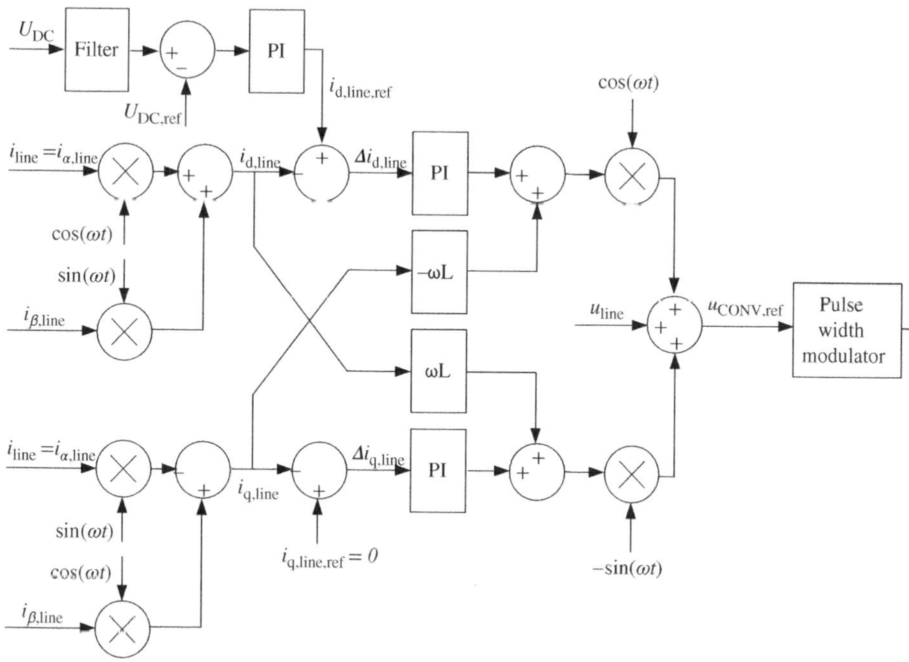

Figure 23.21 Classical single-phase control in virtual *dq* synchronous rotating coordinate system

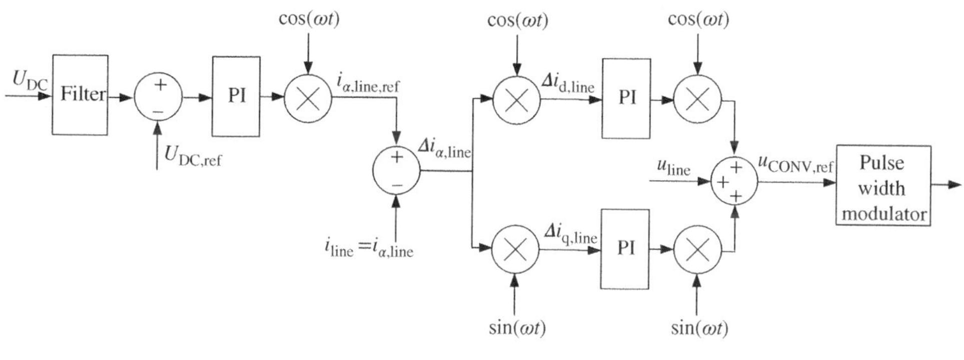

Figure 23.22 Simplified single-phase control in virtual *dq* synchronous rotating coordinate system

and $i_{q,\text{line,ref}}$ and the error is delivered to the PI controllers. Then, the decoupling terms $\omega L i_{q,\text{line}}$ and $\omega L i_{d,\text{line}}$ are applied according to Equation (23.19). Finally, based on Equation (23.17), converter reference voltages in the stationary coordinate system are calculated, where $u_{\alpha,\text{CONV,ref}}$ is used by the modulator and $u_{\beta,\text{CONV,ref}}$ is discarded:

$$u_{\text{CONV,ref}} = u_{\alpha,\text{CONV,ref}} = u_{d,\text{CONV,ref}} \cos(\omega t) - u_{q,\text{CONV,ref}} \sin(\omega t) \tag{23.20}$$

The control presented in Figure 23.21 can be significantly simplified, as shown in Figure 23.22 [42]. Thanks to the simple assumption that for a virtual β-axis error of current control $\Delta i_{\text{d,line}}$ is equal to zero, the following equations:

$$\Delta i_{d,\text{line}} = i_{d,\text{line,ref}} - i_{d,\text{line}} =$$
$$[i_{\alpha,\text{line,ref}} \cos(\omega t) + i_{\beta,\text{line,ref}} \sin(\omega t)] - [i_{\alpha,\text{line}} \cos(\omega t) + i_{\beta,\text{line}} \sin(\omega t)] =$$
$$(i_{\alpha,\text{line,ref}} - i_{\alpha,\text{line}}) \cos(\omega t) + (i_{\beta,\text{line,ref}} - i_{\beta,\text{line}}) \sin(\omega t) \quad (23.21)$$

$$\Delta i_{q,\text{line}} = i_{q,\text{line,ref}} - i_{q,\text{line}} =$$
$$[-i_{\alpha,\text{line,ref}} \sin(\omega t) - i_{\beta,\text{line,ref}} \cos(\omega t)] - [-i_{\alpha,\text{line}} \sin(\omega t) - i_{\beta,\text{line}} \cos(\omega t)] =$$
$$-(i_{\alpha,\text{line,ref}} - i_{\alpha,\text{line}}) \sin(\omega t) + (i_{\beta,\text{line,ref}} - i_{\beta,\text{line}}) \cos(\omega t) \quad (23.22)$$

can be simplified to form:

$$\Delta i_{d,\text{line}} = (i_{\alpha,\text{line,ref}} - i_{\alpha,\text{line}}) \cos(\omega t) \quad (23.23)$$

$$\Delta i_{q,\text{line}} = -(i_{\alpha,\text{line,ref}} - i_{\alpha,\text{line}}) \sin(\omega t) \quad (23.24)$$

Thanks to this, only a simple PLL is needed to obtain the synchronous rotating coordinate system instead of using coordinate transformations from stationary to synchronous rotating $\alpha\beta/dq$ (Equation 23.17), notch filters or storing samples to get a quarter cycle delay.

23.3.3 ABC Natural Reference Frame Current Control – PR-CC

Figure 23.23 presents the block scheme of the PR controller-based current control algorithm, which consists of two cascaded loops. The commanded DC-link voltage $U_{DC,ref}$ is compared with the measured U_{DC} value. The voltage U_{DC} is distorted by 100 Hz AC oscillations derived from the single-phase current i_{line} with 50 Hz fundamental frequency. Because the $U_{DC,ref}$ has a constant DC value, the 100 Hz distortion will be visible in the DC-link voltage error signal σU_{DC}. The σU_{DC} signal is delivered to the PI controller, which generates the reference line current amplitude $i_{line,m,ref}$. As the PI regulator is not able to eliminate phase error, in order to prevent transmission of 100 Hz distortion on the current $i_{line,m,ref}$, a low-pass filter with 30 Hz cutoff frequency is applied on the measured U_{DC} voltage. The delay introduced by the filter causes a reduction in the voltage loop dynamics, and as a consequence the DC-link voltage transient error should be higher. In fact, for a low switching frequency, the step change response of the DC-link PI controller with and without a low-pass filter is similar. The reason is the limited bandwidth of the voltage loop, which is more restrictive than the bandwidth reduction introduced by the filter. However, for higher sampling frequencies, the dynamics of the voltage loop will be reduced. It should also be noted that for a low sampling frequency the time of the step response will be longer than in the case of a high sampling frequency. To improve the dynamics of the DC-link stabilization, the power feed-forward signal (containing information about the load changes) should be added to the output signal of the DC-link voltage controller (see Section 23.3.4). APFF provides

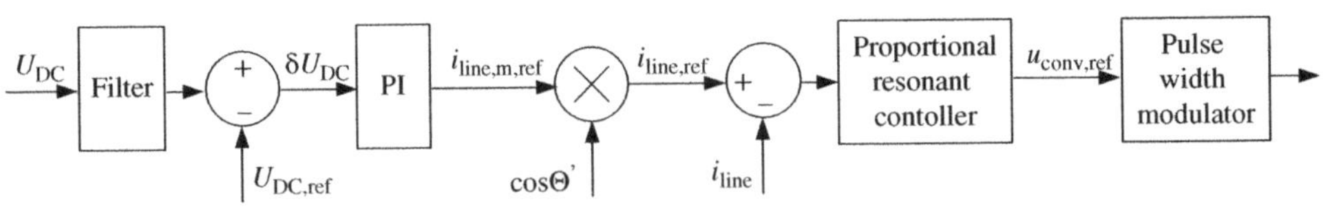

Figure 23.23 Block scheme of PR controller-based current control algorithm

very good stabilization of the DC-link voltage in transient states, and the DC-link voltage overshoot is significantly reduced.

The internal current loop is responsible for system power quality by controlling the grid current. In PR-CC, unity power factor operation can be achieved if the reference line current $i_{\text{line,ref}}$ is given as:

$$i_{\text{line,ref}} = i_{\text{line,m,ref}} \cos(\Theta') \tag{23.25}$$

The cosinus function is used because of the SOGI-PLL phase-angle Θ' generation – $\Theta' = 0$ when the u_{line} amplitude has the maximum positive value. The reference value of $i_{\text{line,ref}}$ is compared with the measured line current i_{line} and the error is delivered to the PR controller. The PR structure shown in Figure 23.24(a) is composed of proportional gain and a resonant integrator [33, 36].

The core of the resonant integrator is the generalized integrator, which achieves a very high gain around the resonant frequency and almost no gain outside this frequency. The transfer function of an ideal PR controller is given by:

$$G_{\text{PR}}(s) = K_{\text{PRp}} + \frac{2K_{\text{PRi}}s}{s^2 + \omega^2} \tag{23.26}$$

where K_{PRp} is the proportional gain and K_{PRi} is the gain of the resonant integrator. The transfer function of the PR controller contains a double imaginary pole adjusted to the fundamental u_{line} frequency ω. Thus, the PR controller is able to track the input phase angle for ω without any steady-state error. Note that the u_{line} voltage feed-forward is not needed. To avoid stability problems with infinite gain, the following transfer function can be used instead of Equation (23.26):

$$G_{\text{PR}}(s) = K_{\text{PRp}} + \frac{2K_{\text{PRi}}\omega_c s}{s^2 + 2\omega_c + \omega^2} \tag{23.27}$$

where cutoff frequency $\omega_c \ll \omega$. For Equation (23.27), the gain is finite (but still high enough to track the input phase-angle Θ' with small steady-state error) and bandwidth can be set by ω_c.

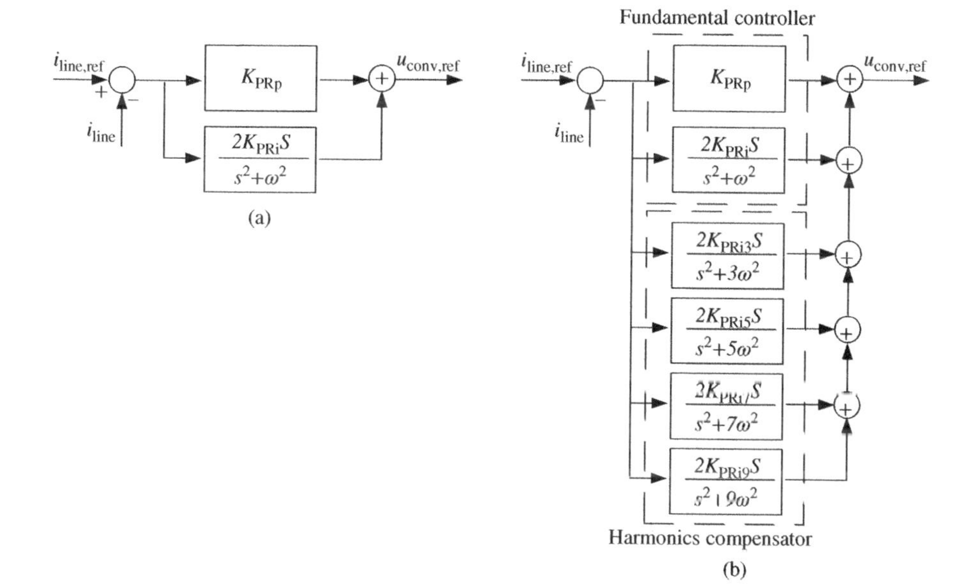

Figure 23.24 Block diagram of: (a) proportional resonant (PR) controller and (b) proportional multiresonant (PMR) controller

The current controller has to be immune for most prominent harmonics in the current spectrum – typically, the third, fifth, and seventh and sometimes the ninth harmonics. In a single-phase system, it means that for each harmonic a separate compensator is needed. For the *abc* natural reference frame, three compensation loops are also needed, where each resonant controller works with the gain of the fundamental PR controller. Still, the individual design of compensator resonant parts should be considered; however, it is possible to use the resonant gain of the fundamental PR controller. As the proportional controller only compensates frequencies very close to the selected harmonics frequencies, the resonant compensator does not affect the dynamics of the fundamental PR controller. Such a structure is called a *proportional multiresonant* (PMR) controller [43–45]. Figure 23.24(b) shows an example of a PMR controller block diagram for the third, fifth, seventh and ninth harmonics compensation, where compensators are parallel connected to the fundamental PR controller. The transfer function of PMR is given by:

$$G_{PR}(s) = K_{PRp} + \frac{2K_{PRi}s}{s^2+\omega^2} + \sum_{h=3,5,7,9} \frac{2K_{PRih}s}{s^2+(h\omega)^2} \tag{23.28}$$

where h is the harmonic order. The harmonics compensator can use the same K_{PRi} gain as the fundamental PR controller for all loops.

The PR controller generates the amplitude of the converter output voltage $u_{CONV,ref}$, which is delivered to the pulse-width modulator.

23.3.4 Controller Design

23.3.4.1 Design of PI-Based Current Control Loop

Figure 23.25 presents a block diagram of a PI-CC loop. The one sampling T_S delay introduced by the control algorithm as well as the statistical $T_S/2$ delay of the PWM generation should also be taken into account. Thus, the time delays in the S&H block of VSC are represented as τ_t – the sum of small time constants:

$$\tau_t = T_S + T_{PWM} = 1.5T_S \tag{23.29}$$

Also, the VSC gain K_{VSC} and dead time D_T should be included. In further considerations, the VSC will be assumed as an ideal amplifier with $K_{VSC} = 1$, with time constant $\tau_0 = 0$.

With the assumption that disturbance u_{line} = const the open-loop transfer function can be written as:

$$G_{iPIo} = \frac{K_{PIp}(1+sT_{PIi})K_{RL}}{sT_{PIi}(1+s\tau_t)(1+sT_{RL})} \tag{23.30}$$

where $T_{RL} = L/R$ is the line choke time constant and $K_{RL} = 1/R$ is the choke gain. With simplification $(1+sT_{RL}) \cong sT_{RL}$ the closed-loop transfer function can be written as [46]:

$$G_{iPIz} = \frac{K_{PIp}(1+sT_{PIi})K_{RL}}{K_{PIp}K_{RL}(1+sT_{PIi}) + s^2T_{PIi}T_{RL} + s^3T_{PIi}T_{RL}\tau_t} \tag{23.31}$$

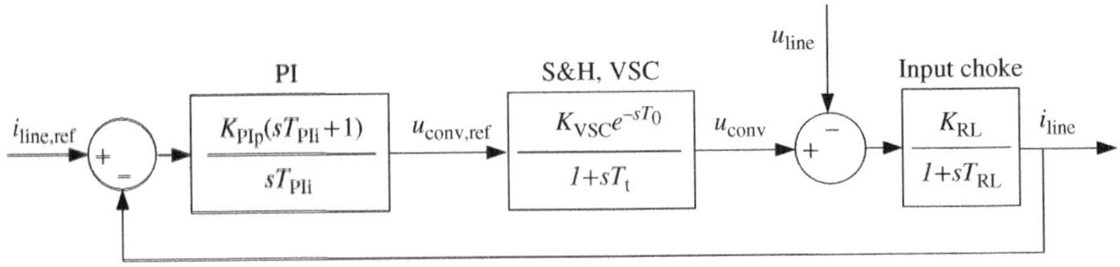

Figure 23.25 Block diagram of proportional integral current control loop

The design procedures for the PI current controllers have been presented in Refs. [47, 48]. For good disturbance rejection performance in transient states, both use the symmetry optimum (SO) [49] design criterion. Therefore, for Equation (23.31), with the assumption that the disturbance u_{line} = const, the PI controller proportional gain K_{PIp} and time constant T_{PIi} can be calculated as:

$$K_{PIp} = \frac{T_{RL}}{2\tau_t K_{RL}} \tag{23.32}$$

$$T_{PIi} = 4\tau_t \tag{23.33}$$

23.3.4.2 Design of PR-Based Current Control Loop

Figure 23.26 presents a block diagram of a PR-CC loop. The one sampling T_S delay introduced by the control algorithm as well as the statistical $T_S/2$ delay of the PWM generation should also be taken into account. Thus, the time delays in the S&H block of VSC are represented as τ_t – the sum of small time constants:

$$\tau_t = T_S + T_{PWM} = 1.5T_S \tag{23.34}$$

Also, the VSC gain K_{VSC} and dead time D_T should be included. In further considerations, the VSC will be assumed as an ideal amplifier with $K_{VSC} = 1$ with time constant $\tau_0 = 0$.

With the assumption that disturbance u_{line} = const the open-loop transfer function can be written as:

$$G_{iPRo} = \frac{K_{PRp}K_{RL}}{(1+s\tau_t)(1+sT_{RL})} + \frac{2K_{PRp}K_{RL}}{(s^2+\omega^2)(1+s\tau_t)(1+sT_{RL})} \tag{23.35}$$

As shown in Ref. [39], the proportional gain of the PR controller K_{PRp} – which determines the dynamics of the system in terms of bandwidth, phase and gain margins – can be tuned in the same way as the PI current controller [47, 48]. According to the SO [49] design criterion, with the assumption that the disturbance u_{line} = const, the PR controller proportional gain can be calculated as:

$$K_{PRp} = \frac{T_{RL}}{2\tau_t K_{RL}} \tag{23.36}$$

To tune the resonant gain of the PR controller K_{PRi}, a graphical frequency response approach can be used. The objective is to guarantee a sufficient phase margin for the system, and to avoid high resonances in the closed loop [50–52]. Figure 23.27 presents an example of a closed-loop frequency response for the designed PMR current control with 3rd, 5th, 7th and 9th harmonic compensation. For the PMR controller, the same gain values can be used as for the PR-based current control with the gain of the harmonics compensator $K_{PRih} = K_{PRi}$.

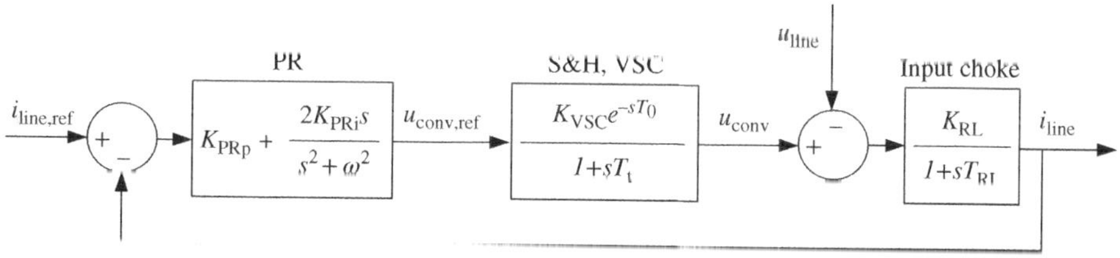

Figure 23.26 Block diagram of proportional resonant current control loop

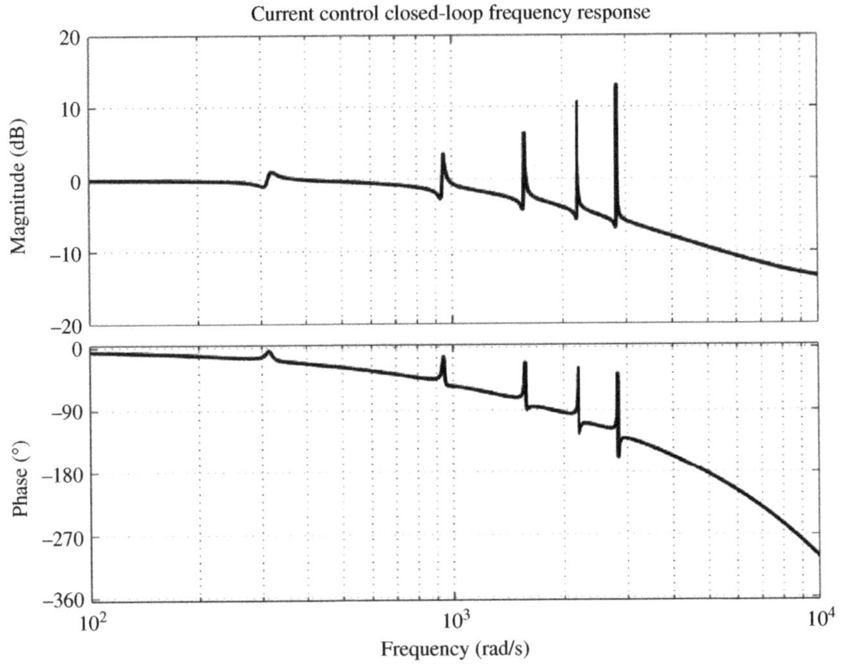

Figure 23.27 Frequency response of current control loop with PMR-based current control and third, fifth, seventh and ninth harmonic compensation

23.3.4.3 Design of DC-Link Voltage Control Loop

Figure 23.28 presents a block diagram of the DC-link voltage control loop with an internal current controller. The inner current control loop with SO design criterion [49] is represented as a first-order transfer function with an equivalent time constant:

$$T_{\mathrm{VSC}} = 4\tau_{\mathrm{t}} \tag{23.37}$$

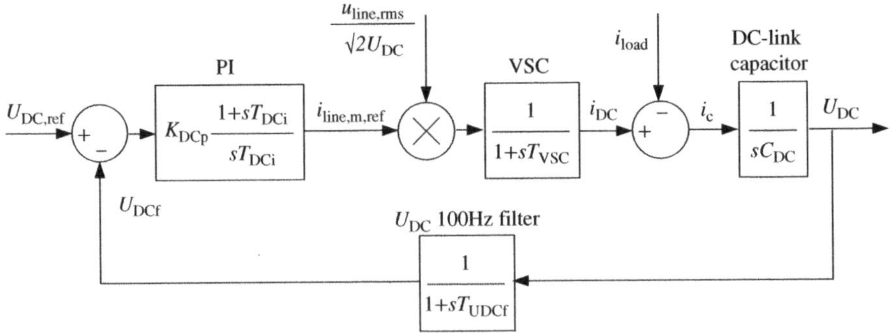

Figure 23.28 Block diagram of DC-link voltage control loop with internal current controller

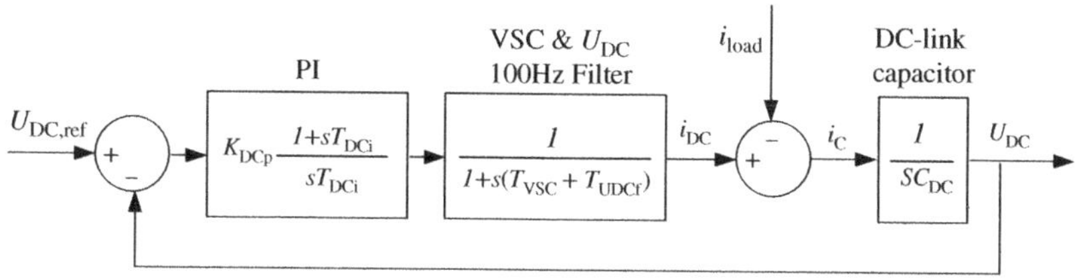

Figure 23.29 Simplified block diagram of DC-link voltage control loop with internal current controller

To prevent 100 Hz distortion transmission on the $i_{line,m,ref}$ reference current, a low-pass filter with a cutoff frequency f_{UDC} is required on the U_{DC} voltage sensor. The time constant of the filter is expressed as:

$$T_{UDCf} = \frac{1}{2\pi f_{UDC}} \tag{23.38}$$

The relation between the amplitude of the referenced current $i_{line,m,ref}$ and DC-link current is described as $\frac{u_{line,RMS}}{\sqrt{2}U_{DC}}$. Taking this and the assumption that the current relation is included in the proportional gain K_{DCp}, Figure 23.29 shows the modified block diagram of the DC-link voltage control loop.

The open-loop voltage control loop transfer function can be written as:

$$G_{uo} = \frac{K_{DCp}(1 + sT_{DCi})}{sT_{DCi}(1 + s(T_{VSC} + T_{UDCf}))sC_{DC}} \tag{23.39}$$

On the basis of the SO design criterion, the proportional gain K_{DCp} and the integral time constant T_{DCi} of the DC-link voltage control loop can be calculated as follows:

$$K_{DCp} = \frac{C_{DC}}{2(T_{VSC} + T_{UDCf})} \frac{u_{line,RMS}}{\sqrt{2}U_{DC}} \tag{23.40}$$

$$T_{DCi} = 4(T_{VSC} + T_{UDCf}) \tag{23.41}$$

As the DC-link referenced voltage $U_{DC,ref}$ is assumed to be constant, no additional pre-filter on $U_{DC,ref}$ is needed.

23.3.5 Active Power Feed-Forward Algorithm

To prevent the transmission of 100 Hz distortion on the line current i_{line}, a low-pass filter with a specified (typically 30 Hz) cutoff frequency is applied on the measured U_{DC} voltage. The delay introduced by the filter causes a reduction in the DC-link voltage loop dynamics. As a consequence, the DC-link voltage transient error is higher and the AC–DC–AC system's dynamic behavior has to be reduced. There are two methods to improve the dynamics of the DC-link stabilization:

- Adding a load feed-forward to the output signal of the DC-link voltage controller from the DC-link load current – in such a case, an additional DC-link current sensor is needed, and the AC–DC converter operates independently of the control of the drive.
- Adding a power feed-forward to the output signal of the DC-link voltage controller from the supplied active load (the DC–AC converter fed motor) – in such a case, the AC–DC converter operates depending on the control of the drive, but no additional sensor is needed.

Thus, for the control of the AC–DC–AC converter, the APFF (based on the control signals) seems to be attractive.

According to Refs. [31, 48], two variants of the APFF based on the power consumed/produced by the DC–AC converter can be listed:

- Active power of the AC–DC converter is calculated from the mechanical speed, commanded torque, and DC–AC converter losses, PF_{Ω}.
- Active power of the AC–DC converter is calculated from the switching states reference converter voltages and actual stator currents, PF_{UI}.

Typically, in HSR, an induction motor (IM) is used. The electromagnetic power of the IM is defined by:

$$P_e = m_e \Omega_m \tag{23.42}$$

where Ω_m is the mechanical speed of the drive. Taking into account the electromagnetic torque m_e equation:

$$m_e = p_b \frac{m_s}{2} \Psi_S I_{Sy} \tag{23.43}$$

where p_b is the number of IM pole pairs, m_s is the number of IM phases, Ψ_S is the IM stator flux and I_{Sy} is the y component of the IM stator current, the electromagnetic power of the IM can be expressed as:

$$P_e = p_b \frac{m_s}{2} \Psi_S I_{Sy} \Omega_m \tag{23.44}$$

The power delivered to the IM also includes the power losses, and the PF_{Ω} should be written as:

$$PF_{\Omega} = P_e + P_{losses} = p_b \frac{m_s}{2} \Psi_S I_{Ey} \Omega_m + P_{losses} \tag{23.45}$$

For nominal torque at mechanical speed $\Omega_m = 0$, the electromagnetic power $P_e = 0$. However, P_{losses} will have a significant value. Thus, P_{losses} cannot be neglected. The estimation of the P_{losses} is difficult, because it requires an exact knowledge of the parameters of the IM. Hence, PF_{UI}, calculated from the reference converter voltages $U_{x,rec}$ and $U_{y,rec}$ (reconstructed from the switching states) and actual stator currents, provides a simple estimation of the active power consumed/produced by the DC–AC converter:

$$PF_{UI} = \frac{3}{2}(I_{Sx} U_{x,rec} + I_{Sy} U_{y,rec}) \tag{23.46}$$

The $U_{x,rec}$ and $U_{y,rec}$ voltages, reconstructed from switching states, contain additional information about the dead time and switching devices voltage drop compensation. Therefore, the PF_{UI} includes the power losses P_{losses}.

If we assume that the losses of the AC–DC converter are neglected, the energy storage variation of the DC-link capacitor will be the integral of the difference between the average AC–DC input power P_{in} and the power delivered to the IM PF_{UI} [31, 48]. Therefore, it can be written as:

$$P_{in} = P_c + PF_{UI} \tag{23.47}$$

where P_c denotes the power of the DC-link voltage feedback control loop:

$$P_c = i_{DC} U_{DC} \tag{23.48}$$

According to the block diagram of the DC-link control loop (Figure 23.29), P_c can be written as:

$$P_c = \frac{K_{DCp}(1 + sT_{DCi})}{sT_{DCi}}(U_{DC,ref} - U_{DCf})U_{DC} \tag{23.49}$$

The DC-link voltage controller generates the reference line current amplitude $i_{line,m,ref}$. If we assume that the AC–DC converter works with the unity power factor (phase shift φ between line current i_{line} and line voltage u_{line} is equal to 0) the instantaneous AC–DC input power can be calculated as:

$$p_{in}(t) = i_{line,m}u_{line,m}\sin^2(\omega t) \tag{23.50}$$

According to Equation (23.50), the average AC–DC input power can be calculated as:

$$P_{in} = \frac{i_{line,m}u_{line,m}}{2} \tag{23.51}$$

Therefore, the reference line current amplitude $i_{line,m,ref}$ should be:

$$i_{line,m,ref} = \frac{2P_{in}}{u_{line,m}} \tag{23.52}$$

If we assume that the power PF_{UI} is delivered to the IM without any DC-link variation (which is the desired situation), then P_c can be assumed to equal 0. In such a situation, Equation (23.46) can be written as:

$$P_{in} = PF_{UI} \tag{23.53}$$

Thus, combining Equation (23.52) with Equation (23.53), the reference line current amplitude from the APFF $i_{PF,line,m,ref}$ is obtained:

$$i_{line,m,ref} = \frac{2P_{in}}{u_{line,m}} \tag{23.54}$$

Figure 23.30 shows the block diagram of the AC–DC–AC converter feeding IM, where $P_{load,ref}$ and P_{load} are the reference and output powers of the DC–AC converter, respectively. Figure 23.31 presents the transient operation of the low-voltage AC–DC–AC converter: AC–DC five-level single-phase FCC and DC–AC three-level three-phase FCC without and with APFF in closed speed control mode under

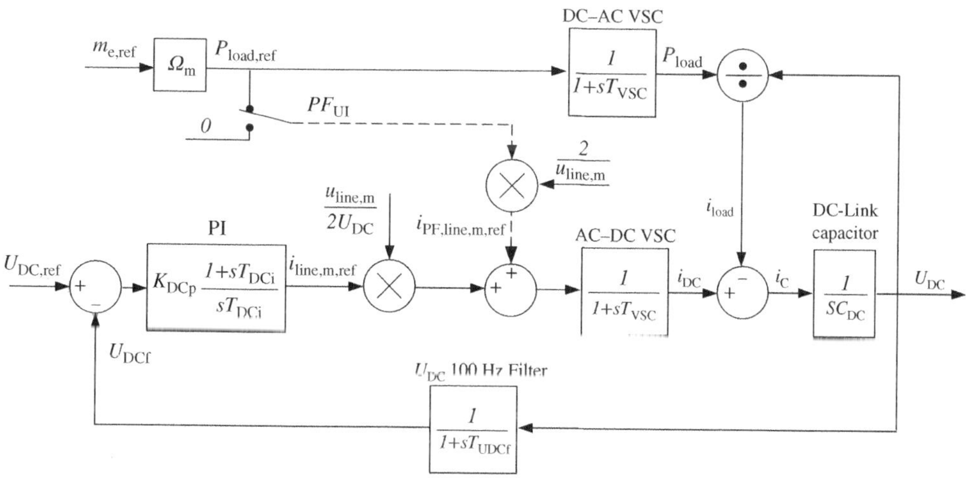

Figure 23.30 Block diagram of the AC–DC–AC converter feeding IM with active power feed-forward – dotted line denotes the influence of the APFF

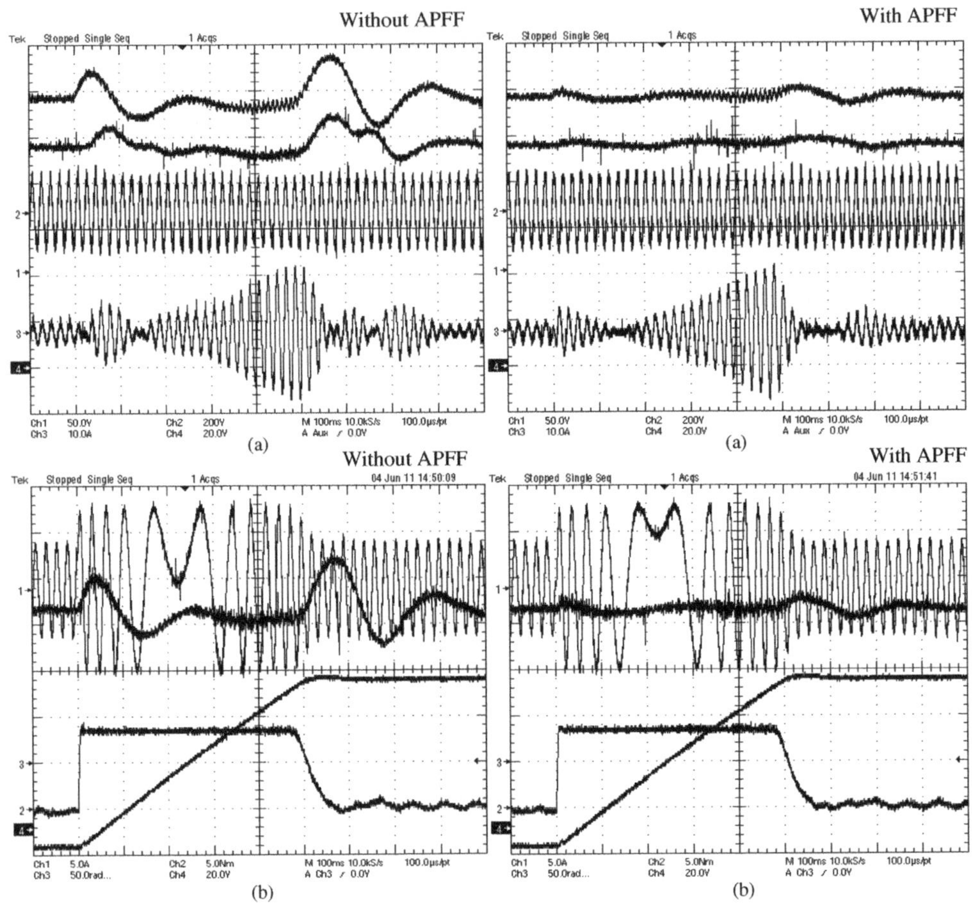

Figure 23.31 Transient operation of the AC–DC–AC FCC without and with APFF: step change in mechanical speed Ω_m. From the top: (a) DC-link voltage U_{DC} (Ch1), rectifier leg *a* flying capacitor voltage U_{FCa} (Ch4), line voltage u_{line} (Ch2) and line current i_{line} (Ch3), (b) IM phase *a* stator current i_{Sa} (Ch1), inverter phase *a* flying capacitor voltage U_{FCa} (Ch4) mechanical speed of IM Ω_m (Ch3) and IM electromagnetic torque m_e (Ch4)

a step change of mechanical speed Ω_m from −85% to 85% of the nominal speed. The AC–DC–AC converter control with the active APFF provides:

- very good stabilization to the DC-link voltage in transient states – the DC-link voltage overshoot is significantly reduced,
- no negative impact on the system's performance in steady states.

Thus, the APFF, by a reduction in the DC-link overvoltage in transients, can efficiently extend the lifetime of the DC-link capacitors.

23.4 Summary

This chapter has reviewed the modulation and control of single-phase grid-side VSD.

The following PWM techniques for single-phase VSC have been presented:

- hybrid (HPWM) and unipolar (UPWM) modulations for parallel-connected H-BCs,
- 1D-N2V and 1D-N3V modulations for H-DCCs,
- 1D-N2V, 1D-N3V and 1D-N(3 + 2R)V modulations for H-FCCs.

Particular emphasis was placed on the impact of the individual modulation techniques and topologies on the quality of the grid current and the harmonic content generated by the converter. The comparison of PWM techniques has shown a clear advantage of parallel-connected H-BC with UPWM and H-FCC with 1D-N(3 + 2R)V modulation for the sake of the highest number of forming states.

Various control techniques for PWM DC/AC converters have been discussed. Among the methods presented, the control of the stationary coordinate system and RC controllers seem to be superior, and is therefore implemented by the industry. Other control methods have high algorithm complexity.

It is believed that thanks to continuing developments in power semiconductor components and digital signal processing, voltage source PWM DC/AC converters will have a strong impact on power conversion, particularly in traction as well as renewable and distributed energy systems.

References

1. Rodriguez, J., Lai, J.S., and Peng, F.Z. (2002) Multilevel inverters: a survey of topologies, controls, and applications. *IEEE Transactions on Industrial Electronics*, **49** (4), 724–738.
2. Rodriguez, R.J., Bernet, S., Wu, B. *et al.* (2007) Multilevel voltage-source-converter topologies for industrial medium-voltage drives. *IEEE Transactions on Industrial Electronics*, **54** (6), 2930–2945.
3. Franquelo, L.G., Rodriguez, J., Leon, J.I. *et al.* (2008) The age of multilevel converters arrives. *IEEE Industrial Electronics Magazine*, **2** (2), 28–39.
4. Rodriguez, J., Franquelo, L.G., Kouro, S. *et al.* (2009) Multilevel converters: an enabling technology for high-power applications. *Proceedings of the IEEE*, **97** (11), 1786–1817.
5. Bose, B.K. (2009) Power electronics and motor drives recent progress and perspective. *IEEE Transactions on Industrial Electronics*, **56** (2), 581–588.
6. Boora, A.A., Zare, F., Ghosh, A., and Ledwich, G. (2007) Applications of power electronics in railway systems. Proceedings of AUPEC 2007, December 2007.
7. Steimel, A. (2008) *Electric Traction – Motive Power and Energy Supply*, Oldenbourg Industrieverlag.
8. Watanabe, T. (1999) Trend of railway technologies and power semiconductor devices. Proceedings of ISPSD 1999, pp. 11–18, May 1999.
9. Shen, J., Taufiq, J.A., and Mansell, A.D. (1997) Analytical solution to harmonic characteristics of traction pwm converters. *IEE Proceedings of Electric Power Applications*, **144** (2), 158–168.
10. Yang, B., Zelaya, H., and Taufiq, J.A. (1990) Computer simulation of a three-phase induction motor traction system fed by single-phase ac/dc pulse converters with current control scheme. Proceedings of IEEE IAS 1990 Annual Meeting, October 1990, pp. 1171–1177.
11. Ryoo, H. J., Kim, J. S., Rim, G.-H. *et al.* (2001) Unit power factor operation of parallel operated ac to dc pwm converter for high power traction application. Proceedings of IEEE PESC 2001, June 2001, Vol. 2, pp. 631–636.
12. Chang, G.W., Hsin Wei, L., and Shin-Kuan, C. (2004) Modeling characteristics of harmonic currents generated by high-speed railway traction drive converters. *IEEE Transactions on Power Delivery*, **19** (2), 766–773.
13. Eini, H.I., Farhangi, S., and Schanen, J.L. (2008) A modular ac/dc rectifier based on cascaded h-bridge rectifier. Proceedings of EPE-PEMC 2008, September 2008, pp. 173–180.
14. Jacobina, C.B., dos Santos, E.C., Rocha, N., and Fabricio, E.L.L. (2010) Single-phase to three-phase drive system using two parallel single-phase rectifiers. *IEEE Transactions on Power Electronics*, **25** (5), 1285–1295.

15. Kouro, S., Malinowski, M., Gopakumar, K. *et al.* (2010) Recent advances and industrial applications of multilevel converters. *IEEE Transactions on Industrial Electronics*, **57** (8), 2553–2580.
16. Akagi, H. (2011) New trends in medium-voltage power converters and motor drives. Proceedings of IEEE ISIE 2011, June 2011, pp. 5–14.
17. Akagawa, E., Kawamoto, S., Tamai, S. *et al.* (1995) Three-level PWM converter-inverter system for next-generation shinkansen. IEE Japan IAS Annual Meeting Rec, pp. 81–82.
18. Wu, C.M., Lau, W.H., and Chung, H. (1999) A five-level neutral-point-clamped h-bridge PWM inverter with superior harmonics suppression: a theoretical analysis. Proceedings of IEEE International Symposium on Circuits and Systems, May/June 1999, pp. 198–201.
19. Cheng, Z. and Wu, B. (2007) A novel switching sequence design for five-level npc/h-bridge inverters with improved output voltage spectrum and minimized device switching frequency. *IEEE Transactions on Power Electronics*, **22** (6), 2138–2145.
20. Etxeberria-Otadui, I., de Heredia, A.L., San-Sebastian, J. *et al.* (2008) Analysis of a h-npc topology for an ac traction front-end converter. Proceedings of EPE-PEMC 2008, September 2008, pp. 1555–1561.
21. Guennegues, V., Gollentz, B., Leclere, L. *et al.* (2009) Selective harmonic elimination PWM applied to h-bridge topology in high speed applications. Proceedings of POWERENG 2009, March 2009, pp. 152–156.
22. Mohan, N., Undeland, T., and Robbins, P.W. (2003) *Power Electronics: Converters, Applications and Design*, John Wiley & Sons, Inc..
23. Ray-Shyang, L. and Ngo, K.D.T. (1995) A PWM method for reduction of switching loss in a full-bridge inverter. *IEEE Transactions on Power Electronics*, **10** (3), 326–332.
24. Holmes, D.G. and Lipo, T.A. (2003) *Pulse Width Modulation for Power Converters, Principles and Practice*, Wiley-IEEE Press.
25. Nabae, A., Takahashi, I., and Akagi, H. (1981) A new neutral-point-clamped PWM inverter. *IEEE Transactions on Industrial Applications*, **17** (5), 518–523.
26. Leon, J.I., Portillo, R., Franquelo, L.G. *et al.* (2007) New space vector modulation technique for single-phase multilevel converters. Proceedings of IEEE ISIE 2007, June 2007, pp. 617–622.
27. Salaet, J. (2006) Contributions to the use of rotating frame control and space vector modulation for multilevel diode-clamped single-phase ac–dc power converters. PhD dissertation, Polytechnic University of Catalonia, Barcelona, Spain.
28. Meynard, T.A. and Foch, H. (1992) Multi-level conversion: high voltage choppers and voltage-source inverters. Proceedings of IEEE PESC 1992, June/July 1992, Vol. 1, pp. 397–403.
29. Lin, B.R. and Hou, Y.L. (2001) High-power-factor single-phase capacitor clamped rectifier. *IEE Proceedings on Electric Power Applications*, **148** (2), 214–224.
30. Meynard, T.A., Foch, H., Thomas, P. *et al.* (2002) Multicell converters: basic concepts and industry applications. *IEEE Transactions on Power Electronics*, **49** (5), 955–964.
31. Stynski, S. (2011) Analysis and control of multilevel AC–DC–AC flying capacitor converter fed from single-phase grid. PhD dissertation, Warsaw University of Technology, Warsaw, Poland.
32. Kazmierkowski, M.P., Krishnan, R., and Blaabjerg, F. (2002) *Control in Power Electronics Selected Problems*, Academic Press.
33. R. Teodorescu, R. Blaabjerg, M. Liserre, and P. C. Loh, "Proportional-resonant controllers and filters for grid-connected voltage-source converters," *IEE Proceedings on Electric Power Applications*, **153**, (5), 750–762, 2006.
34. Dell'Aquila, A., Liserre, M., Monopoli, V.G., and Rotondo, P. (2008) Overview of pi-based solutions for the control of dc buses of a single-phase h-bridge multilevel active rectifier. *IEEE Transactions on Industrial Applications*, **44** (3), 857–866.
35. Blaabjerg, F., Teodorescu, R., Liserre, M., and Timbus, A.V. (2006) Overview of control and grid synchronization for distributed power generation systems. *IEEE Transactions on Industrial Electronics*, **53** (5), 1398–1408.
36. Timbus, A., Liserre, M., Teodorescu, R. *et al.* (2009) Evaluation of current controllers for distributed power generation systems. *IEEE Transactions on Power Electronics*, **24** (3), 654–664.
37. Franklin, G.F., Powell, J.D., and Emami-Naeini, A. (2002) *Feedback Control of Dynamics Systems*, 4th edn, Prentice-Hall.
38. Best, R.E. (2003) *Phase-Locked-Loops: Design, Simulation and Applications*, 5th edn, McGraw-Hill Professional, New-York.
39. Teodorescu, R., Liserre, M., and Rodriguez, P. (2011) *Grid Converters for Photovoltaic and Wind Power Systems*, John Wiley & Sons, Ltd.

40. Salaet, J. (2006) Contributions to the use of rotating frame control and space vector modulation for multilevel diode-clamped single-phase AC–DC power converters. PhD dissertation, Universitat Politecnica de Catalunya, Barcelona, Spain.
41. Salaet, J., Alepuz, S., Gilabert, A., and Bordonau, J. (2004) Comparison between two methods of DQ transformation for single-phase converters control. Application to a 3-level boost rectifier. IEEE PESC 2004, June 2004, pp. 214–220.
42. Miranda, U.A., Aredes, M., and Rolim, L.G.B. (2004) A DQ synchronous reference frame current control for single-phase converters. IEEE PESC 2004, June 2004, pp. 1377–1381.
43. Liserre, M., Blaabjerg, F., and Teodorescu, R. (2006) Multiple harmonics control for three-phase systems with the use of PI-RES current controller in a rotating frame. *IEEE Transactions on Power Electronics*, **21** (3), 836–841.
44. Teodorescu, R., Blaabjerg, F., Liserre, M., and Loh, P.C. (2006) A new breed of proportional-resonant controllers and filters for grid-connected voltage-source converters. *IEE Proceedings on Electric Power Applications*, **153** (5), 750–762.
45. Maknouninejad, A., Simoes, M.G., and Zolot, M. (2009) Single phase and three phase P+ resonant based grid connected inverters with reactive power and harmonic compensation capabilities. Proceeidngs of IEEE Electric Machines and Drives Conference (IEMCD 2009), May 2009, pp. 385–391.
46. Kazmierkowski, M.P. and Tunia, H. (1994) *Automatic Control of Converter-Fed Drives*, Elsevier, Amsterdam, London, New York, Tokyo, PWN Warszawa.
47. Malinowski, M. (2001) Sensorless control strategies for three-phase PWM rectifiers. PhD dissertation, Warsaw University of Technology, Warsaw, Poland.
48. Jasinski, M. (2005) Direct power and torque control of ac–dc–ac converter-fed induction motor drives. PhD dissertation, Warsaw University of Technology, Warsaw, Poland.
49. Levine, W.S. (2000) *Control System Fundamentals*, CRC Press.
50. Lascu, C., Asiminoaei, L., Boldea, I., and Blaabjerg, F. (2009) Frequency response analysis of current controllers for selective harmonic compensation in active power filters. *IEEE Transactions on Industrial Electronics*, **56** (2), 337–347.
51. Castilla, M., Miret, J., Matas, J. *et al.* (2009) Control design guidelines for single-phase grid-connected photovoltaic inverters with damped resonant harmonic compensators. *IEEE Transactions on Industrial Electronics*, **56** (11), 4492–4501.
52. Yepes, A.G., Freijedo, F.D., Lopez, O., and Doval-Gandoy, J. (2011) Analysis and design of resonant current controllers for voltage-source converters by means of nyquist diagrams and sensitivity function. *IEEE Transactions on Industrial Electronics*, **58** (11), 5231–5250.

24

Impedance Source Inverters

Yushan Liu[1,2], Haitham Abu-Rub[1] and Baoming Ge[2,3]
[1]*Department of Electrical and Computer Engineering, Texas A&M University at Qatar, Doha, Qatar*
[2]*School of Electrical Engineering, Beijing Jiaotong University, Beijing, China*
[3]*Department of Electrical Engineering, Texas A&M University, Texas, USA*

24.1 Multilevel Inverters

One of the most suitable power architectures for a photovoltaic (PV) system is the multilevel inverter. Although there are many conventional two-level inverters available in this area, the multilevel inverter provides the following advantages: (1) reduced device voltage stress; (2) negligible total harmonics in the voltage waveforms; (3) smaller output filter size; (4) greater efficiency [1–4]; and (5) an implementation of the so-called *distributed maximum power point tracking* (DMPPT) [5–7]. The fifth advantage extends the MPPT to each panel of a PV system by avoiding series-connected PV arrays, which are often used with the conventional two-level inverter. This minimizes power loss even when mismatching conditions occur. Among the following three main families of multilevel converter: diode-clamped, capacitor-clamped, and cascaded H-bridge, the latter is usually considered in the literature for PV applications [8, 9].

24.1.1 Transformer-Less Technology

To interface the low-voltage (LV) output of an inverter to the grid, a bulky low-frequency transformer is necessary, which involves large size, less efficiency, loud acoustic noise, and high cost [10]. Another choice, instead of a transformer, is to use many PV panels in a string to generate a voltage higher than that of the grid, which will cause power loss of the PV panels in case of mismatching.

Transformer-less topologies are especially deserving of attention because of their higher efficiency, smaller size and weight, and lower price for the PV system [10].

Transformer-less technology is preferable [10] for attaining utility-scale power ratings and medium voltage levels. The cascaded multilevel inverter (CMI) structure is qualified for this purpose and, furthermore, its distributed modules enhance system reliability. At present, three CMI structures for PV power systems have been published [11–15].

The so-called *power electronic transformer* (PET) configurations [11] use a cascaded H-bridge multilevel DC/AC converter on the LV side with a separate DC-link for each section of the PV plant. The

Power Electronics for Renewable Energy Systems, Transportation and Industrial Applications, First Edition.
Edited by Haitham Abu-Rub, Mariusz Malinowski and Kamal Al-Haddad.

configurations allow different voltages on the DC-links; thus, implementation of separate MPPT control algorithms can be carried out for the different PV sections. The high-voltage windings of the employed medium-frequency transformer are interfaced to the medium-voltage grid by means of a multilevel converter comprising a three-phase inverter. The number of voltage levels is selected according to the rated grid voltage and to the characteristics of the switching devices used. However, there are disadvantages: (a) one or three medium frequency transformers cause cost–volume issues, even though isolated from the grid; (b) too many switches cause high cost and loss, are too complex and have low reliability, even though the filter will be small owing to the multilevel voltage, which limits its practical applications to high-power PV systems.

24.1.2 *Traditional CMI or Hybrid CMI*

Traditional CMI or hybrid CMI (Figure 24.1) can connect directly to the grid without a transformer and achieve high efficiency [12–14]. It is an attractive topology because of its modularity, simple layout, fewer components and higher reliability, compared with other multilevel inverters, such as the neutral point clamp inverter and the flying capacitor multilevel inverter. The necessity for isolated DC sources makes this topology an ideal inverter choice for use in PV applications. A 240-kW traditional CMI-based PV inverter is reported in Ref. [12] with efficiency of 98.6% without a transformer. The voltages of hybrid CMI's DC buses are in a geometric sequence, which present several advantages: (a) different kinds of power switches can be used in the H-bridges with different DC bus voltages, switches with small capacity can be used in the H-bridge with lower DC bus voltage, and the on-state loss can be lowered; (b) the cost and complexity of the system is reduced because fewer switches and DC buses are required in the topology; (c) the switching frequency can be reduced significantly. However, this kind of CMI does not have the boost function, which will lead to overrating of the inverter by a factor of two in order to cope with wide (1 : 2) PV voltage changes. For example, a 2 MW inverter is needed for a 1 MW PV power system, which makes the inverter larger, more costly and more difficult for utility-scale applications.

24.1.3 *Single-Stage Inverter Topology*

There are several power converter topologies employed in PV systems, characterized as two-stage or single-stage, transformer or transformer-less and with a two-level or multilevel inverter [1, 11–14]. Single-stage inverters are becoming more attractive compared with two-stage models owing to their compactness, low cost and their reliability [15]. However, the conventional inverter has to be oversized to cope with the wide PV array voltage changes, because a PV panel presents low output voltage with a wide range of variation based on irradiation and temperature, usually with a ratio of 1 : 2.

The two-stage inverter applies a boost DC–DC converter, instead of a transformer, to minimize the required KVA rating of the inverter and to boost the wide range of voltage to a constant desired value. Unfortunately, the switch in the DC–DC converter becomes the killer of the cost and efficiency of the system. For safety reasons, some PV systems have a galvanic isolation, either in the DC–DC boost converter using a high-frequency transformer, or in the AC output side of a line frequency transformer. Both of these added galvanic isolations increase the cost and size of the entire system and decrease the overall efficiency.

24.2 Quasi-Z-Source Inverter

24.2.1 *Principle of the qZSI*

The Z-source inverter (ZSI), as a single-stage power converter with step-up/down function, allows a wide range of PV voltages, and has been reported in applications of PV systems [16]. It can handle the

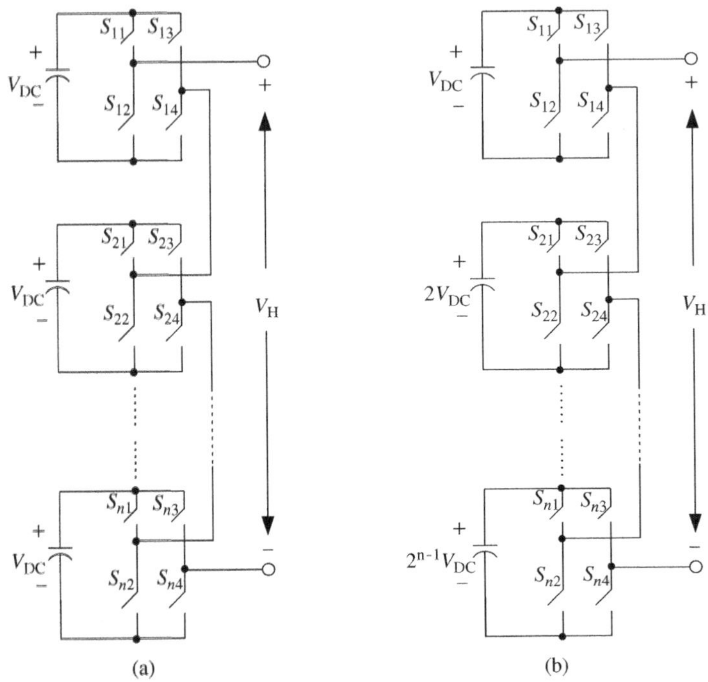

Figure 24.1 (a) Traditional CMI and (b) hybrid CMI

PV DC voltage variation in a wide range without overrating the inverter and implement voltage boost and inversion simultaneously in a single power conversion stage, thus minimizing system cost, reducing component count and cost and improving reliability. Recently proposed quasi-Z-source inverters (qZSI) have some new attractive advantages more suitable for application in PV systems. This will make the PV system much simpler and cheaper because the qZSI draws a constant current from the PV panel, which means that there is no need for extra filtering capacitors. In addition, it features lower component (capacitor) rating and reduces switching ripples to the PV panels [16–23].

The output voltage of the PV panel has a wide variation related to changes of temperature and solar irradiation, which usually presents in a ratio of 1 : 2. It is impossible for the traditional voltage source inverter (VSI) to deal with this wide variation if there is neither overrating of the inverter nor the use of a DC–DC boost converter. For a common single-stage inverter, as shown in Figure 24.2(a) [24, 25], which is used in conventional PV systems to interface the PV array with the utility and/or load, the

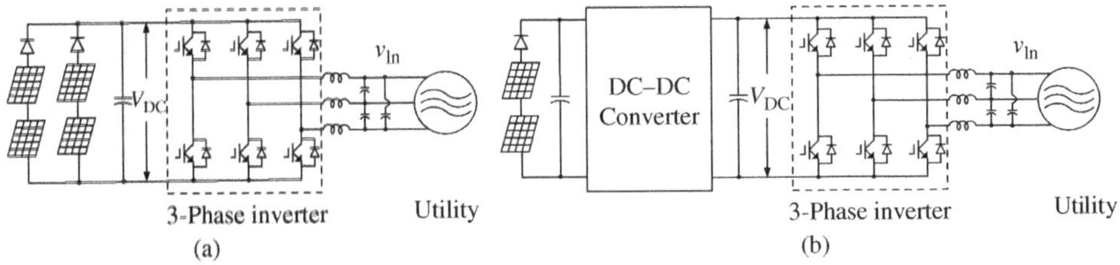

Figure 24.2 Traditional typical configuration of PV system: (a) single-stage VSI and (b) two-stage VSI

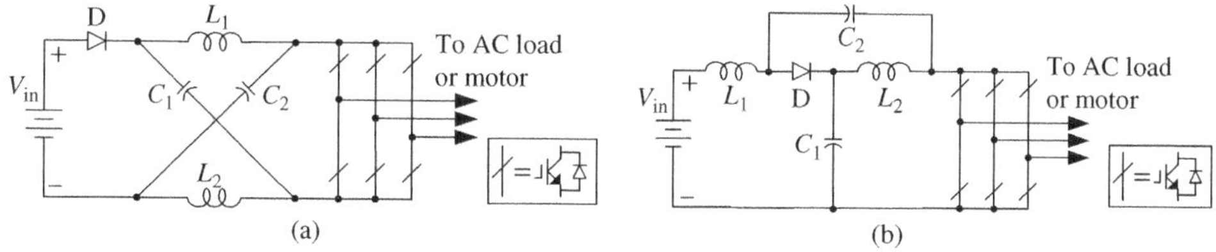

Figure 24.3 Topology of the voltage-fed ZSI and qZSI: (a) voltage-fed ZSI and (b) voltage-fed qZSI

minimum PV DC voltage should be two times the peak value of the AC phase voltage, or $2v_{\text{In}}$. Therefore, considering a PV voltage change range of 1 : 2, the PV array has to produce $2v_{\text{In}}$ to $4v_{\text{In}}$ to feed the inverter. Consequently, the inverter should be designed to block four times the peak value of the AC phase voltage, or $4v_{\text{In}}$. For example, the VSI needs a minimum 340 V DC to produce 208 V AC phase-to-phase voltage for a three-phase system, and the PV voltage has to be 340–680 V considering it changes in a 1 : 2 range. Consequently, 1200 V insulated gate bipolar transistors (IGBTs) are required and the inverter is thereby two times overrated. To overcome this problem, the DC–DC boost circuit is employed, as shown in Figure 1.2(b) [26–31] and 600 V IGBTs can be used for the system. However, the cost will increase and the efficiency will be reduced.

Compared with the configuration of the VSI plus DC–DC boost, the ZSI-based PV system minimizes switching devices, has lower cost and higher reliability [19, 32]. Recently, our group published a class of qZSIs with some new advantages [16, 33], which was from the original ZSI [19]. By using this new quasi-Z source topology, the inverter of a PV system becomes much simpler and its cost can be reduced. This is because the proposed qZSI draws a constant current from the PV panel; thus, there is no need for extra filtering capacitors, and it features lower component (capacitor) rating and reduces switching ripples to the PV panel [17].

Figure 24.3(a and b) shows the traditional voltage-fed ZSI and the recently proposed voltage-fed qZSI, respectively. In the same manner as the conventional ZSI, the qZSI has two general types of operational states at the DC side: the non-shoot-through state (i.e., the six active states and two conventional zero states), and the shoot-through state (i.e., both switches in at least one phase conduct simultaneously). In the non-shoot-through state, the inverter bridge, viewed from the DC side, is equivalent to a current source, whereas in the shoot-through state, the inverter bridge is a short circuit. The equivalent circuits of the two states are shown in Figure 24.4(a and b), respectively. It is well known that the shoot-through state is strictly forbidden in the traditional VSI, because it will cause a short circuit of the voltage source and damage the devices. In the qZSI and ZSI, however, the unique LC and diode network, connected to the inverter bridge, modify the operation of the circuit, allowing the shoot-through states. Furthermore, by using the shoot-through state, the (quasi-) Z-source network boosts the DC-link voltage. This feature will effectively protect the circuit from damage; thus, it improves system reliability significantly. Because of the input inductor L_1, the qZSI draws a continuous constant DC current from the DC source. Compared with the ZSI that draws a discontinuous current, the constant current will reduce significantly the input stress; thus, the qZSI is especially well suited for PV system applications [17].

Assuming that during one switching cycle T, the interval of the shoot-through state is T_0, then the interval of non-shoot-through state is T_1; thus, $T = T_0 + T_1$ and the shoot-through duty ratio $D = T_0/T$. From Figure 24.4(a), during the interval of the non-shoot-through state T_1, there are

$$v_{L_1} = V_{\text{in}} - V_{C_1}, \quad v_{L_2} = -V_{C_2}, \quad v_{\text{PN}} = V_{C_1} - v_{L_2} = V_{C_1} + V_{C_2}, \quad v_{\text{diode}} = 0 \tag{24.1}$$

From Figure 24.4(b), during the interval of the shoot-through state T_0, one can get

$$v_{L_1} = V_{C_2} + V_{\text{in}}, \quad v_{L_2} = V_{C_1} \quad \text{and} \quad v_{\text{PN}} = 0, \quad v_{\text{diode}} = -(V_{C_1} + V_{C_2}) \tag{24.2}$$

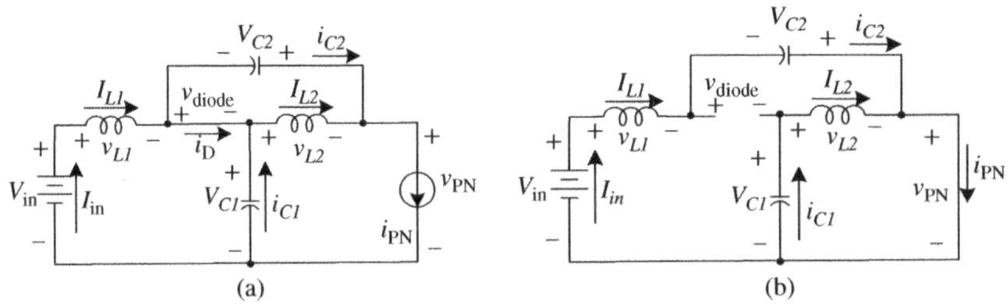

Figure 24.4 Equivalent circuit of the qZSI: (a) non-shoot-through state and (b) shoot-through state

At steady state, the average voltage of the inductors over one switching cycle is zero. From Equations (24.1), (24.2), we have

$$\begin{cases} V_{L_1} = \bar{v}_{L_1} = \dfrac{T_0\left(V_{C_2} + V_{\text{in}}\right) + T_1(V_{\text{in}} - V_{C_1})}{T} = 0 \\ V_{L_2} = \bar{v}_{L_2} = \dfrac{T_0(V_{C_1}) + T_1(-V_{C_2})}{T} = 0 \end{cases} \tag{24.3}$$

Thus,

$$V_{C_1} = \frac{1-D}{1-2D}V_{\text{in}}, \quad V_{C_2} = \frac{D}{1-2D}V_{\text{in}} \tag{24.4}$$

From Equations (24.2) and (24.4), the peak DC-link voltage across the inverter bridge is

$$v_{\text{PN}} = V_{C_1} + V_{C_2} = \frac{T}{T_1 - T_0}V_{\text{in}} = \frac{1}{1-2D}V_{\text{in}} = BV_{\text{in}} \tag{24.5}$$

where B is the boost factor of the qZSI.

The average currents of the inductors L_1 and L_2 can be calculated by the system power rating P

$$I_{L_1} = I_{L_2} = I_{\text{in}} = \frac{P}{V_{\text{in}}} \tag{24.6}$$

According to Kirchhoff's current law and 24.6, we also can get that

$$I_{C_1} = I_{C_2} = I_{\text{PN}} - I_{L_1}; \ I_D = 2I_{L_1} - I_{\text{PN}} \tag{24.7}$$

In summary, the voltage and current stress of the qZSI are shown in Table 24.1, where

- M is the modulation index; v_{ln} is the AC peak phase voltage;
- (2) $m = (1-D)/(1-2D)$; $n = D/(1-2D)$; $B = 1/(1-2D)$

The stress on the ZSI is shown as well for comparison.

From Table 24.1 we can establish that the qZSI inherits all the advantages of the ZSI. It can buck or boost a voltage, cope with a wide range of input voltages and produce a desired voltage for the load or connection to the grid in a single stage. This feature results in the reduced number of switches involved in the PV system and, therefore, the reduced cost and the improved system efficiency. When the voltage of the PV panel is low, it boosts the DC-link voltage, which helps avoid redundant PV panels for higher DC voltage or unessential inverter overrating. As mentioned, it is able to handle the shoot-through state; therefore, it is more reliable than conventional VSI. For the same reason, there is no need to add any dead time into the control schemes, which reduces the output distortion.

In addition, there are some unique merits of the qZSI when compared with conventional ZSI [16, 17]. The ZSI has a discontinuous input current in the boost mode, whereas the input current of the qZSI is

Table 24.1 Voltage and average current of the qZSI and ZSI network

	$v_{L_1} = v_{L_2}$		v_{PN}		v_{diode}		V_{C_1}	V_{C_2}	v_{ln}	$I_{in} = I_{L_1} = I_{L_2}$	$I_{C_1} = I_{C_2}$	I_D
	T_0	T_1	T_0	T_1	T_0	T_1						
ZSI	mV_{in}	$-nV_{in}$	0	BV_{in}	BV_{in}	0	mV_{in}	mV_{in}	$MBV_{in}/2$	P/V_{in}	$I_{PN} - I_{L_1}$	$2I_{L_1} - I_{PN}$
qZSI	mV_{in}	$-nV_{in}$	0	BV_{in}	BV_{in}	0	mV_{in}	nV_{in}	$MBV_{in}/2$	P/V_{in}	$I_{PN} - I_{L_1}$	$2I_{L_1} - I_{PN}$

continuous owing to the input inductor L_1 and this reduces the input stress significantly; thus, it can reduce the capacitance for the output of the PV panels. The two capacitors in the ZSI sustain the same high voltage, whereas the voltage on capacitor C_2 in the qZSI is lower, which allows a lower capacitor voltage rating. For the qZSI, there is a common DC rail between the source and inverter, which is easier to assemble and causes less EMI problems.

24.2.2 Control Methods of the qZSI

24.2.2.1 Buck/Boost Conversion Mode

If the inverter operates entirely in the non-shoot-through state, as shown in Figure 24.4(a), the diode will conduct and the voltage on capacitor C_1 will be equal to the input voltage, whereas the voltage on capacitor C_2 will be zero. Therefore, $v_{PN} = V_{in}$ and the qZSI acts as a conventional VSI:

$$v_{ln} = \frac{v_{PN}}{2}M = \frac{V_{in}}{2}M \tag{24.8}$$

For sinewave pulse width modulation (SPWM), $0 \leq M \leq 1$ and for space vector modulation (SVM), $0 \leq M \leq (2/\sqrt{3})$. Thus, when $D=0$, v_{ln} is always less than $V_{in}/\sqrt{3}$ and this is called the *buck conversion mode of the qZSI.*

When the qZSI operates in boost conversion mode, there are two more modulation references: V_P^* and V_N^*, in addition to the conventional three-phase references: v_a^*, v_b^* and v_c^*. As shown in Figure 24.5, by replacing parts or all of the two conventional zero states with shoot-through states, the non-shoot-through states and shoot-through states will alternate in one switching cycle. Then, the peak DC-link voltage v_{PN} can be boosted by a factor of B; its value is dependent on the shoot-through duty ratio, as defined in Equation (24.5). Notice that the six active states are unchanged and the peak AC voltage becomes

$$v_{ln} = \frac{v_{PN}}{2}M = \frac{V_{in}}{2}BM \tag{24.9}$$

24.2.2.2 Boost Control Methods

All the boost control methods that have been explored for the traditional ZSI, such as simple boost, maximum boost, maximum constant boost, as shown in Figure 24.6 [19, 20, 34], can be applied to the qZSI. It is noticeable that the voltage gain of the qZSI is $G=MB$, whereas the voltage stress across the inverter bridge is BV_{in}. In order to maximize the voltage gain and to minimize the voltage stress on the inverter bridge, one needs to decrease the boost factor B and increase the modulation index M as much as possible.

Figure 24.7 shows the voltage gain versus the modulation index of three boost control methods. All present significantly higher gain than traditional VSI. Among the three boost control methods, the maximum boost control best exploits the conventional zero states; therefore, it has the maximum M and

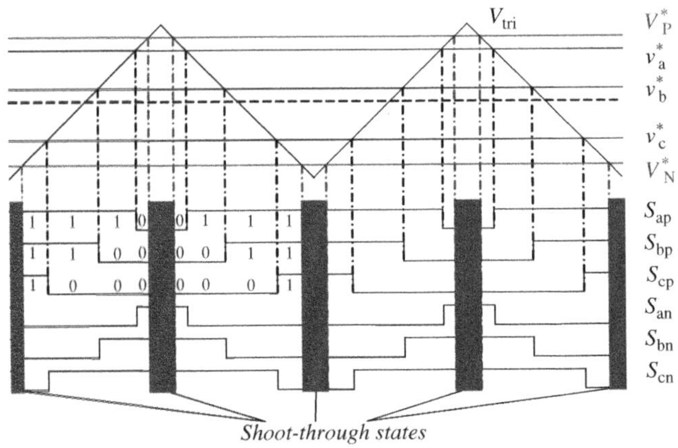

Figure 24.5 SPWM for the qZSI in boost conversion mode

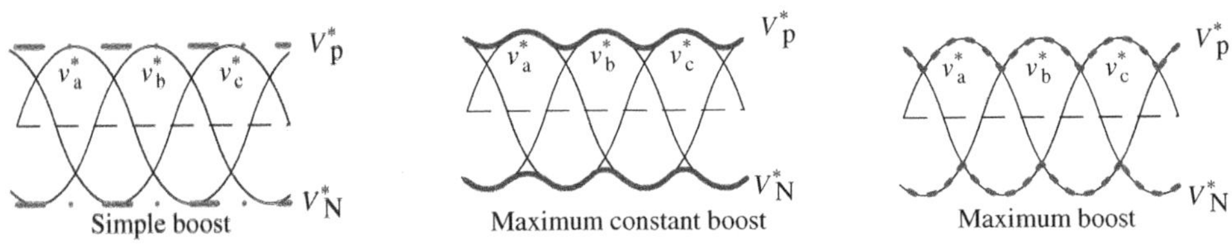

Figure 24.6 Three boost control methods with different V_P^* and V_N^*

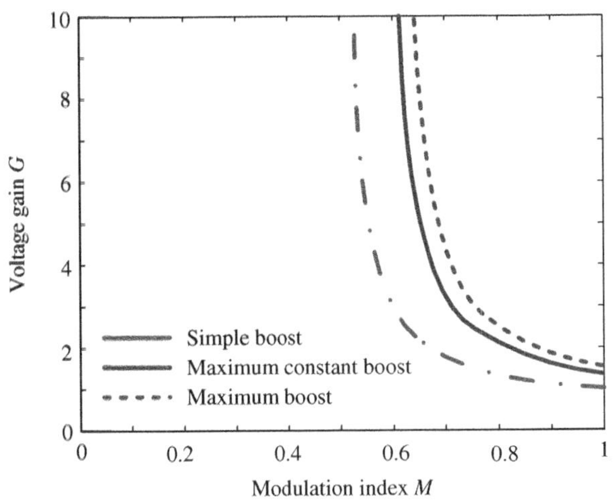

Figure 24.7 Voltage gain versus modulation index

the minimum voltage stress across the inverter bridge with the same voltage gain. However, it has the drawback of low-frequency ripples on the passive components of the qZSI, which requires a larger volume and weight and greater cost of the inductor and capacitor in the qZSI network. The simple boost control has shoot-through states that are spread evenly; thus, it does not involve low-frequency ripples associated with output frequency, but its voltage stress is the largest for a given voltage gain. The maximum constant boost control is a compromise between the other two.

When the third-harmonic injection is combined into the maximum constant boost control, the maximum modulation index can be $M = 2/\sqrt{3}$, and there is lower voltage stress on the inverter bridge. With this method, the shoot-through states are introduced into the switching cycle when the carrier is either greater than V_P^* or less than V_N^*, which is spread evenly in each switching cycle. Thus, the qZSI network does not involve low-frequency ripples and the shoot-through duty ratio is

$$D = \frac{T_0}{T} = 1 - \frac{\sqrt{3}M}{2} \tag{24.10}$$

the boost factor is

$$B = \frac{1}{1 - 2D} = \frac{1}{\sqrt{3}M - 1} \tag{24.11}$$

and the voltage gain is equal to

$$G = MB = \frac{M}{\sqrt{3}M - 1} \tag{24.12}$$

The peak AC phase voltage can be calculated by

$$v_{\ln} = \frac{V_{\text{in}}}{2}G = \frac{MV_{\text{in}}}{2\sqrt{3}M - 2} \tag{24.13}$$

24.2.3 *qZSI with Battery for PV Systems*

Power generated by the solar panel depends on the solar power incident on the panel, the panel temperature and the operating panel voltage. The first two factors are unpredictable because of the weather and seasons. The resultant power of stochastic fluctuations will have a negative effect on the grid. To date, there have been no significant net failures driven by these stochastic fluctuations. Nevertheless, the growing number of solar power plants is forcing us to invent and implement new solutions for this problem. Apart from investments in the extension of the grid and power capacity, as well as selective shutdown of PV systems, the integration of electricity storage systems is a more innovative idea, because the balancing difference between the forecast and real values could minimize this negative effect [35].

In addition, power consumption also presents some characteristics of season and human living habits. In spring and autumn, there are relatively more fine days with a lot of solar irradiation compared with the other seasons. These seasons also have good weather; thus, electric loads such as air conditioners may be used less often. Consequently, increased generation from PV systems and reduced loads cause a voltage rise on a power distribution line. Over weekends, during which the PV systems continue to produce the same amount of power and industrial loads are light, the grid voltage and frequency could easily become high [36]. Overvoltage may exceed the upper tolerance limit at the point of common coupling; usually, a grid overvoltage protection will regulate the output power of the PV system if the AC voltage exceeds the control range. As a result, a significant amount of possible energy will be lost in a clear day. An energy storage unit installed in each PV system can be used for voltage rise avoidance, through charging the excess electric power to the energy storage unit instead of feeding it to the grid. The energy storage unit is similar to an energy buffer, which could be charged using the differential power between the PV power and the output power to the grid, and could maximize the level of power transfer from the

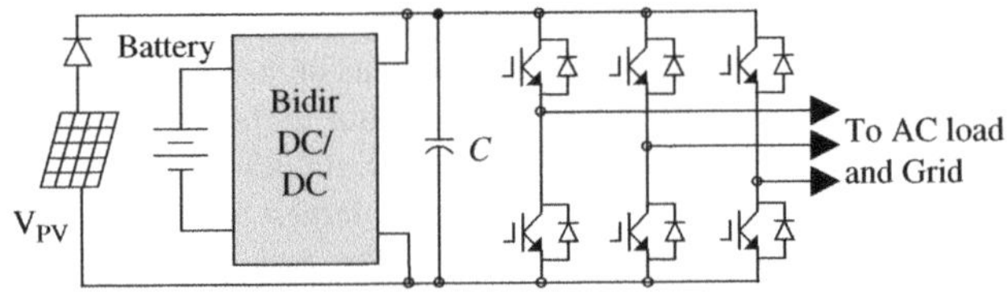

Figure 24.8 Traditional scheme

PV array through the MPPT, resulting in very high efficiency [4]. In addition, PV-grid-connected plants could become more reliable by acquiring the possibility to cope with some important auxiliary services [35]. There is typically a configuration available for this application, shown in Figure 24.8, when the conventional VSI is connected to the grid and/or load, and where a bidirectional DC–DC converter is used to control the battery state of charge (SOC). It is obvious that the DC–DC converter increases both the cost and the system complexity, and reduces reliability and efficiency.

The qZSI has two independent control freedoms: shoot-through duty ratio and modulation index, providing the ability to produce any desired output AC voltage to the grid. It can regulate the battery SOC and control the PV panel output power (or voltage) simultaneously. Second, the qZSI provides the same features of a DC–DC boosted inverter (i.e., buck/boost), yet its single stage is less complex and more cost effective. Third, the qZSI has the benefit of enhanced reliability because momentary shoot-through can no longer destroy the inverter (i.e., both devices of a phase leg can be on for a significant period of time). Furthermore, our proposal includes energy storage (battery) in the qZSI for PV systems without the additional DC–DC converter, as shown in Figure 24.9. This innovative feature can reduce cost and system complexity further and is explained in detail in the next section.

In the system of Figure 24.9, there are three power sources/consumers: the PV panels, battery and the grid/load. As long as we can control the power flow of two of them, the third element automatically matches the power difference. From Equation (24.4) of qZSI, the relationship between the capacitor-C_2 voltage and the voltage of the PV panel is

$$V_{C_2} = \frac{D}{1 - 2D} V_{\mathrm{PV}} \tag{24.14}$$

where D is the shoot-through duty ratio.

In Figure 24.9, the battery voltage V_{b}, equals voltage V_{C_2} of capacitor C_2. The output voltage of a battery is relatively less current dependent because of the much smaller internal resistance. The voltage

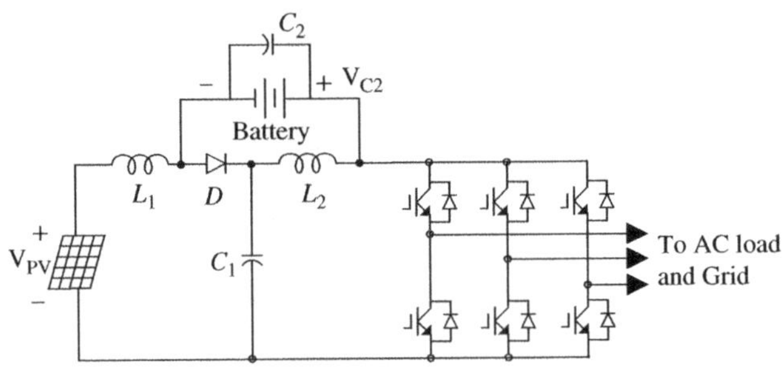

Figure 24.9 Configuration of the qZSI with battery pack for PV system

of a battery changes with the SOC of the battery and will be relatively constant at a certain SOC, and the voltage of the PV panel is highly current dependent; therefore, for a given battery voltage V_b, the voltage of the PV panel is controlled to be

$$V_{PV} = \frac{1-2D}{D} V_b \tag{24.15}$$

At the same time, the output power of qZSI can be controlled by manipulating the modulation index to produce the desired output voltage. The output peak phase voltage of the inverter is

$$v_{ln} = (V_{C_1} + V_b)\left(\frac{M}{2}\right) \tag{24.16}$$

where M is the modulation index based on the reference waveform and the triangular waveform.

The output power to the grid can be expressed as

$$P_{out} = \frac{3pf}{\sqrt{2}} v_{ln} I \tag{24.17}$$

where I is the rms current to the grid and pf is the power factor.

Therefore, the system is able to control the output power of the PV panels and the power injected to the grid at the same time; thus, their difference is the power charging the battery.

In summary, the power of the PV panels is controlled by the shoot-through duty ratio of the qZSI; the output power to the grid is controlled by the output voltage and the current related to the modulation signal. If output power to the grid is higher than the power of the PV panels, the battery is discharged. If the output power to the grid is lower than the power of the PV panels, the battery is charged. If the SOC of the battery becomes too low, the PV panels will provide power to recharge the battery. However, the battery can only be charged when it is not fully charged already.

Let us explain this principle using examples. *Case* 1: for a fixed level of solar irradiance and for a fixed temperature, if the PV panel maintains a maximum output power, the power and voltage of the PV panels will be constant. When power to the grid equals the power of the PV panels, as one would expect, the battery SOC should remain constant with zero average power to the battery. When power to the grid increases to be greater than the power of the PV panels, the battery should supply the additional power requested by the grid; thus, the battery SOC will decrease. When injected grid power falls below the power of the PV panels, the additional power of the PV panel will charge the battery, increasing the SOC. (2) *Case 2*: Power to the grid is kept constant and the power of the PV panels changes because of the variation in solar irradiance. Again, the battery SOC should remain constant while power to the grid matches that of the PV panels. The battery will be charged when the power of the PV panels increases to be greater than the power supplied to the grid, increasing the battery SOC. When the power of the PV panels falls below the power to the grid, the battery will supply the additional power requested by the grid, decreasing the battery SOC.

24.3 qZSI-Based Cascade Multilevel PV System

24.3.1 Working Principle

In grid-connected systems, the panels are usually arranged in strings (series connection) for the required power and voltage levels, where all panels of the string drive the same current. Generally, it is preferable to use the same panels and to keep them away from any shading. However, in residential installations, it is not easy to avoid shading because of the change in sunlight direction throughout the day. Furthermore, obstacles, such as trees, birds and other constructions, can cause partial shading. Some studies have revealed that minor shading can cause a major reduction in solar power output of a PV

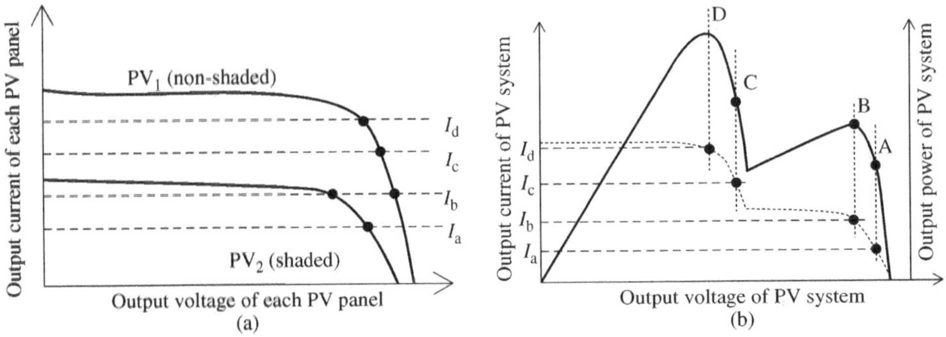

Figure 24.10 Characteristics of a small string consisting of two panels under different solar irradiance levels with the same temperature [5]: (a) current versus voltage of each panel and (b) effects of different solar irradiance levels on a string of PV arrays

array [37]. When a panel is shaded, it generates less optical current, and when connected in series with other panels there is a downgrading of system energy yield, because the low level of current also flows through the non-shaded panels instead of the intrinsic high current [38]. This requires the use of bypass diodes to preserve the PV array voltage and to minimize hot-spot heating and the potential for panel failures when shaded [5, 8, 37]. Figure 24.10(a) shows the I–V characteristics of two panels under different irradiance levels, and Figure 24.10(b) shows the P–V and I–V characteristics of a PV string consisting of two panels with a bypass diode incorporated. The shaded panel becomes short-circuited and the current in the string is that of the non-shaded module, because the bypass diode in parallel with the shaded panel goes into the on state. In this way, the string-generated power and voltage are greater than that in the case of an array with no bypass diodes. Nevertheless, the string power versus voltage becomes a multimodal curve, as indicated in Figure 24.10(b). Its detail is explained in the following [5].

When operating at point A of Figure 24.10(b), both PV panels generate power, but neither one generates maximum power. When operating at point B, the shaded module PV_2 generates maximum power, but the non-shaded module does not generate maximum power. When operating at point C, the non-shaded module generates power, but the shaded module does not generate any power, because the string current flows through its bypass diode. When operating at point D, the non-shaded module generates its maximum power, but the shaded module does not generate any power because the string current flows through its bypass diode. Finally, in the case of both mismatching conditions and the presence of the module bypass diodes, the P–V characteristic curve is a multimodal curve with a number of peaks. The presence of more than one peak in the P–V characteristic of a PV string makes it difficult to implement the absolute maximum power of the PV string. Furthermore, for this case, the maximum output power of the PV string is less than the sum of the maximum generation power of all PV panels. In addition, the reduction of the string voltage at the MPP may not be acceptable according to the system specifications [8].

Reference [37] reported that the shaded panel contributes a 14.06% power loss, and an additional 11.57% loss results from the reduced MPP when the two panels are connected in series. When three or more panels are series-connected, the situation is even worse. The shading condition could be more complicated in practical PV applications, such as more than one shaded panel, making it more difficult to perform MPPT. It is of practical relevance to investigate the possibility of finding a technical and cost-effective solution to such a problem.

If independent control of each panel is possible, to allow different currents and voltages for all panels, the total generated power and voltage will be improved significantly. DMPPT is the best method for this purpose, and it comprises two major schemes. One is to use PV AC-module inverters avoiding the series-connection of panels. When a panel gets shaded, only that inverter will yield lower power, whereas the others continue to perform at their optimum level with their own inverters [38]. However, the conventional two-level inverter outputs LV because of its step-down-only operation and the low input voltage of each panel. It is necessary to step up the voltage suitably for grid connection, by using a low-frequency transformer between the inverter output and the grid, or by a DC–DC boost converter between the PV panel and the inverter input. Both of these configurations will be more costly because of the increased number of devices, and because power losses increase because of adding the DC–DC converter or transformer. Another scheme employs a multilevel inverter to replace its two-level counterpart. In this case, each PV panel is used as a separate power supply for each module of the multilevel inverter [2–4, 8, 39], and the generation point of each PV module can be controlled independently and maximized, that is, the so-called DMPPT, particularly with partial shade covering the PV facility or in the case of the mismatched PV panels. In addition to being able to maximize the power obtained from the PV panels, the multilevel inverter usually presents the advantages of reducing the device voltage stress, being more efficient, generating output AC-voltage with negligible total harmonics and allowing one to operate without transformers to step up the voltage. Among multilevel inverters, the cascade H-bridge inverter is most popular. It uses relatively few power devices and each H-bridge works at very low switching frequency. This offers the possibility of working in high power with low-speed switches and generating low-switching frequency losses, which allows working with a power range similar to the common central inverter topology, but extracting energy in a more efficient way [4].

In our proposal, the qZSI module, shown in Figure 24.9, is used to build the cascade H-bridge inverter as a practical implementation of the aforementioned DMPPT technique. Figure 24.11 presents the proposed three-phase inverter connected directly to the grid. Each phase consists of n qZSI modules, and phases B and C have the same structure as phase A. Normally, this architecture makes each module deliver the same power; however, sometimes this is not so when the output voltages of the qZSI modules are different, even though they have the same rms value of current, which depends on the modulation techniques. However, to extract the instantaneous maximum power from each panel, regardless of the solar irradiance and temperature levels, is the most important goal. The power generated by the PV panel may change owing to variations of solar irradiance and temperature levels, for instance, between summer and winter, a clear day and cloudy day, partial shading and so on. For each module, the stochastic fluctuation of solar power is inconsistent with the desired stable power injected to the grid when using MPPT. The gap between them can be compensated for by inclusion of a battery pack in each qZSI module. This topology allows independent control of the power delivered by each module. The battery pack can charge and discharge to balance the power to the grid. With the DMPPT, power generation of each PV panel is maximized even in the presence of PV panel mismatching conditions, such as supposing the non-shaded, mid-shaded and high-shaded panels of respective modules A1, A2 and An, as shown in Figure 24.11. Each qZSI module contributes to low cost, high efficiency, and reliability and is adaptive to the wide voltage variation range of the PV panel with its special voltage step up/step down function in a single-stage power conversion.

Control of the proposed PV system in Figure 24.11 will take into account the multilevel modulation, independent MPPT of each qZSI module, SOC of the battery pack and the power demand of the grid. There are $3n$ independent shoot-through duty ratios to control $3n$ qZSI modules operating at their own MPPs. The multilevel synthesis is achieved by means of a phase-shift displacement of the carrier waves of the different modules [40], and the shoot-through duty ratio should be combined into the multilevel PWM to form the final driving pulses for all switches. The power demand of the grid is achieved by multilevel modulation. The differential power between the generated maximum power of the PV panels

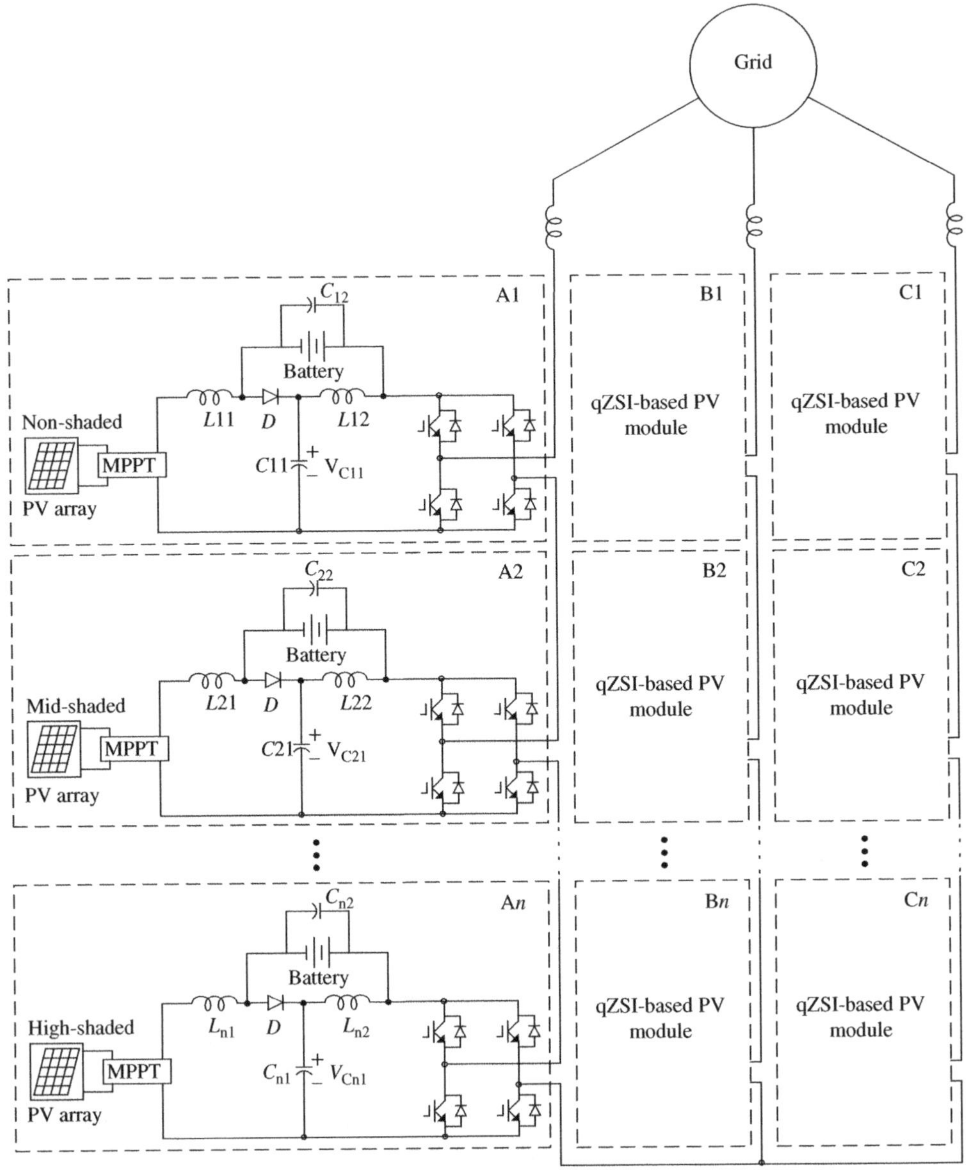

Figure 24.11 qZSI-based cascade multilevel PV power generation system

and the power injected to the grid will charge or discharge the battery pack. The battery SOC of each qZSI module should be monitored during operation and controlled by adjusting the injected power of the grid to maintain the battery SOC within the designed range.

Parameter design of the inductors, capacitors, battery pack and operating voltage of the inverter and modules is a critical step. The optimization of these parameters will be the object of our investigation. The control of each module is elaborated, not only for its own goal but also for the entire PV system.

For the qZS H-bridge legs, the shoot-through zero state does not affect the output voltage, which will not cause harmonic content in the load current. However, the 2ω (ω is fundamental frequency of the grid) voltage ripples are a problem common in the DC-link bus of both traditional H-bridge CMIs and ZS/qZS H-bridge CMIs because the 2ω reactive power flows through each module. For the qZS H-bridge CMI, a small LC filter is sufficient when connected to the grid because the multilevel output voltage is highly sinusoidal. If using the traditional boost converter + H-bridge in the CMI, then the 2ω voltage ripples are still in the DC-link bus. Furthermore, additional switches are needed. The energy storage will require extra circuits for the traditional boost converter VFI system.

24.3.2 Control Strategies and Grid Synchronization

An electrical grid system is complex and dynamic, and is affected by many factors and disturbances. Higher penetration of renewable energy sources is leading to requirements, imposed on power converters that feed the electrical grid system, which are more stringent. Several control strategies are applied to power converters connected to the electrical grid system: voltage-oriented control (VOC), model-predictive control (MPC), direct-power control (DPC), model-predictive direct-power control (MPDPC), direct-power control with space-vector modulation (DPC-SVM) and a control method based on virtual flux (VF) [41–56]. For the VOC and DPC-SVM methods, several modifications have been developed and are reported to involve a change in the voltage modulator to allow the use of multilevel converters. Virtual flux-oriented control (VFOC) for a grid-side PWM rectifier is based on coordinate transformations between stationary α β and synchronous rotating d–q reference systems [41, 42]. This strategy offers fast dynamic responses and a precise, steady-state performance through internal current control loops. Consequently, the final configuration and performance of the system depends largely on the quality of the applied current-control strategy. The simplest technique is hysteresis current control, which provides a fast dynamic response, good accuracy, no DC offset and high robustness. However, the major problem of hysteresis control is the variable switching frequency of the power converter. This makes the switching pattern uneven and random, which results in additional stress on the switching devices and difficulties for the filter design. Therefore, in the literature, several strategies have been reported for improving the performance of current control [41, 43]. Among the several current regulators with constant switching frequency, the most widely used scheme for high-performance current control is the d–q synchronous controller, where the currents being regulated are DC quantities and where it is easy to eliminate steady-state error and offset. DPC is a popular technique employed in grid-connected converters. By selecting appropriate switching states of the power converter, it is possible to control directly the active and reactive powers [44–49]. The technique is similar to the direct torque control, where the optimal switching state is selected by using lookup tables and hysteresis bounds. DPC using the predictive approach is also applied to the active rectifier [54, 55]. MPDPC is an extension of DPC, where the switching lookup table is replaced by an online optimization stage [56]. MPDPC achieves optimal performance by minimizing the switching losses of the converter, while controlling the output of the real and reactive powers.

Hence, the basic and advanced control features should be incorporated for any proper solution. The control technique employed should ensure stability in case of a large grid-impedance change, and be able to ride through during disturbance of the grid voltage. DC-link voltage control, adaptation to the grid voltage variation and grid synchronization, preferably with unity power factor operation, should also be incorporated. The MPPT strategy, and anti-islanding (as required by *IEEE* 1574), should be analyzed, although the monitoring and protection features need to be developed. Other control features are active and reactive power control, harmonic compensation, local voltage control and fault-tolerant operation during converter and grid faults. In addition, a battery energy management system is supposed to be developed and is taken into account in the control system.

24.4 Hardware Implementation

24.4.1 Impedance Parameters

The parameters of the impedance network are one of the important issues for the proper operation of the system. Several works have contributed to pursuing suitable Z-source parameters [17, 57–59]. A design process of a grid-tie PV system isillustrated in the following, according to the grid standard of Qatar.

The desired operating parameters of the converter are presented in Table 24.2. The qZS-CMI system is an 18 kW/7-level cascaded inverter, that is, threecascaded layers with 2 kW for each cell. The input DC voltage ranges from 60 to 120 V (the maximum variation of 1 : 2 is chosen because of the voltage variations of the PV panels).

Each of the quasi-Z-source networks is a combination of two inductors: L_1 and L_2 and two capacitors: C_1 and C_2. As the equivalent circuit shown in Figure 24.4, in shoot-through states, the PV panel and qZS capacitors charge the inductors; in non-shoot-through states, the PV panel and inductors charge the loads and capacitors. Figure 24.12 shows the corresponding charge and discharge waveforms of inductor L_1, L_2 current and capacitor C_1 voltage as an example. The voltage of capacitor C_2 is the same; only the steady-state values are different from capacitor C_1. In the figure, t_1 is the time interval of active states and t_{sh} is the shoot-through time interval, which is equal to (DT_s/n_{sh}). Note that n_{sh} is the section of divided shoot-through duty ratio in each control period. It is four for the designed space-vector modulation and two for the PSSPWM. It can be seen that the purpose of the inductors is to limit the high-frequency current ripple to $r_i\%$ of the maximum inductor current during shoot-through states.

The selection of devices is performed for the worst case, operating with rated power when the PV array voltage is a minimum and all the cascaded qZS-HBI cells have the same working condition. In this scenario, the shoot-through duty cycle reaches its maximum, and the boost ratio of the PV panel voltage is also maximal. With the maximum modulation index M_{max}, the required minimum DC-link voltage v_{DC}

Table 24.2 Desired operating parameters

Parameters	Value
Maximum PV array voltage, $v_{PV,max}$	120 V
Minimum PV array voltage, $v_{PV,min}$	60 V
Rated power of each qZS-HBI, P_{max}	2 kW
Grid voltage	230 V
Grid frequency	50 Hz
Carrier frequency, f_c	2 kHz
Desired peak-to-peak current ripple through the qZS inductors, $r_i\%$	20%
Desired voltage ripple of the qZS capacitors, $r_v\%$	1%

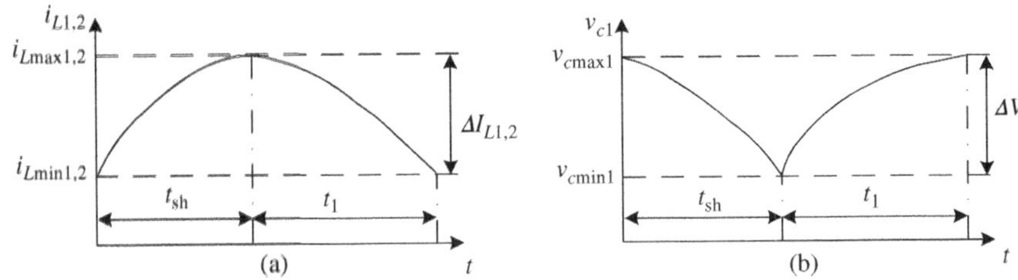

Figure 24.12 Waveforms of (a) inductor current and (b) capacitor voltage

for each cell is

$$\hat{v}_{\mathrm{DC\,min}} = \hat{v}_g / M_{\max} / n \tag{24.18}$$

where v_g with "^" is the amplitude of the grid voltage and n is the number of cascaded cells. The maximum required voltage gain G of the qZSI can be determined by

$$G_{\max} = \frac{(2\hat{v}_g/n)}{v_{\mathrm{PV,min}}} \tag{24.19}$$

Then, the minimum modulation index $M_{\min}$, maximum shoot-through duty ratio D and boost factor B can be determined by

$$M_{\min} = \frac{\pi G_{\max}}{3\sqrt{3}G_{\max} - \pi}, \quad D_{\max} = 1 - \frac{3\sqrt{3}}{2\pi}M_{\min}, \quad B_{\max} = \frac{1}{1 - 2D_{\max}}, \tag{24.20}$$

To avoid the discontinuous conduction mode of the inductor, even at the worst operation point, the inductance can be calculated by

$$L_1 = L_2 = \frac{v_L \Delta T}{\Delta I_L} = \frac{V_{C_1}(T_s D_{\max}/n_{sh})}{(P_{\max}/v_{\mathrm{PV,min}})r_i\%} = \frac{D_{\max}(1 - D_{\max})v_{\mathrm{PV,min}}^2}{n_{sh} f_c P_{\max}(1 - 2D_{\max})r_i\%} \tag{24.21}$$

In the meantime, in non-shoot-through states, two capacitors are in series in the qZSI network. They absorb the current ripple and limit the high-frequency voltage ripple on the inverter bridge to $r_v\%$ of the maximum DC-link voltage, in order to keep the output voltage sinusoidal. Therefore, the capacitance can be obtained as

$$C_1 = C_2 = \frac{2i_C \Delta T}{\Delta v_{DC}} = \frac{2I_{L,\max}(T_s D_{\max}/n_{sh})}{v_{\mathrm{DC,min}} r_v\%} = \frac{2D_{\max}(1 - 2D_{\max})P_{\max}}{n_{sh} f_c v_{\mathrm{PV,min}}^2 r_v\%} \tag{24.22}$$

According to the desired operating parameters listed in Table 24.2 and the commercialized components in the market, the qZS inductance and capacitance can be selected.

24.4.2 *Control System*

The proposed qZS-CMI topology with battery for PV power conversion, which is connected directly to the grid without transformers, has very limited output LC filters. Each phase consists of a series of cascaded qZSI PV modules to reach the grid levels. The qZSI PV modules comprise a group of PV panels and a qZSI with distributed MPPT for maximum energy production, boost, DC–AC inversion, and energy storage, all existing in this one single-stage power conversion unit. The gate drivers, sensors, protection and control platform based on a field programmable gate array (FPGA) or a DSP, and instrumentation circuits, will need to be designed and tested. Then, the entire power converter and PV system can be fabricated and implemented. Sensors will be calibrated and the control algorithms programmed and implemented.

The generic control structure of the proposed overall system is presented in Figure 24.13. The basic functionalities of the PV power system control are illustrated with a block diagram. The basic and advanced control features will be incorporated in the proposed solution. A number of features are required of the control technique employed: (1) ensure stability in case of a large grid-impedance change, (2) be able to ride through during disturbance of the grid voltage, (3) DC-link voltage control, (4) adaptation to grid voltage variation and grid synchronization with unity power factor operation, (5) the MPPT strategy, (6) anti-islanding, as required by *IEEE* 1574 the monitoring and (7) protection features. Other control features are active and reactive power control, harmonic compensation, local voltage control and

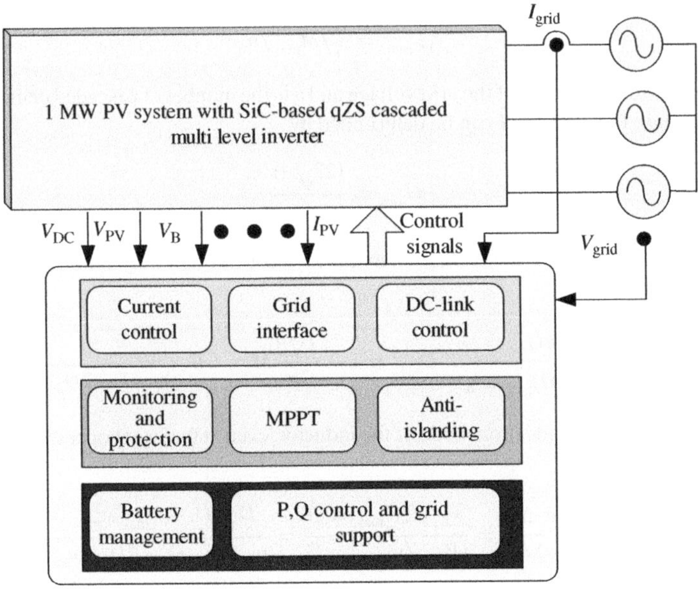

Figure 24.13 Control function diagram of the PV system

fault-tolerant operation during converter and grid faults. In addition, a battery energy management system is required.

With the batteries, each module in the system can output any assigned power no matter what the PV string's MPP, because the power difference can be balanced by the connected battery through charging and discharging. As a result, all modules output balanced power and all three phases feed the balanced power to the grid.

Acknowledgments

This chapter was made possible by NPRP Grant No. 09-233-2-096 and NPRP-EP Grant No. X-033-2-007 from the Qatar National Research Fund (a member of Qatar Foundation). The statements made herein are solely the responsibility of the authors.

References

1. Busquets-Monge, S., Rocabert, J., Rodríguez, P. *et al.* (2008) Multilevel diode-clamped converter for photovoltaic generators with independent voltage control of each solar array. *IEEE Transactions on Industrial Electronics*, **55** (7), 2713–2723.
2. Kerekes, T., Teodorescu, R., Liserre, M. *et al.* (2009) Evaluation of three-phase transformerless photovoltaic inverter topologies. *IEEE Transactions on Power Electronics*, **24** (9), 2202–2211.
3. Negroni, J.J., Guinjoan, F., Meza, C. *et al.* (2006) Energy-sampled data modeling of a cascade H-bridge multilevel converter for grid-connected pv systems. 10th IEEE International Power Electronics Congress, Puebla, October 16–18, 2006, pp. 1–6.
4. Flores, P., Dixon, J., Ortúzar, M. *et al.* (2009) Static var compensator and active power filter with power injection capability, using 27-level inverters and photovoltaic cells. *IEEE Transactions on Industrial Electronics*, **56** (1), 130–138.

5. Carbone, R. (2009) Grid-connected photovoltaic systems with energy storage. Proceedings of International Conference on Clean Electrical Power, Capri, Italy, June 9–11, 2009, pp. 760–767.
6. Femia, N., Lisi, G., Petrone, G. *et al.* (2008) Distributed maximum power point tracking of photovoltaic arrays: novel approach and system analysis. *IEEE Transactions on Industrial Electronics*, **55** (7), 2610–2621.
7. Shimizu, T., Hashimoto, O., and Kimura, G. (2003) A novel high-performance utility-interactive photovoltaic inverter system. *IEEE Transactions on Power Electronics*, **18** (2), 704–711.
8. Busquets-Monge, S., Rocabert, J., Rodríguez, P. *et al.* (2008) Multilevel diode-clamped converter for photovoltaic generators with independent voltage control of each solar array. *IEEE Transactions on Industrial Electronics*, **55** (7), 2713–2723.
9. Negroni, J.J., Guinjoan, F., Meza, C. *et al.* (2006) Energy-sampled data modeling of a cascade H-bridge multilevel converter for grid-connected PV systems. 10th IEEE International Power Electronics Congress, Puebla, October 16–18, 2006, pp. 1–6.
10. Patel, H. and Agarwal, V. (2009) A single-stage single-phase transformer-less doubly grounded grid-connected PV interface. *IEEE Transactions on Energy Conversion*, **24** (1), 93–101.
11. Brando, G., Dannier, A., Del Pizzo, A., and Rizzo, R. (2010) A high performance control technique of power electronic transformers in medium voltage grid-connected PV plants. Proceedings of the XIX International Conference on Electrical Machines – ICEM 2010, Rome, Italy.
12. Lee, J., Min, B., Kim, T. *et al.* (2011) High efficiency grid-connected multi string PV PCS using H-bridge multi-level topology. Proceedings of the 8th International Conference on Power Electronics – ECCE Asia, Shilla-Jeju, Korea, May 30–June 3, 2011, pp. 2557–2560.
13. Xiao, B., Filho, F., and Tolbert, L.M. (2011) Single-phase cascaded H-bridge multilevel inverter with non-active power compensation for grid-connected photovoltaic generators. ECCE2011, Phoenix, AZ, September 17–22, 2011, pp. 2733–2737.
14. Lu, X., Sun, K., Ma, Y. *et al.* (2009) High efficiency hybrid cascade inverter for photovoltaic generation. TENCON 2009, pp. 1–6.
15. Dehbonei, H., Lee, S.R., and Nehrir, H. (2009) Direct energy transfer for high efficiency photovoltaic energy systems part I: concepts and hypothesis. *IEEE Transactions on Aerospace and Electronic Systems*, **45** (1), 31–45.
16. Anderson, J. and Peng, F.Z. (2008) A class of quasi-Z-source inverters. IEEE Industry Applications Society Annual Meeting, IAS '08, Edmonton, Alta, October 5–9, 2008, pp. 1–7.
17. Li, Y., Anderson, J., Peng, F.Z., and Liu, D. (2009) Quasi-Z-source inverter for photovoltaic power generation systems. Twenty-Fourth Annual IEEE Applied Power Electronics Conference and Exposition, APEC2009, Washington, DC, February 15–19, 2009, pp. 918–924.
18. Park, J., Kim, H., Nho, E. *et al.* (2009) Grid-connected PV system using a quasi-Z-source inverter. Twenty-Fourth Annual IEEE Applied Power Electronics Conference and Exposition, APEC2009, Washington, DC, February 15–19, 2009, pp. 925–929.
19. Peng, F.Z. (2003) Z-source inverter. *IEEE Transactions on Industry Applications*, **39** (2), 504–510.
20. Badin, R., Huang, Y., Peng, F.Z., and Kim, H.G. (2007) Grid interconnected Z-source PV system. Proceedings of IEEE PESC'07, Orlando, FL, June 2007, pp. 2328–2333.
21. Park, J.H., Kim, H.G., Chun, T.W. *et al.* (2008) A control strategy for the grid-connected PV system using a Z-source inverter. The 2nd IEEE International Conference on Power and Energy (PECon 08), Johor Baharu, Malaysia, December 1–3, 2008, pp. 948–951.
22. Vilathgamuwa, D.M., Gajanayake, C.J., and Loh, P.C. (2009) Modulation and control of three-phase paralleled Z-source inverters for distributed generation applications. *IEEE Transactions on Energy Conversion*, **24** (1), 173–183.
23. Xu, P., Zhang, X., Zhang, C.W. *et al.* (2006) Study of Z-source inverter for grid-connected PV systems. The 37th IEEE Power Electronics Specialists Conference, PESC'06, Jeju, June 18–22, 2006, pp. 1–5.
24. Calais, M., Myrzik, J., Spooner, T., and Agelidis, V.G. (2002) Inverters for single-phase grid connected photovoltaic systems – an overview. Proceedings of IEEE PESC'02, Vol. 4, pp. 1995–2000.
25. Myrzik, J. and Calais, M. (2003) String and module integrated inverters for single-phase grid connected photovoltaic systems – a review. Proceedings of IEEE Bologna PowerTech Conference, Bologna, Italy, June 2003.
26. Kramer, W., Chakraborty, S., Kroposki, B., and Thomas, H. (2008) Advanced Power Electronic Interfaces for Distributed Energy Systems, Part 1: Systems and Topologies. Technical Report NREL/TP-581-42672, U.S. Department of Commerce, March 2008.
27. Carrasco, J.M., Franquelo, L.G., Bialasiewicz, J.T. *et al.* (2006) Power electronics systems for the grid integration of renewable energy sources: a survey. *IEEE Transactions on Industrial Electronics*, **53** (4), 1002–1016.

28. Li, Q. and Wolfs, P. (2008) A review of the single phase photovoltaic module integrated converter topologies with three different dc link configurations. *IEEE Transactions on Power Electronics*, **23** (3), 1320–1333.
29. Asiminoaei, L., Teodorescu, R., Blaabjerg, F., and Borup, U. (2005) Implementation and test of an online embedded grid impedance estimation technique for PV inverters. *IEEE Transactions on Industrial Electronics*, **52** (4), 1136–1144.
30. Barbosa, P.G., Braga, H.A.C., do Carmo Barbosa Rodrigues, M., and Teixeira, E.C. (2006) Boost current multilevel inverter and its application on single-phase grid-connected photovoltaic systems. *IEEE Transactions on Power Electronics*, **21** (4), 1116–1124.
31. Armstrong, M., Atkinson, D.J., Johnson, C.M., and Abeyasekera, T.D. (2006) Auto-calibrating dc link current sensing technique for transformer-less, grid connected, H-bridge inverter systems. *IEEE Transactions on Power Electronics*, **21** (5), 1385–1396.
32. Peng, F.Z., Shen, M., and Qian, Z. (2005) Maximum boost control of the Z-source inverter. *IEEE Transactions on Power Electronics*, **20** (4), 833–838.
33. Anderson, J. and Peng, F.Z. (2008) Four quasi-Z-source inverters. IEEE Power Electronics Specialists Conference, PESC2008, Rhodes, Greece, June 15–19, 2008, pp. 2743–2749.
34. Shen, M., Wang, J., Joseph, A. *et al.* (2006) Constant boost control of the Z-source inverter to minimize current ripple and voltage stress. *IEEE Transactions on Industry Applications*, **42** (3), 770–778.
35. Bärwaldt, G. and Kurrat, M. (2008) Application of energy storage systems minimizing effects of fluctuating feed-in of photovoltaic systems. CIRED Seminar 2008: SmartGrids for Distribution, Frankfurt, Germany, June 23–24, 2008, pp. 1–3.
36. Ueda, Y., Kurokawa, K., Tanabe, T. *et al.* (2008) Analysis results of output power loss due to the grid voltage rise in grid-connected photovoltaic power generation systems. *IEEE Transactions on Industrial Electronics*, **55** (7), 2744–2751.
37. Xiao, W., Ozog, N., and Dunford, W.G. (2007) Topology study of photovoltaic interface for maximum power point tracking. *IEEE Transactions on Industrial Electronics*, **54** (3), 1696–1704.
38. Rodriguez, C. and Amaratunga, G.A.J. (2008) Long-lifetime power inverter for photovoltaic ac modules. *IEEE Transactions on Industrial Electronics*, **55** (7), 2593–2601.
39. Peng, F.Z., McKeever, J.W., and Adams, D.J. (1997) Cascade multilevel inverters for utility applications. The 23rd International Conference on Industrial Electronics, Control and Instrumentation, IECON97, New Orleans, LA, November 9–14, 1997, pp. 437–442.
40. Kanchan, R.S., Baiju, M.R., Mohapatra, K.K. *et al.* (2005) Space vector PWM signal generation for multilevel inverters using only the sampled amplitudes of reference phase voltages. *IEE Proceedings of Electric Power Applications*, **152** (2), 297–309.
41. Teodorescu, R., Liserre, M., and Rodriguez, P. (2011) *Grid Converters for Photovoltaic and Wind Power Systems*, John Wiley & Sons, Ltd.
42. Wu, B., Lang, Y., Zargari, N., and Kouro, S. (2011) *Power Conversión and Control of Wind Energy Systems*, John Wiley & Sons, Ltd.
43. Blaabjerg, F., Teodorescu, R., Liserre, M., and Timbus, A.V. (2006) Overview of control and grid synchronization for distributed power generation systems. *IEEE Transactions on Industrial Electronics*, **53** (5), 1398–1409.
44. Eloy-Garcia, J., Arnaltes, S., and Rodriguez-Amenedo, J.L. (2007) Extended direct power control for multilevel inverters including DC link middle point voltage control. *Electric Power Applications, IET*, **1** (4), 571–580.
45. Rivera, S., Kouro, S., Cortes, P. *et al.* (2010) Generalized direct power control for grid connected multilevel converters. Proceedings of IEEE International Conference on Industrial Technology (ICIT), Chile, March 14–17, 2010, pp. 1351–1358.
46. Malinowski, M. and Kazmierkowski, M.P. (2003) Simple direct power control of three-phase PWM rectifier using space vector modulation – a comparative study. *EPE Journal*, **13** (2), 28–34.
47. Malinowski, M., Jasiński, M., and Kazmierkowski, M.P. (2004) Simple direct power control of three-phase PWM rectifier using space-vector modulation (DPC-SVM). *IEEE Transaction on Industrial Electronics*, **51** (2), 447–454.
48. Noguchi, T., Tomiki, H., Kondo, S., and Takahashi, I. (1998) Direct power control of PWM converter without power source voltage sensors. *IEEE Transaction on Industrial Electronics*, **34** (3), 473–479.
49. Zhi, D., Xu, L., and Williams, B.W. (2009) Improved direct power control of grid connected DC/AC converter. *IEEE Transaction on Industrial Electronics*, **24** (5), 1280–1292.
50. Malinowski, M., Kazmierkowski, M.P., Hansen, S. *et al.* (2000) Virtual flux based direct power control of three-phase PWM rectifiers. Proceedings of IEEE Thirty-Fifth IAS Annual Meeting and World Conference

on Industrial Applications of Electrical Energy, Roma, Italy, October 8–12, 2000, Vol. 4, pp. 2369–2375 (in English).
51. Malinowski, M., Kazmierkowski, M.P., Hansen, S. *et al.* (2001) Virtual flux based direct power control of three-phase PWM rectifiers. *IEEE Transaction on Industry Application*, **37** (4).
52. Serpa, L.A., Barbosa, P.M., Steimer, P.K., and Kolar, J.W. (2008) Five-level virtual-flux direct power control for the active neutral-point clamped multilevel inverter. Proceedings of the IEEE Power Electronics Specialists Conference PESC, June 15–19, 2008, pp. 1668–1674.
53. Antoniewicz, P. and Kazmierkowski, M.P. (2008) Virtual flux based predictive direct power control of ac/dc converters with on-line inductance estimation. *IEEE Transaction on Industrial Electronics*, **55** (12), 4381–4390.
54. Cortes, P., Rodriguez, J., Antoniewicz, P., and Kazmierkowski, M.P. (2008) Direct power control of an afe using predictive control. *IEEE Transaction on Power Electronics* **23** (5), pp. 2516–2523.
55. Cortes, P., Rodriguez, J., Antoniewicz, P., and Kazmierkowski, M.P. (2008) Direct power control of an AFE using predictive control. *IEEE Transaction on Industrial Electronics*, **23** (5), 2516–2523.
56. Geyer, T. (2010) Model predictive direct current control for multi-level converters. Proceedings of IEEE Energy Conversion Congress and Exposition, Atlanta, GA, September 2010.
57. Rajakaruna, S. and Jayawickrama, L. (2010) Steady-state analysis and designing impedance network of Z-source inverters. *IEEE Transactions on Industrial Electronics*, **57**, 2483–2491.
58. Ellabban, O., van Mier, J., and Lataire, P. (2011) Experimental study of the shoot-through boost control methods for the Z-source inverter. *EPE Journal*, **21**, 18–29.
59. Roasto, L. and Vinnikov, D. (2010) Analysis and evaluation of PWM and PSM shoot-through control methods for voltage-fed qZSI based DC/DC converters. 2010 14th International Power Electronics and Motion Control Conference (EPE/PEMC), September 6–8, 2010, pp. T3-100, T3-105.

Index

Power Electronics for Renewable Energy Systems, Transportation and Industrial Applications, First Edition.
Edited by Haitham Abu-Rub, Mariusz Malinowski and Kamal Al-Haddad.

Printed and bound by CPI Group (UK) Ltd, Croydon, CR0 4YY

17/09/2026

14964300-0008